Discrete-Time Processing of Speech Signals

IEEE Press
445 Hoes Lane, P.O. Box 1331
Piscataway, NJ 08855-1331

Books of Related Interest from IEEE Press . . .

SPEECH COMMUNICATIONS: Human and Machine, Second Edition
Douglas O'Shaughnessy
2000 Hardcover 560 pp ISBN 0-7803-3449-3

THE DIGITAL SIGNAL PROCESSING HANDBOOK
Vijay Madisetti and Douglas Williams
1998 Hardcover 1,776 pp ISBN 0-7803-3475-2

RANDOM PROCESSES FOR IMAGE AND SIGNAL PROCESSING
Edward R. Dougherty
A SPIE Press book published in cooperation with IEEE Press
1999 Hardcover 616 pp ISBN 0-7803-3495-7

THE HANDBOOK OF TABLES AND FORMULAS FOR SIGNAL PROCESSING
Alexander B. Poularikas
A CRC Handbook published in cooperation with IEEE Press
1999 Hardcover 818 pp ISBN 0-7803-4728-5

DSP PROCESSOR FUNDAMENTALS: Architectures and Features
Phil Lapsley et al.
1997 Softcover 224 pp ISBN 0-7803-3405-1

Discrete-Time Processing of Speech Signals

John R. Deller, Jr.
Michigan State University

John H. L. Hansen
University of Colorado at Boulder

John G. Proakis
Northeastern University

IEEE Signal Processing Society, *Sponsor*

The Institute of Electrical and Electronics Engineers, Inc., New York

A JOHN WILEY & SONS, INC., PUBLICATION
New York • Chichester • Weinheim • Brisbane • Singapore • Toronto

First printing (1993) by Macmillan Publishing Co.

For ordering and customer service, call 1-800-CALL-WILEY.
Wiley-Interscience-IEEE **ISBN 0-7803-5386-2**

Printed in the United States of America.
10 9 8 7 6 5 4 3

Library of Congress Cataloging-in-Publication Data

Deller, John R.
Discrete-time processing of speech signals / John R. Deller, Jr., John G. Proakis, John H. L. Hansen.
p. cm.
Originally published: New York : Macmillan, 1993.
Includes bibliographical references and index.
ISBN 0-7803-5386-2
1. Speech processing systems. 2. Discrete-time systems.
I. Proakis, John G. II. Hansen, John H. L. III. Title.
[TK7882.S65D44 2000]
621.382'2—dc21 99-31911
CIP

Dedications

To Joan ... For her boundless love, patience, and support throughout this long project. I would sometimes find her alone in a quiet place reading Charles Dickens or Jane Austen so as not to disturb me while I wrote about speech processing. The juxtaposition was striking. I would imagine some patient spouse 150 years from now reading Deller, Proakis, and Hansen while his or her partner wrote a future great book. The ridiculous fantasy made me smile. When she asked why, I would simply tell her that I was so happy to have such an understanding wife. And so I am. J.D.

To Felia, George, and Elena. J.G.P.

To my wife Holly, Christian and Heather, and in memory of George, a devoted writer.... J.H.L.H.

Contents

Preface to the IEEE Edition

The preface to the original edition of this book remains largely relevant and accurate, since this issue of the book has been changed only in minor ways—the majority of which are corrections to typographical and other errors in the earlier work. We are indebted to many readers of the first edition, too numerous to name, for sending us corrections and suggestions for improvement. Your attention to detail is remarkable and is the source of great and well-deserved embarrassment.

The most significant new information to be noted concerns the means for accessing speech data for use in the end-of-chapter problems. Reflecting the remarkable changes in the Internet since the first edition was published in 1993, a simple and flexible procedure for acquiring these data is now possible. The interested reader can simply access the World Wide Web home pages of either author: Deller (URL:http://www.egr.msu.edu/~deller) or Hansen (URL: http://www.cslu.colorado.edu/rspl), and then follow the link to information about this text. Procedures for downloading data are described there.

The original preface expresses our gratitude to Editor John Griffin for painstakingly nurturing this book into existence in his former editorial life at the original publishing company. The glue on the 1993 binding was not even dry before corporate mergers and reorganizations in the publishing world severed the business relationship between the authors and Mr. Griffin, but certainly not the enduring appreciation for his sometimes heroic efforts that made this book possible. In a fortuitous turn of professional events, the authors were given a second chance to work with Mr. Griffin in his current editorial position at the outstanding IEEE Press. Before the authors allowed him to commit to a second round, it seemed inhumane not to remind him of what we put him through in 1993. Realizing that a good challenge now and then will keep a professional sharp, we simply welcomed the renewed opportunity and continue to appreciate John's support and confidence in this project.

John R. Deller, Jr.
Michigan State University

Preface

Purposes and Scope. The purposes of this book are severalfold. Principally, of course, it is intended to provide the reader with solid fundamental tools and sufficient exposure to the applied technologies to support advanced research and development in the array of speech processing endeavors. As an academic instrument, however, it may also provide the serious student of signal processing an opportunity to strengthen and deepen his or her understanding of the field through the study of one of the most important and interesting contemporary applications of signal processing concepts. Finally, by collecting a large number of contemporary topics with an extensive reference list into a single volume, the book will serve as a convenient resource for those already working in the field.

The book is written by three professors of electrical engineering. This has two implications. First, we view the book as a pedagogical tool. This means we have attempted to keep the student in mind with each sentence and with each chapter. Notation, approach, and general level have been made as uniform as possible across developments and across chapters. Second, the text is written with a clear bias toward the topics and approaches of modern electrical engineering curricula—especially signal processing, systems, and communications. Speech processing is inherently multidisciplinary, and we occasionally indicate to the reader where topics are necessarily treated superficially or not at all, and where the reader can find more information. This occurs principally in areas that would probably be labeled "Speech Science" or "Computer Science."

Level of the Text. A certain level of sophistication with topics in signal processing and stochastic processes is assumed in this book. This background is typical of a solid senior-level course in each discipline at many American universities. Accordingly, the book is intended for use in one or more graduate-level courses. The book could conceivably be used with advanced seniors, but the instructor is urged to consider the degree of maturity that is required in these areas. A good gauge of this factor is available in Chapter 1. Sections 1.1 and 1.2 should be comfortably considered review material for anyone who is to succeed with the text. Sections 1.3 and 1.4 need not be review, but the typical EE senior will have at least some exposure to these topics, even if specific courses in pattern recognition and information theory have not been taken. Nevertheless, Sections 1.3 and 1.4 do provide sufficient background in their respective topics for the remaining chapters, whereas Sections 1.1 and 1.2 are not intended as substitutes for relevant coursework. Section 1.5 is simply a review of some concepts that should be quite familiar to any engineering student, and it is included principally as a means for establishing notation.

Course Planning. The general topical content of the speech processing field as reflected in the book is described in Section 1.6. The instructor might wish to review that section in planning a course around the book (and to have the students read this section as an introduction to the course). Clearly, it will be impossible to cover the entire book in a single quarter or semester. We have found that in a typical semester course (15 weeks), the following can be covered: Chapter 1 (Background material—Brief review of Sections 1.1 and 1.2); Chapter 2 (Speech science topics—Rely heavily on student reading of qualitative material and focus on issues necessary to engineering modeling); Chapter 3 (Modeling—The main goal is the digital model. Highlight the mathematics of the acoustic tube theory and stress the physical significance without extensive in-class formal development of the results); Chapter 4 (Short-term processing—This is often the students' first real exposure to short-time processing. Cover the basics carefully and have the student use the computer.); Chapter 5 (Linear prediction—Cover this central topic thoroughly except for some of the details of the solution methods in Section 5.3.3 that the instructor may choose to omit.); Chapter 6 (Cepstral analysis—The instructor may choose to omit Section 6.3 as explained in the reading notes at the beginning); Chapters 10, 11, and 12 (Recognition basics—Many details will need to be omitted, for example, in Sections 12.2.4–12.2.8); Chapters 13 and 14 (Language modeling and neural network approaches—These topics can be covered only superficially as time permits). Alternatively, the instructor may choose to include Chapters 7, 8, and 9 of Part IV on coding and enhancement, rather than recognition, at the end of the course. These three chapters could be covered in some detail. If time and resources permit, an ideal approach is to thoroughly cover material in Parts I, II, and III in an introductory course, and Parts IV and V in an advanced applications course.

Obtaining Speech Data for End-of-Chapter Problems. It will be noted that many of the problems in this book require real speech files. Although most universities have the facilities to create such files, we offer the following options to instructors who wish to avoid the details of collecting speech files on-site. Several standard databases have been compiled by the U.S. National Institute of Standards and Technology and are available on compact disk from the U.S. Department of Commerce, National Technical Information Service (NTIS). These databases are described in Section 13.8 of the book. For ordering information, call the NTIS at (703) 487-4650, or fax (703) 321-8547. Alternatively, the authors of this book have made available some speech samples that can be downloaded over the electronic mail network using the instructions below.

Speech files can be downloaded to a personal computer (PC) through the WWW internet computer network.[1]

Acknowledgments. We appreciate the support of many people who have contributed to the writing and production of this textbook. We are indebted to a number of colleagues and graduate students who provided critical reviews and suggestions that enhanced and improved the presentation. Among those are Professor Mark Clements, Georgia Tech; Professor Jerry Gibson, University of Texas at Austin; Professor Paul Milenkovic, University of Wisconsin; Professor Larry Paarmann, Wichita State University; Dr. Joseph Picone, Texas Instruments Corporation; Dr. Dale Veeneman, GTE Laboratories, Waltham; and Professor Greg Wakefield, The University of Michigan. We also thank the members of the EE 801 class of Spring Quarter 1991 at Michigan State University, who struggled through the early versions of the problem sets and found (the hard way) the stumbling blocks. We also appreciate the diligent efforts of Mr. Sudhir Kandula in assisting with graphics and problem solutions. We also thank the students of the Robust Speech Processing Laboratory (T. W. Pai, S. Nandkumar, S. Bou-Ghazale, and L. Arslan) for their graphics and proofreading assistance. We also greatly appreciate the encouragement and support of Editor John Griffin of the Macmillan Publishing Company whose patience and good sense of humor were extraordinary. We also wish to thank Senior Production Supervisor Elaine Wetterau of Macmillan for many painstaking hours with the manuscript that contributed greatly to the quality of the finished work.

J. D. would also like to acknowledge research support during the writing of this book from the Whitaker Foundation, the National Science Foundation, and the Office of Naval Research. The research supported by these sponsors enriched the author's understanding of many important topics.

J.R.D.
J.G.P.
J.H.L.H.

[1]See Preface to the IEEE Edition for information on speech data, (pg. xvii).

Acronyms and Abbreviations

Acoustic decoder (AD)
Adaptive comb filtering (ACF)
Adaptive delta modulation (ADM)
Adaptive differential pulse code modulation (ADPCM)
Adaptive noise canceling (ANC)
Adaptive predictive coding (APC)
Adaptive pulse code modulation (APCM)
Adaptive transform coding (ATC)
Additive white Gaussian noise (AWGN)
Advanced Research Projects Agency [of the United States] (ARPA)
American Telephone and Telegraph (AT&T)
Articulation index (AI)
Artificial neural network (ANN)
Augmented transition network (ATN)
Average magnitude difference function (AMDF)
Autoregressive [model] (AR)
Autoregressive—moving average [model] (ARMA)
Back-propagation [algorithm] (BP)
Bahl–Jelinek–Mercer [algorithm] (BJM)
Bellman optimality principle (BOP)
Bits per normalized second (or bits per sample) (bpn)
Bits per second (bps)
Bolt, Beranek, and Newman, Incorporated (BBN)
Bounded input–bounded output [stability] (BIBO)
Carnegie Mellon University (CMU)
Centro Studi e Laboratori Telecomunicazioni (CSELT)
Closed phase (CP)
Cocke–Younger–Kasami [algorithm] (CSY)
Code-excited linear prediction (CELP)
Complex cepstrum (CC)
Connectionist Viterbi training (CVT)
Continuous speech recognition (CSR)
Continuously variable slope delta modulation (CVSD)
Cumulative distribution function (cdf)
Defense Advanced Research Projects Agency [of the United States] (DARPA)
Defense Advanced Research Projects Agency resources management database (DRMD)
Delta modulation (DM)
Diagnostic acceptability measure (DAM)
Diagnostic rhyme test (DRT)
Differential pulse code modulation (DPCM)
Digital signal processing (DSP)
Discrete cosine transform (DCT)
Discrete Fourier series (DFS)
Discrete Fourier transform (DFT)
Discrete time (DT)
Discrete-time Fourier transform (DTFT)
Discrimination information (DI)
Dynamic programming (DP)
Dynamic time warping (DTW)
Estimate–maximize [algorithm] (EM)
Fast Fourier transform (FFT)
Feature map classifier (FMC)
Finite impulse response (FIR)
Finite state automaton (FSA)
Floating point operation (flop)
Forward–backward [algorithm] (F–B)
Frequency weighted segmental signal-to-noise ratio (SNR_{fw-seg})
Grammar-driven connected word recognizer (GDCWR)
Harmonic product spectrum (HPS)
Hear-what-I-mean [speech recognition system] (HWIM)
Hertz (Hz)
Hidden Markov model (HMM)
IEEE International Conference on Acoustics, Speech, and Signal Processing (ICASSP)
Impulse response (IR)

Infinite impulse response (IIR)
International Business Machines Corporation (IBM)
International Phonetic Alphabet (IPA)
ATR Interpreting Telephony Research Laboratories (ATR)
Inverse discrete Fourier transform (IDFT)
Inverse discrete-time Fourier transform (IDTFT)
Inverse fast Fourier transform (IFFT)
Inverse filter (IF)
Inverse sine [parameters] (IS)
Isolated-word recognition (IWR)
Isometric absolute judgment [test] (IAJ)
Learning vector quantizer (LVQ)
Least mean square [algorithm] (LMS)
Left–right [parsing] (LR)
Level building [algorithm] (LB)
Levinson–Durbin [recursion] (L–D)
Linguistics Data Consortium (LDC)
Linguistic decoder (LD)
Line spectrum pair [parameters] (LSP)
Linear constant coefficient differential equation (LCCDE)
Linear prediction [or linear predictive] (LP)
Linear predictive coding (LPC)
Linear, time invariant (LTI)
Linde–Buzo–Gray [algorithm] (LGB)
Log area ratio [parameters] (LAR)
Massachusetts Institute of Technology (MIT)
Maximum *a posteriori* (MAP)
Maximum average mutual information (MMI)
Maximum entropy (ME)
Maximum entropy method (MEM)
Maximum likelihood (ML)
Mean opinion score (MOS)
Mean square error (MSE)
Minimum discrimination information (MDI)
Modified rhyme test (MRT)
Multilayer perceptron (MLP)
National Bureau of Standards [of the United States] (NBS)
National Institute of Standards and Technology [of the United States] (NITS)
Normalized Hertz (norm-Hz)
Normalized radians per second (norm-rps)
Normalized seconds (norm-sec)
One stage [algorithm] (OS)
Orthogonality principle (OP)
Paired acceptability rating (PAR)
Parametric absolute judgment [test] (PAJ)
Partial correlation [coefficients] (parcor)
Perceptron learning [algorithm] (PL)
Power density spectrum (PDS)
Probability density function (pdf)
Pulse code modulation (PCM)
Quadrature mirror filter (QMF)
Quality acceptance rating test (QUART)
Radians per second (rps)
Real cepstrum (RC)
Region of convergence (ROC)
Residual-excited linear prediction (RELP)
Segmental signal-to-noise ratio (SNR_{seg})
Self-organizing feature finder (SOFF)
Short-term complex cepstrum (stCC)
Short-term discrete Fourier transform (stDFT)
Short-term discrete-time Fourier transform (stDTFT)
Short-term inverse discrete Fourier transform (stIDFT)
Short-term memory [model] (STM)
Short-term power density spectrum (stPDS)
Short-term real cepstrum (stRC)
Signal-to-noise ratio (SNR)
Simple inverse filter tracking [algorithm] (SIFT)
Stanford Research Institute (SRI)
Strict sense (or strong sense) stationary (SSS)
Subband coding (SBC)
Systems Development Corporation (SDC)
Texas Instruments Corporation (TI)
Texas Instruments/National Bureau of Standards [database] (TI/NBS)
Time-delay neural network (TDNN)
Time domain harmonic scaling (TDHS)
Traveling salesman problem (TSP)
Vector quantization (VQ)
Vector sum excited linear prediction (VSELP)
Weighted recursive least squares [algorithm] (WRLS)
Weighted-spectral slope measure (WSSM)
Wide sense (or weak sense) stationary (WSS)

PART I

SIGNAL PROCESSING BACKGROUND

CHAPTER 1

Propaedeutic

Read.Me: If you are someone who never reads Chapter 1, please at least read Sections 1.0.2 and 1.0.3 before proceeding!

1.0 Preamble

1.0.1 The Purpose of Chapter 1

If the reader learns nothing more from this book, it is a safe bet that he or she will learn a new word. A *propaedeutic*[1] is a "preliminary body of knowledge and rules necessary for the study of some art or science" (Barnhart, 1964). This chapter is just that—a propaedeutic for the study of speech processing focusing primarily on two broad areas, digital signal processing (DSP) and stochastic processes, and also on some necessary topics from the fields of statistical pattern recognition and information theory.

The reader of this book is assumed to have a sound background in the first two of these areas, typical of an entry level graduate course in each field. It is not our purpose to comprehensively teach DSP and random processes, and the brief presentation here is not intended to provide an adequate background. There are many fine textbooks to which the reader might refer to review and reinforce prerequisite topics for these subjects. We list a considerable number of widely used books in Appendices 1.A and 1.B.

What, then, is the point of our propaedeutic? The remainder of this chapter is divided into four main sections plus one small section, and the tutorial goals are somewhat different in each. Let us first consider the two main sections on DSP and stochastic processes. In the authors' experience, the speech processing student is somewhat more comfortable with "deterministic" DSP topics than with random processes. What we will do in Section 1.1, which focuses on DSP, therefore, is highlight some of the key concepts which will play central roles in our speech processing work. Where the material seems unfamiliar, the reader is urged to seek help in

[1]Pronounced "prō′-pa-doo′-tic."

one or more of the DSP textbooks cited in Appendix 1.A. Our main objective is to briefly outline the essential DSP topics with a particular interest defining notation that will be used consistently throughout the book. A second objective is to cover a few subtler concepts that will be important in this book, and that might have been missed in the reader's first exposure to DSP.

The goals of Section 1.2 on random processes are somewhat different. We will introduce some fundamental concepts with a bit more formality, uniformity, and detail than the DSP material. This treatment might at first seem unnecessarily detailed for a textbook on speech processing. We do so, however, for several reasons. First, a clear understanding of stochastic process concepts, which are so essential in speech processing, depends strongly on an understanding of the basic probability formalisms. Second, many engineering courses rely heavily on stochastic processes and not so much on the underlying probability concepts, so that the probability concepts become "rusty." Emerging technologies in speech processing depend on the basic probability theory and some review of these ideas could prove useful. Third, it is true that the mastery of any subject requires several "passes" through the material, but engineers often find this especially true of the field of probability and random processes.

The third and fourth major divisions of this chapter, Sections 1.3 and 1.4, treat a few topics which are used in the vast fields of statistical pattern recognition and information theory. In fact, we have included some topics in Section 1.3 which are perhaps more general than "pattern recognition" methods, but the rubric will suffice. These sections are concerned with basic mathematical tools which will be used frequently, and in diverse ways in our study, beginning in Part IV of the book. There is no assumption that the reader has formal coursework in these topics beyond the normal acquaintance with them that would ordinarily be derived from an engineering education. Therefore, the goal of these sections is to give an adequate description of a few important topics which will be critical to our speech work.

Finally, Section 1.5 briefly reviews the essence and notation of phasors and steady-state analysis of systems described by differential equations. A firm grasp of this material will be necessary in our early work on analog acoustic modeling of the speech production system in Chapter 3.

As indicated above, the need for the subjects in Sections 1.3–1.5 is not immediate, so the reader might wish to scan over these sections, then return to them as needed. More guidance on reading strategy follows.

1.0.2 Please Read This Note on Notation

The principal tool of engineering is applied mathematics. The language of mathematics is abstract symbolism. This book is written with a conviction that careful and consistent notation is a sign of clear understanding, and clear understanding is derived by forcing oneself to comprehend and use such notation. Painstaking care has been taken in this book to use information-laden and consistent notation in keeping with this philosophy. When we err with notation, we err on the side of excessive notation which is not always conventional, and not always necessary once the topic has been mastered. Therefore, the reader is invited (with your instructor's permission if you are taking a course!) to shorten or simplify the notation as the need for the "tutorial" notation subsides.

Let us give some examples. We will later use an argument m to keep track of the point in time at which certain features are extracted from a speech signal. This argument is key to understanding the "short-term" nature of the processing of speech. The ith "linear prediction" coefficient computed on a "frame" of speech ending at time m will be denoted $\hat{a}(i; m)$. In the development of an algorithm for computing the coefficients, for example, the index m will not be very germane to the development and the reader might wish to omit it once its significance is clear. Another example comes from the random process theory. Numerous examples of sloppy notation abound in probability theory, a likely reason why many engineers find this subject intractable. For example, something like "$f(x)$" is frequently used to denote the probability density function (pdf) for the random variable $\underline{x}$. There are numerous ways in which this notation can cause misunderstandings and even subtle mathematical traps which can lead to incorrect results. We will be careful in this text to delineate random processes, random variables, and values that may be assumed by a random variable. We will denote a random variable, for example, by underscoring the variable name, for example, $\underline{x}$. The pdf for $\underline{x}$ will be denoted $f_{\underline{x}}(x)$, for example. The reader who has a clear understanding of the underlying concepts might choose to resort to some sloppier form of notation, but the reader who does not will benefit greatly by working to understanding the details of the notation.

1.0.3 For People Who Never Read Chapter 1 (and Those Who Do)

To be entitled to use the word "propaedeutic" at your next social engagement, you must read at least some of this chapter.[2] If for no other reason than to become familiar with the notation, we urge you to at least generally review the topics here before proceeding. However, there is a large amount of material in this chapter, and some people will naturally prefer to review these topics on an "as needed" basis. For that reason, we provide the following guide to the use of Chapter 1.

With a few exceptions, most of the topics in Sections 1.1 and 1.2 will be widely used throughout the book and we recommend their review before proceeding. The one exception is the subsection on "State Space Re-

[2]If you have skipped the first part of this chapter, you will be using the word without even knowing what it means.

alizations" in Section 1.1.6, which will be used in a limited way in Chapters 5 and 12. The topics in Sections 1.3 and 1.4, however, are mostly specialized subjects which will be used in particular aspects of our study, beginning in Part IV of the book. Likewise the topic in Section 1.5 is used in one isolated, but important, body of material in Chapter 3.

These latter topics and the "state space" topic in the earlier section will be "flagged" in Reading Notes at the beginning of relevant chapters, and in other appropriate places in the book.

1.1 Review of DSP Concepts and Notation

1.1.1 "Normalized Time and Frequency"

Throughout the book, we will implicitly use what we might call *normalized time and frequency variables.* By this we mean that a discrete time signal (usually speech), say $s(n)$, will be indexed by integers only.[3] Whereas $s(n)$ invariably represents samples of an analog waveform, say $s_a(t)$, at some sample period, T,

$$s(n) = s_a(nT) = s_a(t)\big|_{t=nT} \qquad n = \ldots, -1, 0, 1, 2, \ldots, \tag{1.1}$$

the integer n indexes the *sample number,* but we have lost the absolute time orientation in the argument. To recover the times at which the samples are taken, we simply need to know T.

To understand the "physical" significance of this mathematical convention, it is sometimes convenient to imagine that we have scaled the real-world time axis by a factor of T prior to taking the samples, as illustrated in Fig. 1.1. "Normalized time," say t', is related to real time as

$$t' = \frac{t}{T} \tag{1.2}$$

and the samples of speech are taken at intervals which are exactly[4] "normalized seconds (norm-sec)." In most cases it is perfectly sufficient to refer to the interval between samples as the "sample period," where the conversion to the real-world interval is obvious. However, on a few occasions we will have more than one sampling process occurring in the same problem (i.e., a resampling of the speech sequence), and in these instances the concept of a "normalized second" is useful to refer to the basic sampling interval on the data.

Of course, the normalization of time renders certain frequency quantities invariant. The sample period in normalized time is always unity, and therefore the sample frequency is always unity [dimensionless, but some-

[3]Note that we have referred to $s(n)$ as "discrete time" rather than "digital." Throughout most of this book, we will ignore any quantization of amplitude.

[4]The reader should note that the normalized time axis is actually dimensionless.

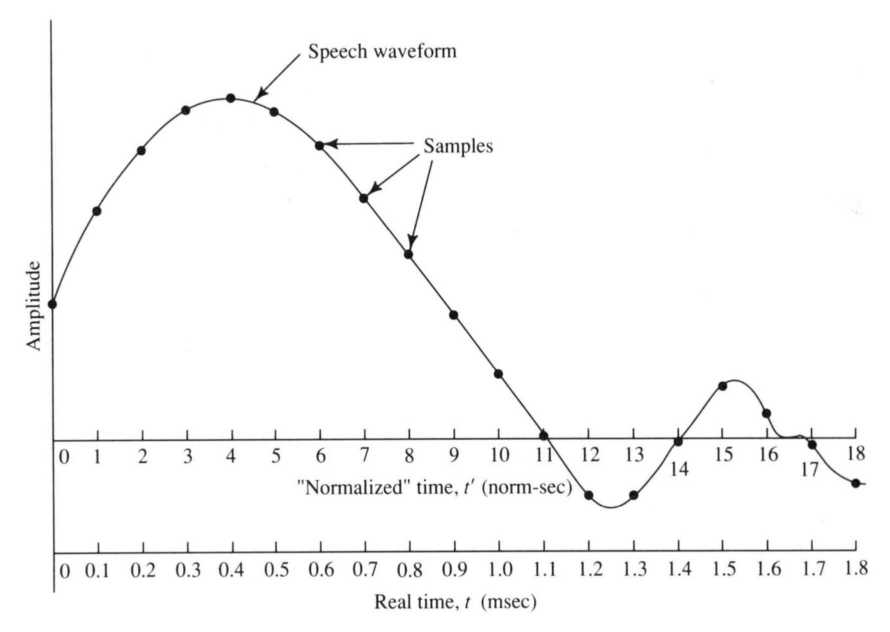

FIGURE 1.1. Segment of a speech waveform used to illustrate the concept of "normalized time." Suppose that samples are to be taken at a rate $F_s = 10$ kHz so that the sample period is $T = 0.1$ msec. The lower time axis represents real time measured in milliseconds, while the upper represents a normalization of the time axis such that the sample times fall at integers. Normalized time, t', is related to real time, t, as $t' = t/T$. We will on a few occasions refer to the sample period in the scaled case as a "normalized second (norm-sec)."

times "normalized Hertz (norm-Hz)"], and the sample radian frequency is always 2π [dimensionless or "normalized radians per second (norm-rps)"]. Accordingly, the Nyquist frequency is always 0.5 norm-Hz, or π norm-rps. In general, the conversions between "real" frequencies, say F (Hz) and Ω (rps) and their normalized counterparts, say f and ω, are given by

$$f = FT \tag{1.3}$$

$$\omega = \Omega T. \tag{1.4}$$

We can easily verify this by examining a single sinusoid at real frequency Ω,

$$x_a(t) = A \sin(\Omega t + \varphi) = A \sin(\Omega T \frac{t}{T} + \varphi). \tag{1.5}$$

The rightmost term can be regarded as a sinusoid at a different frequency, $\omega = \Omega T$, on a different time axis $t' = t/T$,

$$x'_a(t') = A \sin(\omega t' + \varphi). \tag{1.6}$$

Clearly, we obtain the same samples if we sample $x_a(t)$ at $t = nT$, or $x_a'(t')$ at $t' = n$. A magnitude spectrum is plotted against real and normalized frequencies in Fig. 1.2 to illustrate this concept.

In spite of this rather lengthy explanation of "normalized time and frequency," we do not want to overemphasize this issue. Simply stated, we will find it convenient to index speech waveforms by integers (especially in theoretical developments). This is an accepted convention in DSP. The point of the above is to remind the reader that the resulting "normalized" time and frequency domains are simply related to the "real" time and frequency. When the "normalized" quantities need to be converted to "real" quantities, this is very easily accomplished if the sample frequency or period is known. While using the DSP convention for convenience in many discussions, we will always keep the sampling information close at hand so the "real" quantities are known. To do otherwise would be to deny the physical nature of the process with which we are working. In many instances in which practical systems and applications are being discussed, it will make perfect sense to simply work in terms of "real" quantities.

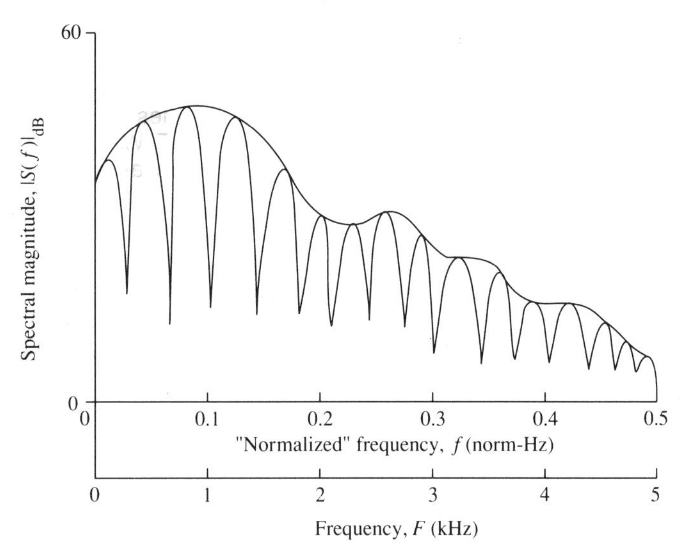

FIGURE 1.2. Magnitude spectrum of a typical speech waveform. This spectrum is based on the DFT of samples of the waveform taken at 10 kHz. The lower frequency axis represents real frequencies measured in kHz, while the upper represents a normalization of the frequency axis concomitant to the time normalization. Normalized frequency, f, is related to real frequency, F, as $f = FT$, where T is the nominal sample period in a given analysis. Accordingly, the "normalized" sampling, say f_s, and Nyquist, f_N, frequencies are invariant with the sample rate, with $f_s = 1$ and $f_N = 1/2$. We will sometimes refer to the units of normalized frequencies as "normalized Hertz (norm-Hz)" or "normalized radians per second (norm-rps)."

1.1.2 Singularity Signals

In the continuous time domain, a *singularity signal* is one for which one or more derivatives do not exist at one or more time points. Although the concept of a derivative is no longer meaningful in discrete time, we often borrow the term *singularity* to describe analogous sequences in sampled time. The two for which we will have the most use are

- The *unit sample sequence* or *discrete-time impulse*, defined by

$$\delta(n) \stackrel{\text{def}}{=} \begin{cases} 1, & \text{if } n = 0 \\ 0, & \text{otherwise} \end{cases}. \tag{1.7}$$

- The *unit step sequence*, defined by[5]

$$u(n) \stackrel{\text{def}}{=} \begin{cases} 1, & \text{if } n \geq 0 \\ 0, & \text{otherwise} \end{cases}. \tag{1.8}$$

The unit step sequence is much more analogous to its analog counterpart than the discrete-time impulse in the sense that $u(n)$ simply represents samples of the analog unit step function (discounting problems that may arise due to different definitions at time zero). On the other hand, recall that the *analog* (Dirac) impulse function, say $\delta_a(t)$, is defined such that it apparently has infinite height, zero width, and unity area. Although the discrete-time impulse plays an analogous role to that played by the analog impulse, it may *not* be interpreted as its samples.

1.1.3 Energy and Power Signals

There are many ways in which a discrete time signal can be classified. One useful grouping is into the categories energy signal, power signal, or neither. Recall that the *energy* of a discrete time signal is defined as[6]

$$E_x \stackrel{\text{def}}{=} \sum_{n=-\infty}^{\infty} |x(n)|^2. \tag{1.9}$$

A signal $x(n)$ is called an *energy signal* if

$$0 < E_x < \infty. \tag{1.10}$$

The *power* in a discrete-time sequence is

[5]The notation $u(n)$ is widely used to indicate the unit step sequence, but $u(n)$ will also refer to a very important waveform, the "glottal volume velocity," throughout the book. Because of the context, there will be no risk of confusion.

[6]The absolute value signs are included because, in general, $x(n)$ is a complex sequence.

$$P_x \stackrel{\text{def}}{=} \lim_{N\to\infty} \frac{1}{2N+1} \sum_{n=-N}^{N} |x(n)|^2. \qquad (1.11)$$

A *power signal* has finite but nonzero power,

$$0 < P_x < \infty. \qquad (1.12)$$

A signal cannot be both a power signal and an energy signal simultaneously, since if $E_x < \infty$, then $P_x = 0$. A signal can, however, be *neither* when $P_x = \infty$ or $E_x = 0$.

For our purposes in speech processing, it is sufficient to associate the energy category with two broad classes of signals. These are

- *Transients*, those which decay (usually exponentially) with time. Examples are

$$x_1(n) = \alpha^n u(n), \qquad |\alpha| < 1 \qquad (1.13)$$

$$x_2(n) = \alpha^{|n|} \cos(n\omega_0 + \psi), \qquad |\alpha| < 1. \qquad (1.14)$$

- *Finite sequences*, those which are zero outside a finite time duration. An example is

$$x_3(n) = e^{\beta n}[u(n+3) - u(n-246)], \qquad |\beta| < \infty. \qquad (1.15)$$

Whereas the energy signals either decay out sufficiently fast or "stop" completely, the power signals neither decay nor increase in their envelopes. The power signals can be associated with three broad classes of signals. These are

- *Constant* signals. An example is

$$x_4(n) = \alpha \qquad -\infty < \alpha < \infty. \qquad (1.16)$$

- *Periodic* signals, those for which $x(n) = x(n+N)$ for some finite N and for all n. Examples are

$$x_5(n) = \alpha \sin(n\omega_0 + \psi), \qquad -\infty < \alpha < \infty \qquad (1.17)$$

$$x_6(n) = [x_3(n)]_{\text{modulo } 512} = \sum_{i=-\infty}^{\infty} x_3(n + i512). \qquad (1.18)$$

- *Realizations of stationary, ergodic stochastic processes* (see Section 1.2.3).

The signals which fall into neither category are the trivial zero signal and those which "blow up" with time. Examples of the latter are $x_1(n)$ and $x_2(n)$ above with the magnitude of α taken to be greater than unity.

1.1.4 Transforms and a Few Related Concepts

At the heart of much of engineering analysis are various frequency domain transforms. Three transforms on discrete-time data will be used extensively throughout this book, and it will be assumed that the reader is familiar with their properties and usage.

The first is the *discrete-time Fourier transform* (DTFT), which, for the sequence $x(n)$, is defined by

$$X(\omega) = \sum_{n=-\infty}^{\infty} x(n)e^{-j\omega n}. \tag{1.19}$$

The *inverse DTFT* (IDTFT) is given by

$$x(n) = \frac{1}{2\pi} \int_{-\pi}^{\pi} X(\omega)e^{j\omega n}\, d\omega. \tag{1.20}$$

The DTFT bears a useful relationship to the continuous-time Fourier transform in the case in which $x(n)$ represents samples of the analog signal[7] $x_a(t')$. In this case $X(\omega)$ will be a periodic (with period 2π), potentially aliased version of $X_a(\omega)$,

$$X(\omega) = \sum_{i=-\infty}^{\infty} X_a(\omega - 2\pi i). \tag{1.21}$$

The existence of the DTFT is not a trivial subject, and we will review only a few important details. A sufficient condition for the DTFT of a sequence $x(n)$ to exist is that the sequence be *absolutely summable*,

$$\sum_{n=-\infty}^{\infty} |x(n)| < \infty. \tag{1.22}$$

This follows immediately from (1.19). Moreover, absolute summability of $x(n)$ is tantamount to *absolute convergence* of the series $\sum_{n=-\infty}^{\infty} x(n)e^{-j\omega n}$ implying that this series *converges uniformly* to a continuous function of ω (Churchill, 1960, Secs. 59 and 60). A sequence that is absolutely summable will necessarily be an energy signal, since

$$E_x = \sum_{n=-\infty}^{\infty} |x(n)|^2 \leq \left[\sum_{n=-\infty}^{\infty} |x(n)| \right]^2. \tag{1.23}$$

There are, however, energy signals that are not absolutely summable (see Problem 1.2). These energy signals will still have DTFTs, but ones whose series converge in a weaker (*mean square*) sense. This can be seen by viewing (1.19) as a conventional Fourier series for the periodic function $X(\omega)$ whose coefficients are $x(n)$. One of the properties of Fourier series is that if the energy in a single period of the function is finite, then the

[7]Note the use of "normalized time" here. If "real" time is used, (1.21) becomes

$$X(\Omega) = \frac{1}{T} \sum_{i=-\infty}^{\infty} X_a\left(\Omega - \frac{2\pi}{T} i\right).$$

series will converge in mean square (Churchill, 1963). In the present case (using the Parseval relation),

$$\int_{-\pi}^{\pi} |X(\omega)|^2 d\omega = 2\pi \sum_{n=-\infty}^{\infty} |x(n)|^2 = 2\pi E_x < \infty, \tag{1.24}$$

so the DTFT will converge in the mean square sense. Practically, this means that the DTFT sum will converge to $X(\omega)$ at all points of continuity, and at points of discontinuity it will converge to the "average" value ("halfway" between the values on either side of the discontinuity).

Properties of the DTFT are detailed in the textbooks cited in Appendix 1.A, and some are reviewed in Problem 1.3. The reader should also recall the numerous symmetry properties of the transform relation which can be useful in simplifying various computations and algorithms.

Whereas the DTFT is most useful for theoretical spectral analysis, it is not computable on a digital computer because it is a function of a continuous argument. In principle, it also works with a sequence of doubly infinite length, which also precludes any practical computation. If we restrict ourselves to the practical situation in which a sequence of finite length is being studied, then the *discrete Fourier transform* (DFT) provides a mapping between the sequence, say

$$x(n), \quad n = 0, 1, 2, \ldots, N-1 \tag{1.25}$$

and a discrete set of frequency domain samples, given by

$$X(k) = \begin{cases} \sum_{n=0}^{N-1} x(n) e^{-j(2\pi/N)kn}, & k = 0, 1, \ldots, N-1 \\ 0, & \text{other } k \end{cases} . \tag{1.26}$$

The *Inverse DFT* (IDFT) is given by

$$x(n) = \begin{cases} \frac{1}{N} \sum_{k=0}^{N-1} X(k) e^{j(2\pi/N)kn}, & n = 0, 1, \ldots, N-1 \\ 0, & \text{other } n \end{cases} . \tag{1.27}$$

The DFT represents exact samples of the DTFT of the finite sequence $x(n)$ at N equally spaced frequencies, $\omega_k = (2\pi/N)k$, for $k \in [0, N-1]$.

The *discrete Fourier series* (DFS) is closely related to the DFT *computationally*, but is quite different philosophically. The DFS is used to represent a *periodic* sequence (hence a *power* signal) with period, say N, using the set of basis functions $e^{j(2\pi/N)kn}$ for $k = 0, \ldots, N-1$. These represent the N harmonic frequencies that may be present in the signal. For a periodic sequence $y(n)$, the expansion is

$$y(n) = \sum_{k=0}^{N-1} C(k) e^{j(2\pi/N)kn}, \tag{1.28}$$

where the coefficients are computed as

$$C(k) = \frac{1}{N} \sum_{n=0}^{N-1} y(n) e^{-j(2\pi/N)kn}. \tag{1.29}$$

[In principle, the $C(k)$'s may be computed over any period of $y(n)$.]

It is occasionally convenient to use an "engineering DTFT" for a periodic signal that technically has no DTFT. The contrived DTFT composed of *analog* impulse functions at the harmonic frequencies weighted by the DFS coefficients is

$$Y(\omega) = 2\pi \sum_{k=-\infty}^{\infty} C(k) \delta_a\left(\omega - k\frac{2\pi}{N}\right). \tag{1.30}$$

Such a construction is not always palatable to a mathematician, but it works for most engineering purposes in the sense that it can be used anywhere that a DTFT is needed for $y(n)$, as long as the rules for *continuous-time* impulse functions are carefully followed. Note that this DTFT correctly asserts, for example, that $y(n)$ has infinite energy at the harmonic frequencies. Consistency with conventional Fourier transform computations is obtained by defining the *magnitude spectrum* of such a DTFT by

$$|Y(\omega)| \stackrel{\text{def}}{=} 2\pi \sum_{k=-\infty}^{\infty} |C(k)| \delta_a\left(\omega - k\frac{2\pi}{N}\right). \tag{1.31}$$

One more contrived quantity is sometimes used. The *power density spectrum* (PDS) for a periodic signal $y(n)$, say $\Gamma_y(\omega)$, is a real-valued function of frequency such that the average power in $y(n)$ on the frequency range ω_1 to ω_2 with $0 \leq \omega_1 < \omega_2 < 2\pi$ is given by

$$\begin{aligned} \text{average power in } y(n) \text{ on } \omega \in [\omega_1, \omega_2] &= \frac{1}{\pi} \int_{\omega_1}^{\omega_2} \Gamma_y(\omega)\, d\omega \\ &= 2 \sum_{k=k_1}^{k_2} |C(k)|^2, \end{aligned} \tag{1.32}$$

where k_1 and k_2 represent the integer indices of the lowest and highest harmonic of $y(n)$ in the specified range.[8] It is not difficult to show that a suitable definition is

$$\Gamma_y(\omega) \stackrel{\text{def}}{=} 2\pi \sum_{k=-\infty}^{\infty} |C(k)|^2 \delta_a\left(\omega - k\frac{2\pi}{N}\right). \tag{1.33}$$

[8] If ω_1 and hence k_1 are zero, then the lower expression in (1.32) should read

$$C(0) + 2\sum_{k=1}^{k_2} |C(k)|^2.$$

[The reader can confirm that this is consistent with (1.32).] By comparing with (1.30), it is clear why some authors choose to write $|Y(\omega)|^2$ as a notation for the PDS. The advantage of doing so is that it gives the contrived DTFT of (1.30) yet another "DTFT-like" property in the following sense: $|X(\omega)|^2$ is properly called the *energy density spectrum* for an *energy* signal $x(n)$ and can be integrated over a specified frequency range to find total energy in that range. $|Y(\omega)|^2$ is thus an analogous notation for an analogous function for a power sequence. The disadvantage is that it introduces more notation which can be easily confused with a more "valid" spectral quantity. We will therefore use only $\Gamma_y(\omega)$ to indicate the PDS of a periodic signal $y(n)$.

The similarity of the DFT to the DFS is apparent, and this similarity is consistent with our understanding that the IDFT, if used outside the range $n \in [0, N-1]$, will produce a periodic replication of the finite sequence $x(n)$. Related to this periodic nature of the DFT are the properties of "circular shift" and "circular convolution" of which the reader must beware in any application of this transform. A few of these notions are reviewed in Problem 1.4.

For interpretive purposes, it will be useful for us to note the following. Although the DTFT does not exist for a periodic signal, we might consider taking the limit[9]

$$\overline{Y}(\omega) = \lim_{N\to\infty} \frac{1}{2N+1} \sum_{n=-N}^{N} y(n)e^{-j\omega n} \tag{1.34}$$

in the hope of making the transform converge. A moment's thought will reveal that this computation is equivalent to the same sum taken over a single period, say

$$\overline{Y}(\omega) = \frac{1}{N} \sum_{n=0}^{N-1} y(n)e^{-j\omega n}. \tag{1.35}$$

We shall refer to $\overline{Y}(\omega)$, $0 \le \omega < 2\pi$, as the *complex envelope spectrum* for a periodic signal $y(n)$. The theoretical significance of the complex envelope is that it can be sampled at the harmonic frequencies to obtain the DFS coefficients for the sequence. The reader will recall that a similar phenomenon occurs in the analog domain where the FT of one period of a periodic signal can be sampled at the harmonics to obtain the FS coefficients.

Finally, with regard to Fourier techniques, recall that the *fast Fourier transform* (FFT) is a name collectively given to several classes of fast algorithms for computing the DFT. The literature on this subject is vast,

[9]The operator notation $\mathcal{L}\{\cdot\}$ will be used consistently in the text to denote a time average of this form. We will formally define the operator when we discuss averages in Section 1.2.3.

but the textbooks cited above provide a general overview of most of the fundamental treatments of the FFT. Some advanced topics are found in (Burris, 1988).

The final transform that will be used extensively in this book is the (two-sided) *z-transform* (ZT), defined by

$$X(z) = \sum_{n=-\infty}^{\infty} x(n)z^{-n}, \tag{1.36}$$

where z is any complex number for which the sum exists, that is, for which

$$\sum_{n=-\infty}^{\infty} \left|x(n)z\right|^{-n} < \infty. \tag{1.37}$$

The values of z for which the series converges comprise the *region of convergence* (ROC) for the ZT. When the series converges, it converges absolutely (Churchill, 1960, Sec. 59), implying that the ZT converges uniformly as a function of z everywhere in the ROC. Depending on the time sequence, the ROC may be the interior of a circle, the exterior of a circle, or an annulus of the form $r_{\text{in}} < |z| < r_{\text{out}}$, where r_{in} may be zero and r_{out} may be infinite. The ROC is often critical in uniquely associating a time sequence with a ZT. For details see the textbooks in Appendix 1.A.

The ZT is formally inverted by contour integration,

$$x(n) = \frac{1}{2\pi j} \oint_{\mathcal{C}} X(z)z^{n-1}\, dz, \tag{1.38}$$

where $\mathcal{C}$ is a counterclockwise contour through the ROC and encircling the origin in the z-plane, but several useful computational methods are well known, notably the *partial fraction expansion* method, and the *residue* method.

The ZT plays a similar role in DSP to that which the Laplace transform does in continuous processing. A good speech processing engineer will learn to "read the z-plane" much the same as the analog systems engineer uses the s-plane. In particular, the reader should be familiar with the correspondence between angles in the z-plane and frequencies, and between z-plane magnitudes and "damping." The interpretation of pole–zero plots in the z-plane is also an essential tool for the speech processing engineer.

Finally, we recall the relationships among the two Fourier transforms and the ZT. From the definitions, it is clear that

$${}^{\text{DTFT}}X(\omega) = {}^{\text{ZT}}X(e^{j\omega}) \tag{1.39}$$

for any ω, so that the DTFT at frequency ω is obtained by evaluating the ZT at angle ω on the unit circle in the z-plane. This is only valid, of course, when the ROC of the ZT includes the unit circle of the z-plane.[10] The periodicity with period 2π of the DTFT is consistent in this regard. Since the DFT represents samples of the DTFT at frequencies ω_k, $k = 0, 1, \ldots, N-1$, it can be obtained by evaluating the ZT at equally spaced angles around the unit circle in the z-plane. Therefore,

$$^{\mathrm{DFT}}X(k) = {}^{\mathrm{DTFT}}X\left(\omega_k = \frac{2\pi}{N}k\right) = {}^{\mathrm{ZT}}X(e^{j(2\pi/N)k}). \tag{1.40}$$

Since we use the same "uppercase," for example, X, notation to indicate all three transforms, it is occasionally necessary in DSP work to explicitly denote the particular transform with, for example, a presuperscript as in (1.40).

1.1.5 Windows and Frames

In all practical signal processing applications, it is necessary to work with *short terms* or *frames* of the signal, unless the signal is of short duration.[11] This is especially true if we are to use conventional analysis techniques on signals (such as speech) with nonstationary dynamics. In this case it is necessary to select a portion of the signal that can reasonably be assumed to be stationary.

Recall that a (time domain) *window*, say $w(n)$, is a real, finite length sequence used to select a desired frame of the original signal, say $x(n)$, by a simple multiplication process. Some of the commonly used window sequences are shown in Fig. 1.3. For consistency, we will assume windows to be *causal* sequences beginning at time $n = 0$. The duration will usually be denoted N. Most commonly used windows are symmetric about the time $(N-1)/2$ where this time may be halfway between two sample points if N is even. Recall that this means that the windows are *linear-phase* sequences [e.g., see (Proakis and Manolakis, 1992)] and therefore have DTFTs that can be written

$$W(\omega) = \left|W(\omega)\right| e^{-j\omega((N-1)/2)}, \tag{1.41}$$

where the phase term is a simple linear characteristic corresponding to the delay of the window that makes it causal.[12]

It will be our convention in this book to use windows in a certain manner to create a frame of the signal. We first reverse the window in time[13] $[w(-n)]$, then shift it so that its leading edge is at a desired time,

[10]The ROC includes the unit circle if and only if $x(n)$ is absolutely summable. Therefore, in keeping with our discussion above, only a uniformly convergent DTFT can be obtained by evaluating the corresponding ZT on the unit circle.

[11]A similar discussion applies to the design of FIR filters by truncation of a desired IIR (see DSP textbooks cited in Appendix 1.A).

[12]If the window were allowed to be centered on $n = 0$, it would have a purely real DTFT and a zero-phase characteristic.

[13]Since we assume windows to be symmetric about their midpoints, this reversal is just to initially shift the leading edge to time zero.

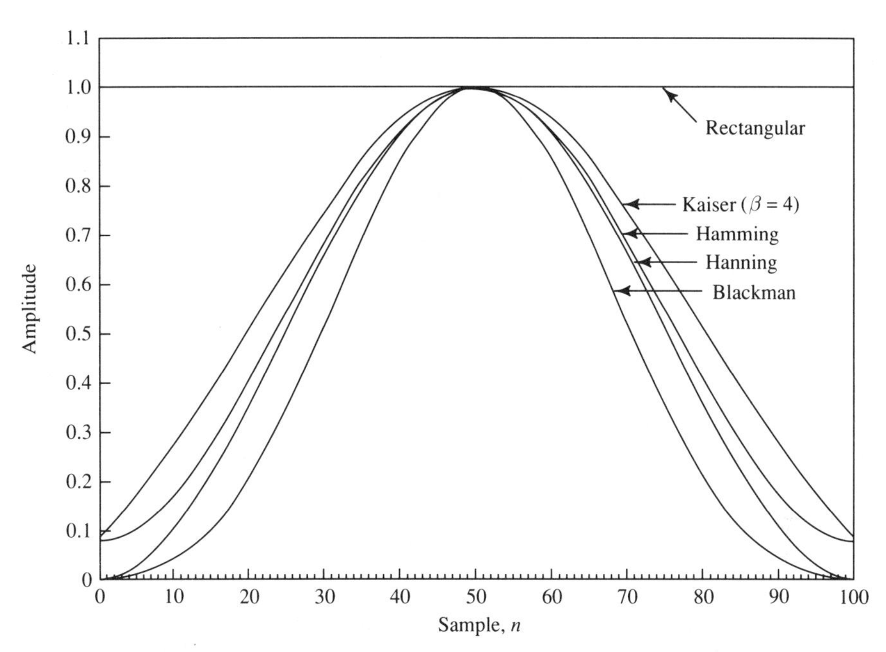

FIGURE 1.3. Definitions and example time plots for the rectangular, Kaiser, Hamming, Hanning, and Blackman windows. All plots are for window lengths $N = 101$, and for the Kaiser window, $\beta = 4$.

m $[w(m-n)]$. A *frame* of the signal $x(n)$ of length N (same as the duration of the window) ending at time m, say $f_x(n;m)$, is obtained as

$$f_x(n;m) = x(n)w(m-n). \tag{1.42}$$

This simple concept will be used extensively in future developments involving frames of speech. In fact, much of the time in this book the frame will be related to a speech sequence denoted $s(n)$ and it will be unnecessary to employ the subscript s because it will be obvious. We will only use a subscript in discussions where frames are being created from more than one signal.

Assume for the moment that $x(n)$ is a stationary signal for all time. Clearly, the temporal properties of $f_x(n;m)$ are distorted with respect to those of $x(n)$ due to the direct modification of the temporal sequence by the window. Correspondingly, the spectral properties also differ as the two transforms are apparently convolved. That is, if $F_x(\omega;m)$ denotes the DTFT of frame $f_x(n;m)$, then

$$F_x(\omega;m) = \frac{1}{2\pi}\int_{-\pi}^{\pi} X(\omega-\theta)W(-\theta)e^{-j\theta m}\,d\theta. \tag{1.43}$$

Now the relationship between $F_x(\omega;m)$ and $X(\omega)$ will only be clear from (1.43) to those who are able to visualize the process of convolving complex functions! Most of us do not have such an imagination. However, we can get some insight into the spectral distortion by assuming with some loss of generality that the reversed and shifted window is centered on time $n = 0$ $[m = (N-1)/2]$. Said another way, this simply means that the true signal transform, $X(\omega)$, against which we are going to compare our frame's transform, $F_x(\omega; m)$, is the one whose signal is assumed to have its time origin in the middle of the window. This, of course, is not always the $X(\omega)$ that represents our standard, but we can use it for insight. In this case (1.41) can be used in (1.43) to yield

$$F_x(\omega; m) = \frac{1}{2\pi} \int_{-\pi}^{\pi} X(\omega - \theta) \left| W(-\theta) \right| d\theta = \frac{1}{2\pi} \int_{-\pi}^{\pi} X(\omega - \theta) \left| W(\theta) \right| d\theta \tag{1.44}$$

where we have replaced $|W(-\theta)|$ by $|W(\theta)|$ since the magnitude spectrum is an even function of θ. In this light it seems that we want our window to have a magnitude spectrum that approximates an (analog) impulse as closely as possible,

$$|W(\theta)| \approx 2\pi \delta_a(\theta), \tag{1.45}$$

since this will imply $F_x(\omega; m) \approx X(\omega)$. When we ponder this for a moment we realize that, in the extreme case, we have concluded the obvious because $|W(\theta)| = 2\pi\delta_a(\theta)$ implies that $w(n) = 1$ for all n. The "best" window in terms of preserving the spectrum is *no* window at all! Of course such a "window" will also preserve the temporal properties of the signal perfectly as well.

For any meaningful window, however, it is the extent to which the approximation (1.45) holds which will determine the preservation of the spectral features of $X(\omega)$. Now all commonly used windows tend to have "lowpass" spectra with one main lobe at low frequencies and various attenuated "sidelobes." This is consistent with the fact that, if viewed as the (usually finite) impulse response of a filter, the window has an averaging effect. Shown in Fig. 1.4, for example, are the magnitude spectra of two commonly used windows, the *rectangular* window, defined as

$$w(n) = \begin{cases} 1, & n = 0, 1, \ldots, N-1 \\ 0, & n \text{ otherwise}, \end{cases} \tag{1.46}$$

and the *Hamming* window,

$$w(n) = \begin{cases} 0.54 - 0.46 \cos(2\pi n/N-1), & n = 0, 1, \ldots, N-1 \\ 0, & n \text{ otherwise}. \end{cases} \tag{1.47}$$

Each is plotted for the case $N = 16$. For any window spectrum to approximate $\delta_a(\omega)$, therefore, there are two desirable features:

- A narrow bandwidth main lobe.
- Large attenuation in the sidelobes.

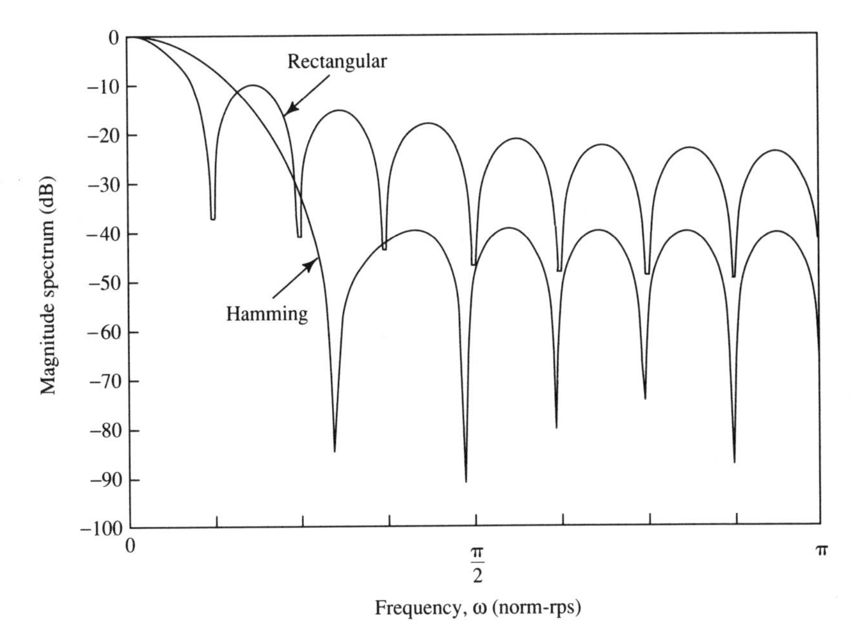

FIGURE 1.4. Magnitude spectra of rectangular and Hamming windows. Window length $N = 16$ is used in each case for clarity. Note that the nominal "bandwidth" (width of main lobe) is $2\pi/N = \pi/8$ for the rectangular case and about twice that for the Hamming. The sidelobe attenuation for the Hamming, however, is 20 dB better outside the passband.

Generally speaking, a narrow main lobe will resolve the sharp details of $|X(\omega)|$ as the convolution (1.44) proceeds, while the attenuated sidelobes will prevent "noise" from other parts of the spectrum from corrupting the true spectrum at a given frequency. As one might expect, an engineering trade-off is encountered in this regard in the choice of a window. The rectangular window, which exactly preserves the temporal characteristics of the waveform over a range of N points, but which abruptly truncates the waveform at the boundaries, has the following spectral characteristics:

- A relatively narrow main lobe (see Fig. 1.4) which decreases with N. (In fact, the width of all the lobes decreases with N, but remember that a very large N begins to defeat the purpose of windowing.)
- The height of all lobes grows with N in such a way that the attenuation in the sidelobes is approximately constant as N grows. This sidelobe attenuation is not good for the rectangular case, typically −20 dB with respect to the main lobe, allowing lots of undesirable spectral energy to be dragged into the resulting spectrum by the convolution (1.44) at a given ω.

Windows with smoother truncations, such as the *Kaiser*, *Hamming*, *Hanning*, and *Blackman* are generally used (see Fig. 1.3). These tend to distort the temporal waveform on the range of N points, but with the benefit of less abrupt truncations at the boundaries. The spectral properties of these windows are generally described as follows:

- For a given N, all will have a wider main lobe than the rectangular. Again, this width decreases with increasing N.
- All have better sidelobe attenuation than the rectangular, typically 10–60 dB better. The popular Hamming window, for example, is -30 dB down in the sidelobes (Fig. 1.4).

Although the choice of windows is somewhat of an art dependent upon experience rather than an exact science, one can use this discussion as a guide to an analytical understanding of the effects of such a choice in the processing of speech or any other signal. Generally, the choice of smoother windows is made because of their preferable sidelobe characteristics.

When analyzing a nonstationary signal like speech, the selection of a window involves another important consideration which is often not treated in introductory textbooks on digital signal processing. From the discussion above, we see that when analyzing a stationary signal, increasing the window length, N, has only beneficial consequences regardless of the type of window used. However, if a window is to be used to sequentially select portions of a nonstationary signal by "sliding" it along in time, a longer window will require a longer period to cross transitional boundaries in the signal and events from different quasi-stationary regions will tend to be blurred together more frequently than if the window were shorter. Therefore another engineering trade-off is encountered in the choice of window *length.* A longer window will tend to produce a better spectral picture of the signal while the window is completely within a stationary region, whereas a shorter window will tend to resolve events in the signal better in time. This trade-off is sometimes called the *spectral-temporal resolution trade-off* and will be discussed further in Chapter 4, where we deal with short-term processing of speech.

1.1.6 Discrete-Time Systems

Elementary Concepts

The following are elementary concepts from discrete time (DT) system theory that will be used intrinsically and extensively throughout the book. It is assumed that the reader has a thorough grounding in these ideas. We list here a number of fundamental topics that will be used without elaboration. If any are unfamiliar, the reader is advised to review them in one of the introductory textbooks indicated in Appendix 1.A.

1. Linearity.
2. Time (shift) invariance.
3. Linear, constant-coefficient difference equation (time domain input–output) description of a linear, time-invariant (LTI) DT system.

4. DT impulse response of an LTI DT system ["$h(n)$"].
5. Convolution sum for an LTI DT system.
6. Bounded input-bounded output (BIBO) stability and relationship to $h(n)$ for an LTI DT system.
7. Causality.
8. System function for an LTI DT system ["$H(z)$"], poles and zeros.
9. Magnitude spectrum and phase spectrum of an LTI DT system and their determination from a pole-zero diagram.
10. Relationship between the linear constant coefficient difference equation and $H(z)$ for an LTI DT system.
11. Relationship between BIBO stability and $H(z)$.
12. Finite impulse response (FIR) and infinite impulse response (IIR) systems and relationships to $H(z)$ and the difference equation.
13. Canonical computational structures for implementing LTI DT systems.

State-Space Realizations of LTI DT Systems

Much of contemporary DT and analog system theory is based upon *state-space* descriptions, rather than input–output descriptions, of systems. In the digital signal processing realm, state-space structures for realizing DT systems have been the subject of intense research and they have been found to have a number of useful numerical properties [e.g., see (Jackson, 1989, Sec. 11.6)]. We have two limited and very specific uses for them in our work, so we review only a few pertinent results here. The reader can refer to a number of other textbooks for further information (see Appendix 1.A).

In the proof of a key result concerning linear prediction analysis in Chapter 5, we will have need of a slight variation of a *Type I* (Proakis and Manolakis, 1992, Sec. 7.5) or *controllable canonical* (Chen, 1984, p. 327) form for a specific LTI DT system with scalar input and output. This form is derived from the input–output description of the system as follows: Consider the system to be governed by the linear constant coefficient difference equation

$$y(n) = \sum_{k=1}^{M} a(k)y(n-k) + \sum_{k=0}^{Q} b(k)x(n-k) \tag{1.48}$$

for which the *direct form II realization* is shown in Fig. 1.5. We assume in that figure and in this discussion that $Q < M$ and we define $b(k) = 0$ for $k > Q$. The (*internal*) *state* of a DT system at time n_0 is defined to be the quantitative information necessary at time n_0 which, together with the input $x(n)$ for $n \geq n_0$, uniquely determines the output $y(n)$ for $n \geq n_0$. The *state variables* of the system are the numerical quantities memorized by the system that comprise the state.

In Fig. 1.5 we have defined the internal variables $v_1(n), \ldots, v_M(n)$.

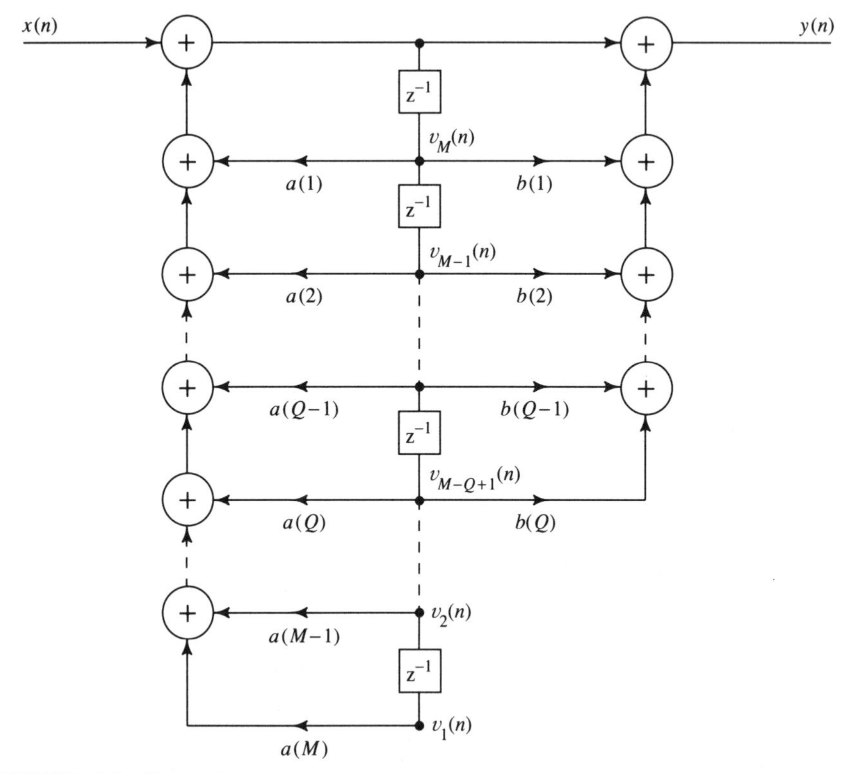

FIGURE 1.5. Direct form II realization of the discrete-time system with input–output description (1.48).

These comprise state variables for this system, as we shall see momentarily. Note that

$$v_i(n+1) = v_{i+1}(n), \qquad i = 1, 2, \ldots, M-1 \tag{1.49}$$

$$v_M(n+1) = x(n) + \sum_{i=1}^{M} a(i) v_{M-i+1}(n). \tag{1.50}$$

These are the *state equations* for the system. Note also that the output can be computed from the state variables at time n using

$$y(n) = b(0)\, v_M(n+1) + \sum_{i=1}^{M} b(i)\, v_{M-i+1}(n) \tag{1.51}$$

$$= b(0)x(n) + \sum_{i=1}^{M} [b(i) + b(0)a(i)]\, v_{M-i+1}(n),$$

which is called simply the *output equation* for the system. It is clear that these state variables do comprise a legitimate state for this system according to the definition. For convenience, the state and output equations can be written in vector-matrix form as

$$\mathbf{v}(n+1) = \mathbf{A}\mathbf{v}(n) + \mathbf{c}x(n) \tag{1.52}$$

$$y(n) = \mathbf{b}^T\mathbf{v}(n) + dx(n), \tag{1.53}$$

in which d is the scalar $d = b(0)$, $\mathbf{A}$ is the $M \times M$ *state transition matrix*

$$\mathbf{A} = \begin{bmatrix} 0 & 1 & 0 & 0 & 0 & \cdots & 0 \\ 0 & 0 & 1 & 0 & 0 & \cdots & 0 \\ 0 & 0 & 0 & 1 & 0 & \cdots & 0 \\ \vdots & \vdots & \vdots & \vdots & \vdots & \cdots & \vdots \\ 0 & 0 & 0 & 0 & 0 & \cdots & 1 \\ a(M) & a(M-1) & a(M-2) & a(M-3) & a(M-4) & \cdots & a(1) \end{bmatrix}, \tag{1.54}$$

and $\mathbf{c}$ and $\mathbf{b}$ are M-vectors [recall the assumption $Q < M$ and the definition $b(k) = 0$ for $k > Q$],

$$\mathbf{c} = \begin{bmatrix} 0 & 0 & 0 & \cdots & 0 & 1 \end{bmatrix}^T \tag{1.55}$$

$$\mathbf{b} = \begin{bmatrix} b(M) + b(0)a(M) \\ b(M-1) + b(0)a(M-1) \\ b(M-2) + b(0)a(M-2) \\ \vdots \\ b(1) + b(0)a(1) \end{bmatrix}. \tag{1.56}$$

Equations (1.52) and (1.53) are very close to the state-space description of an LTI system that will be needed in our work in a limited way. In fact, because of the way we have chosen to define the state variables here, these equations comprise a *lower companion form* state-space model, so named because of the form of the state transition matrix $\mathbf{A}$. A simple redefinition of state variables leads to the *upper companion form* model which we explore in Problem 1.5.

Finally, in our study of hidden Markov models for speech recognition in Chapter 12, we will have need of a state-space description of a system that has a *vector output*. In this case the system will naturally arise in a state-space form and there will be no need for us to undertake a conversion of an input–output description of the system. The system there will have a similar state equation to (1.52) except that the state transition matrix, **A**, will generally not be of a special form like the one above (indicating more complicated dependencies among the states). The output equation will take the form

$$\mathbf{y}(n) = \mathbf{B}\mathbf{v}(n) + \mathbf{d}\mathbf{x}(n) \tag{1.57}$$

in which $\mathbf{y}(n)$ and $\mathbf{d}$ are P-vectors (P outputs) and $\mathbf{B}$ is a $P \times M$ matrix. We will have more to say about this system when its need arises.

1.1.7 Minimum-, Maximum-, and Mixed-Phase Signals and Systems

We have discussed the grouping of signals into energy or power categories. Here we restrict our attention to the subclass of real signals with legitimate DTFTs (those that are absolutely summable) and consider another useful categorization.

The specification of the magnitude spectrum of a discrete-time signal is generally not sufficient to uniquely specify the signal, or, equivalently, the DTFT of the signal. Consider, for example, the magnitude spectrum, $|X(\omega)|$, shown in Fig. 1.6. This spectrum was actually computed for the signal $x_1(n)$ with z-transform,

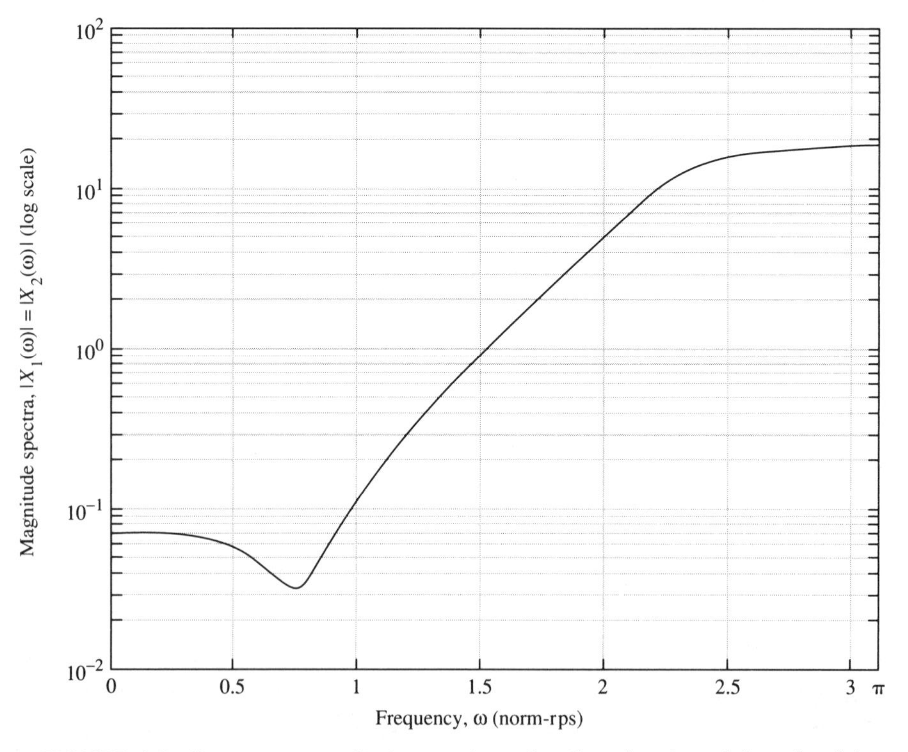

FIGURE 1.6. Common magnitude spectrum for the signals $x_1(n)$ and $x_2(n)$.

$$X_1(z) = \frac{(1-\zeta_1 z^{-1})(1-\zeta_1^* z^{-1})(1-\zeta_2 z^{-1})}{(1-\rho_1 z^{-1})(1-\rho_1^* z^{-1})(1-\rho_2 z^{-1})}, \tag{1.58}$$

with $\zeta_1 = 0.9\angle 45°$, $\zeta_2 = 0.5$, $\rho_1 = 0.7\angle 135°$, and $\rho_2 = -0.5$, and therefore has an analytical description

$$|X_1(e^{j\omega})| = \left| \frac{(1-\zeta_1 e^{-j\omega})(1-\zeta_1^* e^{-j\omega})(1-\zeta_2 e^{-j\omega})}{(1-\rho_1 e^{-j\omega})(1-\rho_1^* e^{-j\omega})(1-\rho_2 e^{-j\omega})} \right|. \tag{1.59}$$

The true phase characteristic is given by

$$\arg\{X_1(e^{j\omega})\} = \arg\left\{ \frac{(1-\zeta_1 e^{-j\omega})(1-\zeta_1^* e^{-j\omega})(1-\zeta_2 e^{-j\omega})}{(1-\rho_1 e^{-j\omega})(1-\rho_1^* e^{-j\omega})(1-\rho_2 e^{-j\omega})} \right\}. \tag{1.60}$$

The magnitude and phase spectra for the signal $x_1(n)$ are found in Figs. 1.6 and 1.7, respectively, and the pole–zero diagram is shown in Fig. 1.8.

If the magnitude spectrum were all that were known to us, however, it would not be possible to deduce this z-transform and corresponding signal with certainty. Indeed, consider the signal $x_2(n)$ with z-transform

$$X_2(z) = \frac{(z^{-1}-\zeta_1)(z^{-1}-\zeta_1^*)(z^{-1}-\zeta_2)}{(1-\rho_1 z^{-1})(1-\rho_1^* z^{-1})(1-\rho_2 z^{-1})}, \tag{1.61}$$

which the reader can confirm has an identical magnitude spectrum to $X_1(z)$ (see Fig. 1.6), but a different phase spectrum that is shown in Fig. 1.7. The pole–zero diagram for the signal $x_2(n)$ is found in Fig. 1.8. Furthermore, there are two other z-transforms which have identical magnitude spectra but different phase spectra. $X_2(z)$ is found from $X_1(z)$ by reflecting both the conjugate zero pair plus the real zero into conjugate reciprocal locations (outside the unit circle) in the z-plane, plus some scaling. The other two magnitude spectrum-equivalent z-transforms are found by reflecting *either* the conjugate pair *or* the real zero.

In general, if a real, causal, absolutely summable signal has a z-transform with C complex pairs of zeros, and R real zeros, then there are $2^{C+R}-1$ other possible signals with identical magnitude spectra but different phase spectra. The signal with all of its zeros inside the unit circle is called a *minimum-phase signal* for reasons explained below. If the signal is the discrete-time impulse response of a system, then the system is said to be a *minimum-phase system or filter.* In the other extreme in which the zeros are completely outside the unit circle, the signal (or system) is called *maximum phase.* All intermediate cases are usually called *mixed phase.*

A little thought about the general relationship between the zero configuration and the phase spectrum (i.e., think about how one deduces a phase plot from the pole–zero diagram) will convince the reader that having all the zeros inside the unit circle will minimize the absolute value of negative phase at a given ω. Conversely, having the zeros outside

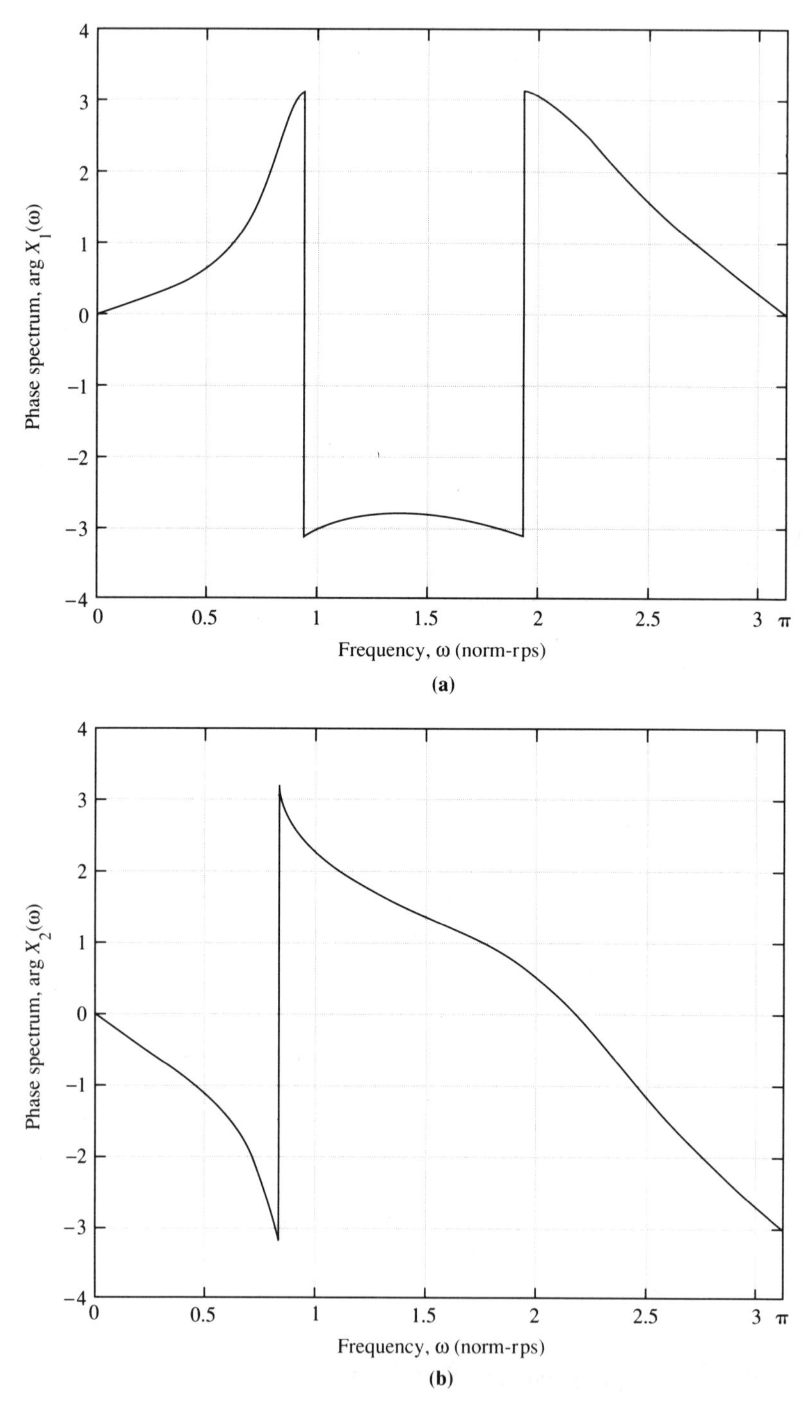

FIGURE 1.7. Phase spectra for the signals (a) $x_1(n)$ and (b) $x_2(n)$.

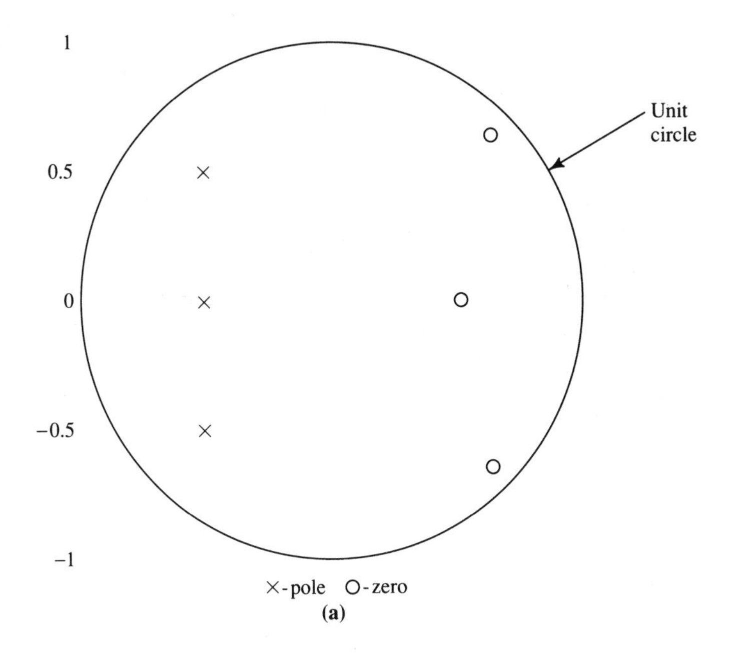

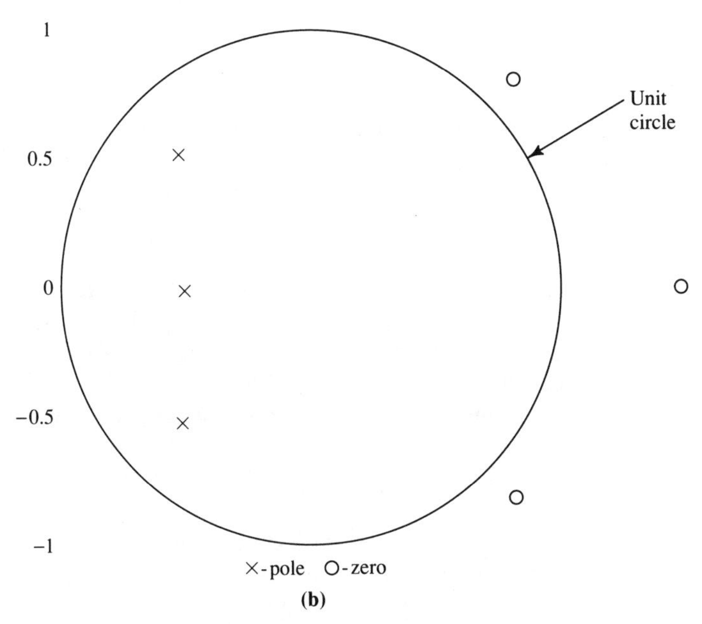

FIGURE 1.8. Pole–zero diagrams for the signals (a) $x_1(n)$ and (b) $x_2(n)$.

the unit circle will maximize the negative phase (see Figs. 1.7 and 1.8). Hence the names minimum and maximum phase are reasonable from this point of view. A more intuitive notion is to found in the time domain, however.

Since the (negative) phase at ω is directly related to the amount of temporal delay of a narrowband component at that frequency, we can infer that the minimum-phase signal is the one which, for a given magnitude spectrum, has a minimum delay of each frequency component in the spectrum. The minimum-phase signal will therefore have the highest concentration of energy near time $n = 0$ of any signal with the same magnitude spectrum. Specifically, if $x_{\min}(n)$ is the minimum-phase signal, and $E_x(m)$ represents the energy in any sequence $x(n)$ in the interval $n \in [0, m]$,

$$E_x(m) \stackrel{\text{def}}{=} \sum_{n=0}^{m} x^2(n), \tag{1.62}$$

then it will be true that[14]

$$E_{x_{\min}}(m) \geq E_x(m) \tag{1.63}$$

for any absolutely summable signal $x(n)$ with the same magnitude spectrum, and for any m. Precisely the opposite holds for the maximum-phase signal, say $x_{\max}(n)$,

$$E_{x_{\max}}(m) \leq E_x(m) \tag{1.64}$$

for any absolutely summable signal $x(n)$ with the same magnitude spectrum, and for any m. The significance of these expressions can be appreciated in Fig. 1.9, where we show the time domain waveforms for $x_1(n)$ above, which we now know is minimum phase, and for $x_2(n)$, which is maximum phase.

Yet another way to view a minimum-phase signal, particularly when it represents the impulse response of a system, is as follows: If $h(n)$ represents a minimum-phase impulse response of a causal stable system, then the z-domain system function, $H(z)$, will have all of its poles and zeros inside the unit circle. Hence there exists a causal, stable *inverse system*, $H^{-1}(z)$, such that

$$H(z)H^{-1}(z) = 1 \tag{1.65}$$

everywhere in the z-plane. If there were even one zero outside the unit circle in $H(z)$, a stable inverse would not exist, since at least one pole in the inverse would be obliged to be outside the unit circle. The existence of a causal stable inverse z-transform for $H(z)$ is therefore a sufficient condition to assure that the signal $h(n)$ (or its corresponding system) is minimum phase.

[14]A proof of this fact is outlined in Problem 5.36 of (Oppenheim and Schafer, 1989).

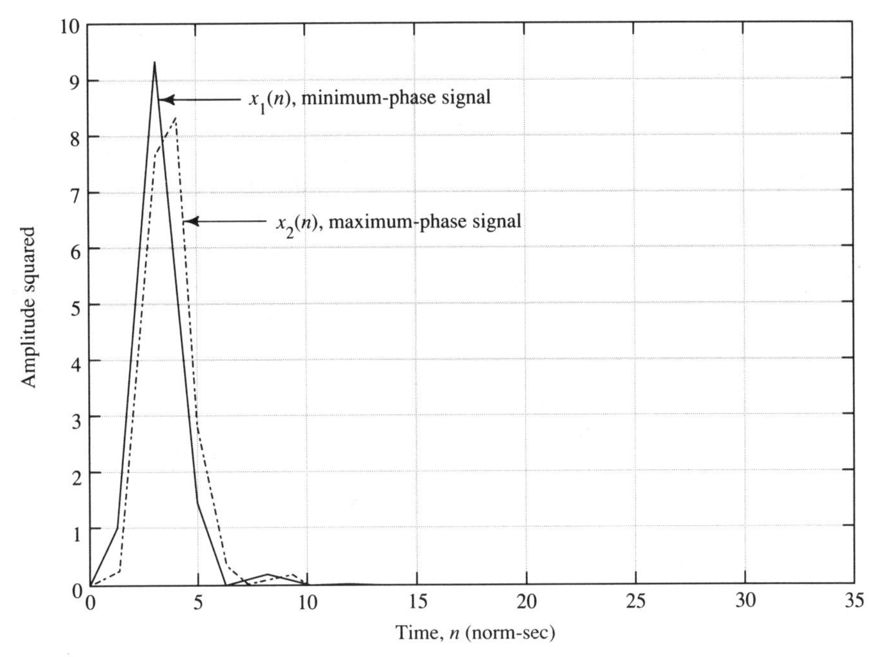

FIGURE 1.9. Time domain plots of minimum-phase signal $x_1(n)$ and maximum-phase signal $x_2(n)$. The signals are squared for convenience.

Finally, we note that we have assumed that signals in this discussion are generally infinite in duration by allowing them to have poles in their z-transforms. (By restricting our discussion to absolutely summable signals, however, we have constrained the poles to be inside the unit circle.) In the case of a real, finite duration ("all zero"), minimum-phase sequence of length N (perhaps the impulse response of an FIR filter), it can be shown that its maximum-phase counterpart is given by

$$x_{\max}(n) = x_{\min}(N-1-n) \tag{1.66}$$

or

$$X_{\max}(z) = z^{-(N-1)}X_{\min}(z^{-1}). \tag{1.67}$$

The concepts of minimum-phase signals and systems will play a key role in the theory of linear prediction and surrounding modeling concepts.

1.2 Review of Probability and Stochastic Processes

We will discover in the next chapter that there are two basic classes of speech sounds, "voiced" and "unvoiced." Generally speaking, the former is characterized by deterministic acoustic waveforms, while the latter corresponds to stochastic waveforms. The difference can be heard in the two sounds present in the word "it," for example. Although random process theory will be necessary to analyze unvoiced signals, we will find that even in the case of voiced sounds it will be very useful to employ analytical techniques which are fundamentally motivated by stochastic process theory, notably the autocorrelation function. In different ways from those used to analyze speech waveforms, we will employ concepts from probability in our study of stochastic models for the coding and recognition of speech. In these and other aspects of our study of speech processing, basic concepts from random process theory will be prerequisite to our pursuits.

As is the case with digital signal processing concepts, it will be necessary for the reader to have a working knowledge of the concepts of probability and stochastic processes, at least at the level of a typical senior or entry-level graduate course. Some of the widely used books in the field are listed in Appendix 1.B, and the reader is encouraged to refer to these textbooks to review concepts as needed.

As noted, one of the central tools of speech processing is the autocorrelation sequence. Several of the more fundamental concepts, in particular stationarity and ergodicity, also play key roles in our work. In the recognition domain, an understanding of basic concepts concerning joint experiments and statistical independence will be essential. It is our purpose here to briefly review these fundamental notions with the autocorrelation sequence and surrounding ideas as a target of this discussion. We will focus on discrete time random processes because of the nature of our application. As was the case in our DSP review, a second objective is to set forth notation for the remainder of the book. This short section is not intended to substitute for a solid course in random processes and will not provide an adequate background for a deep understanding of the stochastic aspects of speech or general signal processing.

1.2.1 Probability Spaces

The science of probability is customarily introduced to engineering students using an axiomatic approach for the sake of mathematical generality and formality. In this context, a formal definition of probability involves the specification of a *sample space*, a *field* or *algebra* of events, and a *probability measure*, which is assumed to conform to some basic axioms. The sample space, say $\mathcal{S}$, is the set of all outcomes of an experiment, plus the null outcome. Each element of $\mathcal{S}$ is called a *sample point*. Collections of sample points (connected by an OR condition) are called *events*. An event may consist of a single sample point.

Although the second component of the probability space is critical to theoretical developments, it is usually of least concern in typical engineering applications. Generally, it is necessary to give some careful thought to which events are to be assigned probabilities. In certain cases, we cannot assign probabilities all possible events, nor can we have too few events, and still have a consistent and meaningful theory of probability. A proper "event space" will turn out to be a *sigma-field* or *sigma-algebra* over $\mathcal{S}$, which is a set of subsets of $\mathcal{S}$ that is closed under complementation, union, and (if $\mathcal{S}$ has an infinite number of elements) countable union. Let us call the algebra $\mathcal{A}$. In typical engineering problems, the algebra of events is often all intervals in some continuum of possible outcomes, or the "power set" of discrete outcomes if $\mathcal{S}$ is finite and discrete. These and other algebras in different situations are naturally used in problems without much forethought.

The third component, probability, is a normalized measure assigned to these "well thought out" sets of events that adheres to four basic axioms. If $P(A)$ denotes the probability of event A, these are

1. $P(\mathcal{S}) = 1$.
2. $P(A) \geq 0, \quad$ for all $A \in \mathcal{A}$.
3. For two *mutually exclusive* events $A, B \in \mathcal{A}$,

$$P(A \cup B) = P(A) + P(B). \tag{1.68}$$

   Mutually exclusive means $A \cap B = \emptyset$, where $\emptyset$ is the null event.
4. For a *countably infinite* set of mutually exclusive events $A_i \in \mathcal{A}$, $i = 1, 2, \ldots$,

$$P\left(\bigcup_{i=1}^{\infty} A_i\right) = \sum_{i=1}^{\infty} P(A_i). \tag{1.69}$$

The first three axioms are very intuitive and reasonable, and indeed are all that are necessary when $\mathcal{A}$ contains a finite set of events. The fourth axiom is necessary for proving certain important convergence results when the sample space is infinite (see the textbooks in Appendix 1.B not labeled "elementary"). The probability measure assigned to an event is usually consistent with the intuitive notion of the relative frequency of occurrence of that event.

The three components of a *probability space* are sufficient to derive and define virtually all important concepts and results in probability theory. Notably, the concepts of statistical independence, and joint and conditional probability, follow from this basic formalism. The two events $A, B \in \mathcal{A}$ are said to be *statistically independent* if[15]

$$P(A \cap B) = P(A)P(B). \tag{1.70}$$

The *joint probability* of events $A, B \in \mathcal{A}$ is defined simply as $P(A \cap B)$, and the *conditional probability* of B given A is

$$P(B|A) \stackrel{\text{def}}{=} \frac{P(B \cap A)}{P(A)}. \tag{1.71}$$

[15]To conserve space in complicated expressions, later in the book we will begin to write $P(A \cap B)$ as $P(A, B)$. That is, the "AND" condition between events will be denoted by a comma. In these introductory sections, however, we use the explicit "$\cap$."

Combined Experiments

This is a good place to review the notion of a *combined experiment.* We will have need of this theory only if such experiments have independent events, so we will restrict our discussion accordingly. Formally, an "experiment" is equivalent to the probability space used to treat that experiment. For example, let experiment 1, $\mathcal{E}_1$, be associated with a probability space as follows:

$$\mathcal{E}_1 = (\mathcal{S}_1, \mathcal{A}_1, P). \tag{1.72}$$

If we wish to combine a second experiment, $\mathcal{E}_2$, we need to have a way of assigning probabilities to combined events. An example will be useful to illustrate the points.

Let $\mathcal{E}_1$ be concerned with a measurement on a speech waveform at a specified time that may take a continuum of values between 0 and 10 volts. Therefore,

$$\mathcal{S}_1 = \{x : 0 \le x \le 10\}. \tag{1.73}$$

The events to which we will assign probabilities consist of all open and closed intervals on this range. Therefore,

$$\mathcal{A}_1 = \{x : x \in (a, b) \text{ or } (a, b] \text{ or } [a, b) \text{ or } [a, b], \text{ where } 0 \le a \le b \le 10\}. \tag{1.74}$$

A second experiment, $\mathcal{E}_2$, consists of a second measurement at a later time, which is assumed to be independent of the first. The voltage in this case ranges from -30 to $+30$ volts, so

$$\mathcal{S}_2 = \{y : -30 \le y \le 30\} \tag{1.75}$$

and $\mathcal{A}_2$ will again consist of open and closed intervals in $\mathcal{S}_2$,

$$\begin{aligned}\mathcal{A}_2 = \{y : y \in (a, b) \text{ or } (a, b] \text{ or } [a, b) \text{ or } [a, b], \\ \text{where } -30 \le a \le b \le 30\}.\end{aligned} \tag{1.76}$$

Now suppose that we want to assign probabilities to joint events such as

$$\begin{aligned}C &= (\text{Event } A \text{ from } \mathcal{E}_1 \cap \text{Event } B \text{ from } \mathcal{E}_2) \\ &= (1 \le x < 5 \cap 15 < y < 25).\end{aligned} \tag{1.77}$$

In this case we simply form a combined experiment or combined probability space that involves a *product sample space* and *product algebra of events,*

$$\mathcal{E} = \mathcal{E}_1 \times \mathcal{E}_2 = (\mathcal{S}, \mathcal{A}, P) = (\mathcal{S}_1 \times \mathcal{S}_2, \mathcal{A}_1 \times \mathcal{A}_2, P). \tag{1.78}$$

This is illustrated in Fig. 1.10. Formally, the event $C \in \mathcal{A}$ is formed by intersecting events $A \times \mathcal{S}_2$ (also in $\mathcal{A}$) with $B \times \mathcal{S}_1$ (also in $\mathcal{A}$) to get

$$C = (A \times \mathcal{S}_2) \cap (B \times \mathcal{S}_1). \tag{1.79}$$

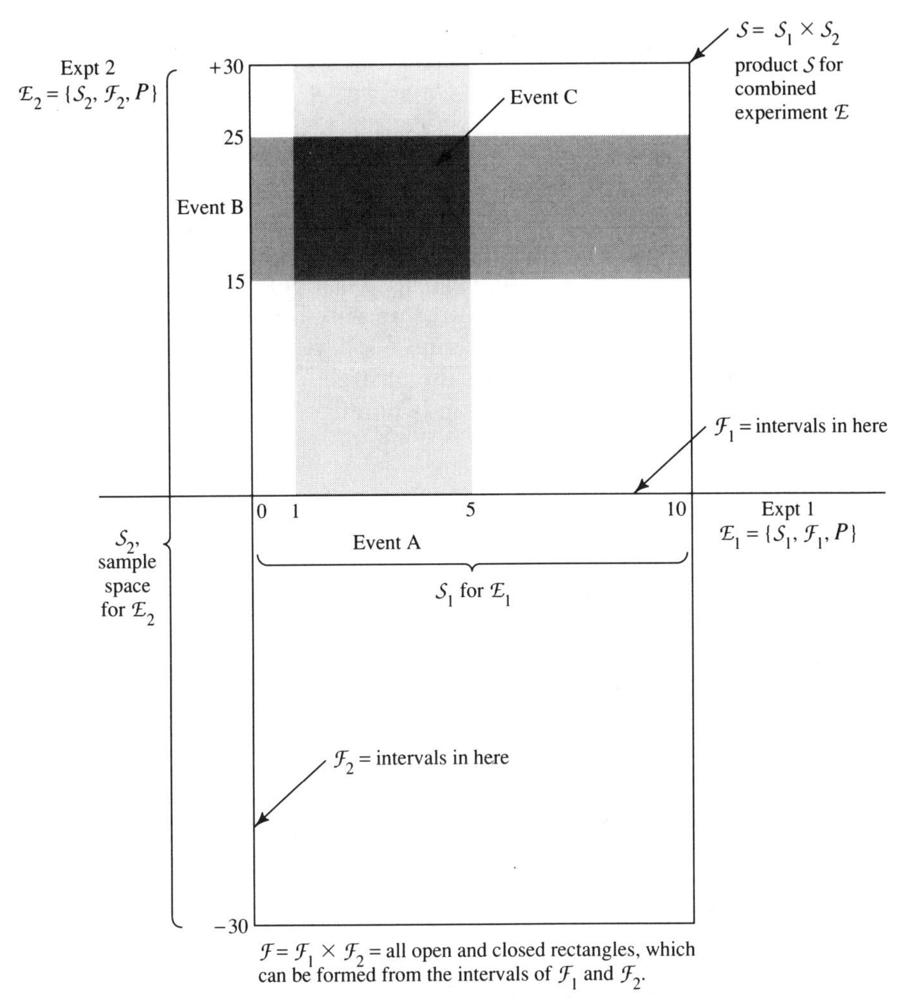

FIGURE 1.10. Combined probability space.

The probability assigned to C will be

$$P(C) = P(A)P(B), \tag{1.80}$$

since we are assuming A and B to be independent. These ideas are easily extended to more than two experiments (see textbooks in Appendix 1.B).

1.2.2 Random Variables

Single Random Variables

Note: *We henceforth use a simple comma to indicate the* AND *condition between two events in the argument of a probability. For example,* $P(A \cap B)$ will be written $P(A, B)$.

Definition of a Random Variable. A (real) *random variable* is the mapping of the sample points in a sample space of an experiment to the real number line. For example, the sample space, $\mathcal{S}$, might be the set of all cities on Earth, and the random variable, say $\underline{x}$, a mapping of the city to its metropolitan population in millions in 1990:

$$\underline{x}(\text{Chicago}) = 6.1. \tag{1.81}$$

Throughout the book, we will follow the convention established here of underscoring a quantity to distinguish it as a random variable (or random vector as discussed below).[16] Later we will use a similar notation to indicate a random *process*. In this context, a lowercase letter which is not underscored is used to indicate, in the abstract, values of the mapping. For example, if the variable σ represents points in $\mathcal{S}$, then we might write something like

$$\underline{x}(\sigma) = x \tag{1.82}$$

to indicate that the random variable $\underline{x}$ maps the outcome σ to the real value x. Note that it is the mapping itself which is the random variable, not the value of the mapping. Nevertheless, we often say "the random variable is 4," when we mean "the random variable has produced a value 4."

It is customary to employ some notation like $\mathcal{S}_{\underline{x}}$ to indicate the *range space* of $\underline{x}$, the intervals and/or points on the real line which constitute the range of the mapping $\underline{x}$. It is also formally convenient to consider an algebra of events in $\mathcal{S}_{\underline{x}}$, say $\mathcal{A}_{\underline{x}}$, to which we might want to assign probabilities.

There are certain conditions that must be met for $\underline{x}$ to be a random variable. Each is ultimately concerned with the concept of *measurability*, a property which allows us to assign probabilities to events in $\mathcal{A}_{\underline{x}}$. Generally speaking, every event in $\mathcal{A}_{\underline{x}}$ must be traceable back to a well-defined event in the original $\mathcal{A}$ probability space so that we know what probability to assign to it. One important criterion is that $\underline{x}$ not be a one-to-many mapping. (A deeper discussion of this issue is found in the textbooks of Appendix 1.B not labeled "elementary.")

The random variable in some engineering problems is only an abstract formality in the sense that the original outcomes of experiments are already real numbers and no mapping is actually necessary. Such will often be the case in this book where the "outcomes of experiments" will be values of a speech sequence at a given point in time. Accordingly, we will encounter no problems with measurability.

A random variable can be either *continuous, discrete*, or *mixed*, referring to whether the mapping produces continua of outcomes, discrete points, or a mixture of the two types. In our speech analysis work, the

[16]Some textbooks use uppercase letters to indicate random variables; still others use boldface. Uppercase letters will have many other significances in this book and boldface quantities are used to indicate vectors or matrices.

random variables will usually represent speech amplitudes; we will primarily work with the continuous case. (Remember that we usually ignore amplitude quantization in this book! The important exception will be in Chapter 7, where we actually consider the issue of quantizing speech for compression purposes.) Later in our recognition work, the random variables will be found to be primarily discrete.

Denoting Probabilities. The outcomes of random variable $\underline{x}$ (elements of $\mathcal{A}_{\underline{x}}$) are formally assigned probabilities using traceback to the original event space $\mathcal{A}$. The "events" in $\mathcal{A}_{\underline{x}}$ with which we will be concerned in this book are points $\underline{x} = x$, and intervals that may be open or closed at either end, for example, $a \leq \underline{x} < b$ or $\underline{x} > 6$. Probabilities of these events will be denoted in the expected way, such as $P(a \leq \underline{x} < b)$ or $P(\underline{x} > 6)$. The generalization to multiple random variables is obvious, for example, $P(\underline{x} \leq \alpha, \beta < \underline{y} \leq \gamma)$. A nuance occurs when describing the probabilities of *point* outcomes. When it is obvious which random variable(s) is (are) involved, we may write $P(x)$ instead of $P(\underline{x} = x)$, or $P(x, y)$ instead of $P(\underline{x} = x, \underline{y} = y)$. Frequently, for absolute clarity, we will retain the random variable in the argument. If the reader finds this unnecessary in certain cases, then he or she should simply use the abbreviated form in any notes or solutions.

As noted above, when a random variable can take only a countable number of values, it is called a *discrete* random variable.[17] In this case the statistical description of, say $\underline{x}$, is modeled entirely by its *probability distribution* $P(\underline{x} = x_i)$, $i = 1, 2, \ldots$. Occasionally, we will want to refer to the probability distribution of $\underline{x}$ in general, and we will write simply[18] $P(\underline{x})$.

cdf and pdf. Associated with a random variable, $\underline{x}$, is a *cumulative distribution function* (cdf), say $F_{\underline{x}}(x)$, defined as

$$F_{\underline{x}}(x) \stackrel{\text{def}}{=} P(\underline{x} \leq x), \tag{1.83}$$

where $P(\underline{x} \leq x)$ means the probability that the random variable $\underline{x}$ produces a value less than or equal to x. Of more use to us is the *probability density function* (pdf),

$$f_{\underline{x}}(x) \stackrel{\text{def}}{=} \frac{d}{dx} F_{\underline{x}}(x). \tag{1.84}$$

We use the derivative in the "engineering" sense in which discontinuities in $F_{\underline{x}}(x)$ (caused by discrete points with nonzero probability) produce impulses in $f_{\underline{x}}(x)$. The continuous part of $F_{\underline{x}}(x)$ might not be differentiable in certain cases which will not concern us [e.g., see (Wong and Hajek,

[17]Carefully note that this term has nothing whatsoever to do with discrete *time*.

[18]A notation which is more consistent with $f_{\underline{x}}(x)$ would be $P_{\underline{x}}(x)$, but this has other obvious disadvantages. For example, how would we denote the probability of the event $\underline{x} \geq x$?

1984, Ch. 1) for details]. Note that a discrete random variable will have a pdf that will consist entirely of impulses at the point outcomes of the random variable. The weighting on the impulse at x_i is $P(\underline{x}=x_i)$.

Returning to (1.84), from the Fundamental Theorem of Calculus, we have

$$P(a<\underline{x}\leq b)=F_{\underline{x}}(b)-F_{\underline{x}}(a)=\int_a^b f_{\underline{x}}(\xi)\,d\xi, \tag{1.85}$$

implying the well-known result that the area under the pdf on the range $(a, b]$ yields the probability that $\underline{x}$ produces a value in that interval.[19]

Some of the commonly used pdf's in speech processing are

1. *Gaussian*:

$$f_{\underline{x}}(x)=\frac{1}{\sqrt{2\pi\sigma_{\underline{x}}^2}}\exp\left\{-\frac{(x-\mu_{\underline{x}})^2}{2\sigma_{\underline{x}}^2}\right\}, \tag{1.86}$$

where $\mu_{\underline{x}}$ is the *average* or *mean* of $\underline{x}$, and $\sigma_{\underline{x}}^2$ is the *variance*, or $\sigma_{\underline{x}}$ is the *standard deviation* (discussed below).

2. *Uniform*:

$$f_{\underline{x}}(x)=\begin{cases}\dfrac{1}{b-a}, & a\leq x<b\\ 0, & \text{otherwise}\end{cases} \tag{1.87}$$

for some $b>a$.

3. *Laplacian*:

$$f_{\underline{x}}(x)=\frac{1}{\sqrt{2\sigma_{\underline{x}}^2}}\exp\left\{-\frac{\sqrt{2}\,|x|}{\sigma_{\underline{x}}}\right\}, \tag{1.88}$$

where $\sigma_{\underline{x}}$ is the standard deviation of $\underline{x}$.

Finally, let us recall the meaning of the *conditional cdf* and *conditional pdf*, which are just natural extensions of the theory,

$$F_{\underline{x}}(x|D)\stackrel{\text{def}}{=}P(\underline{x}\leq x|D)=\frac{P(\underline{x}\leq x, D)}{P(D)} \tag{1.89}$$

and

$$f_{\underline{x}}(x|D)\stackrel{\text{def}}{=}\frac{d}{dx}F_{\underline{x}}(x|D), \tag{1.90}$$

where D is any outcome (event or point) of $\underline{x}$ of nonzero probability.

[19]Care must be taken with impulse functions at the limits of integration if they exist.

Multiple Random Variables

Preliminaries. We are gradually building toward a review of random processes. The next step is to consider relationships among several random variables. We begin by considering relationships between two random variables, noting that many of the concepts we review here have natural generalizations to more than two random variables. At the end of the section, we focus on random *vectors* in which some of these generalizations will arise.

In combining experiments above, we encountered the task of combining two sample spaces at the fundamental level. We assumed that the events in the individual sample spaces were independent. Here we implicitly assume that two random variables, say $\underline{x}$ and $\underline{y}$, map the same $\mathcal{S}$ into two different range spaces, $\mathcal{S}_{\underline{x}}$ and $\mathcal{S}_{\underline{y}}$. The joint range space is simply a product space,

$$\mathcal{S}_{\underline{xy}} = \mathcal{S}_{\underline{x}} \times \mathcal{S}_{\underline{y}}, \tag{1.91}$$

formed in a similar manner to product sample spaces for combined experiments. The joint event algebra, say $\mathcal{A}_{\underline{xy}}$, are events chosen from $\mathcal{S}_{\underline{xy}}$. For most purposes, these will be open and closed rectangles and points in $\mathcal{S}_{\underline{xy}}$. A significant difference between this theory and that of combined experiments is that we do not assume that events in the individual range spaces, $\mathcal{S}_{\underline{x}}$ and $\mathcal{S}_{\underline{y}}$, are independent. We formally assign probabilities to events in $\mathcal{A}_{\underline{xy}}$ by tracing them back to $\mathcal{A}$ to see what event they represent there.

These ideas are readily extended to more than two random variables.

Joint cdf and pdf; Conditional Probability. The *joint cdf* and *joint pdf* are defined formally as

$$F_{\underline{xy}}(x, y) \stackrel{\text{def}}{=} P(\underline{x} \leq x, \underline{y} \leq y) \tag{1.92}$$

and

$$f_{\underline{xy}}(x, y) = \frac{\partial^2}{\partial x \partial y} F_{\underline{xy}}(x, y), \tag{1.93}$$

respectively. Some properties of the these functions are studied in the problems at the end of the chapter. A prevalent joint pdf in engineering is the *joint Gaussian*,

$$f_{\underline{xy}}(x, y) = \frac{1}{2\pi\sigma_{\underline{x}}\sigma_{\underline{y}}\sqrt{1-\rho_{\underline{xy}}^2}} \exp\left\{-\frac{1}{2}Q(x, y)\right\}, \tag{1.94}$$

where

$$Q(x, y) = \frac{1}{1-\rho_{\underline{x}\underline{y}}^2} \left\{ \left(\frac{x-\mu_{\underline{x}}}{\sigma_{\underline{y}}}\right)^2 - 2\rho_{\underline{x}\underline{y}}\left(\frac{x-\mu_{\underline{x}}}{\sigma_{\underline{x}}}\right)\left(\frac{y-\mu_{\underline{y}}}{\sigma_{\underline{y}}}\right) + \left(\frac{y-\mu_{\underline{y}}}{\sigma_{\underline{y}}}\right)^2 \right\}, \tag{1.95}$$

in which $\mu_{\underline{x}}$ and $\mu_{\underline{y}}$ are the means of $\underline{x}$ and $\underline{y}$, $\sigma_{\underline{x}}$ and $\sigma_{\underline{y}}$ are the standard deviations, and $\rho_{\underline{x}\underline{y}}$ is the correlation coefficient. These quantities are special moments, which are reviewed below.

The *conditional probability* of event $A \in \mathcal{A}_{\underline{x}}$ given the occurrence of event $B \in \mathcal{A}_{\underline{y}}$ is defined in the usual way,

$$P(A \in \mathcal{A}_{\underline{x}} | B \in \mathcal{A}_{\underline{y}}) \stackrel{\text{def}}{=} \frac{P(A, B)}{P(B)}, \tag{1.96}$$

where the numerator and denominator are determined by traceback to $\mathcal{A}$. If A is the interval $\underline{x} \le x$, then we have the *conditional cdf*,

$$F_{\underline{x}|\underline{y}}(x|B) \stackrel{\text{def}}{=} P([\underline{x} \le x] \in \mathcal{A}_{\underline{x}} | B \in \mathcal{S}_{\underline{y}}) = \frac{P(\underline{x} \le x, B)}{P(B)} \tag{1.97}$$

and the *conditional pdf*,

$$f_{\underline{x}|\underline{y}}(x|B) \stackrel{\text{def}}{=} \frac{d}{dx} F_{\underline{x}}(x|B). \tag{1.98}$$

All the usual relationships between the cdf and pdf hold with the conditioning information added. For example,

$$F_{\underline{x}|\underline{y}}(x_2|B) - F_{\underline{x}|\underline{y}}(x_1|B) = \int_{x_1}^{x_2} f_{\underline{x}|\underline{y}}(\xi|B)\, d\xi. \tag{1.99}$$

Independence. Two random variables, $\underline{x}$ and $\underline{y}$, are *statistically independent* if and only if for any two events, $A \in \mathcal{A}_{\underline{x}}$ and $B \in \mathcal{A}_{\underline{y}}$,

$$P(A, B) = P(A)P(B). \tag{1.100}$$

It follows immediately for two statistically independent random variables that

$$F_{\underline{x}\underline{y}}(x, y) = F_{\underline{x}}(x)F_{\underline{y}}(y) \tag{1.101}$$

and

$$f_{\underline{x}\underline{y}}(x, y) = f_{\underline{x}}(x)f_{\underline{y}}(y). \tag{1.102}$$

Statistical independence is a very strong condition. It says that outcomes of $\underline{x}$ and $\underline{y}$ tend not to be related in *any* functional way, linear or nonlinear. When two random variables are related linearly, we say that they are *correlated.* To say that $\underline{x}$ and $\underline{y}$ are uncorrelated is to say that there is no *linear* dependence between them. This does not say that they are necessarily statistically independent, for there can still be nonlinear dependence between them. We will see this issue in the topic of vector quantization (Section 7.2.2), where there will be an effort made to extract not only linear dependency (correlation) out of the speech data, but also nonlinear dependency, to produce efficient coding procedures.

Expectation and Moments. The *statistical expectation* or *statistical average* of a scalar function of a random variable, say $g(\underline{x})$, is defined by

$$\mathcal{E}\{g(\underline{x})\} \stackrel{\text{def}}{=} \int_{-\infty}^{\infty} g(x) f_{\underline{x}}(x)\, dx \tag{1.103}$$

assuming that the pdf exists. When $g(\underline{x}) = \underline{x}$, this produces the *average* or *mean value* of $\underline{x}$, $\mu_{\underline{x}}$. Note that when $\underline{x}$ produces only discrete values, say $x_1, x_2, \ldots$, then the pdf consists of impulses and the definition produces

$$\mathcal{E}\{g(\underline{x})\} = \sum_{i=1}^{\infty} x_i P(\underline{x} = x_i). \tag{1.104}$$

The definition is readily generalized to functions of two or more random variables. For example,

$$\mathcal{E}\{g(\underline{x}, \underline{y})\} \stackrel{\text{def}}{=} \int_{-\infty}^{\infty} \int_{-\infty}^{\infty} g(x, y) f_{\underline{xy}}(x, y)\, dx\, dy. \tag{1.105}$$

Particularly useful averages are the *moments* of a random variable. The ith *moment* of the random variable $\underline{x}$ is the number

$$\mathcal{E}\{\underline{x}^i\} = \int_{-\infty}^{\infty} x^i f_{\underline{x}}(x)\, dx. \tag{1.106}$$

Obviously, the first moment is the mean of $\underline{x}$, $\mu_{\underline{x}}$. The ith *central moment* of the random variable $\underline{x}$ is the number

$$\mathcal{E}\{(\underline{x} - \mu_{\underline{x}})^i\} = \int_{-\infty}^{\infty} (x - \mu_{\underline{x}})^i f_{\underline{x}}(x)\, dx. \tag{1.107}$$

A special central moment is the second one ($i = 2$), which we call the *variance* and denote $\sigma_{\underline{x}}^2$. The square root of the variance, $\sigma_{\underline{x}}$, is called the *standard deviation* of $\underline{x}$.

The i, k *joint moment* between random variables $\underline{x}$ and $\underline{y}$ is the number

$$\mathcal{E}\{\underline{x}^i \underline{y}^k\} = \int_{-\infty}^{\infty} \int_{-\infty}^{\infty} x^i y^k f_{\underline{xy}}(x, y)\, dx\, dy \tag{1.108}$$

and the i, k *joint central moment* is the number

$$\mathcal{E}\{(\underline{x}-\mu_{\underline{x}})^i(\underline{y}-\mu_{\underline{y}})^k\} = \int_{-\infty}^{\infty}\int_{-\infty}^{\infty}(x-\mu_{\underline{x}})^i(y-\mu_{\underline{y}})^k f_{\underline{xy}}(x,y)\,dx\,dy. \tag{1.109}$$

When $i = k = 1$, the joint moment is called the *correlation* between $\underline{x}$ and $\underline{y}$, and the joint central moment, the *covariance*. Let us call these numbers $r_{\underline{xy}}$ and $c_{\underline{xy}}$, respectively. A parameter frequently used in the statistical analysis of data (and which appears in the joint Gaussian pdf above) is the *correlation coefficient* given by

$$\rho_{\underline{xy}} = \frac{c_{\underline{xy}}}{\sigma_{\underline{x}}\sigma_{\underline{y}}}. \tag{1.110}$$

We see that the correlation coefficient is the covariance between $\underline{x}$ and $\underline{y}$ normalized to the product of the individual standard deviations.

Correlation and covariance will occur repeatedly in our study of speech, and it is advisable to master their meanings if they are not already very familiar. This is especially true because the terms "autocorrelation" and "covariance" are used in ways that are not consistent with their definitions in some aspects of speech processing. A related pair of somewhat unfortunate terms[20] is the following: $\underline{x}$ and $\underline{y}$ are said to be *orthogonal* if their correlation is zero, and *uncorrelated* if their covariance is zero. Finally, we note that the covariance and correlation are related as

$$c_{\underline{xy}} = r_{\underline{xy}} - \mu_{\underline{x}}\mu_{\underline{y}}. \tag{1.111}$$

The *conditional expectation* of $\underline{y}$, given some event related to random variable $\underline{x}$, say $B \in \mathcal{A}_{\underline{x}}$, is defined as

$$\mathcal{E}\{\underline{y}|B\} \stackrel{\text{def}}{=} \int_{-\infty}^{\infty} y f_{\underline{y}|\underline{x}}(y|B)\,dy. \tag{1.112}$$

It is well known that the best predictor of $\underline{y}$, in the sense of least square error, given some event concerning $\underline{x}$ is given by the conditional expectation. If $\underline{x}$ and $\underline{y}$ are also joint Gaussian, then the conditional expectation also provides the *linear* least square error predictor (see textbooks in Appendix 1.B).

Random Vectors. In discussing more than one random variable at a time, say $\underline{x}_1, \underline{x}_2, \ldots, \underline{x}_N$, it is frequently convenient to package them into a *random vector*,

$$\underline{\mathbf{x}} \stackrel{\text{def}}{=} [\underline{x}_1\underline{x}_2 \cdots \underline{x}_N]^T. \tag{1.113}$$

[20]Speech processing engineers are not responsible for this terminology!

Note that the vector is indicated by a boldface quantity, and the fact that it is a *random* vector is indicated by the line beneath it. The pdf associated with a random vector is very simply the joint pdf among its component random variables,

$$f_{\underline{\mathbf{x}}}(x_1, x_2, \ldots, x_N) \stackrel{\text{def}}{=} f_{\underline{x}_1,\underline{x}_2,\ldots,\underline{x}_N}(x_1, x_2, \ldots, x_N). \tag{1.114}$$

Operations among random vectors follow the usual rules of matrix arithmetic. For example, the operations of inner and outer products of random vectors will be significant in our work. Recall that the *inner product* or l_2 *norm* of a vector (in this case a *random* vector, say $\underline{\mathbf{x}} = [\underline{x}_1 \cdots \underline{x}_N]^T$), is the sum of its squared components. This can be written in a variety of ways,

$$\|\underline{\mathbf{x}}\|^2 = \underline{\mathbf{x}}^T\underline{\mathbf{x}} = \sum_{i=1}^{N} \underline{x}_i^2. \tag{1.115}$$

Note that the inner product of a random vector is itself a *random variable.* The *outer product*, on the other hand, is the product $\underline{\mathbf{x}}\underline{\mathbf{x}}^T$ which creates a *random matrix* whose (i,j) element is the random variable $\underline{x}_i\underline{x}_j$. Of course, the inner and outer products may be computed between two different random vectors.

The *expectation* of a random vector (or matrix) is just the vector (or matrix) of expectations of the individual elements. For example, $\mathcal{E}\{\underline{\mathbf{x}}\}$ is simply the vector of means $[\mu_{\underline{x}_1} \cdots \mu_{\underline{x}_N}]^T$, which we might denote $\boldsymbol{\mu}_{\underline{\mathbf{x}}}$. Another important example is the expectation of the outer product,

$$\mathbf{R}_{\underline{\mathbf{x}}} \stackrel{\text{def}}{=} \mathcal{E}\{\underline{\mathbf{x}}\underline{\mathbf{x}}^T\}, \tag{1.116}$$

which is called the *autocorrelation matrix* for the random vector $\underline{\mathbf{x}}$, since its (i,j) element is the correlation between random variables $\underline{x}_i$ and $\underline{x}_j$. The matrix

$$\mathbf{C}_{\underline{\mathbf{x}}} \stackrel{\text{def}}{=} \mathcal{E}\{(\underline{\mathbf{x}} - \boldsymbol{\mu}_{\underline{\mathbf{x}}})(\underline{\mathbf{x}} - \boldsymbol{\mu}_{\underline{\mathbf{x}}})^T\} \tag{1.117}$$

is called the *covariance matrix* for $\underline{\mathbf{x}}$ for the similar reason.

An example that occurs frequently in engineering problems is the *Gaussian random vector* for which any subset of its random variable components has a joint Gaussian pdf. In particular, the joint pdf among the entire set of N is an N-dimensional Gaussian pdf. That is, if $\underline{\mathbf{x}}$ is a Gaussian random vector, then

$$\begin{aligned} f_{\underline{\mathbf{x}}}(x_1, \ldots, x_N) &= f_{\underline{x}_1,\ldots,\underline{x}_N}(x_1, \ldots, x_N) \\ &= \frac{1}{(2\pi)^{N/2}\sqrt{\det \mathbf{C}_{\underline{\mathbf{x}}}}} \exp\left\{-\frac{1}{2}(\mathbf{x} - \boldsymbol{\mu}_{\underline{\mathbf{x}}})^T \mathbf{C}_{\underline{\mathbf{x}}}^{-1}(\mathbf{x} - \boldsymbol{\mu}_{\underline{\mathbf{x}}})\right\}, \end{aligned} \tag{1.118}$$

where $\mathbf{x}$ denotes the vector of arguments $[x_1 \cdots x_N]^T$ and $\mathbf{C}_{\underline{\mathbf{x}}}$ and $\boldsymbol{\mu}_{\underline{\mathbf{x}}}$ are the covariance matrix and mean vector as defined above. It can be shown that this form reduces to (1.94) in the two-dimensional case.

1.2.3 Random Processes

Basic Concepts

Definition of a Random Process. A (real) discrete[21] *random*, or *stochastic*, *process* is defined as a collection of random variables, each indexed by a point in discrete time. For example, the following set comprises a random process:

$$\{\ldots, \underline{x}(-1), \underline{x}(0), \underline{x}(1), \ldots\} = \{\underline{x}(n), \quad n \in (-\infty, \infty)\}, \tag{1.119}$$

where each random variable represents a model for the generation of values at its corresponding time. There will be many occasions when we will want to refer to a random process by a name. For example, it is too clumsy to write something like "the random process $\{\ldots, \underline{x}(-1), \underline{x}(0), \underline{x}(1), \ldots\}$ is used to model the speech signal. . . ." What shall we call the random process? This is one place where we shall bow to convention and use a less-than-ideal choice. It is common to refer to a random process by the same name as that used for the random variables which constitute it. For example, the random process in (1.119) would be called simply $\underline{x}$,

$$\underline{x} = \{\ldots, \underline{x}(-1), \underline{x}(0), \underline{x}(1), \ldots\} = \{\underline{x}(n), \quad n \in (-\infty, \infty)\}, \tag{1.120}$$

so that we can write "the random process $\underline{x}$ is used to model the speech signal. . . ." Of course, the problem which arises is that $\underline{x}$ may refer to a random variable or a random process. We could further distinguish a random process by using yet another notation, for example, $\underline{\underline{x}}$, but this will turn out to be unnecessary in almost every circumstance. From context, it should always be clear whether an underscored quantity is a random variable or a random process. Note carefully that the notation $\underline{x}$ *never* refers to both. Once it is known that $\underline{x}$ is a random process, then all associated random variables should have time indices, for example, $\underline{x}(n)$. Finally, it should be noted that the random variables in a random process will almost always be indexed by integers in parentheses to indicate their association with discrete time. There will be only limited use for continuous-time random processes in this book.

An example will illustrate how a random process is related to a physical problem. Suppose that we define a simple experiment in which an integer representing one of L digitized speech waveforms is selected at random. For illustrative purposes, we plot segments of all of the waveforms (for $L = 3$) in Fig. 1.11. We can imagine that each time is governed by a random variable, say $\underline{x}(n)$ at time n, and the ordered collection of these random variables is the underlying random process, $\underline{x}$. When the experiment is complete, each random variable will go to work mapping the outcome, for example, "waveform 2," to an amplitude level corresponding to that outcome. For example, $\underline{x}(8)$ maps the outcome "waveform 2" to a value 82 in our figure. For this one experiment, therefore, the totality of all the

[21]We will focus on the discrete case because of our primary interest in discrete signals in this book.

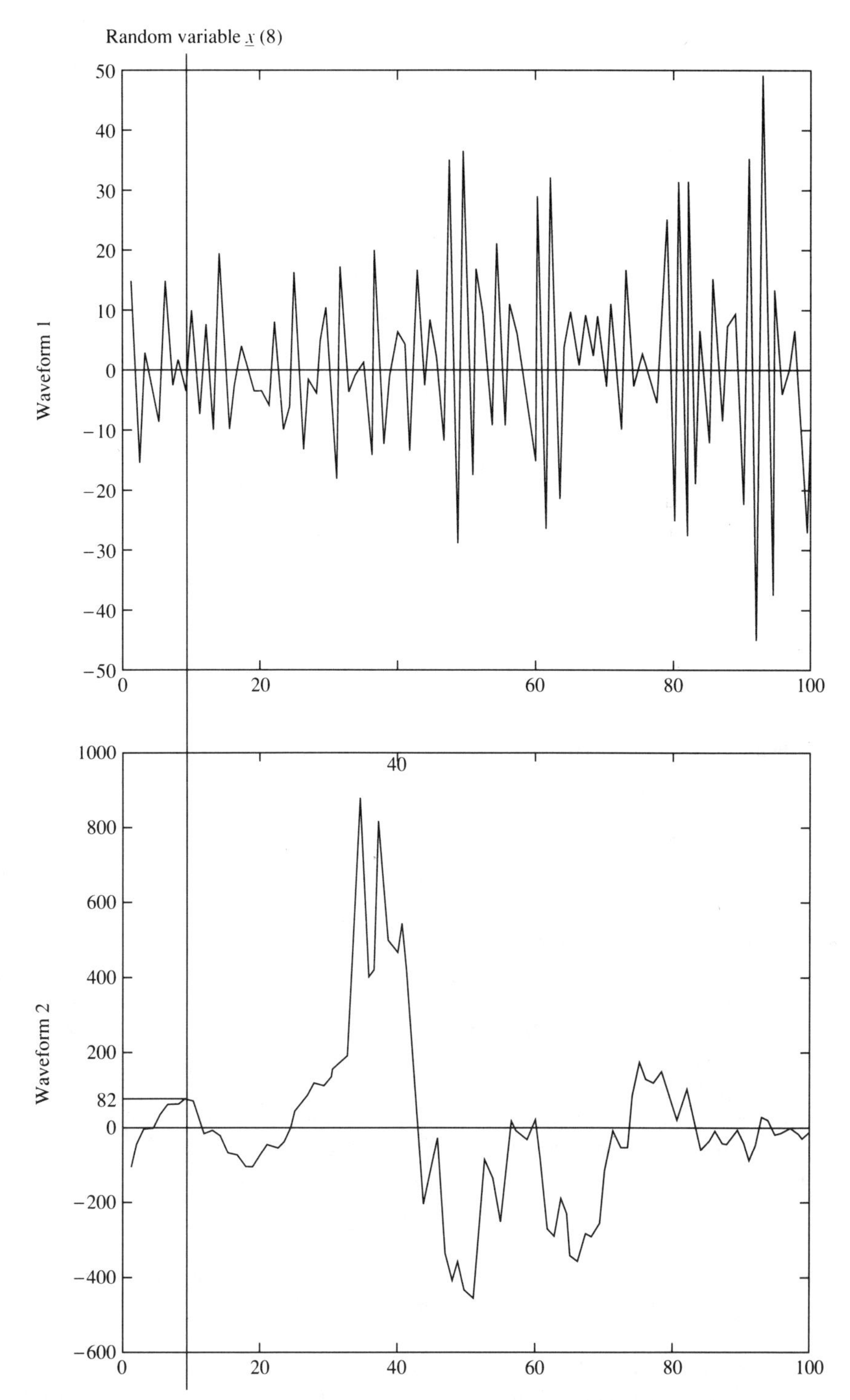

FIGURE 1.11. An ensemble of speech waveforms modeled by random process $\underline{x}$ with random variables $\underline{x}(n)$. (Figure continued on p. 44.)

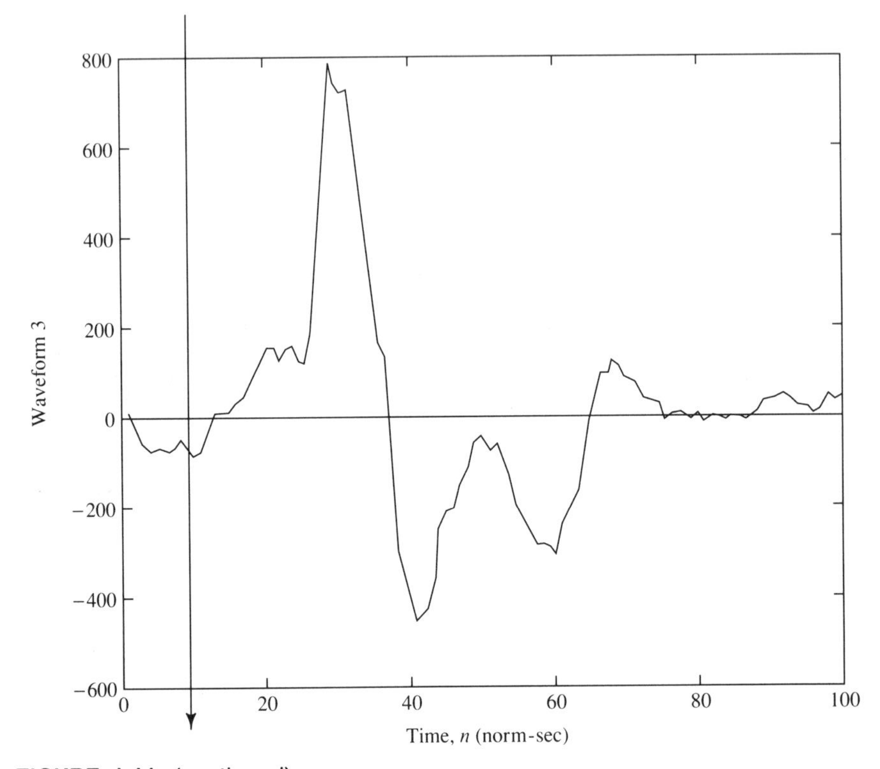

FIGURE 1.11. (continued)

random variables will produce a particular waveform from the experimental outcome, each random variable being responsible for one point. This one waveform is called a *sample function* or *realization* of the random process. The collection of all realizations (resulting from all the experiments) is called an *ensemble.* It should be clear that if we select a time, we will get a random variable. If we select an experimental outcome, we get a realization. If we select both a time and an outcome, we get a *number,* which is the result of the mapping of that outcome to the real line by the random variable at the time we select.

pdf for a Random Process. Associated with any i random variables in a random process is an ith order pdf. For example, for $\underline{x}(n_1)$, $\underline{x}(n_2)$, and $\underline{x}(n_3)$, we have the third-order density

$$f_{\underline{x}(n_1),\underline{x}(n_2),\underline{x}(n_3)}(\xi_1, \xi_2, \xi_3). \tag{1.121}$$

This is consistent with our previous convention of listing all random variables in the joint pdf as subscripts of f.

Independence of Random Processes. We have reviewed the meaning of independence of random variables above. We must also recall the meaning of independent random processes. Two random processes, $\underline{x}$ and $\underline{y}$, are *statistically independent* if, for any times $n_1, n_2, \ldots, n_i$, and $m_1, m_2, \ldots, m_j$, the random variable group $\underline{x}(n_1), \underline{x}(n_2), \ldots, \underline{x}(n_i)$ is independent of $\underline{y}(m_1), \underline{y}(m_2), \ldots, \underline{y}(m_j)$. This, in turn, requires that the joint pdf be factorable as

$$\begin{aligned} &f_{\underline{x}_1(n_1),\ldots,\underline{x}_i(n_i),\underline{y}_1(m_1),\ldots,\underline{y}_j(m_j)}(\xi_1,\ldots,\xi_i,v_1,\ldots,v_j) \\ &\quad = f_{\underline{x}_1(n_1),\ldots,\underline{x}_i(n_i)}(\xi_1,\ldots,\xi_i)f_{\underline{y}_1(m_1),\ldots,\underline{y}_j(m_j)}(v_1,\ldots,v_j). \end{aligned} \tag{1.122}$$

Stationarity. A random process $\underline{x}$ is said to be *stationary to order i* or *i*th-*order stationary* if

$$f_{\underline{x}_1(n_1),\ldots,\underline{x}_i(n_i)}(\xi_1,\ldots,\xi_i) = f_{\underline{x}_1(n_1+\Delta),\ldots,\underline{x}_i(n_i+\Delta)}(\xi_1,\ldots,\xi_i) \tag{1.123}$$

for any times n_1, $n_2, \ldots, n_i$ and any Δ. This means that the joint pdf does not change if we consider any set of i random variables from $\underline{x}$ with the same relative spacings as the original set (which is arbitrary). If $\underline{x}$ is stationary to any order, then it is said to be *strict sense*, or *strong sense*, *stationary* (SSS). We will review a weaker form of stationarity below.

Stationarity has important implications for engineering analysis of a stochastic process. It implies that certain statistical properties of the process are invariant with time, making the process more amenable to modeling and analysis. Consider, for example, the case in which $\underline{x}$ is first-order stationary. Then

$$f_{\underline{x}(n)}(\xi) = f_{\underline{x}(n+\Delta)}(\xi) \tag{1.124}$$

for any n and Δ, from which it follows immediately that every random variable in $\underline{x}$ has the same mean. In this case, it is reasonable to talk about *the* average of the random process, but in general there are as many averages as random variables in a random process. This leads us to the important issue of ergodicity.

Ergodicity and Temporal Averages. Consider a random process, $\underline{x}$, known to be first-order stationary. We might find ourselves in the lab with only one realization of the process, say $x_1(n)$, $n \in (-\infty, \infty)$, wondering whether we could somehow estimate *the* average of $\underline{x}$, say $\mu_{\underline{x}}$. In principle, we should acquire a large number of realizations and use them to compute an empirical average (estimate) of any random variable, say $\underline{x}(n)$ at time n. (It wouldn't matter which n, since the averages should all be the same due to stationarity.) This estimate, obtained by averaging down through the ensemble at a point, is referred to as an *ensemble average.* The ensemble average represents an attempt to estimate $\mathcal{E}\{\underline{x}(n)\}$ at time n, hence to estimate *the* average of process. Since we do not have an ensemble, it would be tempting to estimate $\mu_{\underline{x}}$ by computing a *temporal average* of the realization, $x_1(n)$,

$$\mu_{x_1} = \mathcal{L}\{x_1(n)\} \stackrel{\text{def}}{=} \lim_{N\to\infty} \frac{1}{2N+1} \sum_{n=-N}^{N} x_1(n). \tag{1.125}$$

Note that we have explicitly used a *signal* name, x_1, as a subscript of μ to indicate that it has been computed using the realization rather than an ensemble. Note also the operator $\mathcal{L}$ used to indicate the *long-term time average.* This notation will be used consistently:

$$\mathcal{L}\{\cdot\} \stackrel{\text{def}}{=} \lim_{N\to\infty} \frac{1}{2N+1} \sum_{n=-N}^{N} \{\cdot\}. \tag{1.126}$$

When will μ_{x_1}, the time average, equal $\mu_{\underline{x}}$, the ensemble or statistical average? Generally speaking, a random process is *ergodic* if ensemble averages can be replaced by time averages.[22] In our example, if $\mu_{x_1} = \mu_{\underline{x}}$, $\underline{x}$ is said to be *mean-ergodic,* since this property holds for the mean. Ergodicity will be an important assumption in our work with speech because we frequently will have only one realization with which to compute averages. In particular, second-order ergodicity will play an important role and we will look more carefully at this concept shortly.

Correlation and Covariance Applied to Random Processes

Consider two random variables, say $\underline{x}(n_1)$ and $\underline{x}(n_2)$, taken from a random process $\underline{x}$. Recall that the correlation of these two random variables is $\mathcal{E}\{\underline{x}(n_1)\underline{x}(n_2)\}$. Since the two random variables in this case are drawn from the same random process, we give this the name *autocorrelation* and feature it with a special notation

$$r_{\underline{x}}(n_1, n_2) \stackrel{\text{def}}{=} \mathcal{E}\{\underline{x}(n_1)\underline{x}(n_2)\}. \tag{1.127}$$

Similarly, the *autocovariance* function is given by

$$c_{\underline{x}}(n_1, n_2) \stackrel{\text{def}}{=} \mathcal{E}\{[\underline{x}(n_1) - \mathcal{E}\{\underline{x}(n_1)\}][\underline{x}(n_2) - \mathcal{E}\{\underline{x}(n_2)\}]\}. \tag{1.128}$$

It is a simple matter to show that

$$c_{\underline{x}}(n_1, n_2) = r_{\underline{x}}(n_1, n_2) - \mathcal{E}\{\underline{x}(n_1)\}\,\mathcal{E}\{\underline{x}(n_2)\}. \tag{1.129}$$

It follows immediately from the definition of stationarity that if the random process $\underline{x}$ is at least second-order stationary, then the value of the autocorrelation does not depend on *which* two random variables are selected from $\underline{x}$, but rather their *separation in time.* In this case, we adopt the somewhat sloppy, but very conventional, notation

$$\begin{aligned} r_{\underline{x}}(\eta) &\stackrel{\text{def}}{=} \text{autocorrelation of any two random variables in } \underline{x}, \\ &\quad\;\; \text{which are separated by } \eta \text{ in time} \\ &= \mathcal{E}\{\underline{x}(n)\underline{x}(n-\eta)\} \text{ for any } n. \end{aligned} \tag{1.130}$$

[22]This definition of ergodicity is entrenched in engineering textbooks, but it is not strictly accurate [see (Gray and Davisson, 1986, Ch. 7)].

If a random process is ith-order stationary, it is also $(i-1)$th-order stationary. Therefore a second-order stationary process is also first order and has a constant mean,

$$\mu_{\underline{x}} = \mathcal{E}\{\underline{x}(n)\} \text{ for any } n. \tag{1.131}$$

This leads us to the definition of a weak form of stationarity, which is often sufficient to allow many useful engineering analyses.

A random process $\underline{x}$ is said to be *wide sense*, or *weak sense, stationary* (WSS) if

1. Its autocorrelation is a function of time difference only as in (1.130).
2. Its mean is constant as in (1.131).

We note that

$$\text{SSS} \Rightarrow \text{second-order stationarity} \Rightarrow \text{WSS}, \tag{1.132}$$

but neither of the implications reverses except in the special case of joint Gaussian random variables (see Problem 1.11).

Finally, but very important, note that if $\underline{x}$ is *correlation-ergodic*, then the autocorrelation can be computed using a temporal average

$$r_x(\eta) = \mathcal{L}\{x(n)x(n-\eta)\} = \lim_{N\to\infty} \frac{1}{2N+1} \sum_{n=-N}^{N} x(n)x(n-\eta), \tag{1.133}$$

where $x(n)$ is some realization of $\underline{x}$. This is an extension of the idea of ergodicity discussed above, to a second-order case. Note carefully that the subscript on r is a signal name x, indicating the use of a signal to compute a time average, rather than random variables to compute an ensemble average. We have already introduced this notation, but it is worth reiterating here so that the reader is clear about its significance.

Since speech processing is an applied discipline, we will frequently use temporal, rather than ensemble, averages in our developments. Of course, this is because we have signals, rather than stochastic models, to deal with. On the other hand, there is often much to be gained by modeling speech as a stochastic process. Accordingly, when a speech signal is thought of as a realization of a stochastic process, the underlying process must be *assumed* to have the appropriate stationarity and ergodicity properties to allow the computation of meaningful temporal statistics.[23]

[23]A philosophical point is in order here. A moment's thought will reveal that speech, if thought of as a random process, cannot possibly comprise a *stationary* random process, since speech is a very dynamic phenomenon. This is an indication of the need for "short-term" analytical tools which can be applied to short temporal regions of assumed stationarity. At this point we begin to use formal theory in some rather *ad hoc* and *ad lib* ways. Of course, it is often the case in engineering problems that we use formal theories in rather loose ways in practice. However, the ability to understand the implications of our sloppiness, and the ability to predict and explain success in spite of it, depends entirely on our understanding of the underlying formal principles. In this book, we will stress the dependency of *ad hoc* methods on formal principles.

Multiple Random Processes

We now extend these ideas to the case of two random processes. As a natural extension of the concept of stationarity we have the following: Two random processes, $\underline{x}$ and $\underline{y}$, are said to be *jointly SSS* if

$$f_{\underline{x}_1(n_1),\ldots,\underline{x}_i(n_i),\underline{y}_1(m_1),\ldots,\underline{y}_j(m_j)}(\xi_1,\ldots,\xi_i,v_1,\ldots,v_j)$$
$$= f_{\underline{x}_1(n_1+\Delta),\ldots,\underline{x}_i(n_i+\Delta)\underline{y}_i(m_1+\Delta),\ldots,\underline{y}_j(m_j+\Delta)}(\xi_1,\ldots,\xi_i,v_1,\ldots,v_j) \tag{1.134}$$

for any i random variables from $\underline{x}$, and any j from $\underline{y}$, and for any Δ. It follows that if $\underline{x}$ and $\underline{y}$ are jointly SSS, then each is individually SSS.

From random variables $\underline{x}(n_1)$ and $\underline{y}(n_2)$, chosen from $\underline{x}$ and $\underline{y}$, respectively, we can form the *cross-correlation*,

$$r_{\underline{xy}}(n_1,n_2) \stackrel{\text{def}}{=} \mathcal{E}\{\underline{x}(n_1)\underline{y}(n_2)\}, \tag{1.135}$$

and the *cross-covariance*,

$$c_{\underline{xy}}(n_1,n_2) \stackrel{\text{def}}{=} \mathcal{E}\{[\underline{x}(n_1)-\mathcal{E}\{\underline{x}(n_1)\}][\underline{y}(n_2)-\mathcal{E}\{\underline{y}(n_2)\}]\}. \tag{1.136}$$

Similarly to (1.129), we obtain

$$c_{\underline{xy}}(n_1,n_2) = r_{\underline{xy}}(n_1,n_2) - \mathcal{E}\{\underline{x}(n_1)\}\mathcal{E}\{\underline{y}(n_2)\}. \tag{1.137}$$

As we did in the individual random process case, it will be useful to have a weaker form of stationarity between two random processes. The following conditions are required for $\underline{x}$ and $\underline{y}$ to be declared *jointly WSS*:

1. $\underline{x}$ and $\underline{y}$ are *individually WSS*;
2. $r_{\underline{xy}}(n_1,n_2)$ is a function of $\eta = n_2 - n_1$ only.

It is easy to show that joint SSS $\Rightarrow$ joint WSS (but not the converse). Also, simply by definition, we see that joint WSS $\Rightarrow$ individual SSS, but, again, the converse is not generally true.

As an extension of the concept of ergodicity to the joint random process case, we note that the cross-correlation can be computed using a temporal average over two realizations if the processes are jointly *correlation-ergodic*:

$$r_{\underline{xy}}(\eta) = \mathcal{L}\{x(n)y(n-\eta)\} = \lim_{N\to\infty}\frac{1}{2N+1}\sum_{n=-N}^{N} x(n)y(n-\eta). \tag{1.138}$$

Such a computation, of course, makes no sense unless the two random processes are at least jointly WSS.

Power Density Spectrum

Single Random Process. A general discussion of this important topic is unnecessary for our work with speech and would take us too far afield.

We refer the reader to the textbooks in Appendix 1.B for a general background. For our purposes, it is sufficient to define the *power density spectrum* of a WSS random process $\underline{x}$ as the DTFT of its autocorrelation function,

$$\Gamma_{\underline{x}}(\omega) \stackrel{\text{def}}{=} \sum_{\eta=-\infty}^{\infty} r_{\underline{x}}(\eta) e^{-j\omega\eta}. \tag{1.139}$$

Accordingly, the autocorrelation can be computed from the power density spectrum as

$$r_{\underline{x}}(\eta) = \frac{1}{2\pi} \int_{-\pi}^{\pi} \Gamma_{\underline{x}}(\omega) e^{j\omega\eta}\, d\omega. \tag{1.140}$$

If $\underline{x}$ is also correlation-ergodic and the autocorrelation is computed using time averaging, then, according to our convention, the subscripts will denote realizations. For example,

$$\Gamma_{x}(\omega) = \sum_{\eta=-\infty}^{\infty} r_{x}(\eta) e^{j\omega\eta}. \tag{1.141}$$

The *total*[24] *power* in a second-order stationary real random process is defined as

$$P_{\underline{x}} \stackrel{\text{def}}{=} \mathcal{E}\{\underline{x}^2(n)\} \qquad \text{for any } n. \tag{1.142}$$

To make sense of this definition, we recall the definition of the power in a *signal*, which according to (1.11) is given by[25]

$$P_x = \mathcal{L}\{\,|x(n)|^2\}. \tag{1.143}$$

If $x(n)$ happens to be a realization of $\underline{x}$, and $\underline{x}$ is second-order ergodic, then we see that these two computations are equivalent.

As an aside, we recall that realizations of stationary, ergodic, stochastic processes were listed as a class of power signals in Section 1.2.3. Indeed, we now can appreciate that this is the case. If $x(n)$ is such a realization and is not a power signal, then

$$P_{\underline{x}} = P_x = \{0 \text{ or } \infty\} \tag{1.144}$$

and we encounter a contradiction.

Now that the definition of $P_{\underline{x}}$ makes sense, we note that

$$P_{\underline{x}} = \frac{1}{2\pi} \int_{-\pi}^{\pi} \Gamma_{\underline{x}}(\omega)\, d\omega = \frac{1}{\pi} \int_{0}^{\pi} \Gamma_{\underline{x}}(\omega)\, d\omega = r_{\underline{x}}(0). \tag{1.145}$$

[24]The word *total* is used here to connote that the power in all frequencies is considered.

[25]The absolute value signs appear here because $x(n)$ was assumed to be complex-valued in general in definition (1.11). Since we have focused exclusively upon real random processes, they are superfluous in this discussion.

This result follows immediately from definitions and says that the scaled total area under $\Gamma_{\underline{x}}(\omega)$ yields the total power in $\underline{x}$, making it a sort of density of power on frequency, much like the pdf is a probability density on its variable of interest. In fact, to find the power in any frequency range, say ω_1 to ω_2, for $\underline{x}$, we can compute

$$\text{Power in } \underline{x} \text{ in frequencies } \omega_1 \text{ to } \omega_2 = \frac{1}{\pi}\int_{\omega_1}^{\omega_2} \Gamma_{\underline{x}}(\omega)\, d\omega. \tag{1.146}$$

Finally, we remark that some stochastic processes have all of their power concentrated at discrete frequencies. For example, a process $\underline{x}$ whose random variables are $\{\underline{x}(n) = \cos(\omega_0 n + \underline{\theta}),\ n \in (-\infty, \infty)\}$ with $\underline{\theta}$ a random variable, will have all power concentrated at frequency ω_0. In this case, the autocorrelation (ensemble or temporal) will be periodic with the same frequency, and we must resort to the use of impulses in the PDS much like our work with the PDS for a periodic *deterministic* process.

Two Random Processes. Let us focus here on jointly WSS random processes, $\underline{x}$ and $\underline{y}$, with cross-correlation $r_{\underline{xy}}(\eta)$. In this case the *cross-power spectral density* is given by

$$\Gamma_{\underline{xy}}(\omega) = \sum_{\eta=\infty}^{\infty} r_{\underline{xy}}(\eta)\, e^{-j\omega\eta}. \tag{1.147}$$

We can compute the *cross power* between the two processes,

$$P_{\underline{xy}} = \frac{1}{2\pi}\int_{-\pi}^{\pi} \Gamma_{\underline{xy}}(\omega)\, d\omega, \tag{1.148}$$

which is interpretable as the power that the two random processes generate over and above their individual powers due to the fact that they are correlated.

Noise

Realizations of stochastic processes often occur as unwanted disturbances in engineering applications and are referred to as *noise*. Even when the stochastic signal is not a disturbance, we often employ the term noise. Such will be the case in our speech work, for example, when a noise process appears as the driving function for a model for "unvoiced" speech sounds. (Consider, e.g., the sound that the letter "s" implies.)

One of the most important forms of noise in engineering analysis is (*discrete-time*) *white noise*, defined as a stationary process, say $\underline{w}$, with the property that its power density spectrum is constant over the Nyquist range,

$$\Gamma_{\underline{w}}(\omega) = 2\pi, \qquad \text{for } \omega \in [-\pi, \pi). \tag{1.149}$$

Accordingly, the autocorrelation function for white noise is

$$r_{\underline{w}}(\eta) = \delta(\eta). \tag{1.150}$$

The reader is cautioned to distinguish between *continuous-time* white noise and the phenomenon we are discussing here. Continuous white noise has infinite power and a flat power density spectrum over all frequencies. Just as the discrete-time impulse cannot be considered as samples of the continuous time impulse, so discrete-time white noise should not be considered to be samples of continuous time white noise. In fact, discrete-time white noise may be thought to represent samples of a continuous time stochastic process, which is bandlimited to the Nyquist range and which has a flat power density spectrum over that range.

Random Processes and Linear Systems

It will be useful for us to review a few key results concerning the analysis of LTI discrete time systems with stochastic inputs. Let us restrict this discussion to WSS, second-order ergodic, stochastic processes.

Consider first an LTI system with discrete-time impulse response $h(n)$. Suppose that $x(n)$, a realization of random process $\underline{x}$, is input to the system. The output, say $y(n)$, is given by the convolution sum,

$$y(n) = \sum_{i=-\infty}^{\infty} x(n-i)h(i). \tag{1.151}$$

Of course, the same transformation occurs on the input no matter which realization of $\underline{x}$ it happens to be. We could denote this fact by replacing $x(n-i)$ by its corresponding random variable, $\underline{x}(n-i)$, on the right side of (1.151). Without a rigorous argument,[26] it is believable that the mapping of these random variables by the convolution sum will produce another random variable (for a fixed n), $\underline{y}(n)$, so we write

$$\underline{y}(n) = \sum_{i=-\infty}^{\infty} \underline{x}(n-i)h(i). \tag{1.152}$$

As n varies, a second random process is created at the output, $\underline{y}$. We have assumed $\underline{x}$ to be WSS and second-order ergodic. Let us show that the same is true of $\underline{y}$.

By applying the expectation operator to both sides of (1.152) and interchanging the order of summation on the right, we have

$$\mathcal{E}\{\underline{y}(n)\} = \sum_{i=-\infty}^{\infty} \mathcal{E}\{\underline{x}(n-i)\}\, h(i) \tag{1.153}$$

or

[26]This argument centers on concepts of stochastic convergence that are treated in many standard textbooks (see books in Appendix 1.B not labeled "elementary").

$$\mu_{\underline{y}} = \mu_{\underline{x}} \sum_{i=-\infty}^{\infty} h(i). \tag{1.154}$$

Since this result does not depend on n, we see that $\underline{y}$ is stationary in the mean. A similar result obtains with $\mu_{\underline{y}}$ and $\mu_{\underline{y}}$ replaced by μ_x and μ_y if we begin with (1.151) and use temporal averages, so that $\underline{y}$ is also ergodic in the mean.

In a similar way (see Problem 1.14) we can show that the autocorrelation of $\underline{y}$ is dependent only on the time difference in the arguments and is given by

$$r_{\underline{y}}(\eta) = \sum_{i=-\infty}^{\infty} \sum_{k=-\infty}^{\infty} h(i)h(k)r_{\underline{x}}(\eta + k - i), \tag{1.155}$$

or, in terms of temporal autocorrelations,

$$r_y(\eta) = \sum_{i=-\infty}^{\infty} \sum_{k=-\infty}^{\infty} h(i)h(k)r_x(\eta + k - i). \tag{1.156}$$

We conclude, therefore, that a WSS correlation-ergodic input to an LTI system produces a WSS correlation-ergodic output. This is a fundamental result that will be used implicitly in many places in our work.

Finally, we recall the important relationship between the input and output power spectral densities in the case of LTI systems with WSS inputs,

$$\Gamma_{\underline{y}}(\omega) = |H(\omega)|^2 \Gamma_{\underline{x}}(\omega). \tag{1.157}$$

This result is derived by taking the DTFT of both sides of (1.155).

1.2.4 Vector-Valued Random Processes

At several places in this book, we will encounter random processes that are vector-valued. A *vector-valued random process* $\underline{\mathbf{x}}$ is a collection of random vectors indexed by time,[27]

$$\underline{\mathbf{x}} \stackrel{\text{def}}{=} \{\ldots, \underline{\mathbf{x}}(-1), \underline{\mathbf{x}}(0), \underline{\mathbf{x}}(1), \ldots\}. \tag{1.158}$$

Realizations of these random processes comprise vector-valued signals of the form

$$\{\ldots, \mathbf{x}(-1), \mathbf{x}(0), \mathbf{x}(1), \ldots\}, \tag{1.159}$$

which we customarily denote simply $\mathbf{x}(n)$. (*Note*: We are now employing boldface to indicate vector quantities.)

[27]Again, we will restrict our attention to real processes, but the complex case is a simple generalization.

These sequences will arise in two different ways in our work. In the first case, the elements of each random vector will be random variables representing scalar signal samples, which for some reason are conveniently packaged into vectors. For example, suppose we have a *scalar* signal (random process) $\underline{x} = \{\ldots, \underline{x}(-1), \underline{x}(0), \underline{x}(1), \ldots\}$. We might find it necessary to break the signal into 100-point blocks for coding purposes, thereby creating a vector random process

$$\underline{\mathbf{x}} = \left\{\ldots, \underline{\mathbf{x}}(0) = \begin{bmatrix} \underline{x}(0) \\ \underline{x}(1) \\ \vdots \\ \underline{x}(99) \end{bmatrix}, \underline{\mathbf{x}}(1) = \begin{bmatrix} \underline{x}(100) \\ \underline{x}(101) \\ \vdots \\ \underline{x}(199) \end{bmatrix}, \underline{\mathbf{x}}(2) = \begin{bmatrix} \underline{x}(200) \\ \underline{x}(201) \\ \vdots \\ \underline{x}(299) \end{bmatrix}, \ldots \right\}. \tag{1.160}$$

Note that the "time" indices of the vector random process represent a re-indexing of the sample times of the original random process.

A second type of vector random process will result from the extraction of vector-valued features from frames of speech. We might, for example, extract 14 features from 160-point frames of speech. These frames may be overlapping, as shown in Fig. 1.12. In some cases we might choose to index the resulting random vectors by the end-times of the frames; in others we might reindex the vector sequence using consecutive integers. In either case, it is clear that the vector sequence comprises a vector-valued random process.

For a vector random process, the *mean vector* takes the place of the mean in the scalar case, and the *autocorrelation matrix* plays the role of the autocorrelation. These are

$$\boldsymbol{\mu}_{\underline{\mathbf{x}}(n)} \stackrel{\text{def}}{=} \mathcal{E}\{\underline{\mathbf{x}}(n)\} \tag{1.161}$$

and

$$\mathbf{R}_{\underline{\mathbf{x}}}(n_1, n_2) \stackrel{\text{def}}{=} \mathcal{E}\{\underline{\mathbf{x}}(n_1)\underline{\mathbf{x}}^T(n_2)\}, \tag{1.162}$$

respectively. Note that the mean vector contains the mean of each of the component random variables and the correlation matrix contains the cross-correlations between each component pair in the vectors. We can also speak of the *covariance matrix* of the vector random process $\underline{\mathbf{x}}$, defined as

$$\mathbf{C}_{\underline{\mathbf{x}}}(n_1, n_2) \stackrel{\text{def}}{=} \mathcal{E}\{[\underline{\mathbf{x}}(n_1) - \boldsymbol{\mu}_{\underline{\mathbf{x}}(n_1)}][\underline{\mathbf{x}}(n_2) - \boldsymbol{\mu}_{\underline{\mathbf{x}}(n_2)}]^T\}. \tag{1.163}$$

When the vector random process is WSS, we have a stationary mean vector, and correlation and covariance matrices that depend only on time difference. These are defined, for an arbitrary n, as follows:

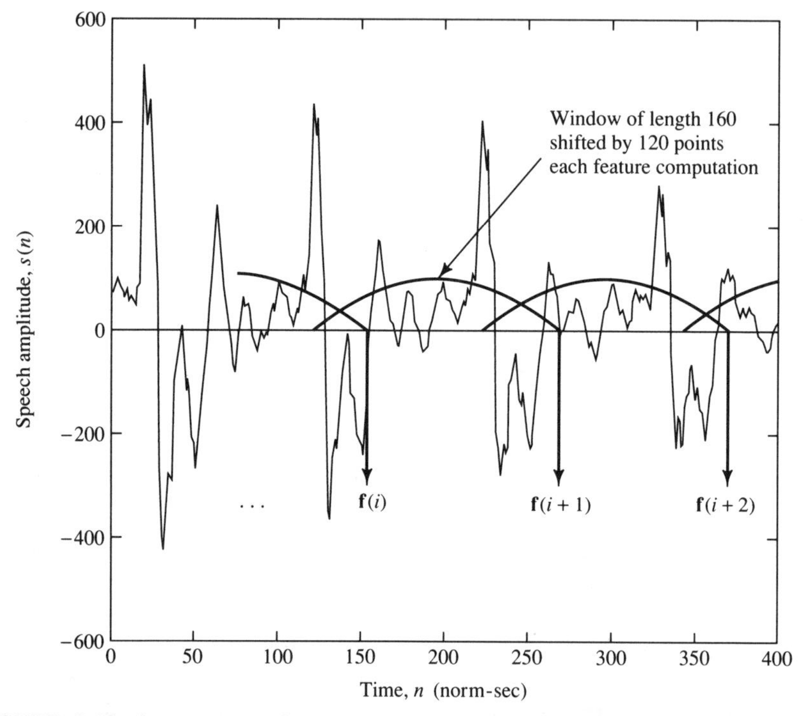

FIGURE 1.12. A vector random process created by extracting vector-valued features from frames of a speech process at periodic intervals. *Note*: Here we index the features by sequential integers. Later we will establish the convention of indexing features by the time of the leading edge of the sliding window.

$$\boldsymbol{\mu}_{\underline{\mathbf{x}}} \stackrel{\text{def}}{=} \boldsymbol{\mu}_{\underline{\mathbf{x}}(n)} = \mathcal{E}\{\underline{\mathbf{x}}(n)\} \tag{1.164}$$

$$\mathbf{R}_{\underline{\mathbf{x}}}(\eta) \stackrel{\text{def}}{=} \mathbf{R}_{\underline{\mathbf{x}}}(n, n-\eta) = \mathcal{E}\{\underline{\mathbf{x}}(n)\underline{\mathbf{x}}^T(n-\eta)\} \tag{1.165}$$

$$\mathbf{C}_{\underline{\mathbf{x}}}(\eta) \stackrel{\text{def}}{=} \mathbf{C}_{\underline{\mathbf{x}}}(n, n-\eta) = \mathcal{E}\left\{[\underline{\mathbf{x}}(n) - \boldsymbol{\mu}_{\underline{\mathbf{x}}}][\underline{\mathbf{x}}^T(n-\eta) - \boldsymbol{\mu}_{\underline{\mathbf{x}}}]^T\right\}. \tag{1.166}$$

Frequently, we are specifically interested in the "zero lag" correlations (or covariance) matrix of a stationary vector random process that plays the role of the variance of the process. For this case, we will write

$$\mathbf{R}_{\underline{\mathbf{x}}} \stackrel{\text{def}}{=} \mathbf{R}_{\underline{\mathbf{x}}}(0) \tag{1.167}$$

and

$$\mathbf{C}_{\underline{\mathbf{x}}} \stackrel{\text{def}}{=} \mathbf{C}_{\underline{\mathbf{x}}}(0) \tag{1.168}$$

for simplicity. The reader should carefully compare these notations with (1.116) and (1.117) and discern the difference in meaning.

Finally, we note that there are temporal versions of these three key statistical matrices that are meaningful when appropriate ergodicity conditions hold. These are, for an arbitrary n,

$$\boldsymbol{\mu}_{\mathbf{x}} \stackrel{\text{def}}{=} \boldsymbol{\mu}_{\mathbf{x}(n)} = \mathcal{L}\{\mathbf{x}(n)\} \tag{1.169}$$

$$\mathbf{R}_{\mathbf{x}}(\eta) \stackrel{\text{def}}{=} \mathcal{L}\{\mathbf{x}(n)\mathbf{x}^T(n-\eta)\} \tag{1.170}$$

$$\mathbf{C}_{\mathbf{x}}(\eta) \stackrel{\text{def}}{=} \mathcal{L}\left\{[\mathbf{x}(n)-\boldsymbol{\mu}_{\mathbf{x}}][\mathbf{x}^T(n-\eta)-\boldsymbol{\mu}_{\mathbf{x}}]^T\right\}. \tag{1.171}$$

We also define

$$\mathbf{R}_{\mathbf{x}} \stackrel{\text{def}}{=} \mathbf{R}_{\mathbf{x}}(0) \tag{1.172}$$

and

$$\mathbf{C}_{\mathbf{x}} \stackrel{\text{def}}{=} \mathbf{C}_{\mathbf{x}}(0). \tag{1.173}$$

1.3 Topics in Statistical Pattern Recognition

Reading Note: Most of the material in this section will not be used until Parts IV and V. The exception is Section 1.3.1, which will first be encountered in Chapter 5.

As in the previous two subsections of this chapter, the material treated here represents a very small sampling of a vast research discipline, with a focus on a few topics which will be significant to us in our speech processing work. Unlike the other two subsections, however, we make no assumption here or in the main text that the reader has a formal background in pattern recognition beyond a casual acquaintence with certain ideas that are inherent in general engineering study. A few example textbooks from this field are listed in Appendix 1.C.

Much of speech processing is concerned with the analysis and recognition of patterns and draws heavily on results from this field. Although many speech processing developments can be successfully understood with a rather superficial knowledge of pattern recognition theory, advanced research and development are not possible without a rigorous understanding. A few advanced speech processing topics in this book will need to be left to the reader's further pursuit, since it is not intended to assume this advanced pattern recognition background, nor is it possible to provide it within the scope of the book.

There are two main branches of pattern recognition—*statistical* and *syntactic*. Generally speaking, the former deals with statistical relationships among features in a pattern, while the latter approaches patterns as structures that can be composed of primitive patterns according to a set of rules. Although these branches are not exactly distinct, they are quite different in philosophy. In our work, the use of the latter is confined to the special problem of language modeling in automatic speech recognition.[28] We therefore defer any discussion of syntactic pattern recognition concepts to Chapter 13. Statistical pattern recognition methods, however, are quite prevalent in many aspects of speech processing, and it will be expedient for us to introduce a few key concepts before starting our study of speech.

We reemphasize that we are only discussing a few small topics in a vast and complex subject. Notably missing from our discussion, for example, is an analysis of how one chooses and evaluates in a rigorous sense the features representing a pattern. Frequently, this is accomplished in a rather *ad hoc* manner in speech processing, but the reader should be aware that a rich theory embracing this issue has been developed. A second example pertains to the covergence of clustering algorithms used to group features into classes. This issue will also be left for further study.

1.3.1 Distance Measures

Given two vectors $\mathbf{x}$ and $\mathbf{y}$ in a multidimensional space, we will frequently be interested in knowing "how far apart" they are. These vectors will often correspond to two time points in a realization of a vector-valued random process, or perhaps vectors drawn from two random processes. For the sake of discussion, let us just refer to $\mathbf{x}$ and $\mathbf{y}$.

It is sufficient for us to be concerned with vectors drawn from Cartesian spaces. The *N-dimensional real Cartesian space*, denoted $\mathbb{R}^N$ is the collection of all N-dimensional vectors with real elements. A *metric*, $d(\cdot\,,\cdot)$, on $\mathbb{R}^N$ is a real-valued function with three properties: For all $\mathbf{x}, \mathbf{y}, \mathbf{z} \in \mathbb{R}^N$,

1. $d(\mathbf{x}, \mathbf{y}) \geq 0$.
2. $d(\mathbf{x}, \mathbf{y}) = 0$ if and only if $\mathbf{x} = \mathbf{y}$.
3. $d(\mathbf{x}, \mathbf{y}) \leq d(\mathbf{x}, \mathbf{z}) + d(\mathbf{z}, \mathbf{y})$.

These properties coincide well with our intuitive notions about a proper measure of distance. Indeed, a metric is often used as a distance measure in mathematics and in engineering.[29]

Any function that meets the properties in the definition above is a legitimate metric on the vector space. Accordingly, there are many metrics, each having its own advantages and disadvantages. Most of the true metrics that we use in speech processing are particular cases of the Minkowski metric, or close relatives. This metric is defined as follows: Let x_k denote the kth component of the N-vector $\mathbf{x}$. Then the *Minkowski metric of order s*, or the *l_s metric*, between vectors $\mathbf{x}$ and $\mathbf{y}$ is

[28] In fact, syntactic pattern recognition has its roots in the theory of formal languages that was motivated by the study of natural languages (see Chapter 13).

[29] We will, however, encounter some distance measures later in the book that are not true metrics.

$$d_s(\mathbf{x}, \mathbf{y}) \stackrel{\text{def}}{=} \sqrt[s]{\sum_{k=1}^{N} |x_k - y_k|^s} . \tag{1.174}$$

Particular cases are

1. The l_1 or *city block* metric,

$$d_1(\mathbf{x}, \mathbf{y}) = \sum_{k=1}^{N} |x_k - y_k| \cdot \tag{1.175}$$

2. The l_2 or *Euclidean* metric,

$$d_2(\mathbf{x}, \mathbf{y}) = \sqrt{\sum_{k=1}^{N} |x_k - y_k|^2} = \sqrt{[\mathbf{x} - \mathbf{y}]^T [\mathbf{x} - \mathbf{y}]} . \tag{1.176}$$

3. The l_∞ or *Chebyshev* metric (corresponds to the Minkowski metric as $s \rightarrow \infty$),

$$d_\infty(\mathbf{x}, \mathbf{y}) = \max_k |x_k - y_k| . \tag{1.177}$$

We should note that the l_s *norm* of a vector $\mathbf{x}$, denoted $\|\mathbf{x}\|_s$, is defined as

$$\|\mathbf{x}\|_s \stackrel{\text{def}}{=} \sqrt[s]{\sum_{k=1}^{N} |x|^s} . \tag{1.178}$$

It follows immediately that the l_s metric between the vectors $\mathbf{x}$ and $\mathbf{y}$ is equivalent to the l_s norm of the difference vector $\mathbf{x} - \mathbf{y}$,

$$d_s(\mathbf{x}, \mathbf{y}) = \|\mathbf{x} - \mathbf{y}\|_s . \tag{1.179}$$

An important generalization of the Euclidean metric is called variously the *weighted Euclidean*, *weighted* l_2, or *quadratic* metric,[30]

$$d_{2w}(\mathbf{x}, \mathbf{y}) \stackrel{\text{def}}{=} \sqrt{[\mathbf{x} - \mathbf{y}]^T \mathbf{W} [\mathbf{x} - \mathbf{y}]} . \tag{1.180}$$

where $\mathbf{W}$ is a positive definite matrix that can be used for several purposes discussed below.

Before proceeding, we should be careful to point out that, in theoretical discussions, we might wish to discuss the distance between two stochastic vectors, say $\underline{\mathbf{x}}$ and $\underline{\mathbf{y}}$. In this case we might write, for example, something like

[30]The quadratic metric is often defined without the square root, but we employ the square root to make the distance more parallel to the Euclidean metric.

$$d_{2w}(\underline{\mathbf{x}}, \underline{\mathbf{y}}) = \sqrt{[\underline{\mathbf{x}} - \underline{\mathbf{y}}]^T \mathbf{W} [\underline{\mathbf{x}} - \underline{\mathbf{y}}]}\,. \tag{1.181}$$

The left side must be interpreted as a random variable that only takes a value when outcomes for $\underline{\mathbf{x}}$ and $\underline{\mathbf{y}}$ are known. The existence of this "random distance" depends upon concepts in stochastic calculus that will not concern us here (see textbooks in Appendix 1.B not labeled "elementary"). For our purposes, we just consider this notation to be a formal way of packaging together all possible outcomes of the distance that depend upon the random values of $\underline{\mathbf{x}}$ and $\underline{\mathbf{y}}$.

1.3.2 The Euclidean Metric and "Prewhitening" of Features

In this section we briefly make some points about the use of the Euclidean distance in engineering, which can have important consequences for performance of resulting algorithms and systems. The concepts discussed here have broader implications for abstract Hilbert spaces, but we will confine the remarks to simple vector spaces. The reader interested in a more formal and comprehensive treatment of these ideas should consult textbooks on linear algebra and functional analysis such as (Hoffman and Kunze, 1961, Ch. 2; Nobel, 1969, Ch. 14; Naylor and Sell, 1971; Lusternik and Sobolev, 1974).

Of the formal metrics in $\mathbb{R}^N$, the Euclidean metric is probably the most widely used in engineering problems. The reason for its popularity is that it fits precisely with our physical notion of distance. When the representation of a vector is based upon an orthonormal basis set, then the Euclidean distance between two vectors in the space conforms exactly to the "natural" distance between them. However, when vector representations are based upon a basis set that is not orthonormal (even if the set is orthogonal), then the Euclidean distance will yield "unnatural" results unless a linear operation is applied which transforms the vector representations to ones based on orthonormal vectors.

These ideas are illustrated in 2-space in Fig. 1.13. The representations of the vectors **a** and **b** with respect to the "natural" basis set $\boldsymbol{\beta}_1$ and $\boldsymbol{\beta}_2$ are

$$\mathbf{x} = \begin{bmatrix} 1 \\ 1 \end{bmatrix} \quad \text{and} \quad \mathbf{y} = \begin{bmatrix} 1 \\ 2 \end{bmatrix}, \tag{1.182}$$

respectively. By this we mean, for example, that

$$\mathbf{x} = \begin{bmatrix} x_1 \\ x_2 \end{bmatrix}, \tag{1.183}$$

where

$$\mathbf{a} = x_1 \boldsymbol{\beta}_1 + x_2 \boldsymbol{\beta}_2. \tag{1.184}$$

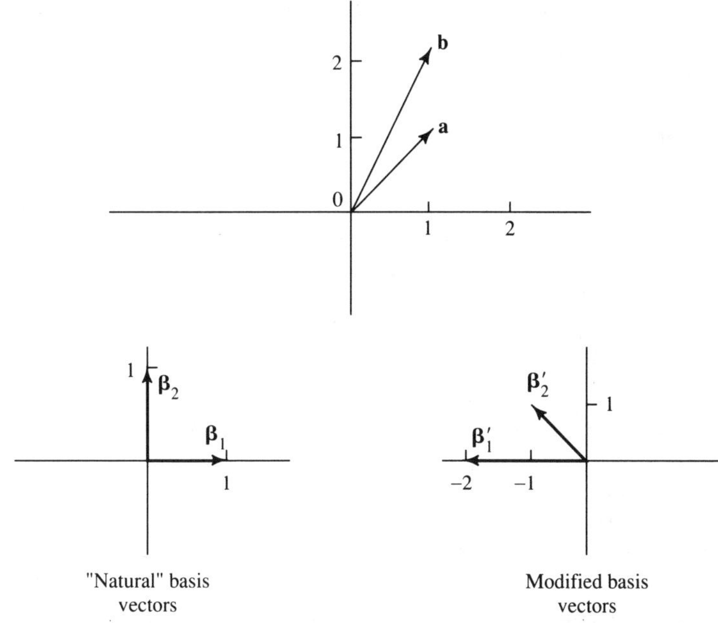

FIGURE 1.13. Vectors used to illustrate concepts of the Euclidean distance metric.

With this construction, everyone would agree that the distance between the vectors **a** and **b** is appropriately given by the Euclidean metric between the representations,

$$d_2(\mathbf{x}, \mathbf{y}) = 1. \tag{1.185}$$

Consistent with our discussion above, it is also true that the distance is given by the l_2, or Euclidean, norm of the difference vector,

$$\|\mathbf{x} - \mathbf{y}\|_2 = 1. \tag{1.186}$$

Suppose, however, that the given basis vectors were $\boldsymbol{\beta}_1'$ and $\boldsymbol{\beta}_2'$. In this case, the representations of **a** and **b** are

$$\mathbf{x}' = \begin{bmatrix} -1 \\ 1 \end{bmatrix} \quad \text{and} \quad \mathbf{y}' = \begin{bmatrix} -\frac{3}{2} \\ 2 \end{bmatrix}, \tag{1.187}$$

respectively. In spite of the fact that **a** and **b** have not moved, the Euclidean distance between these representations is

$$d_2(\mathbf{x}', \mathbf{y}') = \sqrt{\tfrac{5}{4}}. \tag{1.188}$$

We note that the distance would be "incorrect" even if the new basis vectors were orthogonal but not normalized.

What is "wrong" in the second case is the basis upon which the representations are assigned coordinates. The two "coordinates" with the second assignment of basis vectors are not distinct information. Moving in the direction of $\boldsymbol{\beta}_1'$ also includes motion in the direction of $\boldsymbol{\beta}_2'$, and conversely. Only when these bases are made to correspond to distinct (orthonormal) pieces of information does our sense of distance come back into focus and the Euclidean distance become meaningful. Algebraically, if the basis vectors were made to correspond to a proper orthonormal set, then the Euclidean distance would be appropriate. This would require that we transform the vector representations $\mathbf{x}'$ and $\mathbf{y}'$ to their representations on a "proper" set of basis vectors before computing the Euclidean distance. In this case, let us just choose to go back to the orthonormal basis set $\boldsymbol{\beta}_1$ and $\boldsymbol{\beta}_2$, in which case we know that $\mathbf{x}'$ and $\mathbf{y}'$ are transformed back to the original $\mathbf{x}$ and $\mathbf{y}$. Let us call the transformation $\mathbf{V}$. We have, then,

$$\mathbf{x} = \mathbf{V}\mathbf{x}' \qquad \text{and} \qquad \mathbf{y} = \mathbf{V}\mathbf{y}'. \tag{1.189}$$

In this contrived example, $\mathbf{V}$ can be found from (1.189) using simple algebra, because we happen to know what the transformed vectors are. In general, however, finding the transformed representation of a vector corresponding to a change of basis is a simple generalization of the following [see, e.g., (Chen, 1984, p. 17)]. We take the two columns of $\mathbf{V}$ to be: the representation of $\boldsymbol{\beta}_1'$ with respect to basis set $\{\boldsymbol{\beta}_1, \boldsymbol{\beta}_2\}$, and the representation of $\boldsymbol{\beta}_2'$ with respect to $\{\boldsymbol{\beta}_1, \boldsymbol{\beta}_2\}$, respectively.

Now consider computing the Euclidean distance of the transformed vectors to obtain a meaningful measure of their distance apart,

$$\begin{aligned} d_2(\mathbf{V}\mathbf{x}', \mathbf{V}\mathbf{y}') &= \sqrt{[\mathbf{V}\mathbf{x}' - \mathbf{V}\mathbf{y}']^T[\mathbf{V}\mathbf{x}' - \mathbf{V}\mathbf{y}']} \\ &= \sqrt{[\mathbf{x}' - \mathbf{y}']^T \mathbf{V}^T \mathbf{V}[\mathbf{x}' - \mathbf{y}']} \\ &= d_{2w}(\mathbf{x}', \mathbf{y}'). \end{aligned} \tag{1.190}$$

The last line in (1.190) denotes the weighted Euclidean distance with weighting matrix $\mathbf{W} = \mathbf{V}^T\mathbf{V}$. We see that the "meaningful" Euclidean distance for the vectors whose bases are not conducive to proper distance computation can be obtained by using a weighting matrix equivalent to the "square" of the transformation matrix.

It is sometimes desirable that a linear transformation of coordinates not change the rank ordering of distances from some reference vector. If in the above, for example, there were some vector z' such that

$$d_2(\mathbf{x}', \mathbf{z}') < d_2(\mathbf{y}', \mathbf{z}'), \tag{1.191}$$

then it might be desirable that

$$d_2(\mathbf{x}, \mathbf{z}) < d_2(\mathbf{y}, \mathbf{z}). \tag{1.192}$$

Whereas we would want the transformation to make the distance more meaningful, we might not wish to have the rank ordering changed in the new feature space. In general, the (weighted) Euclidean distance does *not* preserve this ordering.

In effect, what we have done in the above example is removed the redundant information in the "bad" vector representations that skews our sense of how naturally far apart they are. This is accomplished by linear transformation of the space, or, equivalently, weighting of the distance metric. This example was meant to build intuition about a more realistic and important problem in pattern recognition. We often encounter (random) vectors of features whose elements are highly correlated, or inappropriately scaled. The correlation and scaling effects will occur, for example, when multiple measurements are made on the same process and mixed in the same feature vector. For example, we might measure the average number of zero crossings[31] per norm-sec in a speech frame, and also the average energy. Clearly, there is no reason to believe that these numbers will have similar magnitudes in a given frame, since they represent quite different measurements on the sequence. Suppose, for example, that in one frame we measure 240 "joules," and 0.1 zero crossing, per norm-sec. In the next, we measure 300 and 0.05. Are these vector representations based on an appropriate orthonormal basis set so that Euclidean distances are meaningful? This answer could be argued either way, but the question is really academic. Our satisfaction with the distance measure here will depend upon how faithfully it reflects the difference in the frames in light of the measurements. So let us explore the question: Do these two frames represent the same sound? If so, we would like the distance to be small.

In answering this question, we should notice two things about the measurements. First, there could be less information in these measurements than we might assume. It could be the case that zero crossings tend to decrease when energy increases (correlation) so that the combination of changes does not make the two frames as different as the outcome might suggest. This point is reminiscent of the nonorthonormal basis case above. Second, note that the zero crossing measure is so relatively small in amplitude that its effect on the distance is negligible. In order for this feature to have more "discriminatory power" (which does not potentially get lost in numerical roundoff errors[32]), the relative scale of the features must be adjusted. (This corresponds to basis vectors of grossly different lengths, orthogonal or not.) An approach to solving this scaling problem is to simply normalize the feature magnitudes so that each has unity variance. Presumably, smaller features will have smaller variances (and conversely) and this will tend to bring the measurements into an appropriate relative scale. The "decorrelation" process is also not difficult; in fact, the scaling can be accomplished simultaneously using the following.

[31]The average number of times the sequence changes sign. This gives a rough measure of frequency content.

[32]Also as a practical matter, the measurement on a "low amplitude" feature is potentially much more susceptible to roundoff error problems in numerical computations, and the presence of grossly misscaled features can cause other numerical problems such as an ill-conditioned covariance matrix (Nobel, 1969, Sec. 8.2).

Suppose that the feature vectors between which we are trying to compute a distance are $\mathbf{x}'$ and $\mathbf{y}'$. Each is an outcome of random vector $\underline{\mathbf{x}}'$ with mean $\boldsymbol{\mu}_{\underline{\mathbf{x}}'}$ and covariance matrix $\mathbf{C}_{\underline{\mathbf{x}}'}$. We would like to transform the original random variable $\underline{\mathbf{x}}'$ to a representation, say $\underline{\mathbf{x}}$, in which all components are uncorrelated and are individually of unity variance. This means that the new covariance matrix $\mathbf{C}_{\underline{\mathbf{x}}}$ should equal $\mathbf{I}$, where $\mathbf{I}$ is the identity matrix. According to the heuristic arguments above, Euclidean distance computed on these vectors will then be intuitively appealing. As in the simple vector example above, we will show that the proper Euclidean distance can be computed using an appropriate weighting matrix in the computation.

The requisite transformation on the feature vectors is easily discovered by focusing on the covariance matrix. Since $\mathbf{C}_{\underline{\mathbf{x}}'}$ is a symmetric matrix, it can be written [see, e.g., (Nobel, 1969, Ch. 10)]

$$\mathbf{C}_{\underline{\mathbf{x}}'} = \boldsymbol{\Phi}\boldsymbol{\Lambda}\boldsymbol{\Phi}^T, \tag{1.193}$$

where $\boldsymbol{\Phi}$ is an orthogonal matrix whose columns are the normalized eigenvectors of $\mathbf{C}_{\underline{\mathbf{x}}'}$, and $\boldsymbol{\Lambda}$ is a diagonal matrix of eigenvalues of $\mathbf{C}_{\underline{\mathbf{x}}'}$. Therefore,

$$\boldsymbol{\Phi}\boldsymbol{\Lambda}\boldsymbol{\Phi}^T = \mathcal{E}\left\{\left[\underline{\mathbf{x}}' - \boldsymbol{\mu}_{\underline{\mathbf{x}}'}\right]\left[\underline{\mathbf{x}}' - \boldsymbol{\mu}_{\underline{\mathbf{x}}'}\right]^T\right\}, \tag{1.194}$$

from which it follows that

$$\mathbf{I} = \mathcal{E}\left\{\boldsymbol{\Lambda}^{-1/2}\boldsymbol{\Phi}^T\left[\underline{\mathbf{x}}' - \boldsymbol{\mu}_{\underline{\mathbf{x}}'}\right]\left[\underline{\mathbf{x}}' - \boldsymbol{\mu}_{\underline{\mathbf{x}}'}\right]^T\boldsymbol{\Phi}\boldsymbol{\Lambda}^{-1/2}\right\}. \tag{1.195}$$

Clearly, therefore, if we transform the feature vectors using the transformation

$$\underline{\mathbf{x}} = \boldsymbol{\Lambda}^{-1/2}\boldsymbol{\Phi}^T\underline{\mathbf{x}}', \tag{1.196}$$

we will be dealing with uncorrelated random vectors for which the Euclidean metric will provide a proper measure of distance. In this case,

$$\begin{aligned} d_2(\boldsymbol{\Lambda}^{-1/2}\boldsymbol{\Phi}^T\underline{\mathbf{x}}', \boldsymbol{\Lambda}^{-1/2}\boldsymbol{\Phi}^T\underline{\mathbf{y}}') &= \sqrt{[\underline{\mathbf{x}}' - \underline{\mathbf{y}}']^T\boldsymbol{\Phi}\boldsymbol{\Lambda}^{-1/2}\boldsymbol{\Lambda}^{-1/2}\boldsymbol{\Phi}^T[\underline{\mathbf{x}}' - \underline{\mathbf{y}}']} \\ &= \sqrt{[\underline{\mathbf{x}}' - \underline{\mathbf{y}}']^T\boldsymbol{\Phi}\boldsymbol{\Lambda}^{-1}\boldsymbol{\Phi}^T[\underline{\mathbf{x}}' - \underline{\mathbf{y}}']} \\ &= \sqrt{[\underline{\mathbf{x}}' - \underline{\mathbf{y}}']^T\mathbf{C}_{\underline{\mathbf{x}}'}^{-1}[\underline{\mathbf{x}}' - \underline{\mathbf{y}}']} \\ &= d_{2w}(\underline{\mathbf{x}}', \underline{\mathbf{y}}'). \end{aligned} \tag{1.197}$$

We see again that a meaningful Euclidean distance between correlated feature vectors can be computed if an appropriate weight is used. It is worth noting that the weighted Euclidean distance which has arisen here is very similar to the Mahalanobis distance that we discuss below.

The linear operation applied to the feature vectors in this procedure is frequently referred to as a *prewhitening transformation*, since it produces feature vectors whose components are uncorrelated and normalized. This terminology is somewhat abusive because the "white" concept applies to (usually scalar) random *processes* that are not being considered here, and also because "white" features would have zero means. This latter point would require, for a better analogy, that $\mathbf{R}_{\underline{\mathbf{x}}}$, rather than $\mathbf{C}_{\underline{\mathbf{x}}}$ be $\mathbf{I}$. Nevertheless, the terminology is widely used and is well understood by signal processing engineers.

Simplifications of the prewhitening procedure are sometimes employed to avoid the computational expense of using the full covariance matrix in the distance expression. The most common simplification is to assume that the features are mutually uncorrelated, but inappropriately scaled relative to one another. In this case $\mathbf{C}_{\underline{\mathbf{x}}'}$ (it is assumed) has the form

$$\mathbf{C}_{\underline{\mathbf{x}}'} = \mathbf{\Lambda}, \tag{1.198}$$

where $\mathbf{\Lambda}$ is a diagonal matrix whose diagonal elements in general are unequal. The transformation that need be done on each incoming vector is represented by $\mathbf{\Lambda}^{-1/2}\underline{\mathbf{x}}$. This amounts to simply normalizing each feature to its standard deviation so that all features may contribute equally to the distance.

1.3.3 Maximum Likelihood Classification

We frequently encounter problems in engineering in which a pattern representation is to be associated with one of a number of classes of patterns. This paradigm will occur in several significant places in our work. The purpose of this section is to explore a few underlying concepts with a particular interest in studying the distance measures that are often used in this endeavor.

Suppose that we have a set of classes, indexed by integers, say $c = 1, 2, \ldots, K$, which are outcomes of the class random variable, $\underline{c}$. Suppose that we also have a feature vector modeled by the random vector $\underline{\mathbf{x}}$. For example, the classes might represent the words in a vocabulary, and the feature vector a list of acoustic features extracted from the utterance of a word to be recognized. Ideally, given a feature vector outcome $\underline{\mathbf{x}} = \mathbf{x}$, we would select the class for which the conditional probability is highest. That is, c^* is the selected class (word) if

$$c^* = \underset{c}{\operatorname{argmax}}\, P(\underline{c} = c \mid \underline{\mathbf{x}} = \mathbf{x}). \tag{1.199}$$

Unfortunately, the training process usually does not permit characterization of the probabilities $P(\underline{c} = c \mid \underline{\mathbf{x}} = \mathbf{x})$. Instead what we learn is the probability that a given class will generate certain feature vectors, rather than the converse. The training process yields conditional probabilities of the form $P(\underline{\mathbf{x}}=\mathbf{x}|\underline{c}=c)$. So we ask whether it makes sense to select

$$c^* = \underset{c}{\operatorname{argmax}}\, P(\underline{\mathbf{x}}=\mathbf{x}|\underline{c}=c). \tag{1.200}$$

By definition

$$P(\underline{c}=c|\underline{\mathbf{x}}=\mathbf{x}) = \frac{P(\underline{c}=c, \underline{\mathbf{x}}=\mathbf{x})}{P(\underline{\mathbf{x}}=\mathbf{x})} \tag{1.201}$$

and

$$P(\underline{\mathbf{x}}=\mathbf{x}|\underline{c}=c) = \frac{P(\underline{c}=c, \underline{\mathbf{x}}=\mathbf{x})}{P(\underline{c}=c)}, \tag{1.202}$$

from which we have

$$P(\underline{c}=c|\underline{\mathbf{x}}=\mathbf{x}) = \frac{P(\underline{\mathbf{x}}=\mathbf{x}|\underline{c}=c)P(\underline{c}=c)}{P(\underline{\mathbf{x}}=\mathbf{x})}. \tag{1.203}$$

Clearly, the choice of c that maximizes the right side will also be the choice of c that maximizes the left side. Therefore,

$$c^* = \underset{c}{\operatorname{argmax}}\; P(\underline{c}=c|\underline{\mathbf{x}}=\mathbf{x}) = \underset{c}{\operatorname{argmax}}\; P(\underline{\mathbf{x}}=\mathbf{x}|\underline{c}=c)P(\underline{c}=c). \tag{1.204}$$

Furthermore, *if the class probabilities are equal,*

$$P(\underline{c}=c) = \frac{1}{K}, \qquad c = 1, 2, \ldots, K, \tag{1.205}$$

then

$$c^* = \underset{c}{\operatorname{argmax}}\; P(\underline{c}=c|\underline{\mathbf{x}}=\mathbf{x}) = \underset{c}{\operatorname{argmax}}\; P(\underline{\mathbf{x}}=\mathbf{x}|\underline{c}=c). \tag{1.206}$$

Therefore, under the condition of equal *a priori* class probabilities, the class decision

$$c^* = \underset{c}{\operatorname{argmax}}\; P(\underline{\mathbf{x}}=\mathbf{x}|\underline{c}=c) \tag{1.207}$$

is equivalent to the more desirable (1.199) for which we do not have probability distributions.

A quantity related to the probability of an event which is used to make a decision about the occurrence of that event is often called a *likelihood measure.* Hence, our decision rule based on given feature vector $\mathbf{x}$ is to choose the class c that maximizes the likelihood $P(\underline{\mathbf{x}}=\mathbf{x}|\underline{c}=c)$. This is called the *maximum likelihood decision.*

There is an implicit assumption in the discussion above that the random feature vector may only assume one of a finite number of outcomes. This is evident in the writing of probability distribution $P(\underline{\mathbf{x}}=\mathbf{x}|\underline{c}=c)$. Where this is not the case, it is frequently assumed that feature vectors associated with a given class are well modeled by a multivariate Gaussian distribution [cf. (1.118)],

$$f_{\underline{\mathbf{x}}|\underline{c}}(x_1, \ldots, x_N|c) = f_{\underline{\mathbf{x}}|\underline{c}}(\mathbf{x}|c)$$
$$= \frac{1}{\sqrt{(2\pi)^N \det \mathbf{C}_{\underline{\mathbf{x}}|c}}} \exp\left\{-\frac{1}{2}(\mathbf{x} - \boldsymbol{\mu}_{\underline{\mathbf{x}}|c})^T \mathbf{C}_{\underline{\mathbf{x}}|c}^{-1}(\mathbf{x} - \boldsymbol{\mu}_{\underline{\mathbf{x}}|c})\right\}, \tag{1.208}$$

where $\mathbf{x}$ denotes the N-vector of arguments (features) $[x_1 \cdots x_N]^T$ and $\mathbf{C}_{\underline{\mathbf{x}}|c}$ and $\boldsymbol{\mu}_{\underline{\mathbf{x}}|c}$ are the class-conditional covariance matrix and mean vector. Without belaboring the issue, it is believable based on our previous discussion that an appropriate likelihood measure for this case is the class-conditional density $f_{\underline{\mathbf{x}}|\underline{c}}(\mathbf{x}|c)$. The class decision is based on maximizing the likelihood,

$$c^* = \underset{c}{\text{argmax}}\ f_{\underline{\mathbf{x}}|\underline{c}}(\mathbf{x}|c). \tag{1.209}$$

We can rid ourselves of the need to compute the exponential by electing instead to maximize $\ln f_{\underline{\mathbf{x}}|c}(\mathbf{x}|c)$. This leads to the decision rule

$$c^* = \underset{c}{\text{argmin}}\left\{[\mathbf{x} - \boldsymbol{\mu}_{\underline{\mathbf{x}}|c}]^T \mathbf{C}_{\underline{\mathbf{x}}|c}^{-1}[\mathbf{x} - \boldsymbol{\mu}_{\underline{\mathbf{x}}|c}] + \ln\left\{\det \mathbf{C}_{\underline{\mathbf{x}}|c}\right\}\right. . \tag{1.210}$$

Note that the maximization has become a *minimization* because we have removed a superfluous minus sign from the computation. Notice also that the first term on the right has the form of a weighted Euclidean distance. Let us further develop this point.

The term on the right side of (1.210) is sometimes considered a *distance* between the given feature vector and the cth class mean, $\boldsymbol{\mu}_{\mathbf{x}|c}$. Accordingly, it provides a measure of "how far $\mathbf{x}$ is from class c." For generality, let us replace the specific outcome of the feature vector, $\mathbf{x}$, with its random variable, $\underline{\mathbf{x}}$, and define the *maximum likelihood distance* as

$$d_{ml}(\underline{\mathbf{x}}, \boldsymbol{\mu}_{\underline{\mathbf{x}}|c}) = [\underline{\mathbf{x}} - \boldsymbol{\mu}_{\underline{\mathbf{x}}|c}]^T \mathbf{C}_{\underline{\mathbf{x}}|c}^{-1}[\underline{\mathbf{x}} - \boldsymbol{\mu}_{\underline{\mathbf{x}}|c}] + \ln\left\{\det \mathbf{C}_{\underline{\mathbf{x}}|c}\right\}. \tag{1.211}$$

We see that for a multiclass, multivariate Gaussian feature problem, choosing the class that minimizes this distance is equivalent to choosing the maximum likelihood class.

A simplification occurs when all classes share a common covariance matrix, say

$$\mathbf{C}_{\underline{\mathbf{x}}} \stackrel{\text{def}}{=} \mathbf{C}_{\underline{\mathbf{x}}|1} = \mathbf{C}_{\underline{\mathbf{x}}|2} = \cdots = \mathbf{C}_{\underline{\mathbf{x}}|K}. \tag{1.212}$$

In this case $\mathbf{C}_{\underline{\mathbf{x}}|c}$ can be replaced by $\mathbf{C}_{\underline{\mathbf{x}}}$ in (1.211) and the final $\ln\{\cdot\}$ can be ignored, since it simply adds a constant to all distances. In this case, we obtain

$$d_M(\underline{\mathbf{x}}, \boldsymbol{\mu}_{\underline{\mathbf{x}}|c}) = [\underline{\mathbf{x}} - \boldsymbol{\mu}_{\underline{\mathbf{x}}|c}]^T \mathbf{C}_{\underline{\mathbf{x}}}^{-1}[\underline{\mathbf{x}} - \boldsymbol{\mu}_{\underline{\mathbf{x}}|c}]. \tag{1.213}$$

This distance is frequently called the *Mahalanobis distance* (Mahalanobis, 1936). We see that for a multiclass, multivariate Gaussian feature problem in which the classes share a common covariance matrix (the way in which features are correlated is similar across classes), choosing the class to which the given feature vector is closest in the sense of the Mahalanobis distance is tantamount to choosing the maximum likelihood class.

Interestingly, we have come full circle in our discussion, for it is apparent that the Mahalanobis distance is nothing more than a "covariance weighted" (squared) Euclidean distance[33] between the feature vector and a special set of deterministic vectors—the means of the classes. Nevertheless, the name Mahalanobis distance is often applied to this distance in this special maximum likelihood problem. Based on our previous discussion, it should be apparent that the Mahalanobis distance represents an appropriate use of the l_2 metric, since the inverse covariance weighting removes the correlation among the features in the vectors.

1.3.4 Feature Selection and Probabilistic Separability Measures

In the preceding section, we discussed a general problem in which a feature vector was associated with one of a number of classes. A subject that we avoided was the selection of features (this process is often called *feature extraction*) and their evaluation in terms of classification performance. These tasks are inseparable, since performance evaluation is often integrated into the search for appropriate features. In this section we make a few brief comments about these issues. One of the objectives is to let the reader know what material is not being covered with regard to this topic, and why. Another is to touch on the subject of probability separability measures and entropy measures, and to explain their specific relationship to speech processing.

Feature selection and evaluation is a vast subject on which much research has been performed and many papers and books written. To attempt to address this subject in any detail would take us too far afield from the main subject of this book. Several excellent textbooks address this field authoritatively and in detail, and we refer the reader to these books and the research literature for detailed study.[34] Second, the importance of feature evaluation procedures is diminished relative to the early days of speech processing. Although statistical pattern recognition techniques are central to the operation and performance of many speech processing tasks (particularly speech recognition), decades of research and development have led to convergence on a few (spectrally based) fea-

[33]Again, we could introduce a square root into the definition to make this distance exactly a Euclidean metric as defined in (1.180), but that would be breaking with convention. The Mahalanobis distance is almost invariably defined without the square root, and it should be clear that for the maximum likelihood problem, whether the distance is squared or not is of no consequence.

[34]For example, see the textbooks in Appendix 1.C and the *IEEE Transactions on Pattern Analysis and Machine Intelligence.*

tures that perform well, and appear to be enduring. This is not to say that new features have not been tried, and that the field is not evolving. Indeed, we have seen, for example, "cepstral" type of features supplant the "LP" type parameters in certain speech recognition tasks during the 1980s. This shift, however, was between two closely related sets of features and was to some extent motivated by computational expediencies. Further, the most frequently cited study behind this shift relies on experimental evidence of improved recognition performance (Davis and Mermelstein, 1980). Although the course of research is very unpredictable, for the foreseeable future, there appears to be no compelling problems that will demand a deep analysis of features.

As if to contradict the statement above, lurking in one little corner of our study (Section 12.2.7) we will mention some directions in speech recognition research that are based on the notions of probabilistic separability and entropy measures. These measures are customarily encountered in the advanced study of feature extraction and evaluation. We conclude this section by broadly discussing the types of feature evaluation, putting the topic of probabilistic separability measures and entropy measures into perspective.

Probabilistic Distance Measures

Ideally, features would be evaluated on their performance in terms of minimizing the rate of classification error. However, error rate is generally a very difficult quantity to evaluate, and other techniques must be employed. Almost all commonly used techniques for feature evaluation involve some attempt to measure the separation of classes when represented by the features.

The simplest techniques for measuring class separation (or *interclass distance*) are based on distance metrics in multidimensional space, especially the Euclidean distance and its variants, which we discussed extensively above. These measures generally do not utilize much of the probabilistic structure of the classes and therefore do not faithfully represent the degree of overlap of the classes in a statistical sense. The *probabilistic separability measures* represent an attempt to capture that information in the evaluation. There are two related types of probabilistic separability measures, the "probabilistic distances" and the "probabilistic dependencies."

To illustrate what is meant by a "probabilistic distance," consider the two-class problem for which class-conditional pdf's are shown for two different features, $\underline{x}$ and $\underline{y}$, in Fig. 1.14. Let us assume that the *a priori* class probabilities are equal, $P(\underline{c}=1)=P(\underline{c}=2)$. In the first case features (scalars, so we can draw a picture in two dimensions) characterized by random variable $\underline{x}$ are employed, and $f_{\underline{x}|\underline{c}}(x|1)$ and $f_{\underline{x}|\underline{c}}(x|2)$ are well separated with respect to the feature values. The classes appear to be almost fully separable based on these densities. On the other hand, when

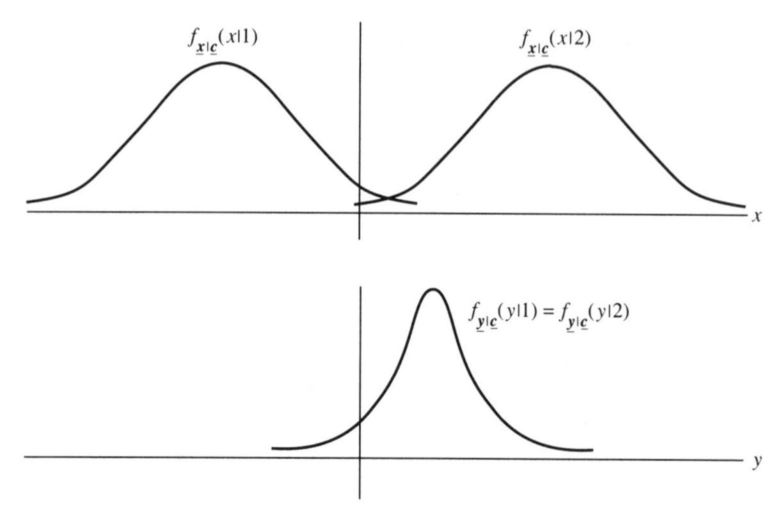

FIGURE 1.14. Class-conditional pdf's for two feature models, $\underline{x}$ and $\underline{y}$, used to introduce the concept of probabilistic distance.

features $\underline{y}$ are used, the separation is extremely poor. In this case $f_{\underline{y}|\underline{c}}(y|1)$ and $f_{\underline{y}|\underline{c}}(y|2)$ are identical and the classes would be completely inseparable based on this feature. That is, this feature would provide no better performance than simply guessing, or random assignment, of the class identity.

Probabilistic distance measures attempt to capture the degree of overlap of the class pdf's as a measure of their distance apart. In general, these measures take the form

$$J = \int_{-\infty}^{\infty} g\left\{f_{\underline{\mathbf{x}}|\underline{c}}(\mathbf{x}|c), P(\underline{c}=c), c = 1, 2, \ldots, K\right\} d\mathbf{x}, \tag{1.214}$$

where $g(\cdot)$ is some function, and $\int_{-\infty}^{\infty} (\cdot)\, d\mathbf{x}$ indicates the integral over the entire N-dimensional hyperplane with N the dimension of the feature vector $\mathbf{x}$. Probabilistic distance measures have the following properties (Devijver and Kittler, 1982):

1. J is nonnegative, $J \geq 0$.
2. J attains a maximum when all classes in the feature space are disjoint, J is maximum if $f_{\underline{\mathbf{x}}|\underline{c}}(\mathbf{x}|c) = 0$ when $f_{\underline{\mathbf{x}}|\underline{c}}(\mathbf{x}|c') \neq 0$ for all $c \neq c'$.
3. $J = 0$ when $f_{\underline{\mathbf{x}}|\underline{c}}(\mathbf{x}|1) = f_{\underline{\mathbf{x}}|\underline{c}}(\mathbf{x}|2) = \cdots = f_{\underline{\mathbf{x}}|\underline{c}}(\mathbf{x}|K)$.

Two examples of probabilistic distance measures for a two-class problem are the *Bhattacharyya distance*,

$$J_B = -\ln \int_{-\infty}^{\infty} \sqrt{f_{\underline{\mathbf{x}}|\underline{c}}(\mathbf{x}|1) f_{\underline{\mathbf{x}}|\underline{c}}(\mathbf{x}|2)}\, d\mathbf{x}, \tag{1.215}$$

and the *divergence*,

$$J_D = \int_{-\infty}^{\infty} [f_{\underline{\mathbf{x}}|\underline{c}}(\mathbf{x}|1) - f_{\underline{\mathbf{x}}|\underline{c}}(\mathbf{x}|2)] \ln \frac{f_{\underline{\mathbf{x}}|\underline{c}}(\mathbf{x}|1)}{f_{\underline{\mathbf{x}}|\underline{c}}(\mathbf{x}|2)} \, d\mathbf{x}, \qquad (1.216)$$

both of which reduce to a Mahalanobis-like distance in the case of Gaussian feature vectors and equal class covariances (see Problem 1.19).

Probabilistic Dependence Measures

Another method for indirectly assessing class pdf overlap is provided by the *probabilistic dependence measures.* These measures indicate how strongly the feature outcomes depend upon their class association. In the extreme case in which the features are independent of the class affiliation, the class conditional pdf's are identical to the "mixture" pdf (pdf of the entire universe of feature vectors),

$$f_{\underline{\mathbf{x}}|\underline{c}}(\mathbf{x}|c) = f_{\underline{\mathbf{x}}}(\mathbf{x}), \qquad \text{for all } c. \qquad (1.217)$$

Conversely, when the features depend very strongly on their class association, we expect $f_{\underline{\mathbf{x}}|\underline{c}}(\mathbf{x}|c)$ to be quite different from the mixture pdf. Therefore, a good indicator of the effectiveness of a set of features at separating the classes is given by the probabilistic dependence measures which quantify the difference between the class conditional pdf's and the mixture pdf. These measures adhere to the same properties noted above for the probabilistic distance measures and are generally of the form

$$J = \int_{-\infty}^{\infty} g\left\{f_{\underline{\mathbf{x}}|\underline{c}}(\mathbf{x}|c), f_{\underline{\mathbf{x}}}(\mathbf{x}), P(\underline{c}=c), c = 1, 2, \ldots, K\right\} d\mathbf{x}, \qquad (1.218)$$

and they adhere to the same properties as those listed above for probabilistic distance measures.

An example of a probabilistic dependence measure that we will encounter in the study of hidden Markov models (Chapter 12) is the *average mutual information,*

$$\overline{M}(\underline{c}, \underline{\mathbf{x}}) = \sum_{c=1}^{K} P(\underline{c}=c) \int_{-\infty}^{\infty} \cdots \int_{-\infty}^{\infty} f_{\underline{\mathbf{x}}|\underline{c}}(\mathbf{x}|c) \log_2 \frac{f_{\underline{\mathbf{x}}|\underline{c}}(\mathbf{x}|c)}{f_{\underline{\mathbf{x}}}(\mathbf{x})} \, d\mathbf{x} \qquad (1.219)$$

where the integral is taken over the multidimensional feature space. We note that if $\mathbf{x}$ takes only a finite number of values, say $\{\mathbf{x}_1, \ldots, \mathbf{x}_L\}$, then (1.219) becomes

$$\overline{M}(\underline{c}, \underline{\mathbf{x}}) = \sum_{c=1}^{K} P(\underline{c}=c) \sum_{l=1}^{L} P(\underline{\mathbf{x}}=\mathbf{x}_l|\underline{c}=c) \log_2 \frac{P(\underline{\mathbf{x}}=\mathbf{x}_l|\underline{c}=c)}{P(\underline{\mathbf{x}}=\mathbf{x}_l)}$$

$$= \sum_{c=1}^{K} \sum_{l=1}^{L} P(\underline{\mathbf{x}} = \mathbf{x}_l, \underline{c} = c) \log_2 \frac{P(\underline{\mathbf{x}} = \mathbf{x}_l, \underline{c} = c)}{P(\underline{\mathbf{x}} = \mathbf{x}_l)\, P(\underline{c} = c)}. \tag{1.220}$$

This measure, which can be seen to be an indicator of the average deviation of $f_{\underline{\mathbf{x}}|\underline{c}}(\mathbf{x}|c)$ from $f_{\underline{\mathbf{x}}}(\mathbf{x})$ [or $P(\underline{\mathbf{x}}|\underline{c})$ from $P(\underline{\mathbf{x}})$] will be given another interpretation when we discuss entropy concepts in Section 1.5.

Entropy Measures

Entropy measures are based on information-theoretic concepts that quantify the amount of uncertainty associated with the outcome of an experiment. In the pattern recognition context, these measures relate how much uncertainty remains about the class membership once a feature measurement is made. This knowledge quantifies the effectiveness of a set of features at conveying information that assists classification. Although we will have no direct use for entropy measures in this book, we will have several occasions to use the concepts of information and entropy. We will therefore address these issues in the next section, and, for completeness, include some comments on entropy measures in pattern recognition at the end of that section.

1.3.5 Clustering Algorithms

The previous discussions were based on the assumption that labeled (according to class) training features were available from which to infer the underlying probability structure of the classes. In some problems, however, information about the class membership of the training vectors is not provided. It is possible that we might not even know the number of classes represented by the training features. The problem of automatically separating training data into groups representing classes is often solved by a *clustering algorithm*.

The process of clustering is part of a more general group of techniques commonly referred to as *unsupervised learning*. As the name would imply, unsupervised learning techniques are concerned with the problem of forming classes from training data without benefit of supervision regarding class membership. Within this group of techniques, clustering algorithms represent a rather *ad hoc* approach to learning classes, which do not attempt to employ deep analysis of the statistical structure of the data. The more formal unsupervised learning methods are called *mode separation techniques* (Devijver and Kittler, 1982, Ch. 10), and we shall not have any use for these methods in our study of speech. Rather, clustering methods are based on the heuristic argument that vectors representing the same class should be "close" to one another in the feature space and "far" from vectors representing other classes. Accordingly, one of the distance metrics discussed above is usually employed in the analysis.

There are two basic classes of clustering algorithms. In *dynamic clustering*, a fixed number of clusters (classes) is used. At each iteration, feature vectors are reassigned according to certain rules until a stable partitioning of the vectors is achieved. We give an important example below. In *hierarchical clustering*, each feature vector is initially a separate cluster, then at each step of the algorithm, the two most similar clusters (according to some similarity criteria) are merged until the desired number of clusters is achieved.

There are a variety of clustering algorithms, but we focus on only one example of an iterative approach which is widely used in speech processing for a number of tasks. This is usually called the *K-means algorithm*, but the "K" simply refers to the number of desired classes and can be replaced by any desired index. The operation of the K-means algorithm is straightforward. Feature vectors are continuously reassigned to clusters, and the cluster centroids updated, until no further reassignment is necessary. The algorithm is given in Fig. 1.15.

The version of K-means given here is sometimes called the *isodata algorithm*. It is different from the original K-means algorithm in that it reassigns the entire set of training vectors before updating the cluster centroids. If means are recomputed after each vector is considered, then the algorithm terminates only after a complete scan of the training set is made without reassignment.

FIGURE 1.15. The K-means algorithm.

Initialization: Choose an arbitrary partition of the training vectors $\{\mathbf{x}\}$ into K clusters, denoted Λ_k, $k = 1, 2, \ldots, K$, and compute the mean vector (centroid) of each cluster, $\overline{\mathbf{x}}_k$, $k = 1, 2, \ldots, K$.

Recursion:

1. For each feature vector, $\mathbf{x}$, in the training set, assign $\mathbf{x}$ to Λ_{k^*}, where

$$k^* = \underset{k}{\operatorname{argmin}}\ d(\mathbf{x}, \overline{\mathbf{x}}_k). \tag{1.221}$$

   $d(\cdot, \cdot)$ represents some distance measure in the feature space.
2. Recompute the cluster centroids, and return to Step 1 if any of the centroids change from the last iteration.

A brief history and more details of the K-means approach from an information theory perspective is given in the paper by Makhoul et al. (1985). In an unpublished 1957 paper [more recently published, see (Lloyd, 1982)], Lloyd, independently of the pattern recognition research efforts, had essentially worked out the isodata algorithm for scalar quantization in pulse code modulation. The generalization of the K-means algorithm to "vector quantization," a technique which we will first encounter in Chapter 7, is sometimes called the *generalized Lloyd algo-*

rithm (Gray, 1984). A further generalization involves the fact that the K-means approach can also be applied to representations of the clusters other than centroids, and to measures of similarities other than distance metrics (Devijver and Kittler, 1982). A measure of similarity which does not necessarily adhere to the formal properties of a distance metric is often called a *distortion measure*. Linde et al. (1980) were the first in the communications literature to suggest the use of vector quantization with K-means and nonmetric distortion measures. Consequently, the K-means algorithm (particularly with these generalizations) is frequently called the *Linde–Buzo–Gray* or *LBG algorithm* in the speech processing and other communications literature.

Generally, the objective of the LBG algorithm is to find a set of, say, K feature vectors (*codes*) into which all feature vectors in the training set can be "quantized" with minimum distortion. This is like adjusting the levels of a scalar quantizer to minimize the amount of distortion incurred when a signal is quantized. This set of code vectors comprises a *codebook* for the feature space. The method is generally described in Fig. 1.16. A slight variation on the LBG method is also shown in Fig. 1.16, which differs in the way in which the algorithm is initialized. In the latter case, the number of clusters is iteratively built up to a desired number (power of two) by "splitting" the existing codes at each step and using these split codes to seed the next iteration.

FIGURE 1.16. The generalized Lloyd or Linde–Buzo–Gray (LBG) algorithm.

Initialization: Choose an arbitrary set of K code vectors, say $\bar{\mathbf{x}}_k$, $k = 1, 2, \ldots, K$.

Recursion:

1. For each feature vector, $\mathbf{x}$, in the training set, "quantize" $\mathbf{x}$ into code $\bar{\mathbf{x}}_{k^*}$, where

$$k^* = \underset{k}{\operatorname{argmin}}\; d(\mathbf{x}, \bar{\mathbf{x}}_k). \tag{1.222}$$

Here $d(\cdot\,,\cdot)$ represents some distortion measure in the feature space.

2. Compute the total distortion that has occurred as a result of this quantization,

$$D = \sum d[\mathbf{x}, Q(\mathbf{x})], \tag{1.223}$$

where the sum is taken over all vectors $\mathbf{x}$ in the training set, and $Q(\mathbf{x})$ indicates the code to which $\mathbf{x}$ is assigned in the current iteration. (This is an estimate of $\mathcal{E}\{d[\mathbf{x}, Q(\mathbf{x})]\}$.) If D is sufficiently small, STOP.

3. For each k, compute the centroid of all vectors $\mathbf{x}$ such that $\bar{\mathbf{x}}_k = Q(\mathbf{x})$ during the present iteration. Let this new set of centroids comprise the new codebook, and return to Step 1.

Alternative LBG algorithm with "centroid splitting."

Initialization: Find the centroid of the entire population of vectors. Let this be the (only) initial code vector.

Recursion. There are I total iterations where 2^I code vectors are desired. Let the iterations be $i = 1, 2, \ldots, I$. For iteration i,

1. "Split" any existing code vector, say $\bar{\mathbf{x}}$, into two codes, say $\bar{\mathbf{x}}(1+\varepsilon)$ and $\bar{\mathbf{x}}(1-\varepsilon)$, where ε is a small number, typically 0.01. This results in 2^i new code vectors, say $\bar{\mathbf{x}}_k^i$, $k = 1, 2, \ldots, 2^i$.
2. For each feature vector, $\mathbf{x}$, in the training set, "quantize" $\mathbf{x}$ into code $\bar{\mathbf{x}}_{k^*}^i$, where $k^* = \underset{k}{\operatorname{argmin}}\, d\left(\mathbf{x}, \bar{\mathbf{x}}_k^i\right)$. Here $d(\cdot,\cdot)$ represents some distortion measure in the feature space.
3. For each k, compute the centroid of all vectors $\mathbf{x}$ such that $\bar{\mathbf{x}}_k^i = Q(\mathbf{x})$ during the present iteration. Let this new set of centroids comprise the new codebook, and, if $i < I$, return to Step 1.

1.4 Information and Entropy

Reading Note: The material in this section will not be needed until Parts IV and V of the text.

The issues discussed here are a few necessary concepts from the field of information theory. The reader interested in this field should consult one of many widely used books on this subject (see Appendix 1.D).

Note that our need for this material in this text will usually occur in cases in which all random vectors (or variables) take discrete values. We will therefore focus on such cases. Similar definitions and developments exist for continuous random vectors [e.g., (Papoulis, 1984)].

1.4.1 Definitions

At a rudimentary level, the field of information theory is concerned with the amount of uncertainty associated with the outcome of an experiment. Once the experiment is performed and the outcome is known, the uncertainty is dispelled. The amount of information we receive when the outcome is known depends upon how much uncertainty there was about its occurrence.

In the pattern recognition problem above, for example, learning which of the K classes (e.g., words) represents the correct answer is informative. How uncertain we were before the answer was revealed (and therefore how much information we receive) depends on the probability distribution of the classes. For example, consider the extreme cases,

$$P(\underline{c} = c) = \frac{1}{K}, \qquad \text{for all } c \tag{1.224}$$

and

$$P(\underline{c}=c)=\begin{cases}1, & c=c' \\ 0, & c\neq c'.\end{cases} \tag{1.225}$$

In the first case in which the class probabilities are uniformly distributed, we have complete uncertainty about the association of a given feature vector, and gain the maximum information possible (on the average) when its true association is revealed. On the other hand, in the second case we have no doubt that true class is c', and no information is imparted with the revelation of the class identity. In either case, the information we receive is in indirect proportion to the probability of the class.[35]

The same intuitive arguments apply, of course, to the outcomes of any random variable—the quantity $\underline{c}$ need not model class outcomes in a pattern recognition problem. Let us therefore begin to view $\underline{c}$ as a general discrete random variable. In fact, for even broader generality, let us begin to work with a random *vector*, $\underline{\mathbf{c}}$, recognizing, of course, that the scalar random variable is a special case. According to the notion that information is inversely proportional to outcome likelihood, Shannon (1948) proposed the following formalism. We define the *information* associated with a particular outcome, $\mathbf{c}$, of a discrete random vector, $\underline{\mathbf{c}}$, to be

$$I(\underline{\mathbf{c}}=\mathbf{c}) \stackrel{\text{def}}{=} \log_2 \frac{1}{P(\underline{\mathbf{c}}=\mathbf{c})} = -\log_2 P(\underline{\mathbf{c}}=\mathbf{c}). \tag{1.226}$$

The information is a measure of uncertainty associated with outcome $\mathbf{c}$-the less likely is the value $\mathbf{c}$, the more information we receive. Although information may be defined using any logarithmic base, usually base two is used, in which case $I(\cdot)$ is measured in *bits*. The sense of this term is as follows: If there are K *equally likely* outcomes, say $\mathbf{c}_1, \ldots, \mathbf{c}_K$, and each is assigned an integer $1, 2, \ldots, K$, then it requires a binary number with $\log_2 K$ bits to identify the index of a particular outcome. In this case, we receive exactly that number of bits of information when it is revealed that the true outcome is $\mathbf{c}$,

$$I(\underline{\mathbf{c}}=\mathbf{c}) = -\log_2 P(\underline{\mathbf{c}}=\mathbf{c}) = \log_2 K. \tag{1.227}$$

$I(\underline{\mathbf{c}}=\mathbf{c})$ can therefore be interpreted as the number of binary digits required to identify the outcome $\mathbf{c}$ if it is one of $2^{I(\underline{\mathbf{c}}=\mathbf{c})}$ equally likely possibilities.

In general, of course, information is a random quantity that depends on the outcome of the random variable. We denote this by writing simply $I(\underline{\mathbf{c}})$. The *entropy* is a measure of *expected* information across all outcomes of the random vector,

[35]According to Papoulis (1981), Planck was the first to describe the explicit relationship between probability and information in 1906.

$$H(\underline{\mathbf{c}}) \stackrel{\text{def}}{=} \mathcal{E}\{I(\underline{\mathbf{c}})\} = -\sum_{l=1}^{K} P(\underline{\mathbf{c}} = \mathbf{c}_l) \log_2 P(\underline{\mathbf{c}} = \mathbf{c}_l). \tag{1.228}$$

Now consider N random vectors, say $\underline{\mathbf{x}}(1), \ldots, \underline{\mathbf{x}}(N)$, each of which produces outcomes from the same finite set,[36] $\{\mathbf{x}_1, \ldots, \mathbf{x}_L\}$. By a natural generalization of the above, the information associated with the revelation that $\underline{\mathbf{x}}(1) = \mathbf{x}_{k_1}, \ldots, \underline{\mathbf{x}}(N) = \mathbf{x}_{k_N}$ is defined as

$$I[\underline{\mathbf{x}}(1) = \mathbf{x}_{k_1}, \ldots, \underline{\mathbf{x}}(N) = \mathbf{x}_{k_N}] \stackrel{\text{def}}{=} -\log_2 P[\underline{\mathbf{x}}(1) = \mathbf{x}_{k_1}, \ldots, \underline{\mathbf{x}}(N) = \mathbf{x}_{k_N}], \tag{1.229}$$

and the entropy associated with these random variables is

$$\begin{aligned} H[\underline{\mathbf{x}}(1), \ldots, \underline{\mathbf{x}}(N)] &\stackrel{\text{def}}{=} \mathcal{E}\{I[\underline{\mathbf{x}}(1), \ldots, \underline{\mathbf{x}}(N)]\} \\ &= -\sum_{l_1=1}^{L} \cdots \sum_{l_N=1}^{L} P[\underline{\mathbf{x}}(1) = \mathbf{x}_{l_1}, \ldots, \underline{\mathbf{x}}(N) = \mathbf{x}_{l_N}] \\ &\quad \times \log_2 P[\underline{\mathbf{x}}(1) = \mathbf{x}_{l_1}, \ldots, \underline{\mathbf{x}}(N) = \mathbf{x}_{l_N}]. \end{aligned} \tag{1.230}$$

$I[\underline{\mathbf{x}}(1), \ldots, \underline{\mathbf{x}}(N)]$ and $H[\underline{\mathbf{x}}(1), \ldots, \underline{\mathbf{x}}(N)]$ are called the *joint information* and *joint entropy*, respectively. If random vectors $\underline{\mathbf{x}}(1), \ldots, \underline{\mathbf{x}}(N)$ are *independent*, then

$$I[\underline{\mathbf{x}}(1), \ldots, \underline{\mathbf{x}}(N)] = \sum_{n=1}^{N} I[\underline{\mathbf{x}}(n)] \tag{1.231}$$

and

$$H[\underline{\mathbf{x}}(1), \ldots, \underline{\mathbf{x}}(N)] = \sum_{n=1}^{N} H[\underline{\mathbf{x}}(n)]. \tag{1.232}$$

In particular, if $\underline{\mathbf{x}}(1), \ldots, \underline{\mathbf{x}}(N)$ are *independent and identically distributed*, then

$$H[\underline{\mathbf{x}}(1), \ldots, \underline{\mathbf{x}}(N)] = NH[\underline{\mathbf{x}}(n)] \qquad \text{for arbitrary } n. \tag{1.233}$$

Intuitively, the information received when we learn the outcome, say $\mathbf{x}_k$, of a random vector, $\underline{\mathbf{x}}$, will be less if we already know the outcome, say $\mathbf{y}_l$, of a correlated random vector, $\underline{\mathbf{y}}$. Accordingly, we define the *conditional information* and *conditional entropy*, respectively, as

$$I(\underline{\mathbf{x}} = \mathbf{x}_k | \underline{\mathbf{y}} = \mathbf{y}) = -\log_2 P(\underline{\mathbf{x}} = \mathbf{x}_k | \underline{\mathbf{y}} = \mathbf{y}) \tag{1.234}$$

[36]This definition is easily generalized to the case in which all random vectors have different sets of outcomes, but we will not have need of this more general case.

and

$$H(\underline{\mathbf{x}}|\underline{\mathbf{y}}) = \mathcal{E}_{\underline{\mathbf{x}},\underline{\mathbf{y}}}\{I(\underline{\mathbf{x}}|\underline{\mathbf{y}})\}$$
$$= \sum_{k=1}^{K}\sum_{l=1}^{L} P(\underline{\mathbf{x}} = \mathbf{x}_l, \underline{\mathbf{y}} = \mathbf{y}_k) \log_2 P(\underline{\mathbf{x}} = \mathbf{x}_l | \underline{\mathbf{y}} = \mathbf{y}_k), \tag{1.235}$$

where $\mathcal{E}_{\underline{\mathbf{x}},\underline{\mathbf{y}}}$ denotes the expectation with respect to both random vectors $\underline{\mathbf{x}}$ and $\underline{\mathbf{y}}$ and where we have assumed that $\underline{\mathbf{y}}$ takes discrete values $\{\mathbf{y}_1, \ldots, \mathbf{y}_K\}$.

Finally, we need to introduce the notion of "mutual information." The pairing of random vector outcomes intuitively produces less information than the sum of the individual outcomes *if the random vectors are not independent.* Upon arriving at the airport, we receive less information when the ticket agent tells us that (1) we missed the plane, and (2) the plane left 15 minutes ago, than either of those pieces of information would provide individually. This is because the two pieces of information are related and contain information about each other. Formally, this means that

$$I(\underline{\mathbf{x}}, \underline{\mathbf{y}}) \leq I(\underline{\mathbf{x}}) + I(\underline{\mathbf{y}}). \tag{1.236}$$

The "shared" information that is inherent in either of the individual outcomes is called the *mutual information* between the random vectors,

$$M(\underline{\mathbf{x}}, \underline{\mathbf{y}}) \stackrel{\text{def}}{=} [I(\underline{\mathbf{x}}) + I(\underline{\mathbf{y}})] - I(\underline{\mathbf{x}}, \underline{\mathbf{y}}). \tag{1.237}$$

It follows from the definitions above that

$$M(\underline{\mathbf{x}}, \underline{\mathbf{y}}) = \log_2 \frac{P(\underline{\mathbf{x}}, \underline{\mathbf{y}})}{P(\underline{\mathbf{x}})P(\underline{\mathbf{y}})}. \tag{1.238}$$

Equation (1.238), in turn, leads to the conclusion that

$$M(\underline{\mathbf{x}}, \underline{\mathbf{y}}) = I(\underline{\mathbf{x}}) - I(\underline{\mathbf{x}}|\underline{\mathbf{y}}) = I(\underline{\mathbf{y}}) - I(\underline{\mathbf{y}}|\underline{\mathbf{x}}). \tag{1.239}$$

This result clearly shows the interpretation of the mutual information as the information that is "shared" by the random vectors.

Like an entropy measure, the *average mutual information,* which we denote $\overline{M}(\underline{\mathbf{x}}, \underline{\mathbf{y}})$, is the *expected* mutual information over all values of the random vectors,

$$\overline{M}(\underline{\mathbf{x}}, \underline{\mathbf{y}}) \stackrel{\text{def}}{=} \mathcal{E}_{\underline{\mathbf{x}},\underline{\mathbf{y}}}\left\{\log_2 \frac{P(\underline{\mathbf{x}}, \underline{\mathbf{y}})}{P(\underline{\mathbf{x}})P(\underline{\mathbf{y}})}\right\}$$
$$= \sum_{k=1}^{K}\sum_{l=1}^{L} P(\underline{\mathbf{x}} = \mathbf{x}_l, \underline{\mathbf{y}} = \mathbf{y}_k) \log_2 \frac{P(\underline{\mathbf{x}} = \mathbf{x}_l, \underline{\mathbf{y}} = \mathbf{y}_k)}{P(\underline{\mathbf{x}} = x_l)P(\underline{\mathbf{y}} = \mathbf{y}_k)}. \tag{1.240}$$

Note from (1.238) and (1.240) that

$$\overline{M}(\underline{\mathbf{x}}, \underline{\mathbf{y}}) = H(\underline{\mathbf{x}}) - H(\underline{\mathbf{x}}|\underline{\mathbf{y}}) = H(\underline{\mathbf{y}}) - H(\underline{\mathbf{y}}|\underline{\mathbf{x}}), \tag{1.241}$$

which immediately leads to the conclusion that if $\underline{\mathbf{x}}$ and $\underline{\mathbf{y}}$ are independent, then there is no average mutual information (no "shared" information on the average).

We will see entropy concepts play a role in several areas of speech coding and recognition in Chapters 7 and 12. The mutual information will be used in an important speech recognition technique in Chapter 12. We illustrate the use of some entropy concepts in pattern recognition in Section 1.4.3.

1.4.2 Random Sources

In several places in our work, we will need to characterize the information conveyed by a (vector) random *process*, say

$$\underline{\mathbf{x}} = \{\ldots, \underline{\mathbf{x}}(-1), \underline{\mathbf{x}}(0), \underline{\mathbf{x}}(1), \ldots\}, \tag{1.242}$$

in which each random vector takes a finite number of discrete outcomes, say $\mathbf{x}_1, \ldots, \mathbf{x}_L$. In communications applications, the random process will often characterize the output of a transmitter where it is given the name *random source*. Nevertheless, from a mathematical point of view, a random source is equivalent to a random process.

How then do we characterize the information from a random source? The usual method is to indicate the entropy per sample (per random vector), which, if the process is[37] *stationary with independent random vectors*, is equivalent to the entropy associated with *any* random vector,

$$H(\underline{\mathbf{x}}) \stackrel{\text{def}}{=} H[\underline{\mathbf{x}}(n)] = -\sum_{l=1}^{L} P[\underline{\mathbf{x}}(n) = \mathbf{x}_l] \log_2 P[\underline{\mathbf{x}}(n) = \mathbf{x}_l]. \tag{1.243}$$

However, if the random vectors are not independent, then we must use[38]

$$H(\underline{\mathbf{x}}) \stackrel{\text{def}}{=} -\lim_{N\to\infty} \sum_{l_1=1}^{L} \cdots \sum_{l_N=1}^{L} P[\underline{\mathbf{x}}(1) = \mathbf{x}_{l_1}, \ldots, \underline{\mathbf{x}}(N) = \mathbf{x}_{l_N}]$$
$$\times \log_2 P[\underline{\mathbf{x}}(1) = \mathbf{x}_1, \ldots, \underline{\mathbf{x}}(N) = \mathbf{x}_{l_N}]. \tag{1.244}$$

If the random vectors are uncorrelated beyond some finite N, then the expression need not contain the limit. Definition (1.244) is useful for theoretical discussions, but it becomes practically intractable for N's

[37]A stationary source with discrete, independent random variables (or vectors) is called a *discrete memoryless source* in the communications field [see, e.g., (Proakis, 1989, Sec. 2.3.2)].

[38]We assume here that the random process starts at $n = 0$.

much larger than two or three. We will see one interesting application of this expression in our study of language modeling in Chapter 13.

1.4.3 Entropy Concepts in Pattern Recognition

Entropy measures are used in pattern recognition problems. To provide an example of the use of the entropy concepts described above, and also to provide closure to our discussion of probabilistic separability measures, we briefly consider that task here. The material here will turn out to be very similar to a development needed in our study of speech recognition. One disclaimer is in order before we begin the discussion. Because we have only studied entropy concepts for the case of *discrete* conditioning vectors, we will only consider the case of discrete feature vectors here. This is consistent with our need for this material in the text, but is at variance with our discussion of features in Section 1.4. The more general case is found in (Devijver and Kittler, 1982, Sec. 5.3.5).

Generalized entropy measures are used to assess the effectiveness of a set of features at pattern classification. The properties of such measures are quite complex, and are described, for example, in (Devijver and Kittler, 1982, App. C). A special case of a generalized entropy measure is what we would simply call the conditional entropy for the set of classes, characterized by random variable $\underline{c}$, conditioned upon knowledge of a feature vector, modeled by random vector $\underline{\mathbf{x}}$. From (1.235)

$$H(\underline{c}|\underline{\mathbf{x}}) = \sum_{c=1}^{K} \sum_{l=1}^{L} P(\underline{c}=c, \underline{\mathbf{x}}=\mathbf{x}_l) \log_2 P(\underline{c}=c|\underline{\mathbf{x}}=\mathbf{x}_l). \tag{1.245}$$

This quantity provides a measure of the average quality of the chosen features over the entire feature space. $H(\underline{c}|\mathbf{x})$ is sometimes called the *equivocation.* We would want the equivocation to be small, meaning that, on the average, the feature vector $\mathbf{x}$ greatly reduces the uncertainty about the class identity.

A related way to view the feature effectiveness is to examine the average mutual information between the random variable $\underline{c}$ and random vector $\underline{\mathbf{x}}$, $\overline{M}(\underline{c}, \underline{\mathbf{x}})$. Ideally, this measure is large, meaning that a given feature outcome contains a significant amount of information about the class outcome. From (1.240) we can write

$$\overline{M}(\underline{c}, \underline{\mathbf{x}}) = \sum_{c=1}^{K} \sum_{l=1}^{L} P(\underline{c}=c, \underline{\mathbf{x}}=\mathbf{x}_l) \log_2 \frac{P(\underline{c}=c, \underline{\mathbf{x}}=\mathbf{x}_l)}{P(\underline{c}=c)P(\underline{\mathbf{x}}=\mathbf{x}_l)}. \tag{1.246}$$

The reader can confirm that this measure is identical to (1.220) discussed above.

It might also be of interest to characterize the average mutual information between two *jointly stationary random sources*, say $\underline{\mathbf{x}}$ and $\underline{\mathbf{y}}$. By this we will simply mean the average mutual information between two random variables at any arbitrary time n. We will write $\overline{M}(\underline{\mathbf{x}}, \underline{\mathbf{y}})$ to emphasize that the random variables are taken from the stationary random sources,

$$\overline{M}(\underline{\mathbf{x}}, \underline{\mathbf{y}}) \stackrel{\text{def}}{=} \sum_{l=1}^{L} \sum_{k=1}^{K} P(\underline{\mathbf{x}} = \mathbf{x}_l, \underline{\mathbf{y}} = \mathbf{y}_k) \log_2 \frac{P(\underline{\mathbf{x}} = \mathbf{x}_l, \underline{\mathbf{y}} = \mathbf{y}_k)}{P(\underline{\mathbf{x}} = \mathbf{x}_l) P(\underline{\mathbf{y}} = \mathbf{y}_k)}. \tag{1.247}$$

Here we have assumed that the sources are vector random processes that take discrete vector values. Versions of (1.247) for other cases, for example, scalar random processes that take continuous values, require obvious modifications [cf. (1.219)].

1.5 Phasors and Steady-State Solutions

In our work with analog acoustic modeling of the speech production system, we will be concerned with the solution of a linear, constant coefficient, differential equation (LCCDE). This short section is intended to remind the reader of some commonly used techniques and notation.

Consider a continuous time system described by an LCCDE

$$\sum_{i=1}^{n} a_i \frac{d^i}{dt^i} y(t) = \sum_{i=1}^{m} b_i \frac{d^i}{dt^i} x(t), \tag{1.248}$$

where $x(t)$ and $y(t)$ are the input and output, respectively. It is often desired to know the *steady-state* response of the system (response after all transients have diminished) to a sinusoidal input, say $x(t) = X \cos(\Omega t + \varphi_x)$. It is frequently convenient to replace the cosine (or sine) by a complex exponential,

$$x(t) = X e^{j(\Omega t + \varphi_x)}, \tag{1.249}$$

recognizing that the solutions to the real (cosine) and imaginary (sine) parts of the exponential will remain separated in the solution because of linearity. Further, it is also frequently useful to rewrite (1.249) as

$$x(t) = \overline{X} e^{j\Omega t}, \tag{1.250}$$

where $\overline{X}$ *is the complex number*

$$\overline{X} = X e^{j\varphi_x}, \tag{1.251}$$

which is called a *phasor* for the exponential signal $x(t)$. Also due to linearity, we know that the input (1.249) will produce an output of the form $y(t) = Y e^{j(\Omega t + \varphi_y)}$, which may also be written in terms of a phasor,

$$y(t) = \overline{Y}e^{j\Omega t}, \tag{1.252}$$

where $\overline{Y} = Ye^{j\varphi_y}$. Putting the forms (1.250) and (1.252) into (1.248), it is found immediately that the terms $e^{j\Omega t}$ cancel and the differential equation solution reduces to one of solving an algebraic equation for $\overline{Y}$ in terms of $\overline{X}$, powers of Ω, and the coefficients a_i and b_i. Engineers often take advantage of this fact and solve algebraic phasor equations directly for steady-state solutions, sidestepping the differential equations completely. We have developed constructs such as "impedance" to assist in these simplified solutions (see below).

In fact, recall that phasor analysis amounts to steady-state frequency domain analysis. In principle, the phasors $\overline{X}$ and $\overline{Y}$ are frequency dependent, because we may enter a variety of inputs (actually an uncountably infinite number!) of the form $x(t) = X\cos(\Omega t + \varphi_x)$, each with a different frequency, Ω; amplitude, X; and phase, φ_x, to produce corresponding outputs of form $y(t) = Y\cos(\Omega t + \varphi_y)$ with frequency-dependent amplitudes and phases. We may reflect this fact by writing the phasors as $\overline{X}(\Omega)$ and $\overline{Y}(\Omega)$. Plugging forms (1.250) and (1.252) into (1.248) with these explicitly frequency-dependent phasors immediately produces the general expression for the output phasor

$$\overline{Y}(\Omega) = \frac{\sum_{i=0}^{m} b_i \Omega^i}{1 + \sum_{i=1}^{n} a_i \Omega^i} \overline{X}(\Omega). \tag{1.253}$$

The ratio $H(\Omega) \stackrel{\text{def}}{=} \overline{Y}(\Omega)/\overline{X}(\Omega)$, is of course the (Fourier) transfer function for the system. Other ratios, in particular, impedances and admittances, result from similar analyses. If, for example, $y(t)$ is a voltage across a discrete electrical component in response to current $x(t)$, then the phasor ratio $Z(\Omega) = \overline{Y}(\Omega)/\overline{X}(\Omega)$ resulting from the (usually simple) differential equation governing the component is the *impedance* (frequency dependent) of that component. The algebraic equations resulting from phasor-based solutions of differential equations mimic the simple "Ohm's law" type relations that arise in DC analysis of resistive circuits. As electrical engineers, we sometimes become so familiar with these simple phasor techniques that we forget their fundamental connection to the underlying differential equation.

In connection with the concepts above, we note that the ratio of phasors is always equivalent to the ratio of *complex* signals they represent,

$$\frac{\overline{Y}(\Omega)}{\overline{X}(\Omega)} = \frac{\overline{Y}(\Omega)e^{j\Omega t}}{\overline{X}(\Omega)e^{j\Omega t}} = \frac{y(t)}{x(t)}. \tag{1.254}$$

This fact is sometimes useful in theoretical discussions in which phasor notations have not been defined for certain signals.

We will make use of these ideas in our early work (Chapter 3) concerning analog acoustic modeling of the speech production system. If necessary, the reader should review these topics in any of a number of engineering textbooks [e.g., (Hayt and Kimmerly, 1971)] or textbooks on differential equations [e.g., (Boyce and DiPrima, 1969)].

1.6 Onward to Speech Processing

Thus ends our review and tutorial of selected background material prerequisite to the study of speech processing. The reader will probably want to refer back to this chapter frequently to recall notational conventions and basic analytical tools. Before beginning our formal study, we make a few introductory comments about the speech processing field, and about the organization of the book.

Brief History. The history of speech processing certainly does not begin with the digital signal processing engineer, nor even with the work of electrical engineers. In an interesting article[39] surveying some of the history of speech synthesis, Flanagan (1972) notes humankind's fascination with speech and voice from ancient times, and places the advent of the scientific study of speech in the Rennaisance when clever mechanical models were constructed to imitate speech. The first well-documented efforts at mechanical speech synthesis occurred in St. Petersburg and Vienna in the late eighteenth century. The 1930s, a century and a half later, are often considered to be the beginning of the modern speech technology era, in large part due to two key developments at Bell Laboratories. The first was the development of pulse code modulation (PCM), the first digital representation of speech (and other waveforms) which helped to pioneer the field of digital communications. The second was the demonstration of the Vocoder (*Vo*ice *Coder*) by Dudley (1939), a speech synthesizer, the design of which first suggested the possibility of parametric speech representation and coding. The subsequent decades have seen an explosion of activity roughly concentrated into decades. We mention a few key developments: intense research on the basic acoustical aspects of speech production and concomitant interest in electronic synthesizers in the late 1940s through the 1960s (Fant, 1960), which was spurred on by the invention of the spectrograph in 1946 (Potter et al., 1966); advances in analysis and coding algorithms (linear prediction, cepstrum) in the 1960s (see Chapters 5 and 6 in this book) made possible by the new digital computing machines and related work in digital signal processing [e.g., (Cooley and Tukey, 1965)]; development of temporally adaptive speech coding algorithms in the 1970s (see Chapter 7); and vast

[39]Also see (Schroeder, 1966). Each of these papers, as well as others describing early work, are reprinted in (Schafer and Markel, 1979).

interest in speech recognition research in the 1970s and 1980s and continuing into the 1990s, grounded in the development of dynamic programming techniques, hidden Markov modeling, vector quantization, neural networks, and significant advances in processor architectures and fabrication (see the chapters of Part V).

Research Areas and Text Organization. There is no precise way to partition the speech processing research field into its component areas. Nevertheless, we offer the following first approximation to a partition that can roughly be inferred from the discussion above:

Speech Science (Speech Production and Modeling) (Part II of this book)
Analysis (Part III)
Coding, Synthesis, Enhancement, and Quality Assessment (Part IV)
Recognition (Part V)

We have organized the book around these themes.

Part II is concerned with providing necessary topics in speech science and with early efforts to model speech production, which are grounded in the physics of the biological system. By *speech science* we mean the use of engineering techniques—spectral analysis, modeling, and so on—in work that is specifically aimed at a better understanding of the physiological mechanisms, anatomy, acoustic, phonetic, and linguistic aspects of normal and abnormal voice and speech production. Naturally, such work is highly interdisciplinary and is least concerned with immediate application of the research results. Needless to say, however, speech science research has been, and continues to be, central to progress in the more applied fields. In Chapter 2, the first chapter in Part II, we examine speech science concepts necessary to "engineer" speech. Our goal is to learn enough about speech to be able to converse with interdisciplinary researchers in various aspects of speech science and speech processing, and to be able to build useful mathematical models of speech production. Chapter 3 begins the quest for a useful mathematical model by building on the science of speech production discussed in Chapter 2. The journey takes us through a discussion of fundamental attempts to model speech production based on the physics of acoustic tubes. These real acoustic models are revealing and provide a firm foundation for the widely used discrete time model, which will be employed throughout the remainder of the book and whose description is the culmination of the chapter.

Speech *analysis* research is concerned with processing techniques that are designed to extract information from the speech waveform. In Part III we take up the most important contemporary tools for analyzing speech by computer. Speech is analyzed for many reasons, including analysis for analysis' sake (basic research into phonetics or better models of speech production), but also to reduce it to basic features for coding, synthesis, recognition, or enhancement. Part III of the book, therefore, comprises the engineering foundation upon which speech processing is built. In the first of these topics (Chapter 4) we examine the general issue of processing short terms of a signal. Most engineering courses ignore the fact that, in the real world, only finite lengths of signals are available for processing. This is particularly true in speech where the signal remains stationary for only milliseconds. The remaining chapters (5 and 6) of Part III introduce the two most important parameterizations of speech in contemporary processing—linear prediction coefficients and cepstral coefficients—their meaning, and the analysis techniques for obtaining them. These parameters are widely used for spectral representations of speech in the areas mentioned above. We shall therefore use them repeatedly as we progress through the material.

Part IV consists of three chapters that cover a rather wide range of topics. This part of the text is concerned with those aspects of speech processing which most directly intersect with the communications technologies. Here we will be concerned with efficient coding for the transmission of speech across channels and its reconstruction at the receiver site. Since the task of synthesis is closely coupled with transmission and reconstruction strategies, we will examine some of the widely used analytical techniques for synthesis in the context of this study. Synthesis for voice response systems, in which a machine is used in place of a human to dispense information, is also an important application domain, and many of the techniques used in communications systems are equally applicable to this problem.

The effectiveness of a coding scheme at preserving the information and the natural quality of the speech can be ascertained by using results from *quality assessment* research. Accordingly, we include this topic in Part IV (Chapter 9). Related to the assessment of quality is the *enhancement* of speech that has been corrupted by any of a number of natural or human-made effects, including coding. This issue will also be addressed in Part IV (Chapter 8).

Speech *recognition* deals with the related problems of designing algorithms that recognize or even understand[40] speech, or which identify the speaker (speech recognition versus speaker recognition).[41] In Part V, we take up the first of these problems, that of recognizing the speech itself. Chapter 10 overviews the problems encountered in trying to recognize speech using a computer. Chapters 11 and 12 introduce the two most widely used techniques for recognizing speech—dynamic time-warping algorithms and the hidden Markov model. The first is a template match-

[40]A speech *recognizer* simply "translates" the message into words, while a speech *understanding* system would be able to ascertain the meaning of the utterance. Speech understanding algorithms can be used as an aid to recognition, by, for example, disallowing nonsensical concatenations of words to be tried, or by "expecting" certain utterances in various conversational contexts.

[41]A slight variation on the latter problem is *speaker verification*, in which the recognizer accepts or rejects the speaker's claim of identity.

ing method following the classical paradigm of statistical pattern recognition with the interesting special problem of time registration of the waveform. The latter is a stochastic method in which statistical characterizations of utterances are automatically learned from training utterances. Chapter 13 introduces the basic principles of language modeling, techniques that reduce entropy by taking advantage of the higher-level structure of spoken utterances to improve recognizer performance. Chapter 14 is a brief introduction to a radically different approach to speech recognition based on massively parallel computing architectures or "artificial neural networks." This field is in its relative infancy compared with techniques based on sequential computing, and it offers interesting challenges and possibilities for future research and development.

Applications. The applications of speech processing are manifold and diverse. In a general way, we have alluded to some of the basic areas above. Among the principal "drivers" of speech processing research in recent years have been the commercial and military support of ambitious endeavors of large scale. These have mainly included speech coding for communications, and speech recognition for an extremely large array of potential applications—robotics, machine data entry by speech, remote control of machines by speech for hazardous or "hands-free" (surgery) environments, communications with pilots in noisy cockpits, and so on. Futuristic machines for human/machine communication and interaction using speech are envisioned (and portrayed in science fiction movies), and in the meantime, more modest systems for recognition of credit card, telephone, and bank account numbers, for example, are in use. In addition, speech processing is employed in "smaller scale" problems such as speaker recognition and verification for military, security, and forensic applications, in biomedicine for the assessment of speech and voice disorders (analysis), and in designing speech and hearing aids for persons with disabilities (analysis and recognition). Inasmuch as speech is the most natural means of communication for almost everyone, the applications of speech processing technology seem nearly limitless, and this field promises to profoundly change our personal and professional lives in coming years.

What Is Not Covered in This Textbook. Speech processing is an inherently interdisciplinary subject. Although the boundaries among academic disciplines are certainly not well defined, this book is written by electrical engineers and tends to focus on topics that have been most actively pursued by digital signal processing engineers.

Significant contributions to this field, especially to speech recognition, have come from research that would usually be classified as computer science. A comprehensive treatment of these "computer science" topics is outside the intended scope of this book. Examples include (detailed discussions of) parsing algorithms for language modeling (see Chapter 13), and knowledge-based and artificial intelligence approaches to recognition[42] [e.g., (Zue, 1985)]. Although we briefly discuss the former, we do not address the latter. Another example concerns the use of "semantic" and "pragmatic" knowledge in speech recognition (see Chapters 10 and 13). Semantics and pragmatics are subjects that are difficult to formalize in conventional engineering terms, and their complexity has precluded a significant impact on speech recognition technology outside the laboratory. We treat these issues only qualitatively in this book.

The speech (and hearing) science domains—anatomy and physiology of speech production, acoustic phonetics, linguistics, hearing, and psychophysics—are all subjects that are fundamentally important to speech processing. This book provides an essential engineering treatment of most of these subjects, but a thorough treatment of these topics obviously remains beyond the scope of the book. The reader is referred to Appendix 1.E for some resources in the area.

Finally, the explosive growth in this field brought about by digital computing has made it impossible for us to provide a thorough account of the important work in speech processing prior to about 1965. Essential elements of the analog acoustic theory of speech, upon which much of modern speech processing is based, are treated in Chapter 3 and its appendix. A much more extensive treatment of this subject is found in the book *Speech Analysis, Synthesis, and Perception* by J. L. Flanagan (1972). This book is a classic textbook in the field and no serious student of speech processing should be unfamiliar with its contents. Other important papers with useful reference lists can be found in the collection (Schafer and Markel, 1979).

Further Information. The appendixes to this chapter provide the reader with lists of books and other supplementary materials for background and advanced pursuit of the topics in this book. In particular, Section 1.E of this appendix is devoted to materials specifically on speech processing. Among the sections are lists of other textbooks, edited paper collections, journals, and some notes on conference proceedings.

1.7 PROBLEMS

1.1. Whereas the unit step sequence, $u(n)$, can be thought of as samples of the continuous time step, say $u_a(t)$, defined as

$$u_a(t) = \begin{cases} 1, & t \geq 0 \\ 0, & t < 0 \end{cases}, \tag{1.255}$$

[42]This and other papers on knowledge-based approaches are reprinted in (Waibel and Lee, 1990).

a similar relationship does not exist between the discrete-time "impulse," $\delta(n)$ and its continuous-time counterpart $\delta_a(t)$.

(a) Consider sampling the signal $u_a(t)$ with sample period T to obtain the sequence $u(n) \stackrel{\text{def}}{=} u_a(nT)$. If we now subject $u(n)$ to the customary ideal interpolation procedure in an attempt to reconstruct $u_a(t)$ (Proakis and Manolakis, 1992, Sec. 6.3), will the original $u_a(t)$ be recovered? Why or why not?

(b) Roughly sketch the time signal, say $\hat{u}_a(t)$, and the spectrum $|\hat{U}_a(\Omega)|$ of the signal that *will* be recovered in part (a).

(c) That $\delta_a(t)$ cannot be sampled fast enough to preserve the information in the time signal is apparent, since the signal has infinite bandwidth, that is, $\Delta_a(\Omega) = 1$. However, to show that any attempt to sample $\delta_a(t)$ results in an anomalous sequence, consider what happens in the frequency domain with reference to (1.21). What is the anomaly in the time sequence that causes this strange frequency domain result?

(d) Carefully sketch and numerically label the time signal, say $\hat{\delta}_a(t)$, and its spectrum $\hat{\Delta}_a(\Omega)$ that results from an ideal interpolation of the unit sample sequence,

$$\delta(n) = \begin{cases} 1, & n = 0 \\ 0, & \text{otherwise} \end{cases}. \tag{1.256}$$

1.2. Consider the following sequences:

(i) $y(n) = \begin{cases} 1/n, & n > 0 \\ 0, & n \leq 0 \end{cases}.$

(ii) $x(n) = [\sin(\omega_c n)]/\pi n,$ restrict ω_c as $0 < \omega_c < \pi$

(a) In each case, classify the sequence according to whether it represents an energy signal, power signal, or neither.

(b) In each case, determine whether the sequence is absolutely summable.

(c) In each case comment on the existence of the DTFT and whether the z-transform ROC includes the unit circle.

1.3. (a) Verify the properties of the DTFT shown in Table 1.1.

(b) Prove Parseval's relation:

$$E_x = \sum_{n=-\infty}^{\infty} |x(n)|^2 = \frac{1}{2\pi} \int_{-\pi}^{\pi} |X(\omega)|^2 \, d\omega. \tag{1.257}$$

1.4. (a) Verify the properties of the DFT shown in Table 1.2. The notation $W \stackrel{\text{def}}{=} e^{-j2\pi/N}$ is used for convenience and all time sequences are assumed to be of length N.

(b) Prove Parseval's relation:

TABLE 1.1. Properties of the DTFT.

Property	Time Domain	Frequency Domain
Linearity	$ax_1(n) + bx_2(n)$	$aX_1(\omega) + bX_2(\omega)$
Delay	$x(n-d)$	$e^{-j\omega d}X(\omega)$
Modulation	$e^{j\omega_0 n}x(n)$	$X(\omega - \omega_0)$
Time reversal	$x(-n)$	$X(-\omega)$
Multiplication	$x(n)y(n)$	$\frac{1}{2\pi}\int_{-\pi}^{\pi} X(\zeta)Y(\omega-\zeta)\,d\zeta = X(\omega) * Y(\omega)$
Convolution	$x(n) * y(n)$	$X(\omega)Y(\omega)$
Conjugation	$x^*(n)$	$X^*(-\omega)$
Differentiation	$nx(n)$	$j\frac{dX(\omega)}{d\omega}$

$$\sum_{n=0}^{N-1} |x(n)|^2 = \frac{1}{N}\sum_{k=0}^{N-1} |X(k)|^2. \tag{1.258}$$

1.5. Suppose that we redefine state variables in Fig. 1.5. The new set, $\{v_i'(n)\}$, is such that $v_1'(n) = v_N(n)$, $v_2'(n) = v_{N-1}(n), \ldots, v_N'(n) = v_1(n)$, where the $v_i(n)$'s are defined in the figure. Develop an upper companion form state space model of the form

$$\mathbf{v}'(n+1) = \mathbf{A}'\mathbf{v}'(n) + \mathbf{c}'x(n) \tag{1.259}$$

$$y(n) = \mathbf{b}'^T\mathbf{v}'(n) + d'x(n), \tag{1.260}$$

in which $\mathbf{A}'$ is obtained from the lower companion form matrix $\mathbf{A}$ of (1.54) by reflecting all elements around the main diagonal.

1.6. (a) How many sequences are there with P nonzero, finite poles and $Z \le P$ nonzero, finite zeros that have identical magnitude spectra? The sequences need *not* be real.

(b) How many of these are causal and "stable"? That is, how many are absolutely summable, meaning that their z-transform ROCs include the unit circle? How many are noncausal and "stable"?

(c) If $Z > P$, how do your answers in part (b) change?

TABLE 1.2. Properties of the DFT.*

Property	Time Domain	Frequency Domain
Linearity	$ax_1(n) + bx_2(n)$	$aX_1(k) + bX_2(k)$
Circular shift	$x(n-d)_{\mathrm{mod}N}$	$W^{kd}X(k)$
Modulation	$W^{ln}x(n)$	$X(k+l)_{\mathrm{mod}N}$
Circular convolution	$x(n)_{\mathrm{mod}N} * y(n)$	$X(k)Y(k)$

*The notation $W \stackrel{\mathrm{def}}{=} e^{-j2\pi/N}$ and all sequences are assumed to be of length N.

(d) Of the causal sequences in part (b), how many are minimum phase? Maximum phase?

1.7. (Computer Assignment) Using a signal processing software package, replicate the experiment of Section 1.1.7, that is, reproduce Figures 1.6–1.8.

1.8. Given the joint probability density function $f_{\underline{xy}}(x, y)$ for two jointly continuous random variables $\underline{x}$ and $\underline{y}$, verify the following using a pictorial argument:

$$P(x_1 < \underline{x} \leq x_2, y_1 < \underline{y} \leq y_2) = \int_{-\infty}^{x_2} \int_{-\infty}^{y_2} f_{\underline{xy}}(x, y)\, dx\, dy$$

$$- \int_{-\infty}^{x_2} \int_{-\infty}^{y_1} f_{\underline{xy}}(x, y)\, dx\, dy - \int_{-\infty}^{x_1} \int_{-\infty}^{y_2} f_{\underline{xy}}(x, y)\, dx\, dy$$

$$+ \int_{-\infty}^{x_1} \int_{-\infty}^{y_1} f_{\underline{xy}}(x, y)\, dx\, dy. \tag{1.261}$$

1.9. Formally verify that, for two jointly continuous random variables $\underline{x}$ and $\underline{y}$,

$$\mathcal{E}\{\mathcal{E}\{h(\underline{y})|\underline{x}\}\} = \mathcal{E}\{h(\underline{y})\} \tag{1.262}$$

where $h(\cdot)$ is some "well-behaved" function of $\underline{y}$. Assume that all relevant pdf's exist.

1.10. For a random process $\underline{x}$ with random variables $\underline{x}(n)$, show that

$$c_{\underline{x}}(n_1, n_2) = r_{\underline{x}}(n_1, n_2) - \mathcal{E}\{\underline{x}(n_1)\}\ \mathcal{E}\{\underline{x}(n_2)\}. \tag{1.263}$$

1.11. In this problem we will show that the implications in (1.132) reverse in the special case in which random variables within a random process are known to be joint Gaussian. Consider a random process $\underline{x}$, known to be WSS. (*Note*: This means that the mean and correlations are time independent, which we denote by writing $\mu_{\underline{x}}$, $\sigma_{\underline{x}}$, and $\rho_{\underline{x}}$.) If two random variables, $\underline{x}(n_1)$ and $\underline{x}(n_2)$ for any n_1 and n_2, in $\underline{x}$ are joint Gaussian,

$$f_{\underline{x}(n_1)\underline{x}(n_2)}(x_1, x_2) = \frac{1}{2\pi\sigma_{\underline{x}}^2 \sqrt{1-\rho_{\underline{x}}^2}}\, e^{-1/2Q(x_1,x_2)}, \tag{1.264}$$

where

$$Q(x_1, x_2) = \frac{1}{1-\rho_{\underline{x}}^2} \left\{ \left(\frac{x_1 - \mu_{\underline{x}}}{\sigma_{\underline{x}}}\right)^2 - 2\rho_{\underline{x}}^2 \left(\frac{x_1 - \mu_{\underline{x}}}{\sigma_{\underline{x}}}\right)\left(\frac{x_2 - \mu_{\underline{x}}}{\sigma_{\underline{x}}}\right) + \left(\frac{x_2 - \mu_{\underline{x}}}{\sigma_{\underline{x}}}\right)^2 \right\}, \tag{1.265}$$

show that the process is also *second-order stationary.* Show, in fact, that the process is SSS.

1.12. For a WSS random process $\underline{x}$, verify that

$$P_{\underline{x}} = \frac{1}{2\pi} \int_{-\pi}^{\pi} \Gamma_{\underline{x}}(\omega)\, d\omega = \frac{1}{\pi} \int_{0}^{\pi} \Gamma_{\underline{x}}(\omega)\, d\omega = r_{\underline{x}}(0). \tag{1.266}$$

1.13. Show that, if a WSS random process $\underline{x}$, which is ergodic in both mean and autocorrelation, is used as input to a stable, linear, time-invariant discrete time system with impulse response $h(n)$, then the output random process $\underline{y}$ is also ergodic in both senses.

1.14. Verify (1.155) and (1.156).

1.15. Verify (1.157).

1.16. Show that (1.210) follows from (1.209).

1.17. Repeat the analysis leading to (1.190) starting with the basis vectors $\boldsymbol{\beta}_1' = [2\ 0]^T$ and $\boldsymbol{\beta}_2' = [0\ 10]^T$. In this case the basis vectors are orthogonal, but grossly out of scale. This corresponds to the case of two features which are uncorrelated but which have widely different variances. What is the form of the eventual weighting matrix, $\mathbf{W}$, for this case? Is it clear what the weighting matrix is doing "physically?" Explain.

1.18. Consider a nonsingular $N \times N$ matrix $\mathbf{W}$ which operates on three N-vectors, $\mathbf{x}'$, $\mathbf{y}'$, and $\mathbf{z}'$, to produce three new vectors $\mathbf{x} = \mathbf{W}\mathbf{x}'$, $\mathbf{y} = \mathbf{W}\mathbf{y}'$, and $\mathbf{z} = \mathbf{W}\mathbf{z}'$. The original vectors are arbitrary except that $\mathbf{x}'$ is closer to $\mathbf{z}'$ when the Euclidean metric is used to measure distance:

$$d_2(\mathbf{x}', \mathbf{z}') < d_2(\mathbf{y}', \mathbf{z}'). \tag{1.267}$$

Show that the linear transformation need *not* preserve the relative distances by finding a vector triplet satisfying (1.267) and a nonsingular $\mathbf{W}$ such that

$$d_2(\mathbf{x}, \mathbf{z}) \geq d_2(\mathbf{y}, \mathbf{z}). \tag{1.268}$$

Can you find conditions on $\mathbf{W}$ so that the relative distances are preserved?

1.19. An example of a probabilistic separability measure for a two-class problem is the Bhattacharyya distance

$$J_B = -\ln \int_{-\infty}^{\infty} \sqrt{f_{\underline{\mathbf{x}}|\underline{c}}(\mathbf{x}|1) f_{\underline{\mathbf{x}}|\underline{c}}(\mathbf{x}|2)\, d\mathbf{x}}. \tag{1.269}$$

Show that this measure reduces to a Mahalanobis-like distance in the case of Gaussian feature vectors and equal class covariances. *Hint*: Use the fact that

$$-\frac{1}{2}\left(\mathbf{x} - \boldsymbol{\mu}_{\underline{\mathbf{x}}|1}\right)^T \mathbf{C}_{\underline{\mathbf{x}}}^{-1} \left(\mathbf{x} - \boldsymbol{\mu}_{\underline{\mathbf{x}}|1}\right) + \frac{1}{2}\left(\mathbf{x} - \boldsymbol{\mu}_{\underline{\mathbf{x}}|2}\right)^T \mathbf{C}_{\underline{\mathbf{x}}}^{-1} \left(\mathbf{x} - \boldsymbol{\mu}_{\underline{\mathbf{x}}|2}\right) = \tag{1.270}$$

(equation continues next page)

$$\frac{(\mu_{\underline{\mathbf{x}}|1} - \mu_{\underline{\mathbf{x}}|2})^T}{2} \mathbf{C}_{\underline{\mathbf{x}}}^{-1} \frac{(\mu_{\underline{\mathbf{x}}|1} - \mu_{\underline{\mathbf{x}}|2})}{2} + \left[\mathbf{x} - \frac{(\mu_{\underline{\mathbf{x}}|1} + \mu_{\underline{\mathbf{x}}|2})}{2}\right]^T \mathbf{C}_{\underline{\mathbf{x}}}^{-1} \left[\mathbf{x} - \frac{(\mu_{\underline{\mathbf{x}}|1} + \mu_{\underline{\mathbf{x}}|2})}{2}\right].$$

1.20. Two stationary binary sources, $\underline{x}$ and $\underline{y}$, are considered in this problem. In each case the random variables of the source, for example, $\underline{x}(n)$, $n = 1, 2, \ldots$, are statistically independent.

(a) Given $P[\underline{x}(n) = 1] = 0.3$ for any n, evaluate the entropy of source $\underline{x}$, $H(\underline{x})$.

(b) In the source $\underline{y}$, the entropy is maximal. Use your knowledge of the meaning of entropy to guess the value $P[\underline{y}(n) = 1]$. Explain the reasoning behind your guess. Formally verify that your conjecture is correct.

(c) Given that $P[\underline{x}(n) = x, \underline{y}(n) = y] = 0.25$ for any n and for any possible outcome, $(x, y) = (0,0)$, $(0,1)$, $(1,0)$, $(1,1)$, evaluate the average mutual information, say $\overline{M}(\underline{x}, \underline{y})$, between the *jointly stationary* random sources $\underline{x}$ and $\underline{y}$.

(d) Find the probability distribution $P[\underline{x}(n), \underline{y}(n)]$ such that the two jointly stationary random sources have no average mutual information.

1.21. Verify (1.231)–(1.233).

APPENDICES: Supplemental Bibliography

1.A Example Textbooks on Digital Signal Processing

Cadzow, J. A. *Foundations of Digital Signal Processing and Data Analysis.* New York: Macmillan, 1987.

Jackson, L. B. *Digital Filters and Signal Processing*, 2nd ed. Norwell, Mass.: Kluwer, 1989.

Kuc, R. *Introduction to Digital Signal Processing.* New York: McGraw-Hill, 1988.

Oppenheim, A. V., and R. W. Schafer. *Discrete Time Signal Processing*, Englewood Cliffs, N.J.: Prentice Hall, 1989.

Proakis, J. G., and D. G. Manolakis. *Digital Signal Processing: Principles, Algorithms, and Applications*, 2nd ed. New York: Macmillan, 1992.

1.B Example Textbooks on Stochastic Processes

Davenport, W. B. *Random Processes: An Introduction for Applied Scientists and Engineers.* New York: McGraw-Hill, 1970. [Elementary]

Gardner, W. A. *Introduction to Random Processes with Applications to Signals and Systems*, 2nd ed. New York: McGraw-Hill, 1990.

Gray, R. M., and L. D. Davisson. *Random Processes: A Mathematical Approach for Engineers.* Englewood Cliffs, N.J.: Prentice Hall, 1986.

Grimmett, G. R., and D. R. Stirzaker. *Probability and Random Processes.* Oxford: Clarendon, 1985.

Helstrom, C. W. *Probability and Stochastic Processes for Engineers*, 2nd ed. New York: Macmillan, 1991.

Leon-Garcia, A. *Probability and Random Processes for Electrical Engineering.* Reading, Mass.: Addison-Wesley, 1989. [Elementary]

Papoulis, A. *Probability, Random Variables, and Stochastic Processes*, 2nd ed. New York: McGraw-Hill, 1984.

Peebles, P. Z. *Probability, Random Variables, and Random Signal Principles*, 2nd ed. New York: McGraw-Hill, 1987. [Elementary]

Pfeiffer, P. E. *Concepts of Probability Theory.* New York: Dover, 1965.

Wong, E., and B. Hajek. *Stochastic Processes in Engineering Systems.* New York: Springer-Verlag, 1984. [Advanced]

1.C Example Textbooks on Statistical Pattern Recognition

Devijver, P. A., and J. Kittler. *Pattern Recognition: A Statistical Approach.* London: Prentice Hall International, 1982.

Fukunaga, K. *Introduction to Statistical Pattern Recognition.* New York: Academic Press, 1972.

Jain, A. K., and R. C. Dubes. *Algorithms for Clustering Data.* Englewood Cliffs, N.J.: Prentice Hall, 1988.

1.D Example Textbooks on Information Theory

Blahut, R. E. *Principles and Practice of Information Theory.* Reading, Mass.: Addison-Wesley, 1987.

Csiszár, I., and J. Körner. *Information Theory.* New York: Academic Press, 1981.

Gallagher, R. G. *Information Theory and Reliable Communication.* New York: John Wiley & Sons, 1968.

Guiasu, S. *Information Theory with Applications.* New York: McGraw-Hill, 1976.

Khinchin, A. Y., *Mathematical Foundations of Information Theory.* New York: Dover, 1957.

McEliece, R. J. *The Theory of Information and Coding.* Reading, Mass.: Addison-Wesley, 1977.

1.E Other Resources on Speech Processing

1.E.1 Textbooks

Flanagan, J. L. *Speech Analysis, Synthesis, and Perception*, 2nd ed. New York: Springer-Verlag, 1972.

Furui, S. *Digital Speech Processing*. New York: Marcel Dekker, 1989.

Furui, S., and M. Sondhi. *Recent Progress in Speech Signal Processing*, New York: Marcel Dekker, 1990.

Markel, J. D., and A. H. Gray. *Linear Prediction of Speech*. New York: Springer-Verlag, 1976.

Morgan, D. P., and C. L. Scofield. *Neural Networks and Speech Processing*. Norwell, Mass.: Kluwer, 1991.

O'Shaughnessy, D. *Speech Communication: Human and Machine*. Reading, Mass.: Addison-Wesley, 1987.

Papamichalis, P. E. *Practical Approaches to Speech Coding*. Englewood Cliffs, N.J.: Prentice Hall, 1987.

Parsons, T. W. *Voice and Speech Processing*. New York: McGraw-Hill, 1986.

Rabiner, L. R., and R. W. Schafer. *Digital Processing of Speech Signals*. Englewood Cliffs, N.J.: Prentice Hall, 1978.

1.E.2 Edited Paper Collections

Dixon, N. R., and T. B. Martin, eds., *Automatic Speech and Speaker Recognition*. New York: IEEE Press, 1979.

Fallside, F., and W. A. Woods, eds., *Computer Processing of Speech*. London: Prentice Hall International, 1985.

Lea, W. A., ed., *Trends in Speech Recognition*. Apple Valley, Minn.: Speech Science Publishers, 1980.

Reddy, R., ed., *Speech Recognition*. New York: Academic Press, 1975.

Schafer, R. W., and J. D. Markel, eds., *Speech Analysis*. New York: John Wiley & Sons (for IEEE), 1979.

Waibel, A., and K. F. Lee, eds., *Readings in Speech Recognition*. Palo Alto, Calif.: Morgan-Kauffman, 1990.

1.E.3 Journals

Among the most widely read journals in English covering the field of speech processing are the following:[43]

[43]IEEE is the Institute of Electrical and Electronics Engineers, the world's largest professional organization whose membership is over 300,000. The Signal Processing Society of the IEEE, the society most directly concerned with speech processing, has a membership exceeding 15,000. Other societies of the IEEE also publish transactions which occasionally contain papers on speech processing. Among them are the *Transactions on Information Theory, Computers, Communications, Pattern Analysis and Machine Intelligence, Automatic Control, Systems Man and Cybernetics, Neural Networks*, and *Biomedical Engineering*. IEE is the Institute of Electronics Engineers, the professional electrical engineering society based in the United Kingdom.

AT&T Technical Journal (Prior to 1985, *Bell System Technical Journal*).
Computer Speech and Language.
IEE Proceedings F: Communications, Radar, and Signal Processing.
IEEE Transactions on Signal Processing (Prior to 1991, *IEEE Transactions on Acoustics, Speech, and Signal Processing*, and prior to 1974, *IEEE Transactions on Audio and Electroacoustics*).
IEEE Transactions on Audio and Speech Processing (initiated in 1993).
Journal of the Acoustical Society of America.[44]
Speech Communication: An Interdisciplinary Journal.

In addition, the *Proceedings of the IEEE* and the *IEEE Signal Processing Magazine* occasionally have special issues or individual tutorial papers covering various aspects of speech processing.

1.E.4 Conference Proceedings

The number of engineering conferences and workshops that treat speech processing is vast—we will make no attempt to list them. However, the most widely attended conference in the field, and the forum at which new breakthroughs in speech processing are often reported, is the annual International Conference on Acoustics, Speech, and Signal Processing, sponsored by the Signal Processing Society of the IEEE. The society publishes an annual proceedings of this conference. By scanning the reference lists in these proceedings, as well as those in the journals above, the reader will be led to some of the other important conference proceedings in the area.

Also see Section 1.G.3 of this appendix.

1.F Example Textbooks on Speech and Hearing Sciences

Borden, G., and K. Harris. *Speech Science Primer: Physiology, Acoustics, and Perception.* Baltimore, Md.: Williams & Wilkins, 1980.

Chomsky, N., and M. Halle. *The Sound Pattern of English.* New York: Harper and Row, 1968.

Daniloff, R., G. Shuckers, and L. Feth. *The Physiology of Speech and Hearing.* Englewood Cliffs, N.J.: Prentice Hall, 1980.

Eimas, P., and J. Miller, eds., *Perspectives on the Study of Speech.* Hillsdale, N.J.: Erlbaum, 1981.

Flanagan, J. L. *Speech Analysis, Synthesis, and Perception*, 2nd ed. New York: Springer-Verlag, 1972.

Ladefoged, P. *A Course in Phonetics.* New York: Harcourt Brace Jovanovich, 1975.

[44]Of the journals listed, this one is most oriented toward the presentation of basic science results in speech and hearing.

LeHiste, I., ed., *Readings in Acoustic Phonetics*. Cambridge, Mass.: MIT Press, 1967.

Lieberman, P. *Intonation, Perception, and Language*, Cambridge, Mass.: MIT Press, 1967.

MacNeilage, P. *The Production of Speech*. New York: Springer-Verlag, 1983.

Minifie, F., T. Hixon, and F. Williams, eds., *Normal Aspects of Speech, Hearing, and Language*. Englewood Cliffs, N.J.: Prentice Hall, 1973.

Moore, B. *An Introduction to the Physiology of Hearing*. London: Academic Press, 1982.

O'Shaughnessy, D. *Speech Communication: Human and Machine*. Reading, Mass.: Addison-Wesley, 1987.

Perkell, J., and D. Klatt, eds., *Invariance and Variability in Speech Processes*. Hillside, N.J.: Lawrence Erlbaum Associates, 1986.

Zemlin, W. *Speech and Hearing Science, Anatomy and Physiology*. Englewood Cliffs, N.J.: Prentice Hall, 1968.

1.G Other Resources on Artificial Neural Networks

1.G.1 Textbooks and Monographs

Kohonen, T. *Self-Organization and Associative Memory*, 2nd ed. New York: Springer-Verlag, 1988.

Kosko, B. *Neural Networks and Fuzzy Systems*. Englewood Cliffs, N.J.: Prentice Hall, 1992.

Morgan, D. P., and C. L. Scofield. *Neural Networks and Speech Processing*. Norwell, Mass.: Kluwer, 1991.

Rumelhart, D. E. *Parallel Distributed Processing*, Vol. 1: *Foundations*, Vol 2: *Psychological and Biological Models*. Cambridge, Mass.: MIT Press, 1986.

Simpson, P. K. *Artificial Neural Systems*. Elmsford, N.Y.: Pergamon Press, 1990.

Zurada, J. M. *An Introduction to Artificial Neural Systems*. St. Paul, Minn.: West Publishing, 1992.

1.G.2 Journals

A few of the widely read journals on ANNs in English are the following:

IEEE Transactions on Neural Networks.
International Journal of Neural Systems.
Neural Computation.
Neural Networks Journal.

In addition, many of the journals listed in Section 1.E.3 of this appendix publish articles on neural network applications to speech processing.

1.G.3 Conference Proceedings

The number of conferences devoted to neural network technology is very large. These are two of the important ones:

IEEE International Conference on Neural Networks.
International Joint Conference on Neural Networks.

Many papers on ANNs related to speech processing are also presented at the IEEE International Conference on Acoustics, Speech, and Signal Processing, which is discussed in Section 1.E.4 of this appendix.

PART II

SPEECH PRODUCTION AND MODELING

CHAPTER 2

Fundamentals of Speech Science

Reading Notes: This chapter treats qualitative concepts and no special reading from Chapter 1 is required.

2.0 Preamble

Fundamentally, of course, speech processing relies on basic research in the speech and hearing sciences, some of which is centuries old, and much of which is ongoing. Few speech processing engineers have the time or opportunity to become expert in these fundamental sciences, so the field of speech processing remains an inherently multidisciplinary one. Nevertheless, the speech processing engineer needs a sound working knowledge of basic concepts from these areas in order to intelligently analyze and model speech, and to discuss findings with researchers in other fields. The purpose of this chapter is to provide the essential background in these allied fields. We touch upon a rather broad array of interdisciplinary subjects. We can only hope to treat elements of these research areas, and the reader in need of deeper study is encouraged to consult the textbooks in Appendix 1.F.

In order for communication to take place, a speaker must produce a speech signal in the form of a sound pressure wave that travels from the speaker's mouth to a listener's ears. Although the majority of the pressure wave originates from the mouth, sound also emanates from the nostrils, throat, and cheeks. Speech signals are composed of a sequence of sounds that serve as a symbolic representation for a thought that the speaker wishes to relay to the listener. The arrangement of these sounds is governed by rules associated with a language. The scientific study of language and the manner in which these rules are used in human communication is referred to as *linguistics*. The science that studies the characteristics of human sound production, especially for the description, classification, and transcription of speech, is called *phonetics*. In this chapter, we deal principally with the latter science. Some material on language, from an analytical point of view, will be found in Chapters 10 and 13.

2.1 Speech Communication

Speech is used to communicate information from a speaker to a listener. Although we focus on the production of speech, hearing is an integral part of the so-called *speech chain*. Human speech production begins with an idea or thought that the speaker wants to convey to a listener. The speaker conveys this thought through a series of neurological processes and muscular movements to produce an acoustic sound pressure wave that is received by a listener's auditory system, processed, and converted back to neurological signals. To achieve this, a speaker forms an idea to convey, converts that idea into a linguistic structure by choosing appropriate words or phrases to represent that idea, orders the words or phrases based on learned grammatical rules associated with the particular language, and finally adds any additional local or global characteristics such as pitch intonation or stress to emphasize aspects important for overall meaning. Once this has taken place, the human brain produces a sequence of motor commands that move the various muscles of the vocal system to produce the desired sound pressure wave. This acoustic wave is received by the talker's auditory system and converted back to a sequence of neurological pulses that provide necessary feedback for proper speech production. This allows the talker to continuously monitor and control the vocal organs by receiving his or her own speech as feedback.[1] Any delay in this feedback to our own ears can also cause difficulty in proper speech production. The acoustic wave is also transmitted through a medium, which is normally air, to a listener's auditory system. The speech perception process begins when the listener collects the sound pressure wave at the outer ear, converts this into neurological pulses at the middle and inner ear, and interprets these pulses in the auditory cortex of the brain to determine what idea was received.

We can see that in both production and perception, the human auditory system plays an important role in the ability to communicate effectively. The auditory system has both strengths and weaknesses that become more apparent as we study human speech production. For example, one advantage of the auditory system is selectivity in what we wish to listen to. This permits the listener to hear one individual voice in the presence of several simultaneous talkers, known as the "cocktail party effect." We are able to reject competing speech by capitalizing on the phase mismatch in the arriving sound pressure waves at each ear.[2] A disadvantage of the auditory system is its inability to distinguish signals that are closely spaced in time or frequency. This occurs when two tones are spaced close together in frequency, one masks the other, resulting in

[1]Loss of this feedback loop contributes significantly to the degradation in speech quality for individuals who have hearing disabilities.

[2]Listeners who are hearing impaired in one ear cannot cancel such interference and can therefore listen to only one speaker at a time.

the perception of a single tone.[3] As the speech chain illustrates, there are many interrelationships between production and perception that allow individuals to communicate among one another. Therefore, future research will not only focus on speech production, hearing, and linguistic structure but will also undoubtedly probe the complex interrelations among these areas.

2.2 Anatomy and Physiology of the Speech Production System

2.2.1 Anatomy

The speech waveform is an acoustic sound pressure wave that originates from voluntary movements of anatomical structures which make up the human speech production system. Let us first give a very brief overview of these structures.

Figure 2.1 portrays a *medium saggital* section of the speech system in which we view the anatomy midway through the upper torso as we look on from the right side. The gross components of the system are the *lungs*, *trachea* (windpipe), *larynx* (organ of voice production), *pharyngeal cavity* (throat), *oral* or *buccal cavity* (mouth), and *nasal cavity* (nose). In technical discussions, the pharyngeal and oral cavities are usually grouped into one unit referred to as the *vocal tract*, and the nasal cavity is often called the *nasal tract*.[4] Accordingly, the vocal tract begins at the output of the larynx, and terminates at the input to the lips. The nasal tract begins at the velum (see below) and ends at the nostrils of the nose. Finer anatomical features critical to speech production include the *vocal folds* or *vocal cords*, *soft palate* or *velum*, *tongue*, *teeth*, and *lips*. The soft tip of the velum, which may be seen to hang down in the back of the oral cavity when the mouth is wide open, is called the *uvula*. These finer anatomical components move to different positions to produce various speech sounds and are known as *articulators* by speech scientists. The *mandible* or *jaw* is also considered to be an articulator, since it is responsible for both gross and fine movements that affect the size and shape of the vocal tract as well as the positions of the other articulators.

As engineers, it is useful to think of speech production in terms of an acoustic filtering operation, so let us begin to associate the anatomy with

[3]An audio compact disc which demonstrates a wide collection of these auditory phenomena, produced by the Institute for Perception Research (IPO), Eindhoven, The Netherlands, 1987, is available from the Acoustical Society of America (Houtsma et al., 1987).

[4]The term *vocal-tract* is often used in imprecise ways by engineers. Sometimes it is used to refer to the combination of all three cavities, and even more often to refer to the entire speech production system. We will be careful in this book not to use "vocal-tract" when we mean "speech production system," but it is inevitable that the term will sometimes be used to mean "vocal-tract and possibly the nasal tract too, depending on the particular sound being considered."

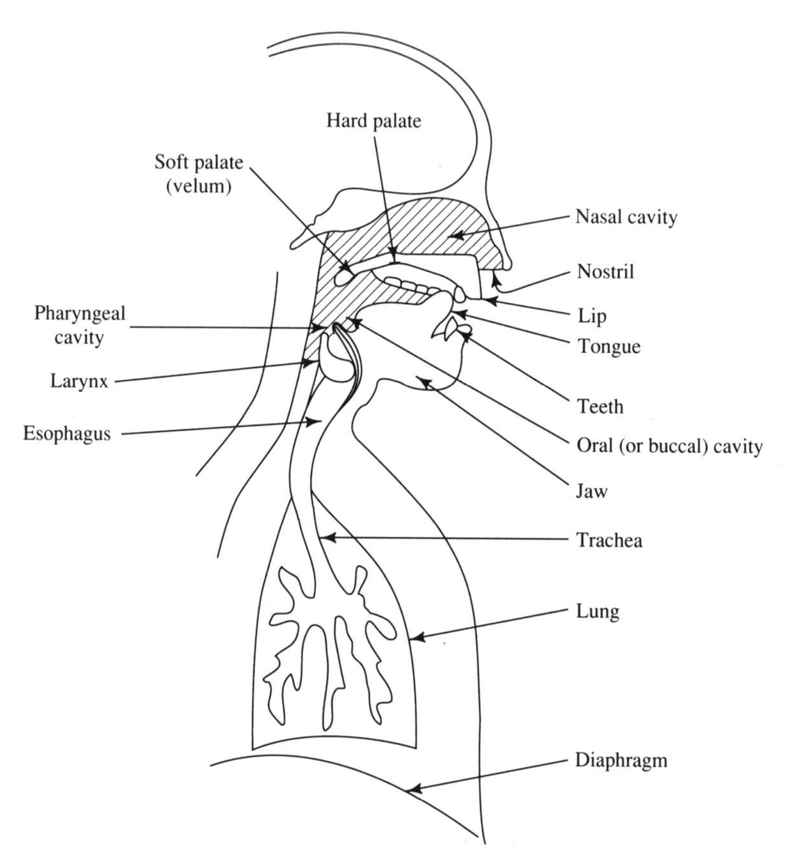

FIGURE 2.1. A schematic diagram of the human speech production mechanism.

such a technical model. The three main cavities of the speech production system (vocal plus nasal tracts) comprise the main acoustic filter. The filter is excited by the organs below it (and in other ways to be described below), and is loaded at its main output by a radiation impedance due to the lips. The articulators, most of which are associated with the filter itself, are used to change the properties of the system, its form of excitation, and its output loading over time. A simplified acoustic model illustrating these ideas is shown in Fig. 2.2.

Let us look more closely at the main cavities (acoustic filter) of the system that contribute to the resonant structure of human speech. In the average adult male (female), the total length of the vocal tract is about 17 (14) cm. The vocal tract length of an average child is 10 cm. Repositioning of the vocal tract articulators causes the cross-sectional area of the vocal tract to vary along its length from zero (complete closure) to greater than 20 cm^2. The nasal tract constitutes an auxiliary path for the transmission of sound. A typical length for the nasal tract in an adult male is 12 cm. Acoustic coupling between the nasal and vocal tracts is controlled by the size of the opening at the velum. In general, nasal

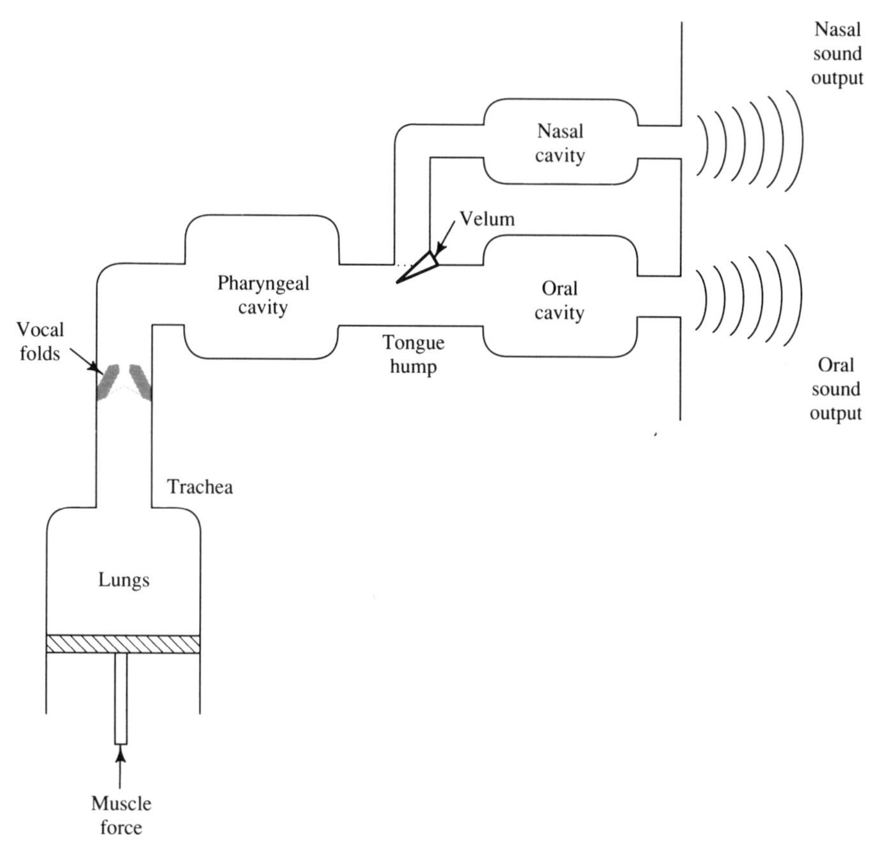

FIGURE 2.2. A block diagram of human speech production.

coupling can substantially influence the frequency characteristics of the sound radiated from the mouth. If the velum is lowered, the nasal tract is acoustically coupled to produce the "nasal" sounds of speech. Velar opening can range from zero to about 5 cm^2 for an average adult male. For the production of nonnasal sounds, the velum is drawn up tightly toward the back of the pharyngeal cavity, effectively sealing off the entrance to the nasal cavity and decoupling it from the speech production system.

Let us now focus on the larynx. From a technical point of view, the larynx has a simple, but highly significant, role in speech production. Its function is to provide a periodic excitation to the system for speech sounds that we will come to know as "voiced." Roughly speaking, the periodic vibration of the vocal folds is responsible for this voicing (more on this below). From an anatomical (and physiological) point of view, however, the larynx is an intricate and complex organ that has been studied extensively by anatomists and physiologists. A diagram showing the main features of the larynx appears in Fig. 2.3. The main framework of

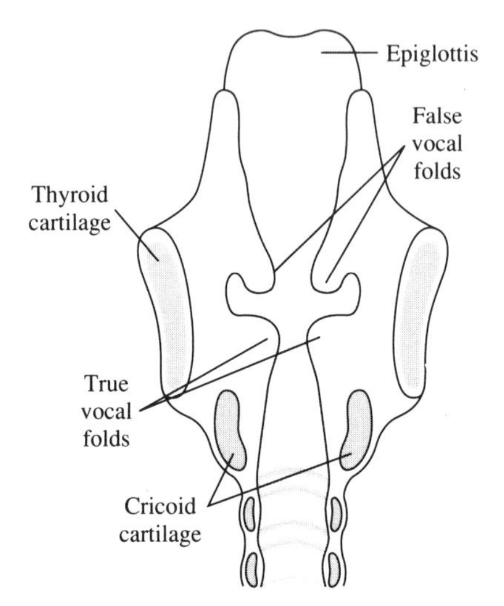

FIGURE 2.3. Cross section of the larynx as viewed from the front. After O'Shaughnessy (1987).

the larynx consists of four cartilages, which are named in the figure. Since the larynx is shown in cross section in Fig. 2.3, most of the largest cartilage, the thyroid cartilage, has been cut away. This cartilage actually consists of two plates that form most of the anterior (front) and lateral (sides) walls of the larynx. They are fused in the front of the larynx at the projection commonly known as the "Adam's apple." This protrusion can be felt in the front of the neck. Another important pair of cartilages which are not visible in Fig. 2.3 are the *arytenoids.* These are significant because the vocal cords attach to them in the posterior part of the larynx. The main cartilages of the larynx are held together by a network of ligaments and membranes that control their positioning during voicing and other functions of the organ. The vocal folds are a pair of elastic bands of muscle and mucous membrane that stretch from the thyroid cartilage in the front to the arytenoids in the rear. Readers interested in more detailed study of this important organ are referred to the books in Appendix 1.F, especially (Zemlin, 1968). Further description of the physiological operation of the larynx is given below.

2.2.2 The Role of the Vocal Tract and Some Elementary Acoustical Analysis

To illuminate the discussion to follow, it will be instructive to digress momentarily to view some elementary characteristics of a real speech waveform. The spectral characteristics of the speech wave are time-varying (or nonstationary), since the physical system changes rapidly over time. As a result, speech can be divided into sound segments that possess similar acoustic properties over short periods of time. We shall discuss these categorizations in some detail later in the chapter. Initially, we note that speech sounds are typically partitioned into two broad categories: (1) *vowels* that contain no major airflow restriction through the vocal tract, and (2) *consonants* that involve a significant restriction and are therefore weaker in amplitude and often "noisier" than vowels. Some of the differences between vowels and consonants are evident in the time waveform for the word "six" spoken by a male speaker in Fig. 2.4. The leading consonant sound "s" is low in amplitude and noiselike. Later we shall adopt a "phonetic alphabet" in which the sound corresponding to the "s" sound will be denoted /s/.) The following vowel portion (denoted /I/) is higher in amplitude and contains strong periodic structure.

As engineers we will have a primary interest in waveforms and what they reveal about the physiological and acoustic aspects of speech. Let us take a moment and view some of the "engineering" aspects of the utterance in Fig. 2.4. It is evident from the figure that time waveform displays of speech can be used to determine signal periodicities, intensities, durations, and boundaries of individual speech sounds. One of the most important properties of speech that is very noticeable in Fig. 2.4 is that speech is not a string of discrete well-formed sounds, but rather a series of "steady-state" or "target" sounds (sometimes quite brief) with intermediate transitions. The preceding and/or succeeding sound in a string can grossly affect whether a target is reached completely, how long it is held, and other finer details of the sound. This interplay between sounds in an utterance is called *coarticulation* (see Section 2.3.4). Changes witnessed in the speech waveform are a direct consequence of movements of the speech system articulators, which rarely remain fixed for any sustained period of time. (This time-varying nature of speech has important implications for speech processing, which we shall explore in Chapter 4 and beyond.) The inability of the speech production system to change instantaneously is due to the requirement of finite movement of the articulators to produce each sound. These articulators are human tissues and/or muscles which are moved from one position to another to produce the desired speech sounds. Unlike the auditory system which has evolved solely for the purpose of hearing, organs used in speech production are shared with other functions such as breathing, eating, and smelling. The multiple role of these organs suggests that their present form may not be optimal for human communication. However, additional information relayed in facial expression or gestures can be used to provide additional input to the listener. For the purposes of human communication, we shall only be concerned with the acoustic signal produced by the talker. In fact, there are many parallels between human and electronic communications. Due to the limitations of the organs for human speech production and the auditory system, typical human speech communication is limited to a bandwidth of 7–8 kHz.

Likewise, engineers can infer significant information about the physical phenomena from frequency domain plots derived from acoustic wave-

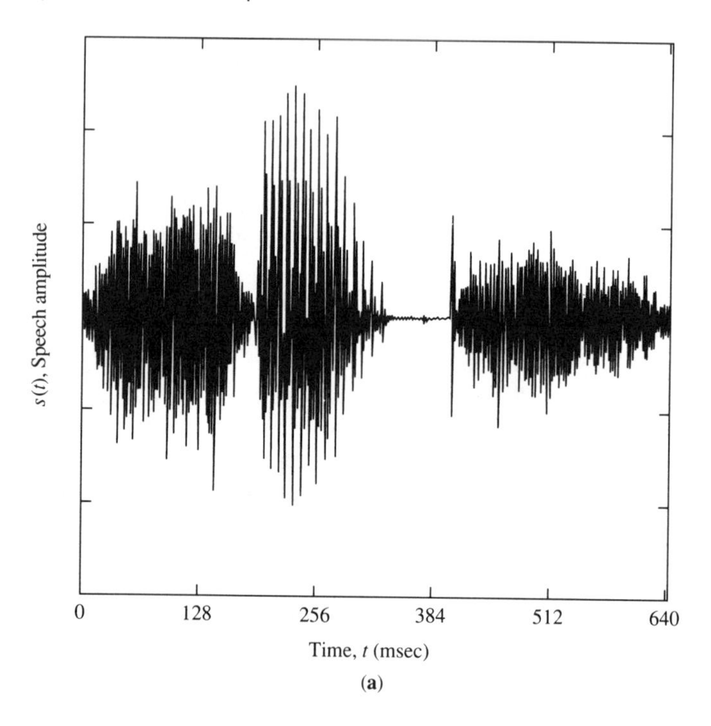

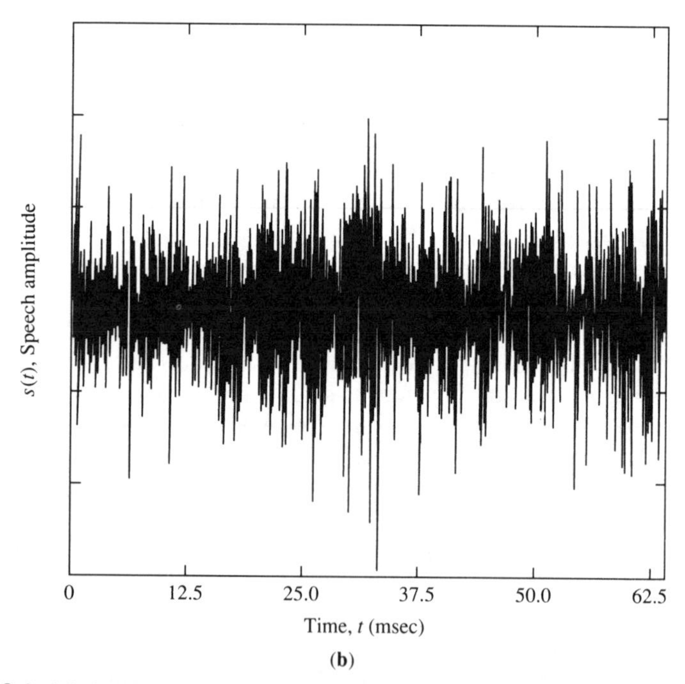

FIGURE 2.4. (a) A speech signal of the word "six" spoken by a male speaker; (b) blowup of a frame of the steady-state region of the initial /s/ sound; (c) blowup of the vowel /I/.

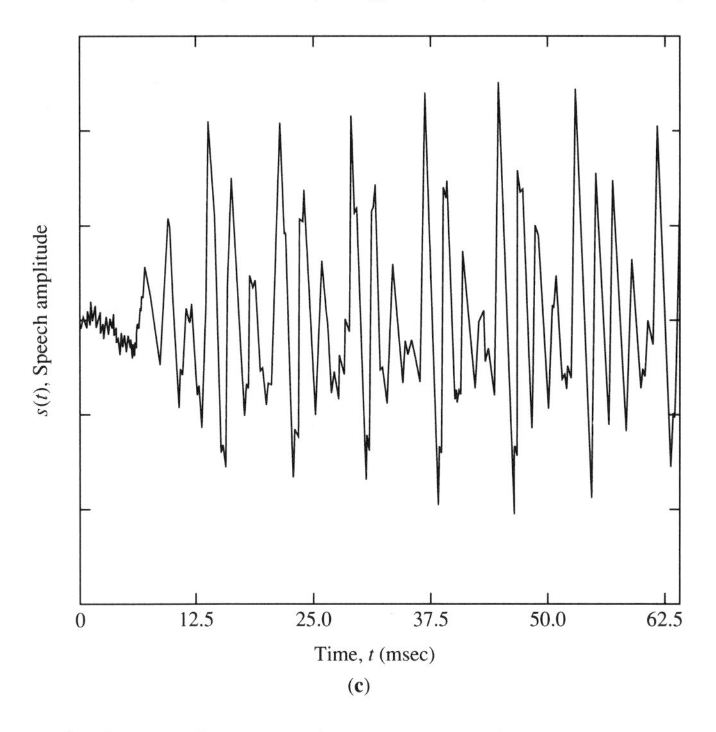

(c)

forms. In Fig. 2.5 we show magnitude spectra of the steady-state portions of /s/ and /I/ in Fig. 2.4, which were computed using a DFT. In the spectrum for /I/, a vowel sound, there is evidence of periodic excitation, while in /s/, there is no such harmonic structure present. These are excitation phenomena which we shall study immediately below. Also note that in each case (and especially for the vowel sound), there are well-defined regions of emphasis (loosely speaking, "resonances") and deemphasis ("antiresonances") in the spectrum. These resonances are a consequence of the articulators having formed various acoustical cavities and subcavities out of the vocal tract[5] cavities, much like concatenating different lengths of organ pipe in various orders. So the locations of these resonances in the frequency domain depend upon the shape and physical dimensions of the vocal tract. Conversely, each vocal tract shape is characterized by a set of resonant frequencies. From a system modeling point of view, the articulators determine the properties of the speech system filter. Since these resonances tend to "form" the overall spectrum, speech scientists refer to them as *formants*. This term is often used to refer to the nominal center frequencies of the resonances, so "formant" may be used interchangeably with *formant frequency*. The formants in the spectrum are usually denoted $F_1, F_2, F_3, \ldots$, beginning with the lowest frequency. In principle, there are an infinite number of formants in a given sound, but in practice, we usually find 3–5 in the Nyquist band after sampling.

[5]Please recall footnote 4.

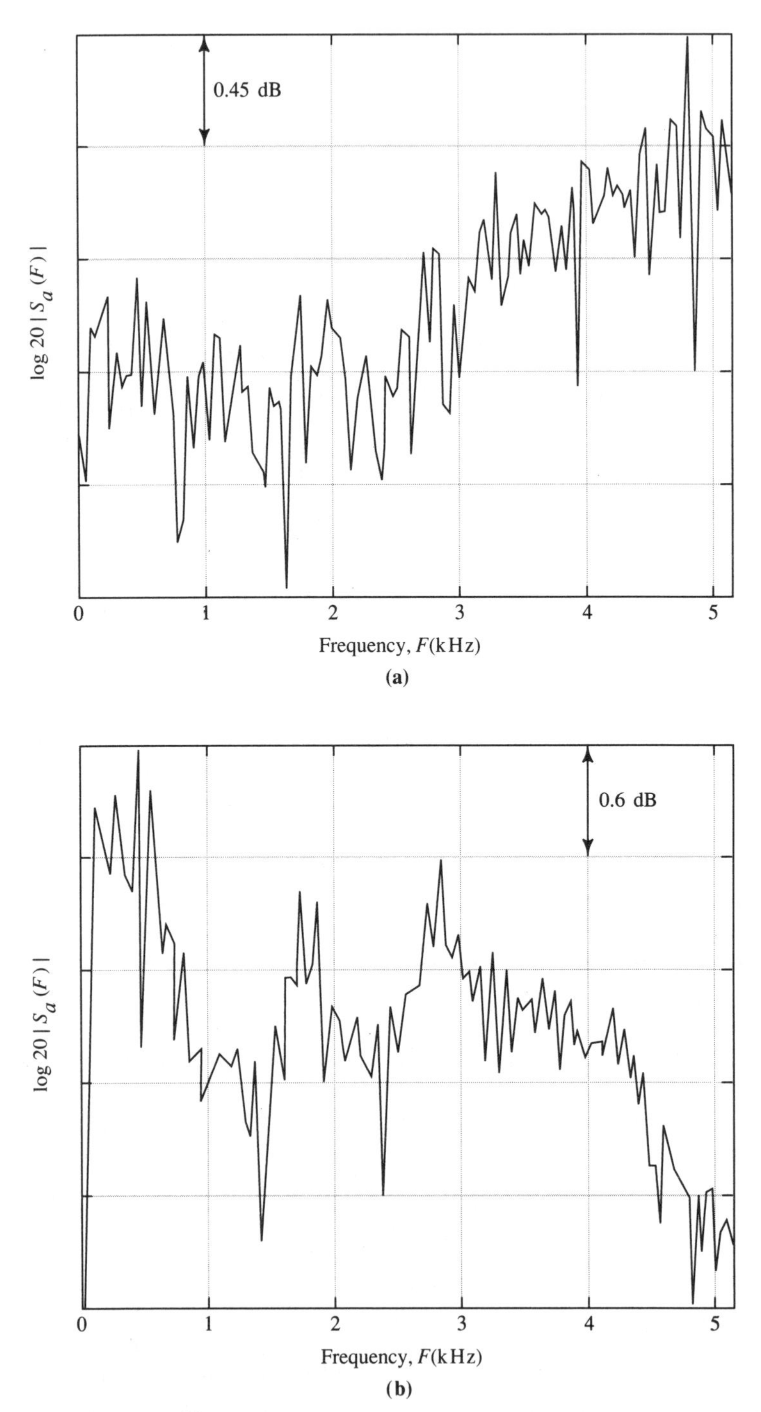

FIGURE 2.5. Magnitude spectra for (a) the /s/ sound, and (b) the /I/ sound, in the utterance of "six" in Fig. 2.4.

Since the vocal tract shape is varied as a function of time to produce the desired speech sounds for communication, so must the spectral properties of the speech signal vary with time. In contemporary signal processing, it is possible to exhibit these changes using a three-dimensional plot of magnitude spectra over time. However, another manner in which to view time-varying spectral characteristics of speech that are of special historical interest is through the use of a speech spectrograph (Flanagan, 1972; Koenig, 1946; Potter et al., 1947). The wideband spectrogram measures the amplitude of sound in a narrow range of frequencies with the use of a fixed filter bank. The speech waveform (time vs. amplitude) is fed into the spectrograph's filter bank, and the respective amplitude of each filter output is recorded on a rotating drum or paper strip chart. Darkness of the output markings reflects the relative amplitude from each filter (i.e., output is time vs. frequency vs. amplitude). Figure 2.6 shows a typical spectrograph system. Peaks in the spectrum such as formant frequencies appear as dark horizontal bands in the spectrogram. Voiced speech is characterized as vertical striations due to the periodic nature of the glottal excitation. Unvoiced speech sounds, due to their noiselike excitation, are characterized by rectangular dark patterns, with somewhat random occurrences of light spots due to sudden variations in energy. Spectrograms can only represent spectral amplitude, with no means of illustrating phase.[6]

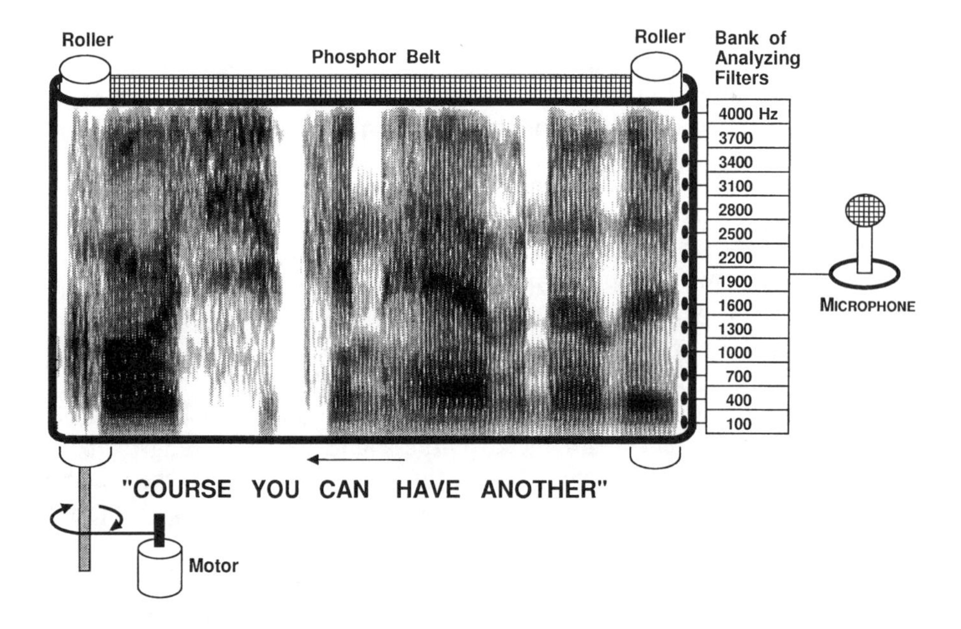

FIGURE 2.6. Bell Laboratories early phosphor belt spectrograph system.

[6]This usually is not a limitation, since phase is not as important as magnitude in a majority of speech applications.

2.2.3 Excitation of the Speech System and the Physiology of Voicing

Types of Excitation

One of the principal features of any speech sound is the manner of excitation. In our work below, we will identify two elemental excitation types: (1) voiced, and (2) unvoiced. Four other types of excitation, which are really just combinations of voiced, unvoiced, and silence, are usually delineated for modeling and classification purposes. These are: (3) mixed, (4) plosive, (5) whisper, and (6) silence.[7] One or more of these excitation types may be blended in the excitation of a particular speech sound or class of sounds. These excitation types pertain to English and many other modern and ancient languages. However, at the end of this section, we mention a few forms of excitation which appear in some of the world's more exotic languages.[8]

Voiced sounds are produced by forcing air through the *glottis* or an opening between the vocal folds. The tension of the vocal cords is adjusted so that they vibrate in oscillatory fashion. The periodic interruption of the subglottal airflow results in quasi-periodic puffs of air that excite the vocal tract. The sound produced by the larynx (or its corresponding signal) is called *voice* or *phonation.* The term *voice* is frequently used to mean "speech" in everyday conversation, but the speech processing engineer should be very careful to discern the difference between the two terms.[9] An example of a voiced sound is the vowel /I/ in the utterance of "six" above.

Unvoiced sounds are generated by forming a constriction at some point along the vocal tract, and forcing air through the constriction to produce turbulence. An example is the /s/ sound in "six." We shall describe this type of excitation further in future discussions.

A sound may be simultaneously voiced and unvoiced (*mixed*). For example, consider the sound corresponding to the letter "z" (phonetic symbol /z/) in the phrase "three zebras." Further, some speech sounds are composed of a short region of silence, followed by a region of voiced speech, unvoiced speech, or both. These *plosive* sounds are formed by making a complete closure (normally toward the front of the vocal tract), building air pressure behind the closure, and suddenly releasing it. A plosive example (silence + unvoiced) is the sound corresponding to /t/ in "pat." Another (silence + voiced) is the /b/ in "boot" in which the voicing for the following "oo" (/u/) vowel is turned on immediately after the re-

[7]Of course, whether "silence" should be called a form of excitation is debatable, but it is useful to include it for modeling purposes.

[8]In this instance, and in many other cases in this book, we restrict the discussion to consideration of American English. To do otherwise would open up broad discussions that would go well beyond the scope of the book. The principles discussed here will provide a solid foundation for the study of other languages.

[9]One often hears a singer described as having a "beautiful voice." This may indeed be the case, but the audience does not attend the concert to hear the singer's *voice*!

lease of pressure. The example of a plosive points out that for modeling purposes, it will be necessary to consider "silence" to be a form of "excitation." Whatever the source of excitation, the vocal tract acts as a filter, amplifying certain frequencies while attenuating others, as we have described above.

Voice Production

Voicing is the most interesting form of excitation in that it involves a special organ for its production. Let us focus momentarily on this excitation type. Voicing is accomplished when the abdominal muscles force the diaphragm up, pushing air out from the lungs into the trachea, then up to the glottis, where it is periodically interrupted by movement of the vocal folds. The repeated opening and closing of the glottis is in response to subglottal air pressure from the trachea. Forces responsible for the glottal pulse cycle affect the shape of the glottal waveform and are ultimately related to its corresponding spectral characteristics. First, subglottal air pressure below the true vocal folds builds up, pushing the vocal folds apart [Fig. 2.7(a),(b)]. The glottal slit begins to open [Fig. 2.7(c)], and air begins to rush out from the trachea through the glottis [Fig. 2.7(d)]. The subglottal air pressure continues to force the glottis to open wider and outward, resulting in increased airflow through the glottis. By using the notion of conservation of energy, the kinetic energy is

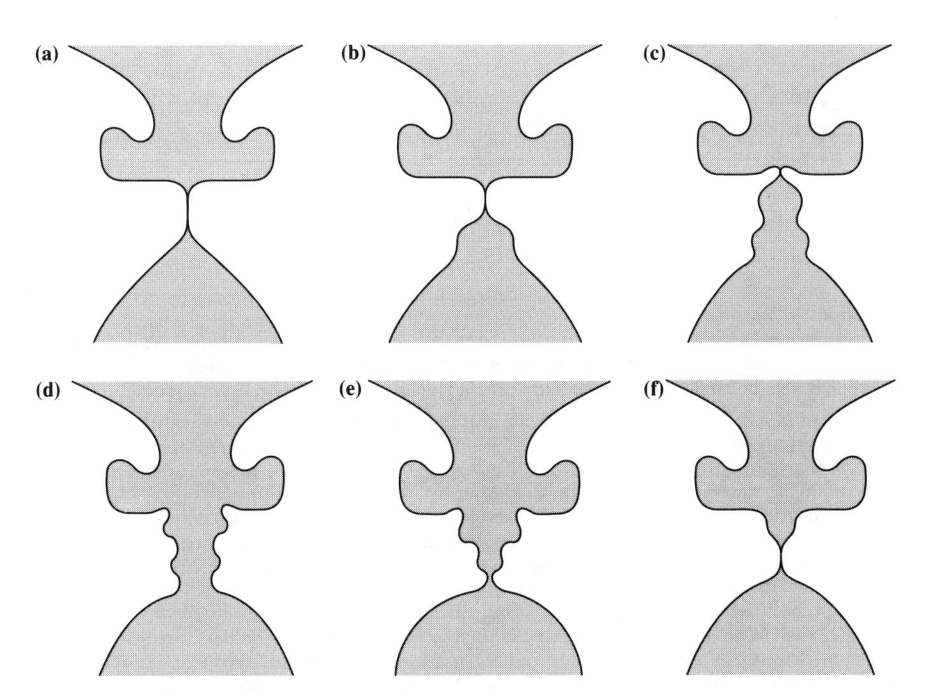

FIGURE 2.7. A sequence of cross sections of the larynx illustrating a complete phonation cycle. After Vennard (1967).

represented as the square of the air velocity; whereas the potential energy is proportional to the air pressure. As the vocal folds spread apart, air velocity increases significantly through the narrow glottis, which causes a local drop in air pressure. Therefore, when the vocal folds are closed, air pressure and potential energy are high. As the glottis opens, air velocity and kinetic energy increase, while pressure and potential energy decrease. The glottis continues to open until the natural elastic tension of the vocal folds equals the separating force of the air pressure. At this point the glottal opening and rate of airflow have reached their maxima. The kinetic energy that was received by the vocal folds during opening is stored as elastic recoil energy, which in turn causes the vocal folds to begin to close [Fig. 2.7(e)]. Inward movement of the vocal folds gathers momentum and a suction effect caused by a Bernoulli force[10] occurs when the glottis becomes narrow enough. Both the elastic restoring force and Bernoulli force act to close the vocal folds abruptly [Fig. 2.7(e)]. The subglottal pressure and elastic restoring forces during closure cause the cycle to repeat.

An example time waveform for the volume velocity[11] (defined carefully later) at the glottis is shown in Fig. 2.8(a). The variation in airflow through the glottis results in a periodic open and closed phase for the glottal or source excitation. The magnitude spectrum of one pulse of the glottal waveform is shown in part (b) of the figure. Note the lowpass nature of this spectrum. This will be significant in our future modeling efforts. In particular, we can measure this lowpass characteristic with the spectral slope $\triangle U_{\text{glottis}}(\Omega)$. Variation in spectral slope is caused primarily through changes in how the puffs of air are produced at the glottis. Its importance will be discussed in greater detail later.

Before leaving the subject of voicing, we should define some vocabulary associated with this phenomenon. The time between successive vocal fold openings is called the *fundamental period* T_o, while the rate of vibration is called the *fundamental frequency* of the phonation, $F_o = 1/T_o$. Of

[10]A Bernoulli force exists whenever there is a difference in fluid pressure between opposite sides of an object. One example of a Bernoulli force in a constricted area of flow occurs when air is blown between parallel sheets of paper held close together. The sheets of paper pull together instead of moving apart because the air velocity is greater, and the pressure is lower, between the sheets than on the outer sides.

[11]Throughout the book, and especially in this chapter and the next, we will be interested in two volume velocity waveforms. These are the volume velocity at the glottis and the volume velocity at the lips. We shall call these $u_{\text{glottis}}(\cdot)$ and $u_{\text{lips}}(\cdot)$, respectively, when both appear in the same discussion. More frequently, the glottal volume velocity will be the waveform of interest, and when there is no risk of confusion, we shall drop the subscript "glottis" and write simply $u(\cdot)$. Further, we shall be interested in both continuous time and discrete time signals in our discussions. To avoid excessive notation, we shall *not* distinguish the two by writing, for example, $u_a(t)$ and $u(n)$ (where subscript "a" means "analog"). Rather, the arguments t and n will be sufficient in all cases to indicate the difference, and we shall write simply $u(t)$ and $u(n)$. Finally, the frequency variable Ω will be used in the consideration of continuous time signals to denote "real-world" frequencies, while "ω" will be used to indicate "normalized" frequencies in conjunction with discrete time signals (see Section 1.1.1).

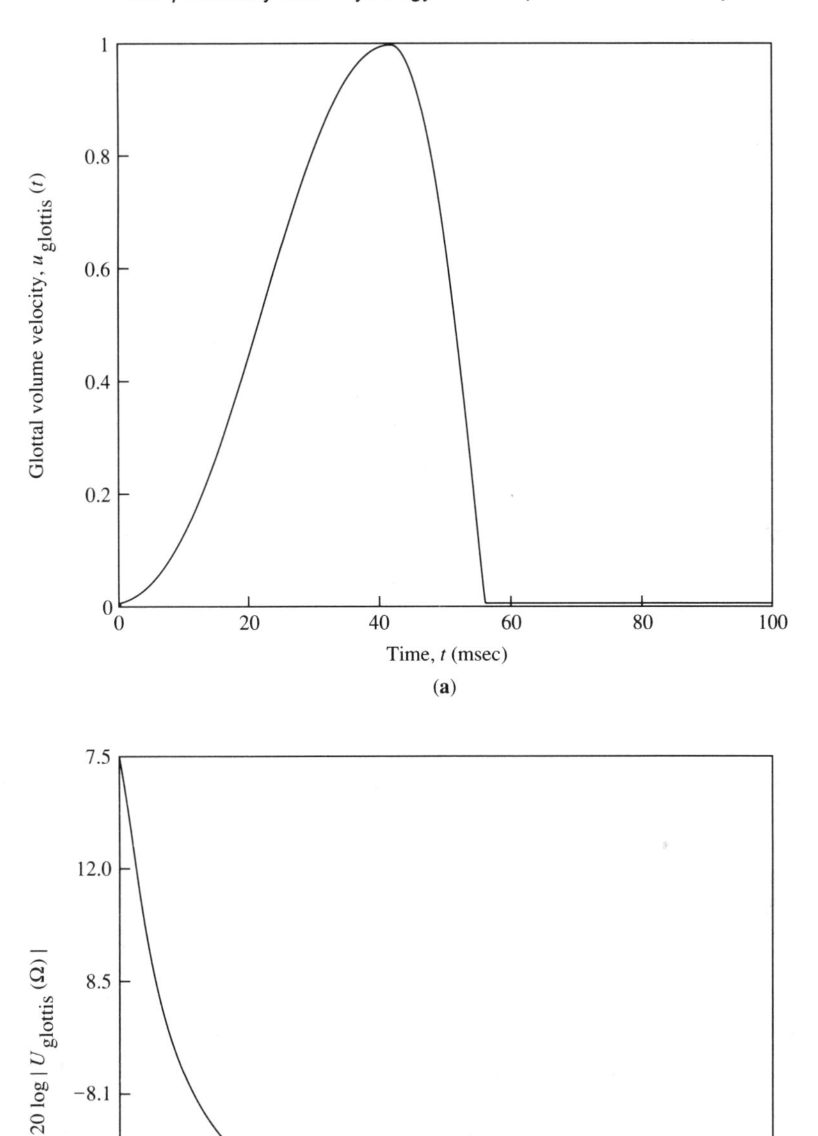

FIGURE 2.8. (a) Time waveform of volume velocity of the glottal source excitation. (b) Magnitude spectrum of one pulse of the volume velocity at the glottis.

course, the fundamental period is evident in the speech waveform as we can see, for example, in Fig. 2.4. The fundamental period is dependent on the size and tension of the speaker's vocal folds at any given instant. Since the average size of the vocal folds in men, for example, is larger than the average in women, the average fundamental frequency of an adult male in speaking a given utterance will often be lower than a female's.

The term *pitch* is often used interchangeably with fundamental frequency. However, there is a subtle difference. Psychoacousticians (scientists who study the perception of sound) use the term *pitch* to refer to the *perceived* fundamental frequency of a sound, *whether or not that sound is actually present in the waveform.* Speech transmitted over the commercial phone lines, for example, are usually bandlimited to about 300–3000 Hz. Nevertheless, a person who is phonating at 110 Hz will be *perceived* as phonating at 110 Hz by the listener, even though the fundamental frequency of the received waveform cannot be less than 300 Hz. In this case, the psychoacoustician would say that the pitch of the received speech waveform is 110 Hz, while the lowest frequency in the signal is 330 Hz. This quirk of the human auditory system requires that we be careful with these terms. Nevertheless, with this caution, we will routinely use the word "pitch" to mean "fundamental frequency" in this book, since it is conventional to do so. Since we will not be concerned with perceptual phenomena, this will not cause ambiguities to arise.

Everyone has a pitch range to which he or she is constrained by simple physics of his or her larynx. For men, the possible pitch range is usually found somewhere between the two bounds 50–250 Hz, while for women the range usually falls somewhere in the interval 120–500 Hz. Everyone has a "habitual pitch level," which is a sort of "preferred" pitch that will be used naturally on the average. Pitch is shifted up and down in speaking in response to factors relating to stress, intonation, and emotion. *Stress* refers to a change in fundamental frequency and loudness to signify a change in emphasis of a syllable, word, or phrase. *Intonation* is associated with the pitch contour over time and performs several functions in a language, the most important being to signal grammatical structure. The markings of sentence, clause, and other boundaries is accomplished through intonation patterns. We shall discuss stress, intonation, and other features related to prosodics in Section 2.3.4.

A More General Look at Excitation Types

The production of any speech sound involves the movement of an airstream. The majority of speech sounds are produced by pushing air from the lungs, through the trachea and pharynx, out through the oral and/or nasal cavities. Since air from the lungs is used, these sounds are called *pulmonic*; since the air is pushed out, they are also labeled as *egressive.* All speech sounds in American English, to which we have nominally restricted our discussion, are pulmonic egressive. Other mechanisms for producing an excitation airstream include *ejectives*, *clicks*, and *implosives*. Clicks and implosive sounds are produced when air is drawn *into* the vocal tract, and are therefore termed *ingressive*. Ejective sounds occur when only air in the oral cavity is pushed out. Ejectives are found in many Native American languages (e.g., Hopi, Apache, Cherokee), as well as in some African and Caucasian languages. Clicks occur in Southern Bantu languages such as Zulu and Xhosa, and are used in the languages spoken by the Bushmen. Implosives occur in the Native American languages, as well as in many languages spoken in India, Pakistan, and Africa.

2.3 Phonemics and Phonetics

2.3.1 Phonemes Versus Phones

Having considered the physical composition of the speech production system and the manner in which we produce speech sounds, we now focus on the collections of sounds that we use to communicate our thoughts. Once a speaker has formed a thought to be communicated to the listener, he or she (theoretically) constructs a phrase or sentence by choosing from a collection of finite mutually exclusive sounds. The basic theoretical unit for describing how speech conveys linguistic meaning is called a *phoneme*. For American English, there are about 42 phonemes* which are made up of vowels, semivowels, diphthongs, and consonants (nasals, stops, fricatives, affricates). Each phoneme can be considered to be a code that consists of a unique set of *articulatory gestures*. These articulatory gestures include the type and location of sound excitation as well as the position or movement of the vocal tract articulators.

We must clearly distinguish between *phonemes* and *phones*, and *phonemics* and *phonetics*. For our purposes, we can think of a phoneme as an ideal sound unit with a complete set of corresponding articulatory gestures. If speakers could exactly and consistently produce (in the case of English) these 42 sounds, speech would amount to a stream of discrete codes. Of course, due to many different factors including, for example, accents, gender, and, most importantly, coarticulatory effects, a given "phoneme" will have a variety of acoustic manifestations in the course of flowing speech. Therefore, any acoustic utterance that is clearly "supposed to be" that ideal phoneme, would be labeled as that phoneme. We see, therefore, that from an acoustical point of view, the phoneme really represents a *class* of sounds that convey the same meaning. Regardless of what is actually uttered for the vowel in "six," if the listener "understands" the word "six," then we would say that the phoneme /I/ was represented in the speech. The phonemes of a language, therefore, comprise a minimal theoretical set of units that are sufficient to convey all meaning in the language. This is to be juxtaposed with the actual *sounds* that are produced in speaking, which speech scientists call *phones*. The study

*The reader might notice that there are 47 symbols shown in Table 2.1 (p. 118) and only 41 phonemes shown in Table 2.3 (p. 120). We intentionally use the phrase "about 42" here because there is considerable variability in enumerating phonemes in the literature. These discrepancies illustrate the point. Similar charts can be found in many texts. If one removes the vocalic /L/, /M/, /N/; the flapped /F/; and the glottal stop /Q/ from Table 2.1, the "basic 42" to which we refer remain.

of the abstract units and their relationships in a language is called *phonemics*, while the study of the actual sounds of the language is called *phonetics*. More specifically, there are three branches of phonetics each of which approaches the subject somewhat differently:

1. *Articulatory phonetics* is concerned with the manner in which speech sounds are produced by the articulators of the vocal system.
2. *Acoustic phonetics* studies the sounds of speech through analysis of the acoustic waveform.
3. *Auditory phonetics* studies the perceptual response to speech sounds as reflected in listener trials.

Our work below will represent a blend of articulatory and acoustic analysis.

If a talker is asked to "speak a phoneme" in isolation, the phoneme will be clearly identifiable in the acoustic waveform. However, when spoken in context, phoneme boundaries become increasingly difficult to label. This is due to the physical properties of the speech articulators. Since the vocal tract articulators consist of human tissue, their positioning from one phoneme to the next is not executed by hard mechanical switches, but by movement of muscles that control articulator movement. Accordingly, there is normally a period of transition between phonemes, which under certain conditions can slightly modify the manner in which a phoneme is produced. Therefore, associated with each phoneme is a collection of *allophones* (variations on phones) that represent slight acoustic variations of the basic unit. Allophones represent the permissible freedom allowed within each language in producing a phoneme, and this flexibility is dependent not only on the phoneme itself, but also on its position within an utterance. Therefore, although we present phonemes in this section as the basic building block for human speech communication, considerable freedom is afforded to the speaker in producing these sounds to convey a thought or concept.

2.3.2 Phonemic and Phonetic Transcription

In our study, there will frequently be a need to denote a phoneme. The process of translating some input form of the language—usually a speech waveform or an *orthographic transcription* (spelling using the conventional alphabet)—into a string of symbols representing the phonemes is called *phonemic transcription*. When the transcription also includes *diacritical marks* on the phonemic symbols that indicate allophonic variations, then the result is a *phonetic transcription*. We will only have need for the former in this book, but will give an example below to illustrate the difference. For a detailed discussion, the reader is referred, for example, to (Ladefoged, 1975).

Much discussion has been devoted to what constitutes an appropriate phonetic alphabet and its use in transcription. In English, for example, the usual orthographic representation using the Roman alphabet obviously makes a poor phonetic device. In 1888, a group of prominent European phoneticians developed what is known as *International Phonetic Alphabet* (IPA) in an effort to facilitate and standardize transcription. The IPA is still widely used and accepted. Part of the IPA is shown in Table 2.1. The complete IPA has sufficient entries to cover phonemes in all the world's languages and not all are used in all languages and dialects.[12]

The IPA is most appropriate for handwritten transcription but its main drawback is that it cannot be typed on a conventional typewriter or a computer keyboard. Therefore, a more recent phonetic alphabet was developed under the auspices of the United States Advanced Research Projects Agency (ARPA), and is accordingly called the *ARPAbet.* There are actually two versions of the ARPAbet, one that uses single-letter symbols, and one that uses all uppercase symbols. The use of all uppercase necessitates some double-letter designators. The two versions of the ARPAbet are given in Table 2.1. Throughout the remainder of this book, we shall consistently use the single-letter ARPAbet symbols for phonetic transcription.

The "raw" symbols shown in Table 2.1 might be more appropriately called a *phonemic* alphabet because there are no diacritical marks indicated to show allophonic variations. For example, a super *h* is sometimes used to indicate *aspiration*, the act of delaying the onset of voicing momentarily while exhaling air through a partially open glottis. The difference is heard in the phonemes /*p*/ (as in "spit") and /p^h/ as in "pit." The difference is subtle, but it is precisely the purpose of diacritical marks to denote subtleties of the phonetic content. In spite of the lack of any significant phonetic information, we shall continue to call the ARPAbet a *phonetic* alphabet, and the transcriptions employing it *phonetic* transcriptions, as is customary in the literature.

As has been our convention so far, we shall place phonetic transcriptions between slashes in this book (e.g., /s/). As some examples of phonetic transcriptions of complete English words, consider the entries in Table 2.2.

2.3.3 Phonemic and Phonetic Classification

Introduction

There is a variety of methods for classifying phonemes. Phonemes can be grouped based on properties related to the

1. Time waveform.
2. Frequency characteristics.

[12]Although there is general agreement on the entries in the IPA, different books use different symbols for the elements. The version shown in Table 2.1 uses principally roman letters. When referring to other books, one should be aware of this possible difference in notation, even within the IPA [e.g., cf. (Ladefoged, 1975)].

TABLE 2.1. Phonetic Alphabets. (We use the single-symbol ARPAbet consistently in this book.)

IPA Symbol	ARPAbet Single-Symbol Version	ARPAbet Upper-case Version	Examples	IPA Symbol	ARPAbet Single-Symbol Version	ARPAbet Upper-case Version	Examples
i	i	IY	heed	v	v	V	vice
ɪ	I	IH	hid	θ	T	TH	thing
e	e	EY	hayed	ð	D	DH	then
ɛ	E	EH	head	s	s	S	so
æ	@	AE	had	z	z	Z	zebra
ɑ	a	AA	hod	ʃ	S	SH	show
ɔ	c	AO	hawed	ʒ	Z	ZH	measure
o	o	OW	hoed	h	h	HH	help
ʊ	U	UH	hood	m	m	M	mom
u	u	UW	who'd	n	n	N	noon
ɝ	R	ER	heard	ŋ	G	NX	sing
ə	x	AX	ago	l	l	L	love
ʌ	A	AH	mud	l̩	L	EL	cattle†
ɑɪ	Y	AY	hide	m̩	M	EM	some†
ɑʊ	W	AW	how'd	n̩	N	EN	son†
ɔɪ	O	OY	boy	ɾ	F	DX	batter‡
ɨ	X	IX	roses	ʔ	Q	Q	§
p	p	P	pea	w	w	W	want
b	b	B	bat	j	y	Y	yard
t	t	T	tea	r	r	R	race
d	d	D	deep	tʃ	C	CH	church
k	k	K	kick	dʒ	J	JH	just
g	g	G	go	ʍ	H	WH	when
f	f	F	five				

† Vocalic l, m, n.

‡ Flapped t.

§ Glottal stop.

TABLE 2.2. Some Examples of Phonetic Transcription.

Orthographic Transcription	Phonetic Transcription Using Single-Symbol ARPAbet
chocolate pudding	CaKxllt pUdG
modern clothing	madRn kloTG
inexcusable	InEkskusxbL
red riding hood	rEd rYdG hUd
nearly over	niRli ovxR
batter up	b@FR Ap

3. Manner of articulation.
4. Place of articulation.
5. Type of excitation.
6. The stationarity of the phoneme.

A phoneme is stationary or *continuant* if the speech sound is produced by a steady-state vocal-tract configuration. A phoneme is *noncontinuant* if a change in the vocal-tract configuration is required during production of the speech sound. The phonemes in Table 2.3 are classified based on continuant/noncontinuant properties. Vowels, fricatives, affricates, and nasals are all continuant sounds. Diphthongs, liquids, glides, and stops all require a vocal-tract reconfiguration during production and hence labeled noncontinuant. Due to their required vocal-tract movement, noncontinuant phonemes are generally more difficult to characterize and model than continuant ones.

Let us now study the different classes of phonemes mentioned above in some detail. We will study both articulatory and acoustic features of these classes.

Vowels and Vowellike Phonemes

Vowels. There are 12 principal vowels in American English. Phoneticians often recognize a thirteenth vowel called a *schwa* vowel, which is a sort of "degenerate vowel" to which many others gravitate when articulated hastily in the course of flowing speech. The phonetic symbol we have adopted for the schwa in this book is /x/. The initial vowel in "ahead" is a schwa vowel. The schwa occurs when the tongue hump does not have time to move into a precise location and assumes a neutral position in the vocal tract, so that the tract approximates a uniform tube. The resulting vowel is short in duration and weak in amplitude. Except for its occurrence as a "lazy" vowel, for our purposes it is not much different from the vowel /A/ occurring in "bud." Therefore, we will speak of a schwa vowel to connote the "unintentional" neutrality of a vowel, but in some of the discussions below will not attempt to distinguish it acoustically from the "proper" vowel /A/.

TABLE 2.3. Phonemes Used in American English.

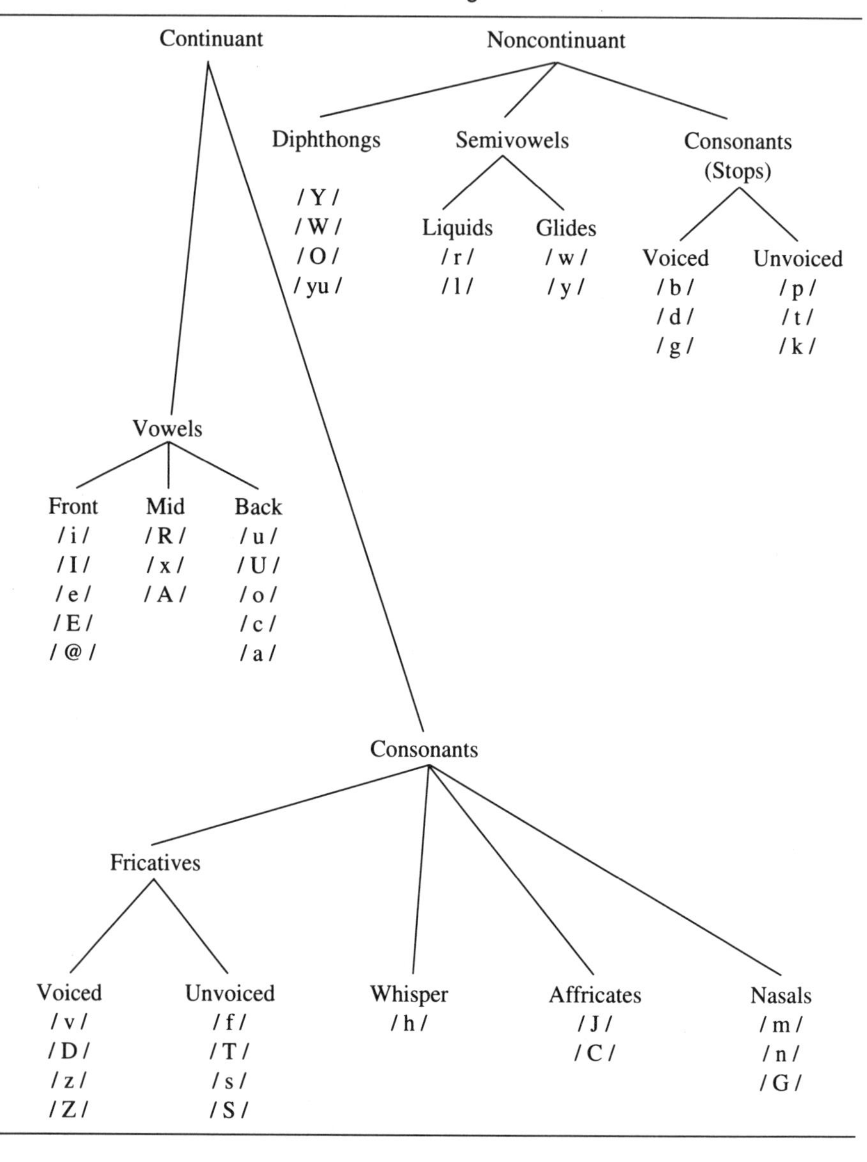

All vowels are phonated and are normally among the phonemes of largest amplitude. Vowels can vary widely in duration (typically from 40–400 msec). The variation in cross-sectional area along the vocal tract determines the formants of the vowel. Vowels are differentiated by the tongue-hump position and the degree of constriction at that position. The position of the hump portion of the tongue (front, central, back) divides the vowels into three groups. The degree to which the hump portion of the tongue is raised toward the palate further delineates each vowel group. Figure 2.9 shows how the vowels are arranged based on these two articulatory features.

The approximate configurations of the vocal-tract articulators for the vowels in Fig. 2.9 are shown in column (a) of Fig. 2.10. The physiological variation between high-medium-low front vowel versus high-medium-low back vowel can be seen by comparing the vocal-tract profiles for /i, I, @/ with /u, o, a/. Also illustrated in Fig. 2.10 are corresponding acoustic waveforms and vocal-tract frequency representations for each vowel. A variety of acoustic features can be seen in the time waveforms and spectral plots. The time waveforms show that vowels are quasi-periodic due to the cyclical vocal-fold movement at the glottis which serves as excitation. The time waveforms also show that the resonant structure of the vocal-tract changes as tongue-hump position and degree of constriction are varied. The changing resonant structure is reflected as shifts in formant frequency locations and bandwidths as pictured in the vocal-tract spectral plots. Vowels can be distinguished by the location of formant frequencies (usually the first three formants are sufficient). As an example, it has been shown through X-ray sketches that the neutral vowel /x/ results in a nearly constant cross-sectional area from the glottis through the lips. The formant frequencies for a male speaker occur near 500, 1500, 2500, 3500 Hz, and so on. F_1 and F_2 are closely tied to the shape of the vocal-tract articulators. The frequency location of the third formant, F_3, is significant to only a few specific sounds. The fourth and higher formants remain relatively constant in frequency regardless of changes in articulation.

Formant frequency locations for vowels are affected by three factors: the overall length of the pharyngeal-oral tract, the location of constrictions along the tract, and the narrowness of the constrictions. A set of rules relating these factors to formants is shown in Table 2.4.

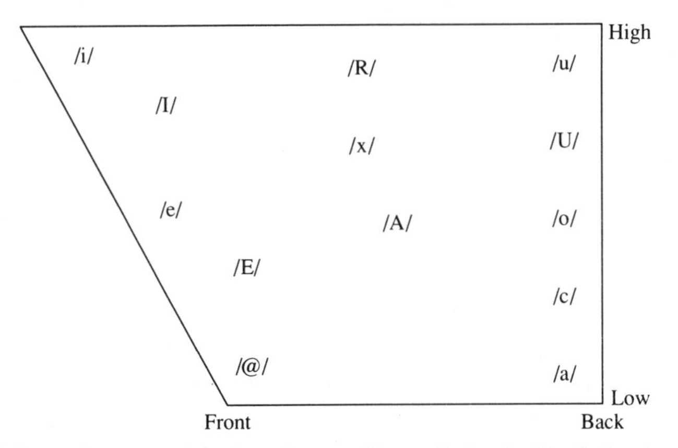

FIGURE 2.9. A diagram showing the position of the bulk of the tongue in the oral cavity during the production of the vowels.

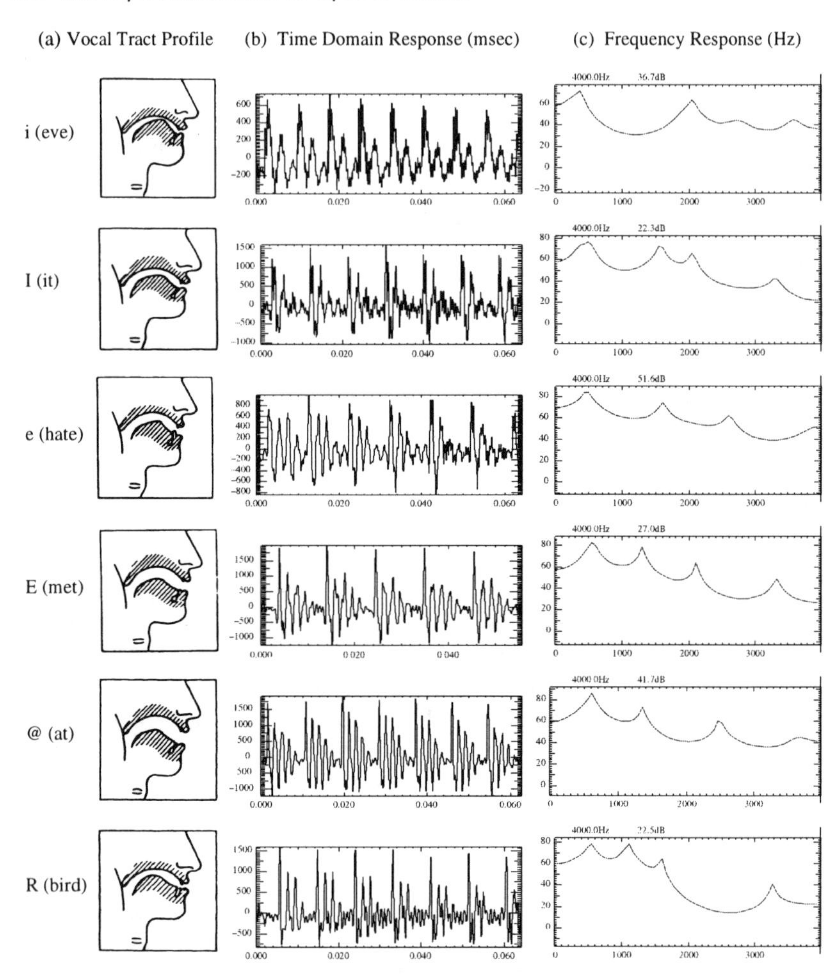

FIGURE 2.10. A collection of features for vowels in American English. Column (a) represents schematic vocal-tract profiles, (b) typical acoustic waveforms, and (c) the corresponding vocal-tract magnitude spectrum for each vowel.

The length of the vocal tract affects the frequency locations of all vowel formants. We see (Table 2.4) that a simple inverse proportionality rule relates the overall vocal-tract length from the glottis to the lips with the location of formant frequencies. However, in general the location and spacing of formants F_3 and above are more closely correlated with vocal-tract length than for F_1 and F_2. F_1 and F_2 formants have been shown to be related to the location and amount of constriction in the vocal tract. Figure 2.10(a) shows diagrams of the vocal tract for different vowels based on an X-ray study by Lindblom and Sandberg (1971). Several observations concerning vocal-tract constriction can be made in these diagrams. The tongue positions for front vowels /i, I, e, E, @/ form a series of vocal configurations that are progressively less constricted at the palate.

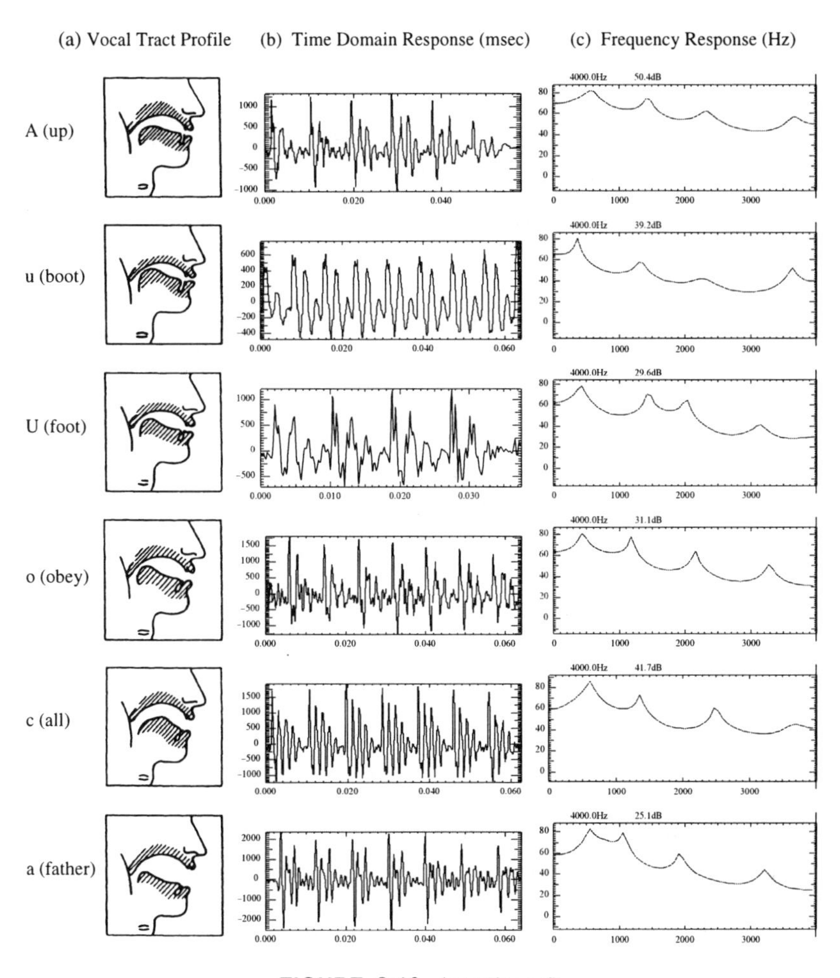

FIGURE 2.10. *(continued)*

Also, the jaw is more open and the pharynx more constricted for the /E, @/ phonemes versus /i, I/. The back vowels /u, U/ differ in the amount of constriction by the tongue toward the back of the palate. As can also be seen, the lips are more rounded for /u/, and less so for /o/. The back vowels /c, a/ have much less lip constriction, but are formed with progressively more pharyngeal constriction going from /o/ to /a/. Generally speaking, there are two rules that relate vocal-tract constriction and F_1. The first is for oral constriction and says that F_1 is lowered by any constriction in the front of the oral cavity. Therefore, if the tongue is at all pushed up toward the front or middle of the palate, F_1 will be lower than for the neutral vowel /x/ ($F_1 = 500$ Hz). The second rule says that F_1 is raised by pharyngeal constriction. Two rules are also seen to relate F_2 to vocal-tract constrictions. The frequency of F_2 is dependent on whether the tongue constriction is near the front of the oral tract (front vowels) or

TABLE 2.4. Rules Relating Formant Frequencies and Vocal-Tract Charateristics for the Vowel Sounds.

Length Rule: The average frequencies of the vowel formants are inversely proportional to the length of the pharyngeal-oral tract (i.e., the longer the tract, the lower its average formant frequencies).

F_1 **Rule—Oral Constriction:** The frequency of F_1 is lowered by any constriction in the front half of the oral section of the vocal tract. The greater the constriction, the more F_1 is lowered.

F_1 **Rule—Pharyngeal Constriction:** The frequency of F_1 is raised by constriction of the pharynx, and the greater the constriction, the more F_1 is raised.

F_2 **Rule—Back Tongue Constriction:** The frequency of F_2 tends to be lowered by a back tongue constriction. The greater the constriction, the more F_2 is lowered.

F_2 **Rule—Front Tongue Constriction:** The frequency of F_2 is raised by a front tongue constriction. The greater the constriction, the more F_2 is raised.

Lip-Rounding Rule: The frequencies of all formants are lowered by lip-rounding. The more the rounding, the more the constriction and subsequently the more the formants are lowered.

near the back (back vowels). F_2 is raised by a front tongue constriction and lowered by a back constriction. The back vowels have the tongue humped up toward the back of the palate to constrict the oral tract. The amount of constriction is largest for /u/, and least for /o/. Therefore, /u/ has a lower F_2 than /o/, with both having a lower F_2 than for neutral /x/. The front vowels are arranged from most tongue constriction to the least as follows: /i, I, e, E, @/.

The final rule relates the effect of lip-rounding on formant locations. It is seen that lip-rounding tends to lower all formants. As the schematic vocal-tract profiles in Fig. 2.10(a) indicate, lip-rounding plays an important part in forming the back vowels. We will also see that lip-rounding affects frequency characteristics of consonants as well. Lip position begins with wide-open lips for /a/ and progresses toward more constricted (i.e., rounded) lip configurations for /c/, /o/, /U/, and /u/.

The central vowels fall between the front and back vowels. These vowels, /R/ and /A/ (or /x/), are formed with constriction in the central part of the oral tract. The /R/ vowel is formed with the central part of the tongue raised mid-high toward the palate. The tip of the tongue is either lifted toward the front part of the palate or pulled backward along the floor of the mouth in a configuration called *retroflexed*. Muscles of the tongue are tensed and the lips are normally not rounded.

The rules in Table 2.4 provide a simple framework for predicting the formant patterns from the vocal-tract shape. They work best when a single constriction is the dominant feature of the vocal tract. When two constrictions operate on the vocal tract, a rule may or may not apply over the entire range of constrictions. As an example, lip-rounding has a more pronounced effect when superimposed on front vowel shapes then for back vowel shapes. During continuous speech, constriction movements of the lips, tongue, and pharynx will usually overlap in time. Early research studies by Stevens and House (1955), and Fant (1960) provide quantitative relations between formant frequencies and degrees of constriction.

The use of vocal-tract rules to identify vowels amounts to an articulatory approach. Since there exists significant variability in vocal-tract shape among the American English-speaking population, distinguishing vowels based on vocal-tract profiles can result in an imprecise characterization. An alternative means is to compare vocal-tract resonances, an acoustical approach. A summary of the first three formant frequencies for vowels is shown in Fig. 2.11. Mean formant frequency location and relative amplitude from a collection of 33 male speakers are shown (Peterson and Barney, 1952). Formant amplitudes are shown in decibels relative to the first formant of /c/. We see that as the degree of constriction decreases from high (/i/) to low (/@/) constricted vowels, the first and second formant locations move progressively closer together (/i/ to /E/ to /@/). For vowels, variation in the first two formant frequencies is more prevalent than for higher formants. With this notion, the first two formants can be used to form a *vowel triangle* (Fig. 2.12). The three point vowels (/i/, /a/, /u/) represent the extreme frequency locations for F_1 and F_2. The movement of F_1–F_2 from /i/ to /@/ is a direct result of increasing tongue height (causing F_1 to increase) and backward movement of the tongue hump (causing F_2 to decrease). This can be observed by noting the tongue position as the following words are spoken: "beet"—"bit"—"bet"—"bat."

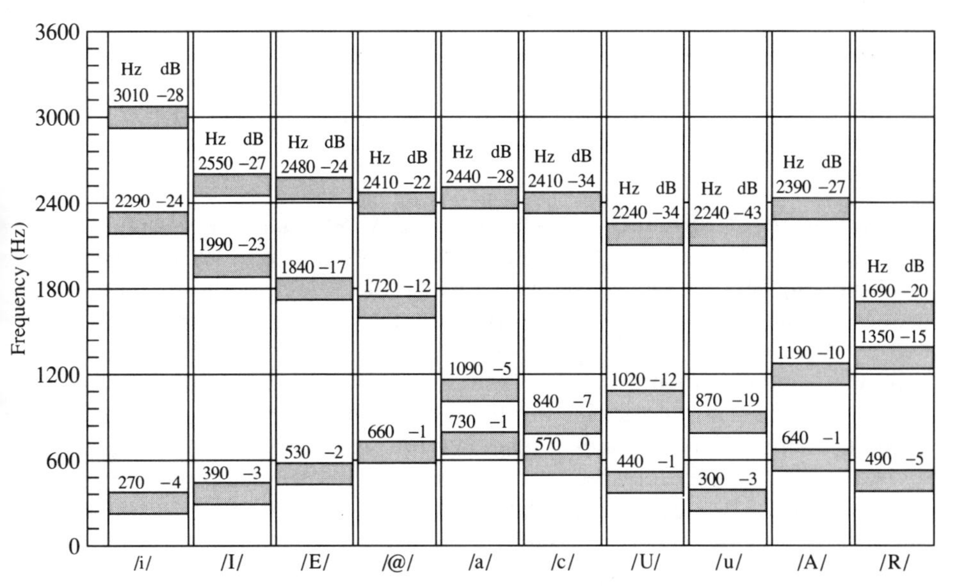

FIGURE 2.11. Average formant locations for vowels in American English (Peterson and Barney, 1952).

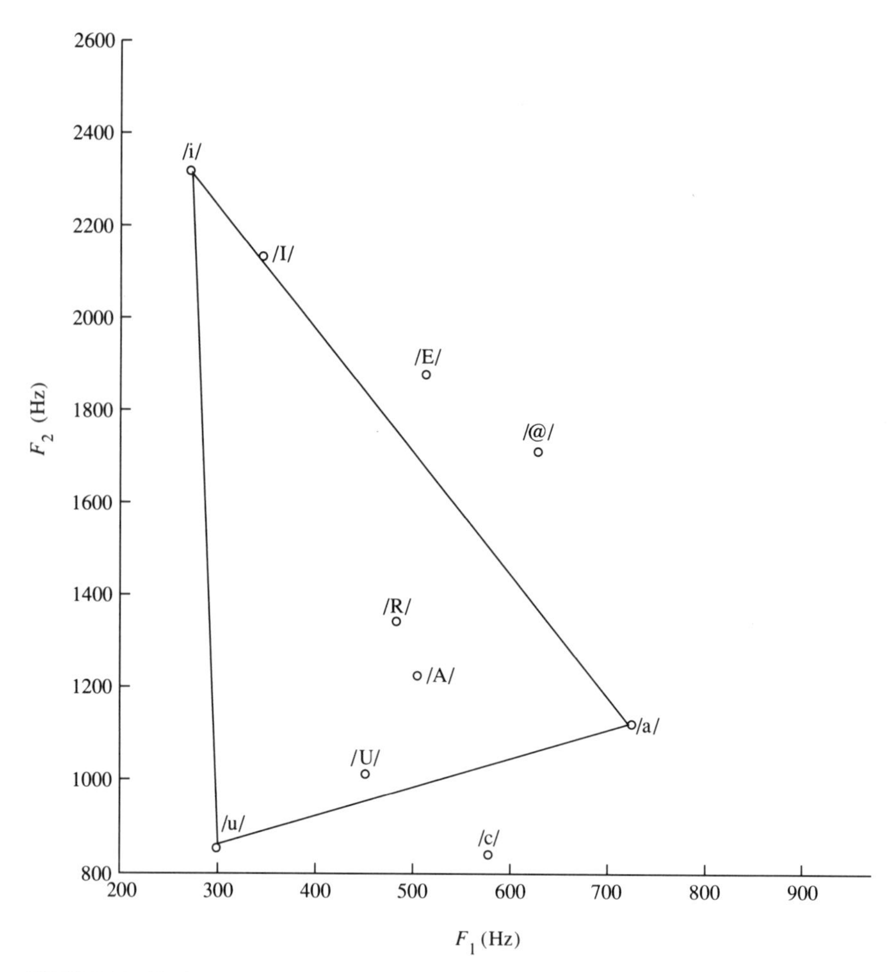

FIGURE 2.12. The vowel triangle. Plot of F_1 versus F_2 for vowels in American English.

Bandwidths of formants can also be used to characterize vowels, although their variation is not as pronounced across vowels as frequency location. Figure 2.13 shows average and extreme values of the first three formant bandwidths from 20 male speakers [each vowel was produced by each speaker twice (Dunn, 1961)]. Formant frequency locations versus bandwidth measurements are shown above the bandwidth results. Vowel bandwidth trends are noted for the three formants (dashed lines for each formant). This graph indicates a tendency for bandwidth of a formant to increase with its center frequency, with this trend much more pronounced for F_3. It is evident that although formant location is of primary importance in characterizing vowels, differences in bandwidth also contribute to the overall vowel character.

Finally, it should be reemphasized that there exists significant variability in vowel formant characteristics (frequencies, spectral magnitudes,

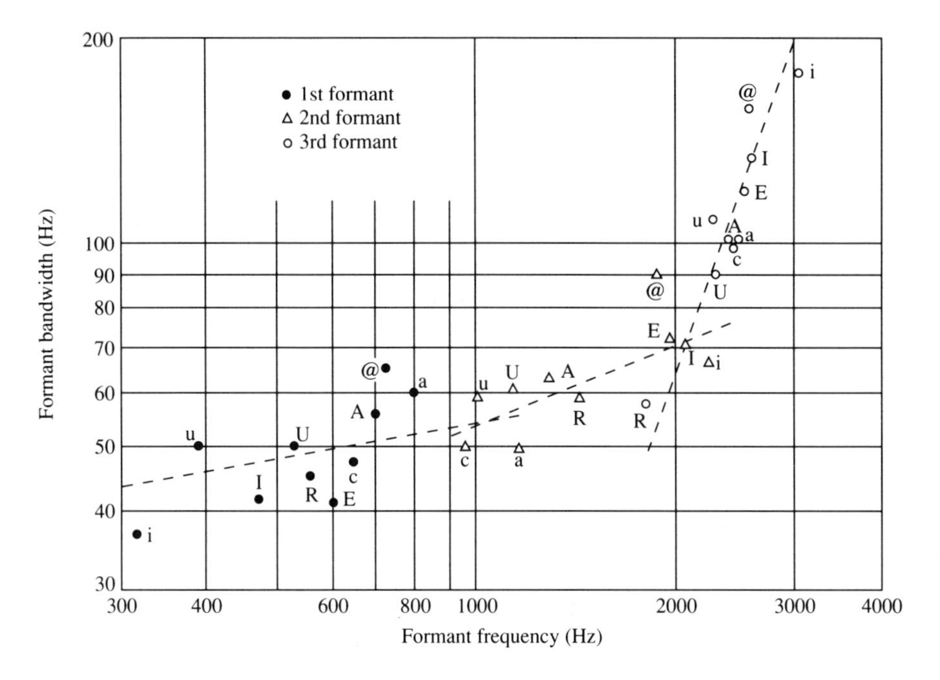

Vowels	F1 Avg.	F1 Extremes		F2 Avg.	F2 Extremes		F3 Avg.	F3 Extremes	
i	38	30	80	66	30	120	171	60	300
I	42	30	100	71	40	120	142	60	300
E	42	30	120	72	30	140	126	50	300
@	65	30	140	90	40	200	156	50	300
a	60	30	160	50	30	80	102	40	300
c	47	30	120	50	30	200	98	40	240
u	50	30	120	58	30	200	107	50	200
U	51	30	100	61	30	140	90	40	200
A	56	30	140	63	30	140	102	50	300
R	46	30	80	59	30	120	58	40	120
Avg.	49.7			64.0			115.2		

FIGURE 2.13. Average formant bandwidths for vowels in American English (Dunn, 1961).

bandwidths) across speakers. However, the data in Figs. 2.11, 2.12, and 2.13 have served as useful guidelines for many purposes over several decades.

Diphthongs. Vowels are voiced speech sounds involving a theoretically constant vocal-tract shape. The only articulator movement occurs during the initial transition to, and the exiting transition from, the nominal vocal-tract configuration for the vowel. A *diphthong* involves an intentional movement from one vowel toward another vowel. There exists some ambiguity as to what constitutes a diphthong, since articulator movement from one vowel toward another might be confused with a sequence of two distinct vowels. A reasonable definition for a diphthong is a vocalic syllable nucleus containing two target vowel positions. The first target vowel is usually longer than the second, but the transition between the targets is longer than either target position (Lehiste and Peterson, 1961). The three diphthongs that are universally accepted are /Y/, /W/, and /O/ (corresponding examples occur in "pie," "out," and "toy"). It should be emphasized that although a diphthong represents a transition from one vowel target to another, frequently it is the case that neither vowel is actually reached. For example, in the diphthong /Y/ ("pie"), the initial vowel position is neither /a/ ("hot") nor /@/ ("had"), and the second is neither /i/ ("heed") nor /I/ ("hid"). The easiest method for determining whether a sound is a vowel or diphthong, is to simply produce the sound. If your vocal tract does not maintain a constant shape, or if the sound cannot be sustained without articulator movement and both vocal targets are vowels, then the sound is a diphthong.

Vocal-tract movement for a diphthong can be illustrated by a plot of the F_1–F_2 transition as in Fig. 2.14. The arrows indicate the direction of motion of the formants versus time. The dashed circles indicate average positions for vowels. Figure 2.14 confirms that the three diphthongs /Y/, /W/, and /O/ move from one vowel target to a second, but in most cases do not achieve either vowel configuration.

The three diphthongs /Y/, /W/, /O/ contain two steady-state vowel target configurations. A second group of dipthongized sounds exists that has only one steady-state target. These sounds are usually referred to as *diphthongized vowels.* They occur in speech because of the tendency to add a "glide" to the beginning or end of long vowels. For example, when we say the word "bay" we usually do not say /be/, but /beI/. The /eI/ sound[13] in this example has a long glide as the first element. Consider, however, /oU/ as in "boat" (/boUt/, not /bot/) has a short steady-state target as the first, followed by a long glide as the second element. Another example is /uw/ as in "who" (/huw/, not /hu/), which has a long steady glide. Finally, the glide-vowel sequence /yu/ as in "you" is sometimes called a diphthong; however, other glide-vowel sequences do exist so that singling this one out can be misleading.

Semivowels. The group of sounds consisting of /l/, /r/, /w/, and /y/ are called *semivowels.* Semivowels are classified as either liquids (/l/ as in "lawn," /r/ as in "ran") or glides (/w/ as in "wet," /y/ as in "yam"). A *glide* is a vocalic syllable nucleus consisting of one target position, with associated formant transitions toward and away from the target. The amount of

[13]Note that single-letter phonetic symbols do not exist for these diphthongized vowels. We create the appropriate sound from two symbols.

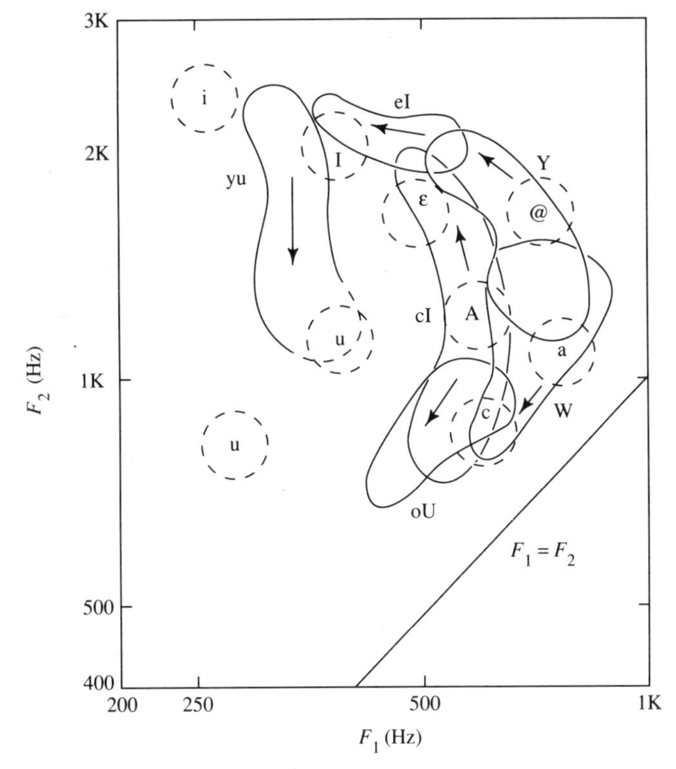

FIGURE 2.14. Movements of F_1 and F_2 for some diphthongs in American English. After Holbrook and Fairbanks (1962).

time spent moving toward and away from the target is comparable to the amount of time spent at the target position. Glides can be viewed as transient sounds as they maintain the target position for much less time than vowels. *Liquids* also possess spectral characteristics similar to vowels, but they are normally weaker than most vowels due to their more constricted vocal tract. Time waveform and vocal-tract profiles for the beginning and ending positions are shown in Fig. 2.15 for liquids and glides.

Consonants

The consonants represent speech sounds that generally possess vocal-tract shapes with a larger degree of constriction than vowels. Consonants may involve all of the forms of excitation that we discussed in Section 2.2.3. Some consonants may require precise dynamic movement of the vocal-tract articulators for their production. Other consonants, however, may not require vocal articulator motion so that their sounds, like vowels, are sustained. Such consonants are classified as continuants. Sounds in which the airstream enters the oral cavity and is completely stopped

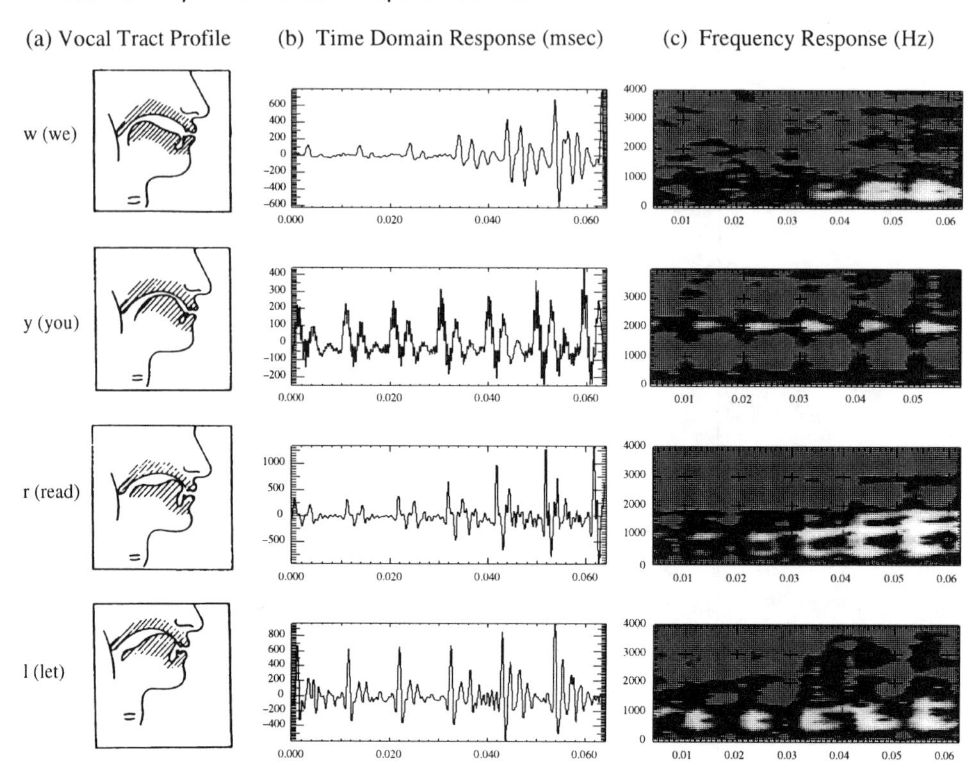

FIGURE 2.15. A collection of features for glides and liquids in American English. Column (a) represents schematic vocal-tract profiles, (b) typical acoustic time waveforms, and (c) the corresponding spectrograms for each glide and liquid.

for a brief period are called *stops* (the term *plosives* is also used to describe nonnasal stops). The continuant, nonplosive consonants are the fricatives /v, D, z, Z, f, T, s, S/, the whisper /h/ (sometimes included as a fricative), the nasals /m, n, G/, and the glides and liquids /w, l, r, y/ (we have chosen to place these semivowels between vowels and consonants, since they require more extensive vocal-tract constriction and articular movement than vowels. These sounds are produced by movements that form partial constrictions of the vocal tract. The stop consonants /b, d, g, p, t, k/ completely inhibit the breath stream for a portion of the articulatory gesture.

Fricatives. Fricatives are produced by exciting the vocal tract with a steady airstream that becomes turbulent at some point of constriction. The point of constriction is normally used to distinguish fricatives as shown in Table 2.5. Locations of constrictions for fricative consonants include *labiodental* (upper teeth on lower lip), *interdental* (tongue behind front teeth), *alveolar* (tongue touching gum ridge), *palatal* (tongue resting on hard or soft palate), and *glottal* (vocal folds fixed and tensed). The constriction in the vocal tract (or glottis) results in an unvoiced excita-

TABLE 2.5. Classification of Nonvowel Speech Sounds by Place and Manner of Articulation.

Place of Articulation / Manner of Articulation	Bilabial	Labiodental	Interdental	Alveolar	Palatal	Velar	Glottal
Nasal stop	/ m / "mom"			/ n / "noon"		/ G / "sing"	
Oral stop	/ p / / b / "pin" "bin"			/ t / / d / "time" "did"		/ k / / g / "kick" "good"	/ Q /
Fricative		/ f / / v / "fine" "vine"	/ T / / D / "thing" "then"	/ s / / z / "sue" "zoo"	/ S / / Z / "shoe" "measure"		/ h / "heat"
Glide	/ w / "wet"				/ y / "yell"		
Liquids				/ l / / r / "like" "run"			

tion. However, some fricatives also have a simultaneous voicing component, in which case they have what we have called mixed excitation. Those with simple unvoiced excitation are usually called *unvoiced fricatives*, while those of mixed excitation are called *voiced fricatives.*

In the unvoiced case, the constriction causes a noise source anterior to the constriction. The location of the constriction serves to determine the fricative sound produced. Unvoiced fricatives include /f/ ("free"), /T/ ("thick"), /s/ ("cease"), /S/ ("mesh"), /h/[14] ("heat"). These are listed in the order in which the point of constriction moves from labiodental to glottal. The constriction also separates the vocal tract into two cavities. The resulting vocal-tract frequency response in general lacks low-frequency energy content, possessing a highpass spectral shape. The frequency cutoff is inversely proportional to the length of the cavity anterior to the constriction. The back cavity acts as an energy trap, which introduces antiresonances in the lower portion of frequency spectrum (Heinz and Stevens, 1961). Vocal-tract profiles, time waveforms, and sample vocal-tract frequency responses for unvoiced fricatives are shown in Fig. 2.16. The major constriction and its effect on low-frequency energy content is evident. Time waveforms also reflect the nonperiodic nature of sound production.

The voiced fricatives /v/ ("vice"), /D/ ("then"), /z/ ("zephyr"), and /Z/ ("measure") are the voiced counterparts of the unvoiced fricatives /f/, /T/, /s/, /S/, respectively. Voiced fricatives also possess the usual frication noise source caused at the point of major constriction, but also have periodic glottal pulses exciting the vocal tract. Since two points of excitation exist for the vocal tract, the vocal-tract spectra of voiced fricatives are expected to differ from unvoiced fricatives. The labiodental /v/ and interdental /D/ voiced fricatives are almost periodic, resulting from a higher degree of excitation at the glottis. The alveolar /z/ and palatal /Z/ voiced fricatives, on the other hand, are more noiselike, giving rise to significant energy in the high-frequency portion of the spectrum. These voiced fricatives also contain a *voice bar* that is a very low-frequency formant (near 150 Hz). The voice bar occurs because the vocal folds are able to vibrate, exciting an occluded oral and nasal cavity. Figure 2.17 illustrates vocal-tract profiles, time waveforms, and sample vocal-tract frequency responses for voiced fricatives. Comparison with Fig. 2.16 reveals the differences in time and frequency characteristics between voiced and unvoiced fricatives. The unvoiced fricatives /f, T, s, S/ have more energy at the middle and high frequencies than at lower frequencies. In contrast, voiced sounds always have more energy in the low frequencies (below 1 kHz) than at high frequencies.

Affricates. Similarly to glides, liquids, and diphthongs, affricates are formed by transitions from a stop to a fricative. The two affricates found

[14]The *whisper* sound /h/ is normally referred to as an unvoiced glottal fricative.

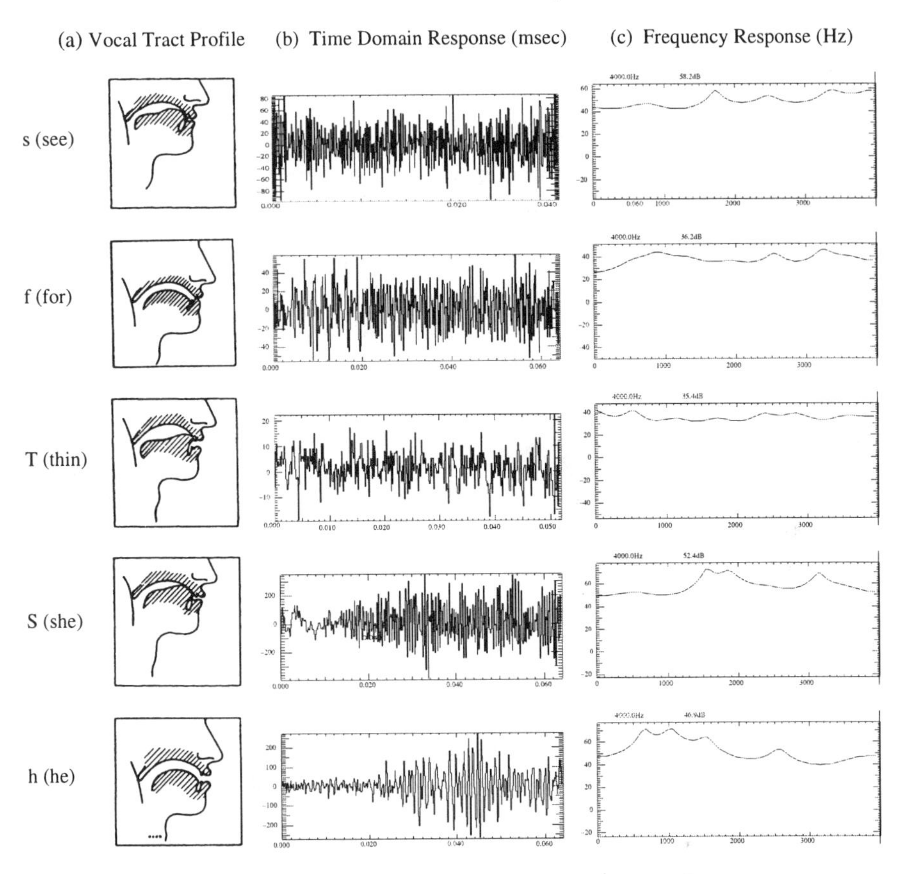

FIGURE 2.16. A collection of features for unvoiced fricatives in American English. Column (a) represents schematic vocal-tract profiles, (b) typical acoustic waveforms, and (c) the corresponding vocal-tract magnitude spectra.

in American English are the unvoiced affricate /C/ as in "change" and the voiced affricate /J/ as in "jam." The unvoiced affricate /C/ is formed by the production of the unvoiced stop /t/, followed by a transition to the unvoiced fricative /S/. The voiced affricate /J/ is formed by producing a voiced stop /d/, followed by a vocal-tract transition to the voiced fricative /Z/.

Stops (or Plosives). The majority of speech sounds in American English can be described in terms of a steady-state vocal-tract spectrum. Some sounds, such as diphthongs, glides, and liquids, possess one or more vocal-tract target shapes, with smooth movement of the vocal-tract articulators either to or from a target position. The *stop* consonants /b, d, g, p, t, k/ are transient, noncontinuant sounds that are produced by building up pressure behind a total constriction somewhere along the vocal tract, and suddenly releasing this pressure. This sudden explosion

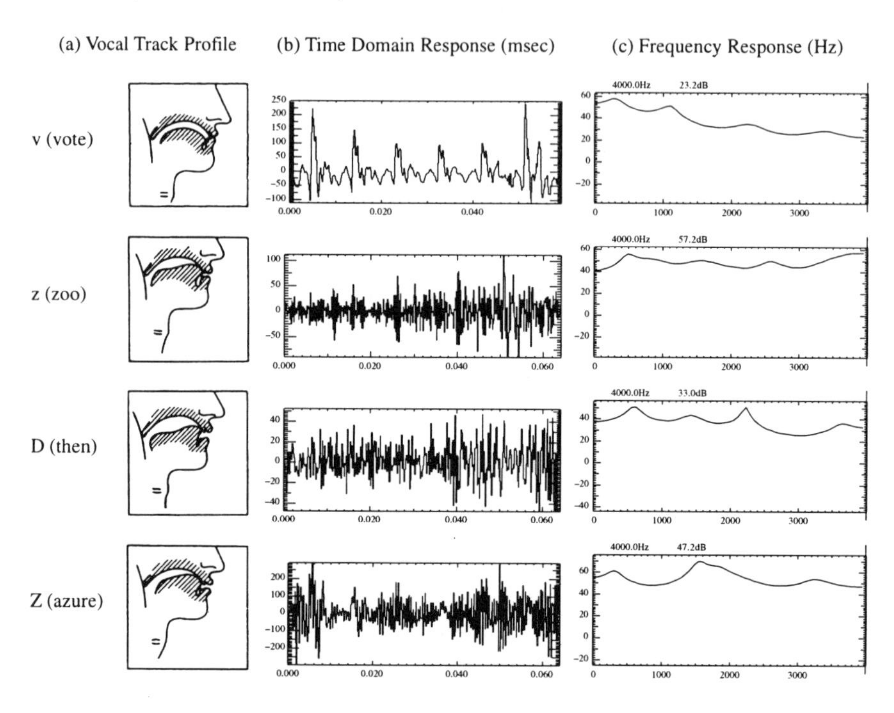

FIGURE 2.17. A collection of features for voiced fricatives in American English. Column (a) represents schematic vocal-tract profiles, (b) typical acoustic waveforms, and (c) the corresponding vocal-tract magnitude spectra.

and aspiration of air characterizes the stop consonants. Stops are sometimes referred to as *plosive* phonemes and their mode of excitation is what we called plosive excitation in Section 2.2.3. As summarized in Table 2.5, the closure can be *bilabial* (/b/ in "be," /p/ in "pea"), *alveolar* (/d/ in "day," /t/ in "tea"), or *velar* (/g/ in "go," /k/ in "key").

The unvoiced stop consonants /p, t, k/ are produced by building up air pressure behind the vocal-tract occlusion, followed by a sudden release. As the air pressure is released, there is a brief period of noiselike frication that is due to the sudden turbulence from the escaping air. The noiselike frication is followed by a steady airflow (aspiration) from the glottis before vocal fold movement is required in the ensuing voiced sound. Unvoiced stops usually possess longer periods of frication than voiced stops. The frication and aspiration is called the *stop release.* The interval of time leading up to the release during which pressure is built up is called the *stop gap.* These features are evident in the time waveforms in Fig. 2.18.

Voiced stops /b, d, g/ are similar to the unvoiced stops, except they include vocal fold vibration that continues through the entire stop or begins after the occlusion release. During the period in which pressure is building behind the oral tract closure, some energy is radiated through the walls of the throat. This leads to the presence of a voice bar in the acoustic analysis.

Characteristics of stops vary depending on their positions within a word or phrase. Most stops are not released if they occur at the end of a syllable (e.g., the /t/ in "cats and dogs," the /k/ in "success"). This is caused by reduced lung pressure that reduces oral pressure behind the

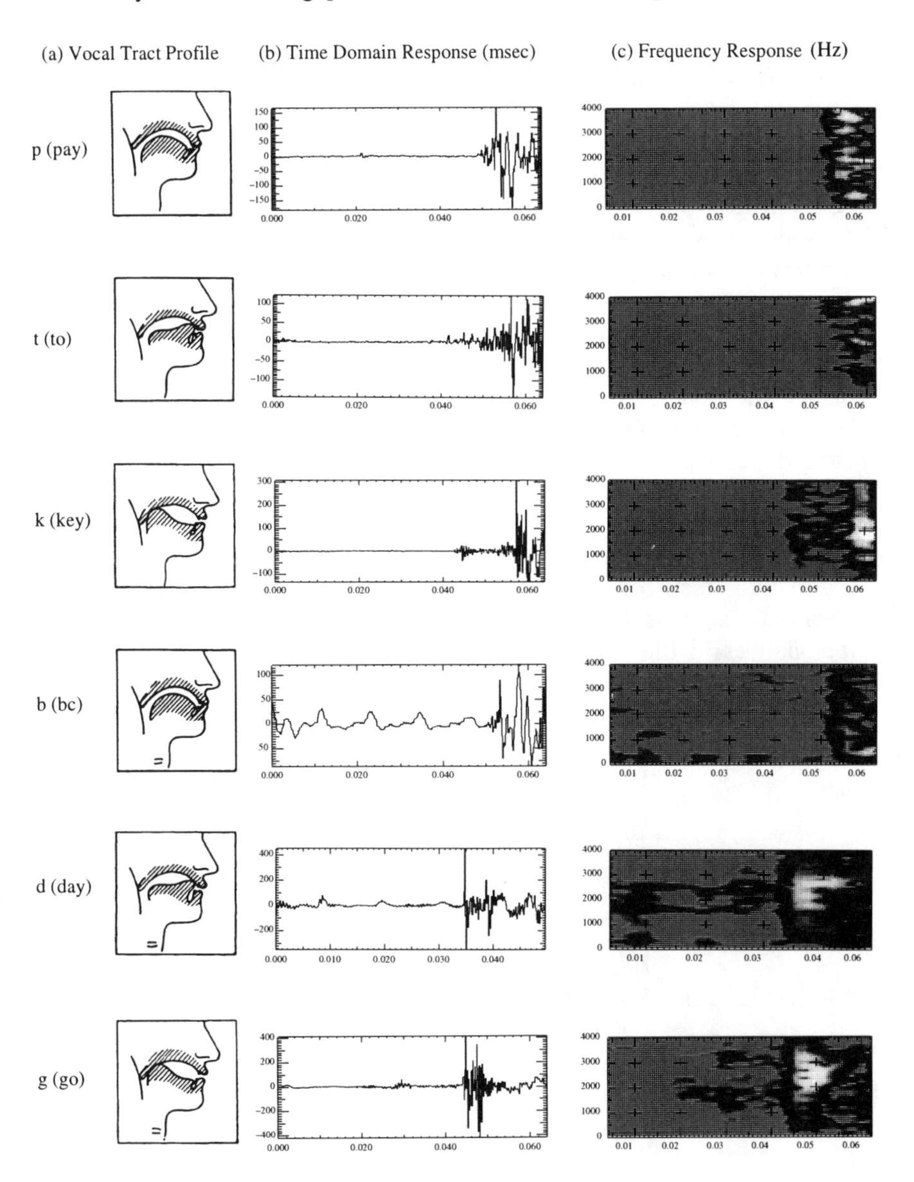

FIGURE 2.18. A collection of features for voiced and unvoiced stops in American English. Column (a) represents schematic vocal-tract profiles just prior to release, (b) typical acoustic waveforms, and (c) the corresponding vocal-tract spectrogram.

stop occlusion. Stops can also be altered if they appear between vowels, resulting in a *tap* or *flap* (Zue and Laferriere, 1979). This variant is produced when one articulator is thrown against another, such as when the tip of the tongue touches the upper teeth or alveolar ridge in "letter." Since stop consonants are transient, their properties are highly influenced by prior or subsequent phonemes (Delattre et al., 1955). Therefore, the waveforms in Fig. 2.18 provide little information for distinguishing the stop consonants. Vocal-tract spectra for stops are sometimes useful; however, a number of allophones can produce radically different spectra at the time of release.

Nasals. The nasal consonants /m, n, G/ are voiced sounds produced by the glottal waveform exciting an open nasal cavity and closed oral cavity. Their waveforms resemble vowels, but are normally weaker in energy due to limited ability of the nasal cavity to radiate sound (relative to the oral cavity). In forming a nasal, a complete closure is made toward the front of the vocal tract, either at the lips (*labial* or *front* nasal /m/ as in "more"), with the tongue resting on the gum ridge (*alveolar* or *mid*-nasal /n/ as in "noon"), or by the tongue pressing at the soft or hard palate (*velar* or *back* nasal /G/ as in "sing"). The velum is opened wide to allow for sound propagation through the nasal cavity.

Figure 2.19 shows vocal-tract profiles, time waveforms, and vocal-tract frequency responses for the three nasals in American English. The closed oral cavity is still acoustically coupled to the pharyngeal and nasal cavities, and will therefore affect the resulting spectral resonances by trapping energy at certain frequencies. This phenomenon gives rise to antiresonances in the overall vocal system. The length of the closed oral

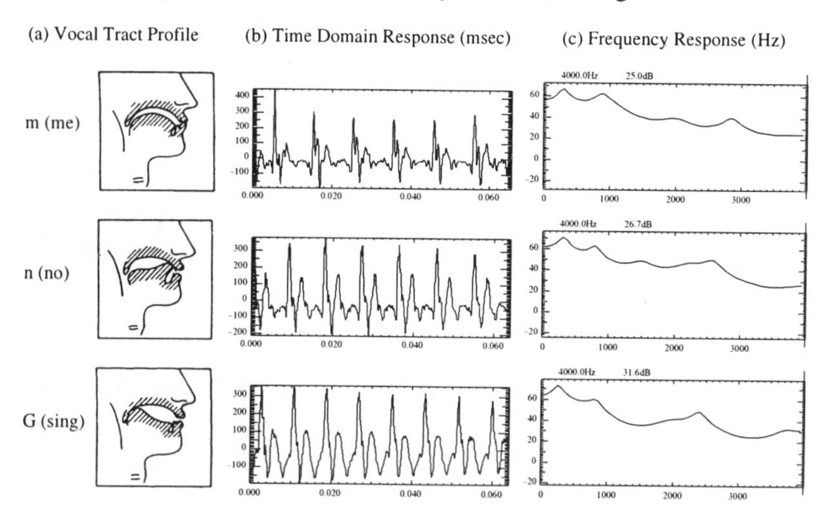

FIGURE 2.19. A collection of features for nasals in American English. Column (a) represents schematic vocal-tract profiles, (b) typical acoustic waveforms, and (c) the corresponding vocal-tract magnitude spectra.

cavity is inversely proportional to the frequency location of the antiresonance. For nasals, formants occur approximately every 850 Hz instead of every 1 kHz. A low formant (F_1) is found near 250 Hz that dominates the frequency spectrum. F_2 is weak, resulting from the antiresonance of the closed oral cavity, and F_3 occurs near 2200 Hz. The antiresonance produces a spectral zero in the frequency response that is inversely proportional to the length of the constricted oral cavity. The spectral zero occurs in the range 750–1250 Hz for /m/, 1450–2200 Hz for /n/, and above 3000 Hz for /G/. Bandwidths of nasal formants are normally wider than those for vowels. This is due to the fact that the inner surface of the nasal cavity contains extensive surface area, resulting in higher energy losses due to conduction and viscous friction. Since the human auditory system is only partially able to perceptually resolve spectral nulls, discrimination of nasals based on the place of articulation is normally cued by formant transitions in adjacent sounds.

Although nasals are the only phonemes that incorporate the nasal cavity to produce their resonant frequency structure, some phonemes become *nasalized* if they precede or follow a nasal sound. This occurs primarily in vowels that precede a nasal, where the velum begins to drop in anticipation of the ensuing consonant. Nasalization produces phonemes that have broader F_1 bandwidths and are less peaked than those without nasal coupling. This is caused by damping of the formant resonance by the loss of energy through the opening into the nasal cavity (House and Stevens, 1956). Other spectral changes include spectral zeros that cancel or trap energy within the vocal tract. The degree to which nasalization affects the spectrum depends on the amount of coupling between the two cavities. If the velum is only slightly lowered, weak nasal cavity coupling produces minor changes in the resulting frequency response. The degree of coupling directly influences the location and strength of the spectral zeros. Changes in spectral characteristics for a nasalized vowel (strongly coupled) are shown in Fig. 2.20. The resulting nasalization produces a zero near 2 kHz and cancels most of the third formant. The position of these zeros should not, however, be considered to hold for weak degrees of nasalization of other vowels, since the vocal-tract shape and amount of velar opening interact to determine the position of spectral zeros.

2.3.4 Prosodic Features and Coarticulation

Prosodics

Our discussion of speech has focused on characterizing speech in terms of articulatory phonetics (the manner or place of articulation) and acoustic-phonetics (frequency structure, time waveform characteristics). Speech production, however, involves a complex sequence of articulatory movements, timed so that vocal-tract shapes occur in the desired pho-

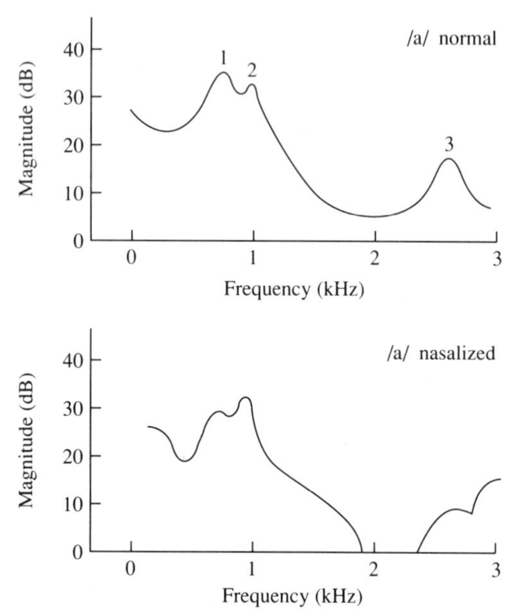

FIGURE 2.20. An example of the effects of nasalization on a vowel spectrum. Note the appearance of spectral nulls in the nasalized case.

neme sequence order. From an acoustic-phonetic point of view, expressive uses of speech depend on tonal patterns of pitch, syllable stresses, and timing to form rhythmic speech patterns. Here we will focus on acoustic-phonetic aspects of rhythm and pitch as a function of their linguistic structure in American English.[15]

Timing and rhythms of speech contribute significantly to the formal linguistic structure of speech communication. The tonal and rhythmic aspects of speech are generally called *prosodic features.* Since they normally extend over more than one phoneme segment, prosodic features are said to be *suprasegmental.* Prosodic features are created by certain special manipulations of the speech production system during the normal sequence of phoneme production. These manipulations are categorized as either source factors or vocal-tract shaping factors. The source factors are based on subtle changes in the speech breathing muscles and vocal folds, while the vocal-tract shaping factors operate via movements of the upper articulators. The acoustic patterns of prosodic features are heard in systematic changes in duration, intensity, fundamental frequency, and spectral patterns of the individual phonemes. The aim of this section is to

[15]It should be noted, however, since foreign languages differ from American English in their basic speech building blocks (phonemes), so too will their tonal patterns of pitch, syllable stresses, and sentence timing. Since most students study a non-native language by first learning vocabulary, it is normally quite easy to identify an early student of a non-native language by his or her incorrect placement of stress and timing.

present a basic explanation of the factors that govern the production of prosodic patterns. A detailed discussion would require a more extensive background in speech science; the interested reader is encouraged to consider textbooks cited in Appendix 1.F, especially (Chomsky and Halle, 1968; Ladefoged, 1967; Lieberman, 1967; Perkell and Klatt, 1986).

Prevailing theories have stated that the time-varying vocal-tract transfer function carries most of the information concerning which phonemes are produced, with glottal source excitation characteristics conveying prosodic cues such as *intonation* and *stress.* Intonation and stress are the two most important prosodic features that convey linguistic information. Stress is used to distinguish similar phonetic sequences or to highlight a syllable or word against a background of unstressed syllables.[16] For example, consider the two phrases "That is insight" and "That is in sight." In the first phrase there is stress on "in" but "sight" is unstressed, while the converse is true in the second phrase. The analytical question that has received a great deal of attention in recent years is how many degrees of stress should be recognized to account for all such contrasts? A structuralist approach[17] suggests that four such degrees are normally distinguishable (from strongest to weakest): (1) primary, (2) secondary, (3) tertiary, and (4) weak. It is only possible to distinguish these stress levels for words in isolation. A good example is the compound word $\overset{1}{e}\overset{4}{le}\overset{3}{va}\overset{4}{tor}\ \overset{2}{o}\overset{4}{pe}\overset{3}{ra}\overset{4}{tor}$, which shows each of the stress levels. One could also rise above the syllables to a *word,* or *lexical, stress* level, as in the examples "Joe writes on a black board" ("black" used as an adjective) and "Joe writes on a blackboard" ("blackboard" is a noun). The second sentence has additional stress placed on "blackboard."

Intonation refers to the distinctive use of patterns of pitch or melody. An analysis of intonation is performed by considering pitch patterns in terms of contours, for which pitch range, height, and direction of change are generally characterized. Intonation performs several useful functions in language, the most important being to signal grammatical structure.[18] Intonation therefore performs a role similar to punctuation in writing. Intonation, however, has much wider contrasts such as marking sentence, clause, or other boundaries, as well as contrasting grammatical sentence structure such as in questions or statements. For example, consider the rising and falling pitch patterns in the second speaker's utterance in the following two pieces of dialogue. In the first, Charles does not know

[16]Stress can refer to syllable, word, phrase, or sentence-level stress. Phonologically, it simply is a means of distinguishing degrees of emphasis or contrast.

[17]A *structuralist* approach to language focuses on the way linguistic features are described in terms of structures and systems. Structure in linguistics can be found in phonology, semantics, and grammar. These concepts will be further discussed in Chapter 13.

[18]For natural sounding text-to-speech systems, it is important to accurately represent intonation.

whether Casey is even trying to come to the meeting, and his intonation (rising pitch at the end) indicates a question:

> *Joan:* "Casey can't be present at the meeting."
> *Charles* (hoping that Casey is at least in transit): "He's on the plane from Dallas, isn't he?"

In this case, "isn't he?" is a question to which Charles expects an answer. In the second, interchange, however, Charles realizes that Casey can't come because he is in transit. His intonation indicates that he just realized why:

> *Joan:* "Casey can't be present at the meeting."
> *Charles* (realizing that Casey is still traveling): "He's on the plane from Dallas, isn't he?"

The "isn't he?" is uttered here with falling pitch indicating that no response is really necessary. A second role of intonation is to relay secondary characteristics such as attitude or emotion (i.e., anger, sarcasm, etc.).[19] In the first interchange above, for example, it is easy to imagine how some anger or annoyance might be expressed by Charles's intonation.

Let us now consider how the glottal source and vocal tract contribute to the production or encoding of prosodic features during speech production. The two principal factors responsible for the variations of glottal source factors for prosodic features are (1) subglottal air pressure and (2) tension of the vocal folds. These factors contribute to variations in fundamental frequency, source spectrum, and source amplitude. If a speaker increases the force on the lungs, a corresponding increase in subglottal air pressure results. This increase in subglottal air pressure in turn causes an increase in the rate at which airflow pulses are produced at the glottis, thus increasing fundamental frequency. Therefore, pitch and loudness both increase when subglottal air pressure increases. The higher the subglottal air pressure, the higher the pitch of the speaker's voice. This relationship approximates a straight line if the log of fundamental frequency is plotted versus subglottal air pressure (Ladefoged, 1963).

Next, we consider an example of how stress and intonation are conveyed by different stress patterns of subglottal pressure, depending on whether the stress is placed on the first or second syllable of a word, and whether the word spoken is from a statement or question. This example is adapted from Ladefoged (1963). In Fig. 2.21 we see the patterns of subglottal air pressure and fundamental frequency for four sentences with the word "digest." Compare the word "digest" spoken as a noun with stress on the first syllable [Fig. 2.21(a) and (b)], with "digest" spoken as a verb with stress on the second syllable [Fig. 2.21(c) and (d)]. It

[19]Many emotions are signaled by constrasts in pitch; however, other prosodic and paralinguistic features also contribute.

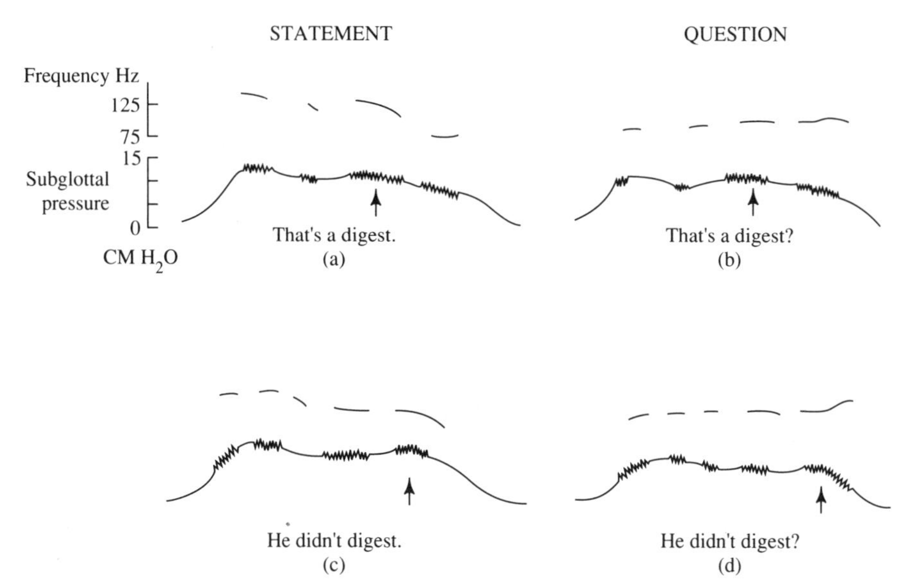

FIGURE 2.21. Relations between fundamental frequency contours and subglottal air pressure for statements and questions with two different word stress patterns. Adapted from Ladefoged (1963).

is clear that the stressed syllable is always spoken with higher subglottal air pressure. In addition, the pressure is normally higher for statements than for questions. Intonation patterns are shown above plots of subglottal air pressure. The pitch frequency is shown during vowel portions (the duration of the pitch contour indicates the duration of each vowel). The general trend for statements is downward pitch contour from the beginning to the end of the sentence. Stressed syllables also possess the longest vowel duration of all the vowels within the sentence. For a statement, if the stressed syllable is a noun, it will have a high pitch on the stressed syllable. For the stressed syllable on the verb, higher pitch occurs on the second syllable, even though this is the last syllable of the sentence. When the same sentences are spoken as questions, the pitch contour generally rises evenly from the beginning to the end of the sentence [Fig. 2.21(b) and (d)]. For the noun in the question, the stressed syllable possesses neither increased duration nor pitch from the following syllable because the rising intonation contour required for a question supersedes lexical stress requirements at the word level. In the final verb form, the stressed syllable will possess increased duration and higher pitch. This example has shown that the amplitude and duration of pitch contours corresponding to stressed and unstressed syllables are dependent on whether the general intonation pattern expresses a statement or a question.

Upon reexamination of the subglottal air pressure patterns in Fig. 2.21, we notice that whether a sentence is spoken as a statement or a question, the overall subglottal air pressure remains approximately constant. What causes the change in pitch to produce the rising intonation for questions and falling intonation for statements? It is currently believed that the major factor controlling pitch is the tension in the vocal folds. Although studies continue in this area, it appears that the strongest stress of a syllable in a phrase is produced by a combination of increased vocal fold tension and a peak in subglottal pressure. Increased vocal fold tension is applied through small muscles in the larynx.

We summarize the major points concerning stress and intonation as follows:

- The phonetic patterns for stress and intonation are based on a blending of changes in vocal fold tension, peaks and valleys of subglottal air pressure, and variation in duration of vowels and consonants.
- Movement of the fundamental frequency (or pitch) is reflected in changes in subglottal air pressure and vocal fold tension. If either (or both) increases, a corresponding increase in fundamental frequency also occurs.
- The general shape of the subglottal air pressure over a phrase is relatively constant and does not reflect the wide variations seen in the pitch contour.

Based on earlier work by Lieberman (1967), a simple *breath group theory* of intonation was proposed to explain how patterns of subglottal pressure and vocal fold tension are encoded to relay stress and intonation. The theory states that the intonation contour will naturally fall toward the end of a breath group due to lower subglottal air pressure, just prior to taking a new breath of air. The breath group can then be marked as a question by simply increasing the vocal fold tension. In its simplest form, the breath group theory assumes a high degree of independence between vocal fold tension and subglottal pressure. Recent studies suggest that intonation contours are primarily controlled by vocal fold tension for high-pitched sections, and by subglottal pressure during lower pitch portions of the contour. It is also thought that the vertical position of the larynx determines whether the vocal fold tension or subglottal pressure is the dominant factor.

Other glottal source factors which are related to stress and intonation are the intensity and spectral balance between high and low frequency regions of the glottal source spectrum. An increase in subglottal pressure will increase the intensity of the open phase of the glottal source waveform (increased amplitude of the air puffs produced at the glottis). If a voiced syllable is stressed, the increased subglottal pressure gives rise to a glottal source spectrum with increased high frequency content, relative to the lower frequencies. Therefore, vowels which are stressed will possess higher magnitudes for the higher-frequency formants than unstressed vowels.

Although these glottal source factors contribute greatly to stress and intonation, variations in duration are also used to encode prosodic features. A syllable is normally longer if it appears in the final portion of a breath group. Unstressed words normally contain vowels that are shorter. Also, the type of consonant that follows a vowel greatly influences the duration of that vowel (e.g., consider the duration of the vowels in "fate" and "fade"). Consonant duration is affected by syllable stress and the position of the syllable in the sentence in a manner similar to those observed for vowels. Phoneticians have described rules for prosodic effects of duration for vowels and consonants to explain the origins of speech motor control. Many of these rules form the basis of natural sounding text-to-speech systems. In general, these rules state that the larger the number of subunits (i.e., phonemes, syllables) in a larger unit, the shorter each subunit becomes up to some limit of compressibility. It is suggested that this duration compression occurs due to the limited number of speech units that can be temporarily held in our speech motor memory. This discussion, though far from complete, has served to relate several factors responsible for prosodic features.

Coarticulation

The production of speech requires movement of the articulators in the vocal tract and the glottal source. Unlike distinct characters found in printed text, phoneme articulations typically overlap each other in time, thereby causing sound patterns to be in transition most of the time. Instead of quick rigid articulator movement between uniform islands of stationary phoneme production, speech is normally produced with smooth movement and timing of the articulators to form the proper vocal-tract shapes to produce the desired phoneme sequence. *Coarticulation* is the term used to refer to the change in phoneme articulation and acoustics caused by the influence of another sound in the same utterance.

Movements of the articulators are fairly independent of one another. The muscle groups associated with each articulator, as well as mass and position, affect the degree and ease of movement. The movements of the tongue and lips are free to overlap in time; however, the actual tongue articulation for a vowel-consonant-vowel sequence can vary depending on the target positions the tongue must attain for the specific vowels and consonants. Because each phoneme has a different set of articulator position goals (not all of which are strictly required), considerable freedom exists during actual speech production. There are two primary factors underlying coarticulation effects:

1. The specific vocal-tract shapes which must be attained.
2. The motor program for production of the sequence of speech units (consonants, vowels, syllables, words, and phrases).

If an articulatory gesture does not conflict with the following phoneme, the given articulator may move toward a position more appropriate for the following phoneme. Such anticipatory coarticulation is classified as *right-to-left* because the vocal system's target to the right induces motion in the present phoneme (to the left). This suggests that some articulators will begin to move during phoneme production earlier than others in anticipation of the next phoneme. This suggests that the motor program needed for performing a sequence of sounds, syllables, and words appears to anticipate the number of remaining speech units (phonemes) to be produced within a breath group, shortens those which simply need to be produced, but retains the reachability of the consonant constrictions and the recognizability of the stressed syllables. If there were no tailoring of the early sounds over a breath group, the later phonemes would have to be crowded in and spoken too quickly for proper listener comprehension. A general theory of motor programming has been proposed based on the *short-term memory* (STM) model. This was proposed by Lindblom, Lyberg, and Holmgren (1977) based on previous principles derived by Klatt (1973, 1976). The model suggests that there is a short-term storage of the instructions for speech articulator movements. The contents of the storage are continuously changing as instructions leave the queue to discharge the actual movements, and new instructions enter for later discharge in their proper sequence. The storage size of this queue is limited; therefore, the available space must be used economically, requiring every speech unit present to be compressed in duration whenever a new segment is submitted. There is a limit, however, on the degree of compressibility, since time is required for articulators to effectively reach target positions as well as to make the transitions between targets. As an example, stop consonants must reach full occlusion to produce audible stop gaps and burst releases. There exists perceptual evidence that listeners actually expect anticipatory (right-to-left) coarticulation (Scharf and Ohde, 1981). In addition to right-to-left coarticulation, there is also left-to-right coarticulation. Left-to-right coarticulation occurs when some of the present phoneme features drift into the following phoneme. Whereas right-to-left coarticulation involves an active forward-looking planning process, left-to-right coarticulation is caused more by articulator momentum or low-level movement constraints than by a higher-level motor control program in the brain. A good example arises in formant transitions in vowels as a result of a preceding consonant (Scharf and Ohde, 1981).

Coarticulation is manifest in a changing acoustic structure caused by movement of articulators as well as changing duration of the phoneme targets once they are reached. The consequence of the limit of compressibility suggests that monosyllabic words are shortened more in anticipation of an increase in the number of later syllables than early bisyllabic words. In a similar fashion, trisyllabic words are shortened less than bisyllabic words. As an example, compare the duration of the word "an" in utterances of "an," and "an official deadline." The duration of the monosyllabic word "an" is significantly reduced when spoken in a phrase. It appears from the STM model that storage in our production queue is in terms of numbers of syllables, and less of that limited capacity is needed for a word with fewer syllables. There is also a continuing readjustment to economize the available storage space dependent on the number of phonemes, syllables, words, or phrases in a sentence. One type of compression occurs when consonants are adjacent between a syllable boundary in the same breath group. Several examples include double consonants like the /kk/ in "black cat" or /pp/ in "stop port," and adjacent differing consonants such as /Gp/ in "Ping–Pong" or /kp/ in "stock pile." If the speaking rate is normal or slow, both consonants can be articulated separately (however, the first consonant may not experience a release). As speaking rate increases, the first consonant merges into the second. At very high speaking rates, only one consonant remains at the beginning of the second syllable. Rapid-utterance speaking rates cause more overlap of articulations than in normal rates with a compression of the duration of consonants and vowels. It appears that very fast utterance production causes vowels to be compressed more than consonants, perhaps due to the requirement that consonants must attain articulator constriction positions that are sufficient to produce perceivable spectral discontinuities.

The smooth flow of speech through movement of the articulators represents a basic area in speech production science that is much broader and more complex than our limited discussion here would indicate. The basic knowledge of speech characteristics in acoustic science has allowed us to specify a series of phenomena that result when certain phonemes appear within a breath group or cluster. Many of these phenomena imply rules that are used to produce more natural sounding synthetic speech for text-to-speech systems. However, since the basic knowledge of how speech movements are exactly organized is not highly developed, we are unable to easily relate the flow of speech articulator movement to timing and spectral characteristics of the acoustic flow. Research continues in this area and at some future time it may be possible to characterize acoustic phonetics properly from the articulator movement flow. Such analysis will also help simplify explanations of how spoken messages are perceived. Until a complete motor theory of speech production has been formulated and accepted, we will have to be satisfied with analytic methods that focus on acoustics (spectral shapes, etc.) and linguistics. Significant progress has been made using such methods in recent years (Lindblom et al., 1977, 1983; MacNeilage, 1970, 1980, 1983; Harris et al., 1977, 1984; Perkell and Klatt, 1986). If an analysis of speech could begin with flow characteristics, then the primary patterns for analysis in digital speech processing would be transitions and effects due to coarticulation or prosodic features, not the fixed vocal-tract shapes discussed earlier in this chapter. This may appear to be a significant limitation; however, we shall see that effective speech modeling, synthesis, and recognition techniques are possible without a complete theory of speech motor production, although future techniques certainly will benefit from such emerging theories.

2.4 Conclusions

The purpose of this chapter has been to present an overview of the fundamentals of speech science. The goal has been to provide a sufficient background to pursue applications of digital signal processing to speech. Speech is produced through the careful movement and positioning of the vocal-tract articulators in response to an excitation signal that may be periodic at the glottis, or noiselike due to a major constriction along the vocal tract, or a combination. Phonemes, the basic units of speech, can be classified in terms of the place of articulation, manner of articulation, or a spectral characterization. Each phoneme has a distinct set of features that may or may not require articulator movement for proper sound production. Language or grammatical structure, prosodic features, and coarticulation are all employed during the production of speech. Although a complete model of such higher level processing does not yet exist, well-defined rules have been collected. Such knowledge is necessary for natural sounding synthesis and text-to-speech systems. A more detailed treatment of the areas of speech science will prove useful in many of the applications in the following chapters, but it will be more useful to explore these questions after becoming familiar with techniques for obtaining general models of speech production and their use in coding, enhancement, or recognition.

2.5 PROBLEMS

2.1 Explain the differences between pitch and fundamental frequency. Which is a function of glottal vocal fold movement? Which is a function of both vocal fold movement and processing by the auditory system?

2.2 Discuss the forces that act to produce a complete phonation cycle during voiced speech production. How do the forces during the phonation vary if the density of air is reduced (e.g., if the speaker is breathing a helium–oxygen mixture)? Suggest a means to remedy this unnatural sounding speech.

2.3 The speech spectrogram has been used for analysis of speech characteristics for many years. Could such a method be used for recognition of speech patterns? What aspects of speech production cannot be resolved from a spectrogram?

2.4 How many phonemes are there in American English? List the attributes which distinguish phonemes into individual speech classes (manner of voicing, place of articulation).

2.5 The following figures are designed to help you become more familiar with vocal-tract shape and speech production. For each drawing, there is only one sound in American English that could be produced with such a

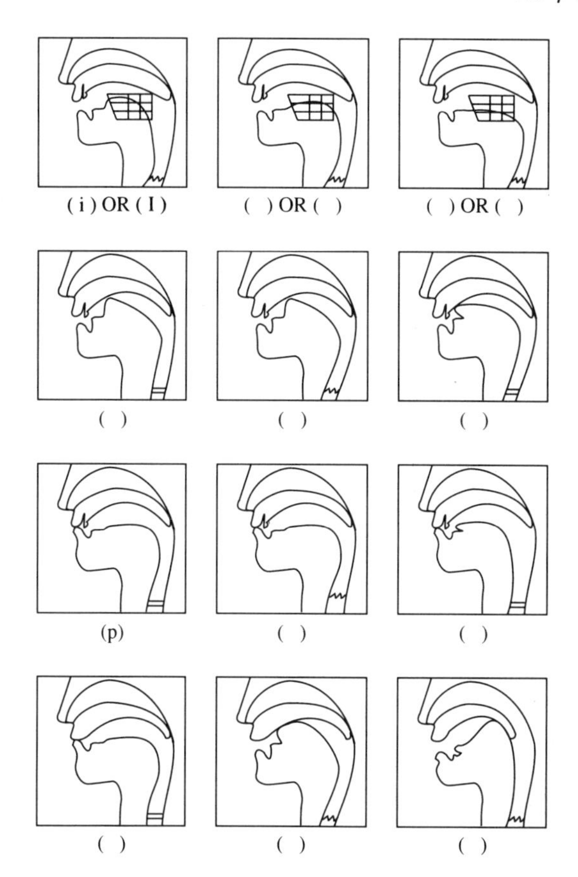

vocal-tract position. For each figure, write the phonetic symbol that corresponds to the shape, and circle the articulator(s) most responsible for this phoneme production.

2.6 Suppose that a typical male speaker of average height (6 feet) possesses a vocal tract of length 17 cm. Suppose also that his pitch period is 8.0 msec. Discuss how voice characteristics would change for a very tall (7 feet) basketball player and an infant (2 feet tall). Assume that their vocal systems are proportioned to their heights.

2.7 The waveform of the sentence "Irish youngsters eat fresh kippers for breakfast," is shown on p. 148. The signal was sampled at 8000 samples/second.

(a) Label all the regions of voiced, unvoiced, and silence.
(b) For the phrase "Irish youngsters," indicate the phonemic boundaries (i.e., mark boundaries for the phonemes /Y/ - /r/ - /I/ - /S/, /y/ - /A/ - /G/ - /s/ - /t/ - /R/ - /z/).
(c) For each the vowel sound, estimate the average pitch period.
(d) Is the speaker an adult male, adult female, or a child?
(e) Find the range for pitch frequency, using estimated minimum and maximum pitch period values from part (c).

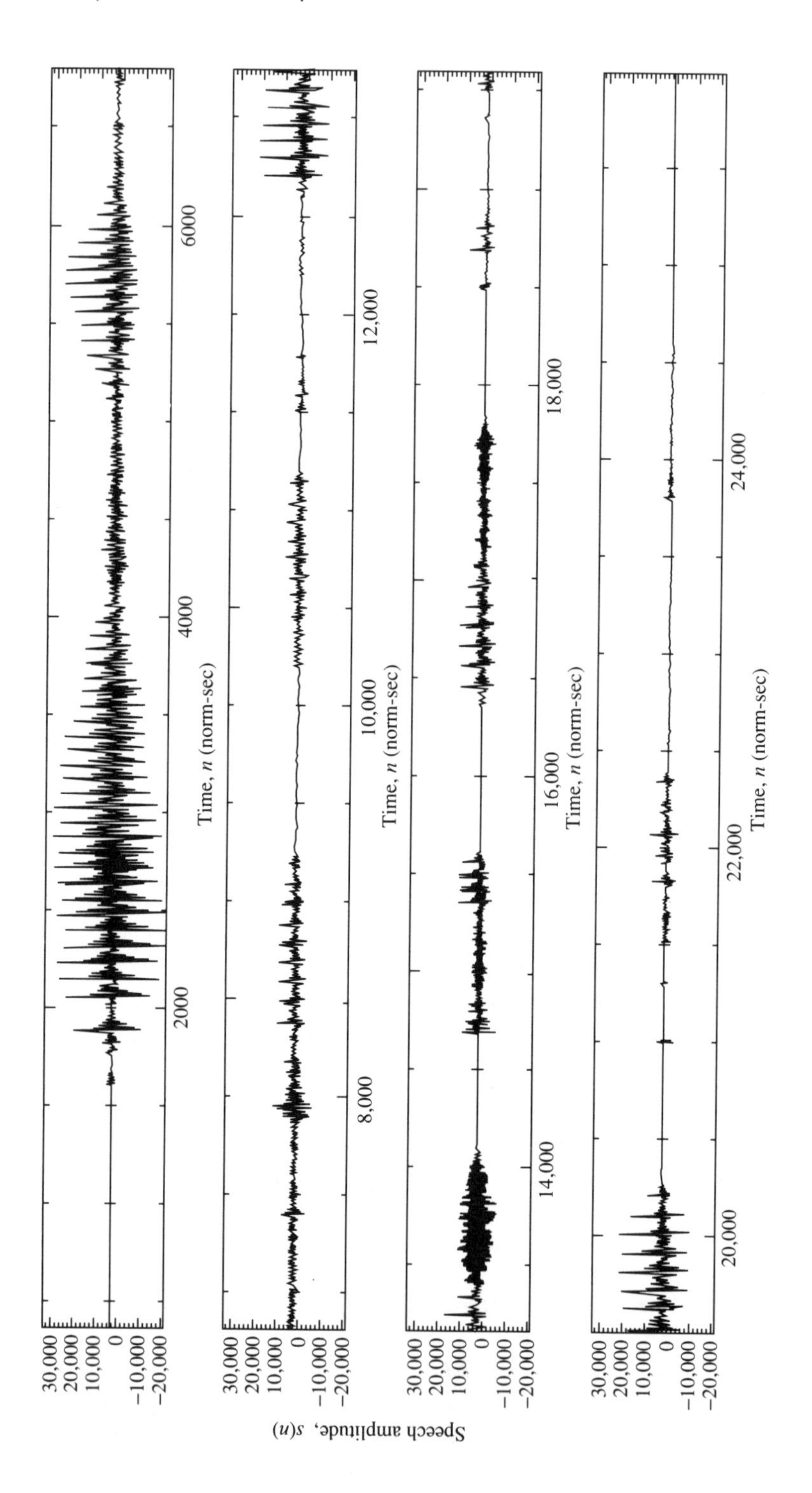

Speech amplitude, s(n)
30,000
20,000
10,000
0
−10,000
−20,000
2000
4000
6000
8,000
10,000
12,000
14,000
16,000
18,000
20,000
22,000
24,000
Time, n (norm-sec)

2.8 It can be shown that vowel formant location varies while producing the following word list: "mat" - "met" - "mit" - "meat." Discuss the differences in vocal-tract articulator position/movement in producing each of these words.

2.9 Vowels are characterized by their formant locations and bandwidths. What spectral features vary across the vowel triangle? How do vocal-tract articulatory movements characterize such a triangle? Does the vowel triangle change for different speakers?

2.10 American English consists entirely of egressive phonemes. List two phonemes that differ only in their type of excitation (i.e., voiced versus unvoiced). Name one language that employs ingressive phonemes.

2.11 Diphthongs are characterized by movement from one vowel target to another. Sketch the initial and final vocal-tract shapes for the /W/ diphthong as in "out." Also sketch the magnitude spectrum for each. Which formants require the greatest articulatory movement?

2.12 What are the differences in vocal-tract articulator movement among diphthongs, liquids, and glides?

2.13 What differentiates the five unvoiced fricatives /f/, /T/, /s/, /S/, /h/? Why are fricatives always lower in energy than vowels?

2.14 Diphthongs, liquids, and glides possess one or more vocal-tract target shapes, with smooth continuant movement of the vocal-tract articulators. Which speech sounds in American English are transient, noncontinuant sounds?

2.15 Nasals differ from other phonemes in that the nasal cavity is part of the speech system's resonant structure. Discuss how the closed oral cavity contributes to the overall magnitude spectrum of a nasal.

2.16 The system of spelling in American English (orthographic transcription) often fails to represent the proper sequence of sounds for word production. For example, the italicized letters in the words below all represent the same sounds:

t*o* tw*o* t*oo*
s*ea* s*ee*
s*ue* sh*oe*
thr*ough* thr*ew*

Also, similar spelling can represent different speech sounds as shown below.

*ch*arter *ch*aracter
d*i*me d*i*m
*th*ick *th*en
*t*op bo*tt*om ca*t*

In other cases, combinations of letters may represent just one sound (e.g., *ph*ysical, *ai*sle), or no sound at all (e.g., *k*now, *p*sychiatry, *p*neumonia).

(a) Below are several groups of words. Pronounce the words and compare the sounds corresponding to the italicized letters. Are the sounds from each group the same phoneme? If so, give one additional word which contains the same speech sound. List the phoneme associated with each italicized letter.

1	2	3
supp*o*se	s*u*ch	*o*pen
y*ou*	p*u*ddle	*ow*l
br*oo*m	t*ou*ch	s*o*ld
t*o*	st*u*d	b*aw*l
s*u*per	p*u*t	c*oa*l

4	5	6
pa*ss*ion	sla*ng*	logg*ed*
fi*sh*ing	fi*ng*er	wait*ed*
a*z*ure	gi*ng*er	baulk*ed*
ca*sh*y	golfi*ng*	bait*ed*
*sh*ift	belo*ng*	trac*t*

(b) Perform a phonetic transcription of the following words. Which letters correspond to a single sound, which are silent, and which combine with another to form a single sound?

unique, license, fizzle, orangutan, autumn, sickle,
tissue, vegetable, quartz

2.17 (Computer Assignment) (a) In the speech files available for this book (see Preface), you will find a file that contains 16-bit integer samples of the vowel /o/, sampled at an 8-kHz sampling rate. Extract 512 samples of this speech sound and plot the time waveform. Next, apply an FFT to the segment and plot the magnitude spectrum.

(i) What features characterize the time waveform?

(ii) From the magnitude spectrum, estimate the fundamental frequency. Is this a male or female speaker?

(iii) From the same magnitude spectrum, estimate the formant locations. How do your measurements compare to typical values for the vowel?

(b) In the same set of computer files, you will find some vowel files whose phonemes are not labeled. For one or more of these files, repeat the steps of part (a), and determine which vowel the file(s) likely represent(s).

CHAPTER 3

Modeling Speech Production

Reading Notes: The discussion and notational conventions established in Section 1.5 will be used extensively in this chapter, beginning in Section 3.1.3.

3.0 Preamble

The general purpose of this chapter is to develop several related models of speech production that will be used throughout the book. We noted at the end of Chapter 1 that much of modern speech processing is based on extensive engineering research into analog acoustic modeling of speech production that was carried out in the middle decades of the twentieth century. A complete description of this work would require one or more volumes, so a thorough account in this book is not possible. Further, many digital speech processing pursuits require only a basic understanding of the central results of the acoustic theory. The purpose of this chapter is to pursue these core ideas. More elaboration on some aspects of the acoustic theory is found in the appendices of this chapter.

3.1 Acoustic Theory of Speech Production

3.1.1 History

The motivation for understanding the mechanism of speech production lies in the fact that speech is the human being's primary means of communication. Through developments in acoustic theory, many aspects of human voice production are now understood. There are areas such as nonlinearities of vocal fold vibration, vocal-tract articulator dynamics, knowledge of linguistic rules, and acoustic effects of coupling of the glottal source and vocal tract that continue to be studied. The continued pursuit of basic speech analysis has provided new and more realistic means of performing speech synthesis, coding, and recognition.

Early attempts to model and understand speech production resulted in mechanical speaking machines. Modern advances have led to electrical analog devices, and ultimately computer-based systems. One of the earliest documented efforts to produce artificial speech was by C. G. Kratzenstein in 1779 (Paget, 1930; Flanagan, 1972) in which he attempted to artificially produce and explain the differences among the five vowels /e, i, Y, o, yu/. He constructed acoustic resonators similar in shape to the human vocal tract and excited them with a vibrating reed, which, like the vocal folds, interrupted an airstream. In roughly the same period, Wolfgang Ritter von Kempelen (1791) demonstrated a much more successful machine for generating connected utterances.[1] In 1791, von Kempelen published a 456-page book describing his observations on human speech production and his experiments during the two decades he had been working on a speaking machine. Von Kempelen's machine used a bellows to supply air to a reed, which, in turn, excited a single resonator cavity. The resonator cavity was constructed of leather and the cross-sectional area was varied by having the operator squeeze the leather tube resonator for voiced speech sounds (Fig. 3.1). Consonant sounds were simulated using four separate constriction mechanisms controlled by the fingers of the operator's second hand. Using von Kempelen's description, Sir Charles Wheatstone built an improved version of the speaking machine which he demonstrated at the Dublin meeting of the British Association for the Advancement of Sciences in 1835 [see Fig. 3.2, from

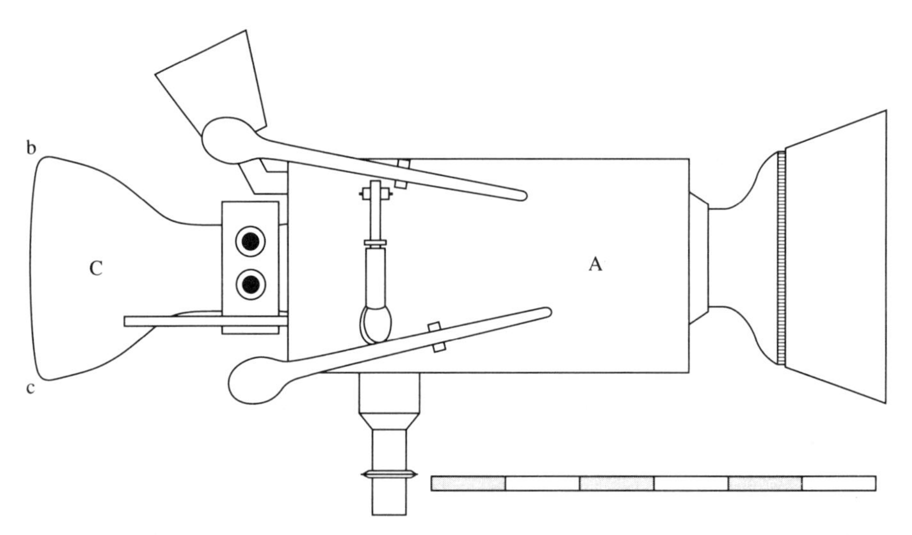

FIGURE 3.1. Von Kempelen's final speaking machine. Adapted from (Dudley and Tarnoczy, 1950).

[1]It has been reported that von Kempelen actually began his early work on vowel production in 1769, but that it was not seriously regarded by his scientific colleagues. Although his work received ample publicity, his research ability had been besmirched by an early deception involving a mechanical chess-playing machine. It seems the primary "mechanical" component of this earlier device was a concealed legless man named Worouski, who was a former commander in the Polish regiment and a master chess player (Flanagan, 1972).

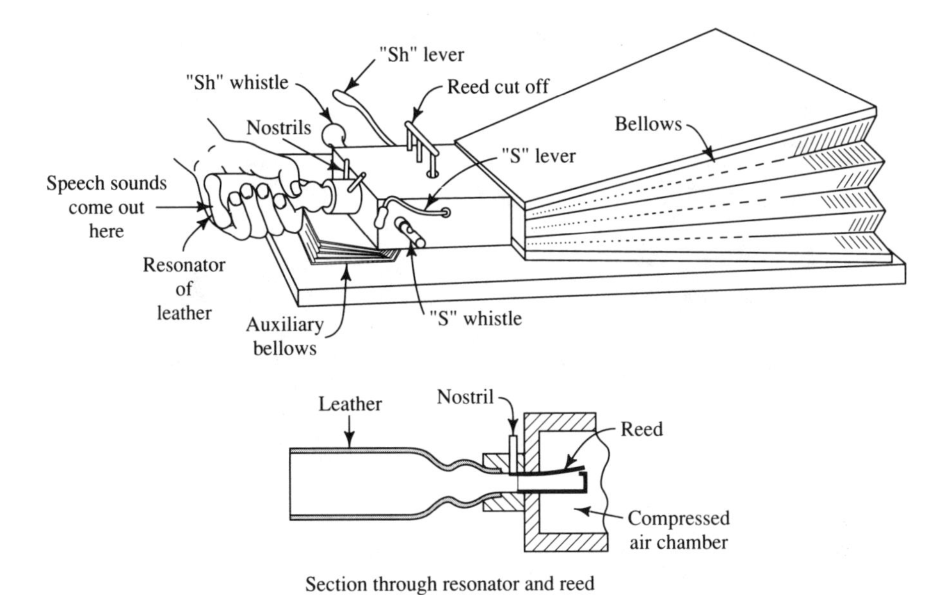

FIGURE 3.2. Wheatstone's reconstruction of von Kempelen's speaking machine. Adapted from (Paget, 1930).

(Dudley and Tarnoczy, 1950; Wheatstone, 1979)].[2] Further developments in mechanical speech modeling and synthesis continued into the 1800s and early 1900s. In 1846, J. Faber demonstrated a speech machine called "Euphonia," which possessed characteristics much like a speech organ. This instrument was said to represent a significant improvement over von Kempelen's, because it allowed for variable pitch, permitting singing. Ordinary, whispered, and conversational speech could also be produced. Other machines include an approach using tuning forks by Helmholtz to produce artificial vowels (Helmholtz, 1875). Paget (1930) built models of plasticene and rubber that could individually produce almost every vowel and consonant sound. Wagner (1936) constructed a vowel-copying electrical circuit that controlled the energy level in each of the first four formant frequency regions.

One of the first all-electrical networks for modeling speech sounds was developed by J. Q. Stewart (1922). In 1939, one of the first all-electrical speech synthesizers known as the "Voder" (from *vo*ice *de*monstrato*r* (Dudley et al., 1939) was demonstrated at the World's Fair (Fig. 3.3). A schematic diagram of the Voder is also shown in Fig. 3.4 (Dudley, 1955). A trained operator was required to "play" the Voder to produce speech.

[2]The efforts of von Kempelen would conceivably have a profound impact on speech modeling. As a boy in Edinburgh, Scotland, Alexander Graham Bell had the opportunity of seeing Wheatstone's construction of von Kempelen's machine. Being greatly impressed by his work, and with assistance from his brother Melville, Bell set out to construct his own speaking machine. His interest ultimately led to his U.S. Patent 174465 of the voice telephone.

FIGURE 3.3. The Voder being demonstrated at the New York World's Fair (Dudley, 1940; Dudley and Tarnoczy, 1950).

The operator manipulated 14 keys with his or her fingers, which controlled vocal-tract resonance structure, a wrist bar which operated excitation carrier (random noise/relaxation oscillator), and a right foot pedal which allowed for variable pitch. Although operator training was quite long (on the order of a year), skilled "players" could produce quite intelligible speech. One of the important aspects of the Voder is the resonance control box, which contained 10 contiguous bandpass filters that span the frequency range of speech. The outputs of the filters were passed through gain controls and were added. Ten fingers were used to operate the gain controls, and three additional keys were provided for transient simulation of stop consonants.

H. K. Dunn (1950) achieved far better quality than that of the Voder with an electrical vocal tract, pictured in Fig. 3.5. The device was based on a vibrating energy source that replaced the vocal folds, and a transmission line model (inductors and capacitors) for the vocal tract using lowpass filter sections that provided the delay experienced by a sound wave traveling along the human vocal tract.

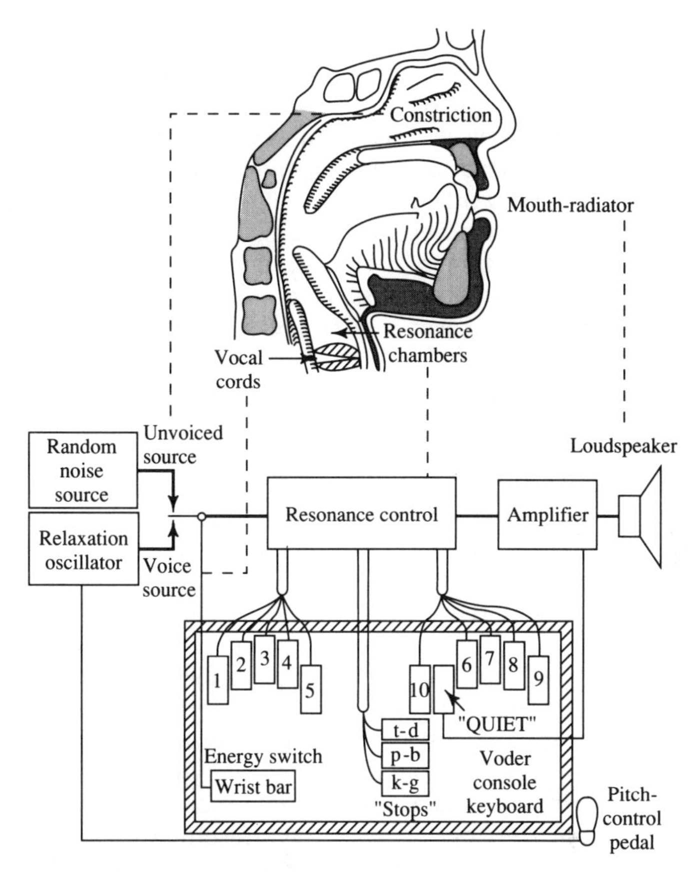

FIGURE 3.4. A schematic diagram of the Voder. Adapted from (H. Dudley, R. R. Riesz, and S. A. Watkins, 1939).

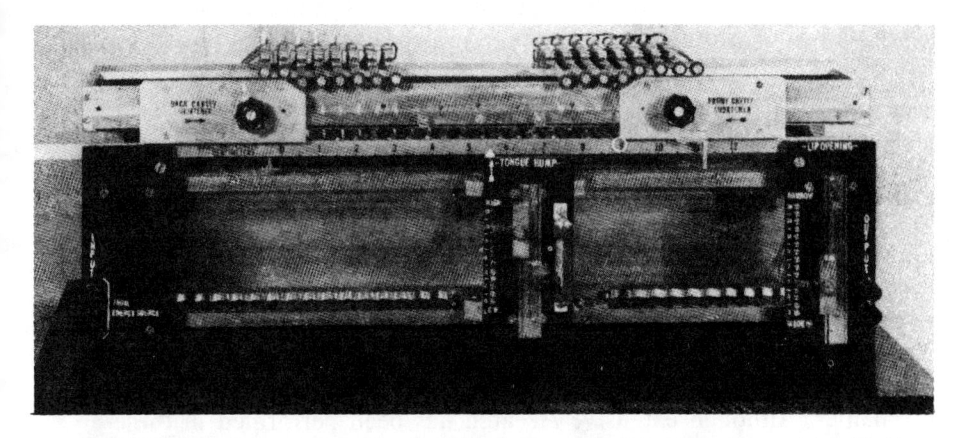

FIGURE 3.5. A front view of the Electrical Vocal Tract developed by H. K. Dunn (1950).

Further developments in speech modeling and synthesis continued with the development of computers. In many regards, advancement in speech modeling led to the development of better speech coding and synthesis methods. Pursuing the historical development of speech modeling further will therefore lead to more recent advances in coding and synthesis, which we shall present in a later chapter. We therefore turn back to the problem of understanding the basic principles of sound propagation and modeling of the human speech production system. Readers interested in further historical discussion of early speech modeling should consider articles by Flanagan (1972) and Klatt (1987).

3.1.2 Sound Propagation

In Chapter 2 we characterized speech sounds in terms of the position and movement of the vocal-tract articulators, variation in their time waveform characteristics, and frequency domain properties such as formant location and bandwidth. This has served to help us understand the differences in how individual phonemes are produced. In this section, we turn to mathematical representations of human speech production to provide a foundation for applications in speech synthesis, coding, enhancement, and recognition. In our brief historical overview of speech modeling, we saw that early attempts at speech modeling and understanding resulted in the construction of mechanical and electrical devices for speech production. The foundations of many of the early attempts to formulate mathematical models are a direct consequence of these early speech production devices. Therefore, the resulting mathematical methods presented here should provide the necessary basis for both analysis and synthesis of speech.

To fully characterize the human speech production system would require a set of partial differential equations that describe the physical principles of air propagation in the vocal system (Beranek, 1954; Morse, 1968; Portnoff, 1973; Rabiner and Schafer, 1978; Sondhi, 1974). Sound generation and propagation requires the characterization of such topics as

1. The time-varying nature of vocal-tract shape.
2. Nasal cavity coupling.
3. The effects of the soft tissue along vocal-tract walls.
4. The effect of subglottal (lungs and trachea) coupling with vocal-tract resonant structure.
5. Losses due to viscous friction along, and heat conduction across, the vocal-tract walls.

Therefore, a complete description requires detailed mathematical analysis and modeling based on acoustic theory and low-viscosity (air) fluid mechanics. Although extensive research has been performed in these areas, a universal theory has not yet emerged. If this is so, we might ask ourselves how the early speech scientists began in their quest to understand and synthesize speech. The answer centers around stationary speech sounds, such as vowels, which offer intriguing parallels with sound generation of early pipe organs. The resulting speech production models presented in this chapter will take on the form of an acoustic-tube model. Since it is assumed that more readers are familiar with electrical circuits than acoustics, electrical analogs of the vocal tract based on transmission lines will be considered. These models are subsequently extended to a discrete-time filter representation for use in processing speech by computer. The development of the discrete-time model is the principal goal of this chapter, as it will serve as the basis for most of the techniques described in Chapter 5 and beyond.

The discussion of human speech production in Chapter 2 reveals three separate areas for modeling. These include the source excitation, vocal-tract shaping, and the effect of speech radiation. For example, a single voiced phoneme such as a vowel, modeled over finite time, can be represented as the product of the following three (Fourier) transfer functions:

$$S(\Omega) = U(\Omega)H(\Omega)R(\Omega) \tag{3.1}$$

with (see footnote 11 in Chapter 2) $U(\Omega) = U_{\text{glottis}}(\Omega)$ representing the voice waveform (source excitation), $H(\Omega)$ the dynamics of the vocal tract, and $R(\Omega)$ the radiation effects. The voice waveform, $u(t)$, is taken to be a *volume velocity*, or "flow," waveform. The transfer function $H(\Omega)$ represents the ratio of the volume velocity at the output of the vocal system to that at the input to the tract. In this sense, $H(\Omega)$ also includes the lips, but in no sense includes the larynx. However, the speech waveform is usually considered to be the sound *pressure* waveform at the output of the vocal system. $H(\Omega)$ does not account for the flow-to-pressure conversion (radiation) function at the lip boundary, which is included in $R(\Omega)$. The reader should appreciate this subtlety because of its significance in many discussions to follow.

As implied by the representation (3.1), the majority of modern speech modeling techniques assumes that these components are linear and separable. Accordingly, the production model for the speech signal is assumed to be the concatenation of acoustical, electrical, or digital models, with no coupling between subsystems. Another assumption made is that of *planar propagation*. Planar propagation assumes that when the vocal folds open, a uniform sound pressure wave is produced that expands to fill the present cross-sectional area of the vocal tract and propagates evenly up through the vocal tract to the lips (see Fig. 3.6).

Assumptions of noncoupling and planar propagation are practical necessities needed to facilitate computationally feasible techniques for speech modeling, coding, and synthesis. Although these conditions are restrictive, they have proved useful in the formulation of the majority of present-day digital speech processing algorithms. It will be useful for us to consider the actual physical mechanism of human speech production

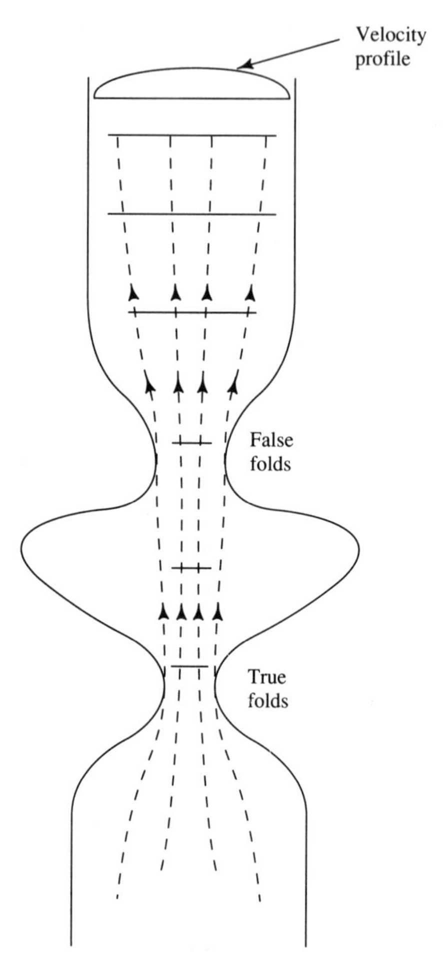

FIGURE 3.6. The classical planar propagation model of sound for speech analysis.

in developing a practical working speech model. However, there are speech scientists who argue that it may not be necessary to model the fine detailed structure and procedures of the physical speech mechanism (i.e., exact articulator movements, characteristics of vocal fold muscle characteristics) if we are merely interested in characterizing broad timing and frequency domain properties of the resulting waveform. Thus it may be best to establish a goal for speech model development before initiating the discussion. For many speech applications such as coding, synthesis, and recognition, good performance can be achieved with a speech model that reflects broad characteristics of timing and articulatory patterns, as well as varying frequency properties. Finer speech modeling methods, based on actual physical traits, are necessary for such areas as analysis of vocal fold movement, effects of pathology on human speech production, and other application areas. In Section 3.2.3 we discuss several experimental studies that seem to suggest the need for alternative speech modeling methods based on physical properties. We begin the discussion by considering a model for the source excitation.

3.1.3 Source Excitation Model

Types of Excitation

Development of an understanding of source characteristics generally requires that we assume independence between source and tract models. Recall from the discussion of speech acoustics in Section 2.2.3 that two basic forms of excitation are possible:

1. *Voiced excitation*—a periodic movement of the vocal folds resulting in a stream of quasi-periodic puffs of air.
2. *Unvoiced excitation*—a turbulent noiselike excitation caused by airflow through a narrow constriction.

Recall also that the following forms of excitation represent important combinations and variations of voiced and unvoiced sources which, for modeling purposes, are often featured as distinct categories:

3. *Plosive excitation*—caused by a buildup of air pressure behind a completely closed portion of the vocal tract, followed by a sudden release of this pressure. The released flow creates an unvoiced or voiced sound depending on which phoneme is being uttered.
4. *Whisper*—a whisper is an utterance created by forcing air through a partially open glottis (glottal fricative) to excite an otherwise normally articulated utterance.
5. *Silence*—included as an excitation form for modeling purposes, since there are short time regions in speech in which no sound occurs. The pause before the burst of a plosive is an example of this phenomenon.

For voiced and unvoiced sounds, the speech signal $s(t)$ can be modeled as the convolution of an excitation signal (see below) and the vocal-tract impulse response $h(t)$.[3] The modeling of the excitation signal and its spectrum represent the goal of this section.

Voiced Excitation

For voiced or sonorant production, airflow from the lungs is interrupted by quasi-periodic vibration of the vocal folds, as illustrated in Fig. 3.7. Details of the forces responsible for vocal fold movement have been discussed in Section 2.2.3. During voiced activity, the vocal folds enter a

[3]For the present, we will ignore radiation effects in the modeling.

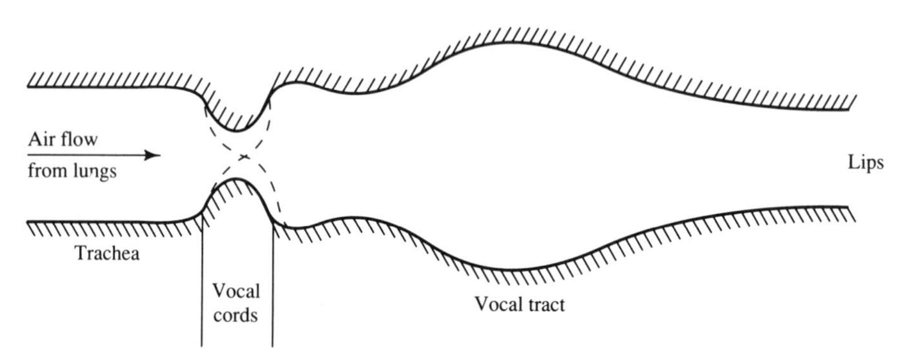

FIGURE 3.7. A schematic representation of the vocal system.

condition of sustained oscillation. The rate at which the vocal folds open and close is determined by subglottal air pressure, the tension and stiffness of the vocal cords, and the area of the glottal opening during periods of rest (i.e., nonspeech activity such as breathing). One means of characterizing vocal fold movement is to measure the volume velocity of air at the glottis and compare it to the sound pressure level at the mouth. Figure 3.8 shows such a comparison. In general, it is difficult to obtain precise measurements of glottal pressure or volume velocity waveforms. In fact, the waveforms shown in Fig. 3.8 were simulated using a method by Ishizaka and Flanagan (1972). However, they do agree with direct observation through high-speed motion photography using mirrors placed at the back of the throat.

The fundamental frequency is one of the most basic and easily measurable features of the glottal waveform. In addition to the fundamental frequency, aspects such as duration of each laryngeal pulse (both open and closed glottal phases), the instant of glottal closure, and the shape of each pulse, play important roles in a talker's ability to vary source characteristics. An analysis of such factors requires the estimation and reconstruction of the excitation signal from the given speech signal. The generally accepted approach for analysis of glottal source characteristics involves direct estimation of the glottal flow wave. Methods that seek to reconstruct the time excitation waveform are referred to as *glottal inverse filtering* algorithms. Due to the high-accuracy requirements of these algorithms, most require large amounts of human interaction with little regard to computational efficiency. These algorithms will be discussed below, and also in Chapter 5 as an application of linear prediction analysis.

If a *discrete-time* linear model for the speech production system is used, the z-domain transfer function for the speech signal $s(n)$ may be written as follows:

$$S(z) = \Theta_0 U(z) H(z) R(z) \tag{3.2}$$

$$= \Theta_0 E(z) G(z) H(z) R(z). \tag{3.3}$$

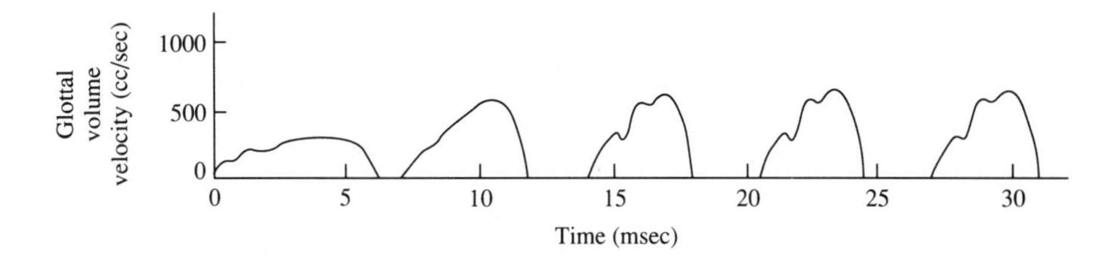

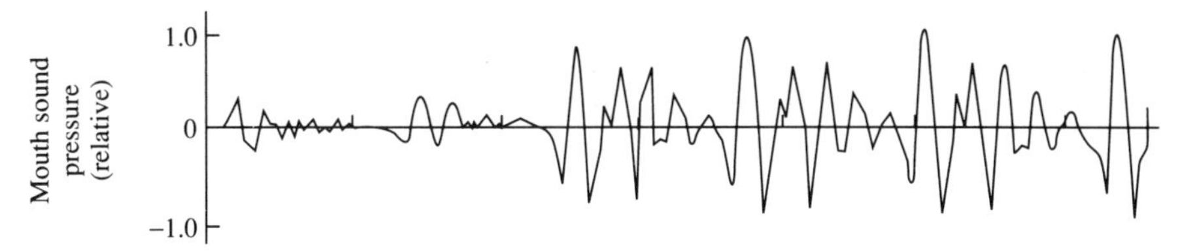

FIGURE 3.8. Example of glottal volume velocity and sound pressure waveforms at the mouth for the vowel /a/. After Ishizaka and Flanagan (1972).

The quantities in (3.2) correspond precisely to those in (3.1) except that we have added a gain constant, Θ_0, for generality. In (3.3) we have refined the source model to include a glottal shaping filter driven by a discrete-time impulse (unit sample) train, $e(n)$, whose pulses are spaced at the pitch period. Consequently, the voiced speech signal $s(n)$ can be thought of as being generated as follows:[4] The pitch pulses $e(n)$ excite the glottal shaping filter $G(z)$, which results in the glottal excitation signal (voice) $u(n)$. Consequently, if $e(n)$ is periodic with period P, then (assuming the voicing lasts forever)

$$u(n) = \sum_{i=-\infty}^{\infty} g(n - iP), \tag{3.4}$$

where $g(n)$ is the impulse response of the glottal shaping filter.[5] This signal is then used to excite the supraglottal system made up of the vocal-tract transfer function $H(z)$ and the output radiation component $R(z)$. The overall amplitude of the system is controlled by Θ_0.

The task of estimating the glottal source requires the formulation of a procedure that inverts the influence of the vocal-tract and radiation components. Such glottal inverse filtering algorithms reconstruct the time excitation waveform $u(n)$. The main component of the inverse filter is the vocal-tract component, $H(z)$. The filter therefore requires estimates of stationary formant locations and bandwidths. Two approaches have been applied in the past for inverse-filter implementation. The first, based in

[4]It should be noted that the pulse train $e(n)$ has no real counterpart in the physical system. The first modeling component with a physical correlate is the voice waveform $u(n)$.

[5]As an aside, from (3.4) one can see the advantage of using an FIR filter to model $G(z)$.

analog technology, involves individual determination of formants requiring user interactive control (Holmes, 1962; Hunt et al., 1978; Lindqvist, 1964, 1965, 1970; Miller, 1959; Nakatsui and Suzuki, 1970; Rothenberg, 1972). The second involves automatic formant measurement by some form of nonstationary linear prediction analysis during the closed glottis interval (Berouti, 1976; Berouti et al., 1977; Childers, 1977; Hedelin, 1981, 1984, 1986; Hunt et al., 1978; Krishnamurthy and Childers, 1986; Larar et al., 1985; Markel and Wong, 1976; Veeneman and BeMent, 1985; Wong et al., 1979). In both approaches, an estimate of the vocal-tract filter $\hat{H}(\omega)$ and a filter representing radiation at the lips $\hat{R}(\omega)$ is used. Given a measurement of the spectral properties of the sound pressure $S(\omega)$, the glottal excitation can, in principle, be estimated to within a gain constant using (3.2),

$$\hat{G}(\omega) = \frac{S(\omega)}{\hat{H}(\omega)\hat{R}(\omega)}. \tag{3.5}$$

[If continuous time analysis is being used, refer to (3.1), which indicates the same result with ω replaced by Ω.] When individual formant determination is used, each formant is canceled individually using a second-order filter with a complex conjugate pair of zeros. The input signal is assumed to be a stationary vowel with high-frequency components and formants that are clearly separable. Miller (1959) was the first to attempt such an approach. The inverse filter formulated contained only one adjustable formant. Lindqvist (1964, 1965, 1970) later developed an approach that contained four tunable filters to cancel the first four formants and a fixed lowpass filter to attenuate higher formants. Holmes (1962) also developed a system with four tunable filters, but also allowed the user to see the excitation pattern of each formant. Fant and Sonesson (1962) compared inverse filtering to waveforms recorded by a photoelectric pitch determination instrument. Results showed good temporal agreement between the two estimates of the instant of glottal closure. The criterion for filter adjustment is based on the postulate that during the closed-glottis interval, the excitation propagated up through the vocal tract approaches zero. The analysis filter therefore models only the vocal-tract characteristics. This requires that the output signal of the inverse filter be maximally flat during this time.

There are some disadvantages and disagreements in the literature concerning inverse filtering techniques. It is generally accepted that at the instant of glottal closure, all frequencies of the vocal tract are excited equally. This can be seen as large impulses in the residual waveform for vocalic speech. Yet, the results by Holmes (1962, 1976) and Hunt et al. (1978) indicate that there may be a secondary excitation at the instant of glottal opening or even during the closed-glottis interval. These results presumably led researchers such as Atal and Remde (1982) and others to look beyond pitch-excited methods to alternative formulations involving "multipulse excitation" (see Chapter 7). Also, it is necessary to select the vowels to be analyzed carefully, since, for example, nasalization can lead to inaccurate determination of formants. Lindqvist (1964) also identified the problem of coupling of the subglottal and supraglottal systems during the open-glottis interval. For these reasons, interactive inverse filter analysis without the use of a pitch detector has almost entirely been confined to male voices, where the closed-glottis interval is well defined and of sufficient duration.

Methods for glottal waveform extraction using linear prediction analysis will be discussed in detail in Chapter 5. Briefly, the techniques are based upon a model of the laryngeal dynamics that includes a closed phase during each cycle. The resulting volume velocity at the glottis $u(n)$ is periodic, ideally resembling a lowpass pulse train as illustrated in Fig. 3.9. The regions of zero volume velocity correspond to closed phases of the glottis. As the vocal folds begin to separate, the volume velocity gradually increases. After reaching their maximum separation, the vocal folds close more rapidly, thus completing the open phase. To extract the glottal waveform, linear prediction analysis is applied during one or more closed phase regions to estimate the vocal-tract dynamics. Ignoring lip radiation effects, during the closed phase, only transient dynamics of the vocal-tract are present in the speech signal. The resulting estimated vocal-tract model, say $\hat{H}(z)$, can be used to recover the glottal waveform according to (3.5).

In addition to the two principal approaches for glottal waveform estimation described above, other methods for modeling the excitation source at the glottis are based on direct waveform models. In these techniques a particular glottal pulse shape is adopted, such as a rising slope leading into a

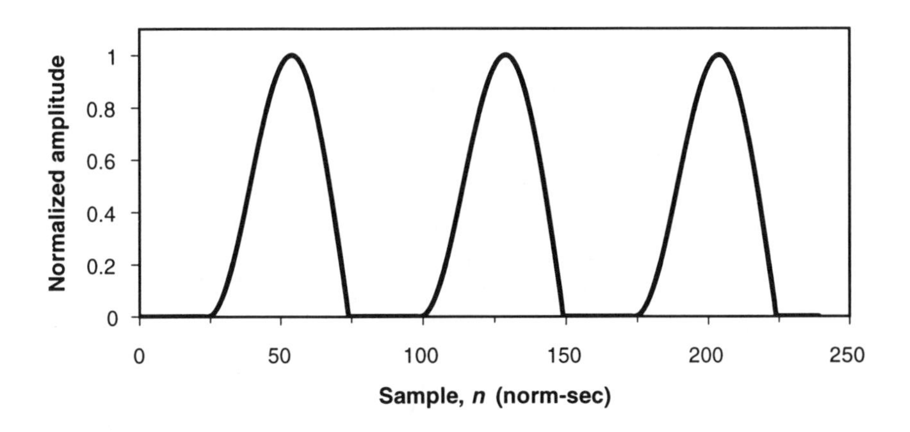

FIGURE 3.9. Example ideal glottal source waveform generated using the Rosenberg pulse model (1971). Three pulses of a sequence with period 75 samples are shown, representing a fundamental frequency $F_0 = 133$ Hz at sample rate $F_s = 10$ kHz.

half-wave rectified sine pulse. The pulse is parameterized by values that determine the precise shape of the waveform (Hedelin, 1986; Rosenberg, 1971). Such a direct modeling approach is generally acceptable for applications such as speech synthesis, where the goal is to produce natural-sounding speech. However, if a model is to be extracted from a given speech signal for coding purposes, for example, the human glottal pulse shape can vary considerably from the assumed model, thereby causing speech of lower quality.

The analytical methods described above are ordinarily used to estimate the glottal waveform. These methods, however, require an accurate and stable model for the vocal-tract and radiation effects, making them less attractive for modeling of speech in some coding and recognition applications. Fixed mathematical models have also been used to model the glottal source. An example is the electrical analog of a vocal fold model developed by Ishizaka and Flanagan (1972) using measurements of subglottal air pressure, volume velocity at the glottis, vocal fold tension, and the resting cross-sectional area of the opening at the vocal folds. Their model is based on a set of nonlinear differential equations. The coupling of these equations with the partial differential equations will be used to model the vocal-tract results in a time-varying acoustic resistance and inductance. These terms are dependent on the cross-sectional area of the glottis as $1/A_{\text{glottis}}(t)$. So if the glottis closes [i.e., $1/A_{\text{glottis}}(t) \to \infty$], the impedances become infinite and the volume velocity becomes zero. In most cases, it is assumed that the coupling between the vocal tract and glottis is weak, so that the glottal impedance can be linearized as (Flanagan, 1972)

$$Z_{\text{glottis}}(\Omega) = R_{\text{glottis}} + j\Omega L_{\text{glottis}} \tag{3.6}$$

with R_{glottis} and L_{glottis} constants during voiced speech activity. The result is the glottal excitation model shown in Fig. 3.10. The term $u_{\text{glottis}}(t)$ represents the volume velocity at the glottis, which has the shape shown in Fig. 3.8. This model will be used to establish boundary conditions once the vocal-tract acoustic tube model has been established.

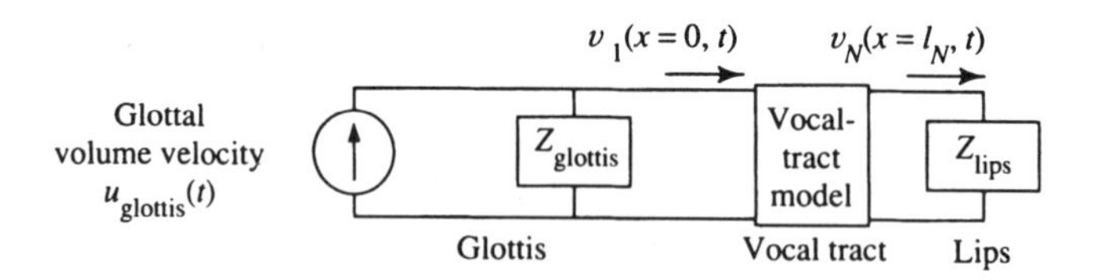

FIGURE 3.10. An approximation of the glottis and lip characteristics. Here, the glottal impedance is represented as a resistance and inductance (see Equation 3.6), which could vary with time or be fixed for production of a given phoneme. The glottal volume velocity $u_{\text{glottis}}(t)$ represents the excitation source. After Flanagan (1972).

Unvoiced Excitation

The discussion of source excitation modeling would not be complete without addressing unvoiced excitation. Unvoiced excitation includes frication at a point of major constriction along the vocal tract or an explosive form during a stop release. Both are normally modeled using white noise. This source of excitation theoretically has no effect on the shape of the speech spectrum, since its power density spectrum is constant over all frequencies. It is also notable that phase is not a meaningful concept in characterizing noise, since the Fourier transform of a noise process does not, strictly speaking, exist. In fact, little attention is paid to the effect of the excitation phase, since good quality unvoiced speech can be obtained for synthesis with random phase and since the spectral amplitude is more important than phase for speech perception (further discussed in Chapter 5). Throughout the history of speech modeling, voiced excitation has always received more research attention than the unvoiced case. This is due to studies in speech perception that suggest that accurate modeling of voiced speech is crucial for natural-sounding speech in both coding and synthesis applications.

In conjunction with this problem of unvoiced excitation and phase considerations, it is worth previewing an issue that will be central to many of the discussions in the future. Suppose that we wish to write an analytical model similar to (3.3) for the unvoiced case. It would be natural to simply omit the glottal shaping filter $G(z)$ and simply let "$E(z)$" represent the driving noise sequence, yielding

$$S(z) = \Theta_0 E(z) H(z) R(z), \tag{3.7}$$

or in terms of the DTFT

$$S(\omega) = \Theta_0 E(\omega) H(\omega) R(\omega). \tag{3.8}$$

The cautious reader may object to this expression on the grounds that a DTFT will generally not exist for the excitation process. However, if we let $\underline{e}$ be the name for the excitation (noise input) process, and $\underline{s}$ the random process characterizing the resulting output (speech), then it is perfectly proper to write

$$\Gamma_{\underline{s}}(\omega) = \Theta_0^2 \Gamma_{\underline{e}}(\omega) |H(\omega) R(\omega)|^2 \tag{3.9}$$

where, recall, $\Gamma_{\underline{x}}(\omega)$ refers to the power density spectrum of the random process $\underline{x}$. However, if we interpret $E(\omega)$ as representative of some *finite time portion* of a realization of $\underline{e}$, and similarly for $S(\omega)$, then (3.7) becomes a valid working expression. Reflecting on (3.3), we realize that a similar problem occurs there because the DTFT does not strictly exist for a stationary pulse train. The issue of short-term processing in speech work is important in both theoretical and practical terms, and we shall pay a great deal of attention to it in the ensuing chapters. For the remainder of this chapter, however, we shall continue to use somewhat sloppy, but readily interpretable, analysis in this regard.

A Plethora of Transfer Functions

Before continuing, we must emphasize a point that is inherent in the discussion above. There are many transfer functions of interest in modeling speech. Of course, each of $G(z)$, $H(z)$, and $R(z)$ represents a transfer function between two important signals; it is hoped that the reader now appreciates the significance of each.[6] The main transfer function variations differ in whether or not the effects of the larynx and/or lips are included. Indeed, therefore, $H(z)$ is one such transfer ratio of interest,

$$H(z) = \frac{U_{\text{lips}}(z)}{U_{\text{glottis}}(z)}. \tag{3.10}$$

Of the other combinations of outputs and inputs that can be formed, the most important for this book is the "complete" transfer function

$$\Theta(z) \stackrel{\text{def}}{=} \frac{S(z)}{E(z)}. \tag{3.11}$$

Depending on whether the sound being modeled is voiced or unvoiced, $\Theta(z)$ may or may not contain the glottal dynamics, $G(z)$. Throughout this chapter and those to follow, it is fundamentally important to understand the nature of the system function under discussion.

3.1.4 Vocal-Tract Modeling

Reading Note: The material on phasors and steady-state solutions in Section 1.5 will be used extensively in the remainder of the chapter. We recommend that the reader review that brief section before continuing.

Pressure and Volume Velocity Relations

A sound wave is produced when the vocal folds vibrate, other articulators move, or by random air particle motion. The propagation follows the laws of physics, which include the conservation of mass, momentum, and energy. Air can be considered a compressible, low-viscosity fluid, which allows the application of laws of fluid mechanics and thermodynamics. If we consider sound propagation in free space, it should be clear that a sound wave propagates radially outward in three dimensions from a point sound source. For human sound production, soft tissue along the vocal-tract prevents radial propagation. Therefore, sound waves normally propagate in only one direction along the vocal tract. To simplify vocal-tract modeling, we assume that sound waves obey planar propagation along the axis of the vocal tract, from the glottis toward the lips. This assumption is strictly valid only for frequencies with wavelengths that are

[6]The argument z can be replaced in this discussion by ω if the DTFT is the transform of interest and by Ω if the "analog" FT is being used.

large compared to the diameter of the vocal tract (less than approximately 4 kHz). To verify this, we calculate the wavelength of a sound wave at 4 kHz as

$$\lambda_{4\,\mathrm{kHz}} = \frac{c}{F} = \frac{340\ \mathrm{m/sec}}{4000\ \mathrm{cycles/sec}} = 8.5\ \mathrm{cm}, \tag{3.12}$$

where c is the speed of sound in air. Here, the wavelength $\lambda_{4\,\mathrm{kHz}}$ of 8.5 cm is much larger than the average diameter of the vocal tract ($\approx$2cm), so that the assumption of planar propagation is reasonable.[7] In addition to planar propagation, we also assume that the vocal tract can be modeled by a hard-walled, lossless series of tubes. Once we have considered the lossless tube-model case, we will introduce the effects of losses due to soft-wall vibration, heat conduction, and thermal viscosity later in this section.

Consider the varying cross-sectional area model of the vocal-tract shown in Fig. 3.11. The two laws that govern sound wave propagation in this tube are the Law of Continuity and Newton's Force Law (force = mass × acceleration),

$$\frac{1}{\rho c^2} \frac{\partial p(x,t)}{\partial t} = -\nabla \cdot \vec{v}_\zeta(x, y, z, t) \tag{3.13}$$

$$\rho \frac{\partial \vec{v}_\zeta(x, y, z, t)}{\partial t} = -\nabla p(x, t), \tag{3.14}$$

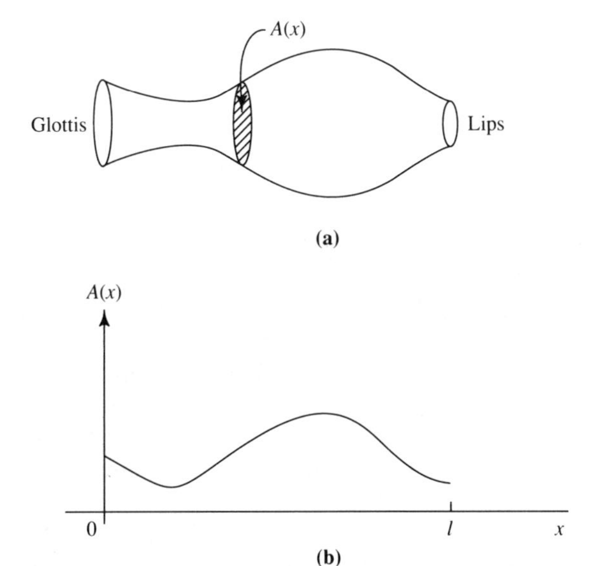

FIGURE 3.11. An ideal, variable cross-sectional area vocal tract.

[7]We shall reconsider this topic in Section 3.2.3.

where ∇ indicates the gradient, $\nabla \bullet$ the divergence, and where

$p(x,t) \stackrel{\text{def}}{=}$ deviant sound pressure from ambient, at location x, time t

$\vec{v}_{\zeta}(x,y,z,t) \stackrel{\text{def}}{=}$ vector velocity of an air particle ζ in the vocal tract at location (x,y,z)

$c \stackrel{\text{def}}{=}$ speed of sound in air (340 m/sec at sea level)

$\rho \stackrel{\text{def}}{=}$ density of air inside the vocal tract.

If one-dimensional planar propagation is assumed, then all particles at a given displacement x will have the same velocity independent of location (y,z) within the cross-sectional area A. Therefore, analysis is more convenient if we examine the velocity of a volume of air rather than the vector velocity of a single particle $\vec{v}_{\zeta}(x,y,z,t)$. The velocity of a volume of air is defined as

$$\vec{v}(x,t) = A(x,t)\,\vec{v}_{\zeta}(x,t), \tag{3.15}$$

where $A(x,t)$ is the cross-sectional area of the vocal tract, and $\vec{v}_{\zeta}(x,t)$ the velocity of any particle ζ in that volume. This assumes that each particle in the volume is moving in the same direction and with the same point velocity (the volume follows planar propagation shown in Fig. 3.6). Since the vector velocity now points in a single direction, *we shall henceforth drop the vector notation over the volume velocity* $\vec{v}(x,t)$ *and write simply* $v(x,t)$. Substituting volume velocity for point velocity in (3.13) and (3.14) results in

$$-\frac{\partial v(x,t)}{\partial x} = \frac{1}{\rho c^2}\frac{\partial[p(x,t)A(x,t)]}{\partial t} + \frac{\partial A(x,t)}{\partial t} \tag{3.16}$$

$$-\frac{\partial p(x,t)}{\partial x} = \rho\frac{\partial[v(x,t)/A(x,t)]}{\partial t}. \tag{3.17}$$

These equations characterize the volume velocity $v(x,t)$ and sound pressure $p(x,t)$ along the vocal tract from glottis $(x=0)$ to lips $(x=l$ or 17.5 cm for a typical male speaker). In general, closed-form solutions for these equations are possible for only a few simple configurations. If, however, the cross-sectional area $A(x,t)$ and associated boundary conditions are specified, numerical solutions are possible. In the case of continuant sounds such as vowels, it is reasonable to assume that the vocal-tract area function $A(x,t)$ does not vary with time. For noncontinuants such as diphthongs, $A(x,t)$ will vary with time. Although detailed measurements of $A(x,t)$ are needed, they are extremely difficult to obtain during production. Even with accurate measurements however, a solution to (3.16) and (3.17) can be difficult. If further simplifications are employed, reasonable solutions can be achieved.

Lossless Tube Model

As noted above, one of the difficulties encountered in solving (3.16) and (3.17) is the inability to characterize the area function $A(x, t)$ along the vocal tract versus time. In order to gain some understanding of the model governed by these coupled differential equations, let us consider a simplified vocal-tract shape consisting of a tube with uniform cross-sectional area. Let the area be fixed in time and space, so that $A(x, t)$ can be replaced by A as shown in Fig. 3.12. The adequacy of such a tube model has been demonstrated by research comparing the sounds produced by physical models with sounds produced by humans. Natural vowel sounds were first produced by speakers and recorded. During production, measurement techniques using X-ray motion pictures were used to sketch the pharyngeal- and oral-tract shapes for each vowel. Using these sketches, tube models were constructed with the same shapes as those observed in the X-rays. Sound energy was then passed through each tube model, and the emerging sound patterns showed agreement with the natural vowel patterns. This work was initiated by Chiba and Kajiyama (1941), and continued by Dunn (1950), Fant (1960, 1968, 1970),

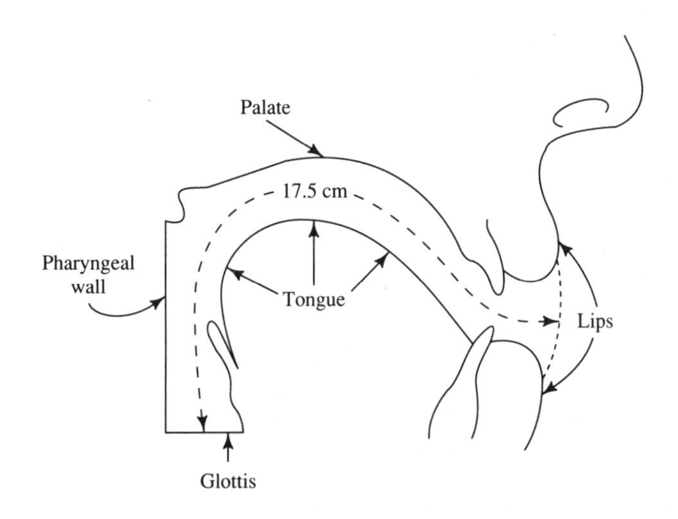

FIGURE 3.12. A diagram of an X-ray trace of the vocal tract and a corresponding uniform tube model for the neutral vowel /x/.

Jakobson et al. (1967), Lindbolm and Sundberg (1971), Perkell (1969), and Stevens and House (1955, 1961).

Figure 3.12 illustrates an X-ray sketch and corresponding uniform tube configuration that approximates the neutral vowel /x/. The ideal source is represented by a piston, which produces sound waves that propagate along the axis of the vocal-tract model. The assumption of a constant cross-sectional area does not ensure the single uniform tube shown in Fig. 3.12, and we need to account for the actual bend in the human vocal tract for this to be a valid model. In a study of the effects of variable angle bends in uniform acoustic tubes, Sondhi (1986) found that curvature does not change the points of resonance by more than a few percent from those of a straight tube model.[8] Since this effect is small, we initially consider the uniform tube case.

Open Termination. First consider the case in which the tube is open at the lips ($x = l$), so that the deviation from ambient pressure at the lips is zero,

$$p(l, t) = p_{\text{lips}}(t) = 0. \tag{3.18}$$

Further, because we are only interested in steady-state analysis, we let the glottal source be modeled by an exponential excitation,

$$v(0, t) = u_{\text{glottis}}(t) = \bar{U}_{\text{glottis}}(\Omega)e^{j\Omega t}, \tag{3.19}$$

where a positive value (of the real part) represents a volume velocity (movement of the plunger in Fig. 3.12) to the right. The complex number $\bar{U}_{\text{glottis}}(\Omega)$ is the phasor for the signal $u_{\text{glottis}}(t)$ (see Section 1.5). With boundary conditions (3.18) and (3.19), and with $A(x, t)$ replaced by a constant A, the steady-state solution to (3.16) and (3.17) for the volume velocity at the lips is shown in Appendix 3.A.1 to be

$$v(l, t) = \frac{\bar{U}_{\text{glottis}}(\Omega)}{\cos(\Omega l/c)} e^{j\Omega t} \stackrel{\text{def}}{=} \bar{U}_{\text{lips}}(\Omega)e^{j\Omega t}. \tag{3.20}$$

The transfer function for the vocal tract can be found by taking the ratio of the phasor volume velocity at the lips to that at the glottis, or equivalently, the ratio of the corresponding complex signals [recall (1.254)],

$$H(\Omega) = \frac{\bar{U}_{\text{lips}}(\Omega)}{\bar{U}_{\text{glottis}}(\Omega)} = \frac{u_{\text{lips}}(t)}{u_{\text{glottis}}(t)} = \frac{1}{\cos(\Omega l/c)}. \tag{3.21}$$

[8]For a 17-cm uniform tube with a 90° bend, variation in formant frequency location from a straight tube was +0.8%, +0.2%, −1.0%, −2.5% for F_1, F_2, F_3, and F_4, respectively.

This function relates the input and output volume velocities for the uniform acoustic tube. The resonant frequencies F_i for this model are found by setting the denominator equal to zero. This occurs when

$$\frac{\Omega_i l}{c} = \frac{\pi}{2}[2i-1] \qquad \text{for} \qquad i = 1, 2, 3, 4, \ldots. \tag{3.22}$$

Since $\Omega_i = 2\pi F_i$, tube resonances occur at frequencies

$$F_i = \frac{c}{4l}[2i-1] \qquad \text{for} \qquad i = 1, 2, 3, 4, \ldots. \tag{3.23}$$

Figure 3.13 shows a plot of the tube transfer function from (3.21), using a tube of length $l = 17.5$ cm and a speed of sound $c = 350$ m/sec.[9] Resonances $F_1, F_2, F_3, \ldots$ occur at $500\,\text{Hz}, 1500\,\text{Hz}, 2500\,\text{Hz}, \ldots$, respectively.

Closed Termination. It is of interest to obtain results for the case in which the oral tract is completely occluded (i.e., closed lip or closed tube condition). Let us reconsider the same uniform acoustic tube terminated by a complete closure. The volume velocity of air at the lips will be zero,

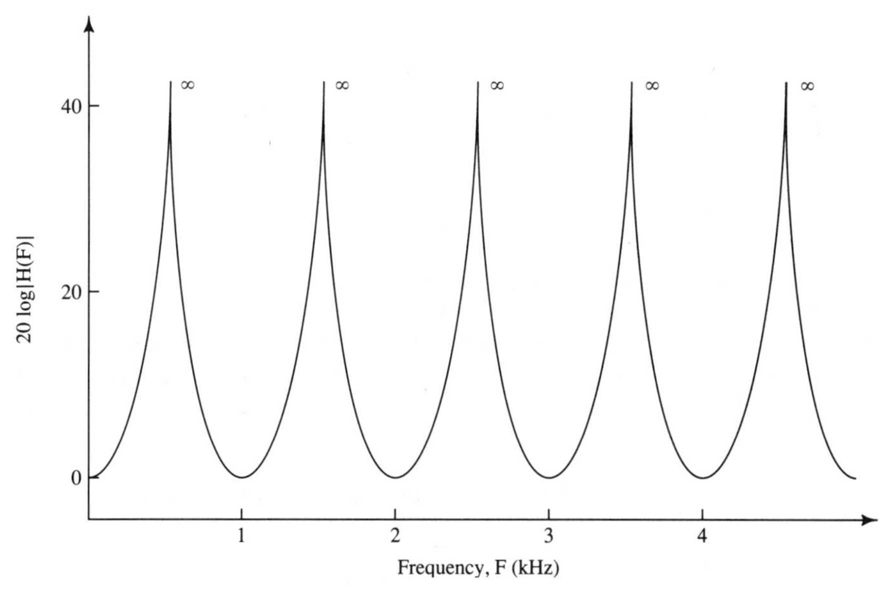

FIGURE 3.13. Frequency response for a uniform acoustic tube.

[9]The speed of sound in air at sea level is 340 m/sec. The speed of sound in moist air at body temperature (37°C) is 350 m/sec.

since there is no air leaving the closed tube. The boundary condition at the lip end is

$$v(l, t) = u_{\text{lips}}(t) = 0. \tag{3.24}$$

Steady-state solutions for volume velocity and pressure are derived in Appendix 3.A.1. Of course, without any detailed mathematics, it is obvious that the transfer function for this case is $H(\Omega) = 0$ for any Ω.

Transmission Line Analogy. It is interesting to note that a uniform cylindrical tube as pictured in Fig. 3.12, which has a plane wave propagating through it, is analogous to a section of a transmission line. The acoustical resistance, mass, and compliance are distributed along the cylinder in the same manner as resistance, inductance, and capacitance are distributed along a transmission line. Acoustic quantities such as sound pressure and volume velocity have their equivalents in transmission line theory as voltage and current (see Table 3.1). A summary of the transmission line–acoustic tube analogies is shown in Table. 3.2.

One key idea that follows from the transmission line analogy is the concept of *acoustic impedance* and *admittance.* Since the sound pressure and volume velocity in the acoustic system are analogous to voltage and current in the transmission line, respectively, we naturally define the acoustic impedance of the tube at distance x from the glottis as

$$Z(x, \Omega) \stackrel{\text{def}}{=} \frac{p(x, t)}{v(x, t)} = \frac{\bar{P}(x, \Omega)}{\bar{V}(x, \Omega)}, \tag{3.25}$$

where $p(x, t)$ and $v(x, t)$ are steady-state solutions as above, and $\bar{P}$ and $\bar{V}$ are the corresponding phasor representations. Also by analogy, we define the *acoustic characteristic impedance* of the uniform tube by

$$Z_o \stackrel{\text{def}}{=} \frac{\rho c}{A}. \tag{3.26}$$

The *acoustic admittance*, $Y(x, \Omega)$, and *acoustic characteristic admittance*, Y_o, are defined as the respective reciprocals.

Several interesting uses are made of the acoustic impedance concept in Appendix 3.A.2. First, the input impedances are characterized for the open and closed termination cases. The results are shown in Table 3.2.

TABLE 3.1. Analogies Between Acoustic and Electrical Quantities for Transmission Line and Acoustic Tube Analysis.

	Acoustic Quantity		Electrical Quantity
$p(x,t)$	Sound pressure	$v(x,t)$	Voltage
$v(x,t)$	Volume velocity	$i(x,t)$	Current
ρ/A	Acoustic inductance	L	Inductance
$A/\rho c^2$	Acoustic capacitance	C	Capacitance

TABLE 3.2. Summary of Useful Acoustic and Electrical Relations for a Uniform Acoustic Tube Analysis.

Uniform Acoustic Tube
Cross-sectional area = A, tube length = l

Open termination		Closed termination	
Acoustic tube	Transmission line	Acoustic tube	Transmission line
$v(0,t) = u_{\text{glottis}}(t)$		$v(0,t) = u_{\text{glottis}}(t)$	
Glottis, Lips, Open tube	Glottis, Lips, Short circuit	Glottis, Lips, Closed tube	Glottis, Lips, Open circuit
sound pressure = $p(x,t)$ volume velocity = $v(x,t)$	voltage = $v(x,t)$ current = $i(x,t)$	sound pressure = $p(x,t)$ volume velocity = $v(x,t)$	voltage = $v(x,t)$ current = $i(x,t)$
$p(l,t) = p_{\text{lips}}(t) = 0$	$v(l,t) = v_{\text{lips}}(t) = 0$	$v(l,t) = u_{\text{lips}}(t) = 0$	$i(l,t) = i_{\text{lips}}(t) = 0$
Characteristic impedance	$Z_0 = \frac{\rho C}{A} = \sqrt{\frac{L}{C}}$	Characteristic impedance	$Z_0 = \frac{\rho C}{A} = \sqrt{\frac{L}{C}}$
Input impedance	$Z_{\text{in open}} = jZ_0 \tan(\Omega\tau)$	Input impedance	$Z_{\text{in closed}} = -jZ_0 \cot(\Omega\tau)$
$v(x,t) = U_{\text{glottis}}(\Omega)\, e^{j\Omega t}\left[\frac{\cos\Omega(\tau - \frac{x}{c})}{\cos(\Omega\tau)}\right]$		$v(x,t) = U_{\text{glottis}}(\Omega)\, e^{j\Omega t}\left[\frac{\sin\Omega(\tau - \frac{x}{c})}{\sin(\Omega\tau)}\right]$	
$p(x,t) = jZ_0 U_{\text{glottis}}(\Omega)\, e^{j\Omega t}\left[\frac{\sin\Omega(\tau - \frac{x}{c})}{\sin(\Omega\tau)}\right]$		$p(x,t) = -jZ_0 U_{\text{glottis}}(\Omega)\, e^{j\Omega t}\left[\frac{\cos\Omega(\tau - \frac{x}{c})}{\cos(\Omega\tau)}\right]$	

Second, just as electrical impedances can be used to study resonance properties of a transmission line, so too can the acoustic impedances be employed to find the resonance frequencies of the acoustic tube. The results are expectedly consistent with those found from the transfer function in (3.21). Finally, the uniform acoustic tube in steady state can be replaced by a T-network of impedances which, in turn, can be used to derive a two-port network of the tube. The resulting network reflects the assumption of no losses in the tube in the occurrence of purely reactive impedances. Details are found in Appendix 3.A.2.

Multitube Lossless Model of the Vocal Tract

Since the production of speech is characterized by changing vocal-tract shape, it might be expected that a more realistic vocal-tract model would consist of a tube that varies as a function of time and displacement along the axis of sound propagation. The formulation of such a time-varying vocal-tract shape can be quite complex. One method of simplifying this model is to represent the vocal tract as a series of concatenated lossless acoustic tubes, as shown in Fig. 3.14. The complete vocal-tract model consists of a sequence of tubes with cross-sectional areas A_k and lengths l_k. The cross-sectional areas and lengths are chosen to approximate the

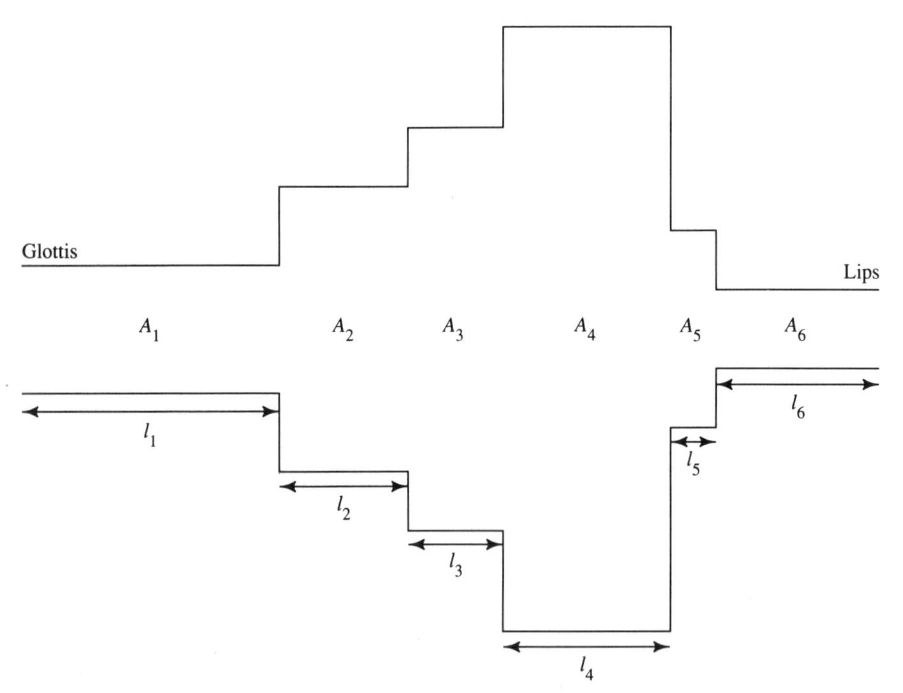

FIGURE 3.14. A vocal-tract model comprised of concatenated acoustic tubes.

vocal-tract area function $A(x)$ (see Fig. 3.11). If a large number of tubes with short lengths is used, then we can expect the formant structure of the concatenated tubes to approach that of a tube with a continuously varying cross-sectional area.

Before beginning the formal study of the multitube model, we note that there are several "intermediate" models between the single-tube case and the general N-tube case that are instructive and useful. However, since the principal goal of our study is the development of a digital model, we shall relegate these intermediate studies to the appendices, where the interested reader can further pursue the details. The two-tube lossless model, discussed in Appendix 3.B, offers an interesting first step toward generality in that it permits the separate modeling of pharyngeal- and oral-tract cavities, while offering an opportunity to exercise the various modeling techniques studied above with tractable mathematics. Three-tube models for nasals and fricatives, discussed in Appendix 3.C, provide the first case in which zeros are present in the model due to the presence of an acoustic "side-cavity" modeling the nasal tract.

We now consider the general N-tube lossless acoustic model, an example of which is illustrated in Fig. 3.14 for $N=6$. The cross-sectional areas $A_1, A_2, \ldots, A_N$ are chosen to approximate the varying area function $A(x,t)$. Here, we assume that a stationary sound is to be modeled, so that the time dependency in $A(x,t)$ can be removed. If an appropriate number of tubes is chosen, the resonant frequencies of this concatenated system of tubes can be made to approximate those of the human vocal tract by approximating $A(x)$. This represents an important step in the development of a computational means to model the vocal tract, since this acoustic tube model will serve as the transition from continuous-time to a discrete-time model based on digital filter theory.

It can be inferred from the discussion in Appendix 3.A.1 that steady-state solutions for volume velocity and sound pressure at point x_k in the kth section of the model have the following forms [see (3.87) and (3.88) in Appendix 3.A.1],

$$v_k(x_k,t)=\frac{A_k}{\rho c}\left[\Psi_k^+(\Omega)e^{j\Omega t}e^{-j\Omega(x_k/c)}-\Psi_k^-(\Omega)e^{j\Omega t}e^{+j\Omega(x_k/c)}\right] \tag{3.27}$$

and

$$p_k(x_k,t)=\Psi_k^+(\Omega)e^{j\Omega t}e^{-j\Omega(x_k/c)}+\Psi_k^-(\Omega)e^{j\Omega t}e^{+j\Omega(x_k/c)}. \tag{3.28}$$

This tube section is characterized by its cross-sectional area A_k and length l_k. The variable x_k represents displacement along the tube measured from left to right (i.e., $0\le x_k \le l_k$), and $\Psi_k^+(\Omega)$ and $\Psi_k^-(\Omega)$ are the complex amplitudes of the rightward (positive direction) and leftward (negative) traveling waves along the kth tube. Suppose that we write

$$v_k(x_k,t)=v_k^+(x_k,t)-v_k^-(x_k,t) \tag{3.29}$$

in which the two terms in the sum designate the respective traveling waves. It is easy to show that, in these terms,

$$v_k(x_k, t) = v_k^+\left(0, t - \frac{x_k}{c}\right) - v_k^-\left(0, t + \frac{x_k}{c}\right) \tag{3.30}$$

and

$$p_k(x_k, t) = Z_{o,k}\left[v_k^+\left(0, t - \frac{x_k}{c}\right) + v_k^-\left(0, t + \frac{x_k}{c}\right)\right], \tag{3.31}$$

where $Z_{o,k}$ is the characteristic impedance of the kth section, defined as in (3.26). Equations (3.30) and (3.31) indicate that the steady-state solutions at any point in the section, say $x_k = x_k'$, can be expressed entirely in terms of the positive- and negative-going volume velocity (or pressure) waves at the left ($x_k = 0$) boundary of the section. This should not be surprising, since there are no losses in the tube, and the volume velocity wave at $x_k = 0$ will therefore appear at $x_k = x_k'$ an appropriate delay or advance in time later. The magnitude of that delay or advance is x_k'/c. Further, as we have discussed above, the pressure and volume velocity are related through the acoustic impedance. Therefore, for convenience in the following, we employ the following abuse of notation:

$$v_k^+(t) \stackrel{\text{def}}{=} v_k^+(0, t) \tag{3.32}$$

$$v_k^-(t) \stackrel{\text{def}}{=} v_k^-(0, t). \tag{3.33}$$

We also define

$$\tau_k \stackrel{\text{def}}{=} \frac{l_k}{c}, \tag{3.34}$$

noting that τ_k is the delay incurred by the sound pressure wave traveling the length of section k.

To determine the interaction of traveling waves between tubes, consider what happens at the juncture between the kth and $(k+1)$st tubes. Since sound wave propagation in a tube obeys the law of continuity and Newton's force law (Halliday and Resnick, 1966), sound pressure and volume velocity must be continuous in both time and space everywhere along the multitube model. Consider the tube juncture shown in Fig. 3.15. Then the law of continuity requires the following boundary conditions to be satisfied at the juncture

$$v_k(x_k = l_k, t) = v_{k+1}(x_{k+1} = 0, t) \tag{3.35}$$

and

$$p_k(x_k = l_k, t) = p_{k+1}(x_{k+1} = 0, t), \tag{3.36}$$

indicating that sound pressure and volume velocity at the end of tube k must equal pressure and velocity at the beginning of tube $k+1$. If (3.27)

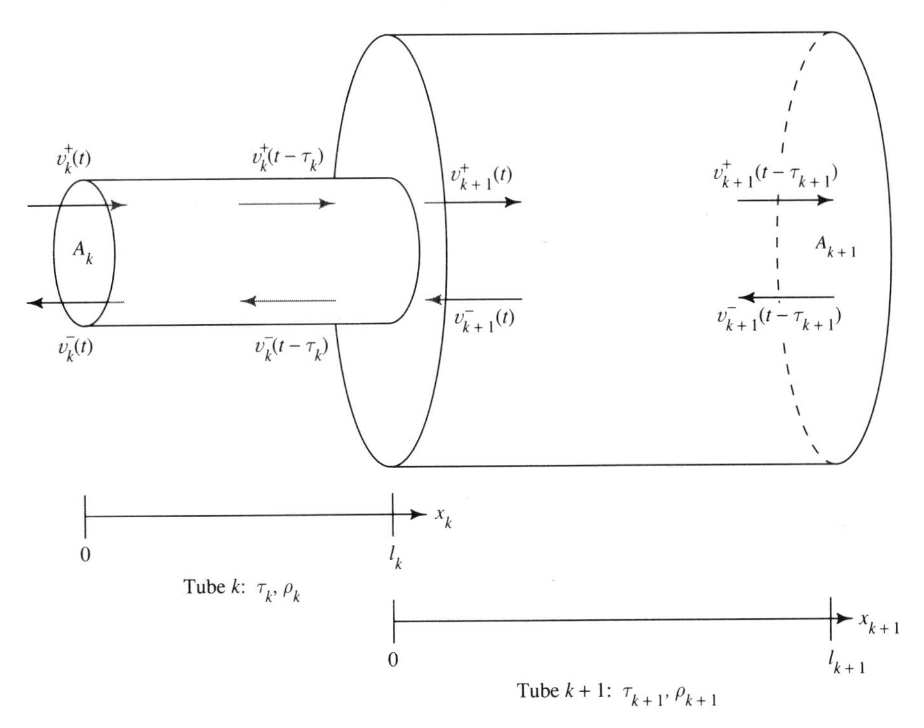

FIGURE 3.15. Sound wave propagation at the juncture between two concatenated acoustic tubes. Each tube is fully characterized by its length and cross-sectional area (l_k, A_k), or its delay and reflection coefficient $[\tau_k = l_k/c, \rho_k = (A_{k+1} - A_k)/(A_{k+1} + A_k)]$.

and (3.28) are substituted into (3.35) and (3.36) and the notation of (3.32), (3.33), and (3.34) employed, the following boundary relations are obtained:

$$v_k^+(t - \tau_k) - v_k^-(t + \tau_k) = V_{k+1}^+(t) - v_{k+1}^-(t) \tag{3.37}$$

$$Z_{o,k}[v_k^+(t - \tau_k) + v_k^-(t + \tau_k)] = Z_{o,k+1}[v_{k+1}^+(t) + v_{k+1}^-(t)]. \tag{3.38}$$

Figure 3.15 illustrates that at the tube juncture, a portion of the positive traveling wave (moving left to right) in tube k is transmitted to tube $k+1$, and a portion is reflected back into tube k. The same occurs for the negative traveling wave (moving right to left) from tube $k+1$, a portion is transmitted into tube k (moving right to left), and a portion is reflected into tube $k+1$ (moving left to right). We can obtain a relation that illustrates transmitted and reflected sound wave propagation at the juncture if (3.37) and (3.38) are solved for the positive wave transmitted into tube $k+1$, $v_{k+1}^+(t)$,

$$v_{k+1}^+(t) = v_k^+(t - \tau_k)\left[\frac{2A_{k+1}}{A_{k+1} + A_k}\right] + v_{k+1}^-(t)\left[\frac{A_{k+1} - A_k}{A_{k+1} + A_k}\right], \tag{3.39}$$

and the negative transmitted wave transmitted into tube k, $v_k^-(t+\tau_k)$ [obtained by subtracting (3.39) from (3.37)],

$$v_k^-(t+\tau_k) = v_{k+1}^-(t)\left[\frac{2A_k}{A_{k+1}+A_k}\right] - v_k^+(t-\tau_k)\left[\frac{A_{k+1}-A_k}{A_{k+1}+A_k}\right]. \tag{3.40}$$

In deriving (3.39) and (3.40), we have used the fact that $Z_{o,k}$ and $Z_{o,k+1}$ contain a common factor ρc that cancels, leaving cross-sectional areas in the results. From (3.39), we see that the volume velocity wave transmitted into tube $k+1$ is composed of a portion transmitted from the forward traveling wave in tube k, $v_k^+(t-\tau_k)$, and a portion reflected from the backward traveling wave in tube $k+1$, $v_{k+1}^-(t)$. Equation (3.40) indicates that the transmitted backward traveling wave into tube k consists of a partially transmitted backward traveling wave from tube $k+1$, $v_{k+1}^-(t)$, and a reflected portion from the forward traveling wave in tube k, $v_k^+(t-\tau_k)$. If we assume that the negative traveling wave $v_{k+1}^-(t)$ in tube $k+1$ is zero, then (3.39) and (3.40) reduce to

$$v_{k+1}^+(t) = v_k^+(t-\tau_k)\rho_k^+ \tag{3.41}$$

$$v_k^-(t+\tau_k) = -v_k^+(t-\tau_k)\rho_k^-, \tag{3.42}$$

where the ratios ρ_k^+ and ρ_k^- are the *transmission coefficient* and *reflection coefficient* between the kth and $(k+1)$st tubes, respectively:

$$\rho_k^+ \stackrel{\text{def}}{=} \frac{2A_{k+1}}{A_{k+1}+A_k} \tag{3.43}$$

$$\rho_k^- \stackrel{\text{def}}{=} \frac{A_{k+1}-A_k}{A_{k+1}+A_k} = \rho_k^+ - 1. \tag{3.44}$$

In fact, since the reflection and transmission coefficients are simply related, it has become conventional to use only the reflection coefficients in analytical and modeling work. Let us henceforth adopt this convention and simplify the notation for the reflection coefficient by omitting the minus sign:[10]

$$\rho_k \stackrel{\text{def}}{=} \rho_k^-. \tag{3.45}$$

Since cross-sectional areas for tubes must be positive, the reflection coefficient at any tube juncture is bounded in magnitude by unity,

$$-1 \le \rho_k \le 1. \tag{3.46}$$

The reflection coefficients become increasingly important as we strive for a digital filter representation.

[10]In Chapter 5 we will encounter the reflection coefficient again in conjunction with the digital model.

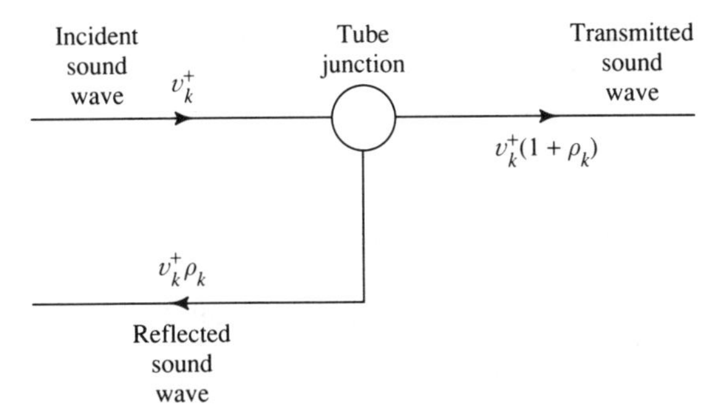

FIGURE 3.16. Signal flow graph of a forward moving sound wave (volume velocity) at the junction between two tubes.

Figure 3.16 presents a signal flow graph at the juncture of the kth and $(k+1)$st tubes for a forward traveling sound wave. Using reflection coefficients, (3.39) and (3.40) can be written as

$$v_{k+1}^+(t) = v_k^+(t-\tau_k)(1+\rho_k) + v_{k+1}^-(t)\rho_k \tag{3.47}$$

$$v_k^-(t+\tau_k) = v_{k+1}^-(t)(1-\rho_k) - v_k^+(t-\tau_k)\rho_k . \tag{3.48}$$

Equations (3.47) and (3.48), which are often called the *Kelly–Lochbaum equations* (Kelly and Lochbaum, 1962), immediately lead to the signal flow graph[11] shown in Fig. 3.17 for the juncture at the kth boundary. This structure was first used by Kelly and Lochbaum to synthesize speech in 1962. Note that this signal flow graph contains equivalent information of the tube diagram in Fig. 3.15, but, whereas the sections were characterized by areas and lengths in the acoustic model, here the sections are more readily characterized by the reflection coefficients and delays.

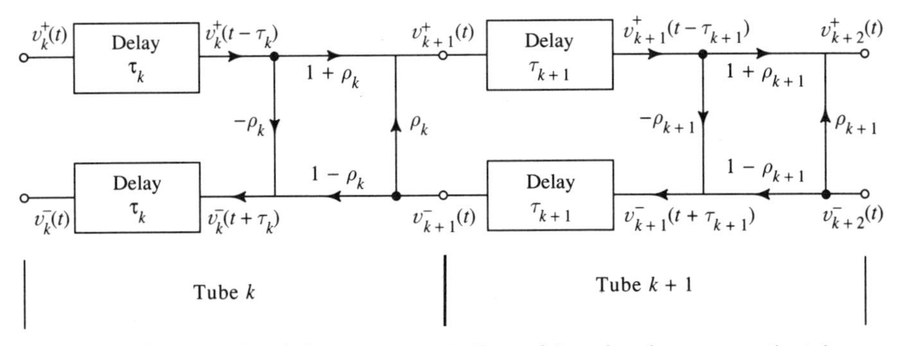

FIGURE 3.17. A signal flow representation of two lossless acoustic tubes.

[11]For an introduction to the use of signal flow graphs in signal processing, see, for example, (Oppenheim and Schafer, 1989).

Figure 3.16 isolates the kth tube junction and illustrates the reflection and transmission of the forward propagating sound wave at the boundary. A propagating sound wave experiences a delay in traveling the length of the each tube, which is represented as a signal delay τ_k in Fig. 3.17 for the kth tube. Since each tube is lossless, the propagating wave is not attenuated. When the sound wave reaches a tube boundary, a portion is transmitted $(1+\rho_k)$, and the remaining portion reflected $(-\rho_k)$. The percentage transmitted/reflected is dependent on the cross-sectional area mismatch at the tube juncture. If the mismatch is small (i.e., $A_k \approx A_{k+1}$), then a larger portion will be transmitted (perfect transmission occurs when $A_k = A_{k+1}$). If a large mismatch is present (i.e., $A_k >> A_{k+1}$), more of the signal will be reflected.

The signal flow representation for two adjoining sections can be extended to N tubes in the obvious way. At each boundary, the flow and pressure dynamics are described in terms of the reflection coefficient and the delay of the section on the left. In order to complete the (lossless) acoustic model of speech production, we need to consider the boundary effects at the lips and glottis. This is the subject of the following section.

Tube Models for the Glottis and the Lips

The formulation of a tube model for either the glottis or the lips requires consideration of the half-infinite acoustic tube in Fig. 3.18. The figure illustrates a tube of cross-sectional area A and infinite length. For such a tube, an incident sound wave $v^+(t)$ [or $p^+(t)$] injected at the source $x=0$ will propagate to the right indefinitely. Since the propagating wave never experiences a tube boundary, no portion will ever be reflected. Now it is shown by letting $l \to \infty$ in (3.106) in Appendix 3.A.2 that the impedance at location x in a lossless open tube of infinite length is simply the characteristic impedance of the tube:

$$\lim_{l \to \infty} Z_{\substack{\text{open}\\ \text{tube}}}(x, \Omega) = Z_o. \tag{3.49}$$

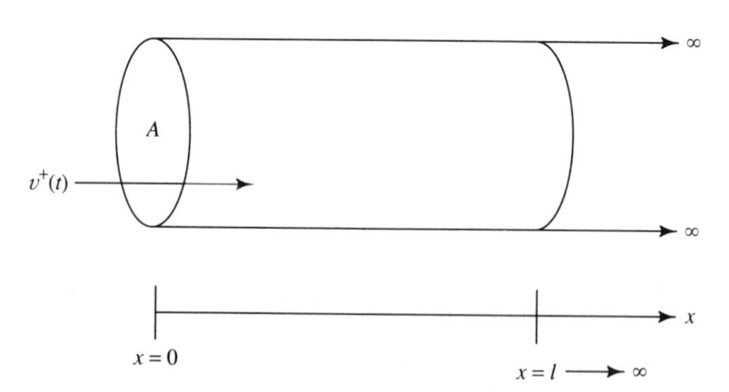

FIGURE 3.18. A half-infinite acoustic tube for modeling glottis and lip radiation.

Since this impedance is assumed to be real, a half-infinite acoustic tube can be used to model a real or resistive load. Under such conditions, the characteristic impedance is sometimes replaced by a resistive load R_o.

Let us now consider the junction between the last vocal-tract tube (call it tube N) and the lips. We assert that the infinite open tube makes an appropriate model for the lip radiation. In any case, let us assume that we can model the lips with a tube section, say section $N+1$. In most situations, real and synthetic, once the sound wave passes the lip boundary, it radiates outward. If the radiating volume velocity wave meets no obstacles, no reflections will propagate back down into the vocal system. Thus the negative traveling wave $v^-_{N+1}(x_{N+1}, t)$ [or $p^-(x_{N+1}, t)$], for any x_{N+1} and any t, should be zero (see Fig. 3.19):

$$v^-_{N+1}(x_{N+1}, t) = p^-_{N+1}(x_{N+1}, t) = 0 \tag{3.50}$$

for all x_{N+1} and t. Therefore, we can model the effects of radiation from the lips using a half-infinite tube. The other property which must hold to make the half-infinite tube appropriate is that the radiation load must be purely real. Assuming that this is the case, and that real impedance Z_{lips} is encountered by the traveling sound wave as it reaches the lips, then the half-infinite tube with cross-sectional area A_{N+1} such that

$$Z_{o,N+1} = \frac{\rho c}{A_{N+1}} = Z_{\text{lips}} \tag{3.51}$$

is a good model.

Let us next derive the signal flow graph for the lip boundary. We use the explicit subscript "lips" to denote the quantities at the lip boundary, so in the present case, $\rho_{\text{lips}} \stackrel{\text{def}}{=} \rho_N$. By definition, the reflection coefficient is given by

$$\rho_{\text{lips}} = \frac{A_{N+1} - A_N}{A_{N+1} + A_N}, \tag{3.52}$$

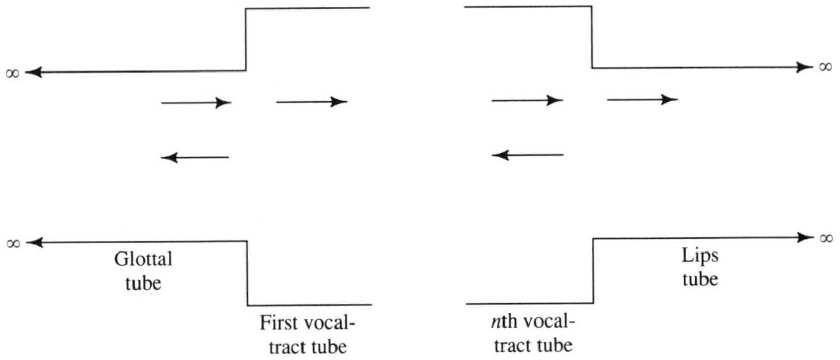

FIGURE 3.19. Wave propagation diagrams of half-infinite acoustic tubes used for modeling the glottis and lips.

which can also be expressed as

$$\rho_{\text{lips}} = \frac{Z_{o,N} - Z_{\text{lips}}}{Z_{o,N} + Z_{\text{lips}}}. \tag{3.53}$$

Now in light of (3.48) and (3.50) we can write

$$v_N^-(t + \tau_N) = -\rho_{\text{lips}} v_N^+(l_N, t - \tau_N). \tag{3.54}$$

Accordingly, the output volume velocity at the lip boundary is

$$\begin{aligned} v_N(l_N, t) = v_N(t - \tau_N) &= v_N^+(t - \tau_N) - v_N^-(t + \tau_N) \\ &= (1 + \rho_{\text{lips}}) v_N^+(t - \tau_N). \end{aligned} \tag{3.55}$$

The signal flow diagram reflecting (3.54) and (3.55) has been appended to the output of the tube model in Fig. 3.20.

A half-infinite tube can also be used to model the termination at the glottis. Consider an incident sound wave at the juncture of the glottis and the first vocal-tract tube as shown in Fig. 3.19.[12] A portion of the volume velocity wave will be transmitted into tube 1, and a portion reflected. The reflected portion travels back down through the subglottal vocal system (trachea and lungs). This backward traveling wave therefore does not contribute appreciably to sound production (it is generally assumed that the soft lung tissue absorbs the majority of this energy). In terms of sound propagation and modeling, this backward traveling wave is typically ignored. The glottal termination is therefore modeled as a volume velocity source $u_{\text{glottis}}(t)$ in parallel with a glottal impedance Z_{glottis} [as in (3.6)]. The glottal impedance

$$Z_{\text{glottis}}(\Omega, t) = R_{\text{glottis}}(t) + j\Omega L_{\text{glottis}}(t) \tag{3.56}$$

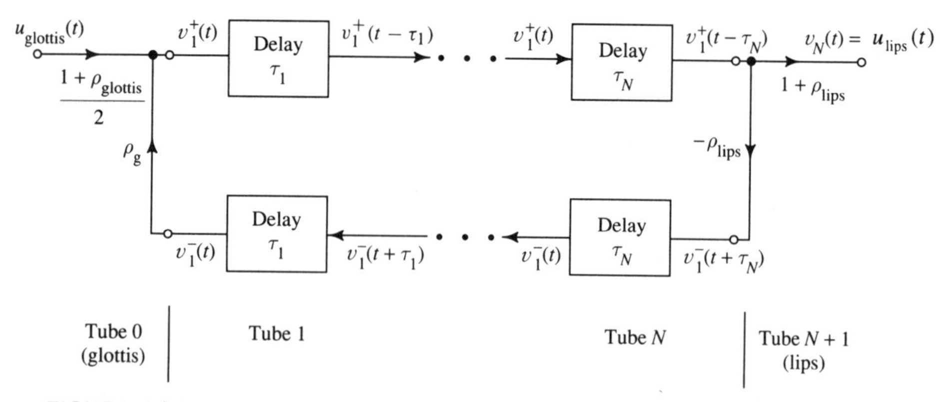

FIGURE 3.20. Signal flow diagrams of acoustic tube junctures for modeling the glottis and lips.

[12] Here we assume that the vocal tract is modeled with a concatenation of N acoustic tubes. The tube closest to the glottis is tube 1.

is a time-varying acoustic impedance which is a function of the inverse cross-sectional area of the glottis, $1/A_{\text{glottis}}(t)$. For example, when the glottis is closed $[A_{\text{glottis}}(t) = 0]$, the glottal impedance becomes infinite. Two equivalent "circuit" models are shown in Fig. 3.21. It is very important to notice that these models are *not* time invariant and that in neither case does the source and impedance vary independently. Consider the current source model, for example. In the case of infinite impedance, the parallel branch containing Z_{glottis} becomes an open circuit, but the source current is also simultaneously reduced to zero. In general, as the glottis opens and closes, the glottal impedance varies between infinity and a finite value, causing the input volume velocity of the vocal tract to vary between zero and a finite value. The variation in the volume velocity wave is exactly the pulselike glottal flow wave shown earlier in Fig. 3.8. One must be careful to remember this time dependence and source-impedance coupling when using the circuit analogies.

If a first-order approximation is used, the time-varying glottal impedance can be approximated by the fixed impedance

$$Z_{\text{glottis}}(\Omega) = R_{\text{glottis}} + j\Omega L_{\text{glottis}}. \tag{3.57}$$

The transmission line circuit model shown in Fig. 3.21 may now be considered a linear, time-invariant circuit. As illustrated in this figure, the net volume velocity into the first vocal-tract tube model is obtained by subtracting that portion lost due to glottal impedance (using a current divider relation[13]),

$$v_1(0, t) = v_0(0, t) - p_1(0, t)\frac{1}{Z_{\text{glottis}}(\Omega)} = u_{\text{glottis}}(t) - p_1(0, t)\frac{1}{Z_{\text{glottis}}(\Omega)}, \tag{3.58}$$

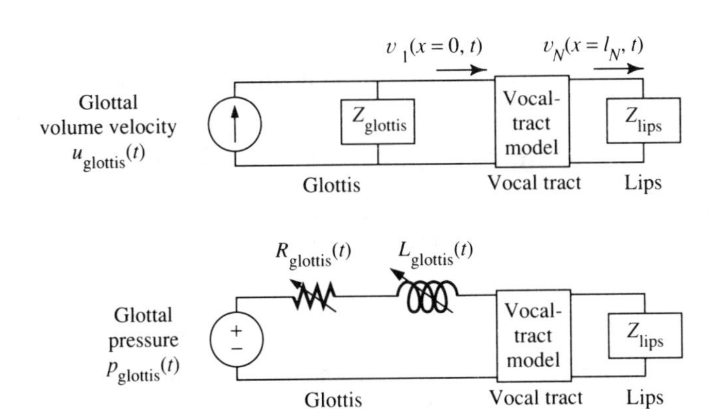

FIGURE 3.21. Transmission line circuit model for the glottis and lips.

[13]For the half-infinite tube used to model the glottis, the location $x = 0$ is taken to be just inside the juncture of the glottis and first vocal-tract tube. The location $x = -\infty$ is the "left end" of the tube.

where the volume velocity $v_0(0,t)$ is taken to be the input glottal volume velocity $u_{\text{glottis}}(t)$, and $p_1(0,t)$ the sound pressure at the entry of the first vocal-tract tube. We seek only steady-state results, so $u_{\text{glottis}}(t)$ is a complex exponential signal of frequency Ω. By using decompositions like (3.30) and (3.31), (3.58) can be written as

$$v_1^+(0,t) - v_1^-(0,t) = u_{\text{glottis}}(t) - \frac{1}{Z_{\text{glottis}}(\Omega)}\left[\frac{\rho c}{A_1}\{v_1^+(t) + v_1^-(0,t)\}\right]. \tag{3.59}$$

Solving for the forward traveling sound wave at the boundary, $v_1^+(0,t)$, we obtain

$$v_1^+(0,t) = \frac{1+\rho_{\text{glottis}}}{2}u_{\text{glottis}}(t) + \rho_{\text{glottis}}v_1^-(0,t), \tag{3.60}$$

where the glottal reflection coefficient is.

$$\rho_{\text{glottis}} = \frac{Z_{\text{glottis}}(\Omega) - (\rho c/A_1)}{Z_{\text{glottis}}(\Omega) + (\rho c/A_1)}. \tag{3.61}$$

The input volume velocity at the glottis is comprised of only that portion transmitted into tube 1, plus that portion from the backward traveling wave in the tube, which is reflected at the glottal tube/tube 1 juncture. A signal flow graph depicting wave propagation from (3.60) is included at the input of Fig. 3.20.

Finally, we notice that, like the lip reflection coefficient, ρ_{lips}, the glottal reflection coefficient ρ_{glottis} is generally frequency dependent because of the involvement of an impedance. However, like Z_{lips}, Z_{glottis} is frequently taken to be real for ease in vocal system modeling. Therefore, the terminal effects of the glottis and lips can be represented in terms of impedance and a volume velocity source for transmission line circuit analysis in Fig. 3.21, or using signal flow diagrams like Fig. 3.20.

Complete Signal Flow Diagram

As an example of a complete representation of sound wave propagation, we consider a two-tube ($N = 2$) acoustic model for the vocal tract. We hasten to point out that the choice $N = 2$ is made for simplicity, and in practice the number N would ordinarily be chosen much larger (typically 10–14; see Chapter 5). Using a two-tube signal flow representation for the vocal tract (from Fig. 3.17), and signal flow representations for the glottis and lips (from Fig. 3.20), we obtain the overall system flow diagram in Fig. 3.22. The overall system transfer function is found by taking the ratio of volume velocity at the lips to that at the glottis. We can use phasors or assume complex exponential signals (see Section 1.5). The

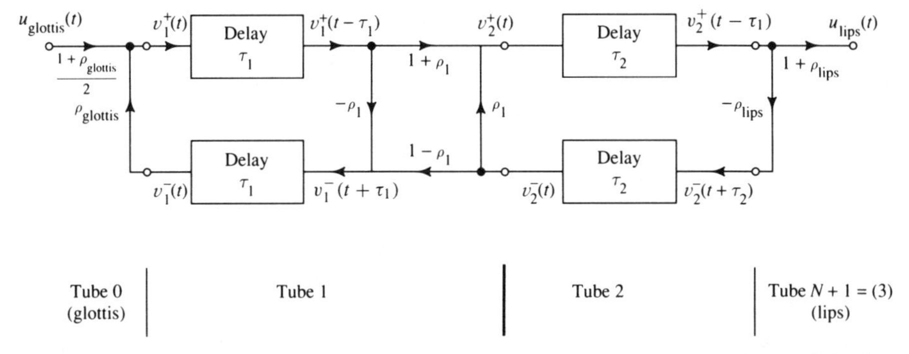

FIGURE 3.22. Overall signal flow diagram for a two-tube acoustic model of speech production.

volume velocity at the lips, $u_{\text{lips}}(t)$, is simply the input sound wave to the half-infinite tube that models the lips,

$$u_{\text{lips}}(t) = v_{N+1}(0, t) = v_3(0, t). \tag{3.62}$$

Accordingly, using Mason's gain rule [see, e.g., (D'Azzo and Houpis, 1966)] on the graph of Fig. 3.22, we have

$$\begin{aligned} H_{\text{two-tube}}(\Omega) &= \frac{\bar{U}_{\text{lips}}(\Omega)}{\bar{U}_{\text{glottis}}(\Omega)} = \frac{u_{\text{lips}}(t)}{u_{\text{glottis}}(t)} \\ &= \frac{[(1+\rho_{\text{glottis}})/2](1+\rho_{\text{lips}})(1+\rho_1)e^{-j\Omega(\tau_1+\tau_2)}}{1+\rho_1\rho_{\text{glottis}}e^{-j\Omega 2\tau_1}+\rho_1\rho_{\text{lips}}e^{-j\Omega 2\tau_2}+\rho_{\text{glottis}}\rho_{\text{lips}}e^{-j\Omega 2(\tau_1+\tau_2)}}. \end{aligned} \tag{3.63}$$

Some features of this transfer function should be noted. First, the magnitude of the numerator is the product of the multiplier terms in the forward signal path, while the phase term $e^{-j\Omega(\tau_1+\tau_2)}$ represents the delay experienced by a signal propagating through the forward path. Second, if $j\Omega$ is replaced by the complex variable s, the poles of the system function $H_{\text{two-tube}}(s)$ represent the complex resonant frequencies of the system. Fant (1970) and Flanagan (1972) have shown that if cross-sectional areas and lengths are chosen properly, this transfer function can approximate the magnitude spectrum of vowels.

Loss Effects in the Vocal-Tract Tube Model

We should return to the starting point of this discussion and recall some simplifying assumptions made regarding the vocal tract. A number of unrealistic constraints have influenced the outcome of this development. In particular, we assumed that the vocal tract can be appropriately modeled by a hard-walled, lossless series of tubes. In fact, energy losses occur during speech due to viscous friction between the flowing air and tract walls, vibration of the walls, and heat conduction through the walls. A proper analysis would require that we modify (3.16) and (3.17) to reflect these losses and proceed accordingly. Of course, some rather intractable mathematics would result, a fact which is compounded by the frequency dependence of the loss effects. The usual approach to this problem taken by speech engineers is learn to anticipate certain frequency domain effects which occur as a result of the tract losses. With a basic understanding of these effects, the speech engineer can usually build successful algorithms and analysis on the simplified lossless model.

The most significant of the losses mentioned above arises due to the tract wall vibration. The walls of the tract are, relatively speaking, rather massive structures and are therefore more responsive to lower frequencies. This fact corresponds to a preferential broadening of the bandwidths of the lower formants. On the other hand, the viscous and thermal losses are most influential at higher frequencies, so that these losses tend to broaden the bandwidths of the higher formants. If we use the analysis in (Flanagan, 1972) and (Portnoff, 1973), it can also be shown that the vibration losses tend to slightly raise the formant center frequencies, while the other losses tend to lower them. The net change is a small shift upward with respect to the lossless tube model.

Whereas the lossless tube model is not, strictly speaking, appropriate for real speech due to the ignored loss effects, the analytical models to which these early "lossless" developments will lead will often be able to "adjust themselves" to handle the aberrations which occur in the spectrum. It is therefore important that we clearly understand the lossless case, and that we be able to anticipate what spectral clues may be present that indicate certain real phenomena.

3.1.5 Models for Nasals and Fricatives

The production of the nasal phonemes /n,m,G/ requires that the velum be open and the lips closed. This creates a very different tube configuration from the vocal-tract case discussed above. In the nasal case, the oral cavity creates an acoustical side cavity branching off the main path from glottis to nostrils. Analytically, the result is a set of three acoustical differential equations, the solution of which, for many practical purposes, is not warranted. As in the case of the vocal-tract losses, it is usually sufficient to be aware of frequency domain effects caused by nasality. In particular, the side cavity introduced by the nasal configuration will tend to trap energy leading to antiresonances (zeros) that appear in addition to the resonances (poles) in the magnitude spectrum. It has also been found that nasal phonemes tend to have somewhat broader bandwidth formants at high frequencies (Fujimura, 1962) due to heat and friction losses along the large walls of the nasal tract.

The modeling of consonants in general, and fricatives in particular, is a complex subject, the details of which would take us too far afield. It has generally been found that a three-tube model of the vocal tract works well to model main front and back cavities connected by a narrow constriction. Each of the tubes produces resonances (formants) whose frequencies are related to the tube lengths. Like nasals, the fricative tract configuration contains cavities that can trap energy leading to antiresonances in the overall spectrum. It has been found that fricatives exhibit very little energy below the first zero in their spectra, which is usually quite high in the Nyquist range.

Like the vocal-tract loss case, in practice we often use a model that was fundamentally derived using the lossless vocal-tract assumption (without side cavities) to analyze nasals and fricatives. To a lesser extent than in the vocal-tract loss case, the ideal model will be able to handle these antiresonance phenomena in the spectra. However, antiresonances are often more problematic than simple broadening of formant bandwidths, because, whereas the latter can be accomplished by a simple shifting of poles in the model, the former requires the presence of spectral zeros for a good representation. However, we shall find that models whose spectra require only poles are to be greatly desired in speech modeling.

3.2 Discrete-Time Modeling

3.2.1 General Discrete-Time Speech Model

We have considered several strategies for modeling the speech production system. Based on early observations of the resonant structure of cylindrical tubes, an acoustic tube model was first considered for speech modeling. An example of a six-tube model is shown in Fig. 3.23(a). We superficially noted the analogy between acoustical systems and the electric transmission line and have occasionally employed that analogy in certain developments. In particular, glottal excitation and load effects from lip radiation were modeled using electric circuit analogies. This analysis and further consideration of properties of the forward and backward traveling sound waves in an acoustic model like Fig. 3.23(a) ultimately led to the signal flow representation in Fig. 3.23(b). A more detailed view of the transmission line analogy in Appendix 3.A.2 can be used to develop the model shown in Fig. 3.23(c) in which a series of T-network impedances is used to model the four vocal-tract tubes found in Fig. 3.23(a). It should be stressed that quantities of cross-sectional area and length for each of the acoustic tubes fully characterize each of the three models.

The resulting signal flow diagram in Fig. 3.23(b) suggests that the lossless tube model exhibits characteristics common to a digital filter model. The final signal flow for the four-tube vocal-tract model contains only additions, multiplications, and delays. These operations are easily implemented in a discrete-time model. The only restriction to satisfy is

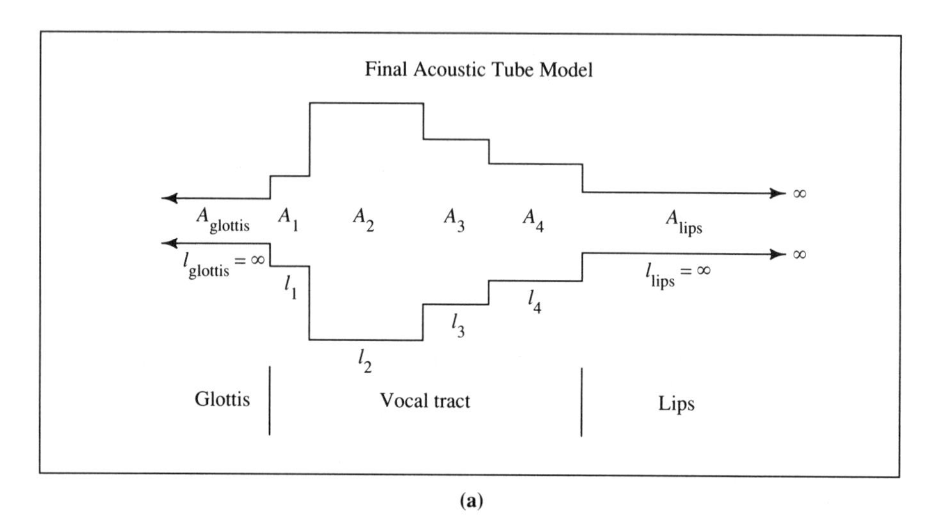

(a)

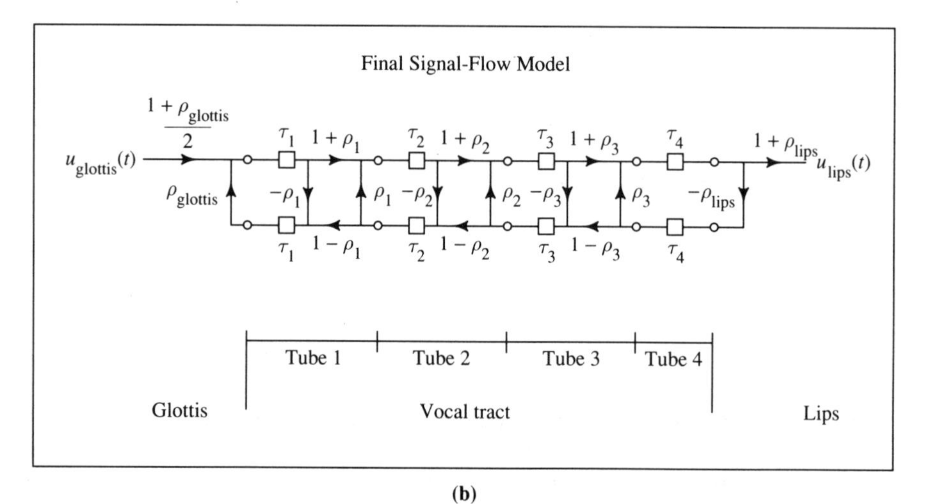

(b)

FIGURE 3.23. A summary of three modeling methods for characterizing speech production. (a) A six-tube acoustic model for speech. Half-infinite acoustic tubes are used to model the glottis and lip radiation effects. (b) A signal flow model for the acoustic tube system pictured in part (a).

that each delay term be a multiple of some unit of time T, corresponding to the sample period in the discrete-time system. To achieve this, consider the cumulative delay experienced by an incident signal in an N-tube system, say

$$\tau_{\text{total}} = \sum_{j=1}^{N} \tau_j = \frac{1}{c} \sum_{j=1}^{N} l_j, \tag{3.64}$$

where τ_{total} represents the amount of time necessary for a sound wave to travel the entire length of vocal-tract model $L = \sum_{j=1}^{N} l_j$. To ensure a

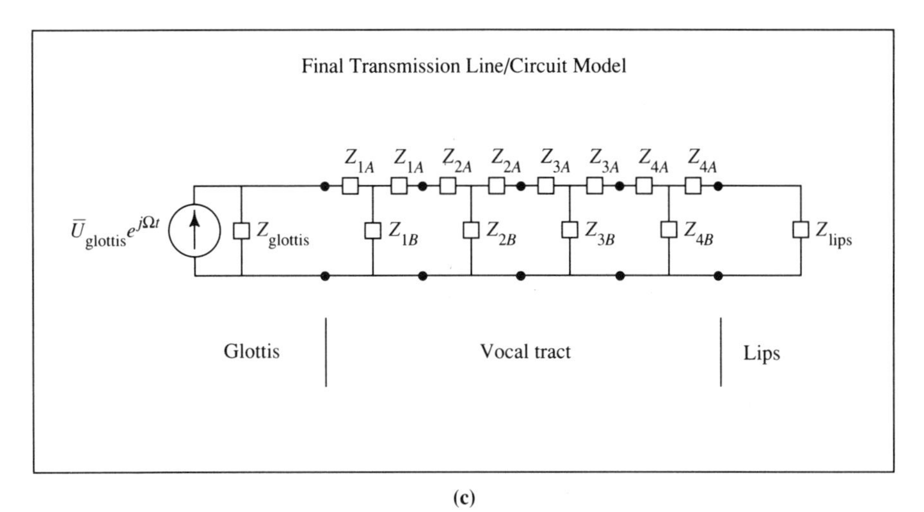

(c)

FIGURE 3.23. (*Cont.*) (c) An equivalent transmission line circuit representation for the acoustic tube model in (a).

smooth transition to the discrete-time domain, let us consider a set of N tubes, each of a fixed common length, say $\Delta_x = L/N$ (see Fig. 3.24). Analysis of wave propagation for this acoustic tube model is equivalent to previous systems, with the exception that here each tube possesses the same delay

$$\tau = \frac{L}{cN} = \frac{\Delta_x}{c}. \tag{3.65}$$

This limits the number of variables available for simulating the vocal-tract cross-sectional area function $A(x, t)$. With this restriction, the signal flow model of Fig. 3.23(b) is replaced by that in Fig. 3.25(a), where each tube delay τ_i has been replaced by the constant delay τ. If a discrete-time impulse is injected into this signal flow model, the earliest an output

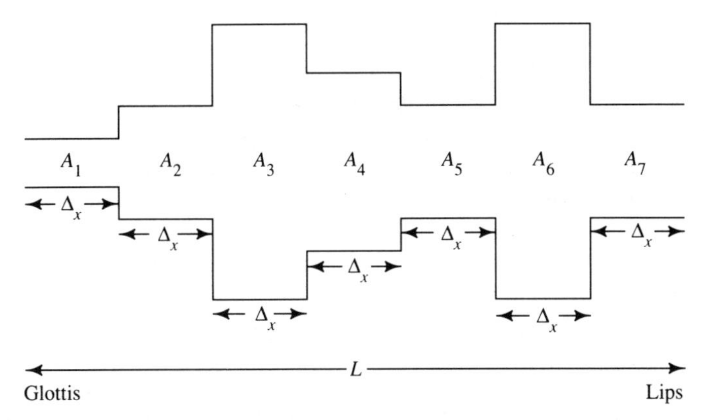

FIGURE 3.24. A concatenation of seven lossless tubes of equal length. Adapted from (Rabiner and Schafer, 1978).

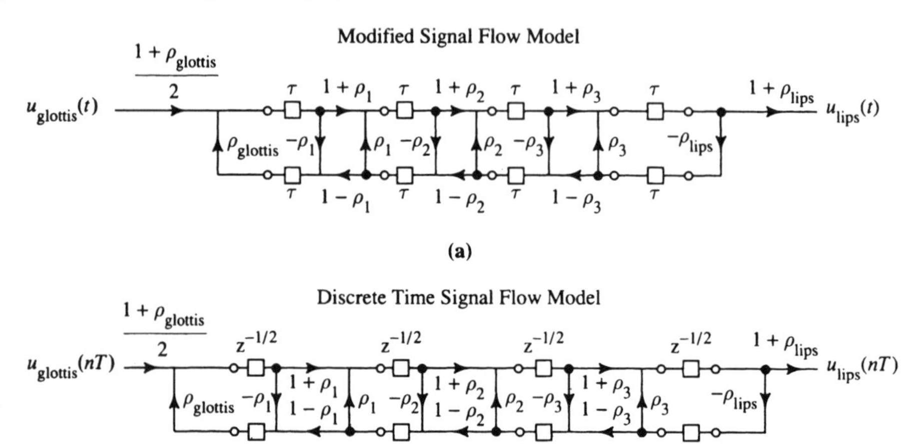

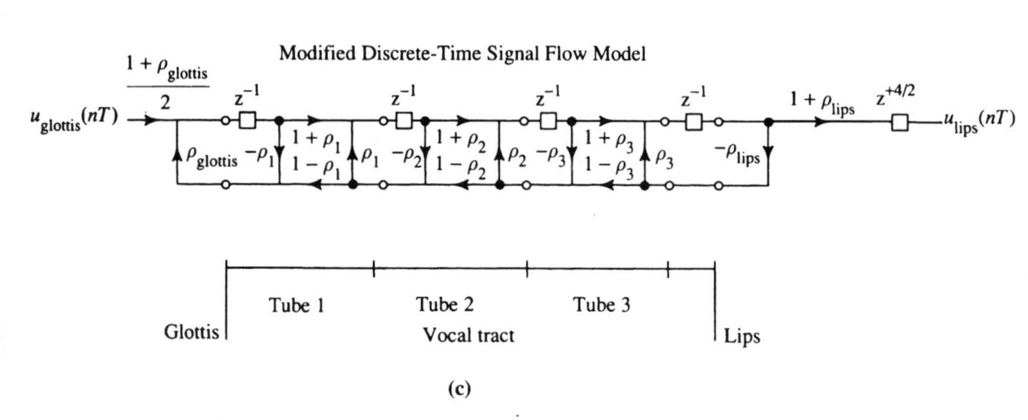

FIGURE 3.25. (a) A modified signal flow model employing lossless tubes of equal length. (b) A discrete-time signal flow diagram for the model shown in (a). (c) A final discrete-time signal flow diagram using whole sample delays for the model shown in (a).

would occur is $N\tau$ units of time (4τ seconds for the model pictured). If a sample period of $T = 2\tau$ is chosen for the discrete-time system, then $N\tau$ seconds corresponds to a shift of $N/2$ samples. For an even number of tubes, $N/2$ is an integer and the output can be obtained by shifting the sequence through the signal flow model. If N is odd, however, output values must be obtained between sample locations, thus requiring an interpolation step. Such a delay in most cases is ignored, since it has little effect for most speech applications.

Therefore, an equivalent discrete-time system can be obtained by substituting sample delays of $z^{-(1/2)}$. The $\frac{1}{2}$ sample delays shown in Fig. 3.25(b) imply that a sample must be interpolated between each input sample, which is infeasible in practice. To address this issue, consider a volume velocity wave traveling the length of a lossless tube section. Figure 3.25 illustrates that the shape of the incident wave at the entrance of tube $j(x_j = 0)$ will not change until the wave has propagated the full length of the tube, meets a tube juncture, is completely or partially reflected, and the reflected wave travels back the entire length of the tube. Therefore, for the volume velocity to change at any boundary location along the signal flow graph requires a round-trip delay of 2τ. This is verified by obtaining the cumulative delay of any closed loop in the signal flow graph of Fig. 3.25(a). For this figure, cumulative delays of $2\tau, 4\tau$, 6τ, and 8τ are possible. If only external signal flow model analysis is considered, equivalent loop equations can be obtained if the half-sample delays in the feedback path are reflected up into the forward path, resulting in Fig. 3.25(c). This final discrete-time signal flow model is equivalent to models (a) and (b) in its ability to characterize speech; however, it has the distinct advantage of employing unit sample delays. A consequence of moving the half-sample delays into the forward path is that an additional delay of $N/2$ samples is introduced. To counteract this effect, a signal "advance" by $N/2$ samples is added after the lip radiation term. Such an advance presents no difficulty if the discrete-time digital filter representation is to be implemented on the computer. The final discrete-time transfer function for a two-tube vocal-tract model can be shown to have the following form [from (3.63)],

$$H_{\text{two-tube}}(z) = \frac{U_{\text{lips}}(z)}{U_{\text{glottis}}(z)} \tag{3.66}$$

$$= \frac{[(1+\rho_{\text{glottis}})/2](1+\rho_1)(1+\rho_{\text{lips}})z^{-1}}{1+(\rho_1\rho_{\text{glottis}}+\rho_1\rho_{\text{lips}})z^{-1}+\rho_{\text{glottis}}\rho_{\text{lips}}z^{-2}}.$$

The two-tube z-transform system function in (3.66) is relatively easy to obtain. However, calculation of transfer functions using signal flow analysis becomes increasingly complex as the number of tubes increases. For example, the four-tube vocal-tract model in Fig. 3.25(c) possesses a transfer function of similar form to (3.66). However, calculation of the coefficients from flow graphs becomes unwieldy as the model order increases to 10 or more. One solution to this problem is to resort to the use of two-port network models for each section of the tube. This general approach is described in Appendix 3.A.2; its application to the fast computation of discrete-time transfer functions for models with a large number of sections is given in Appendix 3.C. The general N-section lossless model is shown there to have a z-domain system function of the form

$$H(z) = \frac{1+\rho_{\text{glottis}}}{2} \cdot \frac{z^{-N/2}\prod_{k=1}^{N}(1+\rho_k)}{1-\sum_{k=1}^{N} b_k z^{-k}}. \tag{3.67}$$

It is interesting to note that this system function is "all pole" in the sense that it has no nontrivial zeros in the z-plane. The poles of $H(z)$ define the resonant or formant structure of the N-tube model. It will be very interesting to return to this form in future discussions of discrete-time models.

One final point is essential. As we have already seen in the early discussions of source modeling, there are various transfer functions that are of interest in speech modeling, and we have urged the reader to understand which transfer function is being considered in a given development. In the present case, $H(z)$ indicates the system function between the volume velocity flow at the output of the glottis and that at the lips. By definition, this transfer ratio is that associated with the vocal tract, and the notation $H(z)$ is consistent with previous developments [see (3.3) and (3.7)]. The reader might wonder, then, about the meaning of tube 0 (glottal tube) and tube $N+1$ (lip tube) in Figs. 3.23 and 3.24. A careful reading of the foregoing will indicate that these "tubes" do not represent the larynx and lips *per se*, but rather are present as a device for modeling the *boundary effects* between the larynx and vocal tract, and vocal tract and lips, respectively. In analytical terms, it is important that the reader not misinterpret tubes 0 and $N+1$ as analog counterparts of models corresponding to $G(z)$ and $R(z)$, respectively, of (3.3). It is interesting, however, that the analog counterpart to $R(z)$, say $R(\Omega)$, *is* involved in the development of tube $N+1$ in the following sense: By definition, the lip radiation transfer function is the transfer function (ratio of phasors) between the lip pressure waveform (speech) and the lip volume velocity

$$R(\Omega) = \frac{\overline{P}_{\text{lips}}(\Omega)}{\overline{U}_{\text{lips}}(\Omega)} = \frac{\overline{S}(\Omega)}{\overline{U}_{\text{lips}}(\Omega)}. \tag{3.68}$$

However, this ratio is precisely the definition of the lip impedance $Z_{\text{lips}}(\Omega)$. It is left as an exercise for the reader to modify the flow diagram of Fig. 3.25 to reflect the same impedance as that in Fig. 3.23c.

3.2.2 A Discrete-Time Filter Model for Speech Production

A reasonably general linear discrete-time model for speech production is shown in Fig. 3.26. This model is called a *terminal-analog model*, meaning that the signals and systems involved in the model are only superficially analogous, if at all, to the true physical system, except at the terminus, where both produce analogous waveforms, speech. In other words, this model attempts to represent the speech production process based on its output signal characteristics. In spite of the fact that there are no provisions for coupling or nonlinear effects between subsystems in the model, this system produces reasonable quality speech for coding purposes. It should be noted that, although it may be possible for different vocal-tract shapes to produce the same speech sound, terminal-analog models ignore this issue.

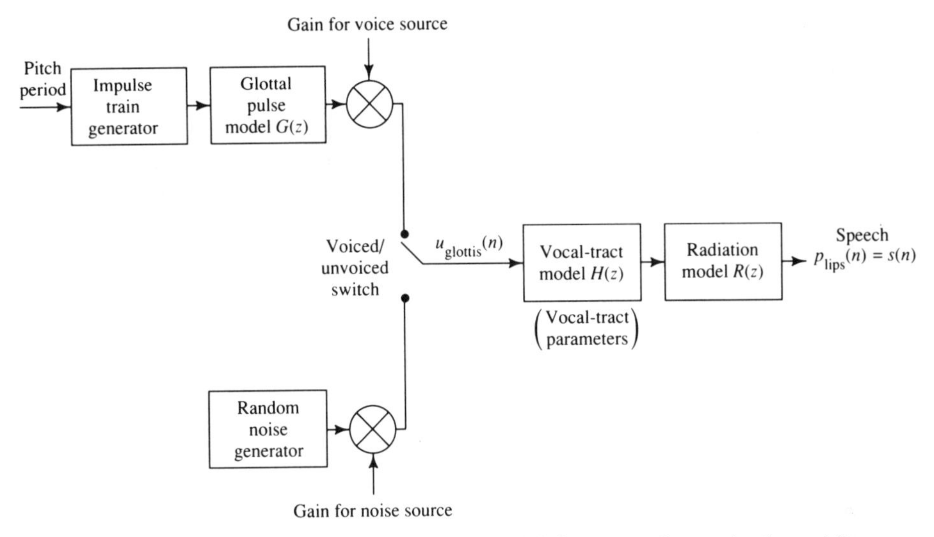

FIGURE 3.26. A general discrete-time model for speech production. After Rabiner and Schafer (1978).

In this general terminal-analog system, a vocal-tract model $H(z)$ and radiation model $R(z)$ are excited by a discrete-time glottal excitation signal $u(n) = u_{\text{glottis}}(n)$. During unvoiced speech activity, the excitation source is a flat spectrum noise source modeled by a random noise generator. During periods of voiced speech activity, the excitation uses an estimate of the local pitch period to set an impulse train generator that drives a glottal pulse shaping filter $G(z)$. This excitation produces a glottal pulse waveform similar in shape to those we have studied in previous discussions. This excitation is certainly flawed, since it cannot address phonemes with more than one source of excitation (e.g., voiced fricatives).

The vocal-tract transfer function $H(z)$ is modeled using a simplification of (3.67),

$$H(z) = \frac{H_o}{1 - \sum_{k=1}^{N} b_k z^{-k}} = \frac{H_o}{\prod_{k=1}^{N} (1 - p_k z^{-1})}, \tag{3.69}$$

where H_o represents an overall gain term and p_k the complex pole locations for the N-tube model. This model is a manifestation of our study of the lossless acoustic tube model that we know to be inadequate for certain classes of phonemes. Nevertheless, this model for $H(z)$ is frequently used to model all speech sounds because of powerful analytical methods that follow from its use. Of course, analytical methods do not justify the use of a model if the performance of resulting techniques is not sufficient. In spite of the fundamental acoustical arguments against the all-pole model, however, it has been very successfully used in many analysis, coding, and recognition tasks, as we shall see in our studies.

Each pair of poles in the z-plane at complex conjugate locations (p_i, p_i^*) roughly corresponds to a formant in the spectrum of $H(z)$. Since $H(z)$ should be a stable system, all poles are inside the unit circle in the z-plane. If the poles in the z-plane are well separated, good estimates for formant frequencies and bandwidths are

$$\hat{F}_i = \left(\frac{F_s}{2\pi}\right)\tan^{-1}\left[\frac{\mathrm{Im}(p_i)}{\mathrm{Re}(p_i)}\right] \tag{3.70}$$

$$\hat{B}_i = -\left(\frac{F_s}{\pi}\right)\ln|p_i|, \tag{3.71}$$

where $\hat{F}_i$ represents the ith formant and p_i its corresponding pole in the upper half of the z-plane. Also, $F_s = 1/T$ denotes the sampling frequency in Hz.

For voiced speech, the digital model must also include a section that models the laryngeal shaping filter, $G(z)$. Depending on the purpose of the model, the glottal filter may also be constrained to be an "all-pole" transfer function like (3.69). It is often suggested that the two-pole signal,

$$g(n) = [\alpha^n - \beta^n]u(n), \qquad \beta < \alpha < 1, \quad \alpha \approx 1, \tag{3.72}$$

in which $u(n)$ is the unit step sequence, is an appropriate choice for the impulse response of the filter. In the sense that this pulse can be made to have a similar *spectral magnitude* to many empirical results, this choice is a good one. Its principal benefit is that it does not require zeros to model $G(z)$. However, an "all-pole" impulse response corresponding to any number of poles is incapable of producing realistic pulse shapes observed in many experiments, because it is constrained to be of minimum phase. In particular, it is not possible to produce pulse shapes for which the "opening phase" is "slower" than the "closing phase" (Problem 3). These features of the pulse have been well documented in many papers in the literature (e.g., Timke et al., 1948; Deller, 1983). Many pulse signals have been suggested in the literature; one of the more popular is due to Rosenberg (1971),

$$g(n) = \begin{cases} \frac{1}{2}\left[1 - \cos\left(\frac{\pi n}{P}\right)\right], & 0 \le n \le P \\ \cos\left[\frac{\pi(n-P)}{2(K-P)}\right], & P \le n \le K \\ 0, & \text{otherwise} \end{cases}, \tag{3.73}$$

where P corresponds to the peak time of the pulse and K the time at which complete closure occurs.

The radiation component, $R(z)$ [or, in terms of impedance and "analog" frequencies, $R(\Omega) = Z_{\text{lips}}(\Omega)$], can be thought of as a low-impedance load that terminates the vocal tract and converts the volume velocity at the lips to a pressure wave in the far field. In the discussion of tube models, we assumed this impedance to be real for convenience, but a more accurate model is given by $Z_{\text{lips}}(\Omega)$ such that (Flanagan, 1972)

$$|Z_{\text{lips}}(\Omega)| = \frac{\Omega K_1 K_2.}{\sqrt{K_1^2 + \Omega^2 K_2^2}} \tag{3.74}$$

and

$$\arg\{Z_{\text{lips}}(\Omega)\} = \frac{\pi}{2} - \arctan\left\{\frac{\Omega K_2}{K_1}\right\}, \tag{3.75}$$

where $K_1 = 128/9\pi^2$ and $K_2 = 8r/3\pi c$, with r the radius of the opening in the lips (assumed circular) and c the velocity of sound. From (3.75) it is clear that $Z_{\text{lips}}(\Omega)$ becomes real only asymptotically as the frequency increases, and tends to a purely imaginary quantity as frequencies decrease. However, similarly to the case of the glottal model above, it is often the objective to model only spectral *magnitude* effects. In this case we note that Z_{lips} has a highpass filtering effect: $|Z_{\text{lips}}(0)| = 0$ and $dZ_{\text{lips}}/d\Omega$ tends to remain positive on most practical Nyquist ranges of frequencies (e.g., 0–4000 Hz). A simple digital filter that has these properties is a differencer,

$$R(z) = Z_{\text{lips}}(z) = 1 - z^{-1}. \tag{3.76}$$

This filter has a single zero at $z_o = 1$ in the z-plane. Since there are occasions in which the inverse of this filter, $R^{-1}(z)$, arises in speech modeling, it is customary to decrease the radius of z_o slightly so that the inverse filter will be stable. In this case the model becomes

$$R(z) = 1 - z_o z^{-1}, \qquad z_o \approx 1, \quad z_o < 1. \tag{3.77}$$

A second reason for moving the zero off the unit circle is that with a microphone approximately 30 cm from the speaker's lips, the analysis has not totally left the acoustic near field; therefore low-frequency preemphasis by a full 6 dB/octave is not fully justified (Flanagan, 1972).

We have gone to some lengths above to assure that both $H(z)$ and $G(z)$ could be modeled in some sense with all poles. The inclusion of the single zero z_o in $R(z)$ will destroy the "all-pole" nature of the total model if we do not find a way to "turn the zero into poles." There are two methods for doing so. We believe that the second is preferable for the reader studying these modeling techniques for the first time. The first method of preserving the all-pole structure of the overall model is to argue that

(3.72) is a good model for the glottal dynamics and that one of its poles will cancel the zero z_o in $R(z)$. Although this argument is satisfactory for the experienced speech processing engineer, it is potentially the source of several misconceptions for the person new to the field. We therefore urge the reader to "decouple" the models for $G(z)$ and $R(z)$ and simply note that the model (3.77) can be written

$$R(z) = 1 - z_o z^{-1} = \frac{1}{\sum_{k=0}^{K} z_o^k z^{-k}}, \tag{3.78}$$

where K is theoretically infinite, but practically finite because $z_o < 1$. Therefore, to the extent that (3.77) is a good model for the lip radiation characteristic, we see that this model can be represented by an all-pole filter with a practically finite number of poles.

Let us summarize these important results. Ignoring the technicalities of z-transform existence [see discussion below (3.7)], we assume that the output (pressure wave) of the speech production system is the result of filtering the appropriate excitation by two (unvoiced) or three (voiced) linear, separable filters. If we ignore the developments above momentarily, let us suppose that we know "exact" or "true" linear models of the various components. By this we mean that we (somehow) know models that will exactly produce the speech waveform under consideration. These models are only constrained to be linear and stable and are otherwise unrestricted. In the unvoiced case

$$S(z) = E(z)H(z)R(z), \tag{3.79}$$

where $E(z)$ represents a partial realization of a white noise process. In the voiced case,

$$S(z) = E(z)G(z)H(z)R(z), \tag{3.80}$$

where $E(z)$ represents a discrete-time impulse train of period P, the pitch period of the utterance. In the above, G, H, and R represent the "true" models. Accordingly, the true overall system function is

$$\Theta(z) = \frac{S(z)}{E(z)} = \begin{cases} H(z)R(z), & \text{unvoiced case} \\ G(z)H(z)R(z), & \text{voiced case} \end{cases}. \tag{3.81}$$

With enough painstaking experimental work, we could probably deduce reasonable "true" models for any stationary utterance of interest. In general, we would expect these models to require zeros as well as poles in their system functions. In fact, there are several arguments above against the appropriateness of an all-pole model. If we were asked to deduce a good speech production model, at this point in our study, we have no grounds for endeavoring to make it all-pole. In fact, the contrary is true. Yet, as we have repeatedly indicated above, an all-pole model is often desirable. An important fundamental point for the reader to understand is that when we focus attention on the all-pole model of speech production, we do so with the understanding that there might be a better model if the objective is to *exactly* generate a given speech waveform using its model.

The preoccupation with an *all-pole* model of the speech production system, however, arises from the fact that a very powerful and simple computational technique, linear prediction analysis (studied in Chapter 5), exists for deriving an all-pole model of a given speech utterance. The extracted model will be optimal in a certain sense, but not in the sense of replicating the waveform from which it is derived. If the identified model cannot necessarily replicate the waveform, it is natural to ask, "In what sense is an all-pole model of the speech production system appropriate?" The answer is inherent in the previous discussions. Tracing back through the discussions for each filter section, it will be noticed that in each case (H, G, and R), an argument is made for the appropriateness of an all-pole filter in the sense of preserving the spectral magnitude of the signal. Apparently, an all-pole model exists that will at least produce a waveform with the correct magnitude spectrum. As we have indicated earlier in the chapter, a waveform with correct spectral magnitude is frequently sufficient for coding, recognition, and synthesis.

We should leave this chapter, then, with the following basic understandings. Computational techniques (to be developed in Chapter 5) exist with which we can conveniently identify (find the filter coefficients, or poles, of) an all-pole model from a speech waveform. This all-pole model will be potentially useful if the objective is to model the magnitude spectral characteristics of the waveform. Our knowledge of acoustics might lead us to believe that there is a better "true" model as formulated in (3.81). However, we will be satisfied to compute a model, say $\hat{\Theta}(z)$, which is accurate in terms of the spectrum. In fact, we will show in Chapter 5 that the "true" model, $\Theta(z)$, and the estimated all-pole model, $\hat{\Theta}(z)$, are clearly related. Ideally, of course, $\hat{\Theta}(z)$ and $\Theta(z)$ will have identical magnitude spectra. It will be shown that if enough poles are included in $\hat{\Theta}(z)$, then the all-pole model will be shown to be the *minimum-phase* part of $\Theta(z)$. This is entirely reasonable, since $\hat{\Theta}(z)$ has all of its singularities inside the unit circle and therefore must be minimum phase.

3.2.3 Other Speech Models

In this chapter, we have discussed several schemes for modeling human speech production. The initial goal was to develop a modeling approach that would match as closely as possible the resonant structure (formant locations) of the human vocal tract that produced the corresponding speech sound. This notion was motivated by earlier scientists who capitalized on their knowledge of the physics of sound propagation in musical instruments such as pipe organs. Early findings prompted the development of mechanical and electromechanical speaking machines. The acoustic tube, transmission line, and digital filter models presented here all assume planar wave propagation along the vocal-tract axis (from glottis to lips). The majority of the discussion has assumed lossless conditions, with a stationary point and type of excitation. The solution for addressing losses normally entails a modification to the output vocal-tract frequency response of a lossless model. The point and type of excitation generally varies across a word or phrase. The general discrete-time speech model presented in Section 3.2.2 reflects a generally accepted means for characterizing the changing speech characteristics for the purposes of coding, synthesis, enhancement, and recognition. Later, in Chapter 5, we will present a fast algorithm for obtaining speech parameters for this model given a short interval of speech. The extraction of these "linear prediction parameters" opens many application areas of interest for the reader. However, we reemphasize that although this is the generally accepted model, it is more of a digital filter representation of the speech waveform than a true model of speech production. Here, we consider several issues that are not adequately addressed in the classical digital filter model, but are submitted as issues for the reader to reflect upon.

The classical digital filter model in Fig. 3.26 provides a switch to distinguish between voiced and unvoiced speech production. This is a serious limitation for several reasons. First, human speech production does not require voicing to turn off immediately prior to an unvoiced phoneme. In addition, several phonemes, such as voiced fricatives (/z/, /v/), possess two sources of excitation, vocal fold movement and a major constriction resulting in both voiced and unvoiced forms of excitation. Since the model is based on acoustic tube theory assuming short-term stationarity, it lacks the ability to characterize rapidly changing excitation properties such as that found in plosive sounds like /t/ and /b/. The method for handling such sounds is to assume short-term stationarity and "hope for the best." Further research is under way to develop a more realistic means of characterizing excitation properties. This work has resulted in several alternative coding schemes, which are discussed in Chapter 7. Unfortunately, these methods have generally been introduced to address the existing limitations of the present digital filter model for speech coding and have only partially addressed the need for formulating improved human speech production models.

In the discussion of speech modeling, this chapter has assumed a linear source/filter model for speech production assuming planar sound propagation [see Fig. 3.27(a)]. This requires excitation at the glottis to be decoupled from the vocal tract. We should expect that coupling of the excitation source and vocal tract influences the resulting speech signal and that such coupling should therefore be incorporated in production modeling. In addition, the linear source/filter model assumes that as the vocal

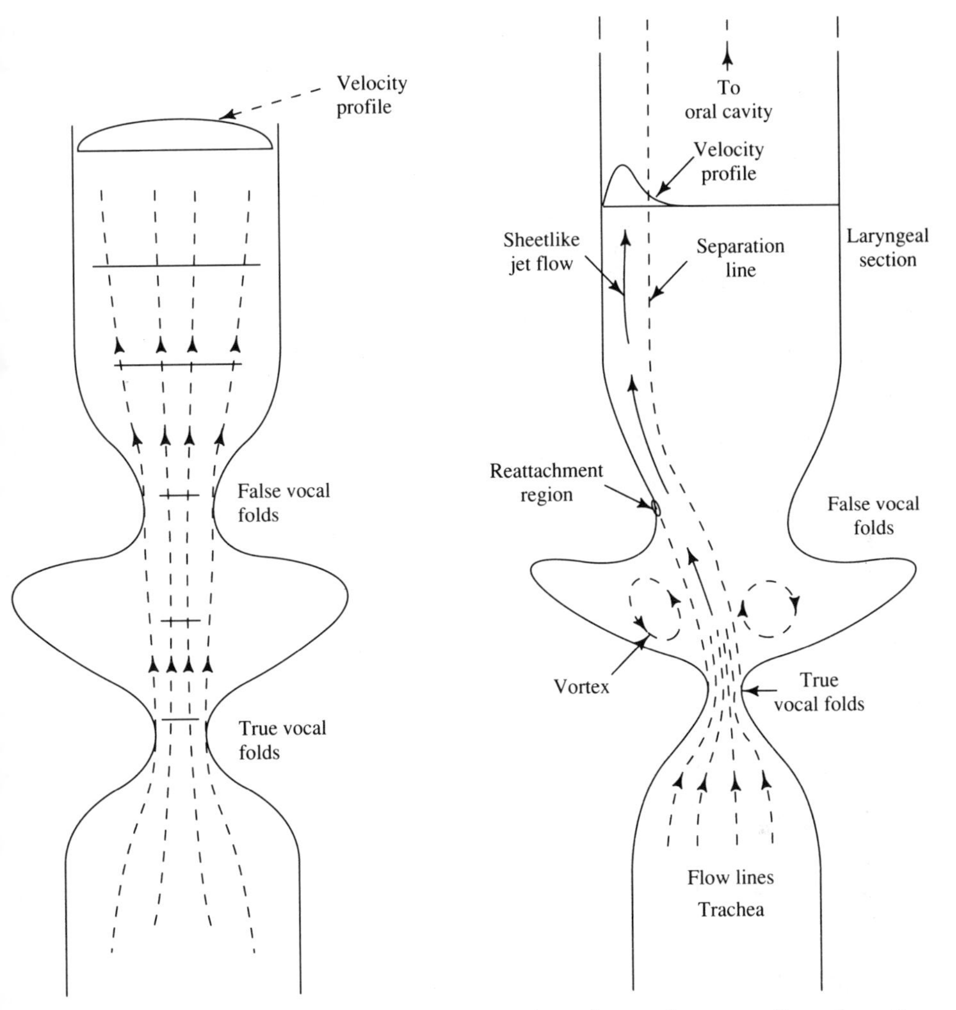

FIGURE 3.27. (a) The classical interpretation of sound propagation along the vocal system. (b) A nonlinear fluid dynamics interpretation of sound propagation along the vocal tract.

folds begin to open, the propagating sound pressure wave expands to fill the cross-sectional area of the vocal tract. However, the physics of the larynx suggest that this cannot occur. Sound pressure and volume velocity measurements within cast models of the human speech system suggest that the nonlinear fluid dynamics as illustrated in Fig. 3.27(b) to be a more realistic means of characterizing sound propagation along the vocal tract. Studies by Teager (1980, 1983, 1989) suggest that the vortices located in the false vocal folds provide the necessary source of excitation during the closed phase of the vocal folds. It has also been suggested that the propagating sound wave adheres to the vocal-tract walls, resulting in laminar flow. A complete solution for such a nonlinear fluid dynamic model requires the solution of the Navier–Stokes equations and has been achieved for a stationary phoneme; however, it may not be computationally feasible to obtain a solution for a time-varying speech model.

The foundation of our developments has been the acoustical analysis of a sequence of acoustic tubes. We have shown that such a system is all-pole and that it is capable of modeling the formant structure of resonant phonemes like vowels. This early discussion, combined with the foreknowledge that the powerful linear prediction method is available to deduce all-pole models of any speech sound, has predisposed our discussion to favor poles and to devote little attention to representing spectral zeros in the model. This is especially important for nasals and useful for addressing model limitations for fricatives and plosives. *Autoregressive—moving average* (ARMA), or pole–zero [e.g., (Proakis and Manolakis, 1992; Johnson, 1985)] methods have been considered for speech. The choice of the numbers of poles and zeros is an important issue in both analysis and synthesis settings. Related to the use of zeros in speech modeling is the characterization of losses in the vocal tract. Losses generally affect formant bandwidth more than location. The use of additional zeros can sometimes provide the necessary modeling effects in the frequency domain.

Although the issues discussed above are important for accurate speech production modeling, in many applications the general digital model in Fig. 3.26 is more than sufficient. In considering a research problem in speech processing, however, the reader should employ this digital model with the knowledge of its limitations. We have alluded to some of these limitations at the end of the preceding section and will return to these important issues in Chapter 5.

3.3 Conclusions

This chapter has covered some of the basic acoustic theory necessary to build important models used pervasively in speech. Decades of research have gone into analog acoustic modeling efforts which have been so central to the technical and quantitative understanding of speech production. Yet, in spite of some lengthy developments, we have only treated this analog research in very superficial terms. Fortunately for our immediate purposes, a deep and detailed study of this rich and varied research field is unnecessary. The reader who becomes seriously involved in the speech processing field will benefit from a more detailed study of original research papers and texts.

Remarkably, a cursory brief overview of the acoustical theory has been sufficient to develop what will emerge in our work as the fundamental basis of much of contemporary speech processing—the discrete-time model of Fig. 3.26. We will find this simple model at the heart of many of the developments in the remaining chapters of the book. One of the most important conclusions taken from this chapter should be an appreciation of how the discrete-time model is a direct descendant of a quite näive physical model, the lossless tube tract model with some simple glottal and lip models at the terminals. In future discussions, it will be useful to remember the model from whence the discrete-time model was derived. Doing so will add much insight into the appropriateness and weaknesses of the model in various applications.

3.4 Problems

3.1. (a) Verify that the reflection coefficient at any tube juncture in an acoustic tube model is bounded in magnitude by unity,

$$-1 \leq \rho_k \leq 1.$$

(b) Suppose that we were to implement a digital filter realization of an N-tube acoustic model. It is also known that high-order digital filter realizations sometimes suffer from issues relating to finite precision arithmetic. Does the result of part (a) suggest a means of testing the N-tube digital filter realization for stability?

3.2. (a) Consider a two-tube lossless vocal-tract model (see Figure 3.17). Draw a signal flow diagram using reflection coefficients and delay elements for the case in which $A_1 = 1\ \text{cm}^2, l_1 = 9\ \text{cm}$, $A_2 = 7\ \text{cm}^2$, and $l_2 = 8\ \text{cm}$. Include glottal and lip radiation effects. What phoneme might this model represent?

(b) Repeat for dimensions $A_1 = 0.9\ \text{cm}^2, l_1 = 9.5\ \text{cm}, A_2 = 0.25\ \text{cm}^2$, $l_2 = 2\ \text{cm}, A_3 = 0.5\ \text{cm}^2, l_3 = 5\ \text{cm}$.

3.3. Consider the following tube model for the vocal tract:

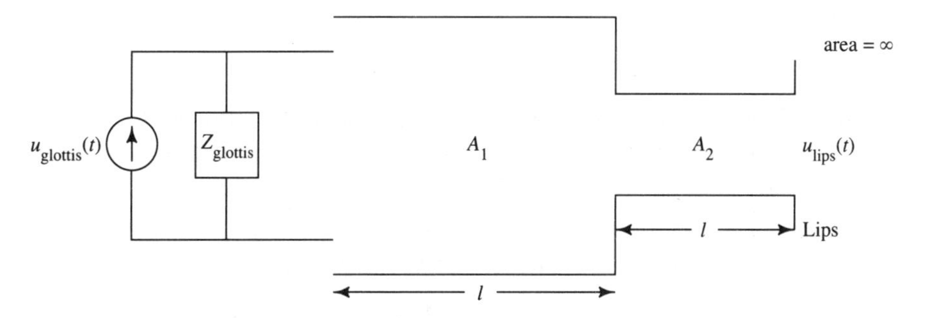

The lips have zero radiation impedance.

$u_{\text{glottis}}(t)$ = volume velocity at the source

Z_{glottis} = real glottal source impedance

A_1, A_2 = area of tubes

This ideal vocal tract is modeled by

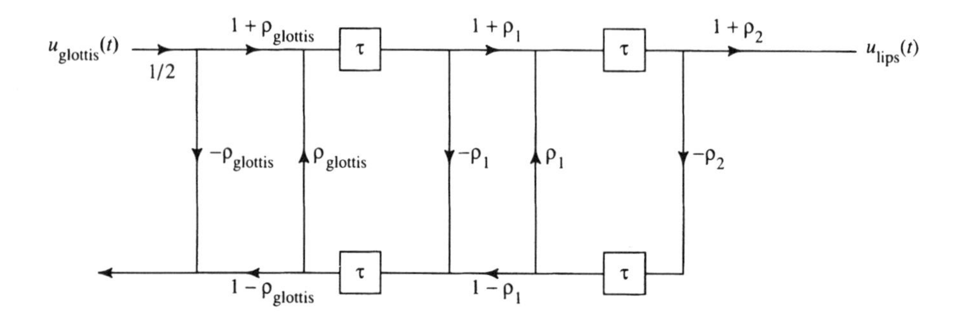

(a) Compute the ideal vocal-tract values ρ_{glottis}, ρ_1, ρ_2, and τ in terms of tube model parameters.

(b) Let this model correspond to a digital system with a sample period of 2τ.

Find the z-transform of the transfer function $\dfrac{U_{\text{lips}}(z)}{U_{\text{glottis}}(z)}$.

(c) Draw a possible set of pole and zero locations of $\dfrac{U_{\text{lips}}(z)}{U_{\text{glottis}}(z)}$ if

1. $Z_{\text{glottis}} = \infty$.

2. $Z_{\text{glottis}} = \dfrac{\rho c}{A}$.

3.4. The effects of lip radiation are often modeled by a single zero system

$$R(z) = 1 - z_o z^{-1}, \qquad z_o \approx 1, \quad z_o < 1.$$

(a) Plot a typical frequency response for such a lip model.

(b) Discuss the importance of such a model as the distance between the speakers' lips and the microphone is varied.

3.5. (Computer Assignment) The glottal shaping filter is often modeled by a two-pole system whose impulse response is given by

$$g(n) = [\alpha^n - \beta^n]u(n), \qquad \beta < \alpha < 1, \quad \alpha \approx 1.$$

(a) Plot several impulse responses for various values of α and β. Make sure that your pulse is practically "closed" at time $n = 64$ in every case.

(b) Based on 128-point DFTs, plot the magnitude and phase responses for two typical examples in part (a). As we discussed in the chapter, more realistic glottal pulse shapes can be obtained using the Rosenberg pulse

$$g(n) = \begin{cases} \frac{1}{2}[1 - \cos(\pi n/P)], & 0 \le n \le P \\ \cos\left([\pi(n-P)]/[2(K-P)]\right), & P \le n \le K. \\ 0, & \text{otherwise} \end{cases}$$

(c) Plot several impulse responses for various values of P and K. Make sure that your pulse is "closed" at time $n = 64$ in every case.

(d) Based on 128-point DFTs, plot the magnitude and phase responses for two typical examples in part (c).

(e) Compare the time-domain and frequency-domain properties of the two glottal pulse models. What realistic time-domain feature of the Rosenberg pulse is apparently unachievable with the two-pole model? How is this feature manifest in the phase spectrum?

APPENDICES

Reading Note: These appendices provide supplemental material on acoustical analysis of speech. In mathematical developments involving steady-state analysis, we omit the argument Ω from phasors and impedances unless required for clarity.

3.A Single Lossless Tube Analysis

3.A.1 Open and Closed Terminations

Open Tube. We derive solutions for the volume velocity and pressure waveforms at distance x in the lossless uniform tube which is open at the termination, and conclude that (3.20) is correct.

Replacing the cross-sectional area function $A(x,t)$ with A in (3.16) and (3.17), and setting $\partial A(x,t)/\partial t = 0$ yields

$$\frac{-\partial v(x,t)}{\partial x} = \frac{1}{\rho c^2} A \frac{\partial p(x,t)}{\partial t} \tag{3.83}$$

$$\frac{-\partial p(x,t)}{\partial x} = \rho \frac{1}{A} \frac{\partial v(x,t)}{\partial t}. \tag{3.84}$$

The boundary conditions (and rationale for them) are given by (3.18) and (3.19) and are repeated here for convenience:

$$v(0,t) = u_{\text{glottis}}(t) = \overline{U}_{\text{glottis}} e^{j\Omega t} \tag{3.85}$$

$$p(l,t) = p_{\text{lips}}(t) = 0. \tag{3.86}$$

Since (3.83) and (3.84) are linear constant coefficient differential equations, their solutions will have the form[14]

$$\begin{aligned} v(x,t) &= v^{+}(x,t) - v^{-}(x,t) \\ &= \frac{A}{\rho c}\left[\Psi^{+}(\Omega)e^{j\Omega t}e^{-j\Omega(x/c)} - \Psi^{-}(\Omega)e^{j\Omega t}e^{+j\Omega(x/c)}\right] \end{aligned} \tag{3.87}$$

and

$$\begin{aligned} p(x,t) &= p^{+}(x,t) + p^{-}(x,t) \\ &= \Psi^{+}(\Omega)e^{j\Omega t}e^{-j\Omega(x/c)} + \Psi^{-}(\Omega)e^{j\Omega t}e^{+j\Omega(x/c)}, \end{aligned} \tag{3.88}$$

where $v^{+}(x,t)$ and $v^{-}(x,t)$ represent traveling waves in the positive and negative directions, respectively, and similarly for $p^{+}(x,t)$ and $p^{-}(x,t)$. The numbers $\Psi^{+}(\Omega)$ and $\Psi^{-}(\Omega)$ are complex numbers that are dependent upon Ω but not x (the effects of x are accounted for in the additional phase factors). Note that since c represents the speed of sound in air, $+x/c$ represents the amount of time needed for a positive moving sound wave to travel x units along the tract (to the right), while $-x/c$ represents the time required for a negative traveling wave to move x units to the left. Hence an additional delay (or phase) factor appears with each term to compensate for the shift along the x dimension. It is also worth noting that for a tube of length l, the delay

$$\tau \stackrel{\text{def}}{=} \frac{l}{c} \tag{3.89}$$

represents the elapsed time for a plane wave to traverse the entire length of the tube.

Applying the boundary conditions to (3.87) and (3.88), or equivalently, writing phasor equations at the boundaries, results in the following relations, which can be solved for the complex amplitudes,

$$\frac{A}{\rho c}\left[\Psi^{+}(\Omega) - \Psi^{-}(\Omega)\right] = \bar{U}_{\text{glottis}} \tag{3.90}$$

$$\Psi^{+}(\Omega)e^{-j\Omega(l/c)} + \Psi^{-}(\Omega)e^{+j\Omega(l/c)} = 0. \tag{3.91}$$

The solutions are

$$\Psi^{+}_{\text{open}}(\Omega) = \bar{U}_{\text{glottis}}\frac{\rho c}{A}\left[\frac{e^{-j\Omega(l/c)}}{2\cos(\Omega l/c)}\right] \tag{3.92}$$

[14]This is just a generalization of the material discussed in Section 3.1.4 to the case in which the additional independent argument x is present. For details the reader may consult any standard textbook on differential equations [e.g., (Boyce and DiPrima, 1969)] or a physics or engineering textbook treating wave equations.

$$\Psi^-_{\text{open}}(\Omega) = -\bar{U}_{\text{glottis}} \frac{\rho c}{A} \left[\frac{e^{-j\Omega(l/c)}}{2 \cos(\Omega l/c)} \right]. \tag{3.93}$$

From (3.87) and (3.88) the steady-state solutions for the volume velocity and pressure at distance x from the origin in the uniform lossless tube of length l are

$$v(x, t) = \bar{U}_{\text{glottis}} \left[\frac{\cos(\Omega(l-x)/c)}{\cos(\Omega l/c)} \right] e^{j\Omega t} \tag{3.94}$$

$$p(x, t) = j \frac{\rho c}{A} \bar{U}_{\text{glottis}} \left[\frac{\sin(\Omega(l-x)/c)}{\cos(\Omega l/c)} \right] e^{j\Omega t}. \tag{3.95}$$

Equation (3.20) follows immediately from (3.94) upon letting $x = l$.

Closed Tube. We derive solutions for the volume velocity and pressure waveforms at distance x in the lossless uniform tube which is closed at the termination.

The boundary conditions are (3.85) and

$$v(l, t) = u_{\text{lips}}(t) = 0. \tag{3.96}$$

Applying these conditions to (3.87) and (3.88) results in (3.90) and

$$\frac{A}{\rho c} \left[\Psi^+(\Omega) e^{-j\Omega(l/c)} - \Psi^-(\Omega) e^{j\Omega(l/c)} \right] = 0. \tag{3.97}$$

Therefore, the complex amplitudes of the traveling waves in the closed-tube case are

$$\Psi^+_{\text{closed}}(\Omega) = \bar{U}_{\text{glottis}} \frac{\rho c}{A} \left[\frac{e^{j\Omega(l/c)}}{j2 \sin(\Omega l/c)} \right] \tag{3.98}$$

$$\Psi^-_{\text{closed}}(\Omega) = \bar{U}_{\text{glottis}} \frac{\rho c}{A} \left[\frac{e^{-j\Omega(l/c)}}{j2 \sin(\Omega l/c)} \right], \tag{3.99}$$

from which the steady-state solutions for volume velocity and sound pressure for the closed tube case follow,

$$v(x, t) = \bar{U}_{\text{glottis}} \left[\frac{\sin(\Omega(l-x)/c)}{\sin(\Omega l/c)} \right] e^{j\Omega t} \tag{3.100}$$

$$p(x, t) = j \frac{\rho c}{A} \bar{U}_{\text{glottis}} \left[\frac{\cos(\Omega(l-x)/c)}{\sin(\Omega l/c)} \right] e^{j\Omega t}. \tag{3.101}$$

3.A.2 Impedance Analysis, T-Network, and Two-Port Network

Input Impedances. The acoustic impedance of the tube at distance x from the glottis is defined as

$$Z(x, \Omega) \stackrel{\text{def}}{=} \frac{p(x,t)}{v(x,t)} = \frac{\bar{P}(x,\Omega)}{\bar{V}(x,\Omega)}, \tag{3.102}$$

where $p(x,t)$ and $v(x,t)$ are steady-state solutions for pressure and volume velocity as described in the main text, and $\bar{P}$ and $\bar{V}$ are the corresponding phasor representations. Also by analogy to electrical impedance, we define the *acoustic characteristic impedance* of the uniform tube by

$$Z_o \stackrel{\text{def}}{=} \frac{\rho c}{A}. \tag{3.103}$$

The *acoustic admittance*, $Y(x, \Omega)$, and *acoustic characteristic admittance*, Y_o, are defined as the respective reciprocals. Using steady-state solutions developed in Appendix 3.A.1, we can immediately derive expressions for the input acoustic impedances of the open and closed tubes. For the open termination, we use (3.20) above and (3.95) in Appendix 3.A.1 to obtain

$$Z_{\substack{\text{in}\\ \text{open}}} = \frac{p(0,t)}{v(0,t)} = jZ_o \tan\left(\frac{\Omega l}{c}\right) = Z_o \tanh(\Omega\tau). \tag{3.104}$$

where τ is defined as the time required for a plane wave to travel the entire length of the tube [see (3.89)]. In a similar manner we find from (3.100) and (3.101) in Appendix 3.A.1 that

$$Z_{\substack{\text{in}\\ \text{closed}}} = \frac{p(0,t)}{v(0,t)} = -jZ_o \cot\left(\frac{\Omega l}{c}\right) = Z_o \coth(\Omega\tau). \tag{3.105}$$

Determination of Resonances Using $Z(x, \Omega)$. One method of determining the resonant modes of a transmission line is to determine the frequencies for which the impedance approaches infinity (or admittance approaches zero). We show first that an equivalent procedure can be carried out with acoustic impedance. In a manner similar to the derivation of (3.104) and (3.105) we find that for a general distance x along the tube, the impedances for the open and closed termination cases are as follows:

$$Z_{\substack{\text{open}\\ \text{tube}}}(x, \Omega) = Z_o \tanh\left[\Omega\left(\tau - \frac{x}{c}\right)\right] \tag{3.106}$$

and

$$Z_{\substack{\text{closed}\\\text{tube}}}(x, \Omega) = Z_o \coth\left[\Omega\left(\tau - \frac{x}{c}\right)\right]. \tag{3.107}$$

For the open tube, resonances of the tube can be obtained by determining frequencies for which the magnitude of the impedance in (3.106) approaches infinity. This occurs when $\Omega[\tau - (x/c)]$ is equal to odd multiples of $\pi/2$,

$$\Omega_i\left(\tau - \frac{x}{c}\right) = \frac{\pi}{2}(2i - 1) \qquad \text{for } i = 1, 2, 3, 4, \ldots. \tag{3.108}$$

Upon substituting for overall tube delay τ, resonances for an open tube occur at the following frequencies:

$$F_i = \frac{c}{4}\frac{2i - 1}{l - x} \qquad \text{for } i = 1, 2, 3, 4, \ldots, \tag{3.109}$$

where x represents the location along the tube. If the resonance structure looking into the tube at the glottis is desired, the position is set to zero ($x = 0$) and constants for tube length are substituted ($l = 17.5$ cm, $c = 350$ m/sec). This results in resonances of 500 Hz, 1500 Hz, 2500 Hz, ... for an open tube, which is equivalent to those determined by setting the denominator equal to zero in (3.21).

T-Network and Two-Port Network Models of the Lossless Tube. The transfer function $H(\Omega)$ in (3.21) was obtained by taking the ratio of volume velocity at the lips to that at the glottis. This is equivalent to a ratio of output to input current for a two-port electrical network. Under steady-state conditions, the uniform tube section can be replaced in terms of external observerable quantities by a T-network of impedances shown in Fig. 3.28. The impedances of this network are described using the following equations and are independent of the tube terminations (i.e, the boundary conditions) (Potter and Fich, 1963; Mason, 1948; Johnson, 1924):

$$Z_1 = Z_o \tanh\left(\Gamma \frac{l}{2}\right) \tag{3.110}$$

$$Z_2 = Z_o \operatorname{csch}(\Gamma l) \tag{3.111}$$

$$Z_o = \left[\frac{R + j\Omega L}{G + j\Omega C}\right]^{1/2} \tag{3.112}$$

$$\Gamma = [(R + j\Omega L)(G + j\Omega C)]^{1/2}. \tag{3.113}$$

In these equations, l represents the length of the transmission line section (or acoustic tube), and $R, G, L,$ and C the distributed resistance, admit-

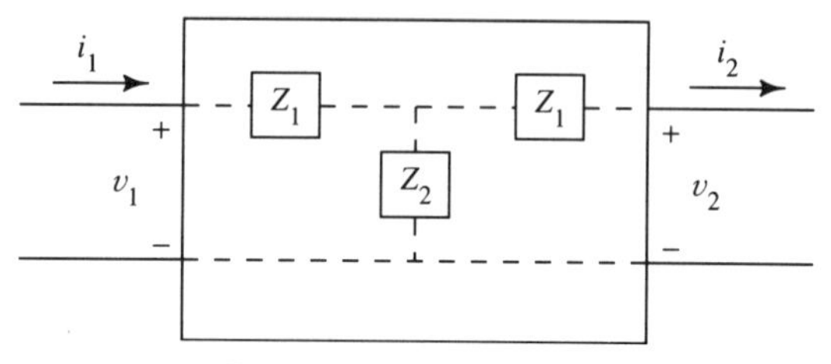

Transmission line T-network

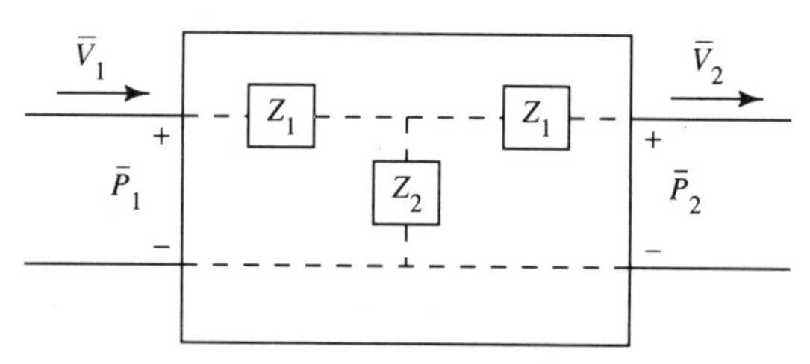

Acoustic tube T-network

FIGURE 3.28. A uniform acoustic tube section with a plane wave that propagates through it can be represented by a T-network with impedances that are hyperbolic functions of frequency.

tance, inductance, and capacitance per unit length of line. Z_o is the characteristic impedance of the line, and Γ is the propagation constant. In this section we neglect dissipative terms R and G representing viscous resistance and energy absorption by the tube walls. Substituting acoustic equivalents for L and C,

$$L = \frac{\rho}{A} \qquad C = \frac{A}{\rho c^2}, \tag{3.114}$$

results in the following impedance relations for the two-port network,

$$Z_o = \frac{\rho c}{A} \qquad \Gamma = \frac{j\Omega}{c} \tag{3.115}$$

$$Z_1 = \frac{\rho c}{A} \tanh\left(\frac{j\Omega l}{2c}\right) = j\frac{\rho c}{A} \tan\left(\frac{\Omega l}{2c}\right) \tag{3.116}$$

$$Z_2 = \frac{\rho C}{A} \operatorname{csch}\left(\frac{j\Omega l}{c}\right) = -j\frac{\rho c}{A} \csc\left(\frac{\Omega l}{c}\right). \tag{3.117}$$

The impedances are therefore reduced to only reactive components, representing a lossless uniform acoustic tube or section of transmission line.

The T-equivalent acoustic circuit employing acoustic quantities is shown in Fig. 3.28, where volume velocity and sound pressure have been substituted for current and voltage.

It is also possible to use the T-network analysis to derive a two-port network model of the lossless tube. To achieve this, it is necessary to reconsider the input–output relations of the acoustic tube T-network of Fig. 3.28. In an assumed steady-state condition, let us denote the phasors associated with the "boundary" volume velocity waveforms as $\bar{V}_1$ and $\bar{V}_2$, meaning that

$$v(0, t) = \bar{V}_1 e^{j\Omega t} \tag{3.118}$$

and

$$v(l, t) = \bar{V}_2 e^{j\Omega t}. \tag{3.119}$$

Similar definitions are given to $\bar{P}_1$ and $\bar{P}_2$ for the pressure waveforms. The T-network is characterized by the following system of equations:

$$\begin{bmatrix} \bar{P}_1 \\ \bar{P}_2 \end{bmatrix} = \begin{bmatrix} Z_{11} & Z_{12} \\ Z_{21} & Z_{22} \end{bmatrix} \begin{bmatrix} \bar{V}_1 \\ \bar{V}_2 \end{bmatrix}. \tag{3.120}$$

The impedances Z_{11}, Z_{21}, Z_{12}, and Z_{22} are found using the two sound pressure equations from the T-network

$$\bar{P}_1 = \bar{V}_1 Z_1 + (\bar{V}_1 - \bar{V}_2) Z_2 \tag{3.121}$$

$$\bar{P}_2 = (\bar{V}_1 - \bar{V}_2) Z_2 - \bar{V}_2 Z_1. \tag{3.122}$$

Substituting T-network impedances $Z_1 = Z_o \tanh(\gamma l/2), Z_2 = Z_o \operatorname{csch}(\gamma l)$, and noting $\gamma = j\Omega/c$ and $\tau = l/c$, results in the following relations:

$$\bar{P}_1 = \bar{V}_1 [Z_1 + Z_2] - \bar{V}_2 Z_2 \tag{3.123}$$

$$= \bar{V}_1 \left[Z_o \tanh\left(\frac{s\tau}{2}\right) + Z_o \operatorname{csch}(s\tau) \right] - \bar{V}_2 Z_o \operatorname{csch}(s\tau) \tag{3.124}$$

$$\bar{P}_2 = \bar{V}_1 Z_2 - \bar{V}_2 [Z_1 + Z_2] \tag{3.125}$$

$$= \bar{V}_1 Z_o \operatorname{csch}(s\tau) - \bar{V}_2 \left[Z_o \tanh\left(\frac{s\tau}{2}\right) + Z_o \operatorname{csch}(s\tau) \right], \tag{3.126}$$

where we have replaced $j\Omega$ by the general complex variable s. The two-port network impedances Z_{12} and Z_{21} are seen to be

$$Z_{12} = -Z_o \operatorname{csch}(s\tau) \tag{3.127}$$

$$Z_{21} = Z_o \operatorname{csch}(s\tau). \tag{3.128}$$

After some simplification (the proof is left as an exercise), the remaining two-port impedances are found to be

$$Z_{11} = Z_o \coth(s\tau) \tag{3.129}$$

$$Z_{22} = -Z_o \coth(s\tau). \tag{3.130}$$

The resulting two-port impedance matrix can be written as

$$\mathbf{Z} = \begin{bmatrix} Z_{11} & Z_{12} \\ Z_{21} & Z_{22} \end{bmatrix} = Z_o \begin{bmatrix} \coth(s\tau) & -\text{csch}(s\tau) \\ \text{csch}(s\tau) & -\coth(s\tau) \end{bmatrix}. \tag{3.131}$$

Due to the defined direction of the output volume velocity $\bar{V}_2$, the two-port impedance matrix $\mathbf{Z}$ is not symmetric. If the output volume velocity were redefined as traveling into the network (as $\bar{V}_1$), then the signs of Z_{12} and Z_{22} would be positive. The present convention was chosen for ease in concatenation of multiple two-port networks.

Thus far, the system of equations in (3.120) has been solved for a given input and output volume velocity. For purposes of concatenation, it will be desirable to rearrange these equations so that output volume velocity and sound pressure can be determined from input velocity and pressure. From the two-port network, we have the following equations:

$$\bar{P}_1 = \bar{V}_1 Z_{11} + \bar{V}_2 Z_{12} \tag{3.132}$$

$$\bar{P}_2 = \bar{V}_1 Z_{21} + \bar{V}_2 Z_{22}. \tag{3.133}$$

Solving for output pressure and velocity, we obtain

$$\bar{P}_2 = \bar{P}_1 \left[\frac{Z_{22}}{Z_{21}} \right] + \bar{V}_1 \left[Z_{21} - \frac{Z_{22} Z_{11}}{Z_{12}} \right] \tag{3.134}$$

$$\bar{V}_2 = \bar{P}_1 \left[\frac{1}{Z_{12}} \right] + \bar{V}_1 \left[-\frac{Z_{11}}{Z_{12}} \right]. \tag{3.135}$$

Substituting the two-port impedances Z_{11}, Z_{12}, Z_{21}, and Z_{22} and simplifying produces the final two-port input–output matrix relation

$$\begin{bmatrix} \bar{P}_2 \\ \bar{V}_2 \end{bmatrix} = \begin{bmatrix} \cosh(s\tau) & -Z_o \sinh(s\tau) \\ (-1/Z_o)\sinh(s\tau) & \cosh(s\tau) \end{bmatrix} \begin{bmatrix} \bar{P}_1 \\ \bar{V}_1 \end{bmatrix}. \tag{3.136}$$

This system of equations provides a means for obtaining characteristics of output pressure and velocity given input quantities and the tube properties of characteristic impedance Z_o and delay τ.

3.B Two-Tube Lossless Model of the Vocal Tract

The two-tube model offers an intermediate step between the single-tube section and the general N-tube model. We can gain greater insight into the interaction of the tract cavities by approximating the vocal tract as a cascade of two tubes with different cross-sectional areas. This is particularly useful for modeling voiced sounds such as vowels. Let us consider a two-tube model configuration for a back vowel (such as /i/ in "heed"). The vocal-tract requires a large or wide open pharyngeal cavity and a narrow oral cavity. The logical acoustic tube model for such a phoneme is to use a large back tube (length l_{ph}, area A_{ph}) for the pharyngeal cavity, followed by a narrow front tube (length l_{oral}, area A_{oral}) that models the oral cavity. Assuming the same sinusoidal volume velocity source (piston in Fig. 3.12), we obtain the two-tube model shown in Fig. 3.29. In order to keep the analysis computationally tractable, we assume that the impedance of the glottis is high compared to the impedance of the vocal tract, and that the impedance outside the lips (load impedance) is negligible when compared to the impedance of the lips and oral cavity. Three methods of analysis will be considered for this two-tube model:

1. Formant location analysis.
2. T-network circuit analysis.
3. Two-port network analysis.

Our goal in each of these analysis methods is to ensure that proper lengths $(l_{\text{ph}}, l_{\text{oral}})$ and cross-sectional areas $(A_{\text{ph}}, A_{\text{oral}})$ are chosen to approximate the desired phoneme's formant locations.

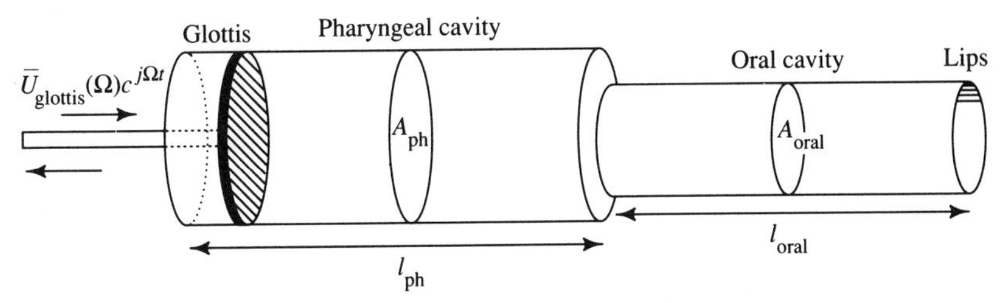

FIGURE 3.29. A two-tube model for the human vocal tract.

Formant Location Analysis. Although an overall transfer function is desirable, there are times when only the knowledge of formant frequencies is necessary in analyzing vocal-tract tube models. We have seen that a single- or multi-tube acoustic model can be viewed as a single- or multi-section transmission line network. The two-tube model in Fig. 3.29 can therefore be regarded as a network of two transmission line circuits, joined at the tube junction in a parallel fashion. If the circuit is broken at this junction, the impedance characteristics of the tubes can be represented by obtaining the input impedance Z_{ph} looking back into the pharyngeal tube, and the impedance Z_{oral} looking forward into the oral cavity tube. The resonances or poles of the general parallel circuit will occur at those frequencies where their admittances sum to zero:

$$\sum_{i=1}^{N} Y_i = \sum_{i=1}^{N} \frac{1}{Z_i} = 0. \tag{3.137}$$

The poles of the parallel circuit can therefore be used to determine the frequencies at which the admittance approaches zero, or impedance becomes infinite. Our analysis continues by determining the input admittance for the pharyngeal and oral cavity tubes (Fig. 3.30). In order to avoid coupling effects between the pharyngeal cavity and subglottal structures (lungs, trachea, larynx), we consider the analysis during the period when the vocal folds are completely closed. The input admittance looking back into the pharyngeal tube is found using the impedance of a closed tube of length l_{ph} [from (3.60)],

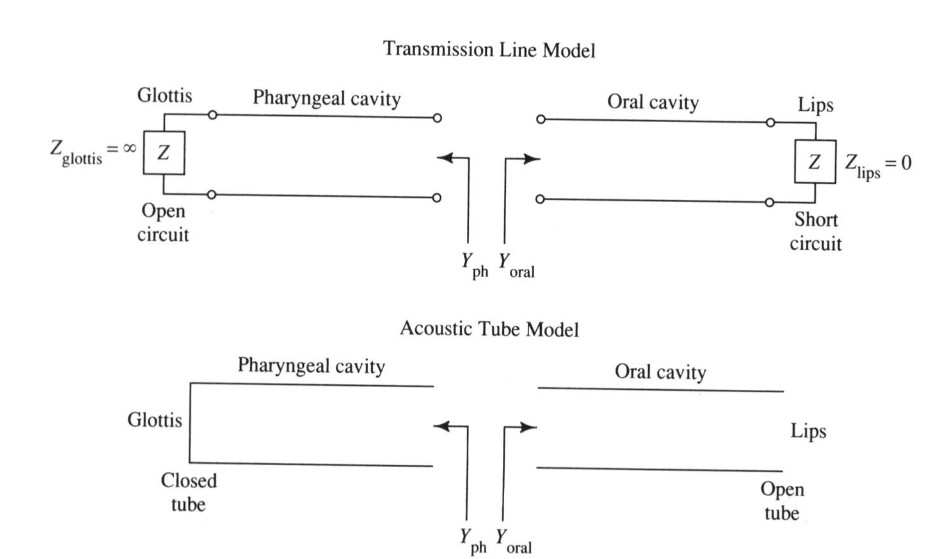

FIGURE 3.30. Impedance analysis of an acoustic two-tube model for the vocal-tract.

$$Y_{\text{ph}} = \frac{1}{Z_{\text{ph}}} = \frac{1}{-jZ_o \cot(\Omega l_{\text{ph}}/c)}$$
$$= j\frac{A_{\text{ph}}}{\rho c}\tan\left(\frac{\Omega l_{\text{ph}}}{c}\right). \tag{3.138}$$

The oral cavity admittance is found using the input impedance of an open tube [from (3.59)],

$$Y_{\text{oral}} = \frac{1}{Z_{\text{oral}}} = \frac{1}{jZ_o \tan(\Omega l_{\text{oral}}/c)}$$
$$= -j\frac{A_{\text{oral}}}{\rho c}\cot\left(\frac{\Omega l_{\text{oral}}}{c}\right). \tag{3.139}$$

From (3.137), the poles occur at frequencies Ω where $Y_{\text{ph}} + Y_{\text{oral}} = 0$, or

$$A_{\text{ph}} \tan\left(\frac{\Omega l_{\text{ph}}}{c}\right) = A_{\text{oral}} \cot\left(\frac{\Omega l_{\text{oral}}}{c}\right). \tag{3.140}$$

A graphical solution for the resonant frequencies can be obtained by plotting $A_{\text{ph}} \tan(\Omega l_{\text{ph}}/c)$ and $A_{\text{oral}} \cot(\Omega l_{\text{oral}}/c)$ and noting the points where the two functions are equal. Figure 3.31 illustrates the graphical solution for a two-tube model for the vowel /i/ with the tube dimensions $l_{\text{ph}} = 9$ cm, $A_{\text{ph}} = 8\ \text{cm}^2$, $l_{\text{oral}} = 6$ cm, and $A_{\text{oral}} = 1\ \text{cm}^2$. This particular model produces resonances at $F_1 = 250$, $F_2 = 1875$, and $F_3 = 2825$ Hz. Typical real vowel formants occur at 270, 2290, and 3010 Hz.

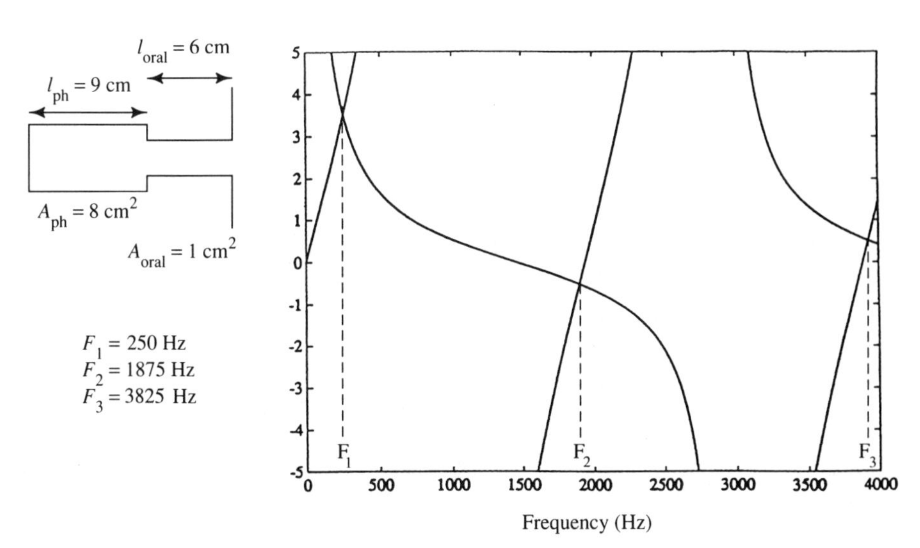

FIGURE 3.31. A two-tube approximation for the vowel /i/ and a corresponding graphical solution for its formant frequencies.

Transmission Line Analysis. Although characterization of formant locations is important in multi-acoustic tube analysis, it does not fully represent the frequency response of the model. If we use the two-tube model in Fig. 3.29, an equivalent electrical circuit consisting of T-network sections from transmission line theory can be constructed. The result is shown in Fig. 3.32. The impedance at the lips is assumed to be zero (open tube, short-circuited line), and that at the glottis to be infinite (closed tube, open-circuited line). Defining $\bar{V}_{12}$ to be the phasor for the steady-state volume velocity at the juncture between the pharyngeal and oral tubes, the two loop equations are,

$$\bar{U}_{\text{lips}} Z_{1A} = Z_{2B}[\bar{V}_{12} - \bar{U}_{\text{lips}}] \tag{3.141}$$

$$[\bar{U}_{\text{glottis}} - \bar{V}_{12}] Z_{1B} = \bar{V}_{12}[Z_{1A} - Z_{2A}] + \bar{U}_{\text{lips}} Z_{2A}. \tag{3.142}$$

Equation (3.141) can then be solved for $\bar{V}_{12}$,

$$\bar{V}_{12} = \bar{U}_{\text{lips}} \left[1 + \frac{Z_{1A}}{Z_{2B}} \right], \tag{3.143}$$

and substituted into (3.142). After simplifying, the following transfer function of output to input volume velocity results,

$$H(\Omega) = \frac{\bar{U}_{\text{lips}}}{\bar{U}_{\text{glottis}}} = \frac{Z_{1B} Z_{2B}}{[Z_{2A} + Z_{2B}][Z_{1A} + Z_{2A} + Z_{1B}] + Z_{2A} Z_{2B}}. \tag{3.144}$$

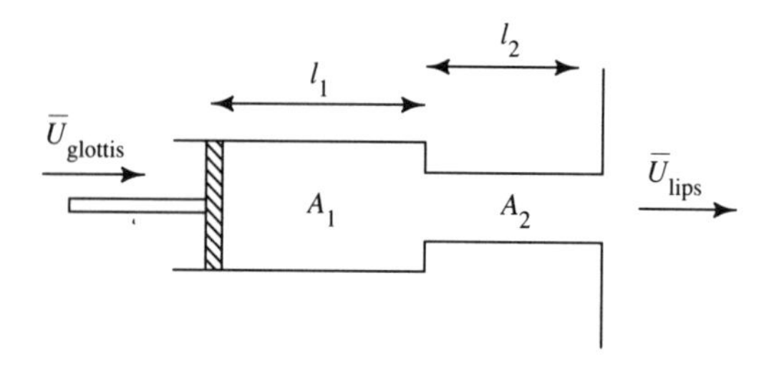

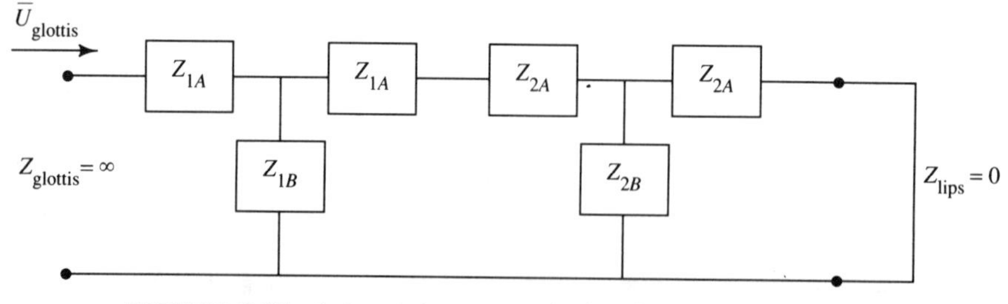

FIGURE 3.32. A two-tube transmission line circuit model.

If the following T-network impedances for the each transmission line section are substituted into (3.144),

$$Z_{1A} = \left(\frac{\rho C}{A_1}\right)\tanh\left(\frac{\Gamma_1 l_1}{2}\right) \qquad Z_{2A} = \left(\frac{\rho C}{A_2}\right)\tanh\left(\frac{\Gamma_2 l_2}{2}\right)$$

$$Z_{1B} = \left(\frac{\rho C}{A_1}\right)\operatorname{csch}(\Gamma_1 l_1) \qquad Z_{2B} = \left(\frac{\rho C}{A_2}\right)\operatorname{csch}(\Gamma_2 l_2), \tag{3.145}$$

and the resulting expression simplified, the final two-tube transfer function is obtained:

$$H(\Omega) = \frac{\bar{U}_{\text{lips}}}{\bar{V}_{\text{glottis}}} = \frac{A_2}{A_1 \sinh(\Gamma_1 l_1)\sinh(\Gamma_2 l_2) + A_2 \cosh(\Gamma_1 l_1)\cosh(\Gamma_2 l_2)}. \tag{3.146}$$

The terms A_1 and l_1 correspond to area and length of the pharyngeal tube, while A_2 and l_2 correspond to dimensions of the oral cavity tube. Under lossless conditions, $\Gamma_1 = \Gamma_2 = j\Omega/c$. From this transfer function, the poles occur at values of $\Omega = 2\pi F$ for which

$$A_1\left[j\sin\left(\frac{\Omega}{c}l_1\right)\right]\left[j\sin\left(\frac{\Omega}{c}l_2\right)\right] = -A_2\left[\cos\left(\frac{\Omega}{c}l_1\right)\right]\left[\cos\left(\frac{\Omega}{c}l_2\right)\right] \tag{3.147}$$

or simplifying,

$$A_1\left[\tan\left(\frac{\Omega}{c}l_1\right)\right] = A_2\left[\cot\left(\frac{\Omega}{c}l_2\right)\right]. \tag{3.148}$$

These resonances occur at the same frequencies as those found using the parallel input impedance calculation from the previous discussion on formant location analysis [see (3.140)]. By employing a two-tube model, a variety of settings for $A_{\text{ph}}, l_{\text{ph}}, A_{\text{oral}}$, and l_{oral} can be used to approximate many articulatory configurations. In Fig. 3.31, an example analysis for one vowel was illustrated. Several additional two-tube models and their corresponding pole structures are shown in Fig. 3.33. For front vowels, Fig. 3.33 confirms that the first and second formants are widely separated, while back vowels possess first and second formants that are closely spaced. Generally speaking, as the ratio of back to front tube areas $(A_{\text{ph}}/A_{\text{oral}})$ increases, the first formant location decreases.

Two-Port Network Analysis. Analysis of the two-tube vocal-tract model was achieved by simplifying the transmission line T-network equations

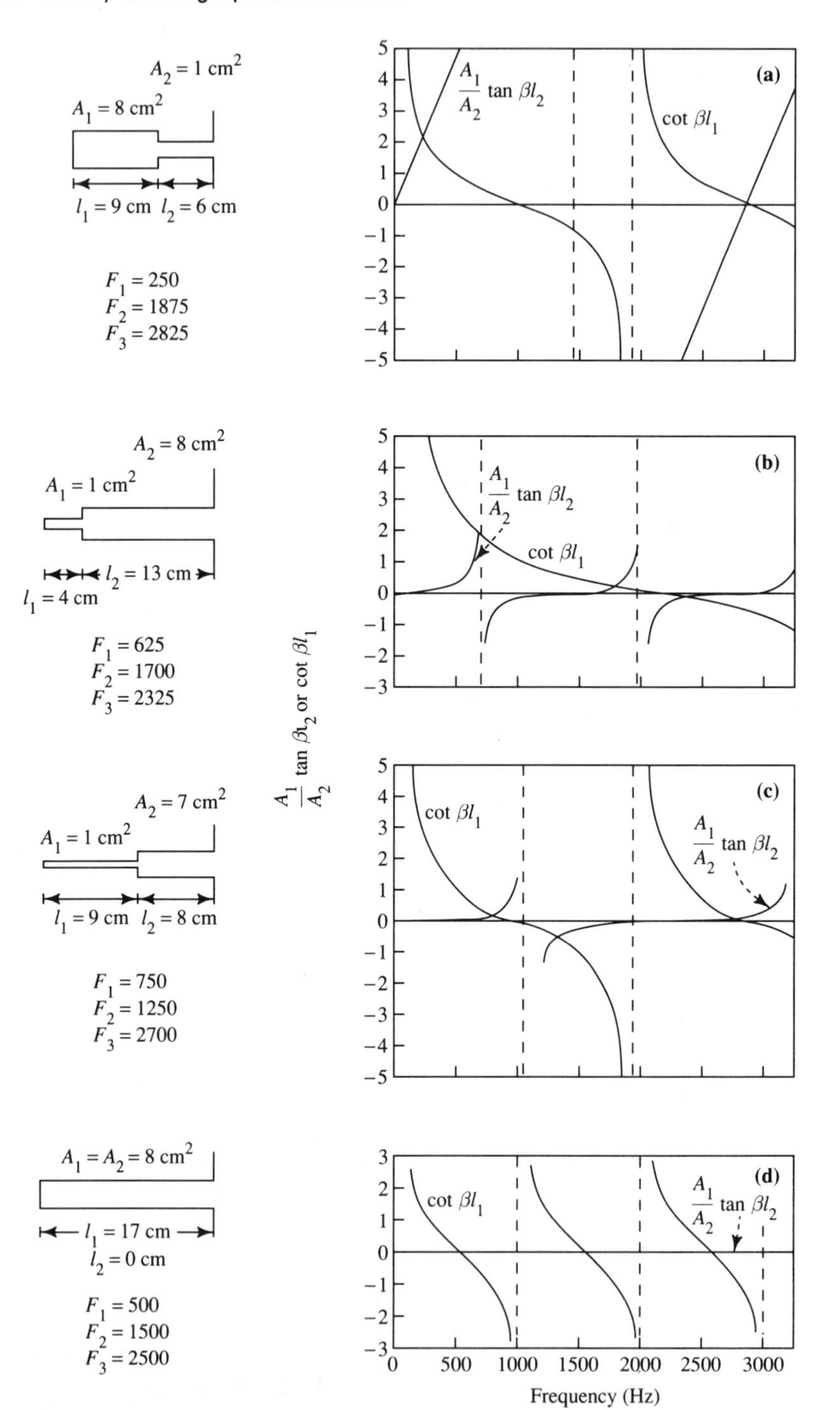

FIGURE 3.33. A collection of two-tube approximations for several phonemes. The solid lines represent cotangent curves. The variable β is the ratio $\Omega/c = 2\pi F/c$. (a) Vowel /i/, "beet"; (b) vowel /@/, "had"; (c) vowel /a/, "hot"; (d) Schwa vowel /x/.

(3.144) to (3.146). Although this is possible in the two-tube case, it becomes increasingly laborious as the number of tubes increases. If the cross-sectional areas at the tube junctions are diverse enough (i.e., sufficient impedance mismatch in the transmission line case), then it can be assumed that no coupling exists between front and back tubes.[15] With the assumption of limited or no coupling, the two-port equations (3.136) immediately yield the overall transfer function,

$$\begin{bmatrix} \bar{P}_{\text{lips}} \\ \bar{U}_{\text{lips}} \end{bmatrix} = \begin{bmatrix} \cosh(j\Omega l_{\text{oral}}/c) & -(\rho c/A_{\text{oral}})\sinh(j\Omega l_{\text{oral}}/c) \\ (-A_{\text{oral}}/\rho c)\sinh(j\Omega l_{\text{oral}}/c) & \cosh(j\Omega l_{\text{oral}}/c) \end{bmatrix} \tag{3.149}$$

$$\cdot \begin{bmatrix} \cosh(j\Omega l_{\text{ph}}/c) & -(\rho c/A_{\text{ph}})\sinh(j\Omega l_{\text{ph}}/c) \\ -(A_{\text{ph}}/\rho c)\sinh(j\Omega l_{\text{ph}}/c) & \cosh(j\Omega l_{\text{ph}}/c) \end{bmatrix} \cdot \begin{bmatrix} \bar{P}_{\text{glottis}} \\ \bar{U}_{\text{glottis}} \end{bmatrix}. \tag{3.150}$$

Again, we consider the period when the glottis is closed and the oral cavity is wide open (i.e., no sound pressure deviation from ambient $\bar{P}_{\text{lips}} = 0$). With these boundary conditions, the following volume velocity ratio is obtained,

$$H(\Omega) = \frac{\bar{U}_{\text{lips}}}{\bar{U}_{\text{glottis}}} = \frac{F(\Omega, l_{\text{ph}}, l_{\text{oral}}, A_{\text{ph}}, A_{\text{oral}})}{A_{\text{oral}}\cot(\Omega l_{\text{ph}}/c) - A_{\text{ph}}\tan(\Omega l_{\text{oral}}/c)}. \tag{3.151}$$

If cross-sectional areas A_{ph} and A_{oral} are diverse enough (i.e., $A_{\text{ph}} >> A_{\text{oral}}$ or $A_{\text{oral}} >> A_{\text{ph}}$), then characteristic impedances for each two-port section are different enough to satisfy the condition for noncoupling. In the majority of cases, tube areas are not so diverse so that all coupling is eliminated. It becomes necessary, however, to continue to rely on the two-port equation (3.136) in order to handle multitube configurations. The result in (3.151) suggests a transfer function with the same pole locations as those found in the parallel impedance formant analysis method, and the T-network transmission line circuit analysis method.

3.C Fast Discrete-Time Transfer Function Calculation

Calculation of vocal-tract transfer functions using signal flow analysis becomes increasingly complex as the number of tubes increases. For example, the four-tube vocal-tract model in Fig. 3.25(c) possesses a transfer function of similar form to (3.66); however, calculation of the coefficients becomes unwieldy as the model order increases to 10 or more. In Appen-

[15]This assumption is similar to the no-coupling assumption of a high output impedance transistor amplifier driving a low-impedance load.

dix 3.A.2, which treats concatenated lossless tubes, we found that it is possible to characterize the input–output relationships of sound pressure and volume velocity using the two-port impedance matrix given in (3.131), where the impedances Z_{11}, Z_{12}, Z_{21}, and Z_{22} were found using T-network impedances from a transmission line circuit [Fig. 3.23(c)]. By following this approach, a similar matrix characterization could be obtained for the signal flow model in Fig. 3.25(c).

For this discussion, consider the single-tube discrete-time flow diagram in Fig. 3.34. Following the approach used for a single transmission line circuit, we obtain a matrix $\mathbf{\Delta}$, which produces forward and backward traveling waves in the $(k+1)$st tube given quantities from the kth tube,[16]

$$\begin{bmatrix} V_{k+1}^{+}(z) \\ V_{k+1}^{-}(z) \end{bmatrix} = \mathbf{\Delta} \begin{bmatrix} V_{k}^{+}(z) \\ V_{k}^{-}(z) \end{bmatrix} = \begin{bmatrix} \delta_{11} & \delta_{12} \\ \delta_{21} & \delta_{22} \end{bmatrix} \begin{bmatrix} V_{k}^{+}(z) \\ V_{k}^{-}(z) \end{bmatrix}. \tag{3.152}$$

Writing the equations implied by the flow diagram, we obtain

$$V_{k+1}^{+}(z) = V_{k}^{+}(z)(1+\rho_k)z^{-1} + V_{k+1}^{-}(z)\rho_k \tag{3.153}$$

$$V_{k+1}^{-}(z) = V_{k}^{-}(z)\frac{1}{1-\rho_k} + V_{k}^{+}(z)\frac{\rho_k z^{-1}}{1-\rho_k}. \tag{3.154}$$

Recall that by omitting the distance argument "x" we implicitly refer to the waveform at the left boundary of the section, $x=0$ [see (3.32) and (3.33)]. From (3.153) and (3.154), we obtain two matrix forms of the solution, first by solving for $V_{k+1}^{+}(z)$ and $V_{k+1}^{+}(z)$ in terms of $V_{k}^{+}(z)$ and

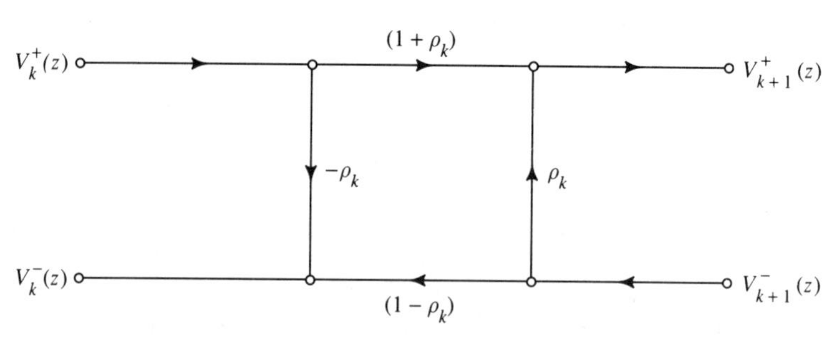

FIGURE 3.34. A single-tube discrete-time flow diagram.

[16]Because we are interested in discrete-time models, the z-transform is used in the following discussion in place of phasors and complex exponential signals that we used for analog steady-state analysis.

$V_k^-(z)$, and second by solving for $V_k^+(z)$ and $V_k^-(z)$ in terms of $V_{k+1}^+(z)$ and $V_{k+1}^+(z)$. These are

$$\begin{bmatrix} V_k^+(z) \\ V_k^-(z) \end{bmatrix} = \begin{bmatrix} \dfrac{z}{1+\rho_k} & -\dfrac{\rho_k z}{1+\rho_k} \\ \dfrac{-\rho_k}{1+\rho_k} & \dfrac{1}{1+\rho_k} \end{bmatrix} \begin{bmatrix} V_{k+1}^+(z) \\ V_{k+1}^-(z) \end{bmatrix}, \tag{3.155}$$

and

$$\begin{bmatrix} V_{k+1}^+(z) \\ V_{k+1}^-(z) \end{bmatrix} = \begin{bmatrix} \dfrac{z^{-1}}{1-\rho_k} & \dfrac{\rho_k}{1-\rho_k} \\ \dfrac{\rho_k z^{-1}}{1-\rho_k} & \dfrac{1}{1-\rho_k} \end{bmatrix} \begin{bmatrix} V_k^+(z) \\ V_k^-(z) \end{bmatrix}. \tag{3.156}$$

We write these two equations in compact form as

$$\mathbf{V}_k(z) = \mathbf{\Delta}_k \mathbf{V}_{k+1}(z) \tag{3.157}$$

and

$$\mathbf{V}_{k+1}(z) = \mathbf{\Delta}_k^{-1} \mathbf{V}_k(z), \tag{3.158}$$

respectively. Evidently,

$$\mathbf{V}_{N+1}(z) = \mathbf{\Delta}_N^{-1}\, \mathbf{\Delta}_{N-1}^{-1} \cdots \mathbf{\Delta}_1^{-1} \mathbf{V}_1(z) = \left[\prod_{k=N}^{1} \mathbf{\Delta}_k^{-1} \right] \mathbf{V}_1(z) \tag{3.159}$$

and

$$\mathbf{V}_1(z) = \mathbf{\Delta}_1\, \mathbf{\Delta}_2 \cdots \mathbf{\Delta}_{N+1} \mathbf{V}_{N+1}(z) = \left[\prod_{k=1}^{N} \mathbf{\Delta}_k \right] \mathbf{V}_{N+1}(z). \tag{3.160}$$

Now the vocal-tract transfer function is the ratio

$$H(z) = \frac{U_{\text{lips}}(z)}{U_{\text{glottis}}(z)} = z^{-N/2} \frac{V_{N+1}(z)}{V_1(z)} \stackrel{\text{def}}{=} z^{-N/2} \frac{V_{N+1}(x=0,z)}{V_1(x=0,z)}, \tag{3.161}$$

where the phase delay $z^{-N/2}$ appears because of the way we have modified the flow diagram in Fig. 3.25(c). It remains, therefore, to apply boundary conditions at the glottis and lips. In (3.160) we assume an N-tube model for the vocal tract, characterized by N signal flow matrices $\mathbf{\Delta}_1, \ldots, \mathbf{\Delta}_N$. However, inspection of the four-tube signal flow diagram in Fig. 3.25(c) reveals no reflection coefficient for the fourth tube, ρ_4. This occurs because the reflection/transmission characteristics of the fourth tube have been lumped with the fifth (half-infinite) tube used to model lip radiation effects. If ρ_{lips} is used in place of ρ_4, (3.160) can be computed. The equation that relates the boundary condition for the lips is written as

$$\mathbf{V}_{N+1}(z) = \begin{bmatrix} U^+_{\text{lips}}(z) \\ U^-_{\text{lips}}(z) \end{bmatrix} z^{-N/2} = \begin{bmatrix} U_{\text{lips}}(z) \\ 0 \end{bmatrix} z^{-N/2}, \tag{3.162}$$

where it is assumed that sound propagating forward into the half-infinite acoustic tube for the lips results in no reflections $[U^-_{\text{lips}}(z) = 0]$.

Using (3.60), we find that the boundary at the glottis is characterized by the relation

$$U_{\text{glottis}}(z) = \mathbf{\Delta}_{\text{glottis}} \mathbf{V}_1(z) \stackrel{\text{def}}{=} \begin{bmatrix} \dfrac{2}{1+\rho_{\text{glottis}}} & \dfrac{-2\rho_{\text{glottis}}}{1+\rho_{\text{glottis}}} \end{bmatrix} \begin{bmatrix} V^+_1(z) \\ V^-_1(z) \end{bmatrix} . \tag{3.163}$$

The use of equations (3.162) and (3.163) in (3.160) yields the ratio $U_{\text{glottis}}(z)/U_{\text{lips}}(z)$, which can then be reciprocated to obtain $H(z)$.

Direct application of equation (3.159) in conjunction with (3.162) and (3.163) is not as straightforward, since (3.163) produces glottal volume velocity $U_{\text{glottis}}(z)$ from input vocal-tract terms $V^+_1(z)$ and $V^-_1(z)$. If the tube matrix product is obtained as follows

$$\begin{bmatrix} \delta_{11} & \delta_{12} \\ \delta_{21} & \delta_{22} \end{bmatrix} = \prod_{k=1}^{N} \mathbf{\Delta}_k^{-1}, \tag{3.164}$$

then, solving for $V^+_1(z)$ and $V^-_1(z)$ in the following system of equations,

$$U_{\text{glottis}}(z) = \begin{bmatrix} \dfrac{2}{1+\rho_{\text{glottis}}} & \dfrac{-2\rho_{\text{glottis}}}{1+\rho_{\text{glottis}}} \end{bmatrix} \begin{bmatrix} V^+_1(z) \\ V^-_1(z) \end{bmatrix} \tag{3.165}$$

$$\begin{bmatrix} U^+_{\text{lips}}(z) \\ 0 \end{bmatrix} = z^{-N/2} \begin{bmatrix} \delta_{11} & \delta_{12} \\ \delta_{21} & \delta_{22} \end{bmatrix} \begin{bmatrix} V_1^{\,+}(z) \\ V_1^{\,-}(z) \end{bmatrix}, \tag{3.166}$$

we can obtain the final vocal-tract transfer function

$$H(z) = \frac{U_{\text{lips}}(z)}{U_{\text{glottis}}(z)} = \left(\frac{1+\rho_{\text{glottis}}}{2} \right) \left[\frac{\delta_{11}\delta_{22} - \delta_{12}\delta_{21} z^{-N/2}}{\delta_{22} z^{-N/2} + \rho_{\text{glottis}} \delta_{21}} \right]. \tag{3.167}$$

In general, it can be seen that for a multitube lossless model, the vocal system transfer function can be expressed as

$$H(z) = \frac{1+\rho_{\text{glottis}}}{2} \frac{z^{-N/2} \prod_{k=1}^{N} (1+\rho_k)}{[1 \quad -\rho_{\text{glottis}}] \cdot \prod_{k=1}^{N} \tilde{\boldsymbol{\Delta}}_k \cdot \begin{bmatrix} 1 \\ 0 \end{bmatrix}}, \tag{3.168}$$

where $\tilde{\boldsymbol{\Delta}}_k$ is the following modified signal flow matrix,

$$\tilde{\boldsymbol{\Delta}}_k = \begin{bmatrix} 1 & -\rho_k \\ -\rho_k z^{-1} & z^{-1} \end{bmatrix}. \tag{3.169}$$

If the denominator of (3.168) is expanded, it is found to have the form

$$D(z) = z^N \left[1 - \sum_{k=1}^{N} b_k z^{-k} \right]. \tag{3.170}$$

Substituting back into (3.168) shows the transfer function for the N-tube lossless vocal-tract system,

$$H(z) = \frac{1+\rho_{\text{glottis}}}{2} \frac{z^{-N/2} \prod_{k=1}^{N} (1+\rho_k)}{1 - \sum_{k=1}^{N} b_k z^{-k}}. \tag{3.171}$$

This important result was discussed in Section 3.2.1 [see material below (3.67)].

PART III

ANALYSIS TECHNIQUES

CHAPTER 4

Short-Term Processing of Speech

Reading Notes: No "advanced" topics from Chapter 1 will be required in this chapter. Basic DSP concepts from Section 1.1 will be used without comment. In particular, the DTFT and DFT will play a significant role in Section 4.3.5. Basic concepts from (scalar-valued) random process theory, which was reviewed in Section 1.2.3 will be used throughout. In particular, Section 4.3.1 will require a solid understanding of the correlation properties of stochastic processes.

4.1 Introduction

Engineering courses frequently ignore the fact that all analysis must be done in finite time. The continuous-time Fourier transform, for example, is a remarkably useful tool for signal analysis. In its fundamental definition, however, the Fourier transform requires our knowledge of the signal for all time, and, further, whatever property or feature we are seeking by use of the Fourier transform (a spectrum, a resonance, bandpass energy, etc.) must remain invariant for all time in the signal. Most of us do not have doubly infinite time in which to process a signal, and most signals do not cooperate by remaining "stationary" with respect to the desired measurement forever. This latter point is particularly true of speech signals, which we can expect to "change" every few milliseconds. In the best case, we as engineers often take what we have learned in the "long term" and apply it to the "short term," counting on our good experience and intuition to compensate for the deviation of reality and theory. In the worst case, we simply ignore the discrepancy and hope for the best. The danger in this latter approach is that when the "best" occurs we do not know why, and we may come to rely on a design or approach that has serious weaknesses that might emerge at a most inopportune time.

Digital signal processing engineers are perhaps more aware of the short-term nature of signal analysis than some other specialists if for no other reason than that our computing machines can only hold and process a finite amount of data at a time. We have also come face to face with the effects of short-term signal analysis in studying topics like the DFT "leakage" phenomenon and FIR filter design. These topics, however, deal with "static" analysis in which a single frame of a signal is operated upon, and the results analyzed with respect to "asymptotic" results. Speech is a dynamic, information-bearing process, however. We as speech processors cannot be content to analyze short-term effects in a single frame. In this chapter we want to build on our DSP background and formally examine short-term processing from the "dynamic" point of view. Our objective here is to learn about analysis of frames of speech as those frames move through time and attempt to capture transient features of the signal. Another goal is to introduce a number of short-term features that are useful in speech processing, some of which will be vital to our work in future chapters.

4.2 Short-Term Measures from Long-Term Concepts

4.2.1 Motivation

Suppose that we wish to extract some information about a *short term* or *frame* of speech spanning the time range $n = m - N + 1, \ldots, m$. We have an intuitive idea that there is a *long-term* concept that could provide the needed information if we could "generalize" it to the short term. For example, suppose that it is desired to know whether a speech sequence is voiced or unvoiced on this short term ending at time $n = m$. We know that voiced speech is generally of higher "power" (average squared value per sample) than unvoiced. The idea, then, would be to employ the concept of average power to assist in the decision. Average power, however, is a long-term concept:[1]

$$P_s = \mathcal{L}\{s^2(n)\}, \tag{4.1}$$

where $\mathcal{L}$ denotes the long-term temporal average[2] introduced in (1.126). Our intuition is telling us here that if $s_1(n)$ is an eternally voiced signal, and $s_2(n)$ eternally unvoiced, then the long-term concept of average power could be useful, since it would be true that

$$P_{s_1} > P_{s_2}, \tag{4.2}$$

and if we could find some "similar" short-term quantities using only the points around m, say $P_{s_i}(m)$, it would be true that

$$P_{s_1}(m) > P_{s_2}(m). \tag{4.3}$$

[Note that once we have narrowed our selection of points to some small region around m, it is inconsequential to the truth of (4.3) whether s_1 and s_2 remain eternally stationary, as long as they have the desired properties on the selected range of points.] $P_s(m)$ (however it is computed)

[1]If the sequence $s(n)$ were complex, $s^2(n)$ should be replaced by $|s(n)|^2$ in the following and in similar expressions [see (1.9) and (1.11)]. We will ignore this generality in our work since the sequences we deal with are almost always real.

[2]If $s(n)$ happens to be a realization of a stochastic process $\underline{s}$ in this discussion, we assume the appropriate stationarity and ergodicity properties to permit the use of temporal averaging.

can be applied to the problem above by determining some threshold on its value below which the signal is classified as unvoiced on the frame ending at m, above which it is deemed voiced.

4.2.2 "Frames" of Speech

Clearly, we are moving toward the practical problem of working with small ranges of the speech sequence. Before continuing down this path, it is important to review the concept of a "frame," which was first introduced in Section 1.1.5. Formally, we define a *frame* of speech to be the product of a shifted window with the speech sequence

$$f_s(n;m) \stackrel{\text{def}}{=} s(n)w(m-n). \tag{4.4}$$

For convenience, we will omit the subscript s when the frame is taken from a speech sequence $s(n)$. This will almost always be the case in our work, and we will include subscripts for clarity when it is not. Although, practically, a frame is just a "chunk" of speech which perhaps has been tapered by the window, formally it is a *new sequence on n* in its own right, which happens to be zero outside the short term $n \in [m-N+1, m]$. Accordingly, we will often find that short-term processing of the speech is tantamount to long-term processing of a frame. The frame created by this process also depends on the end time, m, so that it has a second argument (and also an implicit argument, N). This formality will be very useful to us in the upcoming discussion.

4.2.3 Approach 1 to the Derivation of a Short-Term Feature and Its Two Computational Forms

In general, suppose that $\mathcal{X}_s$ is the long-term feature we have in mind to help us solve a problem. In general, there might be a family of features, each one dependent upon an index λ, so let us write the general long-term feature of the sequence $s(n)$ as $\mathcal{X}_s(\lambda)$. [An example of a feature involving a parameter is the long-term autocorrelation $r_s(\eta)$, which is indexed by the integer η, indicating the lag. A feature may also be parameterized by a *continuous* parameter or by a *vector* of parameters, as we shall see below.] Suppose further that $\mathcal{X}_s(\lambda)$ is computed from $s(n)$ (assuming that s retains the desired property forever) as

$$\mathcal{X}_s(\lambda) = \tilde{\mathfrak{J}}(\lambda)\{s(n)\} = \mathcal{L}\{\mathfrak{J}(\lambda)\{s(n)\}\}, \tag{4.5}$$

where $\tilde{\mathfrak{J}}(\lambda)$ is some operation, generally nonlinear and dependent on λ. For most commonly used long-term features it is found that $\tilde{\mathfrak{J}}(\lambda)$ can be decomposed as $\tilde{\mathfrak{J}}(\lambda) = \mathcal{L} \circ \mathfrak{J}(\lambda)$, where $\mathfrak{J}(\lambda)$ is an operation that produces a new sequence on n, and $\mathcal{L}$ is the temporal average operator. [See the P_s computation above, for example, where $\mathfrak{J}$ is easily seen to be the squaring operation $\mathfrak{J}\{s(n)\} = s^2(n)$.] In light of (4.5), a highly intuitive way to compute a "$\mathcal{X}_s(\lambda)$-like" feature, say, $\mathcal{X}_s(\lambda; m)$, using only the short term, $n \in [m-N+1, m]$ is given in the following "construction principle":

CONSTRUCTION PRINCIPLE 1 *These steps can be used to compute a (family of) short-term feature(s)* $\mathcal{X}_s(\lambda; m)$, *"similar to" the long-term feature(s)* $\mathcal{X}_s(\lambda)$.

1. *Select the desired N-length frame of* $s(n)$ *using a window,* $w(n)$,

$$f(n; m) = s(n)w(m-n). \tag{4.6}$$

2. *Apply a "*$\tilde{\mathfrak{I}}(\lambda)$*-like" operation, say* $\tilde{\mathcal{S}}(\lambda)$, *to the frame:*

$$\begin{aligned}\mathcal{X}_s(\lambda; m) = \tilde{\mathcal{S}}(\lambda)\{s(n)w(m-n)\} &= \frac{1}{N}\sum_{n=-\infty}^{\infty} \mathcal{S}(\lambda)\{s(n)w(m-n)\} \\ &= \frac{1}{N}\sum_{n=-\infty}^{\infty} \mathcal{S}(\lambda)\{f(n; m)\},\end{aligned} \tag{4.7}$$

where it is assumed that $\tilde{\mathcal{S}}(\lambda)$ *can be decomposed as* $(1/N)\sum_{n=-\infty}^{\infty} \circ\, \mathcal{S}(\lambda)$ *just as* $\tilde{\mathfrak{I}}(\lambda) = \mathcal{L} \circ \mathfrak{I}(\lambda)$ *in the long term.*

It should be noted that $\mathcal{S}(\lambda)$ is often the same operation as $\mathfrak{I}(\lambda)$. In fact, let us restrict our discussion to such cases:

$$\mathcal{X}_s(\lambda; m) = \frac{1}{N}\sum_{n=-\infty}^{\infty} \mathfrak{I}(\lambda)\{s(n)w(m-n)\}. \tag{4.8}$$

Clearly, in simple terms, all we have done here is cut out a piece of the speech (created a frame) and applied an operation to it similar to the one that we employ in the long-term version of the feature. This portion of the speech was selected using a window (of length N), which has been shifted in time so that it "ends" at the desired time m.

Applying this principle to the power computation above, we obtain the following *short-term average power* feature over a frame of length N ending at time m:

$$P_s(m) = \frac{1}{N}\sum_{n=-\infty}^{\infty} \{s(n)w(m-n)\}^2. \tag{4.9}$$

We note that no parameter, λ, is involved here, and $\mathfrak{I}\{\cdot\} = \{\cdot\}^2$. Another example that *does* involve a parameter is the autocorrelation. In the long term we have

$$r_s(\eta) = \tilde{\mathfrak{I}}(\eta)\{s(n)\} = \mathcal{L}\{\mathfrak{I}(\eta)\{s(n)\}\} = \mathcal{L}\{s(n)s(n-\eta)\}. \tag{4.10}$$

In this case we see that $\mathfrak{I}$ depends on the parameter η corresponding to the autocorrelation lag, and $\mathfrak{I}(\eta)\{s(n)\} = s(n)s(n-\eta)$. Applying Construction Principle 1, we have the following *short-term autocorrelation* estimator for lag η, over the frame of length N ending at time m,

$$\begin{aligned} r_s(\eta;m) &= \frac{1}{N}\sum_{n=-\infty}^{\infty} \mathcal{T}(\eta)\{s(n)w(m-n)\} \\ &= \frac{1}{N}\sum_{n=-\infty}^{\infty} s(n)w(m-n)s(n-\eta)w(m-n+\eta) \\ &= \frac{1}{N}\sum_{n=m-N+1}^{m} s(n)w(m-n)s(n-\eta)w(m-n+\eta). \end{aligned} \tag{4.11}$$

It is obvious that the operator $\mathcal{T}(\lambda)$ in such problems must have the property that it produces a new sequence on n from the sequence upon which it operates, say,

$$v_{\mathcal{T}}(n;\lambda) = \mathcal{T}(\lambda)\{v(n)\}, \tag{4.12}$$

where $v_{\mathcal{T}}$ will also depend on a parameter. [If $v(n)$ is a *frame*, then both $v(n)$ and $v_{\mathcal{T}}$ will also depend on m.] In addition, $\mathcal{T}(\lambda)$ very often (but not always) has the property that for any two sequences, $x(n), v(n)$,

$$\begin{aligned} \mathcal{T}(\lambda)\{x(n)\,v(n)\} &= \mathcal{T}(\lambda)\{x(n)\}\,\mathcal{T}(\lambda)\{v(n)\} \\ &\stackrel{\text{def}}{=} x_{\mathcal{T}}(n;\lambda)\,v_{\mathcal{T}}(n;\lambda). \end{aligned} \tag{4.13}$$

In the power feature case, for example,

$$[s(n)w(m-n)]^2 = s^2(n)w^2(m-n), \tag{4.14}$$

and in the autocorrelation case,

$$\begin{aligned} \mathcal{T}(\eta)\{s(n)w(m-n)\} &= s(n)w(m-n)s(n-\eta)w(m-n+\eta) \\ &= s(n)s(n-\eta)w(m-n)w(m-n+\eta) \\ &= \mathcal{T}(\eta)\{s(n)\}\mathcal{T}(\eta)\{w(m-n)\}. \end{aligned} \tag{4.15}$$

When the operator $\mathcal{T}$ is separable in this way, (4.8) can be reformulated as follows:

$$\begin{aligned} X_s(\lambda;m) &= \frac{1}{N}\sum_{k=-\infty}^{\infty} \mathcal{T}(\lambda)\{s(k)\}\,\mathcal{T}(\lambda)\{w(m-k)\} \\ &= \frac{1}{N}\sum_{k=-\infty}^{\infty} s_{\mathcal{T}}(k;\lambda)\,w_{\mathcal{T}}(m-k;\lambda). \end{aligned} \tag{4.16}$$

We see that (for any parameter λ), the feature $X_s(\lambda;m)$ can be computed as the *convolution of the sequences* $(1/N)s_{\mathcal{T}}(n;\lambda)$ *and* $w_{\mathcal{T}}(n;\lambda)$ *evaluated at time* $n = m$. To compute the feature at any general time n, we can write

$$X_s(\lambda; n) = \frac{1}{N} s_{\mathcal{T}}(n; \lambda) * w_{\mathcal{T}}(n; \lambda), \tag{4.17}$$

where $*$ denotes convolution. This form, depicted in Fig. 4.1, allows the interpretation of the $X_s(\lambda; n)$ feature as the output of a filter with impulse response $w_{\mathcal{T}}(n; \lambda)$ when the input is $(1/N)s_{\mathcal{T}}(n; \lambda)$.

Thus far we have assumed $w(n)$ and, hence, presumably, $w_{\mathcal{T}}(n; \lambda)$, to be a finite duration [or, if viewed as a filter, finite impulse response (FIR)] window. This "output filter" form for computing $X_s(\lambda; n)$ begs the question as to whether it is actually necessary to employ an FIR filter at the output, particularly in cases in which $X_s(\lambda; n)$ is a feature that is insensitive to the phase of $s(n)$. If we recognize that $w_{\mathcal{T}}(n; \lambda)$ will generally be lowpass in spectral nature, it is possible to generalize (4.17) by substituting $h_{\mathcal{T}}(n; \lambda)$ [any general filter with magnitude spectrum equivalent to that of $w_{\mathcal{T}}(n; \lambda)$] for $w_{\mathcal{T}}(n; \lambda)$ in the computation. It should be noted that the insertion of an infinite impulse response (IIR) output filter into (4.17) will require that $X_s(\lambda; n)$ be computed for each n, as recent past values of the output will be necessary at each time; this is true even though $X_s(\lambda; n)$ might only be desired for a select set of "m's." This does not, however, necessarily imply computational inefficiency. Indeed, even FIR (inherently nonrecursive) forms can sometimes be formulated into a recursive computation for greater efficiency: Consider, for example, the short-term power computation using a rectangular window of length N. Using either (4.9) [from (4.8)] or (4.17) yields

$$P_s(m) = \frac{1}{N} \sum_{n=m-N+1}^{m} s^2(n), \tag{4.18}$$

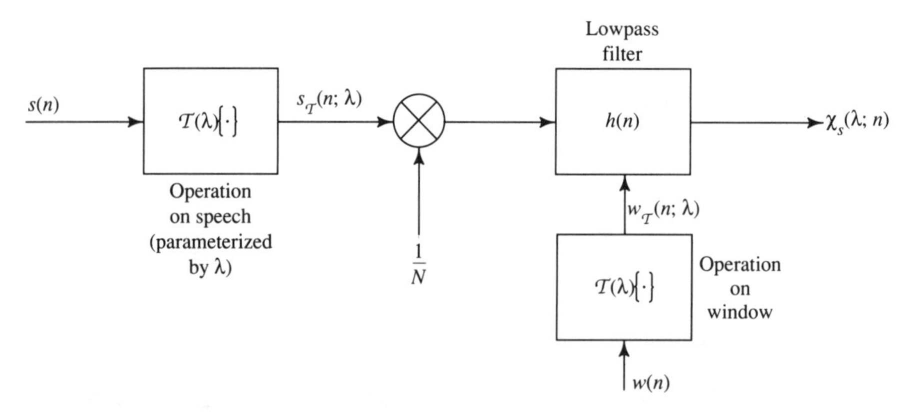

FIGURE 4.1. Computation of the short-term feature $X_s(\lambda; m)$, which is constructed according to Construction Principle 1 and viewed as a convolution.

which can be computed at any m independent of past values of $P_s(n)$. Note, however, that (4.18) can be written as a recursion,

$$P'_s(n) = P'_s(n-1) + s^2(n) - s^2(n-N), \tag{4.19}$$

where $P'_s(n) \stackrel{\text{def}}{=} NP_s(n)$. If $P'_s(n)$ is to be computed at $m = 0, N/4, N/2, \ldots$, for example, the use of (4.18) requires N squaring operations and $(N-1)$ additions every $N/4$ points or four squaring operations and approximately four additions per n (per speech sample or norm-sec); whereas (4.19) requires only two squares and two adds per n once the initial computation is performed.

Finally, it should be noted that the test of merit of a short-term estimator is the accuracy with which it estimates its long-term counterpart [assuming $s(n)$ remains stationary], that is, the degree to which

$$\mathcal{X}_s(\lambda; m) \approx \mathcal{X}_s(\lambda). \tag{4.20}$$

This approximation, in turn, is directly related to the choice of windows in the analysis. A review of this important topic is found in Section 1.1.5. More details on the effects of windows in the design of estimators can be found in the textbooks listed in Appendix 1.A. Let us briefly recall the two considerations in choosing a window: type of window and its length, N. Generally speaking, for a fixed N, two competing factors emerge in the choice of window type (note in the discussion above that the windows act directly on the speech data): the need not to distort the selected points for the waveform versus the need to smooth the abrupt discontinuity at the window boundaries. (A frequency domain interpretation of these factors is found in Section 1.1.5) Generally, the latter consideration emerges as primary, and "smoother" windows such as the Hamming are used. In choosing N, again there are two competitive factors: for a fixed window type, increasing N improves the spectral resolution at a given m by providing more information to the computation. An example arising in voiced speech is shown in Fig. 4.2. However, as the window slides through time (to compute features at various m's), long windows make phonetic boundary straddling more likely and events in time are not resolved as well. This phenomenon is also illustrated in Fig. 4.2. The choice of N is highly problem dependent, but as a rule of thumb, speech can be assumed to remain stationary for frames on the order of 20 msec so that window lengths are chosen accordingly.

Before leaving the generalities of short-term processing to look at some important examples, we examine two more variations on the theme of deducing a short-term feature based on a long-term attribute.

4.2.4 Approach 2 to the Derivation of a Short-Term Feature and Its Two Computational Forms

The long-term temporal average for a speech sequence, $s(n)$, is

$$\mu_s = \mathcal{L}\{s(n)\}. \tag{4.21}$$

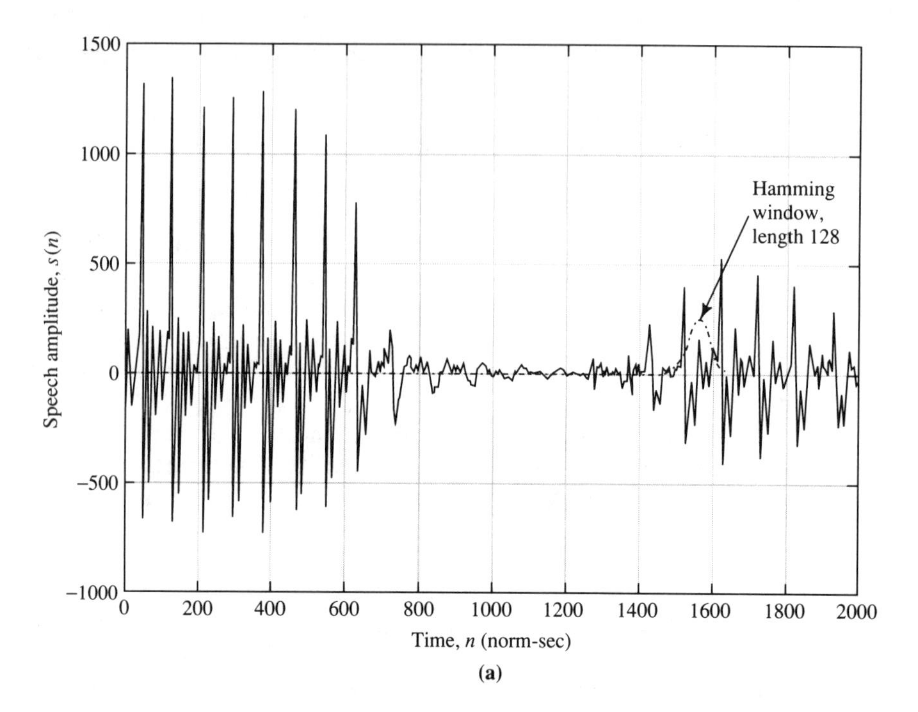

(a)

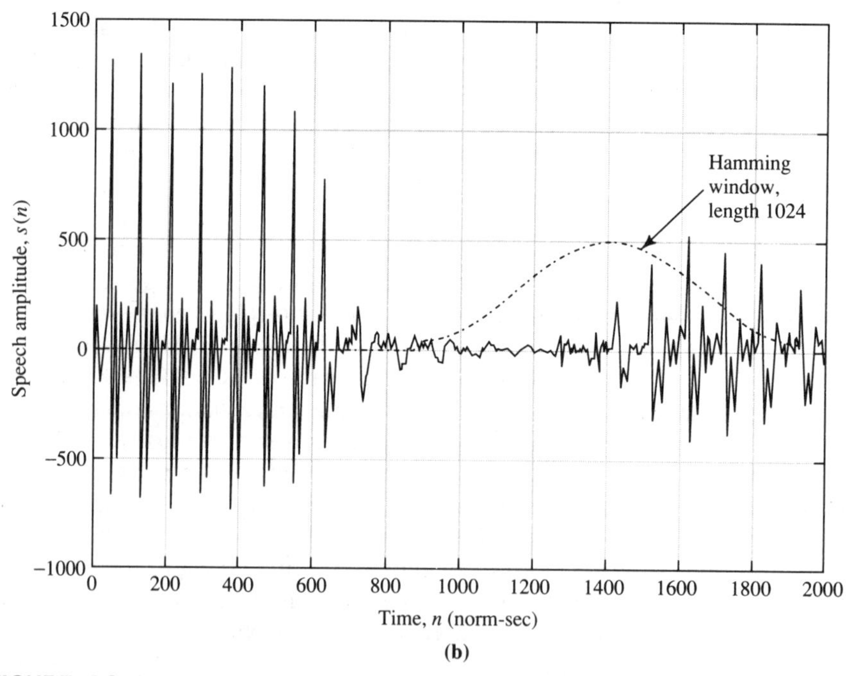

(b)

FIGURE 4.2. Illustration of the time resolution/frequency resolution trade-off encountered in selecting the length of a window. In (a), the window is short and proper frequency information will not be present in the frame. We show an extreme case in which the window is shorter than one pitch period so that the harmonic structure of the excitation will not be resolved by the feature. In (b), where the window is much longer, events will not be well resolved in time since a frame will frequently "straddle" two or more phones.

Using Construction Principle 1, we derive the short-term average of $s(n)$ for the frame ending at m to be

$$\mu_s(m) = \frac{1}{N} \sum_{n=-\infty}^{\infty} s(n) w(m-n) = \frac{1}{N} s(m) * w(m), \tag{4.22}$$

where w is any window [or by the extension below (4.17), any lowpass filter]. Let us define the *short-term average operation,* $\mathcal{A}(m)$ to be such that

$$\mu_s(m) = \mathcal{A}(m)\{s(n)\} = \frac{1}{N} \sum_{n=-\infty}^{\infty} s(n) w(m-n). \tag{4.23}$$

If we note that $X_s(\lambda)$ in (4.5) represents the *long-term temporal average* of the sequence $\mathcal{T}(\lambda)\{s(n)\} = s_{\mathcal{T}}(n;\lambda)$, a second intuitive approach to the construction of a short-term version of $X_s(\lambda)$, namely, $X_s(\lambda;m)$, is to compute the *short-term temporal average* of that same sequence:

CONSTRUCTION PRINCIPLE 2 *If* $X_s(\lambda) = \mathcal{L}\{\mathcal{T}(\lambda)\{s(n)\}\}$, *let*

$$\begin{aligned} X_s(\lambda;m) &\stackrel{\text{def}}{=} \mathcal{A}(m)\{\mathcal{T}(\lambda)\{s(n)\}\} = \mathcal{A}(m)\{s_{\mathcal{T}}(n;\lambda)\} \\ &= \frac{1}{N} \sum_{n=-\infty}^{\infty} s_{\mathcal{T}}(n;\lambda) w(m-n). \end{aligned} \tag{4.24}$$

It is clear that this result is also readily interpreted as a convolution

$$X_s(\lambda;n) = \frac{1}{N} s_{\mathcal{T}}(n;\lambda) * w(n). \tag{4.25}$$

This computation is diagramed in Fig. 4.3.

Comparing the approach of Construction Principle 2 with that of Construction Principle 1, we see that the window, rather than being applied directly to the speech, is applied to the operated-upon speech in Construction Principle 2. The window consequently is unaffected by the operator $\mathcal{T}(\lambda)$ and, where a feature is parametric, the window remains unchanged for varying values of λ. This fact simplifies the computation, which can be seen by comparing Figs. 4.1 and 4.3. In the present case only one output filter need be implemented rather than the one-per-λ required in the former case.

Both Construction Principle 1 and Construction Principle 2 amount to windowed transformations of long-term attributes of the signal, and the two differ principally by whether the window precedes or follows the transformation. In some cases, the two construction principles will produce the same short-term estimator. Most algorithms for computing short-term features in the literature can be classified according to one of these two construction principles.

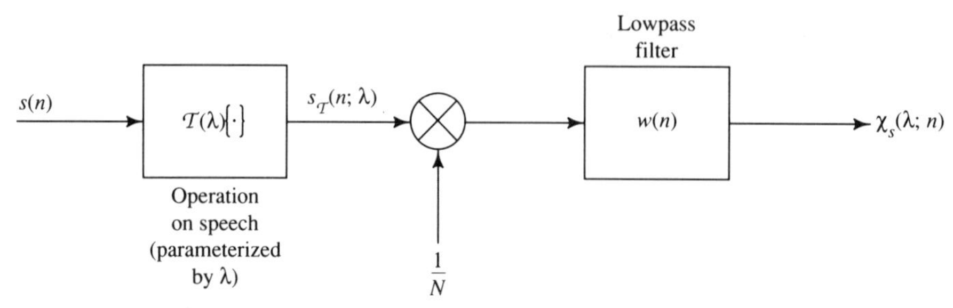

FIGURE 4.3. Computation of the short-term feature $\chi_s(\lambda; m)$, which is constructed according to Construction Principle 2 and viewed as a convolution.

4.2.5 On the Role of "1/*N*" and Related Issues

We have explicitly assumed that the long-term speech signal in the above, $s(n)$, is a *power* signal. Indeed, this is the proper assumption whether we are modeling a voiced or unvoiced phone over infinite time. Recall, from general signal processing theory, however, that if $s(n)$ were an *energy* signal instead, a slightly different set of long-term features would be used. For example, suppose that $x(n)$ is an energy signal. If we attempt to compute the power of $x(n)$ according to (4.1), or its autocorrelation according to (4.10), we will get zero power, or zero for any argument of the autocorrelation. This is not unexpected since, *on the average*, there is no power in an energy signal, nor is there any correlation between points at any lag. It is customary in the energy case to use features which *sum but do not average*,

$$\chi_x(\lambda) = \mathcal{L}'\{\mathcal{T}(\lambda)\{x(n)\}\}, \tag{4.26}$$

where

$$\mathcal{L}'\{\cdot\} \stackrel{\text{def}}{=} \lim_{N\to\infty} \sum_{n=-N}^{N} \{\cdot\}. \tag{4.27}$$

For example, in place of power we would use

$$E_x = \sum_{n=-\infty}^{\infty} x^2(n); \tag{4.28}$$

in place of the "power definition" of autocorrelation, we would use

$$r_x(\eta) = \sum_{n=-\infty}^{\infty} x(n)x(n-\eta). \tag{4.29}$$

In some cases, we have different names for these "energy definition" features, while in others the name remains the same. In the above, for example, since *power is average energy,* when we do not average, the feature is called the "energy" in $x(n)$. (A subtler but analogous situation arises between the discrete-time Fourier series coefficients and the DTFT.) The "autocorrelation" is an example of a "feature" that bears the same name in either case.

We could, of course, create construction principles for frames of energy signals according to the discussion above. It is plausible, if not obvious, that the short-term results would be identical in form to the "power" results in (4.8) and (4.24), except that the "averaging factor" $1/N$ would not appear.

The main point of this discussion is to note the following. The factor $1/N$ in front of the short-term features rarely plays any significant role in short-term analysis. It is therefore often omitted in practice. *This should not be taken to mean that the short-term feature is an estimator of an "energy-type" feature.* On the contrary, the energy feature would not theoretically exist in the long term. For pedagogical purposes, therefore, we usually leave the factor $1/N$ in the features to emphasize the fact that the features are theoretically estimators of power-type quantities.

Another point of confusion that might occur to the reader is that the frame itself is an energy signal even though the signal from which it is drawn in our work is a power signal. As a consequence of this, long-term "energy" analysis on the frame (which is entirely proper) will produce the short-term result of Construction Principle 1 without the scale factor $1/N$. For example, $r_s(\eta;m)$ (without the scale factor) can be obtained as

$$\begin{aligned}\mathcal{L}'\{f(n;m)f(n-|\eta|;m)\} &= \sum_{n=-\infty}^{\infty} f(n;m)f(n-|\eta|;m)\\ &= \sum_{n=m-N+1}^{m} f(n;m)f(n-|\eta|;m),\end{aligned} \tag{4.30}$$

which is easily seen to be equivalent to (4.11) once the scale factor $1/N$ is included. A similar situation can be described for Construction Principle 2.

The fact that an "energy-type" short-term feature is often used in practice, and that the frame itself is an energy signal, should not be allowed to divert the reader's attention from realizing that the short-term feature is fundamentally an estimator of a "power-type" feature of a power signal. Indeed, the counterpart long-term energy feature does not exist in principle. Implications to the contrary can sometimes cause conceptual problems for the student or researcher working on theoretical problems or interpreting practical data. The averaging factor $1/N$, while of no practical significance, is loaded with theoretical meaning, so we leave it in the developments to follow.

We focus now on some important examples of short-term features used in speech processing.

4.3 Example Short-Term Features and Applications

4.3.1 Short-Term Estimates of Autocorrelation

Short-Term Autocorrelation

The autocorrelation sequence will be found to play a central role in many aspects of speech processing. We focus here on several short-term counterparts to the long-term autocorrelation. Recall that we can formalize the autocorrelation as

$$r_s(\eta) = \tilde{\mathcal{J}}(\eta)\{s(n)\} = \mathcal{L}\{\mathcal{J}(\eta)\{s(n)\}\} = \mathcal{L}\{s(n)s(n-\eta)\}. \tag{4.31}$$

Since $r_s(\eta)$ is an even function of η, it is also true that

$$r_s(\eta) = \mathcal{L}\{s(n)s(n-|\eta|)\} \tag{4.32}$$

from which we are free to choose

$$\mathcal{J}(\eta)\{s(n)\} = s(n)s(n-\eta) \tag{4.33}$$

or

$$\mathcal{J}(\eta)\{s(n)\} = s(n)s(n-|\eta|). \tag{4.34}$$

The latter is chosen because it produces an even short-term estimator of autocorrelation. In the material to follow, however, in which the short-term autocorrelation is almost always computed for $\eta \geq 0$, this choice is of no consequence.

Assume that it is desired to compute a short-term autocorrelation estimate for the N points ending at m, for use, for example, in a pitch detection problem (see below). According to Construction Principle 1, one estimator is given by

$$\begin{aligned} r_s(\eta; m) &= \frac{1}{N} \sum_{n=-\infty}^{\infty} \mathcal{J}(\eta)\{s(n)w(m-n)\} \\ &= \frac{1}{N} \sum_{n=-\infty}^{\infty} s(n)w(m-n)s(n-|\eta|)w(m-n+|\eta|). \end{aligned} \tag{4.35}$$

It is clear that this feature can be computed using convolution as in (4.17) with $s_{\mathcal{J}}(n;\eta) = s(n)s(n-|\eta|)$ and $w_{\mathcal{J}}(n;\eta) = w(n)w(n-|\eta|)$. If $w(n)$ is taken to be a rectangular window of length N, then one common estimator emerges,

$$^1r_s(\eta; m) = \frac{1}{N} \sum_{n=m-N+1+|\eta|}^{m} s(n)s(n-|\eta|). \tag{4.36}$$

The following observations are made regarding this estimator:

1. $^1r_s(\eta; m)$ is *biased but asymptotically unbiased.* This means that if $^1r_s(\eta; m)$ is considered to be a random variable (for any η and m), then its expected value is *not* equal to what we are trying to estimate (namely, the long-term autocorrelation) until the window length becomes infinite. Formally, if $s(n)$ is a realization of a wide sense stationary (WSS), correlation-ergodic, random process, $\underline{s}$, then ${}^1r_s(\eta; m)$ can be seen to be one outcome of the random variable ${}^1\underline{r}_s(\eta, m)$, where

$$ {}^1\underline{r}_s(\eta, m) \stackrel{\text{def}}{=} \frac{1}{N} \sum_{n=m-N+1+|\eta|}^{m} \underline{s}(n)\underline{s}(n-|\eta|), \tag{4.37} $$

in which $\underline{s}(n)$ and $\underline{s}(n-|\eta|)$ are random variables drawn from $\underline{s}$. Now,

$$ \begin{aligned} \mathcal{E}\{{}^1\underline{r}_s(\eta, m)\} &= \frac{1}{N} \sum_{n=m-N+1+|\eta|}^{m} \mathcal{E}\{\underline{s}(n)\underline{s}(n-|\eta|)\} \\ &= \frac{N-|\eta|}{N} r_{\underline{s}}(\eta) \\ &= \frac{N-|\eta|}{N} r_s(\eta). \end{aligned} \tag{4.38} $$

(Note that we can replace $r_{\underline{s}}$ by r_s because of the assumed ergodicity.) We conclude that as $|\eta| \to N$ the expected value of the estimator is zero. This is due to the fact that fewer data points are used to compute ${}^1r_s(\eta; m)$ as $|\eta|$ gets close to N. Note, however, that as $N \to \infty$,

$$ \mathcal{E}\{{}^1\underline{r}_s(\eta; m)\} \to r_s(\eta). \tag{4.39} $$

2. ${}^1r_s(\eta, m)$ is a *consistent* estimator (Jenkins and Watts, 1968). This means that both the bias and the variance of the estimator [with respect to $r_{\underline{s}}(\eta) = r_s(\eta)$] vanish as $N \to \infty$.

The bias problem inherent in ${}^1r_s(\eta; m)$ can be eliminated by using an η-dependent rectangular window on the speech,

$$ w(n) = \begin{cases} \sqrt{N/(N-|\eta|)}, & n = 0, 1, 2, \ldots, N-1 \\ 0, & \text{other } n \end{cases}, \tag{4.40} $$

yielding the common estimator,

$$ {}^2r_s(\eta; m) = \frac{1}{N-|\eta|} \sum_{n=m-N+1+|\eta|}^{m} s(n)s(n-|\eta|). \tag{4.41} $$

In this case,[3]

1. ${}^2r_s(\eta; m)$ is unbiased:

$$ \mathcal{E}\{{}^2\underline{r}_s(\eta; m)\} = \frac{N}{N-|\eta|} \mathcal{E}\{{}^1\underline{r}_s(\eta; m)\} = r_{\underline{s}}(\eta) = r_s(\eta). \tag{4.42} $$

[3]As above, ${}^2\underline{r}_s(\eta; m)$ represents the random variable of which ${}^2r_s(\eta; m)$ is an outcome.

2. The variance of ${}^2r_{\underline{s}}(\eta; m)$ with respect to $r_{\underline{s}}(\eta) = r_s(\eta)$ becomes large as $|\eta| \rightarrow N$, making this estimator unreliable for large $|\eta|$ in spite of its unbiasedness (Jenkins and Watts, 1968).
3. ${}^2r_{\underline{s}}(\eta; m)$ is a consistent estimator of $r_s(\eta)$.

Finally, we note another short-term estimator of autocorrelation that avoids the bias problem of ${}^1r_s(\eta; m)$, which can be derived beginning with (4.25):

$$ {}^3r_s(\eta; m) = \frac{1}{N} \sum_{n=-\infty}^{+\infty} \mathcal{T}(\eta)\{s(n)\}\, w(m-n), \tag{4.43}$$

where $w(n)$ is an η-independent window. Letting $w(n)$ be a length N *rectangular* window as above, we obtain

$$ {}^3r_s(\eta; m) = \frac{1}{N} \sum_{n=m-N+1}^{m} s(n)s(n-|\eta|). \tag{4.44}$$

Although developed quite differently, this estimator is closely related to what Rabiner and Schafer (1978, Sec. 4.6) have called the *modified short-term autocorrelation function*.[4]

Short-Term "Covariance"

Consider for a moment that the speech sequence $s(n)$ is a realization of a general stochastic process, $\underline{s}$. In general, the long-term autocorrelation of the random process is a statistical expectation that is a function of two shift parameters, α and β:

$$ r_{\underline{s}}(\alpha, \beta) = \mathcal{E}\{\underline{s}(n-\alpha)\underline{s}(n-\beta)\}, \tag{4.45}$$

where $\underline{s}(n)$ represents the random variable at time n in $\underline{s}$. Only if $\underline{s}$ is at least WSS does $r_{\underline{s}}$ become a function of the (absolute) time difference, $\eta = \beta - \alpha$, and in this case can be written with a single argument. That is, when $\underline{s}$ is WSS,

$$ \begin{aligned} r_{\underline{s}}(\alpha, \beta) &= \mathcal{E}\{\underline{s}(n-\alpha)\underline{s}(n-\beta)\} \\ &= \mathcal{E}\{\underline{s}(n)\underline{s}(n-\eta)\} = r_{\underline{s}}(0, \eta) \end{aligned} \tag{4.46}$$

and it is customary to employ the notation $r_{\underline{s}}(\eta)$ to mean $r_{\underline{s}}(0, \eta)$. When $\underline{s}$ is *nonstationary*, we are forced to employ the two-argument computation, but when $\underline{s}$ is *stationary*, it is still perfectly mathematically correct (even if inconvenient) to work with the two-argument version of $r_{\underline{s}}$. The corresponding two-argument *temporal* autocorrelation function is

[4]Their comments regarding the evenness of the estimator do not apply here.

$$\begin{aligned} r_s(\alpha,\beta) &= \lim_{N\to\infty} \frac{1}{2N+1} \sum_{n=-N}^{N} s(n-\alpha)s(n-\beta) \\ &= \mathcal{L}\{s(n-\alpha)s(n-\beta)\}. \end{aligned} \tag{4.47}$$

Generalizing our discussion leading to (4.25) to allow a two-parameter operator, $\mathcal{J}(\alpha,\beta)$ we see that a short-term version of (4.46) is given by

$$\varphi_s(\alpha,\beta;m) = \frac{1}{N} \sum_{n=-\infty}^{\infty} s_{\mathcal{J}}(n;\alpha,\beta)w(m-n), \tag{4.48}$$

where $s_{\mathcal{J}}(n;\alpha,\beta) = \mathcal{J}(\alpha,\beta)\{s(n)\} = s(n-\alpha)s(n-\beta)$. Letting $w(n)$ be a rectangular window of length N results in the *short-term covariance function,*

$$\varphi_s(\alpha,\beta;m) = \frac{1}{N} \sum_{n=m-N+1}^{m} s(n-\alpha)s(n-\beta). \tag{4.49}$$

We cautioned the reader in Section 1.2.2 to master the correct definitions of the terms *correlation* and *covariance* because they are used in some sloppy ways in signal processing. In fact, there are two pieces of bad jargon associated with the short-term covariance function of which the reader must beware. The first is the name itself, which is accurate only if the average value of the process is zero. For this reason, we avoid the notation $c(\alpha,\beta;m)$, which would imply a sort of parallel to the long-term covariance. The second problem is the frequent reference in the literature to the use of this function in the "nonstationary case." This reference is only an allusion to the need for a two-argument autocorrelation in the case of a stochastic process that is not WSS. Of course, speech is inherently nonstationary. If it were not, we would not bother studying short-term processing. To call this the "nonstationary case" seems to imply that when a single argument estimator is used, the speech is stationary, which of course is not true. On the other hand, with either a one- or two-argument estimator we assume that the signal is long-term stationary and attempt to estimate the long-term properties (autocorrelation) over the short frame. If we did not model the problem in this way, the waveform could not be assumed ergodic and the very use of a temporal average like (4.49) would be nonsense.

Note, finally, that for large N, all of the estimators above become equivalent in the sense that all approach the long-term $r_s(|\eta|)$.

Short-Term Power Spectrum

For a WSS stochastic process $\underline{s}$, we define the *short-term power density spectrum* (stPDS) over a frame ending at m by

$$\Gamma_{\underline{s}}(\omega;m) \stackrel{\text{def}}{=} \sum_{\eta=-\infty}^{\infty} r_{\underline{s}}(\eta;m)e^{-j\omega\eta}, \tag{4.50}$$

where $r_{\underline{s}}(\eta;m)$ is a short-term estimator of autocorrelation. Since $\Gamma_{\underline{s}}(\omega;m)$ is the DTFT of the sequence $r_{\underline{s}}(\eta;m)$, the short-term autocorrelation can be obtained by inverting the transform,

$$r_{\underline{s}}(\eta;m) = \frac{1}{2\pi}\int_{-\pi}^{\pi} \Gamma_{\underline{s}}(\omega;m)e^{j\omega\eta}\,d\omega. \tag{4.51}$$

From (4.51) it is easy to see that the average power in the frame, $P_{\underline{s}} = r_{\underline{s}}(0;m)$, is given by the normalized total area under $\Gamma_{\underline{s}}(\omega;m)$,

$$P_{\underline{s}} = \frac{1}{2\pi}\int_{-\pi}^{\pi} \Gamma_{\underline{s}}(\omega;m)\,d\omega = \frac{1}{\pi}\int_{0}^{\pi} \Gamma_{\underline{s}}(\omega;m)\,d\omega. \tag{4.52}$$

In fact, to find the power in any frequency range, say ω_1 to ω_2, for $\underline{s}$ on the frame, we can compute

$$\text{Power in } \underline{s} \text{ in frequencies } \omega_1 \text{ to } \omega_2 \text{ on the frame} = \frac{1}{\pi}\int_{\omega_1}^{\omega_2} \Gamma_{\underline{s}}(\omega;m)\,d\omega. \tag{4.53}$$

The reader is encouraged to compare these definitions and results with those in Section 1.2.3.

When the signal is deterministic, or when $s(n)$ represents a sample function of an ergodic random process, then the temporal short-term autocorrelation, $r_s(\eta;m)$, is computed and the *temporal short-term power density spectrum* is defined as its DTFT,

$$\Gamma_s(\omega;m) \stackrel{\text{def}}{=} \sum_{\eta=-\infty}^{\infty} r_s(\eta;m)e^{-j\omega\eta}. \tag{4.54}$$

We will give another interpretation of the temporal stPDS in terms of the short-term DTFT in Section 4.3.5.

Short-Term Cross-Correlation and Cross-PDS

From any of the short-term estimators of autocorrelation, we can immediately infer a similar *short-term cross-correlation* function. For example, in a similar manner to the derivation of ${}^1r(\eta;m)$ of (4.36), we could deduce the estimator

$${}^1r_{xy}(\eta;m) = \frac{1}{N}\sum_{n=m-N+1+|\eta|}^{m} x(n)y(n-|\eta|) \tag{4.55}$$

for the two sequences $x(n)$ and $y(n)$. In general, let us simply refer to $r_{xy}(\eta;m)$. Then the *short-term cross-PDS* is defined as

$$\Gamma_{xy}(\omega;m) \stackrel{\text{def}}{=} \sum_{\eta=-\infty}^{\infty} r_{xy}(\eta;m)e^{-j\omega\eta}. \tag{4.56}$$

Application of $r_s(\eta; m)$ to Pitch Detection

The short-term autocorrelation and covariance are among the most useful computations made on the speech waveform. They will play a central role in many upcoming topics. At present, we do not have the tools to describe these advanced applications; so, for the purposes of illustration, we show some relatively simply uses of the short-term feature as it has been applied to the problem of *pitch detection.*

From basic probability theory, we know that we can interpret the autocorrelation as an indicator of the degree of linear relationship that exists between any two random variables spaced η apart in a stationary random process. For an ergodic process, therefore, we infer that the autocorrelation relates the degree of expected linear relationship that exists between any two points that are spaced by η in time in a sample waveform. When we move to the short-term case, by purely heuristic arguments, we expect $r_s(\eta; m)$ to indicate the expected amount of relationship that exists between time points spaced η apart on the window ending at time m.

In Fig. 4.4 we see the waveform for the utterance "three" and plots of $r_s(\eta; m)$ for two values of m. The short-term estimator of autocorrelation is ${}^3r(\eta; m)$ given in (4.44), using a $N = 256$ point window. Note the strong indication of periodicity [large values of $r_s(\eta; 2756)$ for $\eta = iP$, i an integer and P the pitch period] when the window is entirely over the voiced phoneme /i/ ($m = 2756$), and the lack of strong periodicity when the window involves an unvoiced region ($m = 500$). One may get the idea that $r_s(\eta; m)$ would make a good detector and tracker of pitch. Indeed, this idea has been explored, but direct autocorrelation methods are seldom used in practice because they are more error prone than methods that are only slightly more complicated to implement. (Further methods will be discussed as the material becomes accessible later in the book.) The main problem with direct use of $r_s(\eta; m)$ is that the first formant frequency, which is often near or even below the fundamental pitch frequency, can interfere with its detection. If the first formant is particularly strong, this can create a competing periodicity in the speech waveform that is manifest in the autocorrelation.[5] A secondary problem is that the speech is truly only "quasi-periodic," causing the peaks of $r_s(\eta; m)$ to be less prominent, and, in turn, making peak-picking difficult.

Investigators have attempted various signal preprocessing measures to make the autocorrelation focus more intently on the fundamental period of the waveform. Among these are raising the speech waveform to a large odd (to preserve the sign) power (Atal, 1968), and *center clipping.* In the latter, the lowpass filtered[6] time domain waveform is subjected to a nonlinear operation like (Sondhi, 1968)

[5]Recall that there can be energy at frequencies other than the harmonics of the pitch frequency because of the windowing in short-term processing.

[6]In all techniques discussed, it is conventional to remove high-frequency content in the speech waveform first by lowpass filtering.

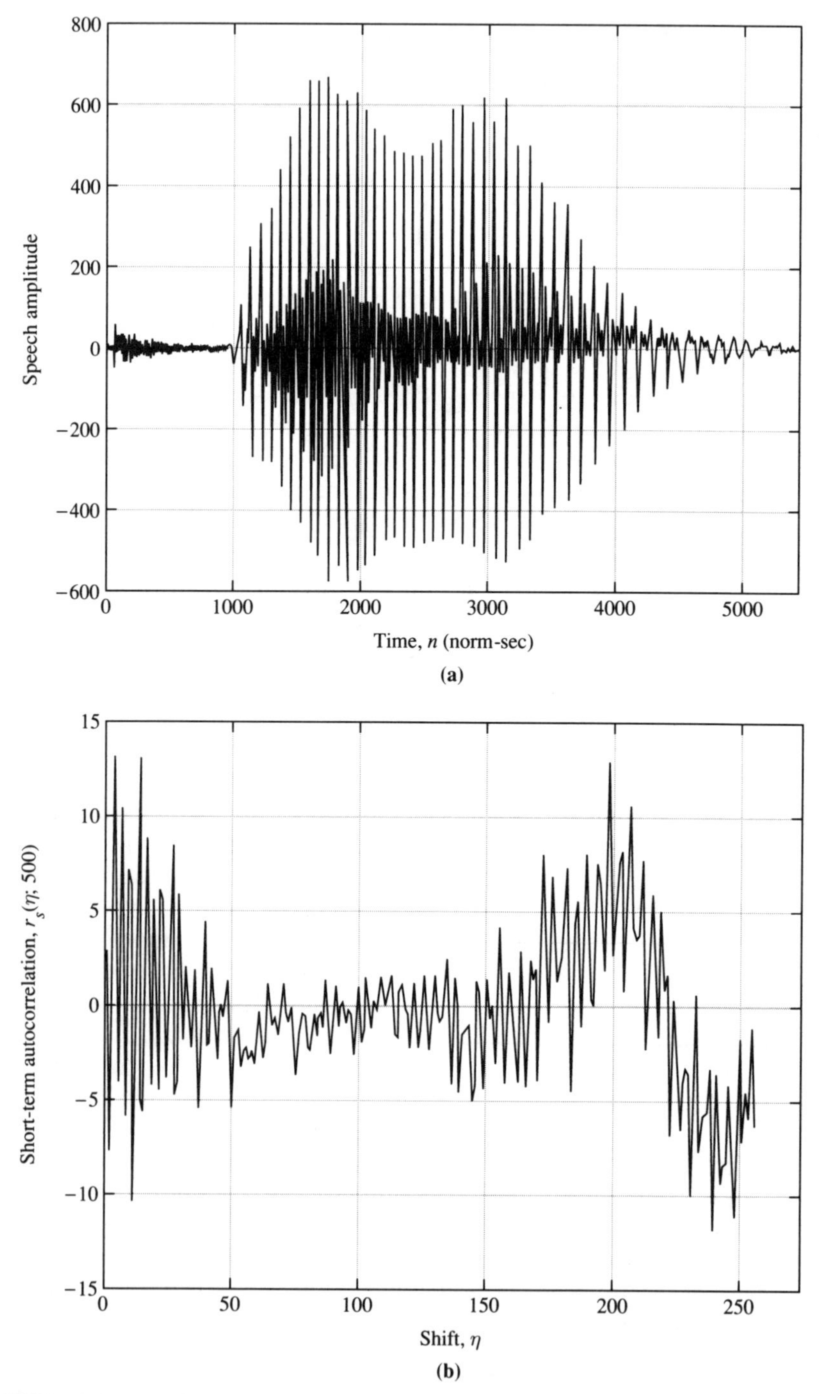

FIGURE 4.4. Utterance of the word "three" and plots of short-term autocorrelation, $r_s(\eta; m)$ versus η based on 256-point Hamming windows ending at $m = 500$ and 2756. These two frames correspond to the fricative /T/ and the vowel /i/, respectively.

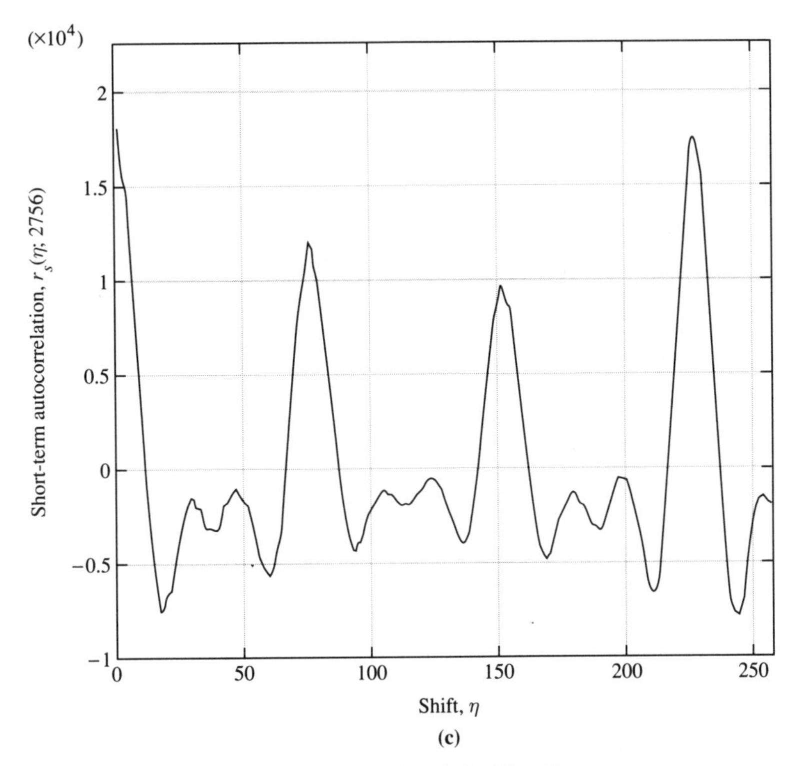

(c)

FIGURE 4.4. *(Cont.)*

$$C\{s(n)\} = \begin{cases} s(n) - C^{+}, & s(n) > C^{+} \\ 0, & C^{-} \le s(n) \le C^{+}. \\ s(n) - C^{-}, & s(n) < C^{-} \end{cases} \tag{4.57}$$

Typically, the clipping limits are set to $\pm 30\%$ of the absolute maximum of the waveform. The clipping operator of (4.57) for this typical case, and its effects on a speech waveform for the word "three" are illustrated in Fig. 4.5. Figures 4.4 and 4.5 can be used to compare the autocorrelation of the unmodified speech and that of the clipped speech for the word "three". Other clipping operators have been investigated by Dubnowski et al. (1975). These are depicted in Fig. 4.6.

Before leaving the issue of center clipping for pitch detection, it should be noted that this procedure constitutes a "whitening" process on the spectrum, since it makes the speech more pulselike (see Problem 4.9). Hence center clipping will tend to emphasize high frequencies in noisy speech and cause pitch detection performance to deteriorate (Paliwal and Aarskog, 1984).

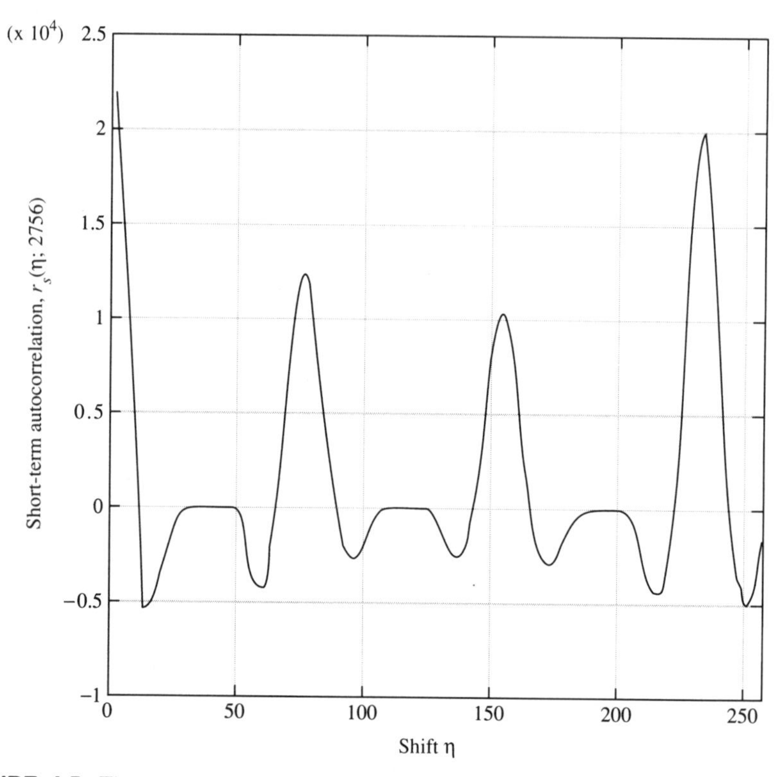

FIGURE 4.5. The experiment of Fig. 4.4 repeated with center clipping applied to the speech prior to computing the autocorrelation. Clipping limits are set to ±30% of the absolute maximum value of the waveform.

4.3.2 Average Magnitude Difference Function

Let us formally define the *average magnitude difference function* (AMDF) for an eternally stationary signal, $s(n)$, by

$$\Delta M_s(\eta) = \mathcal{L}\{|s(n) - s(n-\eta)|\}. \tag{4.58}$$

It should be clear that this family of features, indexed by the time difference parameter η, takes on small values when η approaches the period of $s(n)$ (if any), and will be large elsewhere. Accordingly, it can be used for pitch period estimation. Applying Construction Principle 2 to obtain a realistic family of features, we have

$$\Delta M_s(\eta; m) = \frac{1}{N} \sum_{n=m-N+1}^{m} |s(n) - s(n-\eta)|\, w(m-n). \tag{4.59}$$

The computation of this feature is depicted as a convolution in Fig. 4.7, and its application to a pitch detection problem is illustrated in Fig. 4.8.

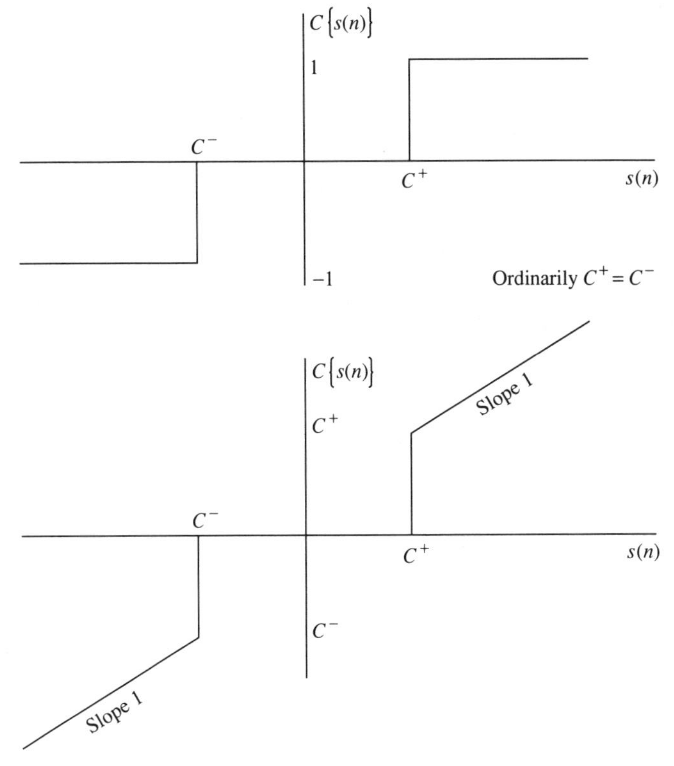

FIGURE 4.6. Dubnowski clipping operators.

4.3.3 Zero Crossing Measure

The number of *zero crossings* (number of times the sequence changes sign) is also a useful feature in speech analysis. Formally defined in the long term, the *zero crossing measure* is

$$Z_s = \mathcal{L}\left\{\frac{|\operatorname{sgn}\{s(n)\} - \operatorname{sgn}\{s(n-1)\}|}{2}\right\}, \tag{4.60}$$

where

$$\operatorname{sgn}\{s(n)\} = \begin{cases} +1, & s(n) \geq 0 \\ -1, & s(n) < 0 \end{cases}. \tag{4.61}$$

Construction Principle 2 can be used to derive a *short-term zero crossing measure* for the N-length interval ending at $n = m$,

$$Z_s(m) = \frac{1}{N} \sum_{n=m-N+1}^{m} \frac{|\operatorname{sgn}\{s(n)\} - \operatorname{sgn}\{s(n-1)\}|}{2} w(m-n). \tag{4.62}$$

Before showing an example of this feature, we will review another and use them together.

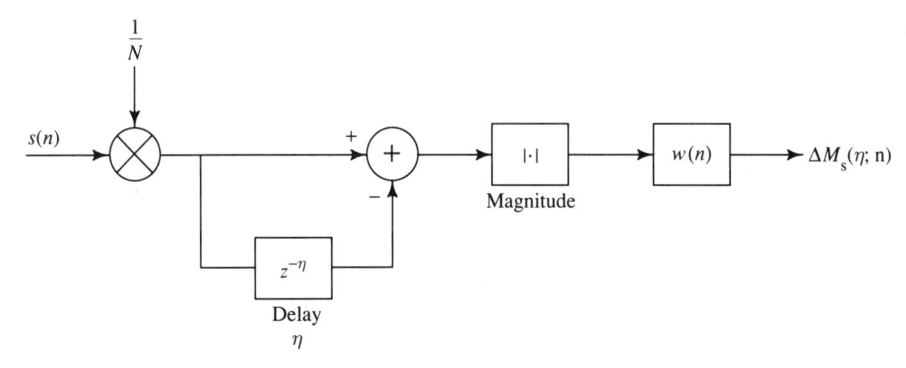

FIGURE 4.7. Short-term AMDF viewed as a convolution.

4.3.4 Short-Term Power and Energy Measures

Definitions

While urging the reader to review the comments in Section 4.2.5, we recall the *short-term power* and *short-term energy* measures for the N-length frame ending at time m,

$$P_s(m) = \frac{1}{N} \sum_{n=m-N+1}^{m} s^2(n) \tag{4.63}$$

and

$$E_s(m) = \sum_{n=m-N+1}^{m} s^2(n), \tag{4.64}$$

respectively. For most practical purposes, these measures provide equivalent information, so that the simpler of the two, $E_s(m)$, is usually preferred.

We now consider an example in which the short-term energy and zero crossing measures are used in combination to segment a signal into speech and "silence" regions.

Endpoint Detection Using Energy and Zero Crossing Measures

The problem of locating the endpoints in time of a discrete utterance is an important problem in many speech processing applications. For ex-

FIGURE 4.8. Application of the short-term AMDF $\Delta M_s(\eta; m)$ to the detection of pitch for the word "seven." (a) Signal for the utterance of "seven." Short-term AMDF computed over rectangular windows of length 256 ending at (b) $m = 2500$, (c) $m = 3000$, (d) $m = 4775$, (e) $m = 5000$. Windows in (b) and (c) are both positioned in "voiced region I" and each produces a similar pitch frequency estimate. Windows in (d) and (e) are both positioned in "voiced region II" and likewise produce similar pitch estimates.

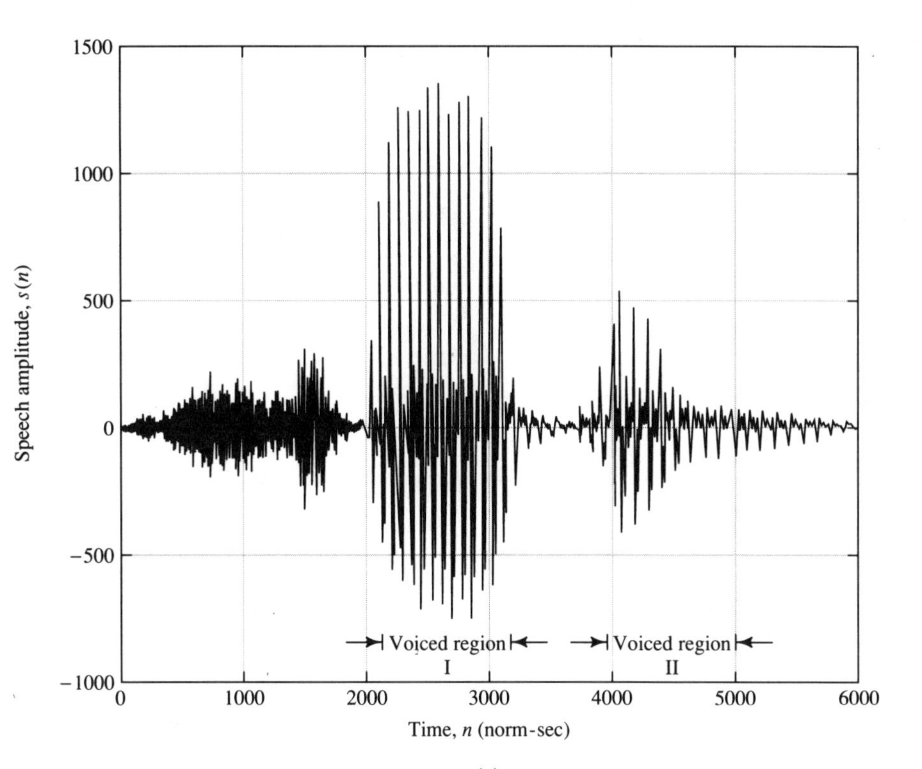
Speech amplitude, $s(n)$
Time, n (norm-sec)
Voiced region I
Voiced region II
1500
1000
500
0
−500
−1000
0
1000
2000
3000
4000
5000
6000
(a)

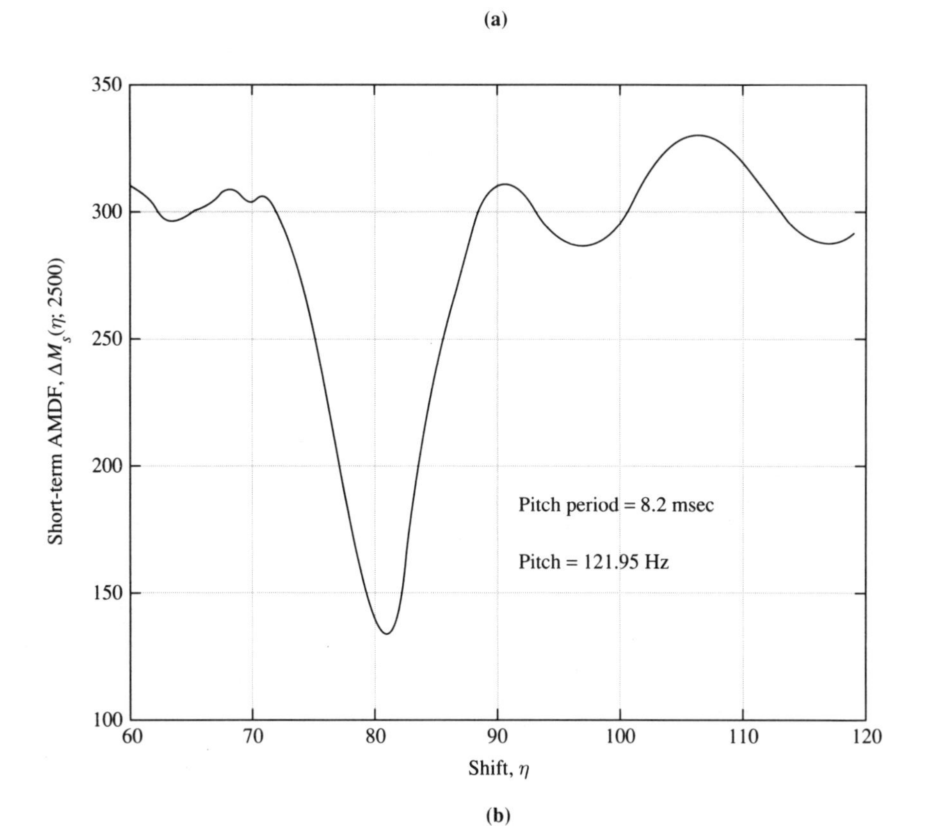
Short-term AMDF, $\Delta M_s(\eta; 2500)$
Shift, η
Pitch period = 8.2 msec
Pitch = 121.95 Hz
350
300
250
200
150
100
60
70
80
90
100
110
120
(b)

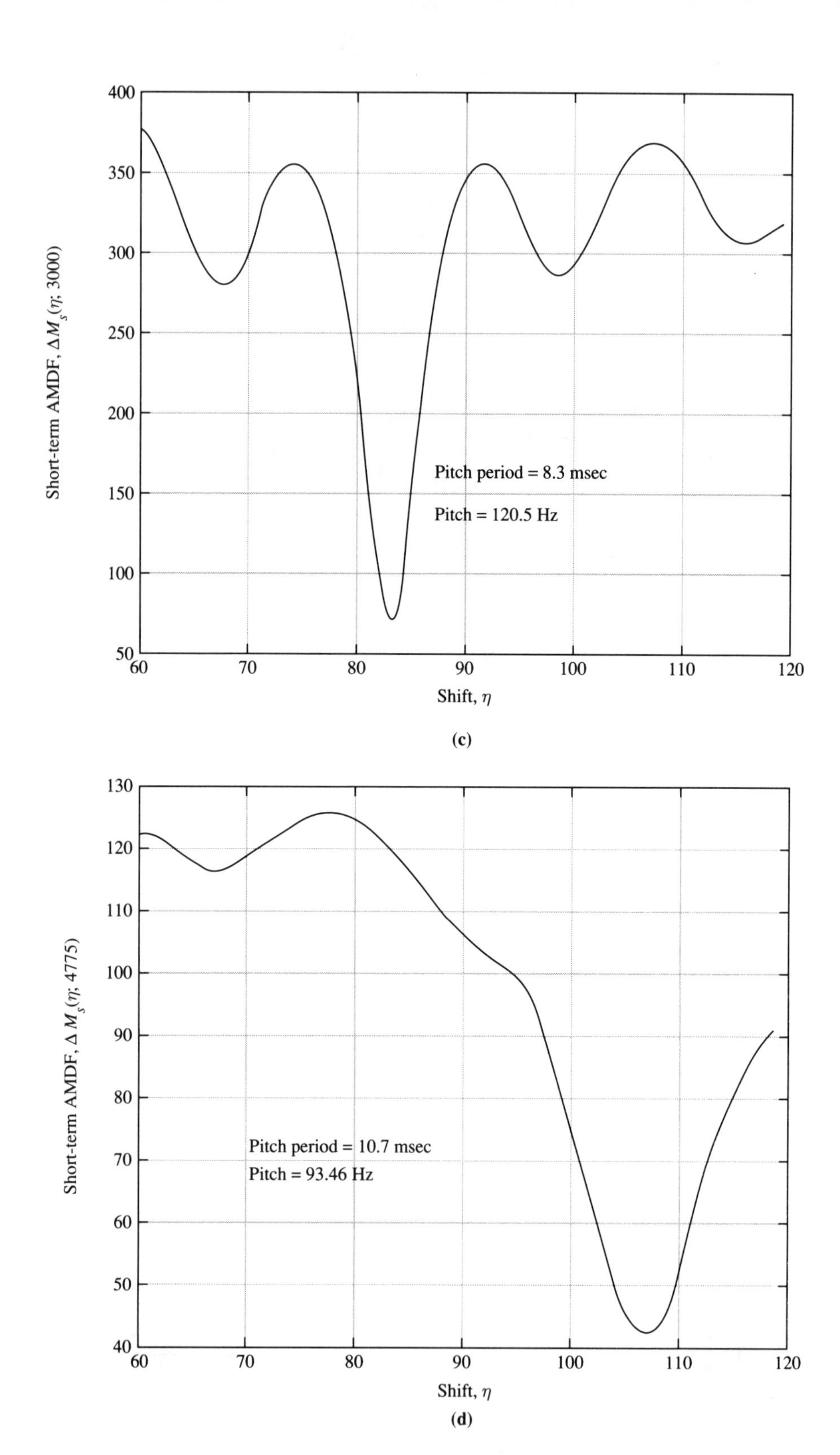

Short-term AMDF, $\Delta M_s(\eta; 3000)$
Pitch period = 8.3 msec
Pitch = 120.5 Hz
Shift, η
400
350
300
250
200
150
100
50
60
70
80
90
100
110
120

(c)

Short-term AMDF, $\Delta M_s(\eta; 4775)$
Pitch period = 10.7 msec
Pitch = 93.46 Hz
Shift, η
130
120
110
100
90
80
70
60
50
40
60
70
80
90
100
110
120

(d)

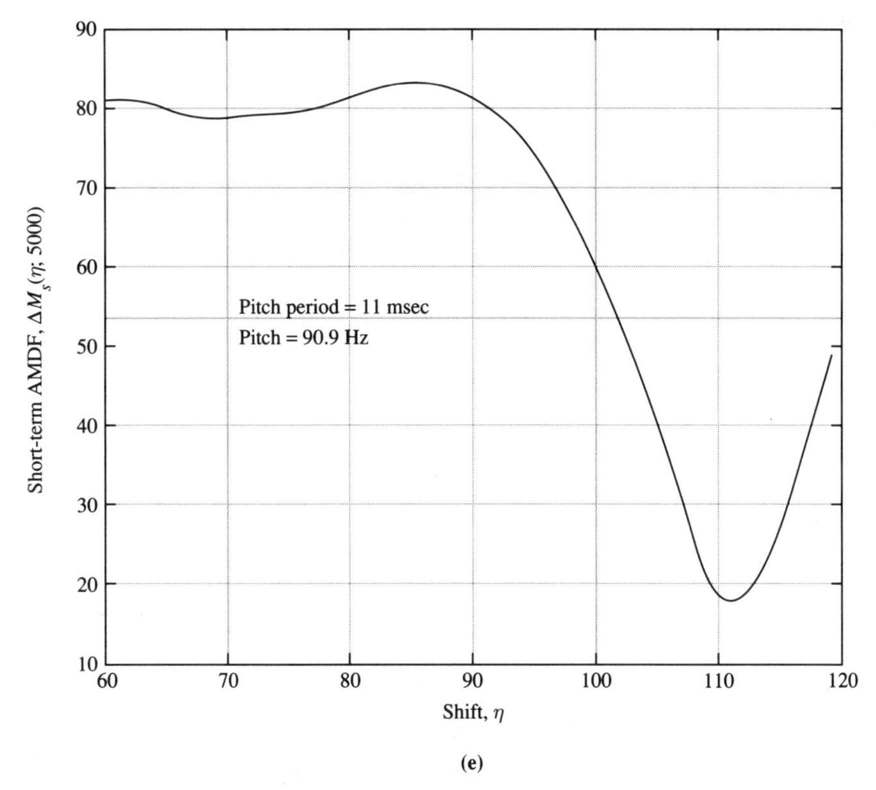

(e)

FIGURE 4.8. (*continued*). Windows in (c) are positioned in "voiced region I and produce a similar pitch frequency estimate. Windows in (d) and (e) are both positioned in voiced region II and likewise produce similar pitch estimates.

ample, in Part V we will study several methods for the recognition of words by comparing their acoustic signals with "template" words in the recognizer. In some methods it is necessary for the incoming word to be as free of "nonspeech" regions as possible to avoid such regions from causing mismatch. The problem of detecting endpoints would seem to be relatively trivial, but, in fact, it has been found to be very difficult in practice, except in cases of very high signal to ("background") noise ratios. Some of the principal causes of endpoint detection failures are weak fricatives (/f/, /T/, /h/) or voiced fricatives that become unvoiced at the end ("has"), weak plosives at either end (/p/, /t/, /k/), nasals at the end ("gone"), and trailing vowels at the end ("zoo"). We will discuss this problem in more detail in the context of speech recognition in Chapter 10.

As a medium for illustrating the use of the short-term zero crossing and energy measures, we briefly discuss the endpoint detection problem. A widely used method for this task was published by Rabiner and Sambur (1975). In Fig. 4.9 we see the short-term zero crossing and en-

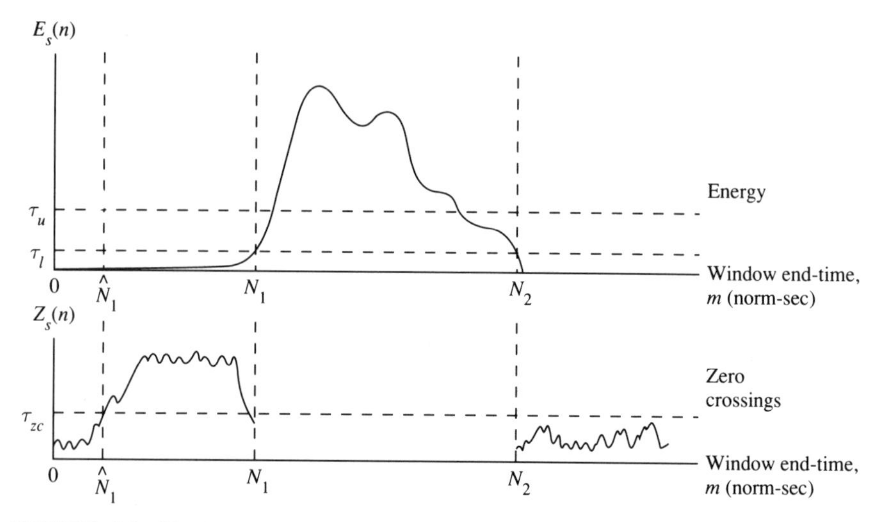

FIGURE 4.9. Short-term energy and zero crossing measures plotted for the word "four." The utterance has a strong fricative at the beginning; hence, the zero crossing level is initially high, and the energy low. The opposite is true as the signal enters the voiced portion of the utterance. After Rabiner and Sambur (1975).

ergy measures[7] plotted for the word "four." These curves are the result of computing each measure every 10 msec on frames of length 10 msec. It is assumed that the first 10 frames are background. They are used to find the mean and variance of each of the features. In turn, these statistics are used to set "upper" and "lower" thresholds, τ_u and τ_l, as shown in the figure. The energy curve is then searched to find the first crossing of the upper threshold τ_u moving toward the middle of the segment from each end. Then we "back down" to the nearest crossing of τ_l in each case. This process yields tenative endpoints N_1 and N_2 in the figure. The double-thresholding procedure prevents the false indication of endpoints by dips in the energy curve. Now we move toward the ends from N_1 and N_2 for no more than 25 frames, examining the zero crossing rate to find three occurrences of counts above the threshold τ_{zc}. If these are not found, the endpoint remains at the original estimate. This is the case with N_2 in Fig. 4.9. If three occurrences are found, then the endpoint estimate is moved backward (or forward) to the time of the first threshold crossing. This is the case for N_1 (moved to $\hat{N}_1$) in the figure.

We can pose the endpoint detection problem as one of discerning

[7]In fact, the "energy" measure here is actually a *magnitude* measure similar to

$$M_s(m) = \sum_{n=m-N+1}^{m} |s(n)|. \tag{4.65}$$

That this measure gives equivalent information to the energy measure discussed above should be apparent.

speech from background noise. A related problem is that of classifying speech into voiced and unvoiced regions. We might, for example, wish to know when a pitch estimate is necessary in a speech coder. We can infer from Fig. 4.9 that the same measures could be used to at least give an initial voiced/unvoiced segmentation. The short-term zero crossing rate tends to be larger during unvoiced regions (indicator of higher frequencies), while the energy will usually be larger during voiced segments.

4.3.5 Short-Term Fourier Analysis

The Short-Term DTFT[8]

The DTFT of an infinitely voiced (or unvoiced) speech sound does not exist because the sequence is not of finite energy. However, we are ultimately going to use an "energy-type" transform on the frame, hence the name "stDTFT." To adhere to our philosophy of beginning with a long-term "power-type" computation, however, we obviously should not begin with the DTFT since the DTFT will[9] "blow up." What *should* we use? To answer this question, and to facilitate the developments below, let us avoid the complications of stochastic processes and focus on the voiced case. We will return to the unvoiced case below. For the voiced case, the discrete Fourier series (DFS) might come to mind as the appropriate long-term starting point, but, looking ahead, we realize that the frame will not generally be a periodic signal. This means that a "short-term DFS" might be less than satisfying because it would not reveal the spectral nature of the frame in between the harmonic frequencies.[10] We seem to need a "compromise" between the Fourier transform and Fourier series with which to begin, and we have the additional requirement that the "transform" accommodate power signals. An excellent choice would be the *complex envelope spectrum*, which we first introduced in (1.35). Indeed a correct interpretation of the stDTFT we are about to develop is that it is a frequency domain transform that serves as an estimator of the complex envelope spectrum of the speech. Accordingly, within a scale factor it is "trying to be" the DFS coefficients at the harmonics of the signal.[11]

[8]Some of the early work on this topic, based on analog signal analysis, is found in papers by Fano (1950) and Schroeder and Atal (1962). More recent work based on discrete-time analysis is reported, for example, in Portnoff (1980). A list of additional references can be found in the tutorial by Nawab and Quatieri (1988).

[9]Actually, for a deterministic signal, we can "create" a DTFT with analog impulse functions at the harmonic frequencies (see Section 1.1.4), but this does not serve any useful purpose here.

[10]Further, in practice, we might not have knowledge of the period of the waveform, and might not have a good means for estimating it.

[11]It will become clear below that another correct interpretation views the stDTFT as the result of an "illegal" long-term DTFT (composed of analog impulse functions at the harmonics) having been convolved with the spectrum of the window creating the frame. Hence the stDTFT is an estimate of the long-term DTFT with potentially severe "leakage" of energy from the harmonic frequencies.

The complex envelope spectrum of the speech is given by [see (1.35)]

$$\bar{S}(\omega) = \frac{1}{P} \sum_{n=0}^{P-1} s(n) e^{-j\omega n}, \tag{4.66}$$

where P is the pitch period of the speech. For formal purposes, recall that this may be written

$$\bar{S}(\omega) = \mathcal{L}\{s(n)e^{-j\omega n}\} = \lim_{N\to\infty} \frac{1}{2N+1} \sum_{n=-N}^{N} s(n) e^{-j\omega n}. \tag{4.67}$$

Now using either Construction Principle 1 or 2 (see Problem 4.2), then ignoring the averaging factor $1/N$, we derive the *short-term DTFT* (stDTFT) (often just called the short-term Fourier transform or short-*time* Fourier transform) for an N-length frame ending at time m,

$$S_s(\omega; m) \stackrel{\text{def}}{=} \sum_{n=m-N+1}^{m} s(n) w(m-n) e^{-j\omega n} = \sum_{n=m-N+1}^{m} f(n;m) e^{-j\omega n}, \tag{4.68}$$

where $w(n)$ is any window of length N. Now the convention of using the uppercase S to indicate a transform of the sequence $s(n)$ makes the subscript unnecessary here, so we will write simply $S(\omega; m)$. Note that if we view $S(\omega; m)$ as a set of features of the speech, then ω plays the role of a *continuous parameter*, so that there is actually an uncountably infinite family of features, one for each ω.

Note that by dropping the factor $1/N$, we have created an "energy-type" computation that has a name, DTFT, which is not used to refer to the similar "power-type" computation. This is similar to what happened when we dropped the $1/N$ from the short-term average power to get short-term energy. For proper interpretation of the short-term energy, however, we encouraged the reader not to think of it as an estimator of "long-term energy" (which theoretically does not exist), but rather as short-term average power (estimating long-term power) with $1/N$ dropped for convenience. Similarly, here we encourage the reader to think of the stDTFT not as an estimator of the DTFT, but rather as an estimator of the complex envelope spectrum with the scale factor omitted for convenience.

For a given $\omega, S(\omega, m)$ can be viewed as the convolution of the complex sequence $s(n)e^{-j\omega n}$ with the real sequence $w(n)$. This computation is diagramed in Fig. 4.10. This view is sometimes called the "filtering interpretation" of the stDTFT. The filtering approach corresponds to the first expression on the right side of the definition in (4.68). The second expression corresponds to the "Fourier transform interpretation," since it involves a conventional DTFT on the frame $f(n; m)$. These interpretations, while suggesting two ways to *compute* the stDTFT, do not really provide clear interpretations of what it *means*. Hence we have been careful to provide this information above.

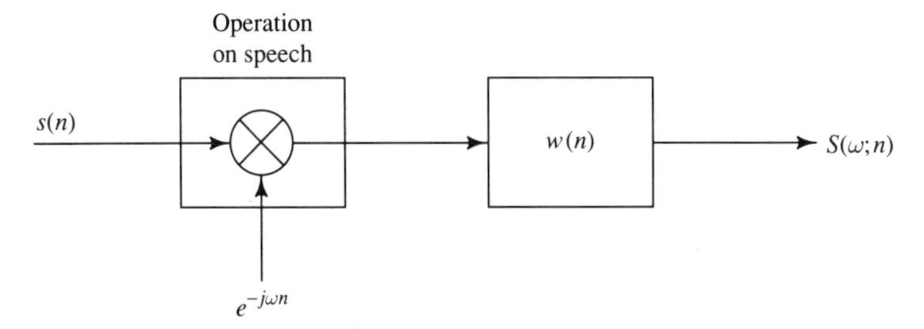

FIGURE 4.10. Computation of $S(\omega; n)$ viewed as a convolution.

Some typical short-term magnitude spectra based on the stDTFT for speech data are shown in Fig. 4.11.[12]

A philosophical remark is in order before leaving the deterministic case. We have interpreted $S(\omega; m)$ as a complex function that is theoretically trying to estimate DFS coefficients (to within a superfluous scale factor $1/N$) when evaluated at the harmonics. This is a correct interpretation, and perhaps not a bad one to keep in mind, but it is probably accurate to say that most DSP engineers do not routinely dwell on the stDTFT in these terms. Instead they just learn to read them, relate them, and accept them as DTFTs of "little signals," which are valuable and informative in their own right. In fact, with experience the reader will probably develop this same intuition about all of the short-term features we have discussed. The formal framework in which we have developed them, however, should contribute significantly to that intuition.

Since there are no "harmonics" present in the unvoiced case, we must also attempt to find a proper interpretation of the stDTFT when $s(n)$ is a stochastic signal. The most meaningful way to view $S(\omega; m)$ for a random process is as the "square root" of the stPDS of the process, that is,[13]

$$\Gamma_s(\omega; m) = \frac{S(\omega; m)S^*(\omega; m)}{N^2} = \frac{|S(\omega; m)|^2}{N^2}, \tag{4.69}$$

where $\Gamma_s(\omega; m)$ is defined in (4.54).

In fact, (4.69) holds whether $s(n)$ is a stochastic or deterministic signal. In a theoretical sense, however, it is more "necessary" for the stochastic case. In the deterministic case, we feel no uneasiness about dealing with a Fourier transform, in this case an stDTFT. We have given a perfectly acceptable interpretation of $S(\omega; m)$ for the deterministic case above. We should feel comfortable that $|S(\omega; m)|^2/N^2$ is a valid stPDS, and that writing it in these terms is acceptable. On the other hand, for a stochastic process for which we have been repeatedly told that Fourier

[12]Of course, these spectra are based on discrete frequency algorithms to be described below.

[13]The factor N^2 appears here because we have dropped it from the stDTFT in the definition.

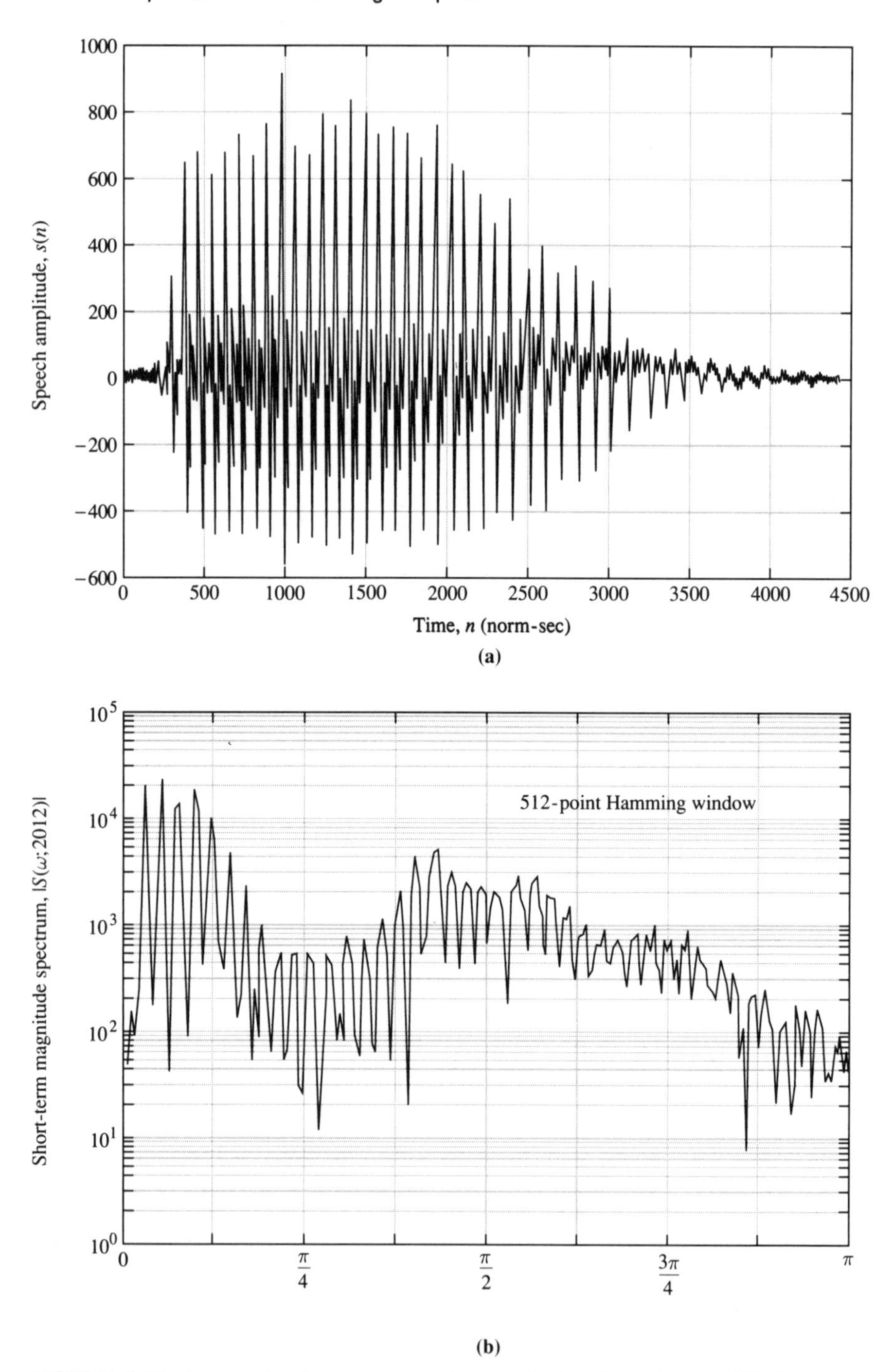

FIGURE 4.11. Some short-term spectral magnitudes based on stDTFTs. (a) Utterance of vowel /a/. (b, c, d) Short-term magnitude spectra based on 512-point Hamming, Hanning, and rectangular windows, respectively.

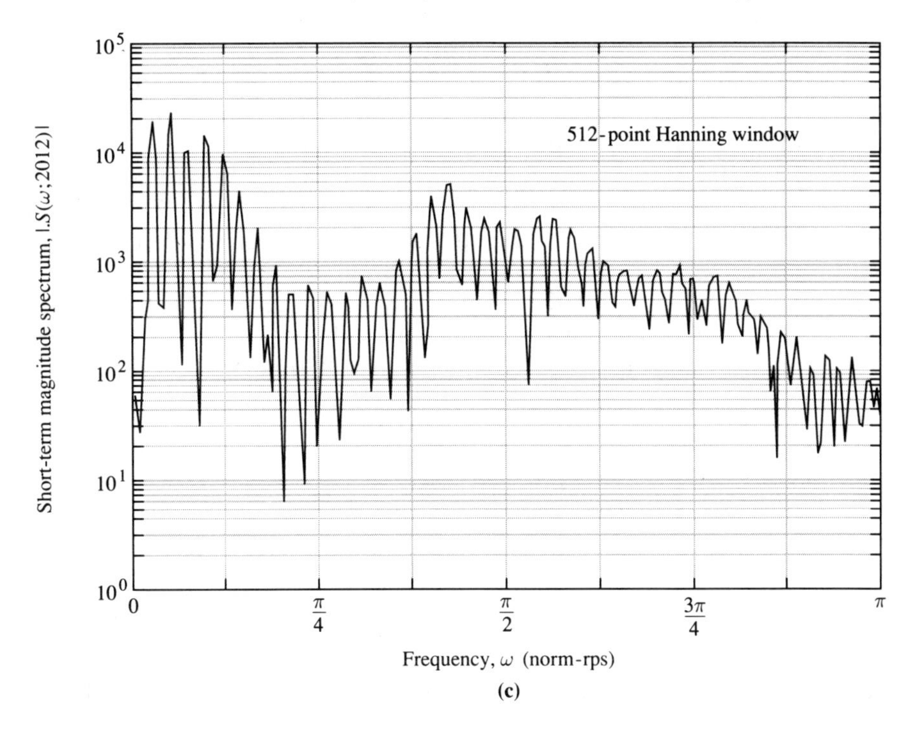

(c)

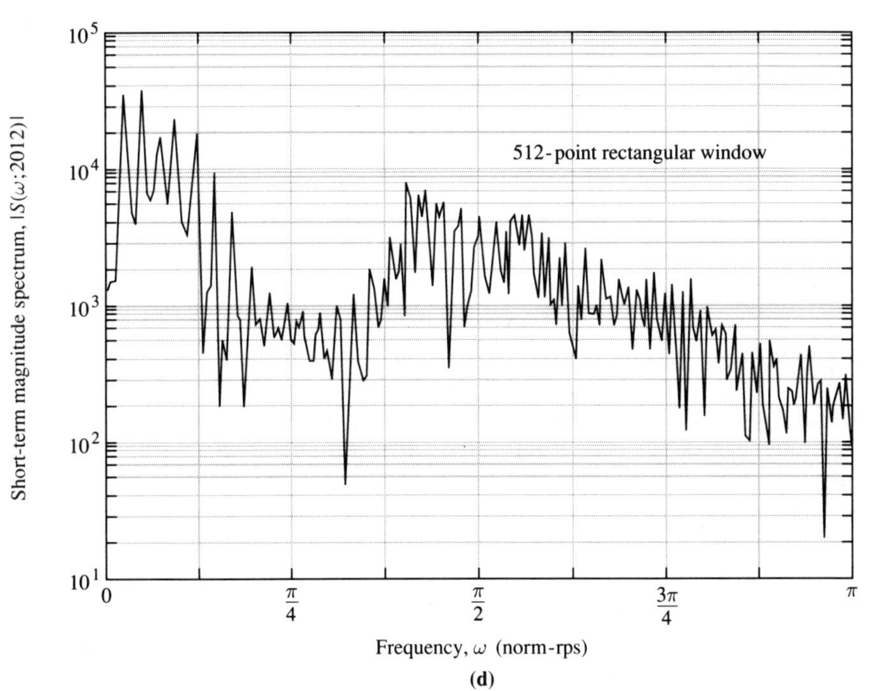

(d)

FIGURE 4.11. (*Cont.*)

transforms make no sense, it is comforting to be able to relate $S(\omega; m)$ to a quantity that "is" meaningful, namely, a power density spectrum. For a deterministic signal, we can view the stDTFT as an estimator of a Fourier transformlike function (the complex envelope spectrum), whereas in the stochastic case, we should only view it as a step toward computing an stPDS, which, in turn, is an estimate of a long-term PDS. In this sense, the phase spectrum of the stDTFT is not meaningful in the stochastic case.

Finally, let us note that the proper way to invert an stDTFT is with the conventional IDTFT inversion formula. The stDTFT is a legitimate DTFT—it just happens to be computed over a time sequence which is zero outside of a specific range. Therefore, we should expect to be able to recover that sequence using the usual IDTFT, and indeed we can:

$$f(n; m) = \frac{1}{2\pi} \int_{-\pi}^{\pi} S(\omega; m) e^{j\omega n} d\omega. \tag{4.70}$$

It is important to note that the stDTFT–IDTFT pair is "delay preserving." By this we mean that the proper phase shift is embodied in the stDTFT so that when it is inverted, the time sequence is properly delayed out to the time range $n = m - N + 1, \ldots, m$. We raise this issue here because it will be important in trying to decide how to formally use the DFT on short terms of speech, a subject which we take up next.

The "Short-Term DFT"

The very title of this little section should seem curious, since the DFT is inherently a short-term entity. However, we need to be cautious, especially in theoretical developments, about how we employ the DFT in short-term analysis. The issue is centered on whether or not we need "delay preservation" in a particular development. By this we mean we must be careful with the DFT *if it is important to keep track of where the frame is located in time.* If so, we will use what we will call the "short-term DFT." If not, we can use the "usual" DFT in the customary way.

What is the "customary way" in which we use DFTs? The DFT is inherently defined for a sequence assumed to have its nonzero portion on the range $n = 0, \ldots, N - 1$. We give little thought to the fact that we effectively shift any N-length sequence down to this time range before computing its DFT (FFT). Of course, the linear phase information corresponding to the proper delay of the signal is lost through this process and will not be reflected in the DFT shift points. Needless to say, if we invert the DFT using the usual IDFT form, the sequence will be returned to the range $n = 0, \ldots, N - 1$. All this is of very little practical consequence, since the user of the algorithm is usually well aware of the appropriate time shift, and, further, the omission of the proper delay from the DFT has no effect on the magnitude spectrum and no practical effect on the phase spectrum.

Occasionally, however, a theoretical problem might arise if we use the DFT in this cavalier fashion on a frame of speech covering the range $n = m - N + 1, \ldots, m$; say $f(n; m)$. Let us consider what happens if we first shift the frame down to the range $n = 0, \ldots, N-1$, and then compute its DFT:

$$f(n; m) \rightarrow f(n + m - N + 1; m) \qquad \text{(shift)} \tag{4.71}$$

$$\sum_{n=0}^{N-1} f(n + m - N + 1; m) e^{-jk(2\pi/N)n} \qquad (\text{DFT for } k = 0, \ldots, N-1). \tag{4.72}$$

Therefore, by a simple change of index on (4.72), we have that the "usual" DFT of the frame is equivalent to

$$\sum_{l=m-N+1}^{m} f(l; m) e^{-jk(2\pi/N)l} e^{jk(2\pi/N)(m-N+1)}, \qquad k = 0, \ldots, N-1. \tag{4.73}$$

The point of this is to note the following: Whereas the DFT is usually asserted to be samples of the DTFT at equally spaced frequencies on the Nyquist range, it is not true that the DFT, if used in the customary manner on a frame of speech, produces samples of the stDTFT at equally spaced frequencies. In fact, it is evident from (4.73) that

$$\begin{aligned} &\text{"usual" DFT of } f(n; m) \\ &= \sum_{n=m-N+1}^{m} f(n; m) e^{-jk(2\pi/N)n} e^{jk(2\pi/N)(m-N+1)}, \qquad k = 0, \ldots, N-1 \\ &= \text{samples of } S(\omega; m) e^{jk(2\pi/N)(m-N+1)} \text{ at } 2\pi k/N, \qquad k = 0, \ldots, N-1. \end{aligned} \tag{4.74}$$

This is because the stDTFT is "delay preserving," and the usual DFT is not.

It is clear that if we want a discrete Fourier transform that preserves the proper delay, and is therefore properly samples of $S(\omega; m)$, we should use

$$S(k; m) = \begin{cases} \displaystyle\sum_{n=m-N+1}^{m} f(n; m) e^{-jk(2\pi/N)n}, & k = 0, \ldots, N-1 \\ 0, & \text{otherwise} \end{cases}, \tag{4.75}$$

which we shall call the *short-term DFT* (stDFT). When the DFT is used in the customary way in which the frame at m is first shifted down in time, we will continue to use the customary term DFT. In this case, for convenience, let us define the shifted frame

$$\overleftarrow{f}(n; m) \stackrel{\text{def}}{=} f(n + m - N + 1; m) \tag{4.76}$$

and write

$$\overleftarrow{S}(k;m) = \begin{cases} \sum_{n=0}^{N-1} \overleftarrow{f}(n;m)e^{-jk(2\pi/N)n}, & k = 0, \ldots, N-1 \\ 0, & \text{otherwise} \end{cases}, \tag{4.77}$$

where the arrows above $\overleftarrow{f}$ and $\overleftarrow{S}$ are just reminders of the shift of the data before the computation.

Finally, we must discover how to invert the stDFT. Since $S(k;m)$ represents equally spaced samples of $S(\omega;m)$ on the Nyquist range, the fundamental theory of the DFT tells us that the use of the "usual" IDFT form on these frequency samples (if not truncated) will produce a periodic (potentially aliased) replication of the time sequence corresponding to $S(\omega;m)$, in this case $f(n;m)$. That is,

$$\frac{1}{N}\sum_{k=0}^{N-1} S(k;m)e^{j(2\pi/N)kn} = \sum_{i=-\infty}^{\infty} f(n+iN;m). \tag{4.78}$$

To obtain an appropriate inverse, we separate out the period corresponding to $i = 0$,

$$f(n;m) = \begin{cases} \frac{1}{N}\sum_{k=0}^{N-1} S(k;m)e^{j(2\pi/N)kn}, & n = m-N+1, \ldots, m. \\ 0, & \text{other } n \end{cases} \tag{4.79}$$

We shall refer to this form as the *short-term IDFT* (stIDFT).

Although it is probably obvious from the discussion above, for completeness we note that to invert the "usual" DFT, $\overleftarrow{S}(k;m)$, we use the "usual" IDFT formula and obtain the version of the frame that has been shifted down to the time origin, $\overleftarrow{f}(n;m) = f(n+m-N+1;m)$.

Use of Short-Term Fourier Analysis in Digital Transmission: Sampling in Time and Frequency

As we have seen, many features used in short-term analysis are actually families of features indexed by some parameter. Accordingly, the feature is essentially a sequence over two dimensions.[14] In the case of $S(\omega;m)$, this has a special significance.

Suppose that we are considering sending $S(\omega;m)$ (for a number of m's) over a digital transmission channel as an alternative to sending the

[14]If the feature parameter is *continuous* as in the case of $S(\omega;n)$, then the feature is a sequence over the time dimension and a *function* over the parameter dimension.

speech. Clearly, we must do some sampling of $S(\omega; m)$ along the ω dimension in order to discretize the (complex-valued) continuous function. This will result in a discrete two-dimensional sequence, with indices over both the frequency and time dimensions. We want to send as few of these numbers as possible in order to preserve channel bandwidth, but we also want to be sure that the receiver is able to reconstruct the speech from the information sent. We pursue here some small set of samples from which $s(n)$ is entirely recoverable.

Let us first fix m and discover how to sample along the ω dimension. From our discussion above, we see that the samples of $S(\omega; m)$, which comprise the stDFT, namely $S(k; m), k = 0, \ldots, N-1$, are a most natural set of samples to use. From the stDFT we can completely recover the frame $f(n; m)$, or, equivalently, the entire stDTFT if desired. Given the frame and knowledge of the window sequence, it is possible to recover $s(n), n = m - N + 1, \ldots, m$.

Now let us consider at what m's we must sample $S(k; m)$. If $w(n)$ is a finite window known to the receiver, then clearly we can just send one set[15] of frequency samples (one stDFT) for each adjacent frame of N speech samples. These can then be used to reconstruct the adjacent frames, and subsequently the N point blocks of speech by removing the window. Consequently, the $S(k; m)$ are computed at times $m = 0, N, 2N, \ldots$. (Note that this represents a resampling of time with respect to the original sampling rate on the speech data.) This method is not consistent with how short-term Fourier-like methods have been used in coding and synthesis, so we need not belabor this discussion. It is interesting to note, however, that no coding efficiency is gained for all this effort, since $N/2$ complex numbers are sent each N norm-sec for a net rate of 1 sample per norm-sec. For the same rate, the original speech waveform can be sent.

Closely related to what we will later call the *filter-bank method* of short-term spectral waveform encoding or synthesis is the following. Consider the computation of $S(k; n)$ at a particular k shown in Fig. 4.12.

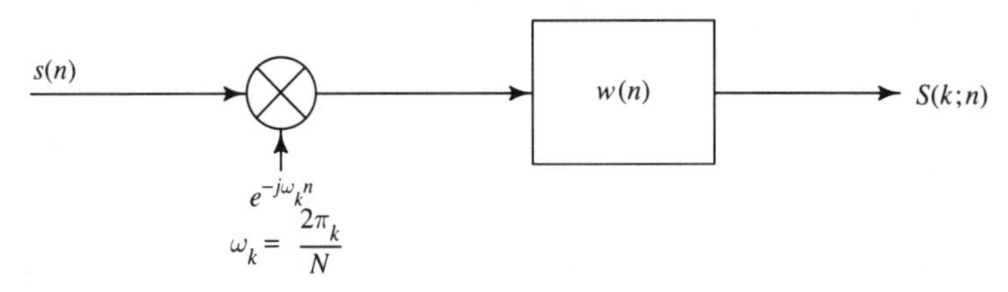

FIGURE 4.12. Computation of the stDFT for a general time n and specific frequency index k representing $\omega_k = 2\pi k/N$.

[15]Note that $N/2$, instead of N, is necessary because of the complex symmetry of the transform.

Note that we have specifically replaced the computation time m by the more general n, indicating that we might wish to compute $S(k;n)$ at every n. In fact, let us assume that it *is* computationally necessary, efficient, or convenient to compute $S(k;n)$ at every n. This will allow for the possibility that $w(n)$ is not a finite-length window and that a recursive computation is necessary. The question remains, however, as to whether it is necessary to *send* $S(k;n)$ for all n even if it is available. The answer is clear when we interpret the system whose impulse response is $w(n)$ as a lowpass filter, which all commonly used windows will be. Consider, for example, the case in which $w(n)$ is a Hamming window. If we define the bandwidth of the lowpass filter to be the positive frequency width of the main spectral lobe, then for the Hamming window this bandwidth is

$$\omega_b = \frac{2\pi}{N} \qquad \text{(norm-rps)}. \tag{4.80}$$

This, therefore, is the nominal Nyquist frequency of $S(k;n)$ when considered as a sequence on n, and the appropriate sampling rate on the sequence is

$$\omega_s = 2\omega_b = \frac{4\pi}{N} \qquad \text{(norm-rps)}. \tag{4.81}$$

From samples (with respect to n) of $S(k;n)$ taken at this rate, the receiver should be able to exactly reconstruct the entire sequence; hence no information will be lost. The sample times correspond to

$$n = i\frac{2\pi}{\omega_s} = i\frac{N}{2}, \qquad i = \ldots, -1, 0, 1, 2, \ldots \qquad \text{(norm-sec)}, \tag{4.82}$$

so that samples of $S(k;n)$ with respect to n need only be sent every $N/2$ computations. Note that this interval corresponds to half the window length. This process in which only a fraction of the sequence values is sent is called *decimation.*

Let us now collect results to determine the coding efficiency. We will need to send about $N/2$ frequency samples (k's) for every time we sample in time (with respect to n). This means $N/2$ complex numbers per $N/2$ norm-sec or one complex number per norm-sec. This represents a *doubling* of required bandwidth with respect to sending the speech waveform itself, since the speech involves only one *real* number per norm-sec.

In spite of the decimation, and in spite of all the trouble of computing the stDFT, we see that no coding efficiency is obtained. In fact, we find that the spectral technique has resulted in less efficient coding than a simple transmission of the speech waveform. This will be found to be quite generally true for commonly used window sequences (see Problem 4.4). The main benefit that obtains from this scheme is a flexible representation of the speech that can be manipulated in both time and frequency. However, this benefit does not outweigh the disadvantage of excessive coding requirements, and simple filter bank systems of this type are used primarily in research. We will return to such systems in Chapter 7 to learn to make them more useful and efficient. The main purpose of this brief preview was to come to terms with $S(k;n)$ as a two-dimensional sequence and to study its sampling.

The Use of $S(\omega;m)$ in Pitch and Formant Estimation

The short-term Fourier transform, in both analog and discrete-time forms, has been the basis for many important developments in speech analysis and synthesis. For a comprehensive overview the reader is referred to the textbooks by Flanagan (1972) and Rabiner and Schafer (1978). Other references are noted in footnote 8. While the stDTFT remains an important tool in some areas of speech processing (e.g., it serves as the basis for some commercial digital spectrographic analyzers), in many contemporary problems spectral features of speech are deduced by other techniques. We will study these methods as we proceed in the book.

Clearly, the stDTFT can serve as a basis for formant analysis of speech, since it very directly contains the formant information in its magnitude spectrum. An example of such a formant analysis system is described by Schafer and Rabiner (1970). The stDTFT has also been used in a number of algorithms for pitch detection. If it is computed with sufficient spectral resolution, then the harmonics of the pitch frequency will be apparent in the short-term spectrum. This idea is the basis for the *harmonic product spectrum* (HPS) (Schroeder, 1968), defined as

$$P(\omega;m) \stackrel{\text{def}}{=} \prod_{r=1}^{R} S(r\omega;m) \tag{4.83}$$

for some relatively small R, typically five. The HPS consists of the product of compressed copies of the original spectrum. The frequency axis is compressed by integer factors, so that the harmonics line up and reinforce the fundamental frequency as shown in Fig. 4.13. The logarithm can also be taken on both sides of (4.83), so that terms are combined additively. This technique has been found to be especially resistant to noise, since the disturbance manifests itself spectrally in ways that do not combine coherently in the HPS. It is also interesting to note that energy need not be present at the fundamental frequency in order to estimate the pitch with the HPS.

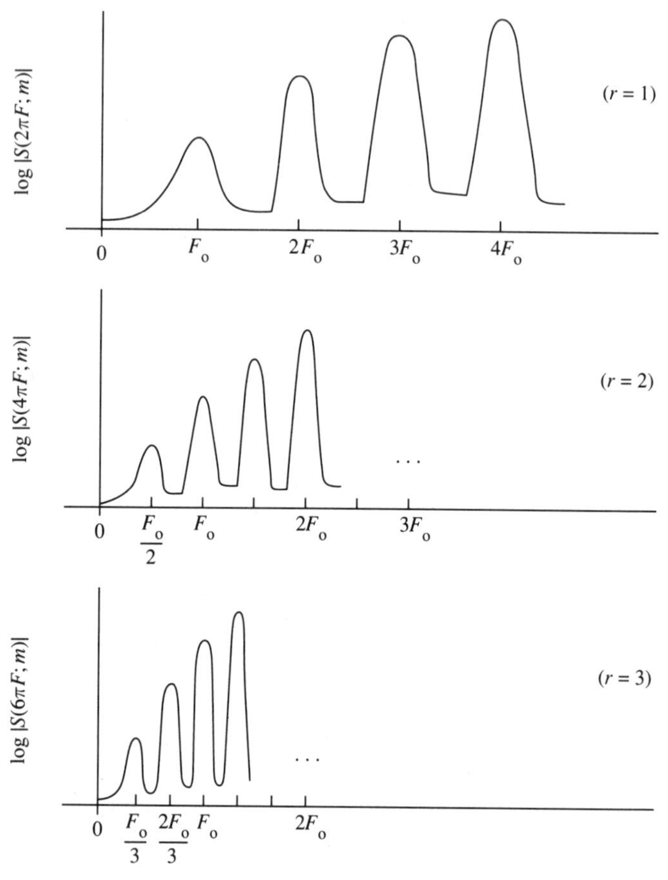

FIGURE 4.13. The harmonic product spectrum is the product of frequency-compressed copies of the original spectrum. In each copy the frequency axis is compressed by a different integer factor so that the harmonics line up and reinforce the fundamental frequency.

4.4 Conclusions

The fundamental purpose of this chapter has been to provide a formal basis for the processing of frames of speech as these frames move though time. Since speech typically remains stationary for ranges of only tens of milliseconds, this issue is of critical importance in the extraction of features from speech. These features, in turn, dynamically characterize the waveform for coding, analysis, or recognition.

Short-term features of speech are generally computed using one of two basic paradigms which we have formalized as "construction principles." These both amount to windowed transformations of long-term attributes of the signal, and the two construction principles differ in essence by whether the window precedes or follows the transformation. Most algorithms in the literature can be classified according to one of these two principles.

A secondary purpose of the chapter has been to introduce some examples of short-term features, both to illustrate the formal developments and because they will be useful in the work to follow. Among the most important of the short-term features we have discussed are the correlation estimators. These will play a central role in linear prediction (LP) analysis, which, in turn, will be at the heart of many future discussions and methods. We take up the subject of LP analysis in the following chapter.

4.5 Problems

4.1. Recall the "voiced" excitation to the digital model of speech production,

$$e(n) = \sum_{q=-\infty}^{\infty} \delta(n - qP). \tag{4.84}$$

(a) Find an expression for the long-term temporal autocorrelation, $r_e(\eta)$.
(b) Find ${}^1r_e(\eta; m)$ of (4.36) for $\eta = 0, 1, \ldots, N-1$ using a rectangular window of length N, where $IP < N < (I+1)P$ for some integer I.
(c) Repeat part (b) for ${}^3r(\eta; m)$ of (4.44).

4.2. (a) Begin with (4.67) and use Construction Principle 1 to derive the (scaled) stDTFT given in (4.68).
(b) Repeat part (a) using Construction Principle 2.
(c) In deriving the stDTFT, both construction principles produce the same result. Can you state general conditions under which the two principles produce the same short-term estimator?

4.3. (a) Determine whether the alleged properties of the stDFT shown in Table 4.1 are correct. The notation $W \stackrel{\text{def}}{=} e^{-j2\pi/N}$ is used for convenience and all frames are of length N.

TABLE 4.1. Alleged Properties of the stDFT.

Property	Time Domain	Frequency Domain
Linearity	$af_1(n; m) + bf_2(n; m)$	$aS_1(k; m) + bS_2(k; m)$
Circular shift	$f(n-d; m)_{\text{mod} N}$	$W^{kd} S(k; m)$
Modulation	$W^{ln} f(n; m)$	$S(k+l; m)_{\text{mod} N}$
Circular convolution	If $f_x(n; m) = \begin{cases} f_s(n; m)_{\text{mod } N} * f_y(n; m), & n = m - N + 1, \ldots, m \\ 0, & \text{other } n \end{cases}$ then $f_x(n; m) \longleftrightarrow S(k; m)Y(k; m)$.	

(b) Determine whether a form of Parseval's relation holds:

$$\sum_{n=m-N+1}^{m} |f(n;m)|^2 = \frac{1}{N} \sum_{k=0}^{N-1} |S(k;m)|^2. \tag{4.85}$$

(c) Determine whether the stDFT is "symmetric" in the sense that $|S(k;m)| = |S(N-k;m)|$ and $\arg S(k;m) = -\arg S(N-k;m)$ for $k = 0, 1, \ldots, [N/2]$, where $[N/2]$ means the largest integer less than or equal to $N/2$.

4.4. What is the effect on coding efficiency of using the rectangular window in the "filter bank" encoder discussed at the end of Section 4.3.5? Express your answer as the ratio of the number of real numbers per norm-sec required to transmit the stDFT versus the number of real numbers required per norm-sec to send the speech waveform itself. The window bandwidth (in terms of N) can be estimated from Fig. 1.4.

4.5. (a) Verify that Construction Principle 2 can be used to derive the short-term zero crossing measure,

$$Z_s(m) = \frac{1}{N} \sum_{n=m-N+1}^{m} \frac{|\operatorname{sgn}\{s(n)\} - \operatorname{sgn}\{s(n-1)\}|}{2} w(m-n) \tag{4.86}$$

from the long-term definition, (4.60).

(b) If the window is rectangular, show that $Z_s(n)$ can be computed recursively (for all n), using a computation of the form

$$Z_s(n) = Z_s(n-1) + \text{other terms}. \tag{4.87}$$

(c) Show the computational structure for your result in part (b).

(d) Assume a rectangular window and suppose that it is desired to have $Z_S(m)$ for $m = 0, N/2, N, 3N/2, 2N, \ldots$. On the average, how many floating point operations (flops) per norm-sec are required to obtain these measures using (i) (4.86) and (ii) the defining relation (4.87)? For simplicity, let one flop be equivalent to one multiplication operation and ignore all additions.

4.6. Repeat parts (b)–(d) of Problem 4.5 for the short-term energy measure defined as

$$E_s(m) = \sum_{n=m-N+1}^{m} s^2(n). \tag{4.88}$$

In tallying flops, consider a squaring operation to be equivalent to one multiplication.

4.7. (a) Recall the expressions for the power and short-term power of the real sequence $s(n)$,

$$P_s = \mathcal{L}\{s^2(n)\} \tag{4.89}$$

and

$$P_s(m) = \frac{1}{N} \sum_{n=m-N+1}^{m} s^2(n), \tag{4.90}$$

respectively. Clearly, $r_s(0) = P_s$ where $r_s(\eta)$ is the long-term autocorrelation sequence. For which short-term autocorrelation estimators, and for which windows, is it true that

$$r_s(0; m) = P_s(m)? \tag{4.91}$$

(b) What is the relation between the short-term autocorrelation, and the short-term energy given in (4.88)?

4.8. (Computer Assignment) Write simple programs for the computation of the short-term AMDF and the general short-term autocorrelation estimator $r(\eta; m)$ given in (4.35). Explore the use of different window types and lengths in estimating the pitch every 1000 samples for the /i/ glissando. (*Note*: The speaker of the glissando (pitch sweep) is an adult male and the sampling rate of the data is 10 kHz. This should help reduce the number of η's considered. This is important to reduce the amount of computation to a reasonable level.)

4.9. (Computer Assignment)

(a) Perform the following operations with a vowel utterance of your choice:

(i) Compute the short-term autocorrelation of (4.35) for a Hamming window of length $N = 512$ for $\eta = 0, 1, \ldots, 256$.

(ii) Compute the $N = 512$-point magnitude spectrum of the waveform based on a Hamming window and an stDFT. (*Note*: The stDFT and conventional DFT are equivalent here because only the magnitude spectrum is required.)

(iii) Repeat steps (i) and (ii) after center clipping the waveform according to (4.57).

(b) Comment on the changes in both the autocorrelation and the spectrum. What do these changes indicate about the effects of the clipping operation on the waveform?

(c) Estimate the pitch using the two autocorrelation results. Which result would provide better performance in an automated procedure?

4.10. (Computer Assignment) Use the short-term AMDF to estimate the pitch in the utterance "seven." Decide upon an appropriate window type and length, frequency of computing $\Delta M_s(\eta; m)$ (i.e., at what m's), and appropriate range of η. The speaker is an adult male and the sampling rate is 10 kHz.

CHAPTER 5

Linear Prediction Analysis

Reading Notes: In addition to the standard material on DSP and stochastic processes, we will have specific uses for the following material in this chapter:

1. *The concepts of minimum-phase systems will play a key role in many developments. A clear understanding of the material in Section 1.1.7 is imperative.*
2. *The reader who intends to study the proof in Appendix 5.A will need to be familiar with the state space concepts in Section 1.1.6.*

5.0 Preamble

Since the earliest papers on the application of linear prediction (LP) to speech analysis (Atal and Schroeder, 1968, 1970; Atal and Hanauer, 1971; Markel, 1971, 1972), thousands of papers, monographs, reports, and dissertations have been written on the subject. The technique has been the basis for so many practical and theoretical results that it is difficult to conceive of modern speech technology without it. Many speech processing engineers initially studied LP concepts in Markel, Gray, and Wakita's early monographs (1971, 1973) or in the Markel and Gray textbook (1976), in Makhoul's (1975) classic tutorial, or in the textbook by Rabiner and Schafer (1978). With all of this, and more recent authoritative tutorial literature on LP, the reader might wonder what remains to be said. We have sought in this treatment to bring a somewhat different perspective to this subject, not, however, just to be different. It has been our experience in teaching this important topic that conventional presentations, although expertly written and comprehensive, do not always convey a complete understanding of this material to the first-time student in two general ways:

1. The student does not perceive a "unified view" of LP due to the introduction by presentation of numerous unclearly related short-term estimation problems, which employ a seemingly unconventional optimization criterion.
2. The student often misunderstands the important phase relationships between the LP model and the signal from which it is derived, resulting in confusion about the temporal properties of the resynthesized speech and the LP residual.[1]

This presentation addresses these problems by making two adjustments to the usual teaching method:

1. The initial focus on the infinite-time-line problem.
2. The careful segregation of the "true" dynamics of the speech system from the all-pole model identified in the LP process.

A detailed emphasis on the classical systems identification/modeling aspect of the LP problem is carried throughout. Once the asymptotic LP solution is well understood by virtue of this initial study, the widely used short-term LP solutions can be presented as simple finite-time estimators.

This chapter details a method for conveying a sound theoretical understanding of LP that proceeds in three sections. In Sections 5.1 and 5.2, the long-term problem is solved and discussed as a classical systems identification problem, unencumbered by the nuances of short-term processing. In Section 5.3, the customary short-term solutions are treated as estimators of three long-term results, and are also interpreted conventionally. Chapter 4 provides the necessary theory of short-term processing, which serves to bridge the long-term and short-term solutions. Many operational topics are treated in the remaining sections of the chapter.

5.1 Long-Term LP Analysis by System Identification

5.1.1 The All-Pole Model

Let us return to the discrete-time model for speech production developed in Chapter 3. We concluded there that during a stationary frame of speech the model would ideally be characterized by a pole–zero system function of the form

$$\Theta(z) = \Theta_0' \frac{1 + \sum_{i=1}^{L} b(i) z^{-i}}{1 - \sum_{i=1}^{R} a(i) z^{-i}}, \tag{5.1}$$

which is driven by an excitation sequence

$$e(n) = \begin{cases} \sum_{q=-\infty}^{\infty} \delta(n - qP), & \text{voiced case} \\ \text{zero mean, unity variance, uncorrelated noise,} & \text{unvoiced case} \end{cases} \tag{5.2}$$

[1]The student often believes, for example, that the LP residual is necessarily a pulse train in the voiced case, or that the two-pole model produces a good temporal representation of the glottal pulse. If you are a first-time student of this material, you probably have no basis for beliefs of any kind along these lines, but come back and reread this footnote after studying the chapter.

The block diagram for this system is shown in Fig. 5.1(a). The objective of an LP algorithm is the identification of the parameters[2] associated with the all-pole system function,

$$\hat{\Theta}(z) = \frac{1}{1 - \sum_{i=1}^{M} \hat{a}(i) z^{-i}}, \tag{5.3}$$

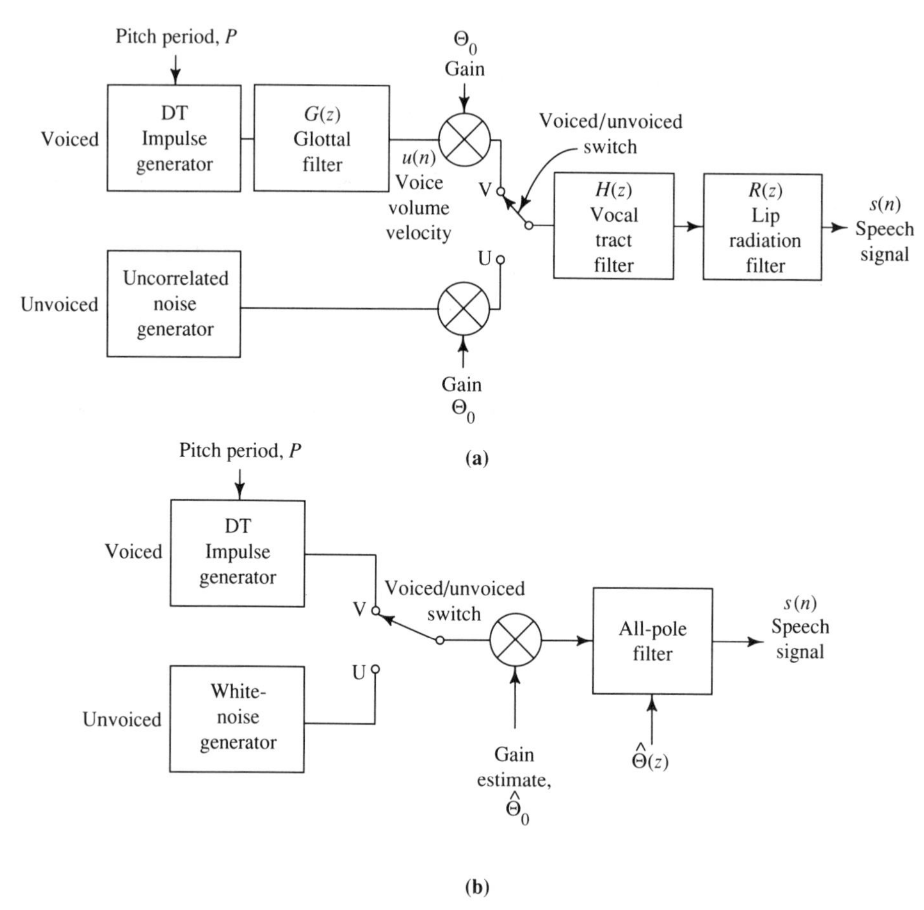

FIGURE 5.1. Discrete-time speech production model. (a) "True" model. (b) Model to be estimated using LP analysis.

[2]In keeping with the notational convention established in earlier chapters, we could attach a subscript such as s to the parameters, $\hat{a}_s(i)$, to indicate that these numbers can be viewed as features of the signal $s(n)$, indexed by a parameter i. This subscript is usually not necessary since it is obvious which signal is related to the features under consideration. We shall therefore omit it for simplicity. Note that we have also begun to index filter coefficients by parenthetical arguments rather than subscripts so that we may use subscripts to indicate signals when necessary.

which is to serve as an estimated model for the true speech production system, $\Theta(z)$. The form of the estimated model is shown in Fig. 5.1(b).

Before initiating a pursuit of these parameters, it is prudent to wonder why an all-pole model is used, in light of the fact that we have seen fit to initially attribute a pole–zero system to the speech. By acoustic analysis of tube models in Chapter 3, we initially built up a simple model of speech production whose system function turned out to be all-pole. We discussed the fact, however, that there are numerous compelling arguments for the inclusion of zeros in the speech production model. It is well known that certain phoneme classes, most notably the vowels, involve vocal-tract configurations that are acoustically resonant, and are therefore appropriately modeled by all-pole structures (Fant, 1956, 1960; Flanagan, 1972). On the other hand, such phoneme classes as nasals and fricatives, and generally any sound that can be modeled by the inclusion of acoustic "side cavities" in the vocal tract, will contain certain spectral nulls, mathematically corresponding to zeros in the system function. Further, it is also known that the glottal pulse waveform, a component of $\Theta(z)$, is better represented by a filter containing zeros[3] (Rosenberg, 1971; Deller, 1983). Ideally, therefore, the method of choice would be to model speech with a pole–zero system function in most cases.

The use of an all-pole model, however, is primarily a matter of analytical necessity. As we shall see, these parameters can be determined using a meaningful strategy (resulting in simple *linear* equations) applied to the very limited information available about the true system (namely, its output).[4] Beyond necessity, however, there is an overriding argument against the need for zeros, if it is the purpose of the LP model to preserve the acoustic information in the speech without regard to temporal relationships: In a practical sense, phase relationships among components of speech have virtually no effect on speech perception.[5] Consider, for example, the ability to clearly understand speech of a stationary talker by a listener moving from room to room within a house. Although the phase relationships among the components of the speech are obviously changing dramatically, the speech "sounds the same" given sufficient amplitude (Carlyon, 1988). With regard to speech perception, therefore, the

[3]Recall the discussion about nonminimum-phase systems in Chapters 1 and 3.

[4]Note that an all-pole model is one that requires delayed values of the *output* only, and minimal information (only the present value) about the input which is unmeasurable in the speech problem.

[5]This is not to say that the "ear" cannot perceive phase differences. The relationships among two or more frequency components within the same "critical band" (see Section 6.2.4) are very significant to the perception of the sound (Carlyon, 1988). There is also some evidence that phase effects can alter the perception of the first formant (hence, vowel perception) (Darwin and Gardner, 1986), but these effects are apparently not very significant. The ultimate test for the "validity" of any engineering assumption is whether the resulting development serves the intended purpose satisfactorily. Whether or not the ignorance of phase is strictly proper, LP analysis based on this assumption has proved to be eminently useful.

human ear is fundamentally "phase deaf" (Milner, 1970, p. 217). Whatever information is aurally gleaned from the speech is extracted from its magnitude spectrum. Further, as we show, a magnitude, but not a phase, spectrum can be exactly modeled with stable poles. Therefore, the LP model can exactly preserve the magnitude spectral dynamics (the "information") in the speech, but might not retain the phase characteristics. In fact, the stable, all-pole nature of the LP representation constrains such a model to be minimum phase quite apart from the true characteristics of the signal being encoded. We will show that (ideally) the model of (5.3) will have the correct magnitude spectrum, but minimum-phase characteristic with respect to the "true" model. If the objective is to code, store, resynthesize, and so on, the magnitude spectral characteristics, but not necessarily the temporal dynamics, the LP model is perfectly "valid" and useful.

5.1.2 Identification of the Model

Preliminaries

To pursue these points more analytically, we need two useful results.

LEMMA 5.1 (SYSTEM DECOMPOSITION LEMMA) *Any causal rational system of form* (5.1) *can be decomposed as*

$$\Theta(z) = \Theta_0 \Theta_{\text{min}}(z) \Theta_{\text{ap}}(z), \tag{5.4}$$

where $\Theta_{\text{min}}(z)$ is minimum phase, and $\Theta_{\text{ap}}(z)$ is all-pass, that is,

$$\left|\Theta_{\text{ap}}(e^{j\omega})\right| = 1 \ \forall \omega.$$

Here Θ_0 *is a constant related to* Θ_0' *and the singularities of* $\Theta(z)$.

Proof.[6] [Based on (Oppenhein and Schafer, 1989, pp. 240–241).] Consider only stable causal systems so that only zeros may be outside the unit circle in the z-plane. Construct a pole–zero diagram for $\Theta(z)$. Any pole or zero inside the unit circle is attributed to $\Theta_{\text{min}}(z)$. Any zero outside the unit circle should be reflected to the conjugate reciprocal location inside the unit circle, and the reflected zeros also become part of Θ_{min}. $\Theta_{\text{ap}}(z)$ contains the reflected zeros in their unreflected positions, plus poles that cancel the reflected ones. Θ_{ap} is normalized by appropriate scaling of the gain term. The pole–zero manipulation is illustrated in Fig. 5.2. The gain scaling is seldom important (as we shall see below), but for completeness, we elaborate using the example system of the figure. The system function is of the form

[6]Although proofs can often be omitted on first reading, it is suggested that you read this one because it leads to an immediate understanding of the lemma.

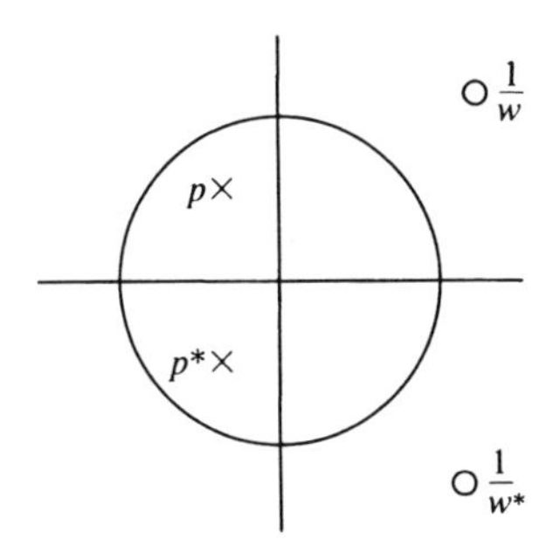

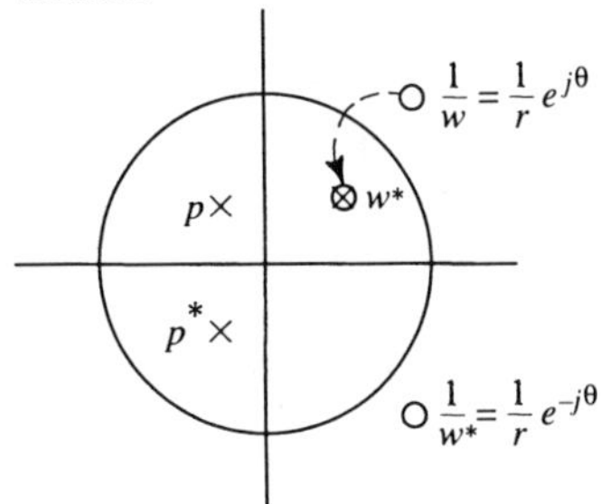

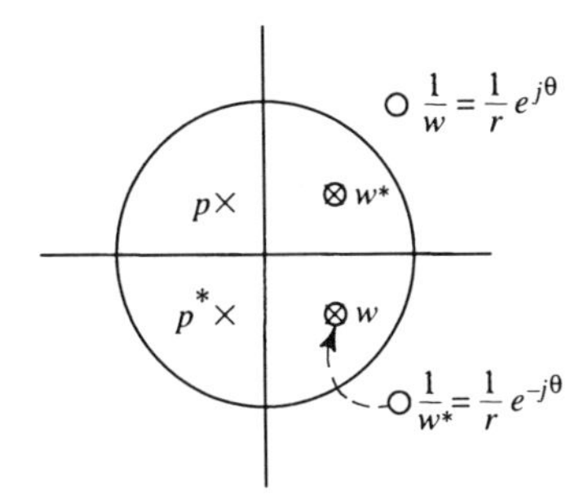

(d) $\Theta_{\min}(z)$ gets all "reflected" zeros plus all original poles. $\Theta_{\mathrm{ap}}(z)$ gets all original zeros plus all "canceling" poles. If $\Theta_{\mathrm{ap}}(z)$ is expressed in the form

$$\Theta_{\mathrm{ap}}(z) = \frac{\prod_i (1 - z_i z^{-1})}{\prod_j (1 - \rho_j z^{-1})},$$

where z_i = original zeros, ρ_j = canceling poles, then it should be multiplied by $\prod_j(-\rho_j)$ to make its magnitude unity. The original gain must be divided by this product.

FIGURE 5.2. Illustration of the proof of the System Decomposition Lemma.

$$\begin{aligned}\Theta(z) &= \Theta_0' \frac{1 + b(1)z^{-1} + b(2)z^{-2}}{1 - a(1)z^{-1} - a(2)z^{-2}} \\ &= \Theta_0' \frac{\left(1 - \frac{1}{w}z^{-1}\right)\left(1 - \frac{1}{w^*}z^{-1}\right)}{\left(1 - pz^{-1}\right)\left(1 - p^*z^{-1}\right)}.\end{aligned} \tag{5.5}$$

Suppose that we choose to first work on the "external" zero at $z = 1/w$. We introduce the conjugate reciprocal zero at $z = w^*$ and the canceling pole at the same location. Grouping these singularities according to the prescription above will cause the ultimate $\Theta_{\min}$ to inherit the term $[1 - w^* z^{-1}]$, and Θ_{ap} will get $(1 - (1/w)z^{-1})/(1 - w^* z^{-1})$. In order that the all-pass term have unity magnitude, we multiply it by $-w$ and let the gain be divided by this term. After each "external" zero has been treated, the original Θ_0' will have been scaled to Θ_0 by absorbing these terms. In the present example, $\Theta_0 = \Theta_0'/(-w)(-w^*) = \Theta_0'/|w|^2$.

LEMMA 5.2 *The minimum-phase component resulting from the decomposition process of Lemma 5.1 can be expressed as an all-pole system*

$$\Theta_{\min}(z) = \frac{1}{1 - \sum_{i=1}^{I} a(i) z^{-i}}, \tag{5.6}$$

where I, while theoretically infinite, can in practice be taken to be a relatively small integer (e.g., 14).

Proof.[7] This result follows readily from the fact that any first-order polynomial of the form $1 - z_0 z^{-1}$ with $|z_0| < 1$ can be written as the ratio

$$\frac{1}{1 + \sum_{k=1}^{\infty} z_0^k z^{-k}}. \tag{5.7}$$

Therefore, by expressing each zero in $\Theta_{\min}(z)$ in the form (5.7), then combining all terms [including the original denominator polynomial of $\Theta_{\min}(z)$] by multiplication, the result is of form (5.6) with I theoretically infinite. However, $a(I)$ can be made arbitrarily small by sufficiently large choice of I. [Ponder the fact that the coefficients in each term like (5.7) are decaying exponentially.] In practice, a sufficient model is obtained by neglecting terms beyond a small integer, typically 8–14.

According to Lemmas 5.1 and 5.2, $\Theta(z)$ can be written

$$\Theta(z) = \Theta_0 \frac{1}{1 - \sum_{i=1}^{I} a(i) z^{-i}} \Theta_{\mathrm{ap}}(z). \tag{5.8}$$

We will show that the LP model will ideally represent the all-pole minimum-phase portion of $\Theta(z)$. The $\hat{a}(i)$'s will be computed in such a way so as to match, if $M = I$, the $a(i)$'s of (5.8). Since $|\Theta_{\mathrm{ap}}(\omega)| = 1$ implies $|\Theta(\omega)| = \Theta_0 |\Theta_{\min}(\omega)|\ \forall \omega$, a correct matching of the $a(i)$ parameters will at least yield a model with a (scaled) correct magnitude spectrum. In case

[7]Some details of this proof are considered in Problem 5.2.

of a truly minimum-phase system, $\Theta_{\text{ap}}(z) = 1$, and an exact match of the parameters yields a model correct in general z-transform.

Linear Prediction Equations

Since a stable all-pole model must be minimum phase, a reasonable goal would be to estimate only the $a(i)$ parameters of $\Theta(z)$ of (5.8). We will show that even if, with ambitious naivete, we attempt to model $\Theta(z)$ "entirely," we will compute what are, in effect, estimates of the $a(i)$'s. We will call these estimates $\hat{a}(i)$ and use them in $\hat{\Theta}(z)$ as our model. These numbers constitute a parametric representation (or "code") for the waveform containing nominally all the magnitude spectral information (except pitch and gain) that is in the waveform itself. It is generally more efficient to transmit, store, analyze, base recognition upon, or synthesize from, these M plus a few parameters (or related numbers) than to work with the much larger number of data points from which the parameters are derived.

Why the name *linear prediction*? Using (5.8), we see that

$$S(z) = \Theta(z)E(z) = \Theta_0 \Theta_{\text{min}}(z)E'(z), \tag{5.9}$$

where $S(z)$ is the z-transform of the output speech sequence, $E(z)$ denotes the z-transform of the input excitation sequence[8], and where we have defined

$$E'(z) \stackrel{\text{def}}{=} E(z)\Theta_{\text{ap}}(z). \tag{5.10}$$

In the time domain, therefore,

$$s(n) = \sum_{i=1}^{I} a(i)s(n-i) + \Theta_0 e'(n) \stackrel{\text{def}}{=} \mathbf{a}^T\mathbf{s}(n) + \Theta_0 e'(n) \tag{5.11}$$

and $s(n)$ can be interpreted as the output of the minimum-phase component of $\Theta(z)$ driven by a phase-altered version of $e(n)$. Except for this input term, the output $s(n)$ can be predicted using a linear combination of its past I values. In statistical terminology, the output is said to *regress* on itself, and the model is often call an *autoregressive* (AR) model in other domains.[9] The $a(i)$'s form the predictor equation coefficients, and their estimates from LP analysis are often called a *linear predictive code*, especially in the communications technologies, where they (or related parameters) are used to encode the speech waveform. For this reason, the process of LP analysis is frequently called *linear predictive coding*, but we will reserve this term for its more proper use in cases where coding is the issue (see Chapter 7). In the literature, the estimates of the numbers $a(i)$,

[8]We have taken the liberty of assuming that these z-transforms exist. When they do not, a similar discussion can be based on the power density spectra.

[9]The notation AR(M) is often used to denote an autoregressive model with M parameters, or, equivalently, M poles. This is typically called an AR model "of order M."

$i = 1, 2, \ldots, I$, namely $\hat{a}(i)$, are variously referred to as "*a*" *parameters*, $\hat{a}(i)$ *parameters*, *LP parameters*, or *LP coefficients*.

Before proceeding, the reader is encouraged to return to (5.11) and carefully note the meaning of the vector notations

$$\mathbf{a} \stackrel{\text{def}}{=} [a(1)a(2)\cdots a(I)]^T \tag{5.12}$$

$$\mathbf{s}(n) \stackrel{\text{def}}{=} [s(n-1)s(n-2)\cdots s(n-I)]^T. \tag{5.13}$$

These and similar vector notations will play a central role in our developments. The superscript T is used to denote the transpose.

With these ideas in mind, we pursue the model. We will find the LP parameters by posing and solving (or leaving it to the reader to solve) four different problems, each of which yields the same solution. We call these "interpretive problems" because each allows us to interpret the resulting model in a different way. The four interpretations of the LP model are

1. System identification interpretation.
2. Inverse filtering interpretation.
3. Linear prediction interpretation.
4. Spectral flattening interpretation.

The result corresponding to the third interpretation will be obtained from two closely related problems.

System Identification Interpretation of the LP Model. In posing and solving the problem by which we obtain the LP parameters, we initially adhere to the theme of system identification established above. Given an unknown system with known input and output as shown in Fig. 5.3, a classic method of deducing a model is to ascribe an impulse response (IR) to the model that minimizes the average squared output error. Let $\hat{\Theta}(z)$ represent the model as shown in Fig. 5.3. The conventional problem is to find $\hat{\theta}(n)$ such that $\mathcal{L}\{\tilde{s}^2(n)\}$ is minimized.[10] Unfortunately, this method fails here, since we are working with a prescribed model form in which the $\hat{a}(i)$ parameters are nonlinearly related to $\hat{\theta}(n)$ (see Problem 5.1). If we "turn the problem around," however, as in part (c) of Fig. 5.3, and match the IR of the inverse model in the same sense, the IR, say $\alpha(n)$, $n = 0, 1, \ldots$, is clearly related to the desired parameters, since

$$\hat{A}(z) = \alpha(0) + \sum_{i=1}^{\infty} \alpha(i) z^{-i} \tag{5.14}$$

[10]We initially assume the speech to be a WSS, second-order ergodic process so that time averages are appropriate here.

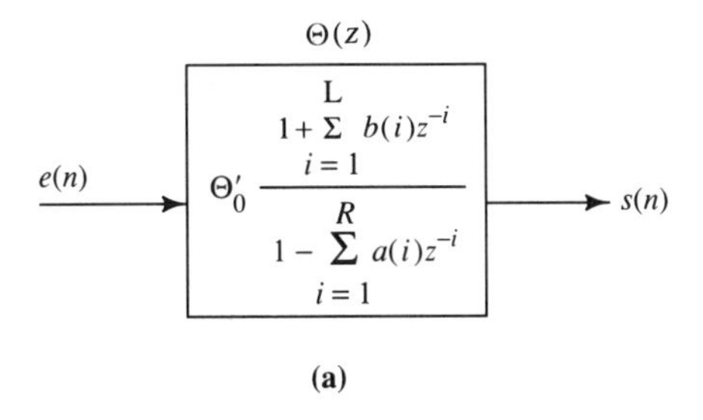

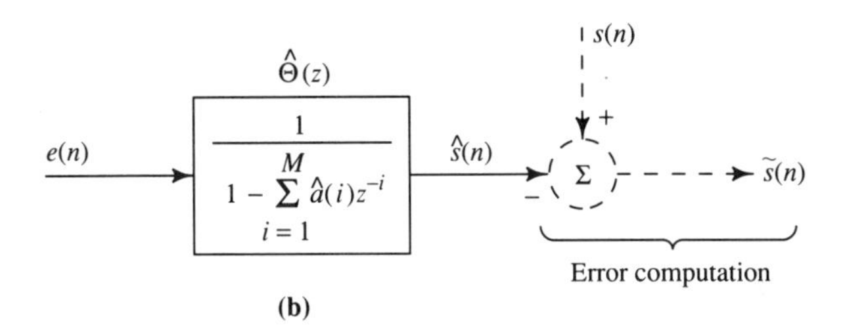

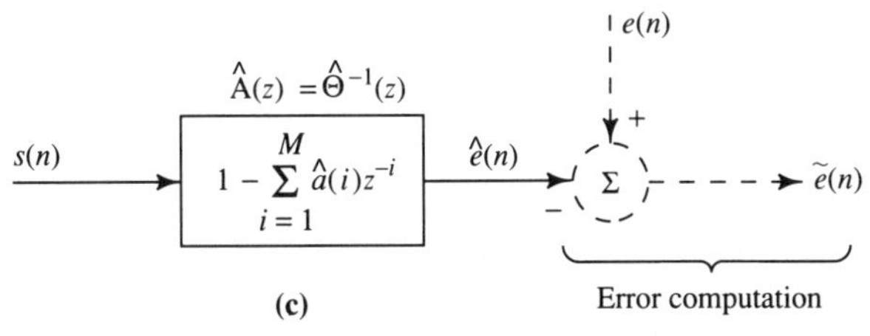

FIGURE 5.3. (a) "True" system model, (b) estimated system, and (c) estimated "inverse" system.

and the desired model form is

$$\hat{A}(z) = 1 - \sum_{i=1}^{M} \hat{a}(i) z^{-i}. \tag{5.15}$$

Comparing (5.14) and (5.15), we see that the model form imposes the following constraints[11] on the IR: $\alpha(0) = 1$ and $\alpha(n) = 0$ for $n > M$. This leads to the following problem:

INTERPRETIVE PROBLEM 5.1 (SYSTEMS IDENTIFICATION) *Design the finite IR (FIR) filter of length $M + 1$, with the constraint $\alpha(0) = 1$, which*

[11]The reason for the constraint $\alpha(0) = 1$ is the lack of knowledge of the gain of the system. Indeed, if we attempt to similarly solve for $\alpha(\eta), \eta = 0, 1, \ldots, M$, (5.18) and (5.20) involve an unsolvable equation (for $\eta = 0$)

$$r_s(0) + \sum_{i=1}^{\infty} \alpha(i) r_s(-i) - \Theta_0' = 0.$$

minimizes $\mathcal{L}\{\tilde{e}^2(n)\}$ *when* $s(n)$ *is input.* [*The LP parameters are given by* $\hat{a}(i) = -\alpha(i), i = 1, 2, \ldots, M$, *and the resulting* $\hat{A}(z)$ *is called the* inverse filter (IF) *in the LP literature.*]

Solution. From part (c) of Fig. 5.3 and (5.14) we have

$$\mathcal{L}\{\tilde{e}^2(n) \mid \alpha(0) = 1\} = \mathcal{L}\left\{\left[\sum_{i=0}^{M} \alpha(i)s(n-i) - e(n)\right]^2 \middle| \alpha(0) = 1\right\}. \tag{5.16}$$

Differentiating with respect to $\alpha(\eta), \eta = 1, 2, \ldots, M$, and setting the results to zero, we obtain

$$2\mathcal{L}\left\{\left[\sum_{i=0}^{M} \alpha(i)s(n-i) - e(n)\right]s(n-\eta) \middle| \alpha(0) = 1\right\} = 0 \tag{5.17}$$

or

$$r_s(\eta) + \sum_{i=1}^{M} \alpha(i)r_s(\eta - i) - r_{es}(\eta) = 0, \qquad \eta = 1, 2, \ldots, M, \tag{5.18}$$

where $r_{es}(\eta) = \mathcal{L}\{e(n)s(n-\eta)\}$ denotes the temporal cross-correlation of the sequences $e(n)$ and $s(n)$, which are either deterministic or realizations of jointly WSS, second-order ergodic random processes; and $r_s(\eta) = \mathcal{L}\{s(n)s(n-\eta)\}$, the temporal autocorrelation of $s(n)$. Now, if we assume $\underline{e}$ to be a unity variance *orthogonal* random process[12], meaning that

$$r_e(\eta) = \mathcal{L}\{e(n)e(n-\eta)\} = \delta(\eta), \tag{5.19}$$

then, using (5.1), it is easy to show that

$$r_{es}(\eta) = \Theta_0'\delta(\eta), \qquad \eta \geq 0, \tag{5.20}$$

which is therefore zero for all positive η. [The violation of (5.19) in the "voiced" case is discussed below.] Finally, therefore,

$$\sum_{i=1}^{M} \alpha(i)r_s(\eta - i) = -r_s(\eta), \qquad \eta = 1, 2, \ldots, M \tag{5.21}$$

and recalling that $\hat{a}(i) = -\alpha(i)$ for $i = 1, 2, \ldots, M$, we have

$$\sum_{i=1}^{M} \hat{a}(i)r_s(\eta - i) = r_s(\eta), \qquad \eta = 1, 2, \ldots, M. \tag{5.22}$$

These M equations, in a short-term form (see Section 5.3), are used to compute the conventional LP parameters. In the statistical and linear algebra literature, they are sometimes called the *normal equations*. For con-

[12]Recall that this is equivalent to saying that $\underline{e}$ is zero mean, uncorrelated.

venience, we will often want to have them packaged as a vector–matrix equation

$$\begin{bmatrix} r_s(0) & r_s(1) & r_s(2) & \cdots & r_s(M-1) \\ r_s(1) & r_s(0) & r_s(1) & \cdots & r_s(M-2) \\ r_s(2) & r_s(1) & r_s(0) & \cdots & r_s(M-3) \\ \vdots & \vdots & \vdots & & \vdots \\ r_s(M-1) & r_s(M-2) & r_s(M-3) & \cdots & r_s(0) \end{bmatrix} \times \begin{bmatrix} \hat{a}(1) \\ \hat{a}(2) \\ \hat{a}(3) \\ \vdots \\ \hat{a}(M) \end{bmatrix} = \begin{bmatrix} r_s(1) \\ r_s(2) \\ r_s(3) \\ \vdots \\ r_s(M) \end{bmatrix}, \tag{5.23}$$

which we will write compactly as

$$\mathbf{R}_s\hat{\mathbf{a}} = \mathbf{r}_s \Rightarrow \hat{\mathbf{a}} = \mathbf{R}_s^{-1}\mathbf{r}_s. \tag{5.24}$$

Inverse Filtering Interpretation of the LP Model. It is customary to employ a subtle truth when formulating the LP problem: If the input is an orthogonal process, then minimizing the average squared *output* of the IF (see Interpretive Problem 5.1) is tantamount to minimizing the average squared *error* in the output. After we have a little experience with LP analysis, this will seem obvious, but, upon first reading, this probably will not seem obvious at all. Let us begin by demonstrating this fact. Note the following regarding the error quantity minimized above:

$$\begin{aligned} \mathcal{L}\{\tilde{e}^2(n) \mid \alpha(0)=1\} &= \mathcal{L}\{[\hat{e}(n)-e(n)]^2 \mid \alpha(0)=1\} \\ &= \mathcal{L}\{\hat{e}^2(n) \mid \alpha(0)=1\} - 2\mathcal{L}\{\hat{e}(n)e(n) \mid \alpha(0)=1\} + \mathcal{L}\{e^2(n)\}. \end{aligned} \tag{5.25}$$

If $\underline{e}$ is a WSS, second-order ergodic, unity variance, orthogonal process, it is not difficult to show (Problem 5.3) that this becomes

$$\mathcal{L}\{\tilde{e}^2(n) \mid \alpha(0)=1\} = \mathcal{L}\{\hat{e}^2(n) \mid \alpha(0)=1\} - 2\Theta_0' + 1, \tag{5.26}$$

which is minimized by minimizing $\mathcal{L}\{\hat{e}^2(n) \mid \alpha(0)=1\}$. We can therefore formulate the problem in four other ways. (We will seek solutions to these in Problems 5.4–5.7.) The first of these interprets the result in terms of an optimal inverse filter:

INTERPRETIVE PROBLEM 5.2 (INVERSE FILTERING) *Design the IF of Fig. 5.3 part* (c), *and* (5.14) *and* (5.15), *which minimizes the average squared output* (*or output power*) *of the filter.*

Linear Prediction Interpretation of the LP Model. Similarly, if

$$\hat{s}(n) = \sum_{i=1}^{M} \hat{a}(i) s(n-i) \tag{5.27}$$

is viewed as a prediction of $s(n)$, and therefore

$$P(z) = \sum_{i=1}^{M} \hat{a}(i) z^{-i} \tag{5.28}$$

as the FIR prediction filter (see Fig. 5.4), then $\hat{e}(n)$ is the *prediction error* (also called the prediction *residual*), and the problem can be posed as follows.

INTERPRETIVE PROBLEM 5.3 (LINEAR PREDICTION) *Find the prediction filter, $P(z)$, which minimizes the average squared prediction error.*

A closely related interpretation is based on a well-known result from the field of mean squared error (MSE) estimation called the *orthogonality principle* (OP). The OP is a general result that applies to many forms of linear and nonlinear estimation problems in which the criterion is to minimize the MSE [see, e.g., (Papoulis, 1984)]. A relevant statement of the OP for the present problem follows. A somewhat more general statement and some elaboration is found in Appendix 5.B.

THEOREM 5.1 (ORTHOGONALITY PRINCIPLE) *Consider the speech sequence $s(n)$ to be a realization of a WSS random process $\underline{s}$ with random variables $\underline{s}(n)$. The error in prediction sequence in Fig. 5.4 will likewise be a realization of a random process $\underline{\hat{e}}$ in this case. The prediction filter $P(z)$ of Fig. 5.4 will produce the minimum mean squared error $\left[\mathcal{E}\{\underline{\hat{e}}^2(n)\}\right.$ will be minimum$\left.\right]$ if and only if the random variable $\underline{\hat{e}}(n)$ is orthogonal to the random variables $\underline{s}(l)$ for $l = n-M, \ldots, n-1$,*

$$\mathcal{E}\{\underline{\hat{e}}(n)\underline{s}(l)\} = 0, \qquad \text{for } l = n-M, \ldots, n-1 \tag{5.29}$$

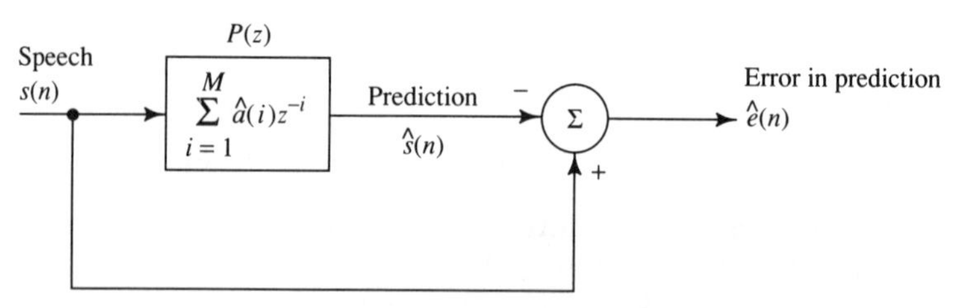

FIGURE 5.4. Prediction error interpretation of the inverse filter.

for any n.

Although we will inherently prove this result in our discussions, explicit proofs are found in many sources [e.g., (Gray and Davisson, 1986, pp. 204–205)].

We can write (5.29) in a slightly different way that will look more familiar,

$$\mathcal{E}\{\underline{\hat{e}}(n)\underline{s}(n-\eta)\} = 0, \qquad \text{for } \eta = 1, \ldots, M \tag{5.30}$$

or

$$r_{\underline{\hat{e}s}}(\eta) = 0, \qquad \text{for } \eta = 1, \ldots, M. \tag{5.31}$$

Further note that if $\underline{s}$ is an ergodic stochastic process, or $s(n)$ a deterministic signal, then condition (5.30) is equivalent to

$$\mathcal{L}\{\hat{e}(n)s(n-\eta)\} = 0, \qquad \text{for } \eta = 1, \ldots, M \tag{5.32}$$

or

$$r_{\hat{e}s}(\eta), = 0, \qquad \text{for } \eta = 1, \ldots, M. \tag{5.33}$$

The reader should compare (5.33) with (5.18). We see that (5.33) is sufficient and necessary to derive the normal equations from (5.18).

This discussion leads us to a variation on Interpretive Problem 5.3:

INTERPRETIVE PROBLEM 5.4 (LINEAR PREDICTION BY ORTHOGONALITY PRINCIPLE) *Design the minimum mean squared error prediction filter of Fig.* 5.4 *using the orthogonality principle.*

Spectral Flattening Interpretation of the LP Model. Finally, we note a frequency domain interpretation that will be useful below.

INTERPRETIVE PROBLEM 5.5 (SPECTRAL FLATTENING) *Design the model of form* (5.3) *that minimizes the integral*

$$\int_{-\pi}^{\pi} \frac{\Gamma_s(\omega)}{|\hat{\Theta}(\omega)|^2} d\omega, \tag{5.34}$$

where $\hat{\Theta}(\omega)$ is the DTFT of the LP model impulse response, and $\Gamma_s(\omega)$ represents the temporal power density spectrum of the signal $s(n)$ [*defined in* (1.141)].

This problem statement is easily verified by showing that the integral is equivalent to $\mathcal{L}\{\hat{e}^2(n)\} = r_{\hat{e}}(0)$ (see Problem 5.7). Interpretive Problem 5.5 is useful for theoretical purposes, but is least useful for design.

Notice carefully that Interpretive Problems 5.2–5.5 yield the customary LP equations, (5.22) or (5.24), whether or not $\underline{e}$ is orthogonal, whereas "getting the right answer" [i.e., (5.22)] to Interpretive Problem

5.1 depends explicitly upon $\underline{e}$ being orthogonal. In this sense, Interpretive Problems 5.2–5.5 are "better" problem statements, but they lack the intuitive quality of Interpretive Problem 5.1. Further, as we shall discuss below, $e(n)$ may usually be considered a realization of a zero-mean uncorrelated process in speech work.

5.2 How Good Is the LP Model?

5.2.1 The "Ideal" and "Almost Ideal" Cases

We now seek to discover how well, and in what sense, the LP model represents the true speech system. Consider first the ideal case in which $M = I$ (model order somehow chosen properly) and $e(n)$ is orthogonal[13] (as assumed). The result is so fundamental that we set it off as a theorem.

THEOREM 5.2 ("IDEAL" IDENTIFICATION CONDITIONS) *The conditions $M = I$ and $e(n)$ orthogonal are sufficient and necessary for the model of* (5.3), *with estimates $\hat{a}(i), i = 1, 2, \ldots, (M = I)$, given by* (5.22), *to exactly represent the minimum-phase component of $\Theta(z)$.*

Proof. A formal proof of this important result is given in Appendix 5.A. The proof is tedious, so we move it out of the reading path and present only a relatively complete sufficiency proof here.

When $M = I$, postmultiplying (5.11) by $s(n)$ and applying the $\mathcal{L}$ operator yields

$$\mathbf{r}_s = \mathbf{R}_s \mathbf{a} + \Theta_0 \mathbf{r}_{e's}, \tag{5.35}$$

where all notation has been defined above except $\mathbf{r}_{e's}$, which is an obvious extension of our conventions. Substituting (5.35) into (5.24), we have

$$\hat{\mathbf{a}} = \mathbf{a} + \Theta_0 \mathbf{R}_s^{-1} \mathbf{r}_{e's} \Rightarrow \tilde{\mathbf{a}} \stackrel{\text{def}}{=} \hat{\mathbf{a}} - \mathbf{a} = \Theta_0 \mathbf{R}_s^{-1} \mathbf{r}_{e's} \tag{5.36}$$

where, $\tilde{\mathbf{a}}$ denotes the error vector. The sufficiency argument is completed by showing that $\mathbf{r}_{e's}$ is the zero vector if $e(n)$ is orthogonal. This is accomplished in Problem 5.8.

Let us next consider the voiced case in which the $e(n)$ orthogonality assumption is violated. For cases of low pitch, however, we will see that $e(n)$ is "almost orthogonal," so that we refer to the case of $M = I$, voiced, low pitch, as "almost ideal." In this case we claim that the LP model very closely approximates the minimum-phase component of $\Theta(z)$, the approximation becoming exact as the pitch frequency of the voice be-

[13]We use the words "$e(n)$ is orthogonal" to mean that $e(n)$ is a realization of a WSS orthogonal random process $\underline{e}$.

comes zero, or $P \to \infty$ (in this case, the excitation becomes a single discrete-time impulse).

This claim is shown as follows: Using definition (5.2), we can show that

$$r_{e'}(\eta) = r_e(\eta) = \frac{1}{P} \sum_{q=-\infty}^{\infty} \delta(\eta - qP). \tag{5.37}$$

Putting this result into the easily demonstrable equation

$$r_{e's}(\nu) = \Theta_0 \sum_{\eta=-\infty}^{\infty} r_{e'}(\eta)\theta_{\min}(\eta - \nu) \tag{5.38}$$

yields

$$r_{e's}(\nu) = \frac{\Theta_0}{P} \sum_{q=-\infty}^{\infty} \theta_{\min}(qP - \nu) \approx \frac{\Theta_0}{P} \theta_{\min}(0)\delta(\nu) = \frac{\Theta_0}{P} \delta(\nu), \tag{5.39}$$

which is valid for large P, since stability assures that the IR of the minimum-phase portion of the system will decay exponentially with n. Practically speaking, the approximation in (5.39) will hold if the IR of the vocal tract has sufficient time to damp out between glottal excitations. This is usually the case for typically lower-pitched "male" phonation, whereas LP techniques often suffer some problems in this regard for higher-pitched female and children's voices (see Fig. 5.5). That the model is only approximate is due to the assumption that (5.20) holds in the derivation. Here we have shown that assumption is nearly justified, clearly leading to a model that accurately estimates the minimum-phase component of the system.

5.2.2 "Nonideal" Cases

In practice, the "true" order of the system, I, is, of course, unknown. An accepted operational rule for the choice of the LP model order is (Markel and Gray, 1976, p. 154):

$$M = \begin{cases} F_s + (4 \text{ or } 5), & \text{voiced} \\ F_s, & \text{unvoiced} \end{cases}, \tag{5.40}$$

where F_s is the sampling frequency of the data in kHz. This rule provides one pole pair per kHz of Nyquist bandwidth, which, in turn, assumes that the spectrum contains one formant per kHz (Fant, 1956). In the voiced case, another 4 or 5 poles are included to model the glottal pulse and to provide "spectral balance."

Let us first consider the "nonideal" case in which M is improperly chosen ($M < I$), but $e(n)$ is properly orthogonal. It is obvious in this case that the resulting estimates, $\hat{a}(i), i = 1, 2, \ldots, M$, cannot be expected to "identify" the $a(i), i = 1, 2, \ldots, I$, of (5.8) in any sense [except, of course,

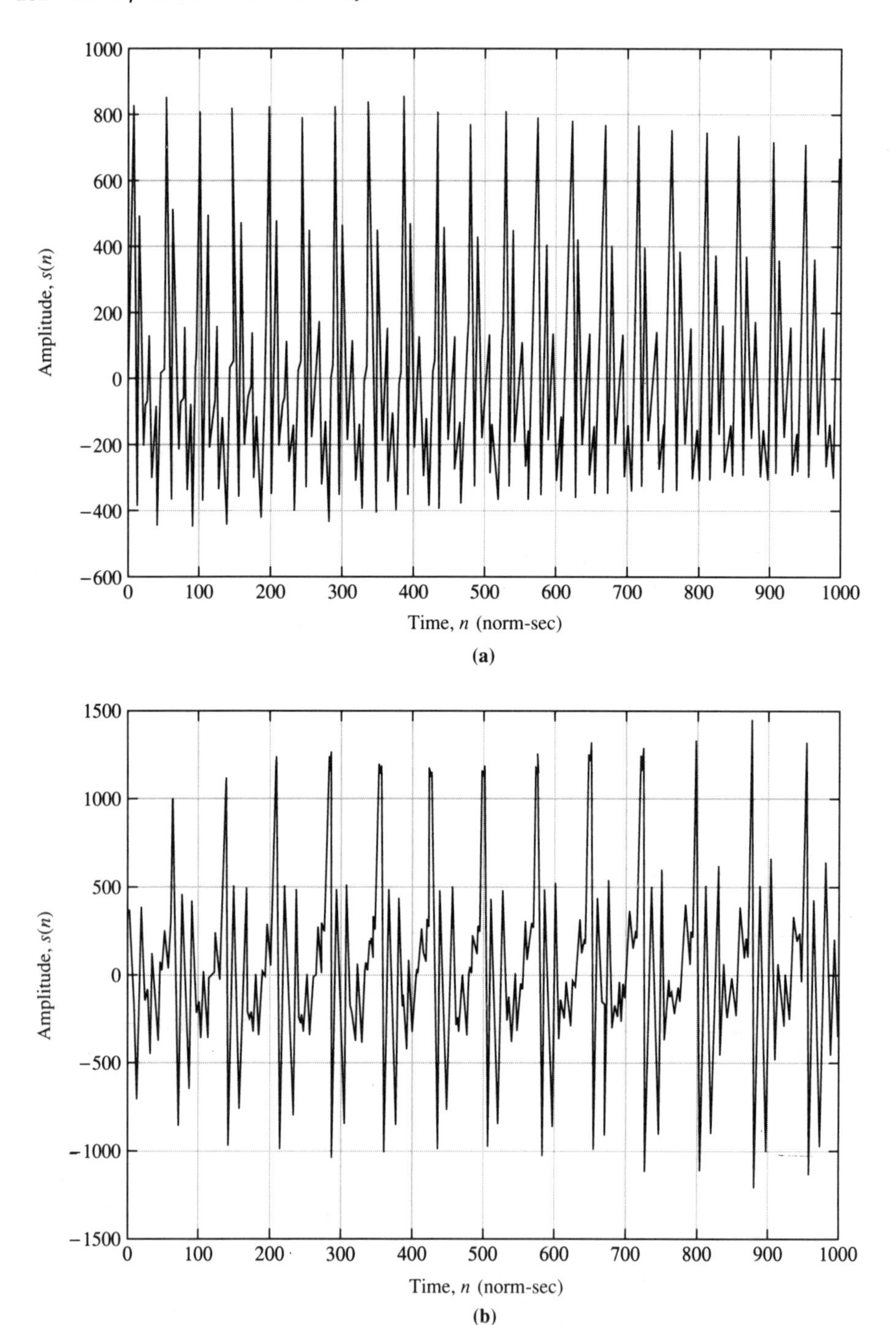

FIGURE 5.5. Typical (a) female and (b) male utterances of the vowel /a/. Based on short-term AMDF analysis in (c) and (d), the estimated pitch frequency for the female voice is 200 Hz, while that for the male is 135 Hz. In the latter case, the waveform is seen to damp out to a large extent between pulses, whereas this is seen not to be true in the former.

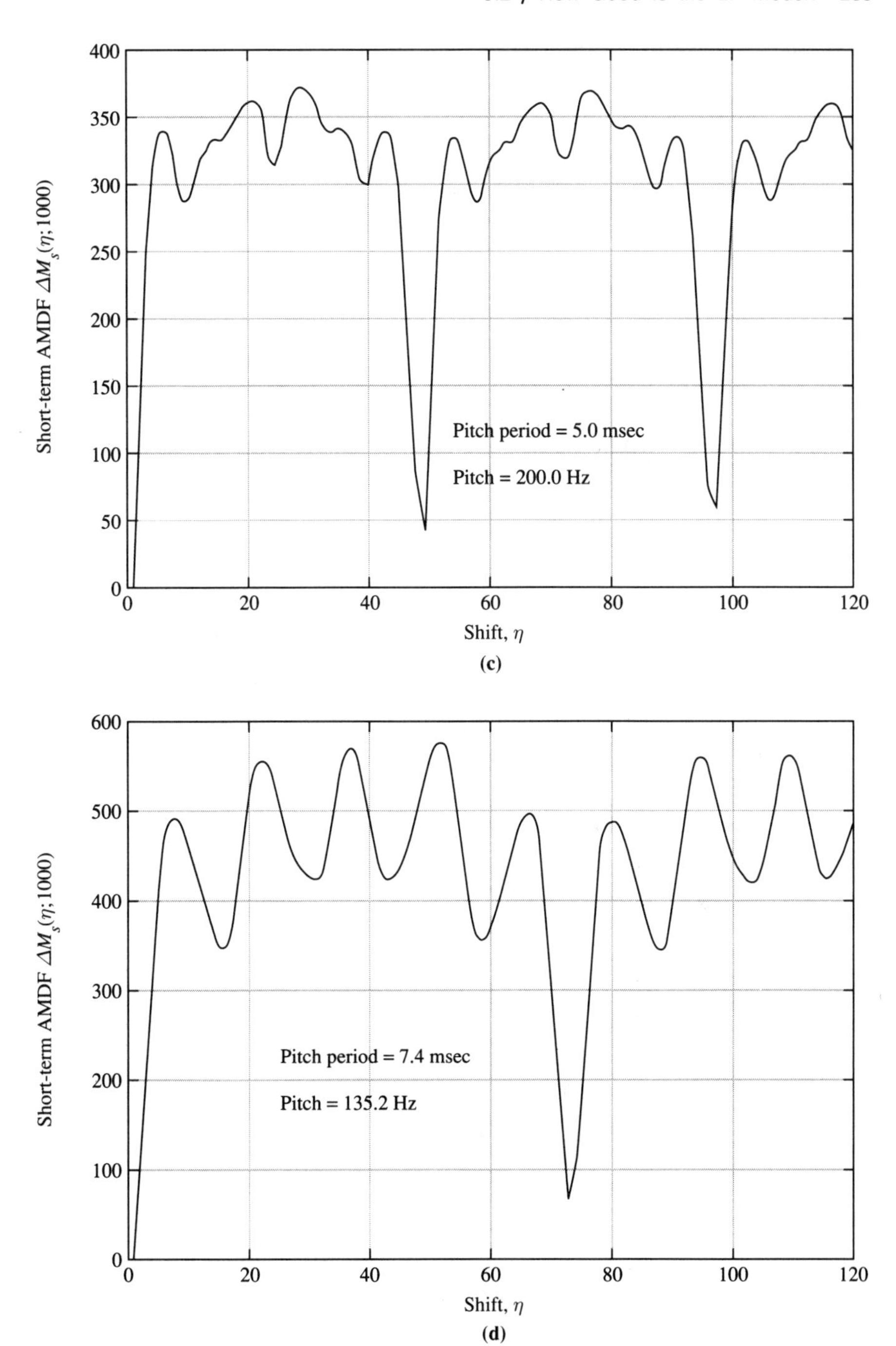

FIGURE 5.5. (*Cont.*)

in the trivial case in which $a(i) = 0, i = M+1, M+2, \ldots, I$]. In theory we can write

$$\Theta_{\min}(z) = \frac{1}{1 - \sum_{i=1}^{I} a(i) z^{-i}}$$

$$= \frac{1}{1 - \sum_{i=1}^{M} a'(i) z^{-i}} \times \frac{1}{1 - \sum_{i=1}^{L} a''(i) z^{-i}} \tag{5.41}$$

$$\stackrel{\text{def}}{=} \Theta_{\min, M}(z) \times \Theta_{\min, L}(z),$$

where $I = M + L$, and interpret the parameters $\hat{a}(i)$ as estimates of the factored subset of coefficients $a'(i), i = 1, 2, \ldots, M$, with errors given by (5.36) with $\mathbf{r}_{e''s}$ replacing $\mathbf{r}_{e's}$, where

$$e''(n) \Leftrightarrow E''(z) \stackrel{\text{def}}{=} \frac{E'(z)}{1 - \sum_{i=1}^{L} a''(i) z^{-i}}. \tag{5.42}$$

As a pertinent aside, note carefully that this same discussion applies to the analysis in which M is "properly chosen" ($M = I$) but $e(n)$ is not orthogonal (and has a z-transform): Diagram the true system as in Fig. 5.6(a), where $e(n)$ is correlated and has a z-transform

$$E(z) = E_{\text{ap}}(z) E_{\min}(z). \tag{5.43}$$

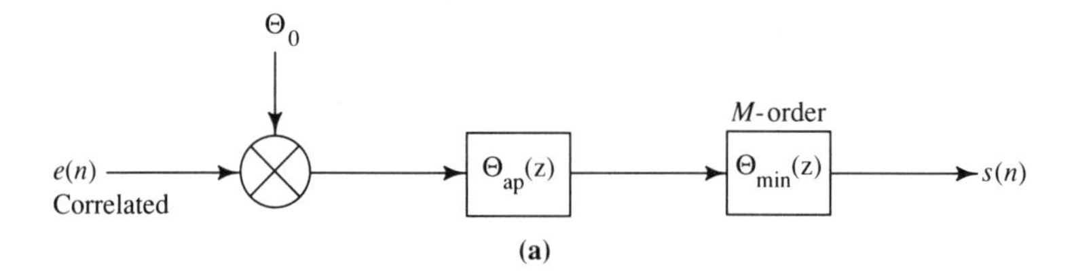

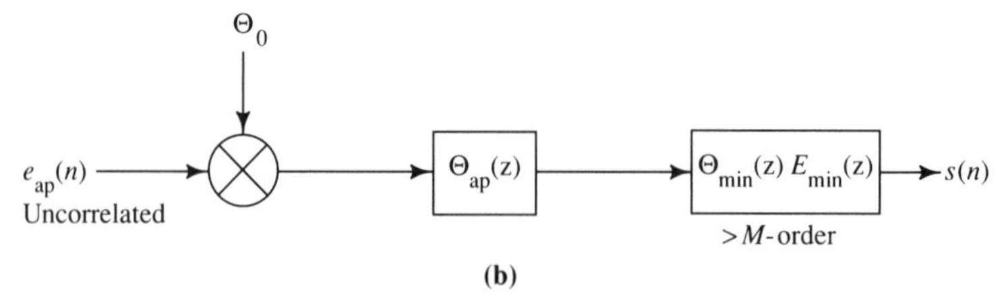

FIGURE 5.6. A correlated input sequence causes an equivalent problem to that caused by an underestimated model order. In (a) the speech is modeled by a correlated input driving an M-order system. This model can be redrawn as in (b), in which an orthogonal input sequence drives a model of order greater than M.

The attempt is to model $\Theta(z)$, which has an M-order minimum-phase component, using an M-order ("properly chosen") analysis. But we can rediagram the system as in Fig. 5.6(b), noting that the system is now driven by an uncorrelated input but has a minimum-phase component that is $M+L=I$ order, where L is the order of $E_{\text{min}}(z)$. We have, therefore, an apparent attempt to identify an M-order subsystem as before.

In either case, this theoretical interpretation provides no practical insight. Exactly which M-order subsystem is identified by such an analysis is indeterminate, and, practically speaking, the meaning of the resulting $\hat{a}(i)$'s (how good or bad, and compared with what?) is unclear. Generally speaking, however, the resulting $\hat{a}(i)$'s will identify the M most resonant poles of the system. This can be seen by recourse to Interpretive Problem 5.5, the spectral flattening interpretation, which asserts that the M-order LP model of (5.3), with coefficients derived using (5.22), will have a magnitude spectrum which minimizes the integral (5.34). It is clear that when $|\hat{\Theta}(\omega)|^2$ cannot match $\Gamma_s(\omega)$ exactly, the task of minimizing the area in (5.34) causes $\hat{\Theta}$ to concentrate on the peaks in the spectrum of $\Gamma_s(\omega)$, or the most resonant frequencies in the speech spectrum. This has the following implications for the discussion at hand: In the case $M<I$ discussed above, since we have assumed $e(n)$ orthogonal, we have

$$\Gamma_e(\omega)=1 \Rightarrow \Gamma_s(\omega)=|\Theta(\omega)|^2 \tag{5.44}$$

and the effort to flatten the peaks in the speech spectrum is tantamount to trying to flatten those in the system spectrum only. Only if $\Theta(\omega)$ has peaks that can be sufficiently flattened by M poles can the LP model be asserted to be a reasonable approximation to $\Theta_{\text{min}}(z)$, in spite of the model underestimation.

On the other hand, in the case in which $e(n)$ is correlated, but M is "large enough," we can see with reference to Fig. 5.6(b) that, once again,

$$\Gamma_{e_{\text{ap}}}(\omega)=1 \Rightarrow \Gamma_s(\omega)=|\Theta(\omega)|^2\,\Gamma_e(\omega) \tag{5.45}$$

and $\hat{\Theta}(z)$ will apparently identify $\Theta(z)$ well (in the usual minimum-phase sense) only if $e(n)$ makes no significant contribution to the spectral peaks of the speech spectrum [which is another way to say that $e(n)$ is "not very correlated"].

An example of the above occurs in speech analysis when it is desired to model only the *vocal-tract* portion of the speech system, which would typically be of order 8 to 10. The entire speech system includes other dynamics, including the glottal system, which typically increase the system order to, say, 14. We may interpret the eighth-order analysis as either the attempt to identify an eighth-order subsystem of a truly fourteenth-order system, or as the attempt to identify a truly eighth-order system (vocal tract) driven by a correlated input (glottal waveform). In either case, the result will be a good minimum-phase approximation to the vocal-tract system, only if the neglected anomaly does not make a significant spectral contribution. Roughly speaking, this will be true for relatively loud, low-pitched phonation in which the vocal waveform is a relatively uncorrelated, low duty-cycle, pulse train.

Our final "nonideal" case is only a slight variation on the above—the case in which $e(n)$ is correlated but has no z-transform. Indeed, this is technically the case with the voiced input to the speech model over the infinite time line. Note, however, that the spectral result of Interpretive Problem 5.5 is still valid, since it requires only that $e(n)$ [hence $s(n)$] have a power density spectrum. As above, we conclude here that only if $\Gamma_e(\omega)$ makes an insignificant contribution to the peakiness of the overall speech spectrum will $\Theta(z)$ be a good representation of $\Theta_{\min}(z)$. Indeed, this is the case as $P \rightarrow \infty$ (pitch frequency becomes small) as discussed above. However, as P becomes small, the pitch pulses begin to significantly distort the spectrum; hence, the identification.

It is instructive to examine this distortion phenomenon from another point of view. It is often pointed out that, if $\hat{\theta}(n)$ represents the IR of the LP model, $\hat{\Theta}(z)$, then

$$r_{\hat{\theta}}(\eta) = r_s(\eta), \qquad \text{for } \eta = 1, 2, \ldots, M. \tag{5.46}$$

In other words, the autocorrelation of the LP model IR matches that of the speech for the first M lags. This is easily shown as follows: According to (5.3),

$$\hat{\theta}(n) = \sum_{i=1}^{M} \hat{a}(i)\hat{\theta}(n-i) + \Theta_0 \delta(n) \tag{5.47}$$

so that, for any $\eta > 0$,

$$\mathcal{L}\{\hat{\theta}(n)\hat{\theta}(n-\eta)\} = \sum_{i=1}^{M} \hat{a}_i \mathcal{L}\{\hat{\theta}(n-i)\hat{\theta}(n-\eta)\} + \Theta_0 \mathcal{L}\{\delta(n)\hat{\theta}(n-\eta)\} \tag{5.48}$$

or

$$r_{\hat{\theta}}(\eta) = \sum_{i=1}^{M} \hat{a}(i) r_{\hat{\theta}}(\eta - i), \qquad \eta = 1, 2, \ldots, M. \tag{5.49}$$

Comparing this result with (5.22) reveals that (5.46) is true.

Now, in the case in which the excitation pitch period, P, is small, it is not difficult to demonstrate that $r_s(\eta)$ [hence $r_{\hat{\theta}}(\eta)$] is a severely aliased version of $r_\theta(\eta)$ (see Problem 5.9). This means that $|\hat{\Theta}(\omega)|^2$ is only mathematically obliged to match $|\Theta(\omega)|^2$ at a few "undersampled" (widely spaced) points in the frequency domain. These points represent the widely spaced harmonic frequencies of the excitation—the only information about $\Theta(\omega)$ that can possibly be represented in the speech sequence.

5.2.3 Summary and Further Discussion

We conclude that when the input to the speech production is an orthogonal sequence (and only then), and when the model order is correctly chosen ($M = I$), then $\hat{\Theta}(z)$ represents the minimum-phase component of $\Theta(z), \Theta_{\min}(z)$, everywhere in the z-plane. The condition of orthogonal $e(n)$ is met for the unvoiced case. We have argued that for sufficiently low pitch, the voiced input may be considered practically orthogonal. Therefore, the LP model will indeed represent the minimum-phase component of the speech.

Some points about the temporal properties of the model signals should be emphasized. When $\hat{\Theta}(z)$ is used for voiced speech reconstruction, the synthesized speech will not be a temporal replication of the speech sequence; rather, it is not hard to demonstrate that if $s'(n)$ represents the synthesized speech, then[14]

$$S'(\omega) = \Theta_0^{-1} S(\omega) e^{-j\varphi_{\mathrm{ap}}(\omega)}, \tag{5.50}$$

where $e^{j\varphi_{\mathrm{ap}}(\omega)}$ is the phase characteristic associated with the all-pass filter. Ideally, the synthesized speech will be a phase-altered, scaled version of $s(n)$. The problem of making synthesized speech sound natural is a complex issue, and it is not the point of this discussion to address that problem. The point is, rather, that speech that is resynthesized using even "ideal" LP parameters will not replicate the original waveform. However, all other factors being "ideal" (perfect transitions among frames, perfect pitch detection, and so on), it is likely the case that the phase scattering would have a negligible effect on the naturalness of the synthetic speech if appropriate scaling is used (see Sec. 5.3.4).

Second, we should take note of the temporal properties of the prediction residual when the speech is inverse filtered. When the IF, $\hat{A}(z)$, is used to filter the speech, the output, $\hat{e}(n)$, is nominally an estimate of the system input, $e(n)$. However, it is easily shown that, in the ideal case,

$$\hat{E}(z) = \Theta_0 E'(z), \tag{5.51}$$

[recall (5.10)] so that the residual will be an estimate of the all-pass filter output, a phase-altered version of the true input. Therefore, the residual might not be a good representation of the expected input pulse train in the voiced case as the various frequency components in the waveform can be "smeared" in time due to the phase delays in $\Theta_{\mathrm{ap}}(z)$ that remain in $\hat{E}(z)$ and are not present in $E(z)$. Algorithms that use the residual directly for pitch detection, for example, do not always yield good performance, at least in part due to this phenomenon.

When M is underestimated [or, equivalently, when $e(n)$ is not orthogonal], phase and temporal relationships between the LP model and the

[14]The DTFT $S(\omega)$ does not theoretically exist here, but we can resort to the use of the "engineering DTFT" described near (1.30).

true system become very unclear. Formally, the problem is the difficult one of determining how well a high-order polynomial is approximated by a polynomial of lower order. In this case we need to resort to spectral arguments alone. We can anticipate that the LP model will approximate the (scaled) magnitude spectrum as well as possible, concentrating on the peaks in the spectrum first. It is also to be noted that, in spite of the total ignorance of the phase throughout these spectral arguments, the resulting model can still be shown to be a minimum-phase filter[15] (Kay and Pakula, 1983; Lang and McClellan, 1980; Burg, 1975). Hence, in spite of our loss of ability to make phase comparisons directly, we can still assert that the LP model is "attempting" to represent the minimum-phase part of the true system.

Finally, it should be noted that, in practice, the choice of M can be made through a study of $r_{\hat{e}}(0)$, the total prediction residual energy [i.e., the value of the integral in (5.34) as a function of M] (see Fig. 5.7) (Chandra and Lin, 1974). If we assume $\hat{e}(n)$ to be orthogonal, when M

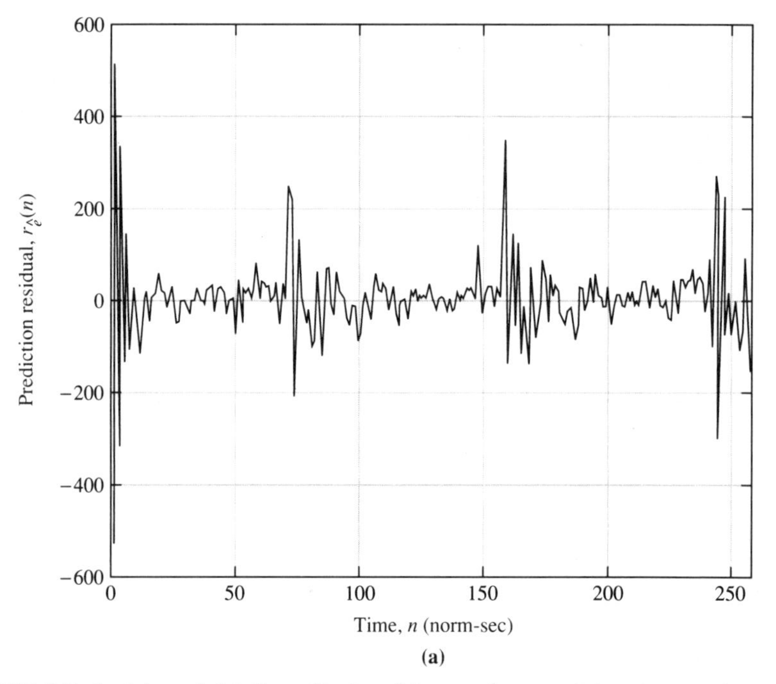

(a)

FIGURE 5.7. In (a) and (b) the effects of increasing model order on the prediction residual are observed for the vowel /a/. In (a) $M = 8$, while in (b) $M = 14$. The increased model order removes more correlation from the signal. In (c), residual energy, $r_{\hat{e}}(0)$ is plotted as a function of M for a voiced and unvoiced phoneme. The LP model is better for voiced sounds, so the asymptotic residual energy is lower.

[15]This is typically done by showing that a pole outside the unit circle can be reflected inside to decrease the mean squared error. Thus the spectrum is preserved and the filter becomes minimum phase.

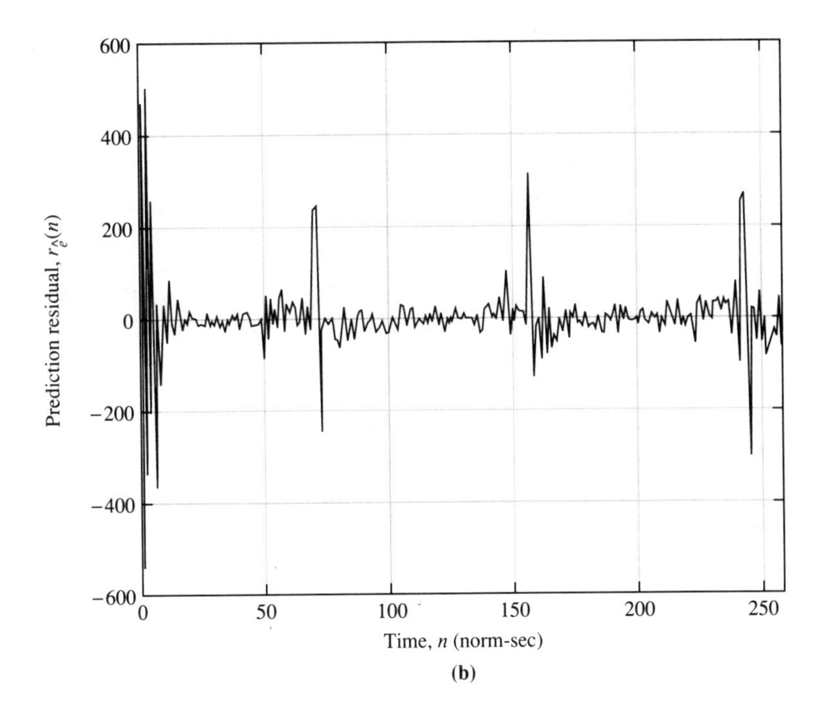

(b)

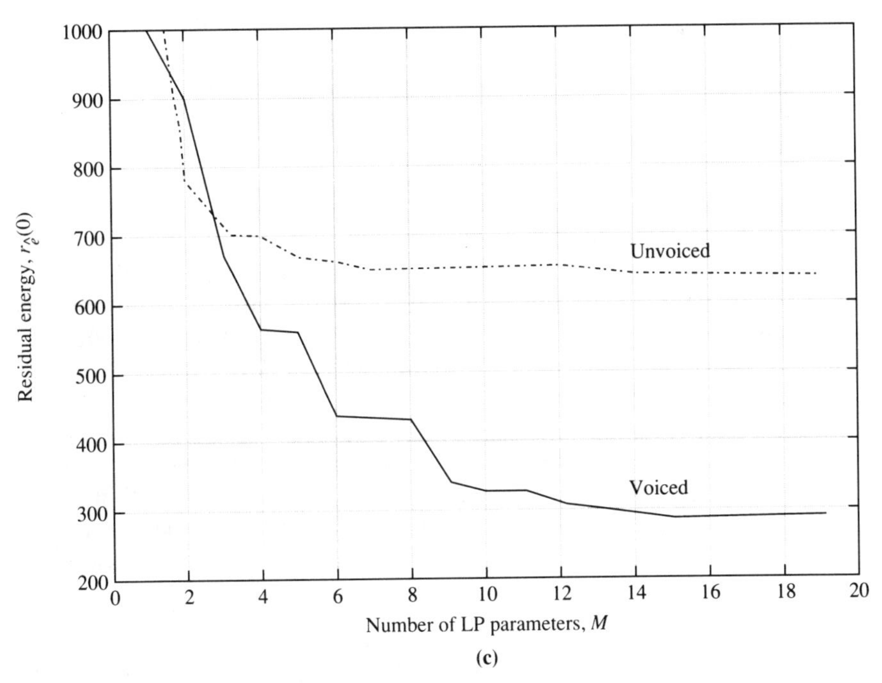

(c)

FIGURE 5.7. *(Cont.)*

reaches I, the integral takes the value Θ_0^2, apparently the minimum possible error energy. For $M < I, r_{\hat{e}}(0)$ is a monotonically nonincreasing function of M, which is easily seen as follows: Suppose that the identified parameters $\hat{a}(i), i = 1, 2, \ldots, M$, result in error energy $r_{\hat{e}}^M(0)$. Then the set $\hat{a}'(i), i = 1, 2, \ldots, M+1$, which are designed to minimize $r_{\hat{e}}^{M+1}(0)$, can always do as well (at minimizing error) as the smaller set, just by letting $\hat{a}'(i) = \hat{a}(i), i \leq M$ and $\hat{a}'(M+1) = 0$. In this case, $r_{\hat{e}}^{M+1}(0) = r_{\hat{e}}^M(0)$.

We must now become realistic about the procedure and consider estimating LP parameters on short intervals of speech data. With a solid understanding of the LP identification process developed through long-term arguments, we must now recall our knowledge of short-term processing from Chapter 4 and bring these concepts to bear on the development of short-term LP analysis. We will discover that our expertise in short-term processing will allow us to quickly develop the needed practical LP estimators.

5.3 Short-Term LP Analysis

5.3.1 Autocorrelation Method

Our objective here will be to use the knowledge and intuition gained from the long-term LP problem, combined with our understanding of short-term analysis, to produce an analogous short-term LP analysis. There are two well-known and widely used short-term LP solutions: the "autocorrelation method" and the "covariance method." Although in many ways the latter solution is more analogous to the long-term problem, we will follow the customary procedure of presenting the autocorrelation method first. This method has two main virtues with respect to the covariance method: It employs a "friendlier" (one argument) estimator for the autocorrelation sequence, and it is always guaranteed (theoretically) to produce a stable LP model (Markel and Gray, 1976).

Desired is an estimate of the LP parameters on the N data points ending at time m: $s(m-N+1), s(m-N+2), \ldots, s(m)$. Let us call the vector of parameter estimates[16] $\hat{\mathbf{a}}(m)$, neglecting the conventional subscript s as above, since it is obvious that these parameters are associated with the sequence $s(n)$. Recall also that $\hat{\mathbf{a}}(m)$ is a vector of M parameters $\hat{a}(i; m)$, $i = 1, 2, \ldots, M$, again omitting the subscript of $\hat{a}_s(i; m)$.

Recall the LP normal equations given in (5.22). Even if we had a complete record of $s(n), n \in (-\infty, \infty)$, the long-term solution would produce a terrible estimate of the parameters related to the interval of interest, since the autocorrelation would contain information from many different phones.

[16]It is important for the reader to think carefully about notation. We will make every effort to keep the notational conventions consistent with those developed in our earlier work.

A natural estimator for the LP parameters that uses only the specified data points is the one that results upon insertion of one of the short-term estimators for autocorrelation discussed in Section 4.3.1 [for generality, say $r_s(\eta; m)$]:

$$\sum_{i=1}^{M} \hat{a}(i;m) r_s(\eta - i; m) = r_s(\eta; m), \qquad \eta = 1, 2, \ldots, M, \tag{5.52}$$

which has a corresponding matrix form

$$\mathbf{R}_s(m)\hat{\mathbf{a}}(m) = \mathbf{r}_s(m) \Rightarrow \hat{\mathbf{a}}(m) = \mathbf{R}_s^{-1}(m)\mathbf{r}_s(m). \tag{5.53}$$

The matrix $\mathbf{R}_s(m)$, called the *(short-term) autocorrelation matrix* in the LP literature, is a Toeplitz operator (since all the elements along any diagonal are equal). The solution of equation (5.52) [or (5.53)] is referred to as the *autocorrelation method* of determining the $\hat{a}(i)$ coefficients.

If the short-term autocorrelation function can be derived from (4.35), then the autocorrelation method has the interesting interpretation as the solution of the problem below. In fact, most tutorial literature presents the autocorrelation method by posing and solving the following problem.

INTERPRETIVE PROBLEM 5.6 (AUTOCORRELATION PROBLEM) *Let $f(n;m)$ be the frame of speech $f(n;m) = s(n)w(m-n)$. Find the linear predictor of $f(n;m)$, which minimizes the total squared error in prediction over all n. (Note that the error is to be minimized over all n, not just the range selected by the window.)*

Solution. First let us define some important notation. We let $\hat{f}(n;m)$ indicate the predicted value of the point $f(n;m)$, and $\varepsilon(n;m)$ be the prediction error[17] at n,

$$\varepsilon(n;m) \stackrel{\text{def}}{=} f(n;m) - \hat{f}(n;m). \tag{5.54}$$

We also let $N\xi(m)$ denote the total squared prediction error over all n related to the analysis of the frame ending at time m. (We have introduced an extra factor of N, the window length, for convenience.) In these terms, we want to minimize

$$\begin{aligned} \xi(m) &= \frac{1}{N} \sum_{n=-\infty}^{\infty} \varepsilon^2(n;m) \\ &= \frac{1}{N} \sum_{n=-\infty}^{\infty} \left[f(n;m) - \sum_{i=1}^{M} \hat{a}(i;m) f(n-i;m) \right]^2. \end{aligned} \tag{5.55}$$

[17]The reader should carefully ponder the fact that $\varepsilon(n;m)$ is not a frame of the long-term error, called $\hat{e}(n)$ above; that is, $\varepsilon(n;m) \neq \hat{e}(n)w(m-n)$.

Differentiating $\xi(m)$ with respect to $\hat{a}(\eta;m)$ and setting the result to zero, we obtain

$$\frac{\partial \xi(m)}{\partial \hat{a}(\eta;m)} = \frac{2}{N}\sum_{n=-\infty}^{\infty}\Bigg[f(n;m)f(n-\eta;m) - \sum_{i=1}^{M}\hat{a}(i;m)f(n-i;m)f(n-\eta;m)\Bigg] = 0. \tag{5.56}$$

Upon dividing by 2, moving the summing operation (including the scale factor $1/N$) over n across terms, and recalling the definition of $f(n;m)$, it is clear that this expression is equivalent to (5.52) if the short-term autocorrelation is computed using (4.35) and the window specified in the problem.

5.3.2 Covariance Method

Covariance Equations

A second estimator for the parameters on the short-term $s(m-N+1)$, $\ldots, s(m)$, is obtained by plugging the short-term covariance function, defined in (4.49), into the long-term normal equations (5.22) as an estimate of autocorrelation:

$$\sum_{i=1}^{M}\hat{a}(i;m)\varphi_s(i,\nu;m) = \varphi_s(0,\nu;m), \qquad \nu = 1,2,\ldots,M. \tag{5.57}$$

This set of equations has the matrix form

$$\mathbf{\Phi}_s(m)\hat{\mathbf{a}}(m) = \boldsymbol{\varphi}_s(m) \Rightarrow \hat{\mathbf{a}}(m) = \mathbf{\Phi}_s^{-1}(m)\boldsymbol{\varphi}_s(m), \tag{5.58}$$

where $\mathbf{\Phi}_s(m)$ is the $M \times M$ matrix with (η,ν) element $\varphi_s(\eta,\nu;m)$; and

$$\boldsymbol{\varphi}_s(m) \stackrel{\text{def}}{=} [\varphi_s(0,1;m)\,\varphi_s(0,2;m)\cdots\varphi_s(0,M;m)]^T. \tag{5.59}$$

The matrix $\mathbf{\Phi}_s(m)$ is called the (*short-term*) *covariance matrix*; it is symmetric but not Toeplitz. Equation (5.57) [or (5.58)] is referred to as the *covariance method* solution.

Note that the covariance method actually makes use of some data points outside the data range $n \in [n-N+1, m]$ in computing the $\varphi_s(\eta,\nu;m)$. The significance of this range, however, is that it comprises the data over which the technique minimizes the prediction error energy, as shown by the following problem.

INTERPRETIVE PROBLEM 5.7 (COVARIANCE PROBLEM) *Find the linear predictor of $s(n)$ that minimizes the mean squared error on the range*[18] $n \in [m-N+1, m]$,

$$\xi(m) = \frac{1}{N} \sum_{n=m-N+1}^{m} [s(n) - \hat{s}(n)]^2 = \frac{1}{N} \sum_{n=m-N+1}^{m} \varepsilon^2(n; m). \tag{5.60}$$

In Problem 5.10, we will show that the solution to Interpretive Problem 5.7 is (5.57) and also explore the relationship to the orthogonality principle discussed above. This problem represents the usual method for introducing the covariance approach.

Note that a frame of speech is *not* created first in this problem. The prediction takes place with respect to the unaltered speech. Whereas the autocorrelation method windows the signal and then seeks to minimize the error in predicting the windowed signal over all n, the covariance method seeks to minimize the error in prediction of the unmodified speech only on the specified range of points.

Parallels with the Long-Term Results

Briefly noted in this section are two results that have practical significance. Both are proven in (Deller, 1984). Each is concerned with the conditions under which we might expect an exact solution using the covariance method. The first is directly parallel to a similar finding in the long-term case. The second, however, has no counterpart in the long-term analysis.

First, in the long-term case, we discovered that the LP solution would be exact if and only if the input to the system, $e(n)$, were orthogonal. A similar result obtains in the short-term covariance case.

THEOREM 5.3 *Let $\Theta(z)$, the speech system model, be minimum-phase. Assume that $\mathbf{\Phi}_s^{-1}(m)$ exists*[19] *and that $\mathbf{s}(m-N+1) = \mathbf{0}$ (recall that this vector represents the M initial conditions on the minimization window). Then the covariance solution is exact if and only if $\varphi_e(0, \nu; m) = K\delta(\nu)$, $\nu = 0, 1, \ldots, N-1$, where K is any finite constant.*

This theorem stipulates a requirement, reminiscent of that in the long-term case, that the input be "locally orthogonal" to produce an exact solution. The reason for the requirement that $\Theta(z)$ be minimum phase is that, in general, $\varphi_e(\eta, \nu; m) \neq \varphi_{e'}(\eta, \nu; m)$, although this is assured as N

[18] We are using the notation $\varepsilon(n; m)$ to indicate the error signal again, but this will not be the same sequence as the one resulting in the autocorrelation method over the same window [c.f. (5.54)].

[19] The necessary and sufficient conditions for the existence of the inverse are found in (Deller, 1984).

gets large. We note that although a strict theorem like this does not exist for the autocorrelation method, intuitively one would expect a practically similar result.

A related result, which is useful practically and has no parallel in the long-term case, is as follows:

THEOREM 5.4 *Let $\Theta(z)$ be a minimum-phase speech system. Assume that $\mathbf{\Phi}_s^{-1}(m)$ exists and that $\mathbf{s}(m-N+1) \neq \mathbf{0}$ (nonzero initial conditions). Then the covariance solution is exact if $e(n)=0, n \in [m-N+1, m]$. This condition is not theoretically necessary, but is practically so.*

According to this theorem there must be no excitation on the minimization window for exact solution in the covariance case when the initial conditions are nonzero.

The Covariance Method and Classical Least Squares Problems

Note: In this section, it will be useful to assume that the analysis window covers the range $n = 1, 2, \ldots, N$, so that $m = N$ and $m - N + 1 = 1$. There is no loss of generality with this choice and it will help us to set up some equations that will be useful in the future when we make this assumption again.

Many useful insights and practical results can be obtained by recognizing the relationship between the covariance method and classical least squares (least MSE) solutions of an overdetermined system of equations. In fact, the covariance method fits exactly the mold of the conventional least squares estimation problem: Given a series of $N \geq M$ observations, $s(1), s(2), \ldots, s(N)$, and a set of related vectors, $\mathbf{s}(1), \mathbf{s}(2), \ldots, \mathbf{s}(N)$, find the linear predictor (of dimension M) relating them, say $\hat{\mathbf{a}}(N)$, which minimizes the mean squared error (sample variance)

$$\xi(N) = \frac{1}{N} \sum_{n=1}^{N} [s(n) - \hat{s}(n)]^2, \tag{5.61}$$

where $\hat{s}(n)$ is the prediction of the nth observation,

$$\hat{s}(n) \stackrel{\text{def}}{=} \sum_{i=1}^{M} \hat{a}(i;N) s(n-i). \tag{5.62}$$

In vector-matrix form we can state the problem: Find the solution, $\hat{\mathbf{a}}(N)$, to the overdetermined system of equations,[20]

$$\mathbf{S}(N)\hat{\mathbf{a}}(N) = \overline{\mathbf{s}}(N), \tag{5.63}$$

[20]The reason for the bar over the vector $\overline{\mathbf{s}}(N)$ is to distinguish it from the notation $\mathbf{s}(N)$, which indicates the vector of past M speech values at time N.

which minimizes

$$\frac{1}{N}\|\mathbf{S}(N)\hat{\mathbf{a}}(N)-\bar{\mathbf{s}}(N)\|^2, \tag{5.64}$$

where

$$\mathbf{S}^T(N) \stackrel{\text{def}}{=} [\mathbf{s}(1)\,\mathbf{s}(2)\cdots\mathbf{s}(N)] \tag{5.65}$$

$$\bar{\mathbf{s}}(N) \stackrel{\text{def}}{=} [s(1)\,s(2)\cdots s(N)]^T \tag{5.66}$$

and $\|\cdot\|$ indicates the l_2 norm. Posed in this sense, the solution is given by [see, e.g., (Golub and Van Loan, 1989)],

$$\frac{1}{N}\mathbf{S}^T(N)\,\mathbf{S}(N)\hat{\mathbf{a}}(N)=\frac{1}{N}\mathbf{S}^T(N)\bar{\mathbf{s}}(N). \tag{5.67}$$

With a little effort, one can show that

$$\mathbf{\Phi}_s(m)=\frac{1}{N}\mathbf{S}^T(N)\mathbf{S}(N) \tag{5.68}$$

and

$$\boldsymbol{\varphi}_s(N)=\frac{1}{N}\mathbf{S}^T(N)\bar{\mathbf{s}}(N), \tag{5.69}$$

so that (5.67) is precisely the covariance solution (see Problem 5.11).

Two points are worth noting. First, statisticians have long used the technique of weighting certain observations more heavily than others in this problem by solving it subject to the constraint that the *weighted* squared error, say,

$$\xi'(N)=\frac{1}{N}\sum_{n=1}^{N}\lambda(n)[s(n)-\hat{s}(n)]^2=\frac{1}{N}\sum_{n=1}^{N}\lambda(n)\varepsilon^2(n;N), \tag{5.70}$$

be minimized, where $\{\lambda(n)\}$ is some set of nonnegative weights. The well-known solution is given by

$$\frac{1}{N}\mathbf{S}^T(N)\mathbf{\Lambda}(N)\mathbf{S}(N)\hat{\mathbf{a}}(N)=\frac{1}{N}\mathbf{S}^T(N)\mathbf{\Lambda}(N)\bar{\mathbf{s}}(N), \tag{5.71}$$

where $\mathbf{\Lambda}(N)$ is a diagonal matrix whose ith diagonal element is $\lambda(i)$. Not surprisingly, this equation is exactly equivalent to the result we would have obtained in our pursuit of the covariance normal equations if we had started with a weighted error criterion, (5.70), rather than (5.60). In this case, we would have concluded with normal equations of the form

$$\mathbf{\Phi}'_s(N)\hat{\mathbf{a}}(N)=\boldsymbol{\varphi}'_s(N)\Rightarrow\hat{\mathbf{a}}(N)=[\mathbf{\Phi}'_s(N)]^{-1}\boldsymbol{\varphi}'_s(m), \tag{5.72}$$

where $\mathbf{\Phi}_s'(N)$ and $\boldsymbol{\varphi}_s'(m)$ are the weighted counterparts to the matrices in (5.58), that is, $\mathbf{\Phi}_s'(N)$ is the matrix with (η, ν) entries

$$\varphi_s'(\eta, \nu; N) = \frac{1}{N} \sum_{n=1}^{N} \lambda(n) s(n-\eta) s(n-\nu) \tag{5.73}$$

and $\boldsymbol{\varphi}_s'(N)$ the M-vector with νth element $\varphi'(0, \nu; N)$. For future reference, let us note explicitly the relations

$$\mathbf{\Phi}_s'(N) = \frac{1}{N} \mathbf{S}^T(N) \mathbf{\Lambda}(N) \mathbf{S}(N) \tag{5.74}$$

and

$$\boldsymbol{\varphi}_s'(N) = \frac{1}{N} \mathbf{S}^T(N) \mathbf{\Lambda}(N) \overline{\mathbf{s}}(N). \tag{5.75}$$

We will call this technique the *weighted* (*short-term*) *covariance method.*

The weighted covariance method offers the advantage of emphasizing a datum that might be desirable in some sense by use of a large $\lambda(n)$, or of rejecting an undesirable point. Although there is no general theory for selecting weights to improve the quality of the LP estimate, the weighted strategy has been used for adaptive identification (Deller and Hsu, 1987) in algorithms for glottal waveform deconvolution and formant estimation (discussed below) (Deller and Picaché, 1989; Deller and Luk, 1987; Larar et al., 1985; Laebens and Deller, 1983; Veeneman and Bement, 1985; Wong et al., 1979) and in algorithms that significantly improve computational efficiency (Deller and Luk, 1989). In turn, these algorithms find application in aspects of coding and recognition, which are subjects of future chapters.[21]

Second, noting this link between the covariance method and the conventional least squares problem opens the door to the use of many conventional methods of solution that could prove useful in speech processing. In particular, the development of two useful temporally recursive forms of the covariance solution based on these ideas will be found in the next section.

Our next order of business is to study some of the commonly used methods for solving the autocorrelation and covariance equations and their practical considerations.

5.3.3 Solution Methods

Whether the autocorrelation or covariance method is employed, we are faced with the solution of a linear vector-matrix problem of the form $\mathbf{Ax} = \mathbf{b}$ in which $\mathbf{A}$ is a square matrix and $\mathbf{x}$ is the vector sought. In prin-

[21]In some cases, "weighting" can simply be "binary" in the sense that points are either included in the estimate or they are not. For other time-selective strategies, the reader is referred to (Steiglitz and Dickinson, 1977) and (Miyoshi et al., 1987).

ciple, one can use any of a number of well-known techniques that can be found in basic linear algebra books [e.g., (Nobel, 1969; Golub and Van Loan, 1989)]. We could even, for example, use some "brute force" method for inverting the matrix **A**, then forming the product $\mathbf{A}^{-1}\mathbf{b}$. A more rational thing to do would be to compute an "**LU**" decomposition of the matrix **A** using Gaussian elimination. Any efficient method for solving a system of equations, however, will take advantage of the special structure inherent in the vector-matrix problem. We have noted that both the covariance and autocorrelation methods produce symmetric matrices, while the autocorrelation matrix is additionally Toeplitz. Another very important property in either case, which is true in all but the rarest of practical circumstances (Deller, 1984), is the positive definiteness of the matrix. Our task here is to take advantage of these properties in creating efficient algorithms for the solution of the two methods. We discuss some basic methods which are popular in speech processing, noting that there are many variations on the solutions presented here. There are also many other approaches from classical numerical algebra that could be applied; new architectures and algorithms might focus attention on one or more of these in the future. We will describe an instance of this latter effect below, which was driven by parallel processing architectures in the 1980s. Intelligent use of any of these methods requires that we be aware of their numerical properties, an issue that would take us too far afield in this discussion. However, the reader might wish to refer to any of a number of excellent books that treat this subject [e.g., (Golub and Van Loan, 1989)] before committing a speech processing project to a particular algorithm.

Levinson–Durbin Recursion (Autocorrelation)

Let us begin with the autocorrelation problem which has the most special structure. In 1947, Levinson (1947) published an algorithm for solving the problem $\mathbf{Ax} = \mathbf{b}$ in which **A** is Toeplitz, symmetric, and positive definite, and **b** is arbitrary.[22] Of course, the autocorrelation equations are exactly of this form, with **b** having a special relationship to the elements of **A**. In 1960, Durbin (1960) published a slightly more efficient algorithm for this special case. Durbin's algorithm is often referred to as the *Levinson–Durbin* (L–D) *recursion* by speech processing engineers.

The L–D recursion is a recursive-in-model-order solution for the autocorrelation equations. By this we mean that the solution for the desired order-M model is successively built up from lower-order models, beginning with the "0th order predictor," which is no predictor at all.

From the vector-matrix form of the autocorrelation equations in (5.53) we can write

$$-\mathbf{R}_s(m)\hat{\mathbf{a}}^M(m) + \mathbf{r}_s(m) = \mathbf{0}. \tag{5.76}$$

[22]A reformulation of Levinson's algorithm is found in (Robinson, 1967).

Be careful to recall that the m's here are time indices indicating the end of the window over which the analysis is being performed. A typical element of $\mathbf{R}_s(m)$, for example, is $r_s(\eta; m)$. A bit of new notation here is the added superscript M on the parameter vector, $\hat{\mathbf{a}}^M(m)$, used to indicate an Mth-order solution. Since the L–D recursion is recursive in the model order, it is necessary in the following development to add superscripts of this form to several variables to keep track of the order of the solution with which they are associated. Quantities without such superscripts are associated with the Mth (final)-order solution. To avoid confusion, any quantity that needs to be raised to a power will be written in brackets first, for example, $[\kappa(3; m)]^2$.

As we did above, let us denote by $\varepsilon(n; m)$ the prediction residual sequence due to the autocorrelation estimate $\hat{\mathbf{a}}(m)$ [see (5.54)]. With this definition, it is not difficult to demonstrate that

$$r_s(0; m) - \sum_{i=1}^{M} \hat{a}^M(i; m) r_s(i; m) = \frac{1}{N} \sum_{n=-\infty}^{\infty} [\varepsilon(n; m)]^2. \tag{5.77}$$

It is clear that the term on the right can be interpreted as the total energy (scaled by a factor $1/N$) in the prediction residual sequence, based on the inverse filter of order M. As we did in (5.55), let us write this quantity as $\xi^M(m)$ (except that this time we have added the superscript) and rewrite (5.77) as

$$r_s(0; m) - \mathbf{r}_s^T(m)\hat{\mathbf{a}}^M(m) = \xi^M(m). \tag{5.78}$$

Now we write (5.76) as

$$\begin{bmatrix} \mathbf{r}_s(m) & \mathbf{R}_s(m) \end{bmatrix} \begin{bmatrix} 1 \\ -\hat{\mathbf{a}}(m) \end{bmatrix} = \mathbf{0}. \tag{5.79}$$

If we stack (5.78) on top of (5.79), we get

$$\begin{bmatrix} r_s(0; m) & \mathbf{r}_s^T(m) \\ \mathbf{r}_s(m) & \mathbf{R}_s(m) \end{bmatrix} \begin{bmatrix} 1 \\ -\hat{\mathbf{a}}(m) \end{bmatrix} = \begin{bmatrix} \xi^M(m) \\ \mathbf{0} \end{bmatrix}. \tag{5.80}$$

Let us denote the augmented matrix on the left by $\tilde{\mathbf{R}}_s^M(m)$. In these terms, we have

$$\begin{aligned} &\tilde{\mathbf{R}}_s^M(m) \times [1 \; -\hat{a}^M(1; m) \; -\hat{a}^M(2; m) \; \cdots \; -\hat{a}^M(M; m)]^T \\ &\quad = [\xi^M(m) \, 0 \, 0 \; \cdots \; 0]^T. \end{aligned} \tag{5.81}$$

The system of equations (5.81) forms the starting point for the L–D recursion. To understand the basic principles of the method, it is sufficient to examine it for the case $M = 3$. In this case, filling in the entries of the $\tilde{\mathbf{R}}_s^3(m)$ matrix, we have

$$\begin{bmatrix} r_s(0;m) & r_s(1;m) & r_s(2;m) & r_s(3;m) \\ r_s(1;m) & r_s(0;m) & r_s(1;m) & r_s(2;m) \\ r_s(2;m) & r_s(1;m) & r_s(0;m) & r_s(1;m) \\ r_s(3;m) & r_s(2;m) & r_s(1;m) & r_s(0;m) \end{bmatrix} \begin{bmatrix} 1 \\ -\hat{a}^3(1;m) \\ -\hat{a}^3(2;m) \\ -\hat{a}^3(3;m) \end{bmatrix} = \begin{bmatrix} \xi^3(m) \\ 0 \\ 0 \\ 0 \end{bmatrix}. \tag{5.82}$$

Note that the augmented matrix, $\tilde{\mathbf{R}}_s^3(m)$, is Toeplitz. Also note that the matrix $\tilde{\mathbf{R}}_s^2(m)$ (associated with the order-two solution) is embedded in $\tilde{\mathbf{R}}_s^3(m)$ in two places. It is the 3×3 matrix obtained by removing either the first row and column, or the fourth row and column, of $\tilde{\mathbf{R}}_s^3(m)$. This consequence of the Toeplitz structure is central to our development.

Now let us assume for the moment that we can write the augmented order-three solution vector in terms of the order-two solution as follows:

$$\begin{bmatrix} 1 \\ -\hat{a}^3(1;m) \\ -\hat{a}^3(2;m) \\ -\hat{a}^3(3;m) \end{bmatrix} = \begin{bmatrix} 1 \\ -\hat{a}^2(1;m) \\ -\hat{a}^2(2;m) \\ 0 \end{bmatrix} - \kappa(3;m) \begin{bmatrix} 0 \\ -\hat{a}^2(2;m) \\ -\hat{a}^2(1;m) \\ 1 \end{bmatrix}. \tag{5.83}$$

[The constant $\kappa(3;m)$ will play a special role in upcoming developments, and will be known as a *reflection coefficient.*] Plugging this result into the left side of (5.82) and remembering the embedded $\tilde{\mathbf{R}}_s^2(m)$'s, we have

$$\begin{bmatrix} r_s(0;m) & r_s(1;m) & r_s(2;m) & r_s(3;m) \\ r_s(1;m) & r_s(0;m) & r_s(1;m) & r_s(2;m) \\ r_s(2;m) & r_s(1;m) & r_s(0;m) & r_s(1;m) \\ r_s(3;m) & r_s(2;m) & r_s(1;m) & r_s(0;m) \end{bmatrix} \begin{bmatrix} 1 \\ -\hat{a}^3(1;m) \\ -\hat{a}^3(2;m) \\ -\hat{a}^3(3;m) \end{bmatrix}$$
$$= \begin{bmatrix} \xi^2(m) \\ 0 \\ 0 \\ q \end{bmatrix} - \kappa(3;m) \begin{bmatrix} q \\ 0 \\ 0 \\ \xi^2(m) \end{bmatrix} = \begin{bmatrix} \xi^3(m) \\ 0 \\ 0 \\ 0 \end{bmatrix}, \tag{5.84}$$

where

$$q \stackrel{\text{def}}{=} r_s(3;m) - \sum_{i=1}^{2} \hat{a}^2(i;m) r_s(3-i;m). \tag{5.85}$$

Equation (5.84) implies two scalar equations:

$$\xi^2(m) - \kappa(3;m)q = \xi^3(m) \tag{5.86}$$

$$q - \kappa(3;m)\xi^2(m) = 0 \Rightarrow \kappa(3;m) = \frac{q}{\xi^2(m)}. \tag{5.87}$$

Eliminating q from the first equation using the second, we obtain

$$\xi^3(m) = \xi^2(m)\{1 - [\kappa(3;m)]^2\}. \tag{5.88}$$

Substituting the definition of q, (5.85), into (5.87) yields

$$\kappa(3;m) = \frac{1}{\xi^2(m)}\left\{r_s(3;m) - \sum_{i=1}^{2} \hat{a}^2(i;m)r_s(3-i;m)\right\}. \tag{5.89}$$

Now we have an expression for $\kappa(3;m)$ in terms of the order-two LP parameters and an expression for $\xi^3(m)$ in terms of $\kappa(3;m)$ and $\xi^2(m)$. Let us now work on finding the order-three LP parameters. If we have $\kappa(3;m)$ (which we do), then everything we need to know is packed in the vector equation (5.83), which easily yields

$$\hat{a}^3(3;m) = \kappa(3;m) \tag{5.90}$$

$$\hat{a}^3(i;m) = \hat{a}^2(i;m) - \kappa(3;m)\hat{a}^2(3-i;m) \qquad \text{for } i = 1, 2. \tag{5.91}$$

FIGURE 5.8. The Levinson-Durbin (L–D) recursion applied to the window $n \in [m-N+1, m]$.

Initialization: $l = 0$

$\xi^0(m)$ = scaled total energy in the "error" from an "order 0" predictor

= average energy in the speech frame $f(n;m) = s(n)w(m-n)$

$= r_s(0;m)$

Recursion: For $l = 1, 2, \ldots, M$,

1. Compute the lth reflection coefficient,

$$\kappa(l;m) = \frac{1}{\xi^{l-1}(m)}\left\{r_s(l;m) - \sum_{i=1}^{l-1} \hat{a}^{l-1}(i;m)r_s(l-i;m)\right\}. \tag{5.92}$$

2. Generate the order-l set of LP parameters,

$$\hat{a}^l(l;m) = \kappa(l;m) \tag{5.93}$$

$$\hat{a}^l(i;m) = \hat{a}^{l-1}(i;m) - \kappa(l;m)\hat{a}^{l-1}(l-i;m),\ i = 1, \ldots, l-1. \tag{5.94}$$

3. Compute the error energy associated with the order-l solution,

$$\xi^l(m) = \xi^{l-1}(m)\{1 - [\kappa(l;m)]^2\}. \tag{5.95}$$

4. Return to Step 1 with l replaced by $l+1$ if $l < M$.

We now have the necessary tools to move from order-two parameters to order-three parameters. Given $\xi^2(m)$, we can compute $r_s(\eta; m)$, $\eta = 1, 2, 3$, from the data, then $\kappa(3; m)$ using (5.89). Then we can compute the order-three LP parameters using (5.90) and (5.91). Finally, we can compute $\xi^3(m)$ using (5.88) in case we want to move on to the next step to compute the order-four LP parameters. The general algorithm is just a straightforward generalization of this specific process and is given in Fig. 5.8.

There are several important features of the L–D recursion to be noted. We know from previous discussion that the sequence of average error energies should be nonincreasing,[23]

$$0 \le \xi^l(m) \le \xi^{l-1}(m) \le \xi^{l-2}(m) \le \cdots \le \xi^0(m). \tag{5.96}$$

If we define the *normalized error energy* at iteration l by the ratio $\xi^l(m)/\xi^0(m)$, then it is clear that

$$0 \le \frac{\xi^l(m)}{\xi^0(m)} \le 1 \tag{5.97}$$

for any l. This quantity can be monitored at each step to detect numerical instabilities. It is also possible to watch either $\xi^l(m)/\xi^0(m)$ or $\xi^l(m)$ to determine when a sufficient model order has been reached.

Similarly, we can use (5.95) and (5.97) to show that

$$\left|\kappa(l; m)\right| \le 1 \qquad \text{for all } l, \tag{5.98}$$

offering another check for numerical stability. As we noted above, these κ parameters are often called *reflection coefficients* because of their close relationship to the reflection coefficients in analog acoustic tube models of the vocal tract (recall Section 3.1.4). We will explore this issue below. Another interpretation, also to be studied in this section, leads to the name *partial correlation*, or *parcor, coefficients* for these parameters. These coefficients play an important role in speech analysis and coding applications, and serve as an alternative to the usual autocorrelation LP coefficients as a parametric representation for the speech. Clearly, the M $\hat{a}(i; m)$ parameters can be obtained from the M reflection coefficients, as we have seen above. It is also not difficult to show that the procedure can be "reversed" to obtain the reflection coefficients from the LP parameters. The algorithm is found in Problem 5.15.

Finally, we note two sets of equivalent parameters that can be computed from the reflection coefficients (and vice versa). These are the *log area ratio* (LAR) *parameters*, another term whose meaning will become clear below, and the *inverse sine* (IS) *parameters*. The LAR and IS param-

[23]The discussion we had about this issue was actually for the "long-term" case in Section 5.2.3. If we replace $r_{\hat{e}}^l(\eta)$ in that discussion by $r_{\varepsilon}^l(\eta; m)$ we get this similar result for the short-term autocorrelation case. Note that our name for $r_{\varepsilon}^l(0; m)$ has been simplified in the present discussion to $\xi^l(m)$.

eters are primarily used in speech coding applications (see Section 7.4.5) and are given, respectively, by

$$g(l;m) = \frac{1}{2}\log\frac{1+\kappa(l;m)}{1-\kappa(l;m)} = \tanh^{-1}\kappa(l;m), \qquad \text{for } l = 1, 2, \ldots, M, \tag{5.99}$$

and

$$\sigma(l;m) = \frac{2}{\pi}\sin^{-1}\kappa(l;\mathrm{m}), \qquad \text{for } l = 1, 2, \ldots, M. \tag{5.100}$$

These parameters are used in place of the reflection coefficients because, when a reflection coefficient has a magnitude near unity, the results have been found to be highly sensitive to quantization errors. These transformations warp the amplitude scale of the parameters to decrease this sensitivity. Note that the IS parameters have the advantage of remaining bounded in magnitude by unity,

$$|\sigma(l;m)| \le 1 \tag{5.101}$$

for any l. Nevertheless, the LAR parameters seem to be more popular, and no significant difference in performance between the two sets has been found. A study of the quantization properties of these transmission parameters is found in (Viswanathan and Makhoul, 1975).

Lattice Structures Based on the L–D Recursion

The L–D Lattice. The L–D recursion leads to a *lattice* formulation of the inverse filter computation. Suppose that we let

$$\hat{A}^l(z;m) \stackrel{\text{def}}{=} 1 - \sum_{i=1}^{l} \hat{a}^l(i;m)z^{-i}. \tag{5.102}$$

Using Step 2 of the L–D recursion, which computes the $\hat{a}^l(i;m)$'s in terms of the $\hat{a}^{l-1}(i;m)$'s and $\kappa(l;m)$, it is not difficult to show that (see Problem 5.16)

$$\hat{A}^l(z;m) = \hat{A}^{l-1}(z;m) - \kappa(l;m)z^{-l}\hat{A}^{l-1}(z^{-1};m). \tag{5.103}$$

We let $\varepsilon^l(n;m)$ denote the error sequence on the window based on the order-l inverse filter,

$$\varepsilon^l(n;m) = f(n;m) - \sum_{i=1}^{l} \hat{a}^l(i;m)f(n-i;m), \tag{5.104}$$

where, as usual, $f(n;m)$ denotes the speech frame ending at time m, $f(n;m) = s(n)w(m-n)$. If we employ the notation

$$\mathcal{E}^l(z;m) \stackrel{\text{def}}{=} \mathcal{Z}\{\varepsilon^l(n;m)\} \tag{5.105}$$

$$F(z;m) \stackrel{\text{def}}{=} \mathcal{Z}\{f(n;m)\}, \tag{5.106}$$

where $\mathcal{Z}$ denotes the z-transform operator, then using (5.102) through (5.104), we can write

$$\begin{aligned}\mathcal{E}^{l}(z;m) &= \hat{A}^{l-1}(z;m)F(z;m) - \kappa(l;m)z^{-l} \\ &= \mathcal{E}^{l-1}(z;m) - z^{-l}\kappa(l;m)\hat{A}^{l-1}(z^{-1};m)F(z;m).\end{aligned} \tag{5.107}$$

It is clear that the first term on the right in this expression corresponds to $\varepsilon^{l-1}(n;m)$, but what about the second? Let us define

$$\mathcal{B}^{l}(z;m) \stackrel{\text{def}}{=} z^{-(l+1)}\hat{A}^{l}(z^{-1};m)F(z;m). \tag{5.108}$$

This is close to the second term except that l has been incremented by unity and we have ignored the reflection coefficient, $\kappa(l;m)$. Noting that

$$\begin{aligned}\beta^{l}(n;m) \stackrel{\text{def}}{=} \mathcal{Z}^{-1}\{\mathcal{B}^{l}(z;m)\} &= f(n-l-1;m) \\ &\quad - \sum_{i=1}^{l} \hat{a}^{l}(i;m)f(n+i-l-1;m),\end{aligned} \tag{5.109}$$

we see that this sequence amounts to the *error in "backward prediction"* of $f(n-l-1;m)$ based on l *future* points. Backward prediction is illustrated in Fig. 5.9.

Now that we understand what $\mathcal{B}^{l}$ means, let us go back to (5.107) and write it in these terms,

$$\mathcal{E}^{l}(z;m) = \mathcal{E}^{l-1}(z;m) - \kappa(l;m)\mathcal{B}^{l-1}(z;m) \tag{5.110}$$

or, in the time domain,

$$\varepsilon^{l}(n;m) = \varepsilon^{l-1}(n;m) - \kappa(l;m)\beta^{l-1}(n;m). \tag{5.111}$$

Also using (5.103) and (5.108), we obtain

$$\begin{aligned}\mathcal{B}^{l}(z;m) &= z^{-(l+1)}\left[\hat{A}^{l-1}(z^{-1};m) - \kappa(l;m)z^{-l}\hat{A}^{l-1}(z;m)\right]F(z;m) \\ &= z^{-1}\left[\mathcal{B}^{l-1}(z;m) - \kappa(l;m)\,\mathcal{E}^{l-1}(z;m)\right]\end{aligned} \tag{5.112}$$

or

$$\beta^{l}(n;m) = \beta^{l-1}(n-1;m) - \kappa(l;m)\varepsilon^{l-1}(n-1;m). \tag{5.113}$$

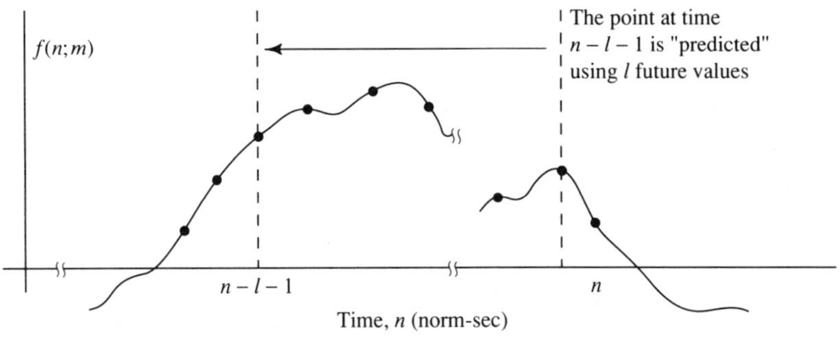

FIGURE 5.9. "Backward prediction" of the speech sequence.

The two time domain equations for the forward and backward prediction errors, (5.111) and (5.113), lead immediately to the *analysis lattice structure* shown in Fig. 5.10. The computations represented by the analysis lattice are used to convert the speech sequence into the prediction error sequence and are equivalent, in theory, to an inverse filter derived using the autocorrelation method. The *synthesis lattice structure* can be used to reverse the process and is derived simply by rewriting (5.111) as

$$\varepsilon^{l-1}(n;m) = \varepsilon^{l}(n;m) + \kappa(l;m)\beta^{l-1}(n;m). \tag{5.114}$$

The structure is shown in Fig. 5.11.

The Itakura–Saito (Parcor) Lattice. The lattice structures derived above are a direct consequence of the L–D recursions and are equivalent to the autocorrelation method. The lattice can, at first glance, be a bit deceiving in this regard because from the diagram it appears that the computations involved in the L–D recursion (in particular, the autocorrelations) have been circumvented. In fact, however, the lattice diagram provides a pictorial view of the evolution of the model order recursions, but it is still quite necessary to compute the L–D type recursions in the process of constructing the lattice. Itakura et al. (1969, 1972), however, demonstrated that the lattice could evolve without explicit computation of the correlations or predictor coefficients. Although their approach is quite novel [see the original papers and (Markel and Gray, 1976, Ch. 2)], we have the information necessary to derive their key result in our developments above. In fact, using the definitions of forward prediction error, $\varepsilon(n;m) = \varepsilon^{M}(n;m)$ [see (5.54) and (5.104)], backward prediction error, $\beta(n;m) = \beta^{M}(n;m)$ [see (5.109)], and total squared (forward) error [see (5.55)], it can be shown that (Marple, 1987, Sec. 7.3.3)

$$\kappa(l;m) = \frac{\sum_{n=m+N-1}^{m} \varepsilon^{l-1}(n;m)\beta^{l-1}(n;m)}{\sqrt{\sum_{n=m+N-1}^{m} [\varepsilon^{l-1}(n;m)]^2 \sum_{n=m+N-1}^{m} [\beta^{l-1}(n;m)]^2}}. \tag{5.115}$$

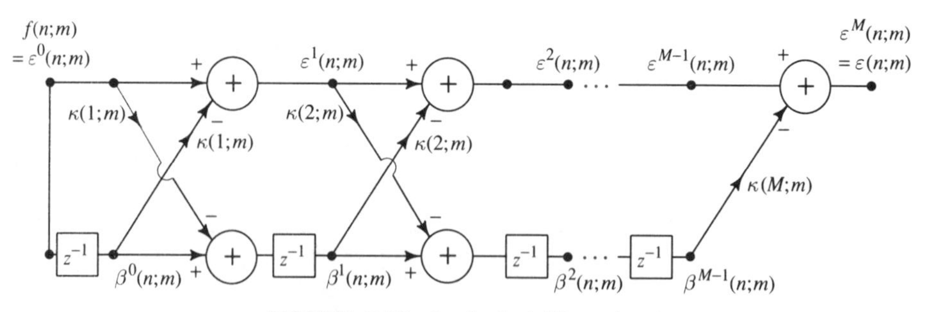

FIGURE 5.10. Analysis lattice structure.

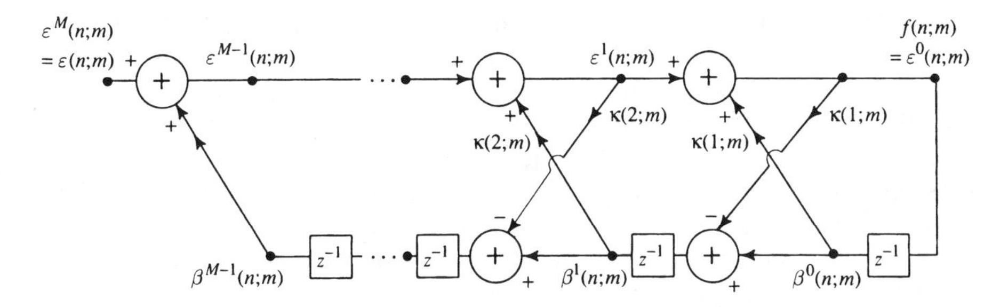

FIGURE 5.11. Synthesis lattice structure.

Although this derivation is a bit tricky, verifying that it is indeed correct is straightforward. We simply need to plug (5.104) and (5.109) into (5.115) to derive

$$\kappa(l;m) = \frac{1}{\xi^{l-1}(m)} \left\{ r_s(l;m) - \sum_{i=1}^{l-1} \hat{a}^{l-1}(i;m) r_s(l-i;m) \right\}, \qquad (5.116)$$

which is the expression used in Step 1 of the L–D recursion above.

At last we can see the reason for calling $\kappa(l;m)$ a *partial correlation* (*parcor*) coefficient. It is evident from (5.115) that $\kappa(l;m)$ represents a normalized cross-correlation between the sequences $\varepsilon^l(n;m)$ and $\beta^l(n;m)$, which statisticians commonly call the correlation coefficient.[24] The word "partial" connotes the lower model orders prior to reaching the "complete" correlation at $l = M$.

The key point is the ability to compute the $\kappa(l;m)$ parameters without recourse to the correlations or prediction parameters. A lattice structure based on this idea is shown in Fig. 5.12.

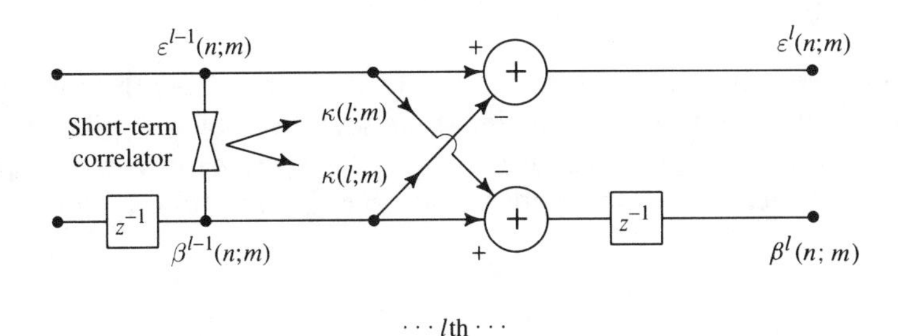

FIGURE 5.12. The parcor lattice of Itakura and Saito. The "correlator" is used to execute equation (5.115) to deduce the $\kappa(l;m)$ coefficient for the lth stage.

[24] Actually, we are being a bit loose with terminology here, since we are using short-term temporal averages rather than statistical expectations employed in the rigorous definition of a correlation coefficient.

Relationship of the Lattice Filter to the Lossless Acoustic Tube Model. A review of Fig. 3.23(b) and Figs. 5.10 and 5.11 will reveal an apparent similarity between the signal flow diagram of the acoustic tube model and the lattice structures developed here. Indeed, the "propagation" of errors through the lattice, and the "transmission" and "reflection" of these errors at the section boundaries, is quite analogous to the behaviors of the forward and backward flow waveforms in the acoustic tube model. In fact, both computational structures are implementing an all-pole transfer function, and each section (in either case) is responsible for the implementation of a pole pair. Further, the acoustic reflection coefficient is sufficient to characterize a given tube section while the lattice reflection coefficient completely characterizes a recursion in the LP computation. Although there are some differences between the tube model and the lattice filter, particularly with regard to details at the inputs and outputs, it is very interesting that such analogous computational structures arise from quite disparate modeling approaches. Whereas the acoustic tube outcome was based on physical considerations, the LP model is based on a system identification approach that makes no overt attempt to model the physics of the system. On the other hand, the fact that the two approaches are either inherently (tube) or explicitly (LP) centered on an all-pole model makes it natural that a similar computational structure could be used to implement the result.

The analogy is manifest in precise terms as follows. Because of the equations we have available without further algebra, it is most convenient to compare the analysis lattice (which produces the excitation from speech) with the dynamics of the *reverse* flow (from lips to glottis) in the acoustic tube. Let us write (5.110) and (5.112) in matrix form,

$$\begin{bmatrix} \mathcal{E}^{l}(z;m) \\ \mathcal{B}^{l}(z;m) \end{bmatrix} = \begin{bmatrix} 1 & -\kappa(l;m) \\ -z^{-1}\kappa(l;m) & z^{-1} \end{bmatrix} \begin{bmatrix} \mathcal{E}^{l-1}(z;m) \\ \mathcal{B}^{l-1}(z;m) \end{bmatrix}. \tag{5.117}$$

Now let us recall (3.155) in Appendix 3.C, which relates the forward and backward volume flows at the *input* of section k of the tube model to those at the *output*. (Recall that section indices increase from lips to glottis in the lattice, while the opposite is true for the tube.) This equation can be written as

$$\begin{bmatrix} V_k^{+}(z) \\ V_k^{-}(z) \end{bmatrix} = \frac{z}{1+\rho_k} \begin{bmatrix} 1 & -\rho_k \\ -z^{-1}\rho_k & 1 \end{bmatrix} \begin{bmatrix} V_{k+1}^{+}(z) \\ V_{k+1}^{-}(z) \end{bmatrix}. \tag{5.118}$$

Except for some scaling and a delay, the computations are seen to be identical in the acoustic tube and lattice sections. The need for the extra factors in the acoustic tube model can be seen by recalling the transfer function for the N-tube acoustic model derived in Appendix 3.C,

$$H(z) = \frac{1+\rho_{\text{glottis}}}{2} \frac{z^{-N/2}\prod_{k=1}^{N}(1+\rho_k)}{\left[1-\sum_{k=1}^{N} b_k z^{-k}\right]}. \tag{5.119}$$

Not surprisingly, if the impedance of the glottis is assumed to be infinite so that $\rho_{\text{glottis}} = 1$, then the denominator coefficients can be generated using the L–D recursion of Fig. 5.10 with the reflection coefficients $-\rho_l$ used in place of $\kappa(l;m)$ (the change of sign is related to the defined directions of flow in the two analyses).

The Burg Lattice

Our main intention here is to show an alternative lattice formulation that is actually identical in form to that in Fig. 5.10, but which has reflection coefficients computed according to a different criterion. Some historical perspective is necessary here before we get into technical details.

J. P. Burg presented a classic paper[25] in 1967 (Burg, 1967) in which he was concerned with the fact that we have no knowledge of the autocorrelation sequence in the estimation procedure outside the range $0 \le \eta \le M$. Effectively, we assume that $r_s(\eta;m) = 0$ outside this range. This is a reflection of the fact that any typical window used on the data will zero out the sequence outside the window. We are therefore imposing a generally incorrect constraint on the solution. Burg thought it made more sense to assume the data sequence to be "maximally uncertain," or to contain the "maximum entropy," on the times that are not included in the estimation process. This, in general, will imply a different autocorrelation sequence outside the "known" range $0 \le \eta \le M$ than the zeros that we have assumed. The power density spectrum that goes with this modified autocorrelation sequence is called the *maximum entropy* (ME) *spectrum*. In general, we would expect the ME spectrum to be different, of course, from the spectrum of the autocorrelation LP model, $|\hat{\Theta}(\omega;m)|^2$. However, in one important case, when the data sequence is a Gaussian random process, Burg shows the ME spectrum to be identical to the autocorrelation LP spectrum, being characterized by an all-pole function with polynomial coefficients that are exactly equivalent to the LP $\hat{a}(i)$ parameters and are computed according to the same normal equations (5.52). Therefore, the autocorrelation method is often called the *maximum entropy method* (MEM), particularly by people working in spectral estimation problems. One should carefully notice that they are equivalent only in the important special case noted.

[25]Burg was actually concerned with spectrum estimation, rather than parameter identification, in his paper.

A year later, Burg published another remarkable paper[26] in which he was concerned with the estimation of the LP parameters, especially under circumstances in which there are very few data in the window upon which to base estimates of autocorrelation. In this paper he developed the basis for the lattice structure we have been discussing above. Burg was the first to notice that, as in the case of the Itakura–Saito lattice, the computation of autocorrelation could be side-stepped altogether and the whole lattice constructed using only reflection coefficients. What makes his method unusual, and the reason we discuss it here at the end, is that his optimization criterion for determining the $\hat{a}(i)$ parameters is somewhat different from the usual autocorrelation criterion. In Burg's work he chooses to find the $\hat{\mathbf{a}}(m)$ vector subject to minimization of the sum of forward and backward prediction error energies. Further, at iteration $l, \hat{a}^l(l;m)$ is computed subject to the minimization of the relevant error quantity, say

$$\tau^l(m) \stackrel{\text{def}}{=} \frac{1}{N} \sum_{n=m+N-1}^{m} [\varepsilon^l(n;m)]^2 + [\beta^l(n;m)]^2 \tag{5.120}$$

with the coefficients, $\hat{a}^l(i;m), i = 1, 2, \ldots, l-1$, remaining fixed at their values from the previous order, $\hat{a}^{l-1}(i;m)$. Since $\hat{a}^l(l;m) = \kappa(l;m)$, this can be interpreted as the pursuit of *reflection coefficients* that are optimized locally with previous reflection coefficients held fixed. Putting (5.111) and (5.113) into (5.120), and differentiating $\tau^l(m)$ with respect to $\kappa(l;m)$, yields

$$\kappa(l;m) = \frac{\displaystyle\sum_{n=m+N-1}^{m} \varepsilon^{l-1}(n;m)\beta^{l-1}(n;m)}{\displaystyle\frac{1}{2}\sum_{n=m+N-1}^{m} \left[\varepsilon^{l-1}(n;m)\right]^2 + \left[\beta^{l-1}(n;m)\right]^2}, \tag{5.121}$$

which are the reflection coefficients used in the Burg lattice. It should be noted that, as in the case of the autocorrelation reflection coefficients, the Burg reflection coefficients have the property that

$$|\kappa(l;m)| \leq 1. \tag{5.122}$$

This fact, along with a demonstration that the solutions are unique and provide minima of the $\tau^l(m)$'s, is found in (Markel and Gray, 1976, Ch. 2).

If we refer to the autocorrelation reflection coefficients as "forward" parcor coefficients (because they are based on minimization of forward error energy), then it is clear that we could also derive a set of "backward" reflection coefficients. The Burg reflection coefficients represent the harmonic mean between the forward and backward coefficients. The

[26]The two bodies of work have become blurred together in the literature, as we discuss below, but each is a distinct contribution to the area.

denominator represents the arithmetic average of the forward and backward error energies, while the autocorrelation-based reflection coefficients contain a geometric mean in the denominator.

The Burg method, when viewed as a spectral estimation technique, has been found to be prone to "line splitting" and also biased estimates for narrowband processes [see (Marple, 1987, Ch. 8) for discussion and references]. Various weighted error strategies have been proposed to alleviate the bias problem (Swingler, 1980; Kaveh and Lippert, 1983; Helme and Nikias, 1985). A specific application to speech involving an exponential forgetting factor is found in (Makhoul and Cosell, 1981).

In summary, after having developed the lattice in Fig. 5.10 that is governed by (5.111) and (5.113), we recognize that, at each step in the recursion, the next stage of the lattice is parameterized by only one quantity, the reflection coefficient. If we therefore solve for $\kappa(l; m)$, which minimizes $\tau^l(m)$ under the constraint of fixed previous coefficients, we will derive (5.121). The Burg lattice is often presented this *ad hoc* way, almost as an afterthought, at the end of discussions on the autocorrelation method. In fact, the Burg method is the predecessor and motivator for the autocorrelation developments.

Finally, it is important to return to the history of the Burg method, and the term MEM. In this "Burg lattice problem," we have essentially solved an "autocorrelation" type problem with a different error minimization criterion. Again under conditions of a Gaussian input and output to the model, the ME considerations yield the same normal equations as if we had just used the standard autocorrelation approach. For this reason, Burg's lattice is sometimes referred to as the MEM, but one should be careful to remember that this is only strictly appropriate in an important special case. Further, the use of the term MEM here tends to blur the distinction between Burg's two important contributions. His 1967 paper was primarily concerned with the presentation of the ME principle, while the 1968 paper showed the ability to decouple the lattice and reflection coefficients from the autocorrelations. Each of these is a profound result on its own merits.

Decomposition Methods (Covariance)

In the covariance case the equations to be solved are of the form (5.58). In this case we still have symmetry and positive definiteness to take advantage of, but the covariance matrix is not Toeplitz.

The most commonly used methods for solving the covariance equations are based on the decomposition of the covariance matrix into lower and upper triangular matrices, say **L** and **U**, such that[27]

$$\mathbf{\Phi}_s(m) = \mathbf{LU}. \tag{5.123}$$

[27]Clearly, **L** and **U** and related quantities depend on time m, but we will suppress this index for simplicity here.

Once this decomposition is achieved, then (5.58) can be solved by sequentially solving the equations

$$\mathbf{L}\mathbf{y} = \boldsymbol{\varphi}_s(m) \tag{5.124}$$

and

$$\mathbf{U}\hat{\mathbf{a}}(m) = \mathbf{y}, \tag{5.125}$$

in each case using simple algorithms to be described below (Golub and Van Loan, 1989).

Since the matrix $\boldsymbol{\Phi}_s(m)$ is symmetric and (assumed) positive definite, the most efficient algorithms for accomplishing the **LU** decomposition are based on the following results from linear algebra:

THEOREM 5.5

1. $\boldsymbol{\Phi}_s(m)$ *can be decomposed as*

$$\boldsymbol{\Phi}_s(m) = \mathbf{L}\boldsymbol{\Delta}\mathbf{L}^T, \tag{5.126}$$

where **L** *is a lower triangular matrix whose diagonal elements are all unity and* $\boldsymbol{\Delta}$ *is a diagonal matrix with all positive elements.*

2. $\boldsymbol{\Phi}_s(m)$ *can be decomposed as*

$$\boldsymbol{\Phi}_s(m) = \mathbf{C}\mathbf{C}^T, \tag{5.127}$$

where **C** *is lower triangular with diagonal elements that are all positive. This is called the Cholesky decomposition of* $\boldsymbol{\Phi}_s(m)$ *and* **C** *the Cholesky triangle.*

For proofs of these results, the reader is referred to (Golub and Van Loan, 1989, Secs. 4.1 and 4.2).

L$\boldsymbol{\Delta}$L^T Decomposition. To develop an algorithm based on the **L$\boldsymbol{\Delta}$L^T** decomposition, we first write scalar equations for the lower triangle elements of $\boldsymbol{\Phi}_s(m)$ in terms of the quantities on the right side of (5.126). For the off-diagonal elements,

$$\varphi_s(i,j;m) = \sum_{k=1}^{j} L(i,k)\Delta(k)L(j,k), \; i = 2, \ldots, M; \quad j = 1, \ldots, i-1, \tag{5.128}$$

and for the diagonal elements,

$$\varphi_s(i,i;m) = \sum_{k=1}^{i} L(i,k)\Delta(k)L(i,k), \tag{5.129}$$

where $L(i, k)$ is the (i, k)-element of $\mathbf{L}$, and $\Delta(k)$ is the kth diagonal element of $\mathbf{\Delta}$. Recalling that the diagonal elements of $\mathbf{L}$ are unity, we can easily convert these equations into solutions for the off-diagonal elements of $\mathbf{L}$ and for the $\Delta(k)$'s. These are as follows:

$$L(i,j) = \frac{\varphi_s(i,j;m) - \sum_{k=1}^{j-1} L(i,k)\Delta(k)L(j,k)}{\Delta(j)}, \tag{5.130}$$

$$i = 2, \ldots, M; j = 1, \ldots, i-1$$

$$\Delta(1) = \varphi_s(1,1;m) \tag{5.131}$$

$$\Delta(i) = \varphi_s(i,i;m) - \sum_{k=1}^{i-1} L(i,k)\Delta(k)L(i,k), \qquad i = 2, \ldots, M. \tag{5.132}$$

These equations, with a small computational trick, comprise an efficient algorithm for computing the $\mathbf{L\Delta L}^T$ decomposition of the covariance matrix shown in Fig. 5.13. The "trick" involves the computation of the quantities $P(i, k) = L(i, k)\Delta(k)$ outside of the inner loop. The savings obtained are explored in Problem 5.18. The algorithm shown requires[28] $O(M^3/6)$ floating point operations (flops).

Once the matrices $\mathbf{L}$ and $\mathbf{U} = \mathbf{\Delta L}^T$ have been found, equations (5.124) and (5.125) are easily solved using the simple algorithms shown in Figs. 5.14 and 5.15, each of which requires about $M(M+1)/2$ flops. In each case the symbol "$\leftarrow$" denotes replacement in memory.

FIGURE 5.13. The $\mathbf{L\Delta L}^T$ decomposition of the covariance matrix.

Initialization: Compute $\Delta(1)$ using (5.131) and recall $L(1, 1) = 1$.

Recursion: For $i = 2, \ldots, M$,
  For $k = 1, \ldots, i-1$,
    $P(i, k) = L(i, k)\Delta(k)$
  Next k
  Compute $\Delta(i)$ using (5.132) with $P(i, k)$ in place of $L(i, k)\Delta(k)$
  For $j = 1, \ldots, i-1$,
    Compute $L(i, j)$ using (5.130) with $P(i, k)$ in place of $L(i, k)\Delta(k)$
  Next j
Next i

---

[28]The notation "$O(\cdot)$" denotes "on the order of" and indicates approximation.

FIGURE 5.14. Forward elimination to solve $\mathbf{L}\mathbf{y} = \boldsymbol{\varphi}_s(m)$.

For $i = 1, 2, \ldots, M$,
  $y(i) \leftarrow \varphi_s(i; m)$
  For $j = 1, 2, \ldots, i-1$,
    $y(i) \leftarrow y(i) - L(i, j)y(j)$
  Next j
  $y(i) \leftarrow y(i)/L(i, i)$
Next i

FIGURE 5.15. Back-substitution to solve $\mathbf{U}\hat{\mathbf{a}}(m) = \mathbf{y}$, where $\mathbf{U} = \boldsymbol{\Delta}\mathbf{L}^T$.

For $i = 1, \ldots, M$,
  $\hat{a}(i; m) \leftarrow y(i)$
  For $j = i+1, \ldots, M$,
    $\hat{a}(i; m) \leftarrow \hat{a}(i; m) - U(i, j)\hat{a}(j; m)$
  Next j
  $\hat{a}(i; m) \leftarrow \hat{a}(i; m)/U(i, i)$
Next i

Before leaving the $\mathbf{L}\boldsymbol{\Delta}\mathbf{L}^T$ decomposition, we note an interesting and useful feature. Similarly to (5.78), we can write the following expression for the average squared error [again let us call it $\xi(m)$] in using the covariance method on the window ending at m:

$$\varphi_s(0, 0; m) - \boldsymbol{\varphi}_s^T(m)\hat{\mathbf{a}}(m) = \xi(m). \tag{5.133}$$

Using the fact that[29] $\hat{\mathbf{a}}(m) = \mathbf{L}^{-T}\boldsymbol{\Delta}^{-1}\mathbf{y}$ [from (5.125)] and the fact that $\mathbf{U} = \boldsymbol{\Delta}\mathbf{L}^T$, we can express this as

$$\xi(m) = \varphi_s(0, 0; m) - \boldsymbol{\varphi}_s^T(m)\mathbf{L}^{-T}\boldsymbol{\Delta}^{-1}\mathbf{y}. \tag{5.134}$$

Now using (5.124), we can write

$$\xi(m) = \varphi_s(0, 0; m) - \mathbf{y}^T\boldsymbol{\Delta}^{-1}\mathbf{y} = \varphi_s(0, 0; m) - \sum_{k=1}^{M} \frac{y^2(k)}{\Delta(k)}. \tag{5.135}$$

This equation can be used to find the prediction error using the vector $\mathbf{y}$ and the diagonal elements of the matrix $\boldsymbol{\Delta}$. It is also interesting to note that if this equation were developed for a smaller model order, say $M' < M$, the result would only differ in the upper limit of the final summation in (5.135), which would then be M'. This means that (5.135) can be used to test for the effect of various model orders on the degree of error minimization.

[29] $\mathbf{L}^{-T}$ denotes the inverse of the transpose of $\mathbf{L}$.

Cholesky Decomposition. The Cholesky decomposition involves the factoring of the covariance matrix as

$$\mathbf{\Phi}_s(m) = \mathbf{C}\mathbf{C}^T, \tag{5.136}$$

where $\mathbf{C}$ is the lower triangular Cholesky factor (see Theorem 5.5). Given that the $\mathbf{L\Delta L}^T$ decomposition exists, it is a simple matter to prove the existence of the Cholesky decomposition for a positive definite $\mathbf{\Phi}_s(m)$ (Problem 5.19). However, an algorithm for computing the Cholesky triangle is better derived by starting from a comparison of matrix elements as we did in the $\mathbf{L\Delta L}^T$ case. Note from (5.127) that

$$\varphi_s(i,j;m) = \sum_{k=1}^{j} C(i,k)C(j,k), \tag{5.137}$$

where the $C(i,j)$'s are the elements of $\mathbf{C}$. Rearranging this equation, we have that

$$C(i,j) = \frac{\varphi_s(i,j;m) - \sum_{k=1}^{j-1} C(i,k)C(j,k)}{C(j,j)}, \quad i > j \tag{5.138}$$

$$C(j,j) = \sqrt{\varphi_s(j,j;m) - \sum_{k=1}^{j-1} C^2(j,k)}\,. \tag{5.139}$$

These equations can be used to compute the lower triangular matrix $\mathbf{C}$ with $O(M^3/6)$ flops according to the algorithm in Fig. 5.16. The solution $\hat{\mathbf{a}}(m)$ can then be obtained using the forward elimination and back-substitution algorithms above. It is to be noted that the Cholesky procedure involves square root taking, which is avoided in the $\mathbf{L\Delta L}^T$ procedure.

The Cholesky decomposition algorithm shown in Fig. 5.16 computes one column of $\mathbf{C}$ at a time. (The computations can also be structured to compute rows.) Note that storage can be saved without computational error by overwriting the memory location assigned to the (i,j) element of $\mathbf{\Phi}_s(m)$ by $C(i,j)$ for any i and j as the algorithm proceeds.

FIGURE 5.16. Cholesky decomposition of $\mathbf{\Phi}_s(m)$.

For $j = 1, \ldots, M$,
  Find $C(j,j)$ using (5.139).
  For $i = j+1, \ldots, M$,
    Find $C(i,j)$ using (5.138).
  Next i
Next j

(Weighted) Recursive Least Squares, Coordinate Rotation Algorithms, and Systolic Processors (Covariance)

The Levinson–Durbin recursion is a recursive-in-model-order solution of the autocorrelation equations. In this material we study methods for solving the covariance equations recursively in time. We will study two widely different algorithms for computing a *weighted recursive least squares* (WRLS) solution for the LP parameters, which is equivalent to a weighted covariance solution.

The interest in a time recursion in LP analysis of speech is naturally motivated, for example, by the need to adaptively change parameter estimates over time. Although the "conventional" form of WRLS was popular for similar identification problems in the control and systems science domains beginning in the 1960s, speech processing engineers did not widely use the method, but instead resorted to more *ad hoc* methods for changing estimates over time. Perhaps the most important reason for this is that WRLS requires $O(M^2)$ flops per speech sample when computed on a sequential machine, whereas when we focus on the computational complexity of various solution methods in this section, we will discover that we are already familiar with other methods that do the job at the expense of $O(M)$ flops. In the 1980s, however, systolic and other forms of parallel processors made it possible to solve this $O(M^2)$ problem at time scales that are many orders of magnitude faster than "real time" in many speech applications. New attention was focused on this algorithm [in particular, for adaptive antenna arrays (McWhirter, 1983)] in that era; from it there promises to arise more elegant solutions to adaptive strategies and discoveries of new uses for weighted covariance estimates in speech.

Here we briefly discuss the conventional form of the WRLS algorithm and then focus on the version that is amenable to systolic processing.

Weighted Covariance and Conventional WRLS. We have thus far been concerned with estimating the LP parameters over a nominal length N window of speech data assumed to begin at time $n = m - N + 1$ and end at $n = m$. In this material we will be performing a recursion through time, effectively increasing the size of our window at each iteration. As we did in the section on classical least squares problems, we will once again find it convenient to assume, without any loss of generality, that *the window begins at time* $n = 1$ *and ends at* $n = N$. We will continue to label our estimates and related quantities by the time of the leading edge of the window, for example, $\hat{\mathbf{a}}(N)$.

In this section we develop a time recursion that if started at time $n = 1$, at any time $N \geq M$ (where M is the number of LP parameters sought), theoretically produces an estimate that is equivalent to that produced on the same window by the covariance method. For generality, we will work with the *weighted* covariance method, since temporal recursions are often the most useful in weighted schemes. If the "unweighted" covariance analysis is desired, it is a simple matter to set all the weights to unity in the following discussion.

Let us proceed to develop the conventional WRLS algorithm. The algorithm will consist of two coupled recursions, one for the (inverse of) the covariance matrix, the other for the LP parameter vector.

We have noted two quite different ways of computing the weighted covariance matrix $\mathbf{\Phi}_s'(N)$. The first consists of computing its elements directly from the speech using the scalar equations (5.73). The second, given in (5.74), arose by comparing the normal equations to a classical linear algebra result. A third, which is immediately evident from the second, is a key equation in the development of the first recursion:

$$\mathbf{\Phi}_s'(N) = \frac{1}{N} \sum_{n=1}^{N} \lambda(n)\mathbf{s}(n)\mathbf{s}^T(n). \tag{5.140}$$

From this form, we can clearly see that

$$N\mathbf{\Phi}_s'(N) = (N-1)\mathbf{\Phi}_s'(N-1) + \lambda(N)\mathbf{s}(N)\mathbf{s}^T(N). \tag{5.141}$$

If we define

$$\mathbf{\Psi}_s(N) \stackrel{\text{def}}{=} \left[N\mathbf{\Phi}_s'(N)\right]^{-1}, \tag{5.142}$$

then we have

$$\mathbf{\Psi}_s(N) = \left[(N-1)\mathbf{\Phi}_s'(N-1) + \lambda(n)\mathbf{s}(N)\mathbf{s}^T(N)\right]^{-1}. \tag{5.143}$$

The following lemma, often called the *matrix inversion lemma*, or *Woodbury's lemma* (Ljung and Söderström, 1983, p. 19; Graupe, 1989), will lead immediately to one key equation.

LEMMA 5.3 (MATRIX INVERSION LEMMA) *Let* **A** *and* **C** *be matrices for which inverses exist, and* **B** *be a matrix such that* $\mathbf{BCB}^T$ *is of the same dimension as* **A**. *Then*

$$\left[A + \mathbf{BCB}^T\right]^{-1} = \mathbf{A}^{-1} - \mathbf{A}^{-1}\mathbf{B}\left[B^T\mathbf{A}^{-1}\mathbf{B} + \mathbf{C}^{-1}\right]^{-1}\mathbf{B}^T\mathbf{A}^{-1}. \tag{5.144}$$

As a consequence of Lemma 5.3, (5.143) can be written in the form (with N replaced by the general n),

$$\mathbf{\Psi}_s(n) = \mathbf{\Psi}_s(n-1) - \lambda(n)\frac{\mathbf{\Psi}_s(n-1)\mathbf{s}(n)\mathbf{s}^T(n)\mathbf{\Psi}_s(n-1)}{1 + \lambda(n)\mathbf{s}^T(n)\mathbf{\Psi}_s(n-1)\mathbf{s}(n)}. \tag{5.145}$$

This equation provides a means for recursively updating the inverse of the weighted covariance matrix.[30]

The second key recursion is used to update the estimate at time n, $\hat{\mathbf{a}}(n)$, once $\mathbf{\Psi}_s(n)$ is found. This equation is as follows:

$$\hat{\mathbf{a}}(n) = \hat{\mathbf{a}}(n-1) + \lambda(n)\mathbf{\Psi}_s(n)\mathbf{s}(n)\varepsilon(n;n-1), \tag{5.146}$$

[30]Actually, the inverse of the covariance matrix without the scale factor $1/N$, which plays no role in our developments, since it is canceled by the same factor on the other side of the normal equations.

where $\varepsilon(n;n-1)$ has precisely the same meaning as always, the prediction error at time n based on the filter designed on the window ending at $n-1$. Equation (5.146) is derived as follows. The auxiliary covariance vector on the right side of (5.72) can be written [similarly to (5.141)]

$$\boldsymbol{\varphi}_s'(N) = \frac{1}{N}\sum_{n=1}^{N}\lambda(n)\mathbf{s}(n)s(n) = \frac{N-1}{N}\boldsymbol{\varphi}_s'(N-1) + \frac{\lambda(N)}{N}\mathbf{s}(N)s(N). \tag{5.147}$$

Combining this with (5.72), we can write

$$\boldsymbol{\Phi}_s'(N)\hat{\mathbf{a}}(N) = \frac{N-1}{N}\boldsymbol{\Phi}_s'(N-1)\hat{\mathbf{a}}(N-1) + \frac{\lambda(N)}{N}\mathbf{s}(N)s(N). \tag{5.148}$$

Adding and subtracting the column vector $[\lambda(N)/N]\mathbf{s}(N)\,\mathbf{s}^T(N)\hat{\boldsymbol{\alpha}}(N-1)$, we have

$$\boldsymbol{\Phi}_s'(N)\hat{\mathbf{a}}(N) = \frac{N-1}{N}\boldsymbol{\Phi}_s'(N-1)\hat{\mathbf{a}}(N-1) + \frac{\lambda(N)}{N}\mathbf{s}(N) \times\left[s(N) - \mathbf{s}^T(N)\hat{\mathbf{a}}(N-1)\right] + \frac{\lambda(N)}{N}\mathbf{s}(N)\mathbf{s}^T(N)\hat{\mathbf{a}}(N-1). \tag{5.149}$$

We recognize the term in brackets as $\varepsilon(N;N-1)$, and the first and third terms on the right side add together to form $\boldsymbol{\Phi}_s'(N)$ according to (5.141). So we have

$$\boldsymbol{\Phi}_s'(N)\hat{\mathbf{a}}(N) = \boldsymbol{\Phi}_s'(N)\hat{\mathbf{a}}(N-1) + \frac{\lambda(N)}{N}\mathbf{s}(N)\varepsilon(N;N-1). \tag{5.150}$$

Premultiplying through by $[\boldsymbol{\Phi}_s'(N)]^{-1}$ yields (5.146) upon recalling (5.142) and replacing N by the more general n.

Before stating the conventional WRLS algorithm, we must note a potential difficulty in initializing it. In principle, at $n=1$, the matrix $\boldsymbol{\Psi}_s(0)$, needed to compute (5.145) at time $n=1$, is the inverse of a matrix of zeros. In practice it is customary to let

$$\boldsymbol{\Psi}_s(0) = \frac{1}{\mu}\mathbf{I}, \tag{5.151}$$

where $\mathbf{I}$ is the identity matrix and μ is a small number, typically 10^{-6}. A discussion of this point is found, for example, in (Ljung and Söderström, 1983; Graupe, 1989). The initialization of $\hat{\mathbf{a}}(0)$ is somewhat arbitrary (Albert and Gardner, 1967, pp. 109–115), although theoretically the algorithm should be initialized with a "correct" value (Hsia, 1977, p. 25; Albert and Gardner, 1967). By this it is meant that the recursion, if started at time n_0, should initialize $\hat{\mathbf{a}}(n_0)$ to the covariance method solution on the interval $[1, n_0]$. This is seldom a concern in practice and the value $\hat{\mathbf{a}}(0) = \mathbf{0}$ is frequently chosen. Lee (1964, pp. 103–106) shows that this choice will result in convergence of the estimator without bias.

The WRLS algorithm is given in Fig. 5.17.

FIGURE 5.17. Conventional weighted recursive least squares algorithm.

Initialization: Set $\hat{\mathbf{a}}(0) = \mathbf{0}$, and $\boldsymbol{\Psi}_s(0)$ according to (5.151).

Recursion: For $n = 1, 2, \ldots,$
Update $\boldsymbol{\Psi}_s(n)$ according to (5.145).
Update $\hat{\mathbf{a}}(n)$ according to (5.146).
Stop at some predetermined point or according to some stopping criterion.

"Systolic" Version of WRLS. We now turn the WRLS algorithm into a version that can be implemented on a systolic processor. It is not our purpose to get into the details of parallel processing architectures, as such a pursuit would take us too far afield. The interested reader can refer to numerous papers on the subject[31] [e.g., (McWhirter, 1983; Rader and Sundaramurthy, 1988)]. Nevertheless, we should reiterate the point that it has been these parallel processing architectures that have driven the interest in this version of the algorithm. Deller and Picaché (1989) have pointed out some additional advantages of this method for speech processing.

"Batch" Givens Rotation Solution. The alternative WRLS method for solving the covariance problem derives from a reexamination of the classical least squares problem noted in Section 5.3.2. Recall that the problem, which is equivalent to the weighted covariance method, is to solve the overdetermined system of equations, (5.63), subject to the weighted error criterion, (5.70). Here we will find it convenient to state the same problem in a slightly different way: Solve the overdetermined system of N equations in M unknowns,

$$\bar{\boldsymbol{\Lambda}}^{1/2}(N)\begin{bmatrix} \mathbf{0}_{M\times M} & \\ \mathbf{s}^T(1) & \longrightarrow \\ \mathbf{s}^T(2) & \longrightarrow \\ \vdots & \\ \mathbf{s}^T(N) & \longrightarrow \end{bmatrix}\hat{\mathbf{a}}(N) = \bar{\boldsymbol{\Lambda}}^{1/2}(N)\begin{bmatrix} \mathbf{0}_{M\times 1} \\ s(1) \\ s(2) \\ \vdots \\ s(N) \end{bmatrix} \tag{5.152}$$

[31]The reader is cautioned that many of the papers on this subject deal with problems in "direction finding" and "beamforming" based on adaptive antenna arrays. In this case, the interest is in directly computing the residual sequence rather than the parameters themselves. Therefore, algorithms and architectures have been developed which, while based on the same basic principles to be discussed here, circumvent the computation of the coefficients altogether and compute the residual directly. A second basic difference that sometimes appears in these papers is that the processing is designed to identify a pole–zero model, rather than an all-pole model as we do in the LP case. We will say a bit more about this problem below. The point to keep in mind is that the LP model is just a special case of the pole–zero model sought in these papers.

subject to the *weighted* error constraint (5.70), where $\bar{\mathbf{\Lambda}}^{1/2}(N)$ is the diagonal matrix,

$$\bar{\mathbf{\Lambda}}^{1/2}(N) \stackrel{\text{def}}{=} \begin{bmatrix} \mathbf{I}_{M\times M} & \mathbf{0} \\ \mathbf{0} & \mathbf{\Lambda}^{1/2}(N) \end{bmatrix} \tag{5.153}$$

with $\mathbf{I}_{M\times M}$ indicating an $M\times M$ identity matrix, and $\mathbf{\Lambda}^{1/2}(N)$ the diagonal matrix with ith diagonal element $\sqrt{\lambda(i)}$, and $\{\lambda(n)\}$ is some set of nonnegative weights. All we have done here is to move the weights out of the error criterion and modified the equations to compensate for the "missing" weights. Another small change with respect to the original statement of the problem is that we have added M "null equations" of the form $\mathbf{0}^T\hat{\mathbf{a}}(N)=0$ to the top of the system. We have arbitrarily assigned unity weights to these equations. This is for convenience in the following discussion; it has absolutely no effect on the solution. Let us denote the system in vector-matrix form by

$$\bar{\mathbf{\Lambda}}^{1/2}(N)\bar{\mathbf{S}}(N)\hat{\mathbf{a}}(N)=\bar{\mathbf{\Lambda}}^{1/2}(N)\bar{\bar{\mathbf{s}}}(N), \tag{5.154}$$

noting that the only difference between $\bar{\mathbf{S}}(N)$ and $\mathbf{S}(N)$ of (5.63) is the zeros at the top. The same is true of $\bar{\bar{\mathbf{s}}}(N)$ and $\bar{\mathbf{s}}(N)$ of (5.63). Using (5.63) and (5.71), we can easily write down a solution to the present system of equations,

$$\frac{1}{N}\bar{\mathbf{S}}^T(N)[\bar{\mathbf{\Lambda}}^{1/2}(N)]^T\bar{\mathbf{\Lambda}}^{1/2}(N)\bar{\mathbf{S}}(N)\hat{\mathbf{a}}(N)=\frac{1}{N}\bar{\mathbf{S}}^T(N)[\bar{\mathbf{\Lambda}}^{1/2}(N)]^T\bar{\mathbf{\Lambda}}^{1/2}(N)\bar{\bar{\mathbf{s}}}(N). \tag{5.155}$$

With a little effort, this result can be shown to be exactly equivalent to (5.71). As promised above, the "null equations" in the system have had no effect on the solution.

Sequential Method. Thus far, we have studied solutions that, in one way or another, have formed the covariance matrix and then used it to obtain the solution for $\hat{\mathbf{a}}(N)$. In this case, we will find the solution without directly computing $\mathbf{\Phi}_s(N)$; instead, we will work directly with the system of equations (5.152) or (5.154). Well-established techniques for finding the solution by *orthogonal triangularization* of the "coefficient" matrix, $\bar{\mathbf{S}}(N)$ are based on the original work of Givens (1958) and Householder (1958). Householder's paper was the first to mention the applicability to the (batch) least squares problem, and the papers of Golub and Businger (1965a, 1965b) were the first to present numerical algorithms. The Givens technique is the focus of our work here.

The "batch" (using all the equations at once) least squares solution of (5.152) by Givens transformation is achieved by "annihilating" elements of $\bar{\mathbf{S}}(N)$ in a row-wise fashion [see, (e.g., Golub and Van Loan, 1989)], "rotating" the information in each element into the upper triangle of the top M rows of the matrix. The *plane rotation* matrix, say $\mathcal{P}$, used to annihilate the (i,j) element of $\bar{\mathbf{S}}(N)$, say $\bar{S}(i,j;N)$, is an $(M+N)\times(M+N)$

matrix differing from the identity matrix only in its $(i, i), (j, j), (i, j)$, and (j, i) elements, which are as follows:[32]

$$\mathcal{P}(i, i) = \mathcal{P}(j, j) = \cos\theta \tag{5.156}$$

$$-\mathcal{P}(i, j) = \mathcal{P}(j, i) = \sin\theta, \tag{5.157}$$

where

$$\theta = \arctan\left[\frac{S(i, j; N)}{S(j, j; N)}\right]. \tag{5.158}$$

Each of these plane rotation matrices is orthonormal, $\mathcal{P}\mathcal{P}^T = \mathbf{I}$. After the rotations, the transformed system is of the form

$$\begin{bmatrix} \mathbf{T}(N) \\ \mathbf{0}_{N \times M} \end{bmatrix} \hat{\mathbf{a}}(N) = \begin{bmatrix} \mathbf{d}_1(N) \\ \mathbf{d}_2(N) \end{bmatrix}, \tag{5.159}$$

where $\mathbf{T}(N)$ is $M \times M$ upper triangular. The system

$$\mathbf{T}(N)\hat{\mathbf{a}}(N) = \mathbf{d}_1(N) \tag{5.160}$$

can be solved using the back-substitution algorithm to obtain the least squares estimate, $\hat{\mathbf{a}}(N)$ (see Problem 5.21).

Next we show how to use this method in a "temporally recursive" mode in which each equation is used to obtain an updated estimate of the LP parameters as it becomes available. If a series of plane rotation matrices is used to annihilate rows of $\overline{\mathbf{S}}(N)$ from left to right (beginning with row $M + 1$, the first real equation below the zeros[33]), each row requires a sequence of M orthonormal plane rotations. Suppose that we denote the product of M plane rotation matrices applied to equation n (row $M + n$) by $\mathcal{R}_n$. Then the overall transformation on the system of equations takes the form

$$\mathcal{R}_N \mathcal{R}_{N-1} \cdots \mathcal{R}_1 \overline{\mathbf{S}}(N)\hat{\mathbf{a}}(N) = \mathcal{R}_N \mathcal{R}_{N-1} \cdots \mathcal{R}_1 \overline{\overline{\mathbf{s}}}(N). \tag{5.161}$$

Let us further define the global transform matrix

$$\mathbf{\Pi}_N \stackrel{\text{def}}{=} \mathcal{R}_N \mathcal{R}_{N-1} \cdots \mathcal{R}_1. \tag{5.162}$$

After the application of $\mathbf{\Pi}_N$, to the system, the result is of the form (5.159).

To move toward a recursive solution, we add one further equation to the system. For the moment, we leave the previous equations unchanged, meaning in particular that their weights cannot change. Suppose, there-

[32]These quantities, of course, depend on N, but we will once again suppress the index for simplicity.

[33]We have avoided the need to process the top M rows in our formulation by filling them with zeros; this makes the following discussion much less cumbersome.

fore, that we were to encounter the same system of N equations plus a new equation representing time $N+1$,

$$\begin{bmatrix} \bar{\boldsymbol{\Lambda}}^{1/2}(N)\overline{\mathbf{S}}(N) \\ \sqrt{\lambda(N+1)}\mathbf{s}^T(N+1) \end{bmatrix} \hat{\mathbf{a}}(N+1) = \begin{bmatrix} \bar{\boldsymbol{\Lambda}}^{1/2}(N)\bar{\bar{\mathbf{s}}}(N) \\ \sqrt{\lambda(N+1)}s(N+1) \end{bmatrix} \tag{5.163}$$

or

$$\bar{\boldsymbol{\Lambda}}^{1/2}(N+1)\overline{\mathbf{S}}(N+1)\hat{\mathbf{a}}(N+1) = \bar{\boldsymbol{\Lambda}}^{1/2}(N+1)\bar{\bar{\mathbf{s}}}(N+1). \tag{5.164}$$

Wishing to find the least square solution to *this* system of equations, we might recognize that we are now faced with an $N+1$ equation version of the problem that we have just solved with N equations, and simply start over, using precisely the same method. We would once again apply the Givens transformations row-wise, this time denoting the row operation matrices by $Q_1, Q_2, \ldots, Q_{N+1}$. Note that the $\mathcal{R}_n$ matrices of the N equation problem are $(M+N)\times(M+N)$, whereas these Q_n matrices are $(M+N+1)\times(M+N+1)$. The key to using the solution in a sequential manner is to recognize that the N equation problem is "embedded" in the $N+1$ equation problem in the following sense: The operations done to the first N equations in the $N+1$ equation problem are identical to those in the N equation problem, as shown in the following lemma.

Lemma 5.4

$$Q_N Q_{N-1} \cdots Q_1 = \begin{bmatrix} \boldsymbol{\Pi}_N & \mathbf{0}_{N\times 1} \\ \mathbf{0}^T_{N\times 1} & 1 \end{bmatrix}, \tag{5.165}$$

where, recall, $\boldsymbol{\Pi}_N \stackrel{\text{def}}{=} \mathcal{R}_N \mathcal{R}_{N-1} \cdots \mathcal{R}_1$.

The proof of this lemma follows immediately from the definitions and some algebraic tedium. By applying Lemma 5.4 to the $N+1$ equations of (5.163), it is seen that after N row transformations in the $N+1$ equation problem, the system is transformed into

$$\begin{bmatrix} \boldsymbol{\Pi}_N\bar{\boldsymbol{\Lambda}}^{1/2}(N)\overline{\mathbf{S}}(N) \\ \sqrt{\lambda(N+1)}\mathbf{s}^T(N+1) \end{bmatrix} \hat{\mathbf{a}}(N+1) = \begin{bmatrix} \boldsymbol{\Pi}_N\bar{\boldsymbol{\Lambda}}^{1/2}(N)\bar{\bar{\mathbf{s}}}(N) \\ \sqrt{\lambda(N+1)}s(N+1) \end{bmatrix}, \tag{5.166}$$

the top M rows of which are exactly (5.160). Therefore, after N row transformations in the $N+1$ equation problem, the estimate $\hat{\mathbf{a}}(N)$ is easily generated and the solution begins to assume a "sequential" flavor.

Notice that in solving the $N+1$ equation problem, it is apparently sufficient to apply the row transformations[34] $Q_{N+1}\mathcal{R}_N\mathcal{R}_{N-1}\cdots\mathcal{R}_1$ if the $\mathcal{R}_n$'s are restricted to operating on the upper N equations of

[34]Because of the restrictions, the matrices in this sequence of operations no longer have compatible dimension and it is not meaningful to interpret this string as a product.

$\bar{\mathbf{\Lambda}}^{1/2}(N+1)\bar{\mathbf{S}}(N+1)$. In fact, by induction, we see that, for an $N+1$ equation problem, the appropriate sequence of operations is given by

$$\mathcal{S}_{N+1}\mathcal{S}_N\mathcal{S}_{N-1}\cdots\mathcal{S}_1, \tag{5.167}$$

where $\mathcal{S}_n$ is defined to be the $(M+n)\times(M+n)$ matrix appropriate for including the nth equation of the system if the system indeed had only n equations, and which is restricted to operating only upon the top n equations of $\bar{\mathbf{\Lambda}}^{1/2}(N+1)\bar{\mathbf{S}}(N+1)$.

Further, in considering, for example, the application of $\mathcal{S}_{N+1}$ to the system of (5.166), we would find that rows and columns $(M+1)$ through N serve only to preserve the rows of zeros (caused by annihilation of equations 1 through N) in the matrix on the left, and the vector $\mathbf{d}_2(N)$ on the right [see also (5.159)]. Since these rows play no further role in the solution, they and the cited rows and columns of $\mathcal{S}_{N+1}$ can be eliminated. Each $\mathcal{S}_n$ operator, therefore, can be formulated as an $(M+1)\times(M+1)$ matrix. This fact is reflected in the algorithm below. Before stating the algorithm, we show how this method is easily modified to be adaptive to the most recent dynamics of the speech signal.

Adaptive Method. One prominent method of causing covariance-like estimators to adapt to the most recent dynamics of the signal and "forget the past" is to include a so-called "forgetting factor" into the recursion. In principle, the covariance method is solved at each N, subject to the error minimization

$$\xi'(N)=\frac{1}{N}\sum_{n=1}^{N}\alpha^{N-n}[s(n)-\hat{s}(n)]^2=\frac{1}{N}\sum_{n=1}^{N}\alpha^{N-n}\varepsilon^2(n;N) \tag{5.168}$$

for some $0<\alpha<1$. Comparing this with (5.70), we see that the weights in this case are $\lambda(n)=\alpha^{N-n}$, and that they become smaller as we look further back into the history of the data. Note that these weights are *time varying* in the sense that if we were to solve the problem on the range $n=1,2,\ldots,N+1$, all weights on the past equations would not remain the same as they were in the N point problem, but instead would be scaled down by a factor of α. At first, this might seem to make a recursion difficult, but in fact the solution is quite simple.

First let us note that, in this case, the equation weighting matrix for the N point problem, $\bar{\mathbf{\Lambda}}^{1/2}(N)$, is as follows:

$$\bar{\mathbf{\Lambda}}^{1/2}(N)=\mathrm{diag}[1\ \cdots\ 1\ \beta^{N-1}\ \beta^{N-2}\ \cdots\ \beta\ 1], \tag{5.169}$$

where $\beta\stackrel{\text{def}}{=}\sqrt{\alpha}$. Consider the system (5.152) or (5.154) with weights given as in (5.169). Suppose that we implement a naive adaptive strategy that introduces a new $\bar{\mathbf{\Lambda}}^{1/2}(N)$ matrix at each N and reuses the Givens approach to obtain $\hat{\mathbf{a}}(N)$. At time N, we can use the sequential technique suggested above. However, it is important to realize that in the course of solving the N equation problem, the intermediate estimates $\hat{\mathbf{a}}(n), n=1, 2,\ldots,N-1$ (if computed), will not correspond to *adaptive* estimates.

Rather, they are just estimates corresponding to the static set of weights at time N. Nevertheless, suppose that we have used the sequential solution at time N, the row transformation matrices being denoted $\mathcal{S}_n$ as in (5.167). Now in the adaptive scheme, the next set of $N+1$ equations to be solved is

$$\begin{bmatrix} \beta\overline{\mathbf{\Lambda}}^{1/2}(N)\overline{\mathbf{S}}(N \\ \mathbf{s}^T(N+1) \end{bmatrix} \hat{\mathbf{a}}(N+1) = \begin{bmatrix} \beta\overline{\mathbf{\Lambda}}^{1/2}(N)\overline{\overline{\mathbf{s}}}(N) \\ s(N+1) \end{bmatrix}. \tag{5.170}$$

A sequential approach becomes apparent given the following:

LEMMA 5.5 *Let the row operators to be applied to this $N+1$ equation system be denoted $\mathcal{S}'_n, n = 1, 2, \ldots, N+1$. Then $\mathcal{S}'_n = \mathcal{S}_n, n = 1, 2, \ldots, N$, where $\mathcal{S}_n$ are the operators that were applied to the N equation step.*

The proof follows readily from the fact that the plane rotations in $\mathcal{S}_n$, $n = 1, 2, \ldots, N$, depend on ratios of elements in $\overline{\mathbf{\Lambda}}^{1/2}(N)\overline{\mathbf{S}}(N)$ which are unaffected by the scalar β.

After N equations are operated upon in the $N+1$ equation step, therefore, we have

$$\begin{bmatrix} \beta\mathbf{T}(N) \\ \mathbf{0}_{N\times M} \\ \mathbf{s}^T(N+1) \end{bmatrix} \hat{\mathbf{a}}(N+1) = \begin{bmatrix} \beta\mathbf{d}_1(N) \\ \beta\mathbf{d}_2(N) \\ s(N+1) \end{bmatrix}. \tag{5.171}$$

To make the algorithm adaptive, therefore, it is not necessary to start over at each step, but only to multiply the upper triangular system by β before rotating in the new equation.

Algorithm. An algorithm to implement either the adaptive or nonadaptive case is given in Fig. 5.18. Further comments on the use of parallel processing are found in the following subsection.

Computational Requirements of the Solution Algorithms

Table 5.1 shows the computational loads, or "complexity," of the five principal classes of algorithms discussed above. In each case, the complexity is given in terms of the approximate number of floating point operations (flops) per speech sample in the analysis frame, where a flop is taken to be one floating point multiplication plus one addition. A summary number is given in the first column along with the method, and the loads of the various subtasks and some typical numbers are shown in the other columns.

FIGURE 5.18. "Systolic" WRLS Algorithm for Adaptive and Nonadaptive Cases.

Dedicate $(M+1)\times M$ memory locations to the elements of the matrix on the left side of the equation, and $(M+1)\times 1$ to the auxiliary vector on the right. These are concatenated into a working matrix, $\mathbf{W}$, of dimension $(M+1)\times(M+1)$. By $\mathbf{W}(n)$ we mean the working matrix at time n (including the nth equation), before the "application of $\mathcal{S}_n$," and by $\mathbf{W}'(n)$, the postrotation matrix.

Initialization:

$$\mathbf{W}(0)=\mathbf{W}'(0)=\left[\mathbf{0}_{(M+1)\times M} \mid \mathbf{0}_{(M+1)\times 1}\right]. \tag{5.172}$$

Recursion: For $n=1,2,\ldots,$

1. (Adaptive Case) Replace the last row of $\mathbf{W}'(n-1)$ by $[\mathbf{s}^T(n) \mid s(n)]$ to form $\mathbf{W}(n)$. (This discards the unnecessary equation left by the $n-1$st row operation.) Multiply the upper M rows of $\mathbf{W}(n)$ by β (neglecting the lower triangle of zeros).
1. (Static Weights Case) Replace the last row of $\mathbf{W}'(n-1)$ by $\left[\sqrt{\lambda(n)}\,\mathbf{s}^T(n) \mid \sqrt{\lambda(n)}\,s(n)\right]$, where $\lambda(n)$ denotes the nth weight, to form $\mathbf{W}(n)$.
2. "Apply $\mathcal{S}_n$." This requires some simple scalar operations:

   For $j=1,2,\ldots,M,$
   $\quad\theta=\arctan[W(M+1,j;n)/W(j,j;n)]$
   $\quad W'(M+1,j;n)=0$
   $\quad$For $k=j,j+1,\ldots,M+1,$
   $\quad\quad W'(j,k;n)=W(j,k;n)\cos\theta+W(M+1,k;n)\sin\theta$
   $\quad$For $k=j+1,j+2,\ldots,M+1,$
   $\quad\quad W'(M+1,k;n)=-W(j,k;n)\sin\theta+W(M+1,k;n)\cos\theta$

3. The first M rows and columns of $\mathbf{W}'(n)$ contain $\mathbf{T}(n)$, and the first M rows of the last column are $\mathbf{d}_1(n)$ [see (5.160)]. Solve for $\hat{\mathbf{a}}(n)$ using back-substitution if desired.
4. Stop at some predetermined point or according to some stopping criterion.

Before discussing the entries in the table, it should be noted that computational expense is only one of a number of factors considered in choosing an algorithm, and this study of complexity should not be interpreted to the contrary. Indeed, as faster processors become available, the difference between the most and least expensive of these techniques might become insignificant in many applications.

The autocorrelation and covariance "batch" methods are seen to be generally the most efficient, with the autocorrelation slightly, but insignificantly, more efficient. Note that various algorithms within these classes exist, some being more efficient than others (Markel and Gray, 1976, Ch. 9); however, all commonly used methods are of $O(M)$ complexity. In choosing between the autocorrelation and covariance methods,

TABLE 5.1. Approximate Number of Floating Point Operations (Flops) Required by Various LP Solutions.[a]

Method and Complexity	Subtasks	Flops per Sample	Typical Number of Flops ($M = 14$, $N = 256$)
Autocorrelation (L–D recursion) $O(M)$	Windowing	1	1
	Autocorrelation update	$O(M)$	14
	Solution	$O(M^2/N)$	0.76
Covariance (Cholesky) $O(M)$	Covariance update	$O(M)$	14
	Solution	$O(M^2/N)$	0.76
Lattice (Itakura or Burg) $O(5M)$	Windowing	1	1
	Solution (parcor coefficents)	$O(5M)$	96
Covariance (W)RLS (conventional) $O(3M^2)$	Covariance update	$O(2M^2)$	392
	Solution	$O(M^2)$	196
	Weights	3	3
	Forgetting factor	3	3
Covariance (W)RLS (QR-decomposition) $O(3M^2/2)$	Covariance update	$O(M^2)$	196
	Solution	$O(M^2/2)$	98
	Weights	3	3
	Forgetting factor	$O(M^2/2)$	98

[a]A flop is taken to be one floating point multiplication plus one addition. Further discussion is found in the text.

the analyst should keep in mind the stability of the solution. The autocorrelation solution is theoretically guaranteed to represent a stable filter if infinite precision arithmetic is used. In practice, finite wordlength computation can cause unstable solutions,[35] but the L–D recursion contains a built-in check for stability embodied in conditions (5.97) and (5.98). Although no such theoretical guarantee or numerical check is available in the covariance case, stability is usually not a problem if the frame is sufficiently large, since the covariance and autocorrelation methods both converge to the same answer as N increases.

[35]A discussion of finite wordlength effects in LP solutions is found in (Markel and Gray 1976, Ch. 9).

The lattice methods are the most computationally expensive of the conventional methods but are among the most popular because of several important properties. First, it should be noted that more efficient lattice structures than those described here are to be found in the work of Makhoul (1977) and Strobach (1991). Second, even with slightly more expensive algorithms, lattice methods offer the inherent ability to directly generate the parcor coefficients (which are often preferable to the LP coefficients, see Section 7.4.5), to monitor stability, and to deduce appropriate model orders "on line."

Finally, if sequential computation is used, the most expensive solutions are the temporal WRLS recursions. However, even on a sequential machine, these methods are advantageous for some applications in providing point-by-point parameter estimates and residual monitoring, as well as convenient adaptation capability. Further, contemporary computing technology has made $O(M^2)$ algorithms not nearly as prohibitive in terms of actual computation time as was the case in the early days of LP analysis. As noted above, however, the primary motivation for the interest in these solutions is the availability of parallel processing architectures that render the Givens rotation-based WRLS method effectively an $O(M)$ process. More details on how parallel computing can be applied to this method is found in the work of Gentleman and Kung (1981); McWhirter (1983); and Deller, Odeh, and Luk (1991, 1989, 1989). It is also to be noted that, in the late 1980s and early 1990s, a parameter estimation strategy known as *set-membership identification* was applied to LP analysis. This is, in fact, the main subject addressed in the papers by Deller et al. noted above. In the paper by Deller and Odeh (1991), it is shown that the set-membership approach has the potential to make even the sequential version of WRLS an $O(M)$ algorithm. Various other "fast" RLS algorithms have been proposed in the literature, notably the *fast transversal filter* (FTF) of Cioffi and Kailath (1984, 1985). The FTF in its most stable form, however, is an $O(8M)$ algorithm, which is close to $O(M^2)$ for speech analysis, and has been shown to be very sensitive to finite precision arithmetic effects (Marshall and Jenkins, 1988).

5.3.4 Gain Computation

Let us return now to the general modeling problem, recalling that we view the total speech production system as consisting of three components, $\Theta(z) = \Theta_0 \Theta_{\min}(z) \Theta_{\mathrm{ap}}(z)$. In the discussions above, we have focused on the problem of modeling the minimum-phase component, $\Theta_{\min}(z)$, the estimate of which yields a sufficient spectral characterization of the speech for many purposes. Thus far, we have said nothing about the possibility of estimating the gain, Θ_0, of the LP model. Indeed, we have discovered implicitly that the LP parameter estimates are insensitive to this quantity. There are many instances, however, in which the relative gains across frames are important, and we seek an estimation procedure here.

Let us recall (5.11), assuming that our model order, M, can be chosen exactly equal to the "true" order of the system, I, so that I will be replaced by M in that equation:

$$s(n) = \sum_{i=1}^{M} a(i)s(n-i) + \Theta_0 e'(n) \stackrel{\text{def}}{=} \mathbf{a}^T \mathbf{s}(n) + \Theta_0 e'(n). \tag{5.173}$$

Recall also that $e'(n)$ represents the phase-scattered version of the driving sequence, $e(n)$, which is a consequence of subjecting $e(n)$ to the all-pass component of $\Theta(z)$. Postmultiplying both sides of (5.173) by $s(n)$ and applying the long-term averaging operator, we have

$$r_s(0) = \sum_{i=1}^{M} a(i) r_s(i) + \Theta_0 r_{e's}(0). \tag{5.174}$$

Now, in the unvoiced case, it is easy to show that

$$r_{e's}(\eta) = \Theta_0 \delta(\eta), \tag{5.175}$$

whereas, in the voiced case, we have from (5.39) that

$$r_{e's}(\eta) \approx \frac{\Theta_0}{P} \delta(\eta), \tag{5.176}$$

where P is the pitch period in samples. This leads us immediately to the following set of estimators for the gain parameter, on the frame ending at time m:

$$\hat{\Theta}_0(m) \approx \sqrt{r_s(0;m) - \sum_{i=1}^{M} \hat{a}(i;m) r_s(i;m)}, \qquad \text{unvoiced case} \tag{5.177}$$

and

$$\hat{\Theta}_0(m) \approx \sqrt{P\left[r_s(0;m) - \sum_{i=1}^{M} \hat{a}(i;m) r_s(i;m)\right]}, \qquad \text{voiced case.} \tag{5.178}$$

These results assume that the autocorrelation method has been used, since the long-term autocorrelations have been replaced by the estimates $r_s(i;m)$. If the covariance method has been used, the estimates $\varphi_s(0;i;m)$ can be inserted instead.

If a short window (less than one pitch period) is used in the voiced case, the short-term autocorrelation sequence, $r_s(i;m)$, has been found to be very sensitive to the placement of the window that defines the frame (Makhoul, 1975), so that longer frames must be used (say 3 or 4 pitch periods). Similarly, it can be inferred from Theorem 5.4 that a short-term covariance window could lead to inaccurate results if a pitch pulse occurs within. Longer windows are generally used in this case as well.[36]

[36]Unless a pitch synchronous estimate is possible. This is ordinarily impractical.

Finally, we show an alternative estimation procedure for the gain. Returning to Fig. 5.4, we can write the following expression for the prediction residual, $\hat{e}(n)$,

$$\hat{e}(n) = -\sum_{i=1}^{M} \hat{a}(i)s(n-i) + s(n). \tag{5.179}$$

If we have done a reasonably good job of selecting the model order, M, then

$$\hat{e}(n) \approx \Theta_0 e'(n). \tag{5.180}$$

Multiplying the left and right sides of (5.179) and (5.180) together, and taking the average, we have

$$r_{\hat{e}}(0) \approx \Theta_0 r_{e's}(0) - \Theta_0 \sum_{i=1}^{M} \hat{a}(i) r_{e's}(i). \tag{5.181}$$

Now using (5.180) and (5.176) leads immediately to the following set of estimators:

$$\hat{\Theta}_0(m) \approx \sqrt{r_{\hat{e}}(0;m)}, \qquad \text{unvoiced case} \tag{5.182}$$

and

$$\hat{\Theta}_0(m) \approx \sqrt{P[r_{\hat{e}}(0;m)]}, \qquad \text{voiced case.} \tag{5.183}$$

5.3.5 A Distance Measure for LP Coefficients

Reading Note: The reader might wish to review Section 1.3.1 before studying this material. The topics are closely related, but not highly dependent upon one another.

In many aspects of speech processing, it is necessary to have a distance measure relating two LP vectors, say $\hat{\mathbf{a}}(m)$ and $\hat{\mathbf{b}}(m')$. In effect, since these parameters bear a close relationship to the short-term spectra of the speech frames from which they were drawn, we are seeking to learn how similar the corresponding spectra are. Recalling the discussion in Section 1.3.1, we know that the unweighted Euclidean distance is not appropriate here because the individual LP parameters in the vectors are highly correlated. A weighted Euclidean distance would be more appropriate, but that would depend on finding an appropriate decorrelating weighting matrix. An estimated covariance matrix for one of the vectors would seem appropriate, and we will return to this possibility below.

The most frequently used distance measure was proposed by Itakura (1975). The easiest way for us to understand this measure is to build on some previous results. Let us recall (5.81), dropping the superscripts M,

$$\begin{aligned} \tilde{\mathbf{R}}_s(m) \times [1 \quad -\hat{a}(1;m) \quad -\hat{a}(2;m) \quad \cdots \quad -\hat{a}(M;m)]^T \\ = [\xi(m) \quad 0 \quad 0 \quad \cdots \quad 0]^T. \end{aligned} \tag{5.184}$$

Remember that $\tilde{\mathbf{R}}_s(m)$ signifies the $(M+1)\times(M+1)$ *augmented* correlation matrix, defined as

$$\tilde{\mathbf{R}}_s(m) \stackrel{\text{def}}{=} \begin{bmatrix} r_s(0;m) & \mathbf{r}_s^T(m) \\ \mathbf{r}_s(m) & \mathbf{R}_s(m) \end{bmatrix}. \tag{5.185}$$

Recall also that $\xi(m)$ denotes the MSE associated with the prediction of the frame $f(n;m)$ over all time using parameters $\hat{\mathbf{a}}(m)$ [see (5.77) and (5.78)]. Let us denote this explicitly by writing $\xi_{\hat{\mathbf{a}}}(m)$. Note that if we define

$$\begin{aligned} \boldsymbol{\alpha}(m) &\stackrel{\text{def}}{=} [1 \quad -\hat{a}(1;m) \quad -\hat{a}(2;m) \quad \cdots \quad -\hat{a}(M;m)]^T \\ &= [1 \quad -\hat{\mathbf{a}}^T(m)]^T, \end{aligned} \tag{5.186}$$

then we can compute the scalar $\xi_{\hat{\mathbf{a}}}(m)$ as

$$\xi_{\hat{\mathbf{a}}}(m) = \boldsymbol{\alpha}^T(m)\tilde{\mathbf{R}}_s(m)\boldsymbol{\alpha}(m). \tag{5.187}$$

This follows immediately upon premultiplying (5.184) through by[37] $\boldsymbol{\alpha}^T(m)$.

Now suppose that we attempt to use the predictor with coefficients $\hat{b}(1;m'),\ldots,\hat{b}(M;m')$ [the elements of $\hat{\mathbf{b}}(m')$] to predict the frame $f(n;m)$. The vector $\hat{\mathbf{b}}(m')$ can be derived from any frame of any signal. It need *not* be associated with the speech sequence $s(n)$ from which we have computed $\hat{\mathbf{a}}(m)$. There will be a certain MSE associated with this prediction, say $\xi_{\hat{\mathbf{b}}}(m)$. With a little manipulation, it can be shown that

$$\xi_{\hat{\mathbf{b}}}(m) = \boldsymbol{\beta}^T(m')\tilde{\mathbf{R}}_s(m)\boldsymbol{\beta}(m'), \tag{5.188}$$

with

$$\boldsymbol{\beta}(m') \stackrel{\text{def}}{=} [1 \quad -\hat{\mathbf{b}}^T(m')]^T. \tag{5.189}$$

We know that

$$\xi_{\hat{\mathbf{a}}}(m) \le \xi_{\hat{\mathbf{b}}}(m) \tag{5.190}$$

because $\xi_{\hat{\mathbf{a}}}(m)$ is the best possible prediction error in the sense of minimizing the average squared prediction error.

How "far" is LP vector $\hat{\mathbf{b}}(m')$ from LP vector $\hat{\mathbf{a}}(m)$? One way of answering this question is to measure how much "better" $\hat{\mathbf{a}}(m)$ is at predicting its "own" frame than $\hat{\mathbf{b}}(m')$ is. A measure of this is the ratio $\xi_{\hat{\mathbf{b}}}(m)/\xi_{\hat{\mathbf{a}}}(m)$, or, taking the logarithm, we define the *Itakura distance* as

$$d_{\mathrm{I}}[\hat{\mathbf{a}}(m), \hat{\mathbf{b}}(m')] \stackrel{\text{def}}{=} \log \frac{\xi_{\hat{\mathbf{b}}}(m)}{\xi_{\hat{\mathbf{a}}}(m)} = \log \frac{\boldsymbol{\beta}^T(m')\tilde{\mathbf{R}}_s(m)\boldsymbol{\beta}(m')}{\boldsymbol{\alpha}^T(m)\tilde{\mathbf{R}}_s(m)\boldsymbol{\alpha}(m)}. \tag{5.191}$$

[37]Actually, we can premultiply by any M-vector with a unity first element, but we choose this form for a reason described below.

Note that this measure will always be positive because of condition (5.190). Also note that while this measure is called a "distance," it is not a true metric because it does not have the required symmetry property, that is

$$d_{\mathrm{I}}[\hat{\mathbf{a}}(m), \hat{\mathbf{b}}(m')] \neq d_{\mathrm{I}}[(\hat{\mathbf{b}}(m'), \hat{\mathbf{a}}(m)] \tag{5.192}$$

(see Section 1.3.1).

Although the Itakura distance was derived using this intuitive interpretation of a ratio of prediction error energies, it was originally derived using a maximum likelihood argument similar to the following. Imagine that the signal $s(n)$ were truly generated by driving the all-pole model with coefficients $\hat{\mathbf{b}}(m')$, by stationary white Gaussian noise. Then, for a frame $f(n; m)$ of length N taken from signal $s(n)$, LP vectors $\hat{\mathbf{a}}(m)$ of order M derived by the autocorrelation method on that frame will tend to be multivariate Gaussian distributed as

$$f_{\underline{\boldsymbol{\alpha}}(m)}[\boldsymbol{\alpha}(m)] = \frac{1}{\sqrt{(2\pi)^N \det \boldsymbol{\Lambda}}} \exp\left\{-\frac{1}{2}[\boldsymbol{\alpha}(m) - \boldsymbol{\beta}(m')]^T \boldsymbol{\Lambda}^{-1}[\boldsymbol{\alpha}(m) - \boldsymbol{\beta}(m')]\right\}, \tag{5.193}$$

where $\boldsymbol{\alpha}(m)$ and $\boldsymbol{\beta}(m')$ are defined above, and where

$$\boldsymbol{\Lambda} \stackrel{\text{def}}{=} \tilde{\mathbf{R}}_s^{-1}(m)[\boldsymbol{\alpha}^T(m)\tilde{\mathbf{R}}_s(m)\boldsymbol{\alpha}(m)]. \tag{5.194}$$

Accordingly, if $\hat{\mathbf{a}}(m)$ and $\hat{\mathbf{b}}(m')$ are related to the same frame (and therefore contain similar spectral information), then intuitively, they should be close in Mahalanobis distance (see Section 1.3.1),

$$d_{\mathrm{M}}[\hat{\mathbf{a}}(m), \hat{\mathbf{b}}(m')] = \frac{[\boldsymbol{\alpha}(m) - \boldsymbol{\beta}(m')]^T \tilde{\mathbf{R}}_s(m)[\boldsymbol{\alpha}(m) - \boldsymbol{\beta}(m')]}{\boldsymbol{\alpha}^T(m)\tilde{\mathbf{R}}_s(m)\boldsymbol{\alpha}(m)}. \tag{5.195}$$

[This expression is often called the *Itakura–Saito distance* in speech processing (Itakura and Saito, 1968).] Further, if $\hat{\mathbf{a}}(m)$ truly is close to $\hat{\mathbf{b}}(m')$, then $\boldsymbol{\alpha}(m) \approx \boldsymbol{\beta}(m')$ and $d_{\mathrm{I}}[\hat{\mathbf{a}}(m), \hat{\mathbf{b}}(m')] \approx d_{\mathrm{M}}[\hat{\mathbf{a}}(m), \hat{\mathbf{b}}(m')]$.

The Itakura distance is probably the most widely used measure of similarity between LP vectors. In Itakura's original paper (1975), it is introduced for use in an isolated word recognition strategy that has since been named "dynamic time warping." The strategy will be the subject of Chapter 11.

5.3.6 Preemphasis of the Speech Waveform

In reading about practical applications of LP analysis, one will often find that the researcher has *preemphasized* the speech waveform prior to computing the LP parameters. By this we mean applying a filter that increases the relative energy of the high-frequency spectrum. Typically, the filter

$$P(z) = 1 - \mu z^{-1} \tag{5.196}$$

is used with $\mu \approx 1$. This filter is identical in form to the filter used to model the lip radiation characteristic. We know that this filter introduces a zero near $\omega = 0$, and a 6-dB per octave shift on the speech spectrum.

The reasons for employing a preemphasis filter are twofold. First, it has been argued that the minimum-phase component of the glottal signal can be modeled by a simple two-real-pole filter whose poles are near $z = 1$ (see Chapter 3). Further, the lip radiation characteristic, with its *zero* near $z = 1$, tends to cancel the spectral effects of one of the glottal poles. By introducing a second zero near $z = 1$, the spectral contributions of the larynx and lips have been effectively eliminated and the analysis can be asserted to be seeking parameters corresponding to the vocal tract only. We know that the speech production model is a greatly simplified analytical model of a complex physical system. Accordingly, we must be careful not to overstate the fact that preemphasis results in an LP spectrum or filter that is free of glottal or lip radiation effects. In the worst case, however, it is clear that the preemphasis will give the higher formants in the vocal tract a better chance to influence the outcome.

The value of μ is taken in the range $0.9 \leq \mu \leq 1.0$, although the precise value seems to be of little consequence. Of course, preemphasis should not be performed on unvoiced speech, in which case $\mu \approx 0$. Both Gray and Markel (1974) and Makhoul and Viswanathan (1974) have worked with an "optimal" value of μ given by

$$\mu = \frac{r_s(1;m)}{r_s(0;m)}, \tag{5.197}$$

where $r_s(\eta;m)$ is the usual short-term autocorrelation sequence for the frame. For unvoiced frames this value is small, whereas for voiced frames it is near unity.

The second reason for preemphasis is to prevent numerical instability. The work on this problem has focused on the autocorrelation method, but the deleterious effects can be expected to be even worse for the covariance case (Markel and Gray, 1976, p. 222). If the speech signal is dominated by low frequencies, it is highly predictable and a large LP model order will result in an ill-conditioned[38] autocorrelation matrix (Ekstrom, 1973). Makhoul (1975) argues that the ill-conditioning of the autocorrelation matrix becomes increasingly severe as the dynamic range of the spectrum increases. If the spectrum has a general "tilt" that is causing the wide dynamic range, then a first-order inverse filter should be able to "whiten" the spectrum. Indeed, the preemphasis filter may be interpreted as such an inverse filter, and μ given by (5.197) is the optimal coefficient in the sense of MSE.

[38]For a general discussion of ill-conditioning, see (Nobel, 1969).

5.4 Alternative Representations of the LP Coefficients

In this short section we remind the reader of two alternative sets of parameters that are theoretically equivalent to the LP parameters, and which can be derived from them, and introduce two others. It is not the purpose here to study alternative models of speech production, although there are many.[39]

In the communications technologies, the LP parameters are rarely used directly. Instead, alternative representations of the LP model are employed. These alternate sets of parameters have better quantization and interpolation properties, and have been shown to lead to systems with better speech quality (Viswanathan and Makhoul, 1975). We have already studied three alternative representations: the reflection coefficients, the log area ratio parameters, and the inverse sine parameters. We will see these systems employed in coding methods in Chapter 7. In this brief section, we introduce another alternative representation to the LP parameters, the line spectrum pair, and preview a second, the cepstral parameters. The latter will be discussed thoroughly in the following chapter, after the necessary background has been presented.

5.4.1 The Line Spectrum Pair

In the 1980s, the *line spectrum pair* (LSP) was introduced as another alternative to the LP parameters. This technology was researched most extensively by the Japanese telephone industry, but some seminal ideas are found in the following papers in English: (Itakura, 1975; Sugamura and Itakura, 1981; Soong and Juang, 1984; Crosmer and Barnwell, 1985).

The LSP is developed by beginning with the z-domain representation of the inverse filter of order M,

$$\hat{A}(z;m) \stackrel{\text{def}}{=} 1 - \sum_{i=1}^{M} \hat{a}(i;m)z^{-i}. \tag{5.198}$$

Now $\hat{A}(z;m)$ is decomposed into two $(M+1)$-order polynomials,

$$P(z;m) = \hat{A}(z;m) + z^{-(M+1)}\hat{A}(z^{-1};m) \tag{5.199}$$

$$Q(z;m) = \hat{A}(z;m) - z^{-(M+1)}\hat{A}(z^{-1};m), \tag{5.200}$$

so that

$$\hat{A}(z;m) = \frac{P(z;m) + Q(z;m)}{2}. \tag{5.201}$$

[39]Indeed, there are many such models, including, for example, models including both poles and zeros (Steiglitz, 1977; Konvalinka and Mataušek, 1979; El-Jaroudi and Makhoul, 1989), sinusoidal models (McAulay and Quatieri, 1986), orthogonal function expansions (Korenberg and Paarmann, 1990), and time-varying LP models and LP models based on alternative formulations [for a review see (McClellan, 1988)].

[The reader is encouraged to take a small-order $\hat{A}(z;m)$ polynomial and work out these three equations.] In light of (5.103), $P(z;m)$ can be interpreted as representing a $(M+1)$-order lattice (analysis) filter with final reflection coefficient $\kappa(M+1;m)=1$. Similarly for $Q(z;m)$ with $\kappa(M+1;m)=-1$. Accordingly, $P(z;m)$ and $Q(z;m)$ correspond to lossless models of the vocal tract with the glottis closed and open, respectively (see Section 5.3.3). In turn, this can be shown to guarantee that all zeros of P and Q lie on the unit circle (Soong and Juang, 1984). In fact, P has a real zero at $z=-1$, Q a zero at $z=1$, and all other zeros are complex and interleaved as shown in Fig. 5.19. These zeros comprise the LSP parameters. The name derives from the fact that each zero pair corresponds to a pole pair in the forward model, which lies on the unit circle. In turn, this pole pair would represent an undamped sinusoid that, in analog terms, would have a line spectrum. Since the zeros occur in complex conjugate pairs for both P and Q, there are only M unique zeros needed to specify the model. The zeros are found by iterative search along the unit circle, taking advantage of the interleaving. Although the zeros are complex, their magnitudes are known to be unity, so that only a single real parameter (the frequency or angle) is needed to specify each one. In fact, coding the frequency differences between zeros has been found to be more efficient than coding the frequencies themselves, leading to a 30% improvement in efficiency over the use of log area ratio parameters. Another strategy involves the use of second-order filter sections to reconstruct the speech from the LSP parameters. In this case each section implements one zero pair, and it is sufficient to know only the cosine of the frequency of the pair. This is another way to reduce the dynamic range of the parameters and improve coding efficiency.

Finally, it is to be noted that the LSP parameters are interpretable in terms of the formant frequencies of the model. Each zero of $\hat{A}(z;m)$ maps into one zero in each of the polynomials $P(z;m)$ and $Q(z;m)$. If the two resulting zeros are close in frequency, it is likely that the "parent" zero in $\hat{A}(z;m)$ represents a formant (narrow bandwidth) in the

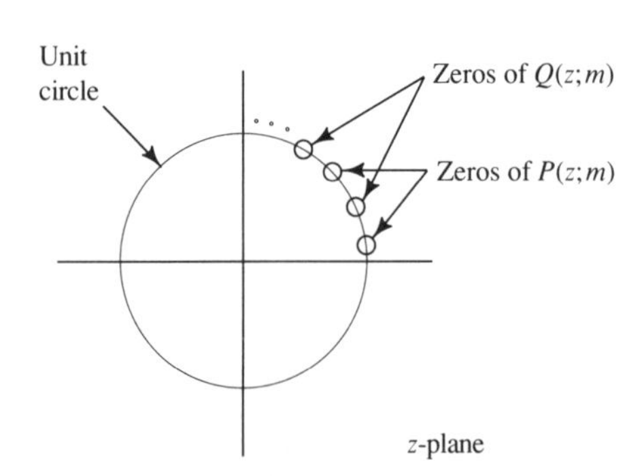

FIGURE 5.19. The interleaved zeros of the LSP polynomials $P(z;m)$ and $Q(z;m)$.

model. Otherwise, the original zero is likely to represent a wide bandwidth spectral feature.

5.4.2 Cepstral Parameters

Because the *cepstral* parameters are so widely used in the speech recognition domain, and because they are frequently derived from the LP parameters, we mention them here for emphasis and completeness. In Chapter 6 we will discuss the relationship between the cepstral and LP parameters. The reader may wish to glance ahead at (6.44) in Section 6.2.4 to preview the conversion formula.

5.5 Applications of LP in Speech Analysis

Throughout the remainder of the book, we will see the LP parameters play a central role in many coding, synthesis, and recognition strategies. In order to give some illustration of the use of LP in the real world while we are still in the analysis part of the book, we focus briefly on the related problems of pitch, formant, and glottal waveform estimation.

5.5.1 Pitch Estimation

In Section 4.3.1 we briefly discussed the possibility of using the short-term autocorrelation as a detector of pitch and indicated that this is seldom done because it is only slightly less expensive than more reliable methods. Some alternative methods were discussed there, including attempts to "prewhiten" the speech by "clipping" before computing the autocorrelation. The *simple inverse filter tracking* (SIFT) algorithm of Markel (1972) follows this basic strategy of prewhitening followed by autocorrelation, but the prewhitening step involves the use of the LP-based IF.

The SIFT algorithm is diagramed in Fig. 5.20. Initially, the digitized speech is lowpass filtered and decimated in order to suppress superfluous high-frequency content and reduce the amount of necessary computation. In Fig. 5.20, for example, a 10-kHz sampling rate is assumed on $s(n)$ and the sequence is lowpass filtered to exclude frequencies above 800 Hz. The sequence is then decimated by a factor 5:1 to create an effective sampling rate of 2 kHz. To create an IF, a low-order analysis ($M \approx 4$) is sufficient, since we would expect no more than two formants in the nominal 1-kHz bandwidth remaining. The short-term LP analysis is typically done on rather small frames of speech (≈ 64 points) for good time resolution. Once the IF is created for a given frame (ending at, say time m), the frame is passed through it to compute the residual, $\varepsilon(n;m)$. Although

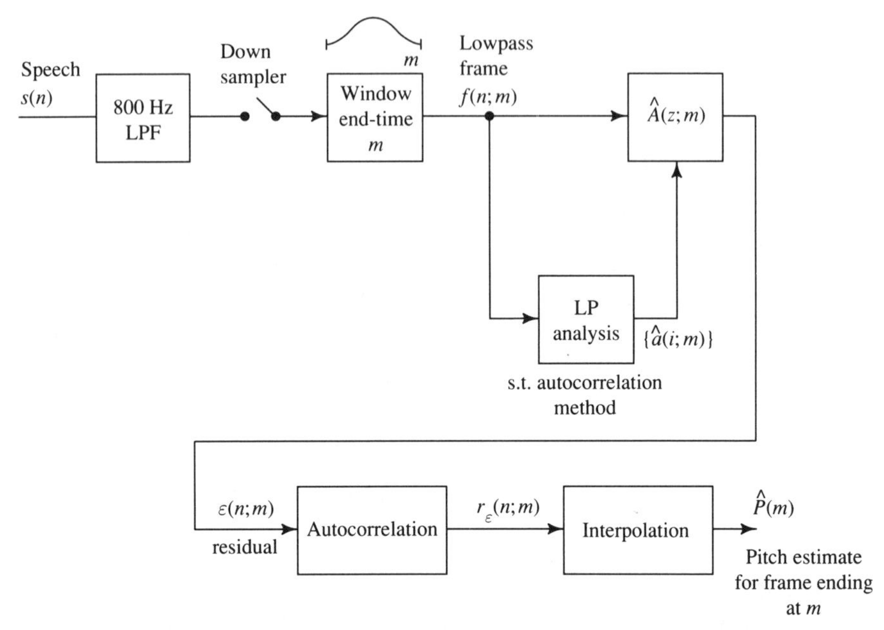

FIGURE 5.20. The SIFT algorithm. After Markel (1972).

it is tempting to believe that $\varepsilon(n; m)$ might exhibit reliable pulses that could be easily selected as excitation times, this is seldom the case. The reader is urged to recall the discussion in Section 5.2.3. In fact, however, the residual can be expected to be a phase-scattered version of the excitation pulse train, and therefore reasonably periodic. The autocorrelation can be used to detect this periodicity; that is the next step in the SIFT algorithm. The pitch period estimate is taken to be[40]

$$\max_{\eta} r_{\varepsilon}(\eta). \tag{5.202}$$

The search is usually restricted to some η's that are in a reasonable neighborhood of the estimate from the last frame (if it was voiced). In fact, since the resolution of this procedure is quite low [in the present example, the temporal spacing between autocorrelation lags is $(2\,\text{kHz})^{-1}$, or 0.5 msec], it is necessary to interpolate between the lags of $r_{\varepsilon}(\eta)$ before performing the peak picking. Hence we see this operation following the autocorrelation in Fig. 5.20. The details of the interpolation are found in (Markel, 1972). Finally, the SIFT procedure is also used to make a voiced/unvoiced decision. This is accomplished by comparing the peak in the present frame to some threshold based on the total residual energy associated with the present frame, for example, $0.4r_{\varepsilon}(0)$.

Some example results from the SIFT algorithm are shown in Fig. 5.21. Details of the procedure are found in (Markel, 1972).

[40]Note that we are, in principle, computing the *long-term* autocorrelation of a finite sequence that has been created by short-term processing.

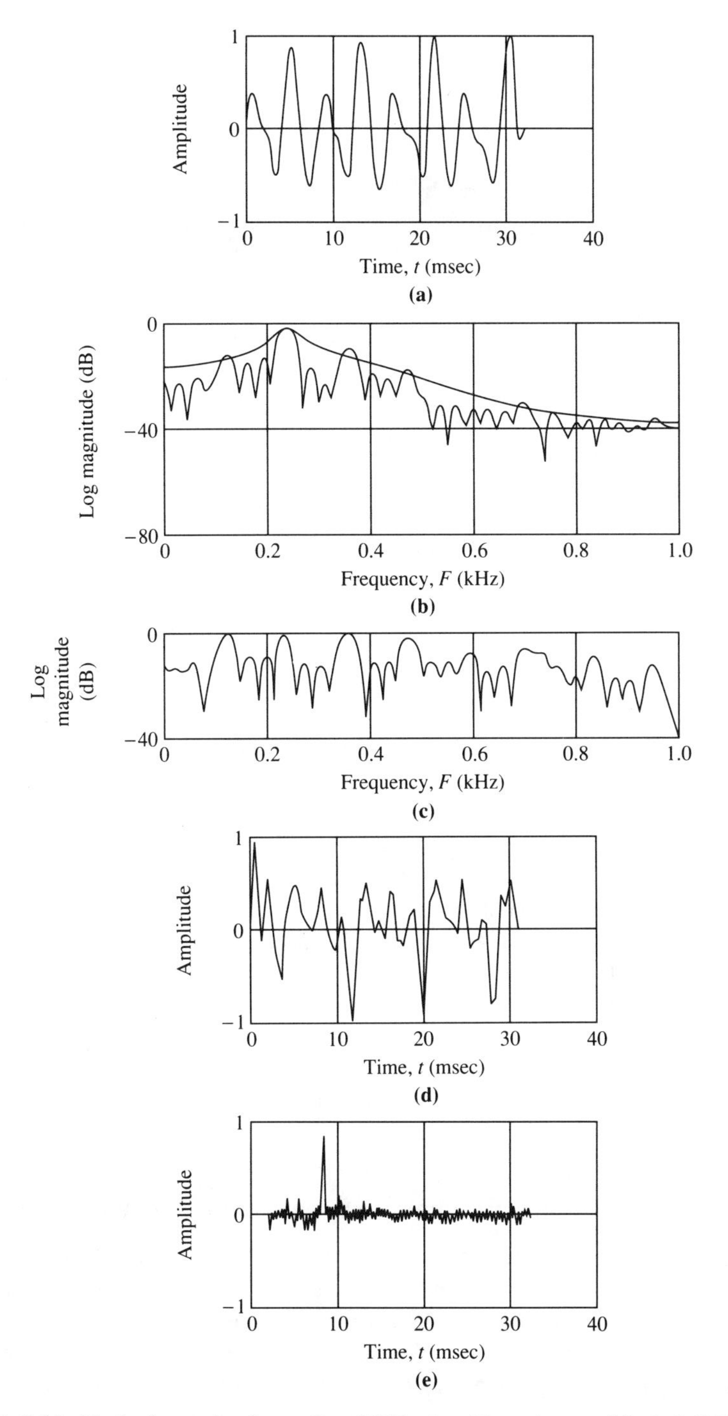

FIGURE 5.21. Typical results from the SIFT algorithm due to Markel (1972) and Markel and Gray (1976). (a) Speech frame. (b) Input spectrum with spectrum of the LP model superposed. (c) Spectrum at the output of the IF. (d) Residual frame at the output of the IF. (e) Autocorrelation of the residual showing a pitch period near 8 msec. (f) Utterance of phrase "linear prediction" and resulting SIFT output.

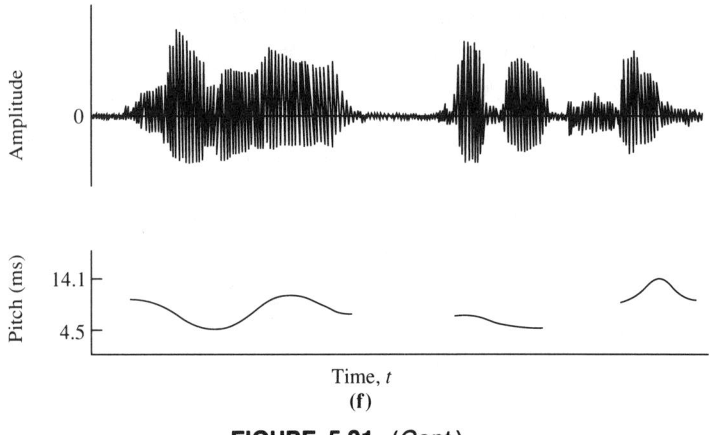

FIGURE 5.21. *(Cont.)*

5.5.2 Formant Estimation and Glottal Waveform Deconvolution

Formant frequencies and bandwidths are principal analytical features of the speech spectrum. Moreover, they are appealing features because they are clearly related to the articulatory act and the perception of the speech. Accordingly, as we shall see later in our work, formant information is used extensively in coding and recognition of speech.

A closely related problem to formant estimation is that of estimating the time-domain glottal waveform. An understanding of the characteristics of the glottal dynamics is important, for example, in speech coding and synthesis (see Chapter 7), and in algorithms for laryngeal pathology detection (see the references below). In this subsection, we briefly examine the application of LP-based techniques to these two related problems of formant and glottal waveform estimation.

Formant Estimation by Spectral Methods

The first of two simple techniques for formant estimation is based on peak finding in an LP-derived magnitude spectrum. Papers of Atal and Hanauer (1971) and Markel (1972) first described a method of this type, although Schafer and Rabiner (1970) had earlier reported a spectral peak picking method based on the cepstrum, which we will study in Chapter 6.

Recall that the short-term IF will approximate the inverse of the minimum-phase component of the speech system,

$$\hat{A}(z;m) = 1 - \sum_{i=1}^{M} \hat{a}(i;m)z^{-i} = \hat{\Theta}^{-1}(z) \approx \Theta_{\min}^{-1}(z). \tag{5.203}$$

Now consider taking the N-point DFT of the sequence

$$\{1, -\hat{a}(1, m), -\hat{a}(2; m), \ldots, -\hat{a}(M; m), 0, 0, 0, \ldots, 0\}, \tag{5.204}$$

where N is typically 256 or 512 and the sequence is accordingly zero-padded. Let us call the DFT $X(k; m)$, so

$$\begin{aligned} X(k; m) &= 1 - \sum_{n=1}^{M} \hat{a}(n; m) e^{-j(2\pi/N)kn} \\ &= \hat{A}(e^{j(2\pi/N)k}; m), \qquad k = 0, 1, \ldots, N-1. \end{aligned} \tag{5.205}$$

Taking the magnitude and reciprocating each point gives

$$|\hat{A}(e^{j(2\pi/N)k}; m)|^{-1} = |\hat{\Theta}(e^{j(2\pi/N)k}; m)| \approx \Theta_0^{-1} |\Theta(e^{j(2\pi/N)k})|, \tag{5.206}$$

which, for a sufficiently large N, yields a high-resolution representation of the (scaled) speech magnitude spectrum. From this spectrum, local maxima are found and those of small bandwidths, and perhaps those restricted to certain neighborhoods for classes of phonemes, are selected as formants.

It is to be noted that an FFT algorithm is employed to obtain (5.205) and that it is not necessary to actually take the reciprocal in (5.206) unless the spectrum is to be displayed. If the IF spectrum is used directly, local *minima* are sought to represent the formants.

Markel (1972) reports that the peak picking of the reciprocal IF spectrum was successful at producing accurate estimates of formant frequencies about 90% of the time in experiments in which he tracked formants in flowing speech. This is a significant improvement over the accuracy that would be expected from an attempt to pick peaks from the unprocessed speech spectrum.

The procedure above involves the computation of the IF spectrum by effectively evaluating $\hat{A}(z; m)$ at equally spaced points on the unit circle. An enhancement to this procedure suggested by McCandless (1974) involves the evaluation of the spectrum on a circle of radius $\rho < 1$. This has the effect of making the valleys in the IF spectrum (peaks in the speech spectrum) more pronounced and easier to discern. This is especially important in cases in which two formants are very closely spaced in frequency. The bandwidths, are, of course, distorted in this case. To carry out this approach in terms of the method described above, it is a simple matter of premultiplying the LP parameters by ρ before computing the FFT. Clearly, the DFT of

$$\{1, -\rho^{-1}\hat{a}(1, m), -\rho^{-2}\hat{a}(2; m), \ldots, -\rho^{-M}\hat{a}(M, m), 0, 0, 0, \ldots, 0\} \tag{5.207}$$

is

$$1 - \sum_{n=1}^{M} \hat{a}(n; m) \rho^{-n} e^{-j(2\pi/N)kn} = \hat{A}(\rho e^{j(2\pi/N)k}; m), \quad k = 0, 1, \ldots, N-1, \tag{5.208}$$

which is the IF spectrum evaluated on the ρ-circle as required.

McCandless (1974) reports that in a study of a large amount of speech material, missing peaks in the reciprocal IF spectrum, due to nasalization or other pole-canceling effects, or due to merger of two separate peaks, were much more common than spurious peaks. Since this is the case, one would expect her enhancement technique to be very useful. In McCandless' test database of 50 sentences, formant analysis was done on voiced frames. The error due to all factors was on the order of a few percent or less with nasality proving to be the most problematic. An algorithmic enhancement was later reported to better handle the rare instances of spurious spectral peaks (McCandless, 1976).

A second variation due to Kang and Coulter (1976), involves moving the zeros of $A(z;m)$ directly onto (or close to) the unit circle before computing the transform. This is done by using the L–D recursion to compute the $\hat{a}(i;m)$ parameters, then setting the last reflection coefficient, $\kappa(M;m)=\hat{a}(M;m)$, equal to -1. Since $-\hat{a}(M;m)$ is easily shown to be the product of all the zeros in the filter, each must lie on the unit circle in this case. The advantage of this approach over McCandless' method is that all singularities become unmistakably prominent, not just those sufficiently close to a ρ-circle. The disadvantage is that the zeros do not move along radial paths to the unit circle resulting in distortion in the formant frequency estimates. Kang and Coulter have referred to these estimates as "pseudoformant" frequencies. As with the McCandless method, of course, the Kang method is unable to directly provide formant bandwidth information.

Pole Extraction Method of Formant Estimation

The early Atal and Hanauer (1971) and Markel (1972) papers also suggested a second approach to using LP parameters for estimating formants. It has been shown by Fant (1956) that appropriate speech spectra can be constructed by ascribing single-pole pairs to each formant. Hence another way to use the information provided by the LP parameters is to extract zeros of the IF polynomial [see (5.203)], which, of course, represent the poles of $\Theta(z)$. Three or four (depending on the Nyquist frequency) resonant pole pairs [zeros of $\hat{A}(z;m)$ near the unit circle] are selected as representative of formants. For a given zero pair, the formant frequency is deduced immediately from the angle of the pair and the bandwidth is related to the pole pair's magnitude.

The zeros are extracted using one of many commonly available routines that are subject to occasional numerical problems. Since these algorithms are of very high computational complexity, this method is primarily used as a research tool and not in real-time systems. Moreover, there is not a simple or predictable relationship between the roots of the IF polynomial and the resonances in the spectrum. Root extraction, therefore, is not as reliable as working with the IF spectrum.

Closed-Phase Deconvolution of the Glottal Waveform

Closed-phase (CP) analysis of voiced speech is based on the assumption that intervals in the signal can be located that correspond to periods in which the glottis is closed (no excitation to the vocal tract). These intervals are then employed to compute estimates of the vocal-tract filter, which, in turn, can be used to inverse filter the speech back to the glottal waveform. Notice that vocal-tract filter, or formant, estimates are a by-product of this analysis. The original work (using digital computing) in this area was done by Berouti, Childers, and Paige (1976, 1977a, 1977b); and Wong, Markel, and Gray (1979). A paper by Deller (1981) unifies the work of Berouti et al. and Wong et al., and it is this unified view that will be taken here as we use the work of these two research groups to illustrate "single-channel" methods (defined below). Related work is to be found in the papers of Mataušek and Batalov (1980), Milenkovic (1986), and Ananthapadmanabha and Yegnanarayana (1979), as well as in earlier papers that describe research based on analog technology (see the references in the papers cited above).

All the papers cited above involve work with a *single-channel* approach in which only the speech signal is used in the analysis. In the 1980s, *two-channel* approaches that supplement the speech signal by an electroglottographic (EGG) trace to locate the CP, were reported by the Childers research group (Krishnamurthy, 1984; Larar et al., 1985; Krishnamurthy and Childers, 1986), and by Veenemann and BeMent (1985). Among other important results, the Childers work used the EGG signal to study variability in Wong's analysis, and Veenemann added some important conceptual developments to the identification process pertaining to experimental usage. We will have more to say about the two-channel method below, but we initially focus on the single-channel methods.

Although elegant in its theoretical simplicity, the single-channel CP method involves some difficult practical issues. We will present only the theoretical rudiments of the methods here; the reader is referred to the original papers for a better appreciation of the implementation considerations.

Let us consider the vocal system model of Fig. 5.22. We assume, for simplicity, that the vocal-tract system is purely minimum phase and can

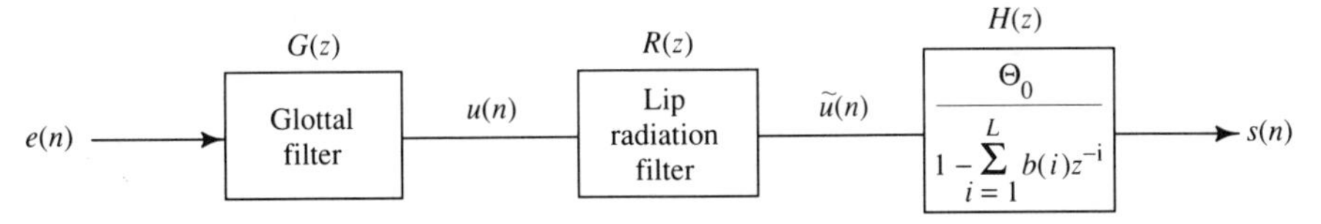

FIGURE 5.22. Speech production model used in the discussion of closed-phase analysis. The conventional speech model is rearranged so that the lip radiation characteristic precedes the vocal-tract model, and the vocal-tract system is assumed minimum phase.

be modeled by an L-order all-pole transfer function. We also assign the overall system gain, Θ_0, to $H(z)$ for convenience,

$$H(z) = \frac{\Theta_0}{1 - \sum_{i=1}^{L} b(i) z^{-i}}. \tag{5.209}$$

We also assume that the lip radiation characteristic can be modeled by a simple differencer,

$$R(z) = 1 - \alpha z^{-1} \tag{5.210}$$

with $\alpha \approx 1$. It is clear from Fig. 5.22 that the speech follows the difference equation

$$s(n) = \sum_{i=1}^{L} b(i) s(n-i) + \Theta_0 u(n) - \Theta_0 \alpha u(n-1), \tag{5.211}$$

where $u(n)$ means the glottal volume velocity, $u_{\text{glottis}}(n)$. For convenience, let us define

$$\tilde{u}(n) = \Theta_0 u(n) - \Theta_0 \alpha u(n-1). \tag{5.212}$$

Single-channel CP glottal deconvolution or inverse filtering (CPIF) involves the estimation of[41] $\Theta_0 u(n)$ given knowledge of $s(n)$ only, under the assumption of a closed phase in the glottal cycle—points for which $u(n) = 0$. It is important to emphasize that voiced speech may not necessarily contain a CP in the glottal cycle, but CP techniques proceed under the assumption that a CP actually exists. Generally speaking, the CP assumption is valid for (at least) the typical "male" pitch range, and for phonation of moderate-to-high intensity. The essence of CP analysis methods is that during a CP, the parameters $b(1), \ldots, b(L)$ can (theoretically) be identified exactly, allowing the construction of a perfect IF with which to deconvolve $\tilde{u}(n)$ from $s(n)$. Then $\Theta_0 u(n)$ can (theoretically) be obtained from $\tilde{u}(n)$ by a simple first-order averaging filter.[42]

All methods for CPIF involve, either implicitly or explicitly, three basic steps:

1. Determination of the CP points in which to compute the CPIF.
2. Computation of the CPIF.
3. Inverse filtering of the speech and postprocessing to obtain an estimate of the glottal waveform.

[41]Of course, we know from our previous work that the numbers Θ_0 and α are generally unknown. The usual assumption is that $\alpha \approx 1$, and seeking $\Theta_0 u(n)$ instead of $u(n)$ removes this problem of indeterminant parameters.

[42]Herein lies one of the practical difficulties, as this process is very sensitive to initial conditions and prone to "drifting" (Wong et al., 1979).

As we have noted, we are chiefly concerned with the first two of these steps. The third step, although theoretically trivial, involves a number of practical difficulties that are treated in the literature.

The single-channel method of Wong, Markel, and Gray for computing the CPIF is formulated in terms of our conventional notation as follows: Let N_w be a window length no longer than the expected size of a CP interval, say, for example, 20 points if the typical cycle period is 100 points. For the N_w-length window ending at time m, compute the L-order covariance solution, letting the average squared error be denoted $\xi(m)$. The procedure is repeated over a range of m's covering at least one cycle of the speech. $\xi(m)$ is monitored as a function of m and the window producing the minimum is assumed to be the best candidate for computing the CPIF (which, note, has already been computed as a by-product of the CP location process), since it offers the "best fit" of an L-order model.

In a sense, the Berouti procedure is a special case of the Wong method with the window size, N_w, chosen as its minimum possible value, $N_w = L$. In this case, it is easily shown (see Problem 5.24) that the covariance equations used on the window, say $n = m - L + 1, \ldots, m$, reduce to the simple solution of the L equations in L unknowns,

$$
\begin{aligned}
s(m) &= \sum_{i=1}^{L} \hat{b}(i)s(m-i) \\
s(m-1) &= \sum_{i=1}^{L} \hat{b}(i)s(m-i-1) \\
&\vdots \\
s(m-L+1) &= \sum_{i=1}^{L} \hat{b}(i)s(m-i-L+1),
\end{aligned}
\tag{5.213}
$$

which are exactly consistent with the model difference equation for each time on this interval if the interval includes only CP points [see (5.211)]. In fact, it was this latter analysis that led Berouti to his method. A major difference in the Berouti method with respect to Wong is in the method of location of the CP interval. Rather than an integrated least squares analysis, Berouti chose the analysis window according to a method for locating the instant of glottal excitation that was developed by Strube (1973).

Intuitively, one would expect the Wong approach to be more robust to noise and model errors as these effects are averaged over more data. A discussion of this point is found in (Deller, 1981).

Although a number of favorable experimental results have been published in the papers cited above, large-scale testing results for these single-channel methods have not been reported. All of the authors acknowledge potential difficulties with the basic assumptions underlying the method, however, and work reported in (Larar et al., 1985) and

(Picaché, 1988) suggests that the technique is not highly reliable without modifications. This is likely due to several basic reasons, all centered on the underlying assumptions. These include:

1. The assumption of a minimum-phase system. The "system" in this case includes any system between the glottis and the samples in the computer memory. This is critical because any significant "smearing" of the temporal relationships of the frequency components of the glottal signal will render the method invalid. Berouti (1976) describes a method for compensating for the phase characteristic of the recording equipment, for example.
2. The basic assumption of a linear filter driven by glottal waveform, including no significant coupling between the dynamics of the vocal tract and the larynx. Any deviation from this model destroys some very fundamental assumptions underlying the method. The work of Teager and Teager (1990) is very relevant to this point.
3. The existence of a closed phase in the glottal cycle. That this is not certain has been borne out by the two-channel studies described below.

One of the most difficult aspects of using the single-channel CPIF method is that the "answer" is unknown, and there is no basis for assessing what is apparently a "reasonable" result. Two-channel methods offer some help with this problem.

In two-channel approaches, the EGG signal is used as an indicator of the closed-phase region. Since most of the computational effort in single-channel approaches is in locating the CP, such approaches greatly increase the efficiency of the analysis. Further, the Larar et al. (1985) two-channel research has cast doubt upon the covariance least square error as an accurate indicator of the CP interval. Although faster and inherently more accurate, the use of an electroglottograph is not always practical, particularly in on-line vocoding schemes in which the glottal waveform estimate is needed. Therefore, the development of reliable single-channel methods remains an interesting problem. Picaché (1988) has described an enhanced covariance-type algorithm that exhibits the potential for more accurate CP location as well as more efficient computation of the estimates. In the paper cited in the opening paragraph of this discussion, Milenkovic (1986) describes a single-channel approach that involves the joint estimation of the vocal-tract model and a linear model of the glottal source. This method is not a CP method and, in fact, shows potential to circumvent many of the problems inherent in assuming the existence of a CP.

5.6 Conclusions

We have been on a long and arduous journey through the concepts, computational methods, and some direct applications of LP analysis. We will continue to see LP analysis play a significant role in the remaining chapters—in endeavors ranging from coding to recognition to enhancement. To say that LP has been important in the development of modern speech processing is to drastically understate its significance. LP analysis has unquestionably become the premiere method for extracting short-term spectral information from speech. In fact, even when LP parameters are not directly used in a design, it is often the case that LP coefficients are first extracted, then converted to another parametric representation. Such is the case with the "cepstral" parameters to which we turn our attention in the following chapter.

5.7 Problems

5.1. In Fig. 5.3(b), suppose that the impulse response, $\hat{\theta}(n)$, which minimizes the mean square output error, $\mathcal{E}\{\tilde{s}^2(n)\}$, has been found. Demonstrate that the desired model parameters, $\hat{a}(i), i = 1, 2, \ldots,$ are nonlinearly related to the impulse response sequence.

5.2. In this problem we provide some supportive details for the proof of Lemma 5.2.

(a) Demonstrate that any first-order polynomial of the form

$$1 - z_0 z^{-1} \qquad \text{with} |z_0| < 1$$

can be written as the ratio

$$\frac{1}{1 + \sum_{k=1}^{\infty} z_0^k z^{-k}}. \tag{5.214}$$

(b) Consider the following polynomial:

$$\begin{aligned} A(z) &= \sum_{j=0}^{\infty} \beta(j) z^{-j} \\ &= \left[1 - \sum_{i=1}^{M} a(i) z^{-i}\right] \prod_{l=1}^{L} \left[1 + \sum_{k_l=1}^{\infty} z_l^{k_l} z^{-k_l}\right], \end{aligned} \tag{5.215}$$

in which $L < \infty$, and $|z_l| < 1$ for all l. Argue heuristically that for any $\varepsilon > 0$, there exists an I such that for all $j \geq I, |\beta(j)| < \varepsilon$. *Hint*: Begin by examining the case $L = 1$.

5.3. Demonstrate (5.26).

5.4. Derive the long-term LP equations (5.22) by solving Interpretive Problem 5.2.

5.5. Derive the long-term LP equations (5.22) by solving Interpretive Problem 5.3.

5.6. Derive the long-term LP equations (5.22) by solving Interpretive Problem 5.4.

5.7. Derive the long-term LP equations (5.22) by solving Interpretive Problem 5.5. (*Note*: The method of solution is suggested under the problem statement.)

5.8. In this problem we complete the sufficiency proof of Theorem 5.2. With reference to (5.36), demonstrate that each element of the vector $\mathbf{r}_{e's}$ is zero if $e(n)$ is a realization of a WSS, correlation ergodic, orthogonal random process $\underline{e}$. Do not use results from Appendix 5.A.

5.9. Consider a voiced speech sequence, $s(n)$, produced by exciting the system with discrete-time impulse response $\theta(n)$ by a discrete-time impulse train, $e(n)$, with period P. Near the end of Section 5.2.2, we asserted that the autocorrelation of the speech, $r_s(\eta)$, is a severely aliased version of the autocorrelation $r_\theta(\eta)$ for small P. (This assertion was then used to interpret the distortion of the LP parameters when P was small.) Verify that this assertion is correct.

5.10. Solve Interpretive Problem 5.7, showing that the solution results in the covariance method equations (5.57).

5.11. Verify that (5.67) is equivalent to the covariance method LP solution, equation (5.58), resulting from the analysis of the short term of speech $s(1), \ldots, s(N)$.

5.12. Return to Interpretive Problem 5.7, the covariance method problem, and solve for the linear predictor that minimizes the weighted squared error

$$\xi(m) = \frac{1}{N} \sum_{n=m-N+1}^{m} \lambda(n)[s(n) - \hat{s}(n)]^2 = \frac{1}{N} \sum_{n=m-N+1}^{m} \lambda(n)\varepsilon^2(n;m), \tag{5.216}$$

where $\{\lambda(n)\}$ represents a set of nonnegative weights. Your solution should be identical to (5.72) and (5.73) if m is taken to be N.

5.13. One of the useful features of the Levinson–Durbin solution of the autocorrelation LP equations is that the reflection coefficient generated at the lth step, $\kappa(l;m)$, is theoretically bounded as

$$|\kappa(l;m)| \leq 1. \tag{5.217}$$

This property allows a check for numerical stability of the computation at each step. Verify that (5.217) is a necessary condition for stability. [For sufficiency, see (Markel and Gray, 1976).]

5.14. Suppose that in using the Levison-Durbin recursion, we find that $|\kappa(l;m)| = 1$. Prove that $\hat{a}^l(1), \ldots, \hat{a}^l(l)$ is a perfect predictor, and that the

L-D algorithm should theoretically be modified so that it will continue to produce the "same" set of LP parameters,

$$\hat{a}^{l+k}(i) = \begin{cases} \hat{a}^l(i), & i = 1, 2, \ldots, l \\ 0, & i = l+1, \ldots, l+k \end{cases} \tag{5.218}$$

for any positive k.

5.15. Verify that the reflection coefficients can be obtained from the LP parameters by the following algorithm:

For $l = M, M-1, \ldots, 1$,
  $\kappa(l; m) = \hat{a}^l(l; m)$
  For $i = 1, 2, \ldots, l-1$,

$$\hat{a}^{l-1}(i; m) = \frac{\hat{a}^l(i; m) + \hat{a}^l(l; m)\hat{a}^l(l-i; m)}{1 - [\kappa(l; m)]^2}$$

  Next i
Next l

All notation is defined in the discussion of the Levinson–Durbin recursion.

5.16. Derive the expression

$$\hat{A}^l(z; m) = \hat{A}^{l-1}(z; m) - \kappa(l; m) z^{-l} \hat{A}^{l-1}(z^{-1}; m), \tag{5.219}$$

which was necessary to deduce the lattice structure from the Levinson–Durbin recursion.

5.17. A key to the Itakura–Saito lattice is the computation of the reflection coefficients that makes them interpretable as parcor coefficients,

$$\kappa(l; m) = \frac{\sum_{n=m+N-1}^{m} \varepsilon^{l-1}(n; m)\beta^{l-1}(n; m)}{\sqrt{\sum_{n=m+N-1}^{m} \left[\varepsilon^{l-1}(n; m)\right]^2 \sum_{n=m+N-1}^{m} \left[\beta^{l-1}(n; m)\right]^2}}. \tag{5.220}$$

Verify that this is a correct expression by employing the method suggested under (5.115).

5.18. Consider the $\mathbf{L\Delta L}^T$ decomposition algorithm used for solving the covariance equations in Section 5.3.3. The algorithm given is structured to avoid the redundant computation of the quantities

$$P(i, k) = L(i, k)\Delta(k).$$

What is the computational savings as a result of this trick with respect to direct use of (5.130)–(5.132)? Express your answer in floating point operations (flops).

5.19. (a) Given that the $\mathbf{L\Delta L}^T$ decomposition exists for a positive definite covariance matrix $\mathbf{\Phi}_s(m)$, show that a Cholesky decomposition exists.
(b) Show that the converse is true.

5.20. Carefully show that the recursion for the inverse weighted covariance matrix, (5.145), follows from (5.143) and the matrix inversion lemma (Lemma 5.3).

5.21. Enroute to developing a "systolic array" version of the WRLS algorithm for solving the covariance equations, we established the upper triangular system of equations (5.160),

$$\mathbf{T}(N)\hat{\mathbf{a}}(N) = \mathbf{d}_1(N) \tag{5.221}$$

by orthogonal triangularization of the original system of equations. We claimed that the solution of this equation for $\hat{\mathbf{a}}(N)$ would produce precisely the covariance solution for the time frame $n = 1, 2, \ldots, N$. Verify that this is the case.

5.22. In the course of developing the Itakura distance measure, we constructed the following expression for the average squared prediction error for the autocorrelation method applied to the frame $f(n; m)$:

$$\xi_{\hat{\mathbf{a}}}(m) = \boldsymbol{\alpha}^T(m)\tilde{\mathbf{R}}_s(m)\boldsymbol{\alpha}(m), \tag{5.222}$$

where

$$\boldsymbol{\alpha}(m) \stackrel{\text{def}}{=} [1 \quad -\hat{a}(1; m) \quad -\hat{a}(2; m) \quad \cdots \quad -\hat{a}(M; m)]^T = [1 \quad -\hat{\mathbf{a}}^T(m)]^T, \tag{5.223}$$

and $\tilde{\mathbf{R}}_s(m)$ is the augmented correlation matrix first encountered in (5.80). Differentiate $\xi_{\hat{\mathbf{a}}}(m)$ with respect to the vector $\hat{\mathbf{a}}(m)$ to derive the autocorrelation solution. (*Note*: The derivative of a scalar function, say f, of a vector $\mathbf{x}$ with respect to the ith element x_i, is defined as the vector $\mathbf{z}$ with ith element $\partial f/\partial x_i$.)

5.23. (Computer Assignment)
Note: In the following problem, it would be useful to first preemphasize the speech signal by filtering it with a simple system with $P(z) = 1 - \mu z^{-1}$ with $\mu \approx 1$. This will tend to emphasize high frequencies, making the estimation of formants easier. For a vowel utterance of your choice, perform the following tasks:
(a) Compute 14 LP parameters using the autocorrelation method over a 256-point frame selected with a Hamming window.
(b) Extract the zeros of the inverse filter polynomial,

$$\hat{A}(z; m) = 1 - \sum_{i=1}^{M} \hat{a}(i; m)z^{-i} \tag{5.224}$$

and estimate the first three formant frequencies based on these zeros.

(c) Now compute the DFT of the LP parameters as suggested in (5.204)–(5.206), and estimate the formant frequencies using the resulting spectrum.

(d) Repeat (c) using the McCandless procedure with $\rho = 0.95$.

(e) Estimate the formant frequencies by simply taking the DFT spectrum of the windowed speech itself.

(f) Discuss your results.

5.24. (a) We introduced the Berouti technique for glottal waveform deconvolution as a special case of the Wong method with the frame size, N_w, chosen as its minimum possible value, $N_w = L$, with L the order of the vocal-tract model. Show that in this case the covariance equations used on the window, say

$$n = m - L + 1, \ldots, m,$$

reduce to the simple solution of the L equations in L unknowns,

$$s(m) = \sum_{i=1}^{L} \hat{b}(i)s(m-i)$$

$$s(m-1) = \sum_{i=1}^{L} \hat{b}(i)s(m-i-1) \tag{5.225}$$

$$\vdots$$

$$s(m-L+1) = \sum_{i=1}^{L} \hat{b}(i)s(m-i-L+1). \tag{5.226}$$

(b) If the vocal-tract system function is truly minimum phase and of order L,

$$H(z) = \frac{\Theta_0}{1 - \sum_{i=1}^{L} b(i)z^{-i}}, \tag{5.227}$$

argue that these equations should provide exact estimates, $\hat{b}(i) = b(i)$ for all i, if the glottis is closed on the interval $m - L + 1, \ldots, m$.

5.25. This problem is concerned with frequency selective LP analysis, which was first developed by Makhoul (1975a, 1975b). Suppose that it is desired to design an LP model

$$\hat{\Theta}(z;m) = \frac{1}{1 - \sum_{i=1}^{M} \hat{a}(i;m)z^{-i}} \tag{5.228}$$

only for the frequency range $0 \le \omega_a \le \omega \le \omega_b \le \pi$. That is, we want all M LP parameters to be dedicated to this region of the spectrum (with the rest of the spectrum simply ignored). Given a frame of speech on the range $n = n - N + 1, \dots, m$, describe the steps of an autocorrelation-like method for estimating these parameters. (*Hint*: Consider computing the autocorrelation sequence, $r_s(1;m), \dots, r_s(M;m)$ using a frequency domain approach.)

5.26. (This problem is for persons who have studied the proof of Theorem 5.2 in Appendix 5.A.) Verify the upper companion form state space model for the speech signal given in (5.231) and (5.232) using the methods of Section 1.1.6.

5.27. (This problem is for persons who have studied the material in Appendix 5.B.) Give a simple proof of Corollary 5.1. The result should follow quite readily from the orthogonality principle.

APPENDIX

5.A Proof of Theorem 5.1

Prerequisite Chapter 1 Reading: *State space concepts in Section 1.1.6.*

When $M = I$, postmultiplying (5.11) by $s(n)$ and applying the $\mathcal{L}$ operator yields

$$\mathbf{r}_s = \mathbf{R}_s\mathbf{a} + \Theta_0\mathbf{r}_{e's}, \tag{5.229}$$

where all notation has been defined above except $\mathbf{r}_{e's}$, which is an obvious extension of our conventions. Substituting (5.229) into (5.24), we have

$$\hat{\mathbf{a}} = \mathbf{a} + \Theta_0\mathbf{R}_s^{-1}\mathbf{r}_{e's} \Rightarrow \tilde{\mathbf{a}} \stackrel{\text{def}}{=} \hat{\mathbf{a}} - \mathbf{a} = \Theta_0\mathbf{R}_s^{-1}\mathbf{r}_{e's}, \tag{5.230}$$

where, $\tilde{\mathbf{a}}$ denotes the error vector. The sufficiency argument is completed by showing that $\mathbf{r}_{e's}$ is the zero vector if $e(n)$ is orthogonal. A convenient framework within which to demonstrate this fact is the upper companion form state space model of the speech production model.

The speech production system has an upper companion form model (see Section 1.1.6)

$$\mathbf{s}(n+1) = \mathbf{A}\mathbf{s}(n) + \mathbf{c}e'(n) \tag{5.231}$$

and

$$s(n) = \mathbf{b}^T\mathbf{s}(n) + de'(n), \tag{5.232}$$

where

$$\mathbf{A} = \begin{bmatrix} a(1) & a(2) & \cdots & a(M) \\ & & & 0 \\ & \mathbf{I}_{(M-1)\times(M-1)} & & 0 \\ & & & \vdots \\ & & & 0 \end{bmatrix} \tag{5.233}$$

and $\mathbf{c} = [\Theta_0 \quad 0 \quad 0 \quad 0 \quad \cdots \quad 0]$, $\mathbf{b}^T$ is equivalent to the top row of $\mathbf{A}$, and $d = \Theta_0$.

Now using this formulation, it is not difficult to show that

$$\mathbf{s}(n) = \sum_{q=0}^{n-q_0-1} \mathbf{A}^q \Theta_0 \mathbf{b} e'(n-q-1) + \mathbf{A}^{n-q_0} s(q_0) \tag{5.234}$$

for all $n > q_0$ in which q_0 is some time for which $e'(q_0) = 0$. Postmultiplying by $e'(n)$ and applying the $\mathcal{L}$ operator yields

$$\mathbf{r}_{e's} = \mathcal{L}\left\{ e'(n) \sum_{q=0}^{n-q_0-1} \mathbf{A}^q \Theta_0 \mathbf{b} e'(n-q-1) \right\} + \mathcal{L}\{\mathbf{A}^{n-q_0} e'(n) s(q_0)\}. \tag{5.235}$$

Now letting $q_0 \to -\infty$ causes $\mathbf{A}^{n-q_0}$ to exponentially approach zero for a stable system [e.g., see (Chen, 1984)], resulting in

$$\mathbf{r}_{e's} = \sum_{q=0}^{\infty} \mathbf{A}^q \Theta_0 \mathbf{b} r_{e'}(q+1). \tag{5.236}$$

The right side is $\mathbf{0}$ when $r_{e'}(\eta) = C\delta(\eta)$ for any constant, C, and it is easy to show that $r_{e'}(\eta) = r_e(\eta)$. This proves sufficiency.

[Necessity of $e(n)$ uncorrelated.] Rewrite (5.236),

$$\begin{aligned} \mathbf{r}_{e's} &= \begin{bmatrix} \mathbf{b} & \mathbf{Ab} & \mathbf{A}^2\mathbf{b} & \cdots \end{bmatrix} \begin{bmatrix} r_e(1) & r_e(2) & \cdots \end{bmatrix}^T \\ &= \begin{bmatrix} \mathbf{C} & \mathbf{A}^M\mathbf{b} & \mathbf{A}^{M+1}\mathbf{b} & \cdots \end{bmatrix} \begin{bmatrix} r_e(1) & r_e(2) & \cdots \end{bmatrix}^T. \end{aligned} \tag{5.237}$$

Now the first M columns of the matrix in (5.237) constitute the *controllability matrix*, $\mathbf{C}$, for the state space system which produces the speech. It is well known that, for the upper companion form model under consideration, $\mathbf{C}$ has full rank, M, and therefore $\mathbf{r}_{e's}$, which is of dimension M, can only equal the zero vector if

$$r_e(\eta) = 0, \qquad \eta = 1, 2, \ldots. \tag{5.238}$$

Further, $r_e(0) \neq 0$ unless $e(n) = 0 \; \forall n$ (a contradiction, since this causes $\mathbf{R}_s$ to be singular), so it is concluded that $\mathbf{r}_{e's} = \mathbf{0}$ only if $r_e(\eta) = C\delta(\eta)$.

(Necessity of $M = I$.) We assume that M cannot be greater than I, since the order of the true system can always be extended by adding parameters equal to zero. If $M < I$, it is obvious that $\hat{\mathbf{a}} \neq \mathbf{a}$, since the vectors are of different dimension.

5.B The Orthogonality Principle

We give a more general statement of the *orthogonality principle* (OP) of linear minimum MSE estimation.

Consider the diagram in Fig. 5.23. Let each sequence be a realization of a WSS stochastic process. We use customary notation; for example, $y(n)$ is a realization of random process $\underline{y}$ with random variables $\underline{y}(n)$. Let us first ignore $x_2(n)$ and $x_1(n)$. The filter represented by impulse response $h(n)$ is to be designed to estimate[43] $y(n)$,

$$\hat{y}(n) = \sum_{i=n_1}^{n_2} h(i)q(n-i), \tag{5.239}$$

where n_1 may be $-\infty$, and n_2 may be $+\infty$. The optimization criterion is that the MSE, $\mathcal{E}\{\underline{e}^2(n)\}$, be minimized. The unique filter that meets this criterion is called $h^\dagger(n)$.

THEOREM 5.6 (ORTHOGONALITY PRINCIPLE) *A linear filter $h(n)$, constrained to be zero except on $n \in [n_1, n_2]$, is the unique minimum MSE filter, $h^\dagger(n)$, iff any random variable in the corresponding error sequence is orthogonal to the random variables in the input used in its computation,*

$$\mathcal{E}\{\underline{e}(n)\underline{q}(n-\eta)\} = 0, \qquad \eta \in [n_1, n_2]. \tag{5.240}$$

A proof of this important theorem can be found in many sources [e.g., (Papoulis, 1984)].

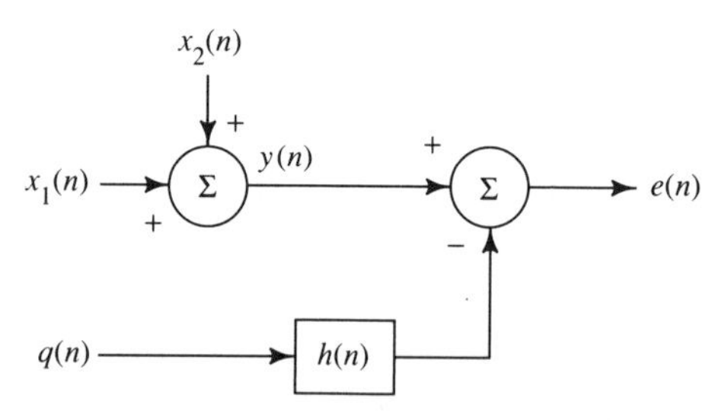

FIGURE 5.23. Diagram of the general linear minimum MSE estimation problem.

[43] The discussion is easily modified to accommodate the prediction of $y(n)$ [meaning that at time n an estimate of $y(n+k), k > 0$ is desired], or the smoothing of $y(n)$ [meaning that at time n an estimate of $y(n-k), k > 0$ is desired].

Note that the OP yields an immediate solution for the optimal filter: Relation (5.240) implies that

$$\mathcal{E}\{[(\underline{y}(n) - \hat{\underline{y}}(n)]\underline{q}(n-\eta)\} = 0, \qquad \eta \in [n_1, n_2]$$

$$\Rightarrow \mathcal{E}\left\{\left[\underline{y}(n) - \sum_{i=n_1}^{n_2} h(i)\underline{q}(n-i)\right]\underline{q}(n-\eta)\right\} = 0, \qquad \eta \in [n_1, n_2]$$

$$\Rightarrow r_{\underline{y}\underline{q}}(\eta) - \sum_{i=n_1}^{n_2} h(i) r_{\underline{q}}(\eta - i) = 0, \qquad \eta \in [n_1, n_2]. \tag{5.241}$$

The last line in (5.241) comprises a set of $n_2 - n_1 + 1$ equations (*normal equations*) that can be solved for the $n_2 - n_1 + 1$ points of the sequence $h^{\dagger}(n)$. The relationship of this result to our work with the speech prediction problem should be apparent.

Let us now bring the sequences $x_2(n)$ and $x_1(n)$ into the problem. With reference to Fig. 5.23, we note that there are occasions in which it is desired to estimate $x_1(n)$, but we have access to only $y(n) = x_1(n) + x_2(n)$. The OP immediately allows us to prove a useful result:

COROLLARY 5.1 *If the random processes $\underline{q}$ and $\underline{x}_2$ are orthogonal, then the MSE filter for estimating $x_1(n)$ is identical to that for estimating $y(n)$.*

The proof is left to the reader as Problem 5.27.

We will encounter a very important application of this result in Chapter 8. In that instance, $q(n)$ and $x_1(n)$ will correspond to noise sequences that are highly correlated, and $x_2(n)$ to a speech sequence that is uncorrelated with either noise signal. In terms of Fig. 5.23, we would like to estimate $x_1(n)$ from $q(n)$ in order to subtract it from $y(n)$. In these terms $e(n)$ can be interpreted as an estimate of $x_2(n)$, the speech sequence. We encounter a problem with this approach, since we do not have access to the sequence $x_1(n)$. It seems, therefore, at first glance, impossible to design the minimum MSE to estimate it. However, we also know (because of Corollary 5.1) that the filter will be identical to that produced by trying to estimate $y(n)$ from $q(n)$. From this point of view, we *do* have sufficient information to design the desired filter. It is interesting that when the problem is approached in this way, we discover that the estimate of the speech is also a signal of minimum mean square (power).

CHAPTER 6

Cepstral Analysis

Reading Notes: DFT spectral analysis (Section 1.1.4) and related phase concepts (Section 1.2.7) play important roles in this material.

6.1 Introduction

According to our usual model, speech is composed of an excitation sequence convolved with the impulse response of the vocal system model. We have access only to the output, yet often find it desirable to eliminate one of the components so that the other may be examined, coded, modeled, or used in a recognition algorithm.

The elimination of one of two combined signals is, in general, a very difficult problem. However, engineers know a great deal about this type of problem when the signals are combined *linearly.* It is worth dwelling on the linear case momentarily because much of the motivation for the techniques to be developed for the more difficult problem will be found in this linear case. Engineers have powerful tools for analyzing signals composed of added (and multiplicatively scaled) components. Indeed, our entire notion of a "frequency domain" is based upon a linear operation (Fourier transform) on signals made up of linearly combined pieces. Suppose, for example, that a low-frequency signal, $x_1(n)$, is corrupted by the addition of high-frequency noise, $w(n)$:

$$x(n) = x_1(n) + w(n). \tag{6.1}$$

Because the Fourier transform is a linear operator, we know that a simple magnitude spectrum of $x(n)$ will allow us to examine the component sequences individually because they occur in different parts of the frequency spectrum. The *spectrum* is the representation of the signal with which we can assess the "separation" of the component parts and perhaps derive needed information about the components. Further, the representations of the component signals are combined linearly in the spectrum.

If it were the objective to assess some properties of the noise, for example, we might be able to derive the desired information from the spectrum and have no need to go any further. If the objective, however, were to remove the noise from the signal, we would presumably design a lowpass filter to remove the undesirable high-frequency component.[1] From a frequency domain point of view, this would consist of construct-

[1]Note that even if the signal and noise were not clearly separable in the spectrum, a filter to remove the noise according to some optimality criterion would still probably be designed using spectral arguments [e.g., Gardner, (1990, Ch. 13)].

ing a system that would remove the unwanted high-frequency spectral energy. The result would then be transformed back into the time domain. Each of the operations taken to produce this filtered result is a linear one, so that the overall operation, say $\mathcal{J}$, is linear. Only because $x_1(n)$ and $w(n)$ are combined linearly can we be confident that putting the signal $x(n)$ into the filter will produce only the low-frequency part $x_1(n)$, that is,

$$\mathcal{J}\{x(n)\} = \mathcal{J}\{x_1(n) + w(n)\} = \mathcal{J}\{x_1(n)\} + \mathcal{J}\{w(n)\} \approx x_1(n). \tag{6.2}$$

If the components were combined in some other way (e.g., convolution), we would generally have no clear idea of the filter's effects on the component parts,

$$\mathcal{J}\{x(n)\} = \mathcal{J}\{x_1(n) * w(n)\} = \text{effects on } x_1(n), w(n)? \tag{6.3}$$

The latter situation above is the case with the speech problem we address. "Cepstral" analysis is motivated by, and is designed for, problems centered on *voiced* speech. According to our speech production model, voiced speech is composed of a convolved combination of the excitation sequence, with the vocal system impulse response,

$$s(n) = e(n) * \theta(n). \tag{6.4}$$

Because the individual parts are not combined linearly, the customary linear techniques provide no apparent help. Like the spectrum, the "cepstrum" will represent a transformation on the speech signal with (ideally) two important properties:

1. The representatives of the component signals will be separated in the cepstrum.
2. The representatives of the component signals will be *linearly combined* in the cepstrum.

If it is the purpose to assess some properties of the component signals, the cepstrum itself might be sufficient to provide the needed information. If it is the purpose to eliminate one of the component signals, then we must press on as we did in the lowpass filtering problem. However, since the representatives of the component signals are linearly combined, we will be able to use *linear filters* to remove undesired cepstral components. The remaining component must then be subjected to the inverse of the transformation that produced the cepstrum in the first place. We will find that the overall operation, say $\mathcal{H}$, will follow a sort of "generalized superposition" principle. For convolution,

$$\mathcal{H}\{s(n)\} = \mathcal{H}\{e(n) * \theta(n)\} = \mathcal{H}\{e(n)\} * \mathcal{H}\{\theta(n)\}. \tag{6.5}$$

If the objective were to eliminate $e(n)$, for example, we would want $\mathcal{H}\{e(n)\} \approx \delta(n)$ and $\mathcal{H}\{\theta(n)\} \approx \theta(n)$. We will come to know these systems that follow such a generalized rule as "homomorphic systems."

Our primary goal in this chapter is to learn to work with the "cepstrum" of a voiced speech signal. As can be inferred from the remarks above, cepstral analysis is a special case within a general class of methods collectively known as "homomorphic" signal processing. Historically, the discovery and use of the cepstrum by Bogert, Healy, and Tukey (1963) and Noll (1967) predate the formulation of the more general homomorphic signal processing approach by Oppenheim (1967, 1969); Schafer (1968); Oppenheim and Schafer (1968); Oppenheim, Schafer, and Stockham (1968); Stockham (1968); and Schafer and Rabiner (1970). In fact, the early cepstrum, which has various similar definitions, turns out to be a simplified version of the one derived as one case of homomorphic processing. The cepstrum derived from homomorphic processing is usually called the *complex cepstrum* (CC) (even though the version of it used is ordinarily *real*), while the Bogert–Tukey–Healy cepstrum is usually just called the "cepstrum." To avoid confusion, we shall refer to the latter as the *real cepstrum* (RC). The definition of the RC used in our work will make it equivalent to the even part of the CC on the region over which the RC is defined. In fact, there are various definitions of the RC, but all are equivalent to the real part of the CC within a scale factor. The reader is encouraged to glance at Table 6.1 for a preview of the nomenclature and notation.

The basic difference between the RC and the CC is that the early cepstrum discards phase information about the signal while the homomorphic cepstrum retains it. Although the CC is more appealing in its formulation, and although the preservation of phase bestows certain properties that are lost with the RC, the CC is often difficult to use in practice and it is the earlier version that is employed most widely in speech analysis and recognition. In fact, however, one of the most important applications of cepstral analysis in contemporary speech processing is the representation of an LP model by cepstral parameters. In this case, the signal parameterized is minimum phase, a condition under which the RC and CC are essentially equivalent. Unless the reader intends to use the cepstrum in phase-sensitive applications (e.g., vocoders), he or she may wish to study the RC carefully, the CC casually, and return to the details of the complex case as needed.

TABLE 6.1. Cepstrum Notation and Terminology Used Throughout the Chapter.

Name	Notation for Signal $x(n)$	Relationship
Complex cepstrum (CC)	$\gamma_x(n)$	
Real cepstrum (RC)	$c_x(n)$	$c_x(n) = \gamma_{x,\text{even}}(n)$
Short-term complex cepstrum (stCC) frame ending at m	$\gamma_x(n;m)$	
Short-term real cepstrum (stRC) frame ending at m	$c_x(n;m)$	$c_x(n;m) = \gamma_{x,\text{even}}(n;m)$

In this chapter, therefore, we shall focus initially upon the RC with some reliance upon future developments concerning the CC to support some of the more formal arguments in our discussions. In keeping with the comments above, the material is presented in such a way that the reader can get a good working knowledge of cepstral analysis of speech using the RC, without a thorough study of the CC. For those wishing a more formal and deeper understanding of cepstral analysis, a complete treatment of the CC is found in Section 6.3. Clearly, anyone who intends to use the cepstrum in practice will ultimately want to master the concepts in Section 6.3, but this is not necessary in the first pass through the material, nor is it essential for a basic understanding of the use of the cepstrum in later material.

Also consistent with the comments above is the fact that most of the formal understanding of the weaknesses of cepstral techniques as they pertain to speech processing is based upon theoretical and experimental work with the CC. Since we have tried to structure this chapter so that the reader can study only the RC if desired, this creates a dilemma. To include this material in the context of the CC study would deprive some readers of some important material. To do otherwise requires that less formality be used in the discussion. We have chosen to sacrifice the formality and describe these findings in a fairly qualitative way in the last section of the chapter. References to the key papers will provide the interested reader the opportunity to further pursue these issues.

6.2 "Real" Cepstrum

6.2.1 Long-Term Real Cepstrum

Definitions and General Concepts

As usual, we will find it convenient to begin our study with a long-term view of the cepstrum, avoiding some of the details of short-term analysis. Once we have laid down the foundations, the transition to the short-term computation will be quick and simple.

The *real cepstrum* (RC) of a speech sequence $s(n)$ is defined as

$$c_s(n) = \mathcal{F}^{-1}\{\log|\mathcal{F}\{s(n)\}|\} = \frac{1}{2\pi}\int_{-\pi}^{\pi} \log|S(\omega)|\, e^{j\omega n}\, d\omega, \tag{6.6}$$

in which $\mathcal{F}\{\cdot\}$ denotes the DTFT. Ordinarily, the natural or base 10 logarithm is used in this computation, but in principle any base can be used. We will assume the natural log throughout. Note carefully that the RC is an even sequence on n, since its DTFT, namely $C_s(\omega) = \log|S(\omega)|$, is real and even. The computation of the RC is shown in block-diagram form in Fig. 6.1.

We have noted above that the cepstrum will be best understood if we focus on voiced speech. Since we are therefore dealing with power (peri-

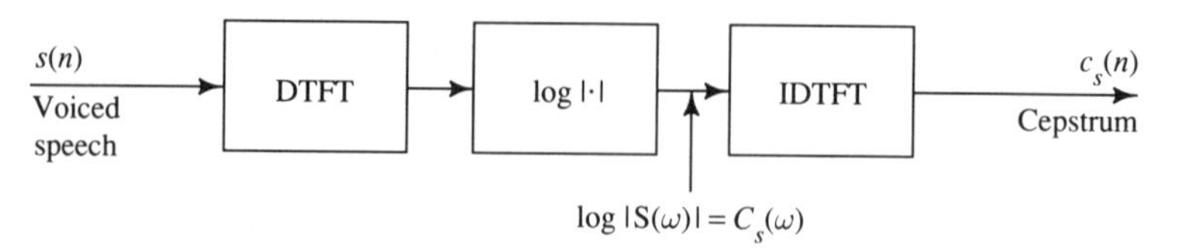

FIGURE 6.1. Computation of the RC.

odic) signals here, the reader might have detected a potential problem in this "long-term" definition. The DTFT of the signal $s(n)$ does not theoretically exist. However, we can make use of the "engineering DTFT" for power signals, which was reviewed near (1.30). In this case "$S(\omega)$" is taken to mean

$$S(\omega) = 2\pi \sum_{k=-\infty}^{\infty} D(k)\delta_a\left(\omega - k\frac{2\pi}{P}\right), \tag{6.7}$$

in which $\delta_a(\cdot)$ is the analog impulse function, the $D(k)$ are the DFS coefficients for the sequence $s(n)$, and P is its period (pitch period). We will verify momentarily that

$$D(k) = \frac{1}{P}\Theta\left(k\frac{2\pi}{P}\right), \tag{6.8}$$

where $\Theta(\omega)$ is the Fourier transfer function of the vocal system model. The use of this "power type" DTFT would seem to be the proper theoretical course of action in this case, but, as sometimes happens, the presence of impulse functions would lead us into some significant theoretical difficulties. Since the point of long-term analysis is to elucidate and motivate, rather than to confuse, we make the following small adjustment, which will circumvent the problems with impulse functions. The source of the impulses is the periodic excitation sequence $e(n)$ for which the DTFT is

$$E(\omega) = \frac{2\pi}{P} \sum_{k=-\infty}^{\infty} \delta_a\left(\omega - k\frac{2\pi}{P}\right). \tag{6.9}$$

Since

$$S(\omega) = E(\omega)\Theta(\omega), \tag{6.10}$$

we see that (6.7) with coefficients given by (6.8) is indeed the correct DTFT for[2] $S(\omega)$. Let us imagine that we apply a very long rectangular window to the speech, say

$$w(n) = \begin{cases} 1, & -\frac{L}{2} < n \leq \frac{L}{2} \\ 0, & \text{other } n, \end{cases} \tag{6.11}$$

for some arbitrarily large, even integer, L. The DTFT of a hypothetical long-term speech sequence and the DTFT of its windowed version are illustrated in Fig. 6.2(a) and (b), respectively. In the windowed spectrum, the amplitude of the spectral peak at the kth harmonic frequency is $[(L+1)/P]|\Theta(2\pi k/P)|$ for any k. This becomes apparent upon convolving the spectrum in Fig. 6.2(c) with the hypothetical long-term spectrum shown in part (a). Therefore, the amplitudes of the spectral peaks grow proportionately with L. It is also clear from the convolution that the width of each "main lobe" in the spectrum (one centered on each harmonic frequency) decreases with L. Further, as L becomes large, it can be shown that the energy (area under the squared magnitude spectrum) associated with a small neighborhood around the kth harmonic frequency becomes $[2\pi(L+1)/P^2]|\Theta(2\pi k/\mathrm{P})|^2$. Therefore, for large L, the DTFT $\tilde{S}(\omega)$ approximates a set of impulses at the harmonics, and the power associated with the kth harmonic is approximately equal to $2\pi|\tilde{S}(2\pi k/P)|^2/(\mathrm{L}+1)$. We can therefore imagine that we have replaced a truly long-term analysis with one that approaches the exact case as closely as desired. This will allow us to work with DTFTs and avoid some cumbersome details.

Now a moment's thought will reveal that, with the possible exception of a few points near the ends of the window, the same windowed speech sequence will be obtained if we window the *excitation* sequence, then drive the vocal system model with the windowed $e(n)$.[3] The model we will use, therefore, for the generation of the long-term speech signal is one of having driven the vocal system with a windowed version of the excitation sequence. The window may be assumed to be arbitrarily large and is only a device to remove the troublesome impulse functions. We therefore will not employ any special notation to denote that there is a long window involved, since the "windowed" versions may be made to arbitrarily closely approximate the truly long-term versions.[4] Rather, we will continue to call the excitation $e(n)$ and the speech $s(n)$. The "tilde" notation in Fig. 6.2 and in the previous paragraph was only a temporary device to distinguish the two spectra.

Let us now return to the computation of the RC. The set of operations leading to $c_s(n)$ is depicted in Fig. 6.1. The first two operations in the figure can be interpreted as an attempt to transform the signal $s(n)$ into a "linear" domain in the sense that the two parts of the signal which are

[2]Recall that a consistent theory of impulse functions requires that $f(\beta)\delta(\alpha)$ be defined to mean $f(\alpha)\delta(\alpha)$ for a continuous function, f, of β.

[3]We will later see that this ability to assume that the window has been applied to the excitation, rather than the speech itself, is critical to some of the short-term spectral arguments.

[4]The reader may feel that we have already entered the realm of short-term processing because a window has been applied. It should be emphasized here that the essence of long-term processing is the assumption that the signal is stationary for all time. This assumption is in effect here, and the window is just a conceptual device to remove some confusing mathematics.

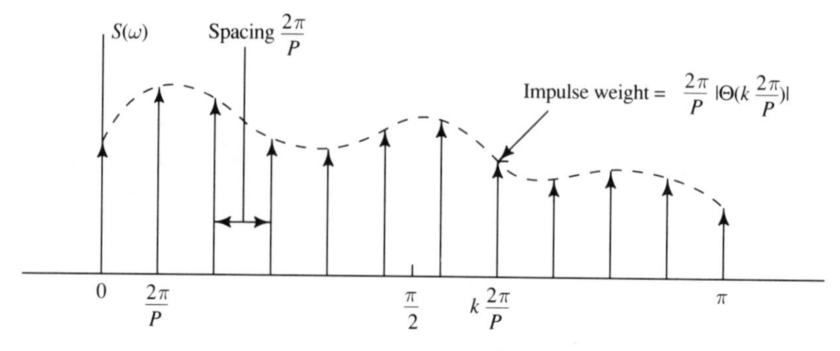

(a) Hypothetical long-term speech spectrum (shown with unrealistically long pitch period, P).

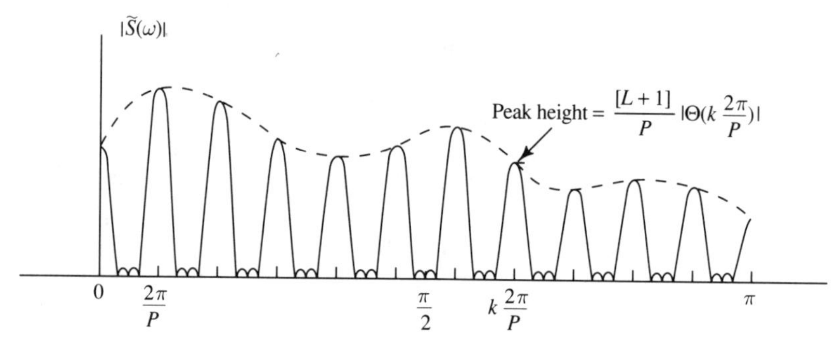

(b) Spectrum of same hypothetical signal after application of a "very long" (large L) rectangular window.

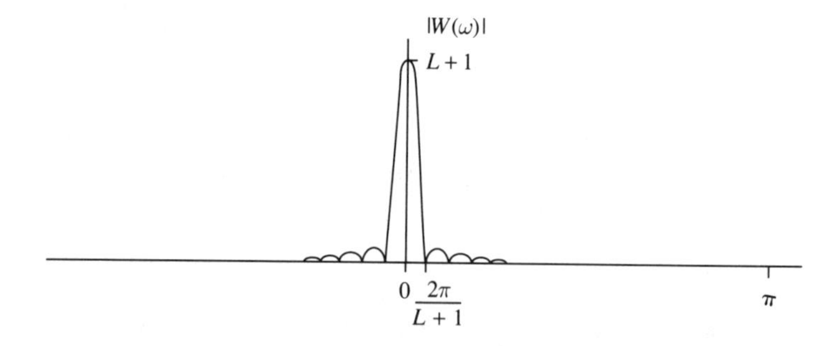

(c) Spectrum of window used in creating part (b).

FIGURE 6.2. (a) DTFT of a hypothetical long-term voiced speech signal. (b) DTFT of the windowed version of the signal in (a). (c) The magnitude spectrum of the rectangular window that is convolved with the long-term speech spectrum to produce the windowed version in (b).

convolved in the unaltered signal now have representatives that are added in the new domain. Let us denote the operation corresponding to the first two boxes by Q_*^{real}, the subscript $*$ denoting that the operator is used to unravel convolution, and the superscript *real* denoting that the log of a real number is taken (this will not be the case with the CC). We have then,

$$\begin{aligned} C_s(\omega) = Q_*^{\text{real}}\{s(n)\} &= \log|S(\omega)| \\ &= \log|E(\omega)\Theta(\omega)| \\ &= \log|E(\omega)| + \log|\Theta(\omega)| \\ &= C_e(\omega) + C_\theta(\omega). \end{aligned} \tag{6.12}$$

Now that we are in a "linear domain," we are free to apply linear techniques to the new "signal," $C_s(\omega)$. In particular, we might wish to apply Fourier analysis to view the "frequency domain" properties of the new "signal." Upon recognizing that $C_s(\omega)$ is a periodic function, the appropriate thing to do is to compute a line spectrum for the "signal," that is, to compute Fourier series coefficients for the "harmonics" of the "signal." These would take the form (see Problem 6.1)

$$\alpha_n = \frac{1}{2\pi}\int_{-\pi}^{\pi} C_s(\omega)e^{-j\omega n}\,d\omega. \tag{6.13}$$

Now, according to the definition,

$$c_s(n) = \frac{1}{2\pi}\int_{-\pi}^{\pi} C_s(\omega)e^{j\omega n}\,d\omega, \tag{6.14}$$

but $C_s(\omega)$ is a real, even function of ω, so that (6.13) and (6.14) produce equivalent results.[5] Therefore, the RC can be interpreted as the Fourier series "line spectrum" of the "signal" $C_s(\omega)$.

We have been careful to put quotation marks around terms that are being used in an unusual manner. The "signal" that is being transformed into the "frequency domain" is, in fact, already in what we consider the frequency domain in usual engineering terms. Therefore, the "new" frequency domain was dubbed the "quefrency domain" by Tukey in the earlier work on the RC, and the "cepstrum" was so named because it plays the role of a "spectrum" in the quefrency domain. The index of the RC (which actually is a discrete-time axis) is called the "quefrency axis." The "harmonic frequencies" of $C_s(\omega)$, which are actually time indices of the cepstrum, are called *rahmonics*. There is an entire vocabulary of amusing

[5]In fact, since $C_s(\omega)$ is even, the IDTFT in the definition can be replaced by a cosine transform,

$$c_s(n) = \frac{1}{2\pi}\int_{-\pi}^{\pi} C_s(\omega)\cos(\omega n)\,d\omega = \frac{1}{\pi}\int_{0}^{\pi} C_s(\omega)\cos(\omega n)\,d\omega.$$

terms that go with cepstral analysis [see the title of the paper by Bogert et al. (1963)], but only the terms *cepstrum* and *quefrency* (and *liftering*, which we discuss below) are widely used by speech processing engineers. We will say more about the origins of the cepstrum below.

The purpose of the RC is to resolve the two convolved pieces of the speech, $e(n)$ and $\theta(n)$, into two additive components, and then to analyze those components with spectral (cepstral) analysis. Clearly, from (6.12),

$$c_s(n) = c_e(n) + c_\theta(n), \tag{6.15}$$

and if the nonzero parts of $c_e(n)$ and $c_\theta(n)$ occupy different parts of the quefrency axis, we should be able to examine them as separate entities—which we were unable to do when they were convolved in $s(n)$.

Intuitive Approach to the Cepstrum (Historical Notes)

We have essentially captured above the thought process that led the inventors of the RC to its discovery. However, their early work is based more on engineering intuition than on mathematical formality. The clause above "if the nonzero parts of $c_e(n)$ and $c_\theta(n)$ occupy different parts of the quefrency axis, we should be able to examine them as separate entities" is the key to the early thinking. Noll (1967) was the first to apply and extend the cepstral notions to speech (pitch detection), but his ideas were based on earlier work on seismic signals by Bogert et al. (1963). The history is quite interesting and is related in Noll's paper in which Tukey is credited with the invention of the cepstrum vocabulary set. With this due credit to the earlier researchers, we will discuss the historical developments from Noll's speech viewpoint, taking some liberties to pose Noll's ideas in our digital signal processing framework.

Viewing the speech spectrum, $|S(\omega)|$, as consisting of a "quickly varying" part, $|E(\omega)|$, and a "slowly varying" part, $|\Theta(\omega)|$ (see Fig. 6.3), Noll simply took the logarithm

$$\log|S(\omega)| = \log|E(\omega)| + \log|\Theta(\omega)| \tag{6.16}$$

to get these two multiplied pieces into additive ones. The reason for wanting to get two additive pieces was to apply a linear operator, the Fourier transform, knowing that the transform would operate individually on two additive components, and further, knowing precisely what the Fourier transform would do to one quickly varying piece and one slowly varying piece. Noll was thinking, as we did above, of the two signals as "time" signals, one "high-frequency," one "low-frequency," knowing that the "high-frequency" signal would manifest itself at big values of frequency in the "frequency domain," and that the "low-frequency" signal would appear at smaller values of "frequency." Since, in fact, the two "signals" were already in the frequency domain, the new vocabulary was employed including the word *quefrency* to describe "frequencies" in this new "frequency domain." Some of the vocabulary is illustrated in Fig. 6.3.

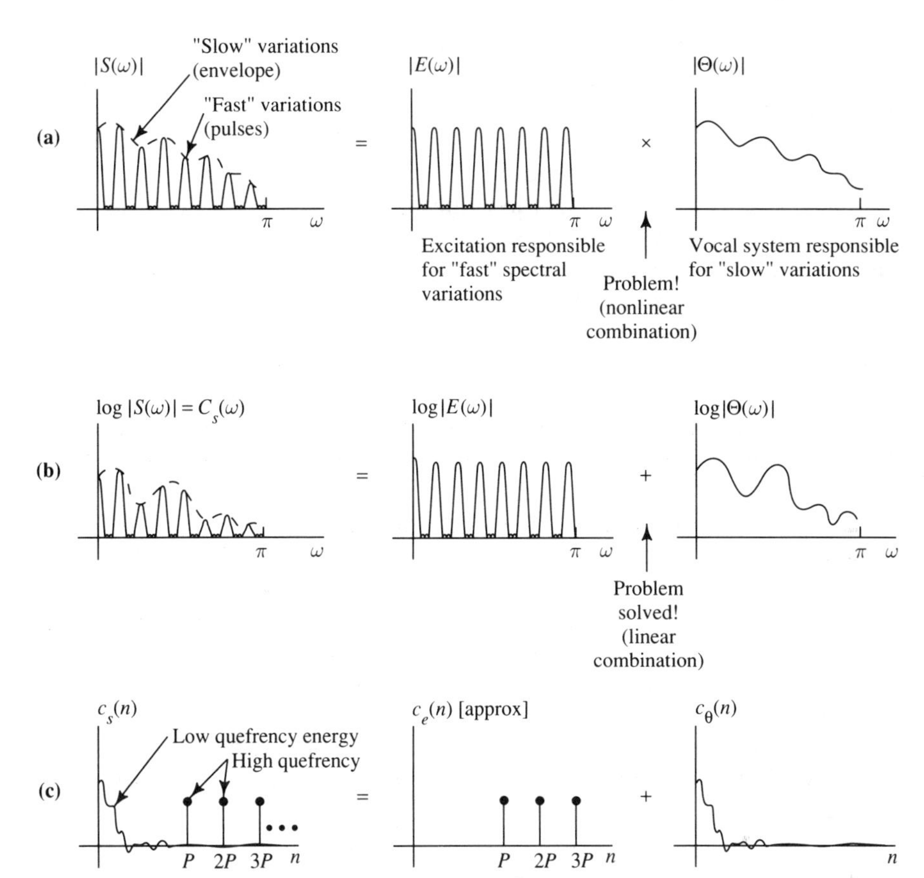

FIGURE 6.3. The motivation behind the RC, and some of the accompanying vocabulary. (a) In the speech magnitude spectrum, $|S(\omega)|$, two components can be identified: a "slowly varying" part (envelope) due to the speech system, $|\Theta(\omega)|$, and a "quickly varying" part due to the excitation, $|E(\omega)|$. These components are combined by addition. Their time domain counterparts, $\theta(n)$ and $e(n)$, are convolved. (b) Once the logarithm of the spectral magnitude is taken, the two convolved signal components, $\theta(n)$ and $e(n)$, have additive correlates in the new "signal," $C_s(\omega)$. The former corresponds to a slowly varying ("low-quefrency") component of $C_s(\omega)$, and the latter to a quickly varying ("high-quefrency") component. (c) When the IDTFT is taken, the slowly varying part yields a "cepstral" component at low quefrencies (smaller values on the time axis), and the component with fast variations results in a "cepstral" component at high quefrencies (larger values on the time axis). The low-quefrency part of the cepstrum therefore represents an approximation to the cepstrum of the vocal system impulse response, $c_\theta(n)$, and the high-quefrency part corresponds to the cepstrum of the excitation, $c_e(n)$.

Before proceeding, we should note a potential source of problems that is often overlooked in discussions of the cepstrum. This "intuitive" level is a good place to flag this issue.[6] If the reader had a sense of uneasiness upon arrival at (6.16), it is probably because of the lack of intuition that most of us have concerning nonlinear operations. Indeed, one moment we had a high-quefrency and a low-quefrency "signal," and the next moment, through the good workings of the logarithm, we see them added together. We must pause, however, and ask: Are these two additive components still high and low quefrency even though we have subjected them to this nonlinear transformation? An affirmative answer is, of course, critical to the proper operation of the technique. The correct answer is, however, that the components do not always remain sufficiently well separated in quefrency even though they may appear to be so before the logarithm is taken. The predictability of this adverse phenomenon is not high because of the nonlinearity.

We will return to the important "separation after the log" issue in the last section of the chapter. In the meantime, the reader is urged to keep in mind that cepstral techniques, from the intuitive conceptual level through the most formal developments, are based on a few tenuous assumptions that can cause unpredictable outcomes. We will point out another hazardous assumption in the discussion of short-term cepstral processing. These assumptions are unavoidable because of the complexity of these interesting techniques. Nevertheless, they are there, and the reader should keep this in mind as he or she applies the methods in practice.

Liftering

Before moving on to a more practical discussion about the RC, let us explore the possibility of doing linear filtering in our new linear domain to select one or the other of the well-separated parts. Since $c_e(n)$ and $c_\theta(n)$ are well-separated in the quefrency domain, we can, in principle, use the RC to eliminate, say, $C_e(\omega)$ from $C_s(\omega)$. This process is called *liftering* (a play on *filtering*), and in this case we use a "low-time lifter" as in Fig. 6.4 (analogous to a lowpass filter in the usual frequency domain). The output of this process in the quefrency domain is an RC, say $c_y(n) \approx c_\theta(n)$. If we now wish to use $c_\theta(n)$ to obtain an estimate of $\theta(n)$ neatly separated from $s(n)$, we need to get out of the quefrency domain, then invert the Q_*^{real} operation. Leaving the quefrency domain is easy—we simply apply a DTFT to the RC. Note that this process results in an estimate of $\log|\Theta(\omega)|$ devoid of any excitation components. This process is called *cepstral smoothing* of the vocal system spectrum, an issue we will study in more detail later. The IDTFT, low-time liftering, DTFT operations comprise a linear filter operation, say $\mathcal{J}$, in the "new" linear domain that was created by the Q_*^{real} operation. When we attempt to return to the

[6]These comments apply equally well to the CC.

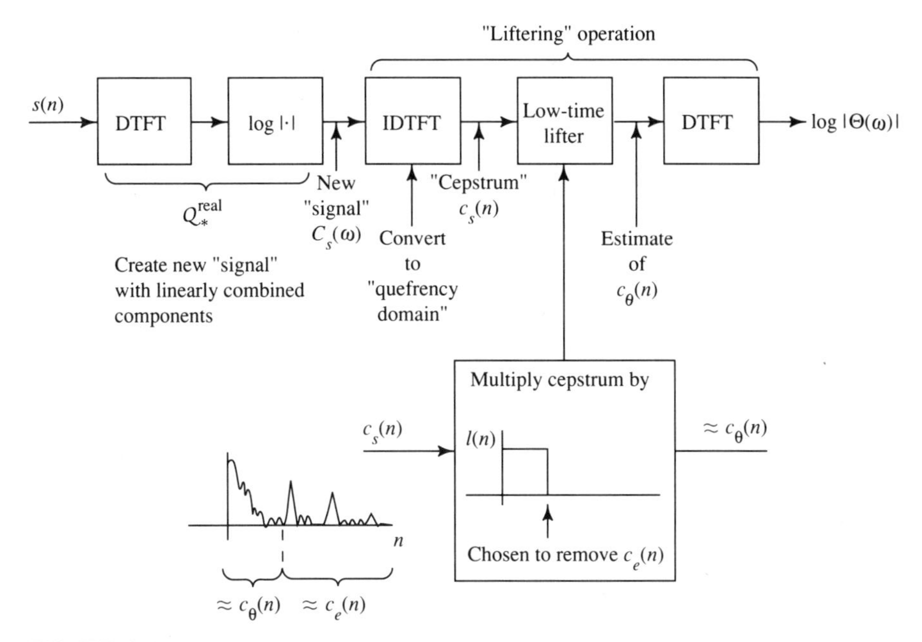

FIGURE 6.4. "Low-time liftering" to remove $C_e(\omega)$ from $C_s(\omega)$. The convolution to linear domain transformation is labeled Q_*^{real}. The IDTFT, low-time windowing, DTFT operations comprise a linear filtering problem in the "new" linear domain. Note that $c_s(n)$ is an even-symmetric sequence and, consequently, $l(n)$ should be even only the positive time parts are illustrated in the figure.

original time domain (invert Q_*^{real}), we encounter a problem. In applying the Q_*^{real} operation, we have discarded the phase spectrum of the original signal. Not surprisingly, therefore, the Q_*^{real} operation in not uniquely invertible, since the phase information is irretrievable. Based on the form of Q_*^{real}, however, the inverse

$$\left[Q_*^{real}\right]^{-1}\{\cdot\} \stackrel{\text{def}}{=} \mathcal{F}^{-1}\{\exp[\cdot]\} \tag{6.17}$$

might be proposed, but it is not difficult to see that this will return a time domain signal with a zero-phase characteristic (which is necessarily noncausal). A second candidate for $\left[Q_*^{real}\right]^{-1}$, which yields a minimum-phase characteristic, will emerge when we discuss the complex cepstrum (also see Problem 6.2).

We conclude from this discussion that liftering is a useful and meaningful process with the RC for obtaining an estimate of the log spectrum of either of the separated components. That is, we can apply a useful linear operation to the RC. However, if the objective is to return to the original time domain with an estimate of the separated signal, the RC will fail, because its "linearizing" operation is not invertible. To complete this task, we would need a phase-preserving linearizing operation. We will find such an operation when we discuss the complex cepstrum. However, we should not leave the reader with the impression that returning to the original time domain with an estimate of the separated signal is an important problem in speech processing. Indeed, although the CC is theoretically interesting, in most practical applications we can get the information we need from the RC directly, or by a liftering operation on it. This is another case among many that we have already discussed, in which magnitude spectral information is sufficient to deduce a useful set of features from speech.

6.2.2 Short-Term Real Cepstrum

In practice we will want to compute the RC on short terms of speech. Formally, the RC is derived from $s(n)$ according to (6.6), which we rewrite with an explicit expression for $S(\omega)$,

$$c_s(n) = \frac{1}{2\pi}\int_{-\pi}^{\pi}\left\{\log\left|\sum_{l=-\infty}^{\infty} s(l)e^{-j\omega l}\right|\right\}e^{j\omega n}\,d\omega \tag{6.18}$$

for all n. By a simple generalization of Construction Principle 1 in Chapter 4, we derive the following *short-term "real" cepstral* (stRC) estimator, $c_s(n;m)$, for the N-length frame of speech ending at time $m, f(n;m) = s(n)w(m-n)$:

$$\begin{aligned} c_s(n;m) &= \frac{1}{2\pi}\int_{-\pi}^{\pi}\left\{\log\left|\sum_{l=-\infty}^{\infty} f(l;m)e^{-j\omega l}\right|\right\}e^{j\omega n}\,d\omega \\ &= \frac{1}{2\pi}\int_{-\pi}^{\pi}\left\{\log\left|\sum_{l=m-N+1}^{m} f(l;m)e^{-j\omega l}\right|\right\}e^{j\omega n}\,d\omega \end{aligned} \tag{6.19}$$

for $n = 0, 1, \ldots$. This amounts to nothing more than using $f(n;m)$ as input to our "usual" algorithm represented by Fig. 6.1. Accordingly, we should replace the DTFT in Fig. 6.1 by the stDTFT. In Fig. 6.5, we have redrawn Fig. 6.1 for this case to feature the fact that the short-term Fourier transform appears in the intermediate computations. This will be important momentarily.

Before proceeding, it is important to reemphasize that the RC, in this case the stRC, makes use of magnitude spectral information only, and disregards all phase information. In particular, the information about the

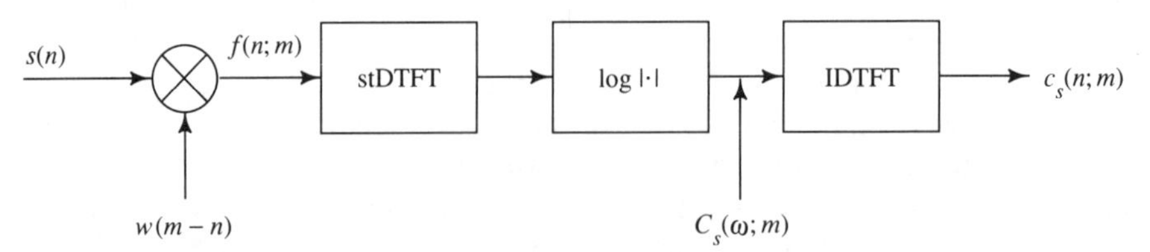

FIGURE 6.5. Computation of the stRC using the DTFT.

phase delay of the window (i.e., the time m) is lost in this analysis. This point will be significant below.

Of course, in practice we will also want to use DFTs in place of the "pencil and paper" DTFTs. This is shown in Fig. 6.6. It is important to note that we have intentionally used the short-term DFT (stDFT) at the input and the "usual" IDFT at the output (review Section 4.3.5). The use of the stDFT at the input is purely for formal convenience, so that the frame need not be reindexed in time. Although the stDFT preserves the phase corresponding to the delay of the window, this information is immediately lost when the spectral magnitude is computed. Along with this phase suppression comes an effective shifting of the input frame down to its zero-phase correspondent, which will be centered on time $n = 0$. The use of the IDFT, rather than the stIDFT, at the output ensures that all computations will produce cepstra on the range $n = 0, \ldots, N-1$, regardless of the window time m. To use the stIDFT would restore the phase information about the window delay and produce a cepstrum on the range $n = m - N + 1, \ldots, m$, but no other phase information about the signal would, of course, be restored. It is conventional to use the stRC on the low time range, so the IDFT is employed. In using the result, we are ordinarily aware of the "m" value associated with a particular computation and we learn to interpret cepstra based at time zero.

By comparing Figs. 6.5 and 6.6 and using our knowledge of the relationship between the IDTFT and IDFT, we see that

$$\tilde{c}_s(n;m) = \begin{cases} \sum_{q=-\infty}^{\infty} c_s(n+qN;m), & n = 0, 1, \ldots, N-1 \\ 0, & \text{other } n \end{cases}, \tag{6.20}$$

where $\tilde{c}_s(n;m)$ denotes the result using DFTs, so that $\tilde{c}_s(n;m)$ is a periodic, aliased, version of the "true" quantity we seek, $c_s(n;m)$. Since the stRC will be of infinite duration (a fact that can be inferred from the material in Section 6.3.1), some aliasing is inevitable. In order that the aliasing not be too severe, one can append zeros onto the speech frame $f(n;m)$ and compute the stDFT, IDFT pair based on more points. It is often necessary to use a significant number of zeros (to extend the effective frame length to, say 512 or 1024 points) to avoid aliasing.[7]

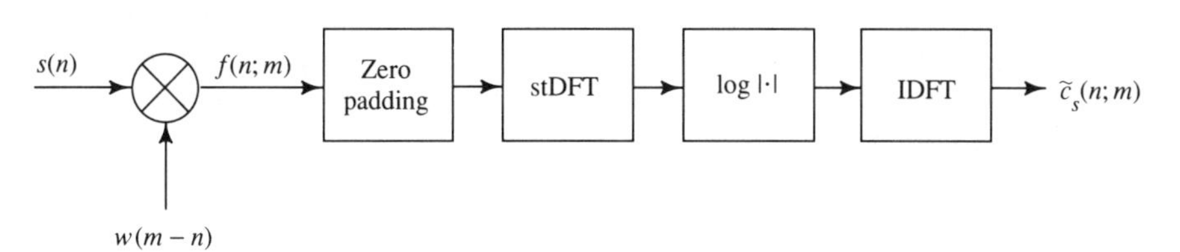

FIGURE 6.6. Computation of the stRC using the DFT.

[7]An alternative method for alias-free computation of the CC has been proposed by Tribolet (1967).

In future discussions, we will not continue to distinguish the RC computed using the DFT from that using the DTFT by the "tilde" notation. In fact, most of our study is based on the more abstract DTFT, although practical implementation requires the use of the DFT. The reader is urged to keep in mind the potential aliasing problem when using the practical computation.

6.2.3 Example Applications of the stRC to Speech Analysis and Recognition

In this section we will study in detail the applications of cepstral analysis to pitch and formant estimation, and then preview some interesting ideas that we will see used repeatedly in the speech recognition chapters of Part V, and also in the material on vocoding in Chapter 7. The third application is among the most important uses of the cepstrum in modern speech processing. In the first two of these applications, the comments made at the end of the "Historical Notes" in Section 6.2.1 are particularly relevant. The reader might wish to review those few paragraphs before reading these applications.

Pitch Estimation

Cepstral analysis offers yet another way to estimate the important fundamental frequency parameter. We have given heuristic arguments above that the RC should do a good job of separating the excitation and vocal system components of the RC in the quefrency domain. In this section we will formalize that argument and show how this information leads to a convenient pitch detection technique.

The voiced speech signal is modeled in the time domain as the convolution of the pulse train excitation and the impulse response of the vocal system,

$$s(n) = e(n) * \theta(n). \tag{6.21}$$

We consider a frame of speech ending at time m:

$$f_s(n;m) = s(n)w(m-n) = \left[e(n) * \theta(n)\right]w(m-n). \tag{6.22}$$

Here we have labeled the frame with the subscript s to indicate that it is derived from the sequence $s(n)$. This is because a second frame will appear in the analysis momentarily. A key to the development is to show that the window can be moved inside the convolution under practical conditions to obtain

$$\begin{aligned} f_s(n;m) &\approx e(n)w(m-n) * \theta(n) \\ &= f_e(n;m) * \theta(n), \end{aligned} \tag{6.23}$$

where $f_e(n;m)$ indicates the frame of $e(n)$ determined by the window ending at time m. An argument in support of (6.23) was first suggested by Oppenheim and Schafer (1968). We elaborate a bit upon their work here. According to (5.2) and (6.21),

$$s(n) = \sum_{q=-\infty}^{\infty} \theta(n - qP), \tag{6.24}$$

where P is the pitch period (in norm-sec) or the spacing of unit samples in $e(n)$. If the time constants of $\theta(n)$ are short compared with pitch period,[8] that is, if

$$\theta(n) \approx 0, \qquad n \geq P' \tag{6.25}$$

for some $P' \leq P$, then, on any period, say $n = rP, rP + 1, \dots, (r+1)P - 1$,

$$s(n) \approx \theta(n - rP). \tag{6.26}$$

Now consider the effect of the window over this period,

$$\begin{aligned} f_s(n; m) &\approx \theta(n - rP)\, w(m - n), \\ n &= rP, rP + 1, \dots, (r+1)P - 1. \end{aligned} \tag{6.27}$$

If the window is long and tapers slowly compared with P', the duration of $\theta(n)$, so that

$$w(m - n) \approx w(m - rP), \qquad n = rP, rP + 1, \dots, rP + P' - 1, \tag{6.28}$$

then we can write

$$\begin{aligned} f_s(n; m) &\approx \theta(n - rP)\, w(m - rP), \\ n &= rP, rP + 1, \dots, (r+1)P - 1. \end{aligned} \tag{6.29}$$

[Note that we have used the fact that $\theta(n - rP) \approx 0$ on the range $n = rP + P', \dots, (r+1)P - 1$ in expanding the time range between (6.28) and (6.29).] Putting the periods back together, we have

$$\begin{aligned} f_s(n; m) &\approx \sum_{q=-\infty}^{\infty} \theta(n - qP)\, w(m - qP) \\ &= e(n)\, w(m - n) * \theta(n), \end{aligned} \tag{6.30}$$

which is the desired result (6.23).

Before proceeding, a very important observation must be made. All of the developments henceforth will rely on a good approximation in (6.30). A good approximation, in turn, depends upon a sufficiently long window in selecting the frame. Indeed, for all but relatively simple applications of short-term cepstral techniques, performance has been found to depend significantly upon the type, length, and placement of the window used to

[8] We have seen this idea before in discussing LP analysis in Section 5.2.1 [cf. (5.39)].

create the frame. The approximation in (6.30) creates one pressure to keep the time domain window as long as possible. However, this is not always consistent with good performance, because there are other considerations that affect the choice of window. We will return to this point in Section 6.4. The reader should keep in mind this potential problem area in the foundation upon which we are building the theory.

Now using the key approximation in (6.23), we seek the stRC of the speech. We do so by noting that the RC of $f_s(n;m)$ will be the sum of the RCs of its two convolved components, $f_e(n;m)$ and $\theta(n)$.[9] First we seek the stRC of the excitation [which is equivalent to the RC of the *frame* $f_e(n;m)$]. If there are Q periods of $e(n)$ inside the window used to create $f_e(n;m)$, and if those periods correspond to indices $q = q_0, q_0 + 1, \ldots, q_0 + Q - 1$, then

$$f_e(n;m) = \sum_{q=q_0}^{q_0+Q-1} w(m - qP)\delta(n - qP). \tag{6.31}$$

Therefore,

$$\begin{aligned} E(\omega;m) &= \sum_{n=m-N+1}^{m} f_e(n;m)e^{-j\omega n} \\ &= \sum_{n=m-N+1}^{m} \sum_{q=q_0}^{q_0+Q-1} w(m - qP)\delta(n - qP)e^{-j\omega n} \\ &= \sum_{q=q_0}^{q_0+Q-1} w(m - qP)e^{-j\omega qP}. \end{aligned} \tag{6.32}$$

Now let us define the sequence

$$\tilde{w}(q) = \begin{cases} w(m - qP), & q = q_0, \ldots, q_0 + Q - 1 \\ 0, & \text{other } q \end{cases}. \tag{6.33}$$

Putting (6.33) into the last line in (6.32), we see immediately that

$$E(\omega;m) = \tilde{W}(\omega P), \tag{6.34}$$

in which $\tilde{W}(\omega)$ denotes the DTFT of the sequence $\tilde{w}(q)$. From (6.34) we conclude that $E(\omega;m)$, hence $\log|E(\omega;m)|$, will be a periodic function of ω with period $2\pi/P$. The precise form of the log spectrum will depend on the numbers $w(m - [q_0 + Q - 1]P), \ldots, w(m - q_0P)$, but the most important point to note is the periodicity of the short-term log spectrum, $\log|E(\omega;m)|$.

[9]There is a subtlety here with which we must be careful. In effect we are seeking the short-term RC of the excitation, and the long-term RC of the impulse response. The stRC of $e(n)$, however, is equivalent to the long-term RC of the frame $f_e(n;m)$ (a common feature extraction occurrence), so that the statement is correct as written.

Let us now examine the final step, IDTFT, in the computation of the stRC for $e(n)$, which takes the form

$$c_e(n;m) = \frac{1}{2\pi}\int_{-\pi}^{\pi} \log\left|E(\omega;m)\right| e^{j\omega n}\, d\omega. \tag{6.35}$$

As in the long-term case [see the discussion surrounding (6.14)], $c_e(n;m)$ can be interpreted as Fourier (cosine) series coefficients for the periodic function $\log\left|E(\omega;m)\right|$. Since the period of $\log\left|E(\omega;m)\right|$ is $2\pi/P$, "harmonics" (or "rahmonics," as Tukey calls them, since they are in the quefrency domain) occur at times $n = i2\pi/(2\pi/P) = iP, i = 0, 1, \ldots$. Therefore, (6.35) will produce a result of the form

$$c_e(n;m) = \sum_{i=-\infty}^{\infty} \alpha_i \delta(n - iP), \tag{6.36}$$

where the numbers α_i depend on $\log\left|E(\omega;m)\right|$, which, in turn, depends on the numbers $w(m-[q_0+Q-1]P), \ldots, w(m-q_0P)$. The important point is that $c_e(n;m)$ is a weighted train of discrete-time impulses spaced at P, the pitch period (see Fig. 6.7). Note also that the weightings on the impulses will decrease as $1/i$ due to the property of the Fourier series coefficients. For future reference, we note in particular that

$$\alpha_0 = c_e(0;m) = \frac{1}{2\pi}\int_{-\pi}^{\pi} \log\left|E(\omega;m)\right| d\omega. \tag{6.37}$$

Let us now focus on the vocal system impulse response and discuss its RC. Recall that because of the theoretical discussion leading to (6.30), we will be seeking the long-term RC for $\theta(n)$ here. First we note, similarly to (6.37), that

$$c_\theta(0) = \frac{1}{2\pi}\int_{-\pi}^{\pi} \log\left|\Theta(\omega)\right| d\omega. \tag{6.38}$$

For $n > 0$, we employ a result that can be inferred from our later work [see (6.85)]: The CC of $\theta(n)$, $\gamma_\theta(n)$, will be bounded on the positive n axis as

$$\left|\gamma_\theta(n)\right| \le (Q^{\text{in}} + P^{\text{in}})\frac{\beta^n}{n} + \frac{Q^{\text{out}}}{n}, \tag{6.39}$$

where Q^{in} (P^{in}) represents the number of zeros (poles) of $\Theta(z)$ inside the unit circle, Q^{out} is the number of zeros outside the unit circle, and β is the maximum magnitude of the set of all zeros and poles inside the unit circle. We can argue that the RC will also be inside this envelope as follows. The RC is equivalent to the even part of the CC. Suppose that for some $n > 0$, the RC exceeds the envelope, for example, in the negative direction. Then the odd part of the CC must be sufficiently large so that the CC at n will be inside the envelope. However, at $-n$, where the RC

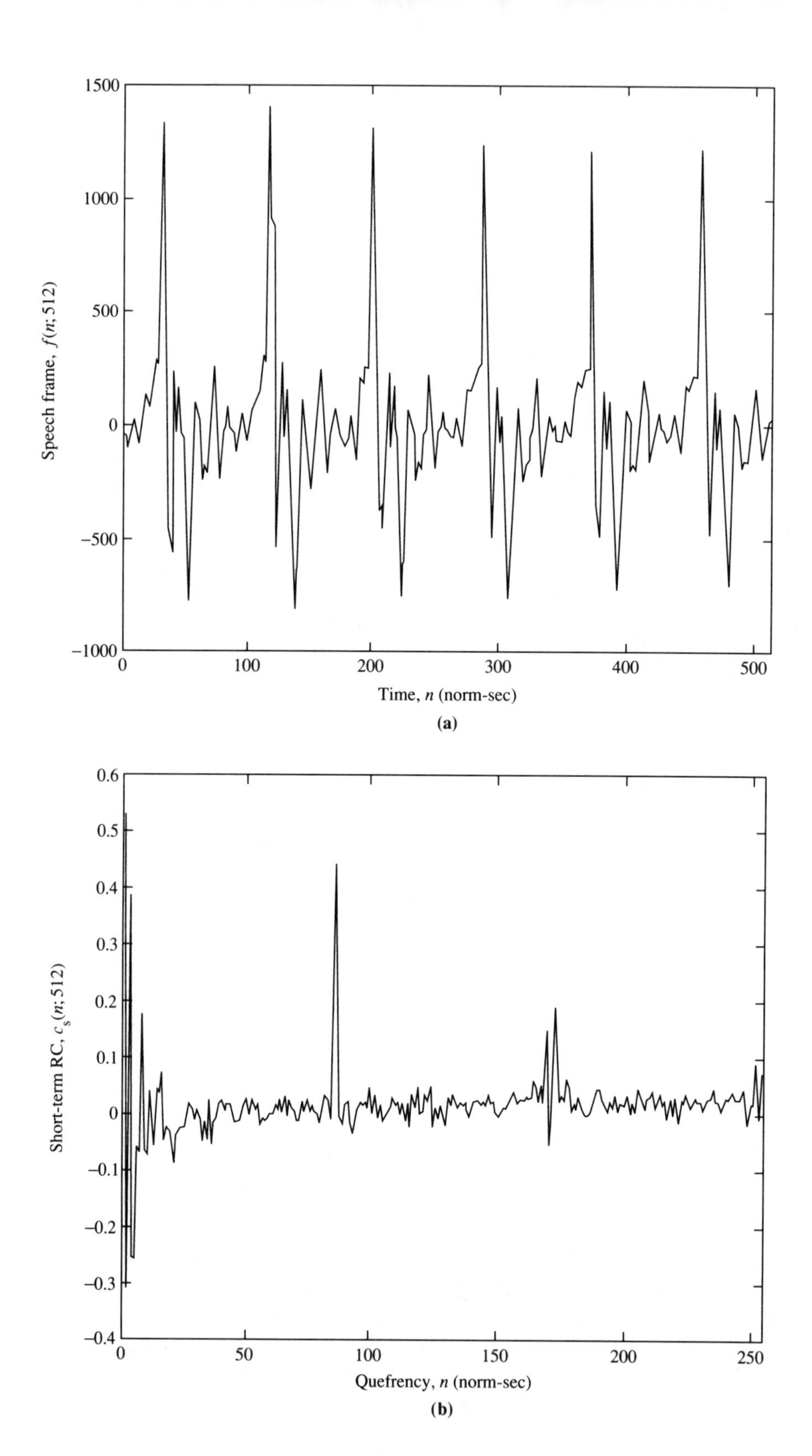
1500
1000
500
0
−500
−1000
Speech frame, $f(n; 512)$
0
100
200
300
400
500
Time, n (norm-sec)
(a)
0.6
0.5
0.4
0.3
0.2
0.1
0
−0.1
−0.2
−0.3
−0.4
Short-term RC, $c_s(n; 512)$
0
50
100
150
200
250
Quefrency, n (norm-sec)
(b)

will also be "too large" in the negative direction, the odd part will also be negative and the sum (CC) will be outside the envelope. From this we conclude that the RC also decays as $1/n$. Clearly, this envelope will usually decay quickly with respect to typical values of the pitch period P.

Putting these results together, we have that

$$c_s(n;m) = c_e(n;m) + c_\theta(n) \tag{6.40}$$

from which we conclude that $c_e(n;m)$ will appear in $c_s(n;m)$ as a pulse train added to the RC of $\theta(n)$. Further, $c_\theta(n)$ usually decays very quickly with respect to P, as is apparent in Fig. 6.7, so that

$$c_s(n;m) \approx \begin{cases} c_e(0;m) + c_\theta(0), & n = 0 \\ c_\theta(n), & 0 < n < P, \\ c_e(n;m) \text{ (weighted pulse train)}, & n \geq P \end{cases} \tag{6.41}$$

where $c_e(0;m)$ and $c_\theta(0)$ are given in (6.37) and (6.38), respectively. To estimate the pitch period, P, on the frame ending at time m, we need only to locate the initial peak in $c_s(n;m)$, which is generally well separated from the vocal system characteristic. The stRC is usually computed at a rate of about once every 10–20 msec, with a typical window size of $N = 256$ points. For a sample rate of 10 kHz, this implies the use of overlapping windows of length 256, which are moved 100–200 points each estimate.

Figure 6.8 shows an example of cepstral pitch detection taken from the work of Noll (1967). The plots on the left (in each case—male and female) are a series of short-term log spectra derived by using an $N = 400$-point Hamming window moved 100 points each computation (so that the frames might be indexed $m = 100, 200, \ldots$). The sampling rate on the original data is $F_s = 10$ kHz. The plots on the right in each case show the stRC according to Noll's definition, which differs from the stRC we have been discussing above by a square and a scale factor (see Problem 6.6). The lack of a clearly defined pitch peak in the first seven frames in the male case, for example, is probably an indication of unvoiced speech. The shape of the spectra for these frames provide some confirmation of this hypothesis. The remaining frames show a strong pitch pulse (of changing fundamental frequency) that could easily be extracted by a peak-picking algorithm.

The example above could lead the reader to the mistaken impression that cepstral pitch detection is a simple problem, free of practical diffi-

FIGURE 6.7. (a) A frame of speech consisting of the phoneme /i/ selected using a rectangular window. (b) stRC of the frame. The voiced excitation manifests itself as a weighted train of discrete-time impulses in the stRC, while the vocal system impulse response is represented by the low-quefrency, quickly decaying portion of the stRC.

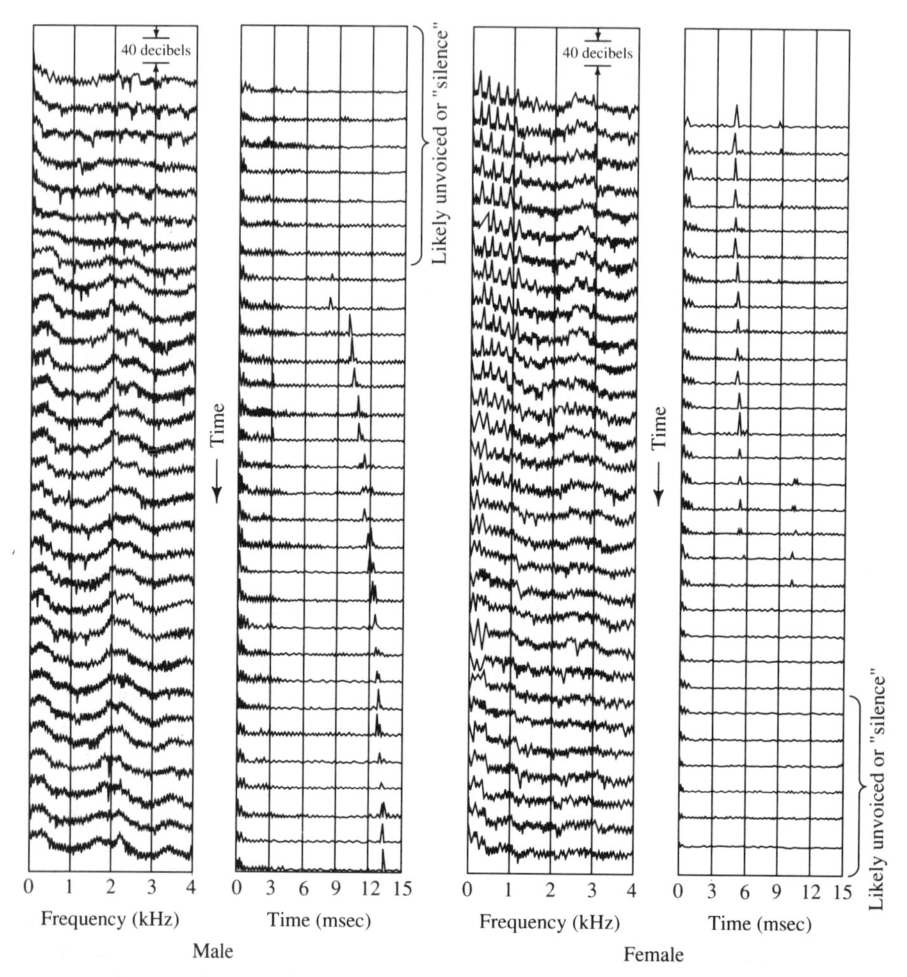

FIGURE 6.8. $\log|S(\omega;m)|$ and $c_s(n;m)$ for a series of m's for a male and a female talker in an experiment performed by Noll. (The definition of the stRC used by Noll differs from our definition by a square and a scale factor, but the general information provided is the same.) After Noll (1967).

culties. In fact, the height of the spectral peak depends upon many factors, notably the size, shape, and placement of the window creating the analysis frame (see Problem 6.3). As an extreme example, suppose that the pitch period is quite long and the window size is chosen so that only one period or less appears in a given frame. In this case $Q = 1$ in the discussion surrounding (6.31) and the reader is encouraged to review the discussion in this light to conclude that $c_e(n;m)$ will no longer consist of a pulse train. Another critical factor is the formant structure of the vocal system. If, for example, the vocal system is essentially a narrowband filter, the periodic component in the spectrum will be masked by the formant filter and no peaks will occur in the stRC. Before attempting the construction of a pitch detection algorithm based on the cepstrum, therefore, the reader is advised to peruse the literature on the subject (Noll, 1967; Rabiner, 1977; Schafer and Rabiner, 1970) to explore the various algorithmic enhancements that have been employed to overcome such difficulties.

Finally, let us recall what we can and cannot do with liftering of the stRC in this regard. (Recall the discussion surrounding Fig. 6.4.) Although there is nothing that prevents liftering of the stRC to remove $c_\theta(n)$ from $c_s(n;m)$, the remaining estimate of $c_e(n;m)$ cannot be used to recover an estimate of $f_e(n;m)$ back in the original time domain. It *can* be used to obtain an estimate of $\log|E(\omega;m)|$, however, but this is seldom the purpose of the analysis. Ordinarily, the purpose is the estimation of pitch for which the stRC itself is adequate.

Formant Estimation

In 1970, Schafer and Rabiner (1970) described the process of cepstral smoothing for formant analysis of voiced speech. Although their work is centered on the more formal CC, the same procedure can be achieved with the RC. In fact, we have already discussed the basic procedure of cepstral smoothing in Section 6.2.1. What we can now add to that discussion is the introduction of the more realistic short-term RC and a more formal discussion of the separability of the cepstral components. We emphasize that this technique will produce an estimate of $\Theta(\omega)$ (the total model transfer function), not just $H(\omega)$ (the vocal tract alone).

To review, to obtain an estimate of $\log|\Theta(\omega)|$, from the speech on a window ending at time m, we execute the following steps:

1. Compute the stRC of the speech $c_s(n;m)$ as above.
2. Multiply $c_s(n;m)$ by a "low-time" window, $l(n)$ to select $c_\theta(n)$:

$$c_\theta(n) \approx c_s(n;m)l(n). \tag{6.42}$$

   (Note that the lifter $l(n)$ should theoretically be an even function of n, or even symmetric about time $(N-1)/2$ if an N-point DFT is used in the next step.)

3. To get the estimate of $\log|\Theta(\omega)|$, DTFT (DFT) the estimate of $c_\theta(n)$.

This entire procedure is diagramed in Figs. 6.9 and 6.10. Figure 6.10 is the completion of the analysis begun in Fig. 6.7.

Recall that the term "cepstral smoothing" refers to the process of removing the high-quefrency effects of the excitation from the spectrum by this analysis. In the next section we will see the cepstrum used to effect further smoothing on a spectrum that is already smoothed by LP analysis.

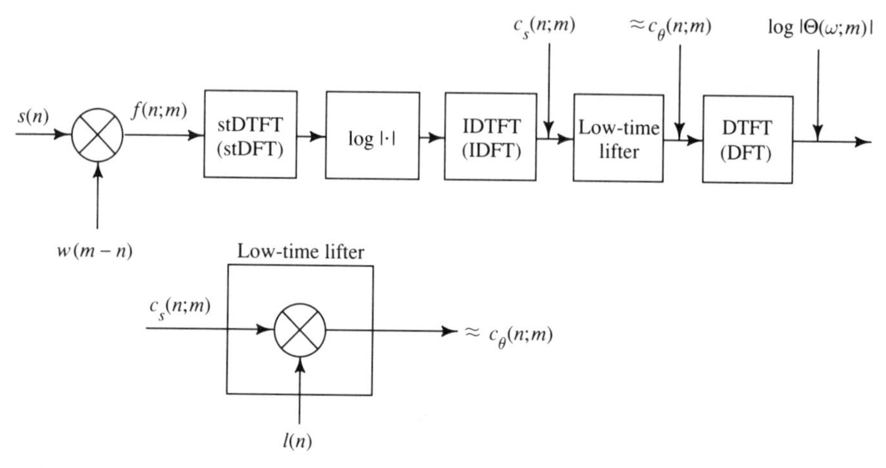

FIGURE 6.9. "Cepstral smoothing" using the stRC. Block diagram of the computations. Note that the processing is almost always terminated after achieving the estimate of $c_\theta(n)$ or $\log|\Theta(\omega)|$. Q_*^{real} is not uniquely defined, and it is therefore not possible to obtain a meaningful estimate of $\theta(n)$.

LP and Cepstral Analysis in Speech Recognition (LP to Cepstrum Conversion and Cepstral Distance Measures)

For many years LP analysis has been among the most popular methods for extracting spectral information from speech. Contributing to this popularity is the enormous amount of theoretical and applied research on the technique, which has resulted in very well-understood properties and many efficient and readily available algorithms. For speech recognition (as well as coding), the LP parameters are a very useful spectral representation of the speech because they represent a "smoothed" version of the spectrum that has been resolved from the *model* excitation.

However, LP analysis is not without drawbacks (Juang et al., 1987). We emphasized the word "model" in the above paragraph because, as we know from our study in Chapter 5, LP analysis does not resolve the vocal-tract characteristics from the glottal dynamics. Since these laryngeal characteristics vary from person to person, and even for within-person utterances of the same words, the LP parameters convey some information to a speech recognizer that degrades performance, particularly for a speaker-independent system.[10] Further, the all-pole constraint

FIGURE 6.10. The cepstral smoothing operation of Fig. 6.9 applied to the real data of Fig. 6.7. In Fig. 6.7 we find the illustration of the input frame for the vowel /i/ and the resulting stRC. (a) By applying a low-time lifter of duration 50 to the stRC, the estimate $c_\theta(n;512)$ results. (b) Computation of the DFT results in an estimate of $\log|\Theta(\omega;512)|$, at, of course, $\omega=2\pi k/M$, where M is the length of the DFT.

[10]A system that recognizes multiple speakers, all or some of whom might not have participated in training the system.

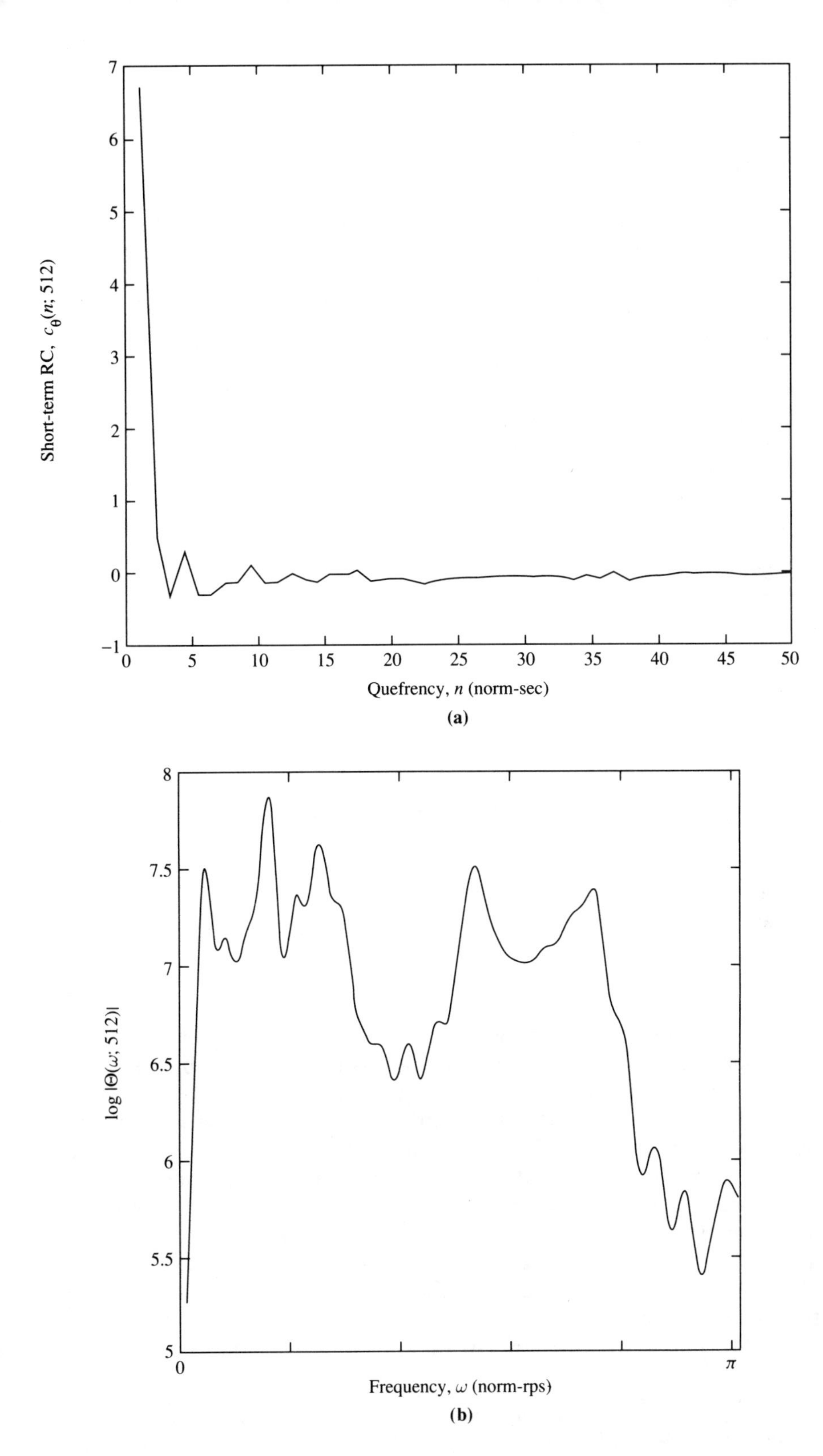
7
6
5
4
3
2
1
0
−1
Short-term RC, $c_\theta(n; 512)$
0
5
10
15
20
25
30
35
40
45
50
Quefrency, n (norm-sec)
(a)
8
7.5
7
6.5
6
5.5
5
log $|\Theta(\omega; 512)|$
0
π
Frequency, ω (norm-rps)
(b)

does not always permit accurate modeling of the spectrum and can lead to spurious effects such as formant splitting or poor modeling of spectral nulls.

In the 1980s, researchers began to improve upon the LP parameters with a cepstral technique. Much of the impetus for this conversion seems to have been a paper by Davis and Mermelstein (1980), which compared a number of parametric representations and found the cepstral method to outperform the "raw" LP parameters in monosyllable word recognition.[11] The method is based on the conversion of the short-term LP parameters to cepstral parameters upon which further smoothing and manipulation is performed.

For theoretical reasons, the conversion formula is more easily derived for converting the LP parameters to the (short-term) complex cepstrum (stCC), which we study in Section 6.3, and the stCC is usually used in this context. As we have noted above, however, there is very little difference between the stCC and the stRC for a minimum-phase signal, like the impulse response of the LP model $\hat{\theta}(n)$. In this case, the stCC, $\gamma_{\hat{\theta}}(n;m)$, is real, causal, and related to the stRC, $c_{\hat{\theta}}(n;m)$, as

$$\gamma_{\hat{\theta}}(n;m) = \begin{cases} c_{\hat{\theta}}(0;m) = \log \hat{\theta}(0) = \log \hat{\mathbf{\Theta}}_0(m), & n = 0 \\ 2c_{\hat{\theta}}(n;m), & n > 0 \\ 0, & 0 \end{cases} . \tag{6.43}$$

The recursion for converting LP to CC parameters is

$$\gamma_{\hat{\theta}}(n;m) = \begin{cases} \log \hat{\Theta}_0(m), & n = 0 \\ \hat{a}(n;m) + \displaystyle\sum_{k=1}^{n-1} \left(\frac{k}{n}\right) \gamma_{\hat{\theta}}(k;m)\hat{a}(n-k;m), & n > 0 \end{cases} , \tag{6.44}$$

in which $\hat{a}(n;m)$ is taken to be zero for $n \notin [1, M]$, and where M denotes the order of the LP model. The proof of this recursion will be the subject of Problem 6.7 after we have studied the CC. Using (6.43), (6.44) is easily modified to compute the stRC rather than the stCC; we will continue our discussion using the RC for consistency with the present material.

Note something very different from the cepstra used previously. In this case, rather than having computed the cepstrum of the speech, we have computed the cepstrum of the impulse response of the LP model. Therefore, we are computing the cepstrum of a sequence that has already been "smoothed" in the sense that the excitation has been removed. The purposes of doing so are generally twofold. First, it is possible to "fine tune" the smoothing operation with several manipulations on $c_{\hat{\theta}}(n;m)$.

[11]In fact, the most successful technique in this study was a cepstral technique, the mel-cepstrum, which is not based on LP analysis, but rather a filter bank spectral analysis. We will study the mel-cepstrum below.

This will be illustrated below. Second, a much more efficient distance computation between parameter sets can be used.

Let us explore the distance computation first, since this issue is central to the smoothing strategy. We denote the system function for an LP model derived on the frame ending at time m by $\hat{\Theta}(z;m)$, or, in terms of the DTFT, $\hat{\Theta}(\omega;m)$. For future reference, let us write*

$$\hat{\Theta}(\omega;m)=\hat{\Theta}_0\hat{A}^{-1}(\omega;m), \tag{6.45}$$

where $\hat{\Theta}_0$ is the estimated model gain, and $\hat{A}(\omega;m)$ represents the inverse filter. Now recall from definition (6.19) that the stRC, $c_{\hat{\theta}}(n;m)$ and $\log|\hat{\Theta}(\omega;m)|$ comprise a DTFT-IDTFT pair. Therefore, using Parseval's relation, we have that the sum of the squared cepstral coefficients is related to the model spectrum as follows,

$$\sum_{n=-\infty}^{\infty} c_{\hat{\theta}}^2(n;m)=\frac{1}{2\pi}\int_{-\pi}^{\pi}\left[\log|\hat{\Theta}(\omega;m)|\right]^2 d\omega. \tag{6.46}$$

If we wish to ignore the gain of the model [see (6.43)], we can omit $c_{\hat{\theta}}(0;m)$ and write

$$\begin{aligned}\sum_{n=1}^{\infty} c_{\hat{\theta}}^2(n;m)&=\frac{1}{4\pi}\int_{-\pi}^{\pi}\left[\log|\hat{A}^{-1}(\omega;m)|\right]^2 d\omega\\ &=-\frac{1}{4\pi}\int_{-\pi}^{\pi}\left[\log|\hat{A}(\omega;m)|\right]^2 d\omega.\end{aligned} \tag{6.47}$$

Accordingly, if we have the cepstral sequences for two LP models, say $\hat{\theta}_1(n;m)$ and $\hat{\theta}_2(n;m)$, then computing the sum of the squared differences yields the mean squared difference in the log spectra,

$$\begin{aligned}&\sum_{n=1}^{\infty}\left[c_{\hat{\theta}_1}(n;m)-c_{\hat{\theta}_2}(n;m)\right]^2\\ &=\frac{1}{2\pi}\int_{-\pi}^{\pi}\left[\log|\hat{A}_1^{-1}(\omega;m)-\hat{A}_2^{-1}(\omega;m)|\right]^2 d\omega.\end{aligned} \tag{6.48}$$

This result indicates that the Euclidean distance between two cepstral sequences is a reasonable measure of spectral similarity in the models. In practice, of course, only finite sequences of cepstral parameters, say $c_{\hat{\theta}_1}(1;m),\ldots,c_{\hat{\theta}_1}(L;m)$ (and similarly for the second model), would be used in the distance computation.[12]

The cepstral parameters may be interpreted as coefficients of a Fourier series expansion of the periodic log spectrum. Accordingly, they are based on a set of orthonormal functions (see Section 1.3.2) and, in a the-

[12]Recall also that these numbers would result from having used discrete frequency transforms, which can potentially induce some cepstral aliasing [see (6.20)]. We will assume that care is taken to prevent this problem.

*The discussion on pages 376–378 of the text, beginning with Eq. (6.45), includes the gain estimate $\hat{\Theta}_0(m)$, as part of the overall LP model $\hat{\Theta}(z;m)$. This was not the convention in Chapter 5.

oretical sense, the Euclidean metric is an appropriate and natural distance measure to use. Nevertheless, several researchers have tried weighted Euclidean distances to determine whether performance improvements would result. Furui (1981), Paliwal (1982), and Nakatsu et al. (1983) have explored the use of various weighting strategies with the Euclidean distance in speech recognition tasks, in each case achieving small improvements in performance. As Tohkura (1987) points out, however, "there is no clear reason why and how [weighting] works, and how to choose an optimal set of weights." Noting that the cepstral coefficients decrease with n (this will be demonstrated below), and that the variances decrease accordingly, Tohkura suggests the use of a diagonal weighting matrix which normalizes the variances of the individual parameters: Let $\underline{\mathbf{c}}_{\hat{\theta}}(m)$ be a random L-vector from which the cepstral vectors to be compared, say $\mathbf{c}_{\hat{\theta}_1}(m)$ and $\mathbf{c}_{\hat{\theta}_2}(m)$, are drawn, where

$$\mathbf{c}_{\hat{\theta}_i}(m) \stackrel{\text{def}}{=} \left[c_{\hat{\theta}_i}(1;m) \cdots c_{\hat{\theta}_i}(L;m)\right]^T. \tag{6.49}$$

In these terms, the Euclidean distance suggested above is

$$d_2[\mathbf{c}_{\hat{\theta}_1}(m), \mathbf{c}_{\hat{\theta}_2}(m)] = \sqrt{[\mathbf{c}_{\hat{\theta}_1}(m) - \mathbf{c}_{\hat{\theta}_2}(m)]^T[\mathbf{c}_{\hat{\theta}_1}(m) - \mathbf{c}_{\hat{\theta}_2}(m)]}. \tag{6.50}$$

Tohkura assumes that the off-diagonal terms of the covariance matrix, $\mathbf{C}_{\hat{\theta}}$, can be ignored so that

$$\mathbf{C}_{\hat{\theta}} = \mathbf{\Lambda}, \tag{6.51}$$

with $\mathbf{\Lambda}$ a diagonal matrix of the variances of the individual coefficients (see Section 1.3.2). The *weighted cepstral distance* is then defined as

$$d_{2w}[\mathbf{c}_{\hat{\theta}_1}(m), \mathbf{c}_{\hat{\theta}_2}(m)] = \sqrt{[\mathbf{c}_{\hat{\theta}_1}(m) - \mathbf{c}_{\hat{\theta}_2}(m)]^T\mathbf{\Lambda}^{-1}[\mathbf{c}_{\hat{\theta}_1}(m) - \mathbf{c}_{\hat{\theta}_2}(m)]}. \tag{6.52}$$

This can be recognized as a standard Euclidean distance weighting strategy as described in Section 1.3.1.

A similar procedure to the above has been tried by Paliwal (1982) in the recognition of vowel sounds. Here the actual covariance of the nth coefficient is replaced by simply n^{-2}. It was later shown by Juang et al. (1987) that under certain reasonable conditions, the cepstral coefficients (for $n > 0$) have zero mean and variance proportional to n^{-2}, thereby providing analytical support for Paliwal's method. The sequence $nc_x(n;m)$ is sometimes called the *root power sums* sequence [see, e.g., (Hanson and Wakita, 1987)], so Paliwal's weighted distance is called the *root power sums measure.*

Several aspects of the weighting strategy were investigated by Tohkura. In general, the weighting was responsible for a significant improvement with respect to unweighted cepstral coefficients, admitting recognition rates of 99% (increased from 95% and 96.5%) on two isolated digit utterance databases, for instance. Among the most significant findings is the fact that the weighting of the lower quefrency coefficients is much more significant than the higher ones. (Note that the effect of the weighting of low-quefrency coefficients is to deemphasize their significance, since their variances are large.) The likely explanation[13] (Juang et al., 1987) is that lower quefrencies correspond to slower changes in the model spectrum, which in turn are related to "spectral tilt" that arises due to the glottal dynamics. The glottal dynamics would generally be expected to interfere with correct recognition. Conversely, weighting of cepstral coefficients beyond about $N = 8$ (this would be approximately the same as the order of the LP analysis) was found to degrade performance. The likely explanation (Juang et al., 1987) is that by normalizing these very small coefficients, one is actually emphasizing features that are prone to numerical errors and artifacts of the computation.

In keeping with Tohkura's findings, which suggest "downweighting" at each end of the cepstral sequence, several lifters have been proposed which taper at both ends (Tohkura, 1987; Juang et al., 1987). The basic forms of these lifters are shown in Fig. 6.11. The analytical form of the raised sine lifter is

$$l(n) = 1 + \frac{L}{2} \sin\left(\frac{\pi n}{L}\right), \tag{6.53}$$

in which L is the length of the cepstral sequence used. Again the lifters should theoretically be made even. Only the positive time parts are shown in the figure.

$$l_1(n) = \begin{cases} 1, & n = 0, 1, \ldots, L. \\ 0, & \text{other } n \end{cases}$$

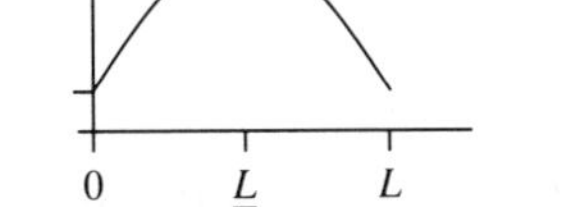

$$l_2(n) = \begin{cases} 1 + \frac{L}{2}\sin\left(\frac{\pi n}{L}\right), & n = 0, 1, \ldots, L. \\ 0, & \text{other } n \end{cases}$$

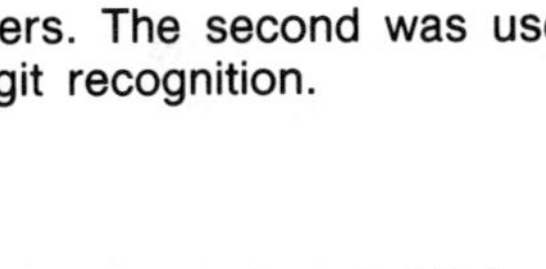

FIGURE 6.11. Some example low-time lifters. The second was used successfully by Juang et al. (1987) for digit recognition.

[13]Tohkura and Juang et al. were working contemporaneously at Bell Laboratories on the related research reported in the two papers cited here.

6.2.4 Other Forms and Variations on the stRC Parameters

The Mel-Cepstrum

Mel-Cepstrum Applied Directly to Speech. In the 1980s, the cepstrum began to supplant the direct use of the LP parameters as the premiere feature in the important "hidden Markov modeling" strategy because of two convenient enhancements that were found to improve recognition rates. The first is the ability to easily smooth the LP-based spectrum using the liftering and weighting processes described above. This process removes the inherent variability of the LP-based spectrum due to the excitation and apparently improves recognition performance (Davis and Mermelstein, 1980).

A second improvement over direct use of the LP parameters can be obtained by use of the so-called "mel-based cepstrum," or simply "mel-cepstrum." In order to explain the significance of this term, we need to briefly delve into the field of *psychoacoustics*, which studies human auditory perception. The discussion here is very superficial and the interested reader is referred to a textbook on the study of hearing for details (see Appendix 1.F).

First we need to understand the term "mel." A *mel* is a unit of measure of *perceived pitch or frequency* of a tone. It does not correspond linearly to the physical frequency of the tone, as the human auditory system apparently does not perceive pitch in this linear manner. The precise meaning of the mel scale becomes clear by examining the experiment by which it is derived. Stevens and Volkman (1940) arbitrarily chose the frequency 1000 Hz and designated this "1000 mels." Listeners were then asked to change the physical frequency until the pitch they perceived was twice the reference, then 10 times, and so on; and then half the reference, $\frac{1}{10}$, and so on. These pitches were labeled 2000 mels, 10,000 mels, and so on; and 500 mels, 100 mels, and so on. The investigators were then able to determine a mapping between the real frequency scale (Hz) and the perceived frequency scale (mels).[14] The mel scale is shown in Fig. 6.12. The mapping is approximately linear below 1 kHz and logarithmic above (Koenig, 1949), and such an approximation is usually used in speech recognition. Fant (1959), for example, suggests the approximation

$$F_{\text{mel}} = \frac{1000}{\log 2}\left[1 + \frac{F_{\text{Hz}}}{1000}\right], \tag{6.54}$$

in which F_{mel} (F_{Hz}) is the perceived (real) frequency in mels (Hz).

Drawing on this idea of a perceptual frequency scale, speech researchers began to investigate the benefits of using a frequency axis which was warped to correspond to the mel scale. The stRC is particularly well

[14]It is interesting to note that the pitch expressed in mels is roughly proportional to the number of nerve cells terminating on the basilar membrane of the inner ear, counting from the apical end to the point of maximal stimulation along the membrane (Stephens and Bate, 1966, p. 238).

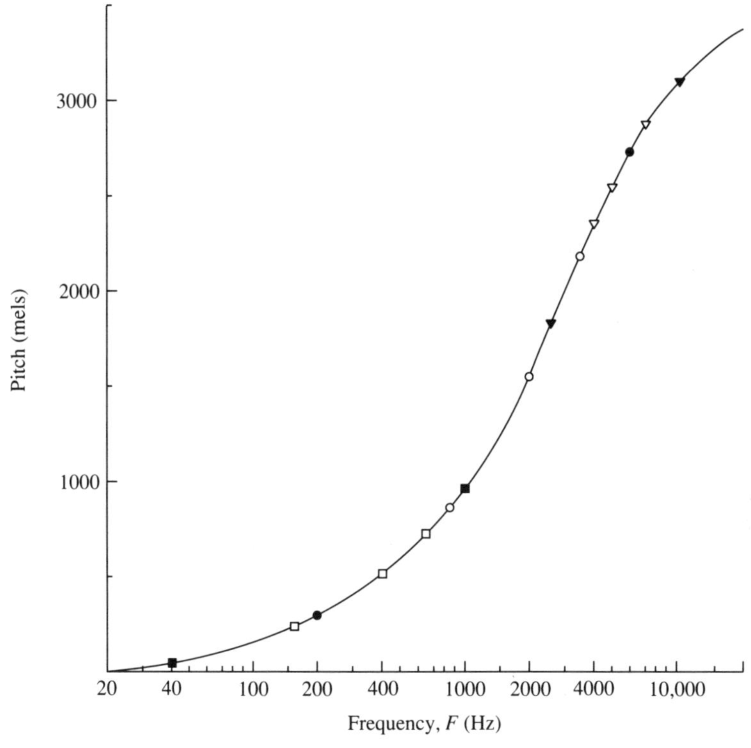

FIGURE 6.12. The mel scale. After Stevens and Volkman (1940).

suited to such a computation, since it works directly with the magnitude spectrum of the speech. (Contrast this, e.g., with the LP parameters, in which the spectral information is not explicit in the computation.) Suppose that it is desired to compute the cepstral parameters at frequencies that are linearly distributed on the range 0–1 kHz, and logarithmically above 1 kHz. Let us refer to Fig. 6.6, in which the DFT is used to compute the stRC. One approach is to "oversample" the frequency axis with the stDFT-IDFT pair by using, say an $N' = 1024$- or 2048-point stDFT, and then selecting those frequency components in the stDFT that represent (approximately) the appropriate distribution. Figure 6.13 illustrates this approach to computing the mel components on a frequency axis assumed to cover a Nyquist range 0–5 kHz. The remaining components can be set to zero, or, more commonly, a second psychoacoustic principle is invoked, which we now discuss.

Loosely speaking, it has been found that the perception of a particular frequency by the auditory system, say Ω_0,[15] is influenced by energy in a *critical band* of frequencies around Ω_0 (Schroeder, 1977; Allen, 1985; O'Shaughnessy, 1987, Ch. 4). Further, the bandwidth of a critical band varies with frequency, beginning at about 100 Hz for frequencies below

[15]Note that we are using uppercase Ω here to designate "real-world" frequencies.

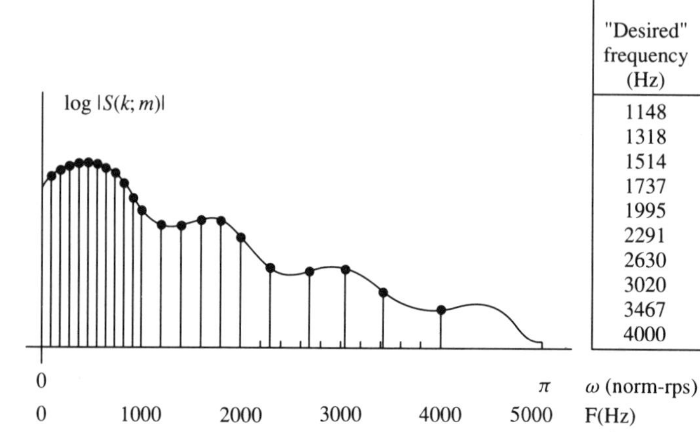

"Desired" frequency (Hz)	"Quantized" DFT frequency (Hz)	"k" value DFT
1148	1152	118
1318	1318	135
1514	1514	155
1737	1738	178
1995	1992	204
2291	2294	235
2630	2627	269
3020	3018	309
3467	3467	355
4000	4004	410

FIGURE 6.13. Using the stDFT to derive the appropriate frequency components for computing the mel-cepstral coefficients. Twenty mel-frequency components are desired on the Nyquist range 0–5 kHz. Ten are committed to the "linear" range 0–1 kHz. These appear at 100, 200, . . . , 1000 Hz. The other 10 are to be at log-distributed frequencies over the range 1–4 kHz. These correspond to "desired frequencies" shown in the table. If we use a 1024-point stDFT, frequency lines at close approximations to these desired frequencies will be available. These are the "quantized frequencies" shown above.

1 kHz, and then increasing logarithmically above 1 kHz. Therefore, rather than simply using the mel-distributed log magnitude frequency components to compute the stRC, some investigators have suggested using the *log total energy* in critical bands around the mel frequencies as inputs to the final IDFT. We use the notation $Y(i)$ to denote the log total energy in the ith critical band. This process is illustrated in Fig. 6.14. The meaning of $Y(i)$ will be understood by examining this figure.

Note how the final IDFT is actually implemented in the figure. We have conceptualized the critical band filters as residing on the Nyquist range only. In principle, to be used as discrete-time filters, their transforms (which are purely real in this case) must be made symmetrical about the Nyquist frequency. Consequently, the critical band filter log energy outputs, designated $Y(i)$ in Fig. 6.14, will be symmetrical about the Nyquist frequency. The integers i index the center frequencies of the critical band filters, each of which is assumed to be centered on one of the frequencies resolved by the original stDFT. In other words, if i indexes center frequency $F_{c,i}$, then

$$F_{c,i} = k\frac{F_s}{N'} \qquad \text{for some } k, \text{ say } k_i, \tag{6.55}$$

for each i, where F_s is the sample frequency, and N' is the number of points used to compute the stDFT (N-length frame of data plus zero padding). Therefore, let us define

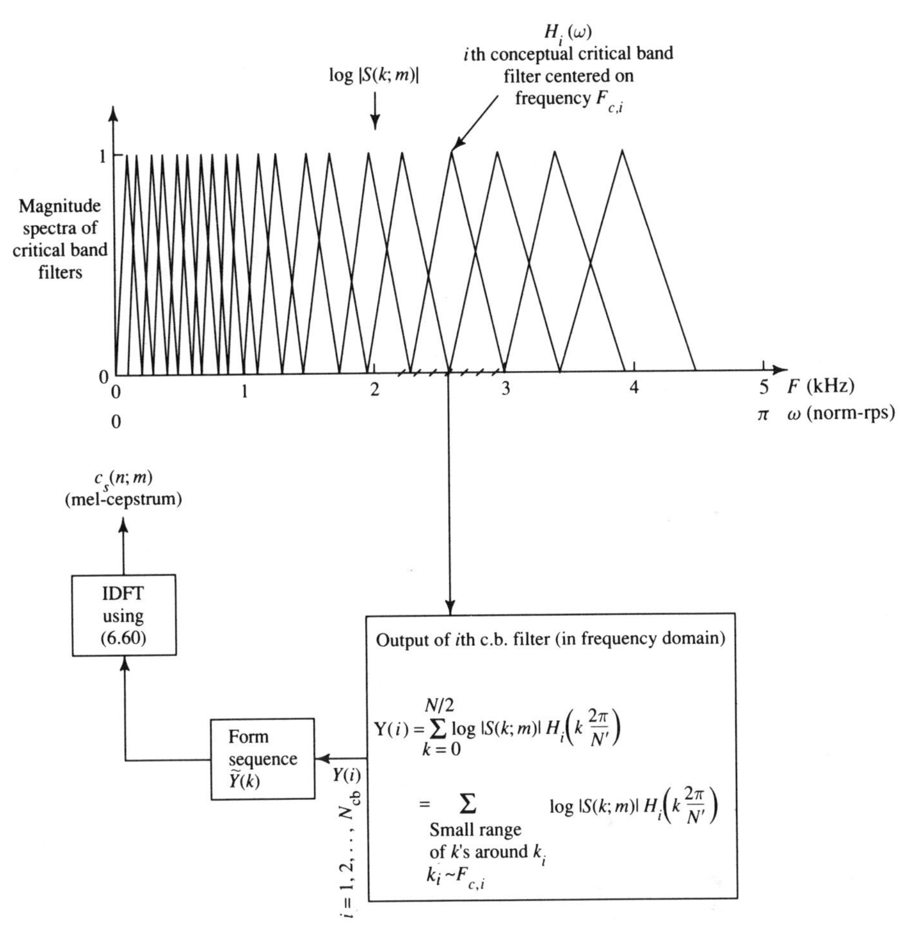

FIGURE 6.14. Use of critical band filters to compute the mel-cepstrum. The filter spectra are labeled "conceptual" critical band filters because, in practice, they must be effectively centered on frequencies corresponding to frequencies sampled by the DFT used to compute $\log|S(k;m)|$. $Y(i)$ denotes the sum of the weighted $\log|S(k;m)|$ terms within the ith critical band filter. $Y(i)$ and similar quantities are sometimes called the (weighted) log energy in the ith critical band. [*Note*: To make this quantity look more like energy, $|S(k;m)|^2$ can be used in place of $|S(k;m)|$ above. Because of the log, however, this would only change $Y(i)$ by a factor of 2.]

$$\tilde{Y}(k) = \begin{cases} Y(i), & k = k_i \\ 0, & \text{other } k \in [0, N'-1] \end{cases} . \tag{6.56}$$

The final IDFT, then, is

$$c_s(n;m) = \frac{1}{N'} \sum_{k=0}^{N'-1} \tilde{Y}(k) e^{jk(2\pi/N')n}. \tag{6.57}$$

[Note that we continue to use the notation $c_s(n;m)$ to denote the mel-cepstrum.] However, since the sequence $\tilde{Y}(k)$ is symmetrical about $N'/2$ ("even"), we can replace the exponential by a cosine,

$$c_s(n;m) = \frac{1}{N'}\sum_{k=1}^{N'} \tilde{Y}(k)\cos\left(k\frac{\pi}{N'}\left(n-\frac{1}{2}\right)\right). \tag{6.58}$$

Again using the symmetry of $\tilde{Y}(k)$, we can write this as

$$\begin{aligned} c_s(n;m) &= \frac{\tilde{Y}(0)}{N'} - \frac{\tilde{Y}(N'/2)}{N'} + \frac{2}{N'}\sum_{k=1}^{(N'/2)-1} \tilde{Y}(k)\cos\left(k\frac{2\pi}{N'}n\right) \\ &= \frac{2}{N'}\sum_{k=1}^{(N'/2)-1} \tilde{Y}(k)\cos\left(k\frac{2\pi}{N'}n\right), \end{aligned} \tag{6.59}$$

where we have assumed N' to be even and $\tilde{Y}(0) = \tilde{Y}(N'/2) = 0$. Of course, if N_{cb} denotes the number of critical band filters used on the Nyquist range, then there are only N_{cb} nonzero terms in the sum of (6.60). Therefore, let us rewrite the sum as

$$\begin{aligned} c_s(n;m) &= \frac{2}{N'}\sum_{\substack{k_i \\ i=1,2,\ldots,N_{\text{cb}}}} \tilde{Y}(k_i)\cos\left(k_i\frac{2\pi}{N'}n\right) \\ &= \frac{2}{N'}\sum_{i=1,2,\ldots,N_{\text{cb}}} \tilde{Y}(k_i)\cos\left(k_i\frac{2\pi}{N'}n\right). \end{aligned} \tag{6.60}$$

The lower form in (6.60) (without the superfluous scale factor $2/N'$) is usually used to compute the final IDFT. This is the form reflected in Fig. 6.14. Note that the use of only the Nyquist range in the final step requires that the stDFT be computed only over this range in the first operation.

Finally, we note that various types of critical band filters have been employed in practice (Davis and Mermelstein, 1980; Dautrich et al., 1983). The set of filters we have used above are essentially those proposed by Davis and Mermelstein. These critical band filters are particularly simple to use computationally, and have been employed in a number of research efforts [see, e.g., (Paul and Martin, 1988)].

Mel-Frequency Warping of LP-based Cepstral Parameters. A mel-frequency warping can also be included in cepstral parameters derived from LP analysis. In this case the cepstral parameters cannot be derived directly from the LP parameters. Rather, it is necessary to compute the log magnitude spectrum of the LP parameters and then warp the frequency axis to correspond to the mel axis. The cepstrum is then computed in the usual way. Shikano (1985), who approximated the mel warping by a bilinear transform of the frequency axis,[16] reported significant recognition improvement with respect to the "unwarped" LP-based cepstral coefficients [see also (Lee et al., 1990)].

Delta, or Differenced, Cepstrum

In addition to the cepstral or mel-cepstral parameters, another popular feature used in contemporary speech recognition is the delta cepstrum.[17] If $c_s(n;m)$ denotes the stRC or mel-cepstrum for the frames of the signal $s(n)$ ending at time m, then the *delta*, or *differenced, cepstrum* at frame m is defined as

$$\Delta c_s(n;m) \stackrel{\text{def}}{=} c_s(n;m+\delta Q) - c_s(n;m-\delta Q) \tag{6.61}$$

for all n. Here Q represents the number of samples by which the window is shifted for each frame. The parameter δ is chosen to smooth the estimate and typically takes a value of 1 or 2 (look forward and backward one or two frames). A vector of such features at relatively low n's (quefrencies) intuitively provides information about spectral changes that have occurred since the previous frame, although the precise meaning of $\Delta c_s(n;m)$ for a particular n is difficult to ascertain.

The delta cepstrum can, of course, also be computed for LP-based cepstral parameters. In this case, all instances of $c_s(n;m)$ above would be replaced by $c_{\hat{\theta}}(n;m)$, where, as usual, this notation refers to the cepstrum of the impulse response of the LP model estimated on the frame ending at time m.

Several researchers have argued that the differencing operation in (6.61) is inherently noisy and should be replaced by a polynomial approximation to the derivative (Furui, 1986; Soong and Rosenberg, 1986); however, many investigators have used the simple differencing operation successfully [e.g., see (Lee et al., 1990)].

Typically, 8–14 cepstral coefficients and their "derivatives" are used for speech recognition in systems that employ cepstral techniques (Juang et al., 1987; Rabiner et al., 1989; Lee et al., 1990). This means, for example, that (6.60) and (6.61) would be computed for $n = 1, 2, \ldots, 8\text{–}14$. From the discussions above, we should feel confident that these "low-time" measures will be associated with the vocal system spectrum and its dynamic changes.

[16]Refer to any of the books in Appendix 1.A for a discussion of the bilinear transform.

[17]Be careful to distinguish this feature from the *differential* cepstrum discussed in Section 6.3.2. Whereas the feature discussed here represents a *time* derivative of the cepstral parameters, the differential cepstrum represents a *frequency* derivative.

Log Energy

The measures $c_x(0; m)$ and $\Delta c_x(0; m)$ (with x standing for either s or $\hat{\theta}$) are often used as relative measures of spectral energy and its change. For the stRC, by definition,

$$c_x(0; m) = \frac{1}{2\pi} \int_{-\pi}^{\pi} \log \left| X(\omega; m) \right| d\omega, \tag{6.62}$$

which differs by a factor of two from the area under the log energy density spectrum. Of course, in the practical case in which the IDFT is used in the final operation, this becomes

$$c_x(0; m) = \frac{1}{N} \sum_{k=0}^{N-1} \log \left| X(k; m) \right|, \tag{6.63}$$

which has a similar interpretation. In the case of the mel-cepstrum,

$$c_s(0; m) = \sum_{i=1, 2, \ldots, N_{\text{cb}}} \tilde{Y}(k_i), \tag{6.64}$$

which clearly is a measure of total spectral energy. We will have more to say about the use of these cepstral features in speech recognition in Part V.

6.3 Complex Cepstrum

6.3.1 Long-Term Complex Cepstrum

Definition of the CC and of Homomorphic Systems

In describing the history of the RC, we suggested that its invention came out of a somewhat less formal line of thinking than we used in introducing it. The notion of moving signals out of a "convolution" domain and into a "linear" domain in a formal way is perhaps better attributed to a retrospective view of the RC in light of the work of Oppenheim, Schafer, and others on the topic of homomorphic signal processing. *Homomorphic signal processing* is generally concerned with the transformation of signals combined in nonlinear ways to a linear domain in which they can be treated with conventional techniques, and then the retransformation of the results to the original nonlinear domain. The general field is quite interesting and is applicable to a number of problems other than speech. We will focus our attention on the cepstrum and leave the general theory to the reader's further pursuit of the literature.

The RC fails the general criterion that a homomorphic technique permit a "round-trip" back to the original nonlinear domain. We discussed this problem near (6.17), where the main problem was found to be the discarding of phase by the operator Q_*^{real}. With a "simple" adjustment to

Q_*^{real}, however, we deduce a *bona fide* homomorphic operation and arrive immediately at the CC. The adjustment is to replace the $\log|S(\omega)|$ operation (see Fig. 6.1) with the complex logarithm of the complete DTFT, phase included. With this change, the *complex cepstrum* (CC) of the signal $s(n)$ is defined similarly to the RC,

$$\gamma_s(n) = \mathcal{F}^{-1}\{\log \mathcal{F}\{s(n)\}\} = \frac{1}{2\pi}\int_{-\pi}^{\pi} \log S(\omega)\, e^{j\omega n}\, d\omega, \qquad (6.65)$$

in which log now denotes the complex logarithm (defined below). The block diagram of the computation, shown in Fig. 6.15, is similar to that for the RC in Fig. 6.1 except that the complex logarithm in the CC case has taken the place of the real log in RC. Consequently, we have labeled the first two operations in Fig. 6.15 Q_*, omitting the superscript *real*. Just as in the RC case, we can interpret this computation as an attempt to move into a linear domain through the operation Q_*, with a subsequent "spectral" (cepstral) analysis through the IDTFT.

By definition, the natural logarithm of a complex number, say z, is

$$\log z = \log |z| + j\arg\{z\}. \qquad (6.66)$$

Therefore,

$$\log S(\omega) = \log |S(\omega)| + j\arg\{S(\omega)\}. \qquad (6.67)$$

In order that $\Upsilon_s(\omega) \stackrel{\text{def}}{=} \log S(\omega)$ be unique,[18] $\arg\{S(\omega)\}$ must be chosen to be an odd continuous function of ω [for details see (Oppenheim and Schafer, 1989, Sec. 12.3)]. This means that when using "canned" routines to find the $\arg\{S(\omega)\}$ in the process of computing $\log S(\omega)$, one must be careful to add multiples of 2π when necessary to make $\arg\{S(\omega)\}$ meet this criterion. This task, which is called *phase unwrapping*, although conceptually simple, is difficult in practice. Algorithms for phase unwrapping are discussed in (Tribolet, 1977).

In light of (6.67), we can redraw the computation of the CC in Fig. 6.15 to explicitly feature the real and imaginary parts of the complex log. This is shown in Fig. 6.16. The upper path of the computation, which treats the real part of the complex log, results in the conjugate symmetric

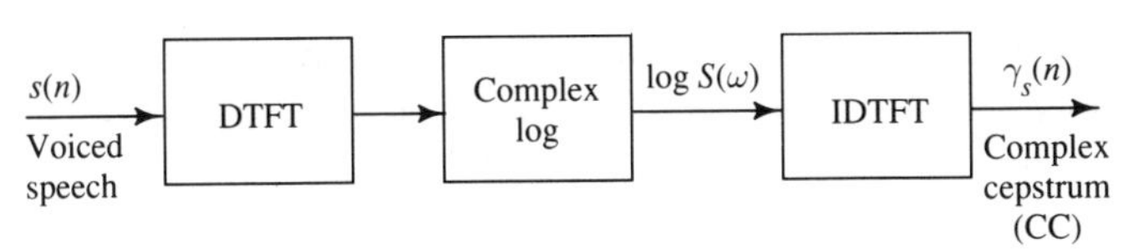

FIGURE 6.15. Computation of the CC.

[18]We avoid writing $\Upsilon_s(\omega)$ to refer to the DTFT of the CC, since this notation would be easily confused with our conventional notation for a power density spectrum. Instead we use $\mathrm{Y}_s(\omega)$ (upsilon).

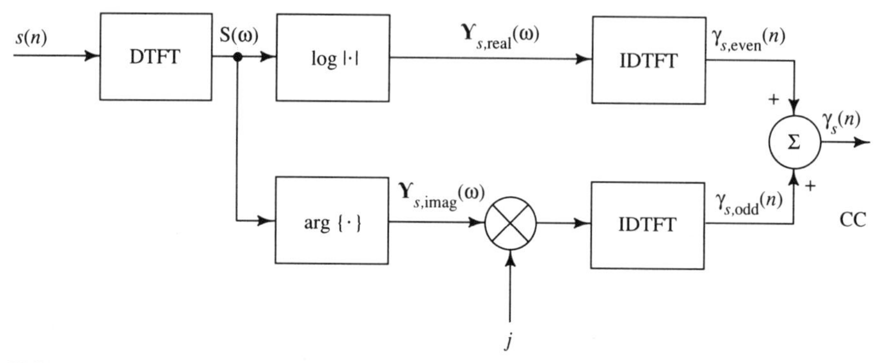

FIGURE 6.16. Computation of the CC showing decomposition into its even and odd parts.

part of the CC, which in this case is real and even. Apparently, the even part of the CC is equivalent to the RC as we have defined it,

$$\gamma_{s,\,\text{even}}(n) = c_s(n). \tag{6.68}$$

The lower path, which involves the imaginary part of the log, results in the complex antisymmetric part of the CC, which will also turn out to be real, and therefore odd. It is also clear from Fig. 6.16 that the even part of the CC is based on the magnitude spectrum of $s(n)$ only (this is, of course, consistent with our understanding of the RC) while the output of the lower path is based on the phase spectrum of $s(n)$. It is this lower path that is missing when we compute the RC and which preserves the phase information that the RC does not.

Before focusing on properties of the CC, we note the construction and meaning of a complete homomorphic system for convolved signals. Just as in the RC case, we can complete a linear operation on the CC by performing a liftering operation,[19] then a forward DTFT. For example, we might wish to "low-time lifter" to remove the effects of the excitation. This is the case shown in Fig. 6.17. Once the linear operation is performed, only an estimate of $\log \Theta(\omega)$ remains, and this "signal" must be reconverted to the original "convolution" domain. This final operation is accomplished through the application of the inverse operation Q_*^{-1}, which in this case exists unambiguously. Q_*^{-1} is illustrated in Fig. 6.17.

In summary, the convolution to linear domain transformation is denoted Q_* and its inverse is Q_*^{-1}. The IDTFT, low-time liftering, DTFT operations comprise a linear filtering operation, say $\mathcal{J}$ in the "new" linear domain. Let us denote the set of operations $[Q_* - \mathcal{J} - Q_*^{-1}]$ by $\mathcal{H}$. Then it is true that the overall system follows a sort of *generalized su-*

[19]This time the lifter can have both magnitude and phase properties.

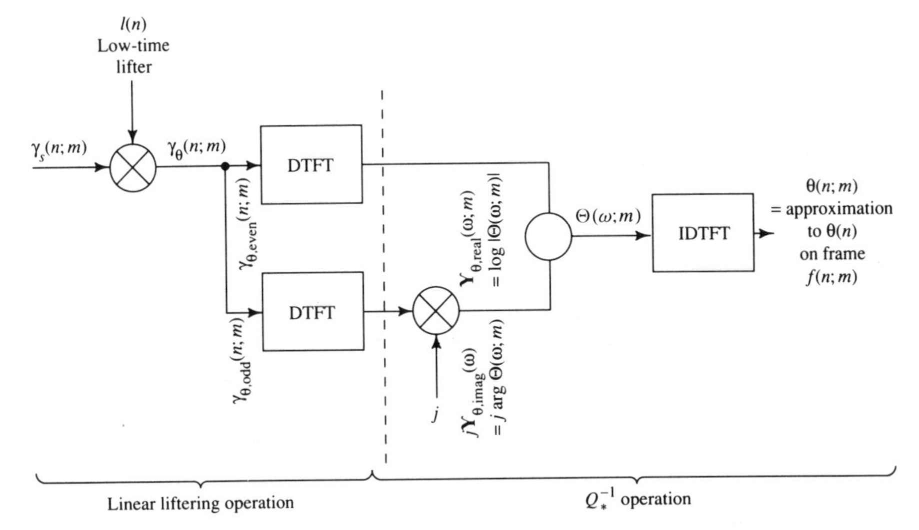

FIGURE 6.17. "Cepstral smoothing" using the CC. In fact, we have anticipated the need for the *short-term* CC and have employed it in this analysis. (The details are a straightforward generalization of our knowledge of the stRC and will be covered below.) This figure shows the block diagram of the computations. Note that the processing can continue all the way back to the original time domain to obtain an estimate of $\theta(n)$ because of the existence of the operation Q_*.

perposition principle where the combination rule[20] "+" has been replaced by "$*$,"

$$\mathcal{H}\{e(n) * \theta(n)\} = \mathcal{H}\{e(n)\} * \mathcal{H}\{\theta(n)\}. \tag{6.69}$$

Because $\mathcal{H}$ retains the same basic algebraic structure as a linear system, we call $\mathcal{H}$ a *homomorphic* ("same shape") system. The $[Q_* - \mathcal{J} - Q_*^{-1}]$ system for $\mathcal{H}$ is called the *canonical form* for the homomorphic system, and the Q_* is called the *characteristic system* for the operation that takes $*$ to $+$. Since the characteristic system is unique, all homomorphic systems that go from $*$ to $*$ differ only in their *linear* parts, $\mathcal{J}$.

Properties of the CC

A study of some of the important analytical properties of the CC will lead us to, among other things, a more formal understanding of some of our intuitive notions about the behavior and usefulness of the RC. We

[20]We have said nothing about rules for scalars here. For details see the discussion in (Oppenheim and Schafer, 1968).

restrict our discussion to (speech) sequences that have rational z-transforms which can be put in the form

$$S(z) = S_0' z^{-D'} \frac{\prod_{k=1}^{Q^{\text{in}}} (1 - \zeta_k^{\text{in}} z^{-1}) \prod_{k=1}^{Q^{\text{out}}} (1 - \zeta_k^{\text{out}} z^{-1})}{\prod_{k=1}^{P^{\text{in}}} (1 - \rho_k^{\text{in}} z^{-1})}, \tag{6.70}$$

where, $\zeta_k^{\text{in}}, k = 1, \ldots, Q^{\text{in}}$ and $\zeta_k^{\text{out}}, k = 1, \ldots, Q^{\text{out}}$ represent the zeros inside and outside the unit circle, respectively, and $\rho_k^{\text{in}}, k = 1, \ldots, P^{\text{in}}$ are the poles inside the unit circle.[21] Note that complex poles and zeros must occur in conjugate pairs. Note also that D' should be nonnegative for causality. The first step toward getting a principal result is to multiply and divide the numerator through by $-\prod_{k=1}^{Q^{\text{out}}} [1/\zeta_k^{\text{out}}] z$, yielding

$$S(z) = S_0 z^{-D} \frac{\prod_{k=1}^{Q^{\text{in}}} (1 - \zeta_k^{\text{in}} z^{-1}) \prod_{k=1}^{Q^{\text{out}}} (1 - \psi_k^{\text{out}} z)}{\prod_{k=1}^{P^{\text{in}}} (1 - \rho_k^{\text{in}} z^{-1})}, \tag{6.71}$$

in which $\psi_k^{\text{out}} \stackrel{\text{def}}{=} 1/\zeta_k^{\text{out}}$ and

$$S_0 = -S_0' \prod_{k=1}^{Q^{\text{out}}} \zeta_k^{\text{out}} \tag{6.72}$$

$$D = D' + Q^{\text{out}}. \tag{6.73}$$

Now, subjecting $s(n)$ [with z-transform as in (6.71)] to the first two operations (Q_*) in Fig. 6.15 yields

$$\begin{aligned}
\log S(\omega) &= \log S_0 + \log e^{-j\omega D} + \sum_{k=1}^{Q^{\text{in}}} \log(1 - \zeta_k^{\text{in}} e^{-j\omega}) \\
&\quad + \sum_{k=1}^{Q^{\text{out}}} \log(1 - \psi_k^{\text{out}} e^{j\omega}) - \sum_{k=1}^{P^{\text{in}}} \log(1 - \rho_k^{\text{in}} e^{-j\omega}) \\
&= \Bigg[\log|S_0| + \sum_{k=1}^{Q^{\text{in}}} \log(1 - \zeta_k^{\text{in}} e^{-j\omega}) + \sum_{k=1}^{Q^{\text{out}}} \log(1 - \psi_k^{\text{out}} e^{j\omega}) \\
&\quad - \sum_{k=1}^{P^{\text{in}}} \log(1 - \rho_k^{\text{in}} e^{-j\omega}) \Bigg] + j(\sigma - \omega D),
\end{aligned} \tag{6.74}$$

[21]It is easy to include poles outside the unit circle in this development, but we have no need for them.

where $\sigma = \pi$ if $S_0 < 0$, and is zero otherwise. We now need to apply the IDTFT operation (see Fig. 6.15) to obtain $\gamma_s(n)$. If we use the fact that

$$\log(1+\alpha) = \alpha - \frac{\alpha^2}{2} + \frac{\alpha^3}{3} - \frac{\alpha^4}{4} + \cdots \tag{6.75}$$

for $|\alpha| < 1$, it is not difficult to show that the IDTFT of (6.74) is

$$\gamma_s(n) = \begin{cases} \log|S_0| + j\sigma, & n = 0 \\ -\sum_{k=1}^{Q^{\text{in}}} \frac{[\zeta_k^{\text{in}}]^n}{n} + \sum_{k=1}^{P^{\text{in}}} \frac{[\rho_k^{\text{in}}]^n}{n} + \frac{D(-1)^{n+1}}{n}, & n > 0. \\ \sum_{k=1}^{Q^{\text{out}}} \frac{[\psi_k^{\text{out}}]^{-n}}{n} + \frac{D(-1)^{n+1}}{n}, & n < 0 \end{cases} \tag{6.76}$$

We need to dwell on the terms $j\sigma$ in the $n = 0$ case, and $D(-1)^{n+1}/n$ in the $n \neq 0$ cases before proceeding. The former is the only imaginary term in the CC and is nonzero only if the gain term S_0 is negative. It is customary to compute only the *real* part of the CC, realizing that we might be throwing away an insignificant piece of phase information corresponding to inversion of the waveform. The "D" term represents another bit of phase information[22] and arises from two sources [see (6.73)] – the presence of nonminimum-phase zeros and the possibility of an initial delay of the waveform. It is customary to eliminate this second contribution by shifting the waveform so that it originates at time zero. (Looking ahead, we see that this will be significant in short-term processing, since, in so doing, we will be giving up the information about the delay of the frame, "m.") With these two assumptions, we have

$$\gamma_s(n) = \begin{cases} \log|S_0|, & n = 0 \\ -\sum_{k=1}^{Q^{\text{in}}} \frac{[\zeta_k^{\text{in}}]^n}{n} + \sum_{k=1}^{P^{\text{in}}} \frac{[\rho_k^{\text{in}}]^n}{n} + \frac{Q^{\text{out}}(-1)^{n+1}}{n}, & n > 0. \\ \sum_{k=1}^{Q^{\text{out}}} \frac{[\psi_k^{\text{out}}]^{-n}}{n} + \frac{Q^{\text{out}}(-1)^{n+1}}{n}, & n < 0 \end{cases} \tag{6.77}$$

In principle, we should go back and modify the definition of the CC so that [for a causal sequence, $s(n)$] the signal is shifted, if necessary, to originate at time zero, and so that the real part of (6.65) is taken:

[22]Consequently, $\frac{D(-1)^{n+1}}{n}$ is an odd function of n.

$$\gamma_s(n) = \text{Real}\left\{\frac{1}{2\pi}\int_{-\pi}^{\pi} \log\left[\sum_{l=-\infty}^{\infty} \overleftarrow{s}(l)e^{-j\omega l}\right] e^{j\omega n}\, d\omega\right\}$$
$$= \text{Real}\left\{\frac{1}{2\pi}\int_{-\pi}^{\pi} \log\left[\overleftarrow{S}(\omega)\right] e^{j\omega n}\, d\omega\right\}, \tag{6.78}$$

where the arrows over $\overleftarrow{s}$ and $\overleftarrow{S}$ are simply reminders that we shift the waveform, if necessary, so that it originates at $n = 0$. It should be clear that this definition will exactly coincide with (6.77).

From this result, the following properties of the CC can be deduced:

1. The CC is *real.* (To convince yourself that the Real$\{\cdot\}$ operation in the new definition does nothing more than strip off a possible "$j\sigma$" term, work with just one conjugate pair of poles, as an example.)
2. $|\gamma_s(n)|$ decays as $|n^{-1}|$. For example, consider $n > 0$:

$$\left| -\sum_{k=1}^{Q^{\text{in}}} \frac{\left[\zeta_k^{\text{in}}\right]^n}{n} + \sum_{k=1}^{P^{\text{in}}} \frac{\left[\rho_k^{\text{in}}\right]^n}{n} + \frac{Q^{\text{out}}(-1)^{n+1}}{n} \right|$$
$$\leq \sum_{k=1}^{Q^{\text{in}}} \frac{\left|\left[\zeta_k^{\text{in}}\right]^n\right|}{n} + \sum_{k=1}^{P^{\text{in}}} \frac{\left|\left[\rho_k^{\text{in}}\right]^n\right|}{n} + \frac{Q^{\text{out}}}{n} \tag{6.79}$$
$$\leq (Q^{\text{in}} + P^{\text{in}})\frac{\beta^n}{n} + \frac{Q^{\text{out}}}{n},$$

   where $\beta \stackrel{\text{def}}{=} \max\left(\max_k |\zeta_k^{\text{in}}|, \max_k |\rho_k^{\text{in}}|\right)$.
3. If $s(n)$ is *minimum phase* (no zeros outside the unit circle), then $\gamma_s(n) = 0, n < 0$ (the CC is causal).
4. Conversely, if $s(n)$ is *maximum phase* (no poles or zeros inside the unit circle), then $\gamma_s(n) = Q^{\text{out}}(-1)^{n+1}/n, n > 0$.
5. The CC is of infinite duration even if $s(n)$ is not. (Recall that a finite duration signal will have only a finite number of zeros and no poles.)

Let us digress for a moment and note an important point about minimum-phase sequences. We do so with the caution that speech and frames of speech (even when they are shifted in time so that they originate at $n = 0$) are generally not minimum-phase sequences, so we must be careful how we apply this information. However, in one very important application of cepstral analysis, we are concerned with the cepstrum of the impulse response of the LP model of a speech frame. In this case the signal considered *is* minimum phase and this point is very useful: For a minimum-phase signal, say $x(n)$, the CC, $\gamma_x(n)$, is completely specified by its even part that, in turn, is precisely the RC, $c_x(n)$. This says that in the case of minimum-phase sequences, the CC and RC comprise equivalent information, and the results obtained for the CC apply equally well to the RC, and vice versa. In fact, in Problem 6.5 we verify that the following relationship exists between $\gamma_x(n)$ and $c_x(n) = \gamma_{x,\text{even}}(n)$ for a minimum-phase sequence:

$$\gamma_x(n) = \begin{cases} 2c_x(n) = \gamma_{x,\text{even}}(n), & n > 0 \\ c_x(n) = \gamma_{x,\text{even}}(n), & n = 0\,. \\ 0, & n < 0 \end{cases} \tag{6.80}$$

6.3.2 Short-Term Complex Cepstrum

It should be clear that the appropriate definition for the *short-term CC* (stCC) for the N-length frame of speech $f(n;m) = s(n)w(m-n)$ is

$$\begin{aligned} \gamma_s(n;m) &= \text{Real}\left\{\frac{1}{2\pi}\int_{-\pi}^{\pi} \log\left[\sum_{l=0}^{N-1} \overleftarrow{f}(l;m)e^{-j\omega l}\right] e^{j\omega n}\, d\omega\right\} \\ &= \text{Real}\left\{\frac{1}{2\pi}\int_{-\pi}^{\pi} \log\left[\overleftarrow{S}(\omega;m)\right] e^{j\omega n}\, d\omega\right\}. \end{aligned} \tag{6.81}$$

Note that the frame has been appropriately shifted down to begin at time zero, and we put an arrow over the frame, $\overleftarrow{f}$, and the stDTFT, $\overleftarrow{S}$, as a reminder. (Recall the discussion in Section 4.3.5.) Accordingly, the index m in the argument of the stCC serves only to catalog the position of the frame. The computation of $\gamma_s(n;m)$ is illustrated in Fig. 6.18. The practical version in which discrete frequency transforms are used is shown in Fig. 6.19. Again the frame is shifted downward so that the "conventional" DFT, rather than the delay-preserving stDFT, is used. As in the case of the RC, a potential for aliasing exists when discrete transforms are used; generally, zero padding of the initial frame is necessary. In this case the DFT-IDFT pair will be based on $N' > N$ points, where N' reflects the zero padding.

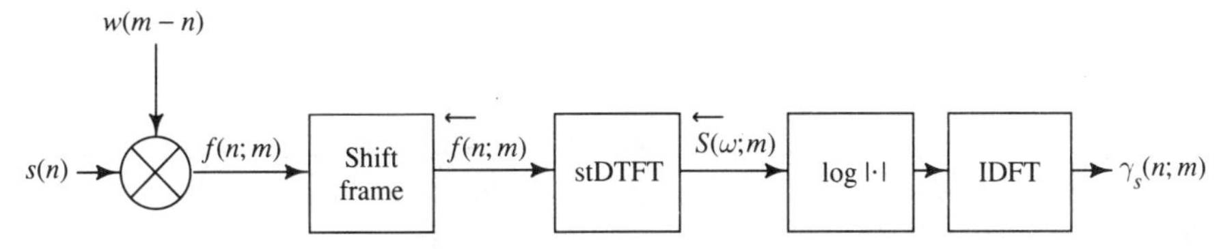

FIGURE 6.18. Computation of the stCC.

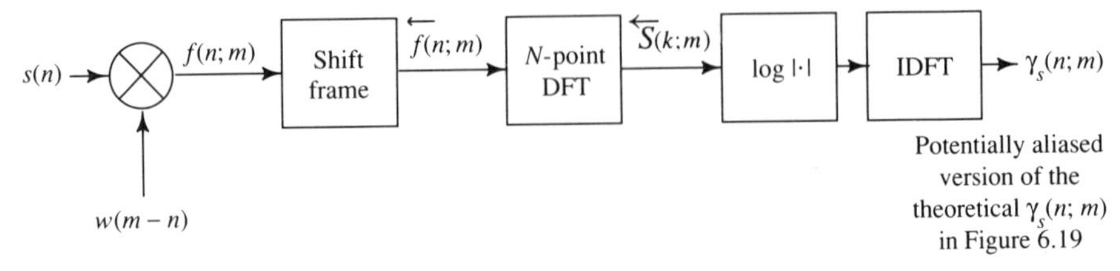

FIGURE 6.19. Computation of the stCC using the DFT-IDFT pair.

6.3.3 Example Application of the stCC to Speech Analysis

As we have repeatedly indicated, much of speech analysis is successfully performed with the simpler RC. However, as we have seen above, the CC is more yielding of theoretical results, which, in turn, provide insight into the behavior of the RC. Indeed, some key points in our discussion of pitch and formant estimation using the stRC were based upon promised future results from the CC. We now look briefly at the related problems of pitch and formant analysis using the stCC, with a primary interest in strengthening the same discussion about the stRC, and a secondary interest in showing that the stCC itself can be used for these purposes.

Of course, as in the stRC case, the stCC of the speech is the sum of the individual cepstra of the excitation and vocal system impulse response:

$$\gamma_s(n;m) = \gamma_e(n;m) + \gamma_\theta(n). \tag{6.82}$$

Note that we have one long-term and one short-term CC on the right side of the equation. Recall that this is a consequence of assuming that the window which creates the frame is slowly varying with respect to the impulse response. The reader is encouraged to review the discussion surrounding (6.23) to refresh his or her memory on this point if it is not clear. As in the case of the RC, the key point is to demonstrate that the two CCs are well separated on the quefrency axis.

In a very similar manner to our work on $c_e(n;m)$, it can be shown that $\gamma_e(n;m)$ takes the form of a pulse train with decaying weights. Whereas we are able to deduce only the form of the pulse train for the stRC [see (6.36)], here a precise expression can be derived. It is (Oppenheim and Schafer, 1968)

$$\gamma_e(n;m) = \sum_{q=0}^{Q-1} \gamma_{v_m}(q)\delta(n-qP), \tag{6.83}$$

where, $v_m(n) \stackrel{\text{def}}{=} w(m-nP)$, and $\gamma_{v_m}(n)$ is its CC. As usual, $w(m-n)$ is the window used to create the frame, and it is assumed that Q pulses of $e(n)$ occur inside the window.

We are also well prepared to make a strong case for the fast decay of $\gamma_\theta(n)$ with respect to P, since we now know that any CC will decay as

$|n^{-1}|$. To be explicit, let us write the z-transform of the vocal system impulse response $\theta(n)$ in the form of (6.71),

$$\Theta(z) = \Theta_0 z^{-Q^{\text{out}}} \frac{\prod_{k=1}^{Q^{\text{in}}} (1 - \zeta_k^{\text{in}} z^{-1}) \prod_{k=1}^{Q^{\text{out}}} (1 - \psi_k^{\text{out}} z)}{\prod_{k=1}^{P^{\text{in}}} (1 - \rho_k^{\text{in}} z^{-1})}. \tag{6.84}$$

Then we have, according to (6.77),

$$\gamma_\theta(n) = \begin{cases} \log|\Theta_0|, & n = 0 \\ -\sum_{k=1}^{Q^{\text{in}}} \frac{[\zeta_k^{\text{in}}]^n}{n} + \sum_{k=1}^{P^{\text{in}}} \frac{[\rho_k^{\text{in}}]^n}{n} + \frac{Q^{\text{out}}(-1)^{n+1}}{n}, & n > 0 \\ \sum_{k=1}^{Q^{\text{out}}} \frac{[\psi_k^{\text{out}}]^{-n}}{n} + \frac{Q^{\text{out}}(-1)^{n+1}}{n}, & n < 0. \end{cases} \tag{6.85}$$

This is the justification for (6.39), which played a key role in the similar discussion about the RC. Finally, therefore, we conclude that

$$\gamma_s(n;m) \approx \begin{cases} \gamma_e(0;m) + \gamma_\theta(0), & n = 0 \\ -\sum_{k=1}^{Q^{\text{in}}} \frac{[\zeta_k^{\text{in}}]^n}{n} + \sum_{k=1}^{P^{\text{in}}} \frac{[\rho_k^{\text{in}}]^n}{n} + \frac{Q^{\text{out}}(-1)^{n+1}}{n}, & 0 < n < P \\ \sum_{k=1}^{Q^{\text{out}}} \frac{[\psi_k^{\text{out}}]^{-n}}{n} + \frac{Q^{\text{out}}(-1)^{n+1}}{n}, & -P < n < 0 \\ \gamma_e(n;m) \text{ (weighted pulse train)}, & |n| \geq P. \end{cases} \tag{6.86}$$

which is similar to (6.41), which we obtained for the stRC.

Although more rigorous than our similar discussion of the stRC, the conclusion is essentially the same. We find that the component excitation and vocal system cepstra are well separated in the quefrency domain, the excitation portion consisting of a high-time pulse train and the significant contribution of the vocal system occurring in the low-time region below quefrency P. In Fig. 6.7 we studied the stRC for the speech frame representing the vowel /i/. In Fig. 6.20(a) we show the stCC for the same speech frame as well as a repetition and extension of the stRC of Fig. 6.7, which appears in Fig. 6.20(b). Note that although the pitch peaks are still present and well separated from the vocal system information in the

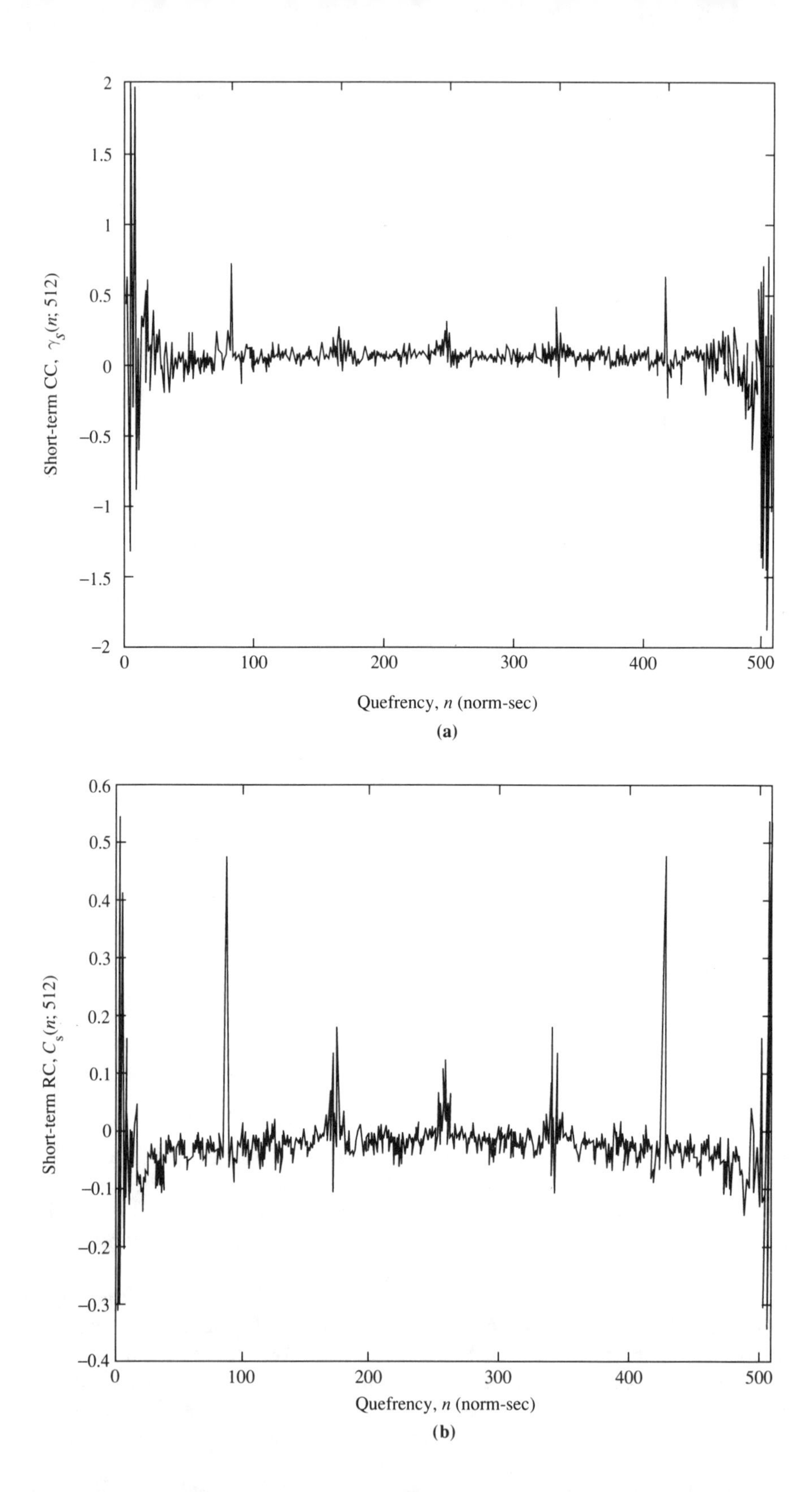

2
1.5
1
0.5
0
−0.5
−1
−1.5
−2
0
100
200
300
400
500
Short-term CC, $\gamma_s(n; 512)$
Quefrency, n (norm-sec)
(a)
0.6
0.5
0.4
0.3
0.2
0.1
0
−0.1
−0.2
−0.3
−0.4
0
100
200
300
400
500
Short-term RC, $C_s(n; 512)$
Quefrency, n (norm-sec)
(b)

stCC, there is a notable difference between the cepstra of Figs. 6.20(a) and (b). The stCC is no longer a purely even function (manifest in symmetry about time 256 in Fig. 6.20(b)). In fact, its even part should be equivalent to the stRC shown in Fig. 6.20(b), but the stCC itself will not in general be even. As we know, the odd part of the stCC is responsible for encoding phase information that is discarded by the stRC.

In principle, the stCC is subject to simple high-time or low-time liftering to remove one or the other of the component cepstra from the speech cepstrum. Recall that it is possible with the liftered stCC to return all the way to the original time domain with an estimate of the excitation or vocal system impulse response because a unique and well-defined Q_*^{-1} operation exists. Although this is seldom the purpose of using cepstral analysis, it is interesting to note this ability, which is lost with the stRC.

6.3.4 Variations on the Complex Cepstrum

At least two interesting variations on the complex cepstrum have appeared in the literature. Each differs from the cepstrum discussed in this chapter in the nonlinear operation applied to the spectrum. In the *spectral root cepstrum* (Lim, 1979) the logarithm is replaced by the operation of raising the DTFT to the power α, say $S^{\alpha}(\omega)$. This system was developed explicitly for the pulse deconvolution problem and has been applied to some simple voiced speech analysis–synthesis experiments in Lim's paper.

A second variation, called the *differential cepstrum* (Polydoros and Fam, 1981), involves the replacement of the logarithm by the derivative of the log. Since

$$\frac{d}{d\omega}\log S(\omega) = \frac{dS(\omega)}{d\omega} S^{-1}(\omega), \tag{6.87}$$

the phase unwrapping problems inherent in the usual cepstrum are not present with this operation. The properties and potential applications of the differential cepstrum are discussed in the paper cited. The technique has not found much application to speech analysis, and the reader should be careful to distinguish the differential cepstrum from the delta, or differenced, cepstrum discussed in Section 6.2.4. The former represents the dynamic behavior of the spectrum with respect to *frequency*, while the latter represents a *time* derivative.

FIGURE 6.20. (a) The stCC of the same frame of speech analyzed in Fig. 6.7. (b) The stRC repeated for convenience and extended to 512 points to show symmetry.

6.4 A Critical Analysis of the Cepstrum and Conclusions

Cepstral techniques have an intriguing history and they have been the subject of much interesting research over the past several decades. Nevertheless, their application to speech processing has been somewhat limited. This is in part due to the computational complexity of cepstral algorithms, and in part due to the mathematical complexity (especially the nonlinear operation) that, on one hand, necessitates certain assumptions that are not always valid, and on the other hand, makes a precise theoretical understanding of the performance difficult.

Among the most important uses of cepstral analysis in recent years has been the conversion of LP parameters to cepstral coefficients for speech recognition. A related use is the computation of mel-cepstral parameters directly from the speech to use as recognition features. Both have been found to improve the performance of speech recognizers, as we shall see in later chapters. In either case, the stRC is used, and in the former case the problem amounts to computing the cepstrum of a minimum-phase signal—a condition under which the stRC and stCC are essentially equivalent. Perhaps not surprisingly, where the cepstrum has had the most successful application record, it has been the relatively simple version of it that has been applied. Also interesting is that this application does not use the cepstrum to separate the excitation from the vocal system spectrum, the general task for which the method was originally developed.

We have mentioned some other areas—pitch and formant estimation—in which either the stRC or stCC could be applied. There are two cautions that have been issued in relation to these general problems. The first is the fact that the nonlinear operation (logarithm) does not always carry with it the assurance that the resulting additive log spectra will be well separated in quefrency, even if the components in the original spectrum apparently are (though combined by multiplication). We will experiment with this idea in Problem 6.10. In this problem, we will take the product of two spectra, one slowly and one quickly varying, such that the "slow" and "fast" components are apparent in the product. By moving the frequencies of variation closer together, we will seek to determine whether there is a point at which the two components are apparent in the spectrum, but not in the log-transformed spectrum. We will find that the logarithm generally preserves the "separability" of the "components." The results of this experiment are generally consistent with experience in speech processing practice and should put to rest some of the uneasiness about the nonlinear operation. Nevertheless, the reader should keep this issue in mind, as it may provide at least a partial explanation for why the cepstrum sometimes fails to resolve the low- and high-quefrency components in practice.

Another "side effect" of the nonlinear log operation that has been of some concern is the potential for improper emphasis of low-level noisy portions of the speech spectrum. This phenomenon can degrade, in unpredictable ways, the performance of coding and recognition strategies based upon the cepstrum.

The other problem area noted in conjunction with the "pitch–formant" application is the importance of the type, length, and placement of the data window used to select the frame for short-term processing.[23] The speech "vocoding" field (see Chapter 7), in which the cepstrum is used in a sort of generalized (and dynamic) pitch and formant analysis, is where this problem has received considerable attention, so we briefly mention some results from this branch of speech processing. In some of the early work on this problem, Quatieri (1979) began with the "approximate" model for short-term cepstral analysis given by (6.23),

$$\begin{aligned} f_s(n;m) &\approx e(n)w(m-n) * \theta(n) \\ &= f_e(n;m) * \theta(n), \end{aligned} \tag{6.88}$$

and used experimental evidence and heuristic arguments to obtain some guidelines for both stRC and stCC analysis. The basic study involved the analysis and resynthesis of synthetic speech using cepstral analysis to perform deconvolution. A general finding of the research is that the stCC, by preserving phase information at significant computational cost and algorithmic overhead, did not significantly outperform the stRC version in terms of speech quality. In either case, by interpreting (6.88) as a smoothing convolution in the frequency domain, Quatieri argues that the spectral nulls (especially due to high pitch) in the speech spectrum represent regions in which the spectral counterpart of (6.88) is least likely to hold. The window length could be adjusted to minimize the spectral distortion—a length of roughly $2P$–$2.5P$, where P is the pitch period, is prescribed. This requires an *adaptive* window size as multiple frames are processed in the utterance. The window should also have large attenuation in its spectral sidelobes, making smoother windows, like the Hamming, an appropriate choice.

The considerations of the data window relationships to phase reconstruction with the stCC are rather detailed and the reader is referred to the original paper for a complete discussion. Briefly, the window alignment with respect to the underlying pulse train is a critical consideration. The window should be aligned to have no linear phase factor with respect to the impulse response $\theta(n)$.

Nearly 10 years after Quatieri's paper was published, Verhelst and Steenhaut (1988) undertook the careful analytical examination of the ap-

[23]Interestingly, the length and shape of the lifter, if spectral deconvolution is the issue, has been found to be less significant than the data window (Verhelst and Steenhaut, 1988).

proximation in (6.88) and published some interesting results. They argue that the findings above are somewhat antithetical to proper theoretical analysis. Indeed, if $s(n)$ is stationary over a long time range, then the analysis frame should be made as long as possible for consistency with the assumption that the window tapers slowly with respect to the pitch period [see (6.28)]. Increasing the frame length should also reduce the sensitivity to the phase of the window with respect to the pulse train. However, experimental work had shown that best performance was obtained with relatively short frames of 2–2.5 pitch periods. Verhelst and Steenhaut show that this paradox is a direct consequence of the approximation (6.88). In fact, assuming that an equality in (6.88) holds is tantamount to assuming that

$$\log \Theta(\omega) = \log \sum_{k=0}^{P-1} e^{\log \Theta(k\omega_P)} \tilde{W}(\omega - k\omega_P), \tag{6.89}$$

where $\tilde{W}(\omega)$ is a normalized version of the window spectrum, and ω_P is the pitch frequency in norm-rps, $2\pi/P$. This rather unlikely equality requires that the vocal system spectrum be an interpolated version of nonlinearly modified versions of its samples at the harmonics! The authors then show that assuming the validity of this result, in turn, is equivalent to assuming that the estimated deconvolved version of $\Theta(\omega)$, say $\hat{\Theta}(\omega; m)$, can be obtained as

$$\hat{\Theta}(\omega; m) \approx \frac{S(\omega)}{\displaystyle\sum_{k=0}^{P-1} W(\omega - k\omega_P)}. \tag{6.90}$$

Clearly, equality in (6.90) is very unlikely in practical problems, but Verhelst and Steenhaut demonstrate in their paper that the approximation is optimized by following the heuristic rules about window length and placement noted above!

Verhelst and Steenhaut go on to show how the nonlinear interpolation process can result in a severe aliasing-like effect in the stCC,[24] $\gamma_\theta(n; m)$, which can be reduced only by reducing the impact of the nonlinearity. They therefore suggest the equalization of the speech spectrum in order to make $\log \Theta(k\omega_p)$ in (6.89) more nearly a constant, thus "linearizing" the interpolation. In effect, the equalization reduces the duration of $\gamma_\theta(n; m)$, thereby reducing the aliasing problem.

In summary, we can appreciate that in addition to the rather large computational complexity required to implement the cepstrum, there is a significant amount of overhead involved in the adaptation of data window sizes and placement, preemphasis, and other *ad hoc* operations to counteract effects of the nonlinear operation. Unmentioned above is the

[24]The reader who has not studied the stCC can simply plug in the stRC here to get the general ideas.

fact that even more overhead is required to appropriately concatenate cepstrally obtained results from frame to frame—in a coding procedure, for example (Quatieri, 1979).

Of course, these considerations become much less important (or even negligible) in applications in which the RC is sufficient. The RC does not preserve phase, but phase is unimportant in some applications, and use of the cepstrum has been principally confined to such applications in practice. The RC, while simpler, is still not immune to anomalies resulting from nonlinearity. However, the RC is simple enough to compute that it can simply be experimented with in deconvolution problems where it seems relevant. A notable example is its use in modeling smoothed speech spectra for recognition. We will see this technique used repeatedly in Part V.

In addition to the speech recognition application, we will also encounter cepstral analysis in one form of speech coder in the next chapter. In fact, having explored the main analysis issues in modern speech processing, we now turn to the study of three related topics that comprise the three chapters of Part IV—coding, enhancement, and quality assessment.

6.5 Problems

6.1. Find the (continuous-time) Fourier series expansion of the periodic "signal" $C_s(\omega)$ of (6.13), in the process showing that (6.13) is the correct expression for the coefficients. Note that $C_s(\omega)$ is a periodic "signal" of "period" $\omega_s = 2\pi$ norm-rps or $f_s = 1$ norm-Hz. Its "fundamental frequency" is therefore $1/f_s = 1$ norm-sec, and its "harmonic frequencies" are $n/f_s = n$ norm-sec.

6.2. (Computer Assignment) The stRC for a signal $s(n)$ on a frame ending at time m is defined as

$$c_s(n) = \mathcal{I}^{-1}\{\log|\mathcal{I}\{f(n;m)\}|\} = \mathcal{I}^{-1}\{Q_*^{\text{real}}\{f(n;m)\}\}, \tag{6.91}$$

in which $f(n;m) = s(n)w(m-n)$, $w(n)$ is the window used to create the frame, and $\mathcal{I}\{\cdot\}$ denotes the DTFT. In the context of introducing liftering in Section 6.2.1, we momentarily pondered whether this operation (in the long-term case) was invertible, but quickly realized that the loss of phase information would render impossible the recovery of the original signal. In this problem we explore two attempts to invert the stRC to reinforce the notion of ambiguous phase. Consider the signal

$$s(n) = \begin{cases} n, & 0 \le n \le 50 \\ 50\cos[0.01\pi(n-50)], & 51 \le n \le 99\,. \\ 0, & \text{other } n \end{cases} \tag{6.92}$$

Create the frame

$$f(n;m) = s(n)w(m-n), \tag{6.93}$$

where $m = 127$ and $w(n)$ is a rectangular window of length $N = 128$. The 64 zeros that are taken along with the main pulse should help prevent aliasing of the cepstrum as the DTFT is replaced by the DFT.

(a) Compute the cepstrum of this frame, $c_s(n;m)$, and then attempt to invert it using the operation

$$f_0(n;m) = \left[Q_*^{\text{real}}\right]^{-1}\{\mathcal{F}\{c_s(n;127)\}\}, \tag{6.94}$$

in which

$$\left[Q_*^{\text{real}}\right]^{-1}\{\cdot\} \stackrel{\text{def}}{=} \mathcal{F}^{-1}\{\exp[\cdot]\}. \tag{6.95}$$

Plot $f(n;m)$ and $f_0(n;m)$, and the phase spectra for each. [*Note*: The phase spectra should be called $\arg\{S(\omega;m)\}$ and $\arg\{S_0(\omega;m)\}$.] Explain what has happened.

[*Note*: It is not actually necessary to compute $c_s(n;127)$. By combining the forward and inverse operations, a very simple computation can be made.]

(b) We can infer from theoretical work with the complex cepstrum (see Problem 6.5) that a causal, minimum-phase correspondent to $s(n)$ [we really mean to $f(n;m)$ here] can be obtained using the following steps:

(i) Create the sequence

$$c_s'(n;m) = c_s(n;m)u_+(n), \tag{6.96}$$

in which

$$u_+(n) \stackrel{\text{def}}{=} 2u(n) - \delta(n), \tag{6.97}$$

where $u(n)$ and $\delta(n)$ are the discrete-time step and impulse sequences, respectively.

(ii) Then compute $f_{\min}(n;m)$ by inserting $c_s'(n;m)$ in place of $c_s(n;m)$ on the right side of (6.94).

Plot $f_{\min}(n;m)$ and its phase spectrum. Comparing these with the time domain and phase plots for $f(n;m)$, is it reasonable to believe that $f_{\min}(n;m)$ is the causal, minimum-phase version of $f(n;m)$? Explain.

(c) Verify theoretically that $f(n;m), f_0(n;m)$, and $f_{\min}(n;m)$ have identical magnitude spectra. [*Hint*: Note that $c_s(n;m)$ is an even sequence with respect to n.]

6.3. Consider the use of the stRC to estimate the pitch described in Section 6.2.3. The theoretical developments underlying this method depend critically upon the assumption that the window used to create the frame varies slowly with respect to the pitch period, P [see the discussion surrounding (6.28)].

(a) Suppose that a speech sequence represents samples of a vowel sound taken at a sampling rate 10 kHz. If a Hamming window is used, how long, N, must the window be so that its amplitude varies by less than 20% over a pitch period in the following cases?
  (i) The pitch frequency is 240 Hz.
  (ii) The pitch frequency is 110 Hz.
  In each case assume that the window is centered on the onset of a pitch period. Note carefully that the worst distortion occurs at the ends of the window, not at the center. Use this fact to make a proper engineering analysis.

(b) In the last section of the chapter, we learned that there are certain situations in which the optimal length of the window is on the order of two pitch periods. Is this constraint adhered to in the cases above?

(c) Generally speaking, do these results suggest more problems will be encountered with higher or lower pitched utterances?

(d) In fact, the height of the cepstral peak indicating the pitch period is dependent upon many factors, including the size, shape, and placement of the window creating the analysis frame. As an extreme example, suppose that the pitch period is quite long and the window size is chosen so that only one period or less appears in a given frame. In this case, $Q = 1$ in the discussion surrounding (6.31). Give an analytical argument assessing the potential for determining the pitch period from the stRC in this case.

6.4. (Computer Assignment)

(a) For a vowel utterance of your choice, plot the log magnitude spectrum based on a stDFT analysis using a Hamming window. Make sure that the window is at least two pitch periods long. Estimate the first three formant frequencies.

(b) Now perform a liftering operation in conjunction with the stRC to produce the log "formant" spectrum and estimate the formants on this basis. Also plot the stRC. Use the same frame of speech as in part (a), but remember to "zero pad" to avoid cepstral aliasing.

(c) Use the stRC derived in part (b) to estimate the pitch of the utterance.

6.5. (a) From the properties of the CC in Section 6.3.1, we can deduce that for a causal, real, minimum-phase sequence $s(n)$, the CC, $\gamma_s(n)$, will also be real and causal. Use this fact to deduce the following relationship between the CC and RC:

$$\gamma_s(n) = c_s(n)[2u(n) - \delta(n)] = \begin{cases} c_s(n), & n = 0 \\ 2c_s(n), & n > 0 \\ 0, & n < 0 \end{cases} \tag{6.98}$$

(b) This fundamental result makes it unnecessary to work with the CC when the signal is minimum phase. On the other hand, note that this result is true for the long-term cepstra and is only relevant to short-term analysis when the frame itself is a minimum-phase sequence. Can you cite an important practical problem in which we take a long-term cepstrum of a minimum-phase signal?

6.6. In one of the first papers to apply cepstral analysis to speech, Noll (1967) defines the cepstrum of $x(n)$ to be the squared cosine transform of the logarithm of the power density spectrum (PDS) of the speech,[25]

$$c_s^{\text{Noll}}(n) \stackrel{\text{def}}{=} [\mathcal{C}\{\log \Gamma_s(\omega)\}]^2. \tag{6.99}$$

$\mathcal{C}\{\cdot\}$ is the cosine transform, the appropriate version of which can be inferred from footnote 5 on page 359.

(a) Find the relationship between Noll's definition of the cepstrum and the definition of the RC, $c_s(n)$, used in this chapter.

(b) Replace the long-term PDS in Noll's definition by the stPDS $\Gamma_s(\omega; m)$ to create the "short-term Noll cepstrum," say $c_s^{\text{Noll}}(n; m)$. Repeat part (a) for $c_s^{\text{Noll}}(n; m)$ and the stRC, $c_s(n; m)$, used in our work.

6.7.* (a) Derive (6.44), which gives a recursion for converting the LP parameters to the stCC of the impulse response of the LP model, $\gamma_{\hat{\theta}}(n; m)$.

(b) With a simple manipulation, derive the recursion for converting the LP parameters to the stRC, $c_{\hat{\theta}}(n; m)$.

6.8. Derive (6.76) from (6.74).

6.9. In a given speech-coding problem, the task is to deduce parametric representations of 256-point frames of speech.

(a) Give an approximate count of the average number of multiplications per frame necessary to compute a set of 14 (real) cepstral coefficients directly from the speech using decimation-in-time FFT algorithms. Assume that before taking FFTs, each frame of 256 points is padded with an additional 256 zeros.

(b) How does the result of part (a) compare with computing 14 LP parameters on each frame using the Levinson–Durbin recur-

[25]Of course, we have taken the liberty of converting Noll's definition into discrete-time terms.

*The discussion on pages 376–378 of the text, beginning with Eq. (6.45), includes the gain estimate $\hat{\Theta}_0(m)$, as part of the overall LP model $\hat{\Theta}(z; m)$. This was not the convention in Chapter 5.

sion, and then converting the LP parameters to log area ratio (LAR) parameters?

(c) Suppose that the LP parameters in part (b) are converted to cepstral parameters rather than LAR parameters. What is the difference between sending the parameters derived in part (a) and those derived here? Would your answer change if 50 cepstral parameters rather than 14 were to be sent?

6.10. (Computer Assignment) This problem is exploratory and has no "correct" answer. The ability to deconvolve two signals using the cepstrum depends critically upon their remaining well separated in "quefrency" once the nonlinear operation of log taking is applied. In a general way, we will explore this issue here.

(a) For simplicity, let us interchange the roles of the time and frequency domains in this problem with respect to cepstral analysis. Take two signals, say $x(n)$ and $y(n)$, which are widely spaced (empirically) in frequency. For example, let us begin with

$$x(n) = 0.1 + |\sin(n)|\, u(n) \tag{6.100}$$

$$y(n) = 0.5 + |\sin(0.1n)|\, u(n) \tag{6.101}$$

Now form the sequence

$$z(n) = x(n)y(n) \tag{6.102}$$

and plot $|z(n)|$. The quickly and slowly varying components should be apparent. It should be obvious that this magnitude function is playing the role of $|S(\omega)|$ in the cepstral analysis.[26]

(b) Now compute $\log|z(n)|$ and plot this result. This sequence is the *sum* of the correspondents of the quickly and slowly varying components,

$$\log|z(n)| = \log|x(n)| + \log|y(n)|. \tag{6.103}$$

Now that the nonlinear operation has been applied, are the effects of the two signals still apparent in the plot?

(c) According to "cepstral thinking," if we compute a DTFT on this sequence, the result should contain two well-separated components. Investigate whether this is the case by taking a 256-point DFT on a windowed frame of $\log|z(n)|$ padded with 256 zeros.

(d) Repeat these steps several times as the frequencies of $x(n)$ and $y(n)$ are brought closer together. Do the frequency domain results stay well separated for each pair of frequencies that seem well separated in terms of their numerical difference? If not, what implication might this have for cepstral analysis?

[26] We will pose this problem in "long-term" quantities and let the reader take care of the "short-term" processing details.

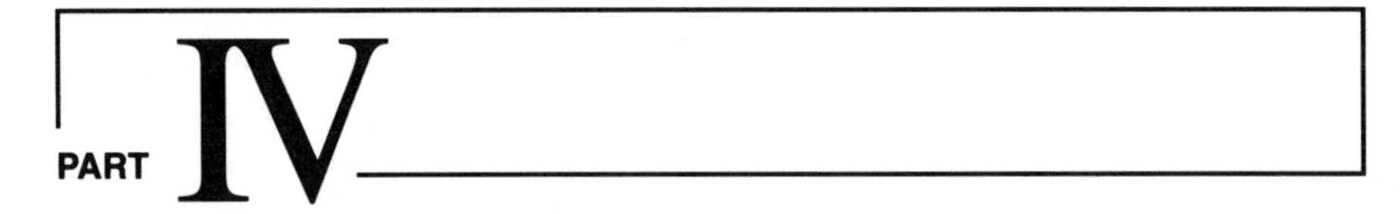
PART IV

CODING, ENHANCEMENT, AND QUALITY ASSESSMENT

CHAPTER 7

Speech Coding and Synthesis

Reading Notes: In addition to the general DSP and stochastic process material in Chapter 1, many developments in this chapter will depend on topics from Sections 1.3 and 1.4.

Generally speaking, there are two fundamental approaches to analyzing signals and systems. The first is to work with "real" time waveforms, using temporal averages and time arguments to obtain desired results. In this case, appropriate stationarity and ergodicity properties are assumed when results from stochastic process theory are required. The second general approach is to work initially with theoretical models, in particular stochastic processes, then move the essential findings into the practical "waveform" world by appropriate stationarity and ergodicity assumptions. The second approach is often more appealing in its formality, generalizability, and rigor, but it does not always lend itself as well to ad hoc *techniques that can be exploited for specific speech tasks.*[1] *Speech processing is inherently an "applied" discipline in which we often have "more waveforms than models" and the "temporal" approach is frequently more appealing or even essential. This approach has been taken throughout most of the book thus far.*

Historically, however, some of the material in this chapter (especially Section 7.2) has been developed and presented in a probabilistic setting. The basic problem addressed, speech coding, is amenable to, and indeed benefits from, a formal modeling approach. Therefore, for the first time in the book, the material will be initially presented in such a framework. The reader is encouraged to review the material in Section 1.2, with special attention to a review of the notation. In particular, it is important to have a clear understanding of the notations used to describe random processes and their key concepts. Recall that we use $\underline{x}$ to indicate a random process. The notation $\{\underline{x}(n), n \in$ (some time interval, say $\mathcal{N}$)$\}$ indicates the set of random variables that comprise $\underline{x}$, that is,

$$\underline{x} = \{\underline{x}(n), n \in \mathcal{N}\}. \tag{7.1}$$

Finally, "$x(n)$" (meaning $\{x(n), n \in \mathcal{N}\}$) is used to indicate a sample function or realization of $\underline{x}$. Vector counterparts to these notations will also be used for vector-valued stochastic processes, and, although our discussions have centered on discrete-time random processes, we will occasionally have reason to work with continuous-time processes as well.

[1] An interesting discussion of the duality between time and ensemble analysis is found in Gardner (1990).

7.1 Introduction

In this chapter we are concerned with the problem of efficiently representing speech signals digitally. The conversion of the analog speech waveform into digital form is usually called *speech coding*, or simply *coding*. A major objective in speech coding is to *compress* the signal, that is, to employ as few bits as possible in the digital representation of the speech signal. The efficient digital representation of the speech signal makes it possible to achieve bandwidth efficiency in the transmission of the signal over a variety of communication channels, or to store it efficiently on a variety of magnetic and optical media, for example, tapes and disks. Since the digitized speech is ultimately converted back to analog form for the user, an important consideration in speech coding is the level of signal distortion introduced by the digital conversion process.

Over the past four or five decades, a variety of speech coding techniques has been proposed, analyzed, and developed. In this chapter we describe the most important of these methods, which may be subdivided into two general categories, *waveform coders* and *voice coders* (*vocoders*).

In waveform encoding we attempt to directly encode speech waveforms in an efficient way by exploiting the temporal and/or spectral characteristics of speech signals. In contrast, voice encoding involves the representation of the speech signal by a set of parameters, the estimation of the parameters from frames of speech, and the efficient encoding of these parameters in digital form for possible transmission or storage.

In our discussion of speech coding techniques, we will assume that the analog speech signal is confined by filtering to a nominal bandwidth of 3–4 kHz. Then, the signal is sampled at a minimum rate of 8000 samples per second to avoid aliasing. In most of the waveform coding techniques to be described, the samples are processed one at a time by quantizing and encoding each sample separately. We call this process *scalar quantization and coding*. In contrast, a block of samples may be quantized as a single entity and the index of the resulting code vector may be transmitted to the receiver. In such a case we call the quantization process *vector quantization*. The latter may be used either for waveform encoding or in a vocoder.

We begin by considering scalar and vector quantization from a general viewpoint. Then we describe waveform encoding techniques in Section 7.3 and vocoder techniques in Section 7.4.

7.2 Optimum Scalar and Vector Quantization

In this section we consider the general problem of coding the output of an analog source from a theoretical viewpoint. In the speech coding problem, the "analog source" may be viewed as the speaker who produces analog acoustic waveforms. Formally, the "analog source" is a continuous-time stochastic process with continuous random variables (see Section 1.4.2), and the spoken waveforms correspond to realizations of the process. Although the work presented in this section is not specific to speech waveform coding, it might be useful to keep the preceding model in mind in order to put the results into the speech coding perspective. The theoretical results in this section relate the source coding rate to the distortion introduced by the quantization process and serve to guide us in the design of efficient speech coding techniques.

To begin, let us suppose that an analog source emits a message waveform $x_a(t)$, which may be considered a sample function of a continuous-time[2] stochastic process $\underline{x}_a$. The subscript a is used to remind the reader that we are temporarily working with an analog waveform drawn from an "analog" stochastic process (source). We assume that $\underline{x}_a$ is a stationary, correlation-ergodic, stochastic process with an autocorrelation function $r_{\underline{x}_a}(\tau) = r_{x_a}(\tau)$ and also a power spectral density function $\Gamma_{\underline{x}_a}(\Omega) = \Gamma_{x_a}(\Omega)$. Furthermore, let us assume that $\underline{x}_a$ is a bandlimited stochastic process, that is,

$$\Gamma_{\underline{x}_a}(\Omega) = 0 \qquad \text{for } |\Omega| > 2\pi W, \tag{7.2}$$

where W is the bandwidth in Hz. From the sampling theorem, we know that any sample function $x_a(t)$ may be represented without loss of information by its samples, say $x(n) = x_a(nT)$ for $-\infty < n < \infty$, as long as $T < 1/2W$. Of course, the sequence $x(n)$ can be considered a realization of the *discrete-time* stochastic process, say $\underline{x}$, which consists of the countable random variables $\underline{x}_a(nT)$ drawn from $\underline{x}_a$.

The sampling process converts the output of an analog source into an equivalent discrete-time sequence of samples. The samples are then quantized in amplitude and encoded as a binary sequence. Quantization of the amplitudes of the sampled signal results in waveform distortion and, hence, a loss in signal fidelity. The minimization of this distortion is considered below from the viewpoint of optimizing the quantizer characteristics.

Our treatment considers two cases, scalar quantization and vector quantization. A scalar quantizer operates on a single sample at a time and represents each sample by a sequence of binary digits. In contrast, a vector quantizer operates on N signal samples ($N > 1$) at a time and thus quantizes the signal vectors in N-dimensional space.

7.2.1 Scalar Quantization

Consider the sequence of samples $x(n)$, a realization of a discrete-time stochastic process $\underline{x}$, which is created by appropriate sampling of a band-

[2]Here we briefly work with a *continuous-time* random process. The concepts used here are analogous to those discussed for the discrete-time case in Section 1.2. The reader is referred to the textbooks listed in Appendix 1.B for details.

limited stochastic analog source as described above. The sequence $x(n)$ is input to the quantizer, which is assumed to have $L = 2^R$ levels. The number of *bits per sample* is therefore

$$R = \log_2 L. \tag{7.3}$$

The units of R may also be designated *bits per normalized second* (bpn), since we consider the interval of time between each sample to be a normalized second. Clearly, the quantity RF_s represents the bit rate per real time, and is measured in *bits per second* (bps). We shall use the abbreviations bpn and bps throughout our discussion.

Now let us denote the output of the quantizer by

$$\tilde{x}(n) = Q[x(n)], \tag{7.4}$$

where $Q[\,\cdot\,]$ represents the mapping (assumed functional) from the sequence $x(n)$ to the L discrete levels. We also assume that the marginal probability density function (pdf) of the stationary stochastic process $\underline{x}_a$ is known and is denoted $f_{\underline{x}_a(t)}(\xi)$ for any t. Because of stationarity, this pdf does not depend on t. Obviously, the sampled process $\underline{x}$ is also stationary and

$$f_{\underline{x}(n)}(\xi) = f_{\underline{x}_a(t)}(\xi) \qquad \text{for arbitrary choices of } t \text{ and } n. \tag{7.5}$$

Since the first-order pdf is the same for any time in either random process, for simplicity we will adopt the notation $f_{\underline{x}}(\xi)$ to mean

$$f_{\underline{x}}(\xi) \stackrel{\text{def}}{=} f_{\underline{x}(n)}(\xi) = f_{\underline{x}_a(t)}(\xi) \qquad \text{for arbitrary } t, n. \tag{7.6}$$

We wish to design the optimum scalar quantizer that minimizes the error in the following sense. Let $\underline{q}$ denote the random process that models the *quantization error sequence*. Realizations of $\underline{q}$ are of the form

$$q(n) \stackrel{\text{def}}{=} \tilde{x}(n) - x(n) = Q[x(n)] - x(n), \tag{7.7}$$

and the random variables in the process are formally described as

$$\underline{q}(n) = \underline{\tilde{x}}(n) - \underline{x}(n) = Q[\underline{x}(n)] - \underline{x}(n). \tag{7.8}$$

In a temporal sense, we desire to find the quantization mapping, Q, that minimizes the average of some function of the error sequence, say $h[q(n)]$. Assuming appropriate stationarity and ergodicity properties, we find Q that minimizes

$$D \stackrel{\text{def}}{=} \mathcal{E}\{h[\underline{q}(n)]\} = \int_{-\infty}^{\infty} h(\xi) f_{\underline{q}(n)}(\xi)\, d\xi \tag{7.9}$$

for an arbitrary n. Using (7.8) and recalling (7.6), we find that the quantity to minimize becomes

$$D = \int_{-\infty}^{\infty} h[Q(\xi) - \xi] f_{\underline{x}}(\xi)\, d\xi. \tag{7.10}$$

In general, an optimum quantizer is one that minimizes D by optimally selecting the output levels and the corresponding input range of each output level. This optimization problem has been considered by Lloyd (1957) and Max (1960), and the resulting optimum quantizer is usually called the *Lloyd–Max quantizer.*

For a *uniform quantizer,* the output levels are specified as

$$\tilde{x}(n) = \tilde{x}_k \stackrel{\text{def}}{=} (2k-1)\frac{\Delta}{2}, \qquad \text{when } (k-1)\Delta \leq x(n) < k\Delta, \tag{7.11}$$

where Δ is the *step size,* as shown in Fig. 7.1. When the uniform quantizer is symmetric with an even number of levels, the average distortion in (7.10) may be expressed as

$$\begin{aligned} D = 2 \sum_{k=1}^{(L/2)-1} \int_{(k-1)\Delta}^{k\Delta} h\left[\frac{(2k-1)\Delta}{2} - \xi\right] f_{\underline{x}}(\xi)\, d\xi \\ + 2 \int_{[(L/2)-1]\Delta}^{\infty} h\left[\frac{(L-1)\Delta}{2} - \xi\right] f_{\underline{x}}(\xi)\, d\xi . \end{aligned} \tag{7.12}$$

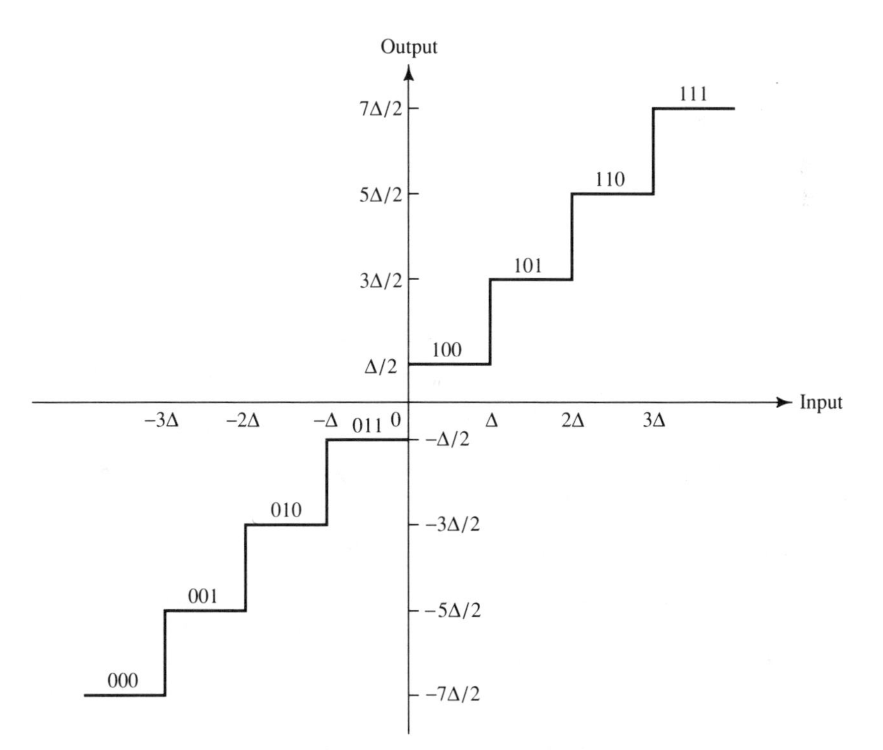

FIGURE 7.1. Input–output characteristic for a uniform quantizer. Shown is the case of a 3-bit quantizer ($2^3 = 8$ levels).

In this case, the minimization of D is carried out with respect to the step size parameter Δ. By differentiating D with respect to Δ, we obtain

$$\sum_{k=1}^{(L/2)-1} (2k-1) \int_{(k-1)\Delta}^{k\Delta} h'\left[\frac{(2k-1)\Delta}{2} - \xi\right] f_{\underline{x}}(\xi)\, d\xi$$
$$+ (L-1) \int_{[(L/2)-1]\Delta}^{\infty} h'\left[\frac{(L-1)\Delta}{2} - \xi\right] f_{\underline{x}}(\xi)\, d\xi = 0, \tag{7.13}$$

where $h'(\cdot)$ denotes the derivative of $h(\cdot)$.

By selecting the error criterion function $h(\cdot)$, the solution of (7.13) for the optimum step size can be obtained numerically on a digital computer for any given pdf $f_{\underline{x}}(\xi)$. For the *mean square error* (MSE) *criterion*, for which

$$h(\alpha) = \alpha^2, \tag{7.14}$$

Max (1960) evaluated the optimum step size Δ_{opt} and the minimum MSE when the pdf is Gaussian, that is,

$$f_{\underline{x}}(\xi) = \frac{1}{\sqrt{2\pi}} e^{-\xi^2/2}. \tag{7.15}$$

Some of these results are given in Table 7.1. We observe that the minimum mean square distortion D_{min} decreases by a little more than 5 dB for each doubling of the number of levels L. Hence each additional bit that is employed in a uniform quantizer with optimum step size Δ_{opt} for a Gaussian-distributed signal amplitude reduces the distortion by more than 5 dB.

By relaxing the constraint that the quantizer be uniform, the distortion can be reduced further. In this case, we let the output level

$$\tilde{x}(n) = y_k, \qquad \text{when } x_{k-1} \le x(n) < x_k. \tag{7.16}$$

TABLE 7.1. Optimum Step Sizes for Uniform Quantization of a Gaussian Random Variable.

Number of Output Levels	Optimum Step Size Δ_{opt}	Minimum MSE D_{min}	$10 \log D_{\text{min}}$ (dB)
2	1.596	0.3634	−4.4
4	0.9957	0.1188	−9.25
8	0.5860	0.03744	−14.27
16	0.3352	0.01154	−19.38
32	0.1881	0.00349	−24.57

For an L-level quantizer, the endpoints are $x_0 = -\infty$ and $x_L = \infty$. The resulting distortion is

$$D = \sum_{k=1}^{L} \int_{x_{k-1}}^{x_k} h(y_k - \xi)\, f_{\underline{x}}(\xi)\, d\xi, \tag{7.17}$$

which is now minimized by optimally selecting the sets of numbers $\{y_k\}$ and $\{x_k\}$.

The necessary conditions for a minimum distortion are obtained by differentiating D with respect to the $\{x_k\}$ and $\{y_k\}$. The result of this minimization is the pair of equations

$$h(y_k - x_k) = h(y_{k+1} - x_k), \qquad k = 1, 2, \ldots, L-1 \tag{7.18}$$

$$\int_{x_{k-1}}^{x_k} h'(y_k - \xi) f_{\underline{x}}(\xi)\, d\xi = 0, \qquad k = 1, 2, \ldots, L. \tag{7.19}$$

As a special case, we again consider minimizing the mean square value of the distortion. In this case, $h(\alpha) = \alpha^2$; hence (7.18) becomes

$$x_k = \frac{y_k + y_{k+1}}{2}, \qquad k = 1, 2, \ldots, L-1, \tag{7.20}$$

which is the midpoint between y_k and y_{k+1}. The endpoints are

$$x_0 = -\infty, \qquad x_L = \infty. \tag{7.21}$$

The corresponding equations determining the numbers $\{y_k\}$ are

$$y_k = \int_{x_{k-1}}^{x_k} \xi f_{\underline{x}}(\xi)\, d\xi, \qquad k = 1, 2, \ldots, L. \tag{7.22}$$

Thus y_k is the centroid (mean value) of $f_{\underline{x}}(\xi)$ between x_{k-1} and x_k. These equations may be solved numerically for any given $f_{\underline{x}}(\xi)$.

Tables 7.2 and 7.3 give the results of this optimization obtained by Max (1960) for the optimum four-level and eight-level quantizers of a

TABLE 7.2. Optimum Four-Level Quantizer for a Gaussian Random Variable.

Level k	x_k	y_k
1	−0.9816	−1.510
2	0.0	−0.4528
3	0.9816	0.4528
4	∞	1.510

$D_{\min} = 0.1175$; $10 \log D_{\min} = -9.3$ dB.

TABLE 7.3. Optimum Eight-Level Quantizer for a Gaussian Random Variable (Max, 1960).

Level k	x_k	y_k
1	−1.748	−2.152
2	−1.050	−1.344
3	−0.5006	−0.7560
4	0	−0.2451
5	0.5006	0.2451
6	1.050	−0.7560
7	1.748	1.344
8	∞	2.152

$D_{\text{min}} = 0.03454$; $10 \log D_{\text{min}} = -14.62$ dB.

Gaussian-distributed signal amplitude having zero mean and unit variance. In Table 7.4 we compare the minimum mean square distortion of a uniform quantizer to that of a nonuniform quantizer for the Gaussian-distributed signal amplitude. From the results of this table, we observe that the difference in the performance of the two types of quantizers is relatively small for small values of R (less than 0.5 dB for $R \leq 3$), but increases as R increases. For example, at $R = 5$, the nonuniform quantizer is approximately 1.5 dB better than the uniform quantizer.

It is instructive to plot the minimum distortion as a function of the bit rate $R = \log_2 L$ (bpn) for both the uniform and nonuniform quantizers. These curves are illustrated in Fig. 7.2. The functional dependence of the distortion D on the bit rate R may be expressed as $D(R)$. This function is called the *distortion-rate function* for the corresponding quantizer. We observe that the distortion-rate function for the optimum nonuniform quantizer falls below that of the optimum uniform quantizer.

TABLE 7.4. Comparison of Optimum Uniform and Nonuniform Quantizers for a Gaussian Random Variable (Max, 1960; Paez and Glisson, 1972).

	$10 \log_{10} D_{\text{min}}$	
R (bpn)	**Uniform (dB)**	**Nonuniform (dB)**
1	−4.4	−4.4
2	−9.25	−9.30
3	−14.27	−14.62
4	−19.38	−20.22
5	−24.57	−26.02
6	−29.83	−31.89
7	−35.13	−37.81

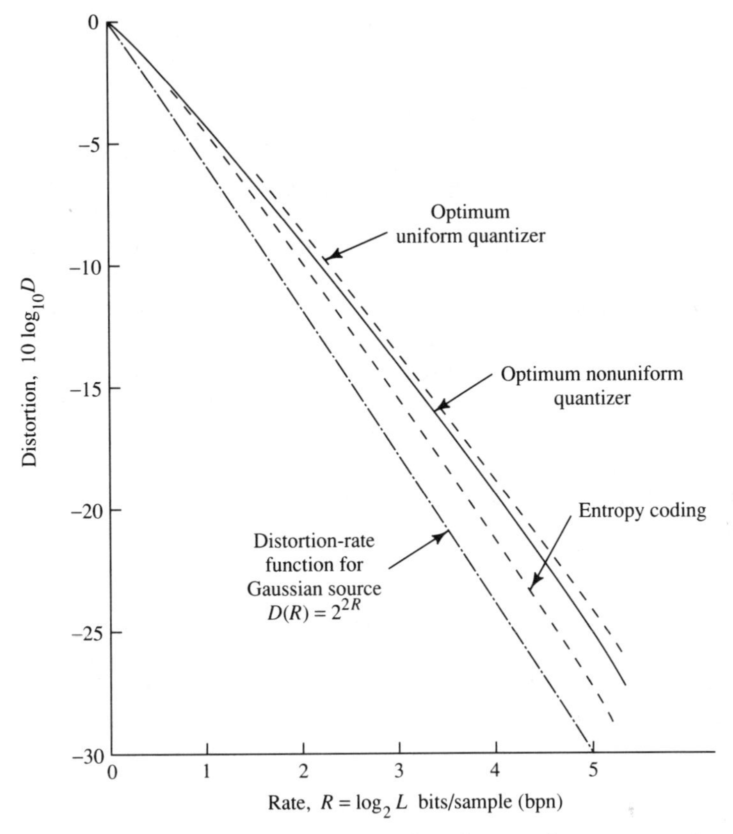

FIGURE 7.2. Distortion versus rate curves for discrete-time, memoryless Gaussian source.

Any quantizer reduces the continuous-amplitude source, $\underline{x}_a$, into a discrete-amplitude source, say $\underline{\tilde{x}}$. As above, suppose that the discrete values taken by the quantized source are

$$\{y_k, 1 \le k \le L\}. \tag{7.23}$$

This set of discrete entities is sometimes called an *alphabet* of the discrete source. Let us denote the probabilities associated with the *symbols* or *letters* from the source by $\{P(y_k) = p_k\}$. If the random variables, $\underline{\tilde{x}}(n)$, from the source are statistically independent, the discrete source is said to be *memoryless*. We know from Section 1.4.2 that such a source has entropy

$$H(\underline{\tilde{x}}) = -\sum_{k=1}^{L} p_k \log_2 p_k. \tag{7.24}$$

An algorithm due to Huffman (1952) provides an efficient method for source encoding based on the notion that the more probable symbols (or blocks of symbols) be assigned fewer bits and the less probable symbols be assigned more bits. The Huffman encoding algorithm yields a variable-length code in which the average number of bits per letter can be made as close to $H(\underline{\tilde{x}})$ as desired. We call this coding method *entropy coding.*

For example, the optimum four-level nonuniform quantizer for the Gaussian-distributed signal amplitude given by (7.15) results in the probabilities $p_1 = p_4 = 0.1635$ for the two outer levels and $p_2 = p_3 = 0.3365$ for the two inner levels. The entropy for this discrete source is $H(\underline{\tilde{x}}) = 1.911$ bits per letter. With entropy coding (Huffman coding) of blocks of output letters, we can achieve the minimum distortion of -9.30 dB with 1.911 bits per letter instead of 2 bits per letter. Max (1960) has given the entropy for the discrete source symbols resulting from quantization. Table 7.5 lists the values of the entropy for the nonuniform quantizer. These values are also plotted in Fig. 7.2 and labeled "entropy coding."

From this discussion we conclude that the quantizer can be optimized when the pdf of the continuous source output $\underline{x}_a$ is known. The optimum quantizer of $L = 2^R$ levels results in a minimum distortion of $D(R)$, where $R = \log_2 L$ bpn. Thus this distortion can be achieved by simply representing each quantized sample by R bits. However, more efficient coding is possible. The discrete source output that results from quantization is characterized by a set of probabilities $\{p_k\}$ that can be used to design efficient variable-length codes (Huffman codes) for the source output (entropy coding). The efficiency of any coding method can be compared, as described below, with the distortion-rate function or, equivalently, the rate-distortion function for the discrete-time, continuous-amplitude source that is characterized by the given pdf.

Rate-Distortion and Distortion-Rate Functions

It is interesting to compare the performance of the optimal uniform and nonuniform quantizers described above with the best achievable performance attained by any quantizer. For such a comparison, we present some basic results from information theory, due to Shannon, which we state in the form of theorems.

TABLE 7.5. Entropy of the Output of an Optimum Nonuniform Quantizer for a Gaussian Random Variable (Max, 1960).

R (bpn)	Entropy (bits/letter)	Distortion, $10 \log_{10} D_{\min}$
1	1.0	−4.4
2	1.911	−9.30
3	2.825	−14.62
4	3.765	−20.22
5	4.730	−26.02

THEOREM 7.1 RATE-DISTORTION FUNCTION FOR A MEMORYLESS GAUSSIAN SOURCE (SHANNON, 1959) *The minimum information rate* (bpn) *necessary to represent the output of a discrete-time, continuous-amplitude, memoryless stationary Gaussian source* [*corresponding to a random process* $\underline{x}$ *with random variables* $\underline{x}(n)$] *based on an* MSE *distortion measure per symbol (single-letter distortion measure) is*

$$R_g(D) = \begin{cases} \frac{1}{2}\log_2(\sigma_{\underline{x}}^2/D), & 0 \leq D \leq \sigma_{\underline{x}}^2 \\ 0, & D > \sigma_{\underline{x}}^2, \end{cases} \tag{7.25}$$

where $\sigma_{\underline{x}}^2$ *is the variance of the Gaussian source output. The function* $R_g(D)$ *is called the rate-distortion function for the source* (*the subscript "g" is used to denote the memoryless Gaussian source*).

We should note that (7.25) implies that no information needs to be transmitted when the distortion $D \geq \sigma_{\underline{x}}^2$. Specifically, $D = \sigma_{\underline{x}}^2$ can be obtained by using zeros in the reconstruction of the signal. For $D > \sigma_{\underline{x}}^2$, we can use statistically independent, zero-mean Gaussian noise samples with a variance of $D - \sigma_{\underline{x}}^2$ for the reconstruction. $R_g(D)$ is plotted in Fig. 7.3.

The rate-distortion function $R(D)$ of a source is associated with the following basic source coding theorem in information theory.

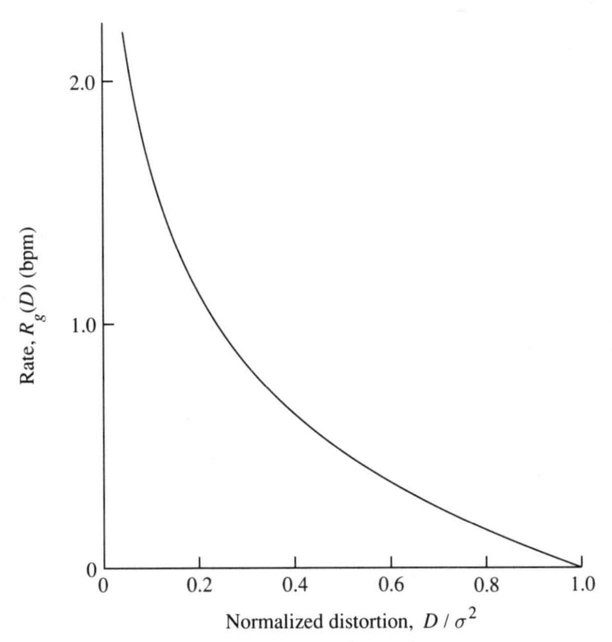

FIGURE 7.3. Rate-distortion function for a continuous-amplitude, memoryless Gaussian source.

THEOREM 7.2 SOURCE CODING WITH A DISTORTION MEASURE (SHANNON, 1959) *There exists a coding scheme that maps the source output into codewords such that for any given distortion D, the minimum rate R(D)* bpn *is sufficient to reconstruct the source output with an average distortion that is arbitrarily close to D.*

It is clear, therefore, that the rate-distortion function $R(D)$ for any source represents (by definition) a lower bound on the source rate that is possible for a given level of distortion.

Let us return to the result in (7.25) for the rate-distortion function of a memoryless Gaussian source. If we reverse the functional dependence between D and R, we may express D in terms of R as

$$D_g(R) = 2^{-2R}\sigma_{\underline{x}}^2. \tag{7.26}$$

This function is called the *distortion-rate function* for the discrete time, memoryless Gaussian source. When we express the distortion in (7.26) in dB, we obtain

$$10\log_{10} D_g(R) = -6R + 10\log_{10}\sigma_{\underline{x}}^2. \tag{7.27}$$

Note that the mean square distortion decreases at a rate of 6 dB per bit. The graph of (7.27) is plotted in Fig. 7.2, and may be used as a basis for comparing the performance of the optimal quantizers against the minimum achievable distortion for the Gaussian source.

If we compare the performance of the optimum nonuniform quantizer to the distortion-rate function, we find, for example, that at a distortion of -26 dB, entropy coding is 0.41 bpn more than the minimum rate given by (7.27) and simple block coding (uniform quantization) of each letter requires 0.68 bpn more than the minimum rate. We also observe that the distortion-rate functions for the optimal uniform and nonuniform quantizers for the Gaussian source approach the slope of -6 dB per bit asymptotically for large R.

Explicit results on the rate-distortion functions for memoryless non-Gaussian sources are not available. However, there are useful upper and lower bounds on the rate-distortion function for any discrete-time, continuous-amplitude, memoryless source. An upper bound is given by the following theorem.

THEOREM 7.3 [UPPER BOUND ON $R(D)$] *The rate-distortion function of a memoryless, continuous-amplitude source with zero mean and finite variance $\sigma_{\underline{x}}^2$ with respect to the* MSE *distortion measure is upper bounded as*

$$R(D) \le \frac{1}{2}\log_2\frac{\sigma_{\underline{x}}^2}{D}, \qquad 0 \le D \le \sigma_{\underline{x}}^2. \tag{7.28}$$

A proof of this theorem is given by Berger (1971). It implies that the Gaussian source requires the maximum rate among all other sources for a specified level of mean square distortion. Thus the rate-distortion $R(D)$ of any continuous-amplitude, memoryless source with zero mean and finite variance $\sigma_{\underline{x}}^2$ satisfies the condition $R(D) \le R_g(D)$. Similarly, the distortion-rate function of the same source satisfies the condition

$$D(R) \le D_g(R) = 2^{-2R}\sigma_{\underline{x}}^2. \tag{7.29}$$

A lower bound on the rate-distortion function also exists. This is called the *Shannon lower bound* for an MSE distortion measure, and is given as

$$R^*(D) = H(\underline{x}) - \frac{1}{2}\log_2 2\pi eD, \tag{7.30}$$

where $H(\underline{x})$ is the *differential entropy* of the continuous-amplitude, memoryless source, defined as

$$H(\underline{x}) \stackrel{\text{def}}{=} -\int_{-\infty}^{\infty} f_{\underline{x}(n)}(\xi)\log_2 f_{\underline{x}(n)}(\xi)\, d\xi \tag{7.31}$$

for arbitrary n.[3] The distortion-rate function corresponding to (7.30) is

$$D^*(R) = \frac{1}{2\pi e}2^{-2[R-H(\underline{x})]}. \tag{7.32}$$

Therefore, the rate-distortion function for any continuous-amplitude, memoryless source is bounded from above and below as

$$R^*(D) \le R(D) \le R_g(D) \tag{7.33}$$

and the corresponding distortion-rate function is bounded as

$$D^*(R) \le D(R) \le D_g(R). \tag{7.34}$$

The differential entropy of the memoryless Gaussian source is

$$H_g(\underline{x}) = \frac{1}{2}\log_2 2\pi e\sigma_{\underline{x}}^2, \tag{7.35}$$

so that the lower bound $R^*(D)$ in (7.30) reduces to $R_g(D)$. Now, if we express $D^*(R)$ in terms of decibels and normalize it by setting $\sigma_{\underline{x}}^2 = 1$ [or dividing $D^*(R)$ by $\sigma_{\underline{x}}^2$], we obtain from (7.32)

$$10\log_{10} D^*(R) = -6R - 6[H_g(\underline{x}) - H(\underline{x})] \tag{7.36}$$

or, equivalently,

$$10\log_{10}\frac{D_g(R)}{D^*(R)} = 6[H_g(\underline{x}) - H(\underline{x})]\,\text{dB} \tag{7.37}$$

$$= 6[R_g(D) - R^*(D)]\,\text{dB}. \tag{7.38}$$

[3]We temporarily ignore the convention established in (7.6) in order to give a precise definition. Also compare (1.243).

The relations (7.37) and (7.38) allow us to compare the lower bound in the distortion with the upper bound, which is the distortion for the Gaussian source. We note that $D^*(R)$ also decreases at -6 dB per bit. We should also mention that the differential entropy $H(\underline{x})$ is upper bounded by $H_g(\underline{x})$, as shown by Shannon (1948).

Table 7.6 lists four pdf's that are models commonly used for source signal distributions. Shown in the table are the differential entropies, the differences in rates in bpn, and the difference in distortion between the upper and lower bounds. Paez and Glisson (1972) have shown that the gamma probability distribution provides a good model for speech signal amplitude. The optimum quantization levels for this amplitude distribution are given in Table 7.7 for $L = 2, 4, 8, 16, 32$. The signal variance has been normalized to unity. From Table 7.6 we note that the gamma pdf shows the greatest deviation from the Gaussian. The Laplacian pdf is the most similar to the Gaussian, and the uniform pdf ranks second of the pdf's shown in Table 7.6. These results provide some benchmarks on the difference between the upper and lower bounds on distortion and rate.

Before concluding this section, let us consider a continuous-time bandlimited Gaussian source with spectral density

$$\Gamma_{\underline{x}_a}(\Omega) = \begin{cases} \sigma^2_{\underline{x}_a}/2W, & |\Omega| \leq 2\pi W \\ 0, & |\Omega| > 2\pi W. \end{cases} \tag{7.39}$$

When the output of this source is sampled at the Nyquist rate, the samples are uncorrelated and, since the source is Gaussian, the samples are also statistically independent. Hence the equivalent discrete-time Gaussian source, $\underline{x}$, is memoryless. The rate-distortion function for each sample is given by (7.25). Therefore, the rate-distortion function for the bandlimited white Gaussian source in bps is

$$R_g(D) = W \log_2 \frac{\sigma^2_{\underline{x}}}{D}, \qquad 0 \leq D \leq \sigma^2_{\underline{x}}, \tag{7.40}$$

The corresponding distortion-rate function is

$$D_g(R) = 2^{-R/W}\sigma^2_{\underline{x}}, \tag{7.41}$$

which, when expressed in decibels and normalized by $\sigma^2_{\underline{x}}$, becomes

$$10 \log \frac{D_g(R)}{\sigma^2_{\underline{x}}} = -\frac{3R}{W}. \tag{7.42}$$

The more general case in which the Gaussian process is neither white nor bandlimited has been treated by Gallager (1968) and Goblick and Holsinger (1967).

TABLE 7.6. Differential Entropies and Rate-Distortion Comparisons of Four Common pdf's for Signal Models.

pdf	$f_{\underline{x}}(x)$	$H(\underline{x})$	$R_g(D) - R^*(D)$ (bpn)	$D_g(R) - D^*(R)$ (dB)
Gaussian	$\frac{1}{\sqrt{2\pi}\sigma_{\underline{x}}} e^{-x^2/2\sigma_{\underline{x}}^2}$	$\frac{1}{2}\log_2\left(2\pi e\sigma_{\underline{x}}^2\right)$	0	0
Uniform	$\frac{1}{2\sqrt{3}\sigma_{\underline{x}}},\ \lvert x\rvert \leq \sqrt{3}\sigma_{\underline{x}}$	$\frac{1}{2}\log_2\left(12\sigma_{\underline{x}}^2\right)$	0.255	1.53
Laplacian	$\frac{1}{\sqrt{2}\sigma_{\underline{x}}} e^{-\sqrt{2}\lvert x\rvert/\sigma_{\underline{x}}}$	$\frac{1}{2}\log_2\left(2e^2\sigma_{\underline{x}}^2\right)$	0.104	0.62
Gamma	$\frac{\sqrt[4]{3}}{\sqrt{8\pi\sigma_{\underline{x}}\lvert x\rvert}} e^{-\sqrt{3}\lvert x\rvert/2\sigma_{\underline{x}}}$	$\frac{1}{2}\log_2\left(4\pi e^{0.423}\sigma_{\underline{x}}^2/3\right)$	0.709	4.25

TABLE 7.7. Optimum Quantizers for Signals with a Gamma Distribution (Paez and Glisson, 1972).

Level	2		4		8		16		32	
	x_i	y_i	x_i	y_i	x_i	y_i	x_i	y_i	x_i	y_i
1	∞	0.577	1.205	0.302	0.504	0.149	0.229	0.072	0.101	0.033
2			∞	2.108	1.401	0.859	0.588	0.386	0.252	0.169
3					2.872	1.944	1.045	0.791	0.429	0.334
4					∞	3.799	1.623	1.300	0.630	0.523
5							2.372	1.945	0.857	0.737
6							3.407	2.798	1.111	0.976
7							5.050	4.015	1.397	1.245
8							∞	6.085	1.720	1.548
9									2.089	1.892
10									2.517	2.287
11									3.022	2.747
12									3.633	3.296
13									4.404	3.970
14									5.444	4.838
15									7.046	6.050
16									∞	8.043
$D_{\min}$	0.6680		0.2326		0.0712		0.0196		0.0052	

7.2.2 Vector Quantization

In Section 7.2.1 we considered the quantization of the output signal from a continuous-amplitude source when the quantization is performed on a sample-by-sample basis, that is, by scalar quantization. In this section we consider the joint quantization of a block of signal samples or a block of signal parameters. This type of quantization is called *block* or *vector quantization* (VQ).

A fundamental result of rate-distortion theory is that better performance can be achieved by quantizing vectors instead of scalars, even if the continuous-amplitude source is memoryless. If, in addition, the signal samples or signal parameters are statistically dependent, we can exploit the dependency by jointly quantizing blocks of samples or parameters, and thus achieve even greater efficiency (lower bit rate) compared with that achieved by scalar quantization.

The general VQ problem may be formulated as follows. Suppose that we have an N-dimensional random vector

$$\underline{\mathbf{x}} = [\underline{x}(1) \quad \underline{x}(2) \quad \cdots \quad \underline{x}(N)]^T, \tag{7.43}$$

where the $\underline{x}(i)$ are real random variables. Later these random variables will correspond to signal samples, or perhaps to parameters such as LP coefficients characterizing a frame of speech. Consequently, we index them by integers in parentheses as a foreshadowing of this fact. The random vector $\underline{\mathbf{x}}$ is governed by a joint pdf

$$f_{\underline{\mathbf{x}}}(\xi_1, \ldots, \xi_N) = f_{\underline{x}(1), \ldots, \underline{x}(N)}(\xi_1, \ldots, \xi_N). \tag{7.44}$$

Formally, the quantizer maps the random vector $\underline{\mathbf{x}}$ to another random vector $\underline{\mathbf{y}}$ of dimension N,

$$\underline{\mathbf{y}} = [\underline{y}(1) \quad \underline{y}(2) \quad \cdots \quad \underline{y}(N)]^T. \tag{7.45}$$

Let us express this mapping as

$$\underline{\mathbf{y}} = Q(\underline{\mathbf{x}}). \tag{7.46}$$

The vector $\underline{\mathbf{y}}$ has a special distribution in that it may only take one of L (deterministic) vector values in $\mathbb{R}^N$. Therefore, its pdf will consist of L impulses over the N-dimensional hyperplane. Let us designate these L values by $\mathbf{y}_1, \ldots, \mathbf{y}_L$.

Basically, the vector quantization of $\underline{\mathbf{x}}$ may be viewed as a pattern recognition problem involving the classification of the outcomes of the random variable $\underline{\mathbf{x}}$, say $\mathbf{x}$, into a discrete number of categories or "cells" in N-space in a way that optimizes some fidelity criterion, such as mean square distortion. For example, consider the quantization of the outcomes of the two-dimensional vectors $\underline{\mathbf{x}} = [\underline{x}(1) \quad \underline{x}(2)]^T$. The two-dimensional space is partitioned into cells as illustrated in Fig. 7.4, where we have arbitrarily selected hexagonal-shaped cells $\{C_k\}$. All input vectors that fall in cell C_k are "quantized" into the vector $\mathbf{y}_k$, which is shown in Fig. 7.4 as the center of the hexagon. In this example, there are

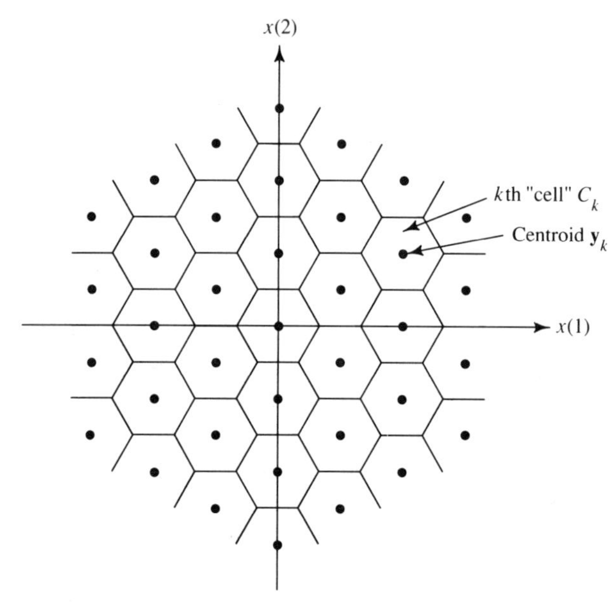

FIGURE 7.4. An example of quantization in two-dimensional space.

$L = 37$ vectors, one for each of the 37 cells into which the two-dimensional space has been partitioned. Under a VQ scheme, if $\mathbf{x}$ is a vector of signal samples to be transmitted or stored, for example, then only the index k of the cell to which $\mathbf{x}$ is assigned is actually transmitted or stored.

In general, quantization of the N-dimensional vector $\mathbf{x}$ into an N-dimensional vector $\mathbf{y}_k$ introduces a quantization error or a distortion $d(\mathbf{x}, \mathbf{y}_k)$. In a statistical sense, the average distortion over the set of input vectors is

$$\begin{aligned} D &= \sum_{k=1}^{L} P(\underline{\mathbf{x}} \in C_k)\mathcal{E}\{d(\underline{\mathbf{x}}, \mathbf{y}_k) | \underline{\mathbf{x}} \in C_k\} \\ &= \sum_{k=1}^{L} P(\underline{\mathbf{x}} \in C_k) \int \cdots \int_{\mathbf{x} \in C_k} d(\underline{\mathbf{x}}, \mathbf{y}_k) f_{\underline{\mathbf{x}}}(\xi_1, \ldots, \xi_N)\, d\xi_1 \cdots d\xi_N, \end{aligned} \tag{7.47}$$

where $P(\underline{\mathbf{x}} \in C_k)$ is the probability that the vector $\underline{\mathbf{x}}$ falls in the cell C_k and $f_{\underline{\mathbf{x}}}$ is the joint pdf defined in (7.44). As in the case of scalar quantization, we can minimize D by selecting the cells $\{C_k\}$ for a given pdf $f_{\underline{\mathbf{x}}}$.

The "pointwise" distortion measures, $d(\underline{\mathbf{x}}, \mathbf{y}_k)$ are typically the distance measures in $\mathbb{R}^N$ that we discussed in Section 1.3.1. These include the l_1, l_2, and l_∞ norms, including weighted versions of the l_2 norm. If LP parameters comprise the feature vector, the Itakura distance or Itakura–Saito distance (see Section 5.3.5) is often used.

Thus far, our discussion has been formulated in terms of abstract (stochastic) quantities. Before proceeding with the theoretical developments, let us point out some of the practical uses of VQ. In practice, of course, we will be faced with the task of associating a real input vector, say $\mathbf{x}$, with one of the vectors $\mathbf{y}_1, \ldots, \mathbf{y}_L$. As noted above, $\mathbf{x}$ is to be thought of as an outcome of the vector-valued random variable $\underline{\mathbf{x}}$. $\mathbf{x}$ might represent an N-length frame of speech that is to be coded. For example, suppose

$$\mathbf{x} = [\text{elements } f(n;m),\ n = m - N + 1, \ldots, m\ ]^T, \tag{7.48}$$

where $f(n;m)$ is our usual notation for a frame of speech ending at time m for a given window. However, VQ is not limited to quantizing a block of signal samples of a source waveform. It can also be applied to quantizing a set of parameters extracted from the data. Examples include linear predictive coefficients, in which case $\mathbf{x}$ is the M-dimensional LP vector

$$\mathbf{x} = [\hat{a}(1;m) \cdots \hat{a}(M;m)]^T. \tag{7.49}$$

These parameters can be considered and quantized as a block by applying some appropriate distortion measure. Ordinarily, the Itakura distance measure would be used as the measure of distortion.

Alternative sets of parameters that may be quantized as a block and transmitted to the receiver are reflection coefficients, log-area ratio parameters, and inverse sine parameters. These were introduced in Section 5.3.3. In these cases the vector $\mathbf{x}$ takes the forms

$$\mathbf{x} = [\kappa(1;m) \cdots \kappa(M;m)]^T, \tag{7.50}$$

$$\mathbf{x} = [g(1;m) \cdots g(M;m)]^T, \tag{7.51}$$

and

$$\mathbf{x} = [\sigma(1;m) \cdots \sigma(M;m)]^T, \tag{7.52}$$

respectively, for coding of the N-length frame ending at time m. The l_2 and l_∞ norms are typically used as measures of distortion for these parameters (Makhoul et al., 1985).

Finally, let us note that although we have been discussing VQ as a means of coding for transmission or storage purposes, it is also widely used as a data compression technique in recognition schemes. We shall see this in Chapter 13. In this case, the cepstral coefficients of a block of speech are frequently vector quantized, and $\mathbf{x}$ takes the form

$$\mathbf{x} = [c_s(1;m) \cdots c_s(M;m)]^T, \tag{7.53}$$

for M cepstral coefficients computed on a frame ending at time m. Appropriate distortion measures for cepstral vectors have been discussed in Section 6.2.3.

Let us return to the mathematical formulation of VQ and consider the partitioning of the N-dimensional space into L cells $\{C_k, 1 \le k \le L\}$

so that the average distortion is minimized over all L-level quantizers. There are two conditions for optimality:

1. The optimal quantizer employs a *nearest neighbor* selection rule, which may be expressed mathematically as follows: Let $\mathbf{x}$ be a vector to be classified (quantized). Then

$$Q(\mathbf{x}) = \mathbf{y}_k \qquad (\mathbf{x} \in C_k) \tag{7.54}$$

if and only if

$$d(\mathbf{x}, \mathbf{y}_k) \leq d(\mathbf{x}, \mathbf{y}_j) \qquad \text{for } k \neq j,\ 1 \leq j \leq L. \tag{7.55}$$

2. Each output vector $\mathbf{y}_k$ is chosen to minimize the average distortion in cell C_k, say D_k. In other words, $\mathbf{y}_k$ is the vector in C_k such that

$$\mathbf{y}_k = \underset{\mathbf{y}}{\text{argmin}}\ D_k = \underset{\mathbf{y}}{\text{argmin}}\ \mathcal{E}\{d(\underline{\mathbf{x}}, \mathbf{y}) | \underline{\mathbf{x}} \in C_k\}$$

$$= \underset{\mathbf{y}}{\text{argmin}} \int \cdots \int_{\mathbf{x} \in C_k} d(\underline{\mathbf{x}}, \mathbf{y}) f_{\mathbf{x}}(\xi_1, \ldots, \xi_n)\, d\xi_1 \ldots d\xi_N. \tag{7.56}$$

The vector $\mathbf{y}_k$ is the centroid of the cell.

Thus these conditions for optimality can be applied to partition the N-dimensional space into cells $\{C_k,\ 1 \leq k \leq L\}$ when the joint pdf $f_{\underline{\mathbf{x}}}(\cdot)$ is known. It is clear that these two conditions represent the generalization of the optimum scalar quantization problem to the N-dimensional vector quantization problem. In general, we expect the code vectors to be closer together in regions where the joint pdf is large and farther apart in regions where $f_{\underline{\mathbf{x}}}(\cdot)$ is small.

As an upper bound on the performance of a vector quantizer, we may use the performance of the optimal scalar quantizer, which can be applied to each component of the vector, as described in the preceding section. On the other hand, the best performance that can be achieved by optimum VQ is given by the rate-distortion function or, equivalently, the distortion-rate function.

The distortion-rate function, which was introduced in the preceding section, may be defined in the context of VQ as follows. In this case we envision a vector-valued input source (stationary stochastic process), say $\underline{\mathbf{x}}$, consisting of a sequence of random vectors $\underline{\mathbf{x}}(m)$. Consider, for example, that each $\underline{\mathbf{x}}(m)$ represents a block (frame) of N speech samples ending at time m, as in (7.48). Each input vector is then quantized to produce an output vector random process, say $\underline{\mathbf{y}}$, with vector random variables $\underline{\mathbf{y}}(m)$. The transformation is of the form

$$\underline{\mathbf{y}}(m) = Q[\underline{\mathbf{x}}(m)] \tag{7.57}$$

and each $\underline{\mathbf{y}}(m)$ has the discrete probability distribution

$$P[\underline{\mathbf{y}}(m) = \mathbf{y}_k] = p_k. \tag{7.58}$$

From (7.47) we have that the average distortion D resulting from representing $\underline{\mathbf{x}}(m)$ by $\underline{\mathbf{y}}(m)$ (for arbitrary m) is $\mathcal{E}\{d[\underline{\mathbf{x}}(m), \underline{\mathbf{y}}(m)]\}$. It is useful to express this on a "per dimension" basis, that is, we define

$$D_N \stackrel{\text{def}}{=} \frac{D}{N} = \frac{\mathcal{E}\{d[\underline{\mathbf{x}}(m), \underline{\mathbf{y}}(m)]\}}{N}. \tag{7.59}$$

Now the information in the output process, which consists exclusively of the vectors $\mathbf{y}_k$, $1 \le k \le L$, can be transmitted at an average bit rate of

$$R = \frac{H(\underline{\mathbf{y}})}{N} \text{ bpn}, \tag{7.60}$$

where $H(\underline{\mathbf{y}})$ is the entropy of the quantized source output,

$$H(\underline{\mathbf{y}}) = -\sum_{k=1}^{L} p_k \log_2 p_k. \tag{7.61}$$

For a given average rate R, the minimum achievable distortion per dimension, $D_N(R)$, is

$$D_N(R) = \min_{Q[\underline{\mathbf{x}}(m)]} \frac{\mathcal{E}\{d[\underline{\mathbf{x}}(m), \underline{\mathbf{y}}(m)]\}}{N} \quad \text{for arbitrary } m, \tag{7.62}$$

where $R \ge H(\underline{\mathbf{y}})/N$ and the minimum in (7.62) is taken over all possible mappings $Q[\,\cdot\,]$. In the limit as the number of dimensions N is allowed to approach infinity, we obtain

$$D(R) = \lim_{N \to \infty} D_N(R), \tag{7.63}$$

where $D(R)$ is the distortion-rate function introduced in Section 7.2.1. It is apparent from this development that the distortion-rate function can be approached arbitrarily closely by increasing the size N of the vectors.

The development above is predicated on the assumption that the joint pdf $f_{\underline{\mathbf{x}}}(\,\cdot\,)$ of the data vector is known. However, in practice, the joint pdf may not be known. In such a case, it is possible to select the quantized output vectors adaptively from a set of *training vectors*, using the K-means algorithm described in Section 1.3.5. This algorithm iteratively subdivides the training vectors into L clusters such that the two necessary conditions for optimality are practically satisfied.

It is appropriate that we digress momentarily to remind the reader of the term *LBG algorithm* that often appears in the present context. In Section 1.3.5 we noted that Lloyd (1957), in considering scalar quantization for pulse code modulation, had essentially developed the K-means algorithm independently of the pattern recognition researchers. Linde et al. (1980) were the first in the communications field to suggest the use of K-means for VQ. (A more complete account of this history is given in Section 1.3.5.) Accordingly, K-means, particularly when used for VQ in the speech processing and other communications literature, is frequently called the *Linde–Buzo–Gray* (LBG) *algorithm*. The LBG algorithm and a slight variation are detailed in Fig. 1.16.

Once we have selected the output vectors $\{\mathbf{y}_k, 1 \leq k \leq L\}$, we have established what is known as a *codebook*. Each input signal vector $\mathbf{x}(m)$ is quantized to the output vector that is nearest to it according to the distortion measure that is adopted. If the computation involves evaluating the distance between $\mathbf{x}(m)$ and each of the L possible output vectors $\{\mathbf{y}_k\}$, the procedure constitutes a *full search*. If we assume that each computation requires F floating point operations (flops),[4] the computational requirement for a full search is

$$C = F \cdot L \tag{7.64}$$

flops per input vector.

If we select L to be a power of two, then $\log_2 L$ is the number of bits required to represent each vector. Now, if R denotes the bit rate per sample [per component or dimension of $\mathbf{x}(m)$], we have $NR = \log_2 L$ and, hence, the computational cost is

$$C = F \cdot 2^{NR}. \tag{7.65}$$

Note that the number of computations grows exponentially with the dimensionality parameter N and the bit rate R per dimension. Because of this exponential increase of the computational cost, VQ has been applied to low bit rate source encoding, such as in coding reflection coefficients or log-area ratios in linear predictive coding.

The computational cost associated with full search can be reduced by slightly suboptimum algorithms [see (Cheng et al., 1984; and Gersho, 1982)]. A particularly simple approach is to construct the codebook based on a binary tree search. Binary tree search is a hierarchical clustering method for partitioning the N-dimensional space in a way that reduces the computational cost of the search to be proportional to $\log_2 L$. This method begins by subdividing the N-dimensional training vectors into two regions using the K-means algorithm with $K = 2$. Thus we obtain two regions and two corresponding centroids, say $\mathbf{y}_1$ and $\mathbf{y}_2$. In the next step, all points that fall into the first region are further subdivided into two regions by using the K-means algorithm with $K = 2$. Thus we obtain two centroids, say $\mathbf{y}_{1,1}$ and $\mathbf{y}_{1,2}$. This procedure is repeated for the second region to yield the two centroids $\mathbf{y}_{2,1}$ and $\mathbf{y}_{2,2}$. Thus the N-dimensional space is divided into four regions, each region having its corresponding centroid. This process is repeated until we have subdivided the N-dimensional space into $L = 2^B$ regions where $B = NR$ is the number of bits per code vector. (Note that B is an integer.) The corresponding code vectors may be viewed as terminal nodes in a binary tree, as shown in Fig. 7.5.

Given a signal (or parameter) vector $\mathbf{x}(m)$, the search begins by comparing $\mathbf{x}(m)$ with $\mathbf{y}_1$ and $\mathbf{y}_2$. If $d[\mathbf{x}(m), \mathbf{y}_1] < d[\mathbf{x}(m), \mathbf{y}_2]$, we eliminate

[4]We again define a flop to be one multiplication and one addition.

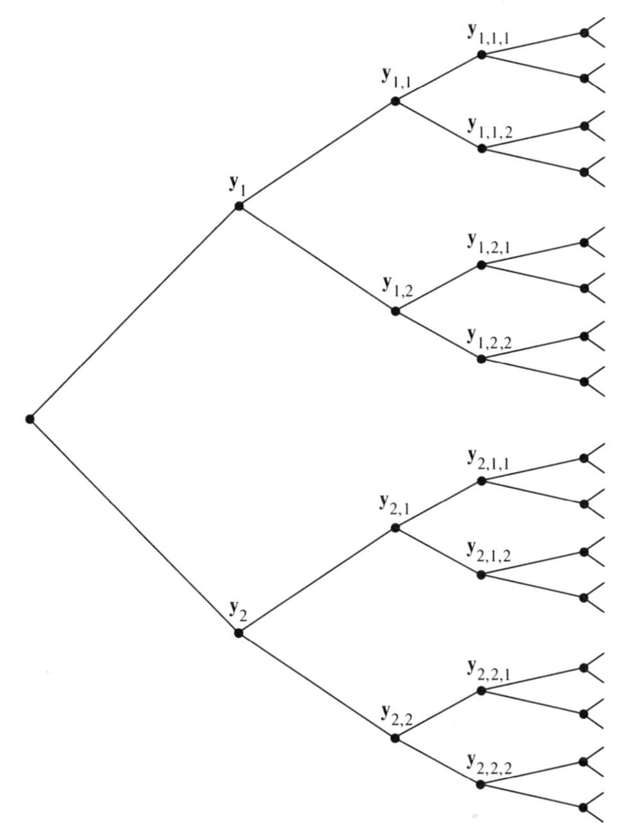

FIGURE 7.5. Uniform tree for binary search vector quantization.

the half of the tree stemming from $\mathbf{y}_2$. Then we compute $d[\mathbf{x}(m), \mathbf{y}_{1,i}]$, $i = 1, 2$. If $d[\mathbf{x}(m), \mathbf{y}_{1,1}] > d[\mathbf{x}(m), \mathbf{y}_{1,2}]$, we eliminate the part of the tree stemming from $\mathbf{y}_{1,1}$ and continue the binary search along $\mathbf{y}_{1,2}$. The search terminates after B steps when we reach a terminal node.

The computational cost of the binary tree search is

$$C = 2N \log_2 L = 2NB \tag{7.66}$$

flops, which is linear in B compared with the exponential cost for full search. Although the cost has been significantly reduced, the memory required to store the vectors has actually increased from NL to approximately $2NL$. The reason for this increase is that we now have to store the vectors at the intermediate nodes in addition to the vectors at the terminal nodes.

This binary tree search algorithm generates a *uniform tree*. In general, the resulting codebook will be suboptimum in the sense that the codewords result in more distortion compared to the codewords generated by the unconstrained method corresponding to a full search. Some improvement in performance may be obtained if we remove the restriction that the tree be uniform. In particular, a codebook resulting in lower distortion is obtained by subdividing the cluster of test vectors having the largest total distortion at each step in the process. Thus, in the first step, the N-dimensional space is divided into two regions. In the second step, we select the cluster with the larger distortion and subdivide it. Now, we have three clusters. The next subdivision is performed on the cluster having the largest distortion. Thus we obtain four clusters and we repeat the process. The net result is that we generate a *nonuniform code tree* as illustrated in Fig. 7.6 for the case $L = 7$. Note that L is no longer constrained to be a power of two. In Section 7.4.6 we compare the distortion of these two suboptimum search schemes with that for full search.

To demonstrate the benefits of vector quantization compared with scalar quantization, we present the following example, due to Makhoul et al. (1985).

EXAMPLE

Let $\underline{\mathbf{x}} = [\underline{x}(1)\ \underline{x}(2)]^T$ be a random vector with uniform joint pdf

$$f_{\underline{\mathbf{x}}}(\xi_1, \xi_2) = f_{\underline{\mathbf{x}}}(\boldsymbol{\xi}) = \begin{cases} 1/ab, & \boldsymbol{\xi} \in \mathbf{C} \\ 0, & \text{otherwise} \end{cases}, \tag{7.67}$$

where $\mathbf{C}$ is the rectangular region illustrated in Fig. 7.7. Note that the rectangle is rotated by 45° relative to the horizontal axis. Also shown in Fig. 7.7 are the marginal densities $f_{\underline{x}(1)}(\xi)$ and $f_{\underline{x}(2)}(\xi)$. If we quantize the outcomes of $\underline{x}(1)$ and $\underline{x}(2)$ separately by using uniform intervals of length Δ, the number of levels needed is

$$L_{x(1)} = L_{x(2)} = \frac{a+b}{\sqrt{2}\Delta}. \tag{7.68}$$

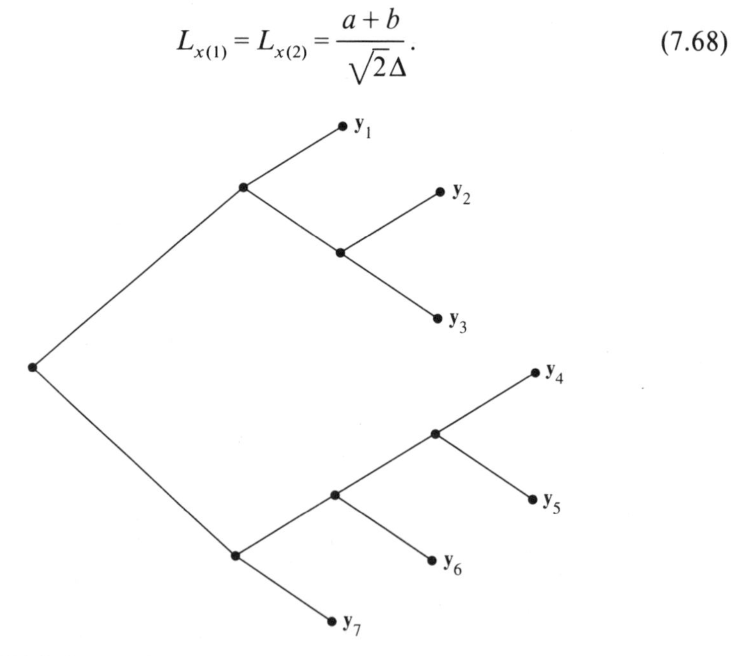

FIGURE 7.6. Nonuniform tree for binary search vector quantization.

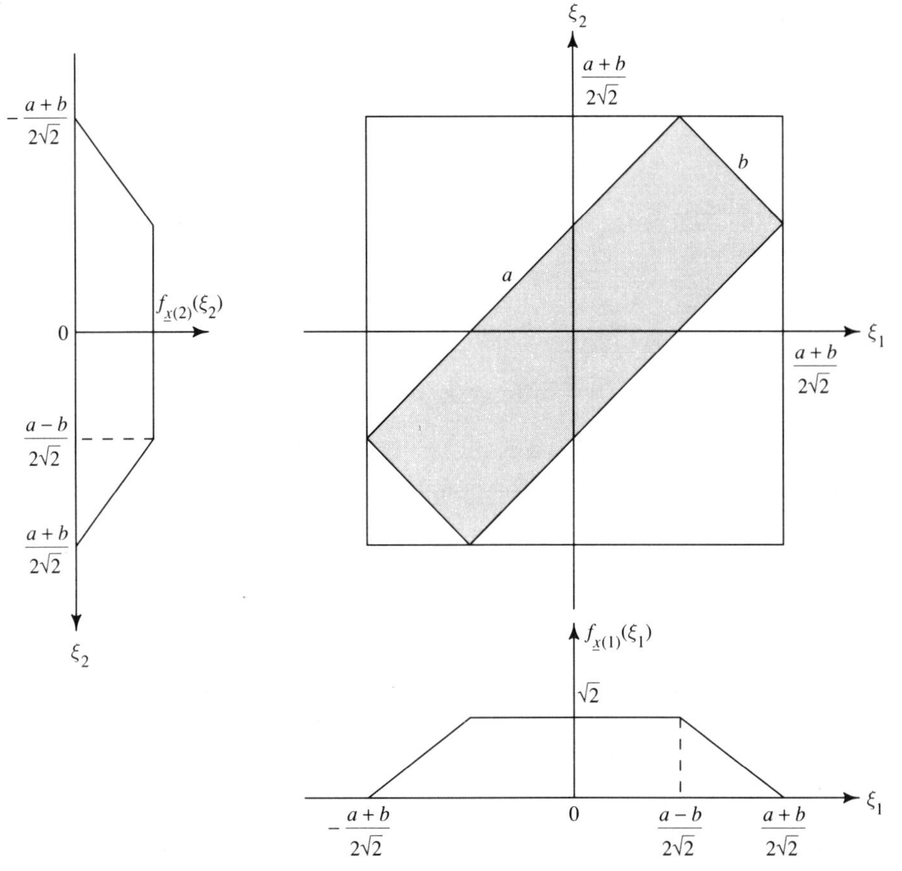

FIGURE 7.7. A uniform pdf in two dimensions (Makhoul et al., 1985).

Hence the number of bits needed for coding the vector outcome $\mathbf{x} = [x(1)\ x(2)]^T$ using scalar quantization is

$$R_{\mathbf{x},\mathrm{SQ}} = R_{x(1)} + R_{x(2)} = \log_2 L_{x(1)} + \log_2 L_{x(2)} = \log_2 \frac{(a+b)^2}{2\Delta^2}. \tag{7.69}$$

Thus the scalar quantization of each component is equivalent to vector quantization with the total number of levels

$$L_{\mathbf{x},\mathrm{SQ}} = L_{x(1)} L_{x(2)} = \frac{(a+b)^2}{2\Delta^2}. \tag{7.70}$$

We observe that this approach is equivalent to covering the large square that encloses the rectangle by square cells, where each cell represents one of the $L_{\mathbf{x},\mathrm{SQ}}$ quantized regions. Since $f_{\underline{\mathbf{x}}}(\boldsymbol{\xi}) = 0$ except for $\boldsymbol{\xi} \in \mathbf{C}$, this encoding is wasteful and results in an increased bit rate.

If we were to cover only the region **C** for which $f_{\underline{\mathbf{x}}}(\xi) \neq 0$ with squares having area Δ^2, the total number of levels that will result is the area of the rectangle divided by Δ^2,

$$L_{\mathbf{x},\mathrm{VQ}} = \frac{ab}{\Delta^2}. \tag{7.71}$$

Therefore, the difference in bit rate between the scalar and vector quantization methods is

$$R_{\mathbf{x},\mathrm{SQ}} - R_{\mathbf{x},\mathrm{VQ}} = \log_2 \frac{(a+b)^2}{2ab}. \tag{7.72}$$

For instance, if $a = 4b$, the difference in bit rate is

$$R_{\mathbf{x},\mathrm{SQ}} - R_{\mathbf{x},\mathrm{VQ}} = 1.64 \text{ bits per vector.} \tag{7.73}$$

Thus vector quantization is 0.82 bpn better for the same distortion.

It is interesting to note that a linear transformation (rotation by 45°) will decorrelate $\underline{x}(1)$ and $\underline{x}(2)$ and render the two random variables statistically independent. Then scalar quantization and vector quantization achieve the same efficiency. Although a linear transformation can decorrelate a vector of random variables, it does not result in statistically independent random variables, in general. Consequently, VQ will always equal or exceed the performance of scalar quantization.

Vector quantization has been applied to several types of speech encoding methods, including both waveform and model-based methods. In model-based methods such as linear predictive coding, VQ has made possible the coding of speech at rates below 1000 bps [see papers by Buzo et al. (1980); Roucos et al. (1982); and Paul (1983)]. When applied to waveform encoding methods, it is possible to obtain good-quality encoded speech at 16,000 bps or, equivalently, at $R = 2$ bpn. With additional computational complexity, it may be possible in the future to implement waveform encoders producing good-quality speech at a rate of $R = 1$ bpn.

7.3 Waveform Coding

7.3.1 Introduction

Methods for digitally representing the temporal or spectral characteristics of speech waveforms are generally called *waveform encoding*. In this section we describe several time domain and frequency domain waveform encoding techniques. These techniques have been widely used in speech telecommunications since the 1950s.

With the exception of Section 7.3.4, the methods considered in Section 7.3 involve scalar quantization of time samples or frequency samples of a speech signal. In contrast to optimum scalar quantization described in Section 7.2.1, which requires knowledge of the pdf of the signal samples, we make few assumptions about specific statistical properties of speech signals. Consequently, the waveform encoding techniques to be described do not achieve the theoretically optimal performance that is achievable when the pdf of the signal samples is known. Nevertheless, the techniques described below are relatively robust and, at sufficiently high bit rates, provide high-quality speech.

7.3.2 Time Domain Waveform Coding

In this section we describe several waveform quantization and encoding techniques that have been applied to speech signals. In particular, we consider pulse code modulation (PCM), differential PCM (DPCM), delta modulation (DM), and several adaptive versions of these methods.

Pulse Code Modulation

Let $s_a(t)$ denote an analog speech waveform and let $s(n)$ denote the samples taken at a sampling rate $F_s \geq 2W$, where W is the highest frequency in the spectrum of $s_a(t)$. In *pulse code modulation* (PCM), each sample of the signal is quantized to one of 2^R amplitude levels, where R is the number of bits used to represent each sample (bpn). Thus the rate from the source is RF_s bps.

The quantized waveform may be modeled mathematically as

$$\tilde{s}(n) = s(n) + q(n), \tag{7.74}$$

where $\tilde{s}(n)$ represents the quantized value of $s(n)$ and $q(n)$ represents the quantization error, which we treat as an additive noise. Assuming that a uniform quantizer is used and the number of levels is sufficiently large, the quantization noise is well characterized statistically as a realization of a stationary random process $\underline{q}$ in which each of the random variables $\underline{q}(n)$ has the uniform pdf

$$f_{\underline{q}(n)}(\xi) = \frac{1}{\Delta}, \qquad -\frac{\Delta}{2} \leq \xi \leq \frac{\Delta}{2}, \tag{7.75}$$

where the step size of the quantizer is $\Delta = 2^{-R}$. The mean square value of the quantization error is

$$\mathcal{E}\{\underline{q}^2(n)\} = \frac{\Delta^2}{12} = \frac{2^{-2R}}{12} \tag{7.76}$$

for arbitrary n. Measured in decibels, the mean square value of the noise is

$$10 \log_{10} \frac{\Delta^2}{12} = 10 \log_{10} \frac{2^{-2R}}{12} = -6R - 10.8 \text{ dB}. \tag{7.77}$$

For example, a 7-bit quantizer results in a quantization noise power of −52.8 dB. We observe that the quantization noise decreases by 6 dB/bit used in the quantizer.

Speech signals have the characteristic that small-signal amplitudes occur more frequently than large-signal amplitudes. Furthermore, human hearing exhibits a logarithmic sensitivity. However, a uniform quantizer provides the same spacing between successive levels throughout the entire dynamic range of the signal. A better approach is to use a nonuniform quantizer that provides more closely spaced levels at the small-signal amplitudes and more widely spaced levels at the large-signal amplitudes. For a nonuniform quantizer with R bits, the resulting quantization error has a mean square value that is smaller than that given by (7.76). A nonuniform quantizer characteristic is usually obtained by passing the signal through a nonlinear device that compresses the signal amplitude, followed by a uniform quantizer. For example, a logarithmic compressor employed in North American telecommunications systems has an input–output magnitude characteristic of the form

$$|y| = \frac{\log(1 + \mu|s|)}{\log(1 + \mu)}, \tag{7.78}$$

where $|s|$ is the magnitude of the input, $|y|$ is the magnitude of the output, and μ is a parameter that is selected to give the desired compression characteristic. Figure 7.8 illustrates this compression relationship for several values of μ. The value $\mu = 0$ corresponds to no compression.

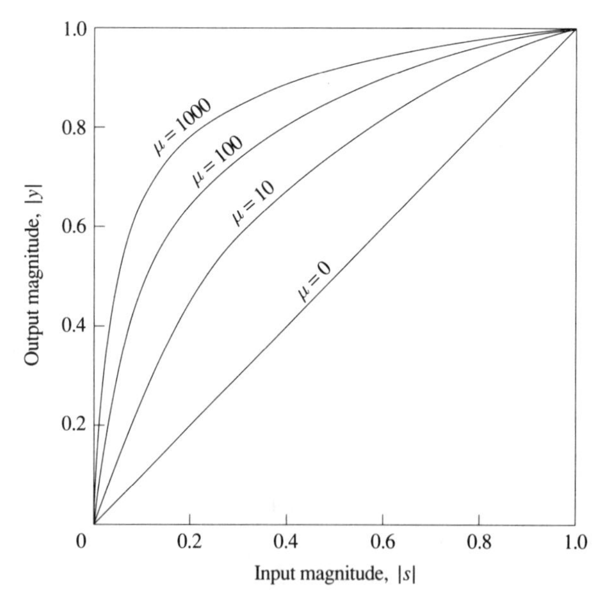

FIGURE 7.8. Input–output magnitude characteristic for a logarithmic compressor.

In the encoding of speech waveforms, the value of $\mu = 255$ has been adopted as a standard in the United States and Canada.[5] This value results in about a 24-dB reduction in the quantization noise power relative to uniform quantization, as shown by Jayant (1974). Consequently, a 7-bit quantizer used in conjunction with a $\mu = 255$ logarithmic compressor produces a quantization noise power of approximately -77 dB compared with -53 dB for uniform quantization.

The logarithmic compressor used in European telecommunications systems is called *A-law* and is defined as

$$|y| = \frac{1 + \log A|s|}{1 + \log A}, \tag{7.79}$$

where A is chosen as 87.56. Although (7.78) and (7.79) are different nonlinear functions, the two compression characteristics are very similar. Figure 7.9 illustrates these two compression functions. At a sampling rate of 8000 samples per second and 8 bpn, the resulting bit rate for the PCM encoded speech is 64,000 bps. Without the logarithmic compressor, a uniform quantizer requires approximately 12 bpn to achieve the same level of fidelity.

In the reconstruction of the signal from the quantized values, the inverse logarithmic relation is used to expand the signal amplitude. The combined compressor-expandor pair is termed a *compandor.*

Differential Pulse Code Modulation

In PCM, each sample of the waveform is encoded independently of all the other samples. However, most source signals including speech sam-

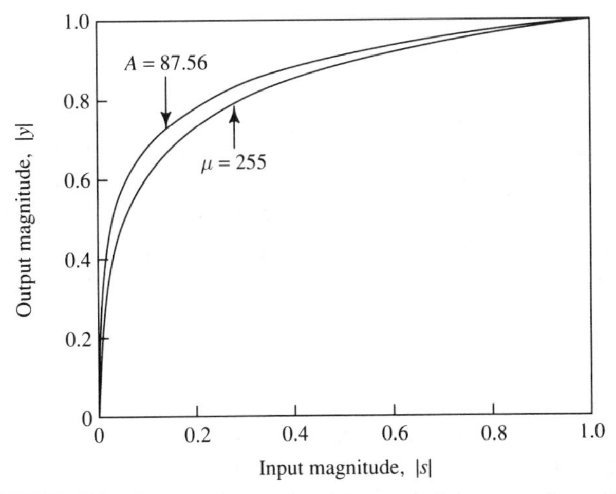

FIGURE 7.9. Comparison of μ-law and A-law nonlinearities.

[5] A piecewise linear approximation to the characteristic for $\mu = 255$ is used in practice.

pled at the Nyquist rate or faster exhibit significant correlation between successive samples. In other words, the average change in amplitude between successive samples is relatively small. Consequently an encoding scheme that exploits the redundancy in the samples will result in a lower bit rate for the source output.

A relatively simple solution is to encode the differences between successive samples rather than the samples themselves. The resulting technique is called *differential pulse code modulation* (DPCM). Since differences between samples are expected to be smaller than the actual sampled amplitudes, fewer bits are required to represent the differences. In this case we quantize and transmit the differenced speech sequence

$$e(n) = s(n) - s(n-1). \tag{7.80}$$

In keeping with the comments in the introductory paragraph, we note that the differencing procedure is a simple attempt to remove redundancy (correlation) from the speech sequence. One way to see this is to think about the spectral "tilt" on the speech spectrum, which is lessened by 6 dB per octave by the differencing operation. We can also easily show by taking long-term averages that the temporal variance (power) of the sequence $e(n)$ is related to that of $s(n)$ as

$$\sigma_e^2 = 2\sigma_s^2\left(1 - \frac{r_s(1)}{\sigma_s^2}\right), \tag{7.81}$$

where we have assumed that the temporal averages of both sequences are zero. Therefore, as long as $r_s(1)/r_s(0) = r_s(1)/\sigma_s^2 > 0.5$ (indicating sufficient correlation), the differencing will reduce the long-term power in the signal. In fact, the long-term temporal autocorrelation ratio $r_s(1)/r_s(0)$ typically exceeds 0.8.

A natural refinement of this general approach is to predict the current sample based on the previous M samples. We know from our experience with linear prediction (LP) in Chapter 5 that LP serves to decorrelate (flatten the spectrum of) the speech signal. In particular, if we use the LP filter in an inverse filter mode, then the residual sequence will be "whitened" with respect to the original sequence. Presumably, this decorrelation procedure will have beneficial consequences for the quantization process. Let us recall the relevant formalisms. We let $s(n)$ denote the current sample from the source and let $\hat{s}(n)$ denote the predicted value of $s(n)$, defined as

$$\hat{s}(n) = \sum_{i=1}^{M} \hat{a}(i)s(n-i). \tag{7.82}$$

Thus $\hat{s}(n)$ is a weighted linear combination of the past M samples and the $\hat{a}(i)$'s are the LP coefficients. If we seek the prediction coefficients that minimize the MSE between $s(n)$ and its predicted value, then the optimal solution is exactly the set of LP coefficients derived in Chapter 5. (In fact, Interpretive Problem 5.2 in Section 5.1.2 is precisely the design problem posed here.) In principle, therefore, the LP parameters are the solution of the normal equations, (5.22) or (5.24). The residual, or error, sequence is the difference[6]

$$e(n) = s(n) - \hat{s}(n) = s(n) - \sum_{i=1}^{M} \hat{a}(i)s(n-i). \tag{7.83}$$

Before proceeding, we should point out that long-term notations are intentionally used in the above. The objective of the LP procedure is not to do an excellent job of prediction on a sample-by-sample basis, but rather to remove correlation in a broad sense for more efficient quantization. Accordingly, a less-than-perfect predictor will suffice. Also, the LP parameters can be computed over a very long corpus of speech data and can be built into the quantizer as static parameters. The case in which the LP parameters are dynamically estimated is discussed below.

Having described the method for determining the predictor coefficients, we now consider the block diagram of a practical DPCM system, shown in Fig. 7.10. In this configuration, the predictor is implemented with a feedback loop around the quantizer. The input to the predictor is denoted as $\tilde{s}(n)$, which represents the signal sample $s(n)$ modified by the quantization process, and the output of the predictor is

$$\hat{\tilde{s}}(n) = \sum_{i=1}^{M} \hat{a}(i)\tilde{s}(n-i). \tag{7.84}$$

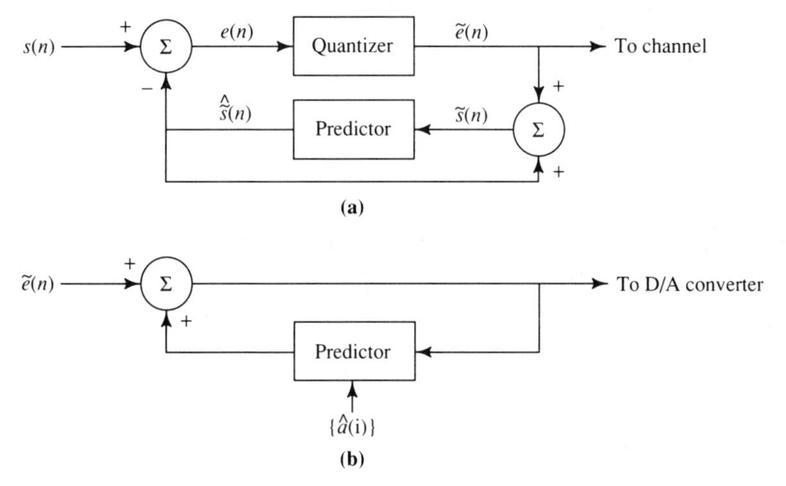

FIGURE 7.10. (a) Block diagram of a DPCM encoder. (b) DPCM decoder at the receiver.

[6]In Chapter 5 this residual signal is called $\hat{e}(n)$ because it is nominally an estimate of the vocal system model input, which was there called $e(n)$. We omit the "hat" in this material for simplicity.

The difference

$$e(n) = s(n) - \hat{\tilde{s}}(n) \tag{7.85}$$

is the input to the quantizer and $\tilde{e}(n)$ denotes the output. Each value of the quantized prediction error $\tilde{e}(n)$ is encoded into a sequence of binary digits and transmitted over the channel to the receiver. The quantized error $\tilde{e}(n)$ is also added to the predicted value $\hat{\tilde{s}}(n)$ to yield $\tilde{s}(n)$.

At the receiver the same predictor that was used at the transmitting end is synthesized and its output $\hat{\tilde{s}}(n)$ is added to $\tilde{e}(n)$ to yield $\tilde{s}(n)$. The signal $\tilde{s}(n)$ is the desired excitation for the predictor and also the desired output sequence from which the reconstructed signal $s_a(t)$ is obtained by filtering, as shown in Fig. 7.10(b).

The use of feedback around the quantizer, as described above, ensures that the error in $\tilde{s}(n)$ is simply the quantization error $q(n) = \tilde{e}(n) - e(n)$ and that there is no accumulation of previous quantization errors in the implementation of the decoder. That is,

$$\begin{aligned} q(n) &= \tilde{e}(n) - e(n) \\ &= \tilde{e}(n) - \left[s(n) - \hat{\tilde{s}}(n)\right] \\ &= \tilde{s}(n) - s(n). \end{aligned} \tag{7.86}$$

Hence $\tilde{s}(n) = s(n) + q(n)$. This means that the quantized sample $\tilde{s}(n)$ differs from the input $s(n)$ by the quantization error $q(n)$ independent of the predictor used. Therefore, the quantization errors do not accumulate.

In the DPCM system illustrated in Fig. 7.10, the estimate or predicted value $\hat{\tilde{s}}(n)$ of the signal sample $\tilde{s}(n)$ is obtained by taking a linear combination of past values $\tilde{s}(n-i)$, $i = 1, 2, \ldots, M$, as indicated by (7.84). An improvement in the quality of the estimate is obtained by including linearly filtered past values of the quantized error. Specifically, the $\hat{\tilde{s}}(n)$ estimate may be expressed as

$$\hat{\tilde{s}}(n) = \sum_{i=1}^{M} \hat{a}(i)\tilde{s}(n-i) + \sum_{i=1}^{L} \hat{b}(i)\tilde{e}(n-i), \tag{7.87}$$

where $\{\hat{b}(i)\}$ are the coefficients of the filter for the quantized error sequence $\tilde{e}(n)$. The block diagram of the encoder at the transmitter and the decoder at the receiver are shown in Fig. 7.11. The two sets of coefficients $\{\hat{a}(i)\}$ and $\{\hat{b}(i)\}$ are selected to minimize some function of the error $e(n) = s(n) - \hat{\tilde{s}}(n)$, such as the MSE.

By using a logarithmic compressor and a 4-bit quantizer for the error sequence $e(n)$, DPCM results in high-quality speech at a rate of 32,000 bps.

Adaptive PCM and Adaptive DPCM

Speech signals are quasi-stationary in nature. One aspect of the quasi-stationary characteristic is that the variance and the autocorrelation

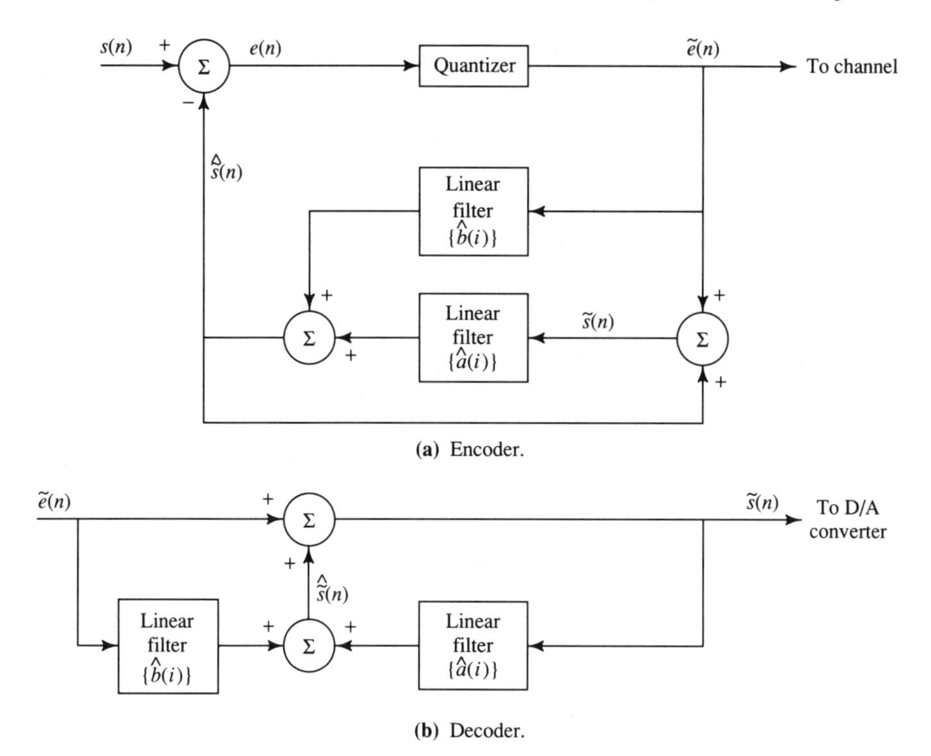

FIGURE 7.11. DPCM modified by the addition of linearly filtered error sequence.

function of the source output vary slowly with time. PCM and DPCM encoders, however, are designed on the basis that the source output is stationary. The efficiency and performance of these encoders can be improved by having them adapt to the slowly time-variant statistics of the speech signal.

In both PCM and DPCM, the quantization error $q(n)$ resulting from a uniform quantizer operating on a quasi-stationary input signal will have a time-variant variance (quantization noise power). One improvement that reduces the dynamic range of the quantization noise is the use of an adaptive quantizer, which may be used in conjunction with PCM to yield *adaptive PCM* (APCM), or with DPCM to yield *adaptive DPCM* (ADPCM).

Adaptive quantizers can be classified as feedforward or feedback. A *feedforward adaptive quantizer* adjusts its step size for each signal sample based on a short-term temporal estimate of the input speech signal variance. For example, the variance can be estimated over a sliding frame of speech (frame length N, sliding end-time m) using the short-term autocorrelation estimator denoted as ${}^3r_s(\eta; m)$ in (4.44). In particular, suppose we are currently at time n and desire an estimate of variance for time $n+1$. Letting $m = n+1$ and $\eta = 0$, we have

$$\sigma_s^2(n+1) \stackrel{\text{def}}{=} {}^3r_s(0; n+1) = \frac{1}{N} \sum_{i=n-N}^{n+1} s^2(i). \tag{7.88}$$

Then, the step size for the quantizer is

$$\Delta(n+1) = \Delta(n)\sigma_s(n+1). \qquad (7.89)$$

In this case, it is necessary to transmit $\Delta(n+1)$ to the decoder in order for it to reconstruct the signal.

A *feedback* (*backward*) *adaptive quantizer* employs the output of the quantizer in the adjustment of the step size. In particular, we may set the step size as

$$\Delta(n+1) = \Delta(n)\,\alpha(n), \qquad (7.90)$$

where the scale factor $\alpha(n)$ depends on the previous quantizer output. For example, if the previous quantizer output is small, we may select $\alpha(n) < 1$ in order to provide for finer quantization. On the other hand, if the quantizer output is large, then the step size should be increased to reduce the possibility of signal clipping. Such an algorithm has been successfully used by Jayant (1974) in the encoding of speech signals. Figure 7.12 illustrates such a (3-bit) quantizer in which the step size is adjusted recursively according to the relation

$$\Delta(n+1) = \Delta(n)M(n), \qquad (7.91)$$

where $M(n)$ is a multiplication factor whose value depends on the quantizer level for the sample $s(n)$, and $\Delta(n)$ is the step size of the

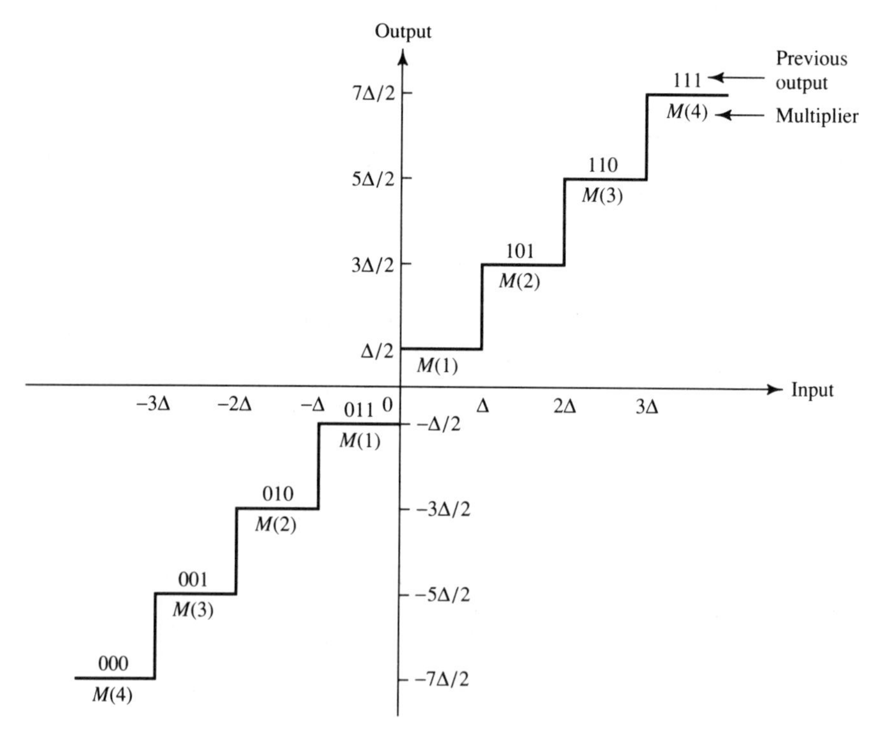

FIGURE 7.12. Example of a quantizer with an adaptive step size (Jayant, 1974).

quantizer for processing $s(n)$. Values of the multiplication factors optimized for speech encoding have been given by Jayant (1974). These values are displayed in Table 7.8 for 2-, 3-, and 4-bit quantization.

In DPCM, the predictor can also be made adaptive. Thus ADPCM with an adaptive predictor takes the configuration shown in Fig. 7.13. The coefficients of the predictor can be changed periodically to reflect the changing signal statistics of the source. In this case the short-term autocorrelation method (see Section 5.3.1) is used to compute estimates of the LP parameters over the current frame, say $f(n;m)$. The resulting error sequence is $\varepsilon(n;m)$ [recall (5.54)]. The predictor coefficients thus determined may be transmitted along with the quantized error, say $\tilde{\varepsilon}(n;m)$, to the receiver, which implements the same predictor. Unfortunately, the transmission of the predictor coefficients results in a higher bit rate over the channel, offsetting in part the lower data rate achieved by having a quantizer with fewer bits (fewer levels) to handle the reduced dynamic range in the error $\varepsilon(n;m)$ resulting from adaptive prediction.

As an alternative to transmitting the prediction coefficients, we can transmit the reflection coefficients, which have a smaller dynamic range and thus result in a lower bit rate (see Sections 5.3.3 and 5.4). To avoid frame edge effects, the reflection (or prediction) coefficients may be interpolated from frame to frame to smooth out large discontinuities in the transmitted filter parameters.

Instead of using the frame processing approach for determining the predictor coefficients as described above, we may adapt these parameter coefficients on a sample-by-sample basis by using a gradient-type algorithm. Similar gradient-type algorithms have also been devised for adapting the filter coefficients $\hat{a}(i)$ and $\hat{b}(i)$ of the DPCM system shown in Fig. 7.11. Details on such algorithms can be found (Jayant and Noll, 1984). Some recursive algorithms for computing the LP parameters are also discussed in Section 5.3.3.

A 32-kbps ADPCM standard has been established by CCITT (Consultative Committee for International Telephone and Telegraph) for interna-

TABLE 7.8. Multiplication Factors for Adaptive Step Size Adjustment (Jayant, 1974).

	PCM			DPCM		
	2	3	4	2	3	4
$M(1)$	0.60	0.85	0.80	0.80	0.90	0.90
$M(2)$	2.20	1.00	0.80	1.60	0.90	0.90
$M(3)$		1.00	0.80		1.25	0.90
$M(4)$		1.50	0.80		1.70	0.90
$M(5)$			1.20			1.20
$M(6)$			1.60			1.60
$M(7)$			2.00			2.00
$M(8)$			2.40			2.40

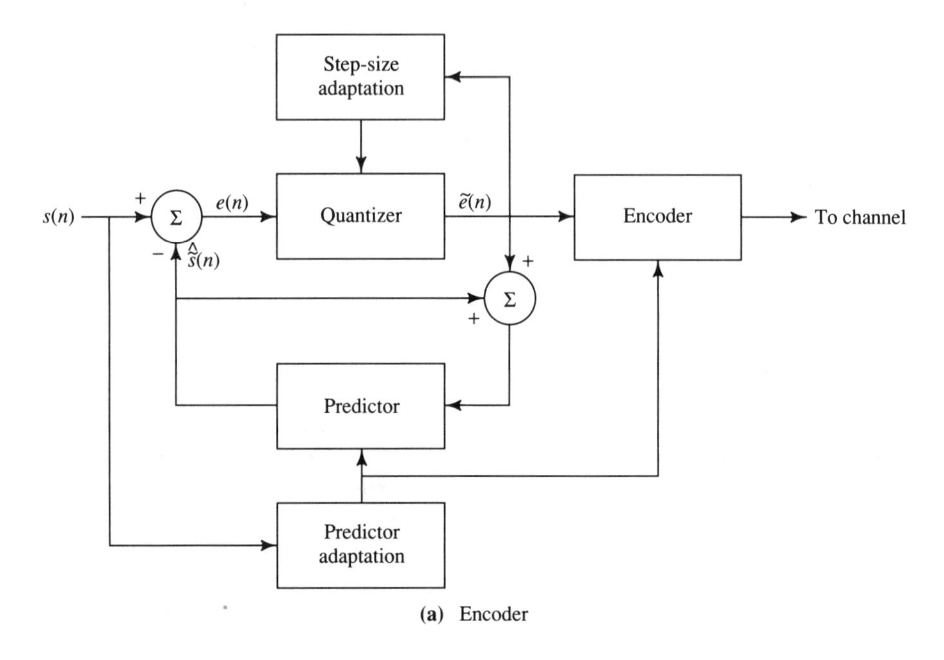

(a) Encoder

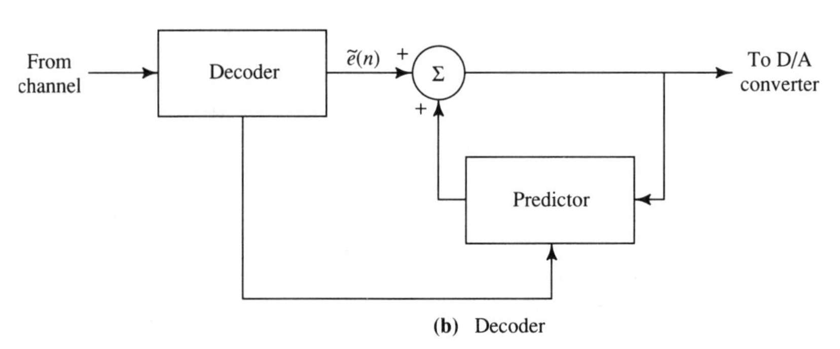

(b) Decoder

FIGURE 7.13. ADPCM with adaptation of the predictor.

tional telephone communications and by ANSI (American National Standards Institute) for North American telephone systems. The ADPCM scheme employs an adaptive feedback quantizer as described above and a pole–zero predictor of the form shown in Fig. 7.13. A gradient algorithm is employed to adaptively adjust the coefficients of the pole–zero predictor. A description of the 32-kbps ADPCM coding standard is given in the paper by Benvenuto et al. (1986).

Delta Modulation

Delta modulation (DM) may be viewed as a simplified form of DPCM in which a two-level (1-bit) quantizer is used in conjunction with a fixed first-order predictor. The block diagram of a DM encoder–decoder is shown in Fig. 7.14(a). We note that

$$\hat{\tilde{s}}(n) = \tilde{s}(n-1) = \hat{\tilde{s}}(n-1) + \tilde{e}(n-1). \tag{7.92}$$

Now, since

$$\begin{aligned} q(n) &= \tilde{e}(n) - e(n) \\ &= \tilde{e}(n) - \left[s(n) - \hat{\tilde{s}}(n)\right], \end{aligned} \tag{7.93}$$

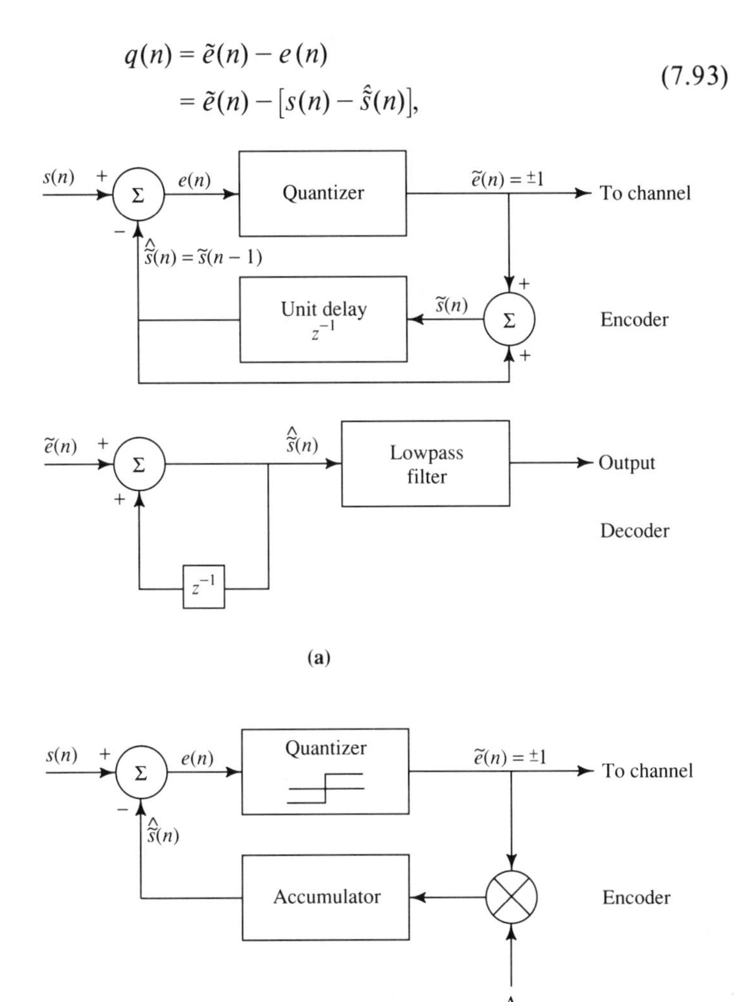

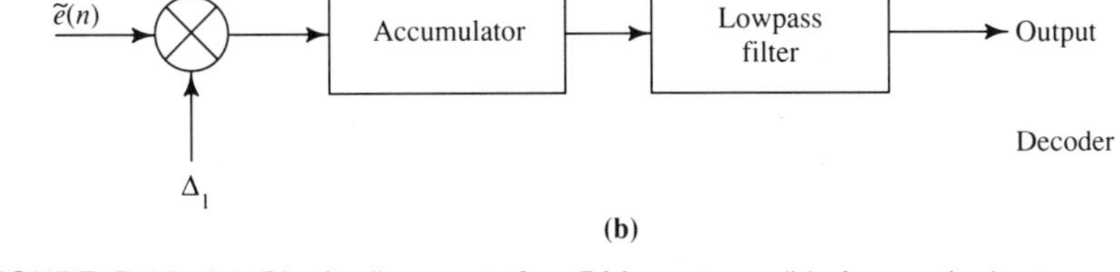

FIGURE 7.14. (a) Block diagram of a DM system. (b) An equivalent realization of a DM system.

it follows that

$$\hat{\tilde{s}}(n) = s(n-1) + q(n-1). \tag{7.94}$$

Thus the estimated (predicted) value $\hat{\tilde{s}}(n)$ is really the previous sample $s(n-1)$ modified by the quantization noise $q(n-1)$. We also note that the difference equation in (7.92) represents an integrator with an input $\tilde{e}(n)$. Hence an equivalent realization of the one-step predictor is an accumulator with an input equal to the quantized error signal $\tilde{e}(n)$. In general, the quantized error signal is scaled by some value, say Δ_1, which is called the *step size.* This equivalent realization is illustrated in Fig. 7.14(b). In effect, the encoder shown in Fig. 7.14(b) approximates a waveform $s_a(t)$ by a linear staircase function. In order for the approximation to be relatively good, the waveform $s_a(t)$ must change slowly relative to the sampling rate. This requirement implies that the sampling rate must be several (at least five) times the Nyquist rate.

Adaptive Delta Modulation

At any given sampling rate, the performance of the DM encoder is limited by two types of distortion, as illustrated in Fig. 7.15. One is called *slope-overload distortion.* This type of distortion is due to the use of a step size Δ_1 that is too small to follow portions of the waveform that have a steep slope. The second type of distortion, called *granular noise,* results from using a step size that is too large in parts of the waveform having a small slope. The need to minimize both of these two types of distortion results in conflicting requirements in the selection of the step size Δ_1. One solution is to select Δ_1 to minimize the sum of the mean square values of these two distortions.

Even when Δ_1 is optimized to minimize the total mean square value of the slope-overload distortion and the granular noise, the performance of the DM encoder may still be less than satisfactory. An alternative solu-

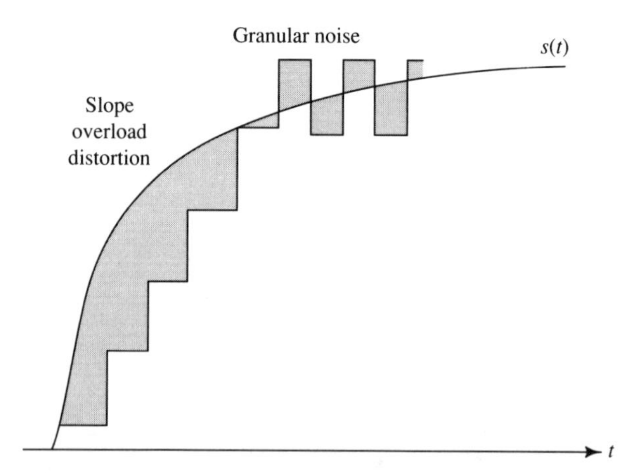

FIGURE 7.15. An example of slope overload distortion and granular noise in a DM encoder.

tion is to employ a variable step size that adapts itself to the short-term characteristics of the source signal. That is, the step size is increased when the waveform has a steep slope and decreased when the waveform has a relatively small slope. This adaptive characteristic is illustrated in Fig. 7.16. This strategy and similar techniques are called *adaptive DM* (ADM).

A variety of methods can be used to set adaptively the step size in every iteration. The quantized error sequence $\tilde{e}(n)$ provides a good indication of the slope characteristics of the waveform being encoded. When the quantized error $\tilde{e}(n)$ changes signs between successive iterations, this indicates that the slope of the waveform in the locality is relatively small. On the other hand, when the waveform has a steep slope, successive values of the error $\tilde{e}(n)$ are expected to have identical signs. From these observations, it is possible to devise algorithms that decrease or increase the step size depending on successive values of $\tilde{e}(n)$. A relatively simple rule devised by Jayant (1970) is to vary adaptively the step size according to the relation

$$\Delta(n) = \Delta(n-1)K^{\tilde{e}(n)\tilde{e}(n-1)}, \qquad n = 1, 2, \ldots, \tag{7.95}$$

where $K \geq 1$ is a constant that is selected to minimize the total distortion. A block diagram of a DM encoder–decoder that incorporates this adaptive algorithm is illustrated in Fig. 7.17.

Several other variations of adaptive DM encoding have been investigated and described in the technical literature. A particularly effective and popular technique first proposed by Greefkes (1970) is called *continuously variable slope delta modulation* (CVSDM). In CVSDM, the adaptive step size parameter may be expressed as

$$\Delta(n) = \alpha\Delta(n-1) + k_1 \tag{7.96}$$

if $\tilde{e}(n)$, $\tilde{e}(n-1)$, and $\tilde{e}(n-2)$ have the same sign; otherwise

$$\Delta(n) = \alpha\Delta(n-1) + k_2. \tag{7.97}$$

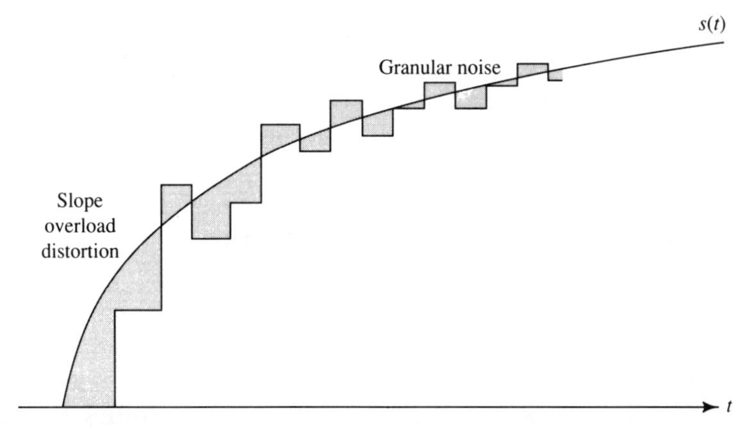

FIGURE 7.16. An example of variable step size DM encoding.

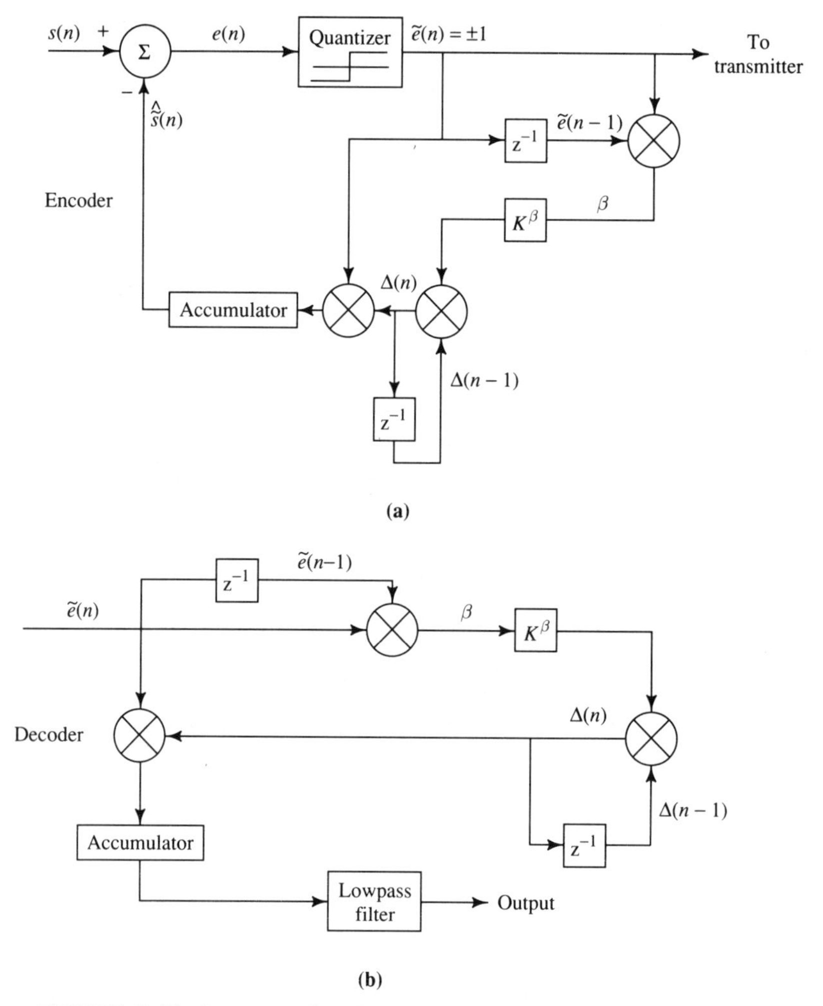

FIGURE 7.17. An example of a DM system with adaptive step size.

The parameters α, k_1, and k_2 are selected such that $0 < \alpha < 1$ and $k_1 > k_2 > 0$. For more discussion on this and other variations of ADM, the interested reader is referred to papers by Jayant (1974) and Flanagan et al. (1979), and to the extensive reference lists contained in these papers.

Adaptive Predictive Coding

Consider a frame of speech to be coded,

$$f(n; m) = s(n)w(m - n). \tag{7.98}$$

If the speech is voiced on this frame, then it is quasi-periodic with an approximate period, P, equal to the pitch period. By exploiting the redundancy inherent in the periodic nature of the signal, it is possible to reduce the bit rate of the waveform encoder. This may be accomplished by estimating the pitch period and using an adaptive predictor of the form

$$\hat{f}(n;m) = \sum_{i=1}^{M} \hat{a}(i;m)f(n-i;m) + \hat{b}(1;m)f(n-P;m). \tag{7.99}$$

The term involving the summation is basically the predictor used in ADPCM. The second term represents the output of a *pitch predictor.* Hence *adaptive predictive coding* (APC) may be viewed as an enhanced version of ADPCM in which the periodicity of voiced speech is used to reduce the size of the error $\varepsilon(n;m) = f(n;m) - \hat{f}(n;m)$. Thus fewer bits are needed to represent the error sequence $\varepsilon(n;m)$.

The prediction coefficients $\hat{a}(i;m)$, $i = 1, 2, \ldots, M$, are determined using short-term LP analysis over the frame, as in ADPCM. For an unvoiced frame, the parameter $\hat{b}(1;m)$ is set to zero. For a voiced frame, the optimum choice of the parameter $\hat{b}(1;m)$ can be determined by using the short-term autocorrelation method for an LP problem in which $f(n;m)$ is to be predicted using only the Pth backlag of the sequence, $f(n-P;m)$. This means that we seek $\hat{b}(1;m)$ such that the energy of the error sequence $[f(n;m) - \hat{b}(1;m)f(n-P;m)]$ is minimized. The answer is easily shown to be

$$\hat{b}(1;m) = \frac{r_s(P;m)}{r_s(0;m)}, \tag{7.100}$$

where $r_s(\eta;m)$ represents a short-term estimator of the autocorrelation of the speech sequence $s(n)$ corresponding to the processing of the frame $f(n;m)$.

In practice, the true pitch period may not be an exact multiple, namely P, of the sampling period $1/F_s$. In such a case, we may use a third-order pitch predictor of the form

$$\begin{aligned}\hat{f}(n;m) = {} & \hat{b}(1;m)f(n-P+1;m) + \hat{b}(2;m)f(n-P;m) \\ & + \hat{b}(3;m)f(n-P-1;m),\end{aligned} \tag{7.101}$$

where the coefficients $\hat{b}(i;m)$, $i = 1, 2, 3$, are selected to minimize the energy in the pitch prediction error sequence $[f(n;m) - \hat{f}(n-P;m)]$. By means of this type of interpolation, we achieve a better estimate of the periodic signal component.

At the receiver, the signal is processed through the linear predictor with coefficients $\hat{a}(i;m)$ and then passed to the pitch predictor, as shown in Fig. 7.18. Gain adaptation may also be incorporated in APC as shown in Fig. 7.18 (also see Section 5.3.4).

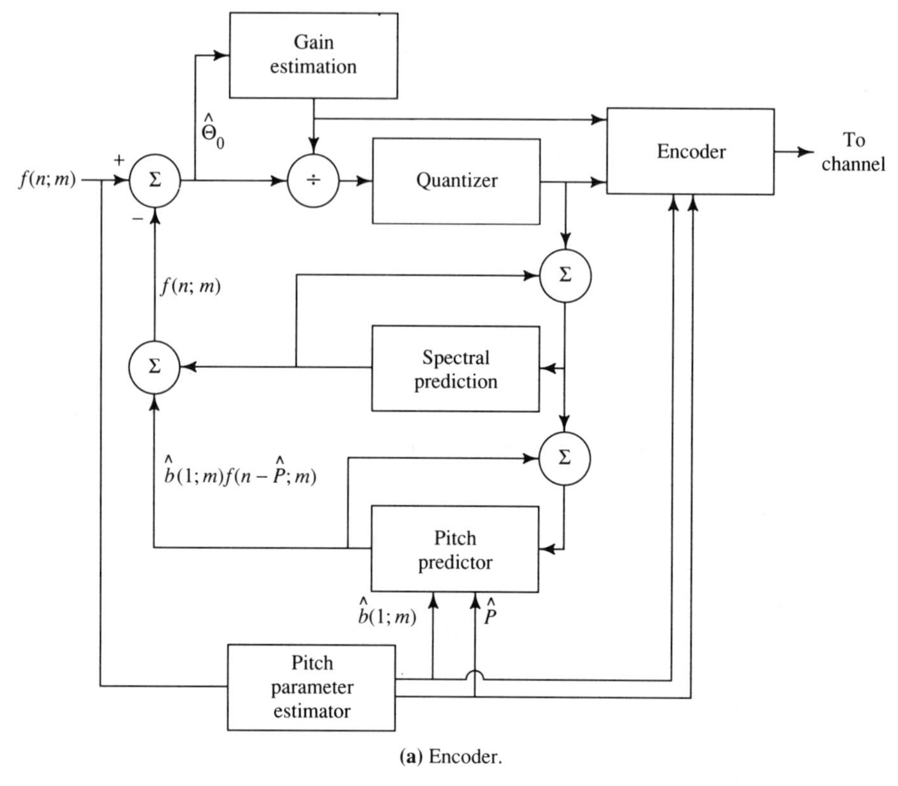

(a) Encoder.

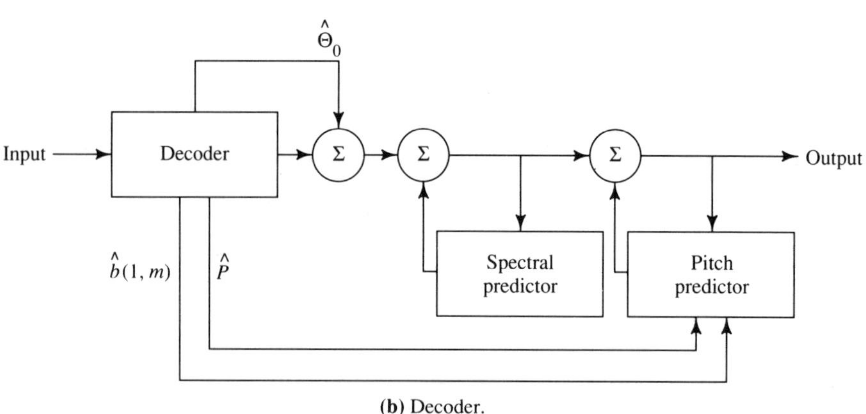

(b) Decoder.

FIGURE 7.18. Block diagram of an APC system.

Adaptive predictive coding has been successfully used to produce communications quality speech at 9600 bps and toll-quality speech at 16,000 bps.

PCM, DPCM, APCM and ADPCM, DM, ADM, and APC are all time domain waveform encoding techniques that attempt to faithfully represent the output waveform from the source. In the next section we consider frequency domain waveform coding techniques.

7.3.3 Frequency Domain Waveform Coding

In this section we consider waveform coding methods that filter a speech signal into a number of frequency bands or subbands and separately encode the signal in each subband. The waveform encoding may be performed either on the time domain waveform in each subband or on the frequency domain representation of the corresponding time domain waveform.

Filter Bank Spectrum Analyzer

In Section 4.3.5 we introduced the *filter bank* method of waveform encoding as an example of the use of the stDFT in speech processing. Let us briefly recall the principles of the system described there.

It is desired to encode the frame of speech

$$f(n; m) = s(n)w(m-n), \tag{7.102}$$

where $w(n)$ is an N-length window. The stDFT can be used to compute samples of the complex spectrum of the frame at normalized frequencies $\omega_k = 2\pi k/N$, $k = 0, 1, \ldots, N-1$,

$$\begin{aligned} S(k; m) &= \sum_{n=m-N+1}^{m} f(n; m) e^{-jk(2\pi/N)n} \\ &= \sum_{n=m-N+1}^{m} s(n)w(m-n)e^{-jk(2\pi/N)n}, \qquad k = 0, \ldots, N-1. \end{aligned} \tag{7.103}$$

We have argued that these samples are sufficient to specify the entire nonzero part of the time domain sequence, $f(n; m)$, or equivalently its short-term Fourier transform $S(\omega; m)$ for any ω. We have also argued that because the window acts as a lowpass filter, it is sufficient to encode the stDFT at frames ending at times

$$m = i\frac{\pi}{\omega_b}, \qquad i = \ldots, -1, 0, 1, 2, \ldots, \tag{7.104}$$

where ω_b represents the bandwidth of the window sequence in norm-rps. For the Hamming window, for example, we found that $\omega_b = 2\pi/N$ and $S(k; m)$ can be reconstructed at any m, given its availability at

$$m = i\frac{N}{2}, \qquad i = \ldots, -1, 0, 1, 2, \ldots. \tag{7.105}$$

Formally, we can compute $S(k; n)$ for each n and then decimate the sequence with respect to n by a factor $N/2$ in this case.

The filter bank implementation assuming a Hamming window is illustrated in Fig. 7.19. Computation of $S(k; n)$ for some k and every n is viewed as the convolution of the complex sequence $s(n)e^{-j\omega_k n}$ with the (FIR) window sequence $w(n)$ [see (7.103)]. Since the bandwidth of the

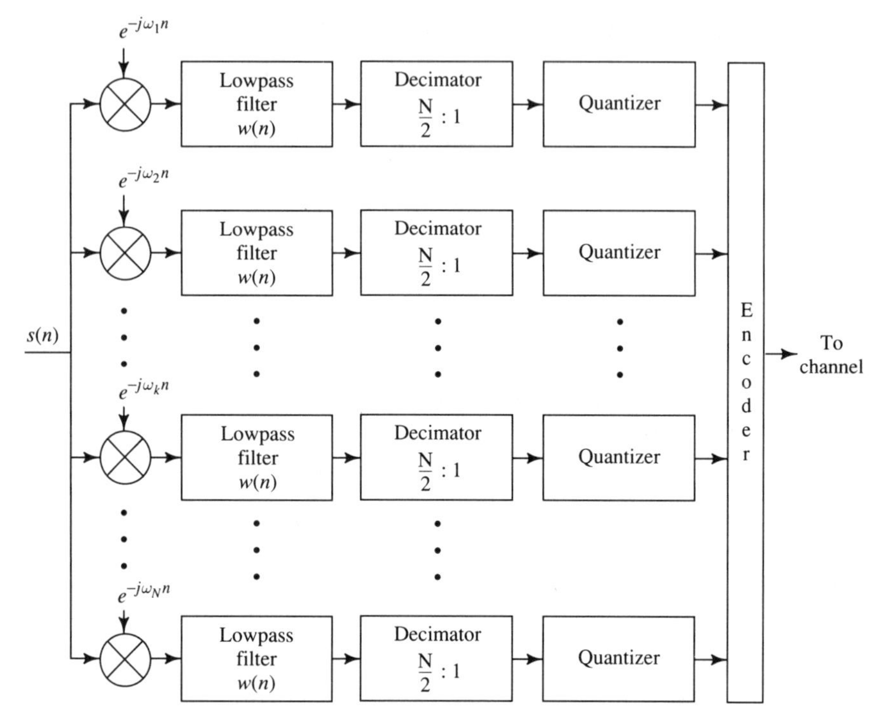

FIGURE 7.19. Filter bank spectrum analyzer.

lowpass filter is relatively narrow, its output is decimated, as discussed above, prior to encoding.

As an aside, we note that this is the first among several systems in which we will see the "short-term" processing "disguised" in the overall operation of the system. In effect, short-term frames of length N are processed for *every* possible frame end-time. The $S(k;n)$'s are therefore computed for each n and become a sequence on the same time index as the original signal. Only selected frames are actually sent, however, by virtue of the decimation of the $S(k;n)$ sequences with respect to the time argument. In such a strategy, we shall see the speech sequence itself, $s(n)$, enter the system, whereas the reader might expect to see frames, $f(n;m)$, as inputs. It is worth pondering where the "frames" are actually created in such a system. It is also worth noting that the decimation is performed for *coding* efficiency rather than computational efficiency. Indeed, the computation of the $S(k;n)$ for each n is wasteful of computation and, in fact, only need be done if the window sequence (lowpass filter) is IIR.

At the receiver, the coded signal samples in each band are interpolated by inserting $(N/2)-1$ zeros between successive received samples; the resulting signals are passed through the same windowing filter as used in the analysis filter bank. The outputs of windowing filters are shifted in frequency and combined as shown in Fig. 7.20.

As concluded in Section 4.3.5, the filter bank implementation provides a conceptual framework for frequency domain waveform coding.

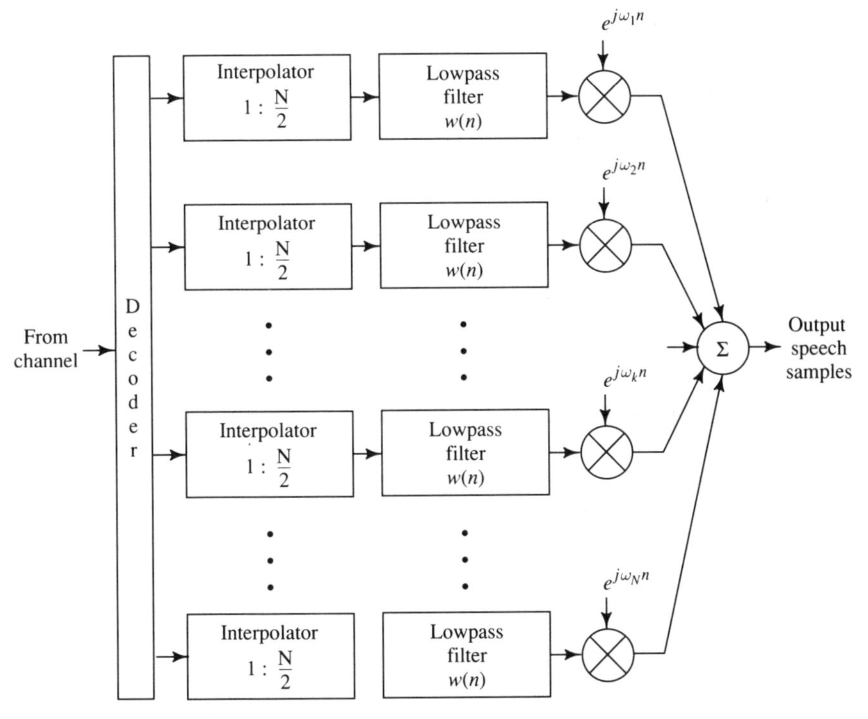

FIGURE 7.20. Filter bank synthesizer.

However, this particular implementation does not provide any coding efficiency compared to the time domain methods such as PCM.

Subband Coding

The filter bank analysis method described above splits the speech signal into N frequency subbands and encodes the signal in each subband separately. In general, the number of signal samples N is large and, consequently, this method does not achieve a low bit rate.

In practice, it is not necessary to split the speech signal into many subbands. In *subband coding* (SBC) the signal is divided into four to eight subbands and the waveform signal in each subband is encoded separately. APCM has been used for the waveform encoding. Since the lower-frequency bands contain most of the spectral energy in voiced speech, more bits are used for the lower-band signals and fewer are used for the signals in the higher-frequency bands.

For example, let us assume that we use two filters to split the signal into two equal subbands as shown in Fig. 7.21. The output of the highpass filter may be translated in frequency to baseband and the signals in the two subbands can be decimated by a factor of two and encoded. Hence, if the original sampling rate is 8 kHz, the rate of the decimated signals is 4 kHz. Suppose that we employ 4-bit APCM for the low-

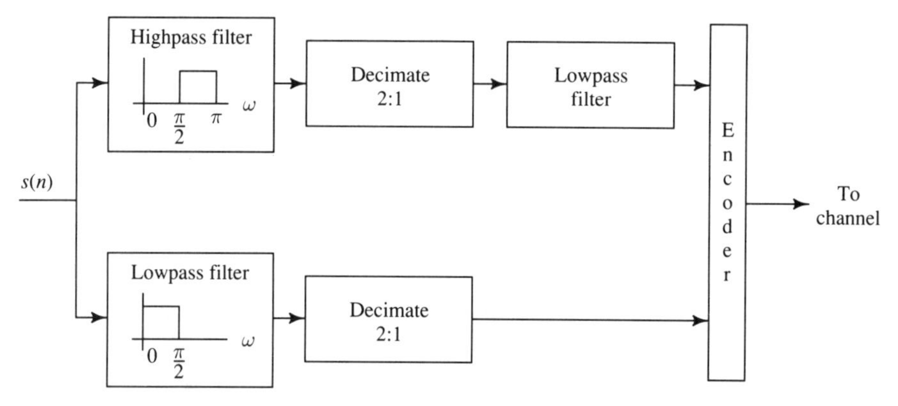

FIGURE 7.21. An example of subband coding.

frequency subband and 2-bit APCM for the high-frequency subband. Then the resulting bit rate is 24,000 bps.

Filter design is particularly important in achieving good performance in SBC. Aliasing resulting from decimation of the subband signals must be negligible. A particularly practical solution to the aliasing problem is to use *quadrature mirror filters* (QMF), which have the frequency response characteristics illustrated in Fig. 7.22. Note that there is aliasing if the outputs of these filters are now decimated by a factor of two. However, the QMF is particularly designed to permit aliasing at the output of the decimator. This aliasing is eliminated by properly designing the synthesis filters such that the images produced by the corresponding interpolators exactly cancel the aliasing. This cancelation of aliasing effects is possible because the QMF analysis/synthesis filter bank is time-variant due to the presence of the decimators and interpolators. A detailed description of the QMF is found in Appendix 7.A.

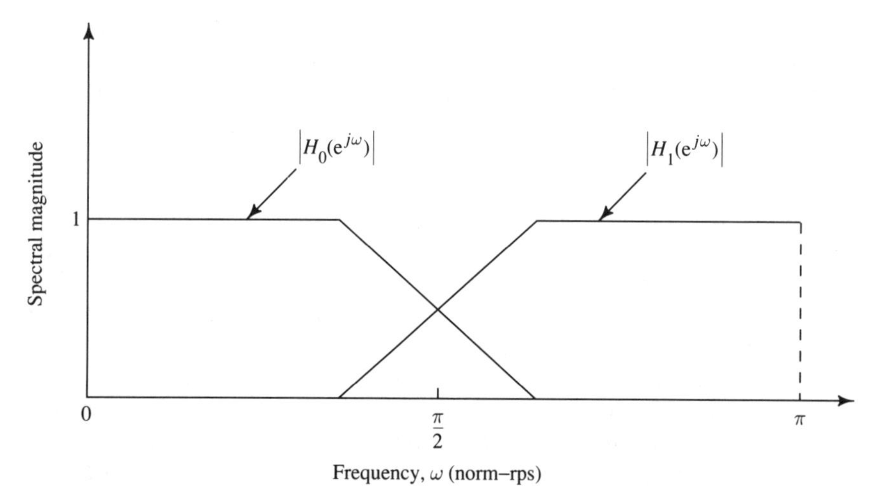

FIGURE 7.22. Ideal frequency response characteristics of a QMF. H_0 and H_1 should be interpreted as z-transforms.

By using QMFs in subband coding, we can repeatedly subdivide the lower-frequency band by factors of two, thus creating octave band filters. For example, suppose that the bandwidth of the speech signal extends to 3400 Hz. Then the first pair of QMFs divides the spectrum into the bands 0–1700 Hz and 1700–3400 Hz. The outputs of these filters can be decimated by a factor of two. A second pair of QMFs can now be used to subdivide the lower-frequency band by two, creating the subbands 0–850 Hz and 850–1700 Hz. Each of these output signals can be decimated by a factor of two. A third subdivision of the lower-frequency band results in two subband signals covering the bands 0–425 Hz and 425–850 Hz, which are then decimated by a factor of two. Thus, with three pairs of QMFs, we can achieve SBC of signals in four frequency bands, 0–425 Hz, 425–850 Hz, 850–1700 Hz, and 1700–3400 Hz, each of which is coded separately by means of APCM. We may allocate 3–4 bpn for the signals in the two lower-frequency bands and 1–2 bpn for the signals in the upper-frequency band. Thus we can achieve bit rates in the range of 16,000–24,000 bps, where the higher rate results in toll-quality speech.

Adaptive Transform Coding

In *adaptive transform coding* (ATC) the analog speech signal is sampled, subdivided into frames of N samples, and transformed into the spectral domain for coding and transmission or storage. At the decoder, each block of spectral samples is transformed back into the time domain and the speech signal is synthesized. To achieve coding efficiency, we assign more bits to the more important spectral coefficients and fewer bits to the less important spectral coefficients. In addition, by using a dynamic allocation in the assignment of the total number of bits to the spectral coefficients, we can adapt to the changing statistics of speech signals.

An objective in selecting the transformation from the time domain to the frequency domain is to achieve uncorrelated spectral samples. In this sense, the Karhunen–Loève transform (KLT) (see Section 1.3.2) is optimal in that it yields spectral values that are uncorrelated. However, this transform is difficult to compute in general. For a frame of N points, $O(N^4)$ flops are necessary to compute the KLT (Wintz, 1972). In addition, the entire autocorrelation matrix of the frame must be sent to the receiver as side information in order that the transform may be inverted. The DFT and the *discrete cosine transform* (DCT) are viable alternatives, although they are suboptimum. Of these two, the DCT yields good performance compared with the KLT and is generally used in practice (Campanella and Robinson, 1971; Zelinski and Noll, 1977; Kuo et al., 1991).

We next wish to define the DCT for a frame of speech $f(n;m) = s(n)w(m-n)$. As with the conventional DFT, the conventional DCT is defined for a finite sequence that begins at the time origin.[7] Accordingly, we first shift the frame down to begin at time $n=0$ [as in (4.76)],

$$\overleftarrow{f}(n;m)=f(n+m-N+1;m). \tag{7.106}$$

The *short-term DCT* (stDCT) is defined as

$$\overleftarrow{C}_s(k;m)=\begin{cases}\overleftarrow{f}(0;m)+\sqrt{2}\displaystyle\sum_{n=1}^{N-1}\overleftarrow{f}(n;m)\cos\left[\frac{(2n+1)\pi k}{N}\right], & k=0,\dots,N-1\\ 0, & \text{otherwise}\end{cases}, \tag{7.107}$$

where the arrows above f and C are simply reminders that this definition includes the shifting of the frame so that it "begins" at time 0. Then the *inverse stDCT* is given by

$$\overleftarrow{f}(n;m)=\begin{cases}\dfrac{1}{N}\left[\overleftarrow{C}(0;m)+\sqrt{2}\displaystyle\sum_{k=1}^{N-1}\overleftarrow{C}(k;m)\cos\left(\frac{(2n+1)\pi k}{N}\right)\right], & k=0,\dots,N-1\\ 0, & \text{otherwise}\end{cases}, \tag{7.108}$$

so that the frame returns to the time range $n=0,\dots,N-1$. Note that the spectral parameters $\overleftarrow{C}(k;m)$ are real-valued. In fact, it is not difficult to show that

$$\overleftarrow{C}(k;m)=\begin{cases}\mathrm{Re}\,\{X(0;m)\}, & k=0\\ \sqrt{2}\,\mathrm{Re}\,\{e^{-(j\pi/2N)k}X(k;m)\}, & k=1,\dots,N-1\end{cases}, \tag{7.109}$$

where $x(n;m)$ is the sequence $\overleftarrow{f}(n;m)$ with N zeros appended, and $X(k;m)$ is its $2N$-point DFT. This equation can be used to compute the DCT using an FFT algorithm.

The trapezoidal window is frequently used in practice to select the frame in computing the DCT. To reduce signal distortion due to edge discontinuities, trapezoidal windows used in adjacent frames may be overlapped by a small amount, typically 10%, such that the sum of the overlapped windows is unity. This condition implies that the trapezoidal ramp length is equal to the amount of overlap.

[7]We could define a "delay-preserving" *short-term* DCT if future theoretical developments required it (see Section 4.3.5), but we shall be content to work with the "conventional" version in describing ATC. The receiver is assumed to be aware of the appropriate m value for each frame.

The bit allocation for encoding the spectral samples $\overleftarrow{C}(k;m)$, $k=0,1,\ldots,N-1$, is done adaptively by monitoring the power as a function of frequency. For this purpose, the N frequency samples are subdivided into L nonoverlapping frequency bands. Each of the N/L spectral values in a frequency band is squared, the logarithm is computed, and the resulting logarithmic values are added together to form an average of a logarithmically scaled power level for the band. Thus we obtain the power level parameters, say $P_s(k;m)$, $k=1,2,\ldots,L$. Typically, L is selected to be in the range of 16–24 frequency bands. From these power measurements in the L bands we can interpolate $P_s(k;m)$ to generate a value corresponding to each of the spectral coefficients. Thus we obtain $P_s'(k;m)$, $k=0,1,\ldots,N-1$.

Based on the values of the $P_s'(k;m)$, the number of bits allocated to coding each of the spectral components $\overleftarrow{C}(k;m)$ is estimated from the formula

$$R_k = \frac{R}{N} + \frac{1}{2}\left[P_s'(k;m) - \frac{1}{N}\sum_{k=0}^{N-1} P_s'(k;m)\right], \tag{7.110}$$

where R is the total number of bits available to encode the entire block of spectral samples. Note that R/N is the average number of bits per spectral component. The second term serves to increase or decrease the number of bits relative to the average, depending on whether the power level of the spectral component is above or below the average power level among the N spectral components.

In general, the R_k will not take integer values and must be rounded off to integers. If $R_k < 0$, it must be set to zero. In addition to coding the $\overleftarrow{C}(k;m)$, the spectral power measurements $P_s(k;m)$, $k=1,\ldots,L$, must also be coded and transmitted to the decoder as side information. Typically, we may use 2–3 bits per parameter. A block diagram of an ATC system is shown in Fig. 7.23.

Some additional bit compression is possible by encoding the differences $\log \overleftarrow{C}(k;m) - [P_s'(k;m)]^2$, $k=1,2,\ldots,N-1$, based on the bit allocation given by (7.110). The spectral parameters $P_s(k;m)$, $k=1,2,\ldots,L$, may also be differentially encoded and transmitted. These are just two of a number of possible variations of the basic ATC method described above.

With ATC it is possible to attain toll-quality speech at 16,000 bps or higher. Communication-quality speech is feasible with ATC at a rate of 9600 bps.

7.3.4 Vector Waveform Quantization

The waveform encoding methods described above are based on scalar quantization. In some of these methods, such as DPCM, we exploit the redundancies in the speech samples to reduce the bit rate. Thus, DPCM

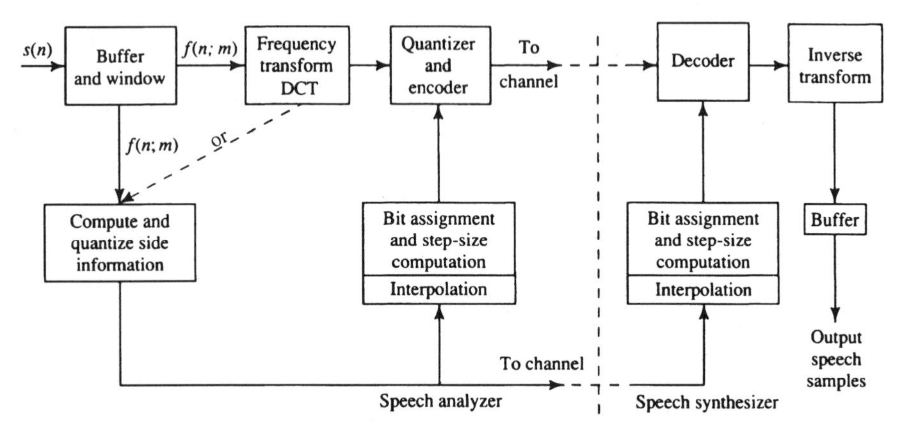

FIGURE 7.23. Block diagram of an adaptive transform coder (Tribolet and Crochiere, 1979).

provides toll-quality speech at about 32,000 bps. With adaptive methods, the bit rate can be reduced further. In particular, APC and ATC provide toll-quality speech at a bit rate of about 16,000 bps.

In order to reduce the bit rate of waveform encoders even further, we can resort to vector waveform quantization, which encodes one frame of speech samples at a time. In particular, we segment the speech waveform samples in frames of N samples and design a codebook containing vectors that are obtained by the K-means algorithm described in Section 1.3.5. Since computational complexity and memory requirements are an important consideration in the design of the codebook, it is necessary to keep the frame size relatively small.

One approach that simplifies the problem to some extent is to extract short-term spectral and pitch information and to code these separately. Therefore, we are left with coding the residual, which tends to be white. Vector quantization is applied to the residual and the index of the quantized vector is transmitted. At the receiver the residual provides the excitation for the synthesis of the speech signal.

Beside the computational complexity and memory costs in the use of vector quantization for waveform encoding, another problem is the potential discontinuity in the synthesis of the waveform due to the quantization of frames of signal samples. With vector waveform quantization it is likely that the end of one quantized vector will not match the beginning of the subsequent quantized vector. This problem is particularly serious at low data rates where code vectors in the codebook are far apart. A remedy is to overlap samples in adjacent frames at the expense of an increase in the bit rate.

In spite of these difficulties, there is much research activity devoted to vector waveform quantization with the objective of achieving toll-quality speech at rates in the range of 8000–10,000 bps.

7.4 Vocoders

The waveform coding techniques described in Section 7.3 are based on either a sample-by-sample, or a frame-by-frame, speech waveform representation either in the time or frequency domain. In contrast, the methods described in this section are based on the representation of a speech signal by an all-pole model of the vocal system upon which we have based numerous results in our work. For convenience, the model is repeated as Fig. 7.24.

Recall that in this model the speech production system is modeled as an all-pole filter. For voiced speech, the excitation is a periodic impulse train with period equal to the pitch period of the speech. For unvoiced speech, the excitation is a white noise sequence. In addition, there can be an estimated gain parameter included in the model. Basically, the differ-

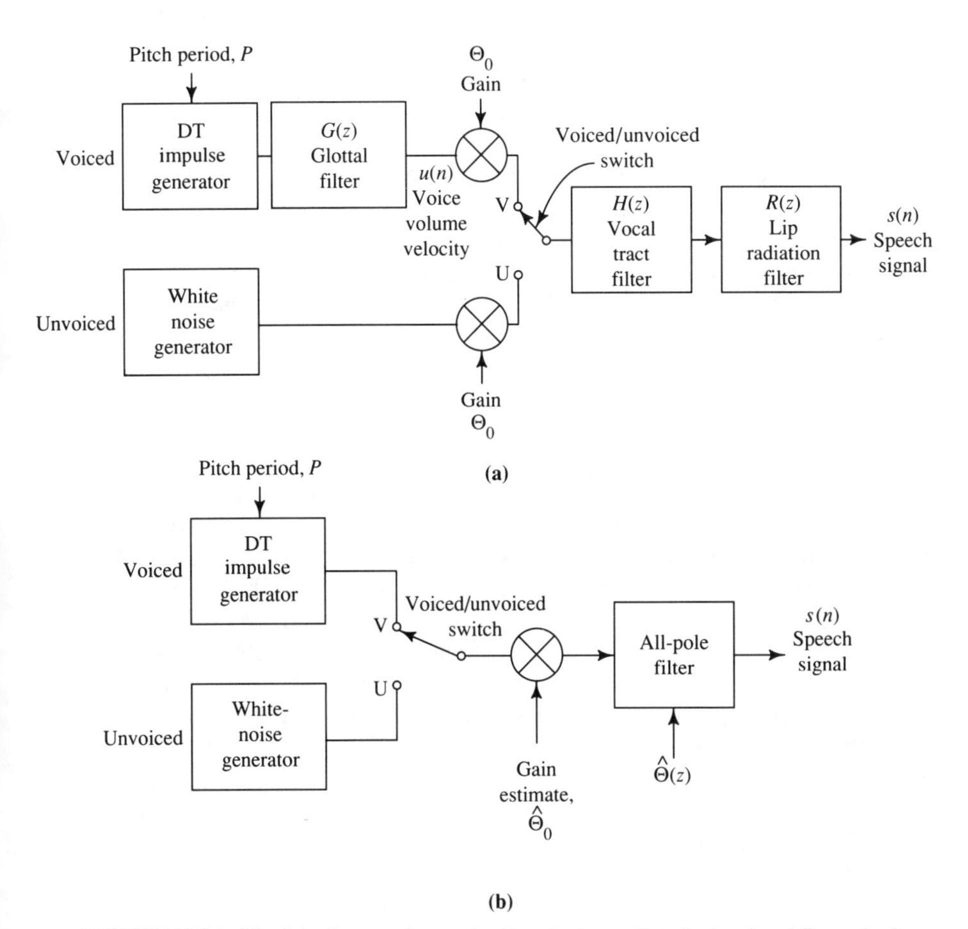

FIGURE 7.24. Model of speech production to be estimated using LP analysis.

ent vocoders described below estimate the model parameters from frames of speech (speech analysis), encode and transmit the parameters to the receiver on a frame-by-frame basis, and reconstruct the speech signal from the model (speech synthesis) at the receiver.

Vocoders usually provide more bandwidth compression than is possible with waveform coding. In particular, the techniques described in this section result in communication-quality speech at data rates in the range of 2400–9600 bps. Our discussion includes the channel vocoder, the cepstral (homomorphic) vocoder, the phase vocoder, the formant vocoder, and the linear predictive coder. The last of these is the most widely used in practice today.

7.4.1 The Channel Vocoder

The channel vocoder employs a bank of analog bandpass filters, each having a bandwidth between 100 Hz and 300 Hz, to estimate the slowly time-varying spectral magnitudes in each band. Typically, 16–20 linear-phase FIR filters are used to cover the audio band 0–4 kHz. Narrower bandwidth filters are employed for the lower-frequency bands and wider bandwidth filters are used in the upper-frequency bands. The output of each filter is rectified and lowpass filtered.

The bandwidth of the lowpass filter is selected to match the time variations in the characteristics of the vocal tract. For example, if the spectral magnitude changes at a 50-Hz rate (a new value every 20 msec), the lowpass filter bandwidth is selected to be in the range of 20–25 Hz. Hence its output may be sampled at a rate of 50 Hz, quantized, and transmitted to the receiver.

In addition to the measurement of the spectral magnitudes, a voicing detector and a pitch estimator are included in the speech analysis. The channel filter bank for speech analysis is shown in Fig. 7.25.

Coding of the voicing detector output requires 1 bit for each speech frame. The representation of the pitch period for voiced speech requires about 6 bits per frame. If we use a 16-channel-bank vocoder and a sampling rate of 50 Hz for each filter, where each sample is coded at 3 to 4 bits, the resulting bit rate is in the range of 2400–3200 bps. Further reductions in bit rate to about 1200 bps can be achieved by exploiting the frequency correlations of the spectral magnitudes. In particular, we may use companded PCM for the first band and DPCM to encode the other spectral samples across the frequency band within each frame.

At the receiver, the speech is synthesized as shown in Fig. 7.26. The signal samples are passed through D/A converters whose outputs are multiplied by the voiced or unvoiced signal sources; the resulting signals are passed through corresponding bandpass filters. The outputs of the bandpass filters are summed to form the output synthesized speech signal.

The channel vocoder is the first and oldest vocoder to have been studied and implemented (Dudley, 1939). When implemented with modern

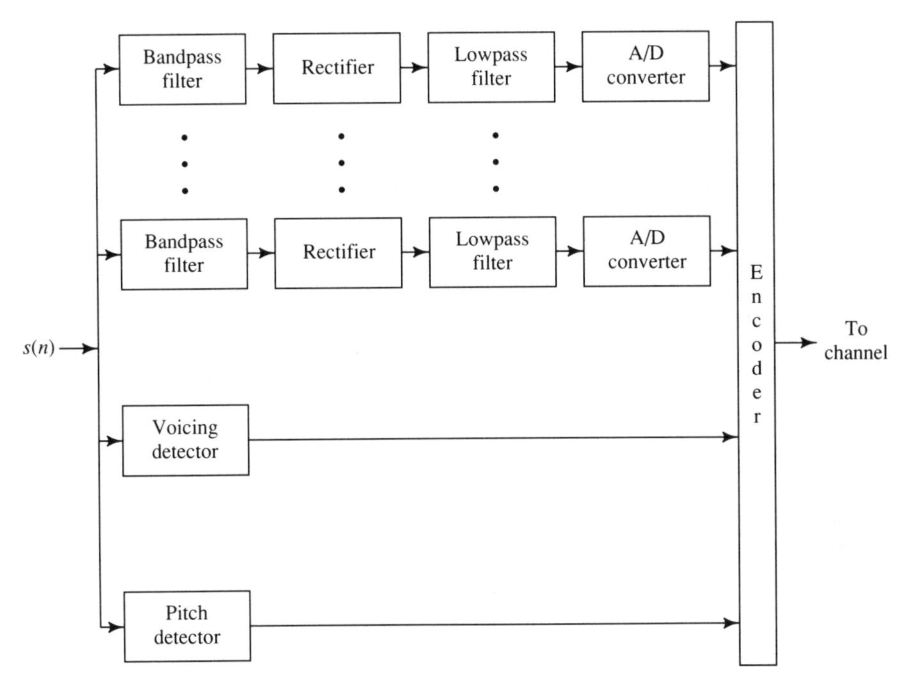

FIGURE 7.25. Block diagram of analyzer for a channel vocoder.

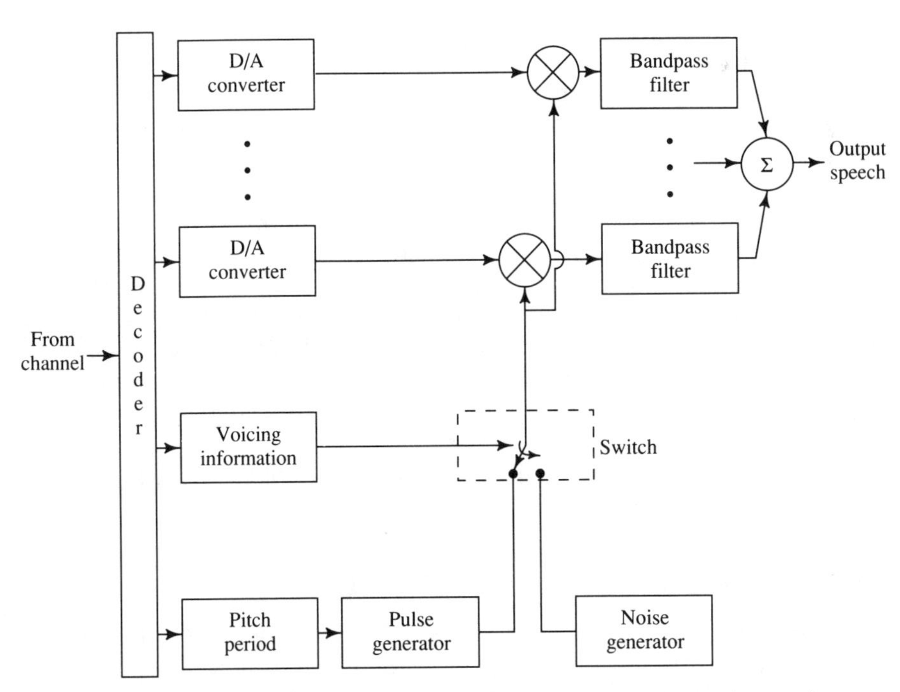

FIGURE 7.26. Block diagram of synthesizer for a channel vocoder.

digital signal processing techniques, it provides communications quality speech at about 2400 bps.

7.4.2 The Phase Vocoder

The phase vocoder is similar to the channel vocoder to the extent that a bank of filters is used to estimate the magnitude spectrum of short-time segments of speech. However, instead of estimating the pitch, the phase vocoder estimates the phase derivative at the output of each filter. By coding and transmitting the phase derivative, this vocoder destroys the relative phase information among the various signal harmonics in the speech signal. By compromising the signal quality contained in the relative phase information, we achieve greater bandwidth compression and, hence, a lower bit rate.

A typical channel of a phase vocoder analyzer is shown in Fig. 7.27. The speech signal in each frequency band is demodulated to baseband and filtered by a lowpass filter having a bandwidth of about 50 Hz. The demodulator frequencies are usually spaced about 100 Hz apart. Consequently, about 25–30 filters are usually employed in the implementation of the phase vocoder. The output of the lowpass filters is used to compute the magnitude and phase derivative of the speech signal in each band. These are sampled at a nominal rate of 50–60 samples per second. (Note that sampling the magnitude and phase of the speech signal at a rate F_s is equivalent to sampling the real-valued bandpass signal at a rate of $2F_s$.) These samples of the magnitude and phase derivative are coded and transmitted to the receiver.

Coding of the spectral magnitude may be done as in the channel vocoder by using log PCM and DPCM, with lower frequencies having greater precision. The phase derivative is usually coded linearly using 2–3 bpn. The resulting bit rate is in the range of 7200 bps.

At the receiver, the signal is synthesized as shown in the block diagram in Fig. 7.28. First, the phase in each channel is integrated. Then the signal magnitude and the resulting phase are used to form two lowpass in-phase and quadrature signal components, which are interpolated and translated in frequency by multiplying each component by $\cos(\omega_k n)$ and $\sin(\omega_k n)$, where ω_k represents the normalized frequency translation. The two signal components are then added to form the bandpass speech signal for each of the frequency bands.

Due to the limitations indicated above, the phase vocoder has not been widely used in practice.

7.4.3 The Cepstral (Homomorphic) Vocoder

In the speech model assumed for vocoder implementation, the voiced speech signal is assumed to be generated by exciting a slowly varying vocal system filter by a periodic sequence. Unvoiced speech is generated by exciting the filter with a white noise sequence. In either case, the

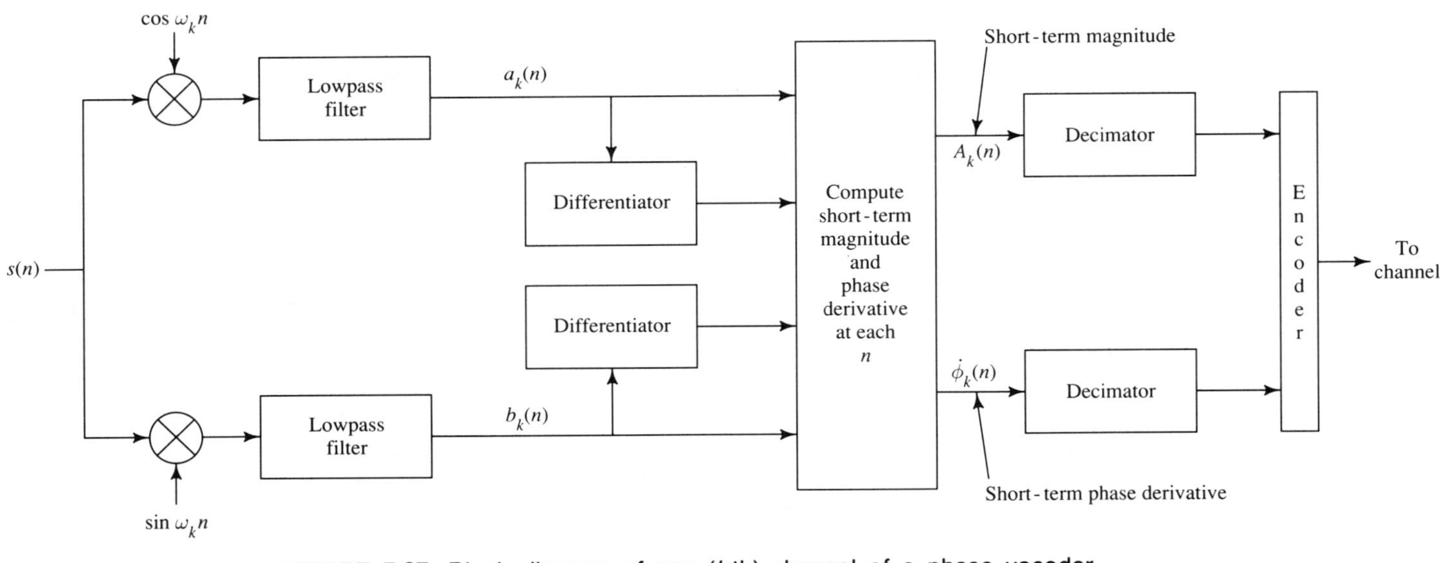

FIGURE 7.27. Block diagram of one (kth) channel of a phase vocoder analyzer. $a_k(n)$ and $b_k(n)$ are bandpass signals (bandwidth typically 50 Hz) at baseband.

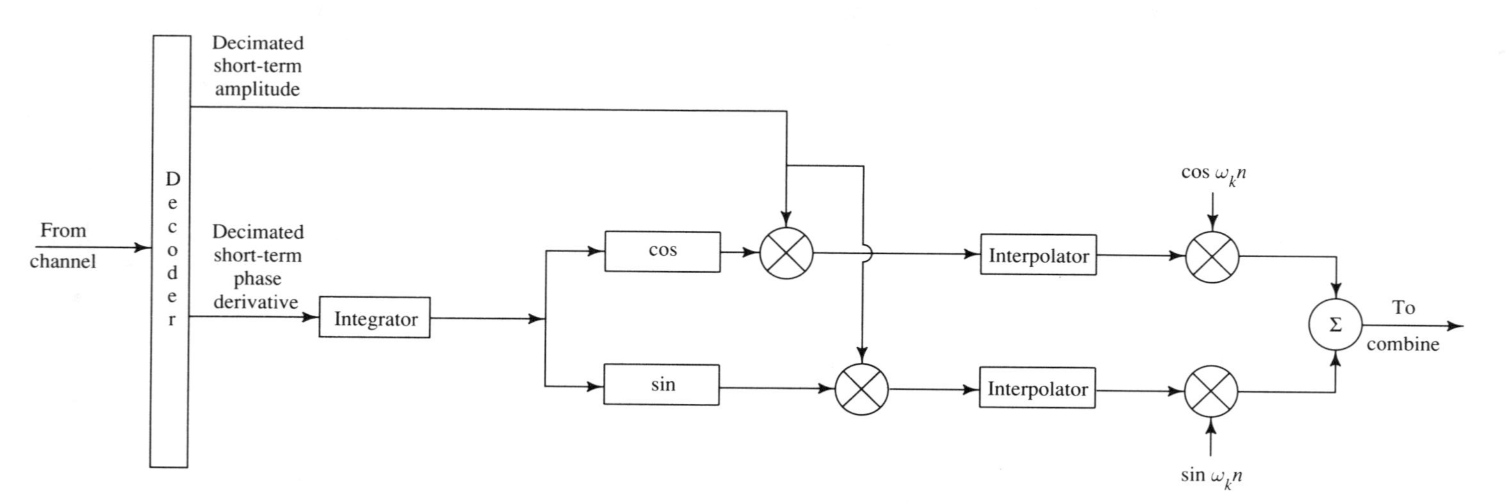

FIGURE 7.28. Block diagram of one (kth) channel of a phase vocoder synthesizer.

vocal system response is slowly varying, whereas the excitation is varying more rapidly.

This basic difference between the excitation and the vocal system response is exploited by the cepstral vocoder. We recall from Chapter 6 that by processing frames of a speech signal in the cepstral domain we can separate the slowly varying vocal system spectrum from the faster varying periodic spectrum due to the pitch. Thus we can separate these two spectra and estimate the characteristics of the vocal system.

As illustrated in Fig. 7.29, the analysis system computes the short-term cepstrum [called the short-term real cepstrum (stRC) in Chapter 6], $c_s(n; m)$, for a given frame $f(n; m)$, and then uses a low-time lifter to extract the component of the cepstrum corresponding to the speech production model impulse response,[8] $c_\theta(n; m)$. Typically, a time window of 2–3 msec is sufficient to exclude the effect of the excitation. Pitch estimation can also be performed by analyzing the cepstral component due to the excitation, $c_e(n; m)$, which is extracted by means of a high-time lifter. Cepstral analysis of a real speech waveform is illustrated in Figs. 6.7, 6.8, and 6.10.

The low-time liftered cepstrum, $c_\theta(n; m)$, is coded using 5–6 bpn, and the resulting bits are transmitted, along with the side information on voiced/unvoiced speech and pitch period, to the receiver. The resulting bit rate is in the range of 6000–10,000 bps.

A block diagram of the synthesizer for the speech signal at the receiver is shown in Fig. 7.30. The cepstral component $c_\theta(n; m)$ is subjected to the customary nonlinear processing that results in the estimate $\hat{\theta}(n; m)$ of the system impulse response corresponding to the frame $f(n; m)$ (compare Section 6.2.3). The speech signal is synthesized by convolving the vocal system response with the appropriate excitation, which is generated at the receiver. Edge effects can be reduced by interpolating or smoothing the cepstral responses from frame to frame. Recall that this series of operations will destroy all phase information [see the discussion surround-

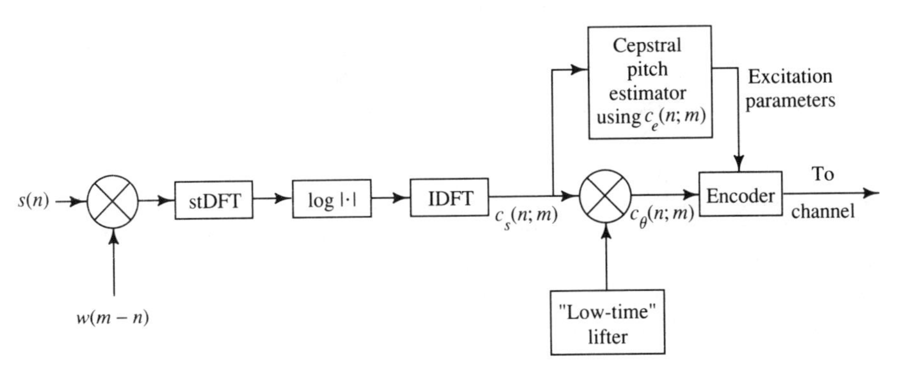

FIGURE 7.29. Block diagram of cepstral vocoder analyzer.

[8]The careful reader will recall that $c_\theta(n; m)$ should not have a second (frame) argument. The reason can be found in the discussion leading to (6.30). However, we include the argument m here to keep track of the frame upon which we are working.

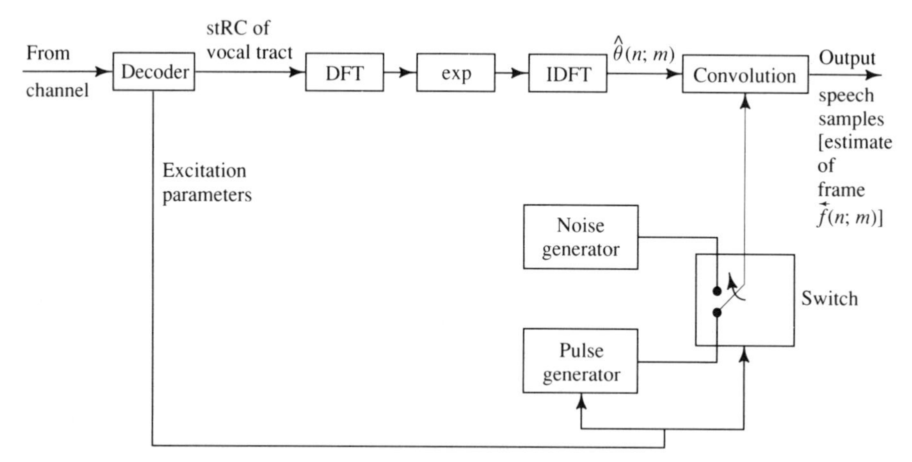

FIGURE 7.30. Block diagram of cepstral vocoder synthesizer.

ing (6.17)]. In particular, the placement of the frame in time is lost when the "real" cepstrum is used, and the formalities here should not be interpreted to the contrary. In practice, this is not a problem, as the receiver will correctly concatenate incoming information.

More significantly, the loss of phase information from the speech spectra degrades speech quality. Phase information can be preserved by computing the complex cepstrum (see Section 6.3). However, the additional computation coupled with the associated phase unwrapping that is necessary at the receiver renders this approach undesirable from a computational viewpoint. Further, one of the general findings of Quatieri's work discussed in Section 6.4 (Quatieri, 1979) is that the complex cepstrum, by preserving phase information at significant computational cost and algorithmic overhead, does not significantly outperform the real cepstrum in terms of speech quality. Even with the use of the real cepstrum, the computational complexity of the cepstral vocoder is its main disadvantage for practical use (see also Section 6.4).

Another type of homomorphic vocoder is based on an *analysis-by-synthesis* method. The stRC is used to estimate the vocal system impulse response at the encoder in each frame, as described above. Then the synthesized speech is generated at the encoder by exciting the vocal system filter model. The difference between the synthetic speech and the original speech signal constitutes an error signal, which is spectrally weighted to emphasize lower frequencies and then minimized by optimizing the excitation signal. Optimal excitation sequences are typically computed over four or five blocks *within the frame duration*, meaning that the excitation is updated more frequently than the vocal system filter.[9]

[9]For clarity in the ensuing discussions, we will use the term *block* to indicate a subframe interval. The *frame* will indicate the usual analysis interval over which the speech is analyzed. Ordinarily this will correspond to the interval over which the vocal-tract characterization is estimated. The characterization of the excitation, however, will often be updated more frequently—over blocks of the analysis frame.

A method for determining the excitation sequence has been described by Chung and Schafer (1990). The excitation signal is determined dynamically every few milliseconds within the frame under analysis, and takes one of three forms. For unvoiced speech, the excitation sequence $e(n)$ is selected from a Gaussian codebook of sequences and has the form

$$e(n) = \beta \rho_k(n), \tag{7.111}$$

where k is the index of the sequence selected from the codebook, and β is a scale factor. For voiced frames, the excitation sequence $e(n)$ consists of two sequences, selected from a time-variant queue of past excitations, of the form

$$e(n) = \beta_1 e(n - d_1) + \beta_2 e(n - d_2), \tag{7.112}$$

where β_1 and β_2 are scale factors and d_1 and d_2 are appropriately selected values of delay. For speech frames in which the excitation signal is classified as mixed, the excitation sequence $e(n)$ is modeled as the sum of a Gaussian codebook sequence $\rho_k(n)$ and a sequence selected from an interval of the past excitation, that is,

$$e(n) = \beta_1 \rho_k(n) + \beta_2 e(n - d), \tag{7.113}$$

where d is the delay.

Spectral weighting is performed on the vocal-tract impulse response and the original speech frame prior to speech synthesis, as shown in Fig. 7.31. Note that the perceptually weighted speech sequence is denoted as $y(n)$, that is,

$$y(n) = f(n; m) * w(n), \tag{7.114}$$

where $f(n; m)$ is the incoming speech frame of length N, $w(n)$ is the impulse response of the weighting filter, and "$*$" denotes convolution. The excitation is applied to the weighted vocal-tract response to produce a

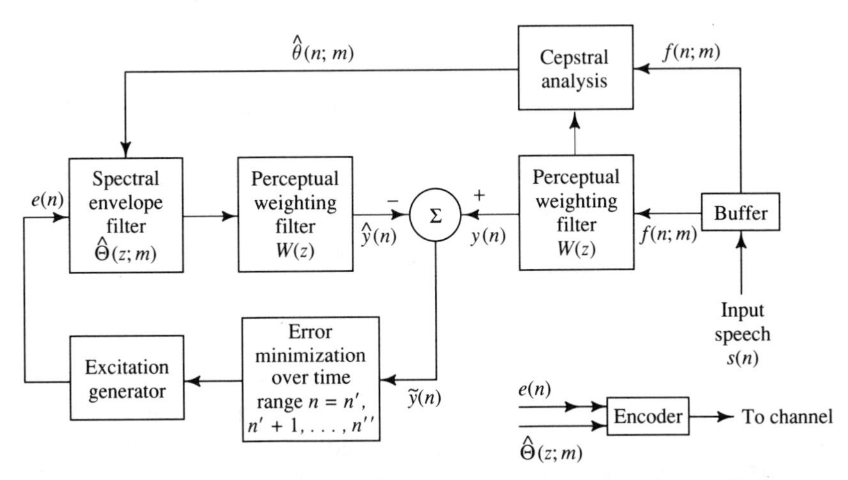

FIGURE 7.31. Analysis-by-synthesis method for obtaining the excitation sequence $e(n)$.

synthetic speech sequence $\hat{y}(n)$. For example, the response of the weighted vocal-tract filter to the mixed excitation may be expressed as

$$\hat{y}(n) = \beta_1 \hat{y}_1(n) + \beta_2 \hat{y}_2(n), \tag{7.115}$$

where $\hat{y}_1(n)$ is the response of the weighted vocal-tract filter to $\rho_k(n)$ and $\hat{y}_2(n)$ is its response to $e(n-d)$. The parameters β_1, β_2, k, and d are selected to minimize the error energy over small blocks of time between the weighted speech $y(n)$ and the synthetic speech $\hat{y}(n)$. For discussion purposes, let us assume that the time range of interest is $n = n', n'+1, \ldots, n''$. To simplify the optimization process, the minimization is performed in two steps. First, β_2 and d are determined to minimize the error energy

$$\xi_2 = \sum_{n=n'}^{n''} [y(n) - \beta_2 \hat{y}_2(n)]^2. \tag{7.116}$$

For a given value of d, the optimum value of β_2, say $\hat{\beta}_2(d)$, is easily shown to be

$$\hat{\beta}_2(d) = \frac{\sum_{n=n'}^{n''} y(n)\hat{y}_2(n)}{\sum_{n=n'}^{n''} \hat{y}_2^2(n)}. \tag{7.117}$$

By restricting the delay d to a small range, the optimization of d is performed by exhaustive search and the resulting $\hat{\beta}_2(d)$ is obtained from (7.117).

Once these two parameters are determined, the optimum choices of β_1 and k are made based on the minimization of the error energy between the residual signal $y_1(n) = y(n) - \beta_2 \hat{y}_2(n)$ and $\beta_1 \hat{y}_1(n)$. Thus β_1 and k are chosen by an exhaustive search of the Gaussian codebook to minimize

$$\xi_1 = \sum_{n=n'}^{n''} [y_1(n) - \beta_1 \hat{y}_1(n)]^2. \tag{7.118}$$

For any given sequence from the codebook, the optimum choice for β_1 is

$$\hat{\beta}_1 = \frac{\sum_{n=n'}^{n''} y_1(n)\hat{y}_1(n)}{\sum_{n=n'}^{n''} \hat{y}_1^2(n)}. \tag{7.119}$$

The optimum excitation parameters and the vocal system impulse response (or its cepstrum sequence) are coded and transmitted to the decoder. The synthesized speech is generated as shown in Fig. 7.32 by exciting the vocal system filter with the excitation signal $e(n)$. The

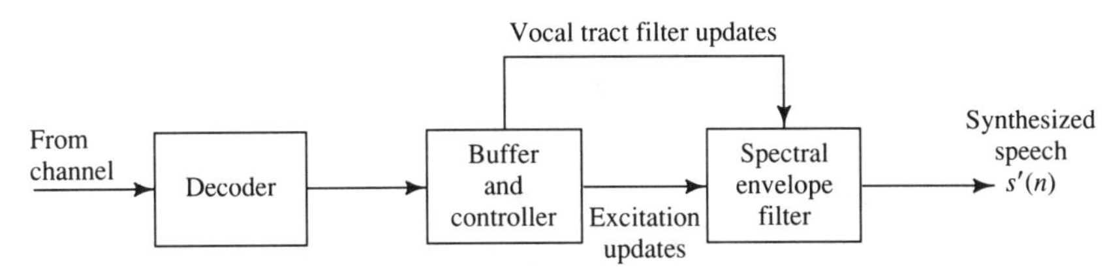

FIGURE 7.32. Synthesizer for the homomorphic vocoder. The controller is included to concatenate and interpolate incoming information from different frames and different excitation blocks within frames.

overlap-add method may be used to combine adjacent output speech records at the receiver.

This analysis-by-synthesis homomorphic vocoder has been implemented by Chung and Schafer (1990). A frame duration of 20 msec was used for the vocal-tract analysis (160 samples of an 8-kHz sampling rate) and 5-msec block duration (40 samples) for determining the excitation. The codebook employed 256 zero-mean Gaussian codewords of 40 samples each. A bit rate of about 3000 bps was achieved with this implementation.

7.4.4 Formant Vocoders

A formant vocoder may be viewed as a type of channel vocoder that estimates the first three or four formants in a segment of speech and their corresponding bandwidths. It is this information plus the pitch period that is encoded and transmitted to the receiver.

For a given frame of speech, each formant may be characterized by a two-pole digital filter of the form

$$\Theta_k(z) = \frac{\Theta_k}{\left(1 - \rho_k e^{j\omega_k} z^{-1}\right)\left(1 - \rho_k e^{-j\omega_k} z^{-1}\right)}, \qquad k = 1, 2, 3, 4, \tag{7.120}$$

where Θ_k is a gain factor, ρ_k is the distance of the complex-valued pole pair from the origin, and $\omega_k = 2\pi F_k T$. Also, F_k is the frequency of the kth formant in Hz, and T is the sampling period in seconds. The bandwidth is determined by the distance of the pole from the unit circle.

The formants can be estimated by linear prediction or by cepstral analysis. The most difficult aspect of the analysis is to obtain accurate estimates of the formants, especially when two formants are very close together. In such a case, the chirp z-transform can provide better frequency resolution and may be used for formant estimation. Nevertheless, this problem has hindered the use of this vocoder in practical applications.

The synthesizer for the formant vocoder may be realized as a cascade of two-pole filters, one filter for each formant, as shown in Fig. 7.33. An alternative realization is a parallel bank of two-pole filters with adjustable gain parameters. In this configuration, the overall filter contains

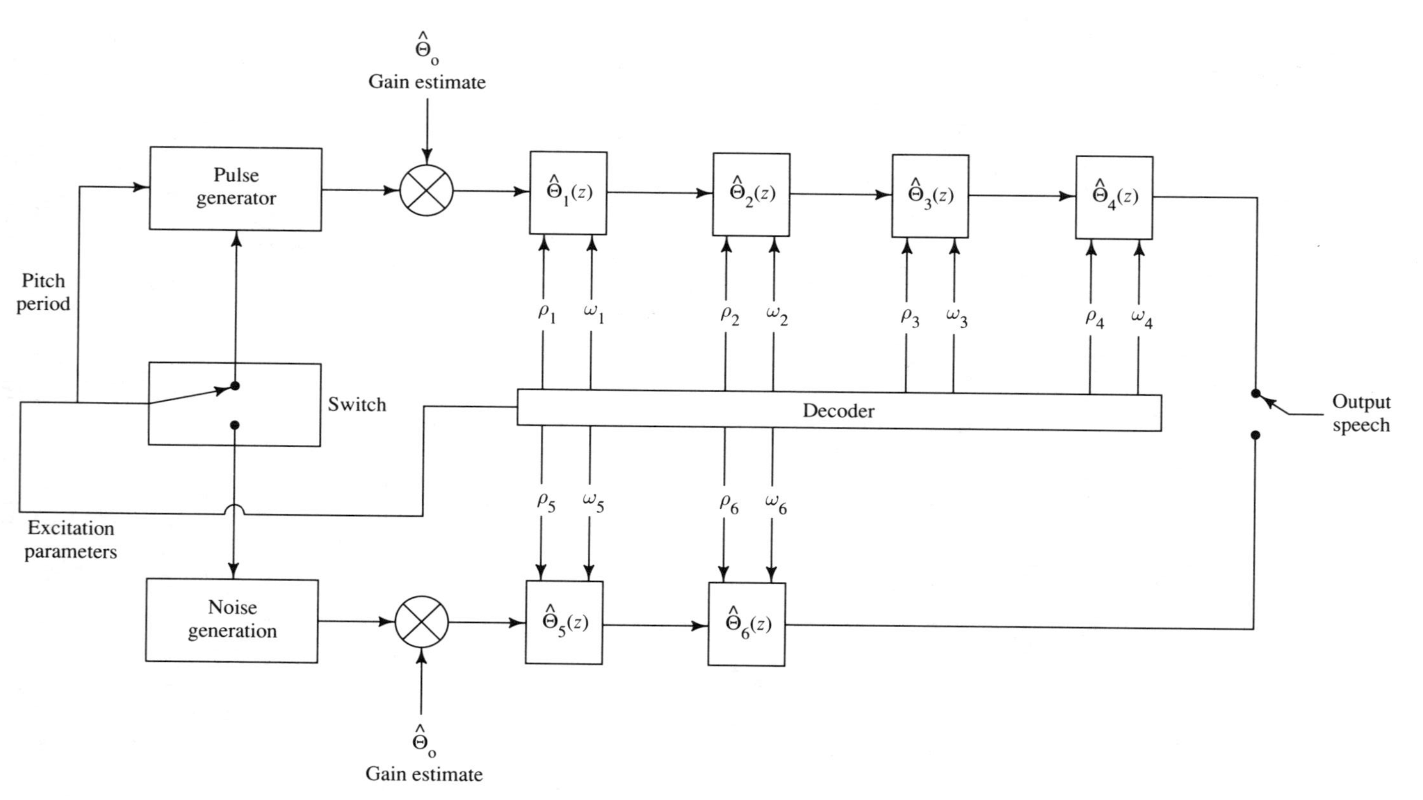

FIGURE 7.33. Block diagram of formant vocoder synthesizer.

zeros whose values depend on the gain parameters. Therefore, the values of the gain parameters must be carefully selected.

For unvoiced speech, the formant vocoder can be simplified to fewer filter sections, typically two. One of the filters may be designed to include a pair of complex conjugate zeros in addition to the poles to provide for a smooth spectrum. Alternatively, linear prediction may be applied to the segments of unvoiced speech.

Since so few parameters are involved in the analysis and synthesis portions of the formant vocoder, significant bandwidth compression is possible. In particular, each formant may be coded into 3–4 bits, the pitch requires 6 bits, and the overall gain factor requires 2–3 bits. Hence it is possible to achieve a bit rate of 1000 bps or less with a formant vocoder. In general, the accuracy of the formant estimates determines the quality of the estimated speech.

7.4.5 Linear Predictive Coding

Generalities

As described in Chapter 5, the objective of LP analysis is to estimate parameters of an all-pole model of the vocal tract. Related problems are to determine the type of excitation and also to estimate the pitch period and the gain parameter.

Suppose that these pieces of information have been determined for a given frame of speech to be transmitted or stored. This is called a *linear predictive coding* (LPC) problem, a name that is frequently applied to LP analysis whether or not *coding* is the issue. Typically, the pitch period requires 6 bits and the gain parameter may be represented by 5 bits after its dynamic range is compressed logarithmically. If the prediction coefficients were to be coded, they would require 8–10 bits per coefficient for accurate representation. The reason for such high accuracy is that relatively small changes in the prediction coefficients result in a large change in the pole positions of the filter model. The accuracy requirements are lessened by transmitting the reflection coefficients, which have a smaller dynamic range, that is, $|\kappa(i;m)| \leq 1$ (see Section 5.3.3). These are adequately represented by 6 bits per coefficient. Thus, for a 10th-order predictor (i.e., five poles), the total number of bits assigned to the model parameters per frame is 72. If the model parameters are changed every 15–30 msec, the resulting bit rate is in the range 2400–4800 bps. Since the reflection coefficients are usually transmitted to the receiver, the synthesis filter at the receiver is implemented as a lattice filter, as shown in Fig. 7.34.

The coding of the reflection coefficients can be improved further by first performing a nonlinear transformation of the coefficients. A problem arises when some of the reflection coefficients are very close to ± 1. In such a case, the quantization error introduced by coding significantly affects the quality of the synthesized speech. By means of an appropriate

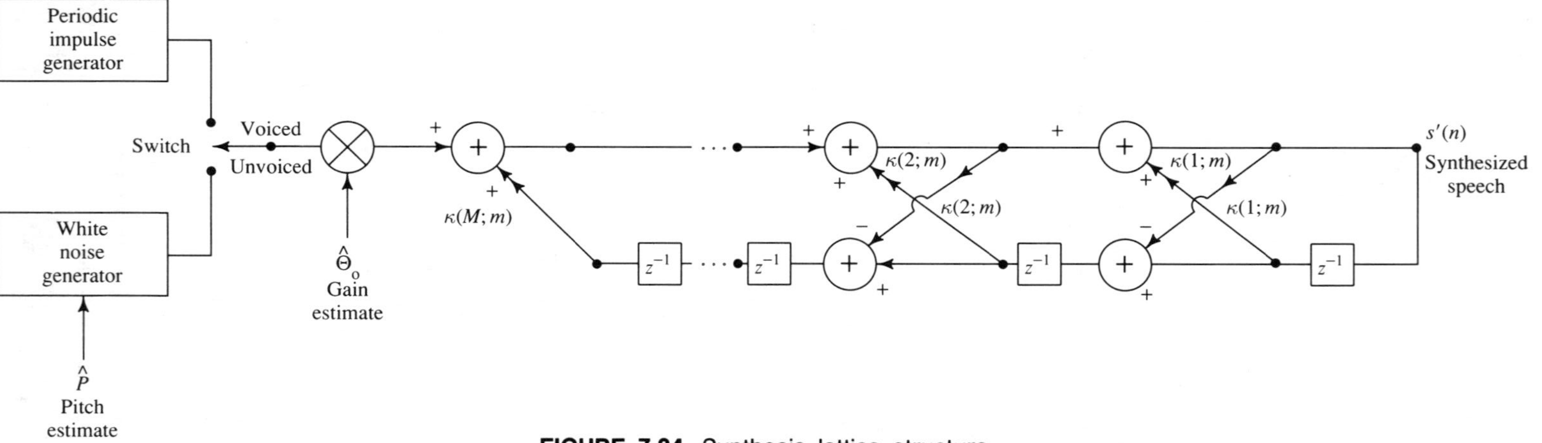

FIGURE 7.34. Synthesis lattice structure.

nonlinear transformation of the reflection coefficients, the scale is warped so that we can apply a uniform quantizer to the transformed coefficients. Thus we obtain an equivalent nonuniform quantizer. The desired transformation should expand the scale near the values of ± 1.

Two nonlinear transformations that accomplish this objective are the inverse sine transform and the inverse hyperbolic tangent [or log-area ratio (LAR)] transform. The inverse sine maps the $\kappa(i;m)$ into

$$\sigma(i;m) = \frac{2}{\pi}\sin^{-1}\kappa(i;m), \qquad 1 \le i \le M, \tag{7.121}$$

and the LAR transform performs the mapping

$$g(i;m) = \tanh^{-1}\kappa(i;m) = \frac{1}{2}\log\frac{1+\kappa(i;m)}{1-\kappa(i;m)}, \qquad 1 \le i \le M. \tag{7.122}$$

These parameters were first introduced in Section 5.3.3. Either of these transformations is effective in reducing the quantization noise inherent in the coding of the reflection coefficients. We should also recall the line spectral pair (LSP) parameters (Section 5.4.1), which may also serve as an efficient alternative to direct use of the LP parameters.

Several methods have been devised for generating the excitation sequence for speech synthesis. Next we describe several LPC-type speech analysis and synthesis schemes that differ primarily in the type of excitation signal that is generated for speech synthesis.

LPC-10 Algorithm

The earliest LPC-based vocoders followed the classic all-pole model of Fig. 5.1 or Fig. 7.24 directly in the use of a pitch-pulse excitation for synthesis of voiced speech, and a noise source excitation for unvoiced. Such a vocoder first caught the public's attention in the educational toy "Speak-and-Spell" produced by Texas Instruments in the 1970s. The basic method is still in use in many vocoding applications and is employed by the U.S. Government as one speech coding standard. The algorithm is usually called *LPC-10*, a reference to the fact that 10 coefficients are typically employed.

One version of LPC-10 is described in the papers by Kang et al. (1979) and Tremain (1982). LPC-10 partitions the speech into 180 sample frames, resulting in a frame rate of 44.44 frames/sec when an 8-kHz sample rate is used on the data. Pitch and a voicing decision are determined by using the AMDF and zero crossing measures discussed in Section 4.3. The covariance method is used to compute the LP solution, and a set of "generalized" reflection coefficients [similar to reflection coefficients arising from the autocorrelation method (Gibson, 1977)] are encoded for storage or transmission. Ten coefficients are computed; the first two are encoded as LARs and the others are encoded linearly. When an unvoiced decision is made, only four coefficients are used. In the stored or transmitted bit stream, 41 bits are used for the reflection coefficients, 7 for pitch and the voiced/unvoiced bit, and 5 for the gain. One additional bit is used for synchronization. Accordingly, a total of 54 bits per frame are sent, yielding a bit rate of 2400 bps.

Residual Excited Linear Prediction Vocoder

Speech quality in LPC can be improved at the expense of a higher bit rate by computing and transmitting a residual error, as done in the case of DPCM. There are various ways in which this can be done. One method is illustrated in the block diagram of Fig. 7.35. Once the LPC model and excitation parameters are estimated from a frame of speech, the speech is synthesized at the transmitter and subtracted from the original speech signal to form a residual error. The residual error is quantized, coded, and transmitted to the receiver along with the model parameters. At the receiver the signal is synthesized by adding the residual error signal to the signal generated from the model. Thus the addition of the residual error improves the quality of the synthesized speech.

Another approach that produces a residual error is shown in Fig. 7.36. In this case, the original speech signal is passed through the inverse (all-zero) filter to generate the residual error signal. This signal has a relatively flat spectrum, but its most important frequency components for

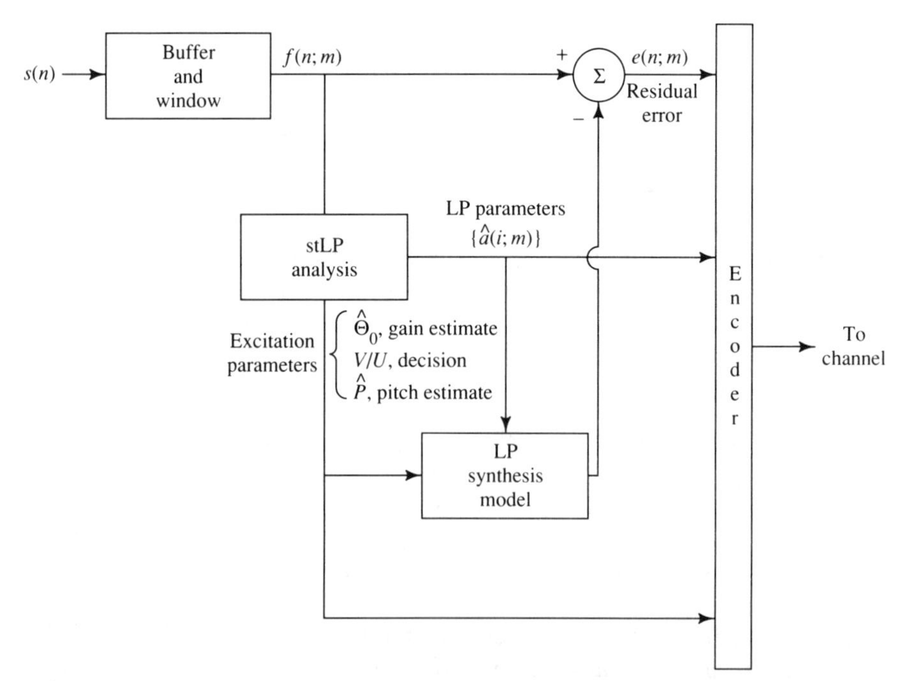

FIGURE 7.35. LPC encoder with residual error transmitted. [Note that the error signal $e(n; m)$ is not the prediction residual].

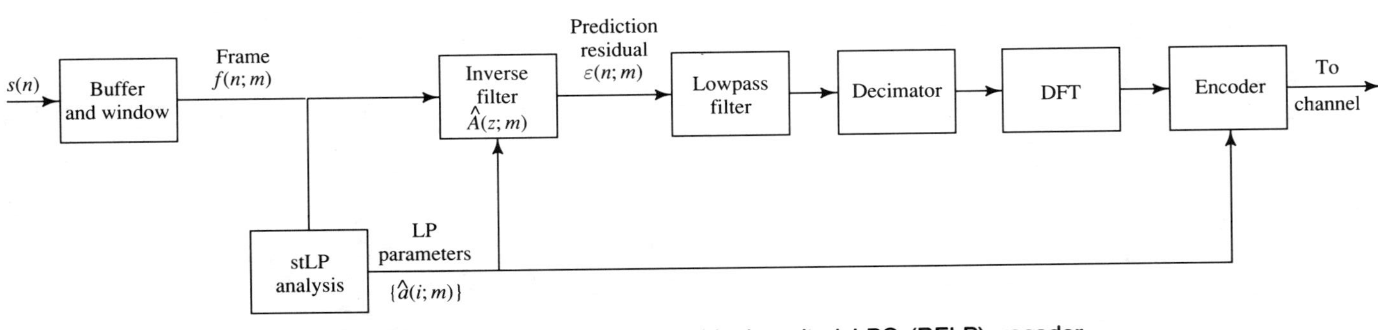

FIGURE 7.36. Analyzer for a residual excited LPC (RELP) vocoder.

improving speech quality are contained in the frequency range below 1000 Hz. Ideally, we would like to transmit the entire residual signal spectrum. However, in order to reduce the bit rate, we pass the residual error through a lowpass filter with bandwidth 1000 Hz, decimate its output, and encode the decimated signal. Usually, the decimated signal is transformed into the frequency domain via the DFT and the magnitude and phase of the frequency components are coded and transmitted to the receiver.

At the receiver, the residual error signal is transformed into the time domain, interpolated, and filtered. The resulting signal contains no high-frequency information, which must be restored in some manner. A simple method for regenerating the high-frequency components is to pass the signal through a full-wave rectifier and then flatten the resulting spectrum by filtering. The lowpass and highpass signals are summed, as shown in Fig. 7.37, and the resulting residual error signal is used to excite the LPC model. We note that this method does not require pitch information and voicing information. The residual error signal provides the excitation to the all-pole LPC model. This LPC vocoder is called a *residual excited linear prediction* (RELP) vocoder; it provides communication-quality speech at about 9600 bps.

Multipulse LPC Vocoder

One of the shortcomings of RELP is the need to regenerate the high-frequency components at the decoder. The regeneration scheme results in a crude approximation of the high frequencies. The *multipulse* LPC method described in this section is a time domain method that results in a better excitation signal for the LPC vocal system filter.

Multipulse LPC is an analysis-by-synthesis method, due to Atal and Remde (1982), which has the basic configuration shown in Fig. 7.38. The LPC filter coefficients are determined from the speech signal samples by the conventional methods described in Chapter 5. Let $\hat{\Theta}(z)$ denote the system function of the all-pole filter, which is usually realized as a lattice filter. This filter is to be understood to have been computed over the N-

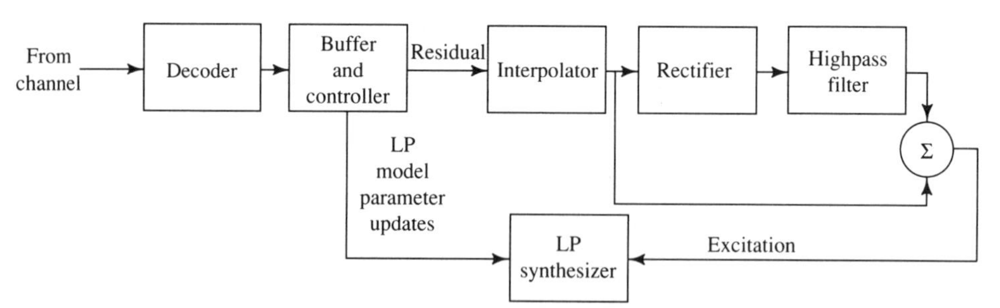

FIGURE 7.37. Synthesizer for a residual excited LPC vocoder. The controller is included to concatenate and interpolate incoming information from different frames.

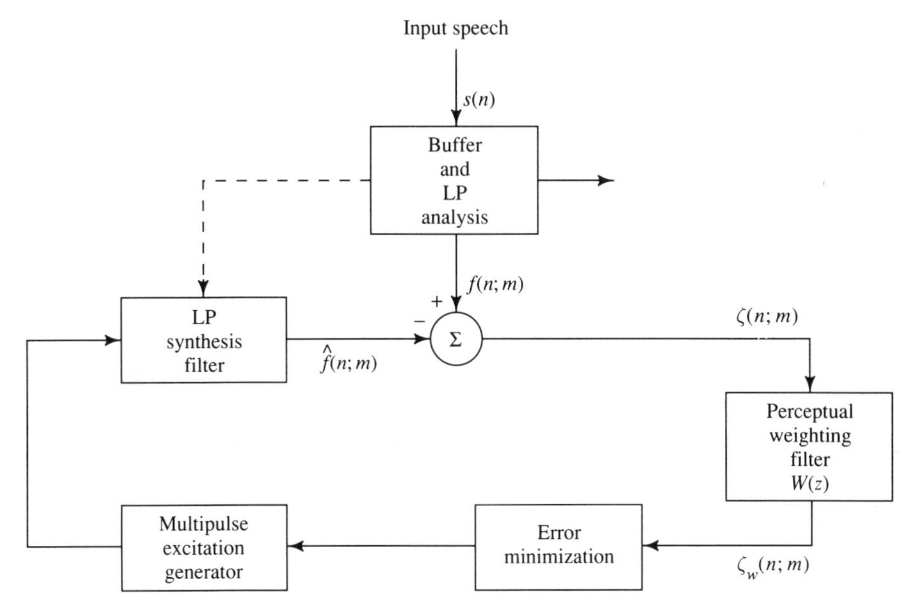

FIGURE 7.38. Analysis-by-synthesis method for obtaining the multipulse excitation.

length speech frame, $f(n; m)$, which ends at time m. The output of the filter is the synthetic speech (frame), say $\hat{f}(n; m)$, which is subtracted from the original speech signal to form the residual error sequence,[10] say

$$\zeta(n; m) = f(n; m) - \hat{f}(n; m). \tag{7.123}$$

The error sequence is passed through a perceptual error weighting filter with system function

$$W(z) = \frac{\hat{\Theta}(z/c)}{\hat{\Theta}(z)} = \frac{\hat{A}(z)}{\hat{A}(z/c)}, \tag{7.124}$$

where c is a parameter, in the range $0 < c \leq 1$, that is used to control the noise spectrum weighting. Note that when $c = 1$, there is no weighting of the noise spectrum and when $c = 0$, $W(z) = \hat{A}(z) = 1/\hat{\Theta}(z)$. In practice, the range $0.7 \leq c \leq 0.9$ has proved effective. Let us denote the perceptually weighted error by $\zeta_w(n; m)$, noting that

$$\zeta_w(n; m) = \zeta(n; m) * w(n). \tag{7.125}$$

The multipulse excitation consists of a short sequence of pulses (discrete-time impulses) whose amplitudes and locations are chosen to minimize the energy in $\zeta_w(n; m)$ over small blocks of the current frame. (For discussion purposes, let us concern ourselves with a block over the time range $n = n', n' + 1, \ldots, n''$.) For simplicity, the amplitudes and locations of the

[10]The reader should appreciate that this is not the residual sequence associated with the LP filter design.

impulses are obtained sequentially by minimizing the error energy for one pulse at a time. In practice only a few impulses, typically 4–8 every 5 msec, are sufficient to yield high-quality synthetic speech.

In particular, let us consider the placement of a pulse of amplitude $\alpha_1(k)$ at location $n = k$. Assuming that this is the initial pulse proposed, prior to its insertion we will have (see Fig. 7.38)

$$\zeta^0(n;m) = f(n;m), \tag{7.126}$$

where the superscript indicates the number of pulses in the excitation. In general, let us write $\zeta^i(n;m)$ and $\zeta^i_w(n;m)$ to denote the residual and perceptually weighted residual sequences with i pulses in the excitation. Further, let us denote the response of the cascaded filters $\hat{\Theta}(z)W(z) = 1/\hat{A}(c^{-1}z)$ to this initial impulse by $\alpha_1(k)\theta_w(n-k)$. It should be clear, therefore, that $\zeta^1_w(n;m)$ can be written

$$\begin{aligned} \zeta^1_w(n;m) &= f(n;m) * w(n) - \alpha_1(k)\theta_w(n-k) \\ &= \zeta^0_w(n;m) - \alpha_1(k)\theta_w(n-k). \end{aligned} \tag{7.127}$$

We seek to minimize

$$\xi_1 = \sum_{n=n'}^{n''} [\zeta^1_w(n;m)]^2 = \sum_{n=n'}^{n''} [\zeta^0_w(n;m) - \alpha_1(k)\theta_w(n-k)]^2 \tag{7.128}$$

by choice of α_1 and k. The minimization with respect to $\alpha_1(k)$ yields

$$\sum_{n=n'}^{n''} [\zeta^0_w(n;m) - \alpha_1(k)\theta_w(n-k)]\theta_w(n-k) = 0 \tag{7.129}$$

or, equivalently,

$$\hat{\alpha}_1(k) = \frac{\rho_{\zeta\theta}(k)}{\rho_{\theta\theta}(k)} \stackrel{\text{def}}{=} \frac{\sum_{n=n'}^{n''} \zeta^0_w(n;m)\theta_w(n-k)}{\sum_{n=n'}^{n''} \theta^2_w(n-k)}. \tag{7.130}$$

We can eliminate $\alpha_1(k)$ from (7.128) by substituting the optimum values $\hat{\alpha}_1(k)$. Thus we obtain

$$\xi_1(k) = \sum_{n=n'}^{n''} [\zeta^0_w(n;m)]^2 - \frac{\rho^2_{\zeta\theta}(k)}{\rho_{\theta\theta}(k)}. \tag{7.131}$$

Hence the optimum location for the first impulse is the value of k, say $\hat{k}$, that minimizes $\xi_1(k)$ or, equivalently, maximizes $\rho^2_{\zeta\theta}(k)/\rho_{\theta\theta}(k)$. Once $\hat{k}$ is obtained, $\hat{\alpha}(k)$ is computed from (7.130).

After the location and the amplitude of the impulse is determined, it is subtracted from $\zeta_w^1(n;m)$ to form a new sequence $\zeta_w^2(n;m)$. The location and amplitude of the second impulse is then determined by minimizing the error energy

$$\xi_2 = \sum_{n=n'}^{n''} [\zeta_w^1(n;m) - \alpha_2(k_2)\theta_w(n-k_2)]^2 \qquad (7.132)$$

in a similar manner. This procedure for determining the locations and amplitudes of the impulses is repeated until the perceptually weighted error is reduced below some specified level or the number of pulses reaches the maximum that can be encoded at some specified bit rate.

There are several variations of the basic multipulse LPC method described above. In particular, Singhal and Atal (1984) have noted that, for voiced speech, the multipulse LPC excitation sequence shows a significant correlation from one pitch period to the next. This observation suggests that the perceptually weighted error can be further reduced by including a long-delay correlation filter in cascade with the speech system $\hat{\Theta}(z)$, as shown in Fig. 7.39. This filter is usually implemented as a pitch predictor with system function

$$\Theta_p(z) = \frac{\Theta_p}{1 - bz^{-\hat{P}}}, \qquad (7.133)$$

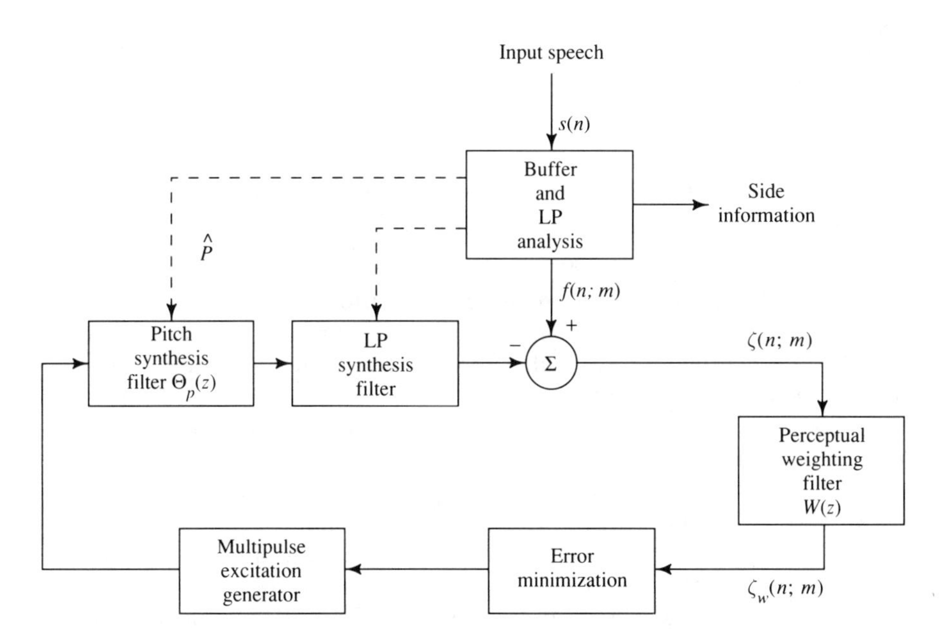

FIGURE 7.39. Analysis-by-synthesis method for obtaining the multipulse excitation with a pitch synthesis filter.

where $0 < b < 1$, $\hat{P}$ is an estimate of the number of samples in the pitch period, and Θ_p is a scale factor. Thus the pitch filter $\Theta_p(z)$ provides the long-term correlation in the excitation and, thus, the correlation in the multipulse excitation is reduced.

Other variations of the basic multipulse LPC method can be devised by adopting different strategies in the optimization of the pulse-signal amplitudes. For example, as the location and amplitude of each new pulse is obtained, one can go back and reoptimize the amplitudes of the previous pulses. Alternatively, we can perform a joint optimization of all the amplitudes of the pulses by solving a set of linear equations after all the pulse locations have been determined. This and several other variations of the basic scheme described above have been suggested in the paper by Lefevre and Passien (1985).

In conclusion, multipulse LPC has proved to be an effective method for synthesizing good-quality speech at 9600 bps. Multipulse LPC vocoders have been implemented for commercial use [see (Putnins et al., 1985)] and have been used for airborne mobile satellite telephone service. The information that is transmitted concerning the excitation sequence includes the locations of the pulses, an overall scale factor corresponding to the largest pulse amplitude, and the pulse amplitudes relative to the overall scale factor. The scale factor is logarithmically quantized, usually to 6 bits. The amplitudes are linearly quantized, usually into 4 bits (one of 16 levels). The pulse locations are usually encoded by means of a differential coding scheme. The excitation parameters are usually updated every 5 msec, while the LPC vocal-tract parameters and the pitch period estimate are updated less frequently, say every 20 msec. Thus a nominal bit rate of 9600 bps is obtained.

Code-Excited Linear Prediction Vocoder

As indicated in the preceding section, multipulse LPC provides good-quality speech signals at 9600 bps. Further reductions in the bit rate can be achieved by better selection of the excitation sequence $e(n)$.

Code-excited linear prediction (CELP) is an analysis-by-synthesis method (see Schroeder and Atal, 1985) in which the excitation sequence is selected from a codebook of zero-mean Gaussian sequences. Hence the excitation sequence is a stochastic signal selected from a stored codebook of such sequences.

The CELP synthesizer is shown in Fig. 7.40. It consists of the cascade of two all-pole filters, with coefficients that are updated periodically. The first filter is a long-delay pitch filter used to generate the pitch periodicity in voiced speech. This filter typically has the form given by (7.133), where its parameters can be determined by minimizing the prediction error energy, after pitch estimation, over a frame duration of 5 msec. The second filter is a short-delay all-pole (vocal-tract) filter used to generate the spectral envelope (formants) of the speech signal. This filter typically

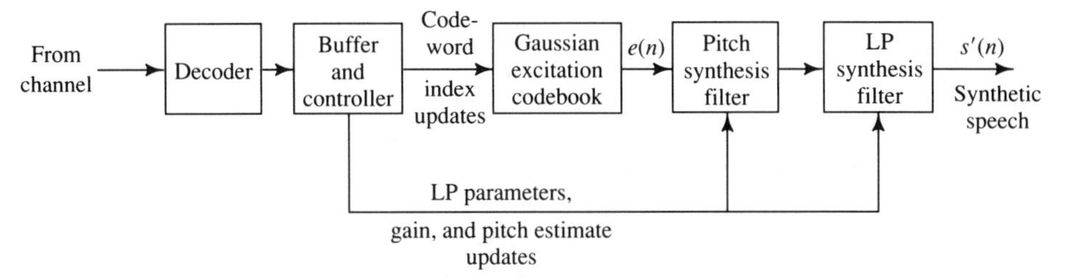

FIGURE 7.40. CELP synthesizer.

has 10–12 coefficients that are determined periodically, as described previously, for example, every 10–20 msec.

A block diagram of the analysis-by-synthesis coder is shown in Fig. 7.41. A stored sequence from a Gaussian excitation codebook is scaled and used to excite the cascade of a pitch synthesis filter and the LPC synthesis filter (computed over the "current" frame). The synthetic speech is compared with the original speech and the difference constitutes the residual error signal, which is perceptually weighted by passing it through a filter that is characterized by the system function in (7.124), as in multipulse LPC. This perceptually weighted error is squared and summed over a subframe block to give the error energy. By performing an exhaustive search through the codebook we find the excitation sequence that minimizes the error energy. The gain factor for scaling the excitation sequence is determined for each codeword in the codebook by minimizing the error energy for the block of samples.

For example, suppose that a speech signal is sampled at a frequency of 8 kHz and the subframe block duration for the pitch estimation and ex-

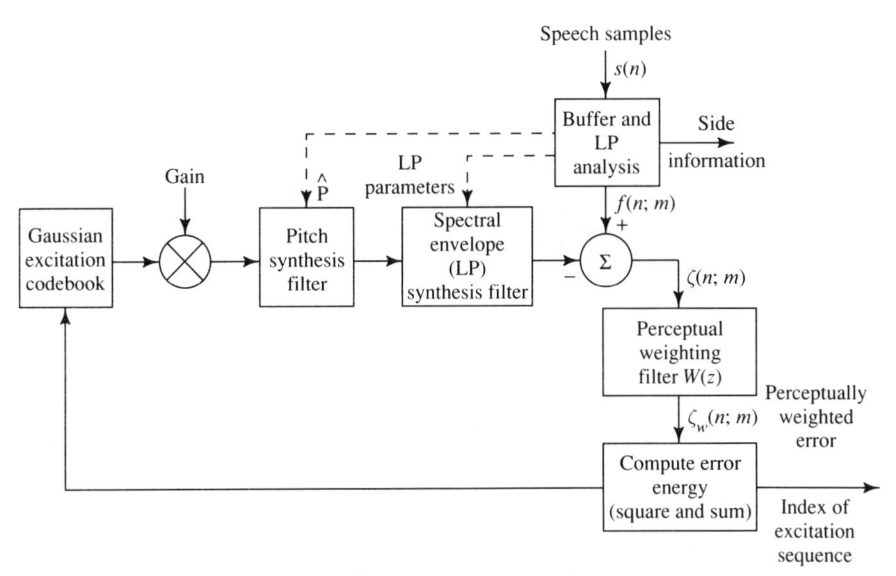

FIGURE 7.41. CELP analysis-by-synthesis coder.

citation sequence selection is performed every 5 msec. Thus we have 40 samples per 5-msec block interval. The excitation sequences consist of 40 samples each, stored in the codebook. A codebook of 1024 sequences has been found to be sufficiently large to yield good-quality speech. For such a codebook size, we require 10 bits to send the codeword index. Hence the bit rate is $\frac{1}{4}$ bpn. The transmission of the pitch predictor parameters and the spectral predictor parameters brings the bit rate to about 4800 bps. Methods for allocating bits dynamically (adaptively) have also been devised; they are described in papers by Kroon and Atal (1988), Yong and Gersho (1988), Jayant and Chen (1989), Taniguchi et al. (1989), and Akamine and Miseki (1990).

CELP has also been used to achieve toll-quality speech at 16,000 bps with a relatively low delay. Although other types of vocoders produce high-quality speech at 16,000 bps, these vocoders typically buffer 10–20 msec of speech samples and encode speech on a frame-by-frame basis. As a consequence, the one-way delay is of the order of 20–40 msec. However, with a modification of the basic CELP, it is possible to reduce the one-way delay to about 2 ms.

The low-delay version of CELP is achieved by using a backward-adaptive predictor with a gain parameter and an excitation vector size as small as 5 samples. A block diagram of the low-delay CELP encoder, as implemented by Chen (1990), is shown in Fig. 7.42. Note that the pitch predictor used in the conventional forward-adaptive coder is eliminated. In order to compensate for the loss in pitch information, the LPC predictor order is increased significantly, typically to an order of about 50. The LPC coefficients are also updated more frequently, typically every 2.5 ms, by performing LPC analysis on previously quantized speech. A 5-sample excitation vector corresponds to an excitation block duration of

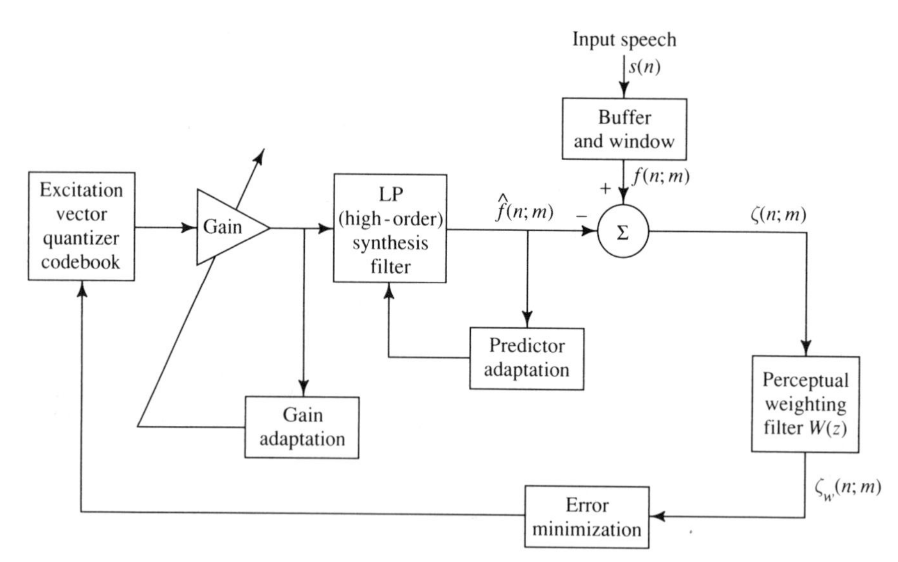

FIGURE 7.42. Low-delay CELP coder.

0.625 msec at an 8-kHz sampling rate. Hence the LPC analysis frame may be four times as long as the excitation block size.

In Chen's implementation, the logarithm of the excitation gain is adapted every subframe excitation block by employing a 10th-order adaptive linear predictor in the logarithmic domain. The coefficients of the logarithmic-gain predictor are updated every four blocks by performing an LPC analysis of previously quantized and scaled excitation signal blocks. The perceptual weighting filter is also 10th order and is updated once every four blocks by employing an LPC analysis on frames of the input speech signal of duration 2.5 msec. Finally, the excitation codebook in the low-delay CELP is also modified compared to conventional CELP. A 10-bit excitation codebook is employed, but the codewords are closed-loop optimized to take into account the effects of predictor adaptation and gain adaptation (Chen, 1990).

Another variation of the conventional CELP vocoder is the *vector sum excited linear prediction* (VSELP) vocoder described in the paper by Gerson and Jasiuk (1990). The VSELP coder and decoder basically differ in the method by which the excitation sequence is formed. To be specific, we consider the 8000-bps VSELP vocoder described by Gerson and Jasiuk.

A block diagram of the VSELP decoder is shown in Fig. 7.43. As we observe, there are three excitation sources. One excitation is obtained from the pitch (long-term) predictor state. The other two excitation sources are obtained from two codebooks, each containing 128 codewords. The outputs from the three excitation sources are multiplied by their corresponding gains and summed. The LPC synthesis filter is implemented as a 10-pole filter and its coefficients are coded and transmit-

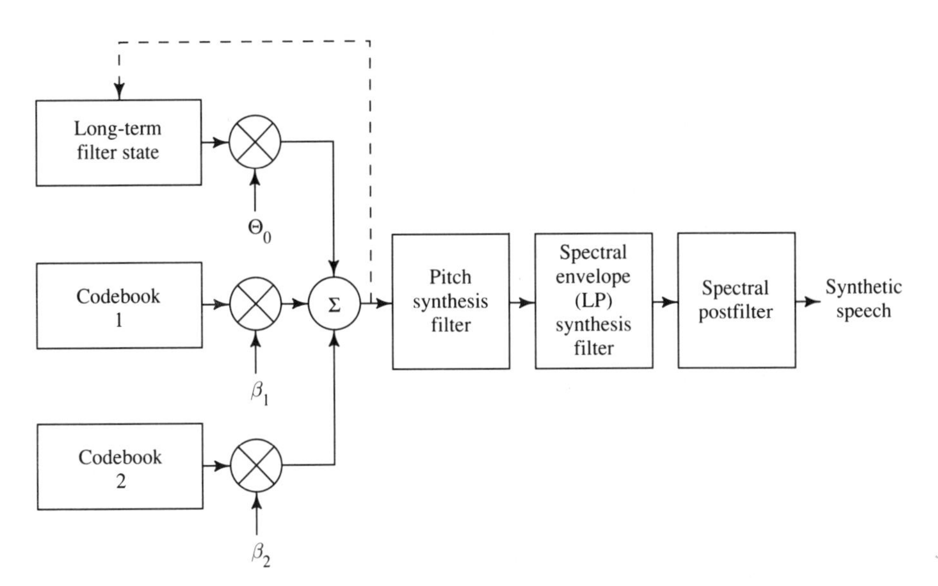

FIGURE 7.43. VSELP decoder.

ted every 20 ms. These coefficients are updated in each 5-ms frame by interpolation. The excitation parameters are also updated every 5 ms.

The 128 codewords in each of the two codebooks are constructed from two sets of seven basis codewords (vectors) by forming linear combinations of the seven basis codewords. Initially, the seven basis codewords in each codebook are selected as zero-mean Gaussian sequences. However, these basis codewords are optimized over a training database by minimizing the total perceptually weighted error.

The long-term filter state is also a codebook with 128 codeword sequences, where each sequence corresponds to a given lag (pitch period) of the filter $\Theta_p(z)$ given by (7.133). In each 5-msec frame, the codewords from this codebook are filtered through the speech system filter $\hat{\Theta}(z)$ and correlated with the input speech sequence. The filtered codeword that has the highest correlation to the speech is used to update the history array and the lag is transmitted to the decoder. Thus the update occurs by appending the best-filtered codeword to the history codebook, and the oldest sample in the history array is discarded. The result is that the long-term state becomes an adaptive codebook.

The three excitation sequences are selected sequentially from each of the three codebooks. Each codebook search attempts to find the codeword that minimizes the total energy of the perceptually weighted error. Then, once the codewords have been selected, the three gain parameters are optimized. Joint gain optimization is sequentially accomplished by orthogonalizing each weighted codeword vector to each of the previously selected weighted excitation vectors prior to the codebook search. These parameters are vector quantized to one of 256 eight-bit vectors and transmitted in every 5-ms frame.

In addition to these gain parameters, the lag in the pitch filter is estimated and transmitted every 5 msec as a 7-bit number. Also, the average speech energy is estimated and transmitted once every 20 msec as a 5-bit number. The 10 LPC coefficients are represented by reflection coefficients and are quantized by use of scalar quantization. In Table 7.9 we

TABLE 7.9. Bit Allocations for 8000-bps VSELP.

Parameter	Bits/5-ms Frame	Bits/20 ms
10 LPC coefficients	—	38
Average speech energy	—	5
Excitation codewords from two VSELP codebooks	14	56
Gain parameters	8	32
Lag of pitch filter	7	28
Total	29	159

summarize the number of bits transmitted to the decoder for the various speech signal parameters and the excitations.

Finally, we observe that an adaptive spectral postfilter is employed in VSELP following the LPC synthesis filter. This postfilter is a pole–zero filter of the form

$$\Theta_s(z) = \frac{B(z)}{\hat{A}(z/c)}, \tag{7.134}$$

where $\hat{A}(z)$ is the denominator of the LPC synthesis filter, and $B(z)$ is the numerator polynomial whose coefficients are determined adaptively to smooth the speech signal spectrum.

The 8000-bps VSELP vocoder described in the paper by Gerson and Jasiuk (1990) has been adopted by the Telecommunications Industry Association (TIA) as the standard speech coder for use in North American digital cellular telephone systems.

7.4.6 Vector Quantization of Model Parameters

In the description of the vocoders given above, VQ is widely used to efficiently represent excitation sequences and other signal parameters and, thus, to reduce the bit rate over the channel. For example, in a channel vocoder, the outputs of the channel bank can be quantized as a vector instead of quantizing each filter output separately. The same applies to the analysis outputs of the phase vocoder, the cepstral vocoder, and the formant vocoder. Thus the bit rate can be significantly reduced as a result of the greater efficiency afforded by VQ relative to scalar quantization. Side information such as pitch and voicing information are usually quantized separately, since they are not highly correlated with the other signal parameters.

Vector quantization has proved to be particularly efficient in LPC, where speech coding rates in the range of 200–800 bps have been achieved [see, e.g., (Wong et al., 1982)]. It is customary to apply VQ to the log-area ratios $g(k; m), k = 1, \ldots, M$, which are obtained directly from the LP coefficients. The commonly used distortion measure for VQ of the LP parameters is the Itakura distance (see Section 5.3.5).

As an example, suppose that the analysis rate in LPC is 50 frames per second. Typically, 10–13 bits per frame are needed for VQ of the log-area ratios. With 6 bits per frame for the pitch and 1 bit per frame for the voicing information, the total rate is about 20 bits per frame and, hence, a bit rate of 1000 bps.

To illustrate the benefits of VQ vis-à-vis scalar quantization for the log-area ratios in LPC, let us refer to Fig. 7.44, taken from the paper by Makhoul et al. (1985). In this case the speech signal was filtered to 5 kHz and sampled at 10 kHz. The LPC model had 14 coefficients. A total of 60,000 frames were used to adaptively train the scalar and vector quantizers. The LPC coefficients were transformed to log-area ratios for scalar and vector coding. The MSE was the distortion measure selected.[11] Shown in Fig. 7.44 is the MSE plotted as a function of the number of bits per vector (hence the number of centroids in the codebook). The performance of two types of scalar quantizers are shown in the figure. The first one, labeled (a), represents the performance of the Lloyd–Max quantizer, where each log-area ratio coefficient is optimally assigned a number of bits, in proportion to its variance, so as to minimize the MSE, as shown by Makhoul et al. (1985). More bits are assigned as the variance grows. The second scalar quantizer, labeled (b), uses optimum bit allocation after performing a rotation operation to remove linear dependence from the vector of log-area ratios. The codebook for the vector quantizer was generated by the K-means algorithm. The bit allocation for the scalar quantizers was based on empirical pdf's obtained from the data.

We note from Fig. 7.44 that VQ requires about 10 bits per vector for $M = 14$ and reduces the bit rate per frame by 5 bits relative to the scalar quantizer (b) and 8 bits per frame relative to the scalar quantizer (a).

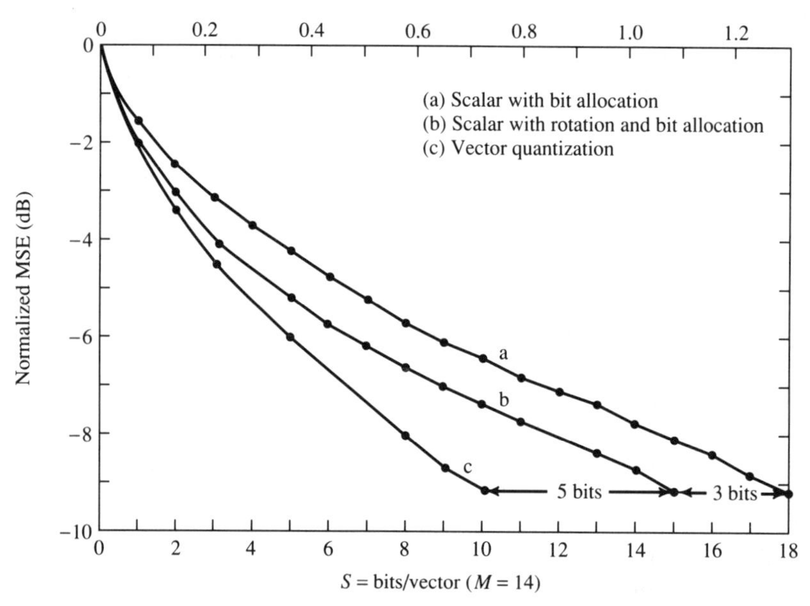

FIGURE 7.44. Normalized MSE in quantizing log-area ratios (LARs) using three methods: (a) scalar quantization with bit allocation; (b) scalar quantization withbit allocation, preceded by eigenvector rotation; and (c) vector quantization. The 3-bit reduction from (a) to (b) takes advantage of linear dependencies (correlation), and the additional 5-bit reduction from (b) to (c) takes advantage largely of nonlinear dependencies (Makhoul et al., 1985).

[11]This means the average squared Euclidean distance between a vector and its assigned centroid over the entire population of vectors.

Codebook design for VQ of the model parameters can be implemented by using the K-means algorithm, as described in Sections 1.4.5 and 7.2.2. An important issue is the search through the codebook. As indicated in Section 7.2.2, a full search through the codebook requires ML multiplications and additions to quantize each input vector. For the example treated above, $M = 14$ and $L = 2^{10}$, so that $ML = 14{,}336$, which is very large. This computational burden is significantly reduced by using the binary tree search method. In such a case, the cost becomes $2M \log_2 L = 280$ multiplications and additions. On the other hand, the memory requirement is now $2ML = 28{,}672$, which is twice as much as for the full search.

We conclude with a comparison of the MSE performance of full search, uniform binary search, and nonuniform binary search for the LPC example with $M = 14$, given above. The results shown in Fig. 7.45, due to Makhoul et al. (1985), illustrate such a comparison. We note that the nonuniform binary tree search method is only slightly inferior to the full search and only slightly better than uniform binary search. Consequently, the increase in distortion for a given bit rate resulting from binary tree search compared to full search is relatively small. Considering the computational savings resulting from the use of the binary tree search algorithm, we find that the compromise in performance is well justified.

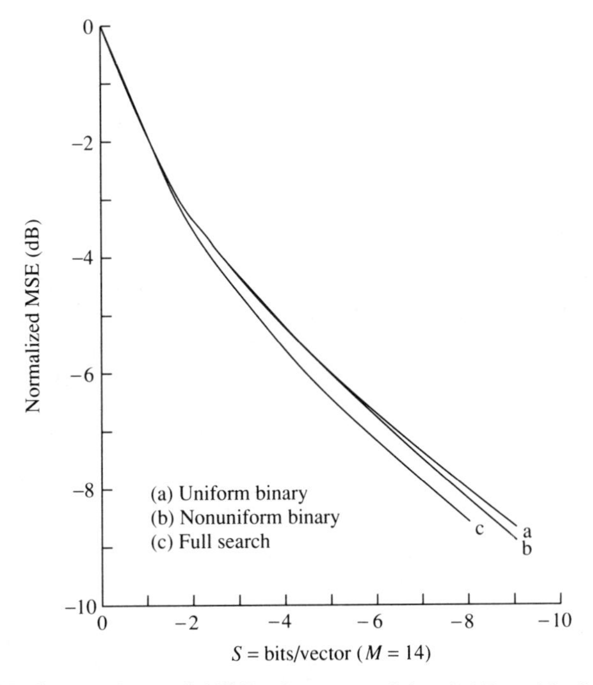

FIGURE 7.45. Comparison of MSE when quantizing LARs with three types of vector quantization: (a) uniform binary search, (b) nonuniform binary search, and (c) full search (Makhoul et al., 1985).

7.5 Measuring the Quality of Speech Compression Techniques

Clearly, a trade-off exists between the efficiency of representation and the fidelity of the resulting speech—bit rate versus speech quality. In essence, the history of speech coding represents an effort to continually expand the envelope of the efficiency-to-quality curve. The relationship between transmission bit rate and quality is shown in Fig. 7.46. Of the two quantities in question, bit rate is highly quantifiable whereas the latter is subject to interpretation. When competing digital speech coding systems are to be evaluated relative to one another, a method that is repeatable, meaningful, and able to reliably measure the sound quality of the speech is needed.

One possible technique for evaluating speech coding algorithms is to use a standardized procedure that employs human listeners to evaluate "goodness." The general test procedure entails having the human evaluation group listen to example coded utterances and recording their opinions of the quality relative to the coded examples and/or known test cases (e.g., standard LPC at 22 kbps, 16 kbps, 10 kbps). Such tests result in subjective measures of speech quality because they are based on the listener group's perception of quality. These tests have been used by industry and the military for more than 30 years to evaluate voice communication systems (Flanagan, 1979; Hecker, 1967; Tribolet, 1978). The tests are known and well understood. Examples include the *diagnostic rhyme test*, the *mean opinion score*, and the *diagnostic acceptability measure* (Voiers, 1977). (These measures are discussed in Chapter 9.) Such tests, however, are expensive in terms of time and resources, requiring, for example, trained listener groups and large amounts of coded data. They are also difficult to administer and are often suspect due to the inherent nonrepeatability of human responses. In general, results from one coder evaluation cannot be compared reliably with another unless the test environment is preserved (same listener group, speech corpus, and presentation order). Many of the subjective tests are based on pairwise comparisons. In order for the listener group to establish statistically significant results, the coding distortion between examples must be fairly

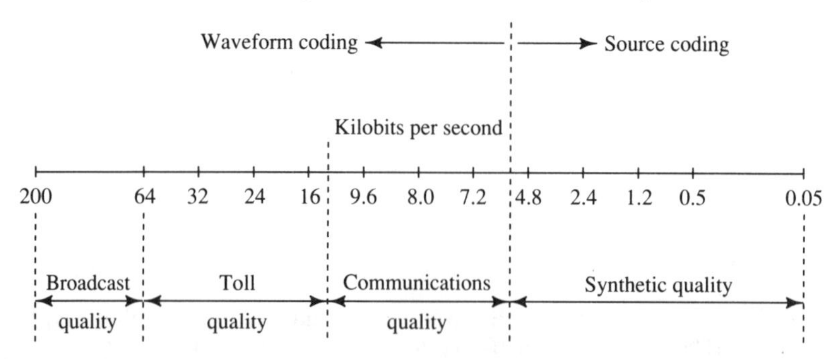

FIGURE 7.46. The range of speech coding transmission rates versus associated quality. After Flanagan et al. (1979).

large. This makes it difficult to use subjective tests to determine optimal vocoder parameter settings, including, for example, bit allocation and codebook size. Since only gross settings can be compared, such tests seldom provide much insight into factors that may lead to improvements in the evaluated systems (Quackenbush, 1988).

An alternative method for evaluating speech coding algorithms is based on a *computable* measure of quality, which is roughly equivalent in accuracy to that of subjective tests. These tests measure the distortion between the input and output signal from coding systems. These objective measures are classified into waveform distortion measures, including, for example, signal-to-noise ratio, and frequency-spectrum distortion measures, such as LPC-based distance measures. The choice of which objective quality measure to use in coder evaluation rests on the measure's ability to predict subjective quality. Some measures are better than others for particular environments (e.g., performance in broadband channel noise versus impulsive noise). In order to determine general coder performance, an objective measure that reliably predicts subjective quality for a broad range of distortions is desirable. In Chapter 9, we discuss a number of subjective and objective intelligibility and quality tests. Before describing them, however, we take up another area in which these quality assessment measures will be useful. Chapter 8 treats the issue of enhancing speech that has been corrupted or distorted by some technical or natural process. Clearly, this area is closely related to the speech coding techniques we have discussed in the present chapter.

7.6 Conclusions

Waveform encoding of speech has been widely used for several decades on telephone channels, and consequently, the various coding methods described in this chapter are well developed. In general, waveform encoding methods yield high (toll)-quality speech at bit rates of 16,000 bps and higher with reasonable implementation complexity. For lower bit rates, complex VQ encoding methods are necessary to achieve high-quality speech. With present VLSI technology, real-time implementation of such highly complex VQ methods are expensive and, hence, impractical.

For speech communication over mobile cellular radio channels, channel bandwidth is more scarce and, hence, bandwidth constraints are more severe than on wire line telephone channels. In this case there is a great need for low bit rate speech coding. This is an application area where LPC-based encoding schemes employing VQ, such as CELP and VSELP, are particularly suitable. It is anticipated that by the mid-1990s, LPC-based VQ techniques will be developed to provide communication-quality speech at rates of 2000–2400 bps. This technology will lead to a quadrupling of the channel capacity for mobile cellular radio communications systems compared with today's CELP and VSELP capabilities.

7.7 Problems

7.1. Let $\underline{x}$ be a stationary first-order Markov source. The random variables $\underline{x}(n)$ denote the output (state) of the source discrete time n. The source is characterized by the state probabilities $P(x_i) \stackrel{\text{def}}{=} P[\underline{x}(n) = x_i]$ for arbitrary n, where $\{x_i,\ i = 1, 2, \ldots, L\}$ is the set of labels for the L possible states; and the transition probabilities,

$$P(x_k|x_i) \stackrel{\text{def}}{=} P[\underline{x}(n) = x_k \,|\, \underline{x}(n-1) = x_i] \text{ for } i, k = 1, 2, \ldots, L$$

and for arbitrary n. The entropy (see Section 1.5) of the Markov source is

$$H(\underline{x}) = \sum_{k=1}^{L} P(x_k)\, H(\underline{x}|x_k), \tag{7.135}$$

where $H(\underline{x}\,|x_k)$ is the entropy of the source conditioned on the source being in state x_k.

(a) Determine the entropy of the binary, first-order Markov source shown in Fig. 7.47, which has transition probabilities

$$P(x_2|x_1) = 0.2 \qquad \text{and} \qquad P(x_1|x_2) = 0.3.$$

Hint: Begin by deriving the expression

$$\begin{aligned} H(\underline{x}|x_1) = \{-\log[1 - P(x_2)]\}[1 - P(x_2|x_1)] \\ +[-\log P(x_2)]P(x_2|x_1) \end{aligned} \tag{7.136}$$

and a similar expression for $H(\underline{x}|x_2)$.

(b) How does the entropy of the Markov source compare with the entropy of a binary discrete memoryless source with the same output probabilities, $P(x_1)$ and $P(x_2)$.

NOTE: Recall that a discrete memoryless source is one for which the output random variables are independent (see Section 1.5). In the present case, this implies that $P(x_k|x_i) = P(x_k)$ for $i, k = 1, 2$.

7.2. This problem will review linear prediction concepts studied in Chapter 5. Consider a discrete-time, stationary, stochastic process $\underline{x}$ with random variables $\underline{x}(n)$, which has mean zero and autocorrelation sequence

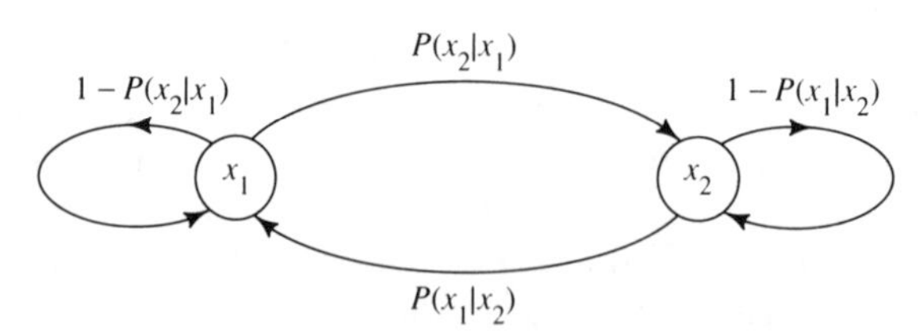

FIGURE 7.47. First-order Markov source.

$$r_{\underline{x}}(\eta) = \begin{cases} 1, & \eta = 0 \\ \frac{1}{2}, & \eta = \pm 1 \\ 0, & \text{otherwise} \end{cases} . \tag{7.137}$$

(a) Determine the prediction coefficient, $\hat{a}(1)$, of the first-order minimum MSE predictor,

$$\hat{x}(n) = \hat{a}(1)x(n-1), \tag{7.138}$$

where the sequence $x(n)$ represents any realization of $\underline{x}$.

(b) Repeat part (a) for the second-order predictor,

$$\hat{x}(n) = \hat{a}(1)x(n-1) + \hat{a}(2)x(n-2). \tag{7.139}$$

7.3. Consider the random variables $\underline{x}_1$ and $\underline{x}_2$, which are characterized by the joint pdf

$$f_{\underline{x}_1,\underline{x}_2}(x_1, x_2) = \begin{cases} \frac{15}{7ab}, & x_1, x_2 \in C \\ 0, & \text{otherwise} \end{cases}, \tag{7.140}$$

where C is the shaded region shown in Fig. 7.48.

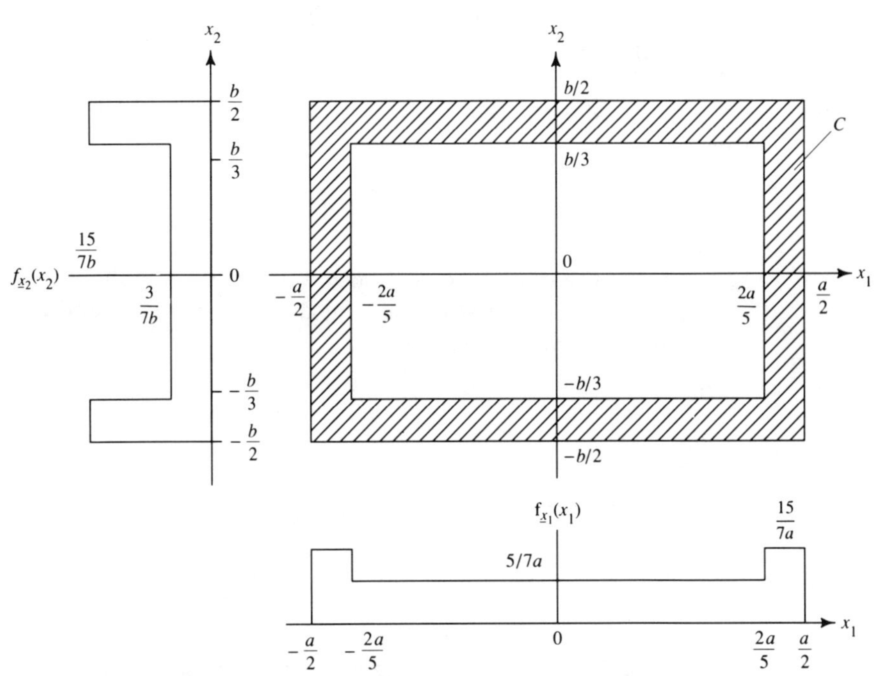

FIGURE 7.48. Joint pdf of random variables $\underline{x}_1, \underline{x}_2$.

(a) Evaluate the bit rates required for uniform quantization of $\underline{x}_1$ and $\underline{x}_2$ separately (scalar quantization), and combined (vector) quantization of $(\underline{x}_1, \underline{x}_2)$.

(b) Determine the difference in bit rate when $a = 4b$.

7.4. In evaluating the performance of a uniform quantizer, it is common practice to model the quantization error as a random process, say $\underline{e}$, which is independent of the random process modeling the continuous-amplitude signal, say $\underline{s}$. This assumption is not valid if the number of quantization levels is small. However, by adding a small dither noise to the signal prior to the quantization, as shown in Fig. 7.49, we can change the statistical character of the error signal.

(a) Suppose that the dither noise sequence, $d(n)$, is modeled as the realization of a zero-mean, white noise random process $\underline{d}$ with pdf

$$f_{\underline{d}(n)}(d) = \begin{cases} 1/\Delta, & |d| \leq \Delta/2 \\ 0, & \text{otherwise} \end{cases} \tag{7.141}$$

for arbitrary n. Let $e_d(n)$ denote the quantization error sequence with dither

$$e_d(n) \stackrel{\text{def}}{=} s(n) - x_q(n), \tag{7.142}$$

where $x_q(n)$ is the quantized sequence with dither. Show that the stochastic process modeling $e_d(n)$, say $\underline{e}_d$, is statistically independent of $\underline{s}$.

(b) Also let $e_u(n)$ denote the quantization error sequence without dither,

$$e_u(n) \stackrel{\text{def}}{=} s(n) - s_q(n), \tag{7.143}$$

where $s_q(n)$ is the quantized sequence without dither. Show that the variances associated with the stochastic processes $\underline{e}_d$ and $\underline{e}_u$, say σ_d^2 and σ_u^2, satisfy the relation

$$\sigma_d^2 > \sigma_u^2. \tag{7.144}$$

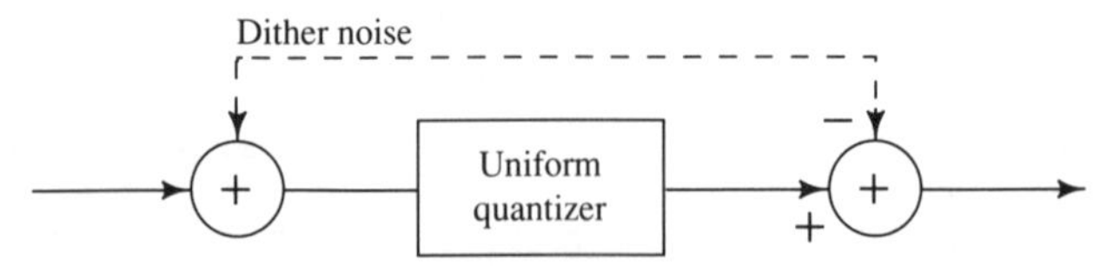

FIGURE 7.49. Addition of dither noise to signal prior to quantization.

(c) Now suppose we subtract the dither noise from $x_q(n)$ (dotted path in Fig. 7.49), so that the resulting error is

$$e(n) = s(n) - x_q(n) - d(n). \tag{7.145}$$

Determine the variance, say σ^2, for the random process $\underline{e}$, and show that

$$\sigma^2 = \sigma_u^2. \tag{7.146}$$

7.5. Let $s(n)$ be a realization of a stationary, zero-mean random process $\underline{s}$, whose autocorrelation sequence $r_{\underline{s}}(\eta)$ is nonzero if and only if $|\eta| \leq N$. The sequence $v(n)$ is defined as

$$v(n) = s(n) - \alpha s(n-D), \tag{7.147}$$

where α is a constant and D is a delay. Here $v(n)$ is modeled by a stationary random process $\underline{v}$.

(a) Determine the variance $\sigma_{\underline{v}}^2$ for $D > N$ and compare it to the variance $\sigma_{\underline{s}}^2$. Which is larger?
(b) Determine the variance $\sigma_{\underline{v}}^2$ for $D < N$ and compare it to the variance $\sigma_{\underline{s}}^2$. Which is larger?
(c) Determine the values of α which minimize the variances $\sigma_{\underline{v}}^2$ for the cases of parts (a) and (b). Find $\sigma_{\underline{v}}^2$ in each case and compare it to $\sigma_{\underline{s}}^2$.

7.6. This problem will review the complex cepstrum (CC) and real cepstrum (RC) treated in Chapter 6. Compute the (long-term) CC and RC for the sequence

$$x(n) = \delta(n) + b\delta(n-D), \tag{7.148}$$

where b is a scale factor and D is a delay.

7.7. This problem is for persons who have studied the material on quadrature mirror filters in Appendix 7.A. By using (7.160), derive the equations corresponding to the structure for the polyphase synthesis section shown in Fig. 7.50.

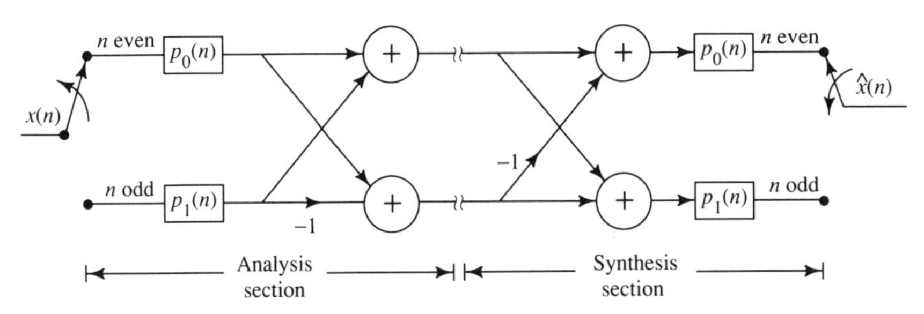

FIGURE 7.50. Polyphase filter structure for the QMF bank.

APPENDIX

7.A Quadrature Mirror Filters

The basic building block in applications of quadrature mirror filters (QMF) is the two-channel QMF bank shown in Fig. 7.51. This is a multirate digital filter structure that employs two decimators in the signal analysis section and two interpolators in the signal synthesis section. The lowpass and highpass filters in the analysis section have impulse responses $h_0(n)$ and $h_1(n)$, respectively. Similarly, the lowpass and highpass filters contained in the synthesis section have impulse responses $g_0(n)$ and $g_1(n)$, respectively.

We know that for an M-fold decimator the relationship between the z-transform of the input and output signals, $X(z)$ and $Y(z)$, respectively, is

$$Y(z) = \frac{1}{M}\sum_{k=0}^{M-1} X\left(z^{1/M}\right)e^{-j2\pi k/M} \tag{7.149}$$

and for an M-fold interpolator we have

$$Y(z) = X\left(z^{M}\right). \tag{7.150}$$

Hence the DTFTs of the signals at the outputs of the two decimators are

$$\begin{aligned} X_{a,0}(\omega) &= \frac{1}{2}\left[X\left(\frac{\omega}{2}\right)H_0\left(\frac{\omega}{2}\right) + X\left(\frac{\omega-2\pi}{2}\right)H_0\left(\frac{\omega-2\pi}{2}\right)\right] \\ X_{a,1}(\omega) &= \frac{1}{2}\left[X\left(\frac{\omega}{2}\right)H_1\left(\frac{\omega}{2}\right) + X\left(\frac{\omega-2\pi}{2}\right)H_1\left(\frac{\omega-2\pi}{2}\right)\right]. \end{aligned} \tag{7.151}$$

If $X_{s,0}(\omega)$ and $\mathrm{X}_{\mathrm{s},1}(\omega)$ represents the two inputs to the synthesis section, the output is simply

$$\hat{X}(\omega) = X_{s,0}(2\omega)G_0(\omega) + X_{s,1}(2\omega)G_1(\omega). \tag{7.152}$$

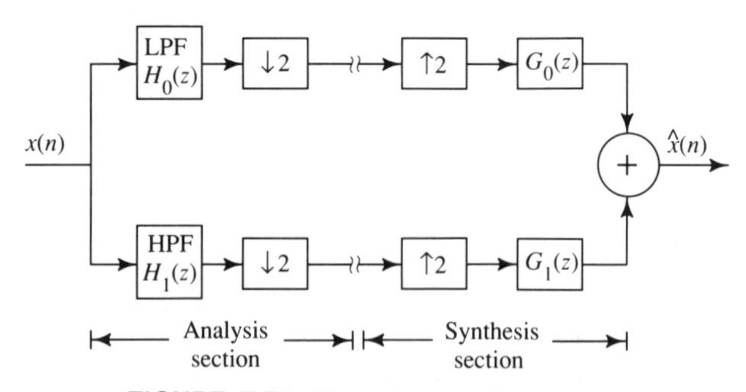

FIGURE 7.51. Two-channel QMF bank.

Now, suppose we connect the analysis filter to the corresponding synthesis filter, so that $X_{a,0}(\omega) = X_{s,0}(\omega)$ and $X_{a,1}(\omega) = X_{s,1}(\omega)$. Then, by substituting from (7.151) into (7.152), we obtain

$$\begin{aligned}\hat{X}(\omega) &= \tfrac{1}{2}\left[H_0(\omega)G_0(\omega) + H_1(\omega)G_1(\omega)\right]X(\omega) \\ &\quad + \tfrac{1}{2}\left[H_0(\omega-\pi)G_0(\omega) + H_1(\omega-\pi)G_1(\omega)\right]X(\omega-\pi).\end{aligned} \tag{7.153}$$

The first term in (7.153) is the desired signal output from the QMF bank. The second term represents the effect of aliasing, which we would like to eliminate. Hence we require that

$$H_0(\omega-\pi)G_0(\omega) + H_1(\omega-\pi)G_1(\omega) = 0. \tag{7.154}$$

This condition can be simply satisfied by selecting $G_0(\omega)$ and $G_1(\omega)$ as

$$\begin{aligned}G_0(\omega) &= H_1(\omega-\pi) \\ G_1(\omega) &= -H_0(\omega-\pi).\end{aligned} \tag{7.155}$$

Thus the second term in (7.153) vanishes.

To elaborate, let us assume that $H_0(\omega)$ is a lowpass filter and $H_1(\omega)$ is a mirror-image highpass filter. Then we may express $H_0(\omega)$ and $H_1(\omega)$ as

$$\begin{aligned}H_0(\omega) &= H(\omega) \\ H_1(\omega) &= H(\omega-\pi),\end{aligned} \tag{7.156}$$

where $H(\omega)$ is the frequency response of a lowpass filter. In the time domain, the corresponding relations are

$$\begin{aligned}h_0(n) &= h(n) \\ h_1(n) &= (-1)^n h(n).\end{aligned} \tag{7.157}$$

As a consequence, $H_0(\omega)$ and $H_1(\omega)$ have mirror-image symmetry about the frequency $\omega = \pi/2$, as shown in Fig. 7.22. To be consistent with the constraint in (7.155), we select the lowpass filter $G_0(\omega)$ as

$$G_0(\omega) = 2H(\omega) \tag{7.158}$$

and the highpass filter $G_1(\omega)$ as

$$G_1(\omega) = -2H(\omega-\pi). \tag{7.159}$$

In the time domain, these relations become

$$\begin{aligned}g_0(n) &= 2h(n) \\ g_1(n) &= -2(-1)^n h(n).\end{aligned} \tag{7.160}$$

The scale factor of two in $g_0(n)$ and $g_1(n)$ corresponds to the interpolation factor that is used to normalize the overall frequency response of the QMF. With this choice of the filter characteristics, the component due to aliasing vanishes. Thus the aliasing resulting from decimation in the analysis section of the QMF bank is perfectly canceled by the image signal spectrum that arises due to interpolation. As a result, the two-channel QMF behaves as a linear, time-invariant system.

If we substitute $H_0(\omega), H_1(\omega), G_0(\omega)$, and $G_1(\omega)$ into the first term of (7.153), we obtain

$$\hat{X}(\omega) = [H^2(\omega) - H^2(\omega - \pi)]X(\omega). \tag{7.161}$$

Ideally, the two-channel QMF bank should have unity gain,

$$|H^2(\omega) - H^2(\omega - \pi)| = 1 \tag{7.162}$$

for all ω, where $H(\omega)$ is the frequency response of a lowpass filter. Furthermore, it is also desirable for the QMF to have linear phase.

Now, let us consider the use of a linear phase filter $H(\omega)$. Hence $H(\omega)$ may be expressed in the form

$$H(\omega) = H_r(\omega)e^{-j\omega(N-1)/2}, \tag{7.163}$$

where $H_r(\omega)$ is the DTFT of the "undelayed" version of $h(n)$ and N is the duration of the filter impulse response. Then

$$\begin{aligned} H^2(\omega) &= H_r^2(\omega)e^{-j\omega(N-1)} \\ &= |H(\omega)|^2 e^{-j\omega(N-1)} \end{aligned} \tag{7.164}$$

and

$$\begin{aligned} H^2(\omega - \pi) &= H_r^2(\omega - \pi)e^{-j(\omega-\pi)(N-1)} \\ &= (-1)^{N-1}|H(\omega - \pi)|^2 e^{-j\omega(N-1)}. \end{aligned} \tag{7.165}$$

Therefore, the overall transfer function of the two-channel QMF that employs linear-phase FIR filters is

$$\frac{\hat{X}(\omega)}{X(\omega)} = \left[|H(\omega)|^2 - (-1)^{N-1}|H(\omega - \pi)|^2\right]e^{-j\omega(N-1)}. \tag{7.166}$$

Note that the overall filter has a delay of $N-1$ samples and a magnitude characteristic

$$A(\omega) = |H(\omega)|^2 - (-1)^{N-1}|H(\omega - \pi)|^2. \tag{7.167}$$

Also note that when N is odd, $A(\pi/2) = 0$, because $|H(\pi/2)| = |H(3\pi/2)|$. This is an undesirable property for a QMF design. On the other hand, when N is even,

$$A(\omega) = |H(\omega)|^2 + |H(\omega - \pi)|^2, \tag{7.168}$$

which avoids the problem of a zero at $\omega = \pi/2$. For N even, the *ideal* two-channel QMF should satisfy the condition

$$A(\omega) = |H(\omega)|^2 + |H(\omega - \pi)|^2 = 1 \tag{7.169}$$

for all ω, which follows from (7.168). Unfortunately, the only filter frequency response function that satisfies (7.169) is one with the trivial magnitude spectrum $H(\omega) = \cos^2(a\omega)$. Consequently, any nontrivial linear-phase FIR filter $H(\omega)$ will introduce some amplitude distortion.

The amount of amplitude distortion introduced by a nontrivial linear-phase FIR filter in the QMF can be minimized by optimizing the FIR filter coefficients. A particularly effective method is to select the filter coefficients of $H(\omega)$ such that $A(\omega)$ is made as flat as possible while simultaneously minimizing (or constraining) the stopband energy of $H(\omega)$. This approach leads to the minimization of the integral squared error

$$\xi = w \int_{\omega_s}^{\pi} |H(\omega)|^2 \, d\omega + (1 - w) \int_0^{\pi/2} [A(\omega) - 1]^2 \, d\omega, \tag{7.170}$$

where w is a weighting factor in the range $0 < w < 1$, and ω_s is the stopband frequency. In performing the optimization, the FIR is constrained to be symmetric (linear phase). This optimization is easily done numerically on a digital computer. This approach has been used by Johnston (1980) and Jain and Crochiere (1984) to design two-channel QMFs. Optimum filter coefficients have been tabulated by Johnston (1980).

As an alternative to the use of linear-phase FIR filters, we may design an IIR filter that satisfies the all-pass constraint given by (7.162). For this purpose, elliptic filters provide especially efficient designs. Since the QMF would introduce some phase distortion, the signal at the output of the QMF can be passed through an all-pass phase equalizer that is designed to minimize phase distortion.

In addition to these two methods for QMF design, one can also design the two-channel QMFs to completely eliminate both amplitude and phase distortion as well as to cancel aliasing distortion. Smith and Barnwell (1984) have shown that such a perfect reconstruction QMF can be designed by relaxing the linear-phase condition of the FIR lowpass filter $H(\omega)$. To achieve perfect reconstruction, we begin by designing a linear-phase FIR half-band filter of length $2N - 1$.

A *half-band filter* is defined as a zero-phase FIR filter whose impulse response $b(n)$ satisfies the condition

$$b(2n) = \begin{cases} b_0, & n = 0 \\ 0, & n \neq 0, \end{cases} \tag{7.171}$$

where b_0 is some constant. Hence all the even-numbered samples are zero except at $n = 0$. The zero-phase requirement implies that $b(n) = b(-n)$. The frequency response of such a filter is

$$B(\omega) = \sum_{n=-K}^{K} b(n) e^{-j\omega n}, \tag{7.172}$$

where K is odd. Also, $B(\omega)$ satisfies the condition that $B(\omega)+B(\pi-\omega)$ be equal to a constant for all frequencies. The typical frequency response characteristic of a half-band filter is shown in Fig. 7.52. We note that the band-edge frequencies ω_p and ω_s are symmetric about $\omega=\pi/2$, and the peak passband and stopband errors are equal. We also note that the filter can be made causal by introducing a delay of K samples.

Now, suppose we design an FIR half-band filter whose impulse response is of length $2N-1$, where N is even, and with frequency response as shown in Fig. 7.52. From $B(\omega)$ we construct another half-band filter with frequency response

$$B_+(\omega)=B(\omega)+\delta e^{j\omega(N-1)}, \tag{7.173}$$

as shown in Fig. 7.53(a). Note that $B_+(\omega)$ is nonnegative and, hence, has the spectral factorization

$$B_+(z)=H(z)H(z^{-1})z^{-(N-1)} \tag{7.174}$$

or, equivalently,

$$B_+(\omega) = |H(\omega)|^2 e^{-j\omega(N-1)}, \tag{7.175}$$

where $H(\omega)$ is the frequency response of an FIR filter of length N with real coefficients. Due to the symmetry of $B_+(\omega)$ with respect to $\omega=\pi/2$, we also have

$$B_+(z)+(-1)^{N-1}B_+(-z)=\alpha z^{-(N-1)} \tag{7.176}$$

or, equivalently,

$$B_+(\omega)+(-1)^{N-1}B_+(\omega-\pi)=\alpha e^{-j\omega(N-1)}, \tag{7.177}$$

where α is a constant. Thus, by substituting (7.174) into (7.176), we obtain

$$H(z)H(z^{-1})+H(-z)H(-z^{-1})=\alpha. \tag{7.178}$$

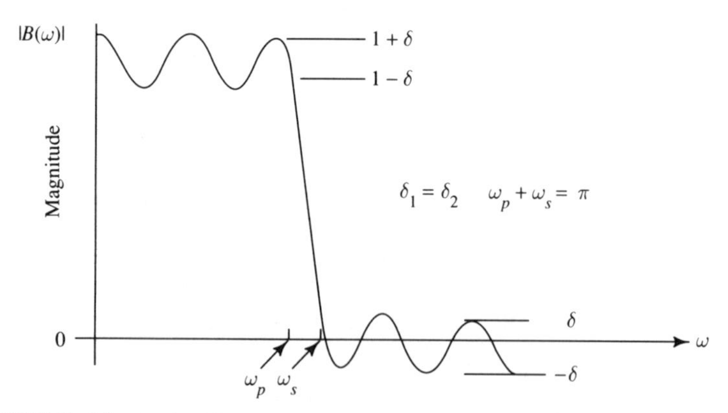

FIGURE 7.52. The typical frequency response characteristic of a half-band filter.

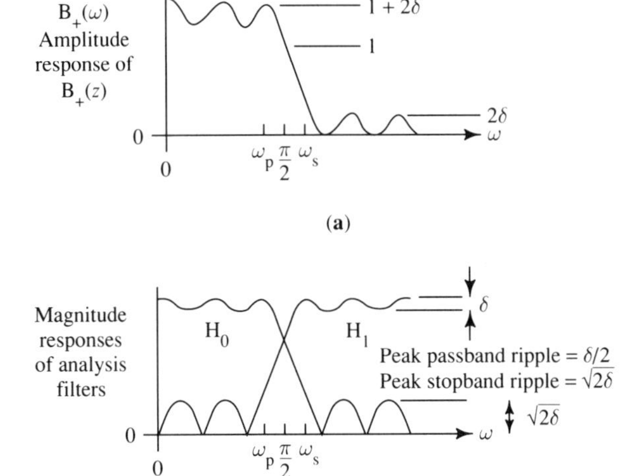

FIGURE 7.53. Perfect reconstruction of QMF constructed from FIR half-band filters.

Since $H(z)$ satisfies (7.178) and since aliasing is eliminated when we have $G_0(z) = H_1(-z)$ and $G_1(z) = -H_0(-z)$, it follows that these conditions are satisfied by choosing $H_1(z), G_0(z)$, and $G_1(z)$ as

$$\begin{aligned} H_0(z) &= H(z) \\ H_1(z) &= -z^{-(N-1)} H_0(-z^{-1}) \\ G_0(z) &= z^{-(N-1)} H_0(z^{-1}) \\ G_1(z) &= z^{-(N-1)} H_1(z^{-1}) = -H_0(-z). \end{aligned} \tag{7.179}$$

Thus aliasing distortion is eliminated and, since $\hat{X}(\omega)/X(\omega)$ is a constant, the QMF performs perfect reconstructon so that $x(n) = ax(n - N + 1)$. However, we note that $H(z)$ is not a linear-phase filter.

The FIR filters $H_0(z), H_1(z), G_0(z)$, and $G_1(z)$ in the two-channel QMF bank are efficiently realized as polyphase filters. Since $U = D = 2$, two polyphase filters are implemented for each decimator and two for each interpolator. However, when we employ linear-phase FIR filters, the symmetry properties of the analysis filters and synthesis filters allow us to simplify the structure and reduce the number of polyphase filters in the analysis section to two and in the synthesis section to another two.

To demonstrate this construction, let us assume that the filters are linear-phase FIR filters of length N (N even), which have impulse responses given by (7.157). Then the outputs of the analysis filter pair, after decimation by a factor of two, may be expressed as

$$
\begin{aligned}
x_{a,k}(n) &= \sum_{n=-\infty}^{\infty} (-1)^{kn} h(n) x(2m-n), \qquad k=0,\ 1 \\
&= \sum_{i=0}^{1} \sum_{l=-\infty}^{\infty} (-1)^{k(2l+1)} h(2l+i) x(2m-2l-i) \qquad (7.180) \\
&= \sum_{l=0}^{N-1} h(2l) x(2m-2l) + (-1)^k \sum_{l=0}^{N-1} h(2l+1) x(2m-2l-1).
\end{aligned}
$$

Now let us define the impulse response of two polyphase filters of length $N/2$ as

$$
p_i(m) = h(2m+i), \qquad i = 0,\ 1. \tag{7.181}
$$

Then (7.180) may be expressed as

$$
\begin{aligned}
x_{a,k}(n) &= \sum_{l=0}^{(N/2)-1} p_0(m) x2(m-l) \\
&\quad + (-1)^k \sum_{l=0}^{(N/2)-1} p_1(m) x(2m-2l-1), \qquad k=0,\ 1.
\end{aligned} \tag{7.182}
$$

This expression corresponds to the polyphase filter structure for the analysis section that is shown in Fig. 7.50. Note that the commutator rotates counterclockwise, and the filter with impulse response $p_0(m)$ processes the even-numbered samples of the input sequence, and the filter with impulse response $p_1(m)$ processes the odd-numbered samples of the input signal.

In a similar manner, by using (7.160), we can obtain the structure for the polyphase synthesis section, which is also shown in Fig. 7.50. This derivation is left as an exercise for the reader (Problem 7.7). Note that the commutator also rotates counterclockwise.

Finally, we observe that the polyphase filter structure shown in Fig. 7.50 is approximately four times more efficient than the direct-form FIR filter realization.

Tutorial treatments of the two-channel QMF and its generalization to multichannel QMF banks are given in two papers by Vaidyanathan (1987, 1990).

CHAPTER 8

Speech Enhancement

Reading Notes: We will make extensive use of concepts from stochastic processes in this chapter, particularly those concerning the autocorrelation and power density spectrum, so that the reader may wish to review the relevant material in Section 1.2, and also Sections 4.2.5, 4.3.1, and 4.3.5. We will also refer to issues relating to speech quality and intelligibility, which are discussed in Chapter 9.

8.1 Introduction

In many speech communication settings, the presence of background interference causes the quality or intelligibility of speech to degrade. When a speaker and listener communicate in a quiet environment, information exchange is easy and accurate. However, a noisy environment reduces the listener's ability to understand what is said. In addition to interpersonal communication, speech can also be transmitted across telephone channels, loudspeakers, or headphones. The quality of speech, therefore, can also be influenced in data conversion (microphone), transmission (noisy data channels), or reproduction (loudspeakers and headphones). The purpose of many enhancement algorithms is to reduce background noise, improve speech quality, or suppress channel or speaker interference. In this chapter, we discuss the general problem of speech enhancement with particular focus on algorithms designed to remove additive background noise for improving speech quality. In our discussion, background noise will refer to any additive broadband noise component (examples include white Gaussian noise, aircraft cockpit noise, or machine noise in a factory environment). Other speech processing areas that are sometimes included in a discussion of speech enhancement include suppression of distortion from voice coding algorithms, suppression of a competing speaker in a multispeaker setting, enhancing speech as a result of a deficient speech production system (examples include speakers with pathology or divers breathing helium–oxygen mixture), or enhancing speech for hearing-impaired listeners. Since the range of possible applications is broad, we will generally limit our discussion to enhancement algorithms directed at improving speech quality in additive broadband noise for speakers and listeners with normal production and auditory systems.

The problem of enhancing speech degraded by additive background noise has received considerable attention in the past two decades. Many approaches have been taken, each attempting to capitalize on specific characteristics or constraints, all with varying degrees of success. The success of an enhancement algorithm depends on the goals and assumptions used in deriving the approach. Depending on the specific application, a system may be directed at one or more objectives, such as improving overall quality, increasing intelligibility, or reducing listener fatigue. The objective of achieving higher quality and/or intelligibility of noisy speech may also contribute to improved performance in other speech applications, such as speech compression, speech recognition, or speaker verification.

As in any engineering problem, it is useful to have a clear understanding of the objectives and the ability to measure system performance in achieving those objectives. When we consider noise reduction, we normally think of improving a signal-to-noise ratio (SNR). It is important to note, however, that this may not be the most appropriate performance criterion for speech enhancement. All listeners have an intuitive understanding of speech quality, intelligibility, and listener fatigue. However, these areas are not easy to quantify in most speech enhancement applications, since they are based on subjective evaluation of the processed signal. Due in part to the efforts of researchers in the speech coding area, testing methods do exist for measuring quality and intelligibility. Although methods for assessing speech quality will be addressed in Chapter 9, we will find it convenient to refer to algorithm performance in terms of these measures later in this chapter. Although the testing methods or measures were originally formulated to quantify distortion introduced by speech coding algorithms, we will find that they can be used to quantify performance for enhancement applications as well.

In this chapter, we will examine an assortment of techniques that attempt to improve the quality or intelligibility of speech. Actual quality improvement will be subject to certain assumptions, such as the type of additive noise, interfering speakers, single or multiple data channels, and available signal bandwidth. Figure 8.1 presents a general speech enhancement framework, illustrating the possible sources of distortion and applications in which an enhanced speech signal is needed. Most speech enhancement techniques focus on reducing the effects of noise introduced at the source. Distortion at the source can be either additive background noise or one or more competing speakers. Speech enhancement can also be useful for reducing distortion introduced in a speech coding algorithm. This distortion is by far the most widely studied of the possible source distortions, due in part to the need for efficient speech-compression techniques by the communications industry. Noise can also be introduced during transmission, as shown in Fig. 8.1. Whatever the origin of the disturbance, the job of the speech enhancement algorithm is to enhance the speech signal prior to processing by the auditory system. Recently, it has also been shown that front-end speech enhancement can also be useful for other speech processing applications, such as processing before coding or recognition.

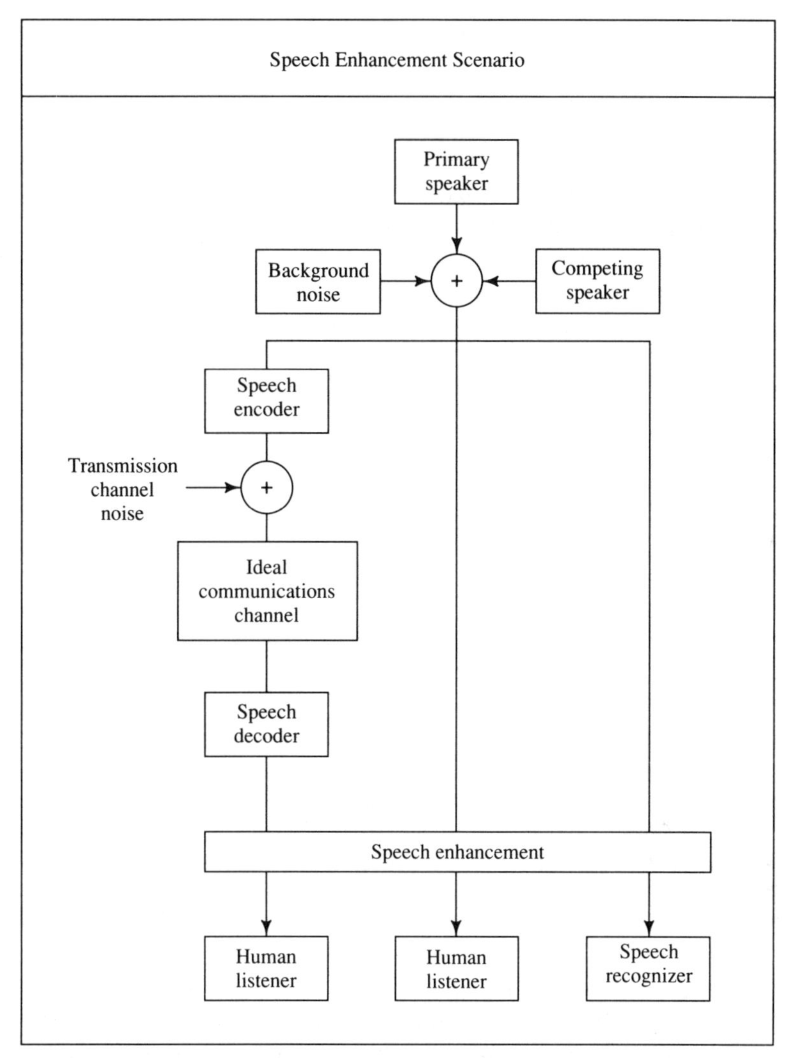

FIGURE 8.1. Typical sources of degradation for speech enhancement applications.

The majority of enhancement techniques seek to reduce the effects of broadband additive noise. Generally speaking, constraints placed on the input speech signal improve the potential for separating speech from background noise.[1] However, such systems also become more sensitive to "deviations" from these constraints. The same reasoning holds for noise assumptions. Confining the noise type improves the chances of removing it, but at the expense of dedicating the technique to a specific interference, such as wideband, narrowband, or competing speaker.

[1]We will see in Part V that constraints on speech will aid in success of recognition.

8.2 Classification of Speech Enhancement Methods

There are a number of ways in which speech enhancement systems can be classified. A broad grouping is concerned with the manner in which the speech is modeled. Some techniques are based on stochastic process models of speech, while others are based on perceptual aspects of speech. Systems based on stochastic process models rely on a given mathematical criterion. Systems based on perceptual criteria attempt to improve aspects important in human perception. For example, one technique may concentrate on improving the quality of consonants, since consonants are known to be important for intelligibility in a manner disproportionate to overall signal energy. Such methods will be discussed in more detail in Section 8.7.

Enhancement algorithms can also be partitioned depending on whether a single-channel or dual-channel (or multichannel) approach is used. For *single-channel* applications, only a single microphone is available. Characterization of noise statistics must be performed during periods of silence between utterances, requiring a stationarity assumption of the background noise. In situations such as voice telephone or radio communications, only a single channel is available. In *dual-channel* algorithms, the acoustic sound waves arrive at each sensor at slightly different times (one is normally a delayed version of the other). Multi- or dual-channel enhancement techniques are based on one of two scenarios. In the first, a primary channel contains speech with additive noise and a second channel contains a sample noise signal correlated to the noise in the primary channel. Normally, an acoustic barrier exists between sensors to ensure that no speech leaks into the noise reference channel. In the second scenario, no acoustic barrier exists, so the enhancement algorithm must address the issue of cross-talk or employ a multisensor beamforming solution. In our discussion, we shall concentrate on methods that assume that (1) the noise distortion is additive, (2) the noise and speech signals are uncorrelated, and (3) only one input channel is available (except for adaptive noise cancelation, where we will assume a dual-channel scenario).

Beyond the classifications based on specific details of the approach, there are four broad classes of enhancement that differ substantially in the general approaches taken. Each of these classes has its own set of assumptions, advantages, and limitations. The first class, addressed in Section 8.3, concentrates on the short-term spectral domain. These techniques suppress noise by subtracting an estimated noise bias (in the power spectral, Fourier transform, or autocorrelation domain) found during nonspeech activity in single-microphone cases, or from a reference microphone in a dual-channel setting. In Section 8.4, we discuss the second class of enhancement techniques, which is based on speech modeling using iterative methods. These systems focus on estimating model parameters that characterize the speech signal, followed by resynthesis of the noise-free signal based on noncausal Wiener filtering. These enhancement techniques estimate speech parameters in noise based on autoregressive, constrained autoregressive, or autoregressive-moving average models. This class of enhancement techniques requires *a priori* knowledge of noise and speech statistics and generally results in iterative enhancement schemes. The third class of systems, discussed in Section 8.5, is based on "adaptive noise canceling" (ANC). Traditional ANC is formulated using a dual-channel time or frequency domain environment based on the "least mean square" (LMS) algorithm. Although other enhancement algorithms can *benefit* from a reference channel, successful ANC *requires* one. The last area of enhancement is based on the periodicity of voiced speech. These methods employ fundamental frequency tracking using either single-channel ANC (a special application) or adaptive comb filtering of the harmonic magnitude spectrum. Fundamental frequency-tracking methods are discussed in Section 8.6.

In this chapter, we consider only a small subset of the possible topics of enhancing speech degraded by noise. Specifically, we address the problem of speech degraded by additive noise as follows,

$$y(n) = s(n) + Gd(n), \tag{8.1}$$

where in general $y(n)$, $s(n)$, and $d(n)$ are realizations of stochastic processes with $s(n)$ representing the original "clean" speech signal, $d(n)$ the degrading noise, and G a gain term that controls SNR. Many practical problems fall into the category of additive distortion, and those that do not can at times be transformed so that they satisfy an additive noise assumption. Consider for example, a convolutional noise degradation,[2]

$$y(n) = s(n) * Gd(n), \tag{8.2}$$

where $*$ represents convolution. If a homomorphic signal transformation is first applied (see Chapter 6), the resulting transformed speech signal and noise component are additive (Oppenheim et al., 1968). Another type of noise distortion is signal-dependent quantization noise found in coding applications such as PCM (see Chapter 7). It has been shown that such noise can be converted to signal-independent noise using a pseudonoise sequence technique (Roberts, 1962).

Another consideration in speech enhancement is that assessment of performance is ultimately related to an evaluation by a human listener. Based on the enhancement context, evaluation may depend on quality, intelligibility, or some other auditory attribute. Therefore, algorithm performance must incorporate aspects of human speech perception. Some techniques are motivated heavily by a mathematical criterion, others focus more on perceptual properties. It is therefore desirable to consider a mathematical criterion that is consistent in some way with human perception. Although no optimum criterion yet exists, some are better than

[2]A similar argument can also be constructed for multiplicative noise.

others. Examples of mathematical criteria that are not particularly well correlated with perceptual quality include mean square error (MSE) and SNR. Performance can also be measured using subjective or objective quality measurement techniques.

We have seen that the vocal tract must vary in order to generate different speech sounds. This movement is reflected in a time-varying linear transfer function. On a short-term basis, it is reasonable to assume that this system is stationary. Therefore, many speech enhancement techniques operate on a frame-by-frame basis and base enhancement on aspects of the slowly varying linear system that reflects properties of speech production. Some of the these aspects include enhancement of the short-term spectral envelope with respect to formant location, amplitude, and bandwidth, as well as the harmonic structure of voiced speech sounds.

8.3 Short-Term Spectral Amplitude Techniques

8.3.1 Introduction

We begin our discussion of speech enhancement by considering methods that focus the processing in the short-term spectral domain. Such techniques seek to enhance a noisy speech signal by subtracting an estimated noise bias. The particular domain in which the subtraction process takes place leads to several alternative formulations. The most popular by far is spectral subtraction, since it was one of the earliest and perhaps the easiest to implement. This has led to many alternative methods involving the "subtraction domain" and variations of a subsequent nonlinear processing step.

Short-term spectral domain methods perform all their processing on the spectral magnitude.[3] The enhancement procedure is performed over frames by obtaining the short-term magnitude and phase of the noisy speech spectrum, subtracting an estimated noise magnitude spectrum from the speech magnitude spectrum, and inverse transforming this spectral amplitude using the phase of the original degraded speech. Since background noise degrades both the spectral magnitude and phase, it is reasonable to question the performance of a technique that does not address noisy phase. We shall explicitly discuss this issue below.

8.3.2 Spectral Subtraction

Generalities

Spectral subtraction is one technique based on direct estimation of the short-term spectral magnitude. In this approach, speech is modeled as a random process to which uncorrelated random noise is added. It is as-

[3]Here we use the term "spectral domain" in a loose sense. The actual domain where processing occurs will vary when we consider refinements to spectral subtraction.

sumed that the noise is short-term stationary, with second-order statistics estimated during silent frames (single-channel) or from a reference channel (dual-channel). The estimated noise power spectrum is subtracted from the transformed noisy input signal.

Let $\underline{s}$, $\underline{d}$, and $\underline{y}$ be stochastic processes representing speech, noise, and noisy speech, respectively. The process $\underline{d}$ is assumed to be uncorrelated, with autocorrelation function

$$r_{\underline{d}}(\eta) = D_0 \delta(\eta), \tag{8.3}$$

where D_0 is some constant. Note that of the three random processes, only $\underline{d}$ can reasonably be assumed stationary. Realizations $s(n)$, $d(n)$, and $y(n)$ are related by

$$y(n) = s(n) + d(n). \tag{8.4}$$

Let us begin by assuming (unrealistically, of course) that $\underline{s}$ and, hence, $\underline{y}$ are stationary processes. Because $\underline{d}$ is an uncorrelated process, it follows immediately that

$$\Gamma_{\underline{y}}(\omega) = \Gamma_{\underline{s}}(\omega) + \Gamma_{\underline{d}}(\omega), \tag{8.5}$$

where $\Gamma_y(\omega)$ is the temporal[4] PDS[5] of the random process $\underline{y}$. Clearly, given $\Gamma_y(\omega)$ and an estimate of $\Gamma_d(\omega)$, say $\hat{\Gamma}_d(\omega)$, it is possible to estimate the PDS of the uncorrupted speech as

$$\hat{\Gamma}_s(\omega) = \Gamma_y(\omega) - \hat{\Gamma}_d(\omega). \tag{8.6}$$

Although theoretically interesting, this analysis has little practical significance, since we deal with real waveforms over short time frames. It does, however, suggest the essence of the spectral subtraction approach to noise elimination. The simplicity of the approach is evident.

Let us now become more realistic and drop the stationarity assumptions on $\underline{s}$ and $\underline{y}$. We are given a signal $y(n)$ (a realization of $\underline{y}$) and the task of estimating the corresponding speech $s(n)$. Recognizing that, at best, $s(n)$ will be "locally stationary" over short time ranges, we select a frame of $y(n)$, say $f_y(n;m)$, using a window of length N ending at time m, $f_y(n;m) = y(n)w(m-n)$. It follows from (8.4) that the selected frame can be expressed in terms of the underlying speech and noise frames,

$$f_y(n;m) = f_s(n;m) + f_d(n;m). \tag{8.7}$$

By analogy to (8.6) we might think to use stPDS[6] and estimate

$$\hat{\Gamma}_s(\omega;m) = \Gamma_y(\omega;m) - \hat{\Gamma}_d(\omega;m), \tag{8.8}$$

[4]The processes are also assumed to be appropriately ergodic so that, for example, $\Gamma_y(\omega) = \Gamma_{\underline{y}}(\omega)$.

[5]In the following, please read the acronym PDS as "power density spectrum" or "power density spectra" as appropriate.

[6]Please recall footnote 5.

where we recall that the stPDS is defined as

$$\Gamma_y(\omega; m) = \frac{1}{N} \sum_{\eta=-\infty}^{\infty} r_y(\eta; m) e^{-j\omega\eta} \tag{8.9}$$

with $r_y(\eta; m)$ the short-term autocorrelation. The careful reader will recognize that this expression is only valid to the extent that $\underline{d}$ remains "locally uncorrelated," meaning that

$$r_d(\eta; m) \approx D_0 \delta(\eta), \tag{8.10}$$

and also uncorrelated with $\underline{s}$,

$$r_{ds}(\eta; m) \approx 0 \tag{8.11}$$

for all η.

Whereas the long-term PDS of (8.5) is part of a mathematical model that is related to time waveforms in only an abstract way, the same is not true of the stPDS of (8.8). In fact, the stPDS is related to the stDTFT in a simple way. For example [see (4.69) in Section 4.3.5],

$$\Gamma_y(\omega; m) = \frac{S_y(\omega; m) S_y^*(\omega; m)}{N^2} = \frac{|S_y(\omega; m)|^2}{N^2}. \tag{8.12}$$

For convenience let us assume that the factor $1/N$ is omitted from the definition (8.9) in this work so that we may write simply

$$\Gamma_y(\omega; m) = |S_y(\omega; m)|^2. \tag{8.13}$$

In effect, therefore, (8.8) offers a way to estimate the short-term magnitude spectrum of the speech, $|S_s(\omega; m)|$. Let us call the estimate $|\hat{S}_s(\omega; m)|$.

In order to estimate the speech frame itself, it is also necessary to obtain an estimate, say $\hat{\varphi}_s(\omega; m) = \arg \hat{S}_s(\omega; m)$, of the short-term phase spectrum of the speech, $\varphi_s(\omega; m) = \arg S_s(\omega; m)$. A moment's thought will indicate that there is no quick way to compute $\hat{\varphi}_s(\omega; m)$. We could, for example, use homomorphic processing, but the complex cepstrum (Section 6.3) is required implying a heavy computational load and overhead. Fortunately, Wang and Lim (1982) have determined that for all practical purposes, it is sufficient to use the noisy phase spectrum, $\varphi_y(\omega; m)$, as an estimate of the clean speech phase spectrum,

$$\hat{\varphi}_s(\omega; m) \approx \varphi_y(\omega; m). \tag{8.14}$$

Further effort to find an improved phase estimate was found to be unwarranted. Therefore, the estimate of the frame of speech resulting from the spectral subtraction method is recovered from the stDTFT estimate

$$\begin{aligned} \hat{S}_s(\omega; m) &= |\hat{S}_s(\omega; m)| e^{j\hat{\varphi}_s(\omega; m)} \\ &= \left[\Gamma_y(\omega; m) - \hat{\Gamma}_d(\omega; m)\right]^{1/2} e^{j\varphi_y(\omega; m)}. \end{aligned} \tag{8.15}$$

$\Gamma_y(\omega; m)$ and $\varphi_y(\omega; m)$ are both obtained from the stDTFT (in practice, the stDFT) of the present noisy speech frame,

$$S_y(\omega; m) = |S_y(\omega; m)| e^{j\varphi_y(\omega; m)} = \Gamma_y^{1/2}(\omega; m) e^{j\varphi_y(\omega; m)}, \tag{8.16}$$

and $\hat{\Gamma}_d(\omega; m)$ can be estimated using any frame of the signal in which speech is not present, or from a reference channel with noise only.

Details, Enhancements, and Applications

Spectral Subtraction Variations and Generalizations. Many variations on the basic strategy above are found in the literature. These are best placed in perspective by presenting a generalized approach due to Weiss and Aschkenasy (1983). Note that the estimator (8.15) can be written

$$\hat{S}_s(\omega; m) = \left[|S_y(\omega; m)|^2 - |\hat{S}_d(\omega; m)|^2 \right]^{1/2} e^{j\varphi_y(\omega; m)}. \tag{8.17}$$

A generalized estimator is given by

$$\hat{S}_s(\omega; m) = \left[|S_y(\omega; m)|^a - |\hat{S}_d(\omega; m)|^a \right]^{1/a} e^{j\varphi_y(\omega; m)} \tag{8.18}$$

where the *power exponent*, a, can be chosen to optimize performance. Regardless of the value of a, these techniques are often just called by the rubric "spectral subtraction," but specific names are sometimes found in the literature. The case $a = 2$, which was used as the motivating case above, is sometimes referred to as *power spectral subtraction* because the noise removal is carried out by subtracting stPDS (squared short-term magnitude spectra). The name *spectral subtraction* is sometimes reserved for the case $a = 1$, in which noise removal is carried out by subtracting *magnitude* spectra. In fact, much of the basis for the ideas above is originally found in papers by Boll (1978; 1979), who employs the $a = 1$ estimator. Techniques using (8.18) directly with other values of a are sometimes called *generalized spectral subtraction.*

Other variations exist in which the "spectral subtraction" is actually implemented in the time domain. A time domain approach corresponding to (8.18) with $a = 2$ is called *correlation subtraction.* When $a = 2$, (8.18) and (8.15) are equivalent, and the magnitude spectral portion of the computation is essentially equivalent to estimating (the square root of)

$$\hat{\Gamma}_s(\omega; m) = \Gamma_y(\omega; m) - \hat{\Gamma}_d(\omega; m) \tag{8.19}$$

or, equivalently,

$$|\hat{S}_s(\omega; m)|^2 = |S_y(\omega; m)|^2 - |\hat{S}_d(\omega; m)|^2. \tag{8.20}$$

Since the stIDTFT is a linear operation, it follows immediately from (8.19) or (8.20) that

$$\hat{r}_s(\eta; m) = r_y(\eta; m) - \hat{r}_d(\eta; m), \tag{8.21}$$

where $\hat{r}_d(\eta; m)$ is an estimate of the short-term autocorrelation of the noise process. Therefore the "spectral subtraction" can also be performed using the autocorrelation. The time domain approach can also be used for values of a other than two, in which case the name *generalized correlation subtraction* is sometimes used. In fact, the INTEL system of Weiss et al. (1974),[7] the first reported "spectral subtraction" technique, is based on correlation subtraction. Finally, we note that the generalized spectral estimator (8.18) was part of an enhanced version of INTEL (Weiss and Aschkenasy, 1983), which also includes generalized cepstral processing. The systems in Fig. 8.2 summarize four variations of spectrum subtraction: (magnitude) spectral subtraction, generalized spectral subtraction, generalized correlation subtraction, and generalized cepstral processing. Although all four of these systems are single-channel, the extension to dual-channel is straightforward and thus provides a recent noise spectral estimate for nonstationary interference.

Negative Spectral Components. In addition to the differences in spectral processing, there is another important aspect of spectral subtraction that is handled differently across various algorithms. From (8.15) [or (8.18)] it is observed that the estimated speech magnitude spectrum is not guaranteed to be positive. Different systems remedy this by performing half-wave rectification or full-wave rectification, or by using a weighted difference coefficient. Most techniques use half-wave rectification (i.e., set negative portions to zero). Forcing negative spectral magnitude values to zero, however, introduces a "musical" tone artifact in the reconstructed speech. This anomaly represents the major limitation of spectral subtraction techniques. In the following material, we pursue details and enhancements to the basic paradigm that have been tried in research and practical systems.

Some Further Enhancements. A variety of techniques centering on the basic principles described above is available for recovering the uncorrupted speech frame, $f_s(n; m)$. We survey a few techniques here with the purpose of pointing out further technical enhancements. Research results for specific spectral subtraction systems are presented here (a comparison with other enhancement techniques can be found in Section 8.7). Basic estimation of the short-term spectral magnitude has resulted in a variety of methods such as spectral subtraction (Boll, 1979), correlation subtraction, and others (Berouti et al., 1977; Curtis and Niederjohn, 1978; Ephraim and Malah, 1983, 1984; Hansen, 1991; Irwin, 1980; Lim, 1978; Un and Choi, 1981).

[7]For historical interest, a discussion of the INTEL system is included in Appendix 8.A.

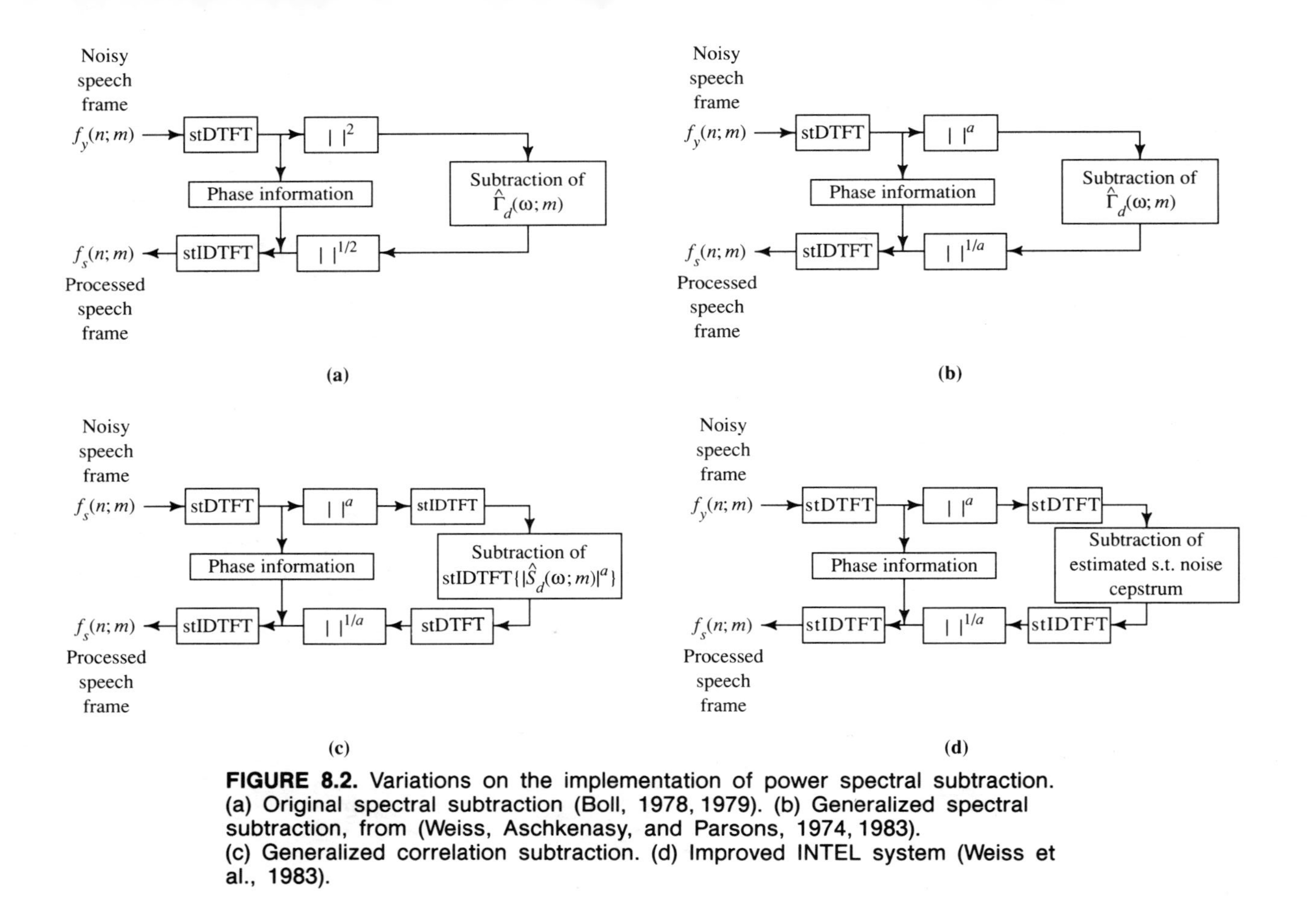

FIGURE 8.2. Variations on the implementation of power spectral subtraction. (a) Original spectral subtraction (Boll, 1978, 1979). (b) Generalized spectral subtraction, from (Weiss, Aschkenasy, and Parsons, 1974, 1983). (c) Generalized correlation subtraction. (d) Improved INTEL system (Weiss et al., 1983).

The system proposed by Boll (1978, 1979) attempts to reduce spectral error by applying three processing steps once the spectral magnitude has been found. The three steps are magnitude averaging, half-wave rectification, and residual noise reduction. The process of magnitude averaging reduces spectral error by performing local averaging of the spectral magnitudes. The magnitude-averaged spectrum is found using the sample mean

$$\overline{|S_y(\omega; m_i)|} \stackrel{\text{def}}{=} \frac{1}{2I+1} \sum_{l=i-I}^{i+I} |S_y(\omega; m_l)|, \tag{8.22}$$

where $m_{i-I}, \ldots, m_{i+I}$ index $2I+1$ frames centered on the "current" frame at m_i. Therefore, the resultant estimator for the speech stDTFT, using the noisy phase $\varphi_y(\omega; m_i)$ from the original distorted speech, is

$$\hat{S}_s(\omega; m_i) = \left[\overline{|S_y(\omega; m_i)|} - |\hat{S}_d(\omega; m_i)|\right] e^{j\varphi_y(\omega; m_i)}, \tag{8.23}$$

where $|\hat{S}_d(\omega; m_i)| = \hat{\Gamma}_d^{1/2}(\omega; m_i)$, an estimate of the magnitude spectrum of the noise frame $f_d(n; m_i)$. The estimator is seen to be of the spectral magnitude type.

The magnitude-averaging method works well if the time waveform is stationary. Unfortunately, the value of I in (8.22) is limited by the short-term stationarity assumption. Therefore, only a few frames of data can be used in averaging. Boll's second processing step is half-wave rectification, which reduces the mean noise level by an amount $|\hat{S}_d(\omega; m)|$. With this rectification, low-variance coherent noise is approximately eliminated. The disadvantage of half-wave rectification is that it is possible for the speech-plus-noise spectrum to be less than $|\hat{S}_d(\omega; m)|$ and consequently, speech information is removed. This step is the major inadequacy of most spectral subtraction techniques, since it is a nonlinear processing step with no mathematical basis other than the requirement that the spectral magnitude be positive.

The last step in Boll's algorithm is residual noise reduction. After half-wave rectification, the spectral bands of speech plus noise above the threshold $|\hat{S}_d(\omega; m)|$ remain, thereby preserving a residual noise component. The argument at this point is that residual noise can be reduced by replacing the present frame value with a minimum value from adjacent frames. The question that arises is, why should such a method work? The answer is that if, for some ω, $|\hat{S}_s(\omega; m)|$ is less than the maximum noise residual, and if it varies from frame to frame, then there is a high probability that the spectrum at that frequency is due to noise. Therefore, the noise can be suppressed by taking the minimum from adjacent frames. If $|\hat{S}_s(\omega; m)|$ is less than the maximum noise residual, but $|\hat{S}_s(\omega; m)|$ is approximately constant between adjacent frames, then a high probability exists that the spectrum at that frequency represents low-energy speech. Therefore, taking the minimum will not affect the information content.

Finally, if $|\hat{S}_s(\omega; m)|$ is greater than the maximum noise residual, then speech is present in the signal at that frequency; therefore, subtracting the noise bias is enough. Boll evaluated this algorithm for speech distorted by helicopter noise. Figure 8.3 shows short-term vocal system spectra of noisy and enhanced helicopter speech. The results showed that spectral subtraction alone does not increase intelligibility as measured by the diagnostic rhyme test (see Chapter 9), but does increase quality, especially in the areas of increased pleasantness and inconspicuousness of noise background. It was also shown that magnitude averaging does reduce the effects of musical tones caused by errors in accurate noise bias estimation.

A further enhancement to spectral processing is to introduce a weighted subtraction term k as

$$\hat{S}_s(\omega; m) = \left[|S_y(\omega; m)|^a - k\,|\hat{S}_d(\omega; m)|^a\right]^{1/a} e^{j\varphi_y(\omega; m)}. \tag{8.24}$$

Berouti et al. (1979) considered such a method with $a = 2$. Their results showed that if the weighted subtraction term k is increased (i.e., overestimating the noise spectrum), musical tone artifacts can be reduced. It was also desirable to adjust k to maintain a minimum and maximum spectral floor based on the estimated input SNR as shown in Fig. 8.4. As with all forms of spectral subtraction, negative values from the subtraction $|S_y(\omega; m)|^a - k\,|\hat{S}_d(\omega; m)|^a$ can be removed by full-wave or half-wave rectification. A frequency-dependent subtraction term $k(\omega)$ was also considered.

Another approach that further modifies spectral subtraction was proposed by McAulay and Malpass (1979). In this method, a spectral decomposition of a frame of noisy speech is performed and a particular spectral line is attenuated based on how much the speech-plus-noise power exceeds an estimate of the background noise. The noise at each frequency component is assumed to be Gaussian, resulting in a maximum likelihood estimate of $|S_s(\omega; m)|$. A further extension, also due to McAulay and Malpass, is to scale the input frequency response $|S_y(\omega; m)|$ by the probability that speech is present in the input signal. Their reasoning is that if the proba-

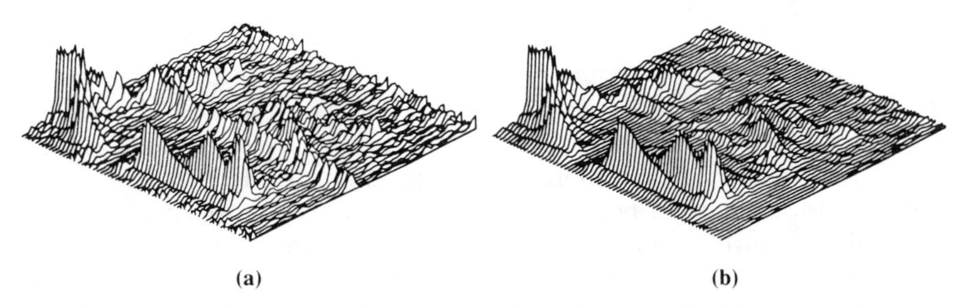

FIGURE 8.3. Examples of (a) noisy and (b) enhanced short-term vocal system spectra of speech degraded with helicopter noise.

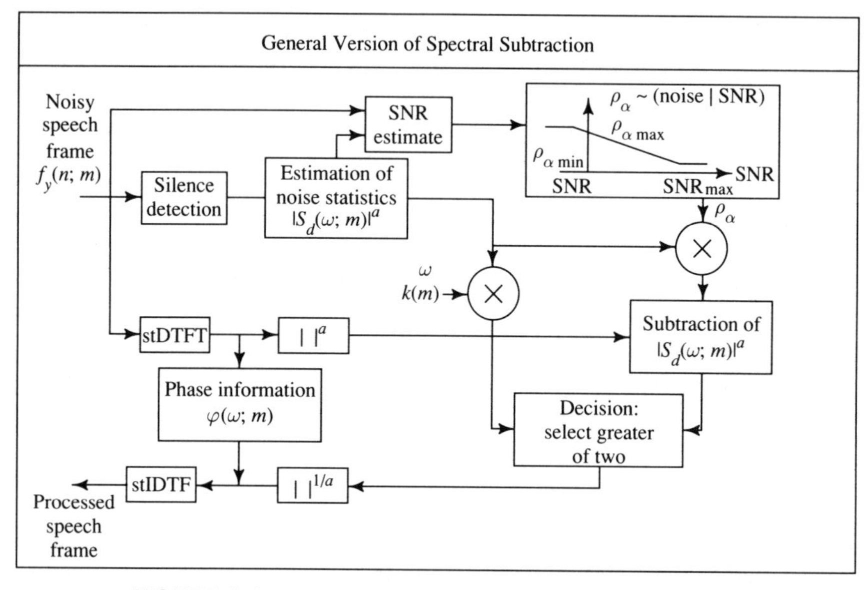

FIGURE 8.4. A general version of spectral subtraction.

bility of noise is high, it would be preferable to further reduce the signal estimate $|\hat{S}_s(\omega; m)|$. Peterson and Boll (1981) considered applying spectral subtraction in separate frequency bands tuned to the loudness components perceived by the auditory system. Other extensions related to those presented here can be found in papers by Curtis and Niederjohn (1978), Preuss (1979), and Un and Choi (1980).

If we combine all of the estimation enhancements above, the most general form of spectral subtraction can be written as

$$\hat{S}_s^{\Sigma,k,a}(\omega; m_i) = \left\{ \frac{1}{2I+1} \sum_{l=-I}^{I} |S_y(\omega; m_l)|^a - k[\omega, P(\text{noise}|\text{SNR}, m_i)]\, |\hat{S}_d(\omega; m_i)|^a \right\}^{1/a} e^{j\varphi_y(\omega; m_i)}, \tag{8.25}$$

where $P(\text{noise}|\text{SNR}, m_i)$ is the probability of only noise being present at the frame ending at m_i given an estimate of the present SNR, and the times m_l index the end-times of the frames used in the magnitude averaging. An extension to this approach was proposed by Hansen (1991), in which a noise-adaptive boundary detector was used to partition speech into voiced/transitional/unvoiced speech sections to allow for a variable noise suppression based on the input speech class, followed by the application of morphological-based spectral constraints to reduce frame-to-frame jitter of speech spectral characteristics. Performance was demonstrated for a variety of speech sound types (vowels, nasals, stops, glides, fricatives, etc.) over a standard spectral subtraction and noncausal Wiener filtering technique.

Single-Channel Spectral Subtraction Evaluation. A variety of single-channel spectral subtraction methods have been discussed. Three factors that greatly influence enhancement performance are (1) the enhancement domain, (2) the power factor term a (also related to the enhancement domain), and (3) processing of negative spectral components. In this section, we briefly summarize two studies that considered these factors. In the first study, Lim (1978) evaluated the correlation subtraction method proposed by Weiss et al. (1974) called INTEL for wideband random noise under varying values of a. Figure 8.5 shows intelligibility scores based on tests involving nonsense sentences. Results with wideband noise show that intelligibility is not improved. If the power exponent is set to $a = 1$, the system reverts to Boll's spectral subtraction technique. This indicates that performance may differ based on the type of distortion (Boll concluded no decrease in intelligibility for helicopter noise). It was also observed that processed speech with $a = 1$ or 0.5 sounded distinctly "less noisy" and of "higher quality" at relatively high SNR.

In a later study, Hansen and Clements (1985, 1987) compared the performance of Boll's spectral subtraction method with that of traditional adaptive Wiener filtering (discussed in the next section). Evaluation was

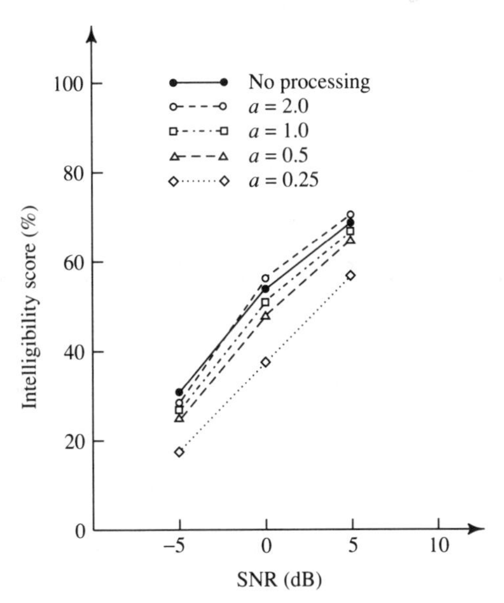

FIGURE 8.5. Intelligibility scores of a spectral subtraction for enhancement of speech degraded by wide-band random noise. Adapted from Lim (1978).

performed for both the half- and full-wave rectification, employing 1–5 frames of magnitude averaging. The evaluation was performed under identical conditions (same distorted utterances, same global SNR estimates). Table 8.1 summarizes the results.[8] Full-wave rectification resulted in improvement over a wider range of SNR, although half-wave rectification had greater improvement over the restricted SNR band of 5–10 dB. In addition, magnitude averaging using frames that look ahead performed poorer than the corresponding equivalent looking back in time. For both rectification approaches, magnitude averaging provided improved quality.

Dual-Channel Spectral Subtraction. The spectral subtraction methods discussed thus far have focused on single-channel techniques. Researchers have also considered dual-channel spectral subtraction methods. Specifically, Hanson et al. (1983), Childers and Lee (1987), and Naylor and Boll (1987) have all considered various forms of spectral subtraction for the purposes of co-talker separation. These methods normally require some *a priori* knowledge of the speaker characteristics (normally fundamental frequency contours) to assist in the enhancement process. The method proposed by Hanson and Wong (1984) considers a power exponent a and the phase difference between speech from two competing speakers. Their results show that magnitude subtraction (i.e., $a = 1$) is preferable at low SNR. Although estimation of pitch and voicing were necessary, they were able to show a significant increase in intelligibility, which has proven to be a difficult task for the competing speaker problem. The greatest improvement occurred for low SNR (−12 dB) with smaller levels of improvement as SNR increased. Finally, Ariki et al. (1986) considered a two-dimensional spectral smoothing and spectral amplitude transformation method. Noise processing is performed in the time versus cepstrum domain, resulting in improved formant characterization with respect to conventional frequency subtraction.

8.3.3 Summary of Short-Term Spectral Magnitude Methods

In this section, we have considered speech enhancement techniques that focus their processing in the short-term spectral domain. These methods are based on subtraction of an estimated noise bias found during nonspeech activity or from a reference channel. The techniques differ in the domain in which subtraction is performed, the power exponent, the presence or absence of the weighted subtraction coefficient based on frequency and/or probability of speech, as well as postprocessing with half- or full-wave rectification, or magnitude averaging.

[8]The Itakura–Saito quality measure is discussed in Chapter 9.

8.4 Speech Modeling and Wiener Filtering

8.4.1 Introduction

Short-term Wiener filtering is an approach in which a frequency weighting for an optimum filter is first estimated from the noisy speech, $y(n)$. The linear estimator of the uncorrupted speech $s(n)$, which minimizes the MSE criterion, is obtained by filtering $y(n)$ with a noncausal Wiener filter. This filter requires *a priori* knowledge of both speech and noise statistics, and therefore must also adapt to changing characteristics. In a single-channel framework, noise statistics must be obtained during silent frames. Also, since noise-free speech is not available, *a priori* statistics must be based upon $y(n)$, resulting in an iterative estimation scheme. The estimation of speech parameters in an all-pole model assuming an additive white Gaussian noise distortion was investigated by Lim and Oppenheim (1978), and later generalized for a colored noise degradation by Hansen and Clements (1985). This approach attempts to solve for the maximum *a posteriori* estimate of a speech waveform in additive white Gaussian noise with the requirement that the signal be the response from an all-pole process. Crucial to the success of this approach is the accuracy of the estimates of the all-pole parameters at each iteration. The estimation procedures that result in linear equations without background noise become nonlinear when noise is introduced. However, by using a suboptimal procedure, an iterative algorithm results in which the estimation procedure is linear at each iteration.

8.4.2 Iterative Wiener Filtering

We begin with the same setup used in the spectral subtraction problem: $\underline{s}, \underline{d}$, and $\underline{y}$ are stochastic processes representing speech, noise, and noisy speech, respectively.[9] The process $\underline{d}$ is assumed to be uncorrelated as in (8.3): $\underline{s}(n)$, $\underline{d}(n)$, and $\underline{y}(n)$ represent random variables from the respective processes, and $s(n)$, $d(n)$, and $y(n)$ denote realizations. Appropriate ergodicity properties are assumed so that time averaging may be used in place of statistical averaging when desirable. Soon we will encounter an estimator for the random process $\underline{s}$, and the estimator itself will be a random process. In anticipation of this estimator, we define the notations $\underline{\hat{s}}$, $\underline{\hat{s}}(n)$, and $\hat{s}(n)$. The noise is additive, so

$$y(n) = s(n) + d(n). \tag{8.26}$$

[9]We shall see below that the results do not change if the uncorrupted speech is deterministic. Therefore, we assume a stochastic process for generality.

TABLE 8.1. Comparison of Enhancement Results for Spectral Subtraction and Traditional Noncausal Wiener Filtering.[a]

	Itakura–Saito Distortion Measure **White Gaussian Noise**								
***SNR* (in dB) (*original measure*)**	**−20 7.22**	**−15 6.97**	**−10 6.49**	**−5 5.79**	**0 4.94**	**5 4.02**	**10 3.08**	**15 2.18**	**20 1.40**
	Spectral Subtraction								
Magnitude Average	*Half-Wave Rectification*								
1	×	×	×	×	×	+0.08	+0.25	+0.20	×
−2	×	×	×	×	+0.03	+0.27	+0.37	×	×
2	×	×	×	×	×	+0.16	+0.25	×	×
3	×	×	×	×	+0.12	+0.41	+0.45	×	×
−4	×	×	×	×	+0.36	+0.67	+0.59	×	×
4	×	×	×	×	+0.21	+0.50	+0.40	×	×
5	×	×	×	+0.10	+0.48	+0.74	+0.52	×	×

Magnitude Average	*Full-Wave Rectification*								
1	+0.05	+0.01	+0.04	+0.10	+0.13	+0.11	+0.05	×	×
−2	×	×	+0.05	+0.20	+0.27	+0.29	+0.19	×	×
2	×	×	+0.08	+0.19	+0.23	+0.26	+0.16	×	×
3	×	×	+0.10	+0.26	+0.38	+0.43	+0.33	×	×
−4	×	×	+0.09	+0.32	+0.48	+0.56	+0.45	×	×
4	×	×	+0.02	+0.24	+0.38	+0.45	+0.32	×	×
5	×	×	+0.04	+0.20	+0.48	+0.56	+0.39	×	×
	Noncausal Wiener Filtering								
Greatest Improvement at Iteration ⇒	−0.04 #1	+0.06 #1	+0.26 #2	+0.52 #3	+0.74 #3	+0.87 #3	+0.92 #3	+0.80 #3	+0.57 #3

[a]Improvements in Itakura–Saito distortion measures of the sentence, "Cats and dogs each hate the other," are shown for the two enhancement classes. The original objective (Itakura–Saito) quality measure across SNR with quality improvement is shown for each type of magnitude averaging. An SNR for which no improvement resulted is indicated with an ×. One to five frames of magnitude averaging were used in the evaluation. The left column in the table indicates the frames used for magnitude averaging. The present frame is identified by a dot at its center, with time increasing toward the right (e.g., magnitude averaging of −2 uses the present and previous past frame; magnitude averaging of 2 uses the present and next future frame).

Our goal is to formulate a linear filter with which to produce an optimal estimate of $s(n)$, say $\hat{s}(n)$, which is optimal in the MSE sense. That is, we desire a filter with impulse response $h^{\dagger}(n)$ such that with input $s(n)$ the output is an estimator $\hat{s}(n)$ for which

$$\xi = \mathcal{E}\{[\underline{s}(n) - \hat{\underline{s}}(n)]^2\} = \mathcal{L}\{[s(n) - \hat{s}(n)]^2\} \tag{8.27}$$

is minimized. For the moment, we allow the filter to be IIR and even noncausal. The reader may recall that this filter, called the *noncausal Wiener filter*, was derived in Problem 5.8 using the orthogonality principle. The result in the frequency domain is

$$H^{\dagger}(\omega) = \left[\frac{\Gamma_s(\omega)}{\Gamma_s(\omega) + \Gamma_d(\omega)}\right], \tag{8.28}$$

where $\Gamma_s(\omega)$ and $\Gamma_d(\omega)$ are the PDS[10] for the processes $\underline{s}$ and $\underline{d}$.

In practice, the filter (8.28) cannot be directly applied to the noisy input speech signal, since $\underline{s}$ is only short-term stationary, and the PDS $\Gamma_s(\omega)$ is generally unknown. One way to approximate the noncausal Wiener filter is to adapt the filter characteristics on a frame-by-frame basis by using the stPDS,

$$H^{\dagger}(\omega; m) = \frac{\hat{\Gamma}_s(\omega; m)}{\hat{\Gamma}_s(\omega; m) + \hat{\Gamma}_d(\omega; m)}. \tag{8.29}$$

The hats over the stPDS are reminders that these spectra must be estimated. For a single-channel enhancement scheme, the noise power spectrum $\Gamma_d(\omega; m)$ is estimated during periods of silence. In dual-channel scenarios, the noise estimate is updated whenever the speech spectrum is reestimated. Estimating the speech stPDS is a more difficult problem, which we address momentarily.

Given the filter response $H^{\dagger}(\omega; m)$, the short-term speech spectrum is then obtained by filtering the noisy speech signal as

$$\hat{S}_s(\omega; m) = H^{\dagger}(\omega; m) S_y(\omega; m) \tag{8.30}$$

either in the time or frequency domain. We should note, however, that $H^{\dagger}(\omega; m)$ has a zero-phase spectrum, so that the output phase of the enhanced speech spectrum $\hat{S}_s(\omega; m)$ is simply the noisy phase from $S_y(\omega; m)$. Therefore, like spectral subtraction methods, adaptive Wiener filtering focuses its processing only in the spectral magnitude domain, but ends up attributing the same phase characteristic to the speech that is used in the spectral subtraction method.

Let us now return to the problem of estimating $\hat{\Gamma}_s(\omega; m)$, which, as we have indicated, is not a trivial problem. Indeed, if we had a good esti-

[10]Again, please read PDS as "power density spectrum" or "spectra" as appropriate.

mate of $\Gamma_s(\omega; m)$ this would imply that we approximately knew the magnitude spectrum of the speech in the frame, since

$$\Gamma_s(\omega; m) = |S_s(\omega; m)|^2. \tag{8.31}$$

However, it is the speech in the frame that we are trying to estimate [see (8.30)] and it is unlikely that we would have an accurate estimate of its spectrum. One approach to the speech spectrum estimation problem is to use an *iterative* procedure in which an ith estimate of $\Gamma_s(\omega; m)$, say $\hat{\Gamma}_s(\omega; m, i)$ [or $|\hat{S}_s(\omega; m, i)|^2$] is used to obtain an $i+1$st filter estimate, say $H^\dagger(\omega; m, i+1)$. In the next sections, we consider several methods for modeling speech in such an iterative framework.

Generalizations of Wiener filtering have been studied in other areas of signal processing. One approach for image restoration employs a noise-scale term k, and a power exponent a, given by

$$H_a^\dagger(\omega) = \left[\frac{\Gamma_s(\omega)}{\Gamma_s(\omega) + k\Gamma_d(\omega)}\right]^a. \tag{8.32}$$

The numbers a and k can be varied to obtain filters with different frequency characteristics. If we were to set $a = 1$ and $k = 1$, (8.32) reverts back to the standard Wiener filter in (8.28). If we set $a = \frac{1}{2}$ and $k = 1$, then (8.32) is equivalent to power spectral filtering. Again, due to the short-term stationarity assumption, (8.32) must be modified for processing on a frame-by-frame basis similarly to (8.29).

8.4.3 Speech Enhancement and All-Pole Modeling

We know from our studies in Chapter 5 that over a given frame of speech, say $f_s(n; m) = s(n)w(m-n)$, an all-pole model of the form

$$\Theta(z; m) = \frac{\Theta_0(m)}{1 - \sum_{i=1}^{M} \hat{a}(i; m) z^{-i}} \tag{8.33}$$

is frequently sufficient to accurately model the magnitude spectrum of the frame. The $\hat{a}(i; m)$'s are the short-term LP coefficients as defined in Chapter 5, where techniques for their estimation in *noise-free* speech are discussed. Techniques for estimating these parameters in noisy speech (which is the case here) have been considered by Magill and Un (1976), Kobatake et al. (1978), and Lim and Oppenheim (1978, 1979).

The method by Lim and Oppenheim is based on *maximum a posteriori* (MAP) estimation of the LP coefficients, gain, and noise-free speech. The method is an iterative one in which the LP parameters and speech frame are repeatedly reestimated. In the following, $f_s(n; m)$ is the (unknown) underlying frame of noise-free speech that we desire to estimate. For simplicity, and without loss of generality, let us take $m = N$, the window length. Here $f_y(n;N)$ is the observed frame of noisy speech. Also, $\hat{\mathbf{a}}(N)$ is our usual notation for the (unknown) M-vector of LP parameters over the frame, and $\hat{\Theta}_0(N)$ is our usual notation for the (unknown) model gain. For simplicity in the discussion to follow, we define:

$$\begin{aligned}
\mathbf{s}_k &\stackrel{\text{def}}{=} k\text{th estimate (from } k\text{th iteration) of the vector} \\
&\qquad \mathbf{s} \stackrel{\text{def}}{=} \begin{bmatrix} f_s(1;N) & f_s(2;N) & \cdots & f_s(N;N) \end{bmatrix}^T \\
\mathbf{s}_I &\stackrel{\text{def}}{=} \text{given or estimated initial conditions for the } \mathbf{s}_k \text{ vector} \\
\mathbf{y} &\stackrel{\text{def}}{=} \text{observable vector of noisy speech} \\
&\qquad \begin{bmatrix} f_y(1;N) & f_y(2;N) & \cdots & f_y(N,N) \end{bmatrix}^T \\
\mathbf{a}_k &\stackrel{\text{def}}{=} k\text{th estimate of the vector } \hat{\mathbf{a}}(N) \\
g_k &\stackrel{\text{def}}{=} k\text{th estimate of the model gain } \hat{\Theta}_0(N).
\end{aligned} \tag{8.34}$$

It is assumed that all unknown parameters are random with *a priori* Gaussian pdf's. The resulting MAP estimator, which maximizes the conditional pdf of the parameters given the observations, corresponds to maximizing[11] $p(\mathbf{a}_k | \mathbf{s}_{k-1})$, which in general requires the solution of a set of nonlinear equations for the additive white Gaussian noise (AWGN) case. In the noisy case, the estimator requires $\mathbf{a}_k, g_k$, and $\mathbf{s}_I$ be chosen to maximize the pdf[12] $p(\mathbf{a}_k, g_k, \mathbf{s}_I | \mathbf{y})$. Essentially, we wish to perform joint MAP estimation of the LP speech modeling parameters and noise-free speech by maximizing the joint density $p(\mathbf{a}_k, \mathbf{s}_k | \mathbf{y}, g_k, \mathbf{s}_I)$, where the terms g_k and $\mathbf{s}_I$ are assumed to be known (or estimated). Lim and Oppenheim consider a suboptimal solution employing sequential MAP estimation of $\mathbf{s}_k$ followed by MAP estimation of $\mathbf{a}_k, g_k$ given $\mathbf{s}_k$. The sequential estimation procedure is linear at each iteration and continues until some criterion is satisfied. With further simplifying assumptions, it can be shown that MAP estimation of $\mathbf{s}_k$ is equivalent to noncausal Wiener filtering of the noisy speech $\mathbf{y}$. Lim and Oppenheim showed that this technique, under certain conditions, increases the joint likelihood of $\mathbf{a}_k$ and $\mathbf{s}_k$ with each iteration. It can also be shown to be the optimal solution in the MSE sense for a white noise distortion [with, say, $\Gamma_d(\omega) = \sigma_d^2$]. The resulting equation for estimating the noise-free speech is simply the optimum Wiener filter (8.28),

$$H^{\dagger}(\omega; N, k) = \frac{\hat{\Gamma}_s(\omega; N, k)}{\hat{\Gamma}_s(\omega; N, k) + \hat{\Gamma}_d(\omega; m)}, \tag{8.35}$$

[11]We omit the conventional subscripts from the pdf's in this discussion because the meaning is clear without them. Further, we use the symbol p rather than f for the pdf to avoid confusion with the frame notation.

[12]The unknowns in this case are the LP model parameters, gain, and initial conditions for the predictor.

where the extra index k is included to indicate the kth iteration (this will become clear below). If $\underline{d}$ is white noise, then $\Gamma_d(\omega)$ can be replaced by σ_d^2. If the Gaussian assumption of the unknown parameters holds, this is the optimum processor in a MSE sense. If the Gaussian assumption does not hold, this filter is the best *linear* processor for obtaining the next speech estimate $\mathbf{s}_{k+1}$. With this relation, sequential MAP estimation of the LP parameters and the speech frame generally follows these steps:

1. Find $\mathbf{a}_k = \underset{\mathbf{a}}{\operatorname{argmax}}\ p(\mathbf{a} | \mathbf{s}_k, \mathbf{y}, g_{k-1}, \mathbf{s}_I)$.
2. Find $\mathbf{s}_k = \underset{\mathbf{s}}{\operatorname{argmax}}\ p(\mathbf{s} | \mathbf{a}_k, \mathbf{y}, g_k, \mathbf{s}_I)$.

The first step is performed via LP parameter estimation and the second step through adaptive Wiener filtering. The final implementation of the algorithm is presented in Fig. 8.6. This approach can also be extended to the colored noise case as shown in (Hansen, 1985). The noise spectral density, or noise variance for the white Gaussian case, must be estimated during nonspeech activity in the single-channel framework.

FIGURE 8.6. Enhancement algorithm based on all-pole modeling/noncausal Wiener filtering; (1) An AWGN distortion, (2) a nonwhite distortion.

Step 1. Estimate $\mathbf{a}_k$ from $\mathbf{s}_k$. Using either:
  a. First M values as the initial condition vector, or
  b. Always assume a zero initial condition $\mathbf{s}_k = 0$.

Step 2. Estimate $\mathbf{s}_k(N)$ given the present estimate $\mathbf{a}_k(N)$.

  a. Using $\mathbf{a}_k$, estimate the speech spectrum: $\Gamma_s(\omega; N, k) = \dfrac{g_k^2}{|1-\mathbf{a}_k^T\mathbf{e}|^2}$, where $\mathbf{e}$ is the vector

$$\left[e^{-j\omega} \quad e^{-j\omega 2} \quad \cdots \quad e^{-j\omega M}\right]^T.$$

  b. Calculate gain term g_k using Parseval's theorem.
  c. Estimate either the degrading

  (1) white noise variance σ_d^2, or (2) colored noise spectrum $\Gamma_d(\omega; N)$

  from a period of silence closest to the utterance.
  d. Construct the noncausal Wiener filter:

$$(1)\ H^\dagger(\omega; N, k) = \left[\frac{\hat{\Gamma}_s(\omega; N, k)}{\hat{\Gamma}_s(\omega; N, k) + \hat{\sigma}_d^2}\right]$$

$$(2)\ H^\dagger(\omega; N, k) = \left[\frac{\hat{\Gamma}_s(\omega; N, k)}{\hat{\Gamma}_s(\omega; N, k) + \hat{\Gamma}_d(\omega; N, k)}\right]$$

  e. Filter the estimated speech $\mathbf{s}_k$ to produce $\mathbf{s}_{k+1}$.
  f. Repeat until some specified error criterion is satisfied.

8.4.4 Sequential Estimation via EM Theory

In this section, we continue to use the simplified notation defined above. The basic sequential MAP estimation procedure above can be formulated in an alternate way.[13] The *estimate-maximize* (EM) algorithm was first introduced by Dempster et al. (1977) as a technique for obtaining maximum likelihood estimation from incomplete data. In the EM algorithm, the observations are considered "incomplete" with respect to some original set (which is considered "complete"). The algorithm iterates between estimating the sufficient statistics of the "complete" data, given the observations and a current set of parameters (E step) and maximizing the likelihood of the complete data, using the estimated sufficient statistics (M step). If the unknown model parameters are distributed in a Gaussian fashion, then it can be shown that the EM approach employing maximum likelihood estimation is equivalent to the original sequential MAP estimation procedure developed by Lim and Oppenheim. To do so, consider a vector of noisy speech data [recall definitions (8.34)]

$$\mathbf{y} = \mathbf{s} + \mathbf{d}, \tag{8.36}$$

where $\mathbf{d}$ has the obvious meaning in light of (8.34), and where the noise is zero mean, Gaussian, with $\Gamma_d(\omega) = \sigma_d^2$. The basic problem, as above, is to estimate $\hat{\mathbf{a}}(N)$ and the speech frame $f_s(n;m)$ (vector $\mathbf{s}$) given the frame $f_y(n;m)$ (vector $\mathbf{y}$). Now, if we view the observed data vector $\mathbf{y}$ as being incomplete and specify some complete data set $\mathbf{s}$ that is related to $\mathbf{y}$ by the relation

$$\mathcal{H}(\mathbf{s}) = \mathbf{y}, \tag{8.37}$$

where $\mathcal{H}(\cdot)$ is a noninvertible (many-to-one) transformation, the EM algorithm (at iteration k) is directed at finding the maximum likelihood estimate of the model parameters, say,

$$\mathbf{a}^* = \underset{\mathbf{a}}{\operatorname{argmax}} \; \log \, p_{\underline{\mathbf{y}},\underline{\mathbf{a}}}(\mathbf{y}, \mathbf{a}), \tag{8.38}$$

with $p_{\underline{\mathbf{y}},\underline{\mathbf{a}}}(\cdot\,,\cdot)$ the pdf for what in the present context may be considered random vectors $\underline{\mathbf{y}}$ and $\underline{\mathbf{a}}$. The algorithm is iterative, with $\mathbf{a}_0$ defined as the initial guess and $\mathbf{a}_k$ defined by induction as follows:

$$\mathbf{a}_k = \underset{\mathbf{a}}{\operatorname{argmax}} \; \mathcal{E}\{\log \, p_{\underline{\mathbf{s}},\underline{\mathbf{a}}}(\mathbf{s}, \mathbf{a}) \,|\, \mathbf{y}, \mathbf{a}_{k-1}\}. \tag{8.39}$$

The basic idea behind this approach is to choose $\mathbf{a}_k$ such that the log-likelihood of the complete data, $\log p_{\underline{\mathbf{s}},\underline{\mathbf{a}}}(\mathbf{s}, \mathbf{a})$ is maximized. However, the joint density function $p_{\underline{\mathbf{s}},\underline{\mathbf{a}}}(\mathbf{s}, \mathbf{a})$ is not available. Therefore, instead of maximizing the log-likelihood, we maximize its expectation given the observed data $\mathbf{y}$ in (8.39). In addition, the current estimate of the parameters $\mathbf{a}_k$ is used rather than the the actual (unknown) $\hat{\mathbf{a}}(N)$. For this reason,

[13] We will encounter this method again in the study of hidden Markov models in Chapter 12.

the conditional expectation is not exact. The algorithm therefore iterates, using each new parameter estimate to improve the conditional expectation on the next iteration cycle (the E step), and then uses this conditional estimate to improve the next parameter estimate (the M step).

The EM approach is similar to the two-step MAP estimation procedure of Lim and Oppenheim; the main difference is that the error criterion here is to maximize the expected log-likelihood function given observed or estimated speech data. Feder et al. (1988, 1989) formulated such a method for dual-channel noise cancelation applications where a controlled level of cross-talk was present. Their results showed improved performance over a traditional least MSE estimation procedure.

8.4.5 Constrained Iterative Enhancement

Although traditional adaptive Wiener filtering is straightforward and useful from a mathematical point of view, there are several factors that make application difficult. Hansen and Clements (1987, 1988, 1991) considered an alternative formulation based on iterative Wiener filtering augmented with speech-specific constraints in the spectral domain. This method was motivated by the following observations. First, the traditional Wiener filter scheme is iterative with sizable computational requirements. Second, and more important, although the original sequential MAP estimation technique is shown to increase the joint likelihood of the speech waveform and all-pole parameters, a heuristic convergence criterion must be employed. This is a disturbing drawback if the approach is to be used in environments requiring automatic speech enhancement. Hansen and Clements (1985) performed an investigation of this technique for AWGN, and a generalized version for additive nonwhite, nonstationary aircraft interior noise. Objective speech quality measures, which have been shown to be correlated with subjective quality (Quackenbush et al., 1985, 1988), were used in the evaluation. This approach was found to produce significant levels of enhancement for white Gaussian noise in three to four iterations. Improved all-pole parameter estimation was also observed in terms of reduced MSE. Only if the pdf is unimodal and the initial estimate for $\mathbf{a}_k$ is such that the local maximum equals the global maximum is the procedure equivalent to the joint MAP estimate of $\mathbf{a}_k$, g_k, and $\mathbf{s}_k$.

Some interesting anomalies were noted that motivated development of the alternative enhancement procedure based on spectral constraints. First, as additional iterations were performed, individual formants of the speech consistently decreased in bandwidth and shifted in location, as indicated in Fig. 8.7. Second, frame-to-frame pole jitter was observed across time. Both effects contributed to unnatural sounding speech. Third, although the sequential MAP estimation technique was shown to increase the joint likelihood of the speech waveform and all-pole parameters, a heuristic convergence criterion had to be employed. Finally, the original technique employs no explicit frame-to-frame constraints, though

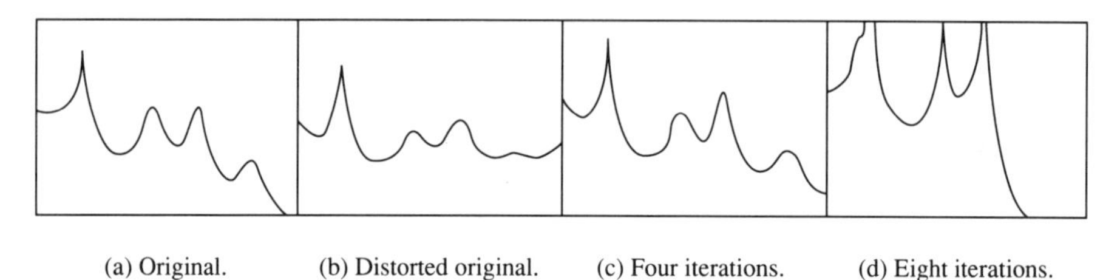

(a) Original. (b) Distorted original. (c) Four iterations. (d) Eight iterations.

FIGURE 8.7. Variation in vocal-tract response across iterations: (a) original, (b) distorted original, (c) 4 iterations, (d) 8 iterations.

it is normally assumed that the characteristics of speech are short-term stationary. The alternative algorithms are based on sequential two-step MAP estimation of the LP parameters and noise-free speech waveform. In order to improve parameter estimation, reduce frame-to-frame pole jitter across time, and provide a convenient and consistent terminating criterion, a variety of spectral constraints were introduced between MAP estimation steps. These constraints are applied based on the presence of perceptually important speech characteristics found during the enhancement procedure. The enhancement algorithms impose spectral constraints on all-pole parameters across time (interframe) and iterations (intraframe), which ensure that

1. The all-pole speech model is stable.
2. The model possesses speech-like characteristics (e.g., poles are not too close to the unit circle causing abnormally narrow bandwidths).
3. The vocal system characteristics do not vary wildly from frame to frame when speech is present.

Due to the imposed constraints, improved estimates $\mathbf{a}_{k+1}$ result.

In order to increase numerical accuracy, and eliminate inconsistencies in pole ordering, the line spectral pair (see Section 5.4.1) transformation was used to implement most of the constraint requirements. The imposition of these constraints helps in obtaining an optimal terminating iteration and improves speech quality by reducing the effects of these anomalies.

The constrained iteration method attempts to bridge the gap between the two broad enhancement philosophies, where the basic sequential MAP estimation procedure serves as the mathematical basis for enhancement while the imposition of constraints between MAP estimation steps attempts to improve aspects important in human perception. Figure 8.8 illustrates results from a single frame of speech for the traditional Wiener filtering method (unconstrained) and constrained approach. Further discussion of quality improvement for iterative speech enhancement methods will be found in Section 8.7. Another speech modeling approach using a dual-channel framework by Nandkumar and Hansen (1992) extends this method by employing auditory-based constraints. Improvement in speech quality was also demonstrated over a portion of the TIMIT database (see Section 13.8).

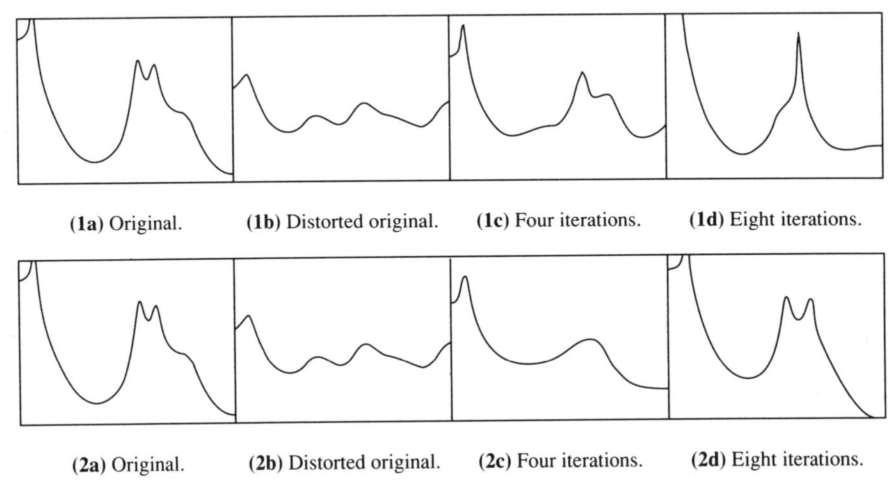

(**1a**) Original. (**1b**) Distorted original. (**1c**) Four iterations. (**1d**) Eight iterations.

(**2a**) Original. (**2b**) Distorted original. (**2c**) Four iterations. (**2d**) Eight iterations.

FIGURE 8.8. Variation in vocal-tract response across iterations for (1a–d) Lim–Oppenheim (1978, 1979) unconstrained enhancement algorithms and (2a–d) Hansen–Clements (1987) constrained enhancement algorithms: (a) original, (b) distorted original, (c) 4 iterations, (d) 8 iterations.

8.4.6 Further Refinements to Iterative Enhancement

All-pole modeling has been shown to be successful in characterizing uncorrupted speech. Techniques have been proposed for estimating all-pole model parameters from noisy observations by Lim and Oppenheim (1978) and Done and Rushforth (1979). Although all-pole modeling of speech has been used in many speech applications, it is known that some sounds are better modeled by a pole–zero system (Flanagan, 1972; Rabiner and Schafer, 1978; O'Shaughnessy, 1988). Musicus and Lim (1979) considered a generalized MAP estimation procedure based on a pole–zero model for speech. Essentially, the procedure requires MAP estimation of the predictor coefficients for both denominator and numerator polynomials, followed by MAP estimation of the noise-free speech through the use of an adaptive Wiener filter. Paliwal and Basu (1987) considered a speech enhancement method based on Kalman filtering. A delayed Kalman filtering method was found to perform better than a traditional Wiener filtering scheme. Another refinement proposed by Gibson et al. (1991) considers scalar and vector Kalman filters in an iterative framework in place of the adaptive Wiener filter for removal of colored noise. Other enhancement techniques based on speech modeling have employed vector quantization and a noisy-based distance metric to determine a more suitable noise-free speech frame for enhancement (Gibson et al., 1988; O'Shaughnessy, 1988). Such methods require a training phase to characterize a speaker's production system. Another speaker-dependent enhancement approach by Ephraim et al. (1988, 1989) employs a hidden Markov model (HMM) to characterize the uncorrupted speech. The parameter set of the HMM is estimated using a K-means clustering algorithm, followed by sequential estimation of the noise-free speech, and HMM state sequences and mixture coefficients. (The HMM is discussed in Chapter 12.) The speech signal estimation process also results in a noncausal Wiener filtering procedure.

The majority of speech-model-based enhancement methods result in iterative procedures. For these methods, a termination criterion is needed. Normally, this is accomplished by listening to successive iterations of processed speech and subjectively determining the iteration with the "best" resulting quality. This knowledge is then used to terminate the procedure at that iteration. Such testing procedures may need to be repeated as noise types or distortion levels vary. Another means of determining the iteration with highest quality is to use objective speech quality measures (Chapter 9).

Finally, we note that noncausal Wiener filtering techniques have also been employed. Lim and Oppenheim (1978) considered such an iterative approach for the AWGN case. Their results showed improvement in speech quality for enhancement at various SNRs. In addition, improvement in all-pole parameter estimation as measured by reduced MSE was also observed. This method was evaluated by Hansen and Clements (1985) for AWGN and slowly varying aircraft cockpit noise. White Gaussian noise results are shown in Table 8.1. This evaluation confirmed that good speech quality can be achieved if the iterative procedure is terminated between three and four iterations. For a colored noise distortion, the method of characterization for the background noise greatly influences enhancement performance. Evaluations with colored noise revealed that a Bartlett spectral estimate (Kay and Marple, 1981) produced higher levels of speech quality compared with other spectral estimation methods (e.g., maximum entropy, Burg, Pisarenko, or periodogram estimates). Further discussion of Wiener filtering performance can be found in Section 8.7.3.

8.4.7 Summary of Speech Modeling and Wiener Filtering

In this section, we have considered speech enhancement techniques that enhance speech by first estimating speech modeling parameters, and then resynthesizing the enhanced speech with the aid of either a noncausal adaptive (over time) Wiener filter or delayed Kalman filter. The techniques differ in how they parameterize the speech model, the criterion for speech enhancement (MSE, MAP estimation, ML estimation, perceptual criteria), and whether they require single- or dual-channel inputs.

8.5 Adaptive Noise Canceling

8.5.1 Introduction

The general technique of *adaptive noise canceling* (ANC) has been applied successfully to a number of problems that include speech, aspects of electrocardiography, elimination of periodic interference, elimination of echoes on long-distance telephone transmission lines, and adaptive antenna theory. The initial work on ANC began in the 1960s. Adaptive noise canceling refers to a class of adaptive enhancement algorithms based on the availability of a primary input source and a secondary reference source. The primary input source is assumed to contain speech plus additive noise,

$$y(n) = s(n) + d_1(n), \tag{8.40}$$

where, as usual, these sequences are realizations of stochastic processes $\underline{y}$, $\underline{s}$, and $\underline{d}_1$. The secondary or reference channel receives an input $d_2(n)$, the realization of a stochastic process $\underline{d}_2$ that may be correlated with $\underline{d}_1$ but not $\underline{s}$ (see Fig. 8.9). All random processes are assumed WSS and appropriately ergodic so that time waveforms can be used in the following analysis.

The adaptive noise canceler consists of an adaptive filter that acts on the reference signal to produce an estimate of the noise, which is then subtracted from the primary input. The overall output of the canceler is used to control any adjustments made to the coefficients of the adaptive filter (often called "tap weights" in this context, see Fig. 8.10). The criterion for adjusting these weights is usually to minimize the mean square energy of the overall output (this might seem odd, but see below). The research area of adaptive filter theory is rich in algorithms and applications. For example, textbooks by Haykin (1991), Messerschmitt (1984), and Proakis et al. (1992) develop an adaptive filter framework and discuss applications in system identification, adaptive channel equalization, adaptive spectral analysis, adaptive detection, echo cancelation, and adaptive beamforming. In this section, we will limit our discussion to the application of ANC for speech enhancement.

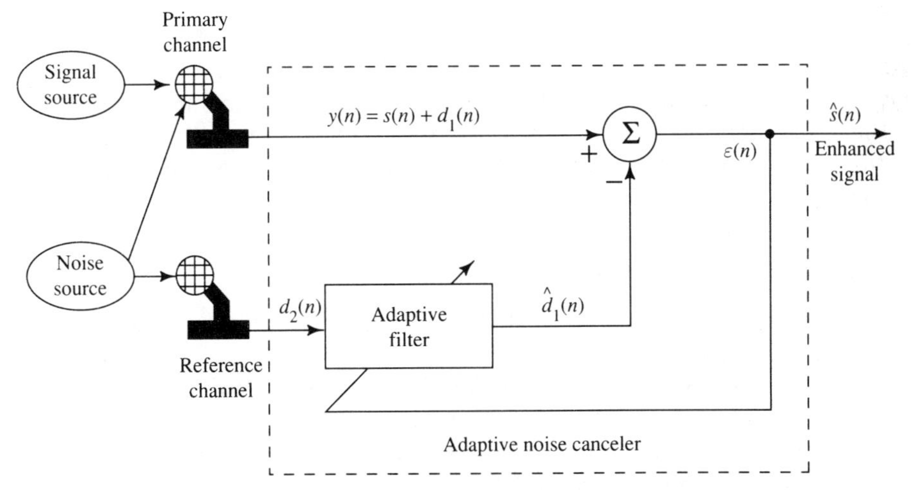

FIGURE 8.9. Flow diagram of adaptive noise canceling.

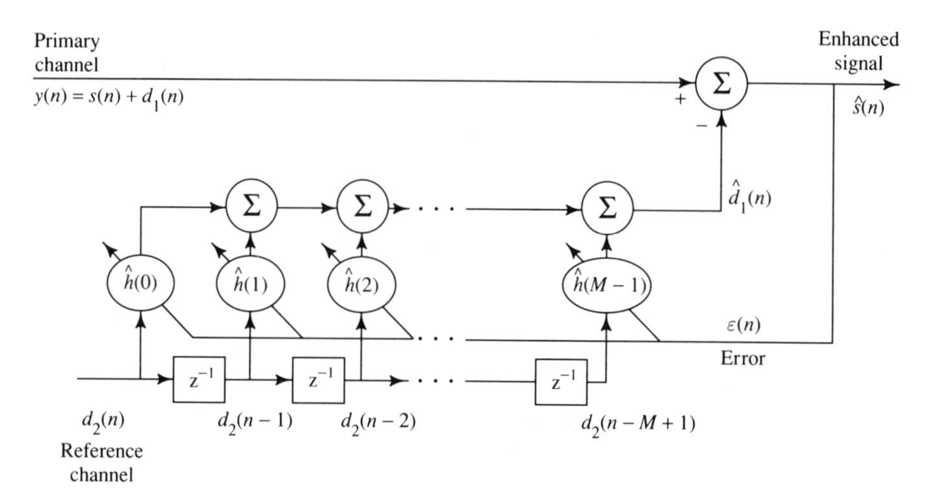

FIGURE 8.10. The LMS adaptive filter.

Most enhancement techniques, such as spectral subtraction and speech-modeling-based approaches, can be generalized to operate in a dual-channel system. However, unlike spectral subtraction and adaptive Wiener filtering, ANC usually *requires* a secondary reference channel. In Section 8.6.2, we discuss a special ANC approach that takes advantage of the periodicity of voiced speech to obviate the second channel.

Initial studies on ANC can be traced to Widrow and his co-workers at Stanford in 1965, and Kelly at Bell Laboratories. In the work by Widrow, an adaptive line enhancer was developed to cancel 60-Hz interference at the output of an electrocardiographic amplifier and recorder. This work was later described in a paper by Widrow et al. (1975). The adaptive line enhancer and its application as an adaptive detector were patented by McCool et al. (1980, 1981) in 1980. The steady-state behavior of the adaptive line enhancer was later studied by Anderson and Satorius (1983) for stationary inputs consisting of finite bandwidth signals embedded in a white Gaussian noise background. Kelly, also in 1965, developed an adaptive filter for echo cancelation that uses the speech signal itself to adapt the filter. This work was later recognized by Sondhi (1967). The echo canceler and its refinements by Sondhi are described in patents by Kelly and Logen (1970) and Sondhi (1970).

8.5.2 ANC Formalities and the LMS Algorithm

The classical approach to dual-channel adaptive filtering, based on a least MSE [the acronym used for "least (or minimum) MSE" in this context is often "LMS"] criterion, was first formulated by Widrow and Hoff (1960, 1975). This technique has the major advantage of requiring no *a priori* knowledge of the noise signal. Figure 8.9 illustrates the basic structure of an adaptive noise canceler.

Our extensive work with LP in Chapter 5 will permit us to get some needed results very quickly. One interesting analytical detail should be pointed out before pursuing these results. In this development it will be entirely sufficient to work exclusively with long-term analysis. Although short-term quantities can be introduced in an obvious place (similar to the transition in the LP developments), this would turn out to be unnecessary here. The reason is that an estimation procedure at the end of the development inherently produces an algorithm that can realistically work in real time frames and in the presence of nonstationary signal dynamics. We will therefore be able to avoid the details of short-term processing in the following discussion with no loss of practical value.

All signals in Fig. 8.9 are assumed to be realizations of WSS stochastic processes with appropriate ergodicity properties so that we may use time waveforms in the analysis. The meaning of each of the signals has been discussed in the introduction. It was explained in the introduction that the objective of the adaptive[14] filter in Fig. 8.9 is to estimate the noise sequence $d_1(n)$ from $d_2(n)$ in order that the noise can be removed from $y(n)$. This seems very reasonable from the diagram. With this interpretation, the output of the noise canceler can be interpreted as an estimate, say $\hat{s}(n)$, of the uncorrupted speech $s(n)$. The filter is FIR with estimated tap weights, say $\hat{h}(i)$, $i = 0, 1, \ldots, M-1$, so that

$$\hat{d}_1(n) = \sum_{i=0}^{M-1} \hat{h}(i) d_2(n-i). \tag{8.41}$$

For convenience, let us define the M-vector of weight estimates

$$\hat{\mathbf{h}} \stackrel{\text{def}}{=} \left[\hat{h}(0) \quad \hat{h}(1) \quad \cdots \quad \hat{h}(M-1)\right]^T. \tag{8.42}$$

Figure 8.10 illustrates the LMS adaptive filter structure.

Now our experience would lead us to discern that a natural optimization criterion is to minimize the MSE between the sequences $d_1(n)$ and $\hat{d}_1(n)$. Unfortunately, the signal $d_1(n)$ is not measurable, so we will be unable to design on this basis. However, a result developed in Appendix 5.B allows us to achieve the same objective from a different viewpoint. It is shown there that attempting to estimate $d_1(n)$ using $d_2(n)$ and a least MSE criterion is equivalent to estimating $d_1(n)$ plus any signal that is orthogonal to $d_2(n)$. In this case, therefore, we may attempt to estimate $y(n)$ from $d_2(n)$ and derive an identical filter to that which would be obtained for estimating $d_1(n)$. It is interesting that in this interpretation the signal $\hat{s}(n)$ is interpreted as an *error* [call it $\varepsilon(n)$], which is to be minimized in mean square. Therefore, the ANC is sometimes described as having been designed by minimizing its output power (or energy in the short-term case).

[14]The filter is not really "adaptive" yet because we are working with a long-term situation in which the relevant properties of all signals are assumed to remain forever stationary.

In keeping with the alternative, but equivalent, optimization criterion, $\hat{\mathbf{h}}$ is chosen such that

$$\begin{aligned}\hat{\mathbf{h}} &= \underset{\mathbf{h}}{\operatorname{argmin}}\, \mathcal{L}\{[y(n) - \hat{y}(n)]^2\} \\ &= \underset{\mathbf{h}}{\operatorname{argmin}}\, \mathcal{L}\left\{\left[y(n) - \sum_{i=0}^{M-1} \hat{h}(i) d_2(n-i)\right]^2\right\}.\end{aligned} \tag{8.43}$$

The result can be derived immediately from (5.241) using the orthogonality principle in Appendix 5.B (using temporal quantities here):

$$r_{yd_2}(\eta) - \sum_{i=0}^{M} \hat{h}(i) r_{d_2}(\eta - i) = 0, \qquad \eta \in [0, M] \tag{8.44}$$

or, in matrix-vector notation similar to previous developments [see (5.23) and (5.24)],[15]

$$\mathbf{R}_{d_2}\hat{\mathbf{h}} = \mathbf{r}_{yd_2}. \tag{8.45}$$

The remaining issue is the solution of (8.45) for the filter tap weights. Indeed, this set of equations differs from those derived for the LP problem only in the presence of cross-correlations in the auxiliary vector on the right side. Accordingly, it possesses all of the symmetry properties of the LP normal equations and can be solved by any of the methods discussed in Chapter 5. However, ANC designs have frequently employed the LMS algorithm, which has been demonstrated to be an effective and practical means for real-time approximation of the filter solution in this application (Koford, 1966; Widrow et al., 1967, 1971). Like the methods in Chapter 5, the LMS algorithm does not require matrix inversion. However, dissimilarly to some of the previous methods, LMS does not require the explicit formulation of a correlation matrix.

Because LMS and its properties have been discussed extensively in the literature [see, e.g., (Proakis et al., 1992; Widrow and Stearns, 1985)], we will only briefly sketch the method here. This development will involve many filter estimates, not just the minimum MSE result of (8.45). Therefore, let us denote a general filter estimate explicitly solving the minimum MSE problem by $\hat{\mathbf{h}}^{\dagger}$. The ith component of $\hat{\mathbf{h}}$ is denoted $\hat{h}(i)$, as in (8.42). Let us denote the MSE for general filter $\hat{\mathbf{h}}$ by $\xi(\hat{\mathbf{h}})$. The *minimum* MSE is therefore $\xi(\hat{\mathbf{h}}^{\dagger})$.

The error $\xi(\hat{\mathbf{h}})$, when considered as a function of the weights as we are doing here, is frequently called an *error surface*. If there are only two

[15]At this point in the development, we could replace the long-term results with short-term estimators relevant to frames of the two observable signals, $f_y(n;m) = y(n)w(n-m)$ and $f_{d_2}(n;m) = d_2(n)w(m-n)$. The work would then proceed similarly to the developments in Section 5.3. We shall see, however, that this will be unnecessary here.

weights, it is possible to imagine a "surface" plotted over the $\hat{h}(0)$–$\hat{h}(1)$ plane. Now, by definition

$$\xi(\hat{\mathbf{h}}) = \mathcal{L}\left\{\left[y(n) - \sum_{i=0}^{M-1} \hat{h}(i) d_2(n-i)\right]^2\right\}. \tag{8.46}$$

Since this expression is a quadratic in $\hat{\mathbf{h}}$, there is a unique minimum (which, of course, occurs at $\hat{\mathbf{h}}^{\dagger}$). Methods for finding the solution discussed thus far have immediately "jumped" to the minimum of the error surface in one step. The LMS algorithm gradually moves toward the minimum by "slowly" moving against the gradient of the error surface. Now from (8.46), the gradient with respect to weight $\hat{h}(\eta)$ is

$$\begin{aligned}\frac{\partial \xi(\hat{\mathbf{h}})}{\partial \hat{h}(\eta)} &= 2\mathcal{L}\left\{y(n) d_2(n-\eta) - \sum_{i=0}^{M-1} \hat{h}(i) d_2(n-i) d_2(n-\eta)\right\} \\ &= 2\left[r_{yd_2}(\eta) - \sum_{i=0}^{M-1} \hat{h}(i) r_{d_2}(\eta - i)\right]\end{aligned} \tag{8.47}$$

or, by differentiating with respect to the entire weight vector at once,

$$\frac{1}{2}\frac{\partial \xi(\hat{\mathbf{h}})}{\partial \hat{\mathbf{h}}} = \mathbf{r}_{yd_2} - \mathbf{R}_{d_2}\hat{\mathbf{h}}. \tag{8.48}$$

If $\hat{\mathbf{h}} = \hat{\mathbf{h}}^{\dagger}$, this gradient will become zero. Now let us denote by $\varepsilon(n)$ the estimation error of $y(n)$, $\varepsilon(n) = y(n) - \hat{y}(n)$. Then it is not difficult to show that the right side of (8.47) is equivalent to an M-vector, say $\mathbf{g}$ (for "gradient"), whose ith component is $2r_{\varepsilon y}(i-1)$. In principle, therefore, we can iteratively compute the filter estimate by moving down the gradient: Let $\hat{\mathbf{h}}^k$ be the estimate from the kth step. Then we take

$$\hat{\mathbf{h}}^k = \hat{\mathbf{h}}^{k-1} - \Delta^k \mathbf{g}, \tag{8.49}$$

where Δ^k, the *step size*, generally varies with k. For the obvious reason, this procedure is called a *steepest-descent algorithm*. This algorithm is guaranteed to converge to $\hat{\mathbf{h}}^{\dagger}$ if $\sum_{k=1}^{\infty} \Delta^k < \infty$ and $\lim_{k\to\infty} \Delta^k = 0$ [see, e.g., (Proakis et al., 1992)].

Unfortunately, the gradient vector $\mathbf{g}$ cannot be computed (i.e., estimated using short-term processing) *a priori* without excessive computational effort, which defeats one of the purposes of using this simple approach. Therefore, we need to estimate $\mathbf{g}$. [To explain clearly, we need to index the iterations, formerly "k," by the time index n (or conversely). There is, of course, no loss of generality in doing so.] At time (iteration) n, we approximate the ith element of $\mathbf{g}$ by $2\varepsilon(n)y(n-i-1)$. From above, we see that this amounts to the approximation

$$r_{\varepsilon y}(i) \approx \varepsilon(n) y(n-i-1). \tag{8.50}$$

Therefore, we have chosen an unbiased estimate of **g**, since the joint wide sense stationarity of $\underline{\varepsilon}$ and $\underline{y}$ follows from original assumptions. The approximation chosen also amounts to using the sample error surfaces due to each incoming point as an estimate of the error surface associated with the entire frame of data. The approximation makes use of the gradient of the MSE, but does not require any squaring or differentiation operations. The resulting recursion is

$$\hat{\mathbf{h}}^n = \hat{\mathbf{h}}^{n-1} - \Delta^n \hat{\mathbf{g}}^n, \tag{8.51}$$

where $\hat{\mathbf{g}}^n$ indicates the estimated gradient associated with time n. In practice, a fixed step size, $\Delta^n = \Delta$, is often used for ease of implementation and to allow for adaptation of the estimate over time as the dynamics of the signal change. This simple algorithm was first proposed by Widrow and Hoff (1960) and is now widely known as the *LMS algorithm.*

Notice that without ever explicitly resorting to short-term analysis, we have an algorithm that is immediately practically applicable. This is because the approximation made in (8.50) is a short-term estimator of the cross-correlation. Because the estimator is very short term (one point), the LMS algorithm has the potential to track time-varying signal dynamics. A discussion of this point is beyond the scope of the present discussion, so we refer the interested reader to one of the cited textbooks.

The convergence, stability, and other properties of LMS have been studied extensively; refer to Widrow and Stearns (1985) or Proakis et al. (1992) for details. It has been shown using long-term analysis (Widrow, 1975) that starting with an arbitrary initial weight vector, the LMS algorithm will converge in the mean and remain stable as long as the following condition on the step-size parameter Δ is satisfied,

$$0 < \Delta < \frac{1}{\lambda_{\max}}, \tag{8.52}$$

where $\lambda_{\max}$ refers to the largest eigenvalue of the matrix $\mathbf{R}_{d_2}$. In practice, the bounds in equation (8.52) are generally modified to ensure a working margin for system stability (Horwitz and Senne, 1981; Tate and Goodyear 1983).

Many alternative approaches for recursive tap filter weight parameter estimation can be found in the literature for controls, system identification, and adaptive filter theory. The interested reader is encouraged to consider texts by Haykin (1991), Messerschmitt (1984), Bellanger (1987), and Proakis et al. (1992).

8.5.3 Applications of ANC

Experimental Research and Development

One of the advantages of dual-channel ANC is that speech with either stationary or nonstationary noise can be processed. In general, the two microphones are required to be sufficiently separated in space, or to contain an acoustic barrier between them to achieve noise cancelation. In this section, we consider several applications of ANC to the problem of enhancing degraded speech.

One of the earlier dual-channel evaluations of ANC for speech was conducted by Boll and Pulsipher (1980). Two adaptive algorithms were investigated: the LMS approach of Widrow et al. (1976) and the gradient lattice approach of Griffiths (1978).[16] Each approach was compared in terms of degree of noise power reduction, algorithm settling time, and degree of speech enhancement. Based on earlier simulation studies (Pulsipher et al., 1979), the typical FIR adaptive filter necessary to estimate the input noise characteristics required 1500 tap weights. Such large filter lengths result in misadjustment, defined by Widrow et al. (1976) as the ratio of excess MSE to minimum MSE. This notion of misadjustment is an important design criterion for dual-channel ANC, since large misadjustment leads to pronounced echo in the resulting speech signal. This occurs because of the adaptive structure of the FIR ANC filter. Fortunately, the echo can be reduced by decreasing the adaptation step size used in updating filter weights, but this increases the settling time of the adaptive filter. Both the LMS and gradient lattice approaches provide comparable noise power reduction. Employing step sizes that correspond to 5% misadjustment, both algorithms converge after 20 seconds of input with a just-noticeable level of echo. The major points from this study suggest that LMS or gradient-lattice-based ANC can provide noise suppression in the time domain, but that a large tap-delay filter is needed. Also, for all of their simulations, Boll and Pulsipher placed the reference microphone directly next to the noise source to eliminate the need for delay estimation caused by noise arriving at each microphone at different time instances.

Although noise cancelation can be achieved using LMS or gradient-lattice ANC, computational requirements become increasingly demanding as adaptive filter lengths grow to as many as 1500 taps. An alternative method of adaptive filtering, based on the complex form of the LMS algorithm, can result in a substantial savings in computation by performing the noise cancelation in the frequency rather than the time domain. The frequency domain LMS adaptive filter is shown in Fig. 8.11. The structure is similar to the conventional time domain filter shown in Fig. 8.10; however input data are frame-processed through input and output N-point FFTs. The filter coefficients are complex and are updated only once per frame using the update equation (Widrow et al., 1975)

[16]The ANC method by Griffiths employs a lattice filter framework, rather than the tap-delay lines (FIR filters) used by the other methods. It has been shown that the successive orthogonalization provided by the lattice offers an adaptive convergence rate that cannot be achieved with tapped-delay lines.

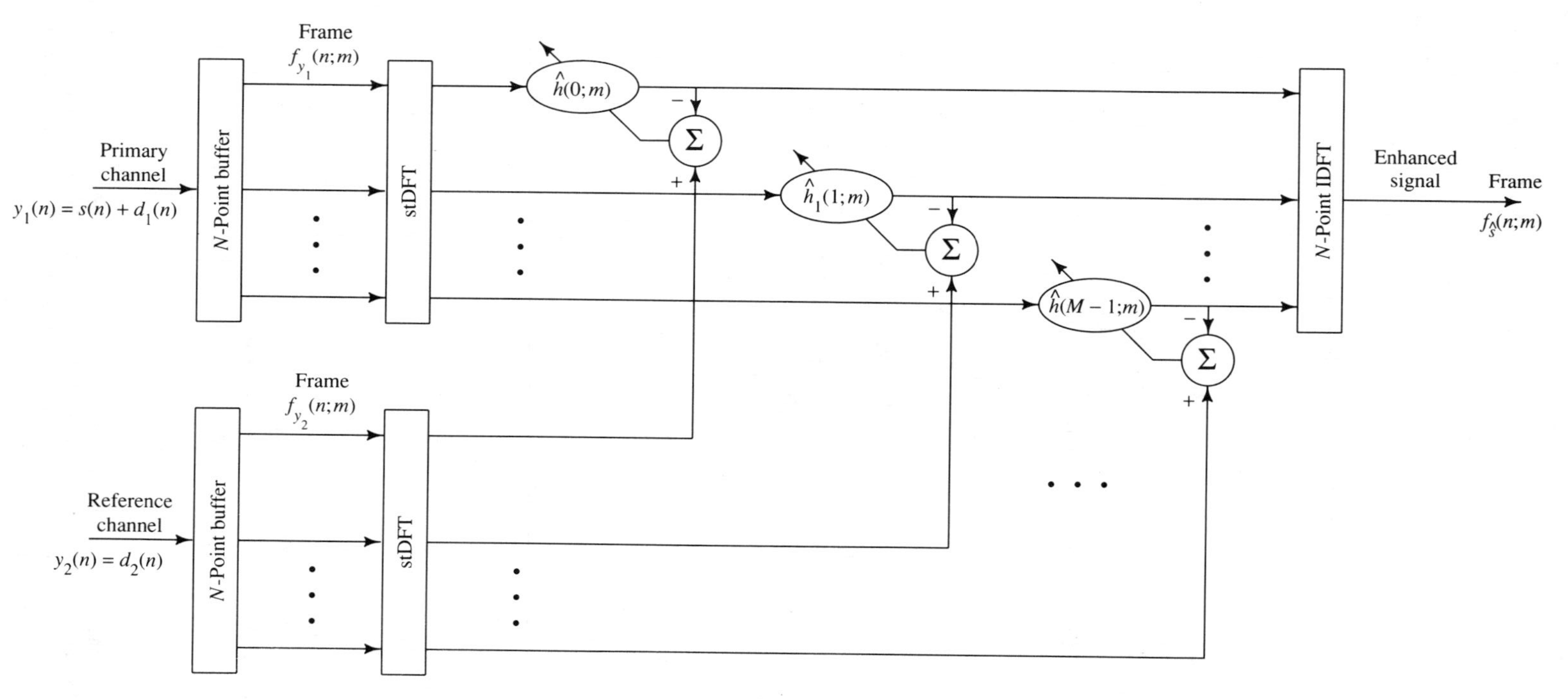

FIGURE 8.11. A frequency domain LMS adaptive filter structure.

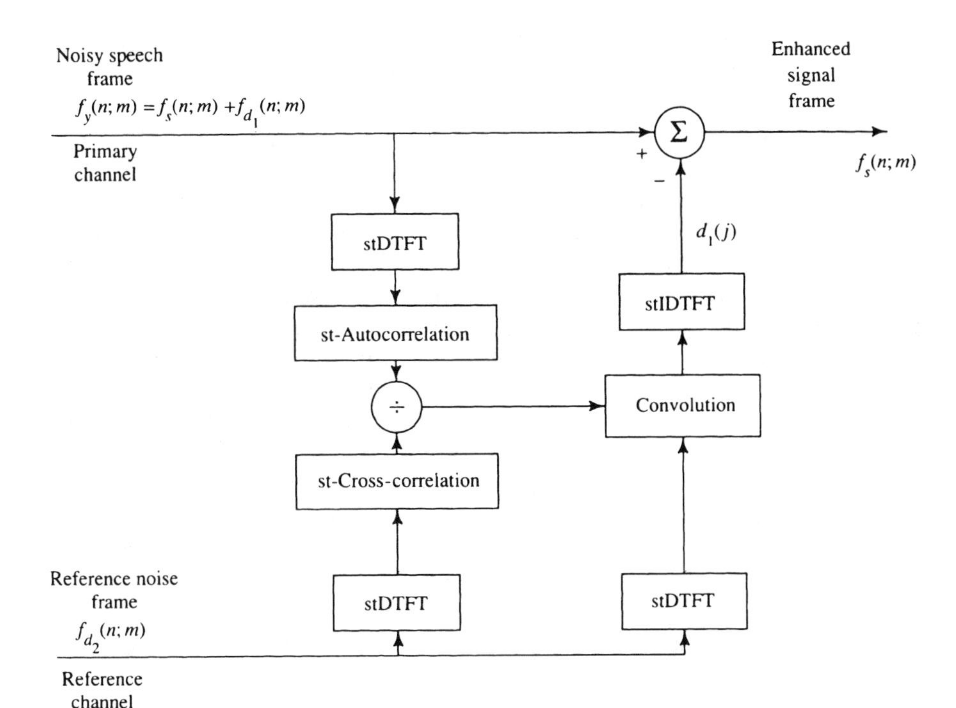

FIGURE 8.12. Adaptive noise cancelation based on the short-term Fourier transform.

$$\hat{\mathbf{h}}(m) = \hat{\mathbf{h}}(m') - \Delta\hat{\mathbf{g}}(m), \tag{8.53}$$

where m' and m index successive frames and $\hat{\mathbf{g}}(m)$ is similar to the long-term vector $\mathbf{g}$ defined above, except its elements are based on short-term cross-correlations over the frame. A comparison of the computational tradeoffs between conventional LMS and frequency domain LMS adaptive filtering is made by Dentino et al. (1978). The ratio of complex multiples required by the frequency domain filter to real multiples required by a conventional LMS filter was found to be

$$\frac{\text{frequency domain complex multiples}}{\text{time domain real multiples}} = \frac{(3M/2)\log_2 M + 2M}{2M^2} = \frac{3\log_2 M + 4}{4M}, \tag{8.54}$$

where M is the number of tap weights. Therefore, if the length required by the time domain LMS adaptive filter exceeds $M = 16$, then the frequency domain approach is computationally superior. In fact, for a typical 1500-tap filter, the computational savings is several orders of magnitude.

An alternative approach for frequency domain ANC can be formulated by explicit estimation of the filter instead of a gradient method such as LMS. Boll (1980) proposed such a method, where the auto and cross-power spectral estimates are used (see Fig. 8.12). Comparable performance to that of LMS and gradient lattice was observed, but with a substantial computational savings. In related work, Reed and Feintuch (1981) considered the statistical behavior of Boll's frequency domain adaptive canceler with white noise inputs. They developed expressions for the mean and variance of the adaptive filter weights, and compared the performance to a time domain canceler. It was shown that the transient responses of both time and frequency domain implementations are the same, but that the inverse transform of the steady-state mean weights of the frequency domain canceler may differ from the steady-state mean weights of the time domain canceler due to frame processing effects of the stDFT. If the signal delay between inputs is small compared with the length of the filter, the steady-state mean weight vector for each canceler is essentially the same. Since time domain approaches can introduce echo in the enhanced speech, such frequency domain adaptive filters serve as a computationally efficient alternative.

Although our discussion of ANC has focused on the dual-channel framework, these systems can be extended to higher dimensions. For example, Ferrara and Widrow (1981) considered the performance of the adaptive line enhancer of Widrow et al. (1975) for enhancing a signal in additive noise with up to I inputs. Although random pulse waveforms were used in place of speech, it is clear that such a method could be generalized for a speech application. It was determined that an $I + 1$ channel ANC was shown to be exactly equivalent in performance to a two-channel ANC whose input SNR is the sum of the SNRs of the I individual adaptive filter inputs. The output SNR density is approximately equal to the sum of the SNR densities of the I individual adaptive filter inputs. As a result, the multichannel adaptive signal enhancer is a generalization of the classic time-delay-and-sum beamforming antenna. Anderson et al. (1983) further evaluated the adaptive line enhancer for finite bandwidth signals embedded in a white Gaussian noise background. Analytical expressions for the weights and output of the LMS adaptive filter were derived as functions of input signal bandwidth and SNR, as well as filter length and bulk delay. Anderson et al. show that there is an optimal filter length whose value depends upon the input signal bandwidth and SNR, for which the broadband gain is maximized.

Adaptive noise canceling has been used in a variety of areas dealing with noisy signals. For example, a general form of ANC can be derived using a signal modeling approach for the purposes of system identification (Friedlander, 1982). Dual-channel ANC has also been used successfully to improve communication for persons with hearing disabilities. In a study by Chabries et al. (1982), the LMS-based ANC was evaluated for hearing-impaired subjects with SNRs ranging from −8 to +12 dB. The interference had the shape of a speech spectrum. At SNRs of −8, −4, and

0 dB, intelligibility as measured using phonetically balanced word lists increased from 0% unprocessing to 30–50% after processing. For SNRs of 4–12 dB, intelligibility increased from a range of 24–46% to 67–75%, thereby suggesting its usefulness for listeners with hearing disabilities.

Dual-Channel ANC in Aircraft Cockpits

One problem with the ANC algorithm is the need for the reference microphone to be well separated (at least acoustically if not physically) from the primary microphone, so that it picks up as little speech as possible. This must be true so that the algorithm does not cancel the speech instead of the noise. In the previously cited study by Boll and Pulsipher (1980), the reference microphone was placed directly next to the noise source, and the primary was placed near the weak signal source but as far as possible from the reference microphone. This satisfies the ANC constraint of a high SNR in the primary channel and a low SNR in the reference. However, it is not always possible to place the reference microphone near the noise source. In fact, for many applications there are multiple noise sources that contribute to noise degradation. In addition, if the two microphones are separated by even a few meters, the adaptive filter must be able to either estimate the delay between the reference and primary signals, or have a long impulse response in order to provide good noise cancelation. This is the cause of the long filter responses often required for successful noise cancelation. As we have noted, however, longer filter lengths require increased computation and tend to introduce reverberation into the processed speech.

An important application of dual-channel ANC in which large microphone spacing is not an issue is in aircraft cockpit environments. This has received considerable interest as a means of improving the performance of existing communication systems: In this case, the pilot's oxygen facemask serves as an acoustic barrier between the two sensors, thereby ensuring that the SNR of the primary sensor be much greater than the SNR of the reference sensor, while permitting close sensor spacing.

Many aspects of the cockpit noise problem have been studied. The interested reader is referred to papers by Harrison et al. (1984, 1986), Darlington et al. (1985), Powell et al. (1987), and Rodriguez et al. (1987).

Cross Talk Within Dual-Channel ANC

In the foregoing discussion of ANC, we enforced a requirement that the primary and reference channels be well separated either physically or by virtue of an acoustic barrier. If the microphones are too close to one another, *cross talk* occurs. A typical adaptive filter will thereby suppress a portion of the input speech characteristics. One means of addressing this problem is to place a second adaptive filter in the feedback loop. Consider the primary $y_1(n)$ and reference $y_2(n)$ channel signals

$$y_1(n) = s_1(n) + d_1(n) \tag{8.55}$$

$$y_2(n) = s_2(n) + d_2(n). \tag{8.56}$$

Here, the primary channel $y_1(n)$ contains speech and a degrading noise component. Due to the close proximity of the secondary microphone, the reference channel contains a low-level speech signal $s_2(n)$ in high-level noise. We assume that the SNR ratio of the primary channel is higher than that for the reference ($\mathrm{SNR}_1 > \mathrm{SNR}_2$). Under such conditions, the low-level speech represents "interference" in the desired noise reference. If a speech reference can be found, then a second adaptive filter can be used to cancel the speech interference in the reference channel. This in turn results in an improved noise reference in which to filter the primary channel. Such a framework for suppressing cross talk using a dual adaptive filter feedback loop is shown in Fig. 8.13. We assume a speech reference $s_{\mathrm{ref}}(n)$ exists as input to adaptive filter I. The tap weights of adaptive filter I are adjusted to produce the best estimate of the low-level speech interference $s_2(n)$ in the MSE sense. This can be done using the standard LMS algorithm or other gradient-descent methods. The estimate $\hat{s}_2(n)$ is subtracted from $y_2(n)$ to produce the estimate $\hat{d}_2(n)$, which is also the MSE estimate assuming uncorrelated speech and noise. The estimated noise signal $\hat{d}_2(n)$ is now used as input to adaptive filter II. A second set of tap weights is adjusted to produce the best estimate of the primary interference $d_1(n)$. The estimate is subtracted from the primary input, resulting in the estimated speech signal $\hat{s}_1(n)$. This represents the enhanced signal as well as the reference for adaptive filter I.

Such a method was considered by Zinser et al. (1985) and Mirchandani et al. (1986) for speech spoken in a helicopter background noise environment. Their evaluations showed increases in SNR in the range 9–11 dB. An increase in intelligibility as measured by the diagnostic

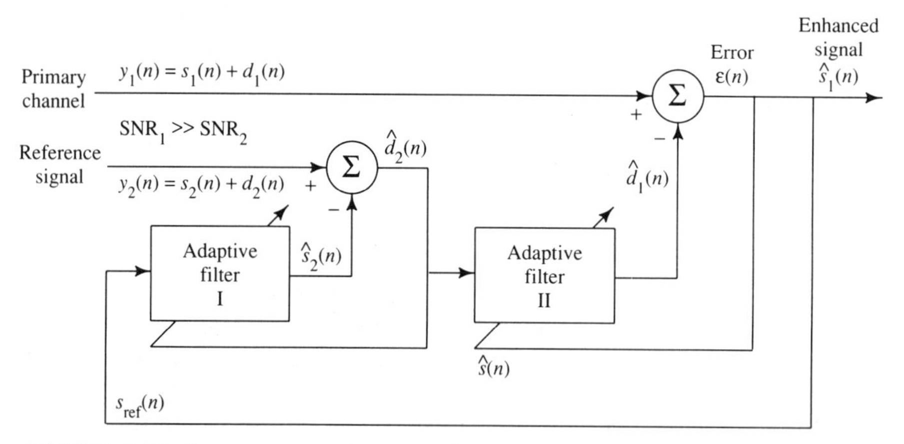

FIGURE 8.13. Suppression of cross talk employing two adaptive noise filters within a feedback loop.

rhyme test[17] of 21 points was also obtained. However, considerable variation in performance was observed using a real-time implementation with different microphones. Implementation issues as well as microphone placement can affect ANC performance. These results do suggest a good potential for effective noise canceling in high-noise environments.

8.5.4 Summary of ANC Methods

In this section, we have considered several methods of dual-channel ANC for enhancing noisy speech. Earlier evaluations that placed the reference microphone directly next to the noise source revealed promising noise cancelation performance. However, further studies that focus on the coherence of the primary and reference microphones suggest that such performance may not be achievable in actual environments. Computational issues from direct-form lattice to gradient-descent techniques based on the LMS algorithm have suggested a variety of ANC implementations. Time versus frequency domain formulations have also been discussed. We found that time domain approaches lend themselves to real-time implementation, but require close microphone placement to avoid long filter lengths and the introduction of echo in the processed speech. Frequency domain approaches offer significant reductions in computational requirements and offer equivalent levels of performance. Finally, while some enhancement methods require accurate characterization or estimation of the noisy speech characteristics, ANC requires a noise reference with no *a priori* knowledge of the input speech characteristics. In high-noise environments, where estimation of such *a priori* knowledge may not be sufficiently accurate, ANC offers a viable means for speech enhancement.

Finally, there is another solution to the problems caused by distant microphone spacing, which has been treated briefly above. This is to deliberately place the microphones as close as possible and to address the issue of speech signal cross talk directly. This subject is discussed further in Appendix 8.B.

8.6 Systems Based on Fundamental Frequency Tracking

8.6.1 Introduction

In this section, we discuss enhancement techniques that are based on tracking the fundamental frequency contour. Such approaches include single-channel ANC, adaptive comb filtering, and enhancement based on harmonic selection or scaling. These techniques capitalize on the prop-

[17]See Chapter 9 for a complete discussion of the diagnostic rhyme test and other intelligibility tests.

erty that waveforms during voiced passages are periodic. This periodicity ideally results in a line spectrum in the frequency domain. Any spectral components between these lines represent noise that can be reduced. Three approaches will be considered. In Section 8.6.2, we discuss the use of single-channel ANC techniques to reduce background noise by capitalizing on the periodicity of voiced speech sections to produce a reference input. In Section 8.6.3, we discuss a method based on comb filtering that passes harmonics of the desired speech and rejects frequency components between harmonics. In Section 8.6.4, we discuss the use of a high-resolution spectrum of noisy speech for scaling and suppression. One useful application area for these techniques has been the competing speaker problem, where the enhancement takes advantage of differences in fundamental frequency contours.

8.6.2 Single-Channel ANC

In the preceding section, we addressed dual-channel ANC. We saw that ANC employing the LMS algorithm requires no *a priori* knowledge of the noise signal, and that classical adaptive filtering by Widrow et al. (1975) assumes a primary channel composed of desired speech plus an uncorrelated noise signal and a second reference signal consisting of noise correlated with noise in the primary channel. Generally speaking, ANC can only be employed when a second channel is available. Suppose, however, that we could *simulate* a reference using data from the primary channel. Under these conditions, traditional ANC can be applied. Sambur (1978) proposed such an approach where, instead of canceling noise in the primary channel, the speech signal is canceled.

In dual-channel ANC, the success of the adaptive filter depends on the availability of a good noise reference input that is free of cross-talk. In most speech enhancement applications, a reference noise channel is not available; therefore, many enhancement techniques must estimate noise characteristics during periods of silence (periods between speech activity) and assume the noise characteristics to be stationary during speech activity. Extracting a noise reference from the input has some disadvantages, including

1. Possible nonstationarity of the noise.
2. A lack of data with which to estimate the noise signal.
3. Silent/noise decision is not error-free.
4. Nonapplicability for some types of distortion that arise from quantization noise.

Although it may be difficult to form a noise reference channel, it is not difficult to obtain a speech reference channel for some classes of speech. Due to the quasi-periodic nature of speech during voiced sections, a reference signal can be formed by delaying the primary data by one or two pitch periods. This reference signal can then be used in the LMS adaptive algorithm, where the criterion for the algorithm is to form the minimum MSE estimate of the uncorrupted speech signal. Consider the noisy speech signal $y_1(n)$,[18]

$$y_1(n) = s(n) + d(n), \tag{8.57}$$

where $s(n)$ is the desired speech and $d(n)$ a realization of the noise. Now, let the reference signal $y_2(n)$ be a delayed version of the primary, $y_2(n) = y_1(n - T_o)$, where T_o represents one pitch period delay in norm-sec. Then under ideal periodic speech conditions, we have

$$y_2(n) = s(n - T_o) + d(n - T_o) = s(n) + d(n - T_o). \tag{8.58}$$

The delayed speech signal $s(n - T_o)$ will be highly correlated with the original speech $s(n)$ (under ideal periodic conditions, they will be perfectly correlated), while the delayed $d(n - T_o)$ and original $d(n)$ noise signals will have low correlation with the speech signal and presumably themselves.[19] This represents the basis for the ANC technique proposed by Sambur (1978). A block diagram of the enhancement system is shown in Fig. 8.14. The primary output is the "enhanced" noise signal $\hat{d}(n)$, but an enhanced speech signal output $\hat{s}(n)$ is also available. The structure of the adaptive filter is illustrated in Fig. 8.15. The FIR filter produces the following output:

$$\hat{s}(n) = \sum_{i=0}^{M-1} \hat{h}(i) y_1(n - T_o - i), \tag{8.59}$$

where $\hat{h}(i)$, $i = 0, \ldots, M-1$, are the FIR filter weights identified in a similar way to methods described in Section 8.5.2.

Sambur investigated this approach for additive white noise and quantization noise. The pitch period was estimated using an average magnitude difference function (Section 4.3.2) and nonlinear smoothing

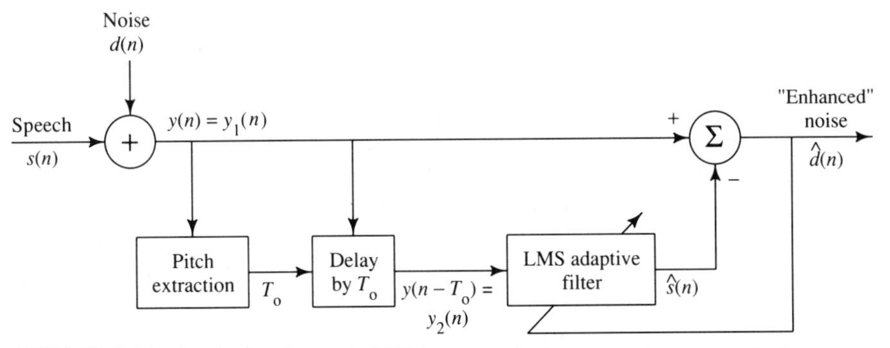

FIGURE 8.14. A single-channel ANC approach for removing speech from noisy speech. Enhanced speech is also a by-product.

[18]We will ignore the details of short-term processing here, since only the familiar problem of pitch estimation requires frame processing.

[19]Of course, this is only approximately true, since the noise process might be nonstationary.

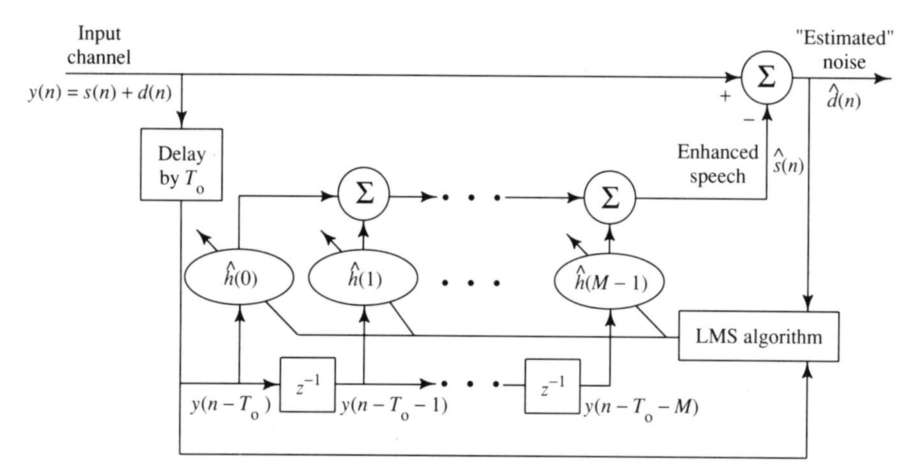

FIGURE 8.15. A single-channel LMS adaptive filter based on fundamental frequency.

(Rabiner et al., 1975). Since this method exploits the periodicity of the input signal, in principle it should only be applied for voiced speech. For unvoiced sections, one of two procedures may be applied. The first approach is to pass the noisy unvoiced speech through the system unprocessed; the second method is to keep the LMS filter response constant and process the unvoiced speech.

In Sambur's work, this approach is shown to improve quality for additive white noise in the SNR range 0–10 dB, with higher levels of improvement as the severity of degrading noise increases. Improved SNR resulted as the LMS filter length M was increased from 6 to 14, especially for lower initial SNRs. Figure 8.16 shows improvements in SNR for varying filter lengths [from equation (8.59)]. It was observed that the more severe the noise, the more dramatic the improvement in SNR. Subjective evaluations were also performed. Listeners concluded that the speech was more pleasant to listen to and "appeared" to have more intelligibility, although no formal tests were performed to determine the level of intelligibility before and after processing. Performance in the presence of quantization noise from a variable-rate delta modulation system was also determined. The LMS adaptive filter removed some of the "granular" quality of the quantized speech. This degradation possesses two types of noise: slope overload (step size too small) and granular noise (hunting due to a too-large step size). ANC removes the granular noise, since it is signal independent and broadband, but leaves slope overload noise unaffected, since it is signal dependent. Sambur also considered this scheme for an LP analysis/synthesis system and found improved all-pole parameter estimation, especially at low SNR.

Although single-channel ANC has been formulated in the time domain, a frequency domain generalization is also possible, following Boll's (1980) dual-channel frequency domain ANC approach.

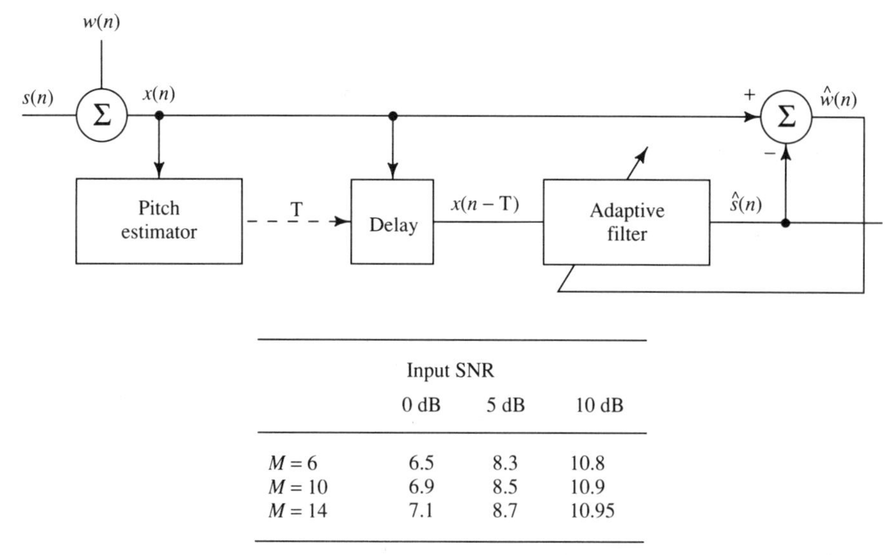

	Input SNR		
	0 dB	5 dB	10 dB
$M = 6$	6.5	8.3	10.8
$M = 10$	6.9	8.5	10.9
$M = 14$	7.1	8.7	10.95

FIGURE 8.16. Improvement in SNR for a single-channel ANC method using the fundamental frequency contour. Adapted from Sambur (1978a).

One of the main limitations of single-channel ANC is the requirement of accurate pitch estimation. Sambur's method was modified by Varner et al. (1983) by removing the pitch estimator and obtaining a reference signal through the use of a low-order DPCM adaptive predictor (model order three). The resulting reference signal contains correlated speech plus uncorrelated noise, which is filtered using an LMS adaptive filter. No quality or intelligibility results were reported, but experiments involving waveforms representing steady-state vowels showed improvement. Finally, another approach (Kim and Un, 1986) attempts to remove the pitch estimator by developing an ANC using both forward- and backward-adaptive filters. The method requires a speech/silence discriminator for narrowband noisy speech and obtains similar levels of enhancement to those resulting in Sambur's work.

8.6.3 Adaptive Comb Filtering

Corrupting noise can take many forms. In some applications speech is degraded by an underlying process that is periodic, resulting in a noise spectrum that also possesses periodic structure. Two methods are available for reducing such noise: *adaptive comb filtering* (ACF) (Lim, 1979) and *time domain harmonic scaling* (TDHS) (Malah et al., 1979). Time domain harmonic scaling will be addressed in the next subsection.

Adaptive comb filtering is similar in its basic assumptions to single-channel LMS-based ANC. Since voiced speech is quasi-periodic, its magnitude spectrum contains a harmonic structure. If the noise is non-periodic, its energy will be distributed throughout the spectrum. The es-

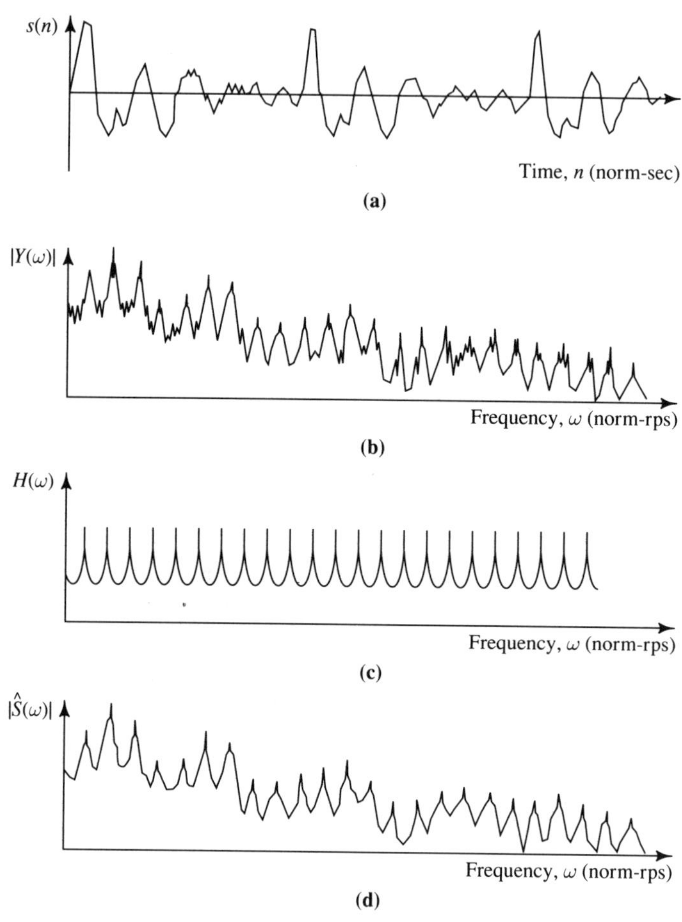

FIGURE 8.17. An example of adaptive comb filtering. (a) A time waveform of a typical section of voiced speech. (b) The amplitude spectrum of noise-corrupted voiced speech. (c) Frequency response of a typical comb filter. (d) Resulting voiced speech spectrum after comb filtering.

sence of comb filtering is to build a filter that passes the harmonics of speech while rejecting noise frequency components between the harmonics. The technique is best explained by considering Fig. 8.17. In Fig. 8.17(a), a periodic time waveform is shown. The noise-corrupted magnitude spectrum of this waveform is displayed in Fig. 8.17(b). The magnitude spectrum indicates that the energy of the periodic signal is concentrated in small energy bands, but that the corrupting noise (white Gaussian in this case) is broadband and distributed throughout the frequency band. It is clear that the magnitude response contains enough energy to allow the accurate estimation of the fundamental frequency component. A filter can be implemented that passes the fundamental frequency plus harmonics while rejecting frequency components between harmonics. Ideally, spacing between each "tooth" in the comb filter should correspond to the fundamental frequency F_o in Hz, and should remain constant throughout the voiced section of speech. Unfortunately, speakers normally vary their pitch and therefore require the comb filter to adapt as data are processed.

A typical block diagram for an adaptive comb filter is shown in Fig. 8.18. The comb filter has large values at the specified fundamental frequency F_o and its harmonics, and low values between. The filter is usually implemented in the time domain as

$$\hat{s}(n) = \sum_{i=-L}^{L} c(i) y(n - iT_o), \tag{8.60}$$

where $c(i)$, $-L \leq i \leq L$, are the $2L+1$ comb filter coefficients (which are positive and symmetrical), $y(n)$ the original noisy speech, T_o the fundamental period in samples, and L a small constant (typically 1–6), which represents the number of pitch periods used forward and backward in time for the filtering process.

Since a comb filter can only be used to enhance noisy voiced speech, a method must be available with which to handle unvoiced speech or silence sections. Two approaches are typical. First, the comb filter can be turned off by setting $c(k) = 0$ for all $k \neq 0$. This has the effect of passing the unvoiced speech through the filter unprocessed. Figure 8.18 shows that a scaling term is used for the unvoiced (or silence) data path. The scaling term (which is typically in the range of 0.3–0.6) is necessary because applying an ACF to voiced sounds reduces the noise energy present. Failure to apply attenuation in unvoiced or silence sections results in unnatural emphasis of unvoiced speech sounds with respect to voiced sounds. The second method for processing unvoiced speech is to maintain a constant set of filter coefficients, obtained from the last voiced speech frame, and process the unvoiced sounds or silence as if they were voiced. This technique has not been as successful as the first.

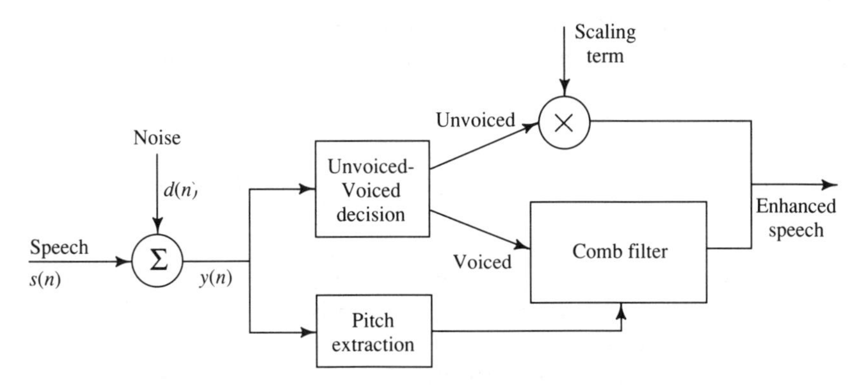

FIGURE 8.18. Block diagram of a typical adaptive comb filter.

The filter of (8.60) is typically implemented in the time domain, whereby the output is an average of delayed and weighted versions of the noisy input. Choosing the delay to correspond to the fundamental period in $y(n)$ results in an averaging process that strengthens all components with that period, and attenuates or cancels others that have no period or a period different from the original fundamental. Clearly, the success of this process is inherently dependent on the accuracy of the estimate of F_o for the desired signal. Certainly, the performance of the comb filter is best when the pitch does not change over the analysis window [e.g., $(2LT_o) + 1$ samples]. The estimation of F_o can in fact be a difficult task, since speech signals often vary from one pitch period to the next. The problem of F_o changing within an analysis window can be approached with the following modification to (8.60):

$$\hat{s}(n) = \sum_{i=-L}^{L} c(i) y(n - iT_o + \zeta_i). \tag{8.61}$$

Here individual pitch period durations are required over the entire data analysis window. These measurements are then used as the timing adjustments ζ_i to align adjacent pitch periods. As the number of pitch periods used in the comb filter, L, increases, so does the need to include the alignment factors ζ_i. Therefore, simple ACFs set $\zeta_i = 0$ and keep the number of pitch periods used for filtering to a minimum (e.g., $L = 1$, resulting in three periods). More sophisticated approaches incorporate larger numbers of pitch periods (e.g., $L = 6$, resulting in 13 periods), and perform pitch alignment using ζ_i timing terms. It is desirable to include as many periods as possible, since the number L is inversely proportional to the bandwidth of each tooth in the comb filter. Larger values of L, therefore, produce more narrow harmonics for the filter and allow for further noise removal. However, due to the short-term stationary property of speech, the number of pitch periods must not represent more than a 30–40-msec duration, thereby limiting the range of L. It has been shown (Lim et al., 1978) that although small increases in SNR occur as the filter length in pitch periods increases from 3 to 7 to 13, intelligibility scores decrease.

Malah and Cox (1982) proposed a generalized comb filtering technique that applies a time-varying weight to each pitch period. The method is similar to normal comb filtering, except that the weights assigned to samples across the pitch period are not fixed. The generalized comb filter has been shown to reduce frame-rate noise for an adaptive transform coder (Cox and Malah, 1981). The method has also shown promise in smearing the structure of simulated periodic interference. Malah and Cox also reported some preliminary experiments for a competing speaker problem (one male and one female). Their initial findings suggest that a tandem system of a fixed comb and generalized comb filters are more effective than a cascade of two fixed comb filters.

Finally, an ACF technique introduced by Frazier et al. (1976) formulates a filter that adjusts itself both globally and locally to the time-varying nature of speech. They showed that classical comb filtering distorts the speech somewhat, but that an approach which adapts itself both globally and locally to the time-varying nature of speech improves performance. Lim et al. (1978), using wideband random noise, and Perlmutter et al. (1977), using a competing speaker, evaluated this adaptive technique for varying filter lengths. Nonsense sentences and exact pitch information (obtained from the noise-free speech) were used in both evaluations. Figure 8.19 illustrates intelligibility scores for both distortions. In both cases, pitch information was obtained from noise-free speech. The competing speaker problem resulted in a decrease in intelligibility. Also, decreases in intelligibility were usually observed for various SNR for wideband random noise. In general, it is not realistic to assume accurate pitch information at low SNR, so intelligibility scores should be lower. Even with decreases in intelligibility, both studies mention that processed speech sounded "less noisy" due to the system's ability to increase the local SNR. No quality tests were performed to verify this hypothesis, however.

8.6.4 Harmonic Selection

If the degrading noise source is a competing speaker, then an enhancement technique similar to comb filtering can be formulated in which spectral harmonics of each speaker are separated based on external pitch estimates. Parsons (1976) proposed such a method in which a high-resolution, short-term spectrum is used to separate competing speakers. All processing is performed in the frequency domain. Speech is generated based on that portion of the spectral content which corresponds to the primary speaker.

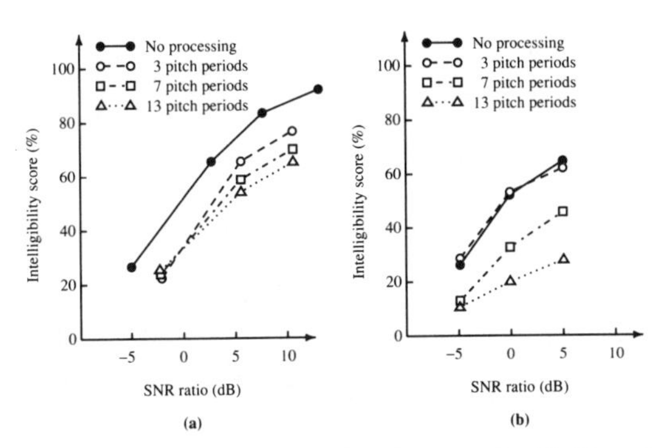

FIGURE 8.19. Intelligibility results for adaptive comb filtering for (a) a competing speaker problem (Perlmutter et al., 1977), and (b) wideband random noise (Lim et al., 1978).

An alternative to frequency domain harmonic selection is *time domain harmonic scaling* (TDHS). Time domain harmonic scaling is a time domain technique that requires pitch-synchronous block decimation and interpolation. The difference between TDHS noise reduction and ACF is that TDHS moves the noise under the pitch harmonics (i.e., masking the background noise), while comb filtering seeks to filter out the noise in the gaps between harmonics. To illustrate the method, consider Fig. 8.20(a), which illustrates the result of a 2:1 decimation procedure for two successive pitch periods to form a single pitch period output. The basic TDHS idea is quite simple. For 2:1 decimation, two input points separated by one pitch period are linearly combined to produce each output point. Window weights are chosen to ensure continuity across the speech waveform. Figure 8.20(b) shows the result of 1:2 interpolation. Again, fundamental frequency, and more specifically accurate pitch period location, is needed to synchronize the windowing operations.

We can also consider the frequency domain consequences of the above decimation and interpolation operations. The spectral characteristics of an ideal voiced section are shown in Fig. 8.21(a). As we have seen in comb filtering, the frequency domain structure appears as "teeth" with an overall spectral shape. Block decimation results in reducing the gaps between these teeth, while block interpolation increases the gap space. The choice of the time domain window will determine the shape of the resulting "bandpass" filter around each pitch tooth. Figure 8.21(b) and (c) illustrates frequency domain consequences of decimation and interpolation.

Time domain harmonic scaling was originally proposed by Malah (1979) for use in perceptually reducing periodic noise in speech. In a later study, Cox and Malah (1981) proposed a hybrid system that uses both ACF and TDHS. An additional benefit of their system is time-scale reduction of input speech for waveform coding and isolated word recognition. Due to its time domain implementation, the choice of an appro-

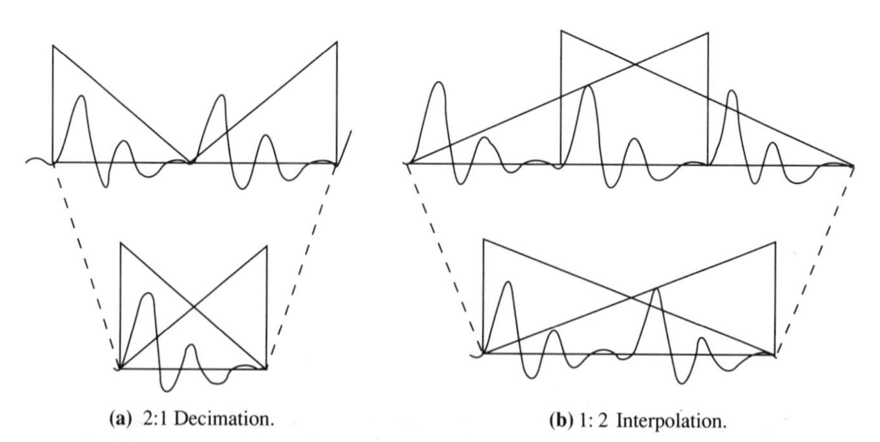

FIGURE 8.20. (a) Decimation and (b) interpolation operations required by TDHS (Malah, 1979; and Cox and Malah, 1981).

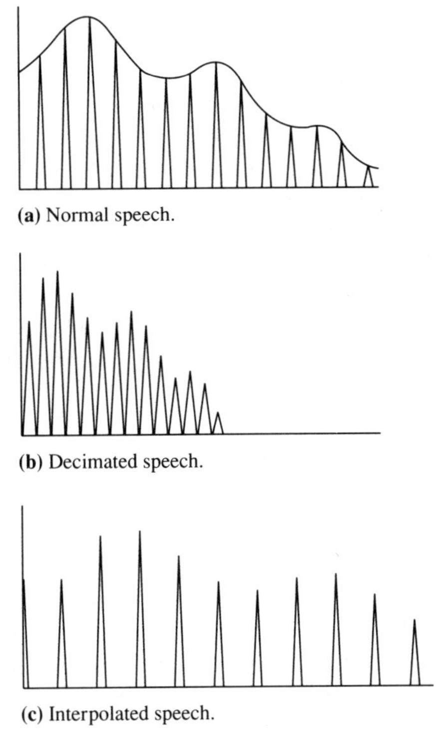

(a) Normal speech.

(b) Decimated speech.

(c) Interpolated speech.

FIGURE 8.21. Frequency interpretation of decimation and interpolation operations required by TDHS.

priate window greatly influences noise cancelation performance. The hybrid system consists of an expansion of the speech spectrum using TDHS, followed by a contraction also using TDHS. This expansion–contraction method does not reduce interference to the same extent as simple TDHS; however, the resulting speech signal is less distorted. In effect, the expansion–compression operation works like a time-varying ACF with the primary difference being that, unlike comb filtering, this technique is guaranteed to be continuous even if pitch characteristics change. Results from adaptive transform coding show that the hybrid enhancement approach can improve speech quality and reduce data transmission rates.

8.6.5 Summary of Systems Based on Fundamental Frequency Tracking

In this section, we have considered techniques based on tracking the fundamental frequency contour. These methods include single-channel ANC, ACF, and enhancement based on harmonic selection or scaling. These techniques exploit the periodicity of the speech signal during voiced passages. Single-channel ANC, ACF, and harmonic scaling all focus on enhancing voiced speech. Therefore, they generally perform poorly for unvoiced speech sections. It is generally known that consonants possess a disproportionate amount of linguistic information when compared with vowels. Since vowels (voiced speech) usually possess larger amounts of energy, broadband noise degradation tends to mask unvoiced sections more than voiced, thus causing decreased intelligibility. Employing a technique that attempts to improve quality in voiced sections may in fact decrease overall speech quality and/or intelligibility. For this reason, these methods are not normally used to attenuate broadband additive noise. Instead, their main area of successful application has been in reducing the effects of a competing speaker, where distinct fundamental frequency contours can be identified (e.g., competing male and female speakers).

8.7 Performance Evaluation

8.7.1 Introduction

A unified performance evaluation of the four areas of speech enhancement considered in this chapter would be a difficult task because of the differences in the assumptions and applications of the varying enhancement algorithms. An algorithm that reduces broadband additive noise may not be appropriate for the competing speaker problem. Generally speaking, a comparative evaluation is only valid if the same test conditions are maintained (same input speech, noise, and quality/intelligibility evaluation methods). At the outset of this chapter, we identified three possible goals for speech enhancement: (1) improving quality, (2) improving intelligibility, and (3) reducing listener fatigue. Since these improvement criteria are based on aspects of human perception, it is necessary to understand how distortion is perceived in noisy speech applications. A careful system evaluation will therefore require the use of either subjective or objective measures of speech quality. In this section, we summarize some performance evaluations of speech enhancement algorithms. In Section 8.7.2 we discuss aspects of perceptual speech enhancement based on various types of highpass, bandpass, or lowpass filtering and/or clipping of the speech waveform. These methods represent operations that greatly influence the intelligibility of speech. In Section 8.7.3, we discuss performance for each of the four classes of enhancement.

8.7.2 Enhancement and Perceptual Aspects of Speech

It is well known that accurate models of speech production in noise can lead to improved speech quality. However, in addition to production aspects, there are also auditory perception aspects that can be exploited in speech enhancement. These are not as well understood as production aspects; however, there are a number of commonly accepted features that contribute to the success of enhancement systems. It is known, for example, that consonants play an important role in speech intelligibility even though they represent a small fraction of the overall signal energy. It is also known that the short-term spectrum is important for speech perception. Formant location is the most important feature, with bandwidth, amplitude, or spectral slope as secondary features. In addition, the second formant is more important perceptually than the first formant (Thomas, 1968; Agrawal and Lin, 1975).

Licklider and Pollack (1948), Thomas (1968), Thomas and Niederjohn (1970, 1972), Thomas and Ohley (1972), and Niederjohn and Grotelueshen (1970a, 1970b, 1970c, 1976) examined the effects of frequency distortion (highpass, bandpass, and lowpass filtering) and amplitude distortion based on infinite-peak clipping on the intelligibility of speech prior to noise degradation. These methods are applicable where the noise-free speech is available for processing, prior to the introduction of noise (e.g., this might occur in voice communication systems). Their results show that removing the first formant (highpass filtering) followed by infinite-peak clipping can increase intelligibility, but with devastating effects on quality (an example of their intelligibility performance is shown in Fig. 8.22). This procedure increases the energy of high-frequency components (most notably the consonants) while decreasing the low-energy frequency components (most notably the vowels). Since this increases the relative SNR, the final noise degradation does not affect the consonants as much without processing. These experiments show that

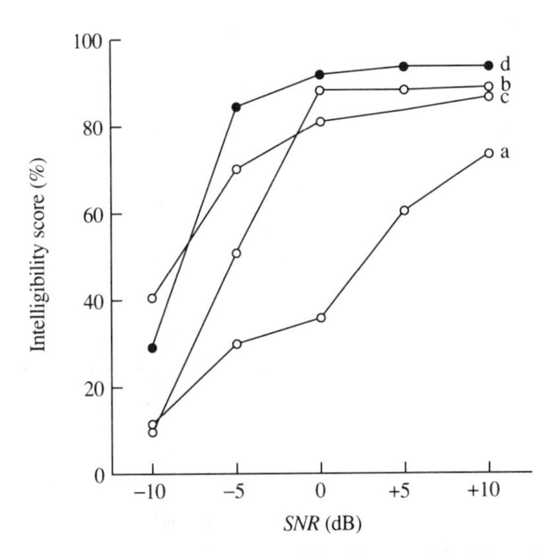

FIGURE 8.22. Highpass filtering and amplitude modification for intelligibility enhancement. Noise level is 90 dB SPL (sound pressure level). Results are shown for (a) normal speech in noise, (b) highpass filtered and clipped speech in noise, (c) highpass filtered speech in noise, and (d) highpass filtered speech with amplitude compression in noise. Adapted from Niederjohn and Grotelueschen (1976).

the first formant is more important in terms of perceived quality and that higher frequencies are more important for intelligibility. It has also been shown that a mild level of highpass filtering alone can increase the crispness of speech by limiting the first formant, without significant loss in speech intelligibility.

As we have seen, speech quality depends upon a good representation of the short-term spectral magnitude, whereas the phase is relatively unimportant. Auditory processing of the spectral magnitude of speech suggests a phenomenon that can be used to suppress narrowband interference. *Auditory masking* is a perceptual characteristic that allows the human auditory system to suppress background noise. The auditory system can sometimes mask one sound with another. For example, consider a narrowband noise source such as a varying sine wave or an artificial noise component. Such distortions can be more annoying, since they fatigue the auditory system faster than broadband noise. An enhancement system can therefore introduce broadband noise in an effort to mask narrowband or artificial noise.

In addition to such methods as lowpass or highpass filtering, one might consider applying a Wiener filter based on the long-term power spectrum of speech. Such processing is optimal in the MSE sense if the signal is stationary and the additive noise is white Gaussian (it is the optimal *linear* filter if the noise is nonwhite). However, such a method is limited, since speech is not truly stationary. Even if the signal were truly stationary, the criterion upon which the Wiener filter is based is not a particularly effective error criterion for speech enhancement. To illustrate this, consider the use of masking to reduce the perceptual effects of a narrowband distortion. By adding broadband noise, we are in effect increasing the MSE while improving the perceived quality of the speech. A second example can be found in modification of the phase characteristics of speech. If speech is filtered with an all-pass filter (with some phase-distorting characteristics), no audible difference is perceived, but a substantial MSE can result. Although the MSE is sensitive to the phase of the speech spectrum, only the spectral magnitude is important for perception.

8.7.3 Speech Enhancement Algorithm Performance

A comparison of speech enhancement algorithms can only be accomplished if evaluation conditions are equivalent. This includes (1) speech material, (2) noise type and level, and (3) quality or intelligibility testing methods. A study by Hansen et al. (1987, 1991) considered an evaluation of the following three enhancement methods: (1) noncausal (unconstrained) Wiener filtering (Lim et al., 1978), (2) spectral subtraction with magnitude averaging (Boll, 1979), and (3) two inter- and intraframe spectral-constrained Wiener filtering methods. Hansen et al. showed that unconstrained Wiener filtering based on an LP all-pole model tends to produce speech with overly narrow bandwidths. However, an alternative iterative speech enhancement scheme that employs spectral constraints placed on redundancies in the human speech production process as represented by the LSP parameters has been shown to overcome this limitation. Figure 8.23 compares quality improvement for each of the three techniques for AWGN. Quality measures for a theoretical limit were obtained by substituting the noise-free LP coefficients into the unconstrained Wiener filter, thereby requiring only one additional iteration to obtain the estimated speech signal. These results show that good quality improvement can be achieved with all three methods. Figure 8.24 shows time versus frequency plots of processed speech from unconstrained and constrained Wiener filtering. A later study (Hansen et al., 1988) compared the performance in colored aircraft cockpit noise. Improvement in speech quality was also demonstrated. Traditional (noncausal) Wiener filtering outperforms spectral subtraction with magnitude averaging for this type of distortion. Although the constrained iterative enhancement algo-

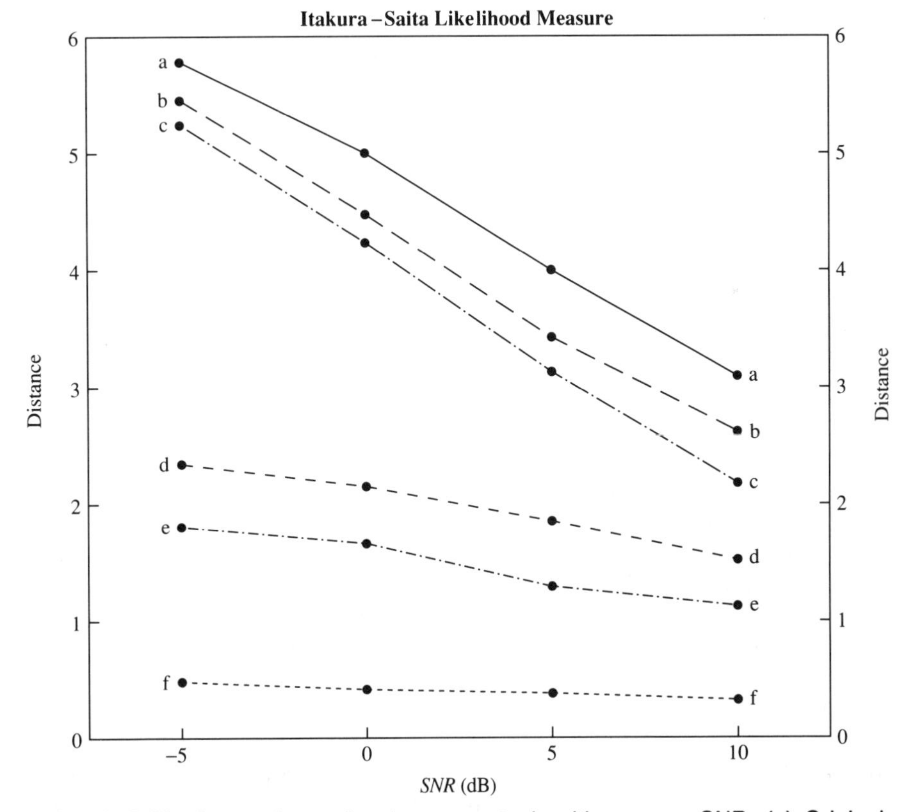

FIGURE 8.23. Comparison of enhancement algorithms over SNR. (a) Original distorted speech. (b) Boll: spectral subtraction, using magnitude averaging. (c) Lim–Oppenheim: unconstrained Wiener filtering. (d) Hansen–Clements: employing interframe constraints (FF-LSP:T). (e) Hansen–Clements: employing inter- and intraframe constraints (FF-LSP:T,Auto:I). (f) Theoretical limit: using undistorted LPC coefficients.

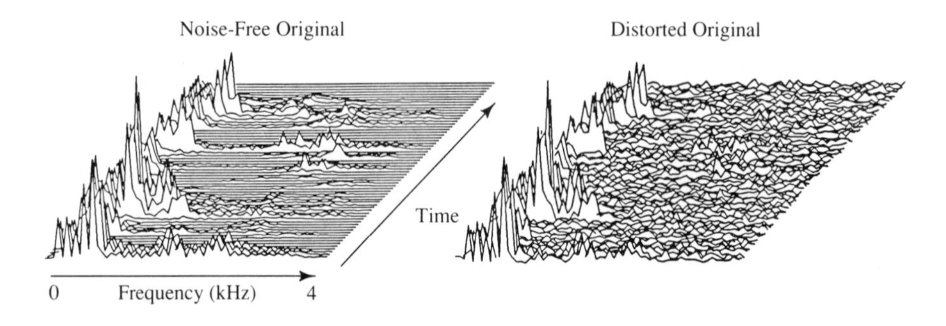

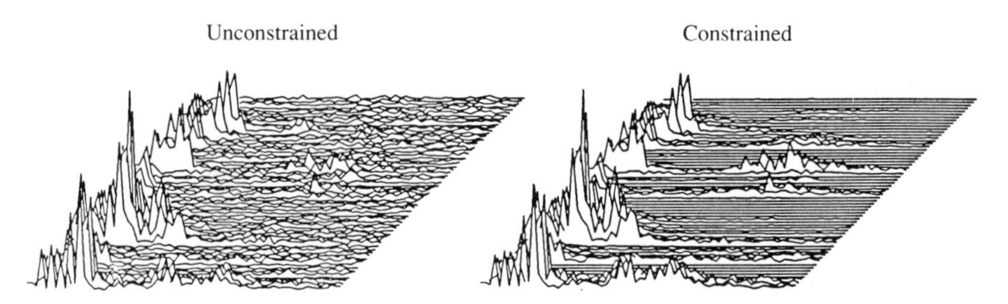

FIGURE 8.24. Time versus frequency plots of the sentence "Cats and dogs each hate the other." The original and distorted original (additive white Gaussian noise, SNR = +5 dB) are shown above. The lower left-hand plot is the response after three iterations of the unconstrained noncausal Wiener filtering approach. The lower right-hand plot is the frequency response after six iterations of an inter- plus intraframe constrained approach.

rithms produced speech of higher quality, computational requirements increased over the unconstrained approach.

8.8 Conclusions

In this chapter, we have considered a variety of approaches for speech enhancement and have incorporated them into a common framework. Due to the large number of applications, assumptions concerning interference, and available input channels, an almost unlimited number of enhancement systems could have been considered.

In conclusion, it should be emphasized that many enhancement systems improve the ratio of speech to noise, and therefore improve quality. This might be all that is important in certain applications in which the context of the material is known to the listener so that intelligibility is not of concern. However, the majority of speech enhancement algorithms actually reduce intelligibility and those that do not generally degrade the quality. This balance between quality and intelligibility suggests that considerable work remains to be done in speech enhancement.

Further, as we have seen here, several enhancement algorithms are designed to improve a mathematical criterion. Although attractive in a mathematical sense, most error criteria are not well correlated with auditory perception. Consequently, the use of both subjective and objective quality measures is necessary to meaningfully compare enhancement algorithm performance. Also, the evaluation of an enhancement algorithm will depend on its ultimate application.

As these concluding comments suggest, speech enhancement continues to be an important research area for improving communications between speaker and listener, speaker and vocoder, or speaker and speech recognizer. In particular, the task of evaluating the effectiveness of an enhancement technique continues to be an important area of investigation. We now turn to this issue in Chapter 9.

8.9 Problems

8.1. (Computer Assignment) In this problem, we will investigate appropriateness of the MSE as a criterion for speech enhancement.

(a) Design a filter which is all-pass with a nonzero phase response. Filter a speech signal with this filter. Play both the original and filtered signals. Can you hear any perceptual differences? If so, list them.

(b) Write a small routine that finds the MSE between two input signals. Find the average MSE between the original and an all-pass filtered speech signal. What can you say about the magnitude of the MSE with respect to any perceptual differences you were able to hear?

(c) Can you suggest an alternate error criterion that might be more appropriate?

8.2. In the discussion of speech enhancement, we addressed methods in the following four areas: (i) short-term spectral amplitude techniques, (ii) speech modeling and Wiener filtering, (iii) adaptive noise canceling, and (iv) fundamental frequency tracking systems. Consider a single algorithm from each area and discuss one advantage or disadvantage of each method. Note possible speaker and/or noise environments in which a technique may be more useful.

8.3. Suppose that we consider the following speech enhancement scenario of hands-off cellular phone dialing in a running automobile (i.e., a speech recognizer is used to automatically dial a desired phone number).

(a) What assumptions would you make on the speech and noise signals in order to formulate a speech enhancement conditioner prior to recognition? What are the design considerations (in terms of mathematical criteria, perceptual criteria, and performance improvement) involved in such an enhancement application? Discuss any trade-offs, restrictions, or limitations in your proposed solution.

(b) Now, suppose that the cellular phone system is based on LP analysis. We wish to extend the speech enhancement conditioner to reduce automobile noise prior to LP encoding within the cellular phone (i.e., output of the enhancement system will be used for automatic recognition when dialing and voice communications when phone communication is established). What additional design considerations may influence your proposed solution in part (a)?*

8.4. In the discussion of spectral subtraction, we identified a problem of resulting "musical tones" that occurs during subtraction.

(a) Discuss why such residual noise persists after spectral subtraction processing and why magnitude averaging can reduce such effects.

(b) Suggest an alternate means of characterizing the background noise interference that might further reduce these tones.

8.5. Consider a speech signal $s(n)$ that is corrupted by an additive sinusoid of the following form:

$$y(n) = s(n) + g \cdot \underline{A} \sin(2\pi f_0 t), \tag{8.62}$$

where g is a fixed term that is adjusted to ensure a desired overall average SNR, and f_0 a fixed sinusoidal frequency.

(a) Assume that $s(n)$ is a steady-state vowel (i.e., deterministic over the time interval of interest) and $\underline{A}$ a uniformly distributed random variable in the range $[0, 1]$. Can a speech enhancement solution be formulated based on spectral subtraction [i.e., $a = 1$ in (8.18)]? If so, find it. If not, why?

(b) Suppose that a random phase is added to the sinusoid as follows

$$y(n) = s(n) + g\,\underline{A} \sin(2\pi f_0 t + \underline{\psi}), \tag{8.63}$$

how would your solution to part (a) change?

(c) Now, let the speech signal $s(n)$ be corrupted by a nonlinear amplitude term as

$$y(n) = \text{sgm}[s(n)] \cdot s(n), \tag{8.64}$$

where

$$\text{sgm}[w] = \begin{cases} -1 & \text{if } w > 0 \\ 1 & \text{if } w \leq 0 \end{cases}. \tag{8.65}$$

What effect does this have on the degraded speech signal $y(n)$? Does a spectral subtraction solution exist? Why or why not?

8.6. One extension to spectral subtraction for reduction of musical tone artifacts is magnitude averaging as discussed in (Boll, 1978). Consider each class of speech (i.e., vowels, stops, etc.) and discuss what trade-offs

*The interested reader may consider recent studies on speech in noisy environments for the GSM (RPE-LTP) voice coder used for the European Digital Mobile Radio system (Hansen and Nandkumar, 1995).

exist for the frame duration choice for magnitude averaging. What speech classes would benefit the most from magnitude averaging and which would be most seriously affected?

8.7. Given the most general form of spectral subtraction [see (8.25)], explain how you would adjust a frequency-dependent weighting factor, $k(\omega)$, in order to obtain improved enhancement for the following speech classes: vowels, nasals, stops, fricatives, silence, or pauses. In selecting $k(\omega)$, consider the broad spectral characteristics of each speech class.

8.8. Consider a signal $s(n)$ that is corrupted by an additive white noise component $\underline{w}$ as,

$$\underline{y}(n) = s(n) + g\underline{w}(n), \tag{8.66}$$

where g is a fixed term that is adjusted to ensure a desired overall average SNR.

(a) Assume that $s(n)$ is a steady-state sinusoid. Derive the optimum noncausal Wiener filter solution for $s(n)$. Can a direct form solution be found, or is there an iterative solution?
(b) Assume $s(n)$ to be a steady-state vowel. How would your answer to part (a) change?
(c) Consider the speech signal $s(n)$ to contain discrete samples of the word "stay." Discuss how the Wiener filter will change as each phoneme is processed.
(d) Suppose that the exponent a in (8.32) is varied. How would the resulting Wiener filter's characteristics change for the word "stay" with $a = 1$ and $a = 2$?

8.9. One method of noncausal Wiener filtering is based on an all-pole model for the speech signal. Assume that speech is degraded by additive white Gaussian noise.

(a) From our discussion of the various LP methods from Chapter 5, discuss the tradeoffs in enhancement using the autocorrelation, covariance, and lattice filter methods of LP analysis. Will stability be an issue?
(b) Suppose that an ARMA (pole–zero) model is used for speech in the Wiener filter. Discuss the sensitivity of the filter to errors in speech model zero location estimation. How does this compare to errors in speech model pole location estimation?

8.10. Consider a simple two-channel processor, consisting of a primary microphone and a reference microphone that are connected. The reference microphone output $y_{\text{ref}}(n)$, is weighted by a value w and then subtracted from the output of the primary microphone, $y_{\text{pri}}(n)$. Show that the mean-square value of the output of this two-channel processor is minimized when the filter weight w achieves the optimum value

$$w_{\text{opt}} = \frac{\mathcal{E}\left[y_{\text{pri}}(n)y^*_{\text{ref}}(n)\right]}{\mathcal{E}\left[|y^*_{\text{ref}}(n)|^2\right]}. \tag{8.67}$$

8.11. Start with the formula for the estimation error for LMS adaptive noise cancellation:

$$e(n) = y_1(n) - \hat{\mathbf{h}}^T \mathbf{y}_2(n), \tag{8.68}$$

where $y_1(n) = s(n) + d(n)$ and $s(n)$ is the desired speech signal, $\hat{\mathbf{h}}$ the tap-weight filter vector, and $\mathbf{y}_2(n)$ the reference channel input vector at time n (assume no cross-talk). Find a relation for the gradient of the instantaneous squared error $|e(n)|^2$ in terms of $s(n)$, $\hat{\mathbf{h}}$, and $\mathbf{y}_2(n)$.

8.12. Consider an ANC scenario as described by (8.40). An estimate of the primary channel noise, $\hat{d}_1(n)$, is given by (8.41). Show that minimizing the MSE between $\underline{d}_1$ and $\hat{\underline{d}}_1$ is equivalent to minimizing the signal estimate $\hat{\underline{s}} = \underline{y} - \hat{\underline{d}}_1$, in a mean-square sense.

8.13. Beginning with the optimization criterion of adaptive noise cancellation from (8.43), derive the minimum MSE solution of (8.45). What assumptions must be made to arrive at this relation?

8.14. Derive a recursive (or lattice) least squares solution to the ANC problem defined in (8.40–8.43). Discuss its advantage or disadvantages with respect to an LMS solution.

8.15. Given the MSE criterion for the ANC scenario as in Problem 8.12, discuss and illustrate the effect of cross-talk (i.e., primary speech component) in the reference channel on the signal estimate, $\hat{\underline{s}}$. Suggest methods you feel might reduce this effect.

8.16. Suppose that two talkers are speaking simultaneously. Speaker A produces the speech signal $s_a(n)$, and speaker B produces the signal $s_b(n)$. Both signals are recorded by a single microphone producing the signal

$$y(n) = s_a(n) + s_b(n). \tag{8.69}$$

It is known that speakers A and B possess vocal-tract lengths of 12 cm and 17 cm, respectively. Suggest a frequency domain scheme to separate the two speakers. What are the trade-offs and assumptions involved in such a strategy.

8.17. (Computer Assignment) We consider the effects of noise on an isolated word utterance. Corrupt a word uttered by a male speaker using three additive noise sources: white Gaussian, a sinusoid, and another isolated word uttered by a female. Adjust the gain of the noise sources so that the SNR is 5 dB in each case.

(a) Listen to each corrupted file and comment on perceived distortion. Noting that the SNR is the same in each case, what can be said about noise effects on vowels, consonants, and upon general intelligibility?

(b) Assuming noise stationarity, find an average power spectrum for each noise source and perform spectral subtraction as illustrated in Fig. 8.2(a). Comment on the results.

8.18. (Computer Assignment) You are given an isolated-word uttered in noise-free conditions. Add white Gaussian noise to this word at an SNR of 5.0 dB.

(a) First, assume that the speech is a stationary process across the entire word. Construct a stationary Wiener filter for the entire word and obtain an estimate of the noise-free signal. Next, perform the same process for a steady-state sound (e.g., vowel) extracted from the given word. Compare the performance in the two cases in terms of output SNR and resulting MSE, and comment on the differences.

(b) Partition the input word utterance into phonemes. Construct a stationary Wiener filter for each phonene and filter each separately. Join the enhanced phonemes and compare its speech quality (waveform characteristics, overall SNR, and spectral representation) with the fixed Wiener filter from the first part of part (a).

8.19. (Computer Assignment) Obtain two degraded speech files by corrupting an isolated-word with white Gaussian noise then white Gaussian noise plus a slowly varying sinusoid. Construct a program for time-domain ANC using LMS. Filter each degraded utterance using correlated noise reference. Discuss the enhancement performance for each noise condition. Can ANC be used to attenuate (i) a slowly varying sinusoid, (ii) white Gaussian noise, or (iii) slowly varying colored noise?

8.A The INTEL System

INTEL,[20] along with other forms of spectral subtraction, fall under the general category of noise suppression prefilters (no assumed *a priori* knowledge of noise statistics). In this approach a spectral decomposition of a frame of noisy speech is performed and a particular spectral line is attenuated depending on how much the measured speech plus noise power exceeds an estimate of the background noise power. One approach in particular, developed by Weiss and Aschkenasy (1974, 1983) implements a real-time audio processor using the INTEL system and a tone component suppression filter.

Weiss and Aschkenasy (1974) originally developed INTEL as a generalization of spectral subtraction. Figure 8.2(a) and (b) illustrates the differences between INTEL and Boll's spectral subtraction. The former involves raising the noisy speech magnitude spectrum to a power a. Later work (1983) resulted in a real-time filter that removes interference from

[20]This section is included for historical purposes.

received or recorded data, termed the *computerized audio processor*. The system is comprised of two processing sections. The first processing operation, called *digital spectral shaping* (DSS), detects and attenuates impulsive and tonal noise. The second is INTEL, which is used to attenuate additive wideband random noise.

For DSS to be effective at tone suppression, three steps must be accomplished. First, the tone must be detected accurately. Next, the system must remove the maximum amount of tone energy once it has been detected, while removing a minimum amount of speech energy. The last step requires that the regenerated speech be maximally free of discontinuities and distortion. The detection process exploits differences between speech and noise. Tone noise is more stationary in both frequency and amplitude compared with the quasi-stationary speech. In the magnitude spectrum of tone noise, peaks result at the frequencies of the tones. In contrast, the speech spectrum is smooth over the entire frequency band with smooth peaks at the formant frequencies and finite nonzero bandwidths. To minimize the speech versus noise overlap in the frequency domain, tone energy should be concentrated into as narrow a spectrum as possible. Figure 8.25 illustrates this processing section.

To achieve minimal overlap between the speech and tone spectra, an appropriate weighting function on the time series must be chosen. Choice of analysis frame length must also be considered in the segmentation portion. The INTEL system uses a frame length of 200 msec, with a Bartlett window overlapped by 50%. This approach for tone removal works well when tone frequencies are different from peaks in the speech spectrum. The greater the difference, the more successful the procedure becomes. If, however, the tone component is random, the success of this approach is limited. This motivated the formulation of the second processing section.

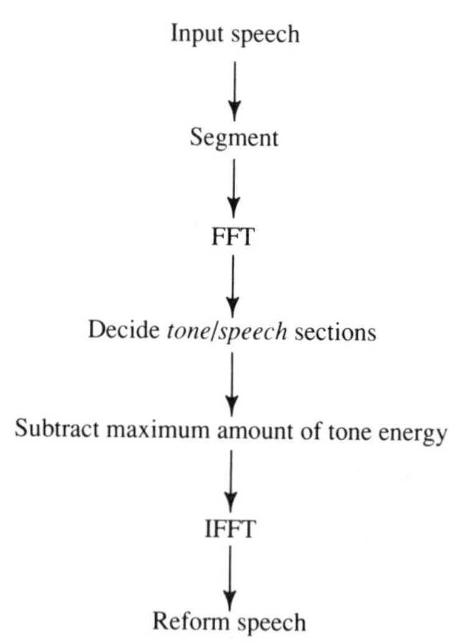

FIGURE 8.25. Tone removal using a computerized audio processor.

The primary use of INTEL is to attenuate additive wideband random noise. The input signal is transformed to a cepstral domain. An estimate of the noise cepstrum is subtracted; the resulting cepstral data are used to reform the enhanced speech. Figure 8.26 illustrates the INTEL procedure. (Note that the power term a has been set to $\frac{1}{2}$ in this implementation.)

Figure 8.26 shows that the difference between INTEL and spectral subtraction is the added transform pair, which results in the subtraction operation being carried out in the cepstrum domain. Improvement over tone subtraction results from differences in cepstral characteristics between the speech and random noise. Above a quefrency of 0.5 msec, the noise energy falls off quickly while speech energy is still present at the pitch period and its harmonics. Therefore, if a noise-only cepstrum can be found, subtracting it from the speech plus noise cepstrum greatly reduces the broadband random noise. How to compute the noise-only

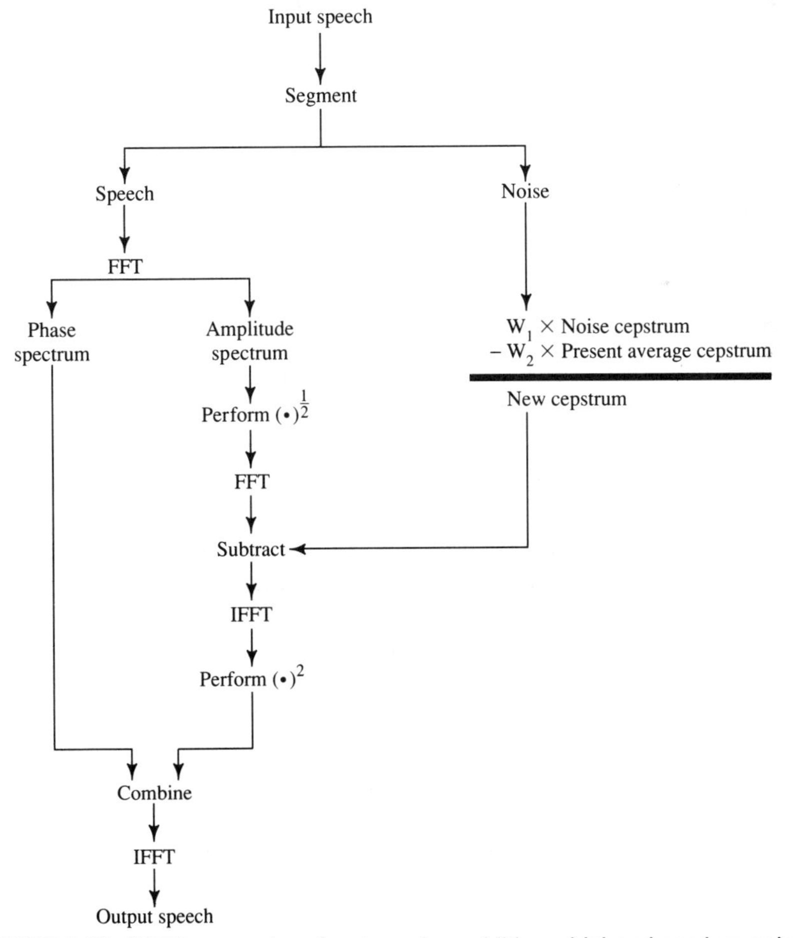

FIGURE 8.26. INTEL procedure for removing additive wideband random noise.

cepstrum must therefore be addressed. To accomplish this, a "lossy moving average" of the noise cepstrum is formed, and the noise cepstrum is then able to follow changes in the noise distribution. The two weights for the present and update noise cepstra must be chosen during processing. Once the algorithm is able to track the noise cepstrum, the choice of scale factor for subtraction between noisy speech and noise-only cepstra must be chosen. Weiss and Aschkenasy concluded that three scale factors were adequate for processing, with the process being carried out as shown in Fig. 8.27. Results indicate that choice of the K_i scalars is dependent on SNR, but somewhat independent of particular noise distribution. Their system used two sets of scale factors: one for use above 6 dB, the second for use below 0 dB. Between 0 and 6 dB, both sets produced similar results.

The INTEL system is useful for reducing wideband random noise, but many disadvantages exist. Since the average noise cepstrum is built up over time, any long sections of silence (1 sec or more) drives the average noise cepstrum to zero. When the signal reappears, a loud noise burst results until the noise cepstrum can be reestablished. Choice of update weights for the noise cepstrum must also be selected. This choice depends on the speed with which the system tracks changes in the noise characteristics. Scale factors for subtraction of the cepstra must also be chosen in some optimal fashion. Although Weiss and Aschkenasy found these to be somewhat independent of noise distribution for the cases analyzed, this may not be true for all types of noise. Although no quantitative results were presented in this investigation, it was suggested that some improvement in intelligibility is possible. This depends on how accurately the noise cepstrum can be updated from silent frames. In most instances, any tones that are encountered will be random. In addition, although INTEL is an improvement over spectral subtraction, it still suffers from the musical tone artifacts found in spectral subtraction.

Speech plus noise cepstrum

$- K_i \times$ [Average noise cepstrum]

Estimated speech only cepstrum

With the three scale factors K_i defined as follows:

K_0 for 0.0, . . . , 0.1 msec

K_1 for 0.1, . . . , 0.5 msec

K_2 for 0.5, . . . , ∞ msec

FIGURE 8.27. "Spectral subtraction" in the INTEL system.

8.B Addressing Cross Talk in Dual-Channel ANC

This appendix presents an expanded ANC framework that incorporates dual-channel cross talk. We present the basic application ideas without concern for the details of short-term processing.

When cross talk is present in an ANC system, a path must exist in the block diagram for speech to enter the reference channel. The speech component in the reference channel could simply be a greatly attenuated speech component, or some filtered version due to propagation, differences in microphone characteristics, or acoustic barrier between the two microphones. Figure 8.28 shows a block diagram of the expanded approach. The filters $H_1(\omega)$ and $H_2(\omega)$ model the frequency-dependent attenuation that the signals $s(n)$ and $d(n)$ experience due to the separation of the primary and reference microphones. The adaptive filter $H(\omega)$ provides an estimate $\hat{d}_1(n)$ of the noise component $d_1(n)$ in the primary channel so that the resulting error term $\varepsilon(n)$ is an estimate of the desired speech $s_1(n)$.

The output from the primary channel can be written as

$$y_1(n) = s_1(n) + d_1(n) = s_1(n) + [d_2(n) * h_2(n)], \tag{8.70}$$

where $d(n) * h_2(n)$ is the degrading noise component resulting from the convolution of the input noise source and impulse response of the noise-shaping filter. Similarly, the reference channel signal can be written as

$$y_2(n) = s_2(n) + d_2(n) = [s_1(n) * h_1(n)] + d_2(n), \tag{8.71}$$

where $h_1(n)$ is the impulse response of the speech-shaping filter. Under typical conditions, the SNR of the primary channel SNR_1 is much higher than that for the reference ($\text{SNR}_1 \gg \text{SNR}_2$). The output of the ANC can be written, assuming an adaptive filter impulse response $h(n)$, as

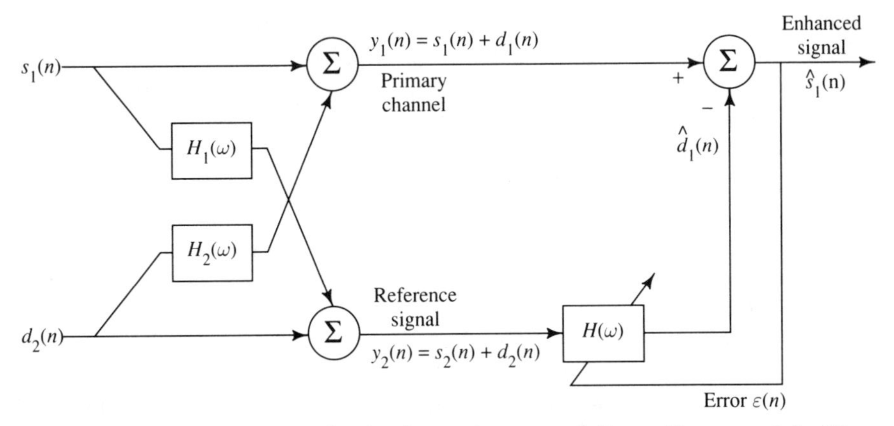

FIGURE 8.28. Dual-channel adaptive noise cancelation with cross talk filter.

$$\begin{aligned} \varepsilon(n) &= y_1(n) - \hat{d}_1(n) \\ &= s_1(n) + [d_2(n) * h_2(n)] - [y_2(n) * h(n)]. \end{aligned} \tag{8.72}$$

After applying a DTFT and substituting the relation for $Y_2(\omega)$, (8.72) becomes

$$\begin{aligned} \varepsilon(\omega) &= S_1(\omega) + [D_2(\omega)H_2(\omega)] - [Y_2(\omega)H(\omega)] \\ &= S_1(\omega) + D_2(\omega)H_2(\omega) - [D_2(\omega)H(\omega) + S_1(\omega)H_1(\omega)H(\omega)]. \end{aligned} \tag{8.73}$$

If we assume that the adaptive filter can adequately approximate the noise-shaping filter [i.e., $H(\omega) \approx H_2(\omega)$] to the extent that the spectral magnitude of $D_2(\omega)H_2(\omega) - D_2(\omega)H(\omega)$ is small when compared with $S_1(\omega) - S_1(\omega)H_1(\omega)H(\omega)$, then the spectral output of the ANC employing a cross-talk filter is

$$\varepsilon(\omega) \approx S_1(\omega)[1 - H_1(\omega)H(\omega)], \tag{8.74}$$

or assuming $H(\omega) \approx H_2(\omega)$,

$$\varepsilon(\omega) \approx S_1(\omega)[1 - H_1(\omega)H_2(\omega)]. \tag{8.75}$$

Therefore, if we could ensure the magnitude of $H_1(\omega)H_2(\omega)$ to be much less than unity across the speech frequency band of interest, the resulting transform $\varepsilon(\omega)$ would represent speech with a minimum of distortion.

Harrison, Lim, and Singer (1984, 1986) considered such a dual-channel ANC approach in a fighter cockpit environment. In their study, the condition $|H_1(\omega)H_2(\omega)| \ll 1$ was satisfied through the use of an oxygen mask, which served as an acoustic barrier between the two microphones. The primary microphone was located inside the mask and the reference microphone was located outside. The oxygen mask provided a barrier that attenuates signals by about 10 dB. This attenuation applies to both the ambient noise-shaping filter $H_2(\omega)$, and the speech-shaping filter $H_1(\omega)$, so that the combined attenuation is approximately 20 dB. It should be noted that these are, in fact, shaping filters, so that the speech component in the primary channel $s_1(n)$ should possess different spectral characteristics than that in reference $s_2(n)$. Since both inputs contain speech, updating the adaptive filter coefficients during speech activity could lead to distorted speech. Instead, Harrison et al. allowed the filter coefficients to adapt only during silence or noise and held them constant during speech activity. This can easily be accomplished by setting the adaptation parameter to zero during speech. This implicitly assumes that the noise-shaping filter $H_2(\omega)$ does not change significantly when speech activity occurs. By directly employing a cross-talk filter, Harrison et al. found a significant reduction in the required filter length for adequate enhancement [from 1500 for Boll and Pulsipher (1980) to 100]. An adaptation time of 120 msec was required for convergence of the LMS algorithm starting from zero initial conditions. Using the exact least squares lattice formulation (Morf and Lee, 1978), convergence from zero initial conditions was achieved in less than 20 msec. In a noise field created using a single source, an improvement of of 11 dB in SNR was obtained.

The improvement obtained by Harrison et al., however, does not occur for all frequencies. In a related study, Darlington, Wheeler, and Powell (1985) attempted to address the practical question of why ANC is successful only for aircraft cockpit noise at low frequencies. The noise levels experienced within fighter cockpit environments typically exceed 90 dB SPL over the entire speech frequency band. For the British helmet and oxygen mask considered in their study, frequency response showed a 30-dB attenuation above 2 kHz, dropping to 0 dB below 100 Hz.[21] The use of ANC in the noise transmission problem across an oxygen mask depends upon the successful identification of the linear filter relating the exterior reference pressure to the noise component detected by the interior mask microphone. Poor coherence in noise transmission across an oxygen mask contributes to the failure of ANC at high frequencies. In addition, the distributed nature of the noise sources within a cockpit environment suggests that a diffuse model could be applied successfully. Darlington et al. point out that in a fully diffuse field, the random incidence of sound at any point causes the coherence between the pressure at two points to decrease as the points become farther apart. Using a notation adopted from Piersol (1978), the *coherence* between two points x and y is defined as

$$\gamma_{xy}^2 = \left[\frac{\sin K_o W}{K_o W}\right], \tag{8.76}$$

where K_o is the wave number and W the spacing between the points. The coherence between the reference noise and primary noise component defines the proportion of the primary noise power that is linearly related to the reference signal. Therefore, the magnitude *coherence* between two WSS random processes, $\underline{x}$ and $\underline{y}$, is defined as

$$\gamma_{\underline{xy}}^2(\omega) = \frac{|\Gamma_{\underline{xy}}(\omega)|^2}{\Gamma_{\underline{x}}(\omega)\Gamma_{\underline{y}}(\omega)}, \tag{8.77}$$

where $\Gamma_{\underline{xy}}(\omega)$ is the cross-PDS of $\underline{x}$ and $\underline{y}$, and $\Gamma_{\underline{x}}(\omega)$ and $\Gamma_{\underline{y}}(\omega)$ the auto-PDS for $\underline{x}$ and $\underline{y}$. We will refer to the magnitude-squared coherence $\gamma_{\underline{xy}}^2(\omega)$ as simply the coherence of $\underline{x}$ and $\underline{y}$. It can be shown that the coherence represents the fraction of $\Gamma_{\underline{y}}$, which is related to $\Gamma_{\underline{x}}$ by a linear filter. For this reason, we can think of the coherence as a frequency-dependent correlation coefficient. Since ANC uses the noise reference signal to estimate the primary noise signal, a large coherence between primary and reference noise signals is necessary if ANC is to be effective.

[21]It would in fact be desirable to increase the noise attenuation characteristics for such passive barriers. It may not be possible to introduce such modifications in present mask/helmet configurations; however, future designs should certainly consider the issue of noise.

CHAPTER 9

Speech Quality Assessment

9.1 Introduction

9.1.1 The Need for Quality Assessment

There are two speech processing areas where speech quality assessment is of primary concern: speech coding or synthesis and speech enhancement. Historically, the majority of quality testing methods have been formulated to evaluate performance of speech coding algorithms. The need for such testing methods became apparent during the development of new analog communication systems for speech data in the 1950s. The advent of digital speech coding algorithms in the 1960s, which led to the fully digital private branch exchange (PBX) speech communication systems of the 1980s, motivated the formulation of new, more sophisticated quality assessment tools for voiceband coding systems. Such tools were essential for optimum coding-system design and effective communications network planning. Recently, many of the same tests or measures have also been successfully applied to quantify improvement for speech enhancement algorithms.

Since this is a rapidly changing field, our purpose in addressing speech quality evaluation is not to make the reader an expert, but rather to expose the reader to useful and reliable methods by which coding or enhancement algorithms can be evaluated. As an example, suppose you were to listen to an old analog record of a speech. If the record has been played many times, the audio may contain hiss, crackle, or other types of background noise. In listening to the recording, you certainly would have some impression of the signal quality. As a listener, you might choose to judge the "processed" speech signal along more than one overall quality scale. For example, on a scale of one to five, how natural sounding is the speech? Does it contain background hiss? Does it sound mechanical? Although the most meaningful judgment of such issues comes from the ultimate human listener, quantifiable means of judging speech quality must be available for testing new coding and enhancement algorithms in research and development laboratories. This is the purpose of quality assessment methods.

To establish a fair means of comparing speech coding or enhancement algorithms, a variety of quality assessment techniques has been formulated. Generally speaking, tests fall into two classes: subjective quality measures and objective quality measures. Subjective measures are based on comparisons of original and processed speech data by a listener or group of listeners, who subjectively rank the quality of speech along a predetermined scale. Objective quality measures are based on a mathematical comparison of the original and processed speech signals. Most objective quality measures quantify quality with a numerical distance measure or a model of how the auditory system interprets quality. Since the distortion introduced by speech coding systems, background noise, and enhancement algorithms varies, a collective body of quality measures and tests has emerged for different applications.

Generally speaking, there are three areas in which it is desirable to measure distortion due to noise. First, as we briefly discussed at the end of Chapter 7, we may want a quality measure for the amount of distortion introduced by a speech compression algorithm. This type of noise can be additive, but in many cases is nonlinear and signal-dependent. Such evaluations can be used to determine parameter settings or bit allocation for parameters, to improve quality, or to further reduce data rates. Second, we may wish to measure the level of distortion introduced by a noisy environment (during speech collection) or during transmission over a noisy channel. In most cases, this noise is additive, broadband, and statistically independent of the speech signal. An exception might be a burst of bit errors over a channel, which may arise, for example, in a speech communication satellite uplink because of sunspot activity. Finally, we may wish to measure the performance of an enhancement algorithm to see if the quality of the processed speech has been improved. Consistency of input speech test data as well as listener groups is essential if the performance of speech algorithms is to be properly compared. Since the noise or distortion introduced in each of the areas mentioned is quite different, the applicability of each quality test will vary. The reader should keep this in mind as we discuss each measure or test.

We begin the main body of this chapter with a discussion of subjective quality measures and illustrate their use in speech coding applications. Next, we consider several objective quality measures and discuss their use in quantifying performance for both coding and enhancement. Finally, the last section briefly discusses the interrelationship between subjective and objective measures. The literature is rich with papers devoted to quality assessment. Several tutorials have been written on both subjective and objective quality measures. One of the most complete treatments is that by Quackenbush et al. (1988). However, as the research continues to mature, measures that are even better able to quantify quality and intelligibility will become available.

Throughout this chapter, any reference to *processed speech* refers to resynthesized speech from a speech coding system, or enhanced speech from an enhancement algorithm. To be entirely consistent with the research literature, this chapter should focus only on quality assessment of speech coding/compression systems, since most if not all quality assessment techniques were formulated for rating such systems. However, we have taken the liberty of expanding this chapter, for several reasons. First, many of the quality tests used for coder evaluation have also been used for enhancement evaluation. Some vocoding algorithms introduce noise or distortion that is similar in spectral content to noise which enhancement algorithms seek to remove. Finally, many quality testing methods used to measure the perceived noise introduced through a speech compression/resynthesis procedure can also be used to measure noise introduced from a surrounding environment during data acquisition.

9.1.2 Quality Versus Intelligibility

In order to evaluate speech processing algorithms, it would be useful to be able to identify the similarities and differences in perceived quality and subjectively measured intelligibility. Engineers have some feeling for the "merit" of their processing systems. This feeling, which is difficult to describe, let alone measure, is referred to as *perceived quality.* The quality of speech addresses "how" a speaker conveys an utterance and may include such attributes as "naturalness," or speaker recognizability. In contrast, speech intelligibility is concerned with *what* the speaker has said—the meaning or information content behind the words. At present we do not clearly understand the interrelationship between perceived quality and intelligibility. Ordinarily, unintelligible speech would not be judged to be high quality; however, the converse need not be true. For example, a very mechanical-sounding synthetic utterance may be highly intelligible. Accordingly, intelligibility can be considered to be one of many "dimensions" of the abstract notion of quality. In the following material, we shall treat intelligibility in this way. When we juxtapose quality and intelligibility, we are really referring to all of the features that contribute to "quality" that are not necessarily required for understanding what is spoken.

The difficulty in separating the notions of quality and intelligibility is due, in part, to the difficulty in isolating and characterizing those acoustic-correlates of quality or intelligibility in speech. However, extensive research has been carried out in developing both subjective and objective tests to ascertain quality and intelligibility. These tests have been used extensively in evaluating speech coding/transmission systems (Flanagan, 1972; Kayser, 1981; Kitawaki et al., 1984; McDermott et al., 1978a, 1978b; Tribolet, 1978). In addition, research on the statistical correlation between objective and subjective measures has been performed in order to formulate good objective measures of quality.

9.2 Subjective Quality Measures

Subjective measures are based on the opinion of a listener or a group of listeners of the quality of an utterance. As suggested by Hecker and Williams (1966), one means of classifying subjective quality measures is to group measures as *utilitarian* or *analytical.* Utilitarian measures employ testing procedures that are both efficient and reliable and that produce a measure of speech quality on a unidimensional scale. The main advantage is that a single number results, which can be used to directly compare speech processing systems.

In contrast, analytical methods seek to identify the underlying psychological components that determine perceived quality. These methods are oriented more toward characterizing speech perception than measuring perceived quality and typically use more than one dimension for reporting results (e.g., rough to smooth, bright to muffled). Studies in this area include those by Kruskal (1964a, 1964b), McGee (1964), McDermott (1969), and Voiers (1964). In this section, we will focus on utilitarian approaches, which can be further divided into those that test for intelligibility and those that measure other aspects of quality. As we have noted above, intelligibility can be viewed as one aspect of quality, since high-quality speech generally implies good intelligibility. However, the converse is not necessarily true. Quality tests (other than those measuring intelligibility) are usually employed to evaluate systems with high intelligibility scores, since low intelligibility is generally a good indicator of poor quality.

The tests and their acronyms to be studied in this chapter appear in Table 9.1. Early subjective measures focused on speech intelligibility, one important aspect of overall speech quality. Several tests have been formulated using rhyme word lists, such as the *modified rhyme test* (MRT) and the *diagnostic rhyme test* (DRT). Here listeners are presented with rhyming words that differ only in their leading consonantal phonemes. The quality of the speech processing system is based on the

TABLE 9.1. Quality Measures Discussed in This Chapter.

Test	Type of Test
Modified rhyme test (MRT)	Subjective intelligibility
Diagnostic rhyme test (DRT)	Subjective intelligibility
Isometric absolute judgment (IAJ)	Subjective quality
Mean opinion score (MOS)	Subjective quality
Paired acceptability rating (PAR)	Subjective quality
Parametric absolute judgment (PAJ)	Subjective quality
Quality acceptance rating test (QUART)	Overall subjective quality
Diagnostic acceptability measure (DAM)	Overall subjective quality
Articulation index (AI)	Objective intelligibility
Signal-to-noise ratio (SNR)	Objective quality
Segmental SNR (SNR_{seg})	Frame-based SNR
Frequency weighted segmental SNR (SNR_{fw-seg})	Frame-based SNR with spectral weighting
Itakura log-likelihood measure	Objective quality
Log-area ratio measure (LAR)	Objective quality
Other LP-based measures	Objective quality
Weighted-spectral slope measure (or Klatt measure) (WSSM)	Objective quality

number and possibly the ease of the listeners' correct responses. Another class of subjective measures that attempts to evaluate more than just speech intelligibility is the *isometric absolute judgment* (IAJ) quality tests. In these tests, listeners are presented with a list of distorted words and are asked to judge their overall quality. In this case, "isometric" means that listeners judge quality without any reference signals. The advantage here is that listeners are free to assign any perceptual meaning to their scale, ranging from excellent to unsatisfactory.

The disadvantage of these tests is that there may be variations or biases among listeners. Therefore, a listener may always rank the distorted speech at some average level lower than another listener. To determine this bias, some researchers present a standard set of distorted utterances to the listener. This bias is assumed constant for the session and is used as an offset for evaluating the required distorted speech. Some of these tests include the "goodness" test, *mean opinion score* (MOS) tests, and the *paired acceptability rating* (PAR) method. Another class of subjective quality measures that addresses the problem of individual preference differences is the *parametric absolute judgment* (PAJ) quality tests. Here, listeners judge distorted speech on several perceptual scales instead of overall quality. For example, these scales might describe the degree of clicking, fluttering, babbling, and so on. These tests are effective, since listeners often agree on the degree of degradation but vary on their dislike for a specific degradation. An overall quality score can then be found by combining the parametric scores. Two of these tests include the *quality acceptance rating test* (QUART), and the very successful *diagnostic acceptability measure* (DAM). Other subjective quality tests are used by speech researchers, but these represent a sampling of the most commonly accepted tests. The remainder of this section discusses these important subjective measures in further detail.

9.2.1 Intelligibility Tests

Rhyme Tests

An early means of testing intelligibility, which is a modification of an articulation test by Fairbanks (1958), is based on constraining the possible listener responses to a set of rhyming words. The Fairbanks test is based on single-syllable words of the form /C-V-C/ (i.e., consonant-vowel-consonant clusters such as "cat," "bat," "tat," "mat," "fat," "sat," etc.). A listener has a response sheet with the leading consonant blank and is required to choose the leading consonant based on his or her interpretation of the test speech. House et al. (1965) made administrative improvements to the Fairbanks test by limiting the possible set of listener responses to a finite set of rhyming words. One of six rhyme words is chosen, greatly simplifying the administration and scoring procedure.

This MRT[1] is more suitable for untrained listeners. Finally, Voiers (1977a, 1977b) refined these rhyme tests to better measure intelligibility in his DRT. The main contribution was to restrict each test to a pairwise comparison, so that the difference in each leading consonant pair differed in just one distinctive feature. Sample word pairs used in the DRT are shown in Table 9.2. Each word pair tests for distinctive features (Jakobson et al., 1967) such as voicing, nasality, sustension, sibilation, graveness, and compactness. A summary of the distinctive features tested in the DRT is shown in Table 9.3. An attribute's presence is indicated by a "+," its absence by a "−," and the neutral condition by an "o." As an example, the test pair "met–net" examines the coders' ability to accurately represent graveness. The overall DRT score is obtained by the correct percentage of responses

$$\mathrm{DRT\%} = \frac{N_{\mathrm{correct}} - N_{\mathrm{incorrect}}}{N_{\mathrm{tests}}} \times 100, \tag{9.1}$$

TABLE 9.2. Word Pairs Used in Diagnostic Rhyme Test (DRT).

veal–feel	bean–peen	zoo–sue	dune–tune
meat–beat	need–deed	moot–boot	news–dues
vee–bee	sheet–cheat	foo–pooh	shoes–choose
zee–thee	cheep–keep	juice–goose	chew–choo
weed–reed	peak–teak	moon–noon	pool–tool
yield–wield	key–tea	coop–poop	you–rue
gin–chin	dint–tint	vole–foal	goat–coat
mitt–bit	nip–dip	moan–bone	note–dote
vill–bill	thick–tick	those–doze	though–dough
jilt–gilt	sing–thing	joe–go	sole–thole
bid–did	fin–thin	bowl–dole	fore–thor
hit–fit	gill–dill	ghost–boast	show–so
zed–said	dense–tense	vault–fault	daunt–taunt
mend–bend	neck–deck	moss–boss	gnaw–daw
then–den	fence–pence	thong–tong	shaw–chaw
jest–guest	chair–care	jaws–gauze	saw–thaw
met–net	pent–tent	fought–thought	bong–dong
keg–peg	yen–wren	yawl–wall	caught–taught
vast–fast	gaff–calf	jock–chock	bond–pond
mad–bad	nab–dab	mom–bomb	knock–dock
than–dan	shad–chad	von–bon	vox–box
jab–gab	sank–thank	jot–got	chop–cop
bank–dank	fad–thad	wad–rod	pot–tot
gat–bat	shag–sag	hop–fop	got–dot

[1] The reader is reminded of Table 9.1 in which all acronyms are summarized.

TABLE 9.3. Consonant Classification Used in Diagnostic Rhyme Test.

	/m/	/n/	/v/	/D/	/z/	/Z/	/Ẑ/	/b/	/d/	/g/	/w/	/r/	/l/	/y/	/f/	/T/	/s/	/S/	/Ŝ/	/p/	/t/	/k/	/h/
Voicing	+	+	+	+	+	+	+	+	+	+	+	+	+	+	−	−	−	−	−	−	−	−	−
Nasality	+	+	−	−	−	−	−	−	−	−	−	−	−	−	−	−	−	−	−	−	−	−	−
Sustention	−	−	+	+	+	+	−	−	−	−	+	+	+	+	+	+	+	+	−	−	−	−	+
Sibilation	−	−	−	−	+	+	+	−	−	−	−	−	−	−	−	−	+	+	+	−	−	−	−
Graveness	+	−	+	−	−	o	o	+	−	o	+	−	o	o	+	−	−	o	o	+	−	o	o
Compactness	−	−	−	−	−	+	+	−	−	+	−	−	o	+	−	−	−	+	+	−	−	+	+
Vowel-like	−	−	−	−	−	−	−	−	−	−	+	+	+	+	−	−	−	−	−	−	−	−	−

where N_{tests} is the number of tests, N_{correct} is the number of correct responses, and $N_{\text{incorrect}}$ is the number of incorrect responses. A typical summary of DRT results is shown in Table 9.4. A system that produces "good"-quality speech should have a DRT score in the range 85–90. As the table shows, application of the DRT will pinpoint exactly why a speech processing system fails, providing valuable insight for further algorithm design. The DRT has enjoyed more widespread use than the MRT and provides very reliable results.

9.2.2 Quality Tests

Intelligibility tests such as the DRT have been widely accepted primarily because they are well defined, accurate, and repeatable. However, they test intelligibility, which is only one facet of the multidimensional space that makes up overall speech quality. Tests that distinguish among speech systems of high intelligibility are usually called speech quality tests. One direct method of measuring speech quality is the MOS. A second, more systematic method is the DAM.

Mean Opinion Score

An opinion rating method can be used to assess the degree of quality for a speech processing system. For the case of voice telephone transmission systems, five grades of quality are distinguished. Although other quality measures evaluate a wider array of speech characteristics that comprise overall quality, the MOS is the most widely used subjective quality measure (IEEE, 1969). In this method, listeners rate the speech under the test on a five-point scale where a listener's subjective impressions are assigned a numerical value (Table 9.5). A training phase is sometimes used before evaluation in order to "anchor" the group of listeners. If a training phase is not used, anchor test phrases with known MOS levels are submitted for listener evaluation. Both procedures normalize listener bias for those who always judge processed speech to be low or high in quality. A standard set of reference signals must be used if the test is to be compared with results from other test sessions. The MOS has been used extensively for evaluation of speech coding algorithms (Daumer and Cavanaugh, 1978; Daumer, 1982; Goodman, 1979; Kitawaki et al., 1984). An advantage of the MOS test is that listeners are free to assign their own meanings of "good" to the processed speech. This makes the test applicable to a wide variety of distortions. At the same time, however, this freedom offers a disadvantage in that a listener's scale of "goodness" can vary greatly (Voiers, 1976). Selection of the subjects, as well as the instructions given to the subjects, can affect opinion scores. Finally, particular attention must be used in maintaining a consistent test condition framework (i.e., the order of presentation, type of speech samples, presentation method, and listening environmental conditions).

TABLE 9.4. Results Form for Diagnostic Rhyme Test. After Papamichalis (1987).

		Test Condition:		EZA-QUIET-ALTEC			Date Tested: 11/11/74	
	Present	S.E.*	Absent	S.E.*	Bias	S.E.*	Total	S.E.*
Voicing	99.0	1.04	98.4	1.10	0.5	1.66	98.7	0.67
Frictional	97.9	2.08	96.9	2.19	1.0	3.32	97.4	1.35
Nonfrictional	100.0	0.00	100.0	0.00	0.0	0.00	100.0	0.00
Nasality	100.0	0.00	99.5	0.52	0.5	0.52	99.7	0.26
Grave	100.0	0.00	100.0	0.00	0.0	0.00	100.0	0.00
Acute	100.0	0.00	99.0	1.04	1.0	1.04	99.5	0.52
Sustention	99.5	0.52	96.9	1.52	2.6	1.35	98.2	0.92
Voiced	100.0	0.00	96.9	2.19	3.1	2.19	98.4	1.10
Unvoiced	99.0	1.04	96.9	2.19	2.1	1.36	97.9	1.57
Sibilation	99.0	0.68	100.0	0.00	−1.0	0.68	99.5	0.34
Voiced	99.0	1.04	100.0	0.00	−1.0	1.04	99.5	0.52
Unvoiced	99.0	1.04	100.0	0.00	−1.0	1.04	99.5	0.52
Graveness	89.1	1.35	94.3	2.22	−5.2	3.22	91.7	0.88
Voiced	99.0	1.04	96.9	2.19	2.1	2.61	97.9	1.11
Unvoiced	79.2	2.23	91.7	4.17	−12.5	6.10	85.4	1.36
Plosive	97.9	1.36	96.9	1.52	1.0	2.46	97.4	0.76
Nonplosive	80.2	2.70	91.7	4.45	−11.5	6.67	85.9	1.56

Compactness	99.0	0.68	100.0	0.00	−1.0	0.68	99.5	0.34
Voiced	97.9	1.36	100.0	0.00	−2.1	1.36	99.0	0.68
Unvoiced	100.0	0.00	100.0	0.00	0.0	0.00	100.0	0.00
Sustained	99.0	1.04	100.0	0.00	−1.0	1.04	99.5	0.52
Interrupted	99.0	1.04	100.0	0.00	−1.0	1.04	99.5	0.52
BK/MD	99.0	1.04	100.0	0.00	−1.0	1.04	99.5	0.52
BK/FR	99.0	1.04	100.0	0.00	−1.0	1.04	99.5	0.52
Experimental†	100.0	0.00	100.0	0.00	0.0	0.00	100.0	0.00

Speaker	LL	CH	RH
List #	302B	307A	310B
DRT score	98.2	97.1	98.3
S.E.*	0.43	0.55	0.52

8 Listeners, crew (2), 576 total words
3 Speaker(s), 192 words per speaker
Standard error for speakers = 0.37
Total voiced score = 98.7
Total unvoiced score = 95.7

Total DRT score = 97.9
Standard error* = 0.33

*Standard errors based on listener means.

†Experimental items are not included in any summary scores.

TABLE 9.5. Mean Opinion Score Five-Point Scale.

Rating	Speech Quality	Level of Distortion
5	Excellent	Imperceptible
4	Good	Just perceptible but not annoying
3	Fair	Perceptible and slightly annoying
2	Poor	Annoying but not objectionable
1	Unsatisfactory	Very annoying and objectionable

In a study by Goodman and Nash (1982), a variety of vocoder configurations were evaluated using MOS tests for seven languages (speech spoken in Britain, Canada, France, Italy, Japan, Norway, and the United States). Listener opinion evaluation using MOS was performed under equivalent test conditions. Two coding algorithms, log-PCM and ADPCM (see Section 7.3.2), were evaluated with various bit error rates and additive white noise to determine if opinion scores are dependent on language. Figure 9.1 summarizes MOS test results for the ADPCM coder versus transmission rate. The results reveal that direct comparison for MOS is difficult, even with carefully controlled experimental conditions.

Studies have shown that reference signals used as part of the evaluation help normalize the MOS so that systems tested at different times and places can be compared in a more reliable manner (Kitawaki et al., 1984; Nakatsui and Mermelstein, 1982; Goodman et al., 1976). One reference signal that is often used is speech degraded by varying amounts of multiplicative white noise, since its distortion with respect to quality is similar to adaptive waveform coder noise (Schroeder, 1968; Law and Seymour, 1962). The speech-to-speech-correlated-noise ratio, referred to as Q in dB, was thereby established as the basic unit of speech quality. In

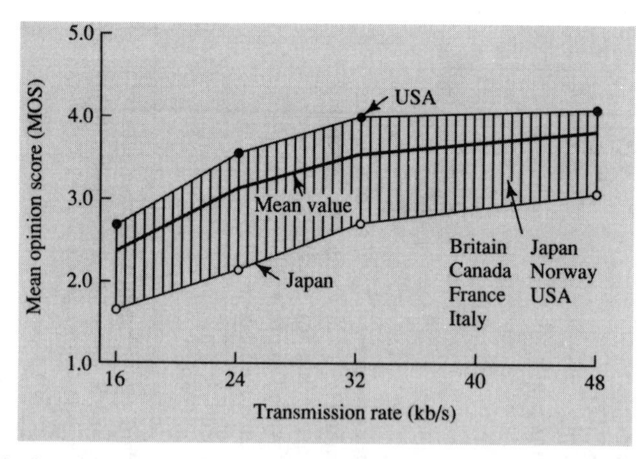

FIGURE 9.1. A summary of listener opinion quality using the MOS for a fixed predictor ADPCM coder with error-free transmission for seven countries. The solid curve is the average MOS measure over all countries. After Goodman and Nash (1982).

coder evaluation, reference signals with a range of Q values are evaluated by the listener group during the MOS test. A plot of MOS versus the Q of the reference signals is obtained (such as that shown in Fig. 9.2). This transforms MOS to an opinion equivalent Q, which can be used to compare quality performance across coding systems such as the waveform coders (APC-AB, ATC, ADPCM, log-PCM) and source coders (LSP, PARCOR) in Fig. 9.3.

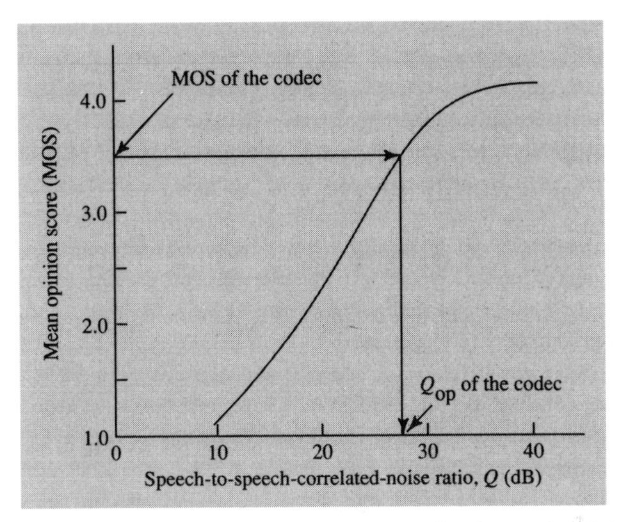

FIGURE 9.2. A plot of speech-to-speech-correlated-noise ratio Q in dB versus mean opinion score.

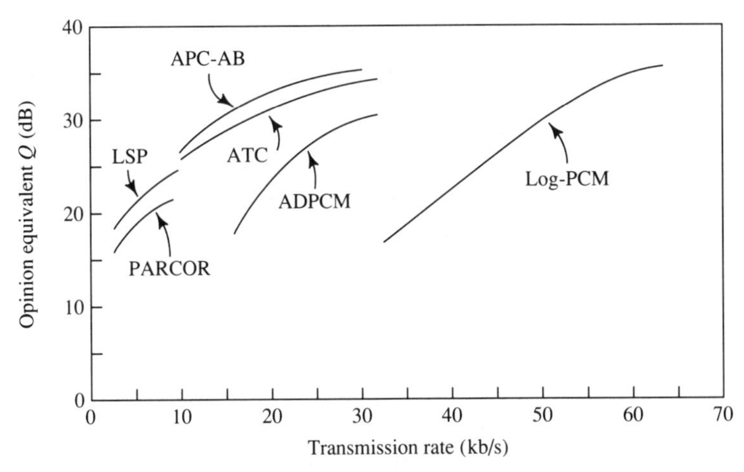

FIGURE 9.3. A summary of quality results versus transmission rate for several waveform and source coders. The opinion equivalent speech-to-speech-correlated-noise ratio Q is shown in decibels.

Diagnostic Acceptability Measure

The DAM (Voiers, 1977) is used for evaluating medium- to high-quality speech. The DAM is unlike other subjective measures in that it incorporates a multidimensional approach. The DAM evaluates a speech signal on 16 separate scales, divided into three categories: signal quality, background quality, and total quality. The multiplicity of scales is an important feature, since it allows the listener to judge signal and background characteristics separately. This fine-grained structure of the DAM ensures that listeners are not required to compromise their subjective impressions. Most listeners agree on the presence of a distortion, but differ on preference. Since the DAM solicits separate reactions from the listener regarding perceived speech signal, background, and total quality, it tends to minimize the sampling error (bias) associated with individual differences in preference found in a single global score approach. Further discussion concerning administration of the DAM test, control of measurement error, and calculation of the DAM scores can be found in the literature (Voiers, 1977; Quackenbush et al., 1988).

An example of the DAM rating form is shown in Fig. 9.4. After listeners judge the processed speech data along signal, background, and overall scales, the responses are processed by weighting each parameter depending on its impact on quality. A sample set of results is shown in Fig. 9.5 for subjective quality versus additive white background noise. It is clear that as the signal-to-noise ratio increases, so does *composite acceptability* of the DAM (an overall measure of quality). The graph clearly shows that additive noise affects the background factor BN (hissing, as in noise-masked speech), while little distortion has been introduced into actual signal quality.

9.3 Objective Quality Measures

Continued research in speech coding and enhancement has necessitated improved quality assessment techniques. Since the goal for coding or enhancement is to produce speech that is perceived by the auditory system to be natural and free of degradation, it is understandable that subjective quality measures be the preferable means of quality assessment. However, as speech compression and enhancement algorithms become increasingly complex, it becomes imperative to be able to distinguish even the most subtle differences in processed speech quality. Further, subjective measures generally serve as a means of obtaining a broad measure of performance. For example, we may wish to investigate the performance of an LPC algorithm for varying types of feedforward or feedbackward predictors. In order for subjective measures to be useful, quality differences must be large enough to be distinguishable in the listener group. If only marginal quality differences exist, it may be difficult to fix algorithm parameters for the resulting quality. Subjective testing requires sig-

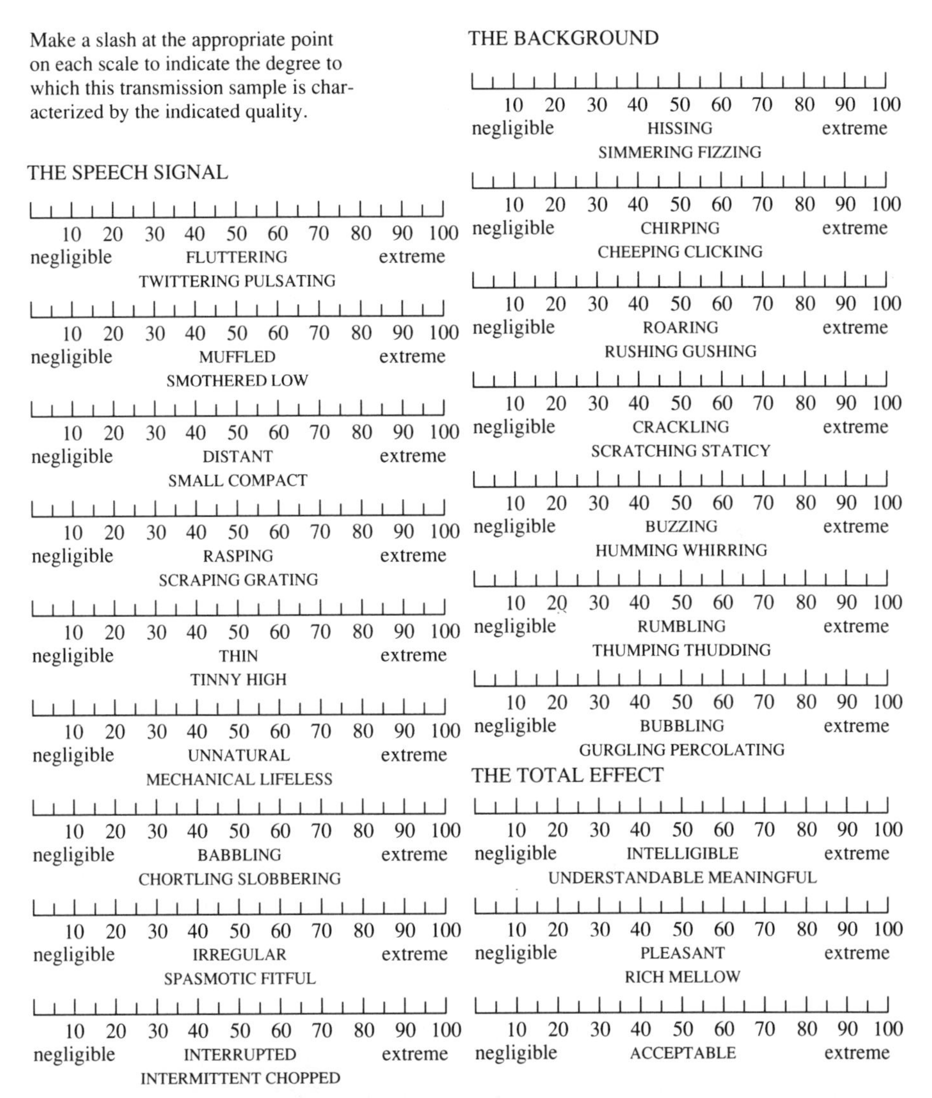

Make a slash at the appropriate point on each scale to indicate the degree to which this transmission sample is characterized by the indicated quality.

THE SPEECH SIGNAL

10 20 30 40 50 60 70 80 90 100
negligible FLUTTERING extreme
TWITTERING PULSATING

10 20 30 40 50 60 70 80 90 100
negligible MUFFLED extreme
SMOTHERED LOW

10 20 30 40 50 60 70 80 90 100
negligible DISTANT extreme
SMALL COMPACT

10 20 30 40 50 60 70 80 90 100
negligible RASPING extreme
SCRAPING GRATING

10 20 30 40 50 60 70 80 90 100
negligible THIN extreme
TINNY HIGH

10 20 30 40 50 60 70 80 90 100
negligible UNNATURAL extreme
MECHANICAL LIFELESS

10 20 30 40 50 60 70 80 90 100
negligible BABBLING extreme
CHORTLING SLOBBERING

10 20 30 40 50 60 70 80 90 100
negligible IRREGULAR extreme
SPASMOTIC FITFUL

10 20 30 40 50 60 70 80 90 100
negligible INTERRUPTED extreme
INTERMITTENT CHOPPED

THE BACKGROUND

10 20 30 40 50 60 70 80 90 100
negligible HISSING extreme
SIMMERING FIZZING

10 20 30 40 50 60 70 80 90 100
negligible CHIRPING extreme
CHEEPING CLICKING

10 20 30 40 50 60 70 80 90 100
negligible ROARING extreme
RUSHING GUSHING

10 20 30 40 50 60 70 80 90 100
negligible CRACKLING extreme
SCRATCHING STATICY

10 20 30 40 50 60 70 80 90 100
negligible BUZZING extreme
HUMMING WHIRRING

10 20 30 40 50 60 70 80 90 100
negligible RUMBLING extreme
THUMPING THUDDING

10 20 30 40 50 60 70 80 90 100
negligible BUBBLING extreme
GURGLING PERCOLATING

THE TOTAL EFFECT

10 20 30 40 50 60 70 80 90 100
negligible INTELLIGIBLE extreme
UNDERSTANDABLE MEANINGFUL

10 20 30 40 50 60 70 80 90 100
negligible PLEASANT extreme
RICH MELLOW

10 20 30 40 50 60 70 80 90 100
negligible ACCEPTABLE extreme

FIGURE 9.4. The DAM rating form. After Quackenbush et al. (1988).

nificant time and personnel resources. For some classes of distortions, it may not always be exactly reproducible. We therefore turn to the class of objective speech quality measures that are reliable, easy to implement, and have been shown to be good predictors of subjective quality.

The performance criterion for an objective speech quality measure is its correlation with subjective quality estimates. To obtain the correlation coefficient, both a subjective and an objective measure must be applied to a database of processed speech. A correlation analysis is applied to determine the ability of the objective quality measure to predict quality as judged by the listeners in the subjective evaluation. Such evaluation procedures have been performed by research laboratories over extensive data-

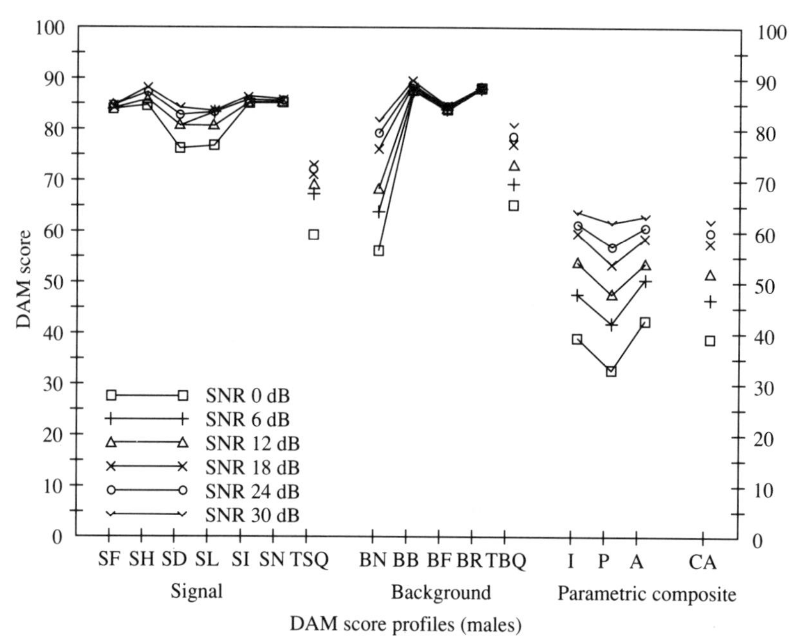

FIGURE 9.5. Subjective quality as measured by the DAM versus presence of additive background noise. After Quackenbush et al. (1988).

bases, incorporating a variety of vocoder distortions (Barnwell and Voiers, 1979; Barnwell et al., 1984; Barnwell, 1985; Quackenbush, 1985; Voiers, 1977). A representative summary is included at the end of this chapter.

Quality assessment via an objective measure provides a quantitative, repeatable, and accurate means of comparing vocoder performance. All objective measures make a direct comparison between an original (or reference) waveform and the processed (resynthesized or enhanced) version. Since a direct comparison is made, it is necessary that the original and processed speech waveforms be synchronized. The common method of implementing an objective measure is to partition speech into frames of typically 10–30 msec in duration, then to compute a distance/distortion measure for each frame. Most measures weigh differences in spectral characteristics between original and processed data. A final measure is formed by combining the distortion measures from each frame. One or more mathematical conditions are normally used in formulating objective measures. These include positive definiteness, symmetry, and the triangle inequality (see Section 1.3.1). If all three conditions are met, the measure is called a metric. In the next sections, we discuss five classes of objective speech quality measures.

9.3.1 Articulation Index

One of the first widely accepted objective speech quality measures is the AI. The AI was originally proposed by French and Steinberg (1947) for quality assessment of analog signals. Other researchers (Flanagan, 1972; House et al., 1965; Kryter, 1962a, 1962b) subsequently developed the AI measure. Although the AI measures only one aspect of quality—intelligibility—it is quite accurate. The AI assumes that the intelligibility of a processed signal is equal to the component intelligibility losses across a set of frequency bands that span the speech spectrum. The frequency limits for each band are normally associated with the critical bands for the human auditory system. In the study by French and Steinberg, the mel scale was used (see Section 6.2.4), although others have also been proposed. Table 9.6 summarizes 20 typical frequency bands used to formulate the AI. The AI assumes that distortion in one band is independent of losses in other bands. Another underlying assumption is that the distortion present in the noisy speech results from either additive noise or signal attenuation. Other processing steps must be incorporated if the AI measure is to be used for other types of distortion (e.g., signal-dependent noise).

The *articulation* within a frequency band is defined as that fraction of the original speech energy perceivable by the listener. Speech is deemed perceivable if it is above the ear's threshold of hearing and below the threshold of pain (see references in Appendix 1.F). If the dynamic range of the processed speech falls entirely within this band in the absence of noise, the AI would have a measure of 1.0. In practice, some residual noise is normally present in the enhanced signal; therefore, the noise spectrum is measured and partitioned into each of the bands and compared to processed signal energy. The final AI is equal to that fraction of the dynamic range of the signal which is below the threshold of pain, above the threshold of hearing, and above the background noise-masking spectrum. Therefore, the bandwidth of each of the filters in Table 9.6 is such that each contributes equally to speech intelligibility. One way to measure AI is to compute the SNR for each band j, for $j = 1, \ldots, 20$, and average the measures. SNR measures for each frequency band must be

TABLE 9.6. Frequency Bands (in Hz) of Equal Contribution to the Articulation Index.

Number	Frequency Limits	Mean	Number	Frequency Limits	Mean
1	200–330	270	11	1600–1830	1740
2	330–430	380	12	1830–2020	1920
3	430–560	490	13	2020–2240	2130
4	560–700	630	14	2240–2500	2370
5	700–840	770	15	2500–2820	2660
6	840–1000	920	16	2820–3200	3000
7	1000–1150	1070	17	3200–3650	3400
8	1150–1310	1230	18	3650–4250	3950
9	1310–1480	1400	19	4250–5050	4650
10	1480–1660	1570	20	5050–6100	5600

limited by the threshold of pain and hearing (in our relation, we limit the SNR to 30 dB). The measure is formulated as

$$\mathrm{AI} = \frac{1}{20} \sum_{j=1}^{20} \frac{\min\{\mathrm{SNR}_j, 30\}}{30}. \tag{9.2}$$

This relation is a "long-term" measure in the sense that each SNR is computed over the entire waveform. Further comments about this point follow in the context of discussing the direct use of the SNR.

It should be emphasized that the AI is a good predictor of intelligibility for the analog communication systems for which it was originally designed. In many digital coding or speech enhancement applications, noise characteristics become signal-dependent and thereby violate the underlying AI assumptions. For example, speech enhancement algorithms such as spectral subtraction with half-wave rectification or nonlinear coders produce noise artifacts that cannot be modeled as additive, or independent across frequency bands. Therefore, care must be taken in using AI measures for speech evaluation. Once a number has been obtained for the AI, it is necessary to relate it to intelligibility. Articulation tests are subject to considerable variability and their results depend strongly on the testing technique, data, and procedure. Usually, it is more relevant to consider differences in intelligibility scores between systems under similar test conditions. As an example, Figure 9.6 illustrates empirical relations between the intelligibility score and the AI for several test conditions.

9.3.2 Signal-to-Noise Ratio

The SNR is the most widely used measure for analog and waveform coding systems and has also been used in the past for assessing enhancement algorithms for broadband noise distortions. Several variations exist, including classical SNR, segmental SNR, and frequency-weighted segmental SNR, to name a few. It is important to note that SNR-based measures are only appropriate for coding or enhancement systems that seek to reproduce the original input waveform. Let $y(n)$ denote a noisy speech signal at time $n, s(n)$ its noise-free equivalent, and $\hat{s}(n)$ the corresponding enhanced or processed signal. These signals are all assumed to be energy signals (Sections 1.1.3, 4.2.5, and 4.3.4). The time error signal can then be written as

$$\varepsilon(n) = s(i) - \hat{s}(i). \tag{9.3}$$

The error energy is then

$$E_\varepsilon = \sum_{n=-\infty}^{\infty} \varepsilon^2(i) = \sum_{n=-\infty}^{\infty} [s(i) - \hat{s}(i)]^2. \tag{9.4}$$

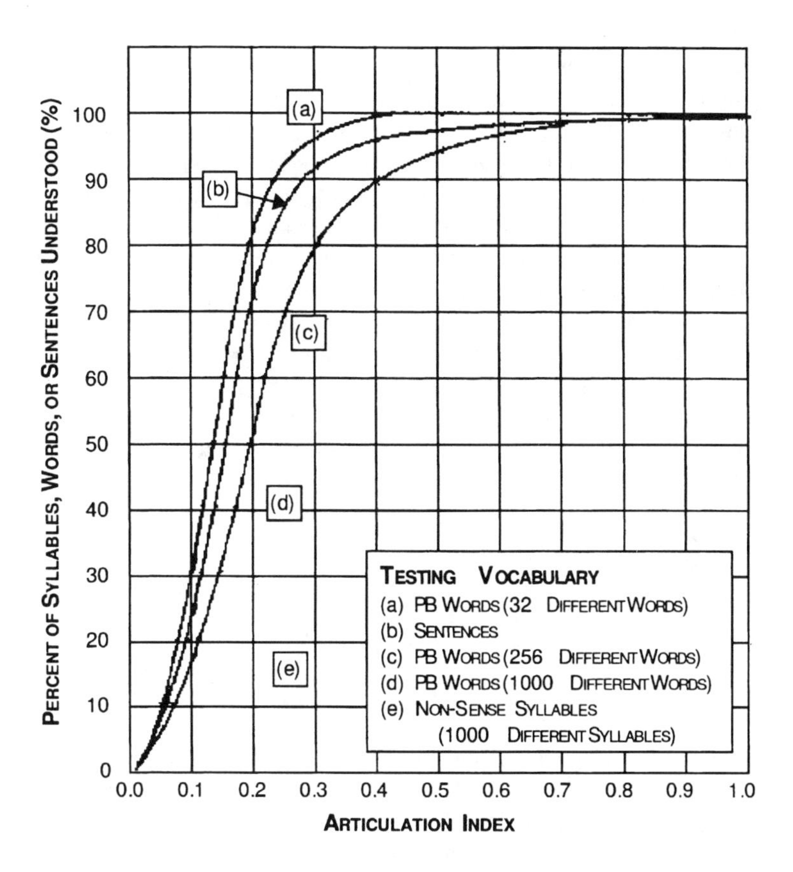

FIGURE 9.6. Several experimental relations between the AI and speech intelligibility (Kryter, 1962a, 1962b).

The energy contained in the speech signal itself is

$$E_s = \sum_{n=-\infty}^{\infty} s^2(n). \tag{9.5}$$

The resulting SNR measure (in dB) is obtained as

$$\mathrm{SNR} = 10 \log_{10} \frac{E_s}{E_\varepsilon} = 10 \log_{10} \frac{\sum_n s^2(n)}{\sum_n [s(n) - \hat{s}(n)]^2}. \tag{9.6}$$

As is the case for many quality measures used in evaluating coding or enhancement systems, the original speech signal $s(n)$ is needed to determine improvement. Therefore, such measures are used primarily in simulation where both degraded and noise-free speech signals are available.

The principal benefit of the SNR quality measure is its mathematical simplicity. The measure represents an average error over time and frequency for a processed signal. It has been well documented, however, that classical SNR is a poor estimator of speech quality for a broad range of speech distortions (McDermott, 1969; McDermott et al., 1978a, 1978b; Tribolet, 1978). This is because SNR is not particularly well related to any subjective attribute of speech quality, and because the SNR weights all time domain errors in the speech waveform equally. The speech energy in general is time varying. If we assume that the noise distortion is broadband with little energy fluctuation, then the SNR measure should vary on a frame-by-frame basis. A deceptively high SNR measure can be obtained if an utterance contains a high concentration of voiced segments, since noise has a greater perceptual effect in low-energy segments (e.g., unvoiced fricatives). A much-improved quality measure can be obtained if SNR is measured over short frames and the results averaged. The frame-based measure is called the *segmental SNR* ($\mathrm{SNR}_{\mathrm{seg}}$), and is formulated as

$$\mathrm{SNR}_{\mathrm{seg}} = \frac{1}{M} \sum_{j=0}^{M-1} 10 \log_{10} \left[\frac{\displaystyle\sum_{n=m_j-N+1}^{m_j} s^2(n)}{\displaystyle\sum_{n=m_j-N+1}^{m_j} [s(n) - \hat{s}(n)]^2} \right], \tag{9.7}$$

where $m_0, m_1, \ldots, m_{M-1}$ are the end-times for the M frames, each of which is length N. For each frame (typically 15-25 msec), an SNR measure is computed and the final measure obtained by averaging these measurements over all segments of the waveform. The segmentation of the SNR permits the objective measure to assign equal weight to loud and soft portions of the speech.

In some cases, problems can arise with the $\mathrm{SNR}_{\mathrm{seg}}$ measure if frames of silence are included, since large negative $\mathrm{SNR}_{\mathrm{seg}}$'s result. This problem is alleviated by identifying silent periods and excluding them from $\mathrm{SNR}_{\mathrm{seg}}$ calculations. Another approach is to set a lower threshold and replace all frames with $\mathrm{SNR}_{\mathrm{seg}}$ measures below it to the threshold (e.g., a 0-dB threshold is reasonable). This prevents the measure from being overwhelmed by a few frames of silence. At the other extreme, frames with $\mathrm{SNR}_{\mathrm{seg}}$ measures greater than 35 dB are not perceived by listeners as being significantly different. Therefore, an upper threshold (normally 35 dB) is used to reset any unusually high $\mathrm{SNR}_{\mathrm{seg}}$ frame measures. The two thresholds thereby prevent the final $\mathrm{SNR}_{\mathrm{seg}}$ measure from being biased in either a positive or a negative direction from a few frames that do not contribute significantly to overall speech quality.

The final SNR measure we will discuss is the *frequency weighted segmental SNR* ($\mathrm{SNR}_{\mathrm{fw-seg}}$) measure. Several variations have been formulated (Tribolet, 1978; Barnwell, 1980) that seek to obtain a segmental SNR within a set of frequency bands normally spaced proportionally to the ear's critical bands (see Section 6.2.4; Scharf, 1970; Davis and Mermelstein, 1980). Such a measure allows for a series of noise-dependent perceptual weights, $w_{j,k}$, to be applied in each band and thereby produces an SNR measure more closely related to a listener's perceived notion of quality. The measure is formed as follows,

$$\mathrm{SNR}_{\mathrm{fw-seg}} = \frac{1}{M} \sum_{j=0}^{M-1} 10 \log_{10} \left[\frac{\sum_{k=1}^{K} w_{j,k} 10 \log_{10} \left[E_{s,k}(m_j) / E_{\varepsilon,k}(m_j) \right]}{\sum_{k=1}^{K} w_{j,k}} \right], \tag{9.8}$$

where M represents the number of speech frames (indexed by $m_0, \ldots, m_{M-1}$), K the number of frequency bands, $E_{s,k}(m_j)$ the short-term signal energy contained in the kth frequency band for the frame of noise-free speech indexed by m_j, and $E_{\varepsilon,k}$ the similar quantity for the noise sequence $\varepsilon(n)$. Studies have shown the $\mathrm{SNR}_{\mathrm{fw-seg}}$ measure to be a better predictor of speech quality than the classical SNR or $\mathrm{SNR}_{\mathrm{seg}}$ measures.

9.3.3 Itakura Measure

As we have discussed in previous chapters, the human auditory system is relatively insensitive to phase distortion (Wang and Lim, 1982). This is particularly important in many speech communication systems (e.g., telephone or radio) and most enhancement algorithms.[2] Therefore, many enhancement and coding systems focus only on the magnitude of the speech spectrum. As a result, the coded or enhanced waveform can be quite different from the original, yet still be perceived similarly by the listener. Measures such as those based on SNR, which obtain a distortion measure based on sample-by-sample differences in the original and processed time waveforms, do not provide a meaningful measure of performance when the two waveforms differ in their phase spectra. Distance measures that are sensitive to variations in the speech spectrum are therefore needed. One of the more successful is the Itakura distance measure, which we discussed in Section 5.3.5 (Itakura, 1975; Crochiere et al., 1980; Chu and Messerschmitt, 1982). This measure is based on the dissimilarity between all-pole models of the reference and enhanced or coded speech waveforms. The distance measure is computed between sets of LP parameters estimated over synchronous frames (typically every 15–30 msec) in the original and processed speech. In Problem 9.8 we demonstrate that the Itakura measure is heavily influenced by spectral dissimilarity due to mismatch in formant locations, whereas errors in matching spectral valleys do not contribute heavily to the distance. This is desirable, since the auditory system is more sensitive to

[2]There are some special noise-reduction applications where an array of microphones is used to perform speech enhancement (such as adaptive noise calceling or some hearing aids). Under these conditions, phase information between microphone inputs is extremely important.

errors in formant location and bandwidth than to the spectral valleys between peaks.

We noted in Section 5.3.5 that the Itakura distance is not a metric because it does not have the required property of symmetry. That is, if $d_I(\cdot,\cdot)$ denotes the Itakura distance, and $\hat{\mathbf{a}}(m)$ and $\hat{\mathbf{b}}(m')$ two LP vectors between which we desire the distance, then

$$d_I(\hat{\mathbf{a}}(m), \hat{\mathbf{b}}(m')) \neq d_I(\hat{\mathbf{b}}(m'), \hat{\mathbf{a}}(m)). \tag{9.9}$$

If a symmetric measure is desired, a combination can be formed such as

$$\tilde{d}_I(\hat{\mathbf{a}}(m), \hat{\mathbf{b}}(m')) = \tfrac{1}{2}\left[d_I(\hat{\mathbf{a}}(m), \hat{\mathbf{b}}(m')) + d_I(\hat{\mathbf{b}}(m'), \hat{\mathbf{a}}(m))\right]. \tag{9.10}$$

In addition to symmetry, this measure also has the property that if the processed spectrum is identical to the original, the resulting distance is zero.

9.3.4 Other Measures Based on LP Analysis

Another class of objective quality measures are the *LPC parametric distances.* These measures are based on the dissimilarity between sets of LP parameters from original and processed speech signals. A variety of coefficients can be used to represent the LP model, some of which include predictor coefficients, reflection or parcor coefficients, log-area ratio (LAR) coefficients, and cepstral parameters. With the exception of the cepstral coefficients (Chapter 6), all of these parameters are discussed in Section 5.3.3 in conjunction with the L–D recursion. It has been shown by Barnwell et al. (1979, 1984, 1985) and Quackenbush et al. (1988) that of all LP-based measures, the LAR measure has the highest correlation with subjective quality. An objective quality measure based on the LAR parameters is obtained by simply computing the Euclidean distance between sets of LARs from original and enhanced speech frames,

$$d_{\text{LAR}}(\mathbf{g}_s(m), \mathbf{g}_{\hat{s}}(m)) = \sqrt{\frac{1}{M}\sum_{l=1}^{M}[g_s(l;m) - g_{\hat{s}}(l;m)]^2}, \tag{9.11}$$

where $g_x(l;m)$ represents the lth LAR for the signal $x(n)$ computed over the frame ending at time m, and $s(n)$ and $\hat{s}(n)$ are the original and processed speech, respectively. Also, $\mathbf{g}_x(m)$ is the vector of LAR parameters for signal $x(n)$ at frame m. Since many coding and enhancement algorithms have been formulated assuming an LP model for speech production, this measure has been useful for evaluating both types of systems.

Since spectral distance measures have been the most widely investigated class of quality measures, many variations exist (Barnwell and Bush, 1978; Barnwell, 1980; Breitkopf and Barnwell, 1981; Gray and Markel, 1976; Gray et al., 1980; Klatt, 1982; Viswanathan et al., 1983).

In fact, any set of LP-based parameters could be used in place of the LAR parameters found in (9.11). A generalized LP spectral distance measure between two frames, ending at times m and m', is based on a weighted Minkowski metric of order β,

$$d_{\text{LP}}(\boldsymbol{\varphi}_s(m), \boldsymbol{\varphi}_{\hat{s}}(m')) = \left| \frac{\sum_{l=1}^{M} w_{l,m,m'} \left[\varphi_s(l;m) - \varphi_{\hat{s}}(l;m') \right]^{\beta}}{\sum_{l=1}^{M} w_{l,m,m'}} \right|^{1/\beta}, \tag{9.12}$$

where $\varphi_x(l;m)$ represents the lth parameter for the signal $x(n)$ computed over the frame ending at time m, and $s(n)$ and $\hat{s}(n)$ are the original and processed speech, respectively. The numbers $w_{l,m,m'}$ are weights added for generality. $\varphi_x(m)$ is the M-vector of parameters $\varphi_x(l;m)$, $l \in [1, M]$. The choice of the weights depends on the specific noise or distortion, choice of LP parameters, and the contribution to overall quality assuming an auditory model.

9.3.5 Weighted-Spectral Slope Measures

The last objective speech quality measures we consider are direct *spectral distance measures.* These measures are based on comparisons of smoothed spectra from the reference and processed speech signals. The smoothed spectrum can be obtained from a variety of methods. For example, one could perform LP analysis and obtain the speech system spectrum (Section 5.5.2), apply a low-time lifter in the cepstral domain (Section 6.2.1), or apply a filter-bank analysis procedure (Section 4.3.5). Of all objective quality measures, the spectral distance measures have been the most widely investigated in recent years (Barnwell and Bush, 1978; Barnwell, 1980; Coetzee and Barnwell, 1989; Gray and Markel, 1976; Gray et al., 1980; Halka, 1991; Klatt, 1976, 1982; Viswanathan et al., 1976, 1983; Wang et al., 1991). Several have gained wide acceptance in quality assessment; of these, we will focus on a specific technique formulated by Klatt (1982) known as the *weighted-spectral slope measure* (WSSM), sometimes referred to as the *Klatt measure.* The WSSM is based on critical-band filter analysis and is therefore more closely related to aspects of listener intelligibility than are measures such as SNR. In this approach, 36 overlapping filters of progressively larger bandwidths are used to estimate the smoothed short-term speech spectra every 12 msec. The filter-bank bandwidths are chosen to be proportional to the ear's critical bands (Section 6.2.4, and Sharf, 1970; Zwicker, 1961) to give equal perceptual weight to each band. Once the filter bank is formed, the measure finds a weighted difference between the spectral slopes in each band. The resulting measure is therefore sensitive to differences in formant location, yet less sensitive to differences in the height of those peaks or differences in spectral valleys.

The procedure for finding the WSSM distortion measure involves three steps. First, the spectral slope in each frequency band of each signal is found using the following computations (the index m refers to the current frame),

$$\Delta|S(k;m)| = |S(k+1;m)| - |S(k;m)|$$
$$\Delta|\hat{S}(k,m)| = |\hat{S}(k+1;m)| - |\hat{S}(k;m)| \tag{9.13}$$

for $k = 1, \ldots, 36$, where $S(k;m)$ is the stDFT of the original reference spectrum evaluated at the center frequency of band k, and $\hat{S}(k;m)$ is the similar quantity for the processed spectrum. The magnitude spectra are expressed in dB even though this is not explicit in (9.13). The second step is to calculate a weight for each band. The magnitude of each weight reflects whether the band is near a spectral peak or valley, and whether the peak is the largest in the spectrum. For a given frame time m, Klatt computes the weight for each spectrum separately, then averages the two sets of weights to obtain $w_{k,m}$, $k = 1, \ldots, 36$. After obtaining the set of band weights, the third step is to form the per-frame spectral distance measure, say

$$d_{\mathrm{WSSM}}(|S(\omega;m)|, |\hat{S}(\omega;m)|) = K + \sum_{k=1}^{36} w_{k,m}\left[|S(k;m)| - |\hat{S}(k;m)|\right]^2, \tag{9.14}$$

where the term K is related to overall sound pressure level of the reference and processed utterances, and also may be adjusted to increase overall performance.

The Klatt measure possesses several properties that are attractive for quality assessment. First, it is not required to identify and time-align speech formants prior to spectral distance computation. Second, no prior knowledge is required for normalizing differences in spectral tilt. Finally, the measure implicitly weights spectral differences due to varying bandwidths in the filters used to estimate the short-time spectrum. This in turn yields a perceptually meaningful frequency weighting.

9.3.6 Global Objective Measures

The objective speech quality measures discussed above each represents a distortion measure between an original and a processed speech frame. The processed speech quality approaches that of the original speech as each frame quality measure approaches zero. The individual frame distances can be averaged over an utterance, or group of utterances, to obtain a "global" distance for the entire signal. Once frame distance measures have been obtained, a variety of other global measures is also possible. If input phonetic label information is available, objective quality measures can be grouped by phonemes (/@/, /f/, /G/), speech classes (vowels, nasals, fricatives), or speakers, to name but a few. Although mean quality is an important measure of performance for a coding or enhancement algorithm, consistency is also important. As an example, let us assume that we have two coding algorithms to compare. Coder A does an outstanding job for all voiced sounds, but does poorly for unvoiced. Coder B performs marginally for both voiced and unvoiced speech. We also assume that the test data used for evaluation have a higher concentration of voiced versus unvoiced speech frames, and that the mean quality for coder A is superior to that of marginal coder B. A listener group might prefer coder B over coder A because the overall quality is more consistent. This aspect, referred to as *listener fatigue*, can be addressed by computing the variance of each quality measure (though few studies include this value). An even better means of representing global speech quality is to estimate a pdf for the resulting measure. This gives a clear indication of algorithm performance.

9.3.7 Example Applications

We close the discussion of objective quality measures with several illustrative examples. Figure 9.7(a) illustrates the accumulated distortion

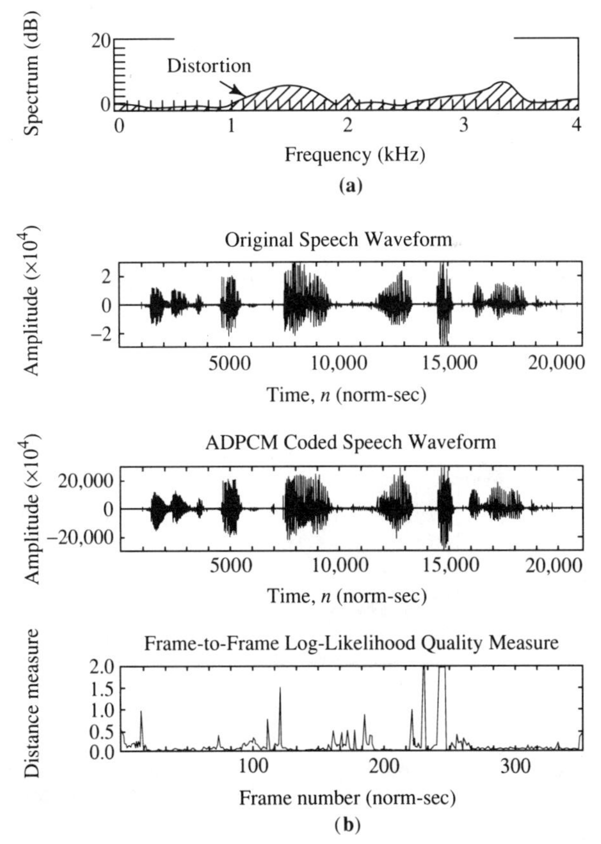

FIGURE 9.7. Example of Itakura quality measure for ADPCM coded speech: (a) Distortion for a frame across frequency; (b) original and coded speech waveforms with frame-to-frame quality measure.

represented by an objective measure. The spectral envelope for a single speech frame before and after ADPCM coding is shown (the plot illustrates distortion versus frequency). Quality can also be seen as a function of time in Fig.9.7(b). Here, the sentence "Only the best players enjoy popularity" spoken by a male is shown before and after ADPCM coding. A frame-to-frame Itakura quality measure d_I is also shown. Essentially, the area under each frame measure corresponds to the accumulated distortion introduced by the coding process. The results show that the coder performs well for steady-state voiced frames, with decreasing performance for dynamically changing speech characteristics (e.g., stop consonants such as /p/ and /t/). Quality measures can also be illustrated in histogram form (Hansen and Nandkumar, 1991, 1992) as shown in Fig. 9.8. Here, an example of the three objective quality measures Itakura d_I, log-area-ratio d_{LAR}, and weighted-spectral slope d_{WSSM}, for a 4.8 Kbit/sec CELP coder (Kemp et al., 1989 Campbell et al., 1991) is shown in histogram form.[3] Deviation from the mean, as well as concentration in

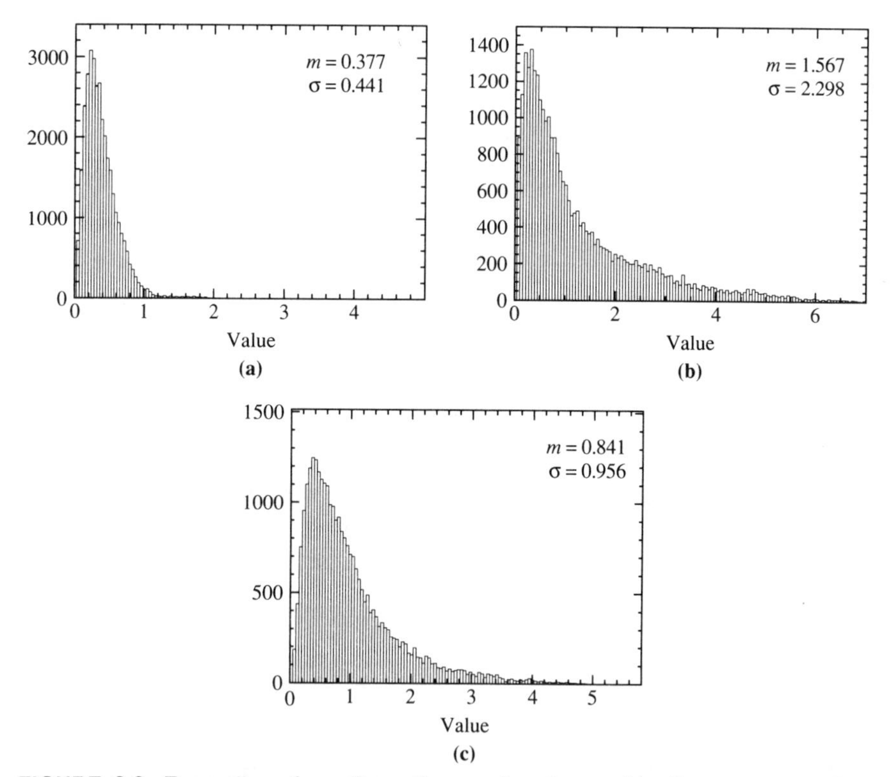

FIGURE 9.8. Examples of quality ratings using three objective measures for CELP coded speech: (a) Itakura measure; (b) LAR measure; and (c) WSSM.

[3] To obtain these histograms, 100 sentences from the TIMIT data base (see Section 13.8) were processed, resulting in approximately 37,000 frame-to-frame measures (Hansen and Nandkumar, 1991).

the distribution tails, help to identify any variability in speech quality from the given coding algorithm.

Finally, an example of objective speech versus time for noisy conditions and speech enhancement is shown in Fig. 9.9. Figure 9.9(a) and (b) are plots of the male speech waveform in noisefree and noisy conditions. The same sentence as in Fig. 9.7 was degraded with 5-dB SNR of additive white Gaussian noise. Figure 9.9(c) illustrates the distortion introduced by additive noise via frame-to-frame Itakura d_I quality measure between Fig. 9.9(a) and (b). Finally, Fig. 9.9(d) shows the result of a single-channel Wiener filter (see Section 8.4) which was used to enhance the speech waveform from Fig. 9.9(b) (three iterations were used). The waveform shows a reduction in noise during periods of silence, as well as a decrease in distortion from the resulting frame-to-frame Itakura quality measure plot in Fig. 9.9(e).

9.4 Objective Versus Subjective Measures

The previous sections have presented a collection of measures or techniques for assessing speech quality for speech coding or enhancement applications. It is prudent to wonder about how one chooses a quality measure for a particular application. The answer depends on the kind of algorithm, the type of noise present in the processed speech signal, and what noise and signal characteristics are deemed acceptable and unacceptable to the listener. Subjective quality measures require careful planning, extensive speaker and listener resources, and normally give little insight as to how to improve an algorithm. Objective quality measures are easy to apply, require few speaker or listener resources, can quantify performance over a variety of scales, and in many cases can point a researcher in the direction of improved performance. The usefulness of an objective quality measure lies in its ability to predict subjective quality. Ideally, a combination of both subjective and objective testing procedures should be employed when evaluating a speech processing system. As a general rule, objective measures are used early in system design to set speech parameters, and subjective measures to assure actual listener preference.

In closing, we summarize the correlation of several objective quality measures with composite acceptability of the DAM in Table 9.7. The results suggest that for waveform coders, the $\text{SNR}_{\text{fw-seg}}$ measure is the best predictor of subjective quality. For other voice coding algorithms, as well as many enhancement algorithms, a number of measures are shown to be well correlated with subjective quality. Points that may be extracted from this table include:

1. The LAR measure, which requires an order of magnitude less computation than other spectral distance measures, demonstrates a competitive correlation coefficient.

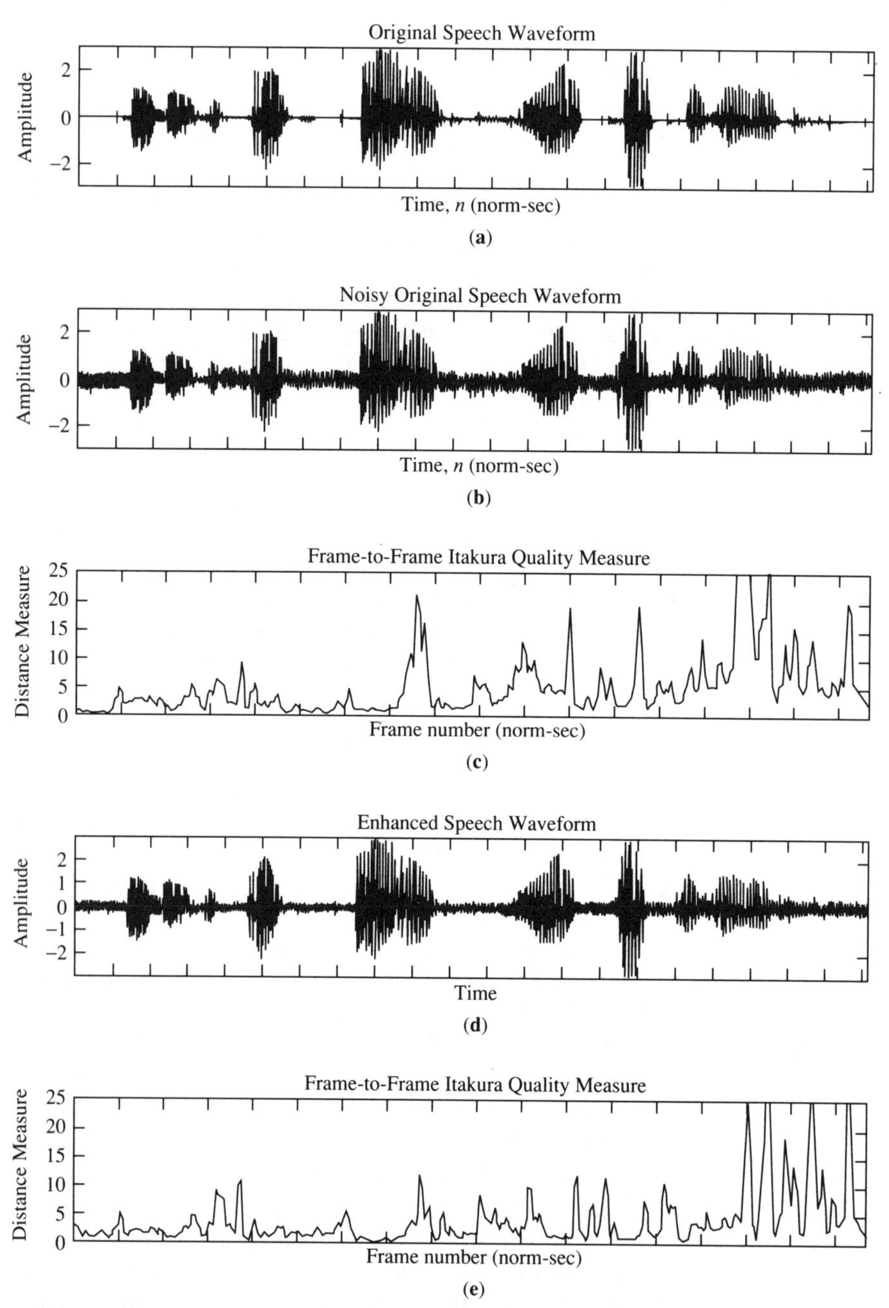

FIGURE 9.9. Examples of the Itakura quality measure for speech degraded by additive noise and enhanced speech.

TABLE 9.7. Comparison of the Average Correlation Coefficient $|\hat{\rho}|$ Between Objective and Subjective Speech Quality (as Measured by Composite Acceptability of DAM).

Objective Quality Measure	$\|\hat{\rho}\|$
SNR	0.24*
SNR_{seg}	0.77*
SNR_{fw-seg}	0.93*
LP-based measures:	
LP coefficients	0.06
Reflection coefficients	0.46
Log predictor coefficients	0.11
Log reflection coefficients	0.11
Linear area ratios	0.24
Log-area ratios	0.62
Itakura distance	0.59
Linear spectral distance	0.38
Inverse linear spectral distance	0.63
Log spectral distance	0.60
Nonlinear spectral distance	0.61
Frequency variant linear spectral distance	0.68
WSSM	0.74
Composite measures:	
Simple and frequency-weighted variant measures	0.86
Parametric objective measures	0.82

*SNR measures are correlated across only waveform coder distortions. After Quackenbush et al. (1988.)

2. Of those measures employing an aural model, the WSSM possesses the highest correlation coefficient with subjective quality.
3. The best predictors of subjective quality are *composite measures*, which are those formed using multiple linear regression on sets of simple measures. The high degree of correlation results by selecting a number of parameters in the composite objective measure that yield maximum correlation. The performance of composite measures can be considered an estimate of the limit of the ability of objective measures to predict subjective quality results.

9.5 Problems

9.1. Discuss the differences between speech quality and speech intelligibility in a medium-bandwidth speech communication system. How does speech coding transmission rate affect speech quality? Define the following terms: synthetic quality, communications quality, toll quality, and broadcast quality.

9.2. We wish to construct a subjective quality-based rhyme test to measure coder performance for unvoiced stop consonants. Using the MRT and DRT as models, devise such a test. Include (i) word list(s), (ii) scoring method, (iii) a test of your procedure, and (iv) a suggested confidence measure.

9.3. Consider speech communications over a typical telephone channel. Under these conditions, it is normally assumed that the transmitted utterance is ideal bandpass filtered with a lower cutoff frequency 200 Hz and an upper cut-off 3600 Hz.

(a) What are the effects on quality and intelligibility due to this processing and why?

(b) Now suppose that the passband of the channel introduces a spectral tilt of 2 dB/octave. How will this affect subjective speech quality?

9.4. Suppose that a narrow-band noise signal $d(n)$ is used to corrupt an input speech signal $s(n)$ that occupies a 4 kHz bandwidth, resulting in a distorted signal $y(n)$. Let the noise signal be uniform in the band $[1.0, 1.2]$ kHz.

(a) Write an expression for the global SNR (i.e., over the entire speech waveform).

(b) Write an expression for a time-domain segmental SNR (time-domain frames of N samples).

(c) Write an expression for a frquency-domain segmental SNR (frequency domain blocks of 200 Hz).

(d) Discuss the trade-offs encountered in the use of each of these measures to predict speech quality. Which is the better predictor of speech quality for this distortion and why?

9.5. (Computer Assignment) A speech-coding strategy occupies a $[0, 4]$ kHz bandwidth. We wish to determine the effect on speech quality of increasing additive white Gaussian noise. Using speech from a male speaker, degrade an input utterance with global SNRs of 5, 10, 20 dB.

(a) Using frames of 128 samples, with a 64-sample overlap from frame to frame, find the segmental SNR measure for the three degraded speech waveforms. Discuss the differences in segmental SNR versus global SNR.

(b) Suppose that the listener has a mild hearing loss so that distortion above 3 kHz is not perceived. Construct a new segmental-based SNR measure that does not include distortion outside the auditory band of this listener.

9.6. Consider an ADPCM vocoder. Assume that a fixed predictor is used with adaptive step size (see Chapter 7).

(a) What is the impact on the LAR quality measure of increasing the data rate from 16 kbit/sec to 32 kbit/sec?

(b) Suppose that the ADPCM vocoder is modified to have fixed step size but adaptive prediction. Assuming a similar increase in transmission rate to that in (a), how will the LAR measure change? Is it better to adapt step size or prediction when considering the LAR quality measure?

9.7. In Fig. 9.8, we see that quality measures for ADPCM are poor for stop consonants, but good for vowels. Discuss why this is true. Is the human auditory system more sensitive to distortion in formant location or formant amplitude? Why?

9.8. (Computer Assignment) Construct a digital simulator for a neutral vowel using four complex pole pairs. For discussion purposes, assume that the Nyquist frequency is 4 kHz. Excite this filter with an 8 msec periodic pulse train and obtain a 1-sec output waveform. Use this as the reference waveform.

(a) Using the original 8-pole filter, decrease the radial pole locations by 5% (i.e., move the eight poles toward the origin in the z-plane). This introduces a distortion in formant amplitude and bandwidth. Obtain 1 sec of the distorted waveform. Using LPA analysis, find the frame-to-frame Itakura quality measure for the formant bandwidth/amplitude distorted signal (use the original signal as reference).

(b) Using the original 8-pole filter, increase the frequency locations of all poles by 5% (e.g., if the first pole-pair is at 500 Hz, the modified pole-pair will be 525 Hz). This introduces a distortion in formant location. Obtain 1 sec of the distorted waveform and find the frame-to-frame Itakura quality measure.

(c) Compare frame-to-frame Itakura quality measures for distortion in formant location versus bandwidth/amplitude. Which distortion introduces a greater loss in speech quality? Repeat the process for a 10% shift in pole location for each distortion.

PART V

Recognition

CHAPTER 10

The Speech Recognition Problem

Reading Note: This chapter is very descriptive; no special reading in Chapter 1 is required.

10.1 Introduction

In this short chapter we introduce the field of speech recognition and discuss the problems that make this endeavor so challenging.

10.1.1 The Dream and the Reality

For most people, speech is the most natural and efficient manner of exchanging information. The goal of speech recognition technology, in a broad sense, is to create machines that can receive spoken information and act appropriately upon that information. Further information exchange from machine to human might then be required using synthetic speech. In these terms, the study of speech recognition is part of a quest for "artificially intelligent" machines that can "hear," "understand," and "act upon" spoken information, and "speak" in completing the information exchange. As anyone who has read or watched science fiction knows, the human capacity for imagining the form and applications of such systems seems unbounded. Our imagination, however, far surpasses our technical abilities in this domain late in the twentieth century. The objective of a robust, intelligent, fluently conversant machine remains a distant goal.

What *is* understood about speech recognition is increasing at a remarkable rate, but present knowledge is amenable to relatively modest tasks and constrained application domains. Roughly speaking, contemporary speech recognition systems with acceptable performance fall into one or more of three broad categories:

1. Those with small vocabularies (~10–100 words).
2. Those in which words are deliberately spoken in isolation from one another (vocabularies can exceed 10,000 words).
3. Those that accept continuous speech but are concerned with relatively constrained "task domains"; for example, messages likely to occur in office correspondence at a particular company (vocabularies typically ~1000–5000 words).

Most systems employed in practical applications are of the small-vocabulary or isolated-word type. Existing systems for more "natural" human–machine communication remain primarily experimental. No existing system, even of those being used in practical applications, is highly robust to environmental noise (office noise, factory noise, extraneous speech, etc.). All perform significantly better if required to recognize only a single speaker who "trains" the system. Even if the system is used to recognize multiple speakers, performance is generally improved if the system users are also the trainers. Whether a single- or a multispeaker system, utterances of cooperative speakers (who articulate clearly, do not use words outside the vocabulary, etc.) are more easily recognized. Although some existing systems take advantage of the grammatical structure of the language, only *experimental* systems have more abstract "cognitive" abilities like discerning *meaning* or learning from mistakes. When such qualities are present, they appear in very primitive measure.

In contrast, most natural application domains that would benefit from speech recognition do not use discrete, clearly articulated, utterances by a single person in a quiet environment, nor is it generally possible to have the system trained by its user population. In fact, speech recognition systems, to be maximally beneficial and universally applicable, must be capable of recognizing continuous speech, and must be able to recognize multiple speakers with possibly diverse accents, speaking styles, and different vocabularies and grammatical tendencies (perhaps even multiple languages); must be able to recognize poorly articulated speech; and must have the ability to recognize speech in noisy environments. Furthermore, all of these capabilities must come in an affordable, sufficiently small system that can operate in real time. Ideally, too, the system should adapt and learn new lexical, syntactic, semantic, and pragmatic information,[1] just as a human can. When placed in this perspective, the field of speech recognition is seen to be in its very early infancy.

The enormity of the problem notwithstanding, great progress has been made in recent decades, and the work that has been accomplished is certainly not without practical value. Small-vocabulary systems, even those requiring discrete word inputs, can be employed in many relatively simple applications to improve the efficiency of entering information to a machine[2]—in manufacturing environments (e.g., for sorting tasks), in applications where the hands are unavailable (in surgery, to assist a person with motor disabilities, in a darkroom, in a cockpit), or in applications where the user must remain in remote contact with the machine (over the phone, in hazardous environments). Systems constrained for use in

[1]These terms are defined below.

[2]Speech input is about twice as fast as information entry by a skilled typist (Kaplan, 1976).

certain "task domains" can be extremely useful in those domains. The possibilities are almost unlimited—one can imagine applications, for example, in library or other information retrieval functions, in air traffic control towers, in emergency response centers, in certain medical environments, or in any situation in which specific tasks tend to restrict the vocabulary and the message content. While application of existing technology proceeds, several promising technologies are contributing to laboratory and commercial systems that attempt to solve one or more of the challenging problems noted above. These methods will be the principal focus of our study.

As we shall discuss, the first step in solving the speech recognition problem was to understand its complexity. Categorizing and discussing the dimensions of the problem is the central issue of this chapter. In light of this overwhelming complexity as it is now understood, it is natural to wonder whether the goal noted in the opening paragraphs is a realistic one. The future will answer that question, and perhaps the reader will contribute to the solution, but a sense of optimism can be derived from assessing recent progress. Very realistic speech recognition systems have been developed in the short period since the invention of the digital computer. In the early 1970s, commercial systems became available that were quite remarkable in their time. These systems addressed a speech recognition problem considered nearly trivial by contemporary standards. They were designed to recognize discrete utterances (usually words) in relatively noise-free environments. The systems employed small vocabularies (10–100 words), and were often used in cases in which the machine was required to recognize only the speaker who trained it. In contrast, near the end of the 1980s researchers at IBM, for example, had developed an experimental system capable of recognizing a vocabulary of 20,000 words when uttered in isolation, or naturally spoken utterances drawn from a 5000-word vocabulary. Not only was such a system almost unimaginable in 1970, but the thought that it would be implemented in a desktop computer the size of a small suitcase probably would have been considered very unrealistic.

Another perspective on modern speech recognition technology is achieved by looking back beyond 1970. Prior to about 1950, the digital computer did not exist. Therefore, what we have called "recent" progress comprises more than half of the period over which speech recognition research has been conducted. Just as the digital computer gave rise to the modern era, so too have advances occurred in proportion to computing speed and memory size.

New interdisciplinary knowledge and improved computing technologies continue to advance the state of the art with each passing year. Tracking this progress is difficult, because the field does not move linearly—there are many different problems involved in the attempt to recognize speech, and each evolving system, whether commercial or research, tends to focus on some aspect of the whole. We will nevertheless attempt to piece together this status and its history as we go along in our study.[3]

10.1.2 Discovering Our Ignorance

Ironically, a realization of the fact that we are in the infancy of the speech recognition field is a consequence of several decades of painstaking research into what was once thought to be a relatively straightforward problem. In the 1950s researchers conceived of a machine that was to be called the "phonetic typewriter" (Fry and Denes, 1958; Dreyfus and Graf, 1962). The goal was to use acoustic features of speech and knowledge of phonetics to turn flowing speech into a phonetic transcription, and eventually into a *graphemic* (conventionally written) transcription of the message. These researchers did not anticipate the extreme difficulty of the task. The relative ease with which we humans communicate using speech obscures the awesome complexity of the task as it is now appreciated; it is a testimony to the remarkable ability of the "human computer." (This latter fact has inspired some researchers to investigate the use of "neural network" architectures in speech recognition. We will have more to say about this in Chapter 14.) In a 1976 article in the *IEEE Spectrum* (Kaplan, 1976), Dr. James Flanagan of Bell Laboratories made the following observation which is a largely accurate reflection of the state of the technology even as we approach the turn of the century: "The problem of speech recognition has not been solved, primarily because the speech communication process is a subtle one. Many of its fundamentals are not well understood. For example, while most researchers recognize that a short-time frequency spectrum of speech bears important information, the human ear and brain are not a laboratory spectrum analyzer. We do not completely understand the inner ear, and what happens beyond the auditory nerve [relaying neural signals from the inner ear to higher auditory centers in the brain] is almost a total mystery." Indeed as with many research endeavors, decades of research have served to point out how little we know about a very complex problem. Many years after these comments were made, the speech process remains fundamentally mysterious, and the engineering view of the brain as a "spec-

[3]Regrettably, we cannot possibly give proper credit to the vast number of researchers at many laboratories and companies around the world who have advanced this field. We can only hope to give a sampling of the systems that represent various concepts. As a matter of policy, we will avoid discussions of specific commercial systems and focus on research developments, except in a small number of cases in which the work represents landmark advances in the field. The reader is encouraged to peruse the *Proceedings of the IEEE International Conferences on Acoustics, Speech, and Signal Processing* (ICASSP), for example, where new results are often first reported. The papers in these *Proceedings* also offer extensive reference lists that will direct the reader to other sources of information. In particular, a wonderfully comprehensive survey of speech recognition advances is given by J. Mariani in the 1989 ICASSP *Proceedings* (Mariani, 1989). Also, in 1990 a collection of papers on the subject of speech recognition was compiled by Waibel and Lee (1990). This collection presents some of the seminal work in the field, focusing principally, but not exclusively, on work in the United States. The papers in this collection also contain many useful reference lists.

tral analyzer" still prevails. We noted in Chapter 3 that Teager and Teager (1990) have urged the speech processing community to consider that analysis techniques based on linear models are quite inappropriate and are hypothetically responsible for hindering greater speech recognition success.

10.1.3 Circumventing Our Ignorance

Although our fundamental understanding of the speech process remains incomplete, a major asset to the field has been the explosive advances in digital computing based on very-large-scale integration of circuit components beginning in the 1980s. Computing speed and abundant memory combined with specialized architectures and signal processors have made it possible to execute enormously complex algorithms that would have been unthinkable in the early days of computing. What is more, researchers can proceed with high confidence that speech recognition strategies that today are implementable only when using vast laboratory systems of networked processors will be run on small systems in the future. It is interesting to note the comments of Dr. Frederick Jelinek of IBM in the same 1976 *IEEE Spectrum* article: "Computers are still too slow and too expensive. Ten years ago [1966] they were even too slow to carry out *research* [emphasis added] in speech recognition. New research is possible, but continuous-speech recognition *products* [emphasis added], by present techniques, would be quite costly. Because programming, even in today's high-level computing languages, is difficult, research is slow. It takes a very long time to test out the simplest experimental idea."

Whereas it is possible to mistake Jelinek's comment for one that might have been made today, relatively speaking, we have come much further in addressing the need for greater computing power than in addressing Flanagan's and Teager and Teager's concern for a more complete view of the speech communication process. In a sense we have used the strength of the former to compensate for the relative weakness of the latter. Some brief discussions of hardware capabilities will be made at appropriate points in the following chapters.

Although advances in hardware have been a major boon to speech recognition technology, certain conceptual advances also underlie high-performance "software." Indeed, in the same era as the concerns above were raised, researchers were beginning to build systems based on *stochastic* models that would, in effect, learn their own representations of the speech process rather than having it deterministically encoded using experts' knowledge. This actually represents another compensation for lack of precise modeling information. Stochastic approaches circumvented the need for extraordinary amounts of complex information necessary to write "deterministic" programs, but at the same time placed heavy demands upon computing systems for both training recognition tasks. [For example, it took more than 13 hours on a DEC-1090 computer to compile the network necessary for one of the first successful large-vocabulary (1011 words) continuous-speech recognizers, HARPY, at Carnegie-Mellon University in 1976 (Reddy, 1976; Lowerre and Reddy, 1980)]. Stochastic approaches (based on linear acoustic models), as we shall see, are now firmly entrenched in most contemporary speech recognition systems with moderate to large vocabularies. Large-vocabulary, continous-speech recognition systems still pose many challenges for computer technologists to create faster hardware and software and much more compact circuitry necessary for these systems to move from the laboratory to the real world. However, it appears likely that continuous-speech recognition systems, and human–machine interaction, will play a significant role in societies of the not-too-distant future.

10.2 The "Dimensions of Difficulty"

In Section 10.1, we described the general goals of the speech recognition task and generally suggested some of the major problems involved. In this section we wish to more formally discuss what Waibel and Lee (1990, p. 2) have called the "dimensions of difficulty" in speech recognition. We address the question of what factors influence the success or failure of a speech recognition system and dictate the degree of sophistication necessary in the design of the system. These factors are enumerated as answers to the following questions:

1. Is the system required to recognize a specific individual or multiple speakers (including, perhaps, all speakers)?
2. What is the size of the vocabulary?
3. Is the speech to be entered in discrete units (usually words) with distinct pauses among them (discrete utterance recognition), or as a continuous utterance (connected or continuous recognition—to be distinguished below).
4. What is the extent of ambiguity (e.g., "know" and "no") and acoustic confusability (e.g., "bee," "see," "pea") in the vocabulary?
5. Is the system to be operated in a quiet or noisy environment, and what is the nature of the environmental noise if it exists?
6. What are the linguistic constraints placed upon the speech, and what linguistic knowledge is built into the recognizer?

We consider each of these questions sequentially in the following subsections.[4]

[4]These "dimensions" tend to focus on the task to be accomplished and accordingly might make the reader think of the correlative difficulty in finding theoretical and algorithmic solutions. However, another facet of this challenge is becoming clear as some of the more difficult algorithmic problems are being solved: the availability of necessary memory resources with which to implement the "solutions." Many existing algorithms require too much memory to even be tested. Therefore, the computing resources necessary to implement a "solution" are becoming a very real part of the difficulty implied by the more extreme answers to these questions.

10.2.1 Speaker-Dependent Versus Speaker-Independent Recognition

Most speech recognition algorithms, in principle, can be used in either a "speaker-dependent" or "speaker-independent" mode, and the designation for a particular system depends upon the mode of training. A *speaker-dependent* recognizer uses the utterances of a single speaker to learn the parameters (or models) that characterize the system's internal model of the speech process. The system is then used specifically for recognizing the speech of its trainer. Accordingly, the recognizer will yield relatively high recognition results compared with a *speaker-independent* recognizer, which is trained by multiple speakers and used to recognize many speakers (who may be outside of the training population). Although more accurate, the apparent disadvantage of a speaker-dependent system is the need to retrain the system each time it is to be used with a new speaker. Beyond the accuracy/convenience trade-off is the issue of necessity. A telephone system [see, e.g., (Wilpon et al., 1990)] that must respond to inquiries from the public is necessarily speaker-independent, while a system used to recognize the severely dysarthric speech of a person with speech disabilities [see, e.g., (Deller et al., 1991)] must be trained to that person's speech. Both types of systems, therefore, are used in practice, and both have been studied extensively in the laboratory.

Before continuing, let us note that some authors distinguish between speaker-independent systems for which the training populations are the same as the users, and those for which the training populations are different from the users. In the former case, the term *multiple speaker* is used while the term "speaker independent" is reserved for the latter. We shall not make this distinction in the following. However, it is important to take note of this issue in comparing the performance of various systems.

10.2.2 Vocabulary Size

Clearly, we would expect performance and speed of a particular recognizer to degrade with increasing vocabulary size. As a rule of thumb, some speech researchers estimate that the difficulty of the recognition problem increases logarithmically with the size of the vocabulary. Memory requirements also increase with increasing vocabulary size, though (as we will be able to infer from the study below) generally not so much as a consequence of the increasing number of words, but rather as a result of the increasing complexity of the recognition task that larger vocabularies imply.

Speech recognition systems or algorithms are generally classified as small, medium, or large vocabulary. There is some variation in the literature on the quantification of these terms, but as a rule of thumb, small-vocabulary systems are those which have vocabulary sizes in the range of 1–99 words; medium, 100–999 words; and large, 1000 words or more.

Since recognizers have been designed for 200,000 words, a 1000-word capability might be called "small" in some contexts, so we need to be careful about the meaning of these loose classifications. Small-vocabulary (as defined here) systems have become routinely available and have been used in tasks such as credit card or telephone number recognition, and in sorting systems (recognizing destinations) for shipping tasks. The focus of the medium-sized vocabulary systems has been experimental laboratory systems for continuous-speech recognition research (driven in part by the availability of standardized databases, to be discussed in Chapter 13). Large-vocabulary systems have been used for commercial products currently aimed at such applications as office correspondence and document retrieval. These systems have been of the isolated-word type in which the speaker must utter each word discretely from the others. It is important to keep in mind that a given-size vocabulary can require far more effort for a speaker-independent system than a speaker-dependent one. Continuous-speech recognition is also much more difficult than discrete utterance recognition (see below); thus, vocabulary size is only one measure of difficulty. It is also true that "linguistic constraints" (see below) can reduce the "per word" difficulty of a fixed-size vocabulary.

As we shall see, for small vocabularies and relatively constrained tasks (e.g., recognizing numerical strings), simple discrete utterance or connected-word recognition strategies[5] can often be employed. In these cases, models for each word in the vocabulary are resident in the system and the list can be exhaustively searched for each word to be recognized. As vocabularies become larger and recognition tasks more complicated, training and storing models for each word is generally impossible and models for subword units (e.g., syllables, phonemes) are employed.[6] Simple exhaustive search of all possible messages (built from these subword units) also becomes unmanageable, and much more sophisticated search algorithms that pare down the number of items searched must be designed. Significant to these algorithms are "linguistic constraints" on the search that eliminate unmeaningful and grammatically incorrect constructions. We discuss this important issue below. Also complicating the recognition task as vocabularies become larger is the potential for an increased number of confusable items in the vocabulary. This issue is also discussed below.

10.2.3 Isolated-Word Versus Continuous-Speech Recognition

Henceforth in this discussion, we will use the term *sentence* to mean any string of words to be recognized that is presumably taken from the vocabulary under consideration. The "sentence" can be what we would

[5]These terms are defined below.

[6]It should be noted that multiple models for each subword unit are frequently necessary to account for coarticulatory and phonological effects. This is yet another effect that is driven upward by increasing vocabularies.

ordinarily think of as a grammatically correct sentence, a simple string of digits, or even, in the "degenerate case," a single word.

Isolated-Word Recognition. *Discrete-utterance* recognizers are trained with discrete renditions of speech units. Since the discrete utterances are usually words, this form of speech recognition is usually called *isolated-word recognition* (IWR). In the recognition phase, it is assumed that the speaker deliberately utters sentences with sufficiently long pauses between words (typically, a minimum of 200 msec is required) so that silences are not confused with weak fricatives and gaps in plosives. Single-word sentences are special cases with "infinite" pauses. The fact that boundaries between words can be located significantly simplifies the speech recognition task. These boundaries are located in various technical ways, including the use of an endpoint detection algorithm to mark the beginning and end (or candidate sets of beginnings and ends) of a word. This is the simplest form of recognition strategy, and it requires a cooperative speaker. It is nevertheless very suitable for certain applications, particularly those in which single-word commands from a small vocabulary are issued to a machine at "lengthy" intervals. A good example of an application with such intervals arises in the sorting machine application noted above, in which the operator utters a destination as each package presents itself on a conveyor.

When the vocabulary size is large, isolated-word recognizers need to be specially constructed and trained using subword models. Further, if sentences composed of isolated words are to be recognized, the performance can be enhanced by exploiting probabilistic (or simply ordering) relationships among words ("syntactic" knowledge) in the sentences. We will be better able to comment on this issue after describing language constraints.

Continuous-Speech Recognition. The most complex recognition systems are those which perform *continuous-speech recognition* (CSR), in which the user utters the message in a relatively (or completely) unconstrained manner. First, the recognizer must be capable of somehow dealing with unknown temporal boundaries in the acoustic signal. Second, the recognizer must be capable of performing well in the presence of all the coarticulatory effects and sloppy articulation (including insertions and deletions) that accompany flowing speech. As an example of cross-word coarticulation effects, the /z/ in "zoo" is pronounced somewhat differently in the utterances of "St. Louis Zoo" and "Cincinnati Zoo." The latter tends to be a true /z/ (voiced fricative) sound, whereas in the former, the voicing tends to be missing. As an example of how intra- as well as interword articulation degenerates in continuous speech, speak the question, "Did you find her?" as discrete words, and then naturally. The latter likely results in "Didjoo (or Didja) finder?" Whereas the CSR problem does not in the extreme case require any cooperation from the speaker, it must compensate for this fact by employing algorithms that are robust to the myriad nuances of flowing speech. CSR systems are the most natural from the user's point of view. They will be essential in many applications in which large populations of naive users interact with the recognizer.

In line with the issue of speaker cooperation, it is worth noting that even IWR systems must be robust to some of the anomalies of continuous speech if used with naive speakers. Often the pause between words by persons who are asked to speak in discrete utterances is not sufficient or even existent. "Pausing" is a very subjective speaking behavior that is sometimes not manifested acoustically. In general, obtaining cooperation from speakers is not simple, and speech recognizers must be robust in handling resulting problems. Pausing is only one such noncooperative behavior. Others include the inclusion of extraneous speech or noise, and use of out-of-vocabulary words (Wilpon et al., 1990; Asadi et al., 1991).

In large-vocabulary CSR speech systems, the same two considerations as in the IWR case apply. Words must be trained as subword units, and interword relationships must be exploited for good performance. There is further pressure in the continuous-speech case to model words in ways that capture the intra- and interword phonological variations, and, perhaps, to learn and exploit probabilistic relationships among subword units ("lexical" and "phonological" knowledge), just as we do with the word relationships ("syntax") at a more macro level of analysis.

"Connected-Speech" Recognition. In small-vocabulary, continuous-speech applications, a recognition technique called *connected-speech recognition* is sometimes used. It is important to note that the term "connected speech" refers to the recognition strategy rather than to the speech itself. In general, the speech is uttered in a continuous manner.

In the connected-speech technique, the sentence is decoded by patching together models built from discrete words and matching the complete utterance to these concatenated models. The system usually does not attempt to model word-boundary allophonic effects, nor sloppy intra- or interword articulation. There is an implicit *assumption* that, while distinct boundaries cannot be located among words, the words are reasonably well articulated. In general, this assumption is violated by the speaker, but the results are improved by speaker cooperation. An example of a "cooperative speaker" application would be the entry of strings of digits representing a credit card number by a sales clerk who has been instructed to "speak slowly and pronounce the digits carefully." An example of an "uncooperative speaker" application would be voice dialing of phone numbers from a public phone. In that case, the average caller (who does not understand speech recognition technology) is not likely to be very cooperative even if asked to be so (Wilpon et al., 1990), and the problem becomes one of recognizing continuous speech. We cannot overemphasize the fact that connected-speech recognition is really recognition of continuous speech, since we intend to use this case to introduce continuous-speech recognition techniques in Chapter 13.

When probabilistic relationships among words (syntax) are known, these can be exploited in the connected-speech recognition approach to improve performance. In the phone-dialing example, we might model the sentences as random strings with equal probabilities of any digit in any time slot, or there might be certain probabilistic relationships among the digits due to higher frequency of calls to one area, for example. In the latter case we could employ this syntactic knowledge to improve performance.

Endpoint Detection. We conclude this subsection by revisiting the problem of endpoint detection, since the proper detection of the onset and termination of the speech amidst background noise is central to the success of many IWR strategies. The problem of endpoint detection has been described as an application of the short-term energy and zero-crossing measures in Section 4.3.4. The approach discussed there was widely used in practice in the 1980s and continues to be useful in a limited number of simpler systems and applications. An even simpler approach that can be used when the speech is more severely bandlimited (to, say, below 3 kHz) relies on threshold settings on the energy only and is described in (Lamel et al., 1981; Wilpon et al., 1984).

More recently, the endpoint detection problem has been addressed using techniques arising from the study of CSR. In this approach, the acoustic signal is modeled as a continuum of silence (or background noise), followed by the desired utterance, then more silence. In this case, the precise location of endpoints is determined in conjunction with the strategy used to actually recognize the words. We shall see how this is accomplished in the next three chapters (see, in particular, Section 12.4.2). With this approach, the endpoint detection stage may be used to provide initial estimates (sometimes crude) or sets of estimates for use in the higher stages.

The somewhat uninspiring problem of endpoint detection would seem to be rather easily solved. In fact, it is often very problematic in practice. Particularly troublesome are words that begin or end in low-energy phonemes like fricatives or nasals, or words that end in unvoiced plosives in which the silence before the release might be mistaken for the end of the word (see Fig. 10.1). Some speakers also habitually allow their words to trail off in energy (see Fig. 10.2). Others tend to produce bursts of breath noise at the ends of words (see Fig. 10.3). Background noise is also an obvious potential source of interference with the correct location of endpoints, with transient noises often causing more of a challenge than stationary noise that can be well modeled.

When the older endpoint technology is used, a short-term energy measure is the principal and most natural feature to use for detection, but each of the problems mentioned above interferes with the effective use of energy measures. Accordingly, safeguards are built into endpoint detectors. The most fundamental problem is the nonstationary nature of the intensity of the speech across words. Thresholds fixed on one relatively

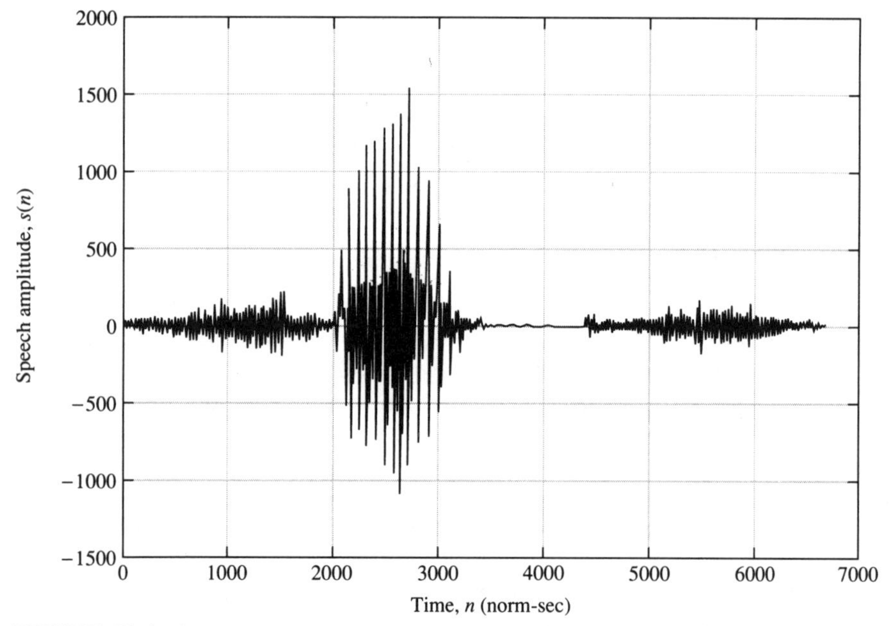

FIGURE 10.1. Acoustic waveform for the word "six" illustrating the two effects in the text.

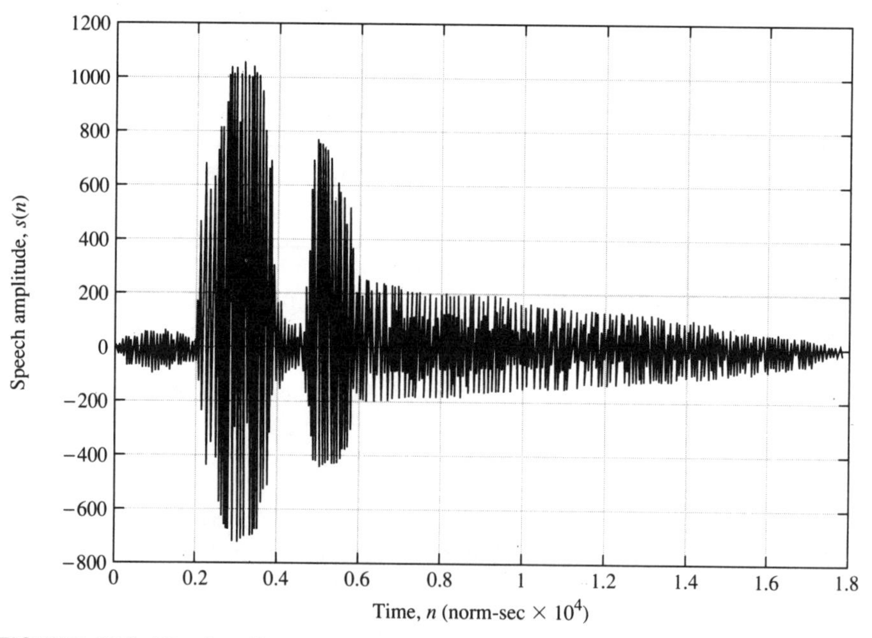

FIGURE 10.2. Word trailing off in intensity. (Waveform for the word "seven.")

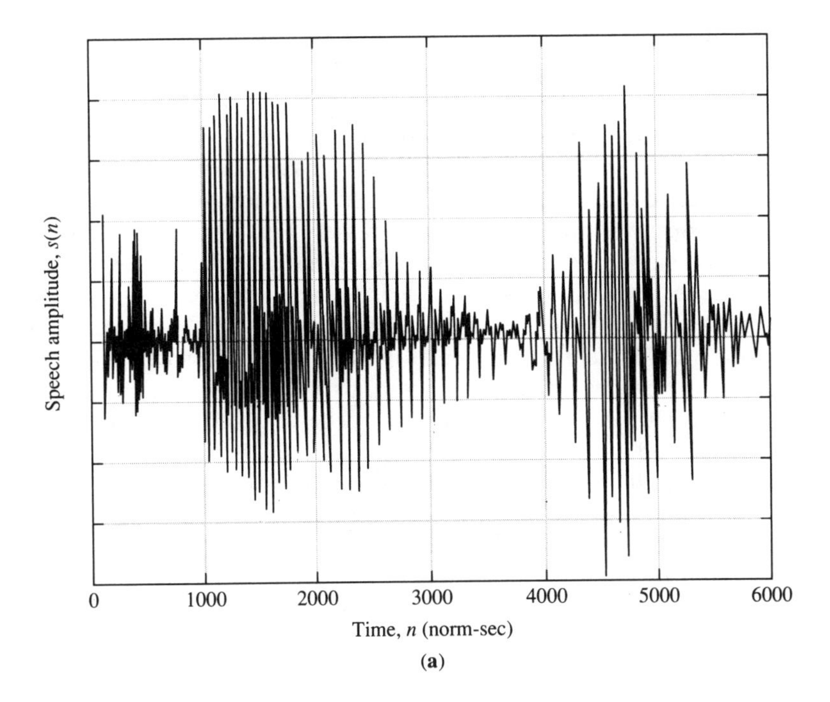

(a)

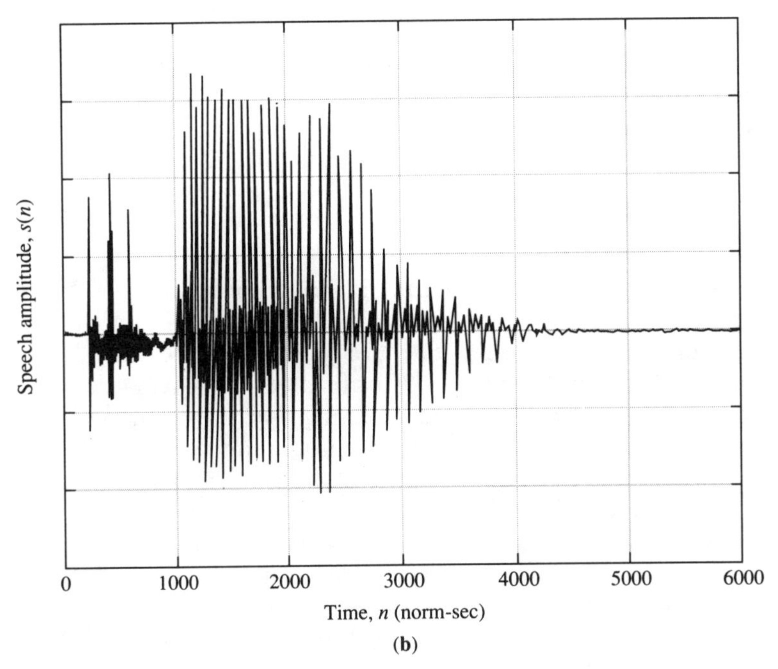

(b)

FIGURE 10.3. Word showing burst of breath noise at end. (Waveform for the word "poof.")

loud word might be entirely inappropriate for endpoint detection on a weaker phonation (even of the same word). Therefore, thresholds are generally normalized to the general energy level of a word, or, alternatively, the intensity of the words can be normalized to use fixed thresholds. Such normalization procedures help to compensate for the fact that when words occur sequentially as part of a spoken sentence, the average energy level tends to decay as the sentence progresses. Further, to a large extent, the energy normalization alleviates the problem noted above with weak sounds at the beginnings or ends of words. The problem with terminal plosives, however, requires that a number of frames of "background" (silence) be determined before the end of the word is declared. Similarly, the problem of transient background sounds being detected as word onsets can be overcome by requiring that a certain number of frames be "above background" in intensity before a word is declared. If this criterion is met, the initial point is the word is found by backtracking to the point of the initial onset of energy. An example technique is described in (Lamel et al., 1981).

The problem of breath noise at the ends of words has been addressed in the paper by Rabiner and Sambur (1975),[7] in which they suggest the preliminary inclusion of the breath noise as part of the word, and then its omission by secondary processing.

Generally, whether the older or newer endpoint detection method is used, the fact that sources of noise can often appear to be valid speech sounds (and vice versa) is also problematic. The solutions to this problem depend very much on the information available about the noise and the technological framework in which the IWR problem is embedded. Some techniques for dealing with the problems of speech in noise were discussed in Chapter 8.

10.2.4 Linguistic Constraints

The most abstract problem involved in speech recognition is endowing the recognizer with the appropriate "language constraints." Whether we view phones, phonemes, syllables, or words as the basic unit of speech, *language* (or *linguistic*) *constraints* are generally concerned with how these fundamental units may be concatenated, in what order, in what context, and with what intended meaning. As we discuss below, this problem is more involved than simply programming the correct grammatical rules for the language. Clearly, the more constrained the rules of language in the recognizer, the less freedom of expression the user has in constructing spoken messages. The challenge of language modeling is to balance the need for maximally constraining the "pathways" that messages may take in the recognizer, while minimizing the degree to which the speaker's freedom of expression is diminished. A measure of the extent to which a given language model constrains permissible discourse in

[7]This is the paper upon which our discussion in Section 4.3.4 was based.

a recognizer is given by the "perplexity" of the language model.[8] This term roughly means the average number of branches at any decision point when the decoding of messages is viewed as the search of paths through a graph of permissible utterances. We will define perplexity more formally in Chapter 13.

Let us begin the consideration of language constraints by posing an abstract model of natural languages. Peirce's model of language (Hartstone and Weirs, 1935) as described by Rabiner and Levinson (1981) includes four components of the natural language code: symbolic, grammatical,[9] semantic, and pragmatic. The *symbols* of a language are defined to be the most fundamental units from which all messages are ultimately composed. In the spoken form of a language, for example, the symbols might be words or phonemes, whereas in the written form, the alphabet of the language might serve as the symbols. Rabiner and Levinson write that "[f]or spoken English, the 40 or so basic sounds or phonemes are a reasonable choice [for the symbols of the spoken form of the language]. Although they are subject to substantial variation, they do correlate highly with measurable spectral parameters." For the purpose of discussion, let us adopt Rabiner and Levinson's suggestion and use the phonemes as the symbols of the language. The *grammar* of the language is concerned with how symbols are related to one another to form ultimate message units. If we consider the sentence to be the ultimate message unit, and we choose phonemes as symbols, then how words are formed from phonemes is properly considered as part of Peirce's grammar, as well as the manner in which words form sentences. How phonemes form words is governed by *lexical* constraints, and how words form sentences by *syntactic* constraints. Lexical and syntactic constraints are both components of the grammar.

Before continuing the discussion of Peirce's model of language, let us view the following "sentences" proposed by Reddy (1976), in light of our definition of grammar. Next to each sentence is a description of its conformity to the linguistic concepts discussed above. These conclusions will be drawn from the discussion below.

1. Colorless paper packages crackle loudly. [grammatically correct]
2. Colorless yellow ideas sleep furiously. [grammatically correct, semantically incorrect]
3. Sleep roses dangerously young colorless. [grammatically (syntactically) incorrect]
4. Ben burada ne yaptigimi bilmiyorum.[10] [grammatically (lexically) incorrect]

[8]Another term used is "habitability".

[9]We have replaced Peirce's word "syntax" and "grammar" for more consistency with our "engineering" formalisms in the following. We will reserve the word *syntax* to refer to the rules that govern how *words* may combine.

[10]According to Reddy, this is a Turkish sentence.

Suppose that in a Peircian model, the grammatical component includes all (lexical) rules that form English words from symbols (phonemes) as well as the classification of these words into parts of speech and the (syntactic) rules by which these parts of speech may be combined to form sentences. Sentence 4 is grammatically incorrect in that the words are not legal concatenations of symbols in English. Sentence 4, therefore, fails at the lowest level (the word or lexical level) of grammatical constraints. Whereas sentence 2 is meaningless, sentences 1 and 2 are both *grammatically* correct in that they consist of proper English words (correct lexically) and the words are correctly concatenated (correct syntactically) according to the rules of English. Although lexically correct, sentence 3 does not obey the rules of English and therefore fails at a higher level of grammar (the syntactic level). Sentences 1 and 2, therefore, are permitted in our language model.

The grammar of a language is, in principle, arbitrary, in the sense that any rule for combining symbols may be posed. We have witnessed this in declaring that sentence 2 is a grammatically correct message. On the other hand, *semantics* is concerned with the way in which symbols are combined to form *meaningful* communication. Systems embued with semantic knowledge traverse the line between speech recognition and *speech understanding*, and draw heavily upon artificial intelligence research on knowledge representation. If our recognizer is semantically constrained so that only meaningful English sentences are permitted, then sentence 2 will clearly fail to be a candidate message. Likewise, sentence 3 is semantically improper. Sentence 4 will presumably fail at the lexical level of grammatical testing and will not be subjected to semantic scrutiny.

Beyond simple "nonsense/meaningful" decisions about symbol strings, semantic processors are often used to impose meaning upon incomplete, ambiguous, noisy, or otherwise hard-to-understand speech. A noisy utterance recognized as "[Indeterminant word], thank you" with the aid of semantic processing might immediately be hypothesized to be either "No, thank you" or "Yes, thank you" with high likelihood. A semantic processor could also choose between "Know, thank you" and "No, thank you," two phrases that might be equally likely without it.

Finally, the *pragmatics* component of the language model is concerned with the relationship of the symbols to their users and the environment of the discourse. This aspect of language is very difficult to formalize. To understand the nature of pragmatic knowledge, consider this sentence: "He saw that gas can burn." Depending on the nature of the conversation, the word "can" might be either a noun (He saw a gas *can* burning, and it was that one.) or a verb (He saw that gas *is able to* burn.). A similar problem occurs with the phrase "rocking chair," which can refer either to a type of chair or a chair that is in the process of tilting back and forth. A source of pragmatic knowledge within a recognizer must be able to discern among these various meanings of symbol strings, and hence find the correct decoding.

The components of Peirce's abstract language model are in essence constraints on the way in which "sound symbols" may form complete utterances in the spoken form of a language. Implicit or explicit banks of linguistic knowledge resident in speech recognizers, sometimes called *knowledge sources*, can usually be associated with a component of Peirce's model. Among the first speech recognition systems to successfully use linguistic constraints on a grand scale was the HARPY system developed at Carnegie–Mellon University (CMU) (Lowerre and Reddy, 1980) as part of the ARPA Speech Understanding Project (Klatt, 1977). Among the principal investigators of the system was Professor D. Raj Reddy of CMU. Reddy, writing in the *Proceedings of the IEEE* in 1976 (Reddy, 1976), gives the following introduction to the use of knowledge sources in speech recognition: "[A] native speaker uses, subconsciously, his knowledge of the language, the environment, and the context in understanding a sentence. . . . [S]ources of knowledge include the characteristics of the speech sounds (*phonetics*), variability in pronunciation (*phonology*), the stress and intonation patterns of speech (*prosodics*), the sound patterns of words (*lexicon*), the grammatical structure of the language . . . (*syntax*),[11] the meaning of the words and sentences (*semantics*), and the context of the conversation (*pragmatics*)." The block diagram of a general speech recognizer showing these sources of knowledge at their appropriate levels of the hierarchy is shown in Fig. 10.4. Note that the "prosodics" knowledge source is shown to interact with both the language and acoustic processors. The acoustic processor is generally considered to be that segment of the recognizer which interfaces the acoustic waveform with the "intelligent" language sector of the recognizer by reducing the waveform to a parametric or feature representation. Since, however, the prosodic content of the utterance is intricately tied to the acoustic content of the waveform, it is difficult to segregate the function of the prosodic knowledge source from the acoustic processing.

We hasten to point out that Fig. 10.4 is a very general system diagram. It is clear from our discussion so far, and issues to be covered later, that speech recognition is an exceedingly complex problem. Accordingly, attempts to solve the problem are manifold and diverse, resulting in numerous and various hardware and software systems. Further, as we have also noted, different speech processing applications require differing degrees of recognition capability, which leads to more diversity among recognizers. Figure 10.4 encompasses most existing systems, but there are probably exceptions. It is certainly the case that not all real-world systems will have all the features shown in the figure.

Existing speech recognizers can be classified into two broad categories, which indicate the direction of the "flow of information" in Fig. 10.4. If the acoustic processing is used to hypothesize various words, phones, and so on, and then these hypotheses are "processed upward" to see if they

[11]In this sentence, Reddy has used the word "grammar" in the more conventional way (learned in primary school), in which it is equivalent to *syntax*.

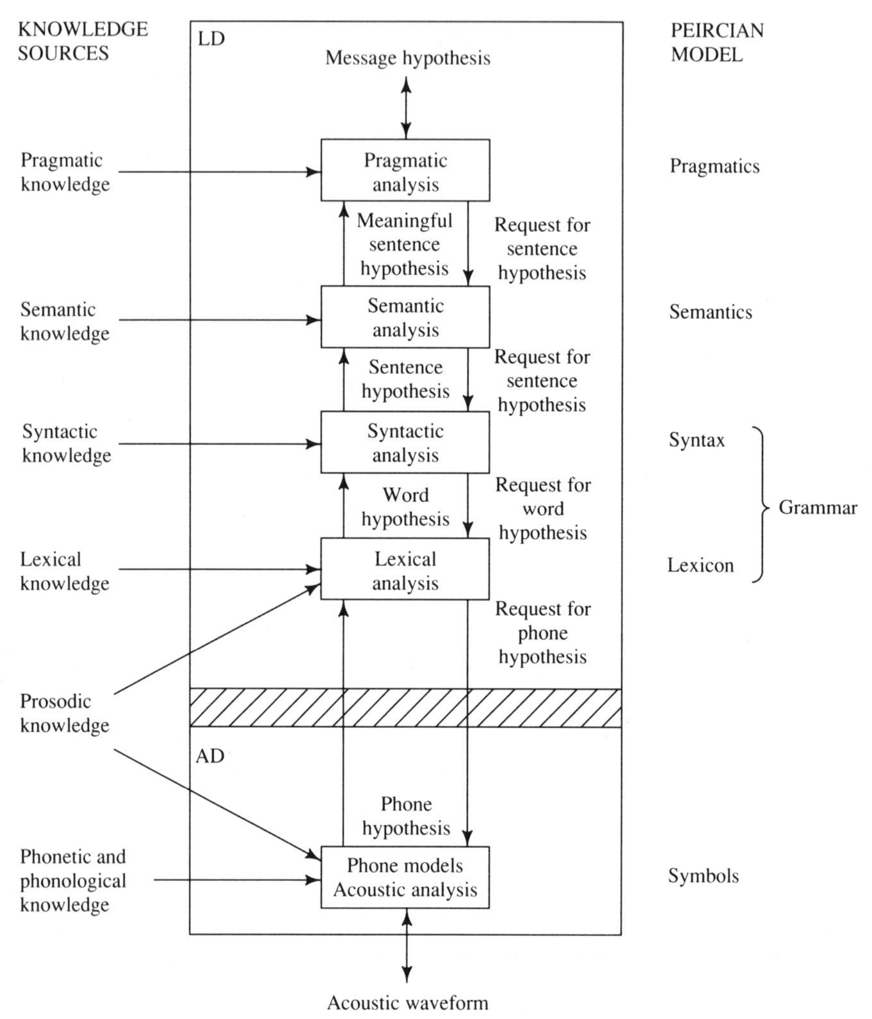

FIGURE 10.4. Block diagram of a general speech recognizer showing the acoustic and linguistic processors. To the left are the knowledge sources placed at the appropriate level of decision-making in the language hierarchy. To the right are the correlate components of the abstract Peircian language model. The illustrated system uses *phones* as the basic symbols of the language. LD = linguistic decoder; AD = acoustic decoder.

can be pieced together in a manner following the "higher level" rules, then the system is said to operate in a *bottom-up* mode. The earlier recognition systems, including HARPY, employ bottom-up processing. Roughly speaking, *top-down* processing begins with sentence hypotheses being posed at the highest levels of the processor. These hypotheses are then scrutinized at each of the lower levels for likelihood of representing the spoken utterance, each level calling on the next lower level to provide information that it uses in its assessment. Ultimately, the acoustic processor is called upon to ascertain whether the acoustics are consistent with the lowest hypothesized abstract symbols; for example, a set of phones. Top-down processing requires much more complex and computationally intensive processing systems than does bottom-up processing. In the early 1990s, researchers began to focus upon techniques that employ a combination of the two types of processing. The theory and practical applications of each of these strategies will be discussed extensively in Chapter 13.

In summary, sources of knowledge will constrain the recognition process and help convert an unmanageably complex decoding process into a tractable one. We will discuss aspects of language modeling at several key points in this part of the book, with a formal introduction to language modeling in Chapter 13. However, our treatment of the highest levels of linguistic constraints, semantics and pragmatics, will be only superficial, since they do not lend themselves well to formal discussion (at least in conventional engineering ways). When we describe certain recognition systems in Chapter 13, we will point out some of their attempts to employ these levels of the hierarchy.

10.2.5 Acoustic Ambiguity and Confusability

Let us now retreat to the lexical level of our problem and note some simpler effects. The extent to which similarities occur in the vocabulary will have an obvious impact on the performance of the recognizer. We identify two types of similarity: "ambiguity" and "confusability" of words. As the term implies, *acoustically ambiguous* words are those that are indistinguishable in their spoken renditions: "know" and "no"; and "two," "to," and "too" are sets of examples. In terms of our formal discussion of language models, we can say that these words consist of the same linguistic symbols. At an acoustic level, therefore, these words are indistinguishable, unless they can be resolved through prosodic subtleties. Ordinarily, higher levels of the recognizer would be called upon to make the correct recognition.

On the other hand, *confusability* refers to the extent to which words can be easily confused because of partial acoustic similarity. The words for the digits 0 through 9 are rather dissimilar acoustically, the most confusable being "one" and "nine" (because of the final nasal) and "five" and "nine" (because of the strong diphthong). The vocabulary consisting of the words for the letters of the alphabet, however, are highly confusable, primarily because of the set *B*, *C*, *D*, *E*, *G*, *P*, *T*, *V*, but also because of the sets *F*, *S*, *X*; *A*, *H*, *J*, *K*; *I*, *Y*; and so on. In each case, the utterances are only discernible by correct recognition of nonvowel phonemes, which are relatively weak in contrast to the vowels. Whereas resolving confusability can be assisted at higher levels of processing in the recognizer, this problem is theoretically solvable at the acoustic level, and there is no substitute for a high-quality acoustic front end in this regard.

10.2.6 Environmental Noise

Real speech recognition systems, of course, will not be used in the "acoustically pristine" environment of the research laboratory using high-quality speech. One of the major challenges of the speech recognition problem is to make the system robust to background noise. This noise can take the form of speech from other speakers; equipment sounds, air conditioners, or fluorescent lighting in the office; heavy equipment noise in a factory environment; or cockpit noise in an aircraft. It might be the consequence of the channel over which the speech reaches the recognizer, ranging from simple microphone variations to complex noise sources over telephone or broadcast channels. The noise might also be created by the speaker himself—lip smacks, breath noises, pops, clicks, coughs, or sneezes. Finally, an unusual form of distortion occurs in speech produced by deep-sea divers breathing a mixture of helium and oxygen. This problem was discussed in Chapter 8.

10.3 Related Problems and Approaches

10.3.1 Knowledge Engineering

We have noted the intention to remain "below" the semantic and pragmatic levels of language processing throughout much of our study of speech recognition. The stated reason is the desire to treat topics that are more readily formulated and quantified in conventional engineering terms. Like the more abstract aspects of language modeling, at the acoustic level there is also a wealth of abstract information that could conceivably be employed to assist in proper recognition. This information includes, for example, prosodic and phonotactic aspects of the speech itself, and perceptual aspects of the human auditory system. As with the more abstract aspects of language, we will see these ideas indirectly enter our study in various and important ways, but there will be no explicit attempt to use them directly. Generally speaking, these sources of information will be used to fine-tune algorithms for better performance.

Attempts to directly employ abstract information in speech recognition systems sometimes fall under the rubric of *knowledge engineering* approaches. Due to the difficulty of expediently quantifying and integrating vast sources of abstract knowledge, such approaches have had only limited impact on the field. Nevertheless, interesting research has been conducted on the subject and the interested reader is encouraged to seek further information in the literature (see Section 1.6). A frequently cited paper on the use of knowledge was written by Zue (1985). Zue's paper also contains a useful reference list, and it has been reprinted in a recent collection (Waibel and Lee, 1990), which also contains some other papers on the subject of knowledge-based approaches.

10.3.2 Speaker Recognition and Verification

For completeness, we also mention the problems of speaker recognition and verification, which are closely related to that of speech recognition. As the terms suggest, *speaker recognition* deals with the problem of identifying *who* among a population is the talker, while *speaker verification* is concerned with ascertaining whether a speaker's claim of identity is correct. In neither case is the objective to recognize the spoken message. Both are interesting problems with many useful applications, but the details of the technologies remain outside the scope of our study. However, the basic technologies studied in the following chapters are the same as those employed in these related fields. The interested reader will therefore be prepared to peruse the research literature for further information (see Section 1.6).

10.4 Conclusions

In this short, descriptive chapter we have introduced the complexities involved in the decidedly nontrivial problem of speech recognition. We have also introduced several concepts that will help us classify different recognizers according to what problem they are attempting to solve. In the remaining chapters of Part V, we will build engineering techniques in the framework of these general ideas.

10.5 Problems

10.1. (Computer Assignment) Write a simple algorithm for computing the short-term energy measure, $E_s(m)$, and use it in conjunction with a threshold setting to detect the onset of the utterances of "*A*," "*B*," and "*C*." Consider various threshold settings and window lengths, and comment on your experiments.

10.2. Consider a "language" for which the symbols are 0, 1, 2, 3, 4, 5, 6, 7, 8, 9, +, −. The formal components of the language are as follows:
Grammatical:

1. Lexical. A "word" consists of any number of digits concatenated by any sequence of + and − signs in any order. The first or last symbol in a word is never a sign (implicitly, therefore, all initial word symbols are nonnegative). The total numerical value of a word must be nine. For example, $4+5-0$ is a word in the language.
2. Syntactic. A "sentence" is any sequence of words such that any word ending in an odd digit must be followed by a word beginning in an even digit, and vice versa. For example,

$$3+6+4-4/7+3-1/2+0-5+8+4 \tag{10.1}$$

is a proper sentence. (The slash is used to show where boundaries between words occur.)

Semantic: To be "meaningful," a sentence must contain words that become increasingly longer. For example,

$$\begin{aligned} &9/2+7+1-1/4+5-5 \\ &+5+0+0+0-0+2-1-1 \end{aligned} \tag{10.2}$$

is a meaningful sentence.

Pragmatic: When "speaking" to a child, each word in a sentence is usually no more than than five digits long. There are no constraints for adult listeners.

(a) Determine whether each of the following sentences is lexically, syntactically, and semantically proper.

(i) $2+3+4/1+8+0+0/7-5-5-5+8+9/0+1-1+2-2+3-3+4-4+5-5+6-6+7-7+8-8+9.$

(ii) $2+3+4/1+8+0/7-5-5-5+8+9/0+1-1+2-2+3-3+4-4+5-5+6-6+7-7+8-8+9.$

(iii) $2+3+4/1+8+0+0/3+7-5-5-5+8+9/0+1-1+2-2+3-3+4-4+5-5+6-6+7-7+8-8+9.$

(iv) $2+3+4/1+8+0+0/7-5-5-5+8+9/1-1+2-2+3-3+4-4+5-5+6-6+7-7+8-8+9.$

(b) Suppose that you are the linguistic analyzer inside of a speech recognizer. An utterance known to have been read from a children's story has the following representation in the recognizer:

$$\begin{aligned} &1+8-1/1+2+7+5/ \\ &3-6+1-1-5+8. \end{aligned} \tag{10.3}$$

The proper representation of this sentence has been corrupted by the possible substitution of an improper digit and sign (or sign and digit) at the beginning and/or end of each word; or by the insertion of an improper digit and sign (or sign and digit) at the beginning and/or end of each word.

10.3. During the course of a radio newscast, the announcer who is discussing the political situation in the Middle East exclaims, "Turkey's role in the Persian Gulf following these messages." Explain how a speech recognizer might deduce an absurd translation of this utterance, and the type of processing that could be used to discern the correct pronouncement. What facts might you program into the processor that would prevent the amusing translation?

10.4. Consider the digits to be symbols in a language of telephone numbers in some geographic region (your city, state, province, country) with which you are familiar with the possible phone numbers. Write a grammar for the language. What is the implication of restricted length(s) of the symbol strings for the success of recognition?

CHAPTER 11

Dynamic Time Warping

Reading Notes: The concept of distances among feature vectors will play an important role in this chapter. The reader might wish to review Sections 1.3.1 and 5.3.5.

11.1 Introduction

In this chapter we begin in earnest our study of the technical methods used in speech recognition. There are two basic classes of methods upon which almost all contemporary speech recognition algorithms using sequential computation are based. The first class that we study in this chapter is based on a form of template matching. These methods draw heavily upon conventional feature-based approaches developed for general statistical pattern recognition problems. Accordingly, we will be able to make use of our general background along these lines from Section 1.3. However, the speech problem has an interesting and important nuance that does not arise in all template-matching problems. This is the need to appropriately temporally align the features of the test utterance with those of the reference utterance before computing a match score. To solve this problem, we will exploit the principles of "dynamic programming," a subject which is taken up first in the chapter. Because one feature string is "warped" (stretched or compressed in time) to fit the other, and because dynamic programming is used to accomplish this task, the class of feature-matching approaches used in speech recognition is often referred to as *dynamic time warping* (DTW).

Dynamic time warping has been successfully employed in simple applications requiring relatively straightforward algorithms and minimal hardware. The technique had its genesis in IWR, but has also been applied to CSR using the connected-speech strategy. Since DTW requires a template (or concatenation of templates) to be available for any utterance to be recognized, the method does not generalize well (to accommodate the numerous sources of variation in speech) and it is not generally used for complex tasks involving large vocabularies. It is also not used for CSR except in the connected-speech paradigm.[1]

[1]These comments apply to DTW as we shall study it in this chapter. They are not true for the "hidden Markov model" that we study in Chapter 12 and which may be considered a stochastic form of DTW.

We will examine the several facets of DTW in this chapter, and then move on to study the second general class of methods based on a "stochastic" approach, the "hidden Markov model," in Chapter 12. We will find that the hidden Markov model can be considered a generalization of DTW; accordingly, it is also heavily based on dynamic programming methods. In turn, many "higher level" problems in speech recognition are based on the theory of hidden Markov modeling. Careful attention to detail in this chapter, therefore, will pay off in much of our future study.

11.2 Dynamic Programming

As implied above, *dynamic programming* (DP) in one of its various forms is at the heart of many commonly used methods for speech recognition. However, DP is a broad mathematical concept for analyzing processes involving optimal decisions, which has been widely applied to many problems outside the speech realm. It behooves us to initially view the DP technique in this broader sense before applying it in specific ways in our speech work.

Dynamic programming has a rich and varied history in mathematics. The interested reader can refer to the article by Silverman and Morgan (1990) for a description of this history and an extensive bibliography. In fact, DP as we discuss it here is actually a subset of the general theory in which we will be concerned with *discrete* sequential decisions. The broader ("analog") theory has wide application in optimal control problems [see, e.g., (Bellman, 1957; Ogata, 1967)]. Silverman and Morgan note that the name of the technique was probably a reference to the fact that its solution algorithm could be programmed on an early digital computer. The late Professor Richard Bellman is generally credited with popularizing DP by applying the technique to the analysis of functional equations in the 1950s, and his aggressive and prolific application of his "principle of optimality" to problems in diverse fields ranging from science and engineering to medicine and economics (Bellman, 1952, 1954, 1957; Bellman and Dreyfus, 1962). The first paper to apply these methods to speech was published by Nagato, Kato, and Chiba (1962), a decade after Bellman's initial work, but it was the paper by Sakoe and Chiba (1978) that seems to have gotten the attention of the speech processing community. The latter paper is frequently referenced as a primary resource on the subject.

We will first view DP in a general framework. Consider the i–j plane shown in Fig. 11.1. The discrete points, or *nodes*, of interest in this plane are, for convenience, indexed by ordered pairs of nonnegative integers as though points in the first quadrant of the Cartesian plane. This indexing of points is only for compatibility with later work; for the purposes of

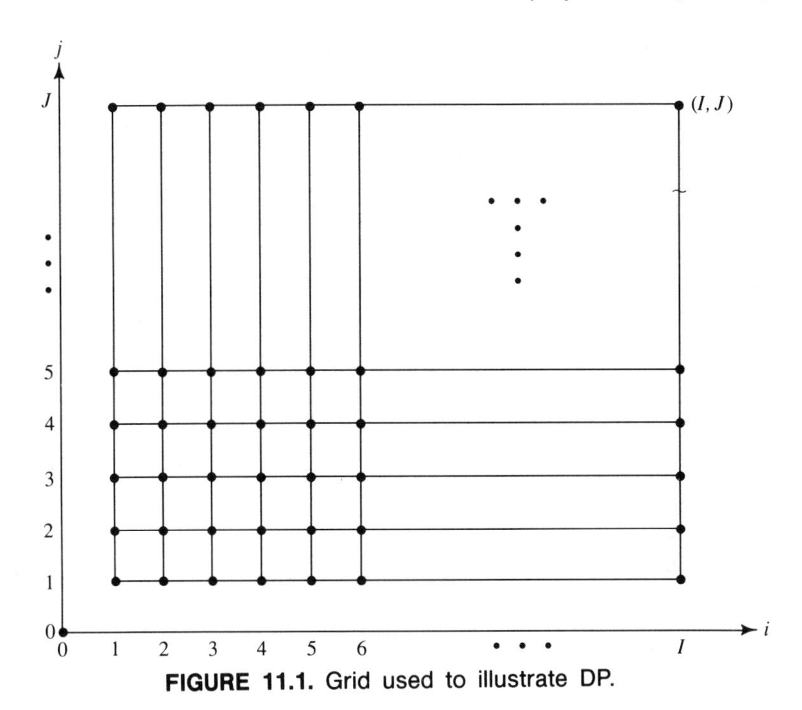

FIGURE 11.1. Grid used to illustrate DP.

discussing DP, they might just as well be labeled with simple integers and displayed arbitrarily in an abstract space. The basic problem is to find a "shortest-distance" or "least-cost" path through the grid, which begins at a designated *original node*, (0, 0), and ends at the designated *terminal node*, (I, J). For future purposes, we need to carefully define what we mean by a "path" through the grid. A *path* from node (s, t) to node (u, v) is an ordered set of nodes (index pairs) of the form

$$(s, t), (i_1, j_1), (i_2, j_2), (i_3, j_3), \ldots, (u, v), \tag{11.1}$$

where the intermediate i_k, j_k's are not, in general, restricted. We will speak of a path as a *complete path* if $(s, t) \equiv (0, 0)$ and $(u, v) \equiv (I, J)$.

Distances, or *costs*, are assigned to paths in one of three natural ways in our work. For ease of discussion, we will refer to these as Types T, N, and B cost assignments (these designations will be seen to refer to Transition, Node, and Both). In a *Type T* case, we have a cost associated with the forward-going *transition* into any node from its predecessor in a path. Suppose that we focus on a node with indices (i_k, j_k). Let us define the notation

$$d_T[(i_k, j_k) \mid (i_{k-1}, j_{k-1})] \stackrel{\text{def}}{=} \textit{transition} \text{ cost from } (i_{k-1}, j_{k-1}) \text{ to node } (i_k, j_k), \tag{11.2}$$

noting that this cost is apparently "Markovian" in its dependence on the immediate predecessor node only.[2] For consistency, we will always assume that $d_T[\cdot]$ is a nonnegative quantity, and that any transition originating at $(0, 0)$ is costless. This latter assumption usually means

$$d_T[(i, j) \mid (0, 0)] = 0, \qquad \text{for all } (i, j), \tag{11.3}$$

although variations may occur.

In the *Type N* case, costs are associated with the *nodes* themselves, rather than with the transitions among them. Let us define the notation

$$d_N(i, j) \stackrel{\text{def}}{=} \text{cost associated with } node \ (i, j) \tag{11.4}$$

for any i and j. In general, we will choose $d_N(\cdot\,,\cdot)$ to be nonnegative, and we will insist that the problem be set up so that node $(0, 0)$ is costless. Usually, this means

$$d_N(0, 0) = 0, \tag{11.5}$$

but variations occur; for example, $d_N(0, 0)$ would be taken to be unity if the node costs were combined by multiplication.

The *Type B* case is that in which *both* transitions and nodes have associated costs. The transition and node costs are usually combined by addition at a given node,

$$d_B[(i_k, j_k) \mid (i_{k-1}, j_{k-1})] \stackrel{\text{def}}{=} d_T[(i_k, j_k) \mid (i_{k-1}, j_{k-1})] + d_N(i_k, j_k). \tag{11.6}$$

The most frequent exception to this case is when they are combined by multiplication,

$$d_B[(i_k, j_k) \mid (i_{k-1}, j_{k-1})] \stackrel{\text{def}}{=} d_T[(i_k, j_k) \mid (i_{k-1}, j_{k-1})] \times d_N(i_k, j_k). \tag{11.7}$$

Note that in the multiplication case we would want $d_T[(i, j) \mid (0, 0)] = 1$ and $d_N(0, 0) = 1$ for a "costless" initiation.

Since most often we will be dealing with "Type B" distance quantities, for convenience we usually drop the subscript B and write simply

$$d[(i_k, j_k) \mid (i_{k-1}, j_{k-1})] \stackrel{\text{def}}{=} d_B[(i_k, j_k) \mid (i_{k-1}, j_{k-1})] \tag{11.8}$$

to refer to such a cost. When we need to specifically use a Type T or Type N distance, we shall use the appropriate subscript for clarity. However, note that there is no loss of generality in always using a Type B distance, since it subsumes the Type T and N varieties as special cases [see (11.6) and (11.7)].

[2] In fact, these transition costs may be non-Markovian in the following sense. Whether the transition into (i_k, j_k) from (i_{k-1}, j_{k-1}) is part of a "legal" path (in accordance with path constraints placed on some problems) may depend on the history of the path leading up to (i_{k-1}, j_{k-1}). As a practical matter, however, we can treat these quantities as Markov, making sure that our path-search algorithms do not seek out paths which are "illegal." Only in this case will we be able to consistently use Markov transition costs without creating theoretical conflicts in later developments. This point will become clearer when we actually discuss such algorithms.

The distance associated with a *complete path* is usually taken as the sum of the costs of these transitions and/or nodes along the path that we can now express as[3]

$$D = \sum_{k=1}^{K} d\left[(i_k, j_k) | (i_{k-1}, j_{k-1})\right], \tag{11.9}$$

in which K is the number of transitions in the path and $i_0 \equiv 0$, $j_0 \equiv 0$, $i_K \equiv I$, and $j_K \equiv J$. The objective, therefore, is to find the path that minimizes D. The most common variations on this problem include cases in which D is found by multiplication of the individual costs,

$$D = \prod_{k=1}^{K} d\left[(i_k, j_k) | (i_{k-1}, j_{k-1})\right], \tag{11.10}$$

and cases in which D is to be *maximized* rather than minimized. The methods described below are applicable with obvious modifications when either or both of these variations occur.

EXAMPLE (Traveling Salespeople)

Before proceeding, let us consider three examples in which the three cost-assignment paradigms would be used. Suppose that each grid point in the i–j plane represents a city on a map. In the first example, suppose that the transitional cost into a city is the time it takes to travel from the previous city. There are no restrictions on which cities may be visited from any given city. The intercity cost is independent of the history of travel prior to the current transition, so that the costs will be Markov. Also suppose that the time it takes to pass through a city is negligible, so that there are no costs assigned directly to the nodes. This, therefore, represents a Type T cost assignment to the grid. If a salesman wants to travel from city $(0, 0)$ to (I, J) at the least cost (shortest time), we would find his path through the grid subject to minimization of D in (11.9) where, in this case, the distances $d_T\left[(i_k, j_k) \mid (i_{k-1}, j_{k-1})\right]$ are the intercity distances.[4]

As an important aside, we should follow up on a point made in footnote 2. It is possible to use the transition costs to formally prevent certain predecessor nodes (or local paths) from being "legal." If, for example, there is no direct route from city (p, q) to (r, s) we could assign

$$d\left[(r, s) \mid (p, q)\right] = \infty \tag{11.11}$$

[3]This total distance appears to omit the cost of the node $(0, 0)$, but recall that we insist that $d_N(0, 0) = 0$.

[4]We should be careful to distinguish this problem, and the one to follow involving a saleswoman's journey through the grid, from what is often called the *Traveling Salesman Problem* (TSP). In the TSP, the salesman must visit *all* cities enroute from $(0, 0)$ to (I, J) and do so over the shortest path. The TSP is a much more difficult problem than those posed here. A further mention of the TSP will be found below.

to formally prevent this transition from occurring on any path. However, we will explicitly avoid the use of these quantities for this purpose, assuming instead that that illegal transitions are handled by the search algorithm without recourse to these costs. [The salesman knows that the bridge is out between cities (p, q) and (r, s) and does not even consider that transition.] To attempt to formally "block" certain transitions using infinite costs will cause us to have to resort to non-Markovian transition probabilities in later work, because a transition that may be "blocked" to a path coming from one "direction" may be permissible to a path coming from another. Indeed, this means that the probabilities must, in fact, *be* non-Markov, but prohibiting assignment (11.11) will allow us to work with the probabilities as though they were Markov.

On the other hand, suppose that a saleswoman has a telephone in her automobile, so that she can transact business while in transit from one city to the next. In this case our model includes no transition costs (time in the car is not "costly"), but it does include costs associated with each city through which she passes. It is known in advance that at each city (i, j) a certain number of units of her product $[d_N(i, j) \geq 0]$, which were sold on a trial basis, will be returned. To find her optimal journey, we would find her least-cost path subject to minimizing D. This problem, of course, represents a Type N assignment of costs to the grid.

An example of the Type B case arises if the number of units returned to the saleswomen at city (i, j) increases linearly with the time it takes to reach (i, j) from her immediate past city.[5] In this case the optimal path must minimize travel time costs (transitional costs) as well as contend with minimizing her losses while in the cities (node costs). In this case, the optimal path will minimize D with individual costs

$$d[(i_k, j_k) \mid (i_{k-1}, j_{k-1})].$$

[Note that in this case, D would be accumulated by addition as in (11.9), but the individual d_B costs would be constructed by multiplication as in (11.7).] We will have an opportunity to use each of these DP problem types in our speech work.

We now seek the powerful DP algorithm for the solution of the shortest-path or least-cost problem. Central to the DP algorithm is the *Bellman optimality principle* (BOP). In order to introduce this concept, let us temporarily refocus our attention on the broader class of paths that need not start at the origin nor end at the terminal node. In general, let the path begin and end at arbitrary nodes (s, t) and (u, v), respectively. We define the notation

$$(s, t) \overset{\rightarrow}{*} (u, v) \tag{11.12}$$

[5] She calls from the previous city and warns of her arrival. The customer has this length of time to ponder the situation.

to be the *best* path (in the sense of minimum cost) leading from (s, t) to (u, v). Also denote by

$$(s, t) \overset{(w, x)}{\underset{*}{\rightarrow}} (u, v) \tag{11.13}$$

the best path segment from (s, t) to (u, v) which also passes through (w, x). In these terms, the BOP can be stated as follows (Bellman, 1957, p. 83).

BELLMAN OPTIMALITY PRINCIPLE

$$(s, t) \overset{(w, x)}{\underset{*}{\rightarrow}} (u, v) = (s, t) \underset{*}{\rightarrow} (w, x) \oplus (w, x) \underset{*}{\rightarrow} (u, v) \tag{11.14}$$

for any $s, t, u, v, w,$ *and* x, *such that* $0 \leq s, w, u \leq I$ *and* $0 \leq t, x, v \leq J$, *where* $\oplus$ *denotes concatenation of the path segments.*

The consequences of this result for efficient algorithm development are quite significant. In particular, it implies that

$$\begin{aligned}(0, 0) &\overset{(i_{k-1}, j_{k-1})}{\underset{*}{\rightarrow}} (i_k, j_k) \\ &= (0, 0) \underset{*}{\rightarrow} (i_{k-1}, j_{k-1}) \oplus (i_{k-1}, j_{k-1}) \underset{*}{\rightarrow} (i_k, j_k),\end{aligned} \tag{11.15}$$

which means that in order to find the best path segment from $(0, 0)$ to (i_k, j_k) that passes through a "predecessor" node (i_{k-1}, j_{k-1}), it is not necessary to reexamine all the partial paths leading from $(0, 0)$ to (i_{k-1}, j_{k-1}) enroute to (i_k, j_k). It is sufficient to simply *extend* $(0, 0) \underset{*}{\rightarrow} (i_{k-1}, j_{k-1})$ over the shortest path segment possible to reach (i_k, j_k). If we define

$$\begin{aligned}D_{\min}(i, j) &\overset{\text{def}}{=} \text{distance from } (0, 0) \text{ to } (i, j) \text{ over the best path} \\ &= \text{distance associated with } (0, 0) \underset{*}{\rightarrow} (i, j)\end{aligned} \tag{11.16}$$

and

$$\begin{aligned}D_{\min}\big[(i_k, j_k) | (i_{k-1}, j_{k-1})\big] &\overset{\text{def}}{=} \text{distance from } (0, 0) \text{ to } (i_k, j_k) \text{ over the best partial path through } (i_{k-1}, j_{k-1}) \\ &= \text{distance associated with } (0, 0) \overset{(i_{k-1}, j_{k-1})}{\underset{*}{\rightarrow}} (i_k, j_k)\end{aligned} \tag{11.17}$$

then, as a direct consequence of (11.15), we can write

$$D_{\min}\big[(i_k, j_k) \mid (i_{k-1}, j_{k-1})\big] = D_{\min}(i_{k-1}, j_{k-1}) + d\big[(i_k, j_k) \mid (i_{k-1}, j_{k-1})\big]. \tag{11.18}$$

This expression describes distance of the best path beginning at $(0,0)$ and (eventually) arriving at (i_k, j_k) from (i_{k-1}, j_{k-1}). The globally optimal path arriving at (i_k, j_k) can be found by considering the set of the best path segments arriving from all possible predecessor nodes and taking the one with minimum distance. The optimal path to (i_k, j_k) therefore has distance

$$\begin{aligned} D_{\min}(i_k, j_k) &= \min_{(i_{k-1}, j_{k-1})} \left\{ D_{\min}\left[(i_k, j_k) \mid (i_{k-1}, j_{k-1})\right] \right\} \\ &= \min_{(i_{k-1}, j_{k-1})} \left\{ D_{\min}(i_{k-1}, j_{k-1}) + d\left[(i_k, j_k) \mid (i_{k-1}, j_{k-1})\right] \right\}. \end{aligned} \tag{11.19}$$

We must be a bit careful in applying (11.19). The minimization is taken over all possible predecessor nodes to (i_k, j_k). This is a key concept, and we shall find in our speech work that the nodes that may precede a particular "(i_k, j_k)" are often highly constrained. We must also be careful with "initialization" of this recursion. By convention, we shall require all paths in our speech work (at least in these initial discussions) to start at $(0,0)$ in the grid. At the beginning of the search, therefore, many "early" points in the grid will have only $(0,0)$ as a legitimate predecessor. For this purpose, we define

$$D_{\min}(0,0) \stackrel{\text{def}}{=} 0. \tag{11.20}$$

Note that (11.19) relates the distance associated with $(0,0) \xrightarrow{*} (i_k, j_k)$, but it does not tell us which nodes are on the path. In some problems it will be sufficient to simply know the distance of the shortest path, but in others we will wish to know precisely *which* path is the shortest one. A simple way to keep track of the nodes on paths is the following: Once the optimal partial path to (i_k, j_k) is found, we simply record the immediate predecessor node on the partial path, nominally at a memory location attached to (i_k, j_k). Suppose that we define

$$\begin{aligned} \Psi(i_k, j_k) &\stackrel{\text{def}}{=} \text{index of the predecessor node to } (i_k, j_k) \\ &\qquad \text{on } (0,0) \xrightarrow{*} (i_k, j_k) \\ &= (i_{k-1}, j_{k-1}). \end{aligned} \tag{11.21}$$

Clearly, if we know the predecessor node to any node on $(0,0) \xrightarrow{*} (i_k, j_k)$, then the entire path segment can be reconstructed by *backtracking* beginning at (i_k, j_k). In particular, if $(i_k, j_k) = (I, J)$, the terminal node in the grid, then the globally optimal path can be reconstructed. If we let (i_k^*, j_k^*) denote the kth index pair on the path $(0,0) \xrightarrow{*} (I, J)$ (assumed to be K nodes long), then

$$
\begin{aligned}
(i_K^*, j_K^*) &= (I, J) \\
(i_{K-1}^*, j_{K-1}^*) &= \Psi(I, J) = \Psi(i_K^*, j_K^*) \\
(i_{K-2}^*, j_{K-2}^*) &= \Psi(\Psi(I, J)) = \Psi(i_{K-1}^*, j_{K-1}^*) \\
&\vdots \\
(i_0^*, j_0^*) &= \cdots = \Psi(i_1^*, j_1^*) = (0, 0).
\end{aligned} \tag{11.22}
$$

We are nearly prepared to provide an example DP algorithm but must consider a very important issue first. The developments above suggest a technique by which we can sequentially extend paths in an optimal manner. We must carefully note, however, that the extension of a "best" path to, say, (i_k, j_k), depends upon the knowledge of optimal paths to all "legal" predecessor nodes of form (i_{k-1}, j_{k-1}). By "legal" we mean all predecessors for which transitions to (i_k, j_k) are possible according to the constraints of the problem. A path search structured so that a set of predecessor nodes with costs already assigned is always available to extend to a given node does not necessarily follow conveniently from the dynamics of the problem. An example of a "difficult" search is shown in Fig. 11.2. The classical Traveling Salesman Problem (TSP) (see footnote 4) is an even more challenging problem, which is discussed in (Papadimitriou and Steiglitz, 1982, Sec. 18.6). A significant amount of thought is some-

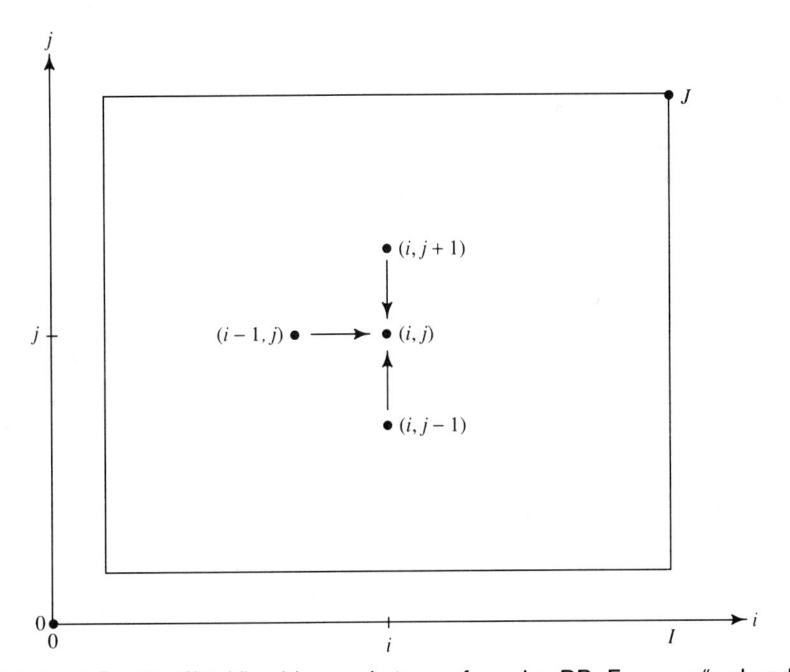

FIGURE 11.2. A "difficult" grid search to perform by DP. For any "ordered search" of the processing of nodes (e.g., bottom to top along columns beginning from the left, right to left along rows beginning at the bottom) the predecessor nodes will not be available for extension.

times required to break the problem down into stages of decisions that follow a pattern yielding a useful recurrence relation.[6] Once found, the memory and computational requirements of the algorithm are sometimes extraordinary as the number of predecessors to keep track of at each stage grows exponentially. However, we do not wish to overstate the problem. There is nothing that invalidates the BOP when the search becomes complicated, and a DP algorithm can be a powerful tool that is often far superior to algorithms based on more naive considerations. [For example, the TSP can be solved using $O(IJ2^{IJ})$ flops, whereas a complete enumeration of all paths in the grid requires $O(IJ-1)!$.]

Although challenges may arise in structuring DP searches, there is often a structure to a sequential decision problem that imposes neat constraints on which nodes may follow others. Fortunately, this will be the case in every attempt to employ DP in our speech work, as we will be dealing with grids that are close to what have been called *layered graphs* (Papadimitriou and Steiglitz, 1982, Sec. 18.6). Generally speaking, this means that the nodes can be lined up in columns (layers) in such a way that there are no "backward" transitions (from right to left) in the grid. These constraints will have a significant impact on the DP algorithm we develop and its computational complexity, in most cases making the algorithm very easy to construct and inexpensive to compute.

EXAMPLE DP ALGORITHM (Traveling Salespeople Revisited) _____

Before proceeding, let us consider an example of how structured paths can yield very simple DP algorithms. We return to the problem of the salesman trying to map a shortest path through the grid of cities. If the salesman is required to drive eastward (in the positive i direction) by exactly one unit with each city transition, this lends a special structure to the problem. In fact, this means that any "legal" predecessor to node (i_k, j_k) will be of the form (i_k-1, j_{k-1}). Only knowledge of a certain set of optimal paths must be available (those in the "column" just to the "west" of i_k) in order to explore extension to (i_k, j_k). Figure 11.3 illustrates this idea. An example DP algorithm that finds the optimal path $(0,0) \overset{*}{\rightarrow} (I,J)$ for this problem is shown in Fig. 11.4. The algorithm proceeds by successively extending, on a column-by-column basis, all possible path segments[7] in the grid according to the inherent prescription above. Each time a path segment is optimally extended, the predecessor node is recorded. Eventually, all paths that lead to (I,J) from predecessor nodes of the form $(I-1,p)$ are found, and the globally optimal path is selected from among them according to (11.19). The best path can then be reconstructed, if desired, by backtracking.

---

[6]One trick that is sometimes useful is to start at the terminal node and work backward. The BOP says that an optimal path to (u, v) through (w, x) must terminate in an optimal path from (w, x) to (u, v). Therefore, one can start at (I, J) and find all predecessor nodes; then find all optimal paths to those predecessors; and so on.

[7]Those that can reach (I, J).

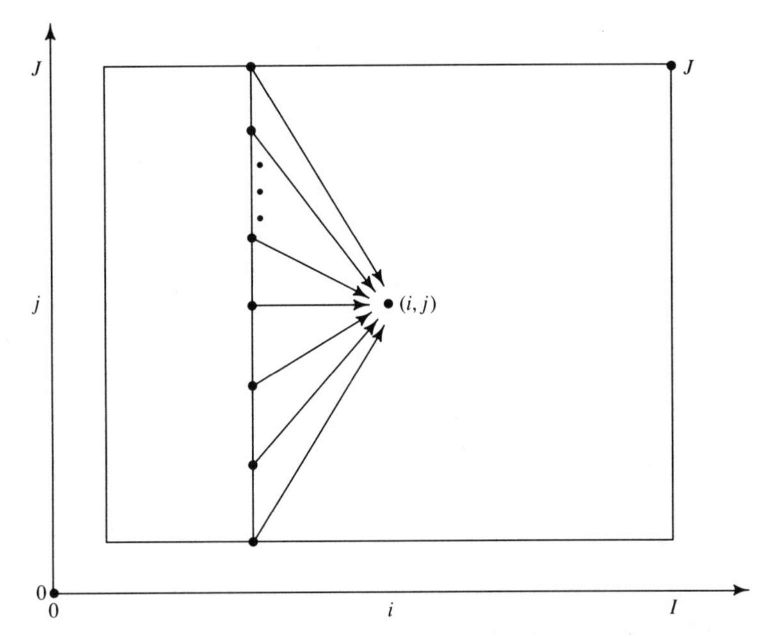

FIGURE 11.3. If all transitions on all paths must move exactly one unit to the "east" in the grid, then only optimal paths to predecessor nodes $(i-1, j')$, for all j', must be known to find the optimal path to (i, j). The search policy should naturally be to complete paths along columns (nodes along a given column can be treated in any order), proceeding from left to right. The only memory requirement is a location to hold the minimum distance, and one for the optimal predecessor, at each (i, j).

FIGURE 11.4. Example dynamic programming algorithm for the "eastbound salesman problem."

Initialization: "Origin" of all paths is node $(0, 0)$.
For $j = 1, 2, \ldots, J$
$D_{\min}(1, j) = d[(1, j) \mid (0, 0)]$
$\Psi(1, j) = (0, 0)$
Next j

Recursion: For $i = 2, 3, \ldots, I$
For $j = 1, 2, \ldots, J$
Compute $D_{\min}(i, j)$ according to (11.19).
Record $\Psi(i, j)$ according to (11.21).
Next j
Next i

Termination: Distance of optimal path $(0, 0) \overset{*}{\rightarrow} (I, J)$ is $D_{\min}(I, J)$.
Best path is found by backtracking as in (11.22).

11.3 Dynamic Time Warping Applied to IWR

11.3.1 DTW Problem and Its Solution Using DP

With a firm grasp of the principles of DP, we are now prepared to apply these ideas to our first speech recognition technique.

Dynamic time warping (DTW) is fundamentally a feature-matching scheme that inherently accomplishes "time alignment" of the sets of reference and test features through a DP procedure. By *time alignment* we mean the process by which temporal regions of the test utterance are matched with appropriate regions of the reference utterance. In this section we will focus on discrete utterances as the unit to be recognized. For the sake of discussion, we will assume the usual case in which these discrete utterances are words, although the reader should keep in mind that there is nothing that theoretically precludes the use of these principles on subword units, or even complete paragraphs (if such is to be considered a "discrete utterance"). The need for time alignment arises not only because different utterances of the same word will generally be of different durations, but also because phonemes within words will also be of different durations across utterances.

Generally speaking, DTW is used in IWR to match an incoming test word (represented by a string of features) with numerous reference words (also represented by feature strings). The reference word with the best match score is declared the "recognized" word. In fairness to the correct reference string, the test features should be aligned with it in the manner that gives the best matching score, to prevent time differences from unduly influencing the match.[8]

Early attempts at compensating for time differences among words consisted of simple linear expansion or compression of the time axis of the test utterance. This procedure is highly dependent upon the correct determination of endpoints and makes no attempt to align intraword phonetic events. In an effort to compensate for this latter deficiency, some researchers tried to line up identifiable events within the test and reference utterances by using energy measures [e.g., (Pruzansky, 1963)]. These techniques were the predecessors to DTW in the sense that they performed a nonlinear mapping of the test utterance onto the reference utterance.

This subject is frequently introduced by asserting that the DTW paradigm offers a systematic method of finding a nonlinear mapping, or "warping," of the time axis of the test utterance onto that of the reference utterance, which effects an optimal match in a certain sense. The name "DTW" derives in part from this apparent quest for an optimal warping function. The problem with placing too much emphasis upon

[8]On the other hand, the reader might wonder about the possibility of warping the test word in such a way that it fits an incorrect reference string quite well. For example, "spills" could be time warped in such a way that it matches "pills" very well. Measures will be taken to prevent such anomalies.

the pursuit of a warping "function" is that frequently the mapping turns out not to be a *function* at all. In this case an *ad hoc* strategy is used to maintain the semblance of a pursuit of mappings. All of this eventually is boiled down to a simple DP algorithm, with the mapping inherent in the optimal path found and serving only as a diversion from the main issues. Therefore, we approach this topic by avoiding the usual starting point and going directly to the DP problem. We begin by looking at a simple version of this technique, then add specific details.

Suppose that we have reduced both a test and a reference utterance of a word to strings of features extracted from the acoustic speech waveform. For example, in each case we might compute 14th-order LP vectors on frames of length $N = 128$ that are shifted by 64 points before each computation. Let us denote the test utterance LP vectors by $\hat{\mathbf{a}}(m)$ and those of the reference utterance by $\hat{\mathbf{b}}(m)$, where m, as usual, denotes the endpoint of the frame in each case. If the test waveform is 1280 samples long, and the reference waveform is 1536 samples long (each assumed to begin at $n = 0$), this procedure will result in the following feature strings:

$$\text{test features:} \quad \hat{\mathbf{a}}(127), \hat{\mathbf{a}}(191), \hat{\mathbf{a}}(255), \ldots, \hat{\mathbf{a}}(1279) \tag{11.23}$$

$$\text{reference features:} \quad \hat{\mathbf{b}}(127), \hat{\mathbf{b}}(191), \hat{\mathbf{b}}(255), \ldots, \hat{\mathbf{b}}(1535). \tag{11.24}$$

In a second example, suppose that we compute 10 cepstral coefficients on the same two signals, in this case using frames of length $N = 256$ that are shifted by 128 points each time. Let us call the vector of test cepstral coefficients $\mathbf{c}(m)$, and the vector of reference coefficients $\mathbf{d}(m)$. In this case we have the following feature strings:

$$\text{test features:} \quad \mathbf{c}(255), \mathbf{c}(383), \mathbf{c}(511), \ldots, \mathbf{c}(1279) \tag{11.25}$$

$$\text{reference features:} \quad \mathbf{d}(255), \mathbf{d}(383), \mathbf{d}(511), \ldots, \mathbf{d}(1535). \tag{11.26}$$

Whether we are dealing with the LP parameters, cepstral parameters, or some other feature strings from some other problem, let us reindex the strings so that they are indexed by simple integers and refer to them as follows:

$$\text{test features:} \quad \mathbf{t}(1), \mathbf{t}(2), \mathbf{t}(3), \ldots, \mathbf{t}(i), \ldots, \mathbf{t}(I) \tag{11.27}$$

$$\text{reference features:} \quad \mathbf{r}(1), \mathbf{r}(2), \mathbf{r}(3), \ldots, \mathbf{r}(j), \ldots, \mathbf{r}(J). \tag{11.28}$$

It is clear that the indices i and j are only related to the original sample times in the acoustic waveform through knowledge of the frame end-times in the analysis. Nevertheless, it is customary to refer to the i and j axes upon which we will lay out our features as "time" axes.

We now develop the formal DTW problem and show how it can be quickly solved using a DP algorithm. Our objective is to match the test and reference features so that they are appropriately aligned. By this procedure we mean that the features are matched pairwise so that the best global matching score for the two strings is obtained. In order to quantify the match score, let us denote the "local" cost of the kth pairing by

$$0 \le d_N(i_k, j_k) = \text{cost of matching the } \mathbf{t}(i_k) \text{ with } \mathbf{r}(j_k). \quad (11.29)$$

This notation, of course, is a foreshadowing of the fact that we are setting up a DP search problem with (at least) Type N cost assignments to the grid. In the example above, in which the features consist of LP vectors, a likely cost function would be

$$d_N(i_k, j_k) = d_I[\mathbf{t}(i_k), \mathbf{r}(j_k)], \quad (11.30)$$

where $d_I(\cdot\,,\cdot)$ represents the Itakura distance; while in the case in which the features are vectors of cepstral coefficients, a candidate cost function would be the Euclidean distance,

$$d_N(i_k, j_k) = d_2[\mathbf{t}(i_k), \mathbf{r}(i_k)] = \|\mathbf{t}(i_k) - \mathbf{r}(j_k)\|_2. \quad (11.31)$$

The global cost of any valid path through the grid, say (i_1, j_1), (i_2, j_2), ..., (i_k, j_k), is simply the sum of these individual match scores,

$$D = \sum_{k=1}^{K} d_N(i_k, j_k). \quad (11.32)$$

Now let us lay the test and reference features out along the abscissa and ordinate of a grid as shown in Fig. 11.5. Clearly associated with any feature pairing of the form

$$\mathbf{t}(i_k) \text{ matched with } \mathbf{r}(j_k), \quad \text{for } k = 1, 2, \ldots, K, \quad (11.33)$$

is the path

$$(i_k, j_k), \quad \text{for } k = 1, 2, \ldots, K. \quad (11.34)$$

In particular, associated with the optimal feature pairing, say,

$$\mathbf{t}(i_k^*) \text{ matched with } \mathbf{r}(j_k^*), \quad \text{for } k = 1, 2, \ldots, K^*, \quad (11.35)$$

is the path

$$(i_k^*, j_k^*), \quad \text{for } k = 1, 2, \ldots, K^*. \quad (11.36)$$

Now to node (i, j) in the grid we assign the cost $d_N(i, j)$ as in (11.29). The cost of any path of form (11.34) is naturally D of (11.32). Therefore, (11.36), which corresponds to a minimum of D, represents the minimum-cost, or shortest-distance, path of form (11.34) through the grid. We have therefore reduced our feature mapping problem into a shortest-distance path search through the i–j plane, and should feel quite confident that we can solve the problem using DP given the right constraints on the search.

An enhancement to the shortest-path problem is appropriate before continuing. Although the Type N problem setup above captures the essence of the DTW problem, it is often desirable to solve a Type B problem instead. The reason is as follows: In making the transition into, say, node (i_k, j_k), we may wish to assign a local cost based not only on the pairing of features $\mathbf{t}(i_k)$ with $\mathbf{r}(j_k)$, *viz.*, $d_N(i_k, j_k)$, but also upon the "tra-

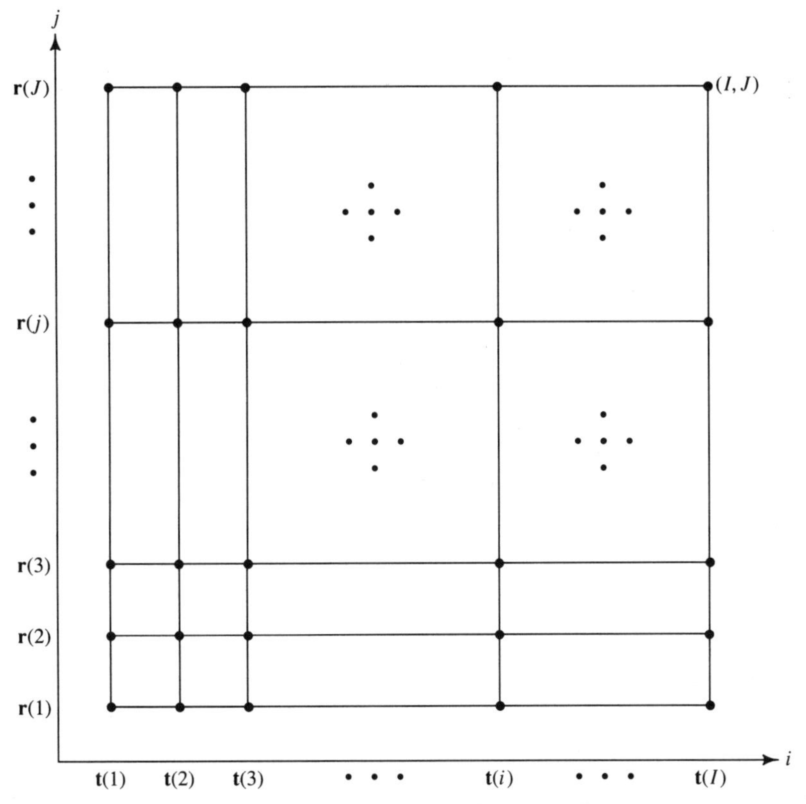

FIGURE 11.5. Test and reference feature vectors associated with the i and j coordinates of the search grid, respectively.

jectory" of the path taken to enter node (i_k, j_k). (Think about the effect of the manner of the transition upon the warping of the two feature strings.) Accordingly, we might wish to use

$$d[(i_k, j_k)|(i_{k-1}, j_{k-1})] = d_T[(i_k, j_k)|(i_{k-1}, j_{k-1})] \times d_N(i_k, j_k) \tag{11.37}$$

as a measure of cost at node (i_k, j_k). In this case (11.32) becomes

$$D = \sum_{k=1}^{K} d[(i_k, j_k)|(i_{k-1}, j_{k-1})]. \tag{11.38}$$

In fact, it is prudent to *normalize* the final distance measure so that paths of different lengths receive an equal opportunity to be optimal. (Imagine, for example, the correct word being penalized simply because its test string of features is longer than that of an incorrect match.) One rational method of normalization is to express $D = D_{\min}(I, J)$ on an "average cost per node" basis. A moment's reflection will indicate that the appropriate calculation is to divide D of (11.38) by the sum of the transition costs,

$$\tilde{D} \stackrel{\text{def}}{=} \frac{\sum_{k=1}^{K} d[(i_k, j_k)|(i_{k-1}, j_{k-1})]}{\sum_{k=1}^{K} d_T[(i_k, j_k)|(i_{k-1}, j_{k-1})]}. \tag{11.39}$$

We will discover later that the ability to incorporate this normalization with a DP solution is not always possible.

Finally, before pursuing the details of the DP problem, let us elaborate on a point made at the outset of this discussion. There we noted that the DTW problem is often introduced in a somewhat different manner in tutorial material and in the early literature on the subject [e.g., (Myers et al., 1980)]. It is often stated that the objective of the DTW problem is to find an optimal warping function, say $\omega(\cdot)$, which maps the i axis onto the j axis in a manner that best aligns the features in the sense of minimizing D of (11.32). The warping function, therefore, nominally creates a relation of the form

$$j = \omega(i). \tag{11.40}$$

We can now appreciate that the difficulty with this approach is that the optimal mapping is not always functional—more than one $\mathbf{r}(j)$ may be associated with a single $\mathbf{t}(i)$. This, for example, will occur if the test string is shorter than the reference string,[9] and it is mandated that every $\mathbf{r}(j)$ must be associated with some $\mathbf{t}(i)$. The analytical method proposed to resolve this problem in (Myers et al., 1980), for example, involves *two* mapping functions, one for each of the i and j axes, onto a third axis. The indices along this third axis may be considered as integers that count the nodes in the path (like k above); hence, in essence, the mappings are simply used to create node pairs. Whether this method or some related interpretation of the "generalized" warping procedure is used [e.g., (Parsons, 1986, p. 298)], the problem quickly boils down to a path-search problem of the type we have described above. Therefore, we have simply chosen to begin with the path-search problem to avoid unnecessary details.

We now proceed to the issue of solving the DTW search by DP.

11.3.2 DTW Search Constraints

In Section 11.2, we discussed in some detail the effects of search constraints on the eventual form of a DP algorithm. Grid searches in DTW problems are usually very highly structured, both to limit the amount of computation and to assure the appropriateness of regions matched be-

[9]It might occur to the reader to simply reverse the roles of the reference and test axes in this case to preserve the functional form of $\omega(\cdot)$. However, this cannot always be done without some effect on the recognition rate (Myers et al., 1980).

tween the test and reference strings. Let us concentrate on this latter issue first, then come back and discuss computational complexity at the end of the section.

The search of the grid is usually subject to four basic types of constraints having to do with physical arguments about the data and reasonable matching. These are discussed in the following subsections.

Endpoint Constraints and "Word Spotting." In some approaches to DTW, the endpoints of the test and reference strings are assumed to match to a reasonable degree. In others, the endpoints are assumed to be virtually unknown and are found inherently in the DTW search.

The strictest form of endpoint constraints in a DTW algorithm requires that the endpoints match exactly:

$$\mathbf{t}(1) \text{ must be paired with } \mathbf{r}(1) \text{ on any candidate path} \quad (11.41)$$

$$\mathbf{t}(I) \text{ must be paired with } \mathbf{r}(J) \text{ on any candidate path.} \quad (11.42)$$

Items (11.41) and (11.42) simply imply that any path we examine must begin at $(1, 1)$ and end at (I, J). Recall that we formally let the DP search originate at a fictitious $(0, 0)$ in the grid. The requirement that the path "begin at $(1, 1)$" here simply means that the only allowable transition out of $(0, 0)$ will be to node $(1, 1)$. It is also important to recall that we insisted in earlier discussions that any Type N assignment of costs to the grid include a zero cost for the node $(0, 0)$. The relevance of this requirement in the present situation should be evident. The reader is encouraged to return to (11.19) and (11.20) and note the effect of these endpoint constraints on the initialization of the recursion.

A much less constrained approach (Bridle and Brown, 1979) uses the DTW search itself to automatically locate endpoints in a feature string by finding the candidate set of points (beginning and end) that yields the best match. This technique is sometimes called *continuous scanning* or simply the *Bridle algorithm. Word spotting* is the process of automatically determining the presence of a word (i.e., the features representing the word) in the context of a longer test string of features that represent, in general, a multiword utterance. However, word spotting can also be used to locate a *single* word whose endpoints are unknown. As illustrated in Fig. 11.6, suppose we lay the test string (length I) out along the abscissa in the customary manner, and place a single-word reference template along the ordinate. For simplicity, let us assume that the endpoints of the reference template are known exactly. Now suppose we allow DTW search paths to begin at any point along the test axis, and to end at any point along the "top" of the search space. Of course, any acceptable path will need to adhere to reasonable constraints such as monotonicity and minimum word length (see below). A moment's thought will indicate that this liberal beginning- and ending-point policy corresponds to highly unconstrained endpoints for the test word. In other words, we do not know *where* in the test string the hypothesized (reference) word might re-

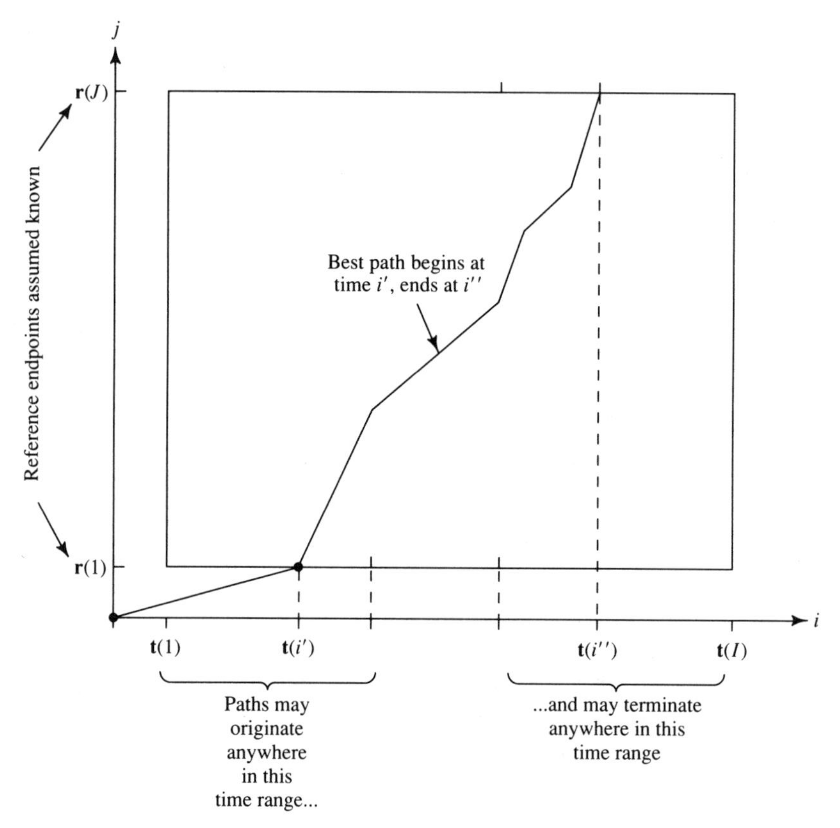

FIGURE 11.6. Word-spotting DTW search.

side, so we are willing to try any reasonable sets of beginning and ending points. A path with a favorable cost, say between test times i' and i'' (see Fig. 11.6), may be considered to be the result of matching the proper segment of the test string (i.e., between its correct endpoints) to the reference string. This results in the reference word becoming a candidate for the recognized word, and also a set of candidate endpoints for the test word. We will see this process used inherently in the "one-stage" algorithm, and more explicitly when we discuss the "grammar-driven connected-word recognition" system, in Section 11.4.4. Many of the formal details of the procedure will be presented in the context of the one-stage algorithm.

Methods exist which are somewhat intermediate between the assumption of known endpoints and the use of word spotting. In Section 10.2.3, we discussed the difficulties encountered in locating the beginning and end of a discrete utterance by direct methods. Relaxing the endpoint constraints on a DTW grid search offers one method for making the algorithm's performance less dependent upon the precise location of these endpoints without complete recourse to the Bridle approach. One method of relaxing endpoint constraints simply consists of "opening up the ends" of the search region of the grid. For example, as illustrated in Fig. 11.7, the initial transition of the path [which for formal purposes is still anchored at $(0, 0)$], is now permitted to arrive at any of the series of nodes $(1, 1)$ to $(1, 1 + \varepsilon)$ in the vertical direction (flexibility in the reference direction), and $(1, 1)$ to $(1 + \varepsilon, 1)$ in the horizontal (flexibility in the test direction). Similar flexibility is also found at the other end of the path.

Another method reported by Rabiner et al. (1978) restricts the search region to a band of points that adaptively follows the path of local minimum cost. The width of the band remains constant, implying multiple potential endpoint nodes at each end of the search. This method has been referred to as UELM for *unrestricted endpoint, local minimum.*

Monotonicity. The path should be *monotonic.* This means that

$$i_{k-1} \le i_k \text{ and } j_{k-1} \le j_k, \tag{11.43}$$

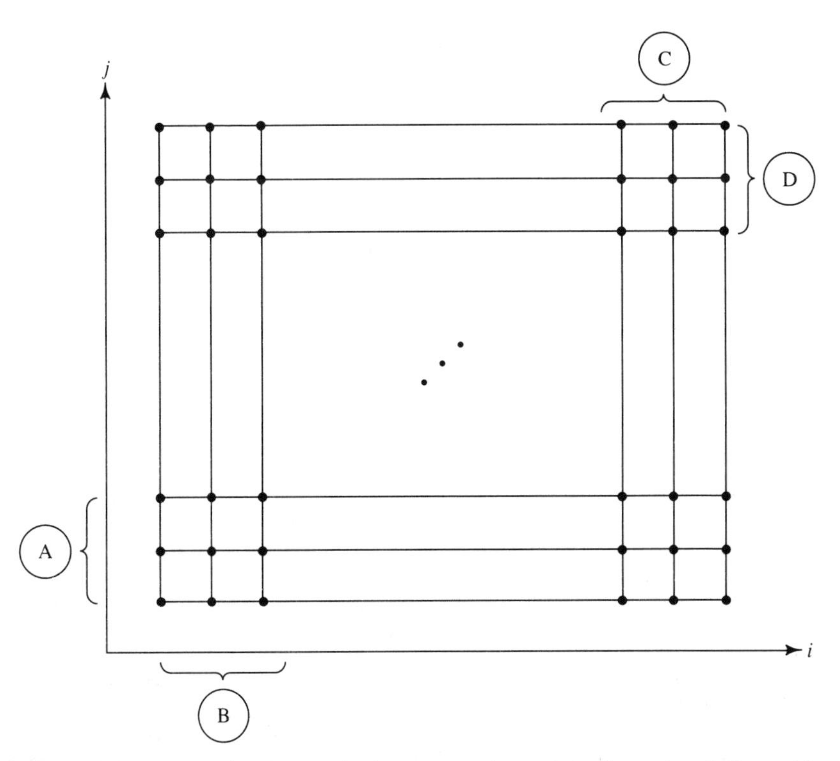

FIGURE 11.7. "Relaxing" the endpoint constraints at both ends of the grid search allows for some uncertainty in the initial and final points in time for both the reference and test waveforms. Regions A and D correspond to uncertainty in the reference string beginning and end, while B and C correspond to uncertainty in the test string endpoints.

which, in turn, require that any candidate path not go "south" or "west" at any time.[10] Physically, this requires, for example, that features of the test waveform must never be matched to features in the reference waveform that are earlier in time than those already matched. This prevention of the path from doubling back on itself is critical to prevent high match scores from very inappropriate warpings. An example is shown in Fig. 11.8.

Global Path Constraints. These constraints follow from a specification of the amount of allowable compression or expansion of "long" stretches of the test string when mapped onto the reference string. They imply regions in the grid outside of which no path must be allowed to go (even if optimal) and significantly pare down the number of grid points that must be considered in the search.

A frequently used global constraint region was suggested by Itakura (1975). An example is shown in Fig. 11.9 for the case in which compres-

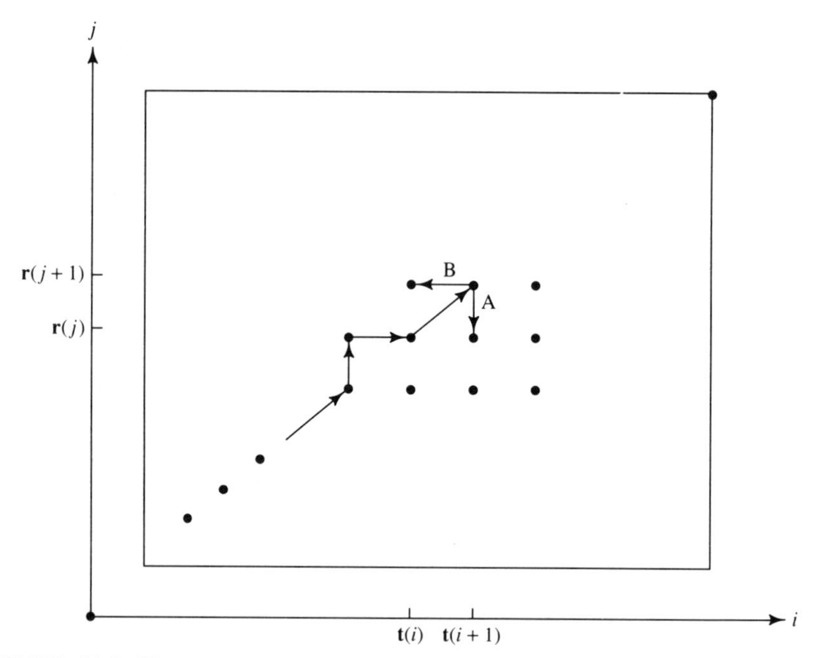

FIGURE 11.8. The monotonicity constraint requires that the search path not make a transition in the negative i or negative j direction at any time. Moving "southward" (transition A in the figure) causes a past reference vector, $\mathbf{r}(j)$, to be reused, and moving "westward" (transition B) causes a past test vector, $\mathbf{t}(i)$, to be reused. Preventing the path from doubling back on itself is critical to preventing high match scores from very inappropriate warping.

[10]The reader should note carefully that there is no implication that $i_k = i_{k-1} + 1$ or that $j_k = j_{k-1} + 1$.

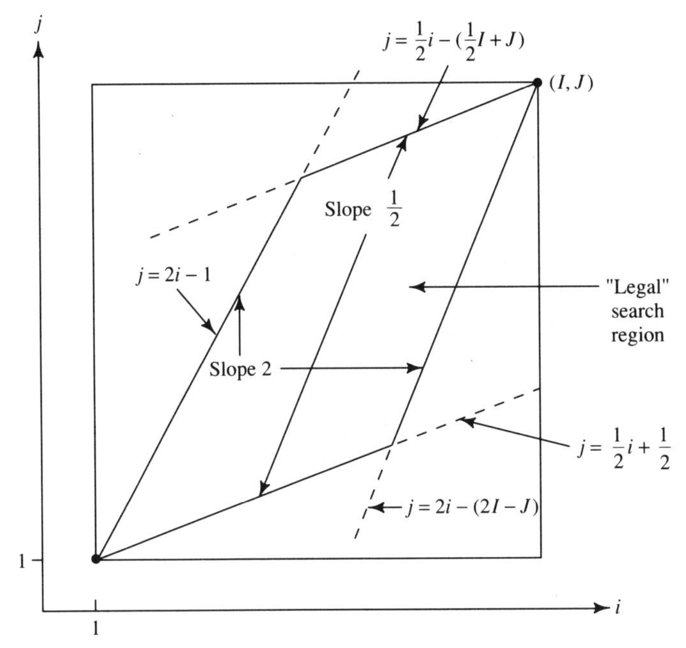

FIGURE 11.9. The Itakura global path search constraints for maximum compression and expansion factors of two. The method for constructing the parallelogram is evident from the figure and is formally described in the text.

sion and expansion factors are each restricted to two. A parallelogram region follows from the following arguments. Worst-case paths (compression factor two, slope $\frac{1}{2}$ and expansion factor two, slope two) are drawn beginning at the tied endpoint $(1, 1)$. It is clear that these paths will generally not intersect with the other obligatory endpoint (I, J) unless they make very sharp "turns" at the top or right boundary of the grid and follow trajectories that are in serious violation of the compression or expansion limits. Therefore, noting that any path which enters (I, J) must not represent compression or expansion of more than two over a long range of test points, worst-case paths are drawn entering (I, J). The interior and boundary nodes of the parallelogram formed by the intersection of the four worst-case paths are deemed appropriate nodes for search.

Note that the Itakura parallelogram degenerates to the single linear path connecting $(1, 1)$ with (I, J) when $I = cJ$ or $cI = J$, where c is the maximum allowable compression or expansion factor. The parallelogram will allow the exploration of the most paths when $I \approx J$. When $c = 2$ and $I \approx J$, then about $I^2/3$ grid points are used, implying that $I^2/3$ costs of the form (11.29) need to be computed. This is to be compared with about I^2 to be computed if the entire grid is searched. We see, then, that search constraints reduce computation as well as place physically reasonable boundaries on the matching process.

A second, simpler type of global search region is imposed by using the constraint that for any node, say (i_k, j_k), on any path to be considered

$$|j_k - i_k| \leq W, \tag{11.44}$$

where W is called the "window width." This constraint generates a simple strip around the purely linear path, as shown in Fig. 11.10. The savings in computation resulting from this search region is explored in Problem 11.2.

Both the Itakura and windowed search regions illustrate the fact that global path constraints are often inseparably related to local path constraints to which we now turn.

Local Path Constraints. Whereas global path constraints are used to restrict the amount of compression or expansion of the test waveform over long ranges of time, local path constraints are used to restrict the local range of a path in the vicinity of a given node in the grid. It is usual to specify the constraints by indicating all legal sets of predecessor nodes to a given node, but this is not universally the case. Myers et al. (1980), for example, have suggested the specification of successor nodes by posing a set of productions in a regular grammar (see Section 13.2.1).

In Fig. 11.11 we show several types of local path constraints in terms of sets of predecessor points that have been studied by various researchers. It is to be noted that in two of the cases, (a) and (e), the immediate predecessor nodes, of the form (i_{k-1}, j_{k-1}) to (i_k, j_k), are unconditionally specified by the constraints. In the other cases, however, whether a particular "(i_{k-1}, j_{k-1})" is a predecessor to (i_k, j_k) depends on the predeces-

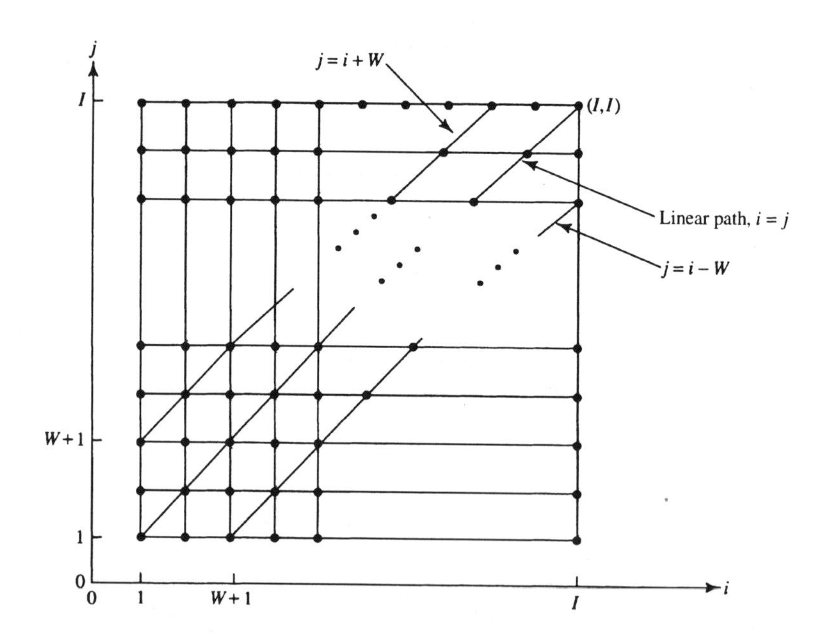

FIGURE 11.10. A simple global search region that restricts the search to a region of "width W" around the purely linear path. Shown: $W = 2$.

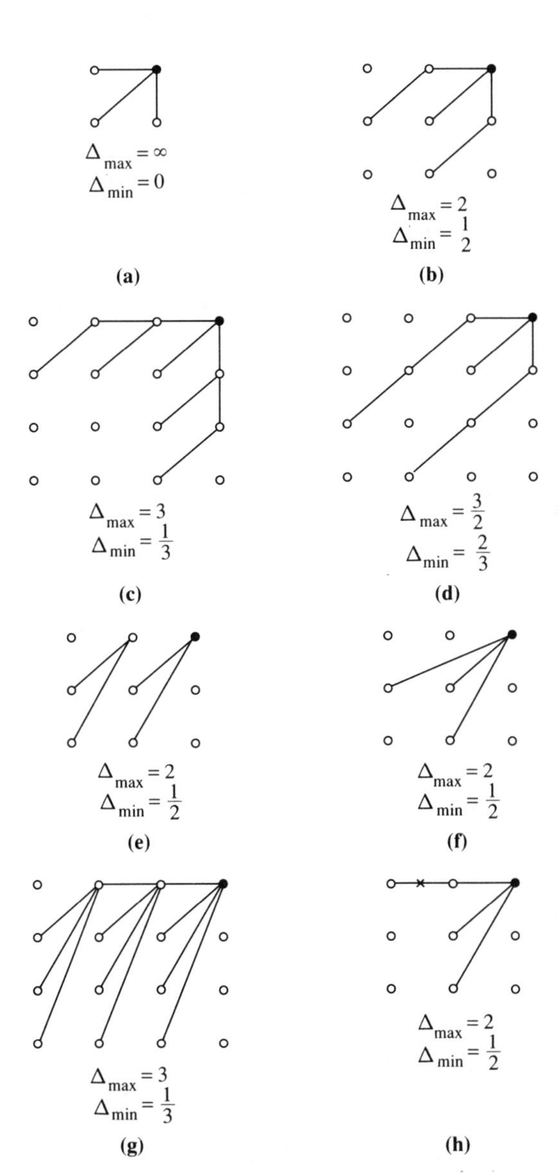

FIGURE 11.11. "Local constraints" on DTW path search. Each case shows the "legal" local paths by which a global path may arrive at the node (l, j). Cases (a) to (d) are considered by Sakoe and Chiba, (h) by Itakura, and (b) and (e) to (h) by Myers et al. in papers cited in the text. In case (h), there is an explicitly forbidden path shown by x. Also shown is the maximum and minimum global expansions of the test with respect to the reference waveform, Δ_{max} and Δ_{min}, which can result under the various local constraints. The relationship between local and global constraints is described in the text.

sor(s) to (i_{k-1}, j_{k-1}). In some cases, this dependence goes back as far as three nodes from the node (i_k, j_k) under consideration. In principle, this means that our Type T cost assignments to the grid are no longer Markov. This is mildly disturbing, since all of our discussions thus far have been based on an assumption that decisions could be made sequentially without recourse to past nodes. The problem is resolved in a manner that preserves the Markov nature of the transition costs as follows: The BOP states that the best path from (s, t) to (u, v), which passes through (w, v), consists of the optimal path $(s, t) \overset{*}{\rightarrow} (w, x)$ concatenated with the optimal path $(w, x) \overset{*}{\rightarrow} (u, v)$. There is no requirement that (w, x) be an immediate predecessor node to (u, v) on the path. As a consequence, we can easily generalize (11.18) as

$$D_{\min}\left[(i_k, j_k) | (i_{k-p}, j_{k-p})\right] = D_{\min}\left[(i_{k-p}, j_{k-p})\right] + \hat{d}\left[(i_k, j_k) | (i_{k-p}, j_{k-p})\right], \tag{11.45}$$

where (i_{k-p}, j_{k-p}) is some legal ("distant") predecessor node to (i_k, j_k) which is p nodes back on the path of interest and where

$$\hat{d}\left[(i_k, j_k) | (i_{k-p}, j_{k-p})\right] \stackrel{\text{def}}{=} \sum_{m=0}^{p-1} d\left[(i_{k-m}, j_{k-m}) | (i_{k-m-1}, j_{k-m-1})\right]. \tag{11.46}$$

Note that local path constraint should be defined so that a unique path exists between any "(i_{k-p}, j_{k-p})" and (i_k, j_k) so that no ambiguity exists in computing $\hat{d}$. This is true, for example, of all constraints shown in Fig. 11.11. To find the optimal path to (i_k, j_k), therefore, we simply take the minimum over all distant predecessors,

$$\begin{aligned} D_{\min}(i_k, j_k) &= \min_{\text{"}(i_{k-p}, j_{k-p})\text{"}} \left\{D_{\min}\left[(i_k, j_k) | (i_{k-p}, j_{k-p})\right]\right\} \\ &= \min_{\text{"}(i_{k-p}, j_{k-p})\text{"}} \left\{D_{\min}\left[(i_{k-p}, j_{k-p})\right] + \hat{d}\left[(i_k, j_k) | (i_{k-p}, j_{k-p})\right]\right\}. \end{aligned} \tag{11.47}$$

We have been a bit sloppy in writing (11.47), since different predecessor nodes may have different values of p, hence the appearance of quotes around the minimization argument (i_{k-p}, j_{k-p}). A simple illustration in Fig. 11.12 will make this point clear.

As usual, we should be careful with the initialization of this recursion. For most points (i_k, j_k), the legal predecessor nodes (i_{k-p}, j_{k-p}) will be determined by the local path constraints. At the outset, however, the initial point of interest will be $(1, 1)$ and its only legal predecessor, according to our formal convention, is $(0, 0)$. Once we recall (11.20), the recursion is ready for use from the outset.

The purpose of a local path constraint is to limit the amount of expansion or compression of the test waveform in a small neighborhood preceding (i_k, j_k). For example, the four Sakoe and Chiba constraints shown in Fig. 11.11 require that any path make no more than m horizontal or vertical transitions without first making n diagonal transitions. n/m ra-

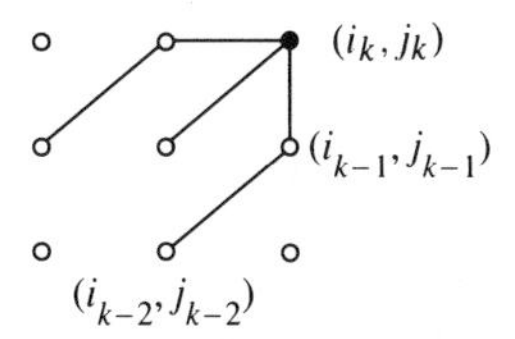

FIGURE 11.12. Illustration of the use of (11.47). Note that the value "p" is generally variable across different local paths within the constraint, as is apparent from this example. $\hat{d}[(i_k, j_k)|(i_{k-p}, j_{k-p})]$ is the accumulated cost along some local path with p transitions. If one of the "outer" paths is used in this example, then $p=2$; if the "inner" path is used, $p=1$.

tios of 0, 1/2, 1, and 2 are found in the figure. We wish to relate these to the global constraints discussed above. Suppose that for a given local path constraint type, there are R possible "local" paths over which to reach (i_k, j_k). For example, for local constraint in Fig. 11.12, there are three possible paths to (i_k, j_k). If Δ_i^r and Δ_j^r represent the total change in the i and j directions, respectively, over local path r, then the maximum and minimum expansion of the test waveform with respect to the reference are given by

$$\Delta_{\max} = \max_{r=1,\ldots,R} \frac{\Delta_j^r}{\Delta_i^r} \tag{11.48}$$

$$\Delta_{\min} = \min_{r=1,\ldots,R} \frac{\Delta_j^r}{\Delta_i^r}, \tag{11.49}$$

respectively. Myers et al. (1980) then give the result for the permissible search region in the i–j plane for the case in which $\Delta_{\min} = \Delta_{\max}^{-1}$. A pair (i, j) is found in the global search space only if both of the following conditions are met:

$$1 + \frac{i-1}{\Delta_{\max}} \leq j \leq 1 + \Delta_{\max}(i-1) \tag{11.50}$$

$$J + \Delta_{\max}(i-I) \leq j \leq J + \frac{i-I}{\Delta_{\max}}. \tag{11.51}$$

There are four implied inequalities on j. By setting each of these to equalities, we obtain the four lines that intersect to form the boundaries of the search region. The reader is encouraged to ponder this issue in light of the discussion surrounding Fig. 11.9.

It should also be noted that the monotonicity constraint, as well as the global path constraints, is (or should be) inherent in the local path constraints. Said another way, local path constraints that violate the requirement of monotonicity should not be chosen.

As we have discussed, a refinement to the local path constraints is the inclusion of transition costs of the form $d_T[(i_{k-m}, j_{k-m})|(i_{k-m-1}, j_{k-m-1})]$

along the various local path sectors. Generally, these weights are used to discourage particular types of deviation away from a linear path through a local neighborhood. Four transition cost types have been proposed by Sakoe and Chiba (1978). These are

$$d_T\big[(i_{k-m},j_{k-m})\big|(i_{k-m-1},j_{k-m-1})\big]=\min\{i_{k-m}-i_{k-m-1},j_{k-m}-j_{k-m-1}\} \tag{11.52}$$

$$d_T\big[(i_{k-m},j_{k-m})\big|(i_{k-m-1},j_{k-m-1})\big]=\max\{i_{k-m}-i_{k-m-1},j_{k-m}-j_{k-m-1}\} \tag{11.53}$$

$$d_T\big[(i_{k-m},j_{k-m})\big|(i_{k-m-1},j_{k-m-1})\big]=i_{k-m}-i_{k-m-1} \tag{11.54}$$

$$d_T\big[(i_{k-m},j_{k-m})\big|(i_{k-m-1},j_{k-m-1})\big]=\big[i_{k-m}-i_{k-m-1}\big]+\big[j_{k-m}-j_{k-m-1}\big]. \tag{11.55}$$

The reader is encouraged to think about the manner in which each of these transition cost strategies influences the path evolution. Note that some cost strategies, when applied to certain path constraints, can result in zero transition costs, a clearly inappropriate result since it makes the matching at the ensuing node "cost-free." An example of this phenomenon occurs if costs of type (11.52) are applied to the "top" local path in Fig. 11.12. To circumvent this problem, Sakoe and Chiba have suggested "smoothing" the transition costs along each local path by replacing each cost by the average cost along the local path. An example is shown in Fig. 11.13.

Finally, let us return to the issue of normalizing the optimal distance in order to express it on a "cost per node" basis. We first introduced this notion when we suggested the normalization (11.39), which is repeated here for convenience,

$$\tilde{D} \stackrel{\text{def}}{=} \frac{\sum_{k=1}^{K} d\big[(i_k,j_k)\big|(i_{k-1},j_{k-1})\big]}{\sum_{k=1}^{K} d_T\big[(i_k,j_k)\big|(i_{k-1},j_{k-1})\big]}. \tag{11.56}$$

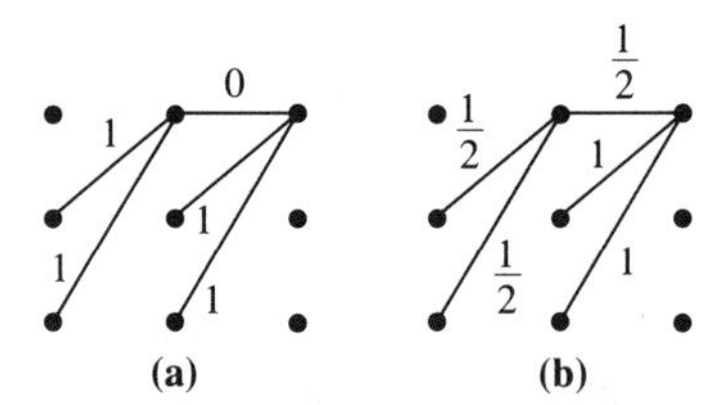

FIGURE 11.13. Adding transition costs to the local constraints of Fig. 11.11. The use of certain general transition costs can sometimes create results that are inappropriate because certain arcs become "cost-free." An example occurs in (a), in which transition cost type (11.52) is applied to the shown local constraints. In (b), "smoothing" is used to prevent this anomaly. Under smoothing, each cost along a local path is replaced by the *average* cost along the local path.

A problem with this normalization is that the normalizing cost (path "length") in the denominator is dependent upon the particular path for transition cost types (11.52) and (11.53). In a DP algorithm, where optimization is done locally, a path-dependent normalization is clearly inappropriate, and we must resort to an arbitrary normalization factor. A natural choice is to simply use I (number of elements in the test string) as a normalization factor (Myers et al., 1980). In this case a DTW algorithm using (11.52) [(11.53)] will be biased toward the use of longer [shorter] paths. In the case of the other two weight types, the normalization is path-independent, since it is easily shown that

$$\sum_{k=1}^{K} d_T\left[(i_k, j_k)\middle|(i_{k-1}, j_{k-1})\right] = \sum_{k=1}^{K} i_k - i_{k-1} = I \tag{11.57}$$

and

$$d_N\left[(i_k, j_k)\middle|(i_{k-1}, j_{k-1})\right] = \left[i_k - i_{k-1}\right] + \left[j_k - j_{k-1}\right] = I + J. \tag{11.58}$$

A detailed study of these various constraint and transition cost strategies has been reported by Myers et al. Their work is based on a 39-word vocabulary (alphabet, digits, and the words "stop," "error," "repeat") with speaker-dependent trials. Local constraints tested were (b) and (e)–(h) in Fig. 11.11. Generally, they find that all constraints perform similarly with respect to accuracy, except (g), which is significantly worse. As expected, restricting path range improves computational efficiency, but at the expense of increased error rate. Interestingly, placing the test sequence on the abscissa improves accuracy, especially if transition cost (11.54) is used. Finally, the DTW algorithms perform best when the test and reference strings are about the same length. The reader is referred to the Myers paper for details.

11.3.3 Typical DTW Algorithm: Memory and Computational Requirements

To summarize the discussion thus far, let us set up a typical DTW algorithm and examine some of the details. The example search grid is illustrated in Fig. 11.14.

Let us first consider the computational requirements of the algorithm. Suppose we choose to work with the Sakoe and Chiba local constraints shown in item (b) of Fig. 11.11. We know from (11.50) and (11.51) that the search region will be restricted in this case to the parallelogram with slopes $\frac{1}{2}$ and 2. Note that we have additionally relaxed the endpoint constraints by a parameter ε. Search over this parallelogram requires that about $IJ/3$ distance measures (Type N costs at the nodes) be computed and that DP equation (11.47) be employed about $IJ/3$ times. This latter figure is often referred to as the "number of DP searches" (Ney, 1984;

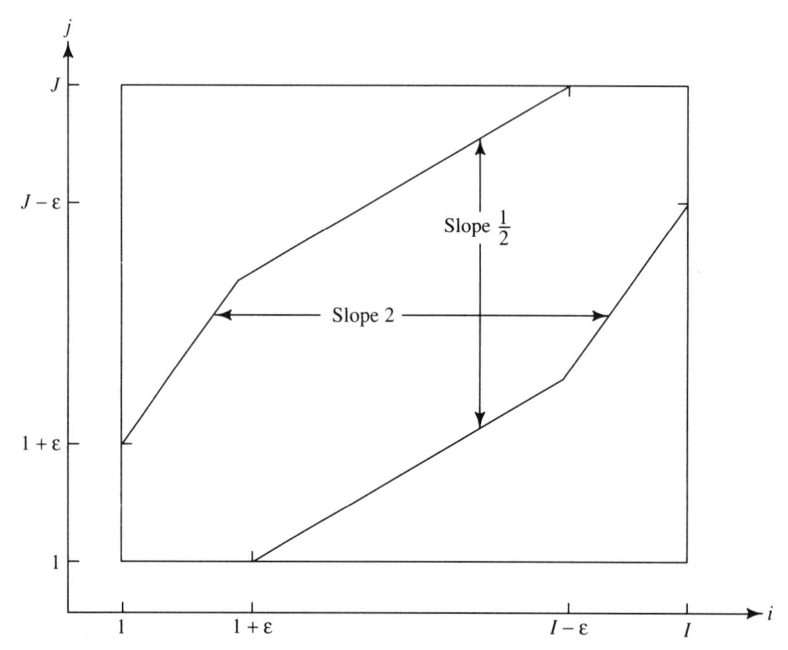

FIGURE 11.14. Illustration of the search space for the example DTW algorithm of Section 11.3.3.

Silverman and Morgan, 1990), and taken together, the number of distance computations and number of DP searches are often used as a measure of computational complexity of a DTW algorithm. We will use these measures in our discussion here and in others to follow, but point out that there are deeper issues involving addressing and computation that should be considered in the serious evaluation of any DTW algorithm [see (Silverman and Morgan, 1990)].

Next, we consider the memory requirements. In light of (11.47), we see that to compute $D_{\min}(i,j)$ for some i, and for any j, it is only necessary that the past quantities $D_{\min}(i-1,j)$ and $D_{\min}(i-2,j)$ be available for all j. Let us therefore employ two $1 \times J$ arrays, say $\boldsymbol{\delta}_1(j)$ and $\boldsymbol{\delta}_2(j)$, $j=1,\ldots,J$, respectively, to record these quantities. The essential memory requirement is therefore modest, totaling only $2J$ locations. Further, note that the computation can be done "in place." We sequentially move through the values of i and, for a given i, compute $D_{\min}(i,j)$ "from top to bottom" — $j=J,J-1,\ldots,2$ (only j's within the parallelogram actually computed). *In-place computation* refers to the fact that $D_{\min}(i,J)$ may replace $\boldsymbol{\delta}_2(J)$, $D_{\min}(i,J-1)$ replace $\boldsymbol{\delta}_2(J-1)$, and so on. Arrays $\boldsymbol{\delta}_2(\cdot)$ and $\boldsymbol{\delta}_1(\cdot)$ are then swapped before proceeding to the next i. If for some reason backtracking to find the optimal path is desired, then additionally, a matrix of size $O(IJ)$ must be allocated to hold the backtracking information.

In Fig. 11.15 we sketch an algorithm for these constraints. Although this example algorithm illustrates the general principles of a DTW solu-

FIGURE 11.15. Typical DTW algorithm for discrete utterance recognition.

Initialization: $D_{\min}(1,j) = d_N(1,j), \quad j = 1, \ldots, 1+\varepsilon$
$D_{\min}(i,1) = d_N(i,1), \quad i = 1, \ldots, 1+\varepsilon$
$\boldsymbol{\delta}_1(j) = D_{\min}(1,j), \quad j = 1, \ldots, J$
$\boldsymbol{\delta}_2(j) = 0, \quad j = 1, \ldots, J$

Recursion: For $i = 2, \ldots, I$
For $j = J, \ldots, 2$
Compute $D_{\min}(i,j)$ using (11.47).
(Note special case for $i = 2$.)
$\boldsymbol{\delta}_2(j) = \boldsymbol{\delta}_1(j)$
$\boldsymbol{\delta}_1(j) = D_{\min}(i,j)$
[Note that $D_{\min}[(i-2,j)]$ is held in $\boldsymbol{\delta}_2(j)$ and
$D_{\min}(i-1,j)]$ is held in $\boldsymbol{\delta}_1(j)$.]
Next j
Next i

Termination: Best path has cost

$$\tilde{D} = \min \begin{cases} D_{\min}(I,j)/I, & j = J-\varepsilon, \ldots, J \\ D_{\min}(i,J)/I, & i = I-\varepsilon, \ldots, I \end{cases}.$$

tion, the details will, of course, change with different constraints on the search. A second example will be considered in Problem 11.4.

11.4 DTW Applied to CSR

11.4.1 Introduction

At the outset of our discussion of DTW, we noted that there is nothing that theoretically precludes the use of the basic DTW approach with any unit of speech declared to be a discrete unit. Practically, however, as template lengths increase, difficulties in both performance and computational expense become significant. In longer utterances, variations in speaking rate, prosodics, and articulation become so vast that local path constraints can no longer be imposed without unacceptable risk of pruning correct paths,[11] often in early stages of the search. The alternative is to greatly relax constraints, but at unacceptable computational cost.

DTW has therefore been applied in a connected-speech paradigm in which individual reference templates representing words are, *in principle*, concatenated to form utterance-length templates that are then compared

[11]This means to terminate an incomplete path due to unacceptable likelihood (see Section 11.4.5).

against the string of feature vectors representing the test utterance. Recall that what distinguishes this strategy as a connected-speech approach is the fact that the reference "model" consists of individually trained pieces. If all concatenations of words are to be tried, then for a V-word vocabulary, with K reference templates per word, and utterances of up to L words in length, $O[(KV)^L]$ reference strings must be matched against each test utterance. Recalling that a search of the DTW grid for a test string of length I and reference string of length J requires $O(\mu IJ)$ distance computations and DP searches (μ is a fraction, typically $\frac{1}{3}$, to account for global path constraints), we see that an exhaustive search over the vocabulary space requires $O[\mu(KV)^L I\bar{J}]$ distance computations, where $\bar{J}$ denotes the average single-word reference string length (in number of frames). This is clearly prohibitive for all but the smallest V's and L's. For example, in a phone-dialing-by-speech problem, Ney (1984) cites the following typical numbers: $V = 10$ (number of digits), $K = 10$, $L = 12$ (maximum number of words in a phone number), $I = 360$ (typical test string length), $\bar{J} = 35$. In this case, we find that exhaustive search requires $O(10^{27})$ distance computations to recognize a phone number. We shall also see that, for at least one method of organizing the computations of an exhaustive search, the number of "global" memory locations required (in addition to the "local" memory required for the basic DP procedure) is $O(3KVLI)$. For our example, this amounts to $O(10^6)$, which, in the early days of DTW (1970s) was in itself prohibitive of exhaustive search. Therefore, much effort has been devoted to finding efficient algorithms requiring less storage for this DTW search. We will generally describe some important results of these efforts in the following sections.

We will discuss three specific algorithms that have been widely used in connected-speech recognition by DTW. These are the "level-building" algorithm (Myers and Rabiner, 1981a, 1981b), the "one-stage" or "Bridle" algorithm (Bridle and Brown, 1979; Bridle et al., 1982), and the "grammar-driven" algorithm (Pawate et al., 1987). The first two of these fall into two general classes of algorithms, which have been called *fixed-memory* and *vocabulary-dependent memory* algorithms, respectively, for reasons that will become apparent (Silverman and Morgan, 1990).

11.4.2 Level Building

Reading Note: This algorithm is of great historical interest, but is no longer widely used in the DTW paradigm. The basic principles will be useful when we study a similar hidden Markov model-based approach in Chapter 12, but the reader may wish to read this section for general information now, and return to the details when needed later.

The Algorithm

Level building (LB) represents one attempt to circumvent the overwhelming number of computations necessary to implement an exhaustive search over all possible reference-word concatenations. This technique was first published by Myers and Rabiner (1981a) and, to some extent, it can be viewed as an attempt to streamline an earlier method reported by Sakoe (1979) called the "two-level" method. Sakoe's method compares reference strings in two passes, one for the individual words, the second for the complete utterance. Level building combines these two tasks and is significantly more efficient.

To understand the operation of LB, first let us picture a large search grid with the test utterance laid out along the i axis, as in Fig. 11.16. The Itakura parallelogram is usually imposed as a global path constraint, and we have shown this region on the grid. Assuming that we know approximately how many, say L, words are present in the test utterance, let us partition the reference axis into L "levels" into which we will "plug" var-

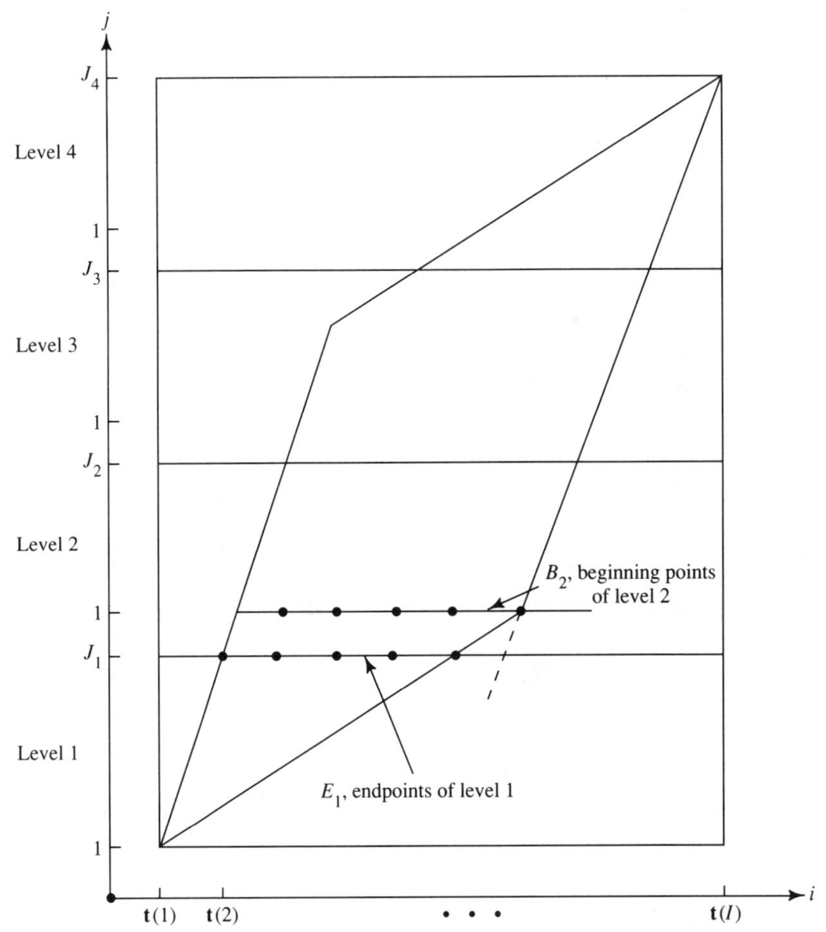

FIGURE 11.16. Large search grid used to introduce the LB algorithm. Shown: four levels. In the text, it is assumed that $J_1 = J_2 = J_3 = J_4 = \bar{J}$.

ious isolated word reference strings.[12] We also make the assumption that all reference templates are approximately the same length, $\bar{J}$, so that each level represents $\bar{J}$ frames along the j axis. This assumption is, of course, not always practical, but remember that the reference strings are preprogrammed into the device and the designer has good control over their structure. Level building becomes difficult to visualize and describe if this assumption is not made. The issue of unequal-length reference templates is briefly discussed in Silverman and Morgan (1990). The inability to conveniently handle reference templates of unequal length is one of the major weaknesses of the LB algorithm.

Rather than introduce a complete utterance along the reference axis, let us simply begin at level 1 and introduce a candidate reference string at that level. For convenience, let us denote the reference string corresponding to word v by $\mathbf{r}^v$,

$$\mathbf{r}^v,\ \text{reference string for word } v = \{\mathbf{r}^v(1), \mathbf{r}^v(2), \ldots, \mathbf{r}^v(\bar{J})\}. \tag{11.59}$$

For $\mathbf{r}^1$, the algorithm proceeds to find the best path to each of the possible ending points of level 1 utilizing conventional DP techniques while adhering to any local path constraints that may apply. Backtracking information is recorded so that paths may be retraced to their origins.[13] Let us assume for simplicity that the initial endpoints of the reference and test strings are tied [(1, 1) must occur on every path], which, recall, means that the formal starting node (0, 0) may only make a transition to (1, 1). This constraint is easily relaxed. It is worth recalling that grid points (in this case within a level) are usually, for a fixed i, evaluated sequentially along the j dimension until the topmost allowable point in the search is reached. Then the search proceeds to the next set of vertical grid points (i is incremented). In this case, this procedure creates vertical "stripes" of evaluated nodes that terminate at the upper boundaries of the levels.

The top of the first level corresponds to $j = J_1$ and a certain range along the i axis, say E_1, to connote "endpoints of level 1." When the top boundary of the first level is reached using $\mathbf{r}^1$, three pieces of information are stored in a three-level array for (i, J_1), for each $i \in E_1$:

1. An identifier of the word associated with $\mathbf{r}^1$ (we assume that this is simply the integer superscript 1);
2. $D^1_{\min}(i, J_1)/i$, where $D_{\min}(\cdot\,,\cdot)$ represents the usual minimum distance computation as in (11.47) with the superscript used to associate this quantity with $\mathbf{r}^1$, and where the division by i represents normalization to path length;
3. The starting point of the path leading to (i, J_1), in this case, (0, 0).

The purposes of these data will become clear as we proceed.

[12]If the test utterance has fewer than L words, we will still be able to find the correct solution. This will be discussed below.

[13]This information is obviously superfluous at level 1 if (0, 0) is the only allowable origin, but it will be essential at higher levels.

Rather than move to the second level, we now insert a second candidate reference string, $\mathbf{r}^2$, into level 1. The path searches to the top boundary are now repeated. If for some $i \in E_1$, say $i', \mathbf{r}^2$ produces a lower-cost path to (i', J_1) than did $\mathbf{r}^1$, that is,

$$D^2_{\min}(i', J_1) < D^1_{\min}(i', J_1), \tag{11.60}$$

then item 1 in the storage array for (i', J_1) is replaced by the integer 2, which indicates that $\mathbf{r}^2$ produced the minimum distance to (i', J_1), and $D^2_{\min}[(i', J_1)]/i'$ replaces the number in item 2. Storage item 3 will not change since the starting point for all paths in level 1 is $(0, 0)$.

This process is repeated for each reference string that can possibly be the first word in the string, $\mathbf{r}^v$, $v = 1, 2, \ldots, V$. At the end of the processing of all reference strings through level 1, the three-part storage array for each node $(i, J_1), i \in E_1$, will contain the following information:

1. $v^*(i, J_1)$, index of the word associated with the best path to (i, J_1);
2. $D^{v^*}_{\min}(i, J_1)$, cost of the best path to (i, J_1);
3. $(0, 0)$, starting node of the best path to (i, J_1).

Moving to level 2 is straightforward. For convenience, let us index the points along the ordinate of level 2 by $j = 1, 2, \ldots, J_2$, rather than $j = J_1 + 1, J_1 + 2, \ldots, J_1 + J_2$, which might seem more consistent with the physical layout of the problem. For formal reasons, we also reindex the important grid points [those formally designated $(i, J_1), i \in E_1$] as $(i, 0)$, $i \in E_1$. Now there are certain grid points in level 2 that represent possible continuations of paths arising out of level 1. These correspond to $j = 1$ and a certain range of i, say B_2, to connote "beginning of level 2." Formally, we can consider the set of points $(i, 0), i \in E_1$, as a set of origins for paths in level 2. Also, for formal record-keeping purposes, we assign the cost in the second level of the storage array at each $(i, 0)$ to that node in the form of a Type N cost. We must remember to append this cost to any path originating from one of these nodes.[14]

At the top boundary of level 2, a three-part storage array is set up to hold the three pieces of information discussed above. This information must be recorded for each node (i, J_2), $i \in E_2$.

For each $\mathbf{r}^v$, a DP search is carried out from each $(i, 0)$, $i \in E_1$, to each (i, J_2), $i \in E_2$. Whenever a lower cost path is found to, say, (i', J_2), the information in its storage array is replaced by the new word index, new cost, and originating point of the superior path.[15]

The procedure described for level 2 is then repeated for successive levels until the final level, say L, is reached. If there is a single tied endpoint, say (I, J_L), then the three-level array will contain the best word in level L, the cost of the global path, and the origin of the best path in

[14]Note that for the first time in our discussions, we have an effective Type N cost assignment to original nodes.

[15]We now see that it is really only necessary to record the appropriate i ($i \in E_1$) value of the originating node, since the j value is known.

level L. The array at this original node, in turn, will indicate the word on the segment of the optimal global path passing through level $L-1$, a cost (not useful), and a pointer to *its* origin. By backtracking in this manner, the word sequence associated with the optimal global path is recovered. It should be apparent that the procedure is essentially the same if the endpoints are relaxed at the top end of level L.[16]

Complexity and Memory Requirements

Dynamic time warping algorithms for connected speech are generally compared with respect to their computational requirements and the amount of necessary storage. With advances in memory technologies, this latter issue is not nearly as critical as it was in the earlier days of the application of DTW (mid-1970s to mid-1980s). To a lesser extent, VLSI fabrication of special-purpose signal processing chips (1980s) has also alleviated the pressures for computationally less-expensive algorithms (more on this below).

As we did with DTW algorithms for discrete utterances, we will use the number of local distance computations and the number of DP searches as important relative measures of the computational complexity of the algorithms. The computational complexity of the LB approach is significantly less than with exhaustive search, and although exhaustive search is not a serious contender for the algorithm of choice, let us recall its computational requirements to get insight into the improvement due to LB. We noted above that exhaustive search requires $O[\mu(KV)^L I\bar{J}]$ distance computations per observation string to be recognized, which in the phone-dialing example tallied to $O(10^{27})$. The vast majority of this load is due to the factor $(KV)^L$, which might be lessened by syntactic constraints. In LB, it is easy to see that the number of distances computed and DP searches required is $O(\mu KVIL\bar{J})$, which, with our typical numbers above is $O(10^7)$. The savings is due to the fact that decisions are effectively made on a level-by-level basis, avoiding the exponential explosion of trials that occurs when decisions are made on concatenated reference templates.

It should be noted that LB has an inherent inefficiency in that many distance computations are repeated. In Fig. 11.17 we see the case in which the same reference template, when used in different levels, requires precisely the same computations to be made. In some cases this redundant computation can be avoided by computing all distances before beginning the LB process. This requires the *a priori* computation and storage of $O(\mu KVI\bar{J})$ distances.

[16]If the endpoints are relaxed at the beginning of level 1, the third level of the storage arrays at the grid points $(i, J_1), i \in E_1$, will still be $(0,0)$ according to our formalism. In any case, the original point of the level 1 path is not useful information in any of the considerations above.

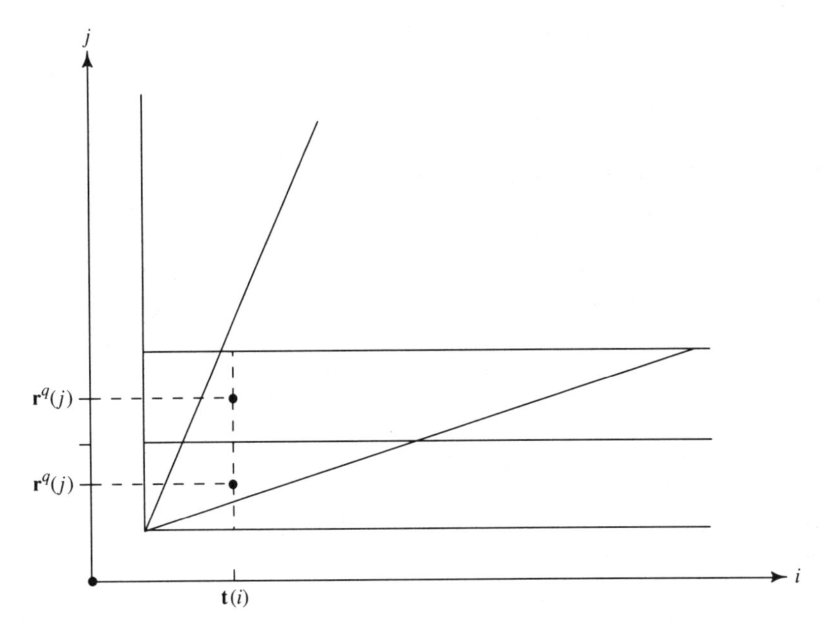

FIGURE 11.17. In the LB algorithm, many distance computations are potentially repeated. Here we see the case in which the same reference template, $\mathbf{r}^q$, when used in different levels, requires many of the same computations to be made.

Memory requirements are a second measure with which connected-speech DTW algorithms are compared. We have discovered that LB requires a three-level storage array associated with the top boundary of each level. Therefore, a $3 \times I$ rectangular array is required for each of L levels,[17] for a total of $O(3LI)$ locations. For our typical numbers above, this amounts to $O(10^3)$ locations, which is relatively insignificant. In addition to this "global" memory, "local" memory is required to run the basic DP procedure. It should be clear that the basic DP search can be carried out by repeatedly using the method described for discrete utterance DTW (see, e.g., Section 11.3.3). The memory required for the memorization of distances is $O(\nu\bar{J})$, where ν is a small integer. The backtracking situation is a bit different here, since the origin of a path through a level is desired, rather than an account of the path itself. It is not difficult to deduce that $O(\nu\bar{J})$ memory locations are also needed for this backtracking task.

Note that once the maximum number of levels and maximum test utterance length is fixed, the total memory requirement of the LB algorithm is fixed. In particular, the amount of memory does not depend on the size of the vocabulary, V. For this reason, Silverman and Morgan (1990) have referred to LB as a *fixed memory* technique.

[17]This is a conservative estimate since storage is required only for endpoints that are admissible according to path constraints.

Enhancements and Further Details

Researchers at AT&T have introduced a number of independently controllable variables that can be used to improve the performance, and in some cases, efficiency, of the LB algorithm (Myers and Rabiner, 1981b; Rabiner and Levinson, 1981; Rabiner et al., 1984). Most of these parameters are illustrated in Fig. 11.18. Parameters like the illustrated δ' and δ'' characterize regions of uncertainty at the beginnings and ends of reference strings in the levels, while δ''' accounts for uncertainty in the end-time of the test string. M_i is a multiplier used to determine the sizes of the beginning-point sets (called B_l for the lth level in the discussion above) at each level. Here ε is a sort of window-width parameter that dictates the amount of permissible local warp along the reference string. This distance is taken around the path that is currently least costly, and accordingly the method is called a *local minimum* search. Such a procedure significantly reduces the allowable search space, thereby decreasing the computational load of the search. Not illustrated are the parameters, say $\tau_{\min}$ and $\tau_{\max}$, that are cost bounds inside of which any accumulated cost must be at a level boundary for the continuation of the associated path. The upper bound comprises a pruning mechanism for ridding the search of unlikely paths, while the lower bound prevents unfair competition by paths that have apparently not been subjected to proper comparisons. The reader is referred to the original papers for details on the use of these parameters.

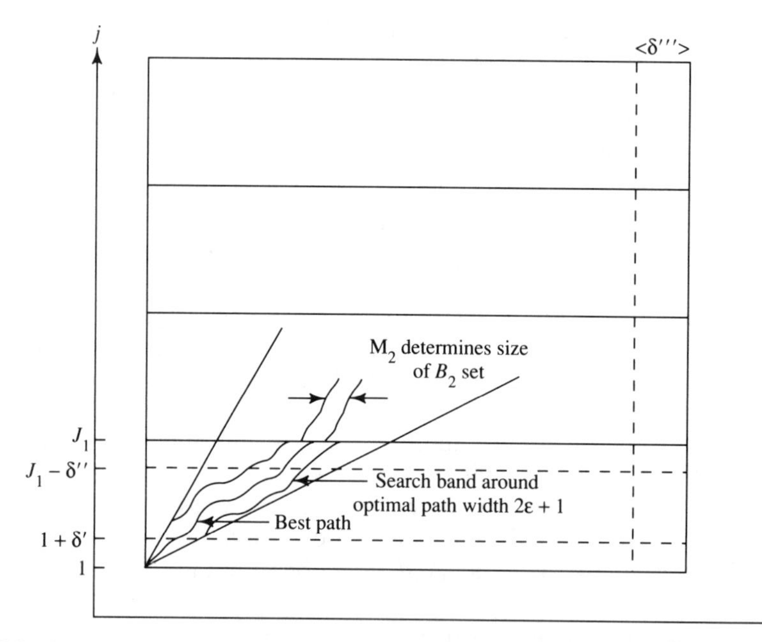

FIGURE 11.18. Illustration of the parameters associated with the "enhanced" LB algorithm.

A second enhancement is concerned with the fact that LB, which inherently segments the test string into component word strings, sometimes inserts incorrect words into the final decoding. Efforts to presegment the test string could potentially alleviate this problem, but at the expense of other errors caused by erroneous boundary decisions. A hybrid approach has been proposed in a paper by Brassard (1985) in which the test string is first segmented into regions approximating syllables; then a local-minimum-search-type DTW algorithm is used to adjust these hypothesized boundaries. Spurious insertions of short words are discouraged by high penalties for path trajectories representing large compression of the test string. Details are found in Brassard's paper.

We noted early in the discussion that an L-level procedure could be used to find the best match to reference strings of length $L-1, L-2, \ldots$. This is accomplished by simply seeking paths that end at grid points (I, J_{L-1}), $(I, J_{L-2}), \ldots$. This idea is illustrated in Fig. 11.19, where the global path constraints have been modified so that the terminal end of the parallelogram is "opened up." Generally, the local and global path constraints can be modified on an *ad hoc* basis. Rabiner and Levinson (1981) point out that such a multiple-path-length approach is particularly useful when several possible path lengths are known *a priori*. This would be the case, for example, in recognizing phone numbers.

Finally, we note that by doubling the storage at the end of each level, it is possible to record a second best path to each endpoint, thus offering the possibility of alternative paths to the "best" path deduced using only one storage array. Notice that by taking combinations of first- and second-best path segments among the levels, 2^L global paths are possible candidates. This procedure can be extended to more than two storage arrays—the limiting case being that for which all candidate reference templates have a ranking at the endpoints of the levels. In this case, LB

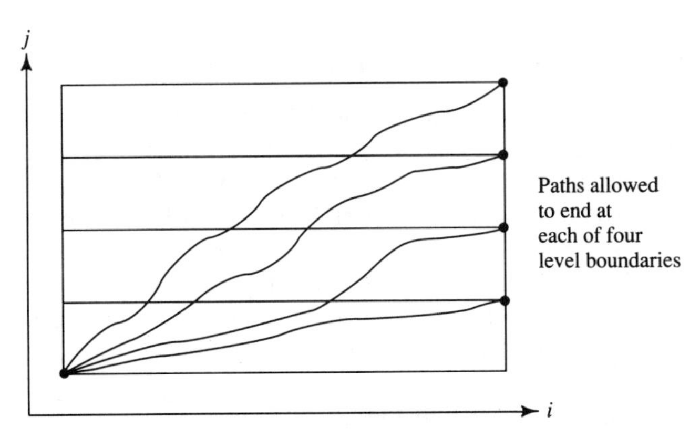

FIGURE 11.19. Level building can be used to recognize a string of words for which the number of words is unknown. This is accomplished by "opening up" the end of the search parallelogram so that the search may end at various level boundaries.

becomes equivalent to an exhaustive search of the grid, $O(3LVI)$ storage locations are required, and the amount of computation is also increased to $O(V^L)$ (cf. the discussion in Section 11.4.1). In this light, we can interpret the LB algorithm (with one level as described above) as an attempt to optimize over the reference string $\mathbf{r}^v$ at each level. The one-stage method to be discussed next will be interpreted as an attempt to optimize over the starting point in the level.

11.4.3 The One-Stage Algorithm

Introduction

The *one-stage* (OS) approach is so-named because the recognition is accomplished by finding an optimal path through a DP grid in "one stage" of computation, rather than by building a series of path "levels" as in the LB technique. The name also contrasts the technique with the earlier "two-level" method of Sakoe (1979), which was briefly described above. The OS method was first described by Vintsyuk (1971), but was not well known until a similar algorithm was reported by Bridle and Brown (1979). A tutorial on the OS method is found in (Ney, 1984). More recently, an enhanced version of the Bridle algorithm has been described by Miller et al. (1987). Ney cites several papers describing systems based on the OS approach, while another is described briefly in (Silverman and Morgan, 1990), and more fully in the paper by Miller et al.

In many ways, the OS approach is much simpler than the LB algorithm and its predecessors. It is also more efficient computationally, and often requires less memory. Because, as we shall see, the memory requirement is proportional to the size of the vocabulary, V, the method has been classified as a *vocabulary-dependent memory technique* (Silverman and Morgan, 1990).

The Algorithm and Its Complexity and Memory Requirements

The simplest way to visualize the OS approach is as a DP problem over a three-dimensional grid as depicted in Fig. 11.20. The test utterance is laid out along the i dimension, as usual. Along the j dimension are the individual word reference strings, $\mathbf{r}^v$, with a different reference word corresponding to each increment along the third axis, which we shall label v. For all legal paths arriving at grid point (i, j, v),[18] we denote the cost of the minimum-cost path by $D_{\min}(i, j, v)$.[19] Of course, the BOP applies equally well to searches over three-dimensional grids, and we can

[18]We specify which paths are legal below.

[19]This quantity will, in many ways, turn out to be similar to what we labeled $D^v_{\min}(i, j)$ in describing the LB algorithm. However, here we specifically use the triplet (i, j, v) to reinforce the notion of a 3-D grid search, whereas the former notation was used to prevent thinking in this way.

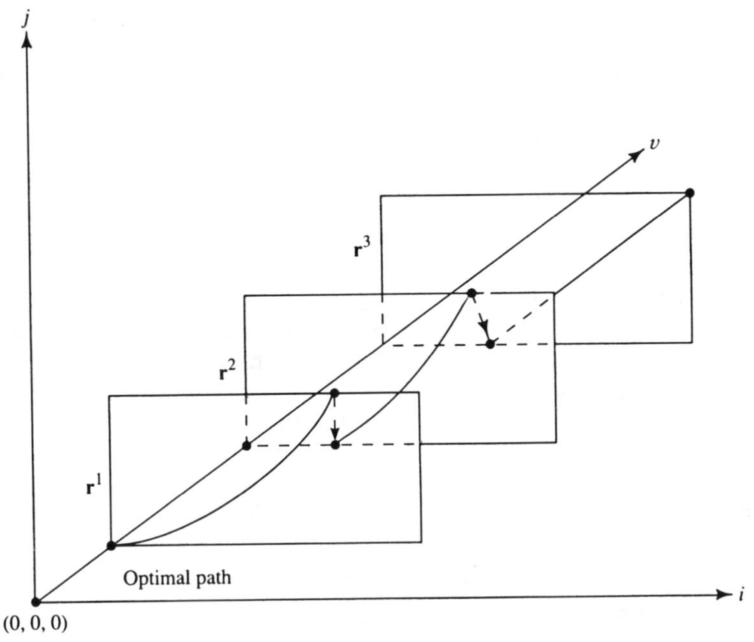

FIGURE 11.20. The OS algorithm can be conceptualized as a search over a three-dimensional grid. Shown is the case of three reference strings. The path indicated in this figure happens to sequentially move through $\mathbf{r}^1, \mathbf{r}^2$, and then $\mathbf{r}^3$. In fact, the optimal path may start at the "southwest" corner of *any* grid and end at the "northeast" corner of any grid, sequentially traversing the grids in any order, including repeated traversals.

write an "upgraded" version of (11.47) [the sentence under (11.47) explains the meaning of the quotes around the minimization argument]:

$$\begin{aligned} D_{\min}(i_k, j_k, v_k) &= \min_{\text{“}(i_{k-p}, j_{k-p}, v_{k-p})\text{”}} \left\{ D_{\min}\left[(i_k, j_k, v_k) \middle| (i_{k-p}, j_{k-p}, v_{k-p})\right]\right\} \\ &= \min_{\text{“}(i_{k-p}, j_{k-p}, v_{k-p})\text{”}} \left\{ D_{\min}(i_{k-p}, j_{k-p}, v_{k-p}) \right. \\ &\qquad \left. + \hat{d}\left[(i_k, j_k, v_k) \middle| (i_{k-p}, j_{k-p}, v_{k-p})\right]\right\}, \end{aligned} \tag{11.61}$$

where

$$\hat{d}\left[(i_k, j_k, v_k) \middle| (i_{k-p}, j_{k-p}, v_{k-p})\right] \stackrel{\text{def}}{=} \sum_{m=0}^{p-1} d\left[(i_{k-m}, j_{k-m}, v_{k-m}) \middle| (i_{k-m-1}, j_{k-m-1}, v_{k-m-1})\right] \tag{11.62}$$

and the paths of the form $(i_{k-p}, j_{k-p}, v_{k-p}), \ldots, (i_k, j_k, v_k)$ represent the allowable local paths to (i_k, j_k, v_k).

A moment's thought will indicate that there should be very strict constraints on the transitions that may be made with respect to the v dimension. Only when a path reaches the upper boundary of one of the grids associated with a particular reference string (particular v) should it be possible to exit that grid. In that case, it should be required that the path continue at the bottom of another grid (possibly the same one to allow the same word to occur consecutively in an utterance). We have, therefore, a set of *within-template* transition rules that govern the path search while the path evaluation is internal to one of the word grids, and a set of *between-template* transition rules that are operative at the top boundaries.

The within-template rules correspond to the usual sorts of local and global constraints to which we have become accustomed in our previous work. For example, the simple Sakoe and Chiba local constraints shown in Fig. 11.11(a), and redrawn here in Fig. 11.21(a), are frequently used. We will henceforth assume these local path constraints in our discussion of the OS algorithm. The results presented here are readily generalized for other path constraints. In the adopted case, if the dimension of the grid corresponding to word v is $I \times J_v$, then all points of the form (i, j, v), $j = 2, 3, \ldots, J_v$, may be preceded by $(i-1, j, v), (i-1, j-1, v)$, and $(i, j-1, v)$ [except, of course, along the left boundary ($i = 1$), where the former two predecessors do not exist]. In this case (11.61) reduces to simply

$$
\begin{aligned}
D_{\min}(i_k, j_k, v) &= \min_{(i_{k-1}, j_{k-1}, v)} \left\{ D_{\min}\left[(i_k, j_k, v) \middle| (i_{k-1}, j_{k-1}, v)\right]\right\} \\
&= \min_{(i_{k-1}, j_{k-1}, v)} \left\{ D_{\min}\left[(i_{k-1}, j_{k-1}, v)\right] + d\left[(i_k, j_k, v) \middle| (i_{k-1}, j_{k-1}, v)\right]\right\},
\end{aligned}
\tag{11.63}
$$

where the $k-1$st node is $(i_k - 1, j_k - 1, v)$, $(i_k - 1, j_k - 1, v)$, or $(i_k, j_k - 1, v)$. Note that the argument v does not change throughout these expressions because the path remains "within a word."

Except at the initialization of path search, which is discussed below, when considering points of the form $(i, 1, v)$ (entry level into grid v), the predecessor points are dictated by the between-template transition rules. A simple set of boundary constraints commonly used is shown in Fig. 11.21(b). We will adopt these constraints for the remaining discussion of the OS algorithm. This formulation is readily generalized. In this case

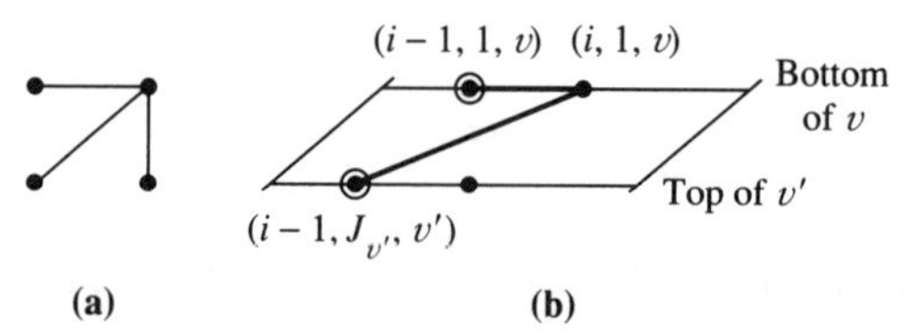

FIGURE 11.21. (a) Sakoe and Chiba local constraints redrawn here for convenient reference. (b) Cross-boundary constraints for the OS algorithm.

$(i, 1, v)$ may be preceded by $(i-1, 1, v)$ and $(i-1, J_{v'}, v')$ for any v', including $v' = v$. In this case, we can write (11.61) as

$$\begin{aligned} D_{\min}(i, 1, v) &= \min\Big\{D_{\min}[(i, 1, v) | (i-1, i, v)], \\ &\quad \min_{v'}\{D_{\min}[(i, 1, v) | (i-1, J_{v'}, v')]\}\Big\} \\ &= \min\Big\{D_{\min}[(i-1, 1, v)] + d[(i, 1, v) | (i-1, 1, v)], \\ &\quad \min_{v'}\{D_{\min}(i-1, J_{v'}, v') + d[(i, 1, v) | (i-1, J_{v'}, v')]\}\Big\}. \end{aligned} \tag{11.64}$$

Associated with each of the transitions in the between-template rules, as well as the within-template rules, is a Type N cost upon arrival at the target node. In general, Type T costs will be incurred on the transitions themselves. All of this information is embodied in (11.63) and (11.64), and in the algorithm to be presented below.

For any v, the node $(1, 1, v)$ may be entered only at the initialization of search. As is our convention, we imagine that all paths originate at a node $(0, 0, 0)$. For all words v that may initiate an utterance, an initial transition is possible from $(0, 0, 0)$ to $(1, 1, v)$ with the only cost being the Type N cost incurred upon arrival at $(1, 1, v)$. Of course, it is possible to loosen the restrictions on the endpoints, in which case further initial transitions will be allowable.

The amount of "local memory" actually necessary to carry out the DP searches is quite small relative to the number of grid points, $O(I\bar{J}V)$, where $\bar{J}$ indicates the average length of the reference strings. Suppose that we focus our attention on the parallel grid corresponding to word v—the search through the grid points (i, j, v) with v fixed. The problem now reverts back to the same search problem encountered above in both IWR and in the LB algorithm. In order to determine $D_{\min}(i, j, v)$ for any i and every $j > 1$, it is only necessary to know the values $D_{\min}(i-1, j, v)$ for $j = 1, 2, \ldots, J_v$. These can be stored in a column vector, say $\boldsymbol{\delta}(j, v)$ (remember that v is fixed at present), of size J_v. We can simply move sequentially through the i values, updating this column vector for each i. The computation can be done in place by moving through the vector from "top to bottom" $(j = J_v, J_v - 1, \ldots, 1)$ and replacing the term $D_{\min}(i-1, j, v)$ [held in $\boldsymbol{\delta}(j, v)$] by $D_{\min}(i, j, v)$ as soon as it is computed. Once again, the computations are performed in "vertical stripes" as we have frequently found to be the case. Since one such column vector is needed for each v, the total amount of local storage necessary is $O(V\bar{J})$. The vth column of the matrix $\boldsymbol{\delta}$ holds the information for word v. Formally and practically, the dimension of $\boldsymbol{\delta}$ is $J_{\max} \times V$, where

$$J_{\max} \stackrel{\text{def}}{=} \max_{v} J_v. \tag{11.65}$$

In general, two or more columns might be necessary to hold the relevant distances for a particular v (see the example in Section 11.3.3). In this case the number of memory locations necessary for the $\boldsymbol{\delta}$ matrix is $O(\nu J_{\max} V)$, where ν is a small integer.

In an effort to discover further memory needs, let us next consider "global memory" required to store the information for recovering the final utterance. As with the LB algorithm, we are not usually interested in recovering the entire node sequence of the optimal path through the 3-D grid. If we were, we could keep a backtracking record at each node in the 3-D grid, and upon reaching the end of the optimal path simply backtrack to discover the sequence of templates used. In this case an $O(I \times \bar{J} \times V)$ matrix of backtracking information must be stored. This is unnecessarily expensive, however, since the boundary-crossing information would suffice to recover the word sequence. What we ultimately need to know is simply this: For any path reaching the top boundary of v [i.e., some grid point, say (i, J_v, v)], what was the last top boundary grid point, say $(i', J_{v'}, v')$, along that path? Knowing these top boundary grid points allows reconstruction of the word sequence. Let us explore how these pieces of information can be stored in a very efficient manner.

Note that if the "preceding" boundary point is $(i', J_{v'}, v')$, then, according to (11.64), the "entry point" to grid v is $(i'+1, 1, v)$. However, also according to (11.64) *any* path with a node of the form $(i'+1, 1, w)$ as an entry point to a word w will also have as its preceding point the node $(i', J_{v'}, v')$. This says that there is only one possible exit point at frame i', the one from word v'. Let us, therefore, create an $I \times 1$ row vector, say $\mathbf{w}(\cdot)$ (to connote best *w*ord), and hold in location i the word with the best exiting path at that frame. When all updating is complete for a particular i, the matrix $\boldsymbol{\delta}(j, v)$ will hold all distances to the top boundaries of words in its locations $\boldsymbol{\delta}(J_v, v)$, $v = 1, 2, \ldots, V$. It should be clear from our discussion above that the information stored in $\mathbf{w}(i)$ is

$$\mathbf{w}(i) = v^*(i) = \underset{v}{\operatorname{argmin}}\ D_{\min}(i, J_v, v) = \underset{v}{\operatorname{argmin}}\ \boldsymbol{\delta}(J_v, v). \qquad (11.66)$$

In order to piece together the appropriate best words, we complement the local memory distance matrix, $\boldsymbol{\delta}(j, v)$, with a $J_{\max} \times V$ matrix, say $\boldsymbol{\beta}(j, v)$, to hold backtracking information. Once we fix v, we then focus on a particular column of this matrix that is associated with word v. Here $\boldsymbol{\beta}$ is used for a very simple purpose. As i is incremented (paths updated in vertical stripes as discussed above), whenever the optimal path to node (i, j, v) is ascertained, the frame, say i', associated with the preceding exit point on that path is recorded in location (j, v) of $\boldsymbol{\beta}$. (Equivalently, frame $i'+1$, associated with the entry point to word v on that path, can be recorded, but we will assume the former.) When all updating is complete at frame i, the best "exiting" path will be at node $[i, J_{v^*(i)}, v^*(i)]$ according to (11.66). We can simply examine location $\boldsymbol{\beta}[J_{v^*(i)}, v^*(i)]$ to discover the frame of the preceding exit node on that path, and hold onto this information in a second "external" array complementing $\mathbf{w}(\cdot)$, say $\mathbf{e}(\cdot)$ (to connote *e*xit points), at location i. That is,

$$\mathbf{e}(i) = \boldsymbol{\beta}[J_{v^*(i)}, v^*(i)], \qquad (11.67)$$

where $v^*(i)$ is given in (11.66). The use of the two external arrays $\mathbf{w}(\cdot)$ and $\mathbf{e}(\cdot)$ is illustrated in Fig. 11.22. Note that, whereas the global memory arrangement is simpler here than in the LB case, the local memory requirement is more complex and the quantity of local memory depends on the vocabulary size, V.

The essential details of the OS method have now been completely discussed, and we are prepared to state the algorithm in a general form. In spite of the fact that the developments above required a bit of thought, the algorithm, which is shown in Fig. 11.23, is a particularly simple one.

Finally, let us note the computational complexity of the OS algorithm. If we employ the Sakoe local constraint assumed throughout our discussion above, there is effectively no global constraint on the search space

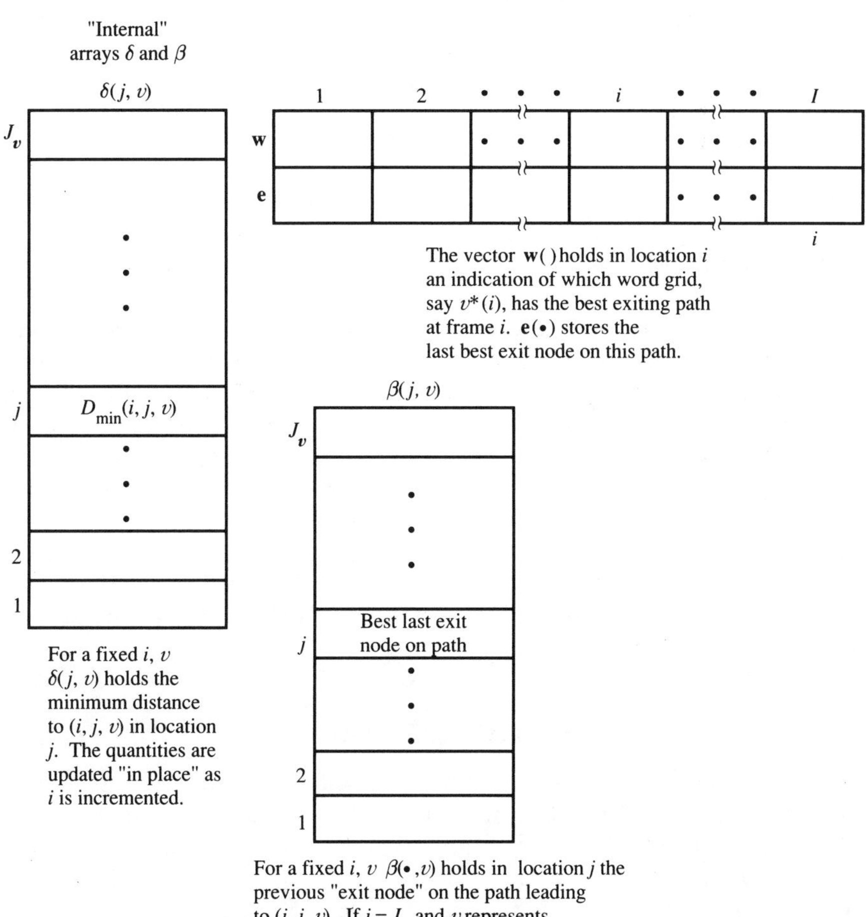

FIGURE 11.22. Illustration of the two external arrays in the OS algorithm.

FIGURE 11.23. The OS algorithm for connected-word recognition.

Initialization: For all v initialize $D_{\min}(1, 1, v) = d[(1, 1, v)|(0, 0, 0)]$.

Recursion:
For $i = 2, \ldots, I$
  For $v = 1, \ldots, V$
    For $j = 2, \ldots, J_v$
    Update $\boldsymbol{\delta}(j, v)$.
    Update $\boldsymbol{\beta}(j, v)$.
    Next j
  Next v
  Set $\mathbf{w}(i)$ according to (11.66).
  Set $\mathbf{e}(i)$ according to (11.67).
Next i

Backtracking: The optimal sequence of words in reverse order is $\mathbf{w}(I), \mathbf{w}(\mathbf{e}(I))$, $\mathbf{w}(\mathbf{e}(\mathbf{e}(I))), \ldots$.

and the number of distances and DP searches required is $VI\bar{J}$. The computational load is a small factor larger in this case than with LB, but the algorithm is simpler and the amount of memory required is significantly less.

Before leaving the OS algorithm, let us seek to understand from whence its main benefit derives with respect to exhaustive search. In the LB algorithm, we found that the reduction in computational effort arose from the willingness to optimize over reference strings at each level of the search. Here it is clear that no such attempt is made, since any reference string may begin a path at any level. However, the benefit derives in this case by optimizing over origins at each boundary so that only one path survives the transition across a boundary at any given frame i. In this sense, the OS algorithm is much more similar to the standard DP search through a grid, and, indeed, the application of the DP method is more straightforward.

Syntactic Information

Although the OS algorithm provides a highly structured way to implement connected-word recognition, it is frequently the case that the computational demands of the algorithm preclude real-time implementation. Consider the phone-dialing task introduced in Section 11.4.1, for example. The vocabulary size is $V = 10$, the number of reference templates per word is $K = 10$, and the typical reference template length is $\bar{J} = 35$. This means that about $KV\bar{J} = 3500$ grid points must be analyzed at each frame of the test string. If the frames are analyzed at 10-msec intervals, this means that 3.5×10^5 grid points must be searched per second. Even for some of the simpler distance measures that might be employed, real-time computation of this magnitude exceeds the capabilities of some of the fastest general-purpose processors available. Memory requirements, although not as restrictive as in the early days of DTW research, can also become problematic in larger tasks. Therefore, recognition problems representing all but the simplest tasks cannot ordinarily be implemented in real-time without further considerations.

Syntactic knowledge is employed in papers by Ney, Mergel, et al. (1987, 1992) to assist in solving the the real-time search problem. Syntactic information frequently takes the form of a graph indicating permissible word sequencing. A transition in the graph represents the production of one word in a sentence, while a complete path through the graph generates an allowable sentence in the language. As expected, syntax ordinarily reduces the computational effort necessary for the search, but one subtle complication must be explained. For various reasons (including the fact that the same word may be pronounced differently in different contexts), if the same word appears in different transitions of the syntax graph, it may have a different set of reference templates for each transition. This fact tends to effectively increase the size of the vocabulary, since each occurrence is effectively treated as a separate word. Whereas the computational load of the unrestricted OS algorithm is $O(VI\bar{J})$, without further consideration, the OS algorithm supplemented by syntax would be $O(V'I\bar{J})$, where V' is the size of the enlarged vocabulary. Of course, the syntax graph has another effect that more than compensates for the small increase in the effective vocabulary: The number of paths, or possible word sequences, is drastically reduced with respect to an unrestricted vocabulary. Let us examine these points more closely in the context of the OS algorithm—in particular using the work of Ney et al. to illustrate these ideas.

In the study by Ney et al., three German speech databases are employed. These databases and the surrounding language (for our purposes, *syntax*) models are described in the paper by Mergel and Paeseler (1987). We will focus on just a few salient details here. In the syntax graph representing the simplest database, for example, there are 3481 allowable transitions (words), each of which is represented by its own set of reference templates. (For our purposes it is sufficient to assume that there is one reference template per transition.) Since there are approximately 900 words in the vocabulary, the syntax graph effectively increases the vocabulary by a factor of four. However, in an unrestricted situation, every word may be followed by 900 other words, meaning that (recall Fig. 11.20) 900 other search grids may be entered upon exiting from each of the word grids. Said another way, paths will traverse almost every point in each of the 900 grids, requiring an evaluation of each of the $900\bar{J}$ points in the "search space." On the other hand, with the benefit of the syntax graph, exiting points from grids will enter only a "few" of the successor grids—in the Ney experiments, frequently 10 or fewer (Mergel and

Paeseler, 1987). As a consequence, although 3481 grids may now be imagined to be stacked as in Fig. 11.20, a vast majority of the search space remains "inactive" and points need not be evaluated in those regions. In addition, a "beam search" is implemented (see below), which further reduces the number of paths. The algorithmic methods for processing only "active" regions of the search space involves a series of list-processing operations. The interested reader is referred to the original papers for details.

Whereas unrestricted word sequences would have resulted in the need to evaluate $3481\bar{J}$ grid points for each test frame, the syntax-directed processing plus beam search was found to reduce this to typically 2% of this number. Even with this very significant reduction, Ney estimates that the execution of 50 million instructions/sec would be necessary to carry out the search in real time if frames are processed each 10 msec. Adjusting the threshold of the beam search results in more or fewer paths being retained, and little gain was found by increasing the number of paths preserved to greater than the nominal 2%.

With the extremely low percentage of paths actually searched in these experiments, it is natural to wonder how often the correct path might be missed. Ney et al. also computed the scores for the correct paths and found that, over six speakers, for one of the databases the correct path was missed only 0.57% of the time.

Some results of the experiments by Ney et al. involving the one database discussed above are shown in Table 11.1. To put into perspective the number of grid points searched (DP searches) per frame, we note that if the average reference template is taken to be $\bar{J} = 50$, then the total number of possible grid points that could be searched each frame is $3481 \times 50 \approx 1.74 \times 10^5$.

We have discussed the issue of syntax in a rather cursory and qualitative way here. Similar ideas concerning language modeling and its effects on recognition will be studied in further detail in future chapters.

TABLE 11.1. Results from Syntax-Driven OS Algorithm Used for Connected-Speech Recognition. After Ney et al. (1992).

Speaker	DP Searches/Frame	Word Error Rate (%)	Missed Paths
F-01	13,900	20.7	2/564
F-02	6,000	9.9	3/376
M-01	7,600	15.6	9/564
M-02	7,400	13.1	0/564
M-03	8,800	7.8	0/564
M-10	8,600	14.7	3/376

Note: The error rate is the number of inserted, deleted, or confused words as a percentage of the total number of correct words. The number of missed paths indicates the number of times the correct path through the search space was certainly missed, as a proportion of the total number of recognized sentences.

11.4.4 A Grammar-Driven Connected-Word Recognition System

To the extent that syntactic information is included in the LB and OS approaches, the incorporation of linguistic information represents a simple form of "top-down" information flow. By this we mean that word strings are hypothesized from "above" and the acoustic processing is used to determine whether the observation sequences would support these hypotheses. Researchers at Texas Instruments have used a different form of recognizer in which the flow of information is "bottom-up" (McMahan and Price, 1986; Pawate et al., 1987; Picone, 1990). In the Pawate paper, this system is referred to as a *grammar-driven connected-word recognizer* (GDCWR). We shall discuss the basic approach here; for details of hardware implementations, the reader is referred to the first two papers cited above.

In the GDCWR, DTW processing is used for word spotting (see Section 11.3.2) to locate words in the test string. These candidate words are then hypothesized (along with their nominal endpoints) to a decoder that evaluates the likelihoods of various concatenations of hypothesized words. This and similar approaches are sometimes called "*N*-best" methods because the *N* (a predetermined number) best hypotheses are submitted for further processing.

As an aside, we note that the word-spotting process is very similar to the LB algorithm in certain ways. The basic difference is that, in word spotting, the paths from searches over different reference strings do not interact at the acoustic processing level. Accordingly, if carried out in the manner suggested here, the acoustic search task of the GDCWR approach has similar complexity and memory requirements for a similar vocabulary. We will therefore not belabor this issue.

The sentence hypothesizer has the task of patching together the hypothesized words in an order indicated by the given endpoints. We will study such linguistic processing systems in much more quantitative detail in Chapter 13. However, the approach considered here is among the earliest language processing systems and can be understood in general terms quite easily. Consider a lattice of permissible word strings as shown in Fig. 11.24. Neglecting the details of endpoint justification, we can imagine that the word-spotting analysis provides costs for traversing a given path in the lattice in the form of likelihoods that the corresponding word will appear in that time slot in a string. With these costs assigned, the lattice can be searched using DP principles and the word string corresponding to the patch with the best overall score can be declared the spoken sentence.

Clearly, there are nontrivial issues involved in matching endpoints of the hypothesized words. Some flexibility in this matching (allowing some overlap or gaps) can be incorporated to allow for uncertainties in the endpoints (McMahan and Price, 1986). Related to this time synchrony issue is the fact that multiple hypotheses (with different boundaries) can occupy the same path through the word lattice. Methods for handling

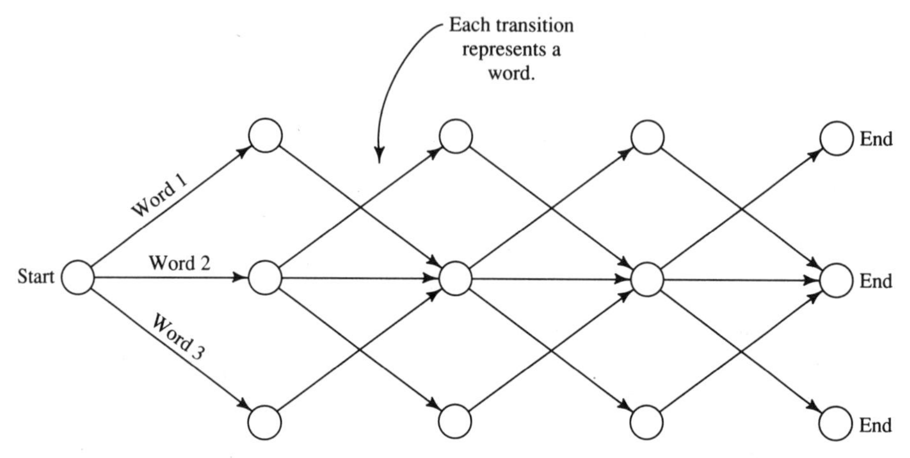

FIGURE 11.24. Lattice of permissible word strings in the language (syntax) processor of the GDCWR.

these synchronization problems will be treated in Chapters 12 and 13. Generally speaking, this issue is handled by variations of the DP search strategy.

11.4.5 Pruning and Beam Search

A final, and very important, constraint that is often imposed in large DP searches is that no path should be extended from a node, say (i,j), for which $D_{\min}(i,j)$ is unacceptably large. The threshold depends, of course, on the measure of cost employed and the nature of the problem. This phenomenon leads to many paths that are terminated early in the search grid, short of their potential for becoming a complete path. This clipping of undesirable paths is called *pruning*.[20]

In searching the DTW grid in the syntax-driven OS method above (or in the case of the GDCWR system, the syntax lattice itself), for example, it is usually the case that relatively few partial paths sustain sufficient probabilities [small enough $D_{\min}(i,j)$ values] to be considered candidates for extension to the optimal path. In order to cut down on what can be an extraordinary number of paths and computations, a pruning procedure is frequently employed that terminates consideration of unlikely paths. Most of the paths will be pruned because they will be so unlikely that their extensions are not warranted. This procedure is often referred to as a *beam search* (Lowerre and Reddy, 1980) since only paths that remain inside a certain acceptable "beam" of likelihoods are retained. Those that fall outside the beam are pruned. A simple beam, for example, would consist of all paths at frame i whose costs fall within, say, $\delta(i)$ of the best path. If, for example, (i,j^*) is the grid point with the best path

[20]The metaphor used here is pruning undesirable branches from plants and trees.

at i, then the path to any (i, j) will be a candidate for extension at frame $i+1$ only if

$$D_{\min}(i, j) \le D_{\min}(i, j^*) + \delta(i), \tag{11.68}$$

where $\delta(i)$ is usually taken to be a constant.

In fact, pruning can be important both for reducing computation and, where memory is allocated as needed, for reducing memory requirements. When memory is statically declared (as in the LB and OS algorithms), there is little memory benefit to pruning. This is one of the considerations that led to the development of the GDCWR system described above, in which pruning could be carried out at the syntax level (Picone, 1992).

We will see this same beam search idea used again in searching the hidden Markov model in Chapter 12, and then again at the linguistic level of processing in Chapter 13.

11.4.6 Summary of Resource Requirements for DTW Algorithms

Table 11.2 provides a convenient summary of computational and memory requirements for various DTW algorithms discussed. In the connected-word cases the numerical values in parentheses indicate requirements for the speech-telephone-number-dialing problem introduced in Section 11.4.1.

TABLE 11.2. Summary of Computational and Memory Requirements for Various DTW Algorithms.

	Typical Memory Requirements		Typical Computational Requirements	
DTW Task	**Local**	**Global**	**Distance Measures**	**DP Searches**
Single word	$2J$ (+IJ if back-tracking used)	—	μIJ	μIJ
LB	$2\bar{J}$ [$O(10^2)$]	$3IL$ [$O(10^3)$]	$O(\mu VIL\bar{J})$ [$O(10^5)$]	$O(\mu VIL\bar{J})$ [$O(10^5)$]
OS	$2V\bar{J}$ [$O(10^3)$]	$2I$ [$O(10^3)$]	$O(VI\bar{J})$ [$O(10^4)$]	$O(VI\bar{J})$ [$O(10^4)$]

Note: In the connected-word cases, the numerical values in square brackets indicate requirements for the speech-telephone-number-dialing problem introduced in Section 11.4.1. μ is a fraction, typically $\frac{1}{3}$, arising from the search constraints, L is the number of levels, $\bar{J}$ is the average single-word reference template length, and V is the vocabulary size. It is assumed that only one reference template is used per word. If K are used, V should be replaced by KV and the numerical estimates modified accordingly. When implemented as suggested in Section 11.4.4, the GDCWR system requires similar amounts of resources. However, a multiprocessor system for this task is described in (Pawate et al., 1987).

Some details should be kept in mind when comparing these numbers. First is that distance metrics are often redundantly computed in the LB scheme (see the discussion surrounding Fig. 11.16). We have noted that in some cases this redundant computation can be avoided by computing all distances before beginning the LB process. This requires the *a priori* computation and storage of $O(\mu VI\bar{J})$ distances. The reader should note how this will change the entries in the LB row of the table. It should also be noted that the "enhanced" LB algorithm of Myers and Rabiner (1981b) results in a savings of about a factor of two in computational load. Also pruning benefits may be obtained from the GDCWR system that are difficult to achieve with the LB and OS algorithms.

11.5 Training Issues in DTW Algorithms

We conclude with a discussion of a very important issue in DTW-based recognition—the construction of appropriate reference templates. That this topic is at the end of the chapter should not be construed as an indication of insignificance. On the contrary, proper "training" of the DTW search is of utmost importance. However, it is also a problem to which there is no simple or well-formulated solution. The appearance of this issue here, therefore, is an instance of saving the worst problem for last. However, its appearance here is also appropriate because it represents one of the central contrasts between the DTW method and the two recognition approaches to follow. Whereas DTW will be found to be very similar in many fundamental ways to the hidden Markov model (HMM) (Chapter 12) and to have many connections to some of the artificial neural network (ANN) approaches (Chapter 14), we will find that these latter techniques (particularly HMM, to which DTW is more closely related) possess vastly superior training methods. The HMM and the ANN can be trained in a "supervised" paradigm in which the model can learn the statistical makeup of the exemplars. The application of the HMM to speech recognition in the 1980s was a revolutionary development, largely because of the supervised training aspect of the model. We will, of course, have more to say about the HMM in the succeeding chapter. Let us focus upon some of the methods used to train the DTW search.

In the simplest DTW task—speaker-dependent IWR—it is frequently sufficient to store unaltered feature strings from one or more utterances of each word. This is the mode that has been more or less implied in the foregoing discussions. This simple training strategy has been called *causal training* (Itakura, 1975; Rabiner and Wilpon, 1980). A major drawback of this strategy is that the quality of any reference template can be established only experimentally. Another problem is that robustness can be increased only by increasing the number of utterances of the same word, which increases computation linearly.

Another technique that has been employed in the IWR problem is that of *averaging*. In this strategy, two or more utterances are time-aligned with respect to one another (by DTW!), then the feature strings are averaged to give a single reference template (Martin, 1975; Sambur and Rabiner, 1976). It is important, of course, to average features for which linear averaging is meaningful. A common choice is the autocorrelation sequence, the average of which may then be converted into another desired feature. Averaging tends to minimize the risk of very spurious and unreliable templates, but at the same time may create a template that is poorer than some of the component strings. Commercial systems have used this method successfully (Martin, 1976).

When DTW is applied in a speaker-independent and/or connected-word recognition system, three further problems emerge. First, different speakers tend to pronounce the same word in different ways. Second, speakers tend to produce isolated words (for training) that are of longer duration than if they were part of continuous speech. Third, in connected-speech applications, the coarticulatory effects between words are not modeled by single-word reference templates. The third problem can only be remedied by painstaking training procedures that attempt to model all such effects. Because more efficient and robust methods have been developed (in particular, the HMM), such procedures have not been explored for DTW. The second problem is important, of course, if multiple reference templates are to be combined. In this case, some form of time normalization is necessary. Several methods of time normalization have been used, including the use of DTW itself to time-align various templates, and techniques for linear compression or expansion of the reference templates (Myers and Rabiner, 1981a, 1981b).

The first problem, that of various pronunciations occurring in the speaker-independent problem, has been addressed through *clustering* techniques. This strategy has produced very good results in both speaker-dependent (Rabiner and Wilpon, 1979) and speaker-independent applications (Levinson et al., 1979). In the clustering approach, multiple, say P (typically 50–100), utterances of a word are reduced to $Q < P$ clusters that, in turn, are represented by one template each. The clusters are achieved by assigning a "feature vector" to each reference template, consisting of its distances from all P exemplars determined by DTW. These feature vectors are then clustered using a method such as the K-means, or isodata, algorithms discussed in Section 1.3.5. There are many details associated with clustering techniques, including the number of clusters to be used, setting criteria and thresholds for separating clusters, and modifying algorithms to operate properly with distances as features. An extensive study of these issues is found in the paper by Rabiner et al. (1979), which addresses the task of speaker-independent recognition of a small vocabulary (digits, alphabet, "stop," "error," "repeat"). The results show a significant improvement of clustered templates with respect to randomly selected ones. The referenced paper also provides a summary of results from several studies of IWR using the various training techniques discussed above.

In summary, training a DTW algorithm is an important task, particularly in speaker-independent systems, which must be given proper attention for successful results. The training issue is one that will contrast greatly with those used in the methods to follow in the remaining chapters.

11.6 Conclusions

The central focus in this chapter has been the DTW algorithm and its various applications and implementations in IWR and CSR. Dynamic time warping is a landmark technology that represents one of the first major breakthroughs in modern speech recognition. Although it is no longer as widely used as technologies to be discussed in later chapters, the fundamental basis for its operation, DP, will be found at the heart of much of what we have yet to study. Therefore, the hard work in this chapter will pay dividends in future study.

The next issue we take up is that of the most prevalent speech-recognition technique, the hidden Markov model (HMM). We will find that the HMM is very similar to a DTW algorithm in certain ways, but that a stochastic component to the HMM provides many useful properties for training, recognition, and general robustness.

In this chapter, we have also seen some glimpses of the usefulness of language processing in speech recognition. It will be interesting to see how some of the same principles we have learned in this chapter will continue to be built upon in Chapter 13.

11.7 Problems

11.1. In this problem we reconsider the algorithm for the "eastbound salesman problem" shown in Fig. 11.4.

(a) In the given algorithm, find the minimum distance path from $(0, 0)$ to $(5, 4)$ ($I = 5$ and $J = 4$) if the distance measure used is the simple Euclidean distance in the plane,

$$\begin{aligned} d[(i,j)|(k,l)] &= d_T[(i,j)|(k,l)] \\ &= \sqrt{(i-k)^2 + (j-l)^2}. \end{aligned} \tag{11.69}$$

In initializing the "first" cities, $(1, j)$, $j = 1, 2, 3, 4$, use (11.69). Indicate which cities are on the optimal path, and give that path's distance. Is the optimal path unique?

(b) Suppose that the salesman's boss gets the idea that better coverage of the territory would be obtained (more cities would be visited) if the salesman were required to take the *maximum* distance path from $(0, 0)$ to (I, J). Modify the algorithm to accommodate this change, and find the maximum distance path.

Indicate the distance of, and the cities on, this path. (*Note:* The strict "eastbound" constraint still applies.)

(c) Was the boss correct about visiting more cities? Is there a less expensive (shorter path) way to visit the same number of (or more) cities?

11.2. In a DTW search, suppose that $I \approx J$, and that we apply a window-width constraint under which no node (i_k, j_k) may be on the optimal path for which

$$|j_k - i_k| > W. \tag{11.70}$$

Approximately how many costs of form (11.29) need to be computed? Your answer will be in terms of I and W.

11.3. One measure of the computational effort of any sequential decision problem solved by DP is the number of times the equation of form (11.47) is used. This is sometimes called the number of *DP searches.*

(a) Consider an LB procedure in which the test utterance is I frames long, the average word-reference template is $\bar{J}$ frames, the vocabulary size is V, the number of levels is L, and continuity constraints are applied which reduce the search space to a factor of μ times the total grid. Give a clear argument showing that the number of DP searches carried out in finding the optimal path through the entire series of levels is $O(\mu VIL\bar{J})$ as claimed in Section 11.4.2.

(b) Using the same figures $I, \bar{J}, V$, and L from part (a), repeat the analysis for the OS algorithm employing the Sakoe and Chiba local path constraints as in Section 11.4.3.

11.4. (a) Modify the DTW algorithm of Section 11.3.3 to accommodate the following:

(i) Initial endpoint $(1, 1)$ and final endpoint (I, J) only;

(ii) The use of local constraint (d) of Fig. 11.11.

(b) What is the minimum and maximum global expansion of the test waveform with respect to the reference waveform with the modified algorithm?

(c) How many distance measures and DP searches will be performed per test waveform, assuming $I \approx J$?

11.5. Modify the algorithm in Section 11.3.3 for use with an LB search. In particular, include the two following backtracking arrays: $\boldsymbol{\beta}_1(j)$ and $\boldsymbol{\beta}_2(j)$, $j = 1, 2, \ldots, \bar{J}$, and explain how they are used.

11.6. In a phone-dialing application, the 10 digits[21] 0(zero), $1, \ldots, 9$ are to be recognized using DTW. There are 30 reference templates (three for each digit) against which each incoming utterance is compared. Suppose

[21]Frequently the utterance "oh" is included in a digit recognition problem, but we ignore it here for simplicity.

that a typical utterance is 0.5 sec long and is sampled at 8 kHz. The sequence is then reduced to a vector of feature parameters over 256-point frames which are shifted by 128 points each computation. This typical length should be assumed for both reference and test strings. Also assume that an Itakura parallelogram with slopes $\frac{1}{2}$ and 2 is applied to the search space. A floating point operation (we will be concerned only with multiplications and divisions here) requires about $\tau\,\mu$sec with the processor used. Log and square root operations require about 25 flops each.

(a) If the feature vector consists of 14 LP parameters and the Itakura distance is used to match vectors at a node, how many digits (spoken discretely) could be recognized in real time with this DTW approach? Your answer will be in terms of τ.

(b) Repeat (a) using 10 cepstral coefficients and the Euclidean distance.

(c) Repeat (a) using a zero-crossing measure and an energy measure as a two-dimensional feature vector, and a weighted Euclidean distance measure. The weighting matrix is simply used to normalize the variances of each feature.

(d) Now suppose that an LB approach is used to recognize complete 7-digit phone numbers. Assume that each utterance requires about 3 sec, but that the same reference strings developed above are used. With the features and distances of (a), (b), and (c), what must be the value of τ for a phone number to be recognized in real time? A typical value of τ for a desktop computer with a floating point processor is 10. Could this recognition problem be implemented on such a machine?

(e) Repeat part (d) for the OS approach.

11.7. (Computer Assignment) Implement the DTW algorithm given in Section 11.3.3 and use it to implement the following experiments.

(a) Reduce each utterance of the 10 digits to one of the sets of feature vectors suggested in Problem 11.6 using the suggested frame length and shift. Now consider a second set of utterances of the 10 digits as test data. Use digit "one" as the reference, and compute optimal distances to all 10 digits in the test set. Repeat for digits "two," "three," and so on, developing a 10×10 matrix of distances. Discuss the results.

(b) With the digit "nine" on the reference axis, use the backtracking facility to determine whether the digits "one" and "five" were temporally aligned with nine in an expected way. Discuss the results.

CHAPTER 12

The Hidden Markov Model

Reading Notes: *Virtually every topic in Chapter 1 will appear in some way in this chapter. Special dependencies are*

1. *In one recognition algorithm encountered in Section 12.2.2, we will briefly use the state space formulation of a discrete-time system described in Section 1.1.6.*
2. *The material in Section 12.2.7 depends strongly on the concepts in Sections 1.3.4 and 1.4, with a special dependency on the concept of mutual information in Section 1.4.3.*

12.1 Introduction

The DTW approach to recognition of isolated words and connected speech underlies a relatively straightforward and mature technology that is the basis for many practical real-time recognition systems. In fact, a number of custom VLSI circuit chips have been developed for DTW solutions (Burr et al., 1984; Ackenhusen, 1984; Jutland et al., 1984; Mann and Rhodes, 1986; Quenot et al., 1986; Takahashi et al., 1986; Gorin and Shively, 1987; Glinski et al., 1987). However, the DTW approach is known to have a number of limitations that restrict it to use in systems with relatively small vocabularies (~100 words) for speaker-independent systems and to moderate vocabularies (~500 words) for speaker-dependent systems. With larger systems, the storage of a sufficient number of templates and the computational cost of search becomes intractable. Whereas it is possible to employ DTW with subword units as reference templates, thereby decreasing the amount of reference information that must be stored (Bahl et al., 1984), accounting for important coarticulatory effects at the boundaries of subword units is difficult or impossible in this case. Coarticulatory effects, which are a problem in any connected-speech recognition strategy, become very significant at subword unit levels because they are naturally more complex at deeper levels of the articulation and because the speaker cannot be asked to control them (as with connected *words*, for example). Once again, we find the general nemesis of speech recognition engineers, variability in the speech, coming to the fore to impede success on a grand scale with DTW.

In the 1970s and especially the 1980s, speech researchers began to turn to stochastic approaches to speech modeling in an effort to address the problem of variability, particularly in large-scale systems. It seems likely that one or more of these methods will be the basis for future large-scale speech understanding systems when such systems finally emerge. The term "stochastic approach" is used to indicate that models are employed that inherently characterize some of the variability in the speech. This is to be contrasted with the straightforward deterministic use of the speech data in template-matching (e.g., DTW) approaches in which no attempt at probabilistic modeling of variability is present. The term "structural methods" is also used to describe the stochastic approaches, since each of these methods is based upon a model with a very important mathematical "structure."

Two very different types of stochastic methods have been researched. The first, the *hidden Markov model* (HMM), is amenable to computation on conventional sequential computing machines. The HMM will be the subject of this chapter. Speech research has driven most of the engineering interest in the HMM in the last three decades, and the HMM has been the basis for several successful large-scale laboratory and commercial speech recognition systems. In contrast, the second class of stochastic techniques in speech recognition to be discussed, that based on the *artificial neural network* (ANN), has been a small part of a much more general research effort to explore alternative computing architectures with some superficial resemblances to the massively parallel "computing" of biological neural systems. Results of the application of ANNs to speech research lag far behind those for HMMs because of the relative youth of the technology. Because of the interest in ANNs and their potential for exciting new technologies, we will give a brief synopsis of ANN research as it applies to speech recognition in Chapter 14.

The history of the HMM precedes its use in speech processing and only gradually became widely known and used in the speech field. The introduction of the HMM into the speech recognition field is generally attributed to the independent work of Baker at Carnegie–Mellon University (Baker, 1975a, 1975b), and Jelinek and colleagues at IBM (Jelinek et al., 1975, 1976). Apparently, work on the HMM was also in progress at the Institute for Defense Analysis in the early 1970s (Poritz, 1988).[1] Interestingly, the field of *syntactic pattern recognition* was independently evolving during this same period, principally due to the research efforts of the late Professor K. S. Fu at Purdue University (Fu, 1982). In the 1980s the close relationships between certain theories of syntactic pattern recognition and the HMM were recognized and exploited.

An exposition of the history of the HMM prior to the speech processing work is related by Levinson (1985) and Poritz (1988). Levinson cites the paper of Dempster et al. (1977), which indicates that the roots of the theory can be traced to the 1950s when statisticians were studying the problem of characterizing random processes for which incomplete observations were available. Their approach was to model the problem as a

[1]The Poritz paper cites unpublished lectures of J. D. Ferguson of IDA in 1974.

"doubly stochastic process" in which the observed data were thought to be the result of having passed the "true" (hidden) process through a "censor" that produced the second process (observed). Both processes were to be characterized using only the one that could be observed. The resulting identification algorithm came to be known as the *estimate-maximize* (EM) *algorithm* (Dempster, 1977). In the 1960s and early 1970s, Baum and colleagues (1966, 1967, 1968, 1970, 1972) worked on a special case of the HMM and developed what can be considered a special case of the EM algorithm, the *forward–backward* (F–B) *algorithm* (also called the *Baum–Welch reestimation algorithm*), for HMM parameter estimation and decoding in time which is linear in the length of the observation string. As we shall see, the F–B algorithm turns an otherwise computationally intractable problem into an easily solvable one. Because the original work is developed in very abstract terms and published in journals not widely read by engineers, it took several years for the methods to come to fruition in the speech recognition problem. Once realized, however, the impact of this technology has been extraordinary.

12.2 Theoretical Developments

12.2.1 Generalities

Introduction

In Chapter 13 we will realize that an HMM is, in fact, a "stochastic finite state automaton"—a type of abstract "machine" used to model a speech utterance. The utterance may be a word, a subword unit, or, in principle, a complete sentence or paragraph. In small-vocabulary systems, the HMM tends to be used to model words, whereas in larger vocabulary systems, the HMM is used for subword units like phones. We will be more specific about this issue at appropriate points in the discussion. In order to introduce the operation of the HMM, however, it is sufficient to assume that the unit of interest is a word. We will soon discover that there is no real loss of generality in discussing the HMM from this point of view.

We have become accustomed to the notion of reducing a speech utterance to a string of features. In discussing DTW, we referred to the feature string representing the word to be recognized as the set of "test features," and denoted it

$$\text{test features:}\quad \mathbf{t}(1), \mathbf{t}(2), \mathbf{t}(3), \ldots, \mathbf{t}(i), \ldots, \mathbf{t}(I). \tag{12.1}$$

The reader is encouraged to review the discussion leading to (11.27) and (11.28) with particular attention to the link between the short-term feature analysis and the resulting test string. In the HMM literature, it is customary to refer to the string of test features as the *observations* or *observables*, since these features represent the information that is "observed" from the incoming speech utterance. We shall adopt this terminology in our discussion, and write

$$\text{observations:}\quad \mathbf{y}(1), \mathbf{y}(2), \mathbf{y}(3), \ldots, \mathbf{y}(t), \ldots, \mathbf{y}(T). \tag{12.2}$$

Note that we have changed the frame index (which we will call *time*) to t (since i is reserved for another purpose in this discussion) and the total number of observations to T. The reader should appreciate, however, that (12.1) and (12.2) are essentially identical.

"Moore" and "Mealy" Forms of the HMM

An HMM, which is always associated with a particular word (or other utterance), is a "finite state machine" capable of generating observation strings. A given HMM is more likely to produce observation strings that would be observed from real utterances of its associated word. During the *training phase*, the HMM is "taught" the statistical makeup of the observation strings for its dedicated word. During the *recognition phase*, given an incoming observation string, it is imagined that one of the existing HMMs produced the observation string. Then, for each HMM, the question is asked: How likely (in some sense) is it that *this* HMM produced this incoming observation string? The word associated with the HMM of highest likelihood is declared to be the recognized word. Note carefully that it is *not* the purpose of an HMM to generate observation strings. We *imagine*, however, that one of the HMMs *did* generate the observation string to be recognized as a gimmick for performing the recognition.

A diagram of a typical HMM with six states is shown in Fig. 12.1. The states are labeled by integers. The *structure*, or *topology*, of the HMM is determined by its allowable state transitions, an important issue to which we return below. The HMM is imagined to generate observation sequences by jumping from state to state and emitting an observation with each jump. Two slightly different forms of the HMM are used. The one usually (but not always) discussed for *acoustic* processing (modeling the signal, as we are doing here) emits an observation upon arrival at each successive state. (It also emits one from the initial state.) The alternative form, generally employed in *language* processing (see Chapter 13), emits an observation during the transition. The "state emitter" form of the model is sometimes called a *Moore machine* in automata theory, while

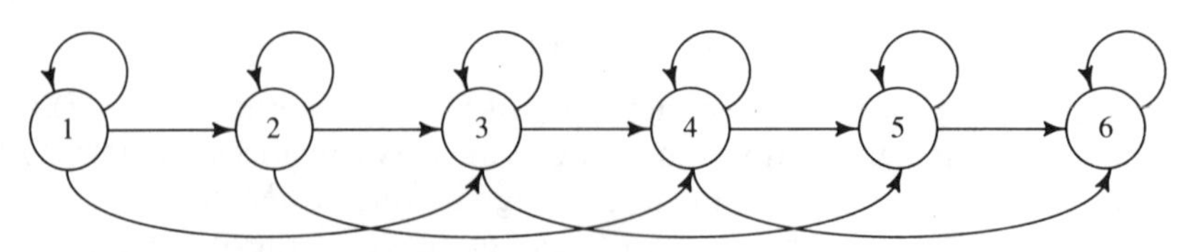

FIGURE 12.1. Typical HMM with six states.

the "transition emitter" form is a *Mealy machine* (Hopcroft and Ullman, 1979, pp. 42–45). Although this terminology is not standard in speech processing, we will use it in our study in order to clearly distinguish the two forms of the HMM. After distinguishing between the two forms in this section, throughout much of the rest of this chapter we will be discussing the Moore form.

The formalities of the transitions are essentially identical for the two forms of the model. At each observation time (corresponding to the times at which we extract observations from the speech utterances to be recognized) a state transition is assumed to occur in the model. The likelihoods of these transitions are governed by the *state transition probabilities*, which appear as labels on the arcs connecting the states. Let us denote the probability of making the transition from state j to state i by[2] $a(i|j)$. The matrix of state transition probabilities is the *state transition matrix*, given by

$$\mathbf{A} = \begin{bmatrix} a(1|1) & a(1|2) & \cdots & a(1|S-1) & a(1|S) \\ & \ddots & & & \\ & & a(i|j) & & \\ & & & \ddots & \\ a(S|1) & a(S|2) & \cdots & a(S|S-1) & a(S|S) \end{bmatrix}, \tag{12.3}$$

where S is the total number of states in the model. The state transition probabilities are assumed to be stationary in time, so that $a(i|j)$ does not depend upon the time t at which the transition occurs. Note that any column of $\mathbf{A}$ must sum to unity, since it is assumed that a transition takes place with certainty at every time.

The sequence of states that occurs enroute to generating a given observation sequence is the first of two random processes associated with an HMM. Suppose that we call this state random process $\underline{x}$. The associated random variables are $\underline{x}(t)$. We then have that

$$a(i|j) \stackrel{\text{def}}{=} P\big(\underline{x}(t) = i \,|\, \underline{x}(t-1) = j\big), \tag{12.4}$$

for arbitrary t. By assumption

$$\begin{aligned} a(i|j_1) = P\big(\underline{x}(t) = i \,|\, &\underline{x}(t-1) = j_1, \\ &\underline{x}(t-2) = j_2, \underline{x}(t-3) = j_3, \ldots\big), \end{aligned} \tag{12.5}$$

meaning that the state transition at time t does not depend upon the history of the state sequence[3] prior to time $t-1$. Under this condition, the

[2]This probability is frequently denoted a_{ji} in the literature, but we denote it more explicitly here to remind the reader of its meaning in terms of its defining probability.

[3]Of course, in theory, nothing precludes dependence upon n past states, but the complexity of training this model and recognition using it increases dramatically with each increment. Further, the memory requirements increase as S^{n+1}.

random sequence is called a *(first-order) Markov process* [see, e.g., (Leon–Garcia, 1989, Ch. 8)]. When the random variables of a Markov process take only discrete values (often integers), then it is called a *Markov chain.* The state sequence $\underline{x}$ in the HMM is a Markov chain, since its random variables assume integer values corresponding to the states of the model. Further, since the state transition probabilities do not depend on t, the Markov chain is said to be *homogeneous* in time.

Although the difference is only formal, in some discussions in the literature involving the Mealy form, the transitions rather than the states will be indexed. That is, each transition will be given a label and the sequence of *transitions*, rather than the sequence of *states*, will be featured. On the few occasions when it is necessary for us to feature the transitions rather than the states in this text, since our states will always be labeled anyway, we will simply refer to the transition between states j and i by $u_{i|j}$. This should be thought of as an integer labeling the transition referenced. Rather than a state sequence, we can discuss the transition sequence modeled by random process $\underline{u}$ with the random variables $\underline{u}(t)$. It should be clear that, for an arbitrary t,

$$P\big(\underline{u}(t) = u_{i|j}\big) = P\big(\underline{x}(t) = i \,|\, \underline{x}(t-1) = j\big) = a(i \,|\, j) \tag{12.6}$$

so that the matrix of transition probabilities is identical to (12.3). This makes perfect sense, since (12.3) is the matrix of probabilities of making the transitions. It is also clear that the transition sequence is a homogeneous Markov chain.

For convenience in the following discussion involving the Moore form, let us define the *state probability vector* at time t to be the S-vector[4]

$$\boldsymbol{\pi}(t) \stackrel{\text{def}}{=} \begin{bmatrix} P(\underline{x}(t) = 1) \\ P(\underline{x}(t) = 2) \\ \vdots \\ P(\underline{x}(t) = S) \end{bmatrix}. \tag{12.7}$$

It should be clear from our discussion above that, for any t,

$$\boldsymbol{\pi}(t) = \mathbf{A}\boldsymbol{\pi}(t-1). \tag{12.8}$$

In fact, given the *initial state probability vector,* $\boldsymbol{\pi}(1)$, it is a simple matter to show by recursion that

$$\boldsymbol{\pi}(t) = \mathbf{A}^{t-1}\boldsymbol{\pi}(1). \tag{12.9}$$

Taken together, therefore, the state transition matrix and the initial state probability vector completely specify the probability of residing in any state at any time.

[4]The symbol $\boldsymbol{\pi}$ is sometimes used in the literature to denote the initial state probability vector which, in our case, is $\boldsymbol{\pi}(1)$.

Let us now turn our attention to the observations. The observation sequence may also be modeled as a discrete-time stochastic process, say $\underline{\mathbf{y}}$, with random variables $\underline{\mathbf{y}}(t)$. For the Moore form, upon entering a state, say state i, at time t, an observation is generated. The generation of the particular observations is governed by the probability density function $f_{\underline{\mathbf{y}}(t)|\underline{x}(t)}(\xi|i)$, which we will call the *observation pdf for state i.* Both $\underline{\mathbf{y}}(t)$ and ξ are, in general, M-dimensional vectors, where M is the dimension of the vector feature extracted from the speech. For mathematical tractability, it is customary to make the unrealistic assumption that the random process $\underline{\mathbf{y}}$ has independent and identically distributed random variables, $\underline{\mathbf{y}}(t)$. In particular, this means that $f_{\underline{\mathbf{y}}(t)|\underline{x}(t)}(\xi|i)$ is not dependent upon t, and we write

$$f_{\underline{\mathbf{y}}|\underline{x}}(\xi|i) \stackrel{\text{def}}{=} f_{\underline{\mathbf{y}}(t)|\underline{x}(t)}(\xi|i) \qquad \text{for arbitrary } t. \tag{12.10}$$

For the Mealy form, the observation densities are slightly different. These are

$$f_{\underline{\mathbf{y}}|\underline{u}}(\xi|u_{i|j}) \stackrel{\text{def}}{=} f_{\underline{\mathbf{y}}(t)|\underline{u}(t)}(\xi|u_{i|j}) \tag{12.11}$$

for arbitrary t. (Remember that $u_{i|j}$ is just an integer indexing the transition from state j to i.)

In a formal sense, then, a Moore-form HMM, say $\mathcal{M}$, is comprised of the set of mathematical entities

$$\mathcal{M} = \left\{S, \boldsymbol{\pi}(1), \mathbf{A}, \left\{f_{\underline{\mathbf{y}}|\underline{x}}(\xi|i), 1 \leq i \leq S\right\}\right\} \tag{12.12}$$

each of which has been described in the discussion above. A similar characterization of the Mealy form is given by

$$\mathcal{M} = \left\{S, \boldsymbol{\pi}(1), \mathbf{A}, \left\{f_{\underline{\mathbf{y}}|\underline{u}}(\xi|u_{i|j}), 1 \leq i,j \leq S\right\}\right\}. \tag{12.13}$$

Henceforth in our discussion, we will focus on the Moore form of the HMM unless otherwise noted.

The Two HMM Problems

Given this formal description of an HMM, we now examine two key issues centered on the training and use of the HMM. These are the following:

1. Given a series of training observations for a given word, how do we train an HMM to represent the word? This amounts to finding a procedure for estimating an appropriate state transition matrix, $\mathbf{A}$, and observation pdf's, $f_{\underline{\mathbf{y}}|\underline{x}}(\xi|i)$, for each state. This represents the HMM *training* problem.
2. Given a trained HMM, how do we find the likelihood that it produced an incoming speech observation sequence? This represents the *recognition* problem.

We begin our pursuit of answers to these questions by examining a case of the HMM with a simple form for the observation pdf's.

12.2.2 The Discrete Observation HMM

Formalities

The *discrete observation HMM* is restricted to the production of a finite set of discrete observations. In this case the naturally occurring observation vectors are quantized into one of the permissible set using the vector quantization (VQ) methods described in Section 7.2.2. Prior to training any of the HMMs for the individual utterance, a set of (continuous) observation vectors from a large corpus of speech is used to derive the codebook. Subsequently, any observation vector used for either training or recognition is quantized using this codebook. If there are K possible vectors (observations) in the codebook, then it is sufficient to assign an observation a single integer, say k, where $1 \le k \le K$. Formally, the vector random process $\underline{\mathbf{y}}$ is replaced by a scalar random process, say $\underline{y}$, where each of the random variables $\underline{y}(t)$ may take only integer values in $[1, K]$. The reader should note that the original observation process is denoted in boldfaced type because, in general, it is a vector, whereas the "quantized" observation process is set in regular typeface, since its random variables take only integers.

More generally, the quantized observation may be considered a distinct "symbol" from an "alphabet" of K symbols. This latter interpretation sometimes gives rise to the name "discrete symbol HMM." In our discussion, it will be sufficient to assume that each observation is assigned an integer.

Note that for the discrete observation HMM, the quantized observation pdf for state i takes the form of K impulses on the real line. In this case it is sufficient to know the probability distribution over the K symbols for each state (weights on the impulses) which we shall denote[5]

$$b(k|i) \stackrel{\text{def}}{=} P(\underline{y}(t) = k | \underline{x}(t) = i). \tag{12.14}$$

These *observation probabilities* are clearly defined to be dependent upon the state, but are assumed independent of time t. In general, we will not know the value assumed by a particular observation, $\underline{y}(t)$, and we will write

$$b(y(t)|i) \stackrel{\text{def}}{=} P(\underline{y}(t) = y(t) | \underline{x}(t) = i). \tag{12.15}$$

For convenience we define the *observation probability matrix*,

[5]Frequently denoted $b_i(k)$ in the literature.

$$\mathbf{B} = \begin{bmatrix} b(1|1) & b(1|2) & \cdots & b(1|S-1) & b(1|S) \\ & \ddots & & & \\ & & b(k|i) & & \\ & & & \ddots & \\ b(K|1) & b(K|2) & \cdots & b(K|S-1) & b(K|S) \end{bmatrix}, \tag{12.16}$$

recalling that K is the number of discrete observations and S is the total number of states in the model.[6] For completeness, let us define the *observation probability vector*,

$$\mathbf{p}(t) \stackrel{\text{def}}{=} \begin{bmatrix} P(\underline{y}(t)=1) \\ P(\underline{y}(t)=2) \\ \vdots \\ P(\underline{y}(t)=K) \end{bmatrix}. \tag{12.18}$$

Then it should be clear that

$$\mathbf{p}(t) = \mathbf{B}\boldsymbol{\pi}(t), \tag{12.19}$$

where $\boldsymbol{\pi}(t)$ is the state probability vector defined in (12.7). Also, using (12.9) in (12.18), we have

$$\mathbf{p}(t) = \mathbf{B}\mathbf{A}^{t-1}\boldsymbol{\pi}(1). \tag{12.20}$$

Equations (12.8) and (12.19) comprise the *state equation* and *observation equation*, respectively, for a *state space* representation of the HMM (see Section 1.1.6). We will have more to say about this representation later.

Finally, we note that the mathematical specification of the HMM is modified somewhat with respect to the general case in (12.12) to reflect the discrete observations:

$$\mathcal{M} = \{S, \boldsymbol{\pi}(1), \mathbf{A}, \mathbf{B}, \{\mathbf{y}_k, 1 \le k \le K\}\}, \tag{12.21}$$

where $\{\mathbf{y}_k, 1 \le k \le K\}$ represents the set of K discrete quantization vectors in the VQ codebook.

Recognition Using the Discrete Observation HMM

Introduction. Of the two problems, training and recognition, the recognition problem is easier, so we examine it first. We assume, therefore, that

[6]Note that in the Mealy form of the HMM given in (12.13), these probabilities would take the form

$$b(y(t)|u_{i|j}) \stackrel{\text{def}}{=} P(\underline{y}(t)=y(t)|\underline{u}(t)=u_{i|j}).$$

How could we create a matrix for these numbers?

fully trained HMMs of the form (12.21) for each word in the vocabulary are available, and that we desire to deduce which of these words a given (quantized) observation sequence $y(1), \ldots, y(T)$ represents. That is, we want to determine the likelihood that each of the models could have produced the observation sequence.

Let us begin by defining some simple notation that will be of critical importance in our discussions. For simplicity, we denote any *partial sequence* of observations in time, say $\{y(t_1), y(t_1+1), y(t_1+2), \ldots, y(t_2)\}$, by

$$y_{t_1}^{t_2} \stackrel{\text{def}}{=} \{y(t_1), y(t_1+1), y(t_1+2), \ldots, y(t_2)\}. \tag{12.22}$$

In particular, the *forward partial sequence* of observations at time t is

$$y_1^t \stackrel{\text{def}}{=} \{y(1), y(2), \ldots, y(t)\} \tag{12.23}$$

and the *backward partial sequence* of observations at time t is

$$y_{t+1}^T \stackrel{\text{def}}{=} \{y(t+1), y(t+2), \ldots, y(T)\}. \tag{12.24}$$

The term "backward" is used here to connote that the sequence can be obtained by starting at the last observation and working backward. Note that the backward partial sequence at time t does not include the observation at time t. It is also useful to note that the forward partial sequence at time T is the complete sequence of observations that, for convenience, we denote simply

$$y \stackrel{\text{def}}{=} y_1^T. \tag{12.25}$$

If we wish to denote a partial sequence of *random variables* in any of the senses above, we will simply underscore the quantity—$\underline{y}_1^t$. This, of course, is an abuse of notation, since $\underline{y}_1^t$ is not a random *variable* but a model for a partial realization of the random *process* $\underline{y}$; however, the meaning should be clear.

A key question is: "What is meant by the 'likelihood' of an HMM?" There are two general measures of likelihood used in the recognition problem. Each leads to its own recognition algorithm, so we must consider them individually.

"Any Path" Method (F–B Approach). A most natural measure of likelihood of a given HMM, say $\mathcal{M}$, would be $P(\mathcal{M} \mid \underline{y} = y)$. However, the available data will not allow us to characterize this statistic during the training process. We will therefore resort to the use of $P(\underline{y} = y \mid \mathcal{M})$, the probability that the observation sequence y is produced, given the model $\mathcal{M}$. The reader is reminded of the discussion in Section 1.3.3 that justifies this substitution. The reason for the name "any path" method is that the likelihood to be computed here is based on the probability that the observations could have been produced using *any* state sequence (path) through the model. The second approach to be considered will insist that the likelihood be based on the *best* state sequence through $\mathcal{M}$.

Let us first consider a "brute force" approach to the computation of $P(y|\mathcal{M})$. Consider a specific state sequence through the HMM of proper length T, say $\mathcal{I} = \{i_1, i_2, \ldots, i_T\}$. The probability of the observation sequence being produced over this state sequence is

$$P(y|\mathcal{I}, \mathcal{M}) = b(y(1)|i_1)b(y(2)|i_2)\cdots b(y(T)|i_T). \tag{12.26}$$

The probability of the state sequence $\mathcal{I}$ is

$$P(\mathcal{I}|\mathcal{M}) = P(\underline{x}(1) = i_1)a(i_2|i_1)a(i_3|i_2)\cdots a(i_T|i_{T-1}). \tag{12.27}$$

Therefore,

$$\begin{aligned} P(y, \mathcal{I}|\mathcal{M}) = b(y(1)|i_1)b(y(2)|i_2)\cdots b(y(T)|i_T) \\ \times P(\underline{x}(1) = i_1)a(i_2|i_1)a(i_3|i_2)\cdots a(i_T|i_{T-1}). \end{aligned} \tag{12.28}$$

In order to find $P(y|\mathcal{M})$, we need to sum this result over all possible paths (mutually exclusive events),

$$P(y|\mathcal{M}) = \sum_{\text{all } \mathcal{I}} P(y, \mathcal{I}|\mathcal{M}). \tag{12.29}$$

Unfortunately, direct computation of (12.28) requires $O(2TS^T)$ flops,[7] since there are S^T possible state sequences, and for each $\mathcal{I}$ [each term in the sum of (12.28)] about $2T$ computations are necessary. This amount of computation is infeasible for even small values of S and T. For example, if $S = 5$ and $T = 100$, then

$$2 \times 100 \times 5^{100} \approx 1.6 \times 10^{72} \tag{12.30}$$

computations are required *per HMM*. This is clearly prohibitive and a more efficient solution must be found.

The so-called *forward-backward* (F–B) *algorithm*[8] of Baum et al. (Baum and Eagon, 1967; Baum and Sell, 1968) can be used to efficiently compute $P(y|\mathcal{M})$. To develop this method, we need to define a "forward-going" and a "backward-going" probability sequence. Let us define $\alpha(y_1^t, i)$ to be the joint probability of having generated the partial forward sequence y_1^t and having arrived at state i at the tth step, given HMM $\mathcal{M}$,

$$\alpha(y_1^t, i) \stackrel{\text{def}}{=} P(\underline{y}_1^t = y_1^t, \underline{x}(t) = i|\mathcal{M}). \tag{12.31}$$

Whereas $\alpha(y_1^t, i)$ accounts for a forward path search ending at a certain state, we will also need a quantity to account for the rest of the search. Let $\beta(y_{t+1}^T|i)$ denote the probability of generating the "backward" partial

[7]Precisely $(2T-1)S^T$ multiplications and $S^T - 1$ additions are required (Rabiner, 1989).

[8]As mentioned in the introduction, this is also called the *Baum–Welch reestimation algorithm*; we will see why in Section 12.2.2 when we use it in the training problem.

sequence y_{t+1}^{T} using model $\mathcal{M}$, given that the state sequence emerges from state i at time t,

$$\beta(y_{t+1}^{T}|i) \stackrel{\text{def}}{=} P(\underline{y}_{t+1}^{T} = y_{t+1}^{T}|\underline{x}(t) = i, \mathcal{M}). \tag{12.32}$$

Actually, we will discover that only the α sequence is necessary to compute $P(y|\mathcal{M})$, which is our present objective, but the β sequence will be very useful in future developments.

Suppose that we now lay the states out in time to form a lattice as shown in Fig. 12.2. At time t, we have arrived at state i and have somehow managed to compute $\alpha(y_1^t, i)$. Suppose further that we wish to compute $\alpha(y_1^{t+1}, j)$ for some state j at the next time. If there were only one path to state j at $t+1$, that arising from i at t (see Fig. 12.2), then clearly

$$\begin{aligned}\alpha(y_1^{t+1}, j) &= \alpha(y_1^t, i)P(\underline{x}(t+1) = j|\underline{x}(t) = i) \\ &\quad \times P(\underline{y}(t+1) = y(t+1)|\underline{x}(t+1) = j) \\ &= \alpha(y_1^t, i)a(j|i)b(y(t+1)|j).\end{aligned} \tag{12.33}$$

Now if there is more than one state "i" at time t through which we can get to j at time $t+1$, then we should simply sum the possibilities,

$$\alpha(y_1^{t+1}, j) = \sum_{i=1}^{S} \alpha(y_1^t, i)a(j|i)b(y(t+1)|j). \tag{12.34}$$

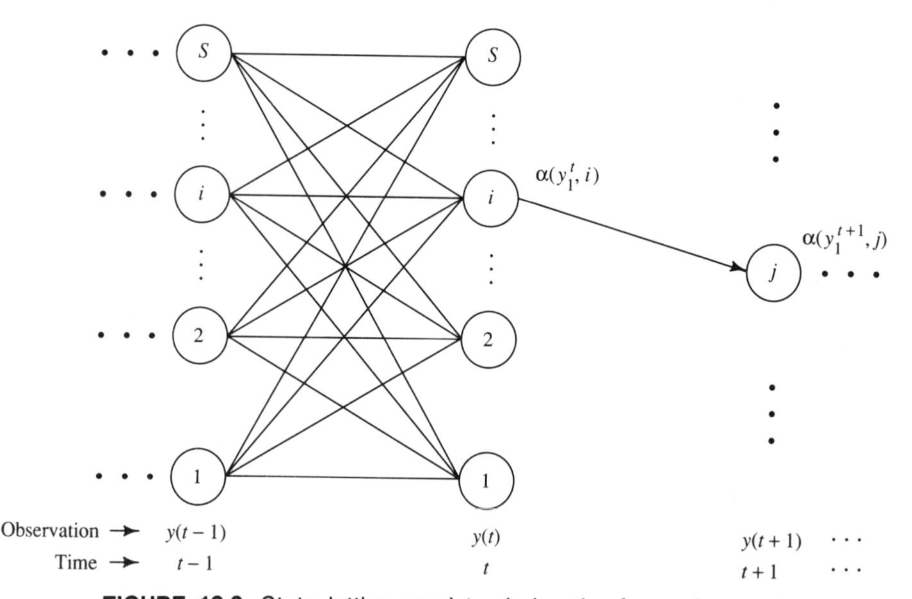

FIGURE 12.2. State lattice used to derive the forward recursion.

This equation suggests a lattice-type computation that can be used to compute the α sequence for each state and for progressively larger t's. This lattice is illustrated in Fig. 12.2. By definition of α, it is clear that the recursion is initiated by setting

$$\alpha(y_1^1, j) = P(\underline{x}(1) = j) b(y(1) | j) \tag{12.35}$$

for each j.

By a similar line of reasoning, the following backward recursion in time can be derived for the β sequence:

$$\beta(y_{t+1}^T | i) = \sum_{j=1}^{S} \beta(y_{t+2}^T | j) a(j | i) b(y(t+1) | j). \tag{12.36}$$

This recursion is initialized by defining y_{T+1}^T to be a (fictitious) partial sequence such that

$$\beta(y_{T+1}^T | i) \stackrel{\text{def}}{=} \begin{cases} 1, & \text{if } i \text{ is a legal final state} \\ 0, & \text{otherwise} \end{cases}, \tag{12.37}$$

where a "legal final state" is one at which a path through the model may end. Note that $\beta(y_{t+1}^T | i)$ will only be used in our developments for $1 \le t \le T-1$, so this last definition is only a convenience to start the recursion.

Now we note that

$$P(y, \underline{x}(t) = i | \mathcal{M}) = \alpha(y_1^t, i) \beta(y_{t+1}^T | i) \tag{12.38}$$

for any state i. Therefore,

$$P(y | \mathcal{M}) = \sum_{i=1}^{S} \alpha(y_1^t, i) \beta(y_{t+1}^T | i). \tag{12.39}$$

The desired likelihood can therefore be obtained at any time slot in the lattice by forming and summing the F–B products as in (12.39). In particular, however, we can work at the final time $t = T$, and by inserting (12.37) into (12.39) obtain

$$P(y | \mathcal{M}) = \sum_{\text{all legal final } i} \alpha(y_1^T, i). \tag{12.40}$$

This expression makes it unnecessary to work with the backward recursion in order to obtain the desired likelihood. Further, a simple algorithm is evident in equations (12.34), (12.35), and (12.40). This is shown in Fig. 12.3.

FIGURE 12.3. Computation of $P(y|\mathcal{M})$ using the forward recursion of the F–B algorithm.

Initialization: Initialize $\alpha(y_1^1, j)$ for $j = 1, \ldots, S$ using (12.35).
Recursion: For $t = 2, \ldots, T$
For $j = 1, \ldots, S$
Update $\alpha(y_1^t, j)$ using (12.34).
Next j
Next t
Termination: Compute likelihood $P(y|\mathcal{M})$ using (12.40).

A careful analysis of this algorithm will show that it requires $O(S^2T)$ flops.[9] For the typical values $S = 5, T = 100$, the likelihood computation will require about

$$100 \times 5^2 = 2500 \tag{12.41}$$

flops, which represents an improvement of 69 orders of magnitude with respect to the same computation carried out directly in (12.30). The key to this reduction is eliminating redundant computation. In turn, this is a consequence of noting that all possible state sequences must merge into one of S states at time t. By summing likelihoods locally at those nodes as we progress through time, the combinatorial explosion of computations that must be performed if paths are considered individually is avoided.

We should note a practical problem here. In the course of computing the $\alpha(\cdot\,,\cdot)$ sequence in the algorithm above, many probabilities are multiplied together [see (12.34)]. This is also true for the computation of the $\beta(\cdot|\cdot)$ sequence [see (12.36)], which will be used in the training procedure to be discussed below. Frequently, these computations cause numerical problems as the results begin to underflow the machine's precision capability. To remedy this problem, a scaling procedure to be described in Section 12.2.5, is employed. The algorithm above should be thought of as only a theoretical result, which must be "scaled" in most cases for practical implementation. The scaling procedure, once understood, is easily added to the algorithm.

Simplified "Any Path" Method (State-Space Approach). The F-B algorithm is generally of $\mathcal{O}(S^2T)$ complexity per model. As claimed in Footnote 9, however, the computation is frequently less due to a model structure (discussed in Section 12.3.2) that does not permit transitions from each state to all others (many of the $a(i|j)$'s are zero). Typically the complexity is $\mathcal{O}(3ST) \approx \mathcal{O}(S^{3/2}T)$. By making a further, albeit theoretically invalid (Mitchell, *et al.*, 1994), simplification, Deller and Snider (1990) suggest an alternative "any path" method of $\mathcal{O}(ST)$

[9]In many HMM applications, the **A** matrix is very sparse (mostly zero elements) and the complexity is typically $O(3ST)$. An alternative "any path" method that assures $O(ST)$ complexity in general is discussed below.

complexity. This method inherently assumes that the observation sequence $\underline{y}(1), \underline{y}(2), \ldots, \underline{y}(T)$ represents a nonstationary, but independent, sequence of random variables. In this case, given probability distributions for each t, the likelihood $P(\underline{y}|\mathcal{M})$ can be computed by forming the product $\prod_{t=1}^{t} P(\underline{y}(t)|\mathcal{M})$.

In fact, the conditional dependence of the observations upon the HMM states is fundamental to the model structure. This, in turn, implies a temporal dependence among the observations $\underline{y}(t)$. In particular, for example, given $\underline{y}(t') = k'$, the observation $\underline{y}(t'') = k''$, $t'' > t'$, is impossible if symbol k' can only be produced in state, say, $i_{t'}$, while symbol k'' can be emitted only by states that are unreachable from state $i_{t'}$. This does not preclude, however, computing the probabilities $P(\underline{y}(t') - k'|\mathcal{M})$ and $P(\underline{y}(t'') - k''|\mathcal{M})$ since each of these outcomes is possible in the absence of any conditioning information. The result of assuming independence of $\underline{y}(t')$ and $\underline{y}(t'')$ in such a circumstance is to assign a nonzero probability to a sequence that cannot be legitimately produced by the model. Even in this bogus situation, however, the computation is not completely blind to the state dynamics of the model, since the probabilities of producing symbols k' and k'' are influenced by the probabilities of residing in the states that can produce them at times t' and t'', respectively. The hidden state structure in such a case, therefore, is more weakly linked to the generation of the outcomes than is prescribed by the HMM. Nevertheless, if the symbols are distributed among the states in such a way that they are effectively "localized in time" (in the above, e.g., the states that produce k'' are very unlikely at times $t > t'$), then the state dependence will be approximately properly accounted for in the computation.

The method proposed by Deller and Snider is based on viewing the HMM as a state-space system for which (12.8) and (12.19) comprise the state equation and observation equation, respectively:

$$\boldsymbol{\pi}(t) = \mathbf{A}\boldsymbol{\pi}(t-1) + \delta(t-1)\mathbf{u}(t-1) \text{ and } \mathbf{p}(t) = \mathbf{B}\boldsymbol{\pi}(t). \qquad (12.42)$$

For formal reasons, an input term, $\mathbf{u}(t)$, is included in the state equation, defined such that $\mathbf{u}(0) = \boldsymbol{\pi}(1)$ and $\mathbf{u}(t)$ is arbitrary otherwise. $\delta(t)$ is the unit sample sequence and $\boldsymbol{\pi}(t)$ is defined to be zero for all $t \leq 0$. $\mathbf{p}(t)$ represents the probabilities of symbols at time t, accounting for the likelihoods of states, but not strictly for the state dependence.

The benefits of this strategy are computational efficiency, numerical stability, and decreased storage requirements. The state matrix can almost always be diagonalized [see e.g., (Chen, 1984)] to yield an equivalent system

$$\overline{\boldsymbol{\pi}}(t) = \overline{\mathbf{A}}\overline{\boldsymbol{\pi}}(t-1) + \delta(t-1)\overline{\mathbf{u}}(t-1) \text{ and } \mathbf{p}(t) = \overline{\boldsymbol{B}}\overline{\boldsymbol{\pi}}(t), \qquad (12.44)$$

where $\overline{\mathbf{A}} = \mathbf{P}\mathbf{A}\mathbf{P}^{-1}$ with $\mathbf{P}$ the matrix of normalized eigenvectors of $\mathbf{A}$, $\overline{\boldsymbol{\pi}}(t) = \mathbf{P}\boldsymbol{\pi}(t)$, $\overline{\mathbf{u}}(t) = \mathbf{P}\mathbf{u}(t)$, and $\overline{\mathbf{B}} = \mathbf{B}\mathbf{P}^{-1}$. Note that the vector $\mathbf{p}(t)$ is not changed by this transformation, so that the transformed system can be used to compute $\mathbf{p}(t)$ at each step, yielding in turn, $P(\underline{y}(t) = y(t)|\mathcal{M})$. The product of these values is the desired likelihood. Note that if $y(t) = k$, then only the kth element of $\mathbf{p}(t)$ need be computed at t. Since $\overline{\mathbf{A}}$ is diagonal, the number of operations necessary to compute the likelihood is $\mathcal{O}(ST)$. In certain cases, it is possible to reduce the average search cost per model to $\mathcal{O}([1-\kappa]NT)$, where $0 \leq \kappa < 1$ by combining essentially redundant computation (Deller and Snider, 1990).

In addition to the computational advantage of the state-space formulation, a numerical advantage is gained by restructuring the computations into this form.

We noted above that the F-B method is subject to serious numerical problems unless a rather elaborate scaling procedure is applied. This problem is avoided here because the accumulation of small numbers by multiplication is concentrated into a single product of $P(y(t)|\mathcal{M})$'s which can be effected by adding negative logarithms. Clearly, $-\log P(y|\mathcal{M})$ serves equally well as a likelihood measure. Note that this likelihood measure becomes more favorable as it becomes smaller.

"Best Path" Method (Viterbi Approach). In the "any path" method taken above, the likelihood measure assigned to an HMM is based on the probability that the model generated the observation sequence using any sequence of states of length T. An alternative likelihood measure, which is slightly less expensive to compute, is based on the probability that the HMM could generate the given observation sequence using the *best* possible sequence of states. Formally, we seek the number $P(y, \mathcal{I}^*|\mathcal{M})$, where

$$\mathcal{I}^* \stackrel{\text{def}}{=} \underset{\mathcal{I}}{\operatorname{argmax}}\, P(y, \mathcal{I}|\mathcal{M}) \tag{12.48}$$

in which $\mathcal{I}$ represents any state sequence of length T. This time we must use the observations in such a way as to compute the probability of their occurrence along the best path through the states. We therefore have the added complication of finding $\mathcal{I}^*$ concurrently with computing the likelihood.

As the reader might have surmised, this problem can be reduced to a sequential optimization problem that is amenable to DP. Consider forming a grid in which the observations are laid out along the abscissa, and the states along the ordinate, as illustrated in Fig. 12.4. (This is nothing more than the "lattice" diagram studied in Fig. 12.2 viewed in a different way.) Each point in the grid is indexed by a time, state pair (t, i). In searching this grid, we impose two simple restrictions:

1. Sequential grid points along any path must be of the form (t, i), $(t+1, j)$, where $1 \le i, j \le S$. This says that every path must advance in time by one, and only one, time step for each path segment.
2. Final grid points on any path must be of the form (T, i_f), where i_f is a legal final state in the model.

The reader is encouraged to ponder these restrictions to determine their reasonableness in relation to the model search.

Suppose that we assign a Type N cost to any node in the grid as follows (the reason for the primes on the following distance quantities will become clear below):

$$d'_N(t, i) = b(y(t)|i) \stackrel{\text{def}}{=} P(\underline{y} = y(t)|\underline{x}(t) = i). \tag{12.49}$$

Further, let us assign a Type T cost to any transition in the grid as

$$d'_T[(t, i)|(t-1, j)] = a(i|j) \stackrel{\text{def}}{=} P(\underline{x}(t) = i|\underline{x}(t-1) = j), \tag{12.50}$$

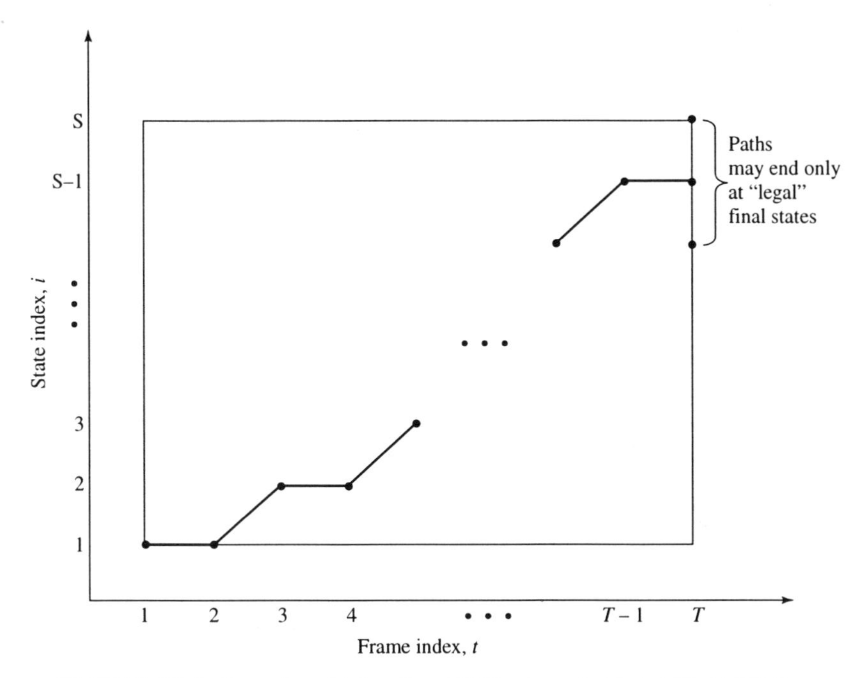

FIGURE 12.4. Search grid for the HMM viewed as a DP problem.

for any i and j and for arbitrary $t > 1$. Also, to account for initial state probabilities, we can allow all paths to originate at a fictitious and costless (Type N) node $(0, 0)$, which makes a transition of cost (Type T) $P(\underline{x}(1) = i)$ to any initial node of the form $(1, i)$. Upon arriving at the initial node, the path will also incur a Type N node cost of the form $b(y(1)|i)$. The accumulated (Type B) cost associated with any transition, say $(t-1, j)$ to (t, i), is therefore

$$\begin{aligned} d'[(t,i)|(t-1,j)] &= d'_T[(t,i)|(t-1,j)]d'_N(t,i) \\ &= a(i|j)b(y(t)|i) \end{aligned} \tag{12.51}$$

for $t > 1$, and

$$\begin{aligned} d'[(1,i)|(0,0)] &= d'_T[(1,i)|(0,0)]d'_N(1,i) \\ &= P(\underline{x}(1)=i)b(y(1)|i) \end{aligned} \tag{12.52}$$

for $t = 1$.

Now let us consider a complete path through the grid of the form

$$(0, i_0),\ (1, i_1),\ (2, i_2), \ldots, (T, i_T) \tag{12.53}$$

with

$$i_0 \stackrel{\text{def}}{=} 0. \tag{12.54}$$

Let the total cost associated with this path be given by the product[10] of the stepwise Type B costs (including the initial one),

$$\begin{aligned} D' &= \prod_{t=1}^{T} d'\left[(t, i_t) \middle| (t-1, i_{t-1})\right] \\ &= \prod_{t=1}^{T} a(i_t | i_{t-1}) b(y(t) | i_t), \end{aligned} \tag{12.55}$$

where, for convenience, we have defined

$$a(i_1 | i_0) = a(i_1 | 0) \stackrel{\text{def}}{=} P(\underline{x}(1) = i_1). \tag{12.56}$$

Then it is clear that D', the "cost" of the path (12.53), is equivalent to its probability of occurrence jointly with the observation sequence y. Formally, since knowing the state sequence $\mathcal{J} = i_1, i_2, \ldots, i_T$ is equivalent to knowing the path (12.53), we can write

$$D' = P(y, \mathcal{J} | \mathcal{M}). \tag{12.57}$$

Therefore, the best path (state sequence), $\mathcal{J}^*$, will be the one of *maximum* cost,

$$[D']^* = P(y, \mathcal{J}^* | \mathcal{M}). \tag{12.58}$$

We have once again reduced our problem to a path search problem that is solvable by DP. The reader will no doubt recall that all of the path searches we have encountered in our work thus far have required solution for the *minimum*-cost path. Here we want the maximum-cost (probability) path. This presents no problem, for the BOP is equally valid when optimization implies maximizing cost. The reader should be able to return to (11.16) and rewrite the ensuing discussion for the case in which maximization is required and return with a ready-made solution for the present problem. However, we circumvent the need to do so by making a small modification to the procedure, which is often used in practice.

In (12.51), (12.52), and (12.55), and at every step along the way to the DP solution suggested above, products of probabilities are formed. These products often become very small, causing numerical problems (often fatal) to arise in the solution. Further, the formation of products is usually more computationally expensive than sums. Therefore, we use the simple trick of taking negative logarithms,[11] which we have used before to turn the expensive, numerically problematic multiplications of proba-

[10]This accumulation of Type B distances by multiplication represents a variation on our customary procedure of adding them [see (11.9)], but multiplication is clearly more appropriate here. Further, we will turn this into a sum momentarily by taking logarithms.

[11]We will also see another justification for the logarithm in studying continuous observation HMMs.

bilities into sums of numerically stable numbers. Note that at the same time the problem is turned into a minimal-cost path search.

Take the negative logarithm of each side of (12.55) to obtain

$$D = \sum_{t=1}^{T} d\big[(t, i_t)\big|(t-1, i_{t-1})\big], \tag{12.59}$$

where

$$D \stackrel{\text{def}}{=} -\log D' \tag{12.60}$$

and

$$\begin{aligned} d\big[(t, i_t)\big|(t-1, i_{t-1})\big] &= -\log d'\big[(t, i_t)\big|(t-1, i_{t-1})\big] \\ &= \big[-\log a(i_t|i_{t-1})\big] + \big[-\log b(y(t)|i_t)\big]. \end{aligned} \tag{12.61}$$

This reveals that we can do the path search by the following assignments: Let the Type N cost at a node (t, i_t) be

$$d_N(t, i_t) = -\log b(y(t)|i_t) = -\log P(\underline{y}(t) = y(t)|\underline{x}(t) = i_t) \tag{12.62}$$

and the Type T cost to any transition in the grid as

$$\begin{aligned} d_T\big[(t, i_t)\big|(t-1, i_{t-1})\big] &= -\log a(i_t|i_{t-1}) \\ &= -\log P(\underline{x}(t) = i_t|\underline{x}(t-1) = i_{t-1}), \end{aligned} \tag{12.63}$$

for any i_t and i_{t-1} and for arbitrary $t > 1$. As above, to account for initial state probabilities, we can allow all paths to originate at a fictitious and costless (Type N) node $(0, 0)$, which makes a transition of cost (Type T) $-\log P(\underline{x}(1) = i_1)$ to any initial node of the form $(1, i_1)$. Upon arriving at the initial node, the path will also incur a Type N node cost of the form $-\log b(y(1)|i_1)$. Further, if we let accumulated (Type B) cost associated with any transition, say $(t-1, i_{t-1})$ to (t, i_t), be obtained by *adding* the Type T and N costs, then we obtain precisely (12.61). Finally, if we also accumulate Type B costs by addition as we move along a path, then we obtain exactly (12.59). Clearly, seeking the path with *minimum* D under this setup is equivalent to seeking the path of *maximum* D', and we have reduced our problem to a shortest-path search with this "log cost" conversion.

Given our vast experience with DP algorithms, developing the steps for this shortest-path problem is quite simple. We first let

$$D_{\min}(t, i_t) \stackrel{\text{def}}{=} \text{distance from } (0,0) \text{ to } (t, i_t) \text{ over the best path.} \tag{12.64}$$

Noting that, for $t > 1$, the only legal predecessor nodes to (t, i_t) are of the form $(t-1, i_{t-1})$, we can write (11.19) as

$$D_{\min}(t, i_t) = \min_{(t-1, i_{t-1})} \big\{D_{\min}(t-1, i_{t-1}) + d\big[(t, i_t)\big|(t-1, i_{t-1})\big]\big\} \tag{12.65}$$

for $t > 1$. Note that, because all predecessor nodes to (t, i_t) must come from "time slot" $t-1$, the right side is really only a minimization over the previous states, and we can write

$$D_{\min}(t, i_t) = \min_{i_{t-1}} \{D_{\min}(t-1, i_{t-1}) + d[(t, i_t)|(t-1, i_{t-1})]\} \tag{12.66}$$

for $t > 1$. Writing the local distance quantity on the right in terms of its model parameters, we have

$$\begin{aligned} D_{\min}(t, i_t) = \min_{i_{t-1}} \{D_{\min}(t-1, i_{t-1}) + [-\log a(i_t|i_{t-1})] \\ + [-\log b(y(t)|i_t)]\} \end{aligned} \tag{12.67}$$

for $t > 1$. This recursion can also be used for $t = 1$ if we recall that $D_{\min}(0, 0) = 0$ [see (11.20)], i_0 means 0, and $a(i_1|0)$ is defined as the initial state probability for state i_1 as in (12.56).

At the end of the search the quantity

$$D^* = \min_{\text{legal } i_T} \{D_{\min}(T, i_T)\}. \tag{12.68}$$

will be the negative logarithm of the probability of joint probability of occurrence of the observation sequence y and the best state sequence, say $\mathcal{I}^* = i_1^*, i_2^*, \ldots, i_T^*$, for producing it. The minimization over "legal i_T" implies that more than one final state might be possible. Clearly, the best of these final states is

$$i_T^* = \underset{\text{legal } i_T}{\operatorname{argmin}} \{D_{\min}(T, i_T)\}. \tag{12.69}$$

We have then that

$$D^* = -\log P(y, \mathcal{I}^* | \mathcal{M}). \tag{12.70}$$

For comparisons across models, this quantity is just as useful as the probability itself, and there is usually no reason to convert it. It should be kept carefully in mind, however, that small logs are favorable because they imply large probabilities. As noted in Section 1.3.3, a measure derived from a probability and used as a measure of likelihood of a certain model or occurrence is often called simply a *likelihood*. We will use this term to refer to the negative log probability in our discussions.

In certain situations, it might be desirable to recover the state sequence associated with the best path through the HMM. This is equivalent to recovering the best path through the grid in the search problem. We are well acquainted with the backtracking procedure to accomplish this task. Let us define $\Psi(t, i_t)$ to be the best last state on the optimal partial path ending at (t, i_t). Then we have

$$\begin{aligned} \Psi(t, i_t) &= \underset{i_{t-1}}{\operatorname{argmin}} \{D_{\min}(t-1, i_{t-1}) + [-\log a(i_t|i_{t-1})] \\ &\quad + [-\log b(y(t)|i_t)]\} \\ &= \underset{i_{t-1}}{\operatorname{argmin}} \{D_{\min}(t-1, i_{t-1}) + [-\log a(i_t|i_{t-1})]\}. \end{aligned} \tag{12.71}$$

We will incorporate this backtracking procedure into the algorithm to follow without further discussion.

We have now developed the essential steps, based on DP principles, for computing the likelihood for a particular HMM in light of the observation sequence $y = y_1^T$. We summarize the steps in a formal algorithm shown in Fig. 12.5. This technique is often called the *Viterbi algorithm*, since it was first suggested by A. J. Viterbi in the context of decoding random sequences (Viterbi, 1967; Viterbi and Omura, 1979; Forney, 1973). The algorithm is also sometimes called the "stochastic form of DP," an allusion to the fact that the Type N costs involved are stochastic quantities.

In conclusion, let us compare the two general approaches to decoding the HMM. The comparison is simple upon recognizing the following. Suppose that we had defined $\alpha(\cdot\,,\cdot)$ in the "any path" approach such that

$$\alpha(y_1^{t+1}, j) = \max_i \alpha(y_1^t, i)a(j|i)b(y(t+1)|j). \tag{12.72}$$

This corresponds to replacing the sum in (12.34) by a maximization and to taking the maximum over all paths entering a node in the lattice of Fig. 12.2. A little thought will convince the reader that this is precisely the distance to grid point $(t+1, i)$ in the Viterbi approach (if we do not use the logarithms). Therefore, the methods are similar but the Viterbi approach requires $(S-1)T$ fewer additions per HMM. Typically, $\frac{1}{10}$ to $\frac{1}{4}$ fewer computations are required with the Viterbi search (Picone, 1992).

Either the Viterbi or F–B algorithm can generate likelihoods for paths that never appeared in the training data. This feature represents an effective "smoothing" of the probability distributions associated with the HMM.

FIGURE 12.5. Computation of $P(y, \mathcal{I}^* | \mathcal{M})$ using the Viterbi algorithm.

Initialization: "Origin" of all paths is node (0, 0).
For $i = 1, 2, \ldots, S$
$D_{\min}(1, i) = a(i|0)b(y(1)|i) = P(\underline{x}(1) = i)b(y(1)|i)$
$\Psi(1, i) = 0$
Next i

Recursion: For $t = 2, 3, \ldots, T$
For $i_t = 1, 2, \ldots, S$
Compute $D_{\min}(t, i_t)$ according to (12.67).
Record $\Psi(t, i_t)$ according to (12.71).
Next i_t
Next t

Termination: Distance of optimal path $(0,0) \xrightarrow{*} (T, i_T^*)$ is given in (12.68).
Best state sequence, $\mathcal{I}^*$, is found as follows:
Find i_T^* as in (12.69).
For $t = T-1, T-2, \ldots, 0$
$i_t^* = \Psi(t, i_{t+1}^*)$
Next t

Self-Normalization. While we are viewing the HMM in a DP framework, a very important distinction should be made between the HMM and the other DP-based technique we have discussed, DTW. Recall that in IWR, for example, when two reference strings of lengths J_1 and J_2 were to be compared with a given test string, exactly J_1 and J_2 distance computations needed to be made per path, respectively. In order to compare the costs, some form of normalization had to be imposed in "fairness" to the longer path. We discovered that time normalization problems in DTW analysis are formidable, and less than ideal *ad hoc* solutions frequently need to be employed.

One of the most useful properties of the HMM is the inherent normalization that occurs. The key to this property can be seen in Fig. 12.4. Note that the ordinate of the search grid is labeled by *state indices* rather than by observations. The number of states in a model is ordinarily much smaller than the length of an observation string to be recognized, T.[12] This means that for every HMM against which we wish to evaluate the test string, the search path will be of exactly length T. Hence, across models, there is no problem of comparing likelihoods resulting from different path lengths. We can view this "self-normalizing" property as a consequence of the doubly stochastic structure of the HMM that, in effect, allows it to always "generate" reference strings of exactly the same length as the test sequence.

Note the implication of this discovery for "higher-level" processing using HMMs. Suppose we were to concatenate several word HMMs together, say $\mathcal{M}_1$, $\mathcal{M}_2$, and $\mathcal{M}_3$, in a similar manner to the building of levels with DTW grids (LB algorithm). Then at any fixed time t in the test sequence, we could compare a path of exactly length t ending at $\mathcal{M}_2$ with one of exactly length t ending at $\mathcal{M}_3$. In language that will be useful to us later, we might say that "hypotheses" (here, hypothesized word strings) of different lengths (two or three words) can be easily and meaningfully compared. This feature of the HMM is profoundly significant, as we shall see in future studies.

Beam Search. In the discussion of DTW grid, we introduced the notion of a beam search, which is used to prune unlikely paths and reduce computation.[13] A similar procedure is employed in the HMM search. Once again we will find that most of the paths will be pruned because they will be so unlikely that their extensions are not warranted. A simple "beam" that is analogous to the example given for DTW search consists of all paths at time t whose costs fall within, say, $\delta(t)$ of the best hypothesis. If, for example, (t, i_t^*) is the grid point with the best path at t, then the path to any (t, i_t) will only be a candidate for extension at time $t+1$ if

[12]This discussion will need to be generalized somewhat when we begin to use HMMs in more sophisticated ways, but the extension of the main ideas will be apparent.

[13]In fact, the term "beam search" was first employed in a system using an HMM-like structure (Lowerre and Reddy, 1980).

$$D_{\min}(t, i_t) \leq D_{\min}(t, i_t^*) + \delta(t). \tag{12.73}$$

It should be noted that beam search can be applied to the F–B search as well as the Viterbi search. It can also be applied to more general forms of HMMs that do not have discrete observations.

Training the Discrete Observation HMM

F–B (Baum–Welch) Reestimation. Next we address the question of training a particular HMM to correctly represent its designated word or other utterance. We assume that we have one or more feature strings of the form $y = y_1^T = \{y(1), \ldots, y(T)\}$ extracted from training utterances of the word (note that this implies that we already have the codebook that will be used to deduce "symbols"), and the problem is to use these strings to find an appropriate model of form (12.21). In particular, we must find the matrices **A** and **B**, and the initial state probability vector, $\boldsymbol{\pi}(1)$. There is no known way to analytically compute these quantities from the observations in any optimal sense. However, an extension to the F–B algorithm (as we know it so far) provides an iterative estimation procedure for computing a model, $\mathcal{M}$, corresponding to a local maximum of the likelihood $P(y|\mathcal{M})$. It is this method that is most widely used to estimate the HMM parameters. Gradient search techniques have also been applied to this problem (Levinson et al., 1983).

The full F–B algorithm as we now present it was developed in the series of papers by Baum and colleagues, which were noted in Section 12.1 (Baum et al., 1966, 1967, 1968, 1970, 1972). The underlying principles described in these papers are beyond the scope of this book. Our objectives here is to clearly describe the algorithm; we leave its original development to the reader's further pursuit of the literature.

To formally describe the F–B procedure, we need to recall some previous notation defined in conjunction with the Mealy form[14] of the HMM [near (12.6)]. We denote by $\underline{u}$ the random process with random variables $\underline{u}(t)$ that model the transitions at time t. Also recall the notation

$$u_{j|i} \stackrel{\text{def}}{=} \text{label for a transition from state } i \text{ to state } j, \tag{12.74}$$

which we extend to include

$$u_{\bullet|i} \stackrel{\text{def}}{=} \text{set of transitions exiting state } i \tag{12.75}$$

$$u_{j|\bullet} \stackrel{\text{def}}{=} \text{set of transitions entering state } j. \tag{12.76}$$

[14]Although we are "borrowing" this notation from the Mealy form, it should be understood that we are seeking parameters for a Moore-form model. It should also be intuitive that a similar procedure could be derived for the Mealy form [see, e.g., (Bahl et al., 1983)].

Let us also define a new set of random processes $\underline{y}_j$, $1 \le j \le S$, which have random variables $\underline{y}_j(t)$ that model the observation being emitted at state j at time t (this may include the "null observation"). The symbol •, which is, in effect, used to indicate an arbitrary *state* above, will also be used to indicate an arbitrary *time* in the following.

Now suppose that we have a model, $\mathcal{M}$, and an observation sequence, $y = y_1^T$. If we are just beginning the process, $\mathcal{M}$ may consist of any legitimate probabilities. We first compute the following numbers:

$$
\begin{aligned}
&\xi(i,j;t) \\
&\quad \stackrel{\text{def}}{=} P\big(\underline{u}(t) = u_{j|i} \big| y, \mathcal{M}\big) \\
&\quad = P\big(\underline{u}(t) = u_{j|i}, y \big| \mathcal{M}\big) / P(y|\mathcal{M}) \\
&\quad = \begin{cases} \dfrac{\alpha(y_1^t, i)a(j|i)b(y(t+1)|j)\beta(y_{t+2}^T|j)}{P(y|\mathcal{M})}, & t = 1, \ldots, T-1 \\ 0, & \text{other } t \end{cases},
\end{aligned}
\tag{12.77}
$$

where the sequences a, b, α, and β are defined in (12.4), (12.14), (12.31), and (12.32), respectively;[15]

$$
\begin{aligned}
\gamma(i;t) &\stackrel{\text{def}}{=} P\big(\underline{u}(t) \in u_{\bullet|i} \big| y, \mathcal{M}\big) \\
&= \sum_{j=1}^{S} \xi(i,j;t) \\
&= \begin{cases} \dfrac{\alpha(y_1^t, i)\beta(y_{t+1}^T|i)}{P(y|\mathcal{M})}, & t = 1, 2, \ldots, T-1 \\ 0, & \text{other } t \end{cases},
\end{aligned}
\tag{12.78}
$$

$$
\begin{aligned}
\nu(j;t) &\stackrel{\text{def}}{=} P\big(\underline{x}(t) = j \big| y, \mathcal{M}\big) \\
&= \begin{cases} \gamma(j;t), & t = 1, 2, \ldots, T-1 \\ \alpha(y_1^T, j), & t = T \\ 0, & \text{other } t \end{cases} \\
&= \begin{cases} \dfrac{\alpha(y_1^t, j)\beta(y_{t+1}^T|j)}{P(y|\mathcal{M})}, & t = 1, 2, \ldots, T \\ 0, & \text{other } t \end{cases},
\end{aligned}
\tag{12.79}
$$

[15]The notation $\underline{u}(t) \in u_{\bullet|i}$ is used below to mean that the random variable $\underline{u}(t)$ takes a value from among those in the set $u_{\bullet|i}$.

and

$$\delta(j,k;t) \stackrel{\text{def}}{=} P(\underline{y}_j(t)=k|y,\mathcal{M})$$
$$= \begin{cases} \nu(j;t), & \text{if } y(t)=k \text{ and } 1\le t\le T \\ 0, & \text{otherwise} \end{cases}$$
$$= \begin{cases} \dfrac{\alpha(y_1^t,j)\beta(y_{t+1}^T|j)}{P(y|\mathcal{M})}, & y(t)=k \text{ and } 1\le t\le T \\ 0, & \text{otherwise} \end{cases} \tag{12.80}$$

Note that we make extensive use of both the forward and backward probability sequences, $\alpha(\cdot\,,\cdot)$ and $\beta(\cdot|\cdot)$, in these computations. Now from these four quantities we compute four related key results:

$$\xi(i,j;\bullet) = P(\underline{u}(\bullet)=u_{j|i}|y,\mathcal{M}) = \sum_{t=1}^{T-1}\xi(i,j;t), \tag{12.81}$$

$$\gamma(i;\bullet) = P(\underline{u}(\bullet)\in u_{\bullet|i}|y,\mathcal{M}) = \sum_{t=1}^{T-1}\gamma(i;t), \tag{12.82}$$

$$\nu(j;\bullet) = P(\underline{u}(\bullet)\in u_{j|\bullet}|y,\mathcal{M}) = \sum_{t=1}^{T}\nu(j;t), \tag{12.83}$$

and

$$\delta(j,k;\bullet) = P(\underline{y}_j(\bullet)=k|y,\mathcal{M}) = \sum_{t=1}^{T}\delta(j,k;t) = \sum_{\substack{t=1\\ y(t)=k}}^{T}\nu(j;t). \tag{12.84}$$

We now give important and intuitive interpretations to the four key quantities (12.81)–(12.84). Let us define the random variables:

$\underline{n}(u_{j|i}) \stackrel{\text{def}}{=}$ number of transitions $u_{j|i}$ for an arbitrary observation sequence of length T and an arbitrary model, (12.85)

$\underline{n}(u_{\bullet|i}) \stackrel{\text{def}}{=}$ number of transitions from the set $u_{\bullet|i}$ for an arbitrary observation sequence of length T and an arbitrary model, (12.86)

$\underline{n}(u_{j|\bullet}) \stackrel{\text{def}}{=}$ number of transitions from the set $u_{j|\bullet}$ for an arbitrary observation sequence of length T and an arbitrary model, (12.87)

$\underline{n}(\underline{y}_j(\bullet)=k) \stackrel{\text{def}}{=}$ number of times observation k and state j occur jointly for an arbitrary observation sequence of length T and an arbitrary model. (12.88)

Now it is easy to show that

$$\xi(i,j;\bullet) = \mathcal{E}\{\underline{n}(u_{j|i})|y, \mathcal{M}\} \tag{12.89}$$

$$\gamma(i;\bullet) = \mathcal{E}\{\underline{n}(u_{\bullet|i})|y, \mathcal{M}\} \tag{12.90}$$

$$\nu(j;\bullet) = \mathcal{E}\{\underline{n}(u_{j|\bullet})|y, \mathcal{M}\} \tag{12.91}$$

$$\delta(j,k;\bullet) = \mathcal{E}\{\underline{n}(\underline{y}_j(t)=k)|y, \mathcal{M}\}. \tag{12.92}$$

For example, since there is either one transition $u_{j|i}$ at any time, or there is none, we have

$$\mathcal{E}\{\underline{n}(u_{j|i})|y, \mathcal{M}\} = [1 \times P(u_{j|i}|y, \mathcal{M})] + [0 \times P(\text{not } u_{j|i}|y, \mathcal{M})] = \xi(i,j;\bullet). \tag{12.93}$$

The interpretations of the other three quantities are verified in a similar manner.

With these interpretations, it is easy to see that the following are reasonable estimates of the model parameters:

$$\bar{a}(j|i) = \frac{\mathcal{E}\{\underline{n}(u_{j|i})|y, \mathcal{M}\}}{\mathcal{E}\{\underline{n}(u_{\bullet|i})|y, \mathcal{M}\}} = \frac{\xi(i,j;\bullet)}{\gamma(i;\bullet)}, \tag{12.94}$$

$$\bar{b}(k|j) = \frac{\mathcal{E}\{\underline{n}(y_j(\bullet)=k)|y, \mathcal{M}\}}{\mathcal{E}\{\underline{n}(u_{j|\bullet})|y, \mathcal{M}\}} = \frac{\delta(j,k;\bullet)}{\nu(j;\bullet)}, \tag{12.95}$$

$$P(\underline{x}(1)=i) = \gamma(i;1). \tag{12.96}$$

Now retrieving results from above, we have

$$\bar{a}(j|i) = \frac{\sum_{t=1}^{T-1} \alpha(y_1^t, i) a(j|i) b(y(t+1)|j) \beta(y_{t+2}^T|j)}{\sum_{t=1}^{T-1} \alpha(y_1^t, i) \beta(y_{t+1}^T|i)}, \tag{12.97}$$

$$\bar{b}(k|j) = \frac{\sum_{\substack{t=1 \\ y(t)=k}}^{T} \alpha(y_1^t, j) \beta(y_{t+1}^T|j)}{\sum_{t=1}^{T} \alpha(y_1^t, j) \beta(y_{t+1}^T|j)}, \tag{12.98}$$

$$P(\underline{x}(1)=i) = \frac{\alpha(y_1^1, i) \beta(y_2^T|i)}{P(y|\mathcal{M})}. \tag{12.99}$$

Having computed (12.97)–(12.99) for all i, j, k, we now have the parameters of a new model, say $\overline{\mathcal{M}}$.

The procedure that we have just been through will probably seem a bit strange and unconvincing after pondering it for a moment. Indeed, what we have done is taken a model, $\mathcal{M}$, and used it in conjunction with the training observation, y, to compute quantities with which to produce a new model, $\overline{\mathcal{M}}$. Recall that our ideal objective is to use the training string, y, in order to find the model, say $\mathcal{M}^*$, such that

$$\mathcal{M}^* = \operatorname*{argmax}_{\mathcal{M}} P(y|\mathcal{M}). \tag{12.100}$$

Now for a given training sequence y, $P(y|\mathcal{M})$ is generally a nonlinear function of the many parameters that make up the model $\mathcal{M}$. This function will accordingly have many *local maxima* in the multidimensional space. This idea is portrayed in two dimensions in Fig. 12.6. The optimal model, $\mathcal{M}^*$, corresponds to the *global maximum* of the criterion function. The significance of what we have done in the foregoing is that repeated reestimation of the model according to these steps is guaranteed to converge to an $\mathcal{M}$ corresponding to a local maximum of $P(y|\mathcal{M})$ (Baum and Sell, 1968). That is, either $\overline{\mathcal{M}} = \mathcal{M}$, or $P(y|\overline{\mathcal{M}}) > P(y|\mathcal{M})$. The model will always improve under the reestimation procedure unless its parameters already represent a local maximum. So this reestimation does not necessarily produce the best possible model, $\mathcal{M}^*$. Accordingly, it is common practice to run the algorithm several times with different sets of initial parameters and to take as the trained model the $\mathcal{M}$ that yields the largest value of $P(y|\mathcal{M})$.

The F–B reestimation algorithm is summarized in Fig. 12.7.

In Section 12.2.2 we noted the numerical problems inherent in the use of this F–B algorithm. We remind the reader that a scaling procedure is

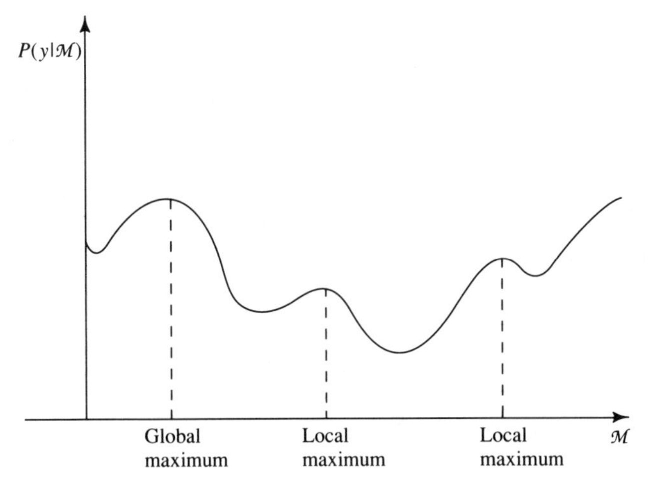

FIGURE 12.6. Conceptualization of the HMM likelihood as a function of model parameters.

FIGURE 12.7. F–B (Baum–Welch) reestimation algorithm.

Initialization: Begin with an arbitrary model $\mathcal{M}$.

Recursion:

1. Use $\mathcal{M} = \{S, \mathbf{A}, \mathbf{B}, \boldsymbol{\pi}(1), \{y_k, 1 \le k \le K\}\}$ and $y = y_1^T$ to compute (12.81)–(12.84).
2. Reestimate the model (call the new model $\overline{\mathcal{M}}$) using (12.94)–(12.96).
3. It will be true that

$$P(y|\overline{\mathcal{M}}) \geq P(y|\mathcal{M}).$$

If

$$P(y|\overline{\mathcal{M}}) - P(y|\mathcal{M}) \geq \varepsilon,$$

return to Step 1 with $\mathcal{M} = \overline{\mathcal{M}}$. Otherwise STOP.

4. Repeat the above steps with several initial models to find a favorable local maximum of $P(y|\mathcal{M})$.

generally necessary for practical implementation of the procedure. This scaling technique will be described as part of our exploration of practical issues in Section 12.2.5.

Finally, we note that the F–B procedure, as well as the Viterbi procedure to follow, are sometimes called *reestimation by recognition* for the obvious reason.

Viterbi Reestimation. Although the F–B algorithm is the most popular algorithm for training the discrete observation HMM, a simpler and equally effective algorithm is available based on the Viterbi decoding approach to recognition (Picone, 1990; Fu, 1982). Suppose, as above, we have a given model, $\mathcal{M}$, and a training sequence $y = y_1^T$ with which to reestimate the model parameters. It is assumed that a particular state is designated as the initial state for the model so that the initial state probabilities need not be estimated. We first evaluate the likelihood, $P(y|\mathcal{M})$, using Viterbi decoding as described in Section 12.2.2. Along the way, we keep track of the following tallies:

$$n(u_{j|i}) \stackrel{\text{def}}{=} \text{number of transitions } u_{j|i} \tag{12.101}$$

$$n(u_{\bullet|i}) \stackrel{\text{def}}{=} \text{number of transitions from the set } u_{\bullet|i} \tag{12.102}$$

$$n(u_{j|\bullet}) \stackrel{\text{def}}{=} \text{number of transitions from the set } u_{j|\bullet} \tag{12.103}$$

$$n(\underline{y}_j(t) = k) \stackrel{\text{def}}{=} \text{number of times observation } k \text{ and state } j \text{ occur jointly.} \tag{12.104}$$

The reestimation equations are

$$\bar{a}(j|i) = \frac{n(u_{j|i})}{n(u_{\bullet|i})}, \tag{12.105}$$

$$\bar{b}(k|j) = \frac{n(\underline{y}_j(t) = k)}{n(u_{j|\bullet})}. \tag{12.106}$$

Having computed (12.105)–(12.106), we now have the parameters of a new model, say $\overline{\mathcal{M}}$. The same, or additional, training strings can be used in further iterations.

The Viterbi reestimation algorithm can be shown to converge to a proper characterization of the underlying observations (Fu, 1982, Ch. 6; Lee and Fu, 1972), and has been found to yield models of comparable performance to those trained by F–B reestimation (Picone, 1990). Further, whereas an HMM corresponds to a particular type of formal grammar (a "regular" or "finite state" grammar), this Viterbi-type training procedure can be used for a broader class of grammars. This issue will be discussed in Chapter 13. Finally, the Viterbi approach is more computationally efficient than the F–B procedure.

12.2.3 The Continuous Observation HMM

Introduction

We now return to the more general case in which the observations are continuous and vector-valued, corresponding to the "unquantized" vectors of features drawn from the speech. In this case the formal description of the HMM contains a multivariate pdf characterizing the distribution of observations within each state. This is given in (12.12), which we repeat here for convenience,

$$\mathcal{M} = \{S, \boldsymbol{\pi}(1), \mathbf{A}, \{f_{\underline{\mathbf{y}}|\underline{x}}(\xi|i), 1 \le i \le S\}\}. \tag{12.107}$$

Recall that we write $f_{\underline{\mathbf{y}}|\underline{x}}(\xi|i)$ rather than $f_{\underline{\mathbf{y}}(t)|\underline{x}(t)}(\xi|i)$ because the process $\underline{\mathbf{y}}$ is assumed to have independent and identically distributed random variables, $\underline{\mathbf{y}}(t)$.

Recognition

We once again need to concern ourselves with the two central problems of training and recognition. The latter is very simple, and the former is solved by variations on the reestimation procedures described above. In the recognition problem, for any incoming observation, say $\mathbf{y}(t)$, we define the *likelihood* of generating observation $\mathbf{y}(t)$ in state j as[16]

$$b(\mathbf{y}(t)|\,j) \stackrel{\text{def}}{=} f_{\underline{\mathbf{y}}|\underline{x}}(\mathbf{y}(t)|\,j). \tag{12.108}$$

With this definition, we can simply resort to the use of any of the recognition methods described above. The resulting measure computed for a model, $\mathcal{M}$, and an observation string, say[17] $\mathbf{y} = \mathbf{y}_1^T$, will be $P(\mathbf{y}|\mathcal{M})$ if the "any path" approach is used, and $P(\mathbf{y}, \mathcal{I}^*|\mathcal{M})$ in the Viterbi case. These measures will no longer be proper probabilities, but it should be clear that they provide meaningful likelihood measures with which appropriate relative comparisons across models can be made. Note, in particular, that taking the negative logarithm of $b(\mathbf{y}(t)|\,j)$ will produce a maximum likelihood (or, almost equivalently, Mahalanobis) distance of the observation $\mathbf{y}(t)$ from the mean of the random vector $\underline{\mathbf{y}}$ if the conditional pdf is assumed to be Gaussian (see Section 1.3.3).

Training

F-B Procedure. To solve the training problem, reestimation formulas have been worked out for a rather broad class of observation pdf's in (Liporace, 1982; Juang, 1985; Juang et al., 1986). The most widely used member of this class is the *Gaussian mixture density*, which is of the form

$$f_{\underline{\mathbf{y}}|\underline{x}}(\xi|i) = \sum_{m=1}^{M} c_{im}\,\mathcal{N}(\xi;\boldsymbol{\mu}_{im}, \mathbf{C}_{im}) \tag{12.109}$$

in which c_{im} is the *mixture coefficient* for the mth component for state i, and $\mathcal{N}(\cdot)$ denotes a multivariate Gaussian pdf with mean $\boldsymbol{\mu}_{im}$ and covariance matrix $\mathbf{C}_{im}$. In order for $f_{\underline{\mathbf{y}}|\underline{x}}(\xi|i)$ to be a properly normalized pdf, the mixture coefficients must be nonnegative and satisfy the constraint

$$\sum_{m=1}^{M} c_{im} = 1, \qquad 1 \le i \le S. \tag{12.110}$$

For a sufficiently large number of mixture densities, M, (12.109) can be used to arbitrarily accurately approximate any continuous pdf. Note that as a special case, a single Gaussian pdf ($M = 1$) may be used to model the observations at any state.

Reestimation formulas have been derived for the three quantities c_{il}, $\boldsymbol{\mu}_{il}$, and $\mathbf{C}_{il}$ for mixture density l in state i (Liporace, 1982; Juang, 1985; Juang et al., 1986). Suppose that we define

[16]How should this likelihood be defined for the Mealy-form HMM?

[17]The notation $\mathbf{y} = \mathbf{y}_1^T$ and similar notations are just the obvious vector versions of (12.22)–(12.25).

$$\nu(i;t,l) \stackrel{\text{def}}{=} P\big(\underline{x}(t)=i \,|\, \mathbf{y}(t) \text{ produced in accordance with mixture density } l\big)$$

$$= \frac{\alpha(\mathbf{y}_1^t, i)\beta(\mathbf{y}_{t+1}^T | i)}{\sum_{j=1}^{S} \alpha(\mathbf{y}_1^t, j)\beta(\mathbf{y}_{t+1}^T | j)} \times \frac{c_{il}\,\mathcal{N}(\xi; \boldsymbol{\mu}_{il}, \mathbf{C}_{il})}{\sum_{m=1}^{M} c_{im}\,\mathcal{N}(\xi; \boldsymbol{\mu}_{im}, \mathbf{C}_{im})}. \tag{12.111}$$

[It is easy to show that this definition is consistent with $\nu(i;t)$ of (12.79) if $M=1$.] Also let

$$\nu(i;\bullet,l) = \sum_{t=1}^{T} \nu(i;t,l). \tag{12.112}$$

Then the mixture coefficients, mean vectors, and covariance matrices are reestimated as follows:

$$\bar{c}_{il} = \frac{\nu(i;\bullet,l)}{\sum_{m=1}^{M} \nu(i;\bullet,m)}, \tag{12.113}$$

$$\bar{\boldsymbol{\mu}}_{il} = \frac{\sum_{t=1}^{T} \nu(i;t,l)\mathbf{y}(t)}{\nu(i;\bullet,l)}, \tag{12.114}$$

$$\bar{\mathbf{C}}_{il} = \frac{\sum_{t=1}^{T} \nu(i;t,l)[\mathbf{y}(t)-\boldsymbol{\mu}_{il}][\mathbf{y}(t)-\boldsymbol{\mu}_{il}]^T}{\nu(i;\bullet,l)}. \tag{12.115}$$

An heuristic interpretation of (12.113)–(12.115) is as follows. Equation (12.113) represents a ratio of the expected number of times the path is in state i and using the lth mixture component to generate the observation to the number of times the path resides in state i. Equation (12.114) is a weighted time average of the observation vectors, weighted according to the likelihood of their having been produced by mixture density l in state i. The computation of the covariance reestimate is a similarly weighted temporal average.

As in the discrete observation case, the use of these reestimation formulas will ultimately lead to a model, $\mathcal{M}$, which represents a *local* maximum of the likelihood $P(\mathbf{y}|\mathcal{M})$. However, finding a *good* local maximum depends rather critically upon a reasonable initial estimate of the vector and matrix parameters $\boldsymbol{\mu}_{il}$ and $\mathbf{C}_{il}$ for each i and l. The reader should be able to discern that this is intuitively correct as he or she ponders the training process. This means that we must somehow be able to use the training data to derive meaningful initial estimates prior to executing the reestimation steps. Below we describe a convenient procedure for deriving good separation of the mixture components for use with a Viterbi procedure. We will discuss this issue further in Section 13.5.4, where the HMMs will be trained in the context of a continuous-speech recognizer. The same general principles discussed there apply to this simpler case assumed here in which each HMM represents a discrete utterance.

Of course, the observation densities comprise only one part of the model description. We still need to find estimates for the state transition and initial state probabilities in the continuous observation case. However, we need not do any further work on this issue, since these probabilities have exactly the same meaning and structure in the continuous observation case and can be found directly from (12.94) and (12.96). The complete training algorithm will therefore be similar to the F–B algorithm given above with (12.95) replaced by (12.113)–(12.115).

Viterbi Procedure. If a Viterbi approach is used, the mean vectors and covariance matrices for the observation densities are reestimated by simple averaging. This is easily described when there is only one mixture component per state. In this case, for a given $\mathcal{M}$, the recognition experiment is performed on the observation sequence. Each observation vector is then assigned to the state that produced it on the optimal path by examining the backtracking information. If the observation $\mathbf{y}(t)$ is produced by state i, let us write "$\mathbf{y}(t) \sim i$." Suppose that N_i observation vectors are so assigned. Then

$$\bar{\boldsymbol{\mu}} = \frac{1}{N_i} \sum_{\substack{t=1 \\ \mathbf{y}(t) \sim i}}^{T} \mathbf{y}(t) \tag{12.116}$$

$$\bar{\mathbf{C}} = \frac{1}{N_i} \sum_{\substack{t=1 \\ \mathbf{y}(t) \sim i}}^{T} [\mathbf{y}(t) - \boldsymbol{\mu}_i][\mathbf{y}(t) - \boldsymbol{\mu}_i]^T. \tag{12.117}$$

When $M > 1$ mixture components appear in a state, then the observation vectors assigned to that state must be subdivided into M subsets prior to averaging. This can be done by clustering, using, for example, the K-means algorithm (see Section 1.3.5) with $K = M$. If there are N_{il} vectors assigned to the lth mixture in state i, then the mixture coefficient c_{il} is reestimated as

$$c_{il} = \frac{N_{il}}{N_i}. \tag{12.118}$$

Some results showing the effect of the number of mixture components per state in a digit recognition experiment (Rabiner et al., 1989) are shown in Fig. 12.8. Further experiments are reported in (Wilpon et al., 1991).

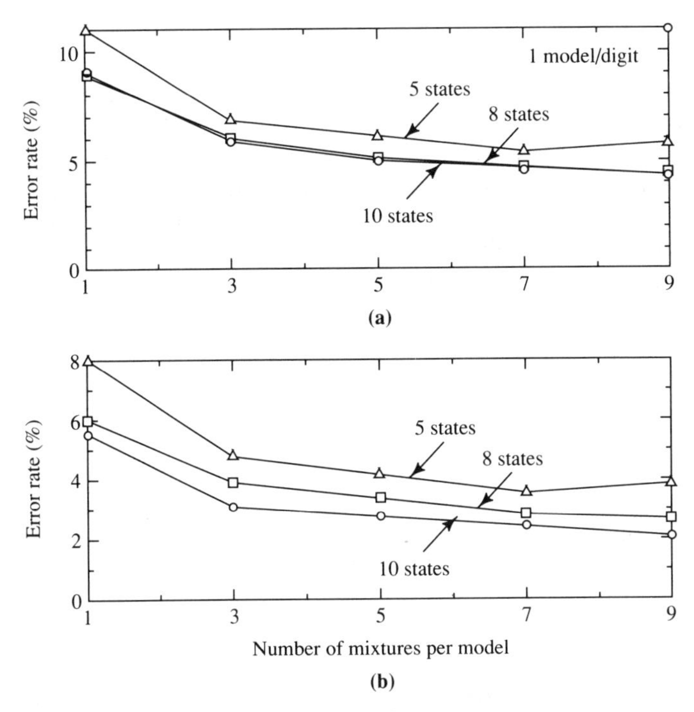

FIGURE 12.8. Error rate versus number of mixture components per state in a speaker-independent digit recognition experiment. Shown also is the effect of various numbers of states in the HMMs. (a) Known length (number of digits uttered) strings. (b) Unknown length strings. After Rabiner et al. (1989).

12.2.4 Inclusion of State Duration Probabilities in the Discrete Observation HMM

One of the benefits of using HMMs is that they obviate a complete *a priori* characterization of the acoustic structure of an utterance to be modeled. If properly "seeded" (see Section 13.5.4), the HMM is capable of "self-organizing" the acoustic data into a meaningful and effective model. Nevertheless, the states of the model are often thought of, to a first approximation, as representing distinct acoustical phenomena in the utterance, such as a vowel sound in a word or a transition between phonemes in a word. The number of states in a model is sometimes chosen to correspond to the expected number of such phenomena. For example, if an HMM is used to model a phoneme (rather than a complete word), then three states might be used—one to capture the transition on either end of the phoneme, and one for the "steady-state" portion. However, the HMM "organizes itself" to maximize an analytic criterion, and not necessarily to correspond to some acoustic structure that the designer may have in mind.

Nevertheless, experimental evidence suggests that states frequently represent identifiable acoustic phenomena. To the extent that this is the case, the conventional HMM as described above has a serious flaw. Acoustic phenomena in speech tend not to be exponentially distributed in duration. One would expect a given phoneme in a given position in a word, for example, to have a normally distributed duration across different renditions of the word. Yet, the durations of states within a conventional HMM have exponential probability distributions. By this we mean the following: Suppose we know that a given HMM, at time t, enters state i. What is the probability that the duration of stay in state i is d frames long? From our knowledge of the transition probabilities, it is easy to show that

$$P\{\underline{x}(t+1)=\underline{x}(t+2)=\cdots=\underline{x}(t+d-1)=i,\\ \underline{x}(t+d)\neq i \,|\, \underline{x}(t)=i, \mathcal{M}\}=[a(i|i)]^{d-1}[1-a(i|i)]. \tag{12.119}$$

Further, because of the stationarity of the model, we know that this result does not depend on t. Therefore, we can simplify (12.119) as follows. Let $\underline{d}_i$ be a random variable modeling the length of stay in state i (from time of entry) for model $\mathcal{M}$. Then,

$$P(\underline{d}_i=d)=[a(i|i)]^{d-1}[1-a(i|i)]. \tag{12.120}$$

For most speech signals such a state duration distribution is inappropriate. One solution to this problem involves the replacement of the self-transition arcs in the model by a duration probability distribution as shown in Fig. 12.9. According to Rabiner (1989) much of the original work on this subject was done by J. Ferguson at the Institute for Defense Analysis, and our discussion of this subject is based largely on Rabiner's description of Ferguson's work. The next five paragraphs are a paraphrasing of the similar material in Rabiner's paper, with a change of notation for consistency with our work.

It is the objective once again to find a reestimation procedure for computing the parameters of the model that maximize the likelihood $P(y|\mathcal{M})$ (training problem). As above, the resulting method will inher-

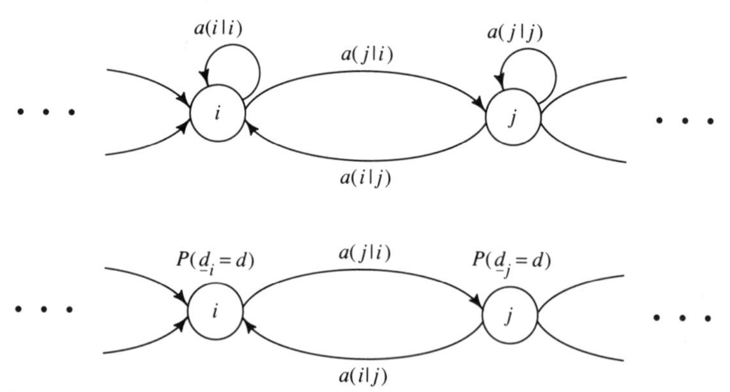

FIGURE 12.9. Replacement of the self-transition arcs in the HMM by a duration probability distribution.

ently contain an F–B procedure for computing $P(y|\mathcal{M})$ for the recognition phase. In this case the HMM includes the duration probability distributions at the states

$$\mathcal{M} = \{S, \boldsymbol{\pi}(1), \mathbf{A}, \mathbf{B}, \{P(\underline{d}_i = d) \text{ for } 1 \le i \le S, 1 \le d \le D\}\}. \tag{12.121}$$

Note that we have included a *maximum allowable duration*, D, for each state. Also note that the diagonal elements of $\mathbf{A}$ need not be estimated.

We begin by redefining the forward-going probability sequence,

$$\alpha(y_1^t, i) \stackrel{\text{def}}{=} P(y_1^t, \underline{x}(t) = i, \underline{x}(t+1) \ne i|\mathcal{M}). \tag{12.122}$$

This is the joint probability of having generated y_1^t, residing at state i at time t, and having the duration of stay in state i end at time t. It is this last event that represents the modification to $\alpha(\cdot\,,\cdot)$ with respect to the previous definition. Let us assume that r states, say $i_1, i_2, \ldots, i_r$, have been visited during the first t observations, and that the durations in these states are $d_1, d_2, \ldots, d_r$, respectively. It is clear that

$$\sum_{s=1}^{r} d_s = t \tag{12.123}$$

and

$$i_r = i. \tag{12.124}$$

By "brute force" reasoning about all possible state sequences, and all possible durations within those states, we can write

$$\begin{aligned}
\alpha(y_1^t, i) = \sum_{i_1, \ldots, i_r} \sum_{d_1, \ldots, d_r} & P(\underline{x}(1) = i_1) P(\underline{d}_{i_1} = d_1) b(y(1)|i_1) \\
& b(y(2)|i_1) \cdots b(y(d_1)|i_1) a(i_2|i_1) \\
& P(\underline{d}_{i_2} = d_2) b(y(d_1+1)|i_2) b(y(d_1+2)|i_2) \\
& \cdots b(y(d_1+d_2)|i_2) \\
& \times \cdots \\
& \times a(i_r|i_{r-1}) P(\underline{d}_{i_r} = d_r)\, b(y(d_1 + \cdots + d_{r-1} + 1)|i_r) \\
& \cdots b(y(t)|i_r),
\end{aligned} \tag{12.125}$$

where the sums $\sum_{i_1,\ldots,i_r} \sum_{d_1,\ldots,d_r}$ indicate that there is a term for every possible sequence of r states, and for every possible set of durations for each state sequence. By induction, we can write (12.125) as

$$\alpha(y_1^t, i) = \sum_{j=1}^{S} \sum_{d=1}^{D} \alpha(y_1^{t-d}, j) a(i|j) P(\underline{d}_j = d) \prod_{s=t-d+1}^{t} b(y(s)|i) \tag{12.126}$$

for $t > D$, where D is the maximum allowable duration in any state. To initialize the recursion, we need the values of $\alpha(y_1^t, j)$ for $t \in [1, D]$, and $j \in [1, S]$. These can be shown to be as follows:

$$\alpha(y_1^1, j) = P(\underline{x}(1) = j) P(\underline{d}_j = 1) b(y(1)|j), \qquad 1 \le j \le S, \tag{12.127}$$

$$\alpha(y_1^2, j) = P(\underline{x}(1) = j) P(\underline{d}_j = 2) \prod_{s=1}^{2} b(y(s)|j) + \sum_{\substack{k=1 \\ k \ne j}}^{S} \alpha(y_1^1, k) a(j|k) P(\underline{d}_j = 1) b(y(2)|j), \qquad 1 \le j \le S, \tag{12.128}$$

$$\alpha(y_1^3, j) = P(\underline{x}(1) = j) P(\underline{d}_j = 3) \prod_{s=1}^{3} b(y(s)|j) + \sum_{d=1}^{2} \sum_{\substack{k=1 \\ k \ne j}}^{S} \alpha(y_1^{3-d}, k) a(j|k) P(\underline{d}_j = d) \prod_{s=4-d}^{3} b(y(s)|j), \quad 1 \le j \le S, \tag{12.129}$$

$$\vdots$$

and so forth, until the $\alpha(y_1^D, j)$ for $1 \le j \le S$ are computed.

As an aside, we note that

$$P(y|\mathcal{M}) = \sum_{i=1}^{S} \alpha(y_1^T, i) \tag{12.130}$$

so that this forward sequence can be used to evaluate the model with respect to an incoming string y for recognition purposes. This is exactly analogous to what happens in the case of the conventional model ("any path" approach), so we need not say anything further about the recognition problem.

Now let us define further forward and backward sequences as follows:

$$\alpha'(y_1^t, i) \stackrel{\text{def}}{=} P(y_1^t, \underline{x}(t) \ne i, \underline{x}(t+1) = i | \mathcal{M}), \tag{12.131}$$

$$\beta(y_{t+1}^T | i) \stackrel{\text{def}}{=} P(y_{t+1}^T | \underline{x}(t) = i, \underline{x}(t+1) \ne i), \tag{12.132}$$

$$\beta'(y_{t+1}^T | i) \stackrel{\text{def}}{=} P(y_{T+1}^t | \underline{x}(t) \ne i, \underline{x}(t+1) = i). \tag{12.133}$$

The reader is encouraged to study each of these definitions to understand its significance. The "unprimed" sequences are concerned with the case in which the residency in state i comes to an end at time t, while the "primed" sequences involve the case in which state i begins at the next observation time, $t+1$. It is not difficult to demonstrate the following relationships among the "primed" and "unprimed" forward and backward sequences:

$$\alpha'(y_1^t, i) = \sum_{j=1}^{S} \alpha(y_1^t, i) a(i|j), \tag{12.134}$$

$$\alpha(y_1^t, i) = \sum_{d=1}^{D} \alpha(y_1^{t-d}, i) P(\underline{d}_i = d) \prod_{s=t-d+1}^{t} b(y(s)|i), \tag{12.135}$$

$$\beta(y_{t+1}^T|i) = \sum_{j=1}^{S} \beta'(y_{t+1}^T|j) a(j|i), \tag{12.136}$$

$$\beta'(y_{t+1}^T|i) = \sum_{d=1}^{D} \beta(y_{t+d+1}^T|i) P(\underline{d}_i = d) \prod_{s=t+1}^{t+d} b(y(s)|i). \tag{12.137}$$

In these terms, the reestimation formulas for the parameters of the model are as follows:

$$\bar{a}(j|i) = \frac{\sum_{t=1}^{T} \alpha(y_1^t, i) a(j|i) \beta'(y_{t+1}^T|j)}{\sum_{j=1}^{S} \sum_{t=1}^{T} \alpha(y_1^t, i) a(j|i) \beta'(y_{t+1}^T|j)}, \tag{12.138}$$

$$\bar{b}(k|i) = \frac{\sum_{\substack{t=1 \\ y(t)=k}}^{T} \left[\sum_{s<t} \alpha'(y_1^s, i) \beta'(y_{s+1}^T|i) - \sum_{s<t} \alpha(y_1^s, i) \beta(y_{s+1}^T|i) \right]}{\sum_{k=1}^{K} \sum_{\substack{t=1 \\ y(t)=k}}^{T} \left[\sum_{s<t} \alpha'(y_1^s, i) \beta'(y_{s+1}^T|i) - \sum_{s<t} \alpha(y_1^s, i) \beta(y_{s+1}^T|i) \right]}, \tag{12.139}$$

$$\bar{P}(\underline{d}_i = d) = \frac{\sum_{t=1}^{T} \alpha'(y_1^t, i) P(\underline{d}_i = d) \beta(y_{t+d+1}^T|i) \prod_{s=t+1}^{t+d} b(y(s)|i)}{\sum_{d=1}^{D} \sum_{t=1}^{T} \alpha'(y_1^t, i) P(\underline{d}_i = d) \beta(y_{t+d+1}^T|i) \prod_{s=t+1}^{t+d} b(y(s)|i)}, \tag{12.140}$$

$$\bar{P}(\underline{x}(1) = i) = \frac{P(\underline{x}(1) = i) \beta(y_1^T|i)}{P(y_1^T|\mathcal{M})}. \tag{12.141}$$

The interpretation of (12.138)–(12.141) is as follows: The formula (12.138) for the reestimation of the transition probabilities is very similar to the usual case except that it "joins" the $\alpha(y_1^t, i)$ term, which corresponds to state i ending at time t, with the $\beta'(y_{t+1}^T | j)$ term, which corresponds to state j starting at time $t+1$, through a transition probability $a(j|i)$. Equation (12.139) represents the ratio of the expected number of times symbol k appears in state i to the expected number of symbols occurring in state i. Equation (12.140) is the ratio of the expected number of times state i occurs with duration d, to the expected number of times state i occurs. Finally, (12.141) is an expression for $P(\underline{x}(1) = i | y_1^T)$.[18]

In spite of the fact that duration densities improve the performance of HMMs in many problems, this improvement is not without costs. First is the increased computational requirements of evaluating the model. A careful analysis of (12.126) and (12.127)–(12.129) will reveal that a D-fold increase in storage, and a factor of $D^2/2$ increase in computation is required with respect to the search of the conventional HMM by forward recursion. Further, D extra parameters must be estimated at each state, thus increasing the amount of information that must be inferred from the training data. This increases the variability of the resulting estimates and thus decreases the reliability of the trained model. One approach to circumventing these problems is to employ parametric probability densities in place of the unspecified probability distributions used above. In this case, only the parameters of the pdf need be estimated during the training, and the computational requirements of the model evaluation are lessened. The Gaussian and gamma pdf's have been successfully used in this manner. For details, the reader is referred to the papers by Levinson (1985, 1986) and Russell and Moore (1985).

Finally, we note that there are simpler alternatives to the F–B-like approach to duration density models taken above. A very different, and much less complex, approach based on clustering techniques is described by Picone (1989). In much earlier work, the "HARPY" system [described in (Lowerre and Reddy, 1980), and in Chapter 13], the states of the model are effectively characterized by uniform probability distributions over some minimum to maximum allowable range, say $d_{\min} \le \underline{d} \le d_{\max}$ for some state. During the recognition phase, a Viterbi search through the model is performed, and a path is penalized for exiting a state too soon or for remaining too long. The estimation of the parameters (nonself-transition) of such a model can be accomplished in a manner similar to the conventional case. We will consider the composition of such a recognition algorithm in Problem 12.8.

[18]At first this might seem inconsistent with our definition of the initial state probability. However, the estimate that we use for this quantity is precisely the probability of being in state i at time $t = 1$ given the training observations. So this estimate is a proper one.

12.2.5 Scaling the Forward–Backward Algorithm

We have cautioned the reader about the numerical problems inherent in using the F–B algorithm for training and recognition. The basic problem is the large numbers of multiplications of numbers less than unity (probabilities) that cause underflow conditions. For example, let us recall (12.34) (and write it for time t instead of $t+1$, and for state i instead of state j),

$$\alpha(y_1^t, i) = \sum_{j=1}^{S} \alpha(y_1^{t-1}, j) a(i|j) b(y(t)|i). \tag{12.142}$$

We see that by recursion we accumulate products of probabilities at each time step, so that the $\alpha(\cdot, \cdot)$ sequence goes exponentially to zero. A remedy for this problem is found in the work of Levinson et al. (1983). Our discussion here is based on a description of the method by Rabiner (1989).

Let us begin by writing the reestimation equation for the state transition probabilities directly in terms of the forward and backward sequences, $\alpha(\cdot, \cdot)$ and $\beta(\cdot|\cdot)$ [recall (12.97)],

$$\bar{a}(j|i) = \frac{\sum_{t=1}^{T-1} \alpha(y_1^t, i) a(j|i) b(y(t+1)|j) \beta(y_{t+2}^T|j)}{\sum_{t=1}^{T-1} \alpha(y_1^t, i) \beta(y_{t+2}^T|j)}. \tag{12.143}$$

Enroute to computing this quantity, suppose we apply a scaling procedure to $\alpha(\cdot, \cdot)$ and $\beta(\cdot|\cdot)$ as follows. After computing $\alpha(y_1^1, i)$ for each state, let us normalize these numbers as follows:

$$\hat{\alpha}(y_1^1, i) \stackrel{\text{def}}{=} \frac{\alpha(y_1^1, i)}{\sum_{j=1}^{S} \alpha(y_1^1, j)}. \tag{12.144}$$

For convenience, let us define $c(1)$ to be the normalization factor,

$$c(1) \stackrel{\text{def}}{=} \frac{1}{\sum_{j=1}^{S} \alpha(y_1^1, j)}, \tag{12.145}$$

and write

$$\hat{\alpha}(y_1^1, i) = c(1) \alpha(y_1^1, i). \tag{12.146}$$

Now to move to the second time slot, we use (12.142) except we insert the normalized values into the right side, and call the result $\tilde{\alpha}(y_1^2, i)$,

$$\tilde{\alpha}(y_1^2, i) = \sum_{j=1}^{S} \hat{\alpha}(y_1^1, j) a(i|j) b(y(t)|i). \tag{12.147}$$

Clearly, from (12.142), (12.146), and (12.147), we have that

$$\tilde{\alpha}(y_1^2, i) = c(1)\alpha(y_1^2, i). \tag{12.148}$$

Once these values are computed for each i, let us scale them by their sum over all states,

$$\hat{\alpha}(y_1^2, i) \stackrel{\text{def}}{=} \frac{\tilde{\alpha}(y_1^2, i)}{\sum_{j=1}^{S} \tilde{\alpha}(y_1^2, j)} = c(2)\tilde{\alpha}(y_1^2, i) = c(2)c(1)\alpha(y_1^2, i). \tag{12.149}$$

Proceeding to $\tilde{\alpha}(y_1^3, i)$, then $\hat{\alpha}(y_1^3, i)$, and so on, we have by induction the relations

$$\tilde{\alpha}(y_1^t, i) = \sum_{j=1}^{S} \hat{\alpha}(y_1^{t-1}, j) a(i|j) b(y(t)|i) \tag{12.150}$$

and

$$\hat{\alpha}(y_1^t, i) = c(t)\tilde{\alpha}(y_1^t, i) = \left(\prod_{\tau=1}^{t} c(\tau)\right)\alpha(y_1^t, i), \tag{12.151}$$

with

$$c(t) = \frac{1}{\sum_{i=1}^{S} \tilde{\alpha}(y_1^t, i)}. \tag{12.152}$$

Now using (12.150)–(12.152) we can show that

$$\hat{\alpha}(y_1^t, i) = \frac{\sum_{j=1}^{S} \hat{\alpha}(y_1^{t-1}, j) a(i|j) b(y(t)|i)}{\sum_{k=1}^{S} \sum_{j=1}^{S} \hat{\alpha}(y_1^{t-1}, j) a(k|j) b(y(t)|k)}. \tag{12.153}$$

This expression makes it possible to compute $\hat{\alpha}(y_1^t, i)$, $i = 1, \ldots, S$, directly from $\hat{\alpha}(y_1^{t-1}, j)$, $j = 1, \ldots, S$, thereby obviating the "intermediate" sequence $\tilde{\alpha}(y_1^t, i)$.

Let us now explore the practical effects of this scaling strategy. Using (12.151) in (12.153), we have

$$\hat{\alpha}(y_1^t, i) = \frac{\sum_{j=1}^{S}\left(\prod_{\tau=1}^{t-1} c(\tau)\right)\alpha(y_1^{t-1}, j)a(i|j)b(y(t)|i)}{\sum_{k=1}^{S}\sum_{j=1}^{S}\left(\prod_{\tau=1}^{t-1} c(\tau)\right)\alpha(y_1^{t-1}, j)a(k|j)b(y(t)|k)} = \frac{\alpha(y_1^t, i)}{\sum_{k=1}^{S}\alpha(y_1^t, k)}. \tag{12.154}$$

We see that, in effect, the procedure we have introduced above scales each $\alpha(y_1^t, i)$ to the sum over all states at time t. However, note that this is accomplished in such a way that small numbers are not encountered in the process.

A similar procedure is used in computing the backward sequence, except that the same scale factor, $c(t)$, is used to scale $\beta(y_{t+1}^T | j)$,[19]

$$\hat{\beta}(y_{t+1}^T | j) = c(t)\tilde{\beta}(y_{t+1}^T | j). \tag{12.155}$$

This is reasonable, since the objective of the scaling is simply to keep the numbers in a useful dynamic range, and since the α's and β's tend to be of the same order of magnitude. With this scaling strategy, it can be shown that

$$\hat{\beta}(y_{t+1}^T | j) = \left(\prod_{\tau=t}^{T} c(\tau)\right)\beta(y_{t+1}^T | j). \tag{12.156}$$

Further, with this choice of scaling, it is not difficult to show that

$$\overline{a}(j|i) = \frac{\sum_{t=1}^{T-1}\hat{\alpha}(y_1^t, i)a(j|i)b(y(t+1)|j)\hat{\beta}(y_{t+2}^T | j)}{\sum_{t=1}^{T-1}\hat{\alpha}(y_1^t, i)\hat{\beta}(y_{t+2}^T | j)} = \frac{\sum_{t=1}^{T-1}\alpha(y_1^t, i)a(j|i)b(y(t+1)|j)\beta(y_{t+2}^T | j)}{\sum_{t=1}^{T-1} a(y_1^t, i)\beta(y_{t+2}^T | j)} \tag{12.157}$$

so that the scaled values can be used in the reestimation equation for the state transition probabilities without modification. In a similar manner, the symbol probabilities can be computed from the "usual" form, (12.98), with the scaled forward and backward sequences inserted.

[19]Recall that $\beta(y_{t+1}^T | j)$ is associated with time t.

Finally, we must determine how to compute the model probability, $P(y|\mathcal{M})$, which, recall, is given by

$$P(y|\mathcal{M}) = \sum_{i=1}^{S} \alpha(y_1^T, i). \tag{12.158}$$

Clearly, it is not appropriate to use the scaled α values in this expression. However, from (12.154) we see that

$$\sum_{i=1}^{S} \hat{\alpha}(y_1^t, i) = 1 \tag{12.159}$$

for any t. In particular, we can write

$$\sum_{i=1}^{S} \hat{\alpha}(y_1^T, i) = \sum_{i=1}^{S} \left(\prod_{\tau=1}^{T} c(\tau) \right) \alpha(y_1^T, i) = 1 \tag{12.160}$$

from which we see immediately that

$$P(y|\mathcal{M}) = \left(\prod_{\tau=1}^{T} c(\tau) \right)^{-1}. \tag{12.161}$$

Since the product of the $c(\tau)$'s is likely to be extremely large, we can compute the logarithm instead,

$$-\log P(y|\mathcal{M}) = \sum_{\tau=1}^{T} \log c(\tau), \tag{12.162}$$

which provides the necessary likelihood measure.

12.2.6 Training with Multiple Observation Sequences

In order to provide a more complete representation of the statistical variations likely to be present across utterances, it is essential to train a given HMM with multiple training utterances. The modification of the Baum–Welsh reestimation equations is straightforward.

First, we assume that we are working with a left-to-right model so that there is no need for estimating the initial state probabilities. Now let us recall (12.94) and (12.95):

$$\bar{a}(j|i) = \frac{\xi(i, j; \bullet)}{\gamma(i; \bullet)}, \tag{12.163}$$

$$\bar{b}(k|j) = \frac{\delta(j, k; \bullet)}{\nu(j; \bullet)}. \tag{12.164}$$

Recall that each numerator and denominator represents an average number of some event related to the model [review (12.89)–(12.92)]. Accordingly, it makes sense to simply sum these events over all observations,

$$\bar{a}(j|i) = \frac{\sum_{l=1}^{L} \xi^{(l)}(i,j;\bullet)}{\sum_{l=1}^{L} \gamma^{(l)}(i;\bullet)}, \tag{12.165}$$

$$\bar{b}(k|j) = \frac{\sum_{l=1}^{L} \delta^{(l)}(j,k;\bullet)}{\sum_{l=1}^{L} \nu^{(l)}(j;\bullet)}, \tag{12.166}$$

where the superscript l indicates the result for the lth observation, say $[y_1^{T_l}]^{(l)} = y^{(l)}$, of which there are a total of L. The length of observation $y^{(l)}$ is denoted T_l. Using the results in (12.77)–(12.84), we can write

$$\bar{a}(j|i) = \frac{\sum_{l=1}^{L} \frac{1}{P(y^{(l)}|\mathcal{M})} \sum_{t=1}^{T_l-1} \alpha^{(l)}(y_1^t, i) a(j|i) b(y(t+1)|j) \beta^{(l)}(y_{t+2}^{T_l}|j)}{\sum_{l=1}^{L} \frac{1}{P(y^{(l)}|\mathcal{M})} \sum_{t=1}^{T_l-1} \alpha^{(l)}(y_1^t, i) \beta^{(l)}(y_{t+1}^{T_l}|i)}, \tag{12.167}$$

$$\bar{b}(k|j) = \frac{\sum_{l=1}^{L} \frac{1}{P(y^{(l)}|\mathcal{M})} \sum_{\substack{t=1 \\ y(t)=k}}^{T_l} \alpha^{(l)}(y_1^t, j) \beta^{(l)}(y_{t+1}^{T_l}|j)}{\sum_{l=1}^{L} \frac{1}{P(y^{(l)}|\mathcal{M})} \sum_{t=1}^{T_l-1} \alpha^{(l)}(y_1^t, j) \beta^{(l)}(y_{t+1}^{T_l}|j)}, \tag{12.168}$$

where the arguments of $\alpha^{(l)}(y_1^T, i)$ or $\beta^{(l)}(y_{t+1}^T|j)$ are assumed to belong to the lth observation, $y^{(l)}$.

Note that scaling these equations is simple. Recall that [see (12.161)]

$$P^{(l)}(y|\mathcal{M}) = \left(\prod_{\tau=1}^{T_l} c^{(l)}(\tau) \right)^{-1}, \tag{12.169}$$

where the $c^{(l)}$ are the scaling factors for the lth observation. This allows us to immediately write

$$\bar{a}(j|i) = \frac{\sum_{l=1}^{L} \sum_{t=1}^{T_l-1} \hat{\alpha}^{(l)}(y_1^t, i) a(j|i) b(y(t+1)|j) \hat{\beta}^{(l)}(y_{t+2}^{T_l}|j)}{\sum_{l=1}^{L} \sum_{t=1}^{T_l-1} \hat{\alpha}^{(l)}(y_1^t, i) \hat{\beta}^{(l)}(y_{t+1}^{T_l}|j)}, \tag{12.170}$$

$$\bar{b}(k|j) = \frac{\displaystyle\sum_{l=1}^{L} \sum_{\substack{t=1 \\ y(t)=k}}^{T_l} \hat{\alpha}^{(l)}(y_1^t, j)\hat{\beta}^{(l)}(y_{t+1}^{T_l}|j)}{\displaystyle\sum_{l=1}^{L} \frac{1}{P(y^{(l)}|\mathcal{M})} \sum_{t=1}^{T_l} \hat{\alpha}^{(l)}(y_1^t, j)\hat{\beta}^{(l)}(y_{t+1}^{T_l}|j)}, \tag{12.171}$$

where $\hat{\alpha}$ and $\hat{\beta}$ indicate the scaled values.

Finally, we note that the use of the Viterbi reestimation approach with multiple observations is very straightforward. In fact, we mentioned this idea when the method was introduced in Section 12.2.2. With a review of (12.105) and (12.106), the enhancement will be apparent.

12.2.7 Alternative Optimization Criteria in the Training of HMMs

Prerequisite Chapter 1 Reading: Sections 1.3.4 and 1.4, especially Subsection 1.4.3.

Thus far, we have adopted a *maximum likelihood* (ML) approach to the design of an HMM. This philosophy asserts that a model is "good" if its parameters are adjusted to maximize the probability $P(\mathbf{y}|\mathcal{M})$ of generating the observation (training) sequences for which it is "responsible." Although the maximum likelihood technique has yielded many encouraging results in practice, there are two fundamental conceptual problems with this approach. First, the signal (as reflected in the observation sequence) might not adhere to the constraints of the HMM, or there might be insufficient data to properly train the HMM even if the model is generally accurate. Second, the ML approach does not include any means of effecting "negative training," in which a model is trained to not only respond favorably to its own class, but to discriminate against productions of other models. These conceptual problems gave rise to research into techniques based on the concept of discrimination in the mid-to-late 1980s.

In the alternative approach, the goal of the training procedure is to find HMM parameters that minimize the so-called *discrimination information* (DI) or *cross entropy* between statistical characterizations of the signal and the model. The technique is therefore referred to as MDI for *minimum DI*. The mathematical details of the general MDI approach are formidable, and we shall not pursue them here. The interested reader is referred to the paper by Ephraim et al. (1989), where a generalized F–B-like algorithm is developed from these considerations. Generally speaking, the MDI method is based on the philosophy that the observations of the signal need not follow the Markovian constraint. Rather, the approach is to match density functions (or probability distributions) of some parametric representations of the signal with those of the model. The DI is a form of probabilistic distance measure (see Section 1.3.4) that indicates the difference in these densities. It has the general form

$$J_{\mathrm{DI}} = \int_{-\infty}^{\infty} f_{\underline{\mathbf{v}}|\underline{c}}(\mathbf{v}|1) \log \frac{f_{\underline{\mathbf{v}}|\underline{c}}(\mathbf{v}|1)}{f_{\underline{\mathbf{v}}|\underline{c}}(\mathbf{v}|2)} d\mathbf{x} \tag{12.172}$$

[cf., e.g., (1.216)], where $\underline{c}=1$ indicates the signal, and $\underline{c}=2$, the HMM, and where $\underline{\mathbf{v}}$ is some vector of parameters that characterizes the signal. Of course, a small DI measure indicates good agreement between the signal and model densities.

In the first of the papers noted above, the ML approach to HMM training is shown to be a special case of the MDI approach. Another special case is the *maximum average mutual information* (MMI) approach which was researched by Bahl et al. (1986) in response to the problems with the ML approach noted above. The MMI approach gives a good sense of the "negative training" aspect of MDI approaches.

Suppose that we have R different HMMs to be trained (e.g., each representing a different word), $\mathcal{M}_1, \ldots, \mathcal{M}_R$. Let us denote by $\underline{\mathcal{M}}$ the random variable indicating a model outcome in an experiment (e.g., $\underline{\mathcal{M}}=1$ indicates that $\mathcal{M}_1$ is chosen). Let us also assume that we have L training observation strings of lengths $T_1, \ldots, T_L$, say $\mathbf{y}^{(1)}, \ldots, \mathbf{y}^{(L)}$. As we have done in the past, let us use the abusive notation "$\underline{\mathbf{y}}=\mathbf{y}^{(k)}$" to mean that realization k has occurred. By a slight generalization of (1.240) to allow for the random vector string, the average mutual information between the random quantities $\underline{\mathbf{y}}$ and $\underline{\mathcal{M}}$ is[20]

$$\overline{M}(\underline{\mathbf{y}}, \underline{\mathcal{M}}) = \sum_{l=1}^{L} \sum_{r=1}^{R} P(\underline{\mathbf{y}}=\mathbf{y}^{(l)}, \underline{\mathcal{M}}=\mathcal{M}_r) \log \frac{P(\underline{\mathbf{y}}=\mathbf{y}^{(l)}, \underline{\mathcal{M}}=\mathcal{M}_r)}{P(\underline{\mathbf{y}}=\mathbf{y}^{(l)})P(\underline{\mathcal{M}}=\mathcal{M}_r)}. \tag{12.173}$$

Note that this can be written

$$\begin{aligned} \overline{M}(\underline{\mathbf{y}}, \underline{\mathcal{M}}) &= \sum_{l=1}^{L} \sum_{r=1}^{R} P(\underline{\mathbf{y}}=\mathbf{y}^{(l)}, \underline{\mathcal{M}}=\mathcal{M}_r) \\ &\quad \times \left[\log P(\underline{\mathbf{y}}=\mathbf{y}^{(l)}|\underline{\mathcal{M}}=\mathcal{M}_r) - \log P(\underline{\mathbf{y}}=\mathbf{y}^{(l)})\right] \\ &= \sum_{l=1}^{L} \sum_{r=1}^{R} P(\underline{\mathbf{y}}=\mathbf{y}^{(l)}, \underline{\mathcal{M}}=\mathcal{M}_r) \\ &\quad \times \left[\log P(\underline{\mathbf{y}}=\mathbf{y}^{(l)}|\underline{\mathcal{M}}=\mathcal{M}_r) \right. \\ &\quad \left. -\log \sum_{m=1}^{R} P(\underline{\mathbf{y}}=\mathbf{y}^{(l)}|\underline{\mathcal{M}}=\mathcal{M}_m)P(\underline{\mathcal{M}}=\mathcal{M}_m)\right]. \end{aligned} \tag{12.174}$$

[20]Recall that any logarithmic base may be used here, so we drop the binary log indicator.

If there is exactly one training string for each HMM ($L = R$) and $\mathbf{y}^{(l)}$ is to be used to train $\mathcal{M}_l$, then, if we assume that $P(\underline{\mathbf{y}} = \mathbf{y}^{(l)}, \underline{\mathcal{M}} = \mathcal{M}_r) \approx \delta(l - r)$, we can approximate (12.174) by

$$\overline{M}(\underline{\mathbf{y}}, \underline{\mathcal{M}}) \approx \sum_{l=1}^{L} \log P(\underline{\mathbf{y}} = \mathbf{y}^{(l)} | \underline{\mathcal{M}} = \mathcal{M}_l) - \log \sum_{m=1}^{L} P(\underline{\mathbf{y}} = \mathbf{y}^{(l)} | \underline{\mathcal{M}} = \mathcal{M}_m) P(\underline{\mathcal{M}} = \mathcal{M}_m). \tag{12.175}$$

From these equations, the "coupling" of each of the models with the remaining set is apparent. Now recall that we want the mutual information shared by random "quantities" $\underline{\mathbf{y}}$ and $\underline{\mathcal{M}}$ to be as large as possible so that the "feature vector" $\mathbf{y}^{(l)}$ is a good indicator of "class" $\mathcal{M}_l$. Hence the name MMI.

During the training phase, the control we have to effect this maximization lies in the parameters of the various models. The minimization therefore takes the form of seeking **A** and **B** matrices, and an initial state probability vector, $\boldsymbol{\pi}(1)$, for the models that maximize $\overline{M}(\underline{\mathbf{y}}, \underline{\mathcal{M}})$. Clearly, this problem involves the solution of a highly nonlinear set of equations with various constraints. There is no known closed-form solution for this maximization, and a standard optimization method such as a gradient-descent algorithm [see, e.g., (Ljung and Söderström, 1983)] must be employed. This method is therefore computationally tedious and it has provided only moderately encouraging results.

A much simpler training strategy that adheres to the basic philosophy of negative training is suggested by Bahl et al. (1988). The method, called *corrective training*, simply involves reinforcing model parameters when the parameters produce correct recognition or correct rejection, and adjusting them when they produce incorrect results or "near misses" (unacceptably favorable likelihoods for incorrect words). The procedure is accomplished using a weighting of counts in the reestimation procedure in accordance with these conditions. Corrective training methods were more recently extended to continuous-speech recognition problems by Lee and Mahajam (1989).

12.2.8 A Distance Measure for HMMs

In the preceding section, we noted the lack of interplay between the models in the ML design process. Upon completion of training, or in attempting to explain recognition performance, it is natural to ask "how different" two models might be, in assessing their discriminatory power. Juang and Rabiner (1985) have studied this problem (see Problem 12.11) and have suggested the following distance measure between models $\mathcal{M}_1$ and $\mathcal{M}_2$,

$$D(\mathcal{M}_1, \mathcal{M}_2) \stackrel{\text{def}}{=} \frac{1}{T_2}\left[\log P(\mathbf{y}^{(2)} | \mathcal{M}_1) - \log P(\mathbf{y}^{(2)} | \mathcal{M}_2)\right], \tag{12.176}$$

where $\mathbf{y}^{(2)}$ is a length T_2 sequence generated by $\mathcal{M}_2$. To create a distance measure that is symmetric with respect to the models, we can use

$$D'(\mathcal{M}_1, \mathcal{M}_2) \stackrel{\text{def}}{=} \frac{D(\mathcal{M}_1, \mathcal{M}_2) + D(\mathcal{M}_2, \mathcal{M}_1)}{2}. \tag{12.177}$$

Definition (12.176) is apparently similar to the mutual information measure, and interpretations of this distance in terms of discrimination information are given in (Juang and Rabiner, 1985).

12.3 Practical Issues

Having completed a review of many of the key theoretical and algorithmic issues underlying the use of HMMs, we now turn to a number of topics pertaining to their use in practice.

12.3.1 Acoustic Observations

Thus far in our discussion, we have denoted the observation vectors $\{\mathbf{y}(1), \ldots, \mathbf{y}(T)\}$ in the abstract, without much regard for what actual quantities are contained in these vectors. Many features have been used as observations, but the most prevalent are the LP parameters, cepstral parameters, and related quantities. These are frequently supplemented by short-term time differences that capture the dynamics of the signal, as well as energy measures such as the short-term energy and differenced energy. A comparative discussion of various techniques is found in (Bocchieri and Doddington, 1986) and (Nocerino et al., 1985).

In a typical application, the speech would be sampled at 8 kHz, and analyzed on frames of 256 points with 156-point overlap. The analysis frame end-times are $m = 100, 200, \ldots, M$ in the short-term analysis. These sample times, m, in turn become observation times $t = 1, 2, \ldots, T$. For a typical word utterance lasting 1 sec, $M = 8000$ and $T = 80$ observations.

On each frame, 8–10 LP coefficients are computed, which are then converted to 12 cepstral coefficients (see Section 6.2.3). Alternatively, 12 mel-cepstral coefficients might be computed directly from the data (see Section 6.2.4). To add dynamic information, 12 differenced cepstral coefficients are also included in the vector (see Section 6.2.4). Finally, a short-term energy measure and a differenced energy measure are included for each frame, for a total of 26 features in the observation.

More practical details on this issue will be presented in Chapter 13 when we discuss the use of HMMs in CSR.

12.3.2 Model Structure and Size

We have alluded to the fact that the HMM is often used at higher levels of a speech recognizer to model linguistic constraints. In the present discussion, we restrict our attention to the use of HMM at the acoustic level as we have tacitly done throughout this chapter.

Another important practical consideration is the structure and size of the model. Recall that by *structure* we mean the pattern of allowable state transitions, and by *size* the number of states to be included. If the HMM is a discrete-observation type, then size also includes the number of levels in the observation codebook, and if continuous-observation, the number of mixtures in the duration densities. At the beginning of Section 12.2.4 we discussed the fact that HMMs are used in part to avoid specific *a priori* statistical and structural characterization of the speech signal. Accordingly, there is not an exact science dictating the size or structure of the model in specific situations. Certain general guidelines are recognized, however.

As we also noted in Section 12.2.4, experimental evidence suggests that states frequently represent identifiable acoustic phenomena. Therefore, the number of states is often chosen to roughly correspond to the expected number of such phenomena in the utterance. If words are being modeled with discrete observations, for example, 5–10 states are typically used to capture the phones in the utterances. Continuous-observation HMMs typically use more states, often one per analysis frame [e.g., (Picone, 1989)]. If HMMs are used to model discrete phones, three states are sometimes used—one each for onset and exiting transitions, and one for the steady-state portion of the phone. An enhancement of this three-state phone model used in the SPHINX system (Lee et al., 1990) is shown in Fig. 12.10, and will be discussed further below. As a cruder measure, the average length of the utterances is sometimes used to determine the number of necessary states. A peculiar version of this idea is embodied in the "fenone" model, which we will discuss in the next subsection. A fenone is an acoustic unit of speech that is nominally one frame long. Accordingly, the HMMs for fenones are very small. Example fenone structures are shown in Fig. 12.11. Both the SPHINX and fenone models as shown use the Mealy form of the HMM.

The relationship of the number of states to the performance of the HMM is very imprecise, and, in practice, it is often necessary to experiment with different model sizes to determine an appropriate number. Figure 12.8, for example, shows a plot of error rate versus the number of states in a digit recognition experiment (Rabiner, 1989). Another study of digit recognition (Picone, 1989) suggests that the number of states should be allowed to vary across models for better results.

The most general structure for the HMM is the so-called *ergodic*[21] model, which allows unconstrained state transitions. In this case none of

[21]For the definition of an ergodic Markov chain, see, for example, (Leon–Garcia, 1989; Grimmett and Stirzaker, 1985).

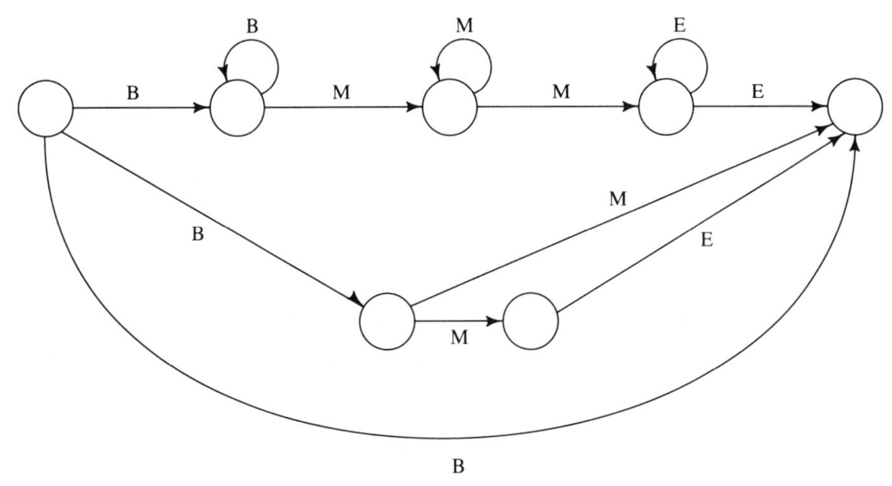

FIGURE 12.10. Phone model used in the SPHINX system. This HMM is of the Mealy type, which generates observations upon transition. The phone model has only three distinct observation distributions, which are shared among the several transitions. These are labeled B, M, and E, to denote beginning, middle, and end.

the elements in the transition probability matrix, $\mathbf{A}$, is constrained to be zero. An example with six states is shown in Fig. 12.12. Such a structure does not coincide well with speech utterances because it does not attempt to model the sequential ordering of events in the signal. Whereas it can be used to provide more flexibility in the generation of observations,[22] this advantage comes at the increased risk of converging on an unsatisfactory local maximum in the training process (see Section 12.2.2). Interestingly, and perhaps not surprisingly, when used with speech, the ergodic model will often train so that it essentially represents a sequential structure (backward transition probabilities turn out zero).

The model structure generally adopted for speech recognition is a *left-to-right* or *Bakis* (Bakis, 1976) model. A typical six-state example is shown in Fig. 12.13. Other models illustrated above are seen to be of the left-to-right variety. The Bakis model has the property that states can be aligned in such a way that only left-to-right transitions are possible. Further, it has a well-defined initial and final state. Such a model naturally suits a signal-like speech that varies in time from left to right. If the states are numbered sequentially from left to right, then the transition probability matrix will be lower diagonal,

$$a(i|j) = 0, \qquad \text{for all } i < j. \tag{12.178}$$

Frequently, the model includes the additional constraint that no more than one or two states may be skipped in any transition. For example, if only one skip is allowed, as shown in Fig. 12.13, then

[22]By this we mean, for example, that a path could "jump backward" to pick up a symbol not found in a particular state.

$$a(i|j) = 0, \qquad \text{for all } i > j + 2. \tag{12.179}$$

Note also that the presence of the initial and final states implies the following:

$$a(i_f|i_f) = 1 \tag{12.180}$$

and

$$P(\underline{x}(1) = i_i) = 1, \tag{12.181}$$

where i_f and i_i represent the final and initial states, respectively. (In Fig. 12.13, $i_f = 6$ and $i_i = 1$.) We will see some variations on this structure when we discuss specific forms of the HMM later in the chapter.

The choice of a "constrained" model structure like the Bakis does not require any modification of the training procedures described above. In the F–B case, it is easily seen that any parameter initially set to zero will remain zero throughout the training [see (12.97)–(12.99)].

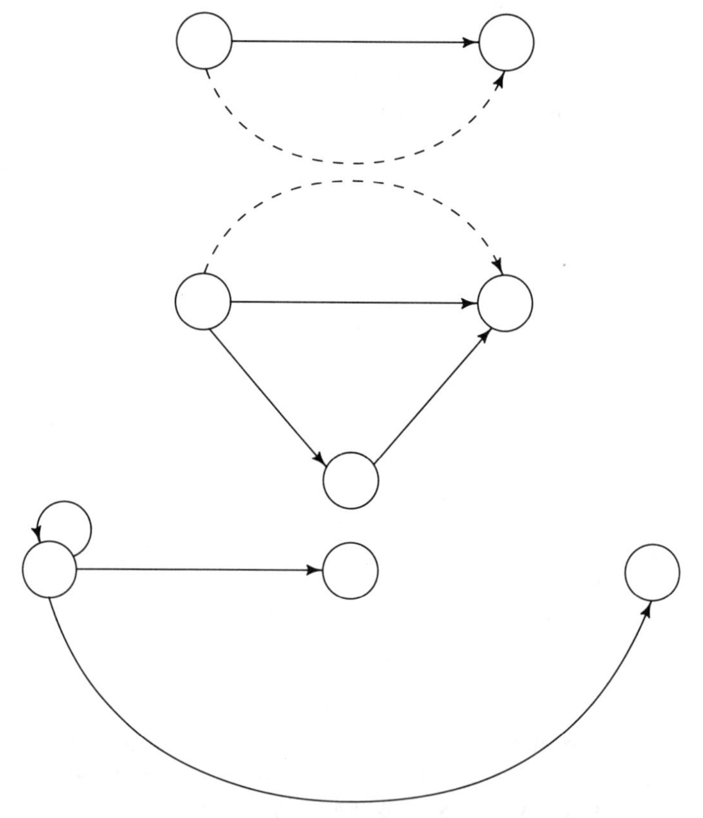

FIGURE 12.11. Example HMM structures for fenones. Each is of Mealy form. The dotted arc represents a transition that generates no output. After Bahl et al. (1988).

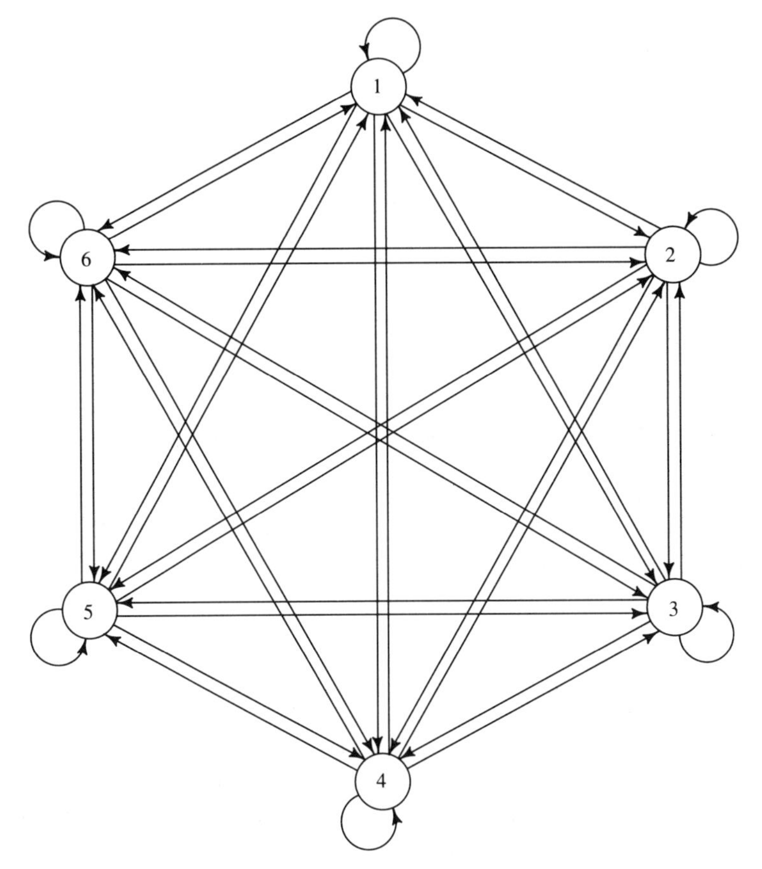

FIGURE 12.12. Ergodic HMM with six states.

The final issue is to decide upon the number of symbols (centroids) in the VQ codebook if a discrete-observation model is used. One measure of the quality of a codebook is the average distance of a vector observation in the training data from its corresponding symbol. This figure is often called the codebook *distortion*. Of course, the smaller the distortion, the more accurately the centroids represent the vectors they replace. A typical plot of distortion versus log codebook size is shown in Fig. 12.14. Although the distortion continues to decrease with codebook size, the benefit per centroid diminishes significantly beyond 32 or 64. Typically, recognizers use codebooks of size 32–256, with the larger sizes

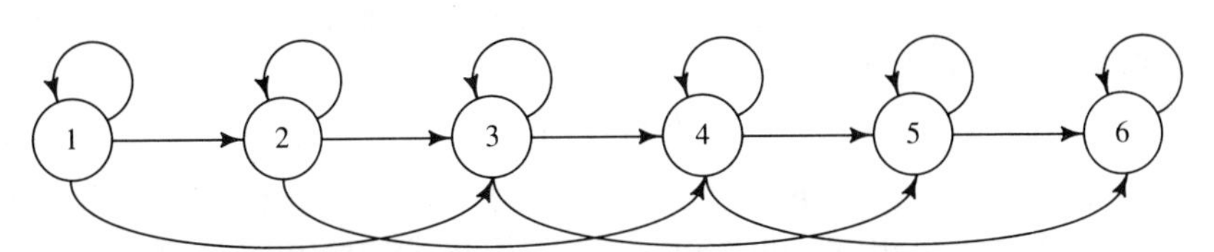

FIGURE 12.13. Example left-to-right, or Bakis, HMM with six states.

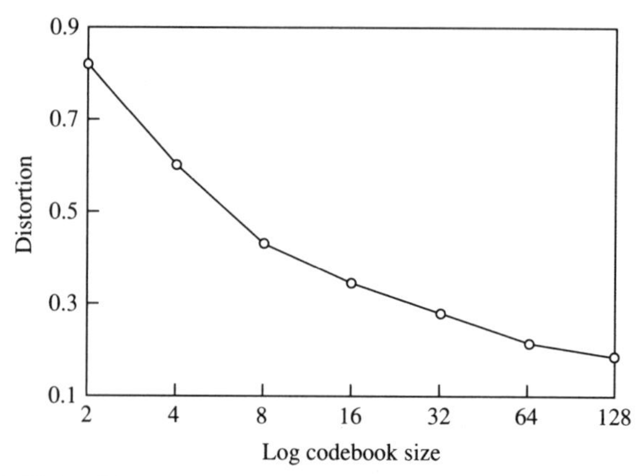

FIGURE 12.14. Typical plot of distortion versus log codebook size.

being more common. Since a larger codebook implies increased computation, there is an incentive to keep the codebook as small as possible without decreasing performance. Like the other parametric decisions that must be made about an HMM, this issue too is principally guided by experimental evidence. Since the centroids loosely correspond to different acoustic phenomena in the speech, some rough guidance is provided by examining the acoustic complexity of the vocabulary. If, for example, the vocabulary is very small, fewer symbols might to be sufficient. In a related, but very unusual circumstance, the *speaker* might be restricted in the number of sounds he or she can reliably produce, for example, because of a speech disability. In this case a smaller codebook might also be preferable, even with a relatively complex vocabulary.

12.3.3 Training with Insufficient Data

From our study of HMM training in Sections 12.2.2 and 12.2.3, it is clear that a large amount of training data is generally necessary to accurately estimate the HMM parameters. As a clear example of the consequences of insufficient data, consider the situation in which a particular observation, say observation j, does not appear in the training of a Moore-form discrete-observation model, say $\mathcal{M}$. Then $b(j|i) = 0$ for any state i. Now consider an observation sequence to be recognized, y_1^T, whose correct model is $\mathcal{M}$. If any of the observations in y_1^T is j, disastrous consequences will occur. This is evident upon reexamining (12.34). One simple remedy is to assure (according to some *ad hoc* strategy) that model parameters are not unacceptably small (Rabiner et al., 1983). For example, we might set

$$b(j|i) = \delta \tag{12.182}$$

in the above situation, in which δ is some small probability. The other observation probabilities in state i must then be readjusted to preserve $\sum_k b(k|i) = 1$.

A more formal approach to supplementing insufficient training data, called *deleted interpolation*, has been suggested by Bahl et al. (1983). To describe this approach, we first need to understand the notion of *tied states*. We discuss this concept from the point of view of the Moore-form HMM, but a parallel development exists for the Mealy form (see the Bahl paper). Two states are said to be *tied* if they share common observation probability distributions (or covariance matrices, etc.). For simplicity, consider the three-state HMM shown in Fig. 12.15(a), which is assumed to be trained using a conventional method like the F–B algorithm in conjunction with a set of training sequences, say $\mathcal{T}$. In Fig. 12.15(b), the "same" HMM is assumed to have been trained with states 1 and 2 tied. To deduce the parameters for the tied model requires only a simple modification of the training algorithms discussed in Section 12.2.2 (see Problem 12.10). At the completion of training $b(k|1) = b(k|2)$ for all observations, k. The advantage of the model with tied states (which is essentially a two-state model) is that the same amount of training data is used to deduce fewer parameters, which will therefore be of lower variance.

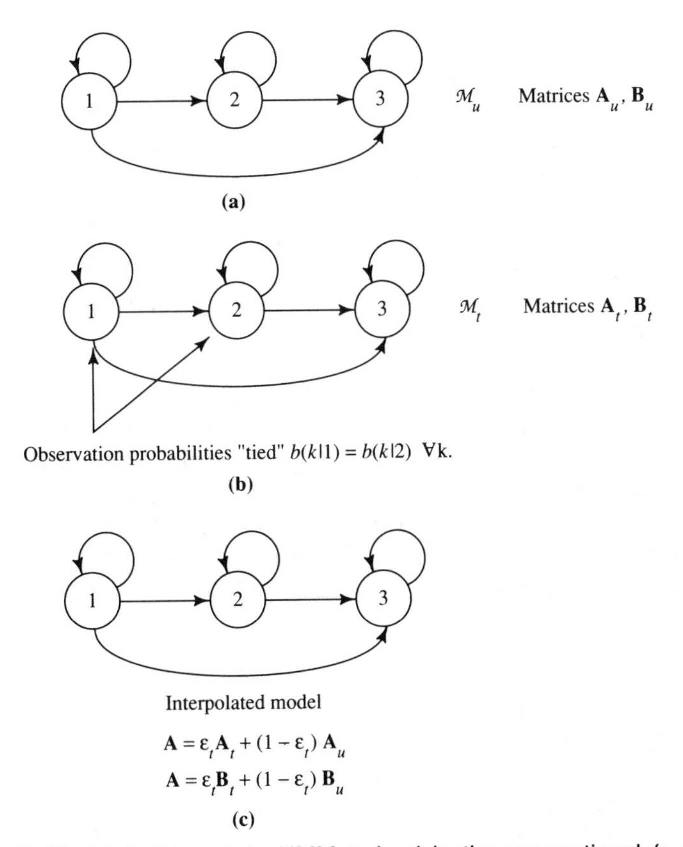

FIGURE 12.15. (a) A three-state HMM trained in the conventional (e.g., F–B algorithm) manner. (b) The "same" three-state HMM with tied states. (c) An "interpolated" model derived from the models of (a) and (b).

Given the HMM trained in the usual way, and the one trained with tied states, which should we use? We might have chosen a three-state model initially due to acoustical or physical considerations. Perhaps, for example, the model represents a word with three phones. Intuitively, three states would seem appropriate for this case. However, if there are insufficient training data to yield reliable parameters using conventional training, then the tied model might be preferable. Deleted interpolation is a method of obtaining a "hybrid" model that automatically includes the proper "proportion" of each of the original models. Suppose that we divide $\mathcal{T}$ into two subsets of training sequences, $\mathcal{T}'$ and $\mathcal{T}''$. Let us use $\mathcal{T}'$ to train both the tied and "untied" models, say $\mathcal{M}_t$ and $\mathcal{M}_u$, respectively. Then we perform recognition experiments on each of the sequences in $\mathcal{T}''$, using both $\mathcal{M}_t$ and $\mathcal{M}_u$ in each experiment. In each case, one of the models will produce a superior likelihood score. Let ε_t indicate the fraction of the strings in $\mathcal{T}''$ for which the tied HMM $\mathcal{M}_t$ gave better recognition scores. If $\mathbf{A}_t$ and $\mathbf{B}_t$ denote the state transition and observation probability matrices for $\mathcal{M}_t$, and $\mathbf{A}_u$ and $\mathbf{B}_u$ the similar matrices for $\mathcal{M}_u$, then the interpolated model has matrices

$$\mathbf{A} = \varepsilon_t \mathbf{A}_t + (1 - \varepsilon_t)\mathbf{A}_u \tag{12.183}$$

$$\mathbf{B} = \varepsilon_t \mathbf{B}_t + (1 - \varepsilon_t)\mathbf{B}_u. \tag{12.184}$$

In fact, the term *deleted interpolation* is usually used to describe a somewhat more complicated procedure than the one discussed above. In this case, the training set $\mathcal{T}$ is iteratively repartitioned and the procedure repeated over each partitioning. There are many methods for constructing the multiple partitions. For example, $\mathcal{T}''$ can be taken to be the first 10% of $\mathcal{T}$ in the first iteration, the second 10% in the second, and so on. For details on this and other issues, the reader is referred to the paper by Bahl et al. (1983).

We should keep in mind that the HMM can be used to model higher-level language structures (Chapter 13) in the speech process, as well as subword phenomena such as phones (see below). The deleted interpolation method is perfectly applicable to HMMs at any level in a recognition system. In fact, the method was developed in conjunction with language modeling (Bahl et al., 1983). For applications to phone models, see, for example, (Schwartz et al., 1985) and (Lee et al., 1990).

12.3.4 Acoustic Units Modeled by HMMs

Early in the chapter, we indicated that it was sufficient for the theoretical developments to imagine that the HMM was being used to model an isolated word. In principle, no changes are necessary in the training or recognition procedures discussed above if some other acoustic unit is modeled by the HMM. We have, for example, mentioned the modeling of phones by the HMM in the last section. In this section we examine the use of HMMs to model a variety of acoustic units in the speech signal.

Let us begin with the most fundamental phonetic unit of speech, the phone. In Chapter 10 we discussed the fact that large-vocabulary systems require the modeling of speech in smaller units than words, since the training of models for a large number of words is generally impractical. In contemporary speech recognition systems, the subword unit chosen is frequently the phone. There are, however, several variations of this approach. The simplest is the use of *context-independent* phone models in which a single model is constructed for each phone. These models are not sensitive to the phonetic context in which the phone occurs; consequently, there are many fewer of them (typically 50) than if context were considered. Use of the example phone model shown in Fig. 12.10 in context-independent mode is discussed in (Lee et al., 1990). Another example of context-independent phone experimentation appears in the early work of the IBM research group (Bahl et al., 1983).

Context-dependent phone models have been used in various systems. Early work on this concept is reported by researchers at IBM (Bahl et al., 1980) and at BBN (Schwartz et al., 1984). Experiments have included models that account for left context, right context, and both left and right. The last case is commonly called a *triphone* model. A major problem with the context-sensitive phone approach is the resulting proliferation of models. For example, for 50 basic phones there are $50^3 = 125{,}000$ possible triphones. Seldom is there sufficient data for training this large number of HMMs. Solutions have included combining certain models based on phonetic knowledge (Derouault, 1987), and combining models based on an automated technique (Lee et al., 1990). This latter technique results in what has been called a *generalized triphone.*

A variation on the idea of context-dependent phones is to separately train phones that are in the context of words which are frequently poorly articulated (*word-dependent* phone models). The so-called "function words" (e.g., "in," "a," "the," "and") comprise 4% of vocabulary of the DARPA Resources Management Database (see Section 13.8), 30% of the spoken words in the database, and accounted for almost 50% of the errors in a recent set of experiments (Lee et al., 1990). Other studies of word-dependent phone models are found in (Chow et al., 1986; Dutoit, 1987).

Other variations on phone models have included *diphones* (phone pairs) (Schwartz et al., 1980), and the inclusion of "stationary" phone models and "transition" models (Cravero et al., 1986). Also, Lee et al. (1988) have constructed HMM-based subword units called *segment models*, which are based solely on acoustic aspects of the signal, without direct concern for the relationships of these models to conventional speech units like phones.[23]

[23]We note in passing that another interesting model based on similar principles is the *stochastic segment model* developed by Roucos and Dunham (1987, 1989). The stochastic segment model is not based on the HMM, however; rather, the acoustic segments are characterized by probability density functions.

As we move up the hierarchy of linguistic units, we next come to the phonemes. Recall that a phoneme is actually an abstract unit that may have multiple acoustic manifestations. From a modeling point of view, the phoneme can be modeled by creating a network of several phone models. This network might provide multiple paths through the phones to account for various pronunciations of the phoneme. An example is shown in Fig. 12.16(a) in which two alternative models for the second phone in the phoneme are apparently possible. In Fig. 12.16(b), the network is portrayed as a Mealy-form HMM whose "observations" are the phone designators. Notice that this network is composed of HMMs embedded inside of a larger HMM. This simple network is a foreshadowing of some of the large recognition networks that will be used in language modeling techniques in Chapter 13.

By extending the idea of the last paragraph, it is possible to build word models out of phoneme models in a similar manner. An example is shown in Fig. 12.17. Again we can interpret this network as a Mealy-form HMM whose observations are abstract units (phonemes). Since the phonemes are, in turn, HMMs with embedded phone models, the resulting network is a triply embedded set of HMMs. In practice, the abstraction of the phoneme is usually omitted, with the word model built directly from the individual phone models. Models of words in terms of their phones are sometimes called *phonetic baseforms* (Bahl et al., 1988).

Of course, the phonetic baseform realized as a network of phone HMMs is an alternative to the use of a whole-word HMM. As we have noted on several occasions, these phone-based models of words are essential for large-vocabulary systems for which training of complete word models is not feasible. However, when the option is available, some experimental evidence suggests that whole-word models outperform word models composed of subword units (Lippmann et al., 1987; Rabiner et al., 1989). This is not, however, an invariant rule.

Above the phoneme level of the acoustic hierarchy are the syllable and demisyllable models. Reports of studies using these units are found in (Hunt et al., 1980), and (Rosenberg et al., 1983), respectively.

Finally, we back down to the bottom of the acoustic hierarchy to discuss an unusual acoustic unit created and named the *fenone* by researchers at IBM. Generally, most contemporary systems using subword models have settled on some variation of phone-based HMMs. We will see this borne out in our discussion of specific systems in Chapter 13. However, the fenone has been used successfully by Bahl et al. (1988). This approach applies to the problem in which the speech waveform for any word is reduced to a vector-quantized observation string, say $y = y_1^T$. Suppose that we take one training word, expressed in terms of its observations, y, as a basis for the model. We know that any other utterance of the same word will produce a similar, but not identical, observation string. We can model the variation by replacing each observation in the original training sequence by a small HMM capable of learning and generating the variability surrounding the original single-frame observation.

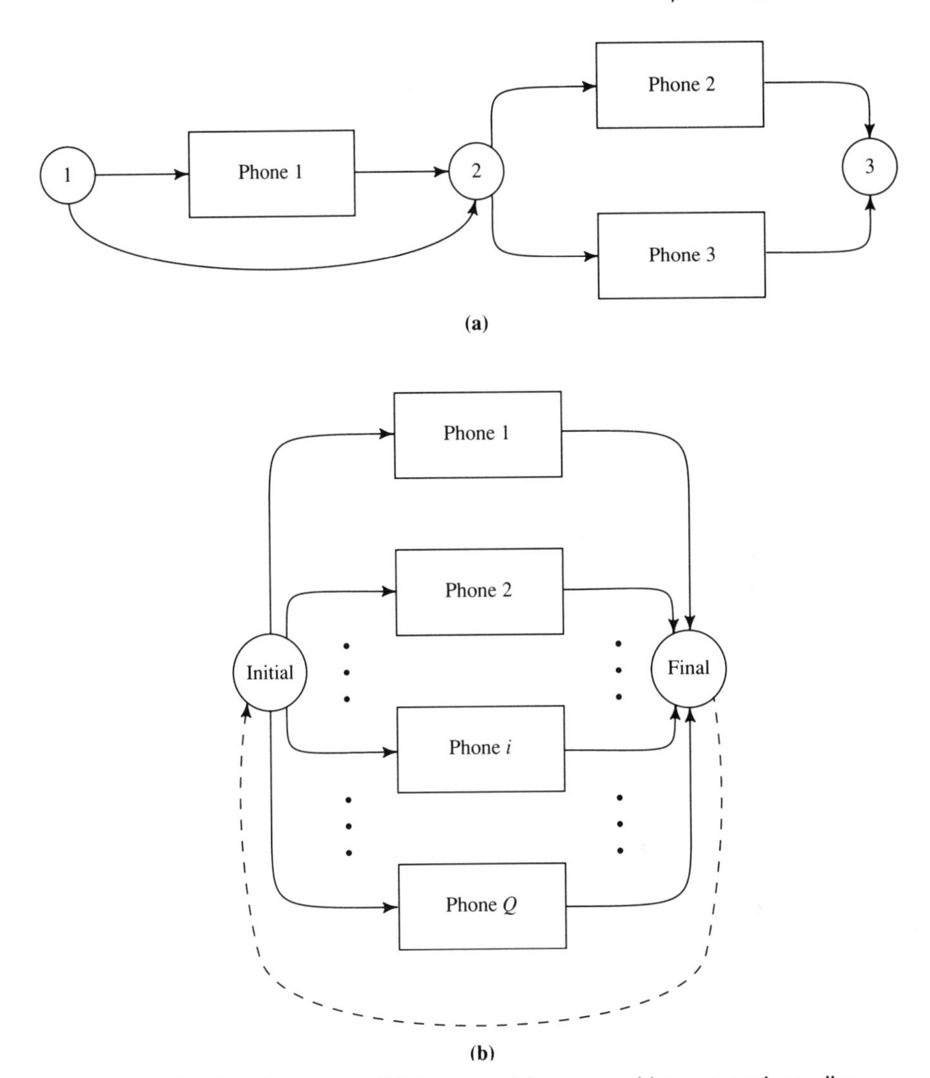

FIGURE 12.16. (a) Phoneme HMM created by networking several smaller phone models into a larger Mealy-form HMM whose output symbols are the phone designators. Inside each box labeled "phone n" is a smaller HMM modeling a particular phone. The dotted arc represents a transition producing no output. (b) An entire language model based on a "looped phonetic model" with Q phones. After Mariani (1989).

Each HMM, then, represents a very small unit of speech corresponding to a small number of frames in the acoustic waveform. This small subphone unit is called a fenone and the representation of the word in these terms is called a *fenone baseform*. Some suggested topologies for fenone models are shown in Fig. 12.11. These are taken from the paper by Bahl et al., and it is to be noted that their work uses Mealy-form HMMs. These fenone-based models of words can be trained and used in recognition in the same manner in which any composite word model is used.

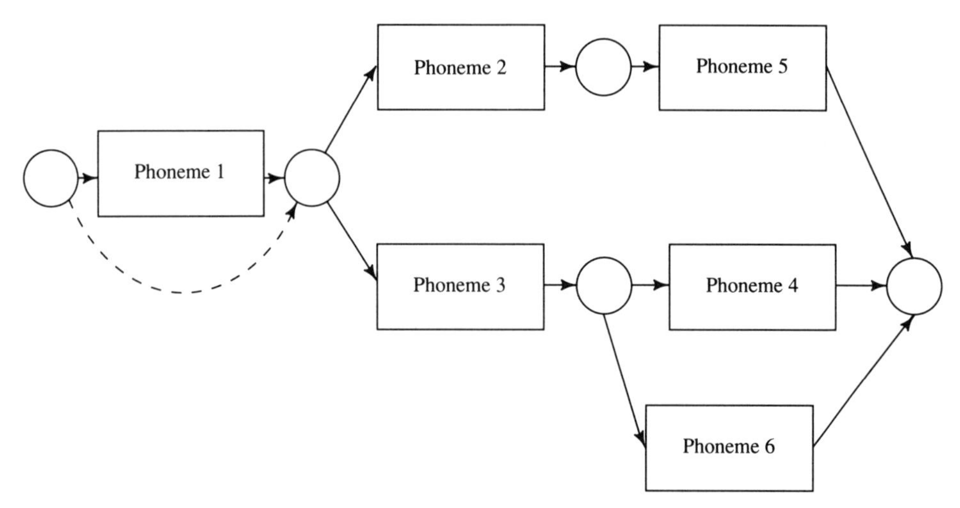

FIGURE 12.17. Word FSA constructed from concatenated phoneme HMMs. This construction is similar to the composition of phonemes from phones in the preceding figure. The phoneme models are, in turn, comprised of phone models. In practice, the intermediate phoneme abstraction is often omitted and the word built "directly" from phone models.

The benefit of these simple models is that they are apparently more robust to speaker-independent training than phonetic models, having reduced the error rate by 28% in an experiment discussed in the Bahl paper. We will see such models play a role in the "TANGORA" system in the following section.

12.4 First View of Recognition Systems Based on HMMs

12.4.1 Introduction

In this section we move closer to the real application domain and review some recognition systems and experimental work that incorporate HMMs. The reason for the phrase "A First View" in the title of this section is that we are prepared at this point in our study to examine only systems that do not employ linguistic constraints. Systems using all but small vocabularies employ such constraints, so we are somewhat restricted in the class of systems we can discuss.

Having said this, however, it should be pointed out that, by studying the HMM, we have inherently learned many important concepts that will contribute directly to our understanding of language modeling. This is because, as we will discover in Chapter 13, the HMM has a one-to-one correspondence with a frequently used language model. We have gotten a small foretaste of this idea in the study of small networks for word modeling. In fact, the HMM-like language model is so prevalent that the concepts of language modeling could almost be covered in this chapter devoted to HMMs. The HMM-like language model is not universally employed in speech recognition systems, however, and as time goes on more and more variations are being developed. Therefore, we treat language models in a separate chapter, and must remain content for the present to examine some work that does not involve language models. In Chapter 13, we will see HMMs used in multiple ways in large-vocabulary systems with linguistic constraints.

Before starting our discussion, we should reiterate a point made in Chapter 10. Speech recognition technology is vast, very detailed, and quickly changing. Accordingly, our goal here and in Section 13.9 will be to present a broad view of some of the approaches to speech recognition, with an emphasis on presenting the fundamental operating principles that will allow the reader to pursue the vast literature in the area. A second objective is to give the reader an appreciation for the performance of HMMs in various strategies. There is most certainly no claim that the few systems and research efforts described here comprise a comprehensive list.

We remind the reader of the abbreviations IWR (isolated-word recognition) and CSR (continuous-speech recognition).

12.4.2 IWR Without Syntax

Let us first consider what is the (relatively) simplest of recognition tasks—the recognition of discrete words spoken in isolation. We explicitly consider the case in which no linguistic information at the syntax level or above is present.[24] Within this task there are two main approaches using HMMs that we have discussed. The first involves the use of single HMMs to represent each word. In the other, subword units are modeled by HMMs and these models are concatenated to form complete word models.

Let us first consider the case in which a single model is dedicated to each word. The methods we have discussed for training and decoding the HMM can be applied directly to the isolated-word models, with one further consideration. We have thus far implicitly assumed that an observation string (whether for training or recognition) represents exactly the speech unit (in this case, a word) corresponding to the model. In practice, however, it is necessary to identify which of the observations represent the word and which correspond to background noise or "silence" at the endpoints. For generality, we will simply refer to "speech" and "nonspeech" segments of the signal. There are three ways to handle this problem. The first is to "manually" (usually impractical) or automatically locate the endpoints of the words and excise them from the nonspeech

[24]This means, for example, that rules for constructing words from phones may be present, but the ordering of words within messages is unknown to the recognizer.

(see Sections 4.3.4 and 10.2.3) in both training and recognition tasks. In this case the model represents only speech, and observation strings to be recognized must be carefully processed to represent only speech. The second approach is to intentionally include nonspeech on both ends of the training samples so that it gets accounted for in each individual model (presumably in the initial and final states). During recognition, the observation strings may optionally contain nonspeech at either end. The third technique is to train separate models for the nonspeech and to concatenate them to either end of the models which represent the words only. In this case, like the second, the incoming observation strings for recognition may have nonspeech at either end that will be properly accounted for by the model.

The third approach to accounting for nonspeech presents us again with the problem of hooking together two or more HMMs in an appropriate way. In our discusion of word models above, we did not worry about how such models would be trained and used. This, of course, is an important issue. Indeed, as we travel up the hierarchy to more complex speech recognition problems, this task will be increasingly more complex and important. Our analysis of this problem here will lay the groundwork for these more sophisticated systems. Figure 12.18 shows the nonspeech model concatenated to the initial and final states of the models for each word in the vocabulary. Note that a transition is present for the path to bypass the initial nonspeech model. The complete network amounts to nothing more than a large HMM that we must now train.

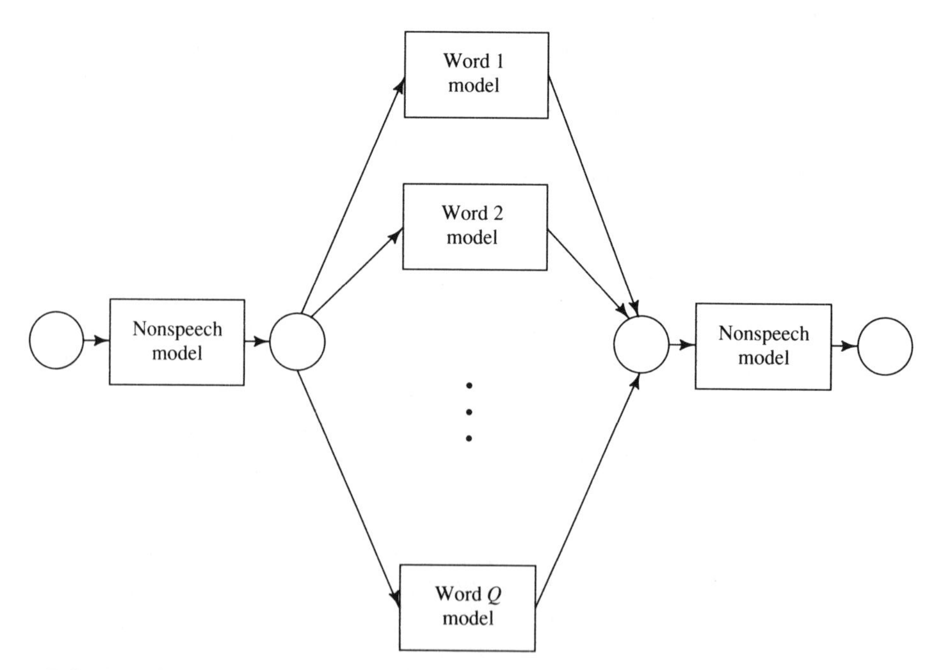

FIGURE 12.18. "Nonspeech" HMMs concatenated to the initial and final states of the models of each word in the vocabulary.

In the training phase, each of individual word models is first "seeded" by a preliminary training pass using an observation string known to be the spoken word only. The nonspeech models are also trained by nonspeech segments. The models are then joined as shown in Fig. 12.18, with some preliminary estimates of transition probabilities on the juncture transitions. A complete training database is then used to retrain the network of models. When training with a particular word, the path is constrained so that it may pass only through that word's model (all other transition probabilities temporarily set to zero at the "boundaries"). The network of HMMs is trained with the entire database in this manner, using the methods for multiple training observations discussed in Section 12.2.6. *After* a complete pass through the training data is completed, the parameters of the network are updated.

A remarkable feature of the procedure described above is that the boundaries between speech and nonspeech need not be known in the training data (except to create the seed models). Although the benefits of this feature are apparent here, we will later see that this ability of HMM networks to automatically locate boundaries is profoundly significant to CSR systems.

Of course, multiple passes through the database are necessary to appropriately train the network. Convergence to a useful model is usually achieved in a few iterations.

In the recognition phase, the large HMM network is searched in the same manner as we search any HMM (see Sections 12.2.2 and 12.2.3). In this case, a beam search is an important tool because of the large number of paths that can be generated through the network. At the completion of the search, the recognized word is discovered by backtracking through the maximum likelihood path.

We know that in large-vocabulary systems, isolated words will likely be modeled by the concatenation of subword HMMs of units such as phones or fenones. In fact, we might want to include various concatenations of subword models to account for several possible phonetic transcriptions of the word.[25] An example of such a "composite" model is shown in Fig. 12.16. While the task is more complicated in terms of recordkeeping, the methods for training and recognition are essentially similar to the case of concatenated "nonspeech–speech–nonspeech" models discussed above. Since this task may be considered a special case of the CSR strategies, we defer specific details until Section 13.7, when we will be better prepared to address this issue.

Some example research on HMM-based IWR has been compiled by Mariani (1989). Speaker-dependent IWR studies using a fenone-based HMM were conducted by IBM on the highly confusable 62-word "key-

[25]Inherent in this network is a set of lexical rules indicating how subword units are combined to form words. These rules, indeed, form a set of linguistic constraints. How these rules were discovered and built into the network will not concern us here, but the problem of learning the linguistic rules from a database will be an important issue in our study of language models in Chapter 13.

board" vocabulary, which includes the alphabet, digits, and punctuation marks (Bahl et al., 1988) as a part of the broader "TANGORA" effort to be described in Chapter 13. These studies yielded a 0.8% error rate. At Lincoln Laboratory, researchers have studied the effects of different types of speaking modes (loud, soft, fast, etc.) and the effects of noise on speaker-dependent HMM-based IWR (Paul et al., 1986). Continuous observation HMMs with "multistyle training" yielded a 0.7% error rate on a 105-word database. This work was later extended to a medium-vocabulary (207 words) speaker-dependent CSR system, which is detailed in (Paul and Martin, 1988).

In the speaker-independent realm, the Centre Nationale des Recherches Scientifique (CNRS) in France has used continuous observation HMMs for the recognition of isolated digits (85% recognition) and signs of the zodiac (89% recognition) over the public phone lines.

12.4.3 CSR by the Connected-Word Strategy Without Syntax

In one widely read paper (Rabiner et al., 1989), researchers at AT&T Bell Laboratories have explored the application of isolated-word HMMs to the recognition of connected digits (connected utterances of the words for the numerals 0–9 plus the word "oh"). Recall that in our earlier work, we have emphasized the fact that the term *connected-speech recognition* is a reference to the decoding technique rather than the manner in which the speaker necessarily utters the message. In general, the speech is uttered in a "continuous" manner without cooperation of the speaker. Such is the case with the digit study discussed here.

As we will see below, knowledge of the syntax of a language (which words may follow which others) significantly improves performance of multiword utterances. Digit strings represent challenging tasks for speech recognizers in that, if the digits are equally likely in each time slot, then there is no syntax.[26] Methods for recognizing digit strings have been based on detailed statistical models and DTW (Bush and Kopec, 1987; Rabiner et al., 1986a, 1986b; Bocchieri and Doddington, 1986). The research described here is based on HMMs used in an LB approach.

Our background will allow us to understand the general principles of HMM-based LB in a few sentences. In discussing DTW-based LB, we learned how to lay out the problem as a sequence of grid searches using Viterbi decoding. The tricky part of the algorithm was setting up the appropriate arrays to "interface" the grid searches at the boundaries of the various levels. Further, we have also learned how to view the HMM recognition problem as a Viterbi search of a grid. LB on HMMs consists of inserting the HMM grid searches in place of the DTW grid searches in

[26]This is an example of what we will call a *Type* 0, or *unrestricted*, grammar in Chapter 13.

the LB algorithm. The other details are essentially identical to those discussed in Section 11.4.2, so we need not belabor the issue.

The AT&T digit study (Rabiner et al., 1989) offers an interesting application of many of the methods we have studied in our work. The individual models are whole-word HMMs of continuous observation type with Gaussian mixture pdf's used in each state to characterize the observations. The main observations consist of cepstral features derived from LP analysis of each frame. Interestingly, a second (energy) feature is also computed in each frame, and the observation likelihood $[b(\mathbf{y}(t)|i)]$ is replaced by the product of this quantity with the probability of the energy observation (see paper for details). State duration probability distributions are also included in the models. The algorithm was tested using single-speaker-trained (same speaker in training and recognition), multispeaker-trained (same speakers in training and recognition), and speaker-independent (different speakers in training from those used in recognition) models. Multiple HMMs were used for each digit. Some typical results are shown in Table 12.1. These research efforts have been applied in various practical domains including speech-telephone dialing, credit card number verification systems, and automatic data entry [see several articles in (*AT&T System Technical Journal*, 1990)]. Another speaker-independent digit recognition application to telephone dialing is found in the work of Jovet et al. (1986).

Many other research centers have contributed to the problem of connected-digit recognition [e.g., (Doddington, 1989)]. The AT&T work was described above because it serves as a good example application of many concepts studied above. We will return to this problem after we have more experience with the concepts of syntax and grammars in Chapter 13.

TABLE 12.1. Typical Results of an AT&T Digit Recognition Study. From (Rabiner, 1989).

	Training Set		Testing Set	
Mode	**Unknown length strings**	**Known length strings**	**Unknown length strings**	**Known length strings**
Speaker trained (50 talkers)	0.39	0.16	0.78	0.35
Multispeaker (50 talkers)	1.74	0.98	2.85	1.65
Speaker independent (112/113 talkers)	1.24	0.36	2.94	1.75

Note: The algorithm was tested using single-speaker-trained (same speaker in training and recognition), multispeaker-trained (same speakers in training and recognition), and speaker-independent (different speakers in training from those used in recognition) models. Multiple HMMs were used for each digit.

12.4.4 Preliminary Comments on Language Modeling Using HMMs

We are finally ready to discuss the most challenging speech recognition problem, that of recognizing utterances from large vocabularies, particularly those uttered as "continuous" speech. It is at this point in our study that we can no longer avoid the issue of linguistic constraints. This is true for two reasons. First, as we already know, large vocabularies require the recognizer to interface with the acoustic signal at a subword level since all word models and variations cannot be stored and searched. This requires the presence of lexical knowledge, which indicates how words are formed from more basic units. (When confronted with this problem above, we simply assumed that the lexical knowledge was somehow provided). Second, even if whole-word models could be used, the presence of grammatical rules above the word level is necessary to reduce the number of word strings searched, and by doing so decreases the entropy of the search, resulting in better performance. We will therefore study the issues of language modeling and CSR together in Chapter 13.

Why all of these remarks in a chapter devoted to HMMs? Systems involving linguistic constraints comprise the most researched area of speech recognition in recent years, and it is in this domain that the powerful "self-organizing" ability of the HMM has had the most significant payoff. As we have noted several times above, we will find that HMMs play a role not only at the acoustic level of the processing, but frequently at the linguistic levels as well. This is because an HMM is in essence a "finite state automaton," an abstract machine that can generate a language produced by a "regular grammar." Because of its simplicity, a regular grammar is often used to model the speech production code. It is when the language is modeled with other than a regular grammar that the recognizer contains some non-HMM aspects.

With this indication that the theory of HMMs that we have painstakingly worked through in this chapter will be central to the interesting problems to follow, we proceed to the next level of our study of speech recognition.

12.5 Problems

12.1. The Baum–Welch F–B algorithm involves both a "forward" probability recursion and a "backward" probability recursion. The forward recursion is given by (12.34) and is initialized by (12.35). This result was carefully developed in the text.

(a) The backward recursion, (12.36), is developed by a similar line of reasoning. Give a careful development of that recursion.

(b) Explain how the backward recursion should be initialized.

12.2. In this problem we seek an analytical relationship between the two "any path" decoding methods. Suppose that you were given the state space model of (12.42) and (12.43). Show how to use this form of the HMM to compute $\alpha(y_1^t, i)$, the forward probability in the F–B algorithm, for $t = 1, 2, \ldots, T$ and for $i = 1, 2, \ldots, S$.

12.3. In developing the Viterbi search, we concluded that the optimal model of length T is the one which maximizes

$$\begin{aligned} D' &= \prod_{t=1}^{T} d'\left[(t, i_t)\middle|(t-1, i_{t-1})\right] \\ &= \prod_{t=1}^{T} a(i_t \mid i_{t-1}) b(y(t) \mid i_t), \end{aligned} \tag{12.185}$$

where

$$a(i_1 \mid i_0) \stackrel{\text{def}}{=} a(i_1 \mid 0) \stackrel{\text{def}}{=} P(\underline{x}(1) = i_1). \tag{12.186}$$

This represents the first DP search encountered for which the maximal-cost path is desired. We circumvented the need to worry about this case by taking the negative logarithm of each side of the cost equation, then minimizing the cost. In this problem, we develop the Viterbi algorithm based directly upon (12.185).

(a) Give a simple argument showing that the BOP works equally well for finding the maximal-cost path.

(b) Modify the Viterbi algorithm in Fig. 12.5 so that it is based upon a maximization of (12.185).

(c) What is wrong with this modified algorithm in practice? Give some rough numbers to support your answer.

12.4. Hidden Markov modeling depends critically upon the unrealistic assumption that the observations in the sequence $\mathbf{y}_1^T = \mathbf{y}$ are statistically independent. (This has often been cited in the literature as a possible explanation for many of the HMM's failings.) One place to simply see how important this assumption is to the methods is in the state space equations (12.42) and (12.43).

(a) Explain where and how the equations would change if the observations within a state were assumed μ-dependent for every state [this means that the random variable $\underline{y}(t)$ is independent of $\underline{y}(u)$ only if $|u - t| > \mu$]. (*Note*: The main point, that the storage requirements and computational complexity increase beyond practicality if $\mu > 0$, can be appreciated by considering $\mu = 1$. You may wish to consider this case, then comment on what happens as μ increases.)

(b) If the state sequence were additionally ν-dependent, how would the equations change? ($\nu = 2$ is sufficient to make the point.)

(c) In either of the cases above, does a simple adjustment to the training process seem likely? Would sufficient training data be available?

12.5. Verify that (12.111) equals (12.79) if $M = 1$.

12.6. (a) It is desired to deduce a continuous observation HMM with scalar observations. The observation densities are to be modeled by a single Gaussian pdf at each state. There are S states in the model. Sketch the part of the training (reestimation) algorithm that infers these densities from the training data.

(b) Repeat (a) for a Viterbi procedure.

12.7. HMM $\mathcal{M}_1$ has been trained on utterances of the word "on." A second HMM, say $\mathcal{M}_2$, is trained on the word "off." A speech-recognition system containing these two models is to be used to control a nuclear power plant.

(a) $\mathcal{M}_1$ has a single Gaussian observation pdf at each state. Following training, it is discovered that, for any state, the feature vectors have approximately independent, mean-zero, unity variance elements. Use (12.67) to show that the likelihood of $\mathcal{M}_1$ having generated any observation sequence is independent of the particular observation sequence. Give an expression for that likelihood. [*Note*: $y(t)$ in (12.67) should be replaced by $\mathbf{y}(t)$ because vector-valued observations are used in this problem.]

(b) $\mathcal{M}_2$ exhibits similar observation pdf's to those of $\mathcal{M}_1$. Give an analytical argument that no matter which word is spoken, the same word (either "on" or "off") will always be recognized.

(c) In practice, of course, utterances of "on" and "off" are equally likely (they occur alternately). Find the long-term recognition rate (probability of correct recognition) using the HMMs $\mathcal{M}_1$ and $\mathcal{M}_2$. Would it be just as effective to randomly select "on" or "off" in response to an utterance? How close to the nuclear power plant would you like to live? (Express your answer in light-years.)

(d) Suppose that for $\mathcal{M}_1$ each state has a single observation pdf similar to that described in (a) except the mean vectors are different. The mean feature vector for state j is $\boldsymbol{\mu}_{\underline{y}|j}$. Show that the Euclidean metric is an appropriate measure of likelihood to use in (12.67),

$$-\log b(\mathbf{y}(t)|\,j) = d_2(\mathbf{y}(t), \boldsymbol{\mu}_{\underline{y}|j}). \tag{12.187}$$

(e) Suppose that $\mathcal{M}_1$ emerged with different state mean vectors [as in (d)] and with approximately equal covariances among the states (elements within feature vectors no longer independent, but covariances do not change much from state to state),

$$\mathbf{C}_{\underline{y}|1} \approx \mathbf{C}_{\underline{y}|2} \approx \cdots \approx \mathbf{C}_{\underline{y}|S}. \tag{12.188}$$

Let $\mathbf{C}_{\underline{y}}$ be the average covariance matrix. Show that the Mahalānobis distance is an appropriate cost to use in (12.67).

12.8. (a) Modify the Viterbi recognition algorithm in Fig. 12.5 so that the self-transition probabilities, $a(i|i), i = 1, \ldots, S$, are effectively replaced by uniform duration distributions,

$$P(\underline{d}_i = d) = \begin{cases} 1/(D_{i,\text{out}} - D_{i,\text{in}} + 1), & D_{i,\text{in}} \leq d \leq D_{i,\text{out}}. \\ 0, & \text{other } d \end{cases} \tag{12.189}$$

(b) A similar approach to that in (a) is to allow the search to remain in state i at no cost for $D_{i,\text{in}}$ to $D_{i,\text{out}}$ observations, but to impose a severe cost penalty for leaving earlier or later. Modify the Viterbi search accordingly.

12.9. Verify (12.157) and derive the similar scaled reestimation formula for the observation probabilities.

12.10. (a) Show how to modify the F–B algorithm for the discrete observation HMM so that the observation probability vectors are shared (tied) between states i_1 and i_2.
(b) Repeat (a) for Viterbi training.

12.11. The following problem is based on an analysis by Juang and Rabiner (1985) used to motivate the need for an HMM distance measure like the one discussed in Section 12.2.8. Consider two discrete observation HMMs, say $\mathcal{M}_1$ and $\mathcal{M}_2$, with the following associated statistics:

$$\begin{aligned} \mathbf{A}_1 &= \begin{bmatrix} p & 1-p \\ 1-p & p \end{bmatrix} \\ \mathbf{B}_1 &= \begin{bmatrix} q & 1-q \\ 1-q & q \end{bmatrix} \\ \boldsymbol{\pi}_1(1) &= \begin{bmatrix} \frac{1}{2} & \frac{1}{2} \end{bmatrix}^T \\ \mathbf{A}_2 &= \begin{bmatrix} r & 1-r \\ 1-r & r \end{bmatrix} \\ \mathbf{B}_2 &= \begin{bmatrix} s & 1-s \\ 1-s & s \end{bmatrix} \\ \boldsymbol{\pi}_2(1) &= \begin{bmatrix} \frac{1}{2} & \frac{1}{2} \end{bmatrix}^T. \end{aligned} \tag{12.190}$$

(a) If there are only two possible observations called y_1 and y_2, show that $\mathcal{E}\{\underline{y}(t)|\mathcal{M}_1\} = \mathcal{E}\{\underline{y}(t)|\mathcal{M}_2\}$ if

$$s = \frac{p + q - 2pq - r}{1 - 2r}. \tag{12.191}$$

If (12.191) holds, therefore, it is reasonable to say that $\mathcal{M}_1$ and $\mathcal{M}_2$ are very similar, since they tend to generate the same observation sequences. However, we show next that this similarity need not be evident in the matrices that comprise the models.

(b) Suppose that $p = 0.6$, $q = 0.7$, and $r = 0.2$. What is s such that (12.191) holds? Are the two corresponding HMMs apparently similar by viewing their matrices?

(c) In vector spaces, vector norms are said to induce a norm upon matrices [see, e.g., (Nobel, 1969, Sec. 13.2)]. For example, the Euclidean norm of a matrix $\mathbf{D}$ is

$$\|\mathbf{D}\|_2 \stackrel{\text{def}}{=} \sqrt{\text{maximum eigenvalue of } \mathbf{D}^T\mathbf{D}}\,. \tag{12.192}$$

Accordingly, the Euclidean distance between matrices $\mathbf{A}_1$ and $\mathbf{A}_2$ is

$$d_2(\mathbf{A}_1, \mathbf{A}_2) \stackrel{\text{def}}{=} \|\mathbf{A}_2 - \mathbf{A}_1\|_2. \tag{12.193}$$

Using the work above, show that the Euclidean distance between the state transition matrices of two HMMs is a very poor indicator of their similarity.

CHAPTER 13

Language Modeling

Reading Notes:

1. *This chapter will not require any of the specialized topics in Chapter 1. The reader might wish to casually review the comments at the beginning of Section 1.3 to help put this material in perspective.*
2. *If it has been a while since the reader has studied Chapter 10, especially Section 10.2.4, it would be worthwhile to review the concepts and terminology.*

13.1 Introduction

In this chapter we discuss the theoretical basis for language models and their practical application, especially to the problem of CSR. It is important to dwell on the phrase "especially to the problem of CSR." Without linguistic processing, we have been able to treat CSR in only fairly superficial ways in our study thus far. The reader might get the idea that now we have reached the most complex problem—CSR—so we need more high-powered tools. In fact, the truth is subtler than that. In a sense, the use of linguistic processing renders all recognition techniques special cases of CSR.[1] Language processing is generally concerned with the attempt to recognize a large pattern (sentence) by decomposing it into small subpatterns according to rules that reduce entropy. We can view the motivations for this chapter as twofold. If the system involves a large vocabulary, the words must be recognized as smaller units. Lexical rules and other subword knowledge can assist with this process. Secondly, whether or not the vocabulary is large (i.e., whether or not the words are recognized as subwords), the recognition of a sentence is benefited by the knowledge of superword (syntactic and "above") knowledge that yields word ordering information. The usual case of a system employing linguistic processing is the more general case in which linguistic rules are applied both "above" and "below" the word level. The word becomes an intermediate entity along the processing path from the sentence to the signal (or vice versa), and knowledge of temporal word boundaries in the acoustic signal, while significant algorithmically and computationally, does not make an IWR problem as distinctly different from CSR as it is in the "no language" case. Having noted this, it is important to point out

[1]Even recognition techniques with no apparent linguistic processing can be framed as special cases.

that IWR is still used in linguistic processing systems where applicable, but the processing fits neatly into the general framework and need not be considered separately.[2] As we will discover, the same comments apply to the "connected-speech" processing technique. It is more effective to view such a system as a special case of general CSR processing. In fact, we will employ the connected-speech approach as a simple way to introduce some of the fundamental principles of linguistic processing.

Of course, all of the comments above will be better understood in retrospect. At the outset, however, it is important to note that language processing tends to unify all speech recognition techniques into a general framework. Although large-vocabulary CSR benefits greatly from linguistic processing, and is essentially impossible without it, language modeling is not just a sophisticated theory employed to assist with the very complex problem of large-vocabulary CSR.

Language modeling research comprises a vast endeavor with a long and interesting history in which numerous and varied systems have been produced. Our goal here will be to provide an overview of some of the basic operating principles, particularly those related to HMMs, and to briefly sketch some of the historical and contemporary developments and systems. Following the study of this chapter, the reader who is interested in specific details of systems will be prepared to pursue the literature on the subject.

Our first task is to learn some formal techniques for the development and study of language models.

13.2 Formal Tools for Linguistic Processing

13.2.1 Formal Languages

Our goals in this section are to gain a working knowledge of some of the concepts of formal languages and automata theory. We will not pursue these topics in depth. The interested reader is referred to one of the standard texts on the subject [e.g., (Fu, 1982; Hopcroft and Ullman, 1979)].

In Chapter 10 we described a model of natural language due to Peirce that included symbolic, grammatical, semantic, and pragmatic components. We then went on to describe knowledge sources due to Reddy that could be embedded in a speech recognizer. These are shown in Fig. 10.4 in relation to the formal components of Peirce's abstract model. Recall

[2]One can correctly argue that IWR, as studied in previous chapters, is a special case of the connected-speech CSR problem when no language model is used. This is, of course, correct, but unlike the present case, attempting to treat IWR as a special case of CSR there would have been pedagogically inappropriate because the known temporal boundaries are very significant in that problem.

that the grammar is a set of rules by which symbols (phonemes in our discussion there) may be properly combined in the natural language. Recall also that the language is the set of all possible combinations of symbols.

The concept of a "language" and a "grammar" can be generalized to any phenomenon that can be viewed as generating structured entities by building them from primitive patterns according to certain rules. Speech is one such patterned phenomenon, but we can also construct grammars that govern the formation of images, strings of binary digits, or computer code. The primitive patterns associated with a "formal language" are called "terminals" of the language. The language itself is defined to be the set of all terminal strings that can be produced by the rules of the "formal grammar." We will make these notions more concrete below.

For our purposes, we can define an *automaton* to be an abstract machine capable of carrying out the rules of a grammar to produce any element of the language. We will discover that an HMM is in essence a "finite state automaton," an abstract machine that can generate a language produced by a "regular grammar." Because of its simplicity, a regular grammar is often used to model the speech production code at all levels of linguistic and acoustic processing. This is why the HMM figures so prominently in the CSR problem. It is when the language is modeled with other than a regular grammar that the recognizer will contain some "non-HMM" aspects. In order to put these issues in perspective, our first task in this chapter will be to learn some principles of formal language modeling.

Formally, a *grammar* of a language is a four-tuple

$$\mathcal{G} = (V_n, V_t, P, S), \tag{13.1}$$

in which V_n and V_t are the *nonterminal* and *terminal vocabularies* (finite sets), P is a finite set of *production rules*, and S is the *starting symbol* for all productions. The sets V_n and V_t are disjoint, and their union, say V, is called the *vocabulary* of the language. We denote by V_t^* the set of all possible strings that can be composed from the elements of V_t. (A similar meaning is given to V_n^* and V^*.) The rules in P are of the form[3]

$$\alpha \rightarrow \beta, \tag{13.2}$$

where α and β are strings over V, with α containing at least one element of V_n.

By definition, V_t^* is the *language* with which the grammar is associated. We can describe the language more formally as follows. Let

$$\gamma \Rightarrow \delta \tag{13.3}$$

[3]In the following, lowercase roman letters will indicate strings over V_t^*, uppercase roman letters strings over V_n^*, and lowercase Greek letters strings over V^*.

denote the fact that there is a rule in P from which δ can be derived from γ. This means that if $\gamma = \omega_1 \alpha \omega_2$ and $\delta = \omega_1 \beta \omega_2$, then the rule $\alpha \rightarrow \beta$ must be in P. The notation

$$\gamma \overset{*}{\Rightarrow} \delta \tag{13.4}$$

means that there are rules in P, by which δ can *eventually* be derived from γ, although not necessarily in one production. In these terms, the language of $\mathcal{G}$ is given by

$$\mathcal{L}(\mathcal{G}) \overset{\text{def}}{=} \left\{x \middle| x \in V_t^* \text{ such that } S \overset{*}{\Rightarrow} x\right\}. \tag{13.5}$$

There are four types of grammars identified by Chomsky (1959a):

1. *Type 0* or *unrestricted.* An unrestricted grammar has no production restrictions. Any string can derive any other string.
2. *Type 1* or *context sensitive.* In a context-sensitive grammar, the production rules must be of the form

$$\omega_1 A \omega_2 \rightarrow \omega_1 \beta \omega_2, \tag{13.6}$$

   where $A \in V_n^*$, and $\omega_1, \omega_2, \beta \in V^*$, with $\beta \neq \emptyset$, the null string. This means that A may be replaced by β *in the context of* ω_1, ω_2.
3. *Type 2* or *context free.* In a context-free grammar, the productions are of the form

$$A \rightarrow \beta, \tag{13.7}$$

   where $A \in V_n^*$, and $\beta \in V^*$, with $\beta \neq \emptyset$. This means that A may be replaced by β *in any context.*
4. *Type 3, finite state,* or *regular.* In a regular grammar, the production rules are of the form

$$A \rightarrow aB \tag{13.8}$$

   or

$$A \rightarrow b, \tag{13.9}$$

   where $A, B \in V_n^*$, and $a, b \in V_t^*$. We see that every production of a regular grammar produces a terminal element of the vocabulary.

Each level of the Chomsky hierarchy contains the one below it as a special case.

A *stochastic grammar* is one whose production rules have probabilities associated with them, for example,

$$\alpha \overset{p}{\rightarrow} \beta, \tag{13.10}$$

where $\overset{p}{\rightarrow}$ indicates that α generates β with probability p. Now each of the strings in the *stochastic language* occurs with a probability that depends on the probabilities of its productions. Otherwise a stochastic grammar or language is the same as a deterministic one. In fact, the grammar that results from removing all probabilities from the productions of a stochastic grammar is called its *characteristic grammar.* A stochastic grammar is classified in the Chomsky hierarchy according to the type of its characteristic grammar.

The formalities described above will allow us to discuss the language constraints in precise terms below. The reader will begin to gain more intuition about these ideas as we apply them to our work.

This is perhaps a good place to highlight the fact that our discussion of language constraints will rarely move above the grammatical level into semantic and pragmatic considerations (see Chapter 10). The details of these topics are beyond the scope of this book because they are very difficult to quantify and reduce to fundamental principles. We will, however, briefly mention these features in certain speech recognition systems when they are described later in the chapter.

13.2.2 Perplexity of a Language

In comparing the performance of speech recognizers, it is important to be able to quantify the difficulty of the recognition task. Language constraints naturally tend to lessen the uncertainty (decrease entropy) of the content of sentences and facilitate recognition. For example, if there are, on average, very few words that may follow any given word in a language, then the recognizer has fewer options to check and recognition performance will be better than if many such words were possible. This example suggests that an appropriate measure of difficulty of any language might involve some measure of the average number of terminals that may follow any given terminal. If the language is viewed as a graph in which terminals are associated with transitions, for example, then this measure would be related to the average "branching factor" at any decision point in the graph. Roughly speaking, this is the quantity measured by "perplexity," which we now seek to formalize.

A formal stochastic language generates terminal strings with certain probabilities. We can view these terminal strings as realizations of a discrete-"time" stationary stochastic process whose random variables take discrete values. These "discrete values" correspond to the individual terminals, and the "time" simply indicates the position of the random terminal in a string. For the sake of discussion, let us assume that the terminals correspond to words and denote this random process by $\underline{w}$. If there are W possible words, say $w_1, \ldots, w_W$, then we know from Section 1.4.2 that the entropy associated with this random process or "source" is[4]

$$\begin{aligned} H(\underline{w}) &= -\mathcal{E}\{\log P(\underline{w}(\cdot) = w_i)\} \\ &= -\sum_{i=1}^{W} P(\underline{w}(\cdot) = w_i) \log P(\underline{w}(\cdot) = w_i), \end{aligned} \tag{13.11}$$

[4]The binary log is used here.

where $\underline{w}(\cdot)$ is an arbitrary random variable in $\underline{w}$ *if* the source has independent and identically distributed random variables. If not, the entropy is given by

$$H(\underline{w}) = -\lim_{N\to\infty} \frac{1}{N}\mathcal{E}\{\log P(\underline{w}_1^N = w_1^N)\}$$
$$= -\lim_{N\to\infty} \frac{1}{N}\sum_{w_1^N} P(\underline{w}_1^N = w_1^N)\log P(\underline{w}_1^N = w_1^N), \tag{13.12}$$

where $\underline{w}_1^N$ denotes the random variable string $\underline{w}(1), \ldots, \underline{w}(N)$, and w_1^N denotes the partial realization $w(1), \ldots, w(N)$, and the sum is taken over all such realizations.[5] Since the words in a language are not likely to be independent, we use (13.12), recognizing that it reduces to (13.11) when they are independent. For an ergodic source, we can compute the entropy using a "temporal" average

$$H(\underline{w}) = -\lim_{N\to\infty} \frac{1}{N}\log P(\underline{w}_1^N = w_1^N). \tag{13.13}$$

In practice, the longer the sentence (larger N) used to estimate H, the better will be the estimate; H represents the average number of bits of information inherent in a word in the language. In turn, this means that $H(\underline{w})$ bits must be extracted by the recognizer from the acoustic data on the average in order to recognize each word.

Of course, the probabilities $P(\underline{w}_1^N = w_1^N)$ are unknown and must be estimated from training data (which can be viewed as example productions of the grammar). Let us call the estimates $\hat{P}(\underline{w}_1^N = w_1^N)$, and the resulting entropy measure $\hat{H}(\underline{w})$,

$$\hat{H}(\underline{w}) = -\lim_{N\to\infty} \frac{1}{N}\log \hat{P}(\underline{w}_1^N = w_1^N). \tag{13.14}$$

It can be shown (Jelinek, 1990) that $\hat{H} \geq H$ if $\underline{w}$ is properly ergodic.

Although the entropy provides a perfectly valid measure of difficulty, speech researchers choose to use the *perplexity*, defined as

$$Q(\underline{w}) \stackrel{\text{def}}{=} 2^{\hat{H}(\underline{w})} \approx 1/\sqrt[N]{\hat{P}(w_1^N)} \tag{13.15}$$

for some large N. To see the sense of this measure, note that if the language has W equally likely words that occur independently of one another in any string, then it follows from (13.11) that the amount of entropy in any string is

$$H(\underline{w}) = \log_2 W. \tag{13.16}$$

Obviously, the size of the vocabulary in this case is related to the entropy as

$$W = 2^{H(\underline{w})}. \tag{13.17}$$

[5] $\underline{w}_1^N$ is similar to the abusive notation we have used before to denote a random model of a partial realization.

Comparing with (13.15), we see that the perplexity of a language can be interpreted as the size of the vocabulary (number of terminals) in another language with equiprobable, independent words, which is equally difficult to recognize. Perplexity therefore indicates an average branching factor for the language modeled by $\underline{w}$. We will see perplexity used to compare performance of various systems later in the chapter.

13.2.3 Bottom-Up Versus Top-Down Parsing

To obtain a more concrete feeling for formal grammatical rules, let us consider an example. In terms of a natural language (Peircian model) we consider the sentence to be a complete utterance, and phonemes to be the basic symbols of the language. Figure 13.1 illustrates the idea of a grammar for this choice of utterance and symbol. For one example utterance, "The child cried as she left in the red plane," the set of rules by which phonemes may ultimately compose this sentence may be inferred from this figure. We see rules for how phonemes form words, how words are classified into parts of speech, how parts of speech form phrases, and how phrases form sentences. In the other direction, we see how this sentence can ultimately be decomposed into its component symbols by a series of rules. The sentence is rewritten in terms of phrases (a noun phrase [NP] and a prepositional phrase [PP]) and parts of speech (verbs [V], a conjunction [CONJ], and a pronoun [PRON]). Then the phrases are decomposed into parts of speech, the parts of speech indicators produce words, and the words are finally decomposed into phonemes.[6]

Although the example above involves a *natural* language, there is nothing preventing us from viewing it as a *formal* language. The terminals are taken to be phonemes and all other quantities but S are taken as nonterminals. Some of the production rules are evident in the example. For example, coming down the right side of the figure, we see the production rules

$$
\begin{aligned}
S &\rightarrow \text{NP, V, CONJ, PRON, V, PP} \\
\text{PP} &\rightarrow \text{PREP, NP} \\
\text{NP} &\rightarrow \text{ART, ADJ, N} \\
\text{ART} &\rightarrow \text{THE} \\
\text{ADJ} &\rightarrow \text{RED} \\
\text{N} &\rightarrow \text{PLANE} \\
\text{THE} &\rightarrow /\text{Dx}/ \\
\text{RED} &\rightarrow /\text{rEd}/ \\
\text{PLANE} &\rightarrow /\text{plen}/,
\end{aligned}
\tag{13.18}
$$

[6]We should note that this is a *phonemic transcription*, meaning that not much of the *phonetic* detail is rendered. For details see (Ladefoged, 1975).

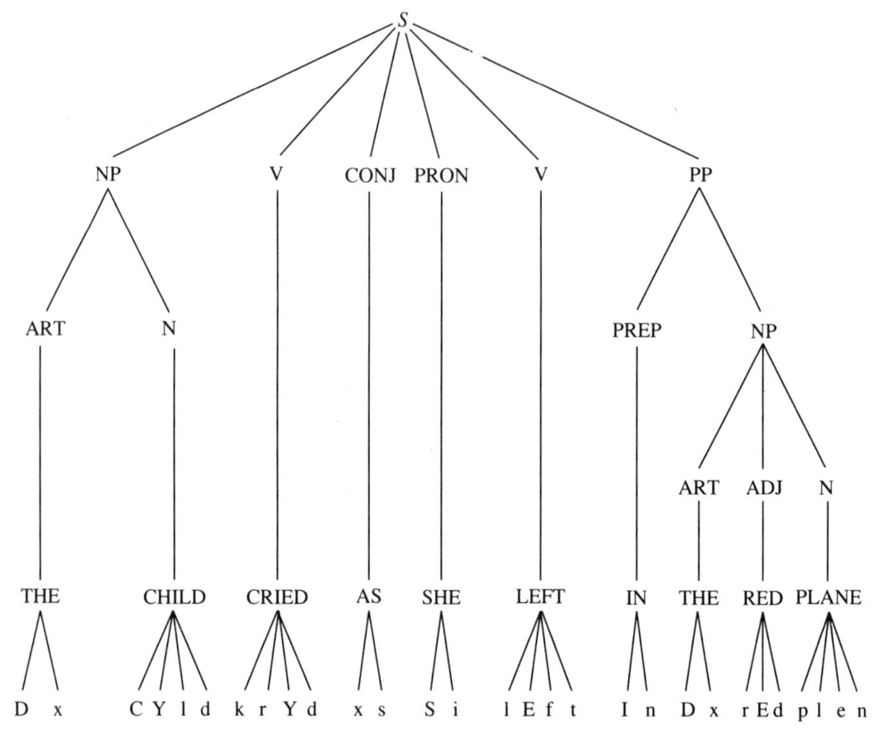

FIGURE 13.1. Production of the utterance "The child cried as she left in the red plane" according to grammatical (syntactic and lexical) rules. The nonterminal vocabulary for this example corresponds to phrases, parts of speech, and words: NP = noun phrase, PP = prepositional phrase, V = verb, CONJ = conjunction, PRON = pronoun, ART = article, PREP = preposition, ADJ = adjective. The terminals are phonemic symbols.

where we have used uppercase to denote the nonterminal quantities and lowercase for the terminals.

We implied above that the grammatical rules (for a particular sentence) may be viewed in either the upward (compose the sentence) or downward (decompose the sentence) direction. Clearly, the information is equivalent. There is a fundamental technical difference, however, inherent in the direction in which the grammatical rules are viewed. The process of determining whether a set of production rules exists in a grammar for composing (from terminals) or decomposing (into terminals) a sentence is called *parsing*. A parsing algorithm that seeks rules for composition, that is, which uses the grammatical rules in the direction

$$\text{terminals} \rightarrow \text{sentence}, \tag{13.19}$$

is called a *bottom-up* parser. On the other hand, if the rules are used for decomposition,

$$\text{sentence} \rightarrow \text{terminals}, \tag{13.20}$$

the algorithm is a *top-down* parser.

Let us examine how parsing with a grammar can be used in the speech recognition problem. We will first look at this issue in broad terms and become more specific as the discussion proceeds. In the bottom-up case, the grammar aids in recognition by disallowing symbol combinations that are not in the language. It can also be used to assign a likelihood to legitimate symbol strings if the grammar is stochastic. The part of the recognizer that converts the acoustic data into linguistic symbols is sometimes called the *acoustic decoder* (AD). Suppose that the AD has processed the utterance and has hypothesized a set of phones (symbols or terminals). The *linguistic decoder* (LD) then goes to work applying the linguistic (in this case grammatical) constraints. Working from the bottom up, the LD can determine whether this set of phones corresponds to a legitimate sentence in the language, and the likelihood of that sentence if appropriate. This can prevent acceptance of erroneous combinations of phones that are *hypothesized* by the AD. More realistically, the AD would probably start hypothesizing phones from the left (*left–right parsing*), guided at each step by the legitimacy and likelihood of the string it is creating. Those hypotheses that are illegal or unlikely would be abandoned before reaching the end of the utterance. This idea will show up as a pruning measure later on. Of course, multiple evolving hypotheses may be considered simultaneously.

In the top-down case, the grammar again prohibits illegal symbol strings, but in this case, none is ever hypothesized. In this case the grammar serves to restrict the number of symbol combinations that must be considered by the recognizer at the acoustic level. The process begins at the top, where the LD hypothesizes a sentence in the language. The rules of the grammar are then used to deduce a possible set of phones corresponding to the hypothesized sentence. This process would also produce an *a priori* likelihood of the derived phones according to the statistical structure of the grammar. The complete likelihood of this phone string given the data (or vice versa) is then computed by the AD. More realistically, the LD would only use the grammar to deduce phones from the left for a given sentence hypothesis and would abandon that sentence if the phone string were turning out to be unlikely.

The main disadvantage of the bottom-up approach is that a sentence cannot be recognized unless each of its symbols is hypothesized by the AD. The bottom-up approach also does not take advantage of the linguistic constraints in decoding the acoustic signal. On the other hand, the language must be highly constrained (with respect to natural discourse) for the top-down approach to be practical, because, in principle (at least the beginning of) every possible sentence must be hypothesized by the LD. However, there are many applications that involve constrained languages with relatively small vocabularies. Some of the tasks that have been explored are data entry, military resource management, archive document retrieval, and office dictation. Most of the research on language modeling since the early 1980s has focused on the top-down approach. Accordingly, most of the systems we will describe are based on top-down linguistic decoding. Nevertheless, before leaving the bottom-up approach entirely, it will be instructive to study an example, which will allow us to contrast the two approaches. Before exploring a bottom-up system, however, we introduce the equivalence between the HMM and the "finite state automaton," and discuss their relationship to regular grammars. These concepts, to which we have alluded several times in previous discussions, will be central to understanding much of the remaining material in the chapter, including the bottom-up and top-down examples to follow.

13.3 HMMs, Finite State Automata, and Regular Grammars

To illustrate some of the basic principles of language modeling, we will use a simple digit recognition problem. The reader should appreciate that this task is trivial in scale relative to many for which contemporary systems have been developed, but it is chosen precisely for its small size. We will, for example, be able to completely illustrate the state diagram representing the language model for this problem, whereas such an illustration would be impossible for a large-vocabulary system. The small size of the vocabulary notwithstanding, many important and quite general principles can be realized with this digit study.

We first encountered the digit recognition problem in the study of DTW-based LB in Chapter 11. We studied the same problem and a similar LB approach using HMMs in Chapter 12. We will return to this LB problem as an example with which to illustrate bottom-up parsing; but first, we will explore some generalities that will serve both the bottom-up and top-down examples.

Clearly, syntactic (word order) information can be added to an attempt to recognize digit strings if it is available. In the digit recognition by LB example in Chapters 11 and 12, we assumed that no syntactic constraints were possible, since every digit could follow every other digit with equal probability. Suppose, however, that we encounter a digit recognition problem where the digits follow one another with distinctly different probabilities. In fact, suppose the probabilistic dependence of digits is as shown in the state diagram of Fig. 13.2. For the moment, let us not worry about how this diagram would be derived, and just accept it as a stochastic model of the digit strings. In this diagram, we have labeled the states with letters so that the state names are not confused with the digits to be recognized. In fact, the states are designated with uppercase letters because they will correspond to nonterminals in a formal language below. The bars over the state designators are simply to distinguish them as particular state names rather than as variables or some other quantities in the following discussion. The digits are spelled out in lowercase letters because they will correspond to terminals in a formal gram-

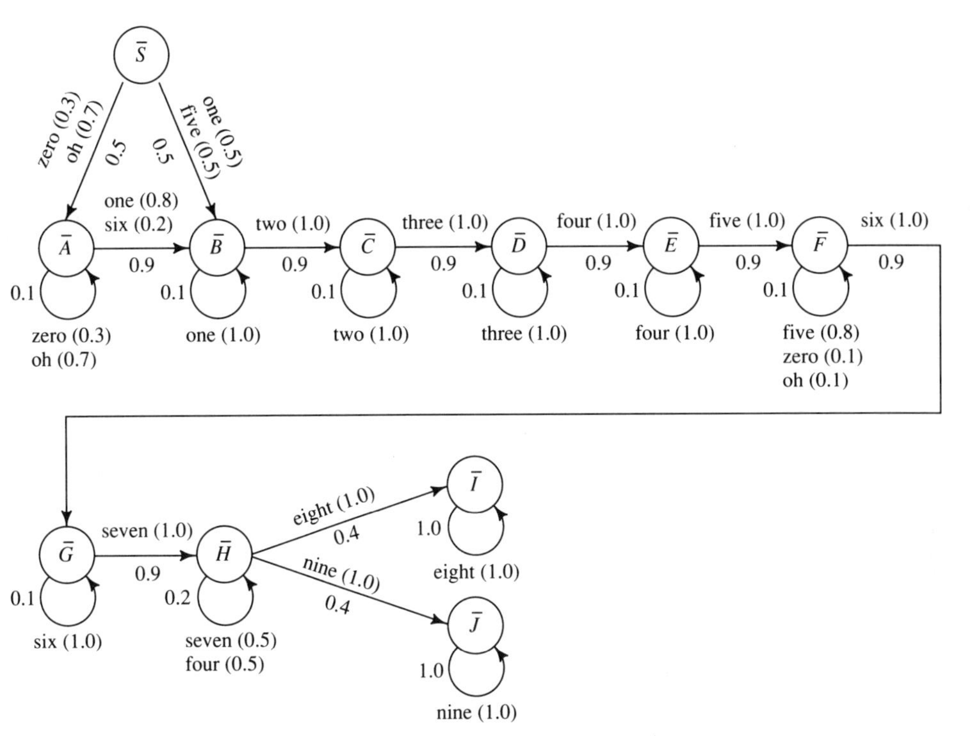

FIGURE 13.2. State diagram showing probabilistic dependencies among the digits in a hypothetical digit recognition problem.

mar below. The diagram is interpreted to mean, for example, that either the digit "one" or the digit "six" may follow an initial string of "zeros" and/or "oh's" ($\bar{S}$ is the initial state). After each initial "zero" or "oh", the probability that a "one" or a "six" will follow is 0.9 (transition from $\bar{A}$ to $\bar{B}$), and the probability that another "zero" or "oh" will follow is 0.1 (transition from $\bar{A}$ to $\bar{A}$). Given that a "one" or "six" appears, "one" will occur with probability (0.8) and "six" with probability (0.2). The numbers in parentheses are the probabilities of emitting the particular digits given the transition that they label. We can also infer that the state sequence (or, equivalently, the transition sequence) is Markov. In fact, it is clear that what we have here is a Mealy version of an HMM whose (discrete) "observations" consist of the names for the digits (words). Accordingly, we can very quickly set up the formalities of this model by analogy to our work with the acoustic-level HMM.

Beside the different forms of the HMMs, one of the main differences between this word-level HMM and the acoustic-level HMM is in the "time" scales of the observations. Let $\underline{w}$ denote the random process with random variables $\underline{w}(l)$, which model the lth digit (word) in any string.[7]

[7]To indicate generality, we will frequently use "word" in place of "digit," even though all words are digits in this example.

These are the "observations" associated with the transitions of the model, and they are analogous to the observations $y(t)$ at the acoustic level. However, l and t are very different indices: t indexes the *acoustic frame* in the speech data and is ordinarily regularly synchronized with the original samples of the data; l, on the other hand, indexes the *word number* in an utterance and is only loosely related to time. Let $\underline{X}$ be a random process with random variables $\underline{X}(l)$, $l = 0, 1, 2, \ldots$, which model the *state sequence* through the model, and $\underline{U}$ be the random process with random variables $\underline{U}(l)$, $l = 1, 2, \ldots$, which model the *transitions* on a path. Also let $U_{I|J}$ be an identifying label of the transition from state J to state I. The state quantities are labeled with uppercase letters because they will have an interpretation as nonterminals in a formal grammar below. The transitions are labeled with uppercase letters because they are closely related to the states, but the transitions will play no direct role in the formal grammar. Now, by direct analogy to the developments at the acoustic level, we can define the (*state*) *transition probabilities*,

$$A(I|J) \stackrel{\text{def}}{=} P\big(\underline{U}(l) = U_{I|J}\big) = P\big(\underline{X}(l) = I \big| \underline{X}(l-1) = J\big) \tag{13.21}$$

for arbitrary l, and the *observation* (*word or digit*) *probabilities*,

$$b\big(w(l) \big| U_{I|J}\big) \stackrel{\text{def}}{=} P\big(\underline{w}(l) = w(l) \big| \underline{U}(l) = U_{I|J}\big) \tag{13.22}$$

also for arbitrary l. Finally, for completeness, we note that the *state probability vector* prior to word l is the vector $\mathbf{\Pi}(l)$ with mth element $P\big(\underline{X}(l) = I_m\big)$ in which I_m, $m = 1, 2, \ldots, M$ represents some ordering of the M states. The most significant of these vectors is the *initial* state probability vector[8] $\mathbf{\Pi}(0)$. Since we always designate the state $\bar{S}$ as the initial state, there is only one nonzero element in $\mathbf{\Pi}(0)$, the one corresponding to $P\big(\underline{X}(0) = \bar{S}\big)$. The reader should be able to extract all of these probabilities for the present example from Fig. 13.2. It should also be clear that all of the training and recognition techniques that we carefully developed for the acoustic HMM apply equally well to this model.

In the parlance of formal language theory, the HMM is known as a *finite state automaton* (FSA). The word *stochastic* or *nondeterministic* might also precede this name to indicate that there are multiple transitions and observations that can be generated by any move in the diagram, and that these are governed by probabilistic rules [for details see, e.g., (Fu, 1982; Hopcroft and Ullman, 1979)]. We will begin to use the term FSA for an HMM that is being used to model *linguistic* information, and reserve the term "HMM" for the acoustic-level model. The reader should appreciate that, for our purposes, the two names refer to abstract models that are equivalent even though they are used to model different phenomena. We should also point out that there is no standard usage of these terms in the speech processing literature. If there is any

[8]For a Mealy-form HMM, it makes the most sense to index the initial state by "time" 0, so that the first *transition* (observation) corresponds to "time" 1.

consistency, it is that the term "HMM" is usually used to indicate the acoustic model, but there are exceptions to this rule as well.

The next point is critical. We show that an FSA has a one-to-one correspondence with a regular grammar. With this knowledge, we will be able to use the FSA as a parser for a regular language. To show this for our present example, we treat the 11 words for the digits as the terminals. Note that we have spelled out the digits in lowercase letters above precisely so they would look like terminals in this part of our discussion: "zero," "oh," "one," ..., "nine." We view the state designators as nonterminals, with the starting state, labeled $\bar{S}$, playing the role of the root of the grammar. The production rules of the grammar are as follows:

$$\bar{S} \xrightarrow{p_1} \text{zero}, \bar{A} \tag{13.23}$$

$$\bar{S} \xrightarrow{p_2} \text{oh}, \bar{A} \tag{13.24}$$

$$\bar{S} \xrightarrow{p_3} \text{one}, \bar{B} \tag{13.25}$$

$$\bar{S} \xrightarrow{p_4} \text{five}, \bar{B} \tag{13.26}$$

$$\bar{A} \xrightarrow{p_5} \text{zero}, \bar{A} \tag{13.27}$$

$$\bar{A} \xrightarrow{p_6} \text{oh}, \bar{A} \tag{13.28}$$

$$\bar{A} \xrightarrow{p_7} \text{one}, \bar{B} \tag{13.29}$$

$$\bar{A} \xrightarrow{p_8} \text{six}, \bar{B} \tag{13.30}$$

$$\bar{B} \xrightarrow{p_9} \text{one}, \bar{B} \tag{13.31}$$

$$\bar{B} \xrightarrow{p_{10}} \text{two}, \bar{C} \tag{13.32}$$

$$\bar{C} \xrightarrow{p_{11}} \text{two}, \bar{C} \tag{13.33}$$

$$\bar{C} \xrightarrow{p_{12}} \text{three}, \bar{D} \tag{13.34}$$

$$\bar{D} \xrightarrow{p_{13}} \text{three}, \bar{D} \tag{13.35}$$

$$\bar{D} \xrightarrow{p_{14}} \text{four}, \bar{E} \tag{13.36}$$

$$\bar{E} \xrightarrow{p_{15}} \text{four}, \bar{E} \tag{13.37}$$

$$\bar{E} \xrightarrow{p_{16}} \text{five}, \bar{F} \tag{13.38}$$

$$\bar{F} \xrightarrow{p_{17}} \text{zero}, \bar{F} \tag{13.39}$$

$$\bar{F} \xrightarrow{p_{18}} \text{oh}, \bar{F} \tag{13.40}$$

$$\bar{F} \xrightarrow{p_{19}} \text{five}, \bar{F} \tag{13.41}$$

$$\bar{F} \xrightarrow{p_{20}} \text{six}, \bar{G} \tag{13.42}$$

$$\bar{G} \xrightarrow{p_{21}} \text{six}, \bar{G} \tag{13.43}$$

$$\bar{G} \xrightarrow{p_{22}} \text{seven}, \bar{H} \tag{13.44}$$

$$\bar{H} \xrightarrow{p_{23}} \text{seven}, \bar{H} \tag{13.45}$$

$$\bar{H} \xrightarrow{p_{24}} \text{four}, \bar{H} \tag{13.46}$$

$$\bar{H} \xrightarrow{p_{25}} \text{eight}, \bar{I} \tag{13.47}$$

$$\bar{H} \xrightarrow{p_{26}} \text{nine}, \bar{J} \tag{13.48}$$

$$\bar{I} \xrightarrow{p'_{27}} \text{eight}, \bar{I} \tag{13.49}$$

$$\bar{I} \xrightarrow{p''_{27}} \text{eight} \tag{13.50}$$

$$\bar{J} \xrightarrow{p'_{28}} \text{nine}, \bar{J} \tag{13.51}$$

$$\bar{J} \xrightarrow{p''_{28}} \text{nine}. \tag{13.52}$$

The significance of the special primed probabilities in (13.49)–(13.52) will be discussed below. We see that this language is governed by a regular grammar because all of its production rules are of the form

$$Q \xrightarrow{p} qR \tag{13.53}$$

or

$$Q \xrightarrow{p} r, \tag{13.54}$$

where $Q, R \in V_n$ and $q, r \in V_t$. Note further that the probabilities associated with the production rules are related to the FSA as follows. Consider, for example, the rule

$$\bar{A} \xrightarrow{p_7} \text{one}, \bar{B}. \tag{13.55}$$

This rule corresponds to a jump from state $\bar{A}$ to $\bar{B}$ in the model with the accompanying generation of observation "one." Accordingly, it is clear from the state diagram that

$$p_7 = (0.9)(0.8) = 0.72. \tag{13.56}$$

With the exception of the rules involving primed probabilities, the rule probabilities are related to the state diagram probabilities in a similar way. In general, the probability, p, associated with the rule

$$Q \xrightarrow{p} qR \tag{13.57}$$

is given in terms of the FSA probabilities as

$$\begin{aligned} p &= P\big(\underline{X}(l) = R \,\big|\, \underline{X}(l-1) = Q\big) P\big(\underline{w}(l) = q \,\big|\, \underline{X}(l) = R, \underline{X}(l-1) = Q\big) \\ &= P\big(\underline{U}(l) = U_{R|Q}\big) P\big(\underline{w}(l) = q \,\big|\, \underline{U}(l) = U_{R|Q}\big), \end{aligned} \tag{13.58}$$

for arbitrary $l > 0$. In turn, we know that this can be written

$$p = A(R|Q)\, b\big(q \,\big|\, U_{R|Q}\big), \tag{13.59}$$

where $Q, R \in V_n$ (equivalently, Q and R are states), and $q \in V_t$ [equivalently, q is a word (digit) in the natural vocabulary].

The one small nuance occurs at the end of the production rule list where we have, for example, both $\bar{J} \xrightarrow{p'_{28}}$ nine, $\bar{J}$ and $\bar{J} \xrightarrow{p''_{28}}$ nine. This is simply so that the grammar may generate a *final* terminal without generating a further nonterminal. The rules (13.49)–(13.52) are therefore necessary for formal reasons. In fact, we could accomplish the same thing in the state diagram by including a phantom final state to which states $\bar{H}, \bar{I}$, and $\bar{J}$ could make final transitions without generating a new state name. Frequently, however, we do not account for the fact that a state transition is a final one in training and using an HMM or FSA. In the present state diagram, for example, the final word is generated by making a final transition into either state $\bar{I}$ or $\bar{J}$, then simply remaining there. We can compensate for this little discrepancy between the FSA and the grammar by allowing the combined production rules

$$\bar{I} \overset{p_{27}}{\longrightarrow} \text{eight}, \bar{I}; \text{ or eight} \tag{13.60}$$

$$\bar{J} \overset{p_{28}}{\longrightarrow} \text{nine}, \bar{J}; \text{ or nine}, \tag{13.61}$$

where $p_k = p'_k + p''_k$ for $k = 27$, 28. With this provision, the state diagram and the grammar correspond precisely.

This example illustrates the equivalence between an FSA or HMM and a regular grammar. Note the following:

1. Traversing a complete path in the FSA (and accumulating probabilities along the path) and generating the corresponding sentence corresponds exactly to the forward or top-down application of a set of production rules to generate the sentence.
2. Given a sentence, exploring the FSA to determine whether a path (or set of productions) exists to generate that sentence corresponds exactly to the bottom-up application of the production rules to determine whether the sentence conforms to the grammar.

Although the FSA offers a simple and highly structured manner to carry out the parsing in either direction, whether it is more convenient to formally view the grammatical structure as a set of production rules, or as an FSA, depends on the nature of the approach taken. The FSA is useful when the linguistic decoding is structured as a dynamic programming search. This will always be the case in the top-down approaches we study that involve regular grammars. On the other hand, if the FSA is just being used as "a graph of the production rules," then the grammatical structure might just as well be thought of in terms of those rules. In the discussions to follow, it will be useful to view the linguistic constraints primarily in terms of the FSA.

13.4 A "Bottom-Up" Parsing Example

In this section, we take a brief look at an example of a bottom-up parsing-based CSR system. We do so by studying a relatively simple example system consisting of the addition of a grammar (linguistic constraints) to the LB approach for digit recognition. The methods described here apply equally well to either DTW- or HMM-based LB, but we will focus on the latter. To put this example in the perspective of the introductory comments of the chapter, note that the LB approach is a "connected-speech" method that does not use grammatical rules below the word level. This is because the *word* can be used as a fundamental unit of speech, since such systems involve small vocabularies. In this sense, the following example involves a simpler system than might ordinarily be embued with a language model, but its simplicity will allow us to focus on the principles rather than the details. After mastering these basic concepts, we will build onto the system by adding grammar at the subword level.

As we proceed through this example, pay careful attention to how the grammar is used. It will be seen that the production rules of the grammar will be used from the "bottom up." This will imply that the LB sector of the system (which comprises the AD) is responsible for a very important task common to all language-based systems, especially CSR systems. This is what we might call the "temporal recordkeeping" function. Because the time boundaries between speech units (in this case words) are unknown, many different regions of the speech must be tried as candidate regions for the first word, second word, and so on. Further, any legitimate solution must be composed of words whose corresponding temporal regions are adjacent and nonoverlapping.[9] As we discuss the example below, the reader is also encouraged to notice how the LB algorithm performs the recordkeeping function. (Actually, we are already familiar with how LB does this, but it is useful to review our understanding in this new situation.) Toward the end of this section, we will discuss how the LD could be made responsible for the temporal recordkeeping in a similar system.

We assume the language model for the digits that was developed in Section 13.3 and illustrated as an FSA in Fig. 13.2. Our objective is to examine now how we can use this linguistic knowledge in assisting in the recognition process in the LB problem. Let us first move down to the acoustic-level processing, and suppose that we have found N candidate word strings, say $w^{[1]}, \ldots, w^{[N]}$, in order of best to worst cost, as a consequence of applying the LB search. (Recall that multiple-sentence hypotheses can be generated by keeping track of the second-, third-, . . . , Nth-best paths through the LB "grid"—see Section 11.4.2.) Let the corresponding cost measures be denoted $D_{\min}^{[1]}(T, J_L), \ldots, D_{\min}^{[N]}(T, J_L)$,[10] where J_L is the indicator for a terminal state in the final HMM. We now intend to subject these strings to "linguistic scrutiny" using the information in the LD's FSA. Before doing so, however, a practical point should be noted.

[9]Actually, some systems have provisions for resolving "overlapping" hypotheses. We will briefly describe one such system at the end of this section.

[10]We have assumed all the strings to be L words long, but this need not be the case.

In Section 11.4.2 we noted the increased memory and computation necessary to seek out multiple-sentence hypotheses using the LB algorithm. In Section 13.2.3 we discussed the possibility of the AD in a bottom-up scheme hypothesizing words (or other units) from the left and receiving "guidance" from the LD as to whether continuation of that string is advisable. Indeed, such an approach can be taken with the LB algorithm. A "linguistic cost" can be integrated with the "acoustic cost" on a "per level" basis in order to find a single-sentence hypothesis consistent with both bodies of information. Although simple in concept, the formal description of such an algorithm is complicated, so this case does not lend itself well to the learning process. Therefore we examine the simpler case in which complete-sentence hypotheses are submitted by the AD to the LD. In Problem 13.3 we return to the enhancement.

Each of the word strings submitted by the AD to the LD is now "recognized" by the LD as though it were an "observation" sequence submitted to an "HMM" for a likelihood score. It should be clear that the decoding of the LD FSA with respect to the observations may proceed according to any of the recognition methods for discrete-observation HMMs described in Chapter 12 (with minor modifications for the Mealy-form FSA). The result for string $w^{[k]}$ is either $P(w^{[k]}|\mathcal{G})$ or $P(w^{[k]}, \mathcal{I}^*|\mathcal{G})$, depending on whether an "any path" or Viterbi approach is used (we assume the former in this discussion). $\mathcal{G}$ (to denote *grammar*) is the name for the FSA model in the LD, and, as usual, $\mathcal{I}^*$ means the best path through the model.

The proposed method for combining the AD and LD costs is very simple. The cost of the kth *best* LB hypothesis at the acoustic level may be written

$$D_{\min}^{[k]}(T, J_L) = -\log P(y, \mathcal{I}^{[k]}|\mathcal{M}^{[k]}), \tag{13.62}$$

where $\mathcal{I}^{[k]}$ represents the globally (over all levels) optimal path through the set of L HMMs $\mathcal{M}^{[k]}$, and J_L is the label of a final state in the final model. Since the HMM string has a unique correspondence to the word string, let us write

$$D_{\min}^{[k]}(T, J_L) = -\log P(y, \mathcal{I}^{[k]}|w^{[k]}). \tag{13.63}$$

Now we alter the acoustic cost for string $w^{[k]}$ by simply adding to it the linguistic cost,[11] say $-\log P(w^{[k]}|\mathcal{G})$, to obtain

$$C_{\min}^{[k]} \stackrel{\text{def}}{=} -\log P(y, \mathcal{I}^{*[k]}|w^{[k]})P(w^{[k]}|\mathcal{G}). \tag{13.64}$$

The globally optimal word string, say w^*, is taken to be the one of minimum acoustic plus linguistic cost,

$$w^* = w^{[k^*]}, \tag{13.65}$$

[11]We add because of the logs. In effect, we are multiplying probabilities.

where

$$k^* = \underset{k}{\operatorname{argmin}}\, C_{\min}^{[k]} = \underset{k}{\operatorname{argmin}} \left\{ -\log P\left(y, \mathcal{J}^{[k]} \middle| w^{[k]}\right) P\left(w^{[k]} \middle| \mathcal{G}\right) \right\}. \tag{13.66}$$

Clearly, the grammatical information has the opportunity to influence the outcome of the final string decision. Consider, for example, a case in which the best acoustic choice, $w^{[1]}$, is a very low probability sentence in the language. With reference to the FSA in Fig. 13.2, for example, we see that the outcome

$$w^{[1]} = \text{zero, one, (two)}^5\text{, three, four, five, six, seven, eight, nine,} \tag{13.67}$$

where (two)^5 means a string of five "two's," is much less likely than, say,

$$w^{[2]} = \text{zero, one, two, three, four, (five)}^5\text{, six, seven, eight, nine.} \tag{13.68}$$

In this case, even though $D_{\min}^{[1]}(T, J_L)$ might be quite small, $C_{\min}^{[1]}$ might be quite large relative to $C_{\min}^{[2]}$.

Let us examine more specifically what the linguistic information is adding to the decision. Ideally, under any strategy we would like to find the word string, w^*, with maximum likelihood in light of the observations; in other words, we want[12]

$$w^* = \underset{w}{\operatorname{argmax}}\, P(w|y). \tag{13.69}$$

Using only acoustic information with no grammar, we must settle for

$$w^\dagger = \underset{w}{\operatorname{argmax}}\, P(y|w). \tag{13.70}$$

It is clear from these developments that the optimal word string to arise from this method is[13]

$$w^* = \underset{w}{\operatorname{argmax}}\, P(y|w)P(w|\mathcal{G}). \tag{13.71}$$

A moment's thought will reveal that $\mathcal{G}$ is simply an alternative way to formalize the unique random process that generates word strings. Therefore,

$$P(w|\mathcal{G}) = P(\underline{w} = w) = P(w), \tag{13.72}$$

where we employed the abusive notation $\underline{w} = w$ to mean that the random process $\underline{w}$ (or $\mathcal{G}$) has produced the partial realization w. Putting (13.72) into (13.71), we have

$$w^* = \underset{w}{\operatorname{argmax}}\, P(y|w)P(w) = \underset{w}{\operatorname{argmax}}\, P(y, w). \tag{13.73}$$

[12]We are ignoring the "best path" dependency of the acoustic decoding, but its inclusion would not change the basic argument here.

[13]Of course, we are only able to select arguments from the strings $w^{[1]}, \ldots, w^{[N]}$, which are provided to the LD.

Apparently, we maximize the joint probability of w and y with the selection of the word string above. Although this seems intuitively reasonable, we can show that this choice is even better than is apparent. Consider the fact that $P(y)$ is a fixed number that does not depend on the choice of word string. Accordingly, we choose w^* to maximize $P(y, w)/P(y)$ and achieve the same answer as in (13.73). However,

$$w^* = \underset{w}{\operatorname{argmax}} \frac{P(y, w)}{P(y)} = \underset{w}{\operatorname{argmax}} P(w \mid y), \tag{13.74}$$

so we have apparently achieved the ideal result of (13.69) with the method suggested above. Finally, note from (13.73) that when every word string is equally likely $[P(w)$ does not depend on $w]$, then the acoustic decision alone is the ideal one. This is entirely expected, since equally likely word strings essentially imply "no grammar." From this formality, therefore, we conclude the obvious, that the language information accounts for the unequal likelihoods of word strings in making the final decision.

As we noted at the outset of this discussion, the LB algorithm, which plays the role of the AD in the above system, is principally responsible for keeping track of timing information. We could, of course, shift this burden to the LD section of the system by recording at the states of the FSA information necessary to piece together hypotheses coming up from the acoustic processor. In this case, the AD is essentially reduced to a hypothesizer that attempts to match HMMs or DTW reference templates to various regions of the speech. For example, the AD might begin by attempting to match a series of candidate first-word HMMs, $\mathcal{M}_1, \ldots, \mathcal{M}_N$, to a range of possible time intervals, say frames $[1, t_1], [1, t_1 + 1], \ldots,$ $[1, t_1 + \tau]$. Any word with an HMM of sufficient likelihood would be hypothesized to the LD (along with the timing information). Hypotheses for the second word would be found by testing a second (perhaps the same) set of models on a different set of intervals. In general, the second set of time intervals will extensively overlap with those of the first set,[14] and the LD is responsible for piecing together an appropriately timed set of hypotheses. Provisions are often made for some overlap at the boundaries of the proposed words. The process carried out by the AD here is what we have previously termed *word spotting* (Section 11.4.4), because it attempts to spot regions corresponding to words in the string of observations.

In essence, the developments above provide the formal details for the GDCWR system described in Section 11.4.4 (McMahan and Price, 1986; Pawate et al., 1987; Picone, 1990). The GDCWR uses an FSA to bottom-up parse words that are hypothesized by a DTW algorithm at the acoustic level. The DTW algorithm is used to implement word spotting

[14]Remember that the time boundaries between the words are not known.

as described above. This system highlights the point that other strategies beside HMMs may be used by the AD to hypothesize words.

Before leaving this example, there are some additional points to be made about the linguistic decoding problem. First, although words were used as the basic unit of speech in this example, similar principles could be used to add grammatical constraints to systems using other terminals, such as phonemes, although this is not common with bottom-up systems, which represent an older technology. Second, we have said very little about the use of "heuristics" in the decoding problem. It might have occurred to the reader that it would be senseless, for example, to try the word "seven" in the second level of the LB algorithm, because there is no possibility of such an occurrence in the language. Of course, algorithms are written to take advantage of such entropy-decreasing information. In fact, for more complex grammars, many heuristic techniques are employed to improve the performance of the parser (Fu, 1982, Ch. 5). A final point, implied by the last comment, is that other grammars may be used in the LD. We will have more to say about this issue in our discussion of top-down parsers, to which we now turn.

13.5 Principles of "Top-Down" Recognizers

13.5.1 Focus on the Linguistic Decoder

Many research groups, particularly beginnning in the early 1980s, have focused on task-oriented speech recognition problems involving large vocabularies ($\sim$1000 words) and constrained grammars. Such problems are amenable to linguistic decoding by top-down parsing approaches. Success with these systems, enhanced by continued research and development of the HMM (FSA), has assured that this trend will continue. Later in the chapter, we will discuss several example systems and provide some historical context to these developments. The goal of the present section is to provide a fundamental understanding of the principles of the top-down approach.

To understand the basic ideas, we return to the digit recognition problem discussed above.[15] Let the LD again be represented by the FSA shown in Fig. 13.2. We again, therefore, focus on a language governed by a regular grammar portrayed by an FSA.

Recall that the most fundamental difference between a top-down and a bottom-up system is the direction in which the hypotheses flow. In the top-down case the LD is responsible for the initiation of downward propagation of hypotheses to the lower level(s) and the reception of measures of likelihood of the hypotheses from below. In the present problem there are only two layers, so the sentence-level decoder will communicate di-

[15]Again, this is a trivial vocabulary and grammar compared with CSR tasks studied in many research centers, but the fundamental points can be made in this context.

rectly with the AD. The linguistic processor is also responsible for the temporal recordkeeping function. We point out that this latter task can be handled in many ways algorithmically, and it is the objective here to indicate one general strategy which will admit understanding of the basic notions.

One of the main advantages of the presence of a regular grammar in a problem is that the search for a solution can be carried out using a DP approach. This is so because, as we have discovered previously, the proposition of a sequence of production rules from a regular grammar corresponds to the navigation through a path in the corresponding FSA. The production rules that best explain the observations correspond to the optimal path through the graph in a certain sense. We will illustrate this using our example, but first it is useful to develop some formalities.

The notation employed in the following discussion is that developed in Section 13.3. In anticipation of the use of DP, we define the following costs. Consider a transition in the FSA corresponding to a production $Q \xrightarrow{p} w_k R$, where w_k represents one of the words (digits) in the natural vocabulary. Suppose this production is producing the lth word in the string. In terms of the FSA, this production involves both the transition $\underline{U}(l) = U_{R|Q}$ and the generation of the observation $\underline{w}(l) = w_k$. Recall that probability p accounts for both the state transition and the word generation. In terms of the various quantities we have defined,

$$
\begin{aligned}
p &= P(Q \rightarrow w_k R) \\
&= P(\underline{X}(l) = R \mid \underline{X}(l-1) = Q) P(\underline{w}(l) = w_k \mid \underline{X}(l) = R, \underline{X}(l-1) = Q) \\
&= P(\underline{U}(l) = U_{R|Q}) P(\underline{w}(l) = w_k \mid \underline{U}(l) = U_{R|Q}) \\
&= A(R|Q) b(w_k | U_{R|Q}).
\end{aligned}
\tag{13.75}
$$

As is suggested by the string of equalities in (13.75), there are many ways to formalize the ensuing discussion, but we will attempt to make the notation as parallel as possible to similar developments in earlier material. We will tend to focus on the language model as an FSA rather than a formal grammar (production rules), but there are a few instances where the production rules are particularly convenient.

Suppose there are many word strings leading to state Q in the FSA which are assumed to be associated with observations $y_1^{t'}$. We (somehow) record the lowest-cost word string and assign that cost to state Q,

$$
D_{\min}(t', Q) \stackrel{\text{def}}{=} \text{cost of best word string to state } Q \text{ associated with observations } y_1^{t'}. \tag{13.76}
$$

Note that the frame time variable "t" does *not* correspond to the operative "time scale" in the upper-level FSA, for this search is to be synchronized with the frame times at the lowest level of the search. Suppose further that this best-path information is available for all states Q that can make a transition to state R, and for all $t' < t$, where t is defined momentarily. Now to the transition corresponding to the production $Q \xrightarrow{p} w_k R$, we assign the "linguistic cost"

$$d_L[Q \xrightarrow{p} w_k R] \stackrel{\text{def}}{=} -\log p. \tag{13.77}$$

Viewing this production as a transition in the FSA, we see that d_L is separable into the conventional Type T and Type N costs that we have associated with an HMM search. From (13.75),

$$d_L[Q \xrightarrow{p} w_k R] = [-\log A(R|Q)] + [-\log b(w_k | U_{R|Q})], \tag{13.78}$$

so we have a cost associated with the transition (Type T) and a cost attached to the generation of the observation (Type N).[16] Embedded within this "linguistic search" is another ongoing search at the acoustic level. For this lower-level search we need to introduce another cost. To any attempt to associate the transition $Q \xrightarrow{p} w_k R$ with the observations $y_{t'+1}^t$, we assign the "acoustic cost"

$$d_A[w_k; t', t] \stackrel{\text{def}}{=} -\log P(y_{t'+1}^t | \mathcal{M}_k), \tag{13.79}$$

where $\mathcal{M}_k$ means the HMM corresponding to word (or, in general, terminal unit) w_k. We have become accustomed to writing $P(y_{t'+1}^t | \mathcal{M}_k)$ as simply $P(y_{t'+1}^t | w_k)$ and we will continue that convention here:

$$d_A[w_k; t', t] \stackrel{\text{def}}{=} -\log P(y_{t'+1}^t | w_k). \tag{13.80}$$

Now we wish to attach some "w_k" to the word string already present at Q by making a transition to R and generating w_k in the process. Note, however, that there will, in general, be many competing transitions coming into R vying for the honor of completing the best word string to state R. There may be multiple strings coming over the same transition in the FSA because of the different observations that can be generated on that transition. There may even be multiple strings coming over the same transition with the same observation generated because of the different

[16]Two points are made here. First, in previous work, we have assigned a Type N cost to an event at a node. Here the cost is associated with the transition because the word is generated during the transition. Second, this definition is essentially the analog of (12.61) used in the search of the HMM in Chapter 12. Accordingly, it might seem to make sense to define d_L with the following notation:

$$d_L[(l, R) | (l-1, Q)] \stackrel{\text{def}}{=} [-\log A(R|Q)] + [-\log b(w(l) | U_{R|Q})].$$

However, there are two subtle reasons for not doing so. The first reason is that the FSA "time" variable l is superfluous in this development, since we will keep track of work ordering by synchronizing the processing to the acoustic-level frames. The other reason is that, unlike the HMM search where the acoustic observations $y(t)$ are known, here the word "observations" $w(l)$ are unknown and must be hypothesized by the FSA. This requires that we speak of the word in a more particular form, w_k.

end-times, t', for previous strings at Q (these will involve different acoustic costs). From the BOP we know that the cost of the best path to R at time t through Q at time t' is obtained by simply adding to the cost of the best path at Q the cost of making the transition to R over the time interval $[t'+1, t]$. That is, for a transition $Q \xrightarrow{p} w_k R$ over the observations $y_{t'+1}^{t}$,

$$\begin{aligned} D_{\min}[(t,R)|(t',Q),w_k] &= D_{\min}(t',Q) + d_A[w_k;t',t] + d_L[Q \xrightarrow{p} w_k R] \\ &= D_{\min}(t',Q) - \log P(y_{t'+1}^{t}|w_k) - \log p \\ &= D_{\min}(t',Q) - \log P(y_{t'+1}^{t}|w_k) \\ &\quad - \log A(R|Q)] - \log b(w_k|U_{R|Q}). \end{aligned} \tag{13.81}$$

Therefore,

$$\begin{aligned} D_{\min}(t,R) &= \min_{\substack{t',\, Q \xrightarrow{p} w_k R \\ t'<t}} D_{\min}[(t,R)|(t',Q),w_k] \\ &= \min_{\substack{t',\, Q \xrightarrow{p} w_k R \\ t'<t}} \{D_{\min}(t',Q) - \log P(y_{t'+1}^{t}|w_k) - \log p\} \\ &= \min_{\substack{t',\, Q \xrightarrow{p} w_k R \\ t'<t}} \{D_{\min}(t',Q) - \log P(y_{t'+1}^{t}|w_k) - \log A(R|Q) \\ &\quad - \log b(w_k|U_{R|Q})\}. \end{aligned} \tag{13.82}$$

It appears that we will be able to implement the linguistic constraints using a Viterbi DP approach to searching the FSA. Before returning to the example to see how this is done, let us take care of one more important detail.

In order to keep track of the best word (terminal) string at (t, R), let us define

$$\Psi(t,R) \stackrel{\text{def}}{=} \text{minimum-cost word string to state } R \text{ associated with observations } y_1^t. \tag{13.83}$$

If the optimal path extension into state R at t was over the transition $Q' \xrightarrow{p'} w_i R$ and over the time interval $[t''+1, t]$, then

$$\Psi(t,R) = \Psi(t'',Q') \oplus w_i, \tag{13.84}$$

where $\oplus$ means concatenation.[17]

We now illustrate this Viterbi search using the digit recognition example. Following the conversion of an acoustic utterance into an observa-

[17]This is an alternative to backtracking for finding the globally optimal word sequence at the end of the search.

tion string, $y = y_1^T$, the LD initiates the process as follows. For some assumed minimum time at which the first word may end in the observation string, say t_1, the LD hypothesizes the presence of all words, say w_k, indicated by production rules of form

$$\bar{S} \xrightarrow{p} w_k R \tag{13.85}$$

to be in the time interval $[1, t_1]$. This means the FSA is searched for all initial transitions (from $\bar{S}$) and hypothesizes the corresponding words to be represented in the observations in the interval $[1, t_1]$. In the present case words "zero," "oh," "one," and "five" would be hypothesized, corresponding to the productions (see Fig. 13.2)

$$\bar{S} \xrightarrow{p_1} \text{zero}, \bar{A} \tag{13.86}$$

$$\bar{S} \xrightarrow{p_2} \text{oh}, \bar{A} \tag{13.87}$$

$$\bar{S} \xrightarrow{p_3} \text{one}, \bar{B} \tag{13.88}$$

$$\bar{S} \xrightarrow{p_4} \text{five}, \bar{B}. \tag{13.89}$$

The request goes down to the AD for a likelihood measure for these words for the observation string $y_1^{t_1}$. The likelihoods $P(y_1^{t_1}|\text{zero})$, $P(y_1^{t_1}|\text{oh})$, $P(y_1^{t_1}|\text{one})$, and $P(y_1^{t_1}|\text{five})$ are reported back. The best costs and words to states $\bar{A}$ and $\bar{B}$ are computed according to (13.82) and (13.84). For example,

$$\begin{aligned} D_{\min}(t_1, \bar{A}) &= \min_{0, \bar{S} \xrightarrow{p} w_k \bar{A}} \{D_{\min}(0, \bar{S}) - \log P(y_1^{t_1}|w_k) - \log p\} \\ &= \min\{-\log P(y_1^{t_1}|\text{zero}) - \log p_1, \\ &\qquad -\log P(y_1^{t_1}|\text{oh}) - \log p_2\}, \end{aligned} \tag{13.90}$$

where $D_{\min}(0, \bar{S}) \stackrel{\text{def}}{=} 0$. If either of the minimum costs to states $\bar{A}$ or $\bar{B}$ is excessive, this means that the corresponding (best) transition into that state does not explain the observations sufficiently well and that path is not started. This amounts to "pruning" the path before it is even originated (an unlikely practical event). If a path is completed to state $\bar{A}$, for example, then the state is flagged as having an *active sentence hypothesis* at time t_1. For simplicity, we will say that "$(t_1, \bar{A})$ is active." Similarly, $(t_1, \bar{B})$ is active if a successful path is initiated there. By definition we say that state $\bar{S}$ has an active sentence hypothesis at time 0; that is, $(0, \bar{S})$ is always active. For activated states the word string (in this case, just one word) is recorded according to (13.83). By definition,

$$\Psi(0, \bar{S}) \stackrel{\text{def}}{=} \varnothing, \text{ the null string in the language.} \tag{13.91}$$

Continuing on to time $t_1 + 1, t_1 + 2, \ldots$ is simple. For each state R, and for successively larger t's, all transitions of form $Q \xrightarrow{p} w_k R$ and all active hypotheses (t', Q) for $t' < t$ compete for extension of their paths to

R. The extension of lowest cost according to (13.82) completes the optimal path to R at time t. For example, let us assume that at the first frame t_1, only state $\bar{A}$ becomes activated in our example FSA. Then at frame $t = t_1 + 1$ this would happen in our example: We see that only states $\bar{A}$ and $\bar{B}$ have predecessor states which have active hypotheses. The paths that could be extended to $\bar{A}$ are the active hypotheses at $(t_1, \bar{A})$ and $(0, \bar{S})$. The active hypotheses that could be extended to $\bar{B}$ are at $(t_1, \bar{A})$ and $(0, \bar{S})$. The operative transitions are associated with rules

$$\begin{aligned} &\bar{S} \xrightarrow{p_1} \text{zero}, \bar{A} \\ &\bar{S} \xrightarrow{p_2} \text{oh}, \bar{A} \\ &\bar{S} \xrightarrow{p_3} \text{one}, \bar{B} \\ &\bar{S} \xrightarrow{p_4} \text{five}, \bar{B} \\ &\bar{A} \xrightarrow{p_5} \text{zero}, \bar{A} \\ &\bar{A} \xrightarrow{p_6} \text{oh}, \bar{A} \\ &\bar{A} \xrightarrow{p_7} \text{one}, \bar{B} \\ &\bar{A} \xrightarrow{p_8} \text{six}, \bar{B}. \end{aligned} \tag{13.92}$$

Note that it is extraordinarily unlikely that either of the last two rules would be used, since this would correspond to the generation of a word using only one acoustic observation.

After all states have been examined at time t, only if $D_{\min}(t, R)$ is sufficiently small does R become activated at t. The best word string to R at t is recorded as in (13.84). The solution is ultimately found in $\Psi(T, R_f^*)$, where T is the length of the observation string and

$$R_f^* \stackrel{\text{def}}{=} \underset{R_f}{\operatorname{argmin}}\, D_{\min}(T, R_f), \tag{13.93}$$

where R_f is any permissible final state in the FSA.

As would be expected, relatively few hypotheses remain active at any given t. This is because relatively few paths involve the appropriate linguistic constraints, HMMs, and time warping to match the incoming acoustic observations. Most of the paths will be pruned because they will be so unlikely that their extensions are not warranted. We saw a similar pruning process take place at the acoustic level to control the number of paths through the acoustic-level HMMs. Similarly to that pruning process, the present procedure is also referred to as a beam search (Lowerre and Reddy, 1980), since only paths that remain inside a certain acceptable beam of likelihoods are retained. Those that fall outside the beam are pruned. A simple beam, for example, would consist of all paths (hypotheses) at time t whose cost fell within, say, $\delta(t)$ of the best hypothesis. If $R^\dagger$ is the active state associated with the best hypothesis at t, then any other state R will only become active at t if

$$D_{\min}(t, R) \le D_{\min}(t, R^\dagger) + \delta(t). \tag{13.94}$$

13.5.2 Focus on the Acoustic Decoder

Thus far we have not paid any attention to the process taking place in the AD. At this level we have (for the digit problem) 11 or more HMMs representing the words in the vocabulary. The AD is receiving requests for likelihoods of the form $P(y_{t'}^{t}|w_k)$ for $t_1 \leq t' < t \leq T$. Recall that $P(y_{t'}^{t}|w_k)$ means $P(y_{t'}^{t}|\mathcal{M}_k)$, where $\mathcal{M}_k$ is the HMM representing word (terminal) w_k. [In fact, if Viterbi decoding is used in the HMM search, the likelihood reported back will be $P(y_{t'}^{t}, \mathcal{J}_k^*|\mathcal{M}_k)$, where $\mathcal{J}_k^*$ is the best state sequence through $\mathcal{M}_k$.] It is important to understand that the HMMs can operate in synchrony with the LD in certain ways.

First, we need to recognize that an entirely parallel search procedure is taking place in each of the HMMs to that occurring in the FSA at the linguistic level. Let us concern ourselves with all requests for $P(y_{t'}^{t}|w_k)$ for a *fixed time* t'. Assuming that $t-t'$ is at least as large as the number of states in $\mathcal{M}_k$, then each state in $\mathcal{M}_k$ has the potential to be active with a viable best path to it at time t. (Ordinarily, for a Bakis model all states would remain active after a sufficient number of observations.) This is nothing new. It is just a different way to view the Viterbi decoding process we discussed in Section 12.2.2. At each new observation frame, all states to which transitions can be made from active states are updated in a similar manner to that which occurs in the FSA. The measure $P(y_{t'}^{t}|w_k)$ is simply the likelihood of the path at the final state at time t when the input observation sequence starts at frame t'. This means that when $P(y_{t'}^{t+1}|w_k)$ is requested, it is simply necessary to update the search in the usual manner and report the likelihood at the final state. There is no need to repeat the entire search for observations $y_{t'}^{t+1}$. In this sense, the operations occurring in each section of the recognizer are occurring in a *frame-synchronous* manner.

When a request is made for a measure $P(y_{t''}^{t}|w_k)$ involving a different observation starting time, it is best to imagine that a new HMM for word w_k is created with which to carry out the search, although, of course, the algorithm can be written to keep track of multiple paths with different starting times propagating through the same HMM.

The differences in the model structure notwithstanding, the FSA at the linguistic level, and the HMMs at the acoustic level are all finite state machines that can be thought of as implementing the rules of regular grammars. The language in the HMMs is much farther removed from a natural language than that in the language level, in that its terminals are the acoustic observations extracted from the speech waveform. Accordingly, the entire speech recognizer (two levels in our current digit recognizer example) can be compiled into a large FSA representing a more complex regular language whose terminals are the acoustic observations, and whose nonterminals are the states of the compiled FSA. A search for a maximum likelihood path through this FSA would result in a corresponding word (digit) string directly, without the need for communication between levels. In fact, in our current example, if we insert the digit HMMs directly into the transitions that produce the corresponding digit in the LD (see Fig. 13.3), it is not difficult to discover how to search this FSA to carry out precisely the same search as was described above. We explore this issue in Problem 13.4.

It can be appreciated from this simple example that one of the major concerns in an algorithm incorporating a language model is the recordkeeping function. Although we have discussed the basic principles of this task above, we have not thought very deeply about practical implementation. A good example of a system that processes continuous speech according to the principles above is found in the papers by Ney et al. (1987, 1992), which also describe the details of recordkeeping. These methods were described in Section 11.4.3. In essence, the Ney papers describe an extension of the one-stage DTW algorithm discussed in Section 11.4.3 to include syntactic information. The recordkeeping is accomplished through a series of list processing operations. This paper represents one of the first reports of the integration of stochastic language models into the CSR problem. A more recent tutorial is found in the paper by Lee and Rabiner (1989).

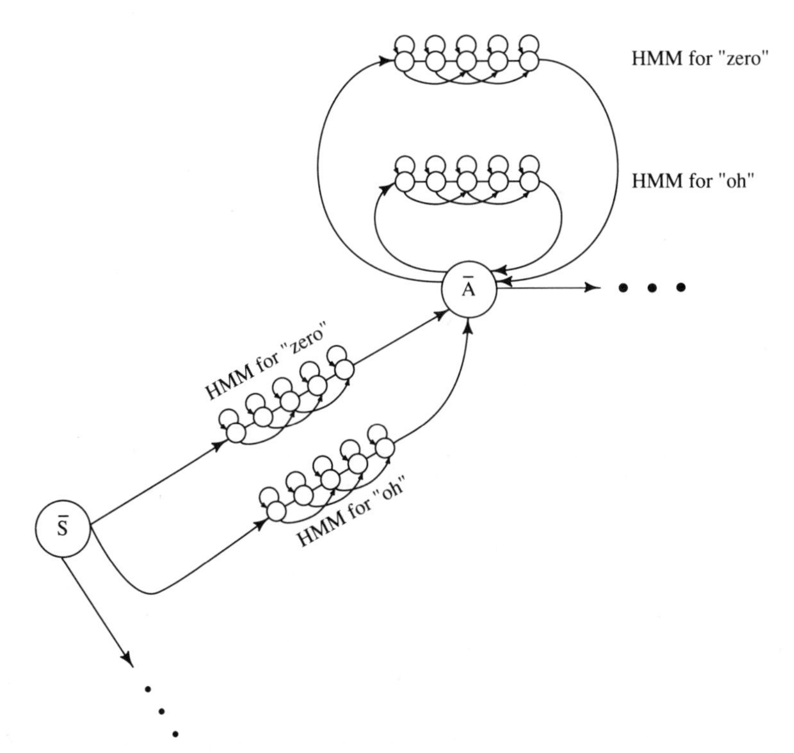

FIGURE 13.3. Acoustic-level HMMs for the digits can be inserted directly into the linguistic-level FSA to create a hybrid FSA that can be searched directly without need for communication between layers. This idea is illustrated here for the initial part of the state diagram of Fig. 13.2.

13.5.3 Adding Levels to the Linguistic Decoder

The general principles of the top-down approach can be understood from our study of the simple digit recognition problem above. In that problem, we employ a two-level approach in which the LD consists of a single FSA. In fact, the basic approach there *is* realistic for the digit problem. A practical recognizer for continuous digits would likely employ two levels as we did in our conceptual discussion. (Sometimes additional layers are added to model noise.)

The reason why two levels are sufficient for the digit problem is that the vocabulary is small and we are able to interface the speech waveform with the recognizer at the word level. Simply put, there are few enough words that we can easily build acoustic models for all of them (even if some require multiple models) and use them as terminals in our formal language.

As we have discussed previously, in larger-vocabulary systems, say 1000–100,000 words [see, e.g., the systems described in Section 13.9 and (Levinson et al., 1988; Deng et al., 1988; Dumouchel et al., 1988)], the collection of data for, and training of, individual word models is prohibitive. In this case, subword units such as phones, diphones, phonemes, or syllables must be employed. This will mean that the AD will work, for example, at the phone level, and the LD will be charged with hypothesizing sentences in terms of phones.

Let us return for a moment to the idea that a two-level system can be compiled into one big FSA by inserting word HMMs into the transitions of the sentence FSA as in Fig. 13.3. In this case, the compiled FSA represents a larger regular grammar with the acoustic observations as terminals. Conversely, we know that this large FSA can be decomposed into a regular grammar with words as terminals and HMMs at the word level representing another regular grammar with acoustic observations as terminals. Now suppose each word were represented by series of appropriately connected *phone* HMM models, as in Fig. 13.4. This representation of a word amounts to a "large" regular grammar with acoustic observations as terminals. By analogy, this word representation could be decomposed into a regular grammar with phones as terminals, and HMMs at the phone level. This process creates a three-level system in which the phone HMM models now comprise the AD, and a new layer is added to the LD. The result for our digit recognition system is illustrated in Fig. 13.5. Of course, further levels could be added by, for example, decomposing words into syllables and then syllables into phones. However, most CSR systems operate with two levels using words at the acoustic level, or three levels with phone models at the acoustic level.

The method for finding a maximum likelihood solution in a three-level system is a natural extension of the methods used with two levels. Again each operation is frame-synchronous across levels. When a request

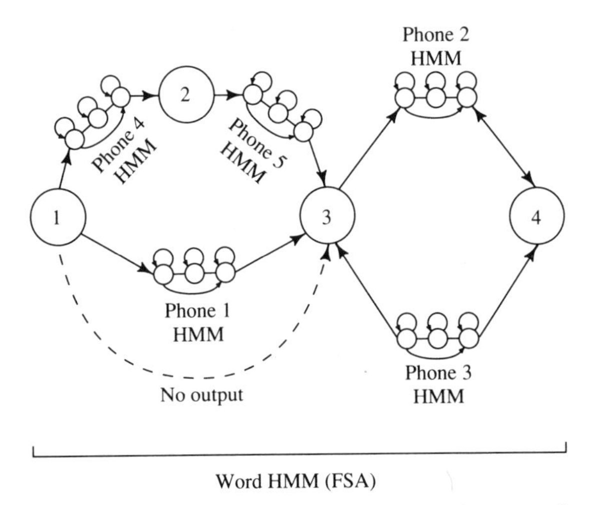

FIGURE 13.4. Word HMM decomposed into a set of appropriately connected phone models.

for $P(y_{t'}^{t}|w_k)$, where w_k is a word, goes down to the word level from the sentence level, the word level must locate an FSA representing word w_k whose first acoustic observation input is $y(t')$, and report back the likelihood of the best path reaching one of its legal final states at time t. In turn, the discovery of this path will be the result of searching this word FSA in the same manner in which we searched the sentence-level FSA in the two-level case. At each frame time, $t', t'+1, \ldots, t$, we will have attempted to extend each active hypothesis using Viterbi decoding. Each of these attempts will have involved requests for likelihoods of phones over certain frame ranges, $P(y_{t''}^{t'''}|z_j)$, where z_j is a phone. Such a request will require locating a phone HMM for z_j that has considered the observations $y_{t''}^{t'''}$ and reporting back the likelihood of its best path to a final state. While the recordkeeping is more complicated with more levels, the basic principles remain the same. A general algorithm for performing the multilevel search is shown in Fig. 13.6.

The discussions above have employed a Viterbi algorithm in which, in principle, all states were considered to see whether extensions could be made from active hypotheses. This is the way in which we are accustomed to using the Viterbi algorithm—moving forward one step by connecting paths from behind. For a large language model in which relatively few states are likely to be active at any time, it is sometimes more efficient to iterate over the active hypotheses at time t rather than over all the states. In this case the hypotheses are pushed forward one step to all possible next states, and then the Viterbi algorithm is performed on all next states receiving more than one extended path. In this case the algorithm above is slightly modified, as shown in Fig. 13.7.

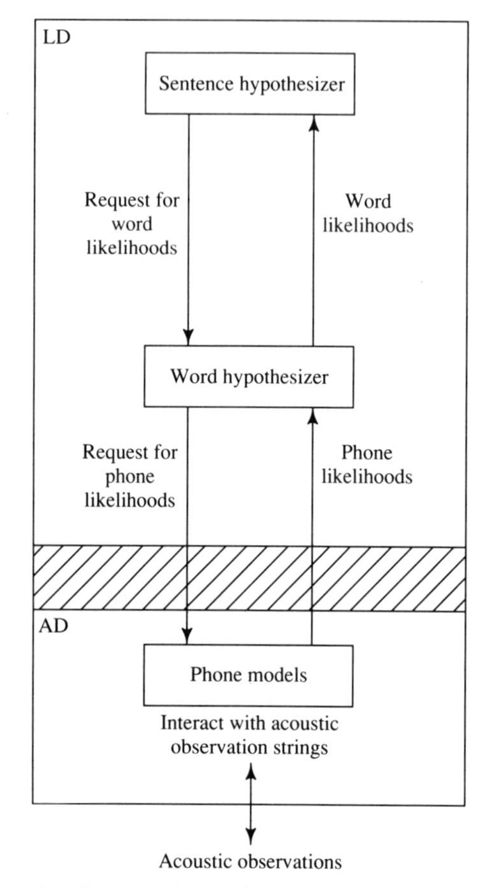

FIGURE 13.5. Three-level recognizer with phone HMMs at the acoustic level (cf. Fig. 10.4). LD = linguistic decoder (corresponds to regular grammar with phone designations as terminals); AD = acoustic decoder (corresponds to regular grammar with acoustic observations as terminals).

As the number of layers, or knowledge sources (KSs), continues to increase (to include, for example, semantics and pragmatics at higher levels and prosodics at the acoustic level), exhaustive search using methods like those described above becomes unrealistic for many applications. In 1990, a method was proposed by Schwartz and Chow (1990) in which the most likely N sentences are located by using selected KSs that are sufficiently informational (entropy reducing) while computationally inexpensive. The remaining KSs are then used to evaluate the hypotheses posed by the first process. For the obvious reason, this paradigm is called *N-best search.* We will be better able to describe some details and results of this form of search after discussing some alternative language models. We will return to it briefly in the context of discussing the BYBLOS recognition system in Section 13.9.

FIGURE 13.6. State-based Viterbi decoding of a CSR system based on a regular grammar.

Initialization: Activate initial state $\bar{S}$ in the sentence-level FSA.

Recursion: For $t = 1, \ldots, T$
For all levels in the LD and AD
Extend paths to all states which can be reached from states active at $t-1$ using the Viterbi algorithm.
Prune
Next level
Next t

Termination: Select hypothesis in sentence-level FSA associated with highest likelihood path to a final state.

FIGURE 13.7. Hypothesis-based Viterbi decoding of a CSR system based on a regular grammar.

Initialization: Activate initial state $\bar{S}$ (hypothesis) in the sentence-level FSA.

Recursion: For $t = 1, \ldots, T$
For all levels in the LD and AD
Extend all active states to their reachable next states.
Perform Viterbi algorithm on all states receiving extensions.
Prune
Next level
Next t

Termination: Select hypothesis in sentence-level FSA associated with highest likelihood path to a final state.

13.5.4 Training the Continuous-Speech Recognizer

In Chapter 12, we discussed in some detail the process of training an HMM to recognize an isolated word. In the sense that a regular grammar-based CSR system represents a "big HMM," we already know much of what we need to know to tackle the problem of training such a system. The isolated-word training is an example of a *supervised* training paradigm in which the model is required to learn an appropriate set of parameters from the presentation of training examples of known classes (words). The same supervised training approach is taken to the problem of training the CSR system, but there are several significant differences that become apparent when one thinks about the task.

These differences emerge when we try to follow up on the suggestion made above that we simply consider the CSR system as one big HMM. Suppose we attempt to place all of the word or phone (or other unit) HMMs in their proper locations in the big network and then train them with example utterances (sentences). The first assumption inherent in this suggestion is that we know the production rules of the grammar! Indeed we might not, and deducing the grammar "manually" from the training data could be an overwhelming task. The first problem, then, is learning the grammar of the language. Then we might recognize another major difference between this big HMM and one that would be used in the isolated-word case. In the isolated-word case, the *entire model* represents a class (word) within the set of classes (vocabulary). In the CSR case, a *path* represents a class (sentence) within the universal set of classes (language).[18] This means that we must find a way to train only one path at a time, which will turn out to be a simple problem. Finally, and perhaps most significantly, in the isolated-word case it is known exactly which observation frames are associated with the HMM, whereas in the CSR case we are faced with a string of observations whose temporal associations with the individual HMMs in the AD are generally unknown.

In order to solve these problems, it is useful to once again decouple the models of the AD from those in the LD. Having done so, let us consider the problem of learning the production rules of the grammar. The technique is very simple. To train the sentence-level FSA, which has words as terminals, we enter orthographic transcriptions of sentences (sentence written out as words) as training "observation" sequences. A version of the F–B algorithm for the Mealy rather than the Moore version of an FSA (Bahl, 1983) is then used to estimate the transition probabilities of the FSA. Alternatively, a Viterbi-like algorithm can be used to estimate the transition probabilities (Fu, 1982, Sec. 6.6). If we were working with Moore forms of the FSA, each of these would be identical to the corresponding approach taken to training isolated-word HMMs. In either case, the entire database of training sequences should be entered before updating the FSA parameters. Note that this procedure is tantamount to "estimating" the production rules of the grammar and their associated probabilities.

As with a "standard" HMM, the structure of the FSA must be prescribed before the probabilities can be assigned in the above procedure. In this case this simply means choosing the number of states (nonterminals) and transitions (terminals) to be included. Initially, each transition should be allowed to generate every observation with nonzero probability, just as each state in a Moore HMM may generate each observation. The F–B algorithm operates in the same fashion, moving from (arbi-

[18]This means that either the F–B approach or the Viterbi approach can be used to decode an isolated-word HMM, but only the Viterbi method can be used for the CSR HMM.

trary) initial transition probabilities to those which represent a local maximum likelihood with respect to the observations.

If there is a second level of the LD, orthographic transcriptions of words in terms of phones, if they are available, can be used to estimate parameters for each of the word FSAs. If these transcriptions are not available, the word FSAs can be estimated along with the acoustic HMMs, as we discuss below.

This is a good point to reemphasize that we are dealing strictly with an LD based on a regular grammar. There are other grammars and linguistic structures that can be used, and each has its own training procedure. These will be discussed below.

Having trained the LD, let us consider the training of the HMMs in the AD. We now return to the problem that a given speech utterance must be used to train only one path of the global FSA. This problem is very easy to solve. For a given training utterance, we temporarily create a small system that recognizes only the corresponding sentence. This is done, in principle, by setting to zero the probabilities of all production rules (transition probabilities) that do not lead to the sentence in its finest known decomposition. By "finest known decomposition" we mean, for example, that if we are dealing with a three-level system and our training sentences are transcribed to the phone level, then presumably the LD grammar is trained all the way through the word-level FSA. In this case we freeze all the production probabilities leading down to the appropriate set of phone models (in the AD) and (in principle) set the rest of the probabilities in the LD to zero. This procedure, in effect, has selected one path through the global FSA. If the sentence is transcribed only in terms of words, and we are training a three-level recognizer, then the probabilities in the sentence-level FSA leading to the appropriate words are frozen, and the rest are set to zero. Assuming that the correct production rules to decompose these words into alternative sets of phones are known,[19] even though the probabilities are not, then the set of acoustic observations can be used to train both the phone models and the word models. In this case we have just constrained the global model to a small set of paths that could conceivably produce the sentence.

We now intend to use an F–B or Viterbi approach to estimate the unknown probabilities on the selected path of the global FSA. We should point out that neither of these procedures will be affected by the fact that some of the probabilities on the path are fixed. For the Viterbi case this is self-evident, and for the F–B case this result is proven by Baum (1972). Now we must face the problem of not knowing where the temporal boundaries are in the acoustic observation string. If these bound-

[19]This just means that all of the phonetic transcriptions of each word are available, even though we do not know which transcription is appropriate for the present sentence. Formally speaking, it means that we know the characteristic grammar for each word even though we do not know the production probabilities.

aries were known, we could simply use the data segments to train the individual phone models. However, the creation of sufficiently large databases marked according to phonetic time boundaries is generally impractical. One of the most remarkable properties of the HMM comes to our aid in this situation. Researchers have discovered that these HMMs can be trained *in context* as long as reasonable "seed" models are used to initiate the estimation procedure. This means that the *entire* observation sequence for a sentence can be presented to the appropriate string of HMMs and the models will tend to "soak up" the part of the observation sequence corresponding to their words or phones, for example. This ability of the HMM has revolutionized the CSR field, since it obviates the time-consuming procedure of temporally marking a database.

However, good initial models must be present in the system before the training begins. This might require some "manual" work to "hand excise" words or phones from some speech samples.[20] Frequently, however, seed models consist of previously trained word or phone models, or are derived from an available marked database[21] [see, e.g., (Lee et al., 1990)]. The seed models need not be excellent representations of the application database, but they must be sufficiently good that they "attract" the proper portions of the observation sequences in the training data. Otherwise, of course, the acoustic-level HMMs will not represent the intended speech unit and recognition performance will be degraded.

As Picone (1990) discusses, seed models are often generated by an iterative process in which a crude model is successively refined into an acceptably good one. He cites, for example, the training of a continuous-observation mixture Gaussian density model in which a hand-excised phone is modeled by a five-state model, one state for each analysis frame in the phone (see Fig. 13.8). Rather than begin with an accurate set of covariance matrices for each state, the procedure begins with a single covariance matrix for all states, and then is iteratively refined.

One point should be emphasized before leaving the issue of training. After the seed models are installed and the training process begins, parameters of the system are not changed until the entire training database has been entered. This is true whether the F–B or Viterbi approach is used. With a little thought about the process, the reason will become evident. The situation is analogous to the use of multiple training sequences in a single HMM.

Finally, we note that in the late 1980s and early 1990s, researchers began to employ a second phase of training based upon the discrimination techniques introduced in Section 12.2.7 for IWR (Ephraim et al., 1989; Bahl et al., 1988). The application of these techniques to CSR was first reported in (Lee and Mahajam, 1989). More recently, Chow (1990) has proposed a method based on the N-best search. These techniques are

[20]This process would likely be accomplished by an expert phonetician using a speech-editing program.

[21]We will discuss standard databases in Section 13.8.

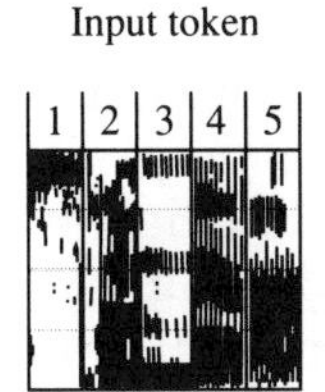

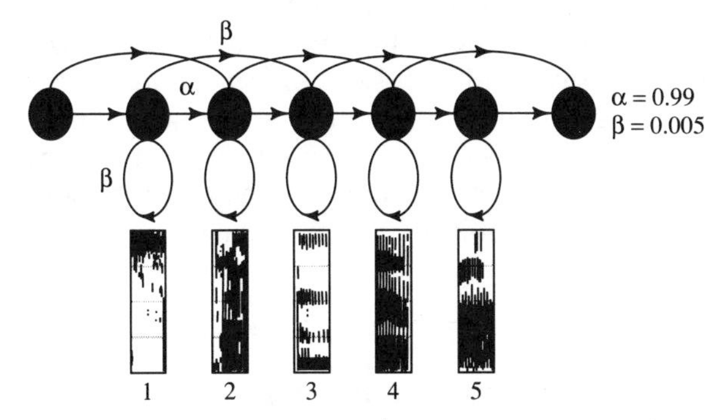

FIGURE 13.8. Training of a continuous-observation mixture Gaussian density model. A hand-excised phone (training token) is modeled by a five-state model, one state for each analysis frame in the phone. In this case, the analysis frames consist of nonoverlapping portions of the utterance chosen by an expert phonetician by examining the spectrographic trace. (The spectrograph gives a frequency versus time record in which the energy in a frequency band is indicated by the intensity of the trace in the corresponding frequency region.) After Picone (1990).

generally based upon the discrimination information or cross-entropy approaches briefly described in Section 12.2.7, but a more *ad hoc* approach follows the *corrective training* approach of Bahl et al. (1988). In fact, the less formal technique was found to outperform the maximum mutual information approach (a form of discrimination information), and some speculations on possible reasons are given in the cited paper.

13.6 Other Language Models

In this section we examine some alternatives to regular grammars for modeling languages.

13.6.1 *N*-Gram Statistical Models

Much of the pioneering research on HMMs in speech processing was carried out at IBM in an effort to develop a large-vocabulary speech recognizer for office dictation. This work has been ongoing since 1972, and in 1986 the IBM group reported on a 5000–20,000-word isolated-word recognizer called the *TANGORA system* (Averbuch et al., 1986). This system will be described in more detail in Section 13.9. Our purpose here is to study the language-modeling technique. This method is described in a frequently cited paper on HMM decoding by Bahl, Jelinek, and Mercer (1983). We shall therefore refer to the method as the *BJM technique* in the following discussion. It is important to recognize that although this method was developed to support an IWR system, it is quite general and applies equally well to the CSR problem.

The BJM technique is developed for the case in which the terminal models in the system correspond to words. In other words, the AD contains word-level HMMs and the LD contains information about how words may be combined to form sentences. Rather than a grammar (at least in the way we have developed it), however, the LD contains a statistical model based on "N-grams," which will be defined shortly. As in systems discussed above, the objective of the recognizer is to find the word sequence, say w^*, such that

$$w^* = \underset{w}{\operatorname{argmax}}\, P(w \mid y), \tag{13.95}$$

where w is any word string and y represents the string of observations [see (13.69) and surrounding discussion].

For visual illustration of the basic concepts, it will be useful to base our discussion on our digit recognition example. We redraw the state diagram of Fig. 13.2 as Fig. 13.9. It is important to recognize that this figure no longer necessarily represents an FSA governed by Markovian transitions. By this we mean that the state diagram represents the allowable paths through words to form sentences, but the statistical dependencies of these words are not necessarily Markov.

One novel feature of the BJM method is that it uses a search method called "stack decoding," which requires subpaths of different length in the LD state diagram to compete. Likelihoods based on direct probability computations become smaller (multiplication of probabilities) as path lengths increase, even for the optimal solution. Since subpaths of lower likelihood have higher risk of being pruned,[22] a path likelihood must be used that provides some degree of fairness in comparing subpaths of unequal length. The likelihood chosen is as follows. We seek the likelihood that the partial word string, w_1^k, begins the optimal path. This is theoretically given by

$$\Lambda(w_1^k) \stackrel{\text{def}}{=} \sum_{t=1}^{T} P(w_1^k, y_1^t)\,\alpha^{T-t} \sum_{w'} P(w', y_{t+1}^T \mid w_1^k, y_1^t), \tag{13.96}$$

where w' represents any word string which can follow w_1^k. The first probability accounts for the likelihood that string w_1^k is associated with observations y_1^t, and the second probability the likelihood that the remaining

[22]This did not happen in the Viterbi decoding methods used for regular grammar systems, in which pruned paths lost the competition with other subpaths of identical length.

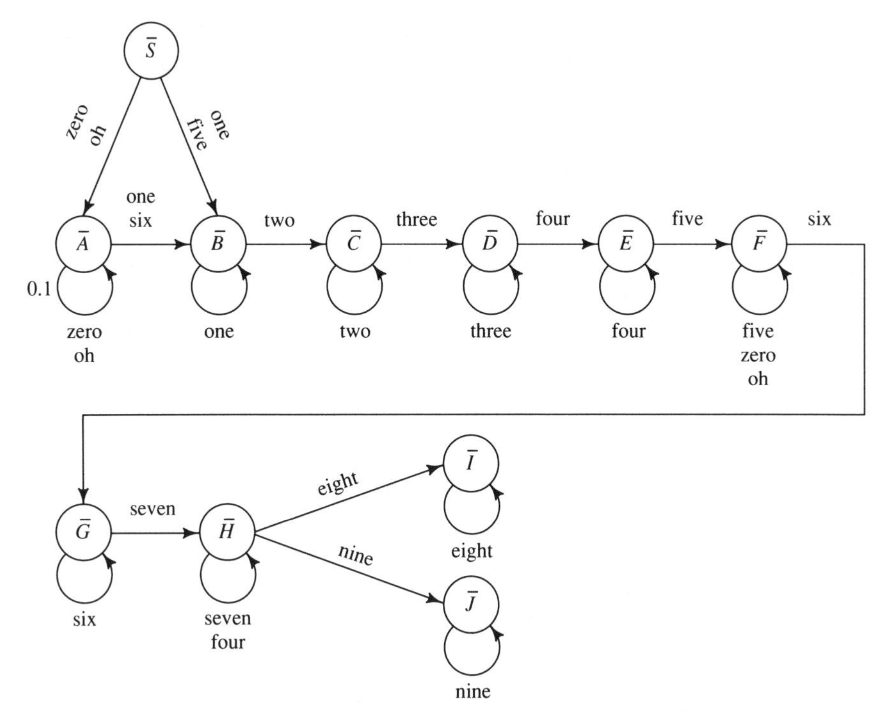

FIGURE 13.9. State diagram of Fig. 13.2 redrawn for convenience. It is important to recognize that this figure no longer necessarily represents an FSA governed by Markovian transitions in the present discussion.

observations are associated with any word string that can legally follow w_1^k. The constant α is chosen to control the rate of growth of the most likely path so that others can compete with it. In other words, it is necessary to control the contribution of the "remaining" sequences "w'" to the likelihood. Bahl et al. suggest that this constant can be found by trial and error, while Levinson (1985) suggests using an estimate for a lower bound on the average likelihood per observation, noting that the optimal path will never be eliminated if α is less than this value.

Some simplifications must be made, since (13.96) contains some impractical quantities. In particular, Bahl et al. suggest approximating $P(w', y_{t+1}^T | w_1^k, y_1^t)$ by ignoring the dependence upon the word string,

$$P(w', y_{t+1}^T | w_1^k, y_1^t) \approx P(y_{t+1}^T | y_1^t) \approx \prod_{\tau=t+1}^{T} P(y(\tau) | y_{\tau-\tau'}^{\tau-1}). \tag{13.97}$$

The latter approximation becomes better as τ' increases, but $\tau' = 1$ is usually adequate. The terms $P(y(\tau) | y_{\tau-\tau'}^{\tau-1})$ can be estimated from training data. Noting that $P(w_1^k, y_1^t) = P(w_1^k)P(y_1^t | w_1^k)$, we can write (13.96) as

$$\Lambda(w_1^k) \approx P(w_1^k) \sum_{t=1}^{T} P(y_1^t | w_1^k) \alpha^{T-t} P(y_{t+1}^T | y_1^t), \tag{13.98}$$

in which the latter conditional probability can be approximated as in (13.97). For a complete path $w = w_1^T$, and observation sequence $y = y_1^T$, this becomes

$$\Lambda(w) = P(w)P(y|w) = P(y, w). \qquad (13.99)$$

The reader is now encouraged to review the discussion surrounding (13.73) to be reminded of the appropriateness of this likelihood measure.

One difference between $\Lambda(w)$ and the measure that would have been computed for w using the Viterbi approach above is that here we attempt to truly maximize $P(y, w)$ (in spite of all the approximations), whereas in the Viterbi approach we maximize the the probability $P(y, \mathcal{I}, w)$, where $\mathcal{I}$ represents a state sequence through the complete network. The difference is similar to that which exists between the Viterbi and "any path" approaches to decoding an HMM.

The BJM technique also involves a second major difference in the manner in which the LD and AD are searched for the optimal solution. The method is a pruned version of the *best-first* search algorithm used in artificial intelligence research [see, e.g., (Nilsson, 1971)]. Let us illustrate this for the digit recognition problem for which the LD network is shown in Fig. 13.9. Suppose we initially pursue all hypotheses arrising from state $\bar{S}$. This will cause us to arrive at states $\bar{A}$ and $\bar{B}$ and to require likelihoods Λ(zero), Λ(oh), Λ(one), and Λ(five). In turn, by examining (13.98), we see that we will need

1. $P(\underline{w}(1) = \text{zero})$, $P(\underline{w}(1) = \text{oh})$, $P(\underline{w}(1) = \text{one})$, and $P(\underline{w}(1) = \text{five})$.
2. $P(y_1^t | \underline{w}(1) = \text{zero})$, $P(y_1^t | \underline{w}(1) = \text{oh})$, $P(y_1^t | \underline{w}(1) = \text{one})$, and $P(y_1^t | \underline{w}(1) = \text{five})$ for $t = 1, \ldots, T$.
3. $P(y_{t+1}^T | y_1^t)$ for $t = 1, \ldots, T$.

Let us assume that the initial probabilities for each word in item 1 are found in a lookup table considered part of the LD. The quantities in item 3 are found using (13.97) with $\tau' = 1$ and a lookup table for the quantities $P(y(t) | y(t-1))$. Repeated calls are made to the AD to compute the quantities $P(y_1^t | w(1))$ for each $w(1)$ and each t. These quantities are stored for a future purpose. Upon completion of these likelihoods, we put each of these single-word partial paths in a *stack* in decreasing order of likelihood. Suppose, for example, the result is

$$\begin{bmatrix} \text{zero} & \Lambda(\text{zero}) \\ \text{one} & \Lambda(\text{one}) \\ \text{five} & \Lambda(\text{five}) \\ \text{oh} & \Lambda(\text{oh}) \end{bmatrix} \qquad (13.100)$$

Now we extend the best partial path, which appears at the top of the stack. This creates several new possibilities: w_1^2 = zero–zero, w_1^2 = zero–oh, w_1^2 = zero–one, and w_1^2 = zero–six. Suppose that we try zero–one first.

To begin to introduce some generality, let us replace the specific word "zero" by $w(1)$ and "one" by $w(2)$. To compute a likelihood for the partial path w_1^2 = zero–one, we will need

1. $P(w_1^2) = P(w(1))P(w(2)|w(1))$.
2. $P(y_1^t|w_1^2)$, for $t = 1, \ldots, T$.
3. $P(y_{t+1}^T|y_1^t)$ for $t = 1, \ldots, T$.

Item 1 requires the quantities $P(w(2)|w(1))$, more linguistic information which is stored in the LD. (In this case, these quantities represent Markov probabilities, but they are not associated with state transitions in the LD state diagram. Rather, they correspond to Markov dependencies between *couples* of state transition arcs in the diagram. This is simply a matter of the way in which the problem is structured in this case.) The quantity $P(w_1^2)$ is stored for the next extension. The item 2 quantities are easy to compute because we have stored the quantities $P(y_1^t|w(1))$ for each $w(1)$ and each t. We have, therefore,

$$P(y_1^t|w_1^2) = P(y_1^{t'}|w(1))P(y_{t'+1}^t|w(2)). \tag{13.101}$$

These quantities are stored for further path extensions. Finally, the item 3 quantities have all been computed in the first iteration and have presumably been stored.

After Λ(zero–one) is computed, this candidate partial path is placed in the stack in its appropriate ranking. If it is no longer at the top, the new top candidate is extended in a similar manner and also placed in the stack in the correct order. Each time, the topmost partial path is extended by one word, and the results placed back in the stack in the proper order. After an iteration is complete, the partial path at the top of the stack is examined to see whether it is a complete path. If it is, then it is declared the optimal path, because extending any partial path below it in the stack cannot result in a path with a better likelihood. It is to be noted that partial paths may fall below some acceptable likelihood and not be placed in the stack even if there is sufficient room. It is also the case that stack size must be limited, and certain partial paths might be lost because of insufficient room. The former has been called *soft pruning* and the latter *hard pruning* (Venkatesh et al., 1991).

It is instructive to move out to the next step in the search. Suppose that the stack resulting from the above is as follows:

$$\begin{bmatrix} \text{zero} & \Lambda(\text{zero}) \\ \text{zero–one} & \Lambda(\text{zero–one}) \\ \text{one} & \Lambda(\text{one}) \\ \text{five} & \Lambda(\text{five}) \\ \text{oh} & \Lambda(\text{oh}) \end{bmatrix}. \tag{13.102}$$

The fact that a single "zero" is at the top of the stack is an indication that there is at least one more extension of this word. Indeed we know that there are three others: zero–zero, zero–oh, and zero–six. After computing the likelihood for each of these partial paths in the next three steps, suppose the stack contains

$$\begin{bmatrix} \text{zero–one} & \Lambda(\text{zero–one}) \\ \text{zero–oh} & \Lambda(\text{zero–oh}) \\ \text{one} & \Lambda(\text{one}) \\ \text{zero–zero} & \Lambda(\text{zero–zero}) \\ \text{zero–six} & \Lambda(\text{zero–six}) \\ \text{five} & \Lambda(\text{five}) \end{bmatrix} \tag{13.103}$$

Now we go to the top and extend the partial path "zero–one" to, say, "zero–one–two." Again, for generality, let us speak of w_1^3, w_1^2, $w(1)$, $w(2)$, and $w(3)$. We need

1. $P(w_1^3) = P(w(1))P(w(2)|w(1))P(w(3)|w_1^2) = P(w_1^2)P(w(3)|w_1^2)$.
2. $P(y_1^t|w_1^3)$ for $t = 1, \ldots, T$.
3. $P(y_{t+1}^T|y_1^t)$ for $t = 1, \ldots, T$.

Again, the item 3 quantities are already available, and the quantities in item 2 are computed similarly to (13.101). It is the first set of quantities on which we need to focus.

In this case the LD needs to have knowledge of the quantities $P(w(3)|w_1^2)$. [Note that $P(w_1^2)$ was stored at the last step.] In general (for a kth extension), we will need $P(w(k)|w_1^{k-1})$. Even for a small vocabulary like the digits, it is easy to see that this requirement quickly becomes prohibitive. It is customary, therefore, to use the approximation

$$P(w_1^k) \approx \prod_{l=1}^{k} P(w(l)|w_{l-N+1}^{l-1}) \tag{13.104}$$

for some small N. The LD contains a statistical characteristic of the language consisting of the probabilities $P(w(l)|w_{l-N+1}^{l-1})$. This is called an *N-gram model* of the language. A 2-gram or *bigram model* assumes a Markovian dependency between words. For most vocabularies the use of N-gram models for $N > 3$ (a 3-gram is a *trigram model*) is prohibitive.

While the BJM technique is sometimes characterized as being based upon a statistical language model as opposed to a grammatical model [see e.g., (Waibel and Lee, 1990, p. 447)], the BJM language can, in fact, be viewed as a regular stochastic language. This point is made in the paper by Levinson (1985). For a trigram model, for example, we see that every word is generated by the rule[23]

$$Q \xrightarrow{p} qR, \tag{13.105}$$

where if q represents the kth word in the string, then $p = P(q|w_{k-2}^{k-1})$. Accordingly, the generation of the words can be modeled as an HMM and the N-gram probabilities can be inferred from training data using an F–B algorithm. In fact, the F–B algorithm reduces to a relatively simple counting procedure in this case. For details see (Bahl et al., 1983; Jelinek, 1990).

As pointed out by Bahl et al., the stack search procedure is not exhaustive and the decoded sentence might not be the most likely one. This can happen, for example, when a poorly articulated word (frequently short "function" words like "a," "the," and "of") causes a poor acoustic match. For this reason, a modified procedure is sometimes used in which all partial paths in the stack with likelihoods within, say Δ, of the maximum likelihood are extended before a final decision is made. A large Δ implies more computation, but lower risk of discounting the correct path.

Finally, we note that Venkatesh et al. (1991) have worked with a multiple stack search that can be used when scattered, rather than sequential left-to-right evaluations, are made on the word models.

13.6.2 Other Formal Grammars

In principle, any formal grammar can be used to model the language in the LD. We will make some brief comments on this issue in this section, but an extensive treatment of the higher levels of the Chomsky hierarchy would take us well beyond the scope of this text. As pointed out by Levinson (1985), any finite language can be generated by a regular grammar, but one motivation for using other grammars is to make the language model conform to a more conventional model of natural language. For example, natural linguistic rules are often presented in context-sensitive form (see, e.g., Fig. 13.1). The disadvantage of higher grammars is the increased complexity encountered in the corresponding parsing algorithms. The number of operations required for Viterbi decoding of a string w using a regular grammar is proportional to $|V_n| \times |w|$, where $|V_n|$ is the size of the nonterminal vocabulary and $|w|$ represents the length of the given sentence. Let us keep this number in mind as we discuss some further methods.

Context-free languages are generally parsed using the *Cocke–Younger–Kasami* (CYK) *algorithm* or *Earley's algorithm*. The CYK algorithm was first developed by Cocke, but independently published by Kasami[24]

[23] Also see the discussion of the Paeseler and Ney (1989) work in Section 13.9.

[24] A more convenient reference to this work is (Kasami and Torii, 1969).

(1965) and Younger (1967). Earley's method, sometimes called the *chart-parsing algorithm*, was published in 1970 (Earley, 1970). The CYK algorithm is essentially a DP approach, whereas Earley's algorithm uses a central data structure called a *chart* to effectively combine intermediate subparses to reduce redundant computation. Each algorithm requires $O(|w|^3)$ operations, but Earley's method reduces to $O(|w|^2)$ if there are no ambiguities in the grammar (Fu, 1982, Sec. 5.5). More recently, Paeseler (1988) has published a modification of Earley's method that uses a beam-search procedure to reduce the complexity to linear in the length of the input string. Example CSR systems based on the Paeseler and CYK algorithms can be found in Paeseler's paper and in (Ney et al., 1987), respectively.

Left–right (LR) *parsing* is an efficient algorithm for parsing context-free languages that was originally developed for programming languages [e.g., (Hopcroft and Ullman, 1979; Fu, 1982)]. A *generalized LR parsing algorithm* has been applied to the CSR problem by Kita et al. (1989) and Hanazawa et al. (1990). The resulting system is called HMM-LR because it is based on HMM analysis of phones driven by predictive LR parsing. The HMM-LR system will be described in Section 13.9.

In discussing the BJM method in Section 13.6.1, we made the point that the trigram model could be posed as a regular grammar and consequently trained using an F–B-type algorithm. It is interesting to note that a system described by Paeseler and Ney (1989) uses a trigram model of *word categories* (triclass model) in a system in which the statistical language model manifests itself as a regular grammar in one form and a context-free grammar in a more efficient form.[25] The set of words (terminals) in this system is partitioned into word classes that share common characteristics. Each word within a given category is equiprobable to all others. In one form of the system, states (nonterminals) represent couples (remember the trigram dependency) of word categories. If the categories are $C_1, C_2, \ldots, C_N$, then any state (nonterminal) is of the form

$$A = \{C_i C_j\}. \tag{13.106}$$

A transition to a state, say $B = \{C_j C_k\}$, occurs with probability $p' = P(C_k | C_i, C_j)$. On the transition, the network generates one of the words (say, nonterminal a) from category C_k with probability

$$p'' = \frac{1}{|C_k|}, \tag{13.107}$$

where $|C_k|$ is the number of words in category C_k. The network can therefore be thought of as a manifestation of productions of the form

$$A \xrightarrow{p} aB, \tag{13.108}$$

[25]A similar idea is the use of a "tri-POS" (parts of speech) statistical model (Derouault and Mérialdo, 1986; Dumouchel et al., 1988).

where $A, B \in V_n$, $a \in V_t$ and $p = p'p''$. This, of course, represents a regular grammar. One transition in the network is illustrated in Fig. 13.10(a). Since every state that may make a transition to B in the above must generate $|C_k|$ observations on the transition, a more efficient set of productions is formed as follows. Let us artificially create two nonterminals, $B_{\text{in}} = \{C_j C_k\}$ and $B_{\text{out}} = \{C_j C_k\}$ from every nonterminal B in the above. Now if we allow rules of the form

$$A_{\text{out}} \xrightarrow{p'} B_{\text{in}} \tag{13.109}$$

and

$$B_{\text{in}} \xrightarrow{p''} aB_{\text{out}}, \tag{13.110}$$

where $A_{\text{out}} = \{C_i C_j\}$, and a, p', and p'' have identical meanings to the above, then the number of necessary transitions in the network is reduced by a factor equal to the number of categories. This revised transition scheme is also illustrated in Fig. 13.10. We see that these rules technically comprise a context-free grammar, although we know it is a thinly disguised regular grammar. It should also be apparent that the latter structure is equivalent to a Moore-form FSA in which the nontermi-

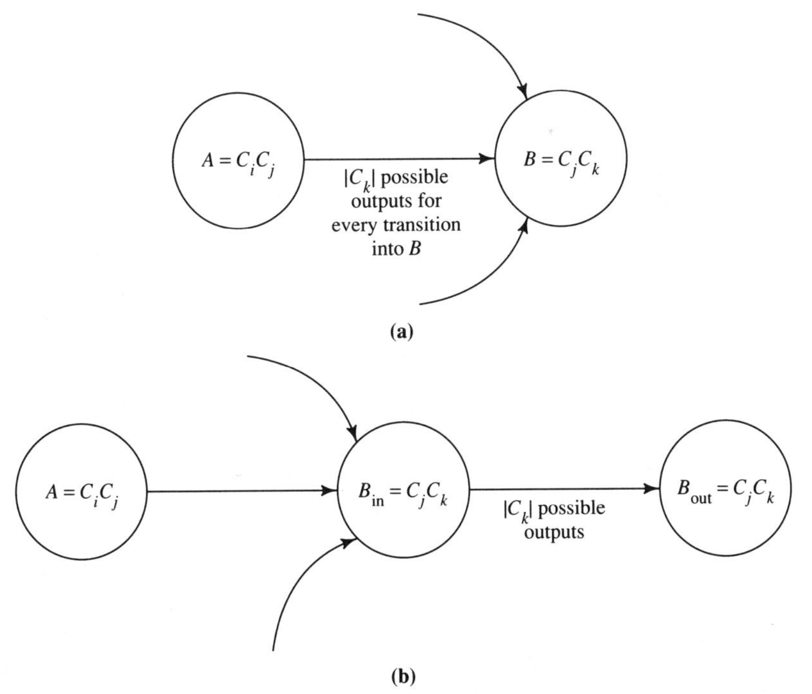

FIGURE 13.10. State model of one production rule in Paeseler and Ney's triclass grammar. (a) "Direct" form corresponding to production rules of form (13.108). (b) Revised transition corresponding to rules of form (13.109) and (13.110).

nals are generated at the states. Rather than split the states apart in the revised network, we could simply allow the states to generate the words.

Specialized grammars have also been used for CSR. *Augmented transition network* (ATN) *grammars* (Woods, 1970, 1983) are similar to context-free grammars but are more efficient due to the merger of common parsing paths. These grammars were developed specifically for natural language processing. An ATN grammar was used in the "HWIM" system (Wolf and Woods, 1977, 1980), discussed further in Section 13.9, in combination with an "island-driven" strategy in which reliable phones, words, or phrases were located using an initial scan, and then built upon using a "middle-out" search. This approach is novel in its divergence from conventional left-to-right parsing. Stochastic *unification grammars* represent generalizations of formal grammars in which features are added to the elements of the formal vocabulary. They have been used in speech processing to model contextual information (Hemphill and Picone, 1989; Shieber, 1986) and to add natural language features (person, number, mood, etc.) to the nonterminal elements of the grammar (Chow and Roucos, 1989). The inclusion of feature information in the grammar represents a step toward speech understanding in its provision of linguistic knowledge beyond the grammatical structure. In their paper, Hemphill and Picone introduce the basic unification grammar formalism and argue that viewing the speech production process as based on a grammar rather than an FSA (in the regular grammar case) has computational advantages when a chart-parsing algorithm is used to generate the hypotheses. In the paper by Chow and Roucos, a speech understanding system is presented that employs an augmented context-free grammar.

As we travel farther up the Chomsky hierarchy, complexity of parsing algorithms increases drastically, and these grammars have not found much application to the speech recognition problem. A DP-type parsing algorithm for context-sensitive grammars has been reported by Tanaka and Fu (1978) of which the complexity is exponential in $|w|$. For unrestricted grammars, there exists no universal parsing algorithm (Fu, 1982, p. 55), although algorithms do exist for special cases (Hopcroft and Ullman, 1979, pp. 267–268). A discussion of this issue and a useful bibliography on the general subject is given by Levinson (1985).

Before leaving this section, we should reiterate a point that has been made several times in earlier discussions: Both F–B-like and Viterbi-like approaches exist for the inference of the probabilities of the production rules of a formal stochastic grammar, given the characteristic grammar (Levinson, 1985; Fu, 1982). Recall that for a regular grammar, this task is equivalent to the problem of finding the probabilities associated with an HMM or FSA with a fixed structure. Consequently, the fact that these training algorithms exist is not surprising. In fact, however, any stochastic grammar may be shown to have a correlative doubly stochastic process. A discussion of this issue and related references are found in Levinson's paper.

13.7 IWR As "CSR"

In Section 12.4.2, we left open for further discussion the task of recognizing isolated words whose models were composed of smaller subword models. We said that this task could be considered a special case of CSR. Indeed, we are now in a position to understand this comment. It should be clear that we could consider the word, rather than the sentence, as the ultimate production in a language, where the word is composed of terminals corresponding to the basic subwords. A formal grammar could be constructed whose production rules would ultimately produce words from the terminals. Now with *sentence* replaced by *word* and *word* replaced by *subword* (whatever its form), any of the discussion above pertaining to a two-level CSR system would apply equally well to this simpler problem.

An accompanying recognizer of isolated words based on these formalities would have an LD consisting of a parser used to hypothesize terminal strings, and an AD that would provide the acoustic matches for the models corresponding to the terminals. If, as would be likely,[26] a regular grammar underlay the linguistic process, then the LD would be representable as an FSA, and the entire discussion of training and recognition algorithms for two-level systems would be applicable to this isolated-word system.

We should also recognize that higher-level linguistic information can be used to assist in the IWR problem when the isolated words comprise sentences or longer messages. A grammar or other model that contains information about how words may comprise sentences is an entropy-reducing device that can assist in improved recognition performance. In a system recognizing sentences comprised of isolated words, therefore, we might have two "coupled" components of the LD, one parsing words into subwords, the other overseeing the order in which words are hypothesized in the first place. This description could, of course, also describe the operation of a CSR recognizer, and a little thought should convince the reader that the present problem is just a special case of the CSR problem.[27] An important example of this approach is the IWR version of the IBM TANGORA recognizer, a system built on the principles of the BJM methods discussed above. We will say more about TANGORA below.

[26]However, context-dependent phone models of words have been found to improve recognition (Bahl et al., 1980; Schwartz et al., 1984).

[27]How would the top-down algorithm discussed in Section 13.5 be modified to accommodate known temporal boundaries in the observation string (corresponding to isolated words)?

13.8 Standard Databases for Speech Recognition Research

A significant boon to the field of speech recognition research has been the availability of standardized databases for system testing, which appeared in the mid- to late 1980s. Among those most frequently cited in the literature are the DARPA Resources Management Database (DRMD) (Price et al., 1988), the TIMIT Acoustic Phonetic Database (Fisher et al., 1986), and the Texas Instruments/National Bureau of Standards (TI/NBS) Database of Connected Digits (Leonard, 1984).

The DRMD is a 1000-word database containing material for speaker-dependent, speaker-independent, and speaker-adaptive recognition. It is based on 21,000 English language utterances about naval resource management, collected from 160 speakers with an array of dialects. The material is partitioned into training and testing data sets. The availability of the DRMD was published in 1988 in the Price et al. paper referenced above, and it has been widely used to test large-vocabulary CSR systems.

The TIMIT database represents another DARPA-supported project. TIMIT is a phonetically transcribed database that was digitally recorded by the Texas Instruments Corporation (TI) and transcribed at the Massachusetts Institute of Technology (MIT). The material for the database was selected by MIT, TI, and the Stanford Research Institute (SRI). It contains the data for 4200 sentences spoken by 630 talkers of various dialects. Data for 420 of the talkers are used as a training database, while the others' data comprise the testing data. Details of the TIMIT database are found in the paper by Fisher et al.

Finally, the TI/NBS[28] database is a collection of digit utterances (0–9, "oh") for use in speaker-independent trials. The data include the speech of 300 men, women, and children recorded in a quiet environment. The material includes strings of digits ranging from one (isolated) to seven long. Details of the TI/NBS database can be found in the paper by Leonard. Also noteworthy is the first industry standard database, TI-46, which was also developed by TI. This collection contains the alphabet, digits, and several command words. It is still being used for neural network research (Chapter 14), among other applications.

Whereas the databases listed above are English-language collections, it is certain that other language databases will be developed in the future. At the time of writing this book, the Acoustic Society of Japan is planning to release a Japanese language database. The SAM project is a European effort to develop large multilingual databases (Mariani, 1989).

Several of the databases discussed above are available on CD-ROM from the U.S. National Institute of Standards and Technology (NITS). Ordering instructions are given in the preface to this book.

[28]The National Bureau of Standards (NBS) is the former name of the NIST mentioned below.

13.9 A Survey of Language-Model-Based Systems

We conclude this chapter with a brief survey of some of the speech recognition systems of historical and contemporary significance. Any such survey will necessarily be incomplete, as we cannot hope to cover the vast array of systems and techniques that have been proposed and implemented over the years. Our objective will be to present a few systems that illustrate various approaches and concepts described in the foregoing material. We will also see some of the higher-level language sources come into play that we have only briefly discussed in this chapter. Finally, this survey will give an indication of the performance capabilities of contemporary speech recognition systems. With one exception, we explicitly focus on systems that are principally research systems and avoid discussion of commercial products. Of course, the ultimate purpose of speech recognition research and development is application to practical problems. Although the speech recognition field is still relatively young, and although many challenging problems remain, many interesting applications have taken place. The reader is referred to (AT&T, 1990), for example, to read about some of these endeavors.

The more recent of the systems described below represent evolving research. Accordingly, we can only provide a brief synopsis of the operating principles extant at the time of completion of this book. The reader is encouraged to consult the literature to discover recent advances in these and other systems. New results are often first reported in the *Proceedings of the IEEE International Conference on Acoustics, Speech, and Signal Processing* (see Appendix 1.E.4). As an example of this continuing evolution, we note that artificial neural network technology, which we take up in the following chapter, has begun to be integrated into some of the contemporary systems described below. In Chapter 14, we shall briefly return to this issue and discuss some of these details.

ARPA Speech Understanding Project. In the United States, the modern era of large-scale automatic speech recognition was ushered in by the Advanced Research Projects Agency (ARPA) of the Department of Defense when, in 1971, it announced a five-year development program with the goal of significantly advancing the field of speech understanding.[29] The ARPA goals for a prototype system are shown in Table 13.1 along with the features of the HARPY system of Carnegie–Mellon University, the only system to exceed all of the stated goals. Klatt (1977) has written a review of the ARPA project that compares and contrasts the architectures and operating principles of four of the systems that resulted from the study.[30]

[29]In spite of the name "speech understanding," only one of the systems described below will be seen to use linguistic knowledge above the syntactic level.

[30]In this paper Klatt also gives a useful list of citations to earlier work in speech recognition. In particular, he recommends the paper by Reddy (1976) from which we quoted in Section 10.2.4.

TABLE 13.1. ARPA 1971 Five-Year Goals for a Prototype Speech Recognition System, Along with the Features of the HARPY System of Carnegie–Mellon University. After Klatt (1977).

ARPA Five-Year Goals (November 1971)	HARPY Characteristics (November 1976)
Accept connected speech	Yes
from many	5 speakers (3 male, 2 female)
cooperative speakers	yes
in a quiet room	computer terminal room
using a good microphone	close-talking microphone
with slight tuning per speaker	20 training sentences per talker
accepting 1000 words	1011 words
using an artificial syntax	average branching factor = 33
in a constraining task	document retrieval
yielding less than 10% semantic error	5% semantic error
in a few times real time	80 times real time
on a 100-MIPS machine	0.4 MIPS PDP-KA10
	using 256K of 36-bit words
	costing $5 per sentence processed.

We give a brief synopsis of the ARPA research with the reminder that these systems, while remarkable achievements in their era, naturally represent early technologies that do not reflect the current state of the art. Accordingly, the reader may wish to casually read through these descriptions for general information only.

The four systems resulting from the ARPA study and their gross performance figures are listed in Table 13.2. Note that the perplexity is used as a measure of the difficulty of the recognition task. As noted by Klatt, given the different branching factors, it is difficult to determine absolute performance differences between the systems. These four systems employed vastly different approaches, but all employed a form of top-down governance of the processing.[31] Briefly, these systems are:

1. The *HARPY* system of Carnegie–Mellon University (CMU) (Lowerre and Reddy, 1980). The basis for HARPY is a massive 15,000-state network that includes lexical representations, syntax, and word boundary rules compiled into a single framework. The resulting FSA is decoded with respect to acoustic measures (see below) using DP and beam search. A predecessor of HARPY is the *DRAGON* system (Baker, 1975), also developed at CMU, which also employed decoding of an FSA using a breadth-first DP search.[32]

[31]In addition to the references cited with each system, Klatt's paper gives a comprehensive description of each of the systems.

[32]This term connotes the fact that all paths are extended in parallel, rather than extending the highest likelihood paths first as is the case, for example, in the BJM technique.

TABLE 13.2. The Four Systems Resulting from the 1971 ARPA Study and Their Gross Performance Figures. After Klatt (1977).

System	Sentences Understood (%)	Perplexity
CMU HARPY	95	33
CMU HEARSAY II	91, 74	33, 46
BBN HWIM	44	195
System Development Corp.	24	105

Note: Percentages are based on more than 100 sentences spoken by several talkers, except for HEARSAY II, which was tested with a smaller data set.

The addition of the beam-search concept in HARPY, however, vastly improved computational efficiency.

The acoustic processing in the HARPY system consists of extracting 14 LP parameters from 10 msec frames of speech. Interestingly, the frames are combined (by summing correlation matrices) if sufficiently similar (according to the Itakura distance) in order to reduce processing time and smooth noise effects. The resulting "acoustic segments" (typically 2–3 frames) are classified into one of 98 groups using the Itakura distance.

2. The *HEARSAY II* system of CMU (Lesser et al., 1975). The HEARSAY system has a radically different architecture from that of HARPY. Information from all sources of acoustic and linguistic knowledge (all below the semantic level) are integrated on a "blackboard" that serves as a controller for the processing. The knowledge processors are quite compartmentalized and relatively easily modified. One component, the "word verifier," is in the form of a HARPY-like FSA for composing words from subword units. The interaction with the acoustic signal occurs in an *island-driven* fashion in which a high-scoring seed word is sought from which to expand to complete sentence hypotheses. A CYK parser is employed in this process in order to consider many hypotheses in parallel.

   Acoustic processing in the HEARSAY II system consists of computing the peak-to-peak amplitudes and zero crossing measures on 10-msec nonoverlapping frames of both a preemphasized speech waveform and a smoothed version of it. Frames of similar measures are grouped into intervals and these intervals, in turn, are classified by manner of articulation using a decision tree and a series of threshold tests on the four parameters.
3. The *HWIM* ("Hear What I Mean") system of BBN, Inc. (Wolf and Woods, 1977, 1980). The HWIM system also uses an island-driven strategy in which complete hypotheses are built from high-scoring seed words. Lexical decoding is accomplished using a sophisticated network of phonological rules applied to a lattice of possible segmentations of the waveform. Word hypotheses are guided from above by syntactic and semantic knowledge, which takes the form

of an ATN grammar (see Section 13.6.2). The complete hypotheses are developed using best-first search.

Acoustic processing in the HWIM system consists of extracting formants (by LP analysis), energy in various frequency bands, zero crossings, and fundamental frequency on 20-msec (Hamming windowed) frames every 10 msec.

4. Systems Development Corporation (SDC) speech understanding system (Ritea, 1975). The SDC system generates a set of alternative phonetic transcriptions from acoustic and phonetic processing that are stored in a "matrix" for processing from above. The system then generally follows the basic paradigm of left-to-right, best-first search using a phonetic "mapper" that interfaces the syntactic and lexical knowledge sources with the phonetic transcription hypotheses from the acoustic processing.

   Acoustic analysis in the SDC system involves computation of energy and zero crossing measures on 10-msec intervals, pitch estimation by a center clipping and autocorrelation technique (Gillman, 1975), and LP analysis on 25.6-msec Hamming windowed frames every 10 msec.

Perhaps one of the most significant findings of the ARPA project is manifest in the HARPY approach—efficiently applied grammatical constraints can comprise a powerful tool in achieving highly accurate performance. This is true in spite of the relatively simple acoustic processing used in the HARPY system. This finding has clearly influenced the course of speech recognition research in the ensuing years.

TANGORA. Concurrent with the ARPA projects, work (continuing today) was in progress at IBM on the application of statistical methods in automatic speech recognition. In particular, early work on the HMM was published by Jelinek (1976) [and independently during the same period by Baker (1975) at CMU]. In 1983, IBM researchers published the paper (Bahl et al., 1983) to which we have frequently referred in our discussions. This paper indicates significantly better performance on the speaker-dependent, constrained-task CSR than was achieved by HARPY. In this paper are many of the seminal ideas on the use of HMMs in the CSR problem. Since we have described the basic technologies in Section 13.6.1, let us just mention here a manifestation of the research. In the early 1980s the IBM group focused their attention on the problem of office dictation. The result, announced in 1984, was a large-vocabulary (5000 words plus a spelling facility), speaker-dependent, isolated-word,[33] near-real-time recognition system built on a vast platform of computing facilities including a mainframe computer and a workstation (Jelinek, 1985). By 1987, the system was scaled down to operate in real time in a

[33]The system uses the techniques of Section 13.6.1, which are applicable to CSR. The IWR problem may be considered a special case in which (word) temporal boundaries are known in the observation sequence.

personal computer with four special-purpose signal processing boards, while having an expanded vocabulary scalable from 5000 to 20,000 words (Averbuch et al., 1987). In 1989, the extension to continuous speaker-dependent utterances of the 5000-word vocabulary was announced (Bahl et al., 1989). The system is called the *TANGORA* system, named for Albert Tangora who is listed in the 1986 *Guinness Book of World Records* as the world's fastest typist (Bahl et al., 1988).

The TANGORA system is based on discrete-observation HMM models using a 200-symbol VQ codebook, and a trigram language model. The concept of fenonic baseforms was added in the 1988 report of the work (Bahl et al., 1988) (see Section 12.3.4) in order to decrease the training time for a new speaker. It is interesting that the training of the 5000-word system requires about 20 minutes to read 100 sentences composed of 1200 words, 700 of which are distinct. Some typical performance results for TANGORA as compiled by Picone (1990) are shown in Table 13.3.

The BYBLOS System. In recent years, ARPA has become DARPA (the Defense Advanced Projects Research Agency), and several systems have been developed under contracts from this agency. Among them are the *BYBLOS* and *SPHINX* systems. The BYBLOS system, developed at BBN, Inc. (Chow et al., 1987; Kubala et al., 1988), is a speaker-dependent CSR system intended for large-vocabulary applications. Its system structure and search mode (decoding is based on the F–B algorithm) closely conform to the general description earlier in this chapter of top-down processors. In fact, the reader might wish to read the paper by Chow et al., because its organization is parallel to our discussions above and will therefore be an easily followed explanation of a real-world CSR system.

One of the unique features of BYBLOS is the inclusion of context-dependent phone models, which were described in Section 12.3.4. This approach allows the models to capture coarticulatory effects. In fact, the name BYBLOS is quite significant in this regard—Byblos is the name of an ancient Phoenecian town (now Jubeil, Lebanon) where the first phonetic writing was discovered. Researchers at BNN chose the name to emphasize the phonetic basis for the system. In 1986, when the system was

TABLE 13.3. Typical Performance Results for TANGORA as Compiled by Picone (1990).

Recognition Task	Word Error Rate (%)
5000-word office correspondence	2.9
20,000-word office correspondence	5.4
2000 most frequent words in office correspondence using phonetic baseforms	2.5
2000 most frequent words in office correspondence using fenonic baseforms	0.7

conceived, there was a widespread belief among the speech processing community that stochastic speech recognition systems based on phonetic units were unfeasible (Makhoul, 1991).

Acoustic processing in BYBLOS consists of the computation of 14 mel-cepstral coefficients every 10 msec using a 20-msec window. The acoustic models are discrete-observation HMMs based on VQ using a 256-symbol codebook.

BYBLOS has been tested under various conditions and with various tasks. In (Kubala et al., 1988), the 1000-word DRMD is used as the test material. Three grammar models were employed for word hypotheses. The first is a regular grammar (FSA) of perplexity 9, the second a *word-pair grammar* (bigram grammar without probabilities) of perplexity 60, and the third a *null grammar* (no grammar) of perplexity 1000 (simply equal to the number of words). Some typical results with data from the DRMD are shown in Table 13.4. Other results are found in (Chow et al., 1987; Kubala et al., 1988).

More recently, N-best search has been incorporated into the BYBLOS system (Schwartz and Chow, 1990; Schwartz et al., 1992); 1990 experiments were performed on the speaker-dependent portion of the DRMD using (in the N-best search) simple language models consisting of no grammar (perplexity 1000), and then a class grammar of perplexity 100. The N-best search, when repeated for $N = 1, 2, \ldots, 100$, produced the correct answer at an average ranking of 9.3 with no grammar, and 1.8 with the relatively unconstrained class grammar. With the class grammar, 99% of the time the correct sentence was found in the top 24 choices.

The SPHINX System. The SPHINX system, developed at Carnegie–Mellon University (Lee et al., 1990), represents another system based on careful phonetic modeling. This system is intended for large-vocabulary CSR in speaker-independent mode.

Like BYBLOS, the SPHINX system generally follows the basic principles of top-down linguistic processing using Viterbi decoding described earlier in the chapter. The most interesting features of the system occur at the lowest levels of the linguistic and acoustic processing.

SPHINX is based on context-dependent, discrete-observation phone models, which in this work are referred to as *triphones*. The basic phone HMM topology used in SPHINX is illustrated in Fig. 12.10. One thousand such phone models were trained on the 1000 most frequent naturally occurring triphones in the DRMD, which was used to test the system (7000 triphones were found in the data). The HMMs are discrete observation, but have an interesting feature that the three 256-symbol codebooks used—cepstral, differential cepstral, and energy features—are derived from LP analysis and are each coded separately. Each transition in the FSA ultimately generates three features and their probabilities are combined. Word duration models are also included in a later version of the system. In addition, measures are taken to account for poorly articulated "function" words such as "a," "the," "of," and so on.

TABLE 13.4. Typical 1988 Performance Results for the BYBLOS System Using Data from the DARPA Resources Management Database and Three Grammars. After Kubala et al. (1988).

	Sentence Pattern				Word Pair				No Grammar	
	Word Error (%)	Word Correct (%)	Sentence Error (%)	Test Set Perplexity	Word Error (%)	Word Correct (%)	Sentence Error (%)	Test Set Perplexity	Word Error (%)	Word Correct (%)
BEF	2.6	98.3	20	8	8.9	93.2	44	62	40.9	62.6
CMR	2.7	99.1	20	7	9.3	94.7	52	66	39.6	65.4
DTB	0.5	100.0	4	10	5.4	96.5	32	64	39.4	63.1
DTD	1.0	99.0	8	8	6.7	94.2	44	54	26.7	75.3
JWS	0.9	99.1	8	9	4.3	96.2	28	59	25.6	75.4
PGH	0.5	99.5	4	9	6.0	96.0	24	56	32.0	70.5
RKM	2.4	98.1	16	10	16.4	89.7	52	64	30.5	71.8
TAB	0.5	100.0	4	9	3.2	97.7	20	67	24.8	76.5
Average	1.4	99.1	10.5	9	7.5	94.8	37.0	62	32.4	70.1

Several language models were derived from the DRMD for testing of SPHINX. These included the same three grammars used to test BYBLOS above plus a bigram model of perplexity 20. Many data indicating performance of SPHINX under various conditions and strategies are given in the paper by Lee et al. Some typical recognition results using the DRMD with different grammars and with various improved versions of the system are shown in Table 13.5.

The LINCOLN System. In Chapter 12, we mentioned the efforts of researchers at Lincoln Laboratories to model speech under various conditions of speed, stress, emotion, and so on. In 1989, Paul (1989) reported on the efforts to expand this research to the large-vocabulary CSR domain for both speaker-dependent and speaker-independent trials. The DRMD was used with a word-pair grammar. Continuous-observation, Gaussian mixture density HMMs are used to model context-sensitive phones. Accordingly, this work involves a system and tasks with similarities to the SPHINX work discussed above. For details, the reader is referred to the paper by Paul.

DECIPHER. The *DECIPHER* system, developed at SRI International, is based on similar principles to systems discussed above (Murveit and Weintraub, 1988; Weintraub et al., 1989; Cohen et al., 1990). This system is notable for its careful attention to modeling of phonological details such as cross-word coarticulatory effects and speaker-specific phonological adaptation. Based on experiments with the DRMD, comparisons to (1988) SPHINX and BYBLOS results in the 1989 Weintraub paper indicate improved performance as a consequence of phonological modeling.

ATR HMM-LR System. Researchers at ATR Interpreting Telephony Research Laboratories (ATR) in Kyoto have developed the *HMM-LR* system, which is based on direct parsing of HMM phone models without any intermediate structures such as phoneme or word models (Kita et al.,

TABLE 13.5. Typical Sentence Recognition Results for the SPHINX System Using the DARPA Resources Management Database with Different Grammars and Various Enhancements to the System. After Lee et al. (1990).

	Grammar		
System Version	**Null**	**Word Pair**	**Bigram**
Baseline	31.1% (25.8%)	61.8% (58.1%)	76.1% (74.8%)
3 codebooks and 4 feature sets	45.6% (40.1%)	83.3% (81.1%)	88.8% (87.9%)
Word duration	55.1% (49.6%)	85.7% (83.8%)	91.4% (90.6%)

Note: Percentages in parentheses represent word recognition rates.

1989; Hanazawa et al., 1990). We have already briefly described LR parsing at the end of Section 13.9. [LR parsing has more recently been incorporated into the SPHINX system (Kita and Ward, 1991).] In HMM-LR, phones are modeled by discrete-observation HMMs employing multiple codebooks including cepstral differences, energy, and an LP-based spectral measure described in (Sugiyama and Shikano, 1981). The training material is a 5456-isolated-word database developed by ATR.

In experiments reported in the 1991 paper by Kita and Ward, the task was to recognize short Japanese phrases (Bunsetsu) uttered by three male speakers and one female speaker. A vocabulary of 1035 words and a grammar of estimated perplexity greater than 100 were used. A phrase recognition rate of 88.4% was achieved with the correct phrase appearing in the top five choices 99% of the time. The results were shown to have benefited from duration models and a speaker adaptation routine.

CSELT. As a part of the European joint research effort ESPRIT, researchers at the Centro Studi e Laboratori Telecomunicazioni (CSELT) and the Universitá di Salerno have developed a 1000-word continuous-speech recognizer that employs a unique word hypothesizer (Fissore et al., 1989). Using a kind of "*N*-best" approach, the CSELT system selects words on the basis of coarse phonetic description, and then refines the hypotheses using more detailed matching. Acoustic decoding is based on HMM phone models and mel-cepstral coefficients. In a language of perplexity 25, experiments involving two speakers uttering 214 sentences produced a word accuracy of 94.5% and correct sentence rate of 89.3%.

Philips Research Laboratory. A natural evolution of the work of Ney et al., which we have discussed extensively in this chapter, is a 10,000-word continuous-speech recognition system described in (Steinbiss et al., 1990; Ney et al., 1992).

Connected-Digit Recognition with Language Models. On a number of occasions we have mentioned the efforts of researchers at AT&T Bell Laboratories to solve the difficult problem of recognition of continuous strings of speaker-independent digits. These efforts have contributed greatly to the understanding of the basic properties of the HMM. In particular, many results on the application of continuous-observation HMMs were developed during the course of this research. We have already discussed the LB algorithm for both DTW- and HMM-based systems, which has also emerged from these efforts. For details of this work, the reader is referred to the paper by Rabiner et al. (1989) and its references. Also noteworthy is an implementation of a continuous-observation HMM-based LB algorithm with a finite state grammar on a systolic processor developed at AT&T (Roe et al., 1989). This latter work represents a conversion of the problem from a "level synchronous" processing scheme, which we saw in the bottom-up parsing example, to one using a "frame synchronous" strategy, as we used in the top-down approach. Some typical results illustrating the effects of various model parameters were given in Chapter 12.

Researchers at Texas Instruments (TI) Corporation have also contributed significantly to the connected-digit recognition problem. A study described in (Doddington, 1989) uses a technique known as "phonetic discriminants" to maximize discrimination information among continuous-observation HMMs representing the digits. In the TI research, the TI/NBS digit database was used as the evaluation material, and a three-level FSA structure—sentence, word, phone—is employed to model the language, except for the digit recognition experiments in which two levels were used. Table 13.6 shows results for four classes of models, explained in the table note.

Very Large Vocabulary Systems. Several research groups have worked on the problem of recognizing very large vocabularies with the aid of language models. At INRS and Bell Northern in Canada, investigators have worked on a 75,000-word speaker-dependent system with several different language models. The best performance was achieved with a trigram model with which 90% recognition was obtained.

At IBM in Paris, experiments have been conducted on a 200,000-word vocabulary in which the entry mode is speaker-dependent syllable-by-syllable utterances (Mérialdo, 1987). Other example systems have been reported by Kimura (1990) and Meisel et al. (1991).

VOYAGER. We have agreed to remain "below" the higher-level knowledge sources, such as semantics, in our study. The interested reader may wish to explore the papers on the *VOYAGER* system developed at MIT as an example of a system with such natural language components (Zue et al., 1990). An interesting feature of this system is the integration of both top-down and bottom-up searching. In a sense, the N-best search proce-

TABLE 13.6. Texas Instruments' Independent Digit Recognition Study. After Picone (1990).

Observation Model	Sentence Error Rate (%)	Word Error Rate (%)
Pooled covariance	3.5	1.3
Diagonal covariance	3.1	1.2
Full covariance	2.1	0.8
Confusion discriminants	1.5	0.5

Note: The four model classes are as follows: (1) The pdf's of states of the model share a common ("pooled") covariance matrix. (2) Each state density is modeled with a diagonal covariance matrix (features assumed uncorrelated). (3) A full covariance matrix is present at each state. (4) "Phonetic discriminants" are employed. The percentage represents string error rate. No grammar is employed.

dures we have discussed several times previously are a primitive form of this integration. As systems become more complex, it is likely that this bidirectional search strategy will become more prevalent.

13.10 Conclusions

We have now completed our study of the attempts to automatically recognize speech using sequential computing machines. The journey has been long and detailed, and it is clear that a tremendous amount of effort and ingenuity has gone into finding solutions for various subproblems. It is equally clear that we still have much to learn, and that the dream of a naturally conversant machine remains a distant goal. The results of the last several decades have been humbling, but have also provided much hope that one or more solutions will eventually be found. Whatever the eventual solution, from our current vantage point it seems likely that dynamic programming, language modeling, and faster computing will be a part of it. Many experts have also argued that, in spite of the vastly significant performance improvements brought about by language models, language models alone will not ultimately yield satisfactory performance. One emerging trend is the use of language models in bidirectional (both top-down and bottom-up) hypothesis formation. More work on the technologies at acoustic level, and more fundamentally on the speech production model itself, will be needed.

Still other researchers have begun to explore the possibility that a radically different computing architecture might hold promise. To this relatively infant speech technology, the artificial neural network, we briefly turn in the next chapter.

13.11 Problems

13.1. (a) A formal grammar $\mathcal{G}_1$ has the terminal set $V_t = \{a, b\}$ and corresponding language

$$\mathcal{L}(\mathcal{G}_1) = \{a^i b^j \mid i, j = 1, 2, \ldots\}. \tag{13.111}$$

Give a possible set of nonterminals and production rules for $\mathcal{G}_1$, in the process showing that $\mathcal{G}_1$ is a finite state grammar.

(b) Suppose that a second grammar, $\mathcal{G}_2$, with the same terminal set, V_t, has corresponding language

$$\mathcal{L}(\mathcal{G}_2) = \{a^i b^i \mid i = 1, 2, \ldots\}. \tag{13.112}$$

Repeat part (a) for this grammar, in the process arguing that $\mathcal{G}_2$ cannot be a finite state grammar.

13.2. (a) Show that if an ergodic language has words that follow a trigram model in which

$$P\big(\underline{w}(l) = w(l)\big|\underline{w}(l-1) = w(l-1), \underline{w}(l-2) = w(l-2), \underline{w}(l-3) = w(l-3)\big) = P\big(\underline{w}(l) = w(l)\big|\underline{w}(l-1) = w(l-1), \underline{w}(l-2) = w(l-2)\big), \tag{13.113}$$

then the entropy of the language is given by

$$H(\underline{w}) = -\lim_{n\to\infty}\frac{1}{n}\left[\sum_{l=3}^{n}\log P\Big(\underline{w}(l) = w(l)\Big|\underline{w}(l-1) = w(l-1), \underline{w}(l-2) = w(l-2)\Big) + \log P\Big(\underline{w}(2) = w(2)\Big|\underline{w}(1) = w(1)\Big) + \log P\Big(\underline{w}(1) = w(1)\Big)\right]. \tag{13.114}$$

[*Hint:* Use (13.13).]

(b) Explain how to estimate the perplexity of the language of part (a) experimentally.

13.3. In a general way, describe a bottom-up recognition system in which the AD hypothesizes words from the left and receives guidance from the LD as to whether continuation of a string is advisable. In particular, describe an LB algorithm in which a linguistic cost is integrated with the acoustic cost on a per level basis in order to find a single sentence hypothesis consistent with both bodies of information. Notice that the LD must remain relatively simple, leaving the recordkeeping function to the AD, if the processing is to remain bottom-up.

13.4. Figure 13.11(a) represents a small linguistic-level FSA that interacts with the acoustic-level word unit HMMs. The search for an optimal path through the utterance proceeds in accordance with the method described in Section 13.5. In Fig. 13.11(b), the two levels have been compiled into a single large FSA network. Carefully describe a Viterbi-like algorithm for searching the compiled FSA with respect to observation string **y**, which will produce the same optimal sentence hypothesis as that produced by the two-level processor. In particular, make clear the means by which your algorithm traverses the boundaries between models in the compiled FSA.

13.5. How would the operation of a top-down CSR system based on a regular grammar change if the endpoints in the word sequence were known? Consider first the case in which the AD contains word models, then the case in which the word models are at an "intermediate" level in the LD. Answer these questions by making modifications to the "state-based" Viterbi algorithm in Fig. 13.6.

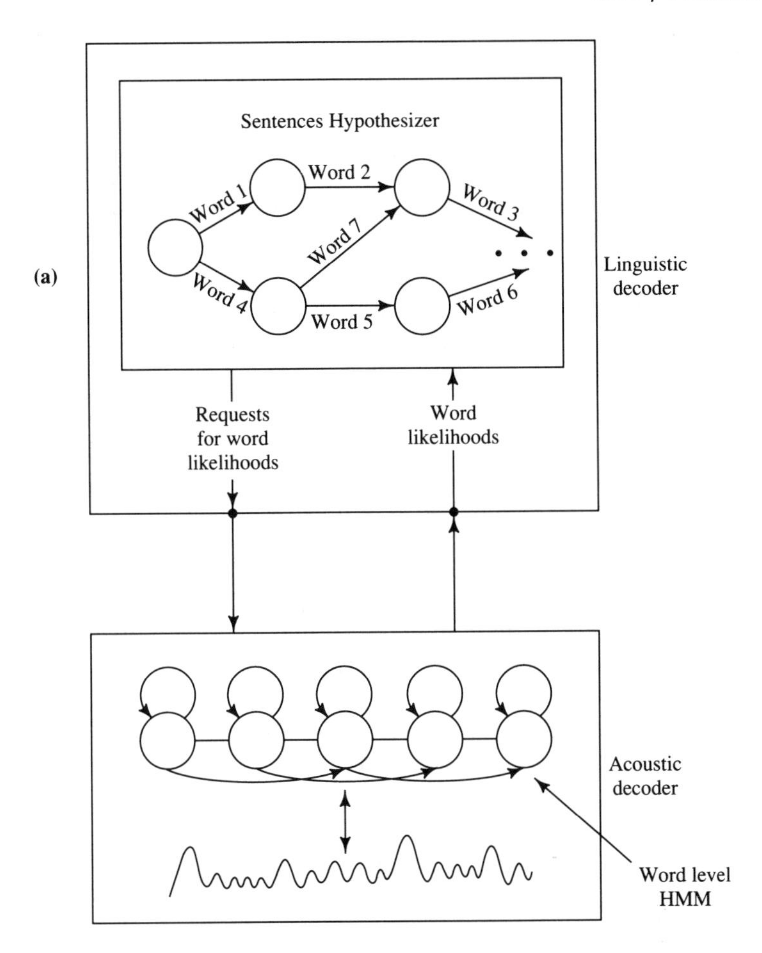

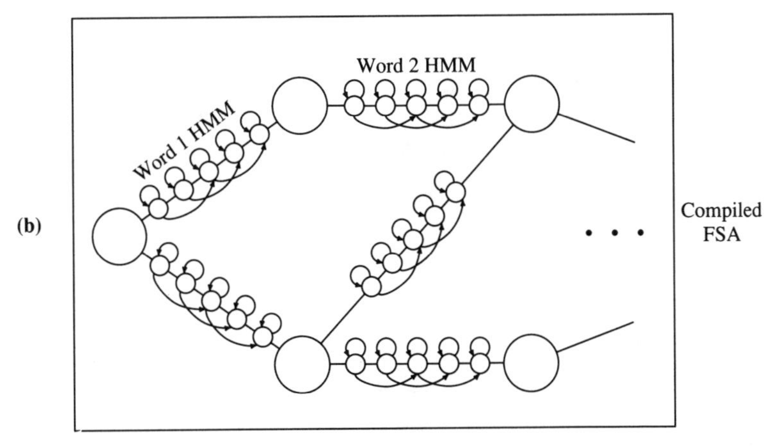

FIGURE 13.11. (a) A small two-level speech recognizer in which the linguistic-level FSA interacts with the acoustic-level word unit HMMs in accordance with the method described in Section 13.5. (b) The two levels compiled into a single large FSA network.

13.6. Give a formal argument that the top-down digit recognizer in Section 13.5 ultimately finds the digit string w^*, where

$$w^* = \underset{w}{\operatorname{argmax}} P(w \mid y). \tag{13.115}$$

13.7. The stack decoding algorithm described in Section 13.6.1 can be greatly simplified if the word boundaries in the observation string (and therefore, the number of words) are known. For simplicity, assume that each utterance is K words long. Suppose t_k is the last observation corresponding to $w(k)$, for $k = 1, 2, \ldots, K$ $(t_K = T)$. In fact, the concern for fairness in comparing paths of different word lengths can be circumvented by providing K stacks and only entering paths of length k in the kth stack.

(a) Without concern for practical algorithm details, generally describe the operation of such a multiple-stack decoding algorithm with a best-first procedure in this known-boundary case. Begin by defining a greatly simplified version of the likelihood $\Lambda(w_1^k)$ found in (13.96).

(b) If memory allocation for the complete set of stacks is limited, so that the kth stack may only hold, say, L_k paths, what is the appropriate relationship among the lengths $L_1, L_2, \ldots, L_K$? Why?

(c) Generally speaking, how does the procedure devised in parts (a) and (b) differ from a "Viterbi" approach to the same problem? Include a comparison of the hard and soft pruning aspects of the two solutions.

13.8. (a) Argue that any N-gram model of a language can be posed as a language governed by a stochastic finite state grammar, $\mathcal{G}$. The N-gram relationship is among the words, and the words are considered the terminal elements in $\mathcal{G}$.

(b) If there are W words in the vocabulary of part (a), show that W terminals, W^N nonterminals, and W^{N+1} productions are required in $\mathcal{G}$. (*Hint*: A partial state diagram might be useful in your argument.)

CHAPTER 14

The Artificial Neural Network

Reading Notes: This chapter requires no special topics from Chapter 1.

14.1 Introduction

In this final chapter we treat an emerging computing technology and its application to speech recognition. The application of *artificial neural networks* (ANNs) to speech recognition is the youngest and least well understood of the recognition technologies. Somewhat unlike other technologies and theories we have discussed for speech recognition—DTW, HMM, language modeling—the speech recognition problem has been an important *application* of ANN technologies but has not been a principal driver of the development of the ANN field. In fact, the application of ANN strategies to speech recognition occupies a small corner in a vast field of theories and applications centering on these computing networks. This research endeavor has not yet matured to the point at which a general technical framework exists, although many of the well-known paradigms have been carefully formalized independently. The field currently consists of numerous and varied architectures and techniques tied together by a common computing "philosophy" that is radically different from that underlying the von Neuman computer. To attempt to describe the totality of this field would take us well beyond the scope of this chapter, and well beyond what is needed for an initial understanding of ANN applications to speech recognition. We will mainly restrict ourselves to the study of a single ANN (called variously a "multilayer perception" or "feedforward neural net," among other names) and enhancements that have been at the center of most ANN applications to automatic speech recognition. Further, we will largely restrict our discussion to technologies that build on familiar ideas from previous chapters.

The ANN research field has a rich, often controversial, history. The work has been highly interdisciplinary, or perhaps more accurately cross-disciplinary, having received attention from physiologists, psychologists, linguists, physicists, mathematicians, computer scientists, and engineers. Research in this field has arisen out of interest in neurology, cognition, perception, vision, speech, linear and nonlinear systems theory, algorithms, and VLSI architectures, among other topics. To a greater or lesser extent, depending on the research aim, attempts have been made to relate ANNs to the operation of the human nervous system and its basic "computing" elements (more on this later). Many of these varied and fascinating aspects of the subject will necessarily be omitted from our discussion here. For a clear and comprehensive treatment of these issues and an extensive bibliography, the reader is encouraged to see the example textbooks and resources in Appendix 1.G. In particular, a comprehensive but concise historical survey of the field is given in the appendix of the Simpson (1990) monograph. Of course, many references cited in the chapter will point to information for further study.

The ANN is based on the notion that complex "computing" operations can be implemented by the massive integration of individual computing units, each of which performs an elementary computation. "Memories" are stored, computations performed, and relations formed through patterns of activity of these simple units, rather than through sequences of logical operations used in conventional von Neuman machines. Motivation for such a computing structure is derived from the human central nervous system (CNS). The CNS consists of an estimated 10^{11}–10^{14} nerve cells, or *neurons*, each of which typically interacts with 10^3–10^4 other neurons as inputs. A simplified anatomical model of a neuron is shown in Fig. 14.1.[1] The cell body, or *soma*, contains the nucleus of the neuron and exhibits protrusions (*dendrites*), which receive inputs from the *axons* of other neurons. The axon is the "transmission line," which transmits an electrical pulse (*action potential*) to neurons farther down the network. Axons in the human body range in diameter from 0.5 μm to 20 μm, and in length from about a millimeter to over a meter. The velocity of propagation of a pulse depends on the diameter of

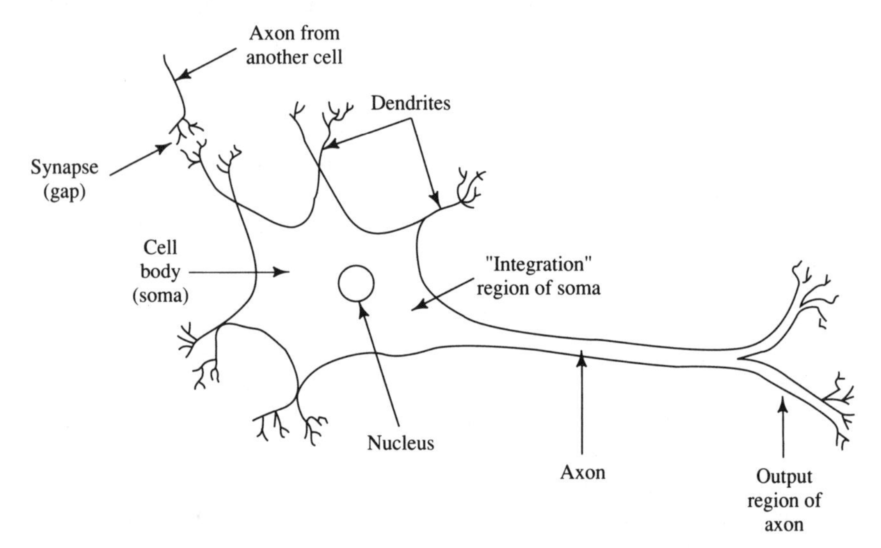

FIGURE 14.1. Simplified anatomical model of the biological neuron.

[1]The following discussion is grossly simplified. For a complete description of the human nervous system, see, for example (Aidley, 1971; Guyton, 1979; Sherwood, 1989).

the axon and other properties of the cell composition, and can range from 0.5 m/sec (small diameters) to 100 m/sec. Whether a receiving neuron will "fire" (send an action potential down its own axon) depends on the inputs received on its dendrites from other neurons. Some of these inputs may be *excitatory* (tending to promote firing, "positive"), while others may be *inhibitory* ("negative"). In most cases the mechanism by which information is transmitted from axon to dendrite (across the gap, or *synapse* of ~200 Angstroms) is actually chemical. *Neurotransmitters*, released by excitatory axons into the synapse, change the electrical properties of the receiving neuron's membrane and promote the transmission of an action potential. Inhibitory axons release substances that hyperpolarize the cell in a manner which prevents action potential formation. Small nerve cells can be stimulated at a rate of about 250 Hz, while large fibers can carry 10 times that many pulses per second.

In a simple model, the response of the receiving neuron is "all or none." That is, the neuron either receives sufficient excitatory input to produce an action potential or it does not. The amplitude of the resulting pulse is not affected linearly by the magnitude of the accumulated inputs. In this sense, the neuron is a nonlinear system whose output represents a nonlinear transformation of its inputs. Note that if the CNS is viewed as a "computer," its only resource for forming computations and relations, or invoking memories or generalizations, resides in the firing patterns of its manifold and richly interconnected simple cells.

Lest we form unrealistic beliefs about our own understanding of the human brain, we must cautiously point out that the portrayal of the neuron given above is a grossly simplified picture of a very complex and varied cell. Physiologists have a significant understanding of the anatomy and chemical and electrical properties of the nerve cell (well beyond the simple description given here), but the operation of the totality of the human brain remains fundamentally mysterious. Whether the concept of the brain as a "computer" that computes with electrical firing patterns has any validity is purely speculative. How such firing patterns could be related to higher cognitive functions such as reasoning, planning, or decision making is unknown. If our model of the brain as a massively powerful computer has any validity, ANNs might already be providing primitive clues to these unknowns. It is critically important to keep in mind, however, that the ANN is motivated by a highly speculative notion of how the brain might function. It is composed of "cells" that are crude, greatly simplified models of the biological neuron. Demonstrations by ANNs of "vision," "speech recognition," or other forms of pattern recognition, or creation of artificial cognitive function by these networks, may ultimately assist in a clearer understanding of brain function. On the other hand, the operation of ANNs might ultimately have almost nothing to do with how the brain functions. Only the future will answer these questions. In the meantime we must be wary of exagerated claims about the relations of ANNs to the human CNS.

14.2 The Artificial Neuron

In this section we introduce the basic processing unit of the ANN and define some useful vocabulary.

The *processing elements* of ANNs—also called *cells*, *neurons*, *nodes*, or *threshold logic units*—are in essence models of the simplified version of the biological neuron given above. A single artificial neuron is shown in Fig. 14.2. Like the simple model of the biological neuron, the artificial neuron has an input region that receives signals from other cells, a "cell body" that integrates incoming signals and determines the output according to a thresholding function, and an output region that carries the cell's response to future cells in the network. Weights are included on each connection to future cells in order to model differing effects of one cell upon several others.

Whereas the thresholding function of the biological neuron was said to be "all or none," the artificial neuron models include a variety of thres-

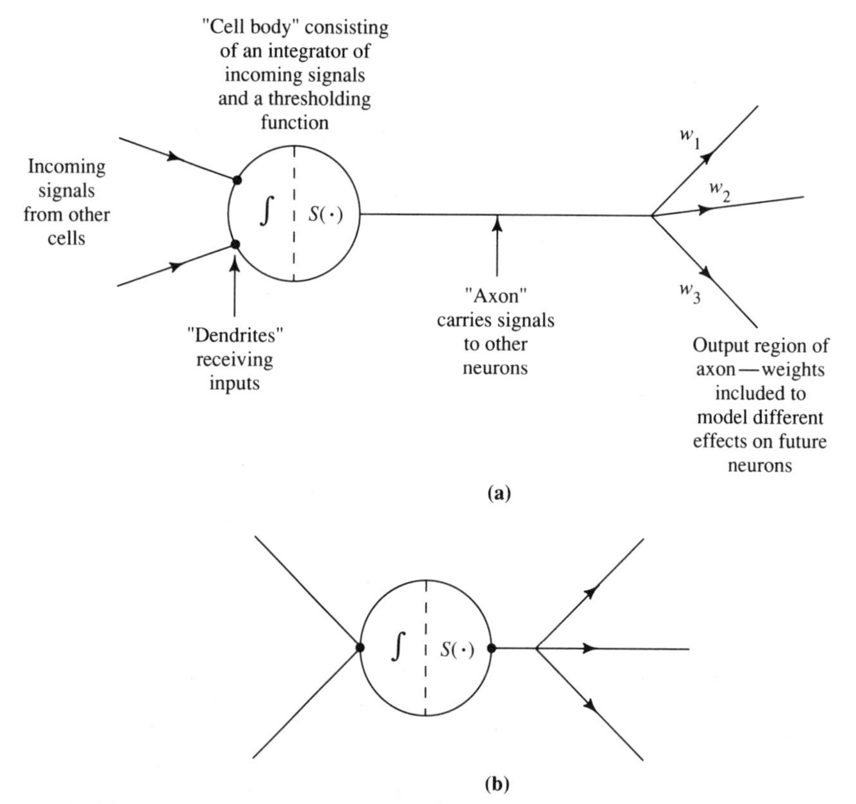

FIGURE 14.2. (a) A single artificial neuron. (b) The axon in the model is superfluous and is frequently omitted. We leave a small axon, which will be later labeled with the cell's output. Also, inputs are usually shown entering a single point (dendrite) on the node.

holding functions, some of which are illustrated in Fig. 14.3. In general, we will say that $S(\cdot)$ is a *thresholding function* if

1. $S(u)$ is a monotonically nondecreasing function of u.
2. $\lim_{u\uparrow+\infty} S(u) = c_+$ and $\lim_{u\downarrow-\infty} S(u) = c_-$ with $|c_+|, |c_-| < \infty$.

Now let us further formalize the artificial neuron. We label some cell, say node k, by n_k. Suppose that there are N inputs arriving at n_k as shown in Fig. 14.4. Frequently, these inputs will be the weighted outputs of predecessor cells, as shown in the figure. For simplicity, let us assume that the N predecessor cells to n_k are n_i, $i = 1, 2, \ldots, N$, and that $k > N$. y'_j denotes the output of n_j for any j (frequently called the *activation* of n_j) and w_{ki} is the weight in the connection to n_k from n_i. Typically, and not surprisingly, the integration region of the cell body formally sums the incoming inputs, so that (ignoring the auxiliary input momentarily)

$$y_k = S\left(\sum_{i=1}^{N} w_{ki} y'_i\right) \stackrel{\text{def}}{=} S(\mathbf{w}_k^T \mathbf{y}'). \tag{14.1}$$

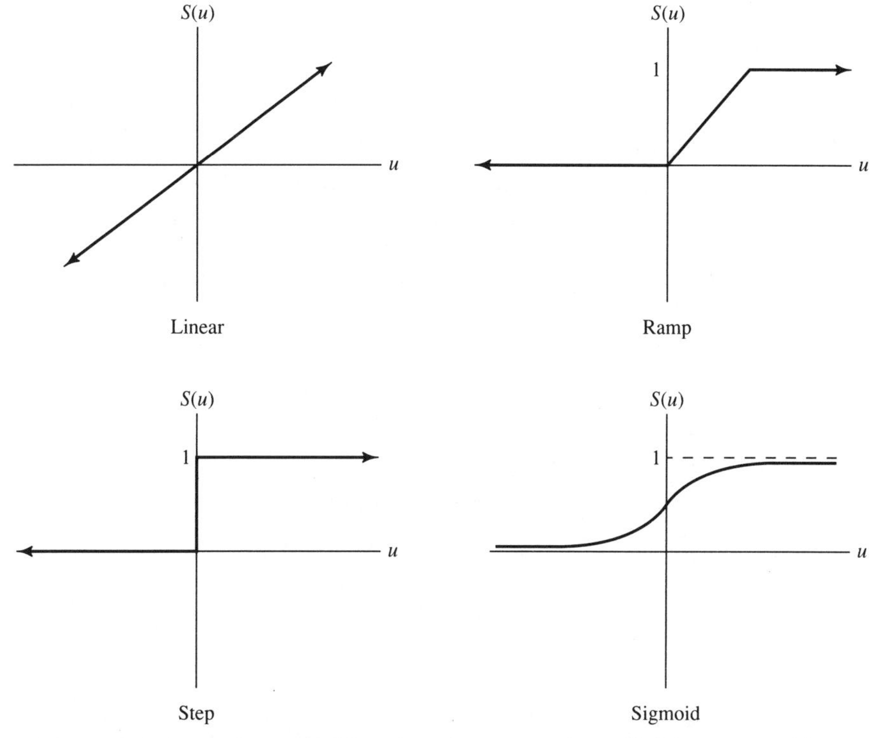

FIGURE 14.3. Typical thresholding functions used in artificial neurons. The sigmoid follows the relation $S(u) = (1 + e^{-u})^{-1}$. Threshold functions are also used in which the lower saturation is -1. A sigmoidlike function with this property is the hyperbolic tangent $S(u) = \tanh(u)$.

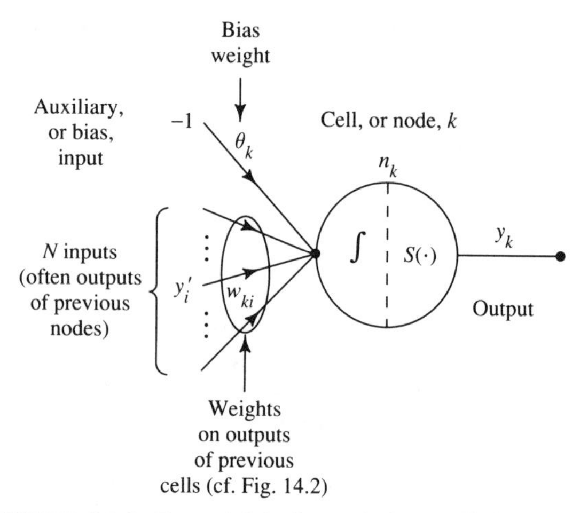

FIGURE 14.4. Formal labeling of the artificial neuron.

If, for example, $S(\cdot)$ is the step threshold function shown in Fig. 14.3, then the output of n_k is as shown in Fig. 14.5. Note that the vector $\mathbf{y}'$ represents the collection of inputs from predecessor cells, and $\mathbf{w}_k$ the vector of weights leading to n_k. Note also the definition

$$u_k \stackrel{\text{def}}{=} \mathbf{w}_k^T \mathbf{y}'. \tag{14.2}$$

This is the scalar quantity that gets passed to the nonlinear thresholding function "inside" the cell. A cell that integrates the incoming weighted inputs by addition, then subjects the result to a nonlinearity (especially the step function) in this fashion, is often called a *perceptron*.

Sometimes an *auxiliary input* is included as an input to the perceptron to serve as a *thresholding bias*. For example, if we include the input θ_k, the activation of n_k becomes

$$y_k = S(\mathbf{w}_k^T \mathbf{y}' - \theta_k) = S(u_k - \theta_k), \tag{14.3}$$

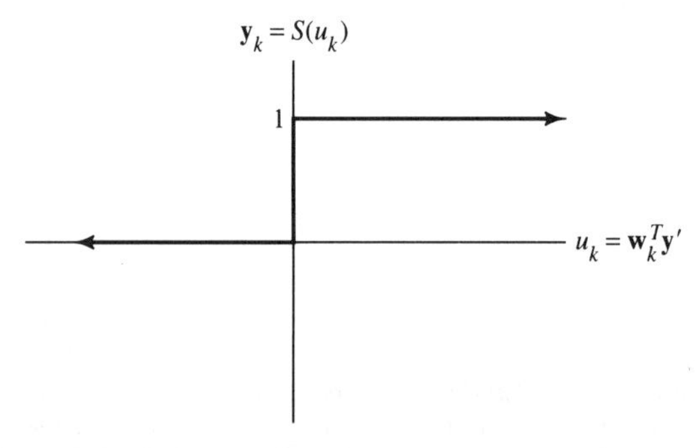

FIGURE 14.5. Computing the output of n_k with inputs $\mathbf{y}'$, input weights $\mathbf{w}_k$, a summing integration, and a step threshold function. Note that u_k indicates the integrated value that is subjected to the nonlinear thresholding.

which is illustrated in Fig. 14.6. Note that in this case the presence or absence of a biasing input need not affect any general formal developments about the cell, since the bias may always be treated as the input from an "$(N+1)$st" cell by letting

$$\begin{aligned} y'_{N+1} &= -1 \\ w_{k,N+1} &= \theta_k. \end{aligned} \tag{14.4}$$

Of course, in this case we assume $k > N+1$. The bias is therefore frequently ignored in deriving formal results about perceptrons.[2]

A second form of "integration" employed in speech processing networks involves the computation of the Euclidean norm between the vector $\mathbf{y}'$ and the connection weights $\mathbf{w}_k$:

$$y_k = S\left(\sqrt{\sum_{i=1}^{N}(w_{ik} - y'_i)^2}\right) = S(\|\mathbf{w}_k - \mathbf{y}'\|_2). \tag{14.5}$$

Such cells are said to implement *radial basis functions* (Broomhead and Lowe, 1988; Bridle, 1988; Moody, 1990) in at least two important practical cases. In the first case we introduce a thresholding bias in the following manner:

$$y_k = S(\|\mathbf{w}_k - \mathbf{y}'\|_2 - \theta_k). \tag{14.6}$$

Now we let $S(\cdot)$ be a "reversed step" threshold function illustrated in Fig. 14.7. Under these circumstances, the cell is activated (produces unity output) if the distance between the input and the connection "weights" is less than θ_k. In other words, the cell is activated if its input vector is found inside a ball of radius θ_k in N-dimensional space. This idea is illustrated for the case $N=2$ in Fig. 14.8.

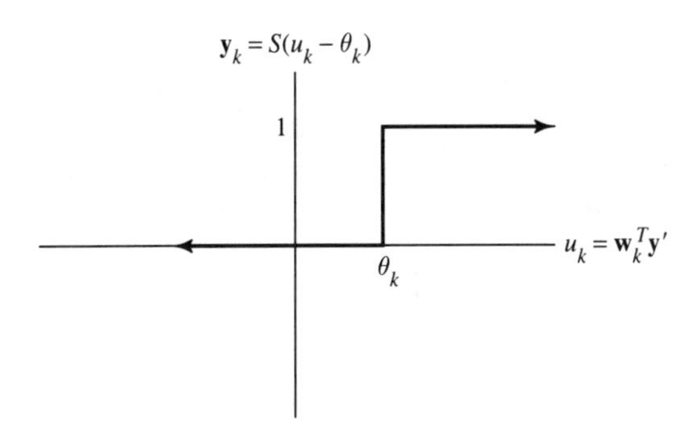

FIGURE 14.6. Effective shift of thresholding function due to auxiliary input. Shown is $\theta_k > 0$.

[2]One instance in which we will need to be careful with this issue is in the drawing of "decision regions" in the feature space, since the bias input is not a real feature. We will have more to say about this issue later.

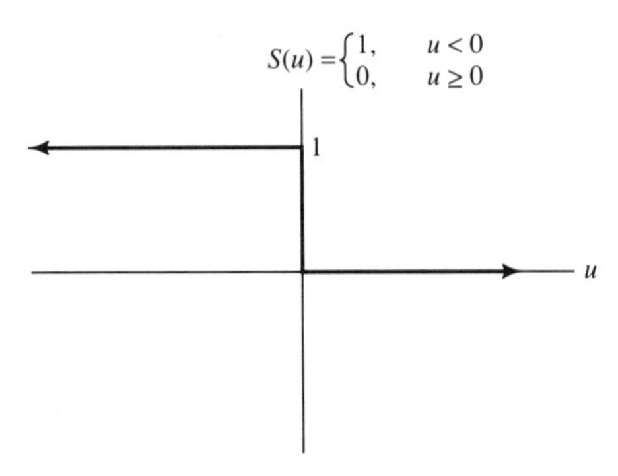

FIGURE 14.7. Reversed step threshold function.

A second case of radial basis function implementation involves no biasing [i.e., output of form (14.5)] and employs a Gaussian nonlinearity

$$S(u) = e^{-u^2}. \tag{14.7}$$

In this case, the activation is constant on any ball centered on $\mathbf{w}_k$. The illustration for $N = 2$ is found in Fig. 14.9. Clearly, the activation level for a Gaussian radial basis function is continuous-valued. However, we might choose to declare the neuron *inactive* if

$$y_k < \rho_k \tag{14.8}$$

for some ρ_k (ρ_k must be less than unity). In this case, it is easily shown that the neuron is active for inputs $\mathbf{y}'$ such that

$$\|\mathbf{w}_k - \mathbf{y}'\| \leq \sqrt{-\log \rho_k} \stackrel{\text{def}}{=} \theta_k. \tag{14.9}$$

In the N-dimensional case, this means that the unit will be activated only if the input resides inside a ball of radius θ_k.

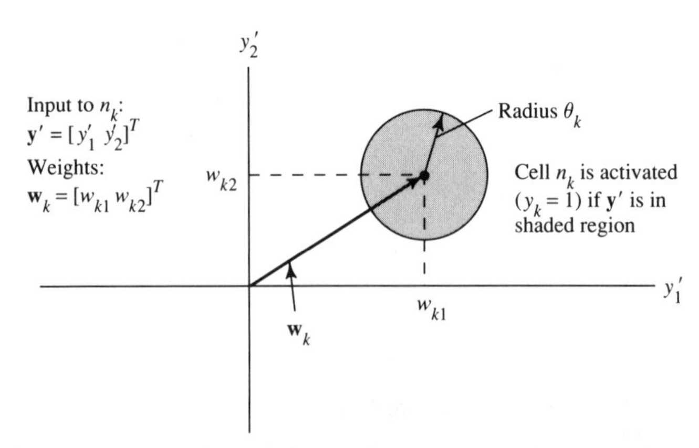

FIGURE 14.8. A radial basis function cell n_k computing a Euclidean distance between its weights $\mathbf{w}_k$ and its inputs $\mathbf{y}'$ and passing that result through a biased reversed step nonlinearity will be activated for inputs within a radius θ_k of its weight vector. The case of two inputs to n_k is illustrated here.

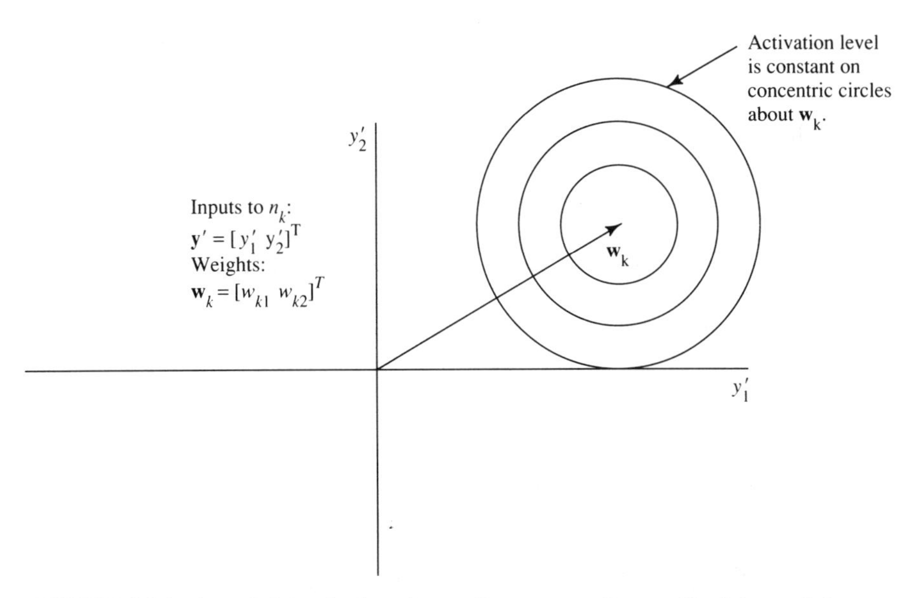

FIGURE 14.9. A radial basis function cell n_k computing a Euclidean distance with no bias between its weights $\mathbf{w}_k$ and its inputs $\mathbf{y}'$ and passing that result through a Gaussian nonlinearity will output constant activation levels for inputs on concentric balls about its weight vector. The case of two inputs to n_k is illustrated here.

In either radial basis function implementation discussed above, there is an effective (or exact) *ball of influence* of radius θ_k inside of which the input will activate the neuron. For this reason, radial basis function neurons are sometimes called *radius-limited perceptrons.*

14.3 Network Principles and Paradigms

14.3.1 Introduction

Having introduced the basic principles of the artificial neuron, we can now begin to connect these units into *networks* of cells comprising computing machines. The resulting ANNs are so named because their topologies consist of "axon-to-dendrite" linkages of the individual cells, reminiscent in a primitive way of the patterns of biological neurons. The computing power of the resulting network derives from the complex interaction of many simple nonlinear elements that perform their operations in parallel. This form of computation stands in stark contrast to the sequential operation of the von Neumann computing machine.

Artificial neural networks have several advantages relative to sequential machines. First, the ability to adapt is at the very center of ANN operations. Adaptation takes the form of adjusting the connection weights in order to achieve desired mappings. Furthermore, ANNs can continue to adapt and learn (sometimes on-line), which is extremely useful in processing and recognizing speech. Adaptation (learning) algorithms continue to be a major focus of research in the ANN field. Second, ANNs tend to be more robust or fault-tolerant than von Neumann machines because the network is composed of many interconnecting neurons, all computing in parallel, and the failure of a few processing units can often be compensated for by redundancy in the network. Similarly, ANNs can often "generalize" from incomplete or noisy data. Finally, ANNs, when used as classifiers, do not require strong statistical characterization or parameterization of data.

Although ANNs can perform many computing functions, they are often used in speech processing to implement pattern recognition—to associate input patterns with classes. Within this function, at least three subtypes of classifiers can be delineated. The first is the most straightforward, in which an output pattern results that identifies the class membership of the input pattern. The second is a *vector quantization* (VQ) function in which vector input patterns are quantized into a class index by the network. In some sense these two functions appear to be about the same task. In ANN discussions, however, the VQ terminology is usually reserved for a particular type of ANN architecture that is trained quite differently (more in keeping with the basic notation of VQ as we know it from past study) than networks for more general types of pattern associators. A third subtype of classifier is the so-called *content-addressable memory* or *associative memory* network. This type of network is used to produce a "memorized" pattern or "class exemplar" as output in response to an input, which might be a noisy or incomplete pattern from a given class. An example of the operation of a content-addressable memory network is shown in Fig. 14.10, which is taken from the article by Lippmann (1987). For discussion of content-addressable memory ANNs, the reader is referred to (Kohonen, 1987). We will have more to say about the first two of these classifier subtypes later.

In addition to pattern recognizers, a second general type of ANN is a *feature extractor.* The basic function of such an ANN is the reduction of large input vectors to small output vectors (features) that effectively indicate the classes represented by the input patterns. In essence, the feature extractor is charged with decreasing the dimension of the representation space by removing nonessential or redundant information. It is also sometimes the case that feature representations will appear as patterns of activation internal to the network rather than at the output [e.g., (Waibel et al., 1989; Elman and Zipser, 1987)].

The provision of a taxonomy of ANN architectures is difficult, since the number of possible interconnections among neurons is unlimited. However, as research and development matures, a few specific preeminent architectures are emerging [e.g., (Simpson, 1990)]. As stated earlier in the chapter, we will concentrate here primarily upon the type of architecture known as "multilayer perceptron." It is upon this architecture that much of the recent research in speech recognition has been based. A second quite different architecture, the "learning vector quantizer", has

Eight exemplar patterns.

(a)

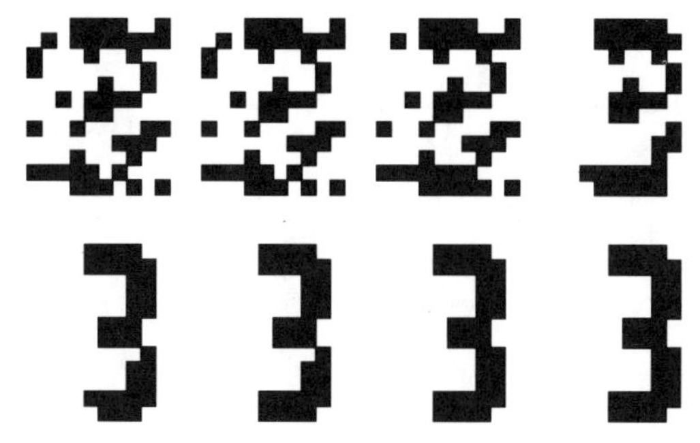

Output patterns for noisy "3" input.

(b)

FIGURE 14.10. Illustration of a content-addressable memory ANN (Lippmann, 1987). Behavior of a Hopfield network [see, e.g., (Simpson, 1990)] when used as a content-addressable memory. A 120-node ANN was trained using the eight shown in (a). The pattern for the digit 3 was corrupted by randomly reversing each bit with a probability of 0.25, then applied to the network at time zero. Outputs at time zero and after the first seven iterations are shown in (b).

been used for VQ of speech and related tasks. We will also briefly describe this architecture in the following.

14.3.2 Layered Networks: Formalities and Definitions

Many ANN architectures can be conveniently viewed as "layers" of cells, as illustrated in Fig. 14.11. A layered structure is one that may be described as follows: A group of N_1 cells designated *layer* 1 receive as their inputs weighted versions of the external inputs of the network. There are N_0 external inputs, one of which may correspond to a bias. The remaining cells in the network (above layer 1) can be grouped into layers $2, 3, \ldots, L$, such that cells in layer l receive as inputs weighted outputs of cells in layer $l-1$. The outputs of the final layer, L, are the external outputs of the network. We use the term "weighted outputs" loosely to mean some combination of the labels on the connections (weights)

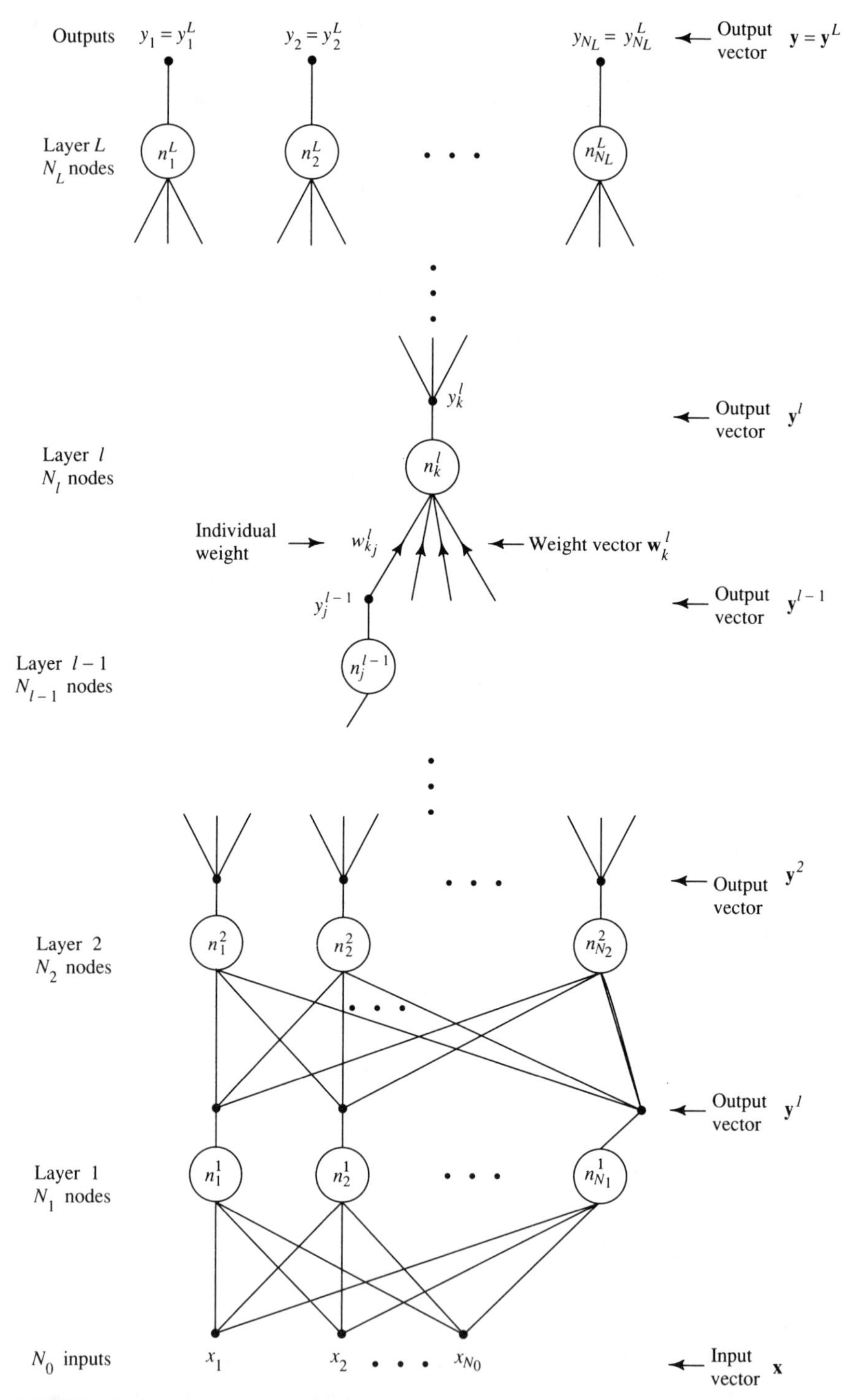

FIGURE 14.11. A layered ANN. In addition to the notation illustrated above, we define $\mathbf{W}^l$ to be the matrix of all weights on connections to layer l, $\mathbf{W}^l \stackrel{\text{def}}{=} [\mathbf{w}_1^l \quad \mathbf{w}_2^l \quad \cdots \quad \mathbf{w}_{N_l}^l]$.

with the outputs of the lower layer. This combination rule depends on the operation of the cell (see Section 14.2). A *hidden layer* is one containing cells whose outputs cannot be measured directly. According to our framework, each layer is hidden except layer L.

To avoid confusion, note that some authors consider the set of inputs to comprise a layer—and might even draw "cells" at the bottom of the network to receive these inputs. These cells are strictly *formal*, performing the mapping $x_i = F(x_i)$ for every x_i, and therefore having no practical significance. We will avoid cells at the bottom of the network. Nevertheless, in reading the literature and comparing some development with our discussion here, the reader should note whether the inputs are counted as a layer in the other work.

A *feedforward*, or *nonrecurrent*, ANN is one for which no cell has a connection path leading from its output back to its input. If such a path can be found, there is feedback in the ANN and the architecture is called *recurrent*. Layering as we have described it above depends not only on nonrecurrence, but also upon *sequential connection*. By this we mean that cells in layer l must be connected to cells in layer $l+m$, where m must not only be positive (no feedback) but it must be exactly unity (sequential). A layered ANN can sustain a few feedback or nonsequential connections without losing its basic layered pattern, but too many such connections erode the fundamental network structure. Such excessive deviations from the layered structure will not arise in our discussion.

A *multilayer perceptron* (MLP), or *feedforward ANN*, is a nonrecurrent layered network in which each of the cells is governed by an activation rule of the form (14.1). A thresholding bias may also be included in each cell of the MLP, and we will assume that it is accounted for as in (14.4) and not show it explicitly. Whereas the perceptron (single cell), as we have defined it above, involves a step nonlinearity, the term "MLP" often refers to networks with less restricted nonlinearities. In fact, let us redefine the (single-cell) perceptron to be an artificial neuron whose activation rule is of the form (14.1) with $S(\cdot)$ any thresholding nonlinearity (see Section 14.2). In this case the MLP may be defined as a nonrecurrent layered network in which each cell is a perceptron. Special cases of the MLP include networks with just one layer of cells (sometimes called a *single-layer perceptron*), and the single-cell perceptron itself.[3]

A second architecture of interest will be the "learning vector quantizer" (LVQ), which is a single-layer network[4] governed by an activation rule similar to the radial basis function rule, (14.6), discussed above. Details of the LVQ will be described below.

[3]We caution the reader that, in addition to the definition we have given for a single-layer perceptron above, the single-cell perceptron is also sometimes called a "single-layer perceptron" because the inputs are sometimes considered to comprise a layer (see the second paragraph of this subsection). To avoid confusion, we will not use the term *single-layer perceptron* in our discussion.

[4]Actually, the LVQ deviates somewhat from the strict definition of a layered network in allowing connections among cells in the single (output) layer.

One of the most interesting and useful aspects of ANNs is their ability to "learn" to implement certain computations. Learning refers to the process of weight adjustment to achieve the desired aim. The MLP and LVQ involve two distinctly different forms of learning paradigms. In *supervised learning*, which is used to train the MLP, a series of training pairs (input vectors and desired output vectors or *targets*) are "shown" to the network and the weights adjusted according to some algorithm. The objective is to reproduce the entire population of target outputs as closely as possible in some sense. Many iterations through the training pairs might be necessary for the learning algorithm to converge on some set of weights (if convergence is possible). On the other hand, some ANN architectures like the LVQ are organized to be trained by *unsupervised learning*. In this case, the network automatically adjusts its own weights so that (training) inputs that are similar in some sense produce similar (or identical) outputs. In effect, the resulting network may be used to classify input data according to the outputs they produce. The training and use of such a network are reminiscent of a clustering procedure used in statistical pattern recognition. Indeed, the one example of a self-organizing ANN that we will consider, the LVQ, is used to achieve VQ. (We will also find that LVQs can also be trained in a supervised mode.)

Since a significant part of our study has been devoted to the very important HMM, it is worth making a few comparisons in the basic technologies. ANNs (in particular MLPs) and HMMs are fundamentally similar in that both have the ability to learn from training data. The process by which HMMs are trained may likewise be considered a form of supervised learning. *What* is learned in each case can be quite dissimilar, both in content and in philosophy, even if both models are being applied to the same problem. The HMM learns the statistical nature of observation sequences presented to it, while the ANN may learn any number of things, such as the classes (e.g., words) to which such sequences are assigned. Although influenced by the statistical nature of the observations, the internal structure of the ANN that is learned is *not* statistical. In its basic form, an ANN requires a fixed-length input, whereas we have discussed the convenient time normalization property of HMMs in Chapter 12. To a greater or lesser degree depending on the design, both systems can be robust to noise, to missing data in the observations, to missing exemplars in the training, and so on. Therefore, although different in philosophy, HMMs and ANNs do have important similarities. However, there is a fundamental difference in the two technologies. Even if we grant that both systems perform mappings—the HMM an observation string to a likelihood, the ANN an input to an output—the fact remains that the dynamics of the HMM are fundamentally linear,[5] whereas the ANN is a nonlinear system. It is from this nonlinear nature that the ANN derives most of its power.

[5]See the state-space interpretation of the HMM in Section 12.2.2.

14.3.3 The Multilayer Perceptron

Some History and Rationale Behind the MLP

In 1957, Frank Rosenblatt, working at Cornell University, created one of the first ANNs with the ability to learn. Rosenblatt developed his network by building on the earlier layered logic concept of McColloch and Pitts (1943). Rosenblatt first worked with the single-cell perceptron (which he so named). In the context of Rosenblatt's early work, the term "perceptron" specifically involves the step threshold nonlinearity. He developed a supervised learning procedure guaranteed to converge to weights that would accurately classify two-class data under a certain condition. For discussion purposes, we have drawn and labeled a perceptron in Fig. 14.12. In anticipation of using more than one cell in the future, we label this cell n_k.

We seek to discover necessary conditions for the convergence of the perceptron weights. Recall the expression for the activation

$$y_k = S(\mathbf{w}_k^T \mathbf{x}) = \begin{cases} 1, & \mathbf{w}_k^T \mathbf{x} \geq 0 \\ 0, & \mathbf{w}_k^T \mathbf{x} < 0 \end{cases}. \tag{14.10}$$

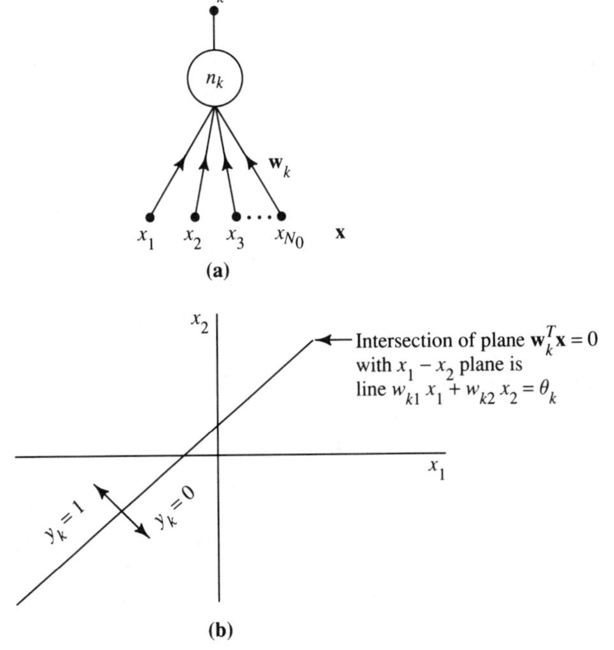

FIGURE 14.12. (a) The perceptron. (b) Decision hyperplane (line) for the case $N_0 = 3$; the third dimension is used for biasing: $x_3 = -1$, $w_{k3} = \theta_k$ (see text).

The boundary between the two "decisions" $y_k = 0$ and $y_k = 1$ in the input vector space is the hyperplane

$$\mathbf{w}_k^T \mathbf{x} = 0. \tag{14.11}$$

If one of the input components, say x_{N_0}, actually corresponds to a bias ($x_{N_0} = -1$ and $w_{k,N_0} = \theta_k$), then, since x_{N_0} takes a constant value -1, (14.11) represents a $(N_0 - 1)$-dimensional hyperplane in the "true" feature space given by

$$\sum_{i=1}^{N_0-1} w_{ki} x_i - \theta_k = 0. \tag{14.12}$$

The inclusion of a bias permits the construction of a decision boundary that is not constrained to pass through the origin (see Problem 14.1). (The bias can be learned like any other weight.) The case of a two-dimensional (both dimensions represent "true" features) space is illustrated in Fig. 14.12(b). The decision hyperplane (in this case, a line) can be placed in any orientation in $\mathbb{R}^2$ by appropriate choice of weights, since a bias term (representing a third dimension) is included in the perceptron.

If the inputs represent two classes whose vectors are separable by a hyperplane (*linearly separable*), then a set of weights can be found to exactly distinguish them via the output of the perceptron. Rosenblatt's *perceptron learning* (PL) algorithm is proven to converge to this set of weights (Block, 1962; Nilsson, 1965). Examples of classes which can and cannot be exactly classified by a perceptron are shown in Fig. 14.13. We note that the bias term can be removed if it is sufficient for the decision boundary to pass through the origin in the feature space. It is unlikely, however, that this would be known *a priori*.

The Rosenblatt PL algorithm is shown in Fig. 14.14. It can be seen that the weights are adjusted (the perceptron learns) only when an error occurs between the actual output and the target (training) output of the network. The parameter η, which takes values $\eta \in [0, 1]$, controls the rate of learning. A trade-off is encountered in the choice of η. This parameter must be chosen large enough to adapt quickly in the presence of errors, yet small enough to allow the weight estimates to stabilize when appropriate values have been reached.

To "teach" the perceptron (adjust its weights) we present the network with a series of training patterns, say $\{(\mathbf{x}(p), \tau_k(p)),\ p = 1, 2, \ldots, P\}$, where $\mathbf{x}(p)$ is the pth input and $\tau_k(p)$ the pth target. Let $y_k(p)$ denote the actual output in response to $\mathbf{x}(p)$ and $\mathbf{w}_k(p)$ be the vector of weights *following* the presentation of training pair p. It is important to understand the meaning of the index p. There will always, of course, be a finite number, P, of training patterns. During training procedures, however, each pattern will be applied multiple times to the network, so that effectively many more than P training pairs will be employed. When an index $p > P$ is encountered, it should be interpreted as follows:

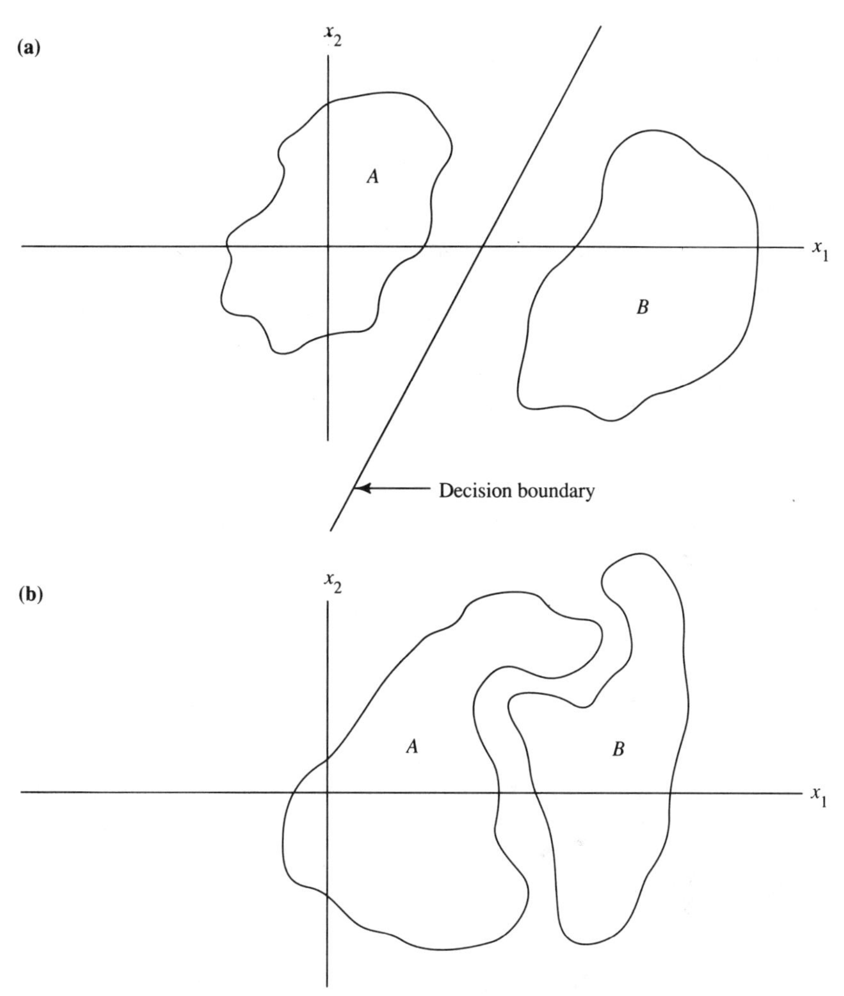

FIGURE 14.13. Regions labeled *A* and *B* represent regions in the feature space from which feature vectors representing classes *A* and *B* may be drawn. (a) The two classes are apparently separable by a line (in general, a hyperplane). (b) The classes are not linearly separable.

$$\text{If } p > P, \text{ then } \begin{cases} \mathbf{x}(p) \text{ means } [\mathbf{x}(\cdot)]_{\text{modulo } P} \text{ evaluated at ``time'' } p \\ \tau_k(p) \text{ means } [\tau_k(\cdot)]_{\text{modulo } P} \text{ evaluated at } p, \end{cases} \tag{14.13}$$

where $[\mathbf{x}(\cdot)]$ represents the original P-length sequence of inputs, and similarly for $[\tau_k(\cdot)]$.

As an aside, we note that the Rosenblatt algorithm becomes exactly the well-known LMS algorithm of Widrow and Hoff (Widrow and Stearns, 1985) when the hard-limiting threshold is replaced by the simple linear mapping $S(u) = u$. The LMS algorithm was described in Chapter 8 and we see it recur in our study of MLPs later in the chapter. For a suffi-

FIGURE 14.14. Perceptron learning algorithm (two-class problem).

Note: Cell is named n_k, and related quantities show subscript k in anticipation of the use of more than one perceptron.

Initialization: Assign small, random, initial weight values $\mathbf{w}_k(0)$.
Select learning constant η.

Recursion: For $p = 1, 2, \ldots$ [continuously iterate through the P training patterns; see (14.13)]
Compute output: $y_k(p) = S(\mathbf{w}_k^T(p)\mathbf{x}(p))$.
Compute error: $\varepsilon_k(p) = \tau_k(p) - y_k(p)$, where

$$\tau_k(p) = \begin{cases} 0, & \text{class 1} \\ 1, & \text{class 2} \end{cases}.$$

Adjust weights: $\mathbf{w}_k(p) = \mathbf{w}_k(p-1) + \eta\varepsilon_k(p)\mathbf{x}(p)$.
Next p

Termination: Stop when weights change negligibly according to some criterion.

ciently small η, the LMS algorithm converges asymptotically to the weights that minimize the total squared error between the target and actual errors in the training population. A perceptron trained with LMS will create a boundary that meets this criterion whether or not it exactly separates the classes. When classes are not linearly separable, LMS avoids the problem of oscillating (nonconverging) weights that can occur with the PL algorithm.

A single layer of perceptrons (as illustrated in Fig. 14.15) can be used to separate multiple classes as long as each class can be separated from all others by a hyperplane. The idea is simply to "assign" a perceptron to each class and to let that cell correspond to a hyperplane separating that class from all others. In Fig. 14.15, for instance, the cell labeled n_A corresponds to the hyperplane (line) labeled A in the weight space. When an element of class A appears in the data, this perceptron will output a "one," and a "zero" otherwise. The perceptrons for classes B and C operate similarly. As noted above, any perceptron corresponding to a hyperplane that does not pass through the origin must contain a bias weight (see Problem 14.1).

The analysis of the perceptron's ability to separate classes was pivotal in the history of ANNs. The proof of convergence of the PL algorithm caused a great deal of excitement and focused a considerable amount of attention on such learning algorithms. Early applications of the perceptron to speech processing are found in the work of Rosenblatt (1962), and other early applications are discussed in Simpson (1990). However, little effort was made to explain the frequently poor classification performance of this simple machine until the first edition of the influential

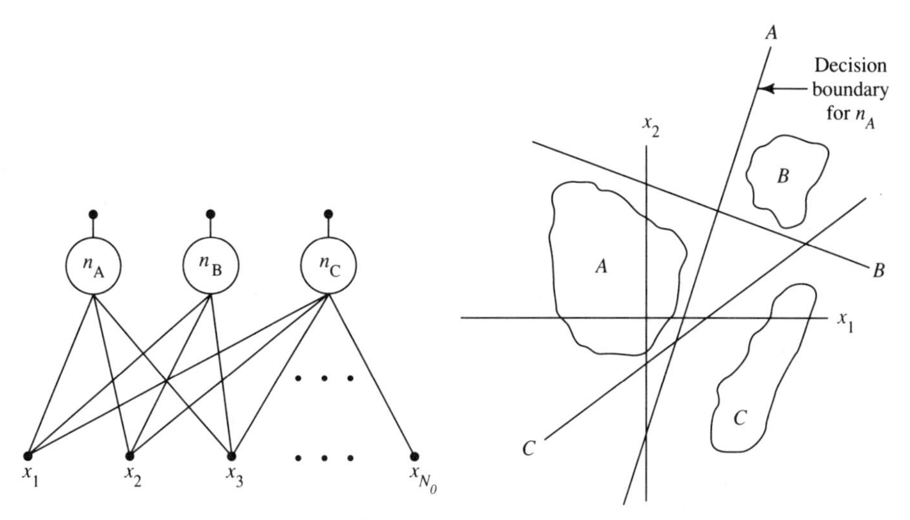

FIGURE 14.15. (a) A layer of three perceptrons can be used to classify three linearly separable classes, A, B, and C. (b) For the case $N_0 = 3$ with the third input $(x_3 = -1)$ corresponding to a bias, the hyperplane (line) boundary formed by perceptron n_A, separates class A from all others.

book *Perceptrons* was published in 1969 (Minsky and Papert, 1969). Exploiting the fact that the perceptron is only capable of linear separability, Minsky and Papert prove a number of theorems that enumerate apparent weaknesses of the perceptron. The exclusive OR (XOR) problem in which classes are not separable by a straight line (see Fig. 14.16) was used to illustrate the perceptron's inadequacies. This problem is still frequently employed today as a test of the efficacy of an ANN architecture

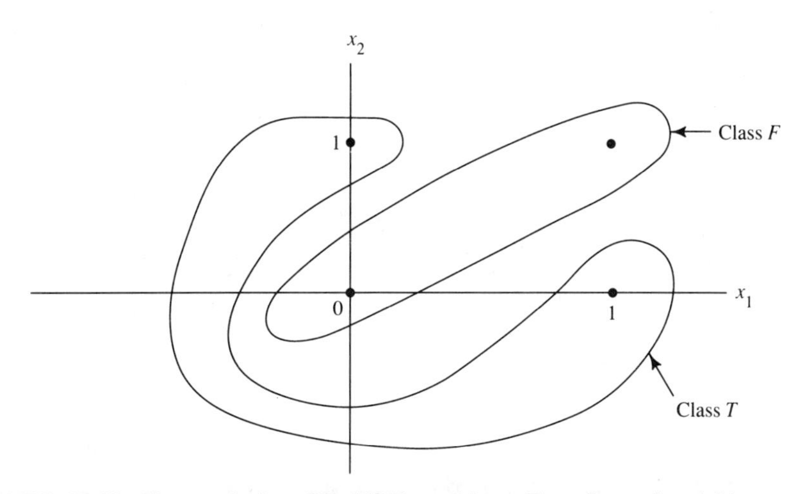

FIGURE 14.16. The exclusive OR (XOR) problem. Two-dimensional binary feature vectors must contain exactly one "one" to correspond to the class T (for "true"). Otherwise, the vector belongs to class F ("false"). The classes are not linearly separable.

or its learning algorithm. Rosenblatt as well as Minsky and Papert were aware of the benefits of using multiple layers in the perceptron architecture. In fact, a two-layer perceptron with a single output cell is capable of distinguishing between any two classes that fall in open or closed convex regions in the feature space. A three-layer perceptron can form arbitrarily complex decision regions if a sufficient number of cells is provided (Lippmann, 1987). These regions are illustrated in Fig. 14.17. The proof of this fact provides some insight into the number of nodes. However, Rosenblatt was unable to find a learning algorithm for a two-layer perceptron,[6] and Minsky and Papert express some doubts in the 1969 version of *Perceptrons* as to whether a learning algorithm would be achievable for the MLP.

In some historical accounts, the book *Perceptrons* is almost blamed for having dealt an unnecessarily devastating blow to ANN research. In others, the book is viewed as an accurate criticism of weaknesses of the perceptron that actually advanced the field by its extensive analysis, and which was misinterpreted by the research community. Block (1970) wrote a review of *Perceptrons* in an attempt to clarify some of these mistaken

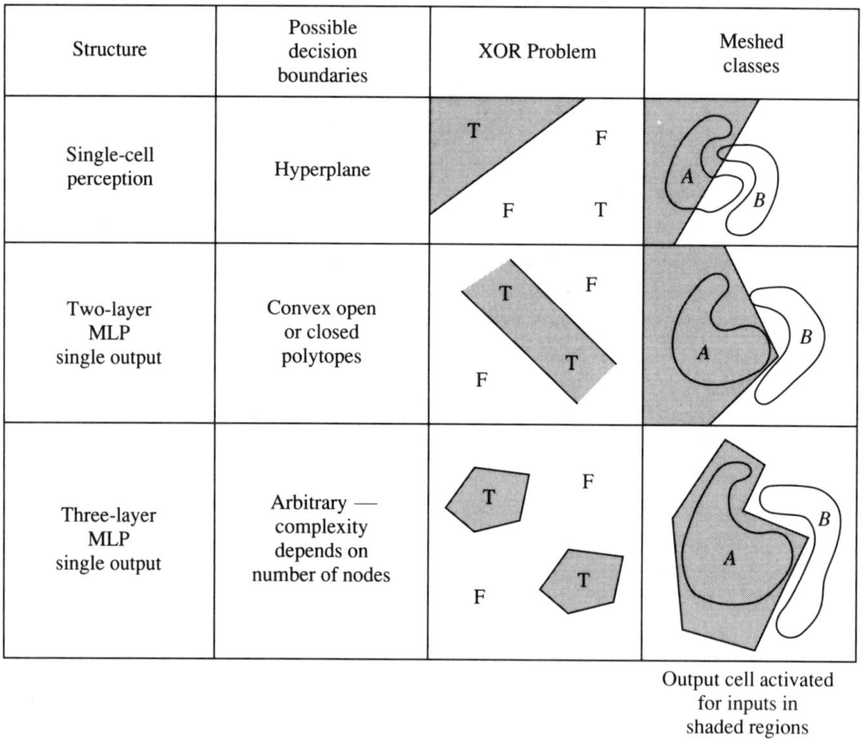

Structure	Possible decision boundaries	XOR Problem	Meshed classes
Single-cell perception	Hyperplane	T F F T	A B
Two-layer MLP single output	Convex open or closed polytopes	T F F T	A B
Three-layer MLP single output	Arbitrary — complexity depends on number of nodes	T F F T	A B

Output cell activated for inputs in shaded regions

FIGURE 14.17. Decision regions formed by MLPs with step threshold nonlinearities and a single output node. Adapted from Lippmann (1987).

[6]However, he had proposed the basic notion of "back-propagation" of error, which we will see used in this task shortly (Rosenblatt, 1962, p. 292).

impressions. Minsky and Papert authored a revised version of *Perceptrons* in the late 1980s (Minsky and Papert, 1988). Whichever view one takes, the 1969 edition of *Perceptions* coincides with the end of the initial period of excitement about the ANN. It was not until the popularizing of the back-propagation algorithm by Rumelhart et al. in the late 1980s (Rumelhart et al., 1986) that the excitement about ANNs began to recur.

Before discussing back-propagation and learning in MLPs, a few details should be elaborated upon. We have noted that a three-layer perceptron is capable of learning arbitrarily complex decision regions in the input feature space given the proper number of nodes in each layer. Of course, by generalization of our previous discussion, it should be intuitively clear that a three-layer output with C outputs could be found to distinguish among C classes comprising arbitrary regions in the input space. However, these results deal with specific problems of class assignment with *binary-valued* outputs. When the step nonlinearities in the network are replaced by smoother thresholding functions like the sigmoid, similar behaviors to those described above are observed if the resulting network is used as a classifier (maximum output indicates the class). Not unexpectedly, however, these networks result in more complex decision regions and are more difficult to analyze. Nevertheless, the smooth nonlinearities make a network amenable to training by contemporary learning algorithms and are in this sense preferable. Also, it should not be construed that the performance of networks will be inferior because of the more complex decision regions. On the contrary, it might be greatly improved (and intuitively one might expect this). However, it is more difficult to predict and analyze performance with the more complex networks.

Beyond the classification problem, in general, we may wish to execute a general mapping $\mathbb{R}^{N_0} \rightarrow \mathbb{R}^{N_L}$ using an ANN, and question whether such a network is possible. In 1989 Hornik et al. proved their so-called *representation theorems*, which show that sigmoidal feedforward architectures can represent any mapping to any degree of accuracy given a sufficient number of *hidden* neurons (Hornik et al, 1989). In fact, the convergence of the mapping to the desired mapping is uniform over the input space with the number of cells. Although theoretically interesting and encouraging, the representation theorems offer no guide to the number of necessary neurons, which may be impractically large. A number of other investigators have attempted to apply Kolmogorov's theorem (Kolmogorov, 1957) to show that a two-layer perceptron with a finite number of cells is capable of executing an arbitrary mapping (Hecht-Nielsen, 1987; Barron and Barron, 1988; Irie and Miyake, 1988; Arai, 1989; Barron, 1989). However, the theorem is not entirely appropriate because, in essence, it does not assume the nonlinearities to be fixed in advance (Irie and Miyake, 1988).

The question as to whether ANNs, and in particular MLPs, can execute arbitrary mappings with practical numbers of cells remains unanswered. Supplemented by exciting experimental findings, the theoretical research, while inconclusive, has encouraged ANN investigators to proceed with good confidence that any mapping can be learned given a sufficiently large network. Most serious, however, is our lack of any well-understood guidelines for selecting the network topology and number of cells. Indeed, as we discuss below, the amount of essential learning time and training data grows exponentially with the number of cells in the network. This is a compelling reason for minimizing the network size. On the other hand, theory (and intuition) suggests that large networks will be necessary to execute complex mappings. For the present, the choice of network size and architecture remains an experimental issue. A compelling reason for the choice of an MLP is the availability of several effective learning algorithms. The premiere among these is the back-propagation algorithm to which we now turn our attention.

Learning in MLPs: The Back-Propagation Algorithm

History. The publication of the two-volume *Parallel Distributed Processing* in 1968 (Rumelhart et al., 1986) helped to create a renaissance of ANN research, in particular by popularizing the *back-propagation* (BP) algorithm (or *generalized delta-rule* as it is called in the book). In this same era, Sejnowski and Rosenberg (1986) used the BP algorithm to successfully train a text-to-speech synthesizer dubbed "NETtalk." The NETtalk experiments were made more dramatic by audio renditions of the synthesized speech at various stages of learning, which were played at conference sessions and for the print and television media. A long-awaited breakthrough in ANN research—the ability to effectively train MLPs—had apparently arrived. [It was not long, however, before similar developments were uncovered in the earlier literature (Werbos, 1974; Parker, 1982). White (1989) has also shown that BP is a special case of stochastic approximation, which has been researched since the 1950s (Tsypkin, 1973). The relative dormancy of the ANN field for 15 years had apparently kept this earlier work from having an impact.] The initial excitement about BP, however, was responsible for many misunderstandings and exaggerated claims about its properties and capabilities. At first many investigators interpreted BP as a means of avoiding the weaknesses of gradient descent (which Minsky and Papert had criticized in *Perceptrons*). It is now appreciated that this is not the case. Further, the algorithm and the architecture were seemingly confused initially. It is now understood that BP cannot embue an architecture with classification properties or other performance capabilities that are not theoretically predictable. For instance, a two-layer perceptron cannot exhibit decision regions other than those shown in Fig. 14.17 by virtue of BP (or any other) training.

Research and practical experience has led to clearer understanding of the BP algorithm's behavior and has moderated performance expectations. In particular, it is now appreciated that BP is a stochastic gradient-descent algorithm subject to the same types of convergence problems as any such algorithm. Nevertheless, BP has proven to be a remarkably successful approach to training MLPs, which has served as an important catalyst in reviving interest in the ANN.

In the following paragraphs, we give a brief development of the BP method and then summarize the algorithm in Fig. 14.19. At the end of the section we comment upon some other algorithms that can also be used to train MLPs.

Algorithm. The network notation of Fig. 14.11 will be used throughout this discussion.

Associated with a given set of training patterns for an MLP, say

$$\{(\mathbf{x}(p), \boldsymbol{\tau}(p)), \quad p = 1, 2, \ldots, P\}, \tag{14.14}$$

and *any* set of network weights, say $\mathbf{W}$, is a total squared output error,

$$\xi(\mathbf{W}) = \frac{1}{2}\sum_{p=1}^{P}[\boldsymbol{\tau}(p) - \mathbf{y}^L(p, \mathbf{W})]^T[\boldsymbol{\tau}(p) - \mathbf{y}^L(p, \mathbf{W})] \tag{14.15}$$

$$= \frac{1}{2}\sum_{p=1}^{P}\sum_{\upsilon=1}^{N_L}[\tau_\upsilon(p) - y_\upsilon^L(p, \mathbf{W})]^2. \tag{14.16}$$

We have included the factor $\frac{1}{2}$ in this expression purely for mathematical convenience, which will become clear below. Note that we have also shown the explicit dependence of the outputs, y_υ^L, on the choice of weights. In general, all inputs and outputs to the nodes in the MLP will depend on the weights in the layers below them. We will explicitly show the dependence of certain of these input and output quantities on weights of interest in the discussion to follow.

The objective of the BP algorithm is to find the weights, say $\mathbf{W}^*$, that minimize $\xi(\mathbf{W})$. If there are, say, N_w weights in the MLP, then a plot of ξ over the N_w-dimensional hyperplane (each dimension representing one weight) is called an *error surface.* Since the MLP implements a nonlinear mapping, in general there will be multiple minima in the error surface[7] and ideally we would like to find $\mathbf{W}^*$ which corresponds to the global minimum. In practice, we must settle for locating weights corresponding to a local minimum, perhaps repeating the procedure several times to find a "good" local minimum. Some measures for analyzing the error surface to assist in this task are described in (Burrascano and Lucci, 1990).

There is no known way to simultaneously adjust all weights in an MLP in a single training step to find a minimum of ξ. In fact, the BP attempts to find a minimum by tackling a much more modest task. Not only does BP consider only one weight at a time (holding all others constant), but it also considers only a single training pattern's error surface

[7]Recall the HMM training problem in Chapter 12.

at a time.[8] By this we mean that in using BP, we (in principle) consider only the error surface, say $\xi(p, \mathbf{W})$, because of a single training pair $(\boldsymbol{\tau}(p), \mathbf{x}(p))$, and repeat the procedure independently for each p. [In fact, as in the Rosenblatt PL algorithm, each training pattern can be introduced many times so that the rules for interpreting the index p shown in (14.13) are in effect here.] The issue of whether such a procedure should be expected to converge to a minimum of the "summed" error surface will be discussed below. Clearly, $\xi(p, \mathbf{W})$ is given by

$$\xi(p, \mathbf{W}) = \frac{1}{2} \sum_{v=1}^{N_L} [\tau_v(P) - y_v^L(p, \mathbf{W})]^2. \tag{14.17}$$

Suppose that we are currently working on weight w_{kj}^l, which has recently been adjusted to value $w_{kj}^l(p-1)$ by processing pattern $p-1$. Back-propagation works by moving w_{kj}^l slightly away from $w_{kj}^l(p-1)$ in the direction that causes $\xi(p, \mathbf{W})$ to decrease along the corresponding dimension. The adjusted value will naturally be called $w_{kj}^l(p)$ in the ensuing discussion. To sense in which direction $\xi(p, \mathbf{W})$ is decreasing along the w_{kj}^l dimension at the value $w_{kj}^l(p-1)$, we evaluate the partial derivative

$$\left. \frac{\partial \xi(p, \mathbf{W})}{\partial w_{kj}^l} \right|_{w_{kj}^l = w_{kj}^l(p-1)} . \tag{14.18}$$

If the gradient is positive at that point, then subtracting a small quantity from $w_{kj}^l(p-1)$ corresponds to moving downhill on the error surface, and vice versa. This means that w_{kj}^l should be adjusted according to the *learning rule*

$$w_{kj}^l(p) = w_{kj}^l(p-1) - \eta(p) \left. \frac{\partial \xi(p, \mathbf{W})}{\partial w_{kj}^l} \right|_{w_{kj}^l = w_{kj}^l(p-1)} , \tag{14.19}$$

where $\eta(p)$ is a small *learning constant* that generally depends on p (more on this sequence below). For the obvious reason, BP is called a *gradient-descent* algorithm.

Finding an expression for the derivative, especially for weights below the output layer L, is made tractable by considering a single weight at a time, as we shall appreciate momentarily. Let us begin with a weight, say

[8] This type of identification problem is often called *stochastic approximation.* The error surface $\xi(p, \mathbf{W})$ for each p may be considered a realization of a random variable, say $\underline{\xi}(\mathbf{W})$, which is parameterized by the matrix $\mathbf{W}$. Ideally, we would like to find the value of $\mathbf{W}$ that minimizes $\mathcal{E}\{\underline{\xi}(\mathbf{W})\}$, but we must be content to work with the realizations. A significant amount of research has been done on this general class of problems. Much of the foundation for the subject, especially as it applies to learning systems, is laid in the classic work of Tsypkin (1973). A rigorous application of stochastic learning theory to the study of the BP algorithm is found in the paper by Stanković and Milosavljević (1991).

w_{kj}^L, in the output layer to be adjusted in response to a training pattern at time p. All other weights are treated as constants held at whatever value they assume at time $p-1$. Beginning with (14.17), it is easy to show that

$$\frac{\partial \xi(p, \mathbf{W})}{\partial w_{kj}^L} = -\left[\tau_k(p) - y_k^L(p, w_{kj}^L)\right] S'\left[u_k^L(p, w_{kj}^L)\right] y_j^{L-1}(p), \tag{14.20}$$

where $S'(\alpha)$ denotes the derivative of the thresholding function evaluated at α,[9] and where all other notation is defined in Fig. 14.18. Let us denote the error at node n_k^L in response to training pattern p by

$$\varepsilon_k^L(p, w_{kj}^L) \stackrel{\text{def}}{=} \tau_k(p) - y_k^L(p, w_{kj}^L) \tag{14.21}$$

so that

$$\frac{\partial \xi(p, \mathbf{W})}{\partial w_{kj}^L} = -\varepsilon_k^L(p, w_{kj}^L) S'\left[u_k^L(p, w_{kj}^L)\right] y_j^{L-1}(p). \tag{14.22}$$

Combining (14.22) with (14.19) provides the necessary mechanism for computing the updated weight w_{kj}^L.

As an aside, we note that if S is taken to be the step threshold function,[10] then the equation used to modify the weights here is identical to the rule used in Rosenberg's PL algorithm. If it was not apparent before, we can now appreciate why Rosenberg's method is called a gradient-descent algorithm as well.

After applying the procedure described above to each weight leading to layer L, we move down to layer $L-1$. Clearly, a similar equation to (14.19) could be used at each node in layer $L-1$ (and lower layers) if "target values" were known for each node and an "error surface" could be formed with respect to these nodes. These target values would somehow need to conspire to minimize the true $\xi(p, \mathbf{W})$ at the output layer. How does one find such target values? The answer becomes evident upon making a "brute force" attempt to compute the gradient of $\xi(p, \mathbf{W})$ with respect to, say, w_{kj}^{L-1}, in layer $L-1$. Note that this procedure is made tractable by focusing on this single weight, even though it is theoretically desirable to change the entire set of weights together. The relevant notation is illustrated in Fig. 14.18. The gradient is

$$\frac{\partial \xi(p, \mathbf{W})}{\partial w_{kj}^{L-1}} = \frac{\partial}{\partial w_{kj}^{L-1}} \left[\frac{1}{2} \sum_{\nu=1}^{N_L} \left[\tau_\nu(p) - y_\nu^L(p, w_{kj}^{L-1})\right]^2\right]. \tag{14.23}$$

[9]BP can only theoretically be used if S is differentiable.

[10]Differentiable everywhere except at the origin.

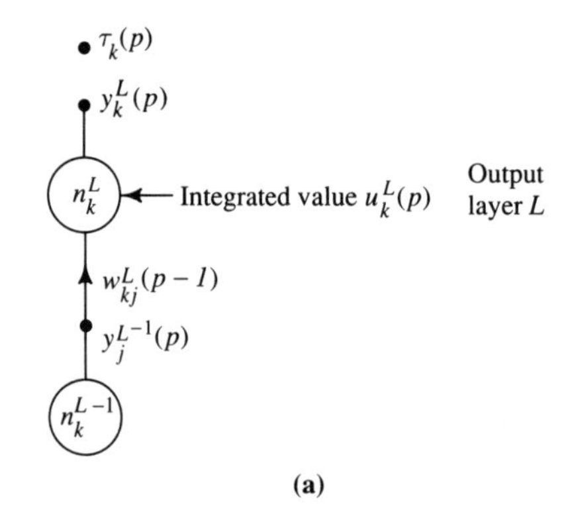

(a)

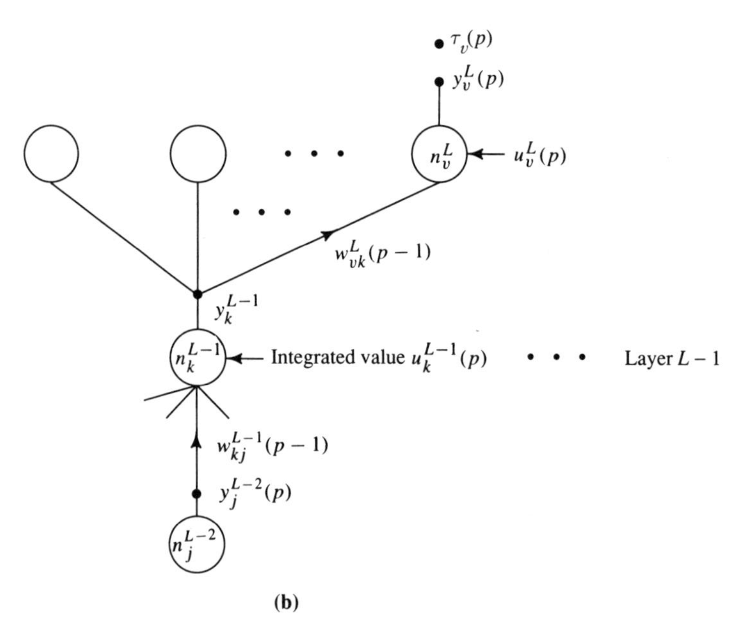

(b)

FIGURE 14.18. Notation used in development of the BP algorithm: (a) Layer L; (b) layer $L-1$.

Recognizing that (see Fig. 14.18)

$$y_v^L(p, w_{kj}^{L-1}) = S(u_v^L(p, w_{kj}^{L-1})) = S[w_{vk}^L y_k^{L-1}(p, w_{kj}^{L-1}) + o_1]$$

$$= S\{w_{vk}^L S[u_k^{L-1}(p, w_{kj}^{L-1})] + o_1\} = S\{w_{vk}^L S[w_{kj}^{L-1} y_j^{L-2}(p) + o_2] + o_1\}, \tag{14.24}$$

where o_1 and o_2 are "other terms" which are not dependent upon w_{kj}^{L-1}, we can apply the chain rule of differential calculus to obtain

$$\frac{\partial \xi(p,\mathbf{W})}{\partial w_{kj}^{L-1}} = -\left[\sum_{v=1}^{N_L} \varepsilon_v^L(p, w_{kj}^{L-1}) S'\left[u_v^L(p, w_{kj}^{L-1})\right] w_{vk}^L(p)\right] \times S'\left[u_k^{L-1}(p, w_{kj}^{L-1})\right] y_j^{L-2}(p). \tag{14.25}$$

The term set off in brackets is called the *back-propagated error* at node n_k^{L-1} in response to pattern p, and we give it the notation

$$\varepsilon_k^{L-1}(p, w_{kj}^{L-1}) \stackrel{\text{def}}{=} \sum_{v=1}^{N_L} \varepsilon_v^L(p, w_{kj}^{L-1}) S'\left[u_v^L(p, w_{kj}^{L-1})\right] w_{vk}^L(p). \tag{14.26}$$

Inserting this notation into (14.25), we have

$$\frac{\partial \xi(p,\mathbf{W})}{\partial w_{kj}^{L-1}} = -\varepsilon_k^{L-1}(p, w_{kj}^{L-1}) S'\left[u_k^{L-1}(p, w_{kj}^{L-1})\right] y_j^{L-2}(p). \tag{14.27}$$

Comparing this expression with (14.22) reveals the reason for the name back-propagated error. In fact, if node n_k^{L-1} were an external node with error equivalent to $\varepsilon_k^{L-1}(p)$, then by the same means we arrived at the "outer layer" expression (14.22), we would have obtained (14.27). In this sense, we can think of node n_k^{L-1} as having a target value

$$\tau_k^{L-1}(p) = y_k^{L-1}\left(p, w_{kj}^{L-1}(p-1)\right) + \varepsilon_k^{L-1}\left(p, w_{kj}^{L-1}(p-1)\right), \tag{14.28}$$

but we have no explicit need for this quantity in the development.

Having now found an expression for the gradient of $\xi(p, \mathbf{W})$ with respect to w_{kj}^{L-1}, we now evaluate it at $w_{kj}^{L-1} = w_{kj}^{L-1}(p-1)$ and use (14.19) to adapt the weight. Note that the significance of the back-propagated error is that of a recordkeeping device. The process is one of computing the required gradient, which could, in principle, be done without recourse to this bookkeeping.

Upon moving down to lower hidden layers in the network, a similar procedure would reveal that, for weight w_{kj}^l,

$$\frac{\partial \xi(p,\mathbf{W})}{\partial w_{kj}^{l}} = -\varepsilon_k^{l}(p, w_{kj}^{l}) S'\left[u_k^{l}(p, w_{kj}^{l})\right] y_j^{l-1}(p), \tag{14.29}$$

where $\varepsilon_k^l(p, w_{kj}^l)$ is the back-propagated error at node n_k^l,

$$\varepsilon_k^{l}(p, w_{kj}^{l}) \stackrel{\text{def}}{=} \sum_{v=1}^{N_{l+1}} \varepsilon_v^{l+1}(p, w_{kj}^{l}) S'\left[u_v^{l+1}(p, w_{kj}^{l})\right] w_{vk}^{l+1}(p). \tag{14.30}$$

With only two exceptions, (14.29) and (14.30) are general expressions that can be used to compute the gradient of the error surface for any weight in the network. The exceptions occur at the top and bottom of the network. For $l = L$, (14.21) must be used to compute ε^L, since this error is not propagated. If $l = 1$, the term $y_j^0(p)$ appears in (14.29). This is to be understood to mean the jth input to the network in pattern p,

$$\mathbf{y}^0(p) \stackrel{\text{def}}{=} \mathbf{x}(p) \quad \text{and} \quad y_j^0(p) \stackrel{\text{def}}{=} x_j(p). \tag{14.31}$$

Equation (14.29) is applied in conjunction with learning rule (14.19) to each node in the network. The process is repeated for each pattern $p = 1, 2, \ldots, P$. Because the weights are adjusted only slightly at each step, many iterations through the training patterns (sometimes an unrealistically large number of them) are usually necessary for the procedure to converge on a minimum-error solution. Before stating the algorithm, therefore, we remind the reader once again of the rules for interpreting the pattern index p given in (14.13). The BP algorithm based on the developments above is shown in Fig. 14.19.

Convergence. General conditions under which the BP algorithm will converge to a useful minimum in a practical number of iterations are unknown. Indeed, because it is such a difficult task, there has been little rigorous analysis of even the asymptotic properties of the method (some studies are cited below). In practice, the behavior of BP has been found to be highly dependent upon the initial choice of weights, and the magnitude of the weights often continues to grow and drift along after the errors at the outputs have apparently been minimized. An analysis of these behaviors is undertaken by Guo and Gelfand (1991). There are two hopeful indicators, however, that the weights will frequently converge satisfactorily in finite time over a wide variety of architectures and mappings. The first is simply experimental evidence that has shown the method to converge for many problems of interest [see, e.g., (Rumelhart et al.,

---

FIGURE 14.19. Back-propagation algorithm.

Initialization: Assign initial weight values $w_{kj}^l = w_{kj}^l(0)$ for all j, k, l.
Select learning constant $\eta(0)$ (may be decreased with time).

Recursion: For $p = 1, 2, \ldots,$ [continuously iterate through patterns; see (14.13)]
  For $l = L, L-1, \ldots, 1$ (layers)
    For all k and j in layer l (weights)
    Compute error $\varepsilon^l[p, w_{kj}^l(p-1)]$ using (14.21) if $l = L$, or (14.26) for other l.
    Compute gradient of $\xi(p, \mathbf{W})$ at $w_{kj}^l(p-1)$ using (14.29).
    Compute updated weight, $w_{kj}^l(p)$, according to (14.19).
    Next weight
  Next l
Next p

Termination: Stop when weights change negligibly according to some criterion.

1986) and Section 14.4]. The second is that BP has been shown to be a form of stochastic approximation (SA)[11] and conditions *are* known for asymptotic convergence for particular forms of SA algorithms (Robbins and Munro, 1951; Blum, 1954; Albert and Gardner, 1967; Eykhoff, 1974; Goodwin and Sin, 1984). Stanković and Milosavljević (1991) and White (1989) have specifically treated the BP algorithm within the framework of SA.[12] It is often noted in this vein that when S is a linear function, $S(\alpha) = K\alpha$ for some constant K, then the BP algorithm for a single-layer network essentially reduces to the well-known LMS algorithm (Widrow and Stearns, 1985), for which the asymptotic convergence properties are well understood.[13]

In spite of the fact that BP frequently exhibits practical convergence, the convergence may require many thousands of iterations through the training patterns. Whereas for a network with N_w weights BP requires only $O(N_w)$ floating point operations (flops) per training pattern, iterating through the training patterns, say, I times, can make this process very expensive. Effectively, $O(IN_w)$ flops per p may be required, where I can be 10^2–10^4 or even higher.

From a heuristic point of view, the difficulty in predicting convergence of the BP algorithm rests in the complicated nature of the error surfaces involved. Recall that the error surface $\xi(\mathbf{W})$ over which the minimum is desired is not considered directly. Rather, trajectories are determined over component error surfaces, $\xi(p, \mathbf{W})$, $p = 1, \ldots, P$, which are summed to yield the complete error surface. Each of these individual surfaces can be quite complex in themselves because of the nonlinearities. Further, finding appropriate initial weight estimates is not easy. Although it is most desirable to initialize the weights near a minimum, the complex nature of $\xi(\mathbf{W})$ does not permit any simple way to estimate where these desirable starting sites might be. Poor initial conditions, combined with a small learning rate, can result in very protracted learning time.

Enhancements and Alternatives. Before leaving the general discussion of the BP algorithm, we note some generalizations and enhancements that have been developed by various researchers. A common modification to the learning rule is to smooth high-frequency variations in the error

[11]See footnote 8.

[12]One of the interesting findings of the Stanković and Milosavljević research is that convergence of the weight matrix is not guaranteed if the popular sigmoid threshold function is used.

[13]In essence, with LMS we need to keep the learning constant, η, sufficiently small so that the weight value does not oscillate about a minimum on the error surface when such a minimum is near. On the other hand, if η is made too small, adaptation will occur at an unacceptably slow rate when the weight value is far from a minimum. Therefore, the learning rate is often decreased over time to balance these competing objectives. Guidelines for setting and adapting the learning rate for LMS are found, for example, in (Widrow and Stearns, 1985; Kosko, 1992, Ch. 5). In the use of BP, η has frequently been set experimentally. We give some example applications below and in Section 14.4 that will illustrate typical choices.

surface by adding a second-order term to the learning rule, which Rumelhart et al. (1986, p. 330) have called a "momentum" term. In this case the learning rule takes the form

$$w_{kj}^{l}(p) = w_{kj}^{l}(p-1) - \Delta w_{kj}^{l}(p), \tag{14.32}$$

in which

$$\Delta w_{kj}^{l}(p) = \eta \frac{\partial \xi(p, \mathbf{W})}{\partial w_{kj}^{l}} \bigg|_{w_{kj}^{l} = w_{kj}^{l}(p-1)} + \alpha \Delta w_{kj}^{l}(p-1), \tag{14.33}$$

where $0 < \alpha < 1$ (typically 0.9). The momentum term allows a larger value of η to be used without causing oscillations in the solution.

Several investigators have explored the possibility of linearizing the dynamics of a feedforward network around the "current" set of weights, then applying some form of linear processing in conjunction with BP to implement the weight modification (Kollias and Anastassiou, 1989; Azimi-Sadjadi et al., 1990; Ghiselli-Crippa and El-Jaroudi, 1991; Hunt, 1992; Deller and Hunt, 1992). The Hunt work is based on a QR decomposition-based RLS algorithm of the type described in Chapter 5. For nodewise training, the fundamental algorithm turns out to be theoretically equivalent to the Azimi-Sadjadi method except for the RLS implementation. A significant improvement in convergence performance was achieved using the method of Hunt and Deller with respect to the BP algorithm, and some results are given in Table 14.1.

Finally, several research groups have been concerned with the extension of the BP algorithm to recurrent networks. Studies are reported in (Almeida, 1987, 1988; Rohwer and Forrest, 1987; Samad and Harper, 1987; Atiya, 1988; Pineda, 1987, 1988; Williams and Zipser, 1988).

14.3.4 Learning Vector Quantizer

The *learning vector quantizer* (LVQ) is a second ANN architecture that is very useful in speech recognition technology. This network was introduced by Kohonen (1981). The LVQ, which is shown in Fig. 14.20, resembles the MLP, but is designed to function in a very different mode. The network consists of a single layer of cells, which differ from a single layer of perceptrons by the interconnections among the top layer. These connections are present in order to implement what is frequently called *competitive learning*, in which the output cells "compete" for the right to respond to a given input. In turn, this learning scheme is central to the unsupervised learning procedure used to train the network.

We have seen the VQ technique employed in various aspects of speech coding and recognition. As the name implies, the LVQ ANN and learning procedure are designed to carry out the VQ task. Analogously to a clustering algorithm (which, in fact, the LVQ implements), the LVQ is presented with training inputs only, and is required to form weights that

TABLE 14.1. Results Comparing Hunt–Deller (H–D) Training with BP and Azimi-Sadjadi (A-S) Training on Several Classical Problems.

	XOR	4-Bit Parity Checker				4-Bit Counter	
	Sigmoid Threshold	Sigmoid Threshold		Step Threshold		Step Threshold	
Training Method	Random Weights	Random Weights	A-S Weights	Random Weights	A-S Weights	Random Weights	A-S Weights
H–D	78	5	5	51	57	1	16
BP	11	0	0	1	53	0	0
A-S	8	0	0	1	37	0	9

Note: Integers indicate the number of times the algorithm converged in 100 trials, each with different initial weights. Convergence is defined very conservatively here. We say the algorithm converges if a solution which classifies the training data without error is achieved. The architectures used for the three problems are XOR: two-layer MLP with three nodes in the hidden layer and a single output; 4-Bit Parity Checker: two-layer MLP with five nodes in the hidden layer and a single output; 4-Bit Counter: two-layer MLP with five nodes in the hidden layer and two outputs. In each case a bias input is added to the architecture. "A-S (initial) weights" are chosen so that the initial output of the networks induce weight updating. Further improvements are discussed in Hunt (1992) and Deller and Hunt (1992).

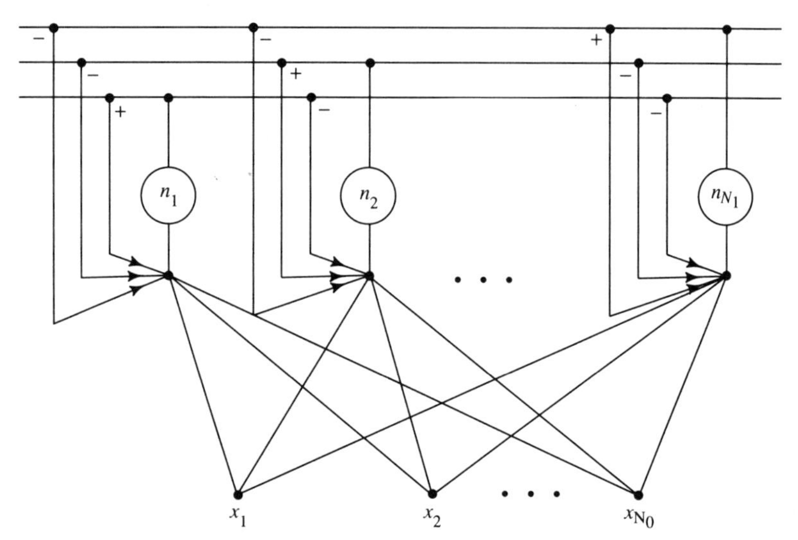

FIGURE 14.20. The LVQ architecture.

produce outputs which effectively classify the input patterns into meaningful groupings.

The so-called *single-winner unsupervised learning* algorithm for the LVQ is shown in Fig. 14.21. The notation used in the algorithm is defined in Fig. 14.20, and the rule for interpreting index p is given in (14.13). Note that all input vectors are normalized to have unity length. This is critical to correct operation in general, and is very important in speech processing. The procedure automatically determines the N_1 best reference vectors (like cluster centers) needed to represent the space spanned by the N_0-dimensional input vectors $\mathbf{x}(p)$, $p = 1, 2, \ldots, P$. Each of these learned reference vectors is used as a weight vector on the connections leading to an output cell. As usual, the weights leading to cell n_k

---

FIGURE 14.21. Single-winner unsupervised learning algorithm for the LVQ.

Initialization: Assign initial weight values randomly on $[0, 1]$.
Select initial learning constant $\eta(0)$ (may be decreased with time).

Recursion: For $p = 1, 2, \ldots$ (cycle through training patterns)

Find $k^* = \underset{k}{\operatorname{argmin}} \left\| \mathbf{w}_k(p-1) - \mathbf{x}(p) \right\|$

Adjust $\mathbf{w}_{k^*}$: $\mathbf{w}_{k^*}(p) = \mathbf{w}_{k^*}(p-1) + \eta(p)[\mathbf{x}(p) - \mathbf{w}_{k^*}(p-1)]$

All $k \neq k^*$: $\mathbf{w}_k(p) = \mathbf{w}_k(p-1)$

Next p

Termination: Stop when weights change negligibly according to some criterion.

---

are denoted $\mathbf{w}_k$. Once the network is trained, an arbitrary input pattern will cause a unity response in output cell n_{k^*}, where $\mathbf{w}_{k^*}$ is the weight vector closest to the input in Euclidean norm. This is analogous to determining which of N_1 cluster centers is closest, and quantizing (classifying) the vector accordingly.

As long as the learning constant, η decays with time [typically, e.g., $\eta(p) = p^{-1}$], the LVQ learning algorithm has been shown to converge to a tessellation of the N_1 space with the weight vectors as centroids (Kohonen, 1987, Sec. 7.5). Extensive learning times are often required.

Learning vector quantizers can also be used in a *multiple-winner unsupervised learning* mode in which a "neighborhood" of cells around the winning cell has its weights reinforced. The learning rule is simply modified in this case so that if a training pattern $\mathbf{x}(p)$ is closest to, say $\mathbf{w}_k$, then any weights associated with any cell in a neighborhood, say $\mathcal{N}_k$, of n_k, are adjusted according to

$$\Delta \mathbf{w}_j = \eta(p)[\mathbf{x}(p) - \mathbf{w}_k], \quad n_j \in \mathcal{N}_k. \tag{14.34}$$

The output layer is often arranged into a planar topology so that the neighborhood of a cell is comprised of all cells within a given radius.

Finally, we note that the LVQ can also be trained in a supervised mode. The training vectors are marked according to one of N_1 classes. When vector $\mathbf{x}(p)$ correctly indicates cell n_k (in the sense of minimum distance to its weights, as above), then $\mathbf{w}_k$ is reinforced,

$$\Delta \mathbf{w}_k = +\eta(p)[\mathbf{x}(p) - \mathbf{w}_k]. \tag{14.35}$$

If class k is incorrectly indicated, then the plus sign in (14.35) is replaced by a negative sign so that n_k is "moved away from $\mathbf{x}(p)$."

Having established the basic principles of the ANN and two important architectures for speech processing, we now turn to some applications of ANN technology in speech recognition.

14.4 Applications of ANNs in Speech Recognition

In this section we describe some example applications of ANNs in speech recognition. The focus here will be upon technologies that employ the MLP and LVQ and that build upon our background in conventional speech recognition approaches. Many other application examples are discussed and cited in (Lippmann, 1989) and in the materials listed in Appendix 1.G.

14.4.1 Presegmented Speech Material

Because the ANN represents a relatively new technology, much of the research into its speech recognition capabilities has focused on the fundamental problem classifying static presegmented speech. Table 14.2,

TABLE 14.2. Example Applications of ANNs to Presegmented Speech. Partially adapted from Lippmann (1989).

Study	Approach/Problem
Elman and Zipser (1987)	MLP—Consonant, vowel recognition
Huang and Lippmann (1988)	MLP and FMC (see text)—Vowel discrimination
Kammerer and Kupper (1988)	MLP and single layer of perceptons—Speaker-dependent and -independent word recognition
Kohonen (1988)	LVQ—Labeled Finish speech
Lippmann and Gold (1987)	MLP—Digit recognition
Peeling and Moore (1987)	MLP—Digit recognition
Ahalt et al. (1991)	MLP and LVQ—Vowel discrimination, gender discrimination, speaker recognition

adapted from the paper by Lippmann (1989), lists a number of representative efforts. Note that all architectures are either MLPs or LVQs, except the *feature map classifier* (FMC) of Huang and Lippmann (1988), which is a hierarchical network consisting of an LVQ-like layer followed by a perceptron-like layer. All MLPs were trained by BP. Where inputs are shown as two-dimensional, this indicates that multiple frames of frequency parameter (e.g., cepstral) vectors were entered simultaneously to form a static two-dimensional pattern.

In reading through the details of these studies, four points emerge. The first is that a sense of cautious optimism about ANN technology is warranted. For digit recognition, Kammerer and Kupper's (1988) single layer of perceptrons outperformed all but one of the commercial recognizers evaluated by Doddington and Schalk (1981). One should appreciate, however, the relatively simple nature of the task undertaken. The work of Peeling and Moore (1987), for example, resulted in very good digit recognition using a two-layer perceptron with 50 hidden nodes. The results are based on simple magnitude spectral measures with random placement of the measured frames within the input grid (the rest were set to zeros) when the word did not fill the entire grid. Rather than indicate a remarkable property of the MLP, Lippmann (1989) speculates that the apparent lack of need for proper time alignment in the grid could be due to the digit corpus, which has been shown to be equally recognizable without time alignment using more conventional approaches (Burton et al., 1985; Shore and Burton, 1983). (Networks that account for the dynamic nature of speech are discussed below.)

Second, even within this small group of experiments, one is impressed that the ANN is not a new class of algorithms, but a new class of computing machines. We have not only to learn the best way to parameterize and process the speech material, but we must determine the very structure of the machine upon which the computing will be performed, and the means by which it is to learn the task (Bodenhausen and Waibel,

1991). Obviously, the data conditioning and architecture problems are not independent. The challenge of deducing general methods and conclusions about these problems from the vast array of ongoing experimental research confronts the ANN researcher.

Third, the problem of long training times for MLPs using the BP algorithm is evident in these experiments. Using array processing, the Kammerer and Kupper (1988) training of a single layer of perceptrons required 25 minutes per talker on the speaker-dependent studies, and 5–9 hours for the speaker-independent training. Elman and Zipser's small database required more than 10^5 iterations to train their MLP, while Huang and Lippmann's training required 5×10^4 trials. These results suggest that appropriate scaling of ANN size for training feasibility might need to be factored against the performance considerations. It is also noteworthy that the computation required for *recognition* using Peeling and Moore's (1987) MLP was typically five times less than that required by an HMM against which performance was compared.

The fourth point concerns the evident advantages of the LVQ-type classifiers over the MLPs for classification problems. Tasks undertaken with static units of speech naturally involve their classification (vowel sounds, words, /b, d, g/ contrasts, etc.). Although the MLP can certainly learn such mappings, LVQ-like networks are specifically designed to perform classification tasks by explicitly computing distance measures. A frequently observed consequence is more efficient training of the LVQ with respect to the MLP. In Huang and Lippmann's (1988) study, for example, a two-layer perceptron required 50,000 iterations for convergence, while the FMC required about 50, a three-order-of-magnitude improvement. Kohonen (1988) witnessed a 15-fold improvement in training time for the LVQ with respect to the MLP in his study of phoneme classification. Along these lines, the reader might also recall the radial basis function (RBF) cells described in Section 14.2. Networks composed of such cells also directly compute distance measures and can be trained efficiently. Recent studies using RBFs appear in the papers of Niranjan and Fallside (1988), Bridle (1988), and Ogelsby and Mason (1991).

14.4.2 Recognizing Dynamic Speech

Other researchers have attempted to account for the dynamically changing nature of speech by building temporal delays into MLPs and integrating information across time in the upper layers of the network. A representative sample of studies is shown in Table 14.3. For illustration, we focus on some of the most-often cited work, which was conducted by Waibel et al. (1989a, 1989b, 1989c).

Waibel's ANN is called the *time-delay neural network* (TDNN). It is so-named because each cell in the network effectively receives not only its nominal input vector but "delayed" values of the input so that it can learn local correlations in the data. A schematic TDNN cell is shown in Fig. 14.22. Sigmoidal nonlinearities are used in the units. The original

TABLE 14.3. Example Applications of ANNs to Dynamic Speech. Partially adapted from Lippmann (1989).

Study	Approach/Problem
Lang and Hinton (1988)	Time-delay MLP—Discrimination of utterances of letters b, d, v, e
Waibel et al. (1989b)	TDNN—/b, d, g/ discrimination
Waibel et al. (1989b)	Enhanced TDNN—Vowel and consonant recognition
Watrous et al. (1987, 1990)	Temporal flow model, recurrent MLP—/b, d, g/ discrimination
McDermott and Katagiri (1988)	Time-delay LVQ—/b, d, g/ discrimination
Komori (1991)	Time state neural network—/b, d, g, m, n/ discrimination
Tebelskis and Waibel (1991)	Predictive networks—CSR
Iso and Wantanabe (1990, 1991)	Predictive network—Large-vocabulary, speaker-dependent IWR
Levin (1990)	Predictive network—Multispeaker connected-digit recognition

version of the TDNN was designed to classify the voiced stops /b, d, g/ in the context of various Japanese words. The TDNN has three layers. The first contains eight cells of the type shown in Fig. 14.22, and the second contains three. At "time p," the layer 1 cells receive the "present" plus two "delays" of the input vector, say $\mathbf{x}(p), \mathbf{x}(p-1), \mathbf{x}(p-2)$, while the layer 2 cells receive the present plus four delays of the output of layer 1, say $\mathbf{y}^1(p), \ldots, \mathbf{y}^1(p-4)$. Each of the cells in layer 2 is assigned to one of the consonants. Let these nodes be labeled n_b^2, n_d^2, n_g^2. Then one can view the output vector from layer 2 as consisting of three components, say

$$\mathbf{y}^2(p) = \left[y_b^2(p) \quad y_d^2(p) \quad y_g^2(p)\right]^T. \tag{14.36}$$

The output layer nodes also contain TDNN units assigned to the individual consonants, but each is responsible for integrating temporal information only. Thus n_b^3, for example, receives (scalar) outputs from n_b^2 only, but taken over the present plus eight delays, $y_b^2(p), \ldots, y_b^2(p-8)$. Fairly extensive preprocessing of the data was required for the experiments performed with the original TDNN. Fifteen frames of 16 mel-scale energy parameters based on the FFT were centered around the hand-labeled onset of the vowel. Frames are effectively computed every 10 msec, and each vector is then normalized. For many contexts and 4000 tokens (2000 used for training, 2000 for testing) taken from three speakers, the TDNN provided a 1.5% error, compared with 6.5% for a discrete-observation HMM approach used on the same feature vectors. The more recent papers by Waibel et al. (1989b, 1989c) describe an approach for merging subnetworks into larger TDNNs for recognizing the complete set of consonants.

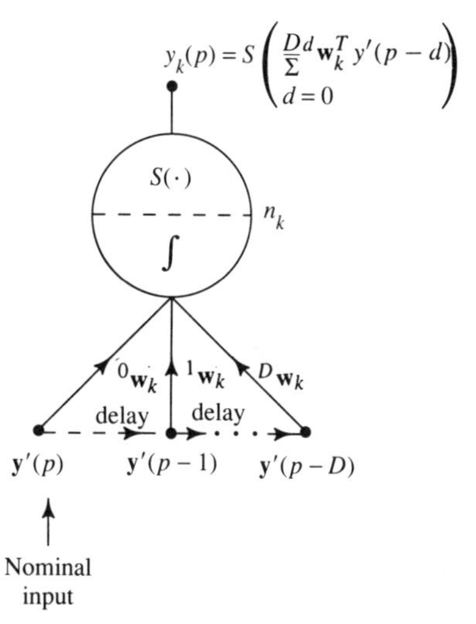

FIGURE 14.22. The TDNN cell. $\mathbf{y}'(p)$ is the nominal vector of inputs to n_k at time p. The D delayed values of this vector are also entered. ${}^d\mathbf{w}_k$ represents the weight vector on the dth delayed copy of the input.

The initial TDNN was trained using BP and provides yet another example of extraordinary times to convergence that may result with this algorithm, especially with larger networks. The authors report that "800 learning samples were used and between 20,000 and 50,000 iterations of the BP loop were run over all training samples." The modular approach to designing the TDNN (Waibel et al., 1989b) was devised in part to remedy the learning time problem.

Other examples of ANNs designed to learn dynamic aspects of speech are found in the papers cited in Table 14.3. In particular, the reader might wish to compare the *time state neural network* of Komori (1991) with the TDNN described above. Another interesting approach uses MLPs to predict patterns rather than to classify them (Iso and Wantanabe, 1990, 1991; Levin, 1990; Tebelskis and Waibel, 1990). The predictive method has some interesting relationships to linear prediction theory and HMMs, which we have studied in detail.

14.4.3 ANNs and Conventional Approaches

Finally, we review some approaches that combine ANN computing with conventional algorithms—in particular, DTW, HMM, and Viterbi search—discussed in earlier chapters. A summary of the approaches to be described here appears in Table 14.4. The ANN contribution to these techniques is principally to serve as an alternative computing structure for carrying out the necessary mathematical operations. The main advantage in this regard is the development of more compact and efficient hardware for real-time implementation. The ANN strategy can also enhance the distance or likelihood computing task by incorporating context

TABLE 14.4. Example Applications Combining ANNs with Conventional Technologies. Partially adapted from Lippmann (1989).

Study	Approach/Problem
Bourlard and Wellekens (1987)	MLPs compute distance scores in DTW
Sakoe et al. (1989)	MLPs compute distance scores in DTW; delayed features used
Lerner and Deller (1991)	Non-MLP preprocessor learns time-frequency representations used in DTW
Franzini et al. (1990)	Connectionist Viterbi training, a hybrid ANN-HMM technique
Various other citations in text	Hybrid ANN-HMM approaches
Lippmann and Gold (1987)	Viterbi net, ANN used to compute HMM probabilities
Niles and Silverman (1990)	HMM network, ANN used to learn and compute HMM probabilities

or by learning which features are most effective. "Back-end" ANNs can also be added to refine the recognition scores and help improve performance. The underlying techniques, however, remain essentially unchanged. Accordingly, each of the methods below will be described in only a general way. To do otherwise would mean getting into fine details of each "machine." The interested reader is referred to the original papers for these details.

ANNs and DTW

Several research groups have described integrations of ANNs into the DTW paradigm. The simplest of these strategies have simply used MLPs to compute the local distance scores used in the DTW search (see Chapter 13). An example due to Bourlard and Wellekens (1987) is shown in Fig. 14.23. The advantage of the MLP computation is that the distance scores include a sensitivity to context as seen in the figure. In one network, 15 feature vectors centered on the "current" feature vector were entered as inputs to the MLP, which had 50 nodes in the second layer and 26 in the third. Each of the 26 outputs corresponded to a single allophone in a German digit vocabulary. Each feature vector consisted of 60 bits, with a single bit set to unity to indicate a codebook correspondent to the relevant frame. The outputs were also binary, so that the distances were actually "all or none." These distance scores were then used in the DTW algorithm in the customary way. The network was trained with hand-excised allophones and exhibited perfect recognition of 100 tokens from a single speaker. More challenging recognition problems with the same basic design are also described in the Bourlard and Wellekens paper.

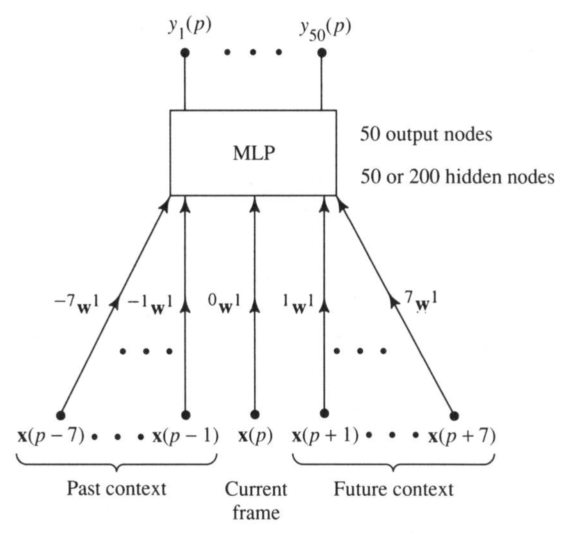

FIGURE 14.23. A feedforward ANN used to compute allophone distance scores for DTW search in a study by Bourlard and Wellekens (1987).

Sakoe et al. (1989) describe a procedure similar to the Bourlard technique, except that delayed values of the input frames are also fed to the network, somewhat akin to the TDNN. The Sakoe group calls their system the *dynamic programming neural network.* The delays in the network further compensate for spectral variability in the feature strings, which is not accounted for in the direct distance computations used by conventional DTW. With feature vectors of 16 mel-cepstral coefficients computed each 10 msec, a two-layer perceptron incorporating four delays per input was used to obtain a 99.3% recognition performance in a speaker-independent Japanese digit recognition experiment.

Lerner and Deller (1991) describe a non-MLP DTW preprocessing network called the *self-organizing feature finder* (SOFF). This network is interesting in its ability to, in an unsupervised strategy, learn features in time-frequency representations of speech. The feature detectors, which are the cells of the network, learn to compute a score that indicates the direction of any test feature vector with respect to its corresponding feature. In this sense the network is similar to an LVQ, but a predefined architecture and number of cells is not specified. Several layers of feature vectors are used to integrate spectral information into small feature vectors, which are then subjected to DTW. An array of speaker-dependent and -independent and multispeaker experiments are reported with average error rates less than 2% on the digit database.

ANNs and HMMs

Likewise, several approaches combining ANNs and HMM-like techniques have been suggested. These have ranged from several hybrid approaches that combine the ANN and HMM paradigms (Cheng et al., 1992; Franzini et al., 1990; Franzini et al., 1989; Huang and Lippmann, 1988; Morgan and Bourlard, 1990; Ramesh et al., 1992; Singer and Lippmann, 1992; Sun et al., 1990) to networks that directly implement the computations required of the HMM (Lippmann and Gold, 1987; Niles and Silverman, 1990). In addition, the network developed by Niles and Silverman is capable of learning the probability structure of the model.

As an example of the hybrid approach, we describe the work of Franzini et al. (1990) because it is closely related to research studied in Chapter 12. The so-called *connectionist Viterbi training* (CVT) is designed to recognize continuous speech and was tested on the 6000 digit strings in the TI/NBS database (see Section 13.8). One thousand of the strings were used for training. Discrete-symbol (cepstral vector) HMMs similar to those used in the SPHINX system were employed as word-dependent phone models, which, in turn, were concatenated into word models and then digit sentence models. The sentence network was trained using the customary F–B procedure (see Section 12.2.2). The symbol strings of the training utterances were then segmented along the various paths using Viterbi backtracking so that their corresponding speech samples could be associated with the various arcs in the network. Each arc was then assigned a recurrent MLP. Each MLP was then trained using BP to compute the output probabilities in response to the original speech in frames of 70 msec surrounded by three left and three right frames. Iterative realignment and retraining was used to improve performance. For unknown-length strings, the word recognition rate was 98.5% using the CVT and the string accuracy was 95.0%, while for known-length strings, these tallies were 99.1 and 96.1%. It is to be noted that these results have poorer error rates than some of the best HMM-based digit results by a factor of about three (Doddington, 1989). However, improvements to the CVT system were reported in 1991 (Haffner et al., 1991) that reduced the error rate on the same task by more than 50% [see also (Haffner, 1992)].

One of the primary advantages of the CVT system is that it obviates the use of VQ and the concomitant distortion of the feature vectors. The output distributions in the HMM are also represented in the MLP without unsubstantiated statistical assumptions. There is also evidence that ANNs might be superior to HMMs at static pattern classification (Waibel et al., 1989), and thus it might be beneficial to replace low-level HMM processing with ANNs when possible.

It is noteworthy that the CVT presented a formidable BP training problem. The researchers were able to scale the problem to about 10^{12} floating point operations by taking several measures to speed the BP algorithm. For details see (Franzini et al., 1990).

The *Viterbi net* reported by Lippmann and Gold (1987) was among the first approaches to integrating HMM-like technology and ANNs. In response to an input feature vector, the network computes a quantity that is proportional to the log probability computed by a Viterbi decoder (HMM). The weights of the network cannot be learned, but rather must be downloaded from a conventional training algorithm (like the F–B algorithm). In isolated-word tests with the Lincoln Laboratory Stress Style database (Lippmann et al., 1987), the system performed very comparably to robust HMM models.

The *HMM network* of Niles and Silverman (1990) is a fully HMM-equivalent network in its ability to learn and compute with the probabilistic structure. The network has recurrent connections, as one might expect, due to the inherent feedback in the HMM dynamics [see (12.42) and (12.43)]. One of the interesting aspects of this study is the demonstrated relationship between the BP algorithm used to train MLPs and the F–B algorithm for HMMs. Hochberg et al. (1991) have reported results of recognition experiments for the HMM network. For a vocabulary consisting of the alphabet, digits, and two control words, models were trained with vector-quantized cepstral coefficients, delta cepstral coefficients, and energy and delta energy features computed every 10 msec over 40-msec frames. About three minutes of speech per 38 talkers required three hours of training (using five workstations operating in parallel) for an F–B-like procedure, and 12 hours for a gradient-ascent maximum likelihood procedure. Continuous utterances were segmented using a Viterbi backtracking procedure as in the CVT. Experiments were performed with both bigram and null grammars with a number of different training strategies. Results are reported in (Hochberg et al., 1991). The findings are somewhat inconclusive and are discussed in detail in the paper. Nevertheless, they demonstrate potential for this interesting approach.

14.4.4 Language Modeling Using ANNs

Several research groups have explored the possibility of using ANNs to model language information. Tasks have included N-gram word category prediction (Nakamura and Shikano, 1989), modeling of a regular grammar (Liu et al., 1990), modeling of a context-free grammar (Sun et al., 1990), and the integration of TDNNs with a parsing strategy (Sawai, 1991). Semantic and other more abstract information has also been modeled in relatively simple experiments using ANNs. A particularly interesting study is reported by Gorin et al. (1991) on the adaptive acquisition of language using a connectionist network. For a review of other selected systems, the reader is referred to (Morgan and Scofield, 1991, Ch. 8), and to the general references cited in Appendix 1.G.

14.4.5 Integration of ANNs into the Survey Systems of Section 13.9

Not surprisingly, ANN technology has been integrated into some of the recent systems surveyed in Section 13.9, or has been used to develop alternative systems. The research group at BNN responsible for the

BYBLOS system has developed a new system based on "segmental neural nets" and HMMs integrated through the use of the N-best search approach. The ANNs are used for improved phonetic modeling in this system (Austin et al., 1992). ANNs have been employed in the DECIPHER system to estimate the output probabilities of the HMMs (Renals et al., 1992). In another recent study, the LR parsing approach developed at ATR has been combined with a TDNN (Sawai, 1991). The trend in these and similar studies has been to use the ANN technology to implement a specialized function that it performs well. Given the complexity of training very-large-scale ANNs, this trend seems likely to continue until major breakthroughs in training methods take place.

14.5 Conclusions

We began our study of speech recognition in Chapter 10 with an enumeration of the challenges faced by speech recognition engineers. Whereas tremendous progress has been achieved in addressing these problems in the past several decades, the performance and capabilities of solutions on sequential machines remains far short of human recognition. The ANN represents an opportunity to explore new and unconventional approaches to these difficult problems. ANN solutions can potentially add massively parallel computing and alternative strategies for adaptation to the techniques upon which speech processing engineers can draw. The current state of ANN research and development for speech recognition, however, lags far behind that of conventional methods, and the ultimate impact of this relatively immature field is uncertain.

We have explored in this chapter some basic principles underlying the ANN concept and two general types of ANN architectures—the MLP and LVQ—that have natural application to speech recognition. The renaissance of interest in ANN has been made possible in large part by the discovery of a training algorithm, BP, for the more difficult of the two, the MLP. The challenge of finding such a training method, and the arduous task of understanding its convergence properties, are both reflective of a central difference between the ANN and more conventional engineering approaches: The ANN is generally a nonlinear system. Herein lies its power and much of its mystery, and with the encouraging results reported here and elsewhere come new challenges for speech processing engineers to explain and unify these results and help to build a general theory of ANN computing. It is important to keep in mind, however, that learning all about ANNs (or any other technology) is not necessarily the key to building large-scale, robust, continuous-speech recognizers. In focusing on the technology and not the basic problems, we might once again advance the technical power underlying the systems without gaining much understanding of the deep and complex problems we are trying to solve. In this sense, recent efforts to model various quantifiable aspects of audition [for reviews, see (Lippmann, 1989; Greenberg, 1988)] might emerge as more fruitful than the more popular ANN approaches described in this chapter.

We have noted that many other ANN architectures have been explored, and have encouraged the interested reader to pursue the literature in the field. It should also be pointed out that ANN applications to speech have included tasks other than recognition, and recognition tasks other than those discussed here. Some example applications include the following:

1. Keyword spotting (Morgan et al., 1990; Anderson, 1991).
2. Synthesis (Sejnowski and Rosenberg, 1986; Scordilis and Gowdy, 1989; Rahim and Goodyear, 1989).
3. Articulatory modeling (Xue et al., 1990).
4. Enhancement and noise robustness (Tamura and Waibel, 1988; Tamura, 1989; Paliwal, 1990; Barbier and Chollet, 1991; Mathan and Miclet, 1991).
5. Voiced-unvoiced-silence discrimination (Ghiselli-Crippa and El-Jaroudi, 1991).
6. Speaker recognition and verification (Bennani et al., 1990; Morgan et al., 1989; Ogelsby and Mason, 1990, 1991).

14.6 Problems

14.1. Consider the simple perceptron shown in Fig. 14.24(a). The thresholding function is the customary step threshold defined in Fig. 14.3. Also θ represents a biasing weight and the input to its connection is always -1. Three classification problems are posed below. In each case, two-dimensional input training vectors, $\{\mathbf{x}(p),\ p=1,\ldots,P\}$, come from one of two classes, A or B. The training vectors are drawn from populations that are uniformly distributed over the regions shown in Fig. 14.24(b).

(a) Find a set of weights, $\mathbf{w}=[w_1 \quad w_2 \quad \theta]^T$, to which the Rosenblatt PL algorithm might converge for the class regions in (i). Is the bias connection to the perceptron necessary in this case?

(b) Repeat part (a) for the class regions (ii).

(c) For the class regions (iii), will the PL algorithm be able to find an exact classification boundary in the feature space? Can you speculate on what boundary the LMS algorithm might deduce? Assume that all classification errors are equally costly.

14.2. In order to better understand the development of the BP algorithm, do the following:

(a) Verify the gradient expression (14.20).

(b) Find explicit expressions for o_1 and o_2 in (14.24).

(c) Verify the gradient expression (14.25).

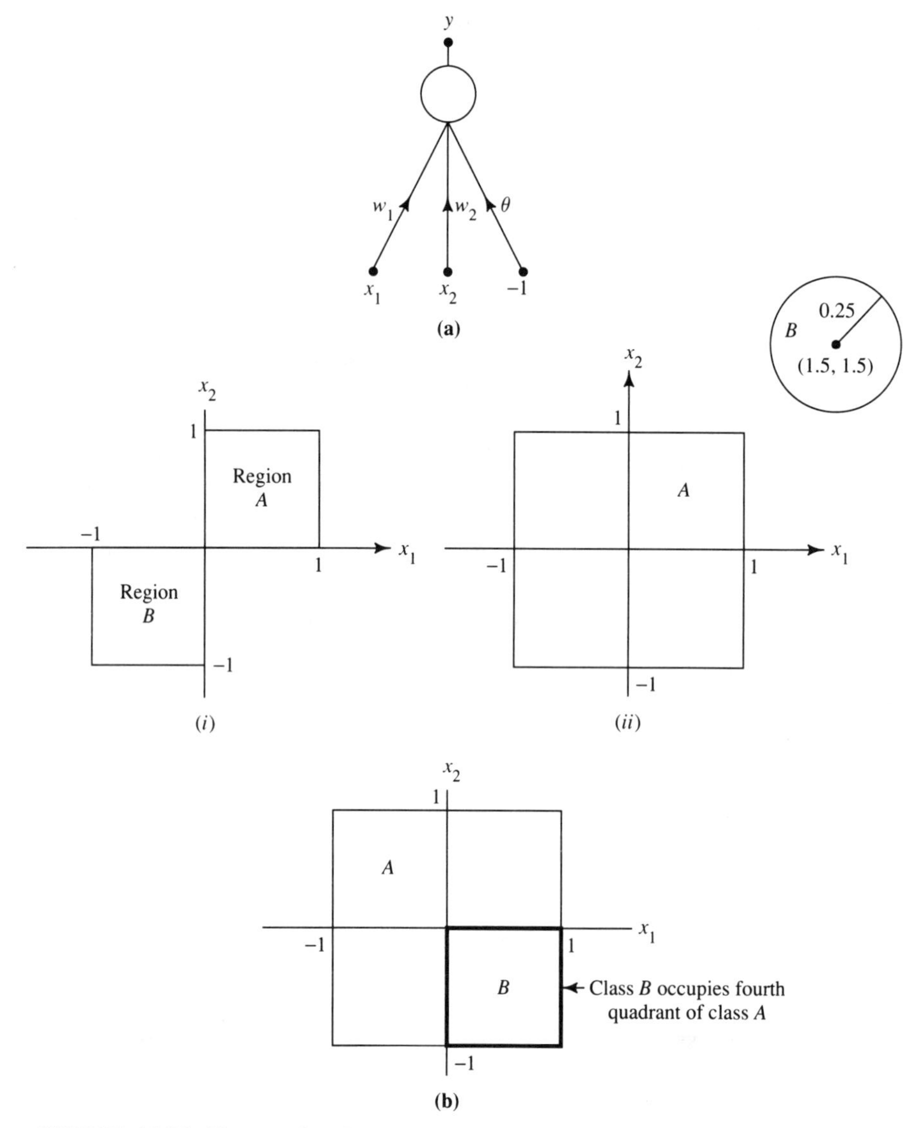

FIGURE 14.24. Figures for Problem 14.1. (a) Perceptron. (b) Feature vectors are uniformly distributed over the regions shown in three situations.

14.3. (Computer Exercise) By a means of your choice, compute the first two formant frequencies for 10 utterances of each of the vowels /a/, /i/, /u/. We consider three networks for classifying vowels sounds:

(a) Consider a single layer of three perceptrons, each with three inputs. The training inputs to the perceptrons are the vectors (augmented with a bias),

$$\mathbf{x}(p) = [F_1(p) \quad F_2(p) \quad -1]^T, \qquad p = 1, 2, \ldots, 30, \tag{14.37}$$

where $F_i(p)$ is the ith formant frequency, expressed in kHz, for the pth vowel training utterance. Let the weights for the kth perception be denoted.

$$\mathbf{w}_k = [w_{k,1}(p) \quad w_{k,2}(p) \quad \theta]^T, \qquad k = 1, 2, 3. \tag{14.38}$$

(i) Use Rosenblatt's PL algorithm to learn the weights for each perceptron. Experiment with different learning rates, η, and note the number of iterations to satisfactory convergence in each case.

(ii) Plot the 30 training vectors (excluding the constant -1) in two dimensions, along with the three decision boundaries learned by the network:

$$w_{k,1}F_1 + w_{k,2}F_2 = \theta_k, \qquad k = 1, 2, 3. \tag{14.39}$$

(iii) Comment on your results. How many training vectors are misclassified by the decision boundaries?

(b) (i) Train the weights of a three-output LVQ to achieve the classification problem above.

(ii) Plot the "cluster centers" (indicated by the weight vectors of the LVQ) on the same plot with the feature vectors from part (a).

(iii) Comment on your results. How many training vectors are misclassified by the LVQ?

(c) Experiment with various MLP structures with two inputs and three outputs, using different numbers of layers and different thresholding functions. Using the BP algorithm, can you find at least one structure that will provide superior performance to the perceptron and LVQ networks designed above in classifying the training vectors?

Bibliography

ACKENHUSEN, J. G. "The CDTWP: A programmable processor for connected word recognition," *Proceedings of the IEEE International Conference on Acoustics, Speech, and Signal Processing*, San Diego, Calif., vol. 3, paper 35.9, 1984.

AGRAWAL, A., and W. C. LIN. "Effect of voiced speech parameters on the intelligibility of /p, b/ words," *Journal of the Acoustical Society of America*, vol. 57, no. 1, pp. 217–222, Jan. 1975.

AHALT, S. C., T.-P. JUNG, and A. K. KRISHNAMURTHY. "A comparison of FSCL-LVQ neural networks for vowel classification." Chapter 6 in vol. III of P. Antognetti and V. Milutinović, eds., *Neural Networks: Concepts, Applications, and Implementations*. Englewood Cliffs, N.J.: Prentice Hall, 1991.

AIDLEY, D. J. *Physiology of Excitable Cells*. New York: Cambridge University Press, 1971.

AKAMINE, M., and K. MISEKI. "CELP coding with an adaptive density pulse excitation model," *Proceedings of the IEEE International Conference on Acoustics, Speech, and Signal Processing*, pp. 29–32. 1990.

ALBERT, A. E., and L. A. GARDNER. *Stochastic Approximation and Nonlinear Regression*. Cambridge, Mass.: M.I.T. Press, 1967.

ALLEN, J. "Cochlear modeling," *IEEE Acoustics, Speech, and Signal Processing Magazine*, vol. 2, pp. 3–29, 1985.

ALMEIDA, L. "A learning rule for asynchronous perceptrons with feedback in a combinatorial environment," *Proceedings of the* 1st *IEEE International Conference on Neural Networks*, San Diego, Calif., vol. II, pp. 609–618, 1987.

———. "Back-propagation in perceptrons with feedback." In R. Eckmiller and C. v.d. Malsberg, eds., *NATO ISI Series, Vol. F41: Neural Computers*. New York: Springer-Verlag, pp. 199–208, 1988.

ANANTHAPADMANABHA, T. V., and B. YEGNANARAYANA. "Epoch extraction from linear prediction residual for identification of the closed glottis interval," *IEEE Transactions on Acoustics, Speech, and Signal Processing*, vol. 27, pp. 309–319, Aug. 1979.

ANDERBERG, M. R. *Cluster Analysis for Applications*. New York: Academic Press, 1973.

ANDERSON, C., and E. SATORIUS. "Adaptive enhancement of finite bandwidth signals in white Gaussian noise," *IEEE Transactions on Acoustics, Speech, and Signal Processing*, vol. 31, pp. 17–28, Feb. 1983.

ANDERSON, T. "Speaker-independent phoneme recognition with an auditory model and a neural network: A comparison with traditional techniques," *Proceedings of the IEEE International Conference on Acoustics, Speech, and Signal Processing*, Toronto, Canada, vol. 1, pp. 149–152, 1991.

ARAI, M. "Mapping abilities of three-layer neural networks," *Proceedings of the International Joint Conference on Neural Networks*, Washington, D.C., vol. I, pp. 419–423, June 1989.

ARIKI, Y., K. KAJIMOTO, and T. SAKAI. "Acoustic noise reduction method by two dimensional spectrum smoothing and spectral amplitude transformation," *Proceedings of the IEEE International Conference on Acoustics, Speech, and Signal Processing*, Tokyo, Japan, pp. 97–100, 1986.

ASADI, A., R. SCHWARTZ, and J. MAKHOUL. "Automatic modelling for adding new words to a large-vocabulary continuous speech recognition system," *Proceedings of the IEEE International Conference on Acoustics, Speech, and Signal Processing*, Toronto, Canada, vol. 1, pp. 305–308, 1991.

ATAL, B. S. "Automatic speech recognition based on pitch contours." Ph.D. dissertation, Polytechnic Institute of Brooklyn, New York, 1968.

ATAL, B. S., and L. S. HANAUER. "Speech analysis and synthesis by linear prediction of the speech wave," *Journal of the Acoustical Society of America*, vol. 50, pp. 637–655, 1971.

ATAL, B. S., and J. R. REMDE. "A new model of LPC excitation for producing natural-sounding speech at low bit rates," *Proceedings of the IEEE International Conference on Acoustics, Speech, and Signal Processing*, Paris, pp. 614–617, May 1982.

ATAL, B. S., and M. R. SCHROEDER. "Predictive coding of speech signals." In Y. Konasi, ed., *Report of the* 6th *International Congress on Acoustics*, Tokyo, Japan, 1968.

———. "Adaptive predictive coding of speech signals," *Bell System Technical Journal*, vol. 49, pp. 1973–1986, 1970.

ATIYA, A. "Learning on a general network." In D. Anderson, ed., *Proceedings of the 1987 IEEE Conference on Neural Processing Systems—Natural and Synthetic*. New York: American Institute of Physics, pp. 22–30, 1988.

AT&T System Technical Journal, vol. 69, Sept.–Oct. 1990.

AUSTIN, S., G. ZAVALIAGKOS, J. MAKHOUL et al. "Speech recognition using segmental neural nets," *Proceedings of the IEEE International Conference on Acoustics, Speech, and Signal Processing*, San Francisco, Calif., vol. I, pp. I-625–I-628, 1992.

AVERBUCH, A., L. BAHL, R. BAKIS et al. "An IBM-PC based large-vocabulary isolated-utterance speech recognizer," *Proceedings of the IEEE International Conference on Acoustics, Speech, and Signal Processing*, Tokyo, Japan, vol. 1, pp. 53–56, 1986.

AVERBUCH, A., L. BAHL, R. BAKIS et al. "Experiments with the TANGORA 20,000 word speech recognizer," *Proceedings of the IEEE International Conference on Acoustics, Speech, and Signal Processing*, Dallas, Tex., vol. 2, pp. 701–704, 1987.

AZIMI-SADJADI, M., S. CITRIN, and S. SHEEDVASH. "Supervised learning process of multilayer perceptron neural networks using fast least squares," *Proceedings of the IEEE International Conference on Acoustics, Speech, and Signal Processing*, Albuquerque, N.M., vol. 3, pp. 1381–1384, 1990.

BAHL, L. R., R. BAKIS, J. BELLAGARDA et al. "Large vocabulary natural language continuous speech recognition," *Proceedings of the IEEE International Conference on Acoustics, Speech, and Signal Processing*, Glasgow, Scotland, vol. 1, pp. 465–467, 1989.

BAHL, L. R., R. BAKIS, P. S. COHEN et al. "Recognition results with several experimental acoustic processors," *Proceedings of the IEEE International Conference on Acoustics, Speech, and Signal Processing*, Washington, D.C., vol. 1, pp. 249–251, 1979.

———. "Further results on the recognition of a continuously read corpus," *Proceedings of the IEEE International Conference on Acoustics, Speech, and Signal Processing*, Denver, Colo., vol. 2, pp. 872–875, 1980.

BAHL, L. R., P. F. BROWN, P. V. DESOUZA et al. "Maximum mutual information estimation of hidden Markov model parameters for speech recognition," *Proceedings of the IEEE International Conference on Acoustics, Speech, and Signal Processing*, Tokyo, Japan, vol. 1, pp. 49–52, 1986.

———. "A new algorithm for the estimation of hidden Markov model parameters," *Proceedings of the IEEE International Conference on Acoustics, Speech, and Signal Processing*, New York, vol. 1, pp. 493–496, 1988.

BAHL, L. R., P. F. BROWN, P. V. DESOUZA et al. "Acoustic Markov models used in the TANGORA speech recognition system," *Proceedings of the IEEE International Conference on Acoustics, Speech, and Signal Processing*, New York, vol. 1, pp. 497–500, 1988.

BAHL, L. R., S. K. DAS, P. V. DESOUSA et al. "Some experiments with large-vocabulary isolated word sentence recognition," *Proceedings of the IEEE International Conference on Acoustics, Speech, and Signal Processing*, San Diego, Calif., vol. 2, paper 26.5, 1984.

BAHL, L. R., F. JELINEK, and R. L. MERCER. "A maximum likelihood approach to continuous speech recognition," *IEEE Transactions on Pattern Analysis and Machine Intelligence*, vol. 5, pp. 179–190, Mar. 1983.

BAKER, J. K. "Stochastic modeling for automatic speech understanding." In D. R. Reddy, ed., *Speech Recognition*. New York: Academic Press, pp. 521–542, 1975. Reprinted in (Waibel and Lee, 1990).

———. "The DRAGON system—An overview," *IEEE Transactions on Acoustics, Speech, and Signal Processing*, vol. 23, pp. 24–29, Feb. 1975. Reprinted in (Dixon and Martin, 1979).

BAKIS, R. "Continuous speech word recognition via centisecond acoustic states," *Proceedings of the 91*st *Annual Meeting of the Acoustical Society of America*, Washington, D.C., 1976.

BARBIER, L., and G. CHOLLET. "Robust speech parameters extraction for word recognition in noise using neural networks," *Proceedings of the IEEE International Conference on Acoustics, Speech, and Signal Processing*, Toronto, Canada, vol. 1, pp. 145–148, 1991.

BARNHART, C. L., ed. *The American College Dictionary.* New York: Random House, 1964.

BARNWELL, T. P. "Correlation analysis of subjective and objective measures for speech quality," *Proceedings of the IEEE International Conference on Acoustics, Speech, and Signal Processing*, Denver, Colo., pp. 706–709, 1980.

———. "A comparison of parametrically different objective speech quality measures using correlation analysis with subjective quality results," *Proceedings of the IEEE International Conference on Acoustics, Speech, and Signal Processing*, Denver, Colo., pp. 710–713, 1980.

———. "Improved objective quality measures for low bit speech compression," National Science Foundation, Final Technical Report ECS-8016712, 1985.

BARNWELL, T. P., and A. M. BUSH. "Statistical correlation between objective and subjective measures for speech quality," *Proceedings of the IEEE International Conference on Acoustics, Speech, and Signal Processing*, Tulsa, Okla., pp. 595–598, 1978.

BARNWELL, T. P., M. A. CLEMENTS, S. R. QUACKENBUSH et al. "Improved objective measures for speech quality testing," DCA Final Technical Report, no. DCA100-83-C-0027, Sept. 1984.

BARNWELL, T. P., and W. D. VOIERS. "An analysis of objective measures for user acceptance of voice communication systems," DCA Final Technical Report, no. DCA100-78-C-0003, Sept. 1979.

BARRON, A. R. "Statistical properties of artificial neural networks," *Proceedings of the IEEE Conference on Decision and Control*, Tampa, Fla., vol. 1, pp. 280–285, 1989.

BARRON, A. R., and R. L. BARRON. "Statistical learning networks: A unifying view," *Proceedings of the Symposium on the Interface Between Statistics and Computing Science*, Reston, Va., Apr. 1988.

BAUM, L. E. "An inequality and associated maximization technique in statistical estimation for probabilistic functions of Markov processes," *Inequalities*, vol. 3, pp. 1–8, 1972.

BAUM, L. E., and J. A. EAGON. "An inequality with applications to statistical estimation for probabilistic functions of Markov processes and to a model for ecology," *Bulletin of the American Mathematical Society*, vol. 73, pp. 360–363, 1967.

BAUM, L. E., and T. PETRIE. "Statistical inference for probabilistic functions of finite state Markov chains," *Annals of Mathematical Statistics*, vol. 37, pp. 1554–1563, 1966.

BAUM, L. E., T. PETRIE, G. SOULES et al. "A maximization technique in the statistical analysis of probabilistic functions of Markov chains," *Annals of Mathematical Statistics*, vol. 41, pp. 164–171, 1970.

BAUM, L. E., and G. R. SELL. "Growth functions for transformations on manifolds," *Pacific Journal of Mathematics*, vol. 27, pp. 211–227, 1968.

BELLANGER, M. G. *Adaptive Digital Filters and Signal Analysis.* New York: Marcel Dekker, 1987.

BELLMAN, R. "On the theory of dynamic programming," *Proceedings of the National Academy of Sciences*, vol. 38, pp. 716–719, 1952.

———. "The theory of dynamic programming," *Bulletin of the American Mathematical Society*," vol. 60, pp. 503–516, 1954.

———. *Dynamic Programming.* Princeton, N.J.: Princeton University Press, 1957.

BELLMAN, R., and S. E. DREYFUS. *Applied Dynamic Programming.* Princeton, N.J.: Princeton University Press, 1962.

BENNANI, Y., F. FOGELMAN, and P. GALLINARI. "A neural net approach to automatic speaker recognition," *Proceedings of the IEEE International Conference on Acoustics, Speech, and Signal Processing,* Albuquerque, N.M., vol. 1, pp. 265–268, 1990.

BENVENUTO, N., G. BERTOCCI, and W. R. DAUMER. "The 32-kbs ADPCM coding standard," *AT&T Technical Journal,* vol. 65, pp. 12–22, Sept.–Oct. 1986.

BERANEK, L. L. *Acoustics.* New York: McGraw-Hill, 1954.

BERGER, T. *Rate Distortion Theory.* Englewood Cliffs, N.J.: Prentice Hall, 1971.

BEROUTI, M. G. "Estimation of the glottal volume velocity by the linear prediction inverse filter." Ph.D. dissertation, University of Florida, 1976.

BEROUTI, M. G., D. G. CHILDERS, and A. PAIGE. "Glottal area versus glottal volume velocity," *Proceedings of the IEEE International Conference on Acoustics, Speech, and Signal Processing,* Hartford, Conn., vol. 1, pp. 33–36, 1977.

BEROUTI, M., R. SCHWARTZ, and J. MAKHOUL. "Enhancement of speech corrupted by acoustic noise," *Proceedings of the IEEE International Conference on Acoustics, Speech, and Signal Processing,* Washington, D.C., pp. 208–211, 1979.

BLOCK, H. "The perceptron: A model for brain functioning I, Analysis of a four-layer series coupled perceptron II," *Review of Modern Physics,* vol. 34, pp. 123–142, 1962.

———. "A review of *Perceptrons,*" *Information and Control,* vol. 17, pp. 501–522, 1970.

BLUM, J. R. "Multidimensional stochastic approximation procedure," *Annals of Mathematical Statistics,* vol. 25, pp. 737–744, 1954.

BOCCHIERI, E. L., and G. R. DODDINGTON. "Frame specific statistical features for speaker-independent speech recognition," *IEEE Transactions on Acoustics, Speech, and Signal Processing,* vol. 34, pp. 755–764, Aug. 1986.

BODENHAUSEN, U., and A. WAIBEL. "Learning the architecture of neural networks for speech recognition," *Proceedings of the IEEE International Conference on Acoustics, Speech, and Signal Processing,* Toronto, Canada, vol. 1, pp. 117–120, 1991.

BOGERT, B. P., M. J. R. HEALY, and J. W. TUKEY. "The quefrency alanysis of time series for echoes: Cepstrum, pseudo-autocovariance, cross-cepstrum and saphe cracking." In M. Rosenblatt, ed., *Proceedings of the Symposium on Time Series Analysis.* New York: John Wiley & Sons, pp. 209–243, 1963.

BOLL, S. F. "Suppression of noise in speech using the SABER method," *Proceedings of the IEEE International Conference on Acoustics, Speech, and Signal Processing,* Tulsa, Okla., pp. 606–609, 1978.

———. "Suppression of acoustic noise in speech using spectral subtraction," *IEEE Transactions on Acoustics, Speech, and Signal Processing,* vol. 27, pp. 113–120, Apr. 1979.

———. "Adaptive noise canceling in speech using the short-time transform," *Proceedings of the IEEE International Conference on Acoustics, Speech, and Signal Processing*, Denver, Colo., pp. 692–695, 1980.

BOLL, S. F., and D. C. PULSIPHER. "Suppression of acoustic noise in speech using two microphone adaptive noise cancellation," *IEEE Transactions on Acoustics, Speech, and Signal Processing*, vol. 28, pp. 751–753, Dec. 1980.

BOURLARD, H., and C. J. WELLEKENS. "Speech pattern discriminations and multi-layer perceptrons," *Computer Speech and Language*, Dec. 1987.

BOYCE, W. E., and R. C. DIPRIMA. *Elementary Differential Equations and Boundary Value Problems.* New York: John Wiley & Sons, 1969.

BRASSARD, J.-P. "Integration of segmenting and nonsegmenting approaches in continuous speech recognition," *Proceedings of the IEEE International Conference on Acoustics, Speech, and Signal Processing*, Tampa, Fla., vol. 3, pp. 1217–1220, 1985.

BREITKOPF, P., and T. P. BARNWELL. "Segmentation preclassification for improved objective speech quality measures," *Proceedings of the IEEE International Conference on Acoustics, Speech, and Signal Processing*, Atlanta, Ga., pp. 1101–1104, 1981.

BRIDLE, J. "Neural network experience at the RSRE Speech Research Unit," *Proceedings of the ATR Workshop on Neural Networks and Parallel Distributed Processing*, Osaka, Japan, 1988.

BRIDLE, J. S., and M. D. BROWN. "Connected word recognition using whole word templates," *Proceedings of the Institute for Acoustics, Autumn Conference*, pp. 25–28, Nov. 1979.

BRIDLE, J. S., R. M. CHAMBERLAIN, and M. D. BROWN. "An algorithm for connected word recognition," *Proceedings of the IEEE International Conference on Acoustics, Speech, and Signal Processing*, Paris, vol. 2, pp. 899–902, 1982.

BROOMHEAD, D. S., and D. LOWE. "Radial basis functions, multivariable functional interpolation, and adaptive networks," Technical Report RSRE Memorandum No. 4148, Royal Speech and Radar Establishment, Malvern, Worcester, England, 1988.

BURG, J. P. "Maximum entropy spectral analysis," *Proceedings of the 37*th *Meeting of the Society of Exploration Geophysicists*, 1967.

———. "Maximum entropy spectral analysis." Ph.D. dissertation, Stanford University, 1975.

BURR, B. J., B. D. ACKLAND, and N. WESTE. "Array configurations for dynamic time warping," *IEEE Transactions on Acoustics, Speech, and Signal Processing*, vol. 32, pp. 119–128, 1984.

BURR, D. J. "Experiments on neural net recognition of spoken and written text," *IEEE Transactions on Acoustics, Speech, and Signal Processing*, vol. 36, pp. 1162–1168, July 1988.

BURRASCANO, P., and P. LUCCI. "A learning rule eliminating local minima in multi-layer perceptrons," *Proceedings of the IEEE International Conference on Acoustics, Speech, and Signal Processing*, Albuquerque, N.M., vol. 2, pp. 865–868, 1990.

BURRIS, C. S. "Efficient Fourier transform and convolution." Chapter 4 in J. S. Lim and A. V. Oppenheim, eds. *Advanced Topics in Signal Processing.* Englewood Cliffs, N.J.: Prentice Hall, 1988.

BURTON, D. K., J. E. SHORE, and J. T. BUCK. "Isolated-word speech recognition using multi-section vector quantization codebooks," *IEEE Transactions on Acoustics, Speech, and Signal Processing,* vol. 33, pp. 837–849, Aug. 1985.

BUSH, M. A., and G. E. KOPEC. "Network-based connected digit recognition," *IEEE Transactions on Acoustics, Speech, and Signal Processing,* vol. 35, pp. 1401–1413, 1987.

BUSINGER, P. A., and G. H. GOLUB. "Linear least squares solutions by Householder transformations," *Numerical Mathematics,* vol. 7, pp. 269–276, 1965.

BUZO, A., A. H. GRAY, JR., R. M. GRAY et al. "Speech coding based upon vector quantization," *IEEE Transactions on Acoustics, Speech, and Signal Processing,* vol. 28, pp. 562–574, Oct. 1980.

CAMPANELLA, S. J., and G. S. ROBINSON. "A comparison of orthogonal transformations for digital speech processing," *IEEE Transactions on Communications,* vol. 19, part 1, pp. 1045–1049, Dec. 1971.

CARLYON, R. Personal communication, 1988.

CHABRIES, D. M., R. W. CHRISTIANSEN, R. H. BREY, et al. "Application of the LMS adaptive filter to improve speech communication in the presence of noise," *IEEE International Conference on Acoustics, Speech, and Signal Processing,* Paris, pp. 148–151, 1982.

CHANDRA, S., and W. C. LIN. "Experimental comparisons between stationary and non-stationary formulations of linear prediction applied to speech," *IEEE Transactions on Acoustics, Speech, and Signal Processing,* vol. 22, pp. 403–415, 1974.

CHEN, C. T. *Linear System Theory and Design.* New York: Holt, Rinehart and Winston, 1984.

CHEN, J. H. "High-quality 16 KBPS speech coding with a one-way delay less than 2 ms," *Proceedings of the IEEE International Conference on Acoustics, Speech, and Signal Processing,* Albuquerque, N.M., vol. 1, pp. 453–456, 1990.

CHENG, D. Y., A. GERSHO, B. RAMAMURTHI et al. "Fast search algorithms for vector quantization and pattern matching," *Proceedings of the IEEE International Conference on Acoustics, Speech, and Signal Processing,* San Diego, Calif., paper 9.11, 1984.

CHENG, Y., D. O'SHAUGHNESSY, V. GUPTA et al. "Hybrid segmental-LVQ/HMM for large vocabulary speech recognition," *Proceedings of the IEEE International Conference on Acoustics, Speech, and Signal Processing,* San Francisco, vol. I, pp. I-593–I-596, 1992.

CHIBA, T., and KAJIYAMA, M. *The Vowel, Its Nature and Structure.* Tokyo: Tokyo-Kaiseikan Pub. Co., 1941.

CHILDERS, D. G. "Laryngeal pathology detection," *CRC Reviews in Bioengineering,* vol. 2, pp. 375–424, 1977.

CHILDERS, D. G., and C. K. LEE. "Co-channel speech separation," *Proceedings of the IEEE International Conference on Acoustics, Speech, and Signal Processing,* Dallas, Tex., pp. 181–184, 1987.

CHILDERS, D. G., D. P. SKINNER, and R. C. KEMERAIT. "The cepstrum: A guide to processing," *Proceedings of the IEEE*, vol. 65, pp. 1428–1443, Oct. 1977.

CHOMSKY, N. "Three models for the description of language," *IEEE Transactions on Information Theory*, vol. 2, pp. 113–124, 1956.

———. "On certain formal properties of grammars," *Information and Control*, vol. 2, pp. 137–167, 1959.

———. "A note on phrase structured grammars," *Information and Control*, vol. 2, pp. 393–395, 1959.

CHOMSKY, N., and G. A. MILLER. "Finite state languages," *Information and Control*, vol. 1, pp. 91–112, 1958.

CHOW, Y. L. "Maximum mutual information estimation of HMM parameters for continuous speech recognition using the *N*-best algorithm," *Proceedings of the IEEE International Conference on Acoustics, Speech, and Signal Processing*, Albuquerque, N.M., vol. 2, pp. 701–704, 1990.

CHOW, Y. L., M. O. DUNHAM, O. A. KIMBALL et al. "BYBLOS: The BBN continuous speech recognition system," *Proceedings of the IEEE International Conference on Acoustics, Speech, and Signal Processing*, Dallas, Tex., vol. 1, pp. 89–92, 1987.

CHOW, Y. L., and S. ROUCOS. "Speech understanding using a unification grammar," *Proceedings of the IEEE International Conference on Acoustics, Speech, and Signal Processing*, Glasgow, Scotland, vol. 2, pp. 727–730, 1989.

CHOW, Y. L., R. M. SCHWARTZ, S. ROUCOS et al. "The role of word-dependent coarticulatory effects in a phoneme-based speech recognition system," *Proceedings of the IEEE International Conference on Acoustics, Speech, and Signal Processing*, Tokyo, Japan, vol. 3, pp. 1593–1596, 1986.

CHU, P. L., and D. G. MESSERSCHMITT. "A weighted Itakura–Saito spectral distance measure," *IEEE Transactions on Acoustics, Speech, and Signal Processing*, vol. 30, pp. 545–560, Aug. 1982.

CHUNG, J. H., and R. W. SCHAFER. "Excitation modeling in a homomorphic vocoder," *Proceedings of the IEEE International Conference on Acoustics, Speech, and Signal Processing*, Albuquerque, N.M., vol. 1, pp. 25–28, 1990.

CHURCHILL, R. V. *Complex Variables and Applications*, 2nd ed. New York: McGraw-Hill, 1960.

———. *Fourier Series and Boundary Value Problems*, 2nd ed. New York: McGraw-Hill, 1963.

CIOFFI, J. M., and T. M. KAILATH. "Fast recursive least squares transversal filters for adaptive filtering," *IEEE Transactions on Acoustics, Speech, and Signal Processing*, vol. 32, pp. 304–337, Apr. 1984.

———. "Windowed fast transversal filter adaptive algorithms with normalization," *IEEE Transactions on Acoustics, Speech, and Signal Processing*, vol. 33, pp. 607–625, June 1985.

COHEN, M., H. MURVEIT, J. BERNSTEIN et al. "The DECIPHER speech recognition system," *Proceedings of the IEEE International Conference on Acoustics, Speech, and Signal Processing*, Albuquerque, N.M., vol. 1, pp. 77–80, 1990.

COOLEY, J. W., and J. W. TUKEY. "An algorithm for the machine computation of the complex Fourier series," *Mathematical Computation*, vol. 19, pp. 297–301, 1965.

COX, R. V., and D. MALAH. "A technique for perceptually reducing periodically structured noise in speech," *Proceedings of the IEEE International Conference on Acoustics, Speech, and Signal Processing*, Atlanta, Ga., pp. 1089–1092, 1981.

CRAVERO, M., R. PEIRACCINI, and F. RAINERI. "Definition and evaluation of phonetic units for speech recognition by hidden Markov models," *Proceedings of the IEEE International Conference on Acoustics, Speech, and Signal Processing*, Tokyo, Japan, vol. 3, pp. 2235–2238, 1986.

CROCHIERE, R. E., J. E. TRIBOLET, and L. R. RABINER. "An interpretation of the log likelihood ratio as a measure of waveform coder performance," *IEEE Transactions on Acoustics, Speech, and Signal Processing*, vol. 28, pp. 367–376, Aug. 1980.

CROSMER, J. R., and T. P. BARNWELL. "A low bit rate segment vocoder based on line spectrum pairs," *Proceedings of the IEEE International Conference on Acoustics, Speech, and Signal Processing*, Tampa, Fla., vol. 1, pp. 240–243, 1985.

CURTIS, R. A., and R. J. NIEDERJOHN. "An investigation of several frequency-domain processing methods for enhancing the intelligibility of speech in wide-band random noise," *Proceedings of the IEEE International Conference on Acoustics, Speech, and Signal Processing*, Tulsa, Okla., pp. 602–605, 1978.

DARLINGTON, P., P. D. WHEELER, and G. A. POWELL. "Adaptive noise reduction in aircraft communication systems," *Proceedings of the IEEE International Conference on Acoustics, Speech, and Signal Processing*, Tampa, Fla., pp. 716–719, 1985.

DARWIN, C. J., and R. B. GARDNER. "Mistuning a harmonic of a vowel: Grouping and phase effects on vowel efficiency," *Journal of the Acoustical Society of America*, vol. 79, pp. 838–845, Mar. 1986.

DAUMER, W. R. "Subjective comparison of several efficient speech coders," *IEEE Transactions on Communications*, vol. 30, pp. 655–662, Apr. 1982.

DAUMER, W. R., and J. R. CAVANAUGH. "A subjective comparison of selected digital coders for speech," *Bell System Technical Journal*, vol. 57, pp. 3109–3165, Nov. 1978.

DAUTRICH, B., L. R. RABINER, and T. B. MARTIN. "On the effects of varying filter bank parameters on isolated word recognition," *IEEE Transactions on Acoustics, Speech, and Signal Processing*, vol. 31, pp. 793–807, 1983.

DAVIS, S. B., and P. MERMELSTEIN. "Comparison of parametric representations for monosyllabic word recognition in continuously spoken sentences," *IEEE Transactions on Acoustics, Speech, and Signal Processing*, vol. 28, pp. 357–366, Aug. 1980.

D'AZZO, J. J., and C. H. HOUPIS. *Feedback Control System Analysis and Synthesis*, 2nd ed. New York: McGraw-Hill, 1966.

DELATTRE, P. C., A. M. LIBERMAN, and F. S. COOPER, "Acoustic loci and transitional cues for consonants," *Journal of Acoustical Society of America*, vol. 27, no. 4, pp. 769–773, July 1955.

DELLER, J. R. "Some notes on closed phase glottal inverse filtering," *IEEE Transactions on Acoustics, Speech, and Signal Processing*, vol. 29, pp. 917–919, Aug. 1981.

____. "On the time domain properties of the two-pole model of the glottal waveform and implications for LPC," *Speech Communication: An Interdisciplinary Journal*, vol. 2, pp. 57–63, 1983.

———. "On the identification of autoregressive systems excited by periodic signals of unknown phase," *IEEE Transactions on Acoustics, Speech, and Signal Processing*, vol. 32, pp. 638–641, 1984.

DELLER, J. R., and D. HSU. "An alternative adaptive sequential regression algorithm and its application to the recognition of cerebral palsy speech," *IEEE Transactions on Circuits and Systems*, vol. 34, pp. 782–787, July 1987.

DELLER, J. R., D. HSU, and L. J. FERRIER. "On the use of hidden Markov modelling for recognition of dysarthric speech," *Computer Methods and Programs in Biomedicine*, vol. 2, pp. 125–139, June 1991.

DELLER, J. R., and S. D. HUNT. "A simple 'linearized' learning algorithm which outperforms back-propagation," *Proceedings of the International Joint Conference on Neural Networks*, Baltimore, Md., vol. III, pp. 133–138, 1992.

DELLER, J. R., and T. C. LUK. "Set-membership theory applied to linear prediction analysis of speech," *Proceedings of the IEEE International Conference on Acoustics, Speech, and Signal Processing*, Dallas, Tex., vol. 2, pp. 653–656, 1987.

———. "Linear prediction analysis of speech based on set-membership theory," *Computer Speech and Language*, vol. 3, pp. 301–327, 1989.

DELLER, J. R., and S. F. ODEH. "Implementing the optimal bounding ellipsoid algorithm on a fast processor," *Proceedings of the IEEE International Conference on Acoustics, Speech, and Signal Processing*, Glasgow, Scotland, vol. 2, pp. 1067–1070, 1989.

———. "Adaptive set-membership identification in $O(m)$ time for linear-in-parameters models," *IEEE Transactions on Acoustics, Speech, and Signal Processing*, May 1993.

DELLER, J. R., and G. P. PICACHÉ. "Advantages of a Givens rotation approach to temporally recursive linear prediction analysis of speech," *IEEE Transactions on Acoustics, Speech, and Signal Processing*, vol. 37, pp. 429–431, Mar. 1989.

DELLER, J. R., and R. K. SNIDER. "'Quantized' hidden Markov modelling for efficient recognition of cerebral palsy speech," *IEEE International Symposium on Circuits and Systems*, New Orleans, La., vol. 3, pp. 2041–2044, 1990.

DEMPSTER, A. P., N. M. LAIRD, and D. B. RUBIN. "Maximum likelihood from incomplete data via the EM algorithm," *Journal of the Royal Statistical Society*, vol. 39, pp. 1–88, 1977.

DENG, L., M. LENNIG, V. GUPTA et al. "Modeling acoustic-phonetic detail in an HMM-based large vocabulary speech recognizer," *Proceedings of the IEEE International Conference on Acoustics, Speech, and Signal Processing*, New York, vol. 1, pp. 509–512, 1988.

DENTINO, M., J. MCCOOL, and B. WIDROW. "Adaptive filtering in the frequency domain," *Proceedings of the IEEE*, vol. 66, pp. 1658–1659, Dec. 1978.

DEROUAULT, A.-M. "Context-dependent phonetic Markov models for large vocabulary speech recognition," *Proceedings of the IEEE International Conference on Acoustics, Speech, and Signal Processing*, Dallas, Tex., vol. 1, pp. 360–363, 1987.

DEROUAULT, A.-M., and B. MÉRIALDO. "Natural language modeling for phoneme-to-text transcription," *IEEE Transactions on Pattern Analysis and Machine Intelligence*, vol. 8, pp. 742–749, Nov. 1986.

DEVIJVER, P. A., and J. KITTLER. *Pattern Recognition: A Statistical Approach.* London, England: Prentice Hall International, 1982.

DIDAY, E., and J. C. SIMON. "Cluster analysis." In K. S. Fu, ed., *Digital Pattern Recognition.* New York: Springer-Verlag, 1976.

DIXON, N. R., and T. B. MARTIN, eds. *Automatic Speech and Speaker Recognition.* New York: IEEE Press, 1979.

DODDINGTON, G. R. "Phonetically sensitive discriminants for improved speech recognition," *Proceedings of the IEEE International Conference on Acoustics, Speech, and Signal Processing,* Glasgow, Scotland, vol. 1, pp. 556–559, 1989.

DODDINGTON, G. R., and T. B. SCHALK. "Speech recognition: Turning theory into practice," *IEEE Spectrum,* pp. 26–32, Jan. 1981.

DONE, W. J., and C. K. RUSHFORTH. "Estimating the parameters of a noisy all-pole process using pole-zero modeling," *Proceedings of the IEEE International Conference on Acoustics, Speech, and Signal Processing,* Washington, D.C., pp. 228–231, 1979.

DREYFUS-GRAF, J. "Phonetograph und Schwallellen-Quantelung, *Proceedings of the Stockholm Speech Communication Seminar,* Stockholm, Sweden, Sept. 1962.

DUBNOWSKI, J. J., R. W. SCHAFER, and L. R. RABINER. "Real time digital hardware pitch detector," *IEEE Transactions on Acoustics, Speech, and Signal Processing,* vol. 24, pp. 2–8, Feb. 1976.

DUDLEY, H. "The vocoder," *Bell Labs Record,* vol. 17, pp. 122–126, 1939. Reprinted in (Schafer and Markel, 1979).

———. "The carrier nature of speech," *Bell System Technical Journal,* vol. 19, pp. 495–515, 1940.

———. "Fundamentals of speech synthesis," *Journal of the Audio Engineering Society,* vol. 3, pp. 170–185, 1955.

DUDLEY, H., R. R. REISZ, and S. S. A. WATKINS. "A synthetic speaker," *Journal of the Franklin Institute,* vol. 227, pp. 739–764, 1939.

DUDLEY H., and T. H. TARNOCZY. "The speaking machine of Wolfgang von Kempelen," *Journal of the Acoustical Society of America,* vol. 22, pp. 151–166, 1950.

DUMOUCHEL, P., V. GUPTA, M. LENNIG et al. "Three probabilistic language models for a large-vocabulary speech recognizer," *Proceedings of the IEEE International Conference on Acoustics, Speech, and Signal Processing,* New York, vol. 1, pp. 513–516, 1988.

DUNN, H. K. "The calculation of vowel resonances, and an electrical vocal tract," *Journal of the Acoustical Society of America,* vol. 22, pp. 740–753, 1950.

———. "Methods of measuring vowel formant bandwidths," *Journal of the Acoustical Society of America,* vol. 33, pp. 1737–1746, Dec. 1961.

DURBIN, J. "Efficient estimation of parameters in moving-average models," *Biometrika,* vol. 46, parts 1 and 2, pp. 306–316, 1959.

———. "The fitting of time series models," *Review of the Institute for International Statistics,* vol. 28, pp. 233–243, 1960.

DUTOIT, D. "Evaluation of speaker-independent isolated-word recognition systems over telephone network," *Proceedings of the European Conference on Speech Technology*, Edinburgh, Scotland, pp. 241–244, 1987.

EARLEY, J. "An efficient context-free parsing algorithm," *Communications of the Association for Computing Machinery*, vol. 13, pp. 92–102, 1970.

EKSTROM, M. P. "A spectral characterization of the ill-conditioning in numerical deconvolution," *IEEE Transactions on Audio and Electroacoustics*, vol. 21, pp. 344–348, Aug. 1973.

EL-JAROUDI, A., and J. MAKHOUL. "Speech analysis using discrete spectral modelling," *Proceedings of the 32*nd *Midwest Symposium on Circuits and Systems*, Champaign, Ill., vol. 1, pp. 85–88, 1989.

ELMAN, J. L., and D. ZIPSER. "Learning the hidden structure of speech," ICS Report 8701, University of California at San Diego, 1987.

EPHRAIM, Y., A. DEMBO, and L. R. RABINER. "A minimum discrimination information approach for hidden Markov modeling," *IEEE Transactions on Information Theory*, vol. 35, pp. 1001-1013, Sept. 1989.

EPHRAIM, Y., and D. MALAH. "Speech enhancement using optimal non-linear spectral amplitude estimation," *Proceedings of the IEEE International Conference on Acoustics, Speech, and Signal Processing*, Boston, pp. 24.1.1–4, 1983.

———. "Speech enhancement using a minimum mean-square error short-time spectral amplitude estimator," *IEEE Transactions on Acoustics, Speech, and Signal Processing*, vol. 32, pp. 1109–1121, Dec. 1984.

EPHRAIM, Y., D. MALAH, and B. H. JUANG. "On the application of hidden Markov models for enhancing noisy speech," *Proceedings of the IEEE International Conference on Acoustics, Speech, and Signal Processing*, New York, pp. 533–536, 1988.

———. "Speech enhancement based upon hidden Markov modeling," *Proceedings of the IEEE International Conference on Acoustics, Speech, and Signal Processing*, Glasgow, Scotland, pp. 353–356, May 1989.

EYKHOFF, P. *System Identification*. New York: John Wiley & Sons, 1974.

FAIRBANKS, G. "Test of phonetic differentiation: The rhyme test," *Journal of the Acoustical Society of America*, vol. 30, pp. 596–600, July 1958.

FANO, R. M. "Short-time autocorrelation functions and power spectra," *Journal of the Acoustical Society of America*, vol. 22, pp. 546–550, Sept. 1950.

FANT, C. G. M. "Analysis and synthesis of speech processes." In B. Malmberg, ed., *Manual of Phonetics*. North-Holland: Amsterdam, The Netherlands, pp. 173–277, 1968.

———. "On the predictability of formant levels and spectrum envelopes from formant frequencies." In *For Roman Jakobson*. The Hague, The Netherlands: Mouton, 1956.

———. "Acoustic description and classification of phonetic units," *Ericsson Technics*, no. 1, 1959. Reprinted in (Fant, 1973).

———. *Acoustic Theory of Speech Production*. The Hague, The Netherlands: Mouton, 1960.

———. *Acoustic Theory of Speech Production*. The Hague, The Netherlands: Mouton, 1970.

———. *Speech Sounds and Features*. Cambridge, Mass.: M.I.T. Press, 1973.

FANT, G., and B. SONESSON. "Indirect studies of glottal cycles by synchronous inverse filtering and photo-electrical glottography," *Quarterly Progress & Status Report*, Speech Transmission Laboratory, eds. Royal Institute of Technology, Stockholm, Sweden, vol. 4, 1962.

FEDER, M., and A. V. OPPENHEIM. "A new class of sequential and adaptive algorithms with applications to noise cancellation," *Proceedings of the IEEE International Conference on Acoustics, Speech, and Signal Processing*, New York, pp. 557–560, 1988.

FEDER, M., A. OPPENHEIM, and E. WEINSTEIN. "Maximum-likelihood noise cancellation in microphones using estimate-maximize algorithm," *IEEE Transactions on Acoustics, Speech, and Signal Processing*, vol. 37, pp. 1846–1856, Feb. 1989.

FERRARA, E. R., and B. WIDROW. "Multichannel adaptive filtering for signal enhancement," *IEEE Transactions on Acoustics, Speech, and Signal Processing*, vol. 29, pp. 766–775, June 1981.

FISHER, W. M., G. R. DODDINGTON, and K. M. GOUDIE-MARSHALL. "The DARPA speech recognition research database: Specifications and status," *Proceedings of the DARPA Speech Recognition Workshop*, pp. 93–99, 1986.

FISSORE, L., P. LAFACE, G. MICCA et al. "A word hypothesizer for a large vocabulary continuous speech understanding system," *Proceedings of the IEEE International Conference on Acoustics, Speech, and Signal Processing*, Glasgow, Scotland, vol. 1, pp. 453–456, 1989.

FLANAGAN, J. L. *Speech Analysis, Synthesis, and Perception*, 2nd ed. New York: Springer-Verlag, 1972.

———. "Voices of men and machines," *Journal of the Acoustical Society of America*, vol. 51, pp. 1375–1387, Mar. 1972.

FLANAGAN, J. L. "Speech coding," *IEEE Transactions on Communication Theory*, vol. 27, pp. 710–736, Apr. 1979.

FORNEY, G. D. "The Viterbi algorithm," *Proceedings of the IEEE*, vol. 61, pp. 268–278, Mar. 1973.

FRANZINI, M. A., K.-F. LEE, and A. WAIBEL. "Connectionist Viterbi training: A new hybrid method for continuous speech recognition," *Proceedings of the IEEE International Conference on Acoustics, Speech, and Signal Processing*, Albuquerque, N.M., vol. 1, pp. 425–428, 1990.

FRANZINI, M. A., M. J. WITBROCK, and K.-F. LEE. "A connectionist approach to continuous speech recognition," *Proceedings of the IEEE International Conference on Acoustics, Speech, and Signal Processing*, Glasgow, Scotland, vol. 1, pp. 424–428, 1989.

FRAZIER, R. H., S. SAMSON, L. D. BRAIDA et al. "Enhancement of speech by adaptive filtering," *Proceedings of the IEEE International Conference on Acoustics, Speech, and Signal Processing*, Philadelphia, pp. 251–253, Apr. 1976.

FRENCH, N. R., and J. C. STEINBERG. "Factors governing the intelligibility of speech sounds," *Journal of the Acoustical Society of America*, vol. 19, pp. 90–119, Jan. 1947.

FRIEDLANDER, B. "System identification techniques for adaptive noise canceling,"

IEEE Transactions on Acoustics, Speech, and Signal Processing, vol. 30, pp. 699–709, Oct. 1982.

FRY, D. B., and P. DENES. "The solution of some fundamental problems in mechanical speech recognition," *Language and Speech*, vol. 1, pp. 35–58, 1958.

FU, K. S. *Syntactic Pattern Recognition and Applications.* Englewood Cliffs, N.J.: Prentice Hall, 1982.

FURUI, S. "Cepstral analysis technique for automatic speaker verification," *IEEE Transactions on Acoustics, Speech, and Signal Processing*, vol. 29, pp. 254–272, Apr. 1981.

———. "Speaker-independent isolated word recognition using dynamic features of the speech spectrum," *IEEE Transactions on Acoustics, Speech, and Signal Processing*, vol. 34, pp. 52–59, Feb. 1986.

GABEL, R. A., and R. A. ROBERTS. *Signals and Linear Systems*, 2nd ed. New York: John Wiley & Sons, 1980.

GABRIEL, C. M., "Machine parlant de M. Faber," *Journale de Physique*, vol. 8, pp. 274–275, 1879.

GALLAGHER, R. G. *Information Theory and Reliable Communication.* New York: John Wiley & Sons, 1968.

GARDNER, W. A. *Introduction to Random Processes with Applications to Signals and Systems*, 2nd ed. New York: McGraw-Hill, 1990.

GENTLEMAN, W. M., and H. T. KUNG. "Matrix triangularization by systolic arrays," *Proceedings of the Society of Photooptical Instrumentation Engineers* (*Real Time Signal Processing IV*), San Diego, Calif., vol. 298, pp. 19–26, 1981.

GERSHO, A. "On the structure of vector quantizers," *IEEE Transactions on Information Theory*, vol. 28, pp. 157–166, Mar. 1982.

GERSON, I. A., and M. A. JASIUK. "Vector sum excited linear prediction (VSELP) speech coding at 8 kbps," *Proceedings of the IEEE International Conference on Acoustics, Speech, and Signal Processing*, Albuquerque, N.M., vol. 1, pp. 461–464, 1990.

GHISELLI-CRIPPA, T., and A. EL-JAROUDI. "A fast neural network training algorithm and its application to voiced-unvoiced-silence classification of speech," *Proceedings of the IEEE International Conference on Acoustics, Speech, and Signal Processing*, Toronto, Canada, vol. 1, pp. 441–444, 1991.

GIBSON, J. D., "On reflection coefficients and the Cholesky decomposition," *IEEE Transactions on Acoustics, Speech, and Signal Processing*, vol. 25, pp. 93–96, Feb. 1977.

GIBSON, J. D., T. R. FISHER, and B. KOO. "Estimation and vector quantization of noisy speech," *Proceedings of the IEEE International Conference on Acoustics, Speech, and Signal Processing*, New York, pp. 541–544, 1988.

GIBSON, J. D., B. KOO, and S. D. GRAY. "Filtering of colored noise for speech enhancement and coding," *IEEE Transactions on Acoustics, Speech, and Signal Processing*, vol. 39, pp. 1732–1744, Aug. 1991.

GILLMAN, R. A. "A fast frequency domain pitch algorithm" (abstract), *Journal of the Acoustical Society of America*, vol. 58, p. S63(A), 1975.

GIVENS, W., "Computation of plane unitary rotations transforming a general matrix to triangular form," *Journal of the Society for Industrial and Applied Math*, vol. 6, pp. 26–50, 1958.

GLINSKI, S., T. M. LALUMIA, D. CASSIDAY et al. "The graph search machine (GSM): A programmable processor for connected word speech recognition and other applications, *Proceedings of the IEEE International Conference on Acoustics, Speech, and Signal Processing*, Dallas, Tex., vol. 1, pp. 519–522, 1987.

GOBLICK, T. J., JR., and J. L. HOLSINGER. "Analog source digitization: A comparison of theory and practice," *IEEE Transactions on Information Theory*, vol. 13, pp. 323–326, Apr. 1967.

GOLUB, G. H. "Numerical methods for solving least squares problems," *Numerical Mathematics*, vol. 7, pp. 206–216, 1965.

GOLUB, G. H., and C. F. VAN LOAN. *Matrix Computations*, 2nd ed. Baltimore, Md.: Johns Hopkins University Press, 1989.

GOODMAN, D. J., B. J. MCDERMOTT, and L. H. NAKATANI. "Subjective evaluation of PCM coded speech," *Bell System Technical Journal*, vol. 55, pp. 1087–1109, Oct. 1976.

GOODMAN, D., and R. D. NASH. "Subjective quality of the same speech transmission conditions in seven different countries," *IEEE Transactions on Communications*, vol. 30, pp. 642–654, Apr. 1982.

GOODMAN, D., C. SCAGLIA, R. E. CROCHIERE et al. "Objective and subjective performance of tandem connections of waveform coders with an LPC vocoder," *Bell System Technical Journal*, vol. 58, pp. 601–629, Mar. 1979.

GOODWIN, G. C., and K. S. SIN. *Adaptive Prediction, Filtering, and Control.* Englewood Cliffs, N.J.: Prentice Hall, 1984.

GOREN, A. L., and R. SHIVELY. "The ASPEN parallel computer, speech recognition and parallel dynamic programming," *Proceedings of the IEEE International Conference on Acoustics, Speech, and Signal Processing*, Dallas, Tex., vol. 2, pp. 976–979, 1987.

GORIN, A., S. LEVINSON, and A. GERTNER. "Adaptive acquisition of spoken language," *Proceedings of the IEEE International Conference on Acoustics, Speech, and Signal Processing*, Toronto, Canada, vol. 2, pp. 805–808, 1991.

GRAUPE, D. *Time Series Analysis, Identification, and Adaptive Filtering*, 2nd ed. Malabar, Fla: Krieger, 1989.

GRAY, A. H., and J. D. MARKEL. "A spectral flatness measure for studying the autocorrelation method of linear prediction of speech analysis," *IEEE Transactions on Acoustics, Speech, and Signal Processing*, vol. 22, pp. 207–217, 1974.

———. "Distance measures for speech processing," *IEEE Transactions on Acoustics, Speech, and Signal Processing*, vol. 24, pp. 380–391, 1976.

GRAY, R. M., A. BUZO, A. H. GRAY et al. "Distortion measures for speech processing," *IEEE Transactions on Acoustics, Speech, and Signal Processing*, vol. 28, pp. 367–376, Aug. 1980.

GRAY, R. M., and L. D. DAVISSON. *Random Processes: A Mathematical Approach for Engineers.* Englewood Cliffs, N.J.: Prentice Hall, 1986.

GREEFKES, J. A. "A digitally companded delta modulation modem for speech transmission," *Proceedings of the IEEE International Conference on Communications*, pp. 7.33–7.48, June 1970.

GREENBERG, S. "The ear as a speech analyzer," *Journal of Phonetics*, vol. 16, pp. 139–149, 1988.

GRIFFITHS, L. J. "An adaptive lattice structure for noise-canceling applications," *Proceedings of the IEEE International Conference on Acoustics, Speech, and Signal Processing*, Tulsa, Okla., vol. 1, pp. 87–90, 1978.

GRIMMETT, G. R., and D. R. STIRZAKER. *Probability and Random Processes*. Oxford, England: Clarendon, 1985.

GUO, H., and S. B. GELFAND. "Analysis of gradient descent learning algorithms for multilayer feedforward networks," *IEEE Transactions on Circuits and Systems*, vol. 38, pp. 883–894, Aug. 1991.

GUYTON, A. C. *Physiology of the Human Body*. Philadelphia: Saunders, 1979.

HAFFNER, P. "Connectionist word level classification in speech recognition," *Proceedings of the IEEE International Conference on Acoustics, Speech, and Signal Processing*, San Francisco, vol. I, pp. I-621–I-624, 1992.

HAFFNER, P., M. FRANZINI, and A. WAIBEL. "Integrating time alignment and neural networks for high performance continuous speech recognition," *Proceedings of the IEEE International Conference on Acoustics, Speech, and Signal Processing*, Toronto, Canada, vol. 1, pp. 105–108, 1991.

HALLIDAY, D., and R. RESNICK. *Physics* (Parts I and II). New York: John Wiley & Sons, 1966.

HANAZAWA, T., K. KITA, S. NAKAMURA et al. "ATR HMM-LR continuous speech recognition system," *Proceedings of the IEEE International Conference on Acoustics, Speech, and Signal Processing*, Albuquerque, N.M., vol. 1, pp. 53–56, 1990.

HANSEN, J. H. L. "A new speech enhancement algorithm employing acoustic endpoint detection and morphological based spectral constraints," *Proceedings of the IEEE International Conference on Acoustics, Speech, and Signal Processing*, Toronto, Canada, pp. 901–904, 1991.

HANSEN, J. H. L., and O. N. BRIA. "Lombard effect compensation for robust automatic speech recognition in noise," *Proceedings of 1990 International Conference on Spoken Language Processing*, Kobe, Japan, pp. 1125–1128, Nov. 1990.

HANSEN, J. H. L., and M. A. CLEMENTS. "Enhancement of speech degraded by non-white additive noise," Final Technical Report submitted to Lockheed Corp., DSPL-85-6, Georgia Institute of Technology, Atlanta, Ga., Aug. 1985.

———. "Constrained iterative speech enhancement with application to automatic speech recognition," *Proceedings of the IEEE International Conference on Acoustics, Speech, and Signal Processing*, New York, pp. 561–564, 1988.

———. "Constrained iterative speech enhancement with application to speech recognition," *IEEE Transactions on Signal Processing*, vol. 39, pp. 795–805, Apr. 1991.

———. "Iterative speech enhancement with spectral constraints," *Proceedings of the IEEE International Conference on Acoustics, Speech, and Signal Processing*, Dallas, Tex., vol. 1, pp. 189–192, Apr. 1987.

———. "Objective quality measures applied to enhanced speech," *Proceedings of the Acoustical Society of America*, 110th Meeting, Nashville, Tenn., p. C11, Nov. 1985.

———. "Stress compensation and noise reduction algorithms for robust speech recognition," *Proceedings of the IEEE International Conference on Acoustics, Speech, and Signal Processing*, Glasgow, Scotland, vol. 1, pp. 266–269, 1989.

———. "Use of objective speech quality measures in selecting effective spectral estimation techniques for speech enhancement," *Proceedings of the IEEE 32*nd *Midwest Symposium on Circuits and Systems*, Champaign, Ill., pp. 105–108, 1989.

HANSON, B. A., and H. WAKITA. "Spectral slope distance measures with linear prediction analysis for word recognition in noise," *IEEE Transactions on Acoustics, Speech, and Signal Processing*, vol. 35, pp. 968–973, July 1987.

HANSON, B. A., and D. Y. WONG. "The harmonic magnitude suppression (HMS) technique for intelligibility enhancement in the presence of interfering speech," *Proceedings of the IEEE International Conference on Acoustics, Speech, and Signal Processing*, San Diego, Calif., pp. 18A.5.1–4, 1984.

HANSON, B. A., D. Y. WONG, and B. H. JUANG. "Speech enhancement with harmonic synthesis," *Proceedings of the IEEE International Conference on Acoustics, Speech, and Signal Processing*, Boston, pp. 24.2.1–4, 1983.

HARRISON, W. A., J. S. LIM, and E. SINGER. "Adaptive noise cancellation in a fighter cockpit environment," *Proceedings of the IEEE International Conference on Acoustics, Speech, and Signal Processing*, San Diego, Calif., pp. 18A.4.1–4, 1984.

———. "A new application of adaptive noise cancellation," *IEEE Transactions on Acoustics, Speech, and Signal Processing*, vol. 34, pp. 21–27, Feb. 1986.

HARTSTONE, C., and P. WEIRS, eds. *Collected Papers of Charles Sanders Peirce.* Cambridge, Mass.: Harvard University Press, 1935.

HAYKIN, S. *Adaptive Filter Theory*, 2nd ed. Englewood Cliffs, N.J.: Prentice Hall, 1986.

HAYT, W. H., and J. E. KEMMERLY. *Engineering Circuit Analysis*, 2nd ed. New York: McGraw-Hill, 1971.

HECHT-NIELSEN, R. "Kolmogorov's mapping neural network existence theorem," *Proceedings of the 1*st *IEEE International Conference on Neural Networks*, San Diego, Calif., vol. III, pp. 11–15, 1987.

HECKER, M. H. L., and C. E. WILLIAMS. "Choice of reference conditions for speech preference tests," *Journal of the Acoustical Society of America*, vol. 39, pp. 946–952, Nov. 1966.

HEDELIN, P. "QD—An algorithm for non-linear inverse filtering," *Proceedings of the IEEE International Conference on Acoustics, Speech, and Signal Processing*, Atlanta, Ga., vol. 1, pp. 366–369, 1981.

———. "A glottal LPC-vocoder," *Proceedings of the IEEE International Conference on Acoustics, Speech, and Signal Processing*, San Diego, Calif., pp. 1.6.1–4, 1984.

———. "High quality glottal LPC-vocoding," *Proceedings of the IEEE International Conference on Acoustics, Speech, and Signal Processing*, Tokyo, Japan, pp. 465–468, 1986.

HEINZ, J. M., and K. N. STEVENS, "On the properties of voiceless fricative consonants," *Journal of the Acoustical Society of America*, vol. 33, pp. 589–596, 1961.

HELME, B., and C. L. NIKIAS. "Improved spectrum performance via a data-adaptive weighted Burg technique," *IEEE Transactions on Acoustics, Speech, and Signal Processing*, vol. 33, pp. 903–910, Aug. 1985.

HELMHOLTZ, H. L. F. von. *Sensations of Tone*. Translated by A. J. Ellis (1875). New York: Dover, 1954.

HEMPHILL, C., and J. PICONE. "Speech recognition in a unification grammar framework," *Proceedings of the IEEE International Conference on Acoustics, Speech, and Signal Processing*, Glasgow, Scotland, vol. 2, pp. 723–726, 1989.

HOCHBERG, M. M., J. T. FOOTE, and H. F. SILVERMAN. "Hidden Markov model/neural network training techniques for connected alpha-digit speech recognition," *Proceedings of the IEEE International Conference on Acoustics, Speech, and Signal Processing*, Toronto, Canada, vol. 1, pp. 109–112, 1991.

HOFFMAN, K., and R. KUNZE. *Linear Algebra*. Englewood Cliffs, N.J.: Prentice Hall, 1961.

HOLMES, J. N. "An investigation of the volume velocity waveform at the larynx during speech by means of an inverse filter." In G. Fant, ed., *Proceedings of the Speech Communication Seminar*, Speech Transmission Laboratory, Royal Institute of Technology, Stockholm, Sweden, p. B4, 1962.

———. "An investigation of the volume velocity waveform at the larynx during speech by means of an inverse filter," *Congress Report: 4th International Congress on Acoustics*, Copenhagen, Denmark, 1962.

———. "Formant excitation before and after glottal closure," *Proceedings of the IEEE International Conference on Acoustics, Speech, and Signal Processing*, Philadelphia, vol. 1, pp. 39–42, 1976.

HOPCROFT, J. E., and J. D. ULLMAN. *Introduction to Automata Theory, Languages, and Computation*. Reading, Mass.: Addison-Wesley, 1979.

HORNIK, K., M. STINCHCOMBE, and H. WHITE. "Multilayer feedforward networks are universal approximators," *Neural Networks*, vol. 2, pp. 359–366, 1989.

HOROWITZ, K. D., and K. D. SENNE, "Performance advantage of complex LMS for circuits and systems, *IEEE Transactions on Circuits and Systems*, vol. 28, pp. 562–576, June 1981.

HOUSE, A. S., and K. N. STEVENS. "Analog studies of the nasalization of vowels." *Journal of Speech and Hearing Disorders*, vol. 21, pp. 218–232, 1956.

HOUSE, A. S., C. E. WILLIAMS, M. H. L. HECKER et al. "Articulation testing methods: Consonantal differentiation with a closed response set," *Journal of the Acoustical Society of America*, vol. 37, pp. 158–166, 1965.

HOUSEHOLDER, A. S. "The approximate solution of matrix problems," *Journal of the Association for Computing Machinery*, vol. 5, pp. 204–243, 1985.

HOUTSMA, A. J. M., T. D. ROSSING, and W. M. WAGENAARS. *Auditory Demonstrations*, Institute for Perception Research (IPO), Eindhoven, Netherlands, 1987. Available from the Acoustical Society of America.

HSIA, T. C. *Identification: Least Squares Methods*. Lexington, Mass.: Heath, 1977.

HUANG, W. M., and R. P. LIPPMANN. "Neural networks and traditional classifiers." In D. Anderson, ed., *Proceedings of the 1987 IEEE Conference on Neural Processing Systems—Natural and Synthetic*. New York: American Institute of Physics, pp. 387–396, 1988.

HUFFMAN, D. A. "A method for the construction of minimum redundancy codes," *Proceedings of the IRE*, vol. 40, pp. 1098–1101, Sept. 1952.

HUNT, S. D. "Layer-wise training of feedforward neural networks based on linearization and selective data processing." Ph.D. dissertation, Michigan State University, 1992.

HUNT, M. J., J. S. BRIDLE, and J. N. HOLMES. "Interactive digital inverse filtering and its relation to linear prediction methods," *Proceedings of the IEEE International Conference on Acoustics, Speech, and Signal Processing*, Tulsa, Okla., vol. 1, pp. 15–19, 1978.

HUNT, M. J., M. LENNIG, and P. MERMELSTEIN. "Experiments in syllable-based recognition of continuous speech," *Proceedings of the IEEE International Conference on Acoustics, Speech, and Signal Processing*, Denver, Colo., vol. 2, pp. 880–883, 1980.

"IEEE recommended practice for speech quality measurements," *IEEE Transactions on Audio and Electroacoustics*, pp. 227–246, Sept. 1969.

IRIE, B., and S. MIYAKE. "Capabilities of three-layered perceptrons," *Proceedings of the* 2nd *IEEE International Conference on Neural Networks*, San Diego, Calif., vol. 1, pp. 641–648, 1988.

IRWIN, M. J. "Reduction of broadband noise in speech by spectral weighting," *Proceedings of the IEEE International Conference on Acoustics, Speech, and Signal Processing*, Denver, Colo., pp. 1045–1051, 1980.

ISHIZAKA, K., and J. FLANAGAN. "Synthesis of voiced sounds from a two mass model of the vocal cords," *Bell System Technical Journal*, pp. 1233–1268, Aug. 1972.

ISO, K., and T. WANTANABE. "Speaker-independent word recognition using a neural prediction model," *Proceedings of the IEEE International Conference on Acoustics, Speech, and Signal Processing*, Albuquerque, N.M., vol. 1, pp. 441–444, 1990.

———. "Large vocabulary speech recognition using neural prediction model," *Proceedings of the IEEE International Conference on Acoustics, Speech, and Signal Processing*, Toronto, Canada, vol. 1, pp. 57–60, 1991.

ITAKURA, F. "Minimum prediction residual principle applied to speech recognition," *IEEE Transactions on Acoustics, Speech, and Signal Processing*, vol. 23, pp. 67–72, Feb. 1975. Reprinted in (Dixon and Martin, 1979) and (Waibel and Lee, 1990).

———. "Line spectrum representation of linear prediction coefficients of speech signals" (abstract), *Journal of the Acoustical Society of America*, vol. 57, p. 535, 1975.

ITAKURA, F., and S. SAITO. "Analysis-synthesis telephone based on the maximum-likelihood method," *Proceedings of the* 6th *International Congress on Acoustics*, Japan, pp. C17–C20, 1968.

———. "Speech analysis-synthesis system based on the partial autocorrelation coefficient," *Proceedings of the Acoustical Society of Japan Meeting*, 1969.

———. "On the optimum quantization of feature parameters in the PARCOR speech synthesizer," *Record of the IEEE Conference on Speech Communication and Processing*, New York, pp. 434–437, 1972.

ITAKURA, F., S. SAITO, Y. KOIKE et al. "An audio response unit based on partial correlation," *IEEE Transactions on Communication Theory*, vol. 20, pp. 792–796, 1972.

JACKSON, L. B. *Digital Filters and Signal Processing*, 2nd ed. Norwell, Mass.: Kluwer, 1989.

JAIN, V. K., and R. E. CROCHIERE. "Quadrature mirror filter design in the time domain," *IEEE Transactions on Acoustics, Speech, and Signal Processing*, vol. 32, pp. 353–361, Apr. 1984.

JAIN, A. J., A. WAIBEL, and D. S. TORETZKY. "PARSEC: A structured connectionist parsing system for spoken language," *Proceedings of the IEEE International Conference on Acoustics, Speech, and Signal Processing*, San Francisco, vol. I, pp. I-205–I-208, 1992.

JAKOBSON, R., C. G. FANT, and M. HALLE. *Preliminaries to Speech Analysis: Distinctive Features and Their Correlates*. Cambridge, Mass.: M.I.T. Press, 1967.

JAYANT, N. S. "Adaptive delta modulation with a one-bit memory," *Bell System Technical Journal*, pp. 321–342, Mar. 1970.

———. "Digital coding of speech waveforms: PCM, DPCM, and DM quantizers," *Proceedings of the IEEE*, vol. 62, pp. 611–632, May 1974.

———. *Waveform Quantization and Coding*. New York: IEEE Press, 1976.

JAYANT, N. S., and J. H. CHEN. "Speech coding with time-varying bit allocation to excitation and LPC parameters," *Proceedings of the IEEE International Conference on Acoustics, Speech, and Signal Processing*, Albuquerque, N.M., vol. 1, pp. 65–68, 1989.

JAYANT, N. S., and P. NOLL. *Digital Coding of Waveforms*. Englewood Cliffs, N.J.: Prentice Hall, 1984.

JELINEK, F. "Continuous speech recognition by statistical methods," *Proceedings of the IEEE*, vol. 64, pp. 532–556, Apr. 1976.

———. "Development of an experimental discrete dictation recognizer," *Proceedings of the IEEE*, vol. 73, pp. 1616–1624, Nov. 1985.

———. "Self-organized language modeling for speech recognition." In (Waibel and Lee, 1990).

JELINEK, F., L. R. BAHL, and R. L. MERCER. "Design of a linguistic statistical decoder for the recognition of continuous speech," *IEEE Transactions on Information Theory*, vol. 21, pp. 250–256, May 1975.

JENKINS, G. M., and D. G. WATTS. *Spectral Analysis and Its Applications*. San Francisco, Calif.: Holden-Day, 1968.

JOHNSON, C. R. *Lectures on Adaptive Parameter Estimation*. Englewood Cliffs, N.J.: Prentice Hall, 1988.

JOHNSTON, J. D. "A filter family designed for use in quadrature mirror filter banks," *Proceedings of the IEEE International Conference on Acoustics, Speech, and Signal Processing*, San Diego, Calif., vol. 1 pp. 291–294, 1980.

JOUVET, D., J. MONNÉ, and D. DUBOIS. "A new network-based, speaker-independent connected-word recognition system," *Proceedings of the IEEE International Conference on Acoustics, Speech, and Signal Processing*, Tokyo, Japan, vol. 2, pp. 1109–1112, 1986.

JUANG, B.-H. "Maximum likelihood estimation for mixture multivariate stochastic observations of Markov chains," *AT&T System Technical Journal*, vol. 64, pp. 1235–1249, July–Aug. 1985.

JUANG, B.-H., S. E. LEVINSON, and M. M. SONDHI. "Maximum likelihood estimation for multivariate mixture observations of Markov chains," *IEEE Transactions on Information Theory*, vol. 32, pp. 307–309, Mar. 1986.

JUANG, B.-H., and L. R. RABINER. "A probabilistic distance measure for hidden Markov models," *AT&T System Technical Journal*, vol. 64, pp. 391–408, Feb. 1985.

JUANG, B.-H., L. R. RABINER, and J. G. WILPON. "On the use of bandpass liftering in speech recognition," *IEEE Transactions on Acoustics, Speech, and Signal Processing*, vol. 35, pp. 947–954, July 1987.

JUTLAND, F. C., G. CHOLLET, and N. DEMASSIEUX. "VLSI architectures for dynamic time warping using systolic arrays," *Proceedings of the IEEE International Conference on Acoustics, Speech, and Signal Processing*, San Diego, Calif., vol. 2, paper 34A.5, 1984.

KAISER, J. F. "Reproducing the cocktail party effect" (abstract), *Journal of the Acoustical Society of America*, vol. 32, p. 918, July 1960.

KAMMERER, B., and W. KUPPER. "Experiments for isolated-word recognition with single and multilayer perceptrons," *Abstracts of the 1st Annual International Neural Network Society*, Boston, in *Neural Networks*, vol. 1, p. 302, 1988.

KANG, G. S., and D. C. COULTER. "600 bits per second voice digitizer (linear predictive formant vocoder)," Naval Research Laboratory Report, 1976.

KANG, G. S. et al. "Multirate processor for digital voice communications," Naval Research Laboratory Report 8295, 1979.

KAPLAN, G. "Words into action I," *IEEE Spectrum*, vol. 17, pp. 22–26, June 1980.

KASAMI, T. "An efficient recognition and syntax algorithm for context-free languages," Scientific Report AFCRL-65-758, Bedford, Mass.: Air Force Cambridge Research Laboratory, 1965.

KASAMI, T., and K. TORII. "A syntax analysis procedure for unambiguous context-free grammars," *Journal of the Association for Computing Machinery*, vol. 16, pp. 423–431, 1969.

KAVEH, M., and G. A. LIPPERT. "An optimum tapered Burg algorithm for linear prediction and spectral analysis," *IEEE Transactions on Acoustics, Speech, and Signal Processing*, vol. 31, pp. 438–444, Apr. 1983.

KAY, S., and L. PAKULA. "Simple proofs of the minimum phase property of the prediction error filter," *IEEE Transactions on Acoustics, Speech, and Signal Processing*, vol. 31, p. 510, Apr. 1983.

KAY, S. M., and S. L. MARPLE, "Spectrum analysis—A modern perspective," *Proceedings of the IEEE*, vol. 69, pp. 1380–1419, Nov. 1981.

KAYSER, J. A. "The correlation between subjective and objective measures of coded speech quality and intelligibility following noise corruptions." M.S. thesis Air Force Institute of Technology, Wright–Paterson Air Force Base, Ohio, Dec. 1981.

KELLY, J. L., and C. C. LOCHBAUM, "Speech synthesis," *Proceedings of the 4th International Congress on Acoustics*, vol. G42, pp. 1–4, 1962. Also appears in *Proceedings of the Stockholm Speech Communications Seminar*, Royal Institute of Technology, Stockholm, Sweden, 1962.

KELLY, J. L., and R. F. LOGAN. *Self-Adaptive Echo Canceller.* U.S. Patent 3,500,000, Mar. 10, 1970.

KIM, J. W., and C. K. UN, "Enhancement of noisy speech by forward/backward adaptive digital filtering," *Proceedings of the IEEE International Conference on Acoustics, Speech, and Signal Processing*, Tokyo, Japan, vol. 1, pp. 89–92, 1986.

KIMURA, S. "100,000-word recognition system using acoustic segment networks," *Proceedings of the IEEE International Conference on Acoustics, Speech, and Signal Processing*, Albuquerque, N.M., vol. 1, pp. 61–64, 1990.

KITA, K., T. KAWABATA, and H. SAITO. "HMM continuous speech recognition using predictive LR parsing," *Proceedings of the IEEE International Conference on Acoustics, Speech, and Signal Processing*, Glasgow, Scotland, vol. 2, pp. 703–706, 1989.

KITA, K., and W. H. WARD. "Incorporating LR parsing into SPHINX," *Proceedings of the IEEE International Conference on Acoustics, Speech, and Signal Processing*, Toronto, Canada, vol. 1, pp. 269–272, 1991.

KITAWAKI, N., M. HONDA, and K. ITOH. "Speech quality assessment methods for speech coding systems," *IEEE Communications Magazine*, vol. 22, pp. 26–33, Oct. 1984.

KLATT, D. "A digital filter bank for spectral matching," *Proceedings of the IEEE International Conference on Acoustics, Speech, and Signal Processing*, Philadelphia, pp. 573–576, 1976.

———. "Prediction of perceived phonetic distance from critical-band spectra: A first step," *Proceedings of the IEEE International Conference on Acoustics, Speech, and Signal Processing*, Paris, pp. 1278–1281, 1982.

———. "Review of the ARPA speech understanding project," *Journal of the Acoustical Society of America*, vol. 62, pp. 1324–1366, Dec. 1977. Reprinted in (Dixon and Martin, 1979) and (Waibel and Lee, 1990).

———. "Review of text-to-speech conversion for English," *Journal of the Acoustical Society of America*, vol. 82, pp. 737–793, Sept. 1987.

KOBATAKE, H., J. INARI, and S. KAKUTA, "Linear predictive coding of speech signals in a high ambient noise environment." *Proceedings of the IEEE International Conference on Acoustics, Speech, and Signal Processing*, pp. 472–475, Apr. 1978.

KOENIG, W. "A new frequency scale for acoustic measurements," *Bell Telephone Laboratory Record*, vol. 27, pp. 299–301, 1949.

KOENIG, W., H. K. DUNN, and L. Y. LACY. "The sound spectrograph," *Journal of the Acoustical Society of America*, vol. 17, pp. 19–49, July 1946.

KOFORD, J., and G. GRONER. "The use of an adaptive threshold element to design a linear optimal pattern classifier," *IEEE Transactions on Information Theory*, vol. 12, pp. 42–50, Jan. 1966.

KOHONEN, T. "Automatic formation of topological maps in a self-organizing system." In E. Oja and O. Simula, eds. *Proceedings of the 2*nd *Scandinavian Conference on Image Analysis*, pp. 214–220, 1981. See also (Kohonen, 1987).

———. *Content Addressable Memories*, 2nd ed. New York: Springer-Verlag, 1987.

———. "An introduction to neural computing," *Neural Networks*, vol. 1, pp. 3–16, 1988.

KOLLIAS, S., and D. ANASTASSIOU. "An adaptive least squares algorithm for the efficient training of artificial neural networks," *IEEE Transactions on Circuits and Systems*, vol. 36, pp. 1092–1101, Aug. 1989.

KOLMOGOROV, A. N. "On the representation of continuous functions of many variables by superposition of functions of one variable and addition," *Dokl. Acad. Nauk USSR*, vol. 144, pp. 953–956, 1957.

KOMORI, Y. "Time state neural networks (TSNN) for phoneme identification by considering temporal structure of phonemic features," *Proceedings of the IEEE International Conference on Acoustics, Speech, and Signal Processing*, Toronto, Canada, vol. 1, pp. 125–128, 1991.

KONVALINKA, I. S., and M. R. MATAUŠEK. "On the simultaneous estimation of poles and zeros in speech analysis, and ITIF: Iterative inverse filtering algorithm," *IEEE Transactions on Acoustics, Speech, and Signal Processing*, vol. 27, pp. 485–492, Oct. 1979.

KORENBERG, M. J., and L. D. PAARMANN. "An orthogonal ARMA identifier with automatic order estimation for biological modelling," *Annals of Biomedical Engineering*, vol. 17, pp. 571–592, 1989.

KOSKO, B. *Neural Networks and Fuzzy Systems*. Englewood Cliffs, N.J.: Prentice Hall, 1992.

KRISHNAMURTHY, A. K. "Two channel analysis for formant tracking and inverse filtering," *Proceedings of the IEEE International Conference on Acoustics, Speech, and Signal Processing*, San Diego, Calif., vol. 3, pp. 36.3.1–36.3.4, 1984.

KRISHNAMURTHY, A. K., and D. G. CHILDERS. "Two channel speech analysis," *IEEE Transactions on Acoustics, Speech, and Signal Processing*, vol. 34, pp. 730–743, Aug. 1986.

KROON, P., and B. S. ATAL. "Strategies for improving the performance of CELP coders at low bit rates," *Proceedings of the IEEE International Conference on Acoustics, Speech, and Signal Processing*, New York, vol. 1, pp. 151–154, Apr. 1988.

KRUSKAL, J. B. "Multidimensional scaling by optimizing goodness of fit to a numerical hypothesis," *Psychometrika*, vol. 29, pp. 1–27, 1964.

———. "Nonmetric multidimensional scaling: A numerical method," *Psychometrika*, vol. 29, pp. 115–129, 1964.

KRYTER, K. D. "Methods for the calculation of the articulation index," *Journal of the Acoustical Society of America*, vol. 34, pp. 1689–1697, Nov. 1962.

———. "Validation of the articulation index," *Journal of the Acoustical Society of America*, vol. 34, pp. 1698–1702, Nov. 1962.

KUBALA, F., Y. CHOW, A. DERR et al. "Continuous speech recognition results of the BYBLOS system on the DARPA 1000-word resource management database," *Proceedings of the IEEE International Conference on Acoustics, Speech, and Signal Processing*, New York, vol. 1, pp. 291–294, 1988.

KUO, C. J., J. R. DELLER, and A. K. JAIN. "Transform encryption coding of images," submitted to *IEEE Transactions on Image Processing*, Aug. 1992.

LADEFOGED, P. *A Course in Phonetics*. New York: Harcourt Brace Jovanovich, 1975.

LAEBENS, J. L., and J. R. DELLER. "'SISSI'—A silent input selective sequential identifier for AR systems," *Proceedings du Neuvième Colloque sur le Traitment du Signal et ses Applications*, Nice, France, vol. 2, pp. 989–994, 1983.

LAMEL, L. F., L. R. RABINER, A. E. ROSENBERG et al. "An improved endpoint detector for isolated word recognition," *Proceedings of the IEEE International Conference on Acoustics, Speech, and Signal Processing*, Atlanta, Ga., vol. 29, pp. 777–785, 1981.

LANG, S. W., and G. E. HINTON. "The development of the time-delay neural network architecture for speech recognition," Technical Report No. CMU-CS-88-152, Carnegie–Mellon University, 1988.

LANG, S. W., and J. H. MCCLELLAN. "Frequency estimation with maximum entropy spectral estimators," *IEEE Transactions on Acoustics, Speech, and Signal Processing*, vol. 28, pp. 716–724, Dec. 1980.

LARAR, J. N., Y. A. ALSAKA, and D. G. CHILDERS. "Variability in closed phase analysis of speech," *Proceedings of the IEEE International Conference on Acoustics, Speech, and Signal Processing*, Tampa, Fla., vol. 2, pp. 1089–1092, 1985.

LAW, H. B., and R. A. SEYMOUR. "A reference distortion system using modulated noise," *Proceedings of the IEE*, pp. 484–485, Nov. 1962.

LEE, C.-H., and L. R. RABINER. "A frame synchronous network search algorithm for connected word recognition," *IEEE Transactions on Acoustics, Speech, and Signal Processing*, vol. 37, pp. 1649–1658, Nov. 1989.

LEE, C.-H., F. K. SOONG, and B.-H. JUANG. "A segment model based approach to speech recognition," *Proceedings of the IEEE International Conference on Acoustics, Speech, and Signal Processing*, New York, vol. 1, pp. 501–504, 1988.

LEE, H. C., and K. S. FU. "A stochastic syntax analysis procedure and its application to pattern classification," *IEEE Transactions on Computers*, vol. 21, pp. 660–666, July 1972.

LEE, K.-F., H.-W. HON, and D. R. REDDY. "An overview of the SPHINX speech recognition system," *IEEE Transactions on Signal Processing*, vol. 38, pp. 35–45, Jan. 1990.

LEE, K.-F., and S. MAHAJAM. "Corrective and reinforcement learning for speaker-independent continuous speech recognition," Technical Report No. CMU-CS-89-100, Carnegie–Mellon University, Jan. 1989.

LEE, R. K. C. *Optimal Estimation, Identification, and Control*. Cambridge, Mass.: M.I.T. Press, 1964.

LEFEVRE, J. P., and O. PASSIEN. "Efficient algorithms for obtaining multipulse excitation for LPC coders," *Proceedings of the IEEE International Conference on Acoustics, Speech, and Signal Processing*, Tampa, Fla., vol. 2, pp. 957–960, 1985.

LEHISTE, I., and G. E. PETERSON. "Transitions, glides, and diphthongs," *Journal of the Acoustical Society of America*, vol. 33, pp. 268–277, Mar. 1961.

LEONARD, R. G. "A database for speaker independent digit recognition," *Proceedings of the IEEE International Conference on Acoustics, Speech, and Signal Processing*, San Diego, Calif., vol. 3, paper 42.11, 1984.

LEON-GARCIA, A. *Probability and Random Processes for Electrical Engineering*. Reading, Mass.: Addison-Wesley, 1989.

LERNER, S. Z., and J. R. DELLER. "Speech recognition by a self-organizing feature finder," *International Journal of Neural Systems*, vol. 2, pp. 55–78, 1991.

LESSER, V. R., R. D. FENNELL, L. D. ERMAN et al. "Organization of the HEARSAY II speech understanding system," *IEEE Transactions on Acoustics, Speech, and Signal Processing*, vol. 23, pp. 11–23, Feb. 1975.

LEVIN, E. "Word recognition using hidden control neural architecture," *Proceedings of the IEEE International Conference on Acoustics, Speech, and Signal Processing*, Albuquerque, N.M., vol. 1, pp. 433–436, 1990.

LEVINSON, N. "The Weiner RMS (root mean square) error criterion in filter design and prediction," *Journal of Mathematical Physics*, vol. 25, pp. 261–278, 1947. Also appears as Appendix B in (Weiner, 1949).

LEVINSON, S. E. "Structural methods in automatic speech recognition," *Proceedings of the IEEE*, vol. 73, pp. 1625–1650, Nov. 1985.

———. "Continuously variable duration hidden Markov models for automatic speech recognition," *Computer Speech and Language*, vol. 1, pp. 29–45, Mar. 1986.

LEVINSON, S. E., A. LJOLJE, and L. G. MILLER. "Large vocabulary speech recognition using a hidden Markov model for acoustic/phonetic classification," *Proceedings of the IEEE International Conference on Acoustics, Speech, and Signal Processing*, New York, vol. 1, pp. 505–508, 1988.

LEVINSON, S. E., L. R. RABINER, A. E. ROSENBERG et al. "Interactive clustering techniques for selecting speaker-independent reference templates for isolated word recognition," *IEEE Transactions on Acoustics, Speech, and Signal Processing*, vol. 27, pp. 134–141, Apr. 1979.

LEVINSON, S. E., L. R. RABINER, and M. M. SONDHI. "An introduction to the application of the theory of probabilistic functions of a Markov process to automatic speech recognition," *Bell System Technical Journal*, vol. 62, pp. 1035–1074, Apr. 1983.

LICKLIDER, J. C. R., and I. POLLACK. "Effects of differentiation, integration, and infinite peak clipping upon the intelligibility of speech," *Journal of the Acoustical Society of America*, vol. 20, pp. 42–51, Jan. 1948.

LIM, J. S. "Evaluation of a correlation subtraction method for enhancing speech degraded by additive white noise," *IEEE Transactions on Acoustics, Speech, and Signal Processing*, vol. 26, pp. 471–472, Oct. 1978.

———. "Spectral root homomorphic deconvolution system," *IEEE Transactions on Acoustics, Speech, and Signal Processing*, vol. 27, pp. 223–232, June 1979.

LIM, J. S., and A. V. OPPENHEIM. "All-pole modeling of degraded speech," *IEEE Transactions on Acoustics, Speech, and Signal Processing*, vol. 26, pp. 197–210, June 1978.

———. "Enhancement and bandwidth compression of noisy speech," *Proceedings of the IEEE*, vol. 67, pp. 1586–1604, Dec. 1979.

LIM, J. S., A. V. OPPENHEIM, and L. D. BRAIDA. "Evaluation of an adaptive comb filtering method for evaluating speech degraded by white noise addition," *IEEE Transactions on Acoustics, Speech, and Signal Processing*, vol. 26, pp. 354–358, Aug. 1978.

LINDBOLM, B. E. F., and J. E. F. SUNDBERG, "Acoustic consequences of lip, tongue, jaw, and larynx movement," *Journal of the Acoustical Society of America*, vol. 50, pp. 1166–1179, 1971.

LINDE, Y., A. BUZO, and R. M. GRAY. "An algorithm for vector quantizer design," *IEEE Transactions on Communications*, vol. 28, pp. 84–95, Jan. 1980.

LINDQVIST, J. "Inverse filtering. Instrumentation and techniques," *Quarterly Progress and Status Report*, Speech Transmission Laboratory, Royal Institute of Technology, Stockholm, Sweden, vol. 4, 1964.

———. "Studies of the voice source by means of inverse filtering," *Quarterly Progress and Status Report*, Speech Transmission Laboratory, Royal Institute of Technology, Stockholm, Sweden, vol. 2, 1965.

———. "The voice source studies by means of inverse filtering," *Quarterly Progress and Status Report*, Speech Transmission Laboratory, Royal Institute of Technology, Stockholm, Sweden, vol. 1, 1970.

LIPORACE, L. A. "Maximum likelihood estimation for multivariate observations of Markov sources," *IEEE Transactions on Information Theory*, vol. 28, pp. 729–734, Sept. 1982.

LIPPMANN, R. P. "An introduction to computing with neural nets," *IEEE Acoustics, Speech, and Signal Processing Magazine*, vol. 4, pp. 4–22, Apr. 1987.

———. "Review of research on neural networks for speech recognition," *Neural Computation*, vol. 1, Mar. 1989. Reprinted in (Waibel and Lee, 1990).

LIPPMANN, R. P., and B. GOLD. "Neural-net classifiers useful for speech recognition," *Proceedings of the 1st IEEE International Conference on Neural Networks*, San Diego, Calif., vol. IV, pp. 417–425, 1987.

LIPPMANN, R. P., E. A. MARTIN, and D. P. PAUL. "Multi-style training for robust isolated-word speech recognition," *Proceedings of the IEEE International Conference on Acoustics, Speech, and Signal Processing*, Dallas, Tex., vol. 2, pp. 705–708, 1988.

LIU, Y. D., G. Z. SUN, H. H. CHEN et al. "Grammatical inference and neural network state machines," *Proceedings of the International Joint Conference on Neural Networks*, Washington, D.C., vol. I, pp. 285–288, Jan. 1990.

LJUNG, L., and T. SÖDERSTRÖM. *Theory and Practice of Recursive Identification*. Cambridge, Mass.: M.I.T. Press, 1983.

LLOYD, S. P. "Least squares quantization in PCM," Bell Laboratories Technical Note, 1957. Reprinted in *IEEE Transactions on Information Theory*, vol. 28, pp. 129–137, Mar. 1982.

LOWERRE, B. T., and D. R. REDDY. "The HARPY speech understanding system." In W. A. Lea, ed., *Trends in Speech Recognition*. Englewood Cliffs, N.J.: Prentice Hall, 1980.

LUCKE, H., and F. FALLSIDE. "Expanding the vocabulary of a connectionist recognizer trained on the DARPA resource management corpus," *Proceedings of the IEEE International Conference on Acoustics, Speech, and Signal Processing*, San Francisco, vol. I, pp. I-605–I-608, 1992.

LUSTERNIK, L. A., and V. J. SOBOLEV. *Elements of Functional Analysis*. New York: Halsted (Wiley), 1974.

MCAULAY, R. J., and M. L. MALPASS. "Speech enhancement using a soft-decision noise suppression filter," *IEEE Transactions on Acoustics, Speech, and Signal Processing*, vol. 28, pp. 137–145, Apr. 1980.

MCAULAY, R. J., and T. F. QUATIERI. "Speech analysis-synthesis based on a sinusoidal representation," *IEEE Transactions on Acoustics, Speech, and Signal Processing*, vol. 34, pp. 744–754, Aug. 1986.

MCCANDLESS, S. S. "An algorithm for automatic formant extraction using linear prediction spectra," *IEEE Transactions on Acoustics, Speech, and Signal Processing*, vol. 22, pp. 135–141, Apr. 1974.

MCCLELLAN, J. H. "Parametric signal modelling." Chapter 1 in J. S. Lim and A. V. Oppenheim, eds., *Advanced Topics in Signal Processing*. Englewood Cliffs, N.J.: Prentice Hall, 1988.

MCCOLLOCH, W. S., and W. PITTS. "A logical calculus of the ideas imminent in nervous activity." *Bulletin of Mathematical Biophysics*, vol. 5, pp. 115–133, 1943.

MCCOOL, J. M. et al. *Adaptive Line Enhancer*, U.S. Patent 4,238,746, Dec. 9, 1980.

MCCOOL, J. M. et al. *An Adaptive Detector*. U.S. Patent 4,243,935, Jan. 6, 1981.

MCDERMOTT, B. J. "Multidimensional analysis of circuit quality judgments," *Journal of the Acoustical Society of America*, vol. 45, pp. 774–781, Mar. 1969.

MCDERMOTT, B. J., C. SCAGLIOLA, and D. J. GOODMAN. "Perceptual and objective evaluation of speech processed by adaptive differential PCM," *Proceedings of the IEEE International Conference on Acoustics, Speech, and Signal Processing*, Tulsa, Okla., pp. 581–585, Apr. 1978.

_____. "Perceptual and objective evaluation of speech processed by adaptive differential PCM," *Bell System Technical Journal*, vol. 57, pp. 1597–1619, May 1978.

MCDERMOTT, E., and S. KATAGIRI. "Shift-invariant, multi-category phoneme recognition using Kohonen's LVQ2," *Proceedings of the IEEE International Conference on Acoustics, Speech, and Signal Processing*, Glasgow, Scotland, vol. 1, pp. 81–84, 1989.

MCGEE, V. E. "Semantic components of the quality of processed speech," *Journal of Speech and Hearing Research*, vol. 7, pp. 310–323, 1964.

MCMAHAN, M., and R. B. PRICE. "Grammar driven connected word recognition on the TI-SPEECH board," *Proceedings of Speech Tech*, New York, pp. 88–91, 1986.

MACNEILAGE, P. F. "Motor control of serial ordering of speech," *Psychological Review*, vol. 77, pp. 182–196, 1970.

McWhirter, J. G. "Recursive least squares solution using a systolic array," *Proceedings of the Society of Photooptical Instrumentation Engineers* (*Real Time Signal Processing* IV), San Diego, Calif., vol. 431, pp. 105–112, 1983.

Mahalanobis, P. C. "On the generalized distance in statistics," *Proceedings of the National Institute of Science* (India), vol. 12, pp. 49–55, 1936.

Makhoul, J. "Linear prediction: A tutorial review," *Proceedings of the IEEE*, vol. 63, pp. 561–580, Apr. 1975.

———. "Spectral linear prediction: Properties and applications," *IEEE Transactions on Acoustics, Speech, and Signal Processing*, vol. 23, pp. 283–296, June 1975.

———. "Stable and efficient lattice methods for linear prediction," *IEEE Transactions on Acoustics, Speech, and Signal Processing*, vol. 25, pp. 423–428, Oct. 1977.

———. Personal communication, 1991.

Makhoul, J., and L. Cosell. "Adaptive lattice analysis of speech," *IEEE Transactions on Acoustics, Speech, and Signal Processing*, vol. 29, pp. 654–659, June 1981.

Makhoul, J., S. Roucos, and H. Gish. "Vector quantization in speech coding," *Proceedings of the IEEE*, vol. 73, pp. 1551–1588, Nov. 1985.

Makhoul, J., and R. Viswanathan. "Adaptive preprocessing for linear predictive speech compression systems" (abstract), *Journal of the Acoustical Society of America*, vol. 55, p. 475, 1974.

Malah, D., and R. V. Cox. "Time-domain algorithms for harmonic bandwidth reduction and time scaling of speech signals," *IEEE Transactions on Acoustics, Speech, and Signal Processing*, vol. 27, pp. 121–133, Apr. 1979.

———. "A generalized comb filtering technique for speech enhancement," *Proceedings of the IEEE International Conference on Acoustics, Speech, and Signal Processing*, Paris, vol. 1, pp. 160–163, 1982.

Mann, J. R., and F. M. Rhodes. "A wafer-scale DTW multiprocessor," *Proceedings of the IEEE International Conference on Acoustics, Speech, and Signal Processing*, Tokyo, Japan, vol. 3, pp. 1557–1560, 1986.

Mariani, J. "Recent advances in speech recognition," *Proceedings of the IEEE International Conference on Acoustics, Speech, and Signal Processing*, Glasgow, Scotland, vol. 1, pp. 429–440, May 1989.

Markel, J. D. *Formant Trajectory Estimation from a Linear Least Squares Inverse Filter Formulation*. Monograph No. 7, Santa Barbara, Calif.: Speech Communication Research Laboratory, Inc., 1971.

———. "The Prony method and its application to speech analysis" (abstract), *Journal of the Acoustical Society of America*, vol. 49, p. 105, Jan. 1971.

———. "Digital inverse filtering: A new tool for formant trajectory estimation," *IEEE Transactions on Audio and Electroacoustics*, vol. 20, pp. 129–137, June 1972.

———. "The SIFT algorithm for fundamental frequency estimation," *IEEE Transactions on Audio and Electroacoustics*, vol. 20, pp. 367–377, Dec. 1972.

Markel, J. D., and A. H. Gray. *Linear Prediction of Speech*. New York: Springer-Verlag, 1976.

MARKEL, J. D., A. H. GRAY, and H. WAKITA. *Linear Prediction of Speech—Theory and Practice.* Monograph No. 10, Santa Barbara, Calif.: Speech Communication Research Laboratory, Inc., 1973.

MARKEL, J. D., and D. Y. WONG. "Considerations in the estimation of glottal volume velocity waveforms," *Proceedings of the Acoustical Society of America,* 91st Meeting, RR6, 1976.

MARPLE, S. L. *Digital Spectral Analysis with Applications.* Englewood Cliffs, N.J.: Prentice Hall, 1987.

MARSHALL, D. F., and W. K. JENKINS. "A fast quasi-Newton adaptive filtering algorithm," *Proceedings of the IEEE International Conference on Acoustics, Speech, and Signal Processing,* New York, pp. 1377–1380, 1988.

MARTIN, T. B. "Practical applications of voice input to machine," *Proceedings of the IEEE,* vol. 64, pp. 487–501, Apr. 1976.

MATAUŠEK, M. R., and V. S. BATALOV. "A new approach to the determination of the glottal waveform," *IEEE Transactions on Acoustics, Speech, and Signal Processing,* vol. 28, pp. 616–622, Dec. 1980.

MATHAN, L., and L. MICLET. "Rejection of extraneous input in speech recognition applications using multi-layer perceptrons and the trace of HMMs," *Proceedings of the IEEE International Conference on Acoustics, Speech, and Signal Processing,* Toronto, Canada, vol. 1, pp. 93–96, 1991.

MAX, J. "Quantizing for minimum distortion," *IRE Transactions on Information Theory,* vol. 6, pp. 7–12, Mar. 1960.

MEISEL, W. S., M. T. ANIKST, S. S. PIRZADEH et al. "The SSI large-vocabulary speaker-independent continuous speech recognition system," *Proceedings of the IEEE International Conference on Acoustics, Speech, and Signal Processing,* Toronto, Canada, vol. 1, pp. 337–340, 1991.

MERGEL, D., and A. PAESELER. "Construction of language models for spoken database queries," *Proceedings of the IEEE International Conference on Acoustics, Speech, and Signal Processing,* Dallas, Tex., vol. 2, pp. 844–847, 1987.

MÉRIALDO, B. "Speech recognition using a very large size dictionary," *Proceedings of the IEEE International Conference on Acoustics, Speech, and Signal Processing,* Dallas, Tex., vol. 1, pp. 364–367, 1987.

MESSERSCHMITT, D. G. *Adaptive Filters.* Norwell, Mass.: Kluwer, 1984.

MILENKOVIC, P. "Glottal inverse filtering by joint estimation of an AR system with a linear input model," *IEEE Transactions on Acoustics, Speech, and Signal Processing,* vol. 34, pp. 28–42, Feb. 1986.

MILLER, L. G., and A. L. GORIN. "A structured network architecture for adaptive language acquisition," *Proceedings of the IEEE International Conference on Acoustics, Speech, and Signal Processing,* San Francisco, Calif., vol. I, pp. I-201–I-204, 1992.

MILLER, R. L. "Nature of the vocal cord wave," *Journal of the Acoustical Society of America,* vol. 31, pp. 667–677, 1959.

MILLER, S. M., D. P. MORGAN, H. F. SILVERMAN et al. "Real-time evaluation system for a real-time connected speech recognizer," *Proceedings of the IEEE International Conference on Acoustics, Speech, and Signal Processing,* Dallas, Tex., vol. 2, pp. 801–804, 1987.

MILNER, P. M. *Physiological Psychology*. New York: Holt, Rinehart and Winston, 1970.

MINSKY, M. L., and S. PAPERT. *Perceptrons*. Cambridge, Mass.: M.I.T. Press, 1969.

_____. *Perceptrons*, 2nd ed. Cambridge, Mass.: M.I.T. Press, 1988.

MIRCHANDANI, G., R. C. GAUS, and L. K. BECHTEL. "Performance characteristics of a hardware implementation of the cross-talk resistant adaptive noise canceller," *Proceedings of the IEEE International Conference on Acoustics, Speech, and Signal Processing*, Tokyo, Japan, pp. 93–96, Apr. 1986.

MITCHELL, C., M. HARPER, and L. JAMIESON, "Comments on 'Reducing computation in HMM evaluation'" *IEEE Transactions on Speech and Audio Processing*, vol. 2, pp. 542–543, Oct. 1994.

MIYOSHI, Y., K. YAMAMOTO, R. MIZOGUCHI et al. "Analysis of speech signals of short pitch period by sample selective linear prediction," *IEEE Transactions on Acoustics, Speech, and Signal Processing*, vol. 35, pp. 1233–1240, Sept. 1987.

MOODY, J. "Fast learning in multi-resolution hierarchies." In D. S. Touretzky, ed., *Advances in Neural Information Processing Systems*. San Mateo, Calif.: Morgan–Kaufmann, 1989.

MORGAN, D. P., L. RIEK, D. P. LOCONTO et al. "A comparison of neural networks and traditional classification techniques for speaker identification," *Proceedings of the Military and Government Speech Technology Conference*, Washington, D.C., pp. 238–242, 1989.

MORGAN, D. P., and C. L. SCOFIELD. *Neural Networks and Speech Processing*. Norwell, Mass.: Kluwer, 1991.

MORGAN, D. P., C. L. SCOFIELD, T. M. LORENZO et al. "A keyword spotter which incorporates neural networks for secondary processing," *Proceedings of the IEEE International Conference on Acoustics, Speech, and Signal Processing*, Albuquerque, N.M., vol. 1, pp. 113–116, 1990.

MORGAN, N., and H. BOURLARD. "Continuous speech recognition using multilayer perceptrons with hidden Markov models," *Proceedings of the IEEE International Conference on Acoustics, Speech, and Signal Processing*, Albuquerque, N.M., vol. 1, pp. 413–416, 1990.

MORSE, P. M., and K. U. INGARD. *Theoretical Acoustics*. New York: McGraw-Hill, 1968.

MURVEIT, H., and M. WEINTRAUB. "1000-word speaker-independent continuous-speech recognition using hidden Markov models," *Proceedings of the IEEE International Conference on Acoustics, Speech, and Signal Processing*, New York, Vol. 1, pp. 115–118, 1988.

MUSICUS, B. R. "An iterative technique for maximum likelihood parameter estimation on noisy data." M.S. thesis, Massachusetts Institute of Technology, June 1979.

MUSICUS, B. R., and J. S. LIM. "Maximum likelihood parameter estimation on noisy data," *Proceedings of the IEEE International Conference on Acoustics, Speech, and Signal Processing*, Washington, D.C., pp. 224–227, Apr. 1979.

MYERS, C. S., and L. R. RABINER. "A level building dynamic time warping algorithm for connected word recognition," *IEEE Transactions on Acoustics, Speech, and Signal Processing*, vol. 29, pp. 284–296, Apr. 1981.

_____. "Connected digit recognition using a level building DTW algorithm,"

IEEE Transactions on Acoustics, Speech, and Signal Processing, vol. 29, pp. 351–363, June 1981.

MYERS, C. S., L. R. RABINER, and A. E. ROSENBERG. "Performance tradeoffs in dynamic time warping algorithms for isolated word recognition," *IEEE Transactions on Acoustics, Speech, and Signal Processing*, vol. 28, pp. 622–635, Dec. 1980.

NAGATA, K., Y. KATO, and S. CHIBA. "Spoken digit recognizer for the Japanese language," *Proceedings of the 4th International Conference on Acoustics*, 1962.

NAKAMURA, M., and K. SHIKANO. "A study of English word category prediction based on neural networks," *Proceedings of the IEEE International Conference on Acoustics, Speech, and Signal Processing*, Glasgow, Scotland, vol. 2, pp. 731–734, 1989.

NAKATSU, H., H. NAGASHIMA, J. KOJIMA et al. "A speech recognition method for telephone voice," *Transactions of the Institute of Electronics, Information, and Computer Engineers* (in Japanese), vol. J66-D, pp. 377–384, Apr. 1983.

NAKATSUI, M., and P. MERMELSTEIN. "Subjective speech-to-noise ratio as a measure of speech quality for digital waveform coders," *Journal of the Acoustical Society of America*, vol. 72, pp. 1136–1144, Oct. 1982.

NAKATSUI, M., and J. SUZUKI. "Method of observation of glottal-source wave using digital inverse filtering in the time domain," *Journal of the Acoustical Society of America*, vol. 47, pp. 664–665, 1970.

NANDKUMAR, S., and J. H. L. HANSEN. "Dual-channel iterative speech enhancement with constraints based on an auditory spectrum," submitted to *IEEE Transactions on Signal Processing*, 1992.

———. "Dual-channel speech enhancement with auditory spectrum based constraints," *Proceedings of the IEEE International Conference on Acoustics, Speech, and Signal Processing*, San Francisco, pp. 36.7.1–4, 1992.

National Institute of Standards and Technology (NIST). "Getting started with the DARPA TIMIT CD-ROM: An acoustic phonetic continuous speech database" (prototype), Gaithersburg, Md., 1988.

NAWAB, S. H., and T. F. QUATIERI. "Short time Fourier transform." Chapter 6 in J. S. Lim and A. V. Oppenheim, eds., *Advanced Topics in Signal Processing.* Englewood Cliffs, N.J.: Prentice Hall, 1988.

NAYLOR, A. W., and G. R. SELL. *Linear Operator Theory.* New York: Holt, Rinehart and Winston, 1971.

NAYLOR, J. A., and S. F. BOLL. "Techniques for suppression of an interfering talker in co-channel speech," *Proceedings of the IEEE International Conference on Acoustics, Speech, and Signal Processing*, Dallas, Tex., pp. 205–208, 1987.

NEY, H. "The use of a one stage dynamic programming algorithm for connected word recognition," *IEEE Transactions on Acoustics, Speech, and Signal Processing*, vol. 32, pp. 263–271, Apr. 1984.

NEY, H., R. HAEB-UMBACH, B. TRAN et al. "Improvements in beam search for 10,000-word continuous speech recognition," *Proceedings of the IEEE International Conference on Acoustics, Speech, and Signal Processing*, San Francisco, vol. I, pp. I-9–I-12, 1992.

NEY, H., D. MERGEL, A. NOLL et al. "A data-driven organization of the dynamic programming beam search for continuous speech recognition," *Proceedings of the IEEE International Conference on Acoustics, Speech, and Signal Processing*, Dallas, Tex., vol. 2, pp. 833–836, 1987.

———. "Data-driven search organization for continuous speech recognition," *IEEE Transactions on Signal Processing*, vol. 40, pp. 272–281, Feb. 1992.

NIEDERJOHN, R. J., and J. H. GROTELUESCHEN. "The enhancement of speech intelligibility in high noise levels by high-pass filtering followed by rapid amplitude compression," *IEEE Transactions on Acoustics, Speech, and Signal Processing*, vol. 24, pp. 277–282, Aug. 1976.

NILES, L. T., and H. F. SILVERMAN. "Combining hidden Markov models and neural network classifiers," *Proceedings of the IEEE International Conference on Acoustics, Speech, and Signal Processing*, Albuquerque, N.M., vol. 1, pp. 417–420, 1990.

NILSSON, N. J. *Learning Machines*. New York: McGraw-Hill, 1965.

———. *Problem Solving Methods in Artificial Intelligence*. New York: McGraw-Hill, 1971.

NIRANJAN, M., and F. FALLSIDE. "Neural networks and radial basis functions in classifying static speech patterns," Technical Report CUED/F-INFENG/TR 22, Cambridge University, Cambridge, England, 1988.

NOBEL, B. *Applied Linear Algebra*. Englewood Cliffs, N.J.: Prentice Hall, 1969.

NOCERINO, N., F. K. SOONG, L. R. RABINER et al. "Comparative study of several distortion measures for speech recognition," *Proceedings of the IEEE International Conference on Acoustics, Speech, and Signal Processing*, Tampa, Fla., vol. 1, pp. 25–28, 1985.

NOLL, A. M. "Cepstrum pitch determination," *Journal of the Acoustical Society of America*, vol. 41, pp. 293–309, Feb. 1967.

OGATA, K. *State Space Analysis of Control Systems*. Englewood Cliffs, N.J.: Prentice Hall, 1967.

OGELSBY, J., and J. S. MASON. "Optimization of neural models for speaker identification," *Proceedings of the IEEE International Conference on Acoustics, Speech, and Signal Processing*, Albuquerque, N.M., vol. 1, pp. 261–264, 1990.

———. "Radial basis function networks for speaker recognition," *Proceedings of the IEEE International Conference on Acoustics, Speech, and Signal Processing*, Toronto, Canada, vol. 1, pp. 393–396, 1991.

OPPENHEIM, A. V. "Generalized superposition," *Informatics and Control*, vol. 11, pp. 528–536, Nov.–Dec. 1967.

———. "A speech analysis-synthesis system based on homomorphic filtering," *Journal of the Acoustical Society of America*, vol. 45, pp. 458–465, 1969.

OPPENHEIM, A. V., and R. W. SCHAFER. "Homomorphic analysis of speech," *IEEE Transactions on Audio and Electroacoustics*, vol. 16, pp. 221–226, June 1968.

———. *Discrete Time Signal Processing*. Englewood Cliffs, N.J.: Prentice Hall, 1989.

OPPENHEIM, A. V., R. W. SCHAFER, and T. G. STOCKHAN, JR. "Nonlinear filtering of multiplied and convolved signals," *Proceedings of the IEEE*, vol. 56, pp. 1264–1291, Aug. 1968.

O'SHAUGHNESSY, D. *Speech Communication: Human and Machine.* Reading, Mass.: Addison-Wesley, 1987.

———. "Speech enhancement using vector quantization and a formant distance measure," *Proceedings of the IEEE International Conference on Acoustics, Speech, and Signal Processing,* New York, pp. 549–552, 1988.

OSTENDORF, M., and S. ROUCOS. "A stochastic segment model for phoneme-based continuous speech recognition," *IEEE Transactions on Acoustics, Speech, and Signal Processing,* vol. 37, pp. 1857–1869, Dec. 1989.

PAESELER, A. "Modification of Earley's algorithm for speech recognition." In H. Niemann et al., eds., *Recent Advances in Speech Understanding and Dialog Systems.* New York: Springer-Verlag, 1988.

PAESELER, A., and H. NEY. "Continuous speech recognition using a stochastic language model," *Proceedings of the IEEE International Conference on Acoustics, Speech, and Signal Processing,* Glasgow, Scotland, vol. 2, pp. 719–722, 1989.

PAEZ, M. D., and T. H. GLISSON. "Minimum mean squared error quantization in speech PCM and DPCM systems," *IEEE Transactions on Communication Theory,* vol. 20, pp. 225–230, Apr. 1972.

PAGET, SIR RICHARD. *Human speech.* London and New York: Harcourt, 1930.

PALIWAL, K. K. "On the performance of the quefrency-weighted cepstral coefficients in vowel recognition," *Speech Communication: An Interdisciplinary Journal,* vol. 1, pp. 151–154, May 1982.

———. "Evaluation of various linear prediction parametric representations in vowel recognition," *Signal Processing,* vol. 4, pp. 323–327, July 1982.

———. "Neural net classifiers for robust speech recognition under noisy environments," *Proceedings of the IEEE International Conference on Acoustics, Speech, and Signal Processing,* Albuquerque, N.M., vol. 1, pp. 429–432, 1990.

PALIWAL, K. K., and A. AARSKOG. "A comparative performance evaluation of pitch estimation methods for TDHS/sub-band coding of speech," *Speech Communication: An Interdisciplinary Journal,* vol. 3, pp. 253–259, 1984.

PALIWAL, K. K., and A. BASU. "A speech enhancement method based on Kalman filtering," *Proceedings of the IEEE International Conference on Acoustics, Speech, and Signal Processing,* Dallas, Tex., vol. 1, pp. 177–180, 1987.

PAPADIMITRIOU, C., and K. STEIGLITZ. *Combinatorial Optimization: Algorithms and Complexity.* Englewood Cliffs, N.J.: Prentice Hall, 1982.

PAPAMICHALIS, P. E. *Practical Approaches to Speech Coding.* Englewood Cliffs, N.J.: Prentice Hall, 1987.

PAPOULIS, A. *Probability, Random Variables, and Stochastic Processes,* 2nd ed. New York: McGraw-Hill, 1984.

PARKER, D. B. "Learning logic," Invention Report S81-64, File 1, Office of Technology Licensing, Stanford University, 1982.

PARSONS, T. W. "Separation of speech from interfering speech by means of harmonic selection," *Journal of the Acoustical Society of America,* vol. 60, no. 4, pp. 911–918, Oct. 1976.

———. *Voice and Speech Processing.* New York: McGraw-Hill, 1986.

PAUL, D. B. "An 800 bps adaptive vector quantization vocoder using a perceptual distance measure," *Proceedings of the IEEE International Conference on Acoustics, Speech, and Signal Processing*, Boston, vol. 1, pp. 73–76, 1983.

PAUL, D. B. "The LINCOLN robust continuous speech recognizer," *Proceedings of the IEEE International Conference on Acoustics, Speech, and Signal Processing*, Glasgow, Scotland, vol. 1, pp. 449–452, 1989.

PAUL, D. B., R. P. LIPPMANN, R. P. CHEN et al. "Robust HMM-based techniques for recognition of speech produced under stress and in noise," *Proceedings of Speech Tech*, New York, 1986.

PAUL, D. B., and E. A. MARTIN. "Speaker stress-resistant continuous speech recognition," *Proceedings of the IEEE International Conference on Acoustics, Speech, and Signal Processing*, New York, vol. I, pp. 283–286, 1988.

PAWATE, B. I., M. L. MCMAHAN, R. H. WIGGINS et al. "Connected word recognizer on a multiprocessor system," *Proceedings of the IEEE International Conference on Acoustics, Speech, and Signal Processing*, Dallas, Tex., vol. 2, pp. 1151-1154, 1987.

PEELING, S. M., and R. K. MOORE. "Experiments in isolated digit recognition using the multi-layer perceptron," Technical Report No. 4073, Royal Speech and Radar Establishment, Malvern, Worcester, England, 1987.

PERKELL, J. *Physiology of Speech Production.* Research Monograph, no. 53, Cambridge, Mass.: M.I.T. Press, 1969.

PERLMUTTER, Y. M., L. D. BRAIDA, R. H. FRAZIER et al. "Evaluation of a speech enhancement system," *Proceedings of the IEEE International Conference on Acoustics, Speech, and Signal Processing*, Hartford, Conn., pp. 212–215, 1977.

PETERSON, G. E., and H. L. BARNEY, "Control methods used in a study of the vowels." *Journal of the Acoustical Society of America*, vol. 24, 175–184, 1952.

PETERSON, T. L., and S. T. BOLL. "Acoustic noise suppression in the context of a perceptual model," *Proceedings of the IEEE International Conference on Acoustics, Speech, and Signal Processing*, Atlanta, Ga., pp. 1086–1088, 1981.

PICACHÉ, G. P. "A Givens rotation algorithm for single channel formant tracking and glottal waveform deconvolution." M.S. thesis, Northeastern University, Boston, 1988.

PICONE, J. "On modeling duration in context in speech recognition," *Proceedings of the IEEE International Conference on Acoustics, Speech, and Signal Processing*, Glasgow, Scotland, vol. 1, pp. 421–424, 1989.

———. "Continuous speech recognition using hidden Markov models," *IEEE Acoustics, Speech, and Signal Processing Magazine*, vol. 7, pp. 26–41, July 1990.

———. Written communication, 1992.

PINEDA, F. "Generalization of back-propagation to recurrent neural networks," *Physical Review Letters*, vol. 18, pp. 2229–2232, 1987.

———. "Generalization of back-propagation to recurrent and high-order neural networks." In D. Anderson ed., *Proceedings of the 1987 IEEE Conference on Neural Processing Systems—Natural and Synthetic.* New York: American Institute of Physics, pp. 602–611, 1988.

POLYDOROS, A., and A. T. FAM. "The differential cepstrum: Definition and properties," *Proceedings of the IEEE International Symposium on Circuits and Systems*, vol. 1, pp. 77–80, 1981.

PORITZ, A. M. "Hidden Markov models: A guided tour," *Proceedings of the IEEE International Conference on Acoustics, Speech, and Signal Processing*, New York, vol. 1, pp. 7–13, 1988.

PORTNOFF, M. R. "Representations of signals and systems based on the short-time Fourier transform," *IEEE Transactions on Acoustics, Speech, and Signal Processing*, vol. 28, pp. 55–69, Feb. 1980.

PORTNOFF, M. R., and R. W. SCHAFER, "Mathematical considerations in digital simulations of the vocal tract" (abstract), *Journal of the Acoustical Society of America*, vol. 53, p. 294, Jan. 1973.

POTTER, R. G., G. A. KOPP, and H. G. KOPP. *Visible Speech*. New York: Van Nostrand, 1947. Reprinted: New York: Dover, 1966.

POWELL, G. A., P. DARLINGTON, and P. D. WHEELER. "Practical adaptive noise reduction in the aircraft cockpit environment," *Proceedings of the IEEE International Conference on Acoustics, Speech, and Signal Processing*, Dallas, Tex., pp. 173–176, 1987.

PREUSS, R. D. "A frequency domain noise cancellation preprocessor for narrowband speech communications systems," *Proceedings of the IEEE International Conference on Acoustics, Speech, and Signal Processing*, Washington, D.C., pp. 212–215, 1979.

PRICE, P. J., W. FISHER, J. BERNSTEIN et al. "A database for continuous speech recognition in a 1000-word domain," *Proceedings of the IEEE International Conference on Acoustics, Speech, and Signal Processing*, New York, vol. 1, pp. 651–654, 1988.

PROAKIS, J. G. *Digital Communications*, 2nd ed. New York: McGraw-Hill, 1989.

PROAKIS, J. G., and D. G. MANOLAKIS. *Digital Signal Processing: Principles, Algorithms, and Applications*, 2nd ed. New York: Macmillan, 1992.

PROAKIS, J. G., C. RADER, F. LING et al. *Advanced Topics in Signal Processing*. New York: Macmillan, 1992.

PRUZANSKY, S. "Pattern-matching procedure for automatic talker recognition," *Journal of the Acoustical Society of America*, vol. 35, pp. 354–358, 1963.

PULSIPHER, D. C., S. F. BOLL, C. K. RUSHFORTH et al. "Reduction of nonstationary acoustic noise in speech using LMS adaptive noise cancelling," *Proceedings of the IEEE International Conference on Acoustics, Speech, and Signal Processing*, Washington, D.C., pp. 204–208, 1979.

PUTNINS, Z. A., G. A. WILSON, I. KOMAR et al. "A multi-pulse LPC synthesizer for telecommunications use," *Proceedings of the IEEE International Conference on Acoustics, Speech, and Signal Processing*, Tampa, Fla., pp. 989–992, 1985.

QUACKENBUSH, S. R. "Objective measures of speech quality." Ph.D. dissertation, Georgia Institute of Technology, 1985.

QUACKENBUSH, S. R., T. P. BARNWELL, and M. A. CLEMENTS. *Objective Measures of Speech Quality*. Englewood Cliffs, N.J.: Prentice Hall, 1988.

QUATIERI, T. F. "Minimum and mixed phase speech analysis—synthesis by adaptive homomorphic deconvolution," *IEEE Transactions on Acoustics, Speech, and Signal Processing*, vol. 27, pp. 328–335, Aug. 1979.

QUENOT, G., J. L. GAUVAIN, J. J. GANGOLF et al. "A dynamic time warp VLSI processor for continuous speech recognition," *Proceedings of the IEEE International Conference on Acoustics, Speech, and Signal Processing*, Tokyo, Japan, vol. 3, pp. 1549–1552, 1986.

RABINER, L. R. "On the use of autocorrelation analysis for pitch detection," *IEEE Transactions on Acoustics, Speech, and Signal Processing*, vol. 26, pp. 24–33, Feb. 1977.

———. "On creating reference templates for speaker independent recognition of isolated words," *IEEE Transactions on Acoustics, Speech, and Signal Processing*, vol. 26, pp. 34–42, Feb. 1978.

———. "A tutorial on hidden Markov models and selected applications in speech recognition," *Proceedings of the IEEE*, vol. 77, pp. 257–285, Feb. 1989.

RABINER, L. R., and S. E. LEVINSON. "Isolated and connected word recognition: Theory and selected applications," *IEEE Transactions on Communications*, vol. 29, pp. 621–659, May 1981.

RABINER, L. R., S. E. LEVINSON, A. E. ROSENBERG et al. "Speaker-independent recognition of isolated words using clustering techniques," *IEEE Transactions on Acoustics, Speech, and Signal Processing*, vol. 27, pp. 336–349, Aug. 1979.

RABINER, L. R., S. E. LEVINSON, and M. M. SONDHI. "On the application of vector quantization and hidden Markov models to speaker-independent isolated word recognition," *Bell System Technical Journal*, vol. 62, pp. 1075–1105, 1983.

RABINER, L. R., A. E. ROSENBERG, and S. E. LEVINSON. "Considerations in dynamic time warping algorithms for discrete utterance recognition," *IEEE Transactions on Acoustics, Speech, and Signal Processing*, vol. 26, pp. 575–582, Dec. 1978.

RABINER, L. R., and M. R. SAMBUR. "An algorithm for determining the endpoints of isolated utterances," *Bell System Technical Journal*, vol. 54, pp. 297–315, Feb. 1975.

RABINER, L. R., M. SAMBUR, and C. SCHMIDT. "Applications of a non-linear smoothing algorithm to speech processing," *IEEE Transactions on Acoustics, Speech, and Signal Processing*, vol. 23, pp. 552–557, Dec. 1975.

RABINER, L. R., and R. W. SCHAFER. *Digital Processing of Speech Signals*. Englewood Cliffs, N.J.: Prentice Hall, 1978.

RABINER, L. R., and J. G. WILPON. "Applications of clustering techniques to speaker-trained isolated word recognition," *Bell System Technical Journal*, vol. 58, pp. 2217–2231, 1979.

———. "A simplified, robust training procedure for speaker trained, isolated word recognition systems," *Journal of the Acoustical Society of America*, vol. 68, pp. 1271–1276, 1980.

RABINER, L. R., J. G. WILPON, and B.-H. JUANG. "A segmental *k*-means training procedure for connected word recognition based on whole word reference patterns," *AT&T System Technical Journal*, vol. 65, pp. 21–31, May–June 1986.

———. "A model based connected-digit recognition system using either hidden

Markov models or templates," *Computer Speech and Language*, vol. 1, pp. 167–197, Dec. 1986.

RABINER, L. R., J. G. WILPON, A. QUINN et al. "On the application of embedded digit training to speaker independent connected digit training," *IEEE Transactions on Acoustics, Speech, and Signal Processing*, vol. 32, pp. 272–280, Apr. 1984.

RABINER, L. R., J. G. WILPON, and F. K. SOONG. "High performance connected digit recognition using hidden Markov models," *IEEE Transactions on Acoustics, Speech, and Signal Processing*, vol. 37, pp. 1214–1225, Aug. 1989.

RADER, C. M., and S. SUNDAMURTHY. "Wafer scale systolic array for adaptive antenna processing," *Proceedings of the IEEE International Conference on Acoustics, Speech, and Signal Processing*, New York, vol. IV, pp. 2069–2071, 1988.

RAHIM, M., and C. GOODYEAR. "Articulatory synthesis with the aid of a neural net," *Proceedings of the IEEE International Conference on Acoustics, Speech, and Signal Processing*, Glasgow, Scotland, vol. 1, pp. 227–230, 1989.

RAMESH, P., S. KATAGIRI, and C.-H. LEE. "A new connected word recognition algorithm based on HMM/LVQ segmentation and LVQ classification," *Proceedings of the IEEE International Conference on Acoustics, Speech, and Signal Processing*, Toronto, Canada, vol. 1, pp. 113–116, 1991.

REDDY, D. R. "Speech recognition by machine: A review," *Proceedings of the IEEE*, vol. 64, pp. 501–531, Apr. 1976. Reprinted in (Dixon and Martin, 1989) and (Waibel and Lee, 1990).

———. "Words into action II: A task oriented system," *IEEE Spectrum*, vol. 17, pp. 26–28, June 1980.

REED, F. A., and P. L. FEINTUCH. "A comparison of LMS adaptive cancellers implemented in the frequency domain and the time domain," *IEEE Transactions on Acoustics, Speech, and Signal Processing*, vol. 29, pp. 770–775, June 1981.

RENALS, S., N. MORGAN, M. COHEN et al. "Connectionist probability estimation in the DECIPHER speech recognition system," *Proceedings of the IEEE International Conference on Acoustics, Speech, and Signal Processing*, San Francisco, vol. I, pp. I-601–I-604, 1992.

RICHARDS, D. L., and J. SWAFFIELD, "Assessment of speech communication links," *Proceedings of the IEE*, vol. 106, pp. 77–89, Mar. 1959.

RITEA, B. "Automatic speech understanding systems," *Proceedings of the 11th IEEE Computer Society Conference*, Washington, D.C., pp. 319–322, 1975.

ROBBINS, H., and S. MUNRO. "A stochastic approximation method," *Annals of Mathematical Statistics*, vol. 22, pp. 400–407, 1951.

ROBERTS, L. G. "Picture coding using pseudo-random noise," *IRE Transactions on Information Theory*, vol. 8, pp. 145–154, Feb. 1962.

ROBINSON, A. J., and F. FALLSIDE. "A recurrent error propagation network speech recognition system," *Computer Speech and Language*, vol. 5, July 1991.

ROBINSON, E. A. *Statistical Communication and Detection*. New York: Hafner, 1967.

RODRIGUEZ, J. J., J. S. LIM, and E. SINGER. "Adaptive noise reduction in aircraft communication systems," *Proceedings of the IEEE International Conference on Acoustics, Speech, and Signal Processing*, Dallas, Tex., pp. 169–172, 1987.

ROE, D. B., A. L. GORIN, and P. RAMESH. "Incorporating syntax into the level-building algorithm on a tree-structured parallel computer," *Proceedings of the IEEE International Conference on Acoustics, Speech, and Signal Processing*, Glasgow, Scotland, vol. 2, pp. 778–781, 1989.

ROHWER, R., and B. FORREST. "Training time dependence in neural networks," *Proceedings of the 1st IEEE International Conference on Neural Networks*, San Diego, Calif., vol. II, pp. 701–708, 1987.

ROSENBERG, A. E. "Effects of glottal pulse shape on the quality of natural vowels," *Journal of the Acoustical Society of America*, vol. 49, pp. 583–590, Feb. 1971.

ROSENBERG, A. E., L. R. RABINER, J. G. WILPON et al. "Demisyllable-based isolated word recognition system," *IEEE Transactions on Acoustics, Speech, and Signal Processing*, vol. 31, pp. 713–726, June 1983.

ROSENBLATT, F. "The perceptron: A perceiving and recognizing automaton," Cornell Aeronautical Laboratory Report 85-460-1, 1957.

———. *Principles of Neurodynamics.* Washington, D.C.: Spartan Books, 1962.

ROTHENBERG, M. "The glottal volume velocity waveform during loose and tight voiced glottal adjustment," *Actes du 7ème Congrès International des Sciences Phonétiques*, Montreal, Canada, pp. 380–388, 1972.

ROUCOS, S., and M. L. DUNHAM. "A stochastic segment model for phoneme-based continuous speech recognition," *Proceedings of the IEEE International Conference on Acoustics, Speech, and Signal Processing*, Dallas, Tex., vol. 1, pp. 73–76, 1987.

ROUCOS, S., R. SCHWARTZ, and J. MAKHOUL. "Segment quantization for very-low-rate speech coding," *Proceedings of the IEEE International Conference on Acoustics, Speech, and Signal Processing*, Paris, pp. 1565–1569, 1982.

RUMELHART, D. E., G. E. HINTON, and R. J. WILLIAMS. "Learning internal representations by error propagation." Chapter 8 in D. E. Rumelhart and J. L. McClelland, eds., *Parallel Distributed Processing*—Vol. 1: *Foundations.* Cambridge, Mass: M.I.T. Press, 1986.

RUSSELL, M. J., and R. K. MOORE. "Explicit modelling of state occupancy in hidden Markov models for automatic speech recognition," *Proceedings of the IEEE International Conference on Acoustics, Speech, and Signal Processing*, Tampa, Fla., vol. 1, pp. 5–8, 1985.

SAKOE, H. "Two-level DP matching: A dynamic programming based pattern recognition algorithm for connected word recognition," *IEEE Transactions on Acoustics, Speech, and Signal Processing*, vol. 27, pp. 588–595, Dec. 1979.

SAKOE, H., and S. CHIBA. "Dynamic programming algorithm optimization for spoken word recognition," *IEEE Transactions on Acoustics, Speech, and Signal Processing*, vol. 26, pp. 43–49, Feb. 1978.

SAKOE, H., R. ISOTANI, K. YOSHIDA et al. "Speaker-independent word recognition using dynamic programming neural networks," *Proceedings of the IEEE International Conference on Acoustics, Speech, and Signal Processing*, Glasgow, Scotland, vol. 1, pp. 29–32, 1989.

SAMAD, T., and P. HARPER. "Associative memory storage using a variant of the generalized delta rule," *Proceedings of the 1st IEEE International Conference on Neural Networks*, San Diego, Calif., vol. III, pp. 173–184, 1987.

SAMBUR, M. R. "LMS adaptive filtering for enhancing the quality of noisy speech," *Proceedings of the IEEE International Conference on Acoustics, Speech, and Signal Processing*, Tulsa, Okla., pp. 610–613, Apr. 1978.

———. "Adaptive noise canceling for speech signals," *IEEE Transactions on Acoustics, Speech, and Signal Processing*, vol. 26, no. 5, pp. 419–423, Oct. 1978.

SAMBUR, M. R., and L. R. RABINER. "A statistical decision approach to recognition of connected digits," *IEEE Transactions on Acoustics, Speech, and Signal Processing*, vol. 24, pp. 550–558, 1976.

SAWAI, H. "TDNN-LR continuous speech recognition system using adaptive incremental TDNN training," *Proceedings of the IEEE International Conference on Acoustics, Speech, and Signal Processing*, Toronto, Canada, vol. 1, pp. 53–56, 1991.

SCHAFER, R. W. "Echo removal by generalized linear filtering." Ph.D. dissertation, Massachusetts Institute of Technology, 1968.

SCHAFER, R. W., and J. D. MARKEL, eds. *Speech Analysis*. New York: John Wiley & Sons, 1979.

SCHAFER, R. W., and L. R. RABINER. "System for automatic formant analysis of voiced speech," *Journal of the Acoustical Society of America*, vol. 47, pp. 634–648, Feb. 1970.

SCHARF, B. "Critical bands." In J. V. Tobias, ed., *Foundations of Modern Auditory Theory*. New York: Academic Press, pp. 157–202, 1970.

SCHROEDER, M. R. "Vocoders: Analysis and synthesis of speech," *Proceedings of the IEEE*, vol. 54, pp. 720–734, May 1966.

———. "Period histogram and product spectrum: New methods for fundamental frequency measurement," *Journal of the Acoustical Society of America*, vol. 43, pp. 829–834, Apr. 1968.

———. "Recognition of complex acoustic signals," *Life Science Research Reports*, vol. 55, pp. 323–328, 1977.

SCHROEDER, M. R., and B. S. ATAL. "Generalized short-time power spectra and autocorrelation," *Journal of the Acoustical Society of America*, vol. 34, pp. 1679–1683, Nov. 1962.

———. Code-excited linear prediction (CELP): High-quality speech at very low bit rates," *Proceedings of the IEEE International Conference on Acoustics, Speech, and Signal Processing*, Tampa, Fla., pp. 937–940, 1985.

SCHWARTZ, R., S. AUSTIN, F. KUBALA et al. "New uses for the *N*-best sentence hypotheses within the BYBLOS speech recognition system," *Proceedings of the IEEE International Conference on Acoustics, Speech, and Signal Processing*, San Francisco, vol. 1, pp. I-1–I-4, 1992.

SCHWARTZ, R. M., and Y. L. CHOW. "The *N*-best algorithm: An efficient and exact procedure for finding the *N* most likely sentence hypotheses," *Proceedings of the IEEE International Conference on Acoustics, Speech, and Signal Processing*, Albuquerque, N.M., vol. 1, pp. 81–84, 1990.

SCHWARTZ, R. M., Y. L. CHOW, O. A. KIMBALL et al. "Context dependent modeling for acoustic-phonetic recognition of continuous speech," *Proceedings of the IEEE International Conference on Acoustics, Speech, and Signal Processing*, Tampa, Fla., pp. 1205–1208, 1985.

SCHWARTZ, R. M., Y. L. CHOW, S. ROUCOS et al. "Improved hidden Markov modelling of phonemes for continuous speech recognition," *Proceedings of the IEEE International Conference on Acoustics, Speech, and Signal Processing*, San Diego, Calif., vol. 3, paper 35.6, 1984.

SCHWARTZ, R. M., J. KLOVSTAD, J. MAKHOUL et al. "A preliminary design of a phonetic vocoder based on a diphone model," *Proceedings of the IEEE International Conference on Acoustics, Speech, and Signal Processing*, Denver, Colo., pp. 32–35, 1980.

SCORDILIS, M. S., and J. N. GOWDY. "Neural network based generation of fundamental frequency contours," *Proceedings of the IEEE International Conference on Acoustics, Speech, and Signal Processing*, Glasgow, Scotland, vol. 1, pp. 219–222, 1989.

SEJNOWSKI, T. J., and C. R. ROSENBERG. "NETtalk: A parallel network that learns to read aloud," Technical Report JHU/EECS-86/01, Johns Hopkins University, 1986.

SENEFF, S. "Modifications to formant tracking algorithm of April 1974," *IEEE Transactions on Acoustics, Speech, and Signal Processing*, vol. 24, pp. 192–193, Apr. 1976.

SHANNON, C. E. "A mathematical theory of communication," *Bell System Technical Journal*, vol. 27, pp. 379–423 and 623–656, 1948.

_____. "A coding theorem for a discrete source with a fidelity criterion," *IRE National Convention Record*, part 4, pp. 142–163, Mar. 1959.

SHERWOOD, L. *Human Physiology*. St. Paul, Minn.: West Publishing, 1989.

SHIEBER, S. M. "An introduction to unification-based approaches to grammar," CSLI Lecture Notes no. 4, Center for the Study of Language and Information, Stanford University, 1986.

SHIKANO, K. "Evaluation of LPC spectral matching measures for phonetic unit recognition" (technical report), Computer Science Department, Carnegie–Mellon University, May 1985.

SHORE, J. E., and D. K. BURTON. "Discrete utterance speech recognition without time alignment," *IEEE Transactions on Information Theory*, vol. 29, pp. 473–491, 1983.

SILVERMAN, H. F., and D. P. MORGAN. "The application of dynamic programming to connected speech recognition," *IEEE Acoustics, Speech, and Signal Processing Magazine*, vol. 7, pp. 6–25, July 1990.

SIMPSON, P. K. *Artificial Neural Systems*. New York: Pergamon Press, 1990.

SINGER, E., and R. LIPPMANN. "A speech recognizer using radial basis function neutral networks in an HMM framework," *Proceedings of the IEEE International Conference on Acoustics, Speech, and Signal Processing*, San Francisco, vol. I, pp. I-629–I-632, 1992.

SINGHAL, S., and B. S. ATAL. "Improving performance of multi-pulse LPC coders at low bit rates," *Proceedings of the IEEE International Conference on Acoustics, Speech, and Signal Processing*, San Diego, Calif., vol. 1, pp. 131–134, 1984.

SMITH, M. J. T., and T. P. BARNWELL. "A procedure for designing exact reconstruction filter banks for tree structured subband coders," *Proceedings of the IEEE International Conference on Acoustics, Speech, and Signal Processing*, San Diego, Calif., pp. 27.1.1–2.7.1.4, Mar. 1984.

SONDHI, M. M. "An adaptive echo canceller," *Bell System Technical Journal*, vol. 46, pp. 497–511, 1967.

_____. *Closed Loop Adaptive Echo Canceller Using Generalized Filter Networks*. U.S. Patent 3,499,999, Mar. 10, 1970.

_____. "Model for wave propagation in a lossy vocal tract," *Journal of the Acoustical Society of America*, vol. 55, pp. 1070–1075, May 1974.

_____. "New methods of pitch extraction," *IEEE Transactions on Audio and Electroacoustics*, vol. 16, pp. 262–268, June 1968.

_____. "Resonances of a bent vocal tract," *Journal of the Acoustical Society of America*, vol. 79, pp. 1113–1116, Apr. 1986.

SOONG, F. K., and B.-H. JUANG. "Line spectrum pair and speech compression," *Proceedings of the IEEE International Conference on Acoustics, Speech, and Signal Processing*, San Diego, Calif., vol. 1, pp. 1.10.1–4, 1984.

SOONG, F. K., and A. E. ROSENBERG. "On the use of instantaneous and transitional spectral information in speaker recognition," *Proceedings of the IEEE International Conference on Acoustics, Speech, and Signal Processing*, Tokyo, Japan, vol. 2, pp. 877–890, 1986.

STANKOVIĆ, S. S., and M. M. MILOSAVLJEVIĆ. "Training of multi-layer perceptrons by stochastic approximation." Chapter 7 in vol. IV of P. Antognetti and V. Milutinović, eds., *Neural Networks: Concepts, Applications, and Implementations*, Englewood Cliffs, N.J.: Prentice Hall, 1991.

STEIGLITZ, K. "On the simultaneous estimation of poles and zeros in speech analysis," *IEEE Transactions on Acoustics, Speech, and Signal Processing*, vol. 25, pp. 429–433, Oct. 1977.

STEIGLITZ, K., and B. DICKINSON. "The use of time domain selection for improved linear prediction," *IEEE Transactions on Acoustics, Speech, and Signal Processing*, vol. 25, pp. 34–39, Jan. 1977.

STEINBISS, V., A. NOLL, A. PAESELER et al. "A 10,000-word continuous speech recognition system," *Proceedings of the IEEE International Conference on Acoustics, Speech, and Signal Processing*, Albuquerque, N.M., vol. 1, pp. 57–60, 1990.

STEPHENS, R. W. B., and A. E. BATE. *Acoustics and Vibrational Physics*. New York: St. Martin's Press, 1966.

STEVENS, K. N., and A. S. HOUSE. "An acoustical theory of vowel production and some implications," *Journal of Speech and Hearing Research*, vol. 4, p. 303, 1961.

_____. "Development of a quantitative description of vowel articulation," *Journal of the Acoustical Society of America*, vol. 27, pp. 484–493, 1955.

STEVENS, S. S., and J. VOLKMAN. "The relation of pitch to frequency," *American Journal of Psychology*, vol. 53, p. 329, 1940.

STEWART, J. Q. "An electrical analogue of the vocal cords," *Nature*, vol. 110, p. 311, 1922.

STOCKHAM, T. G. "The application of generalized linearity to automatic gain control," *IEEE Transactions on Audio and Electroacoustics*, vol. 16, pp. 828–842, June 1968.

STROBACH, P. "New forms of Levinson and Schur algorithms," *IEEE Signal Processing Magazine*, pp. 12–36, Jan. 1991.

STRUBE, H. W. "Determination of the instant of glottal closure from the speech wave," *Journal of the Acoustical Society of America*, vol. 56, pp. 1625–1629, 1974.

SUGAMURA, N., and F. ITAKURA. "Speech data compression by LSP analysis-synthesis technique," *Transactions of the Institute of Electronics, Information, and Computer Engineers*, vol. J64-A, pp. 599–606, 1981.

SUGIYAMA, M., and K. SHIKANO. "LPC peak weighted spectral matching measure," *Transactions of the Institute of Electronics, Information, and Computer Engineers*, vol. J64-A, pp. 409–416, 1981.

SUN, G. Z., H. H. CHEN, Y. C. LEE et al. "Recurrent neural networks, hidden Markov models, and stochastic grammars," *Proceedings of the International Joint Conference on Neural Networks*, San Diego, Calif., vol. I, pp. 729–734, June 1990.

SWINGLER, D. N. "Frequency errors in MEM processing," *IEEE Transactions on Acoustics, Speech, and Signal Processing*, vol. 28, pp. 257–259, Apr. 1980.

TAKAHASHI, J. I., S. HATTORI, T. KIMURA et al. "A ring array processor architecture for highly parallel dynamic time warping," *IEEE Transactions on Acoustics, Speech, and Signal Processing*, vol. 34, pp. 1302–1309, Oct. 1986.

TAMURA, S. "An analysis of a noise reduction neural network," *Proceedings of the IEEE International Conference on Acoustics, Speech, and Signal Processing*, Glasgow, Scotland, vol. 3, pp. 2001–2004, 1989.

TAMURA, S., and A. WAIBEL. "Noise reduction using connectionist models," *Proceedings of the IEEE International Conference on Acoustics, Speech, and Signal Processing*, New York, vol. 1, pp. 553–556, 1988.

TANAKA, E., and K. S. FU. "Error correcting parsers for formal languages," *IEEE Transactions on Computers*, vol. 27, pp. 605–615, July 1978.

TANIGUCHI, T., S. UNAGAMI, and R. M. GRAY. "Multimode coding: Applications to CELP," *Proceedings of the IEEE International Conference on Acoustics, Speech, and Signal Processing*, Glasgow, Scotland, vol. 1, pp. 156–159, 1989.

TATE, C. N., and C. C. GOODYEAR. "Note on the convergence of linear predictive filters, adapted using the LMS algorithm," *IEE Transactions*, vol. 130, pp. 61–64, Apr. 1983.

TEAGER, H. M. "Some observations on oral air flow during phonation," *IEEE Transactions on Acoustics, Speech, and Signal Processing*, vol. 28, pp. 599–601, Oct. 1980.

TEAGER, H. M., and S. M. Teager. "A phenomenological model for vowel production in the vocal tract." In R. G. Daniloff, ed., *Speech Sciences: Recent Advances*. San Diego, Calif.: College-Hill Press, pp. 73–109, 1983.

———. "Evidence for nonlinear production mechanisms in the vocal tract," *NATO Advanced Study Institute, Speech Production and Modelling*, Chateau Bonas, France, July 17-29. 1989. Also in W. J. Hardcastle and A. Marchal, eds. *Proceedings of the NATO ASI*, Norwell, Mass.: Kluwer, 1990.

TEBELSKIS, J., and A. WAIBEL. "Large vocabulary recognition using linked predictive neural networks," *Proceedings of the IEEE International Conference on Acoustics, Speech, and Signal Processing*, Albuquerque, N.M., vol. 1, pp. 437–440, 1990.

THOMAS, I. B. "The influence of first and second formants on the intelligibility of clipped speech," *Journal of the Audio Engineering Society*, vol. 16, pp. 182–185, Apr. 1968.

THOMAS, I. B., and R. J. NIEDERJOHN. "Enhancement of speech intelligibility at high noise levels by filtering and clipping," *Journal of the Audio Engineering Society*, vol. 16, pp. 412–415, Oct. 1968.

———. "The intelligibility of filtered-clipped speech in noise," *Journal of the Audio Engineering Society*, vol. 18, pp. 299–303, June 1970.

THOMAS, I. B., and W. J. OHLEY. "Intelligibility enhancement through spectral weighting," *IEEE Conference on Speech Communications and Processing*, pp. 360–363, 1972.

THOMAS, I. B., and A. RAVINDRAN. "Intelligibility enhancement of already noisy speech signals," *Journal of the Audio Engineering Society*, vol. 22, pp. 234–236, May 1974.

TIMKE, R. H., H. VONLEDEN, and P. MOORE. "Laryngeal vibrations: Measurement of the glottic wave," *American Medical Association Archives of Otolaryngology*, vol. 68, pp. 1–19, July 1948.

TOHKURA, Y. "A weighted cepstral distance measure for speech recognition," *IEEE Transactions on Acoustics, Speech, and Signal Processing*, vol. 35, pp. 1414–1422, Oct. 1987.

TREMAIN, T. E. "The government standard linear predictive coding algorithm: LPC-10," *Speech Technology*, vol. 1, pp. 40–49, Apr. 1982.

TRIBOLET, J. M. "A new phase unwrapping algorithm," *IEEE Transactions on Acoustics, Speech, and Signal Processing*, vol. 25, pp. 170–177, Apr. 1977.

TRIBOLET, J. M., P. NOLL, B. J. MCDERMOTT et al. "A study of complexity and quality of speech waveform coders," *Proceedings of the IEEE International Conference on Acoustics, Speech, and Signal Processing*, Tulsa, Okla., pp. 586–590, 1978.

TSENG, H. P., M. J. SABIN, and E. A. LEE. "Fuzzy vector quantization applied to hidden Markov modeling," *Proceedings of the IEEE International Conference on Acoustics, Speech, and Signal Processing*, Dallas, Tex., vol. 2, pp. 641–644, 1987.

TSYPKIN, YA. Z. *Foundations of the Theory of Learning Systems.* Translated by Z. J. Nikolic. Orlando, Fla.: Academic Press, 1973.

UN, C. K., and K. Y. CHOI. "Improving LPC analysis of noisy speech by autocorrelation subtraction method," *Proceedings of the IEEE International Conference on Acoustics, Speech, and Signal Processing*, Atlanta, Ga., pp. 1082–1085, 1981.

VAIDYANATHAN, P. P. "Quadrature mirror filter banks, *M*-band extensions and perfect reconstruction techniques," *IEEE Acoustics, Speech, and Signal Processing Magazine*, vol. 4, pp. 4–20, July 1987.

———. "Multirate digital filters, filter banks, polyphase networks and applications," *Proceedings of the IEEE*, vol. 78, pp. 56–93, Jan. 1990.

VARNER, L. W., T. A. MILLER, and T. E. EGER. "A simple adaptive filtering technique for speech enhancement," *Proceedings of the IEEE International Conference on Acoustics, Speech, and Signal Processing*, Boston, pp. 24.3.1–4, 1983.

VEENEMAN, D. E., and S. L. BEMENT. "Automatic glottal inverse filtering of speech," *IEEE Transactions on Acoustics, Speech, and Signal Processing*, vol. 33, pp. 369–377, Apr. 1985.

VENKATESH, C. G., J. R. DELLER, and C. C. CHIU. "A graph partitioning approach to signal decoding" (technical report), Speech Processing Laboratory, Department of Electrical Engineering, Michigan State University, Aug. 1990.

VERHELST, W., and O. STEENHAUT. "A new model for the complex cepstrum of voiced speech," *IEEE Transactions on Acoustics, Speech, and Signal Processing*, vol. 34, pp. 43–51, Feb. 1986.

———. "On short-time cepstra of voiced speech," *Proceedings of the IEEE International Conference on Acoustics, Speech, and Signal Processing*, New York, vol. 1, pp. 311–314, 1988.

VINTSYUK, T. K. "Element-wise recognition of continuous speech composed of words from a specified dictionary," *Kibernetika*, vol. 7, pp. 133–143, Mar.–Apr. 1971.

VISWANATHAN, R., and J. MAKHOUL. "Quantization properties of the transmission parameters in linear predictive systems," *IEEE Transactions on Acoustics, Speech, and Signal Processing*, vol. 23, pp. 309–321, June 1975.

VISWANATHAN, V. R., J. MAKHOUL, and W. H. RUSSELL. "Towards perceptually consistent measures of spectral distance," *Proceedings of the IEEE International Conference on Acoustics, Speech, and Signal Processing*, Philadelphia, pp. 485–488, 1976.

VISWANATHAN, V. R., W. H. RUSSELL, and A. W. HUGGINS. "Objective speech quality evaluation of medium and narrowband real-time speech coders," *Proceedings of the IEEE International Conference on Acoustics, Speech, and Signal Processing*, Boston, pp. 543–546, 1983.

VITERBI, A. J. "Error bounds for convolutional codes and an asymptotically optimal decoding algorithm," *IEEE Transactions on Information Theory*, vol. 13, pp. 260–269, Apr. 1967.

VITERBI, A. J., and J. K. OMURA. *Principles of Digital Communication*. New York: McGraw-Hill, 1979.

VOIERS, W. D. "Perceptual bases of speaker identity," *Journal of the Acoustical Society of America*, vol. 36, pp. 1065–1073, June 1964.

———. "Diagnostic acceptability measure for speech communication systems," *Proceedings of the IEEE International Conference on Acoustics, Speech, and Signal Processing*, Hartford, Conn., pp. 204–207, 1977.

———. "Diagnostic evaluation of speech intelligibility." In M. E. Hawley, ed., *Speech Intelligibility and Speaker Recognition*. Stroudsburg, Pa.: Dowden, Hutchenson, and Ross, pp. 374–387, 1977.

———. "Methods of predicting user acceptance of voice communications systems," Final Report, Dynastat, Inc., DCA100-74-C-0056, July 1976.

———. "Interdependencies among measures of speech intelligibility and speech quality," *Proceedings of the IEEE International Conference on Acoustics, Speech, and Signal Processing*, Denver, Colo., pp. 703–705, 1980.

WAGNER, K. W. "Ein neues elektrisches Sprechgerät zur Nachbildung der menschlichen Vokale," *Abhandl. d. Preuss. Akad d. Wissenschaft*, 1936.

WAIBEL, A., T. HANAZAWA, G. HINTON et al. "Phoneme recognition using time-delay neural networks," *IEEE Transactions on Acoustics, Speech, and Signal Processing*, vol. 37, pp. 328–339, Mar. 1989. Reprinted in (Waibel and Lee, 1990).

WAIBEL, A., and K.-F. LEE, eds. *Readings in Speech Recognition*. Palo Alto, Calif.: Morgan–Kauffmann, 1990.

WAIBEL, A., H. SAWAI, and K. SHIKANO. "Consonant recognition by modular construction of large phonemic time-delay neural networks," *Proceedings of the IEEE International Conference on Acoustics, Speech, and Signal Processing*, Glasgow, Scotland, vol. 1, pp. 112–115, 1989. Reprinted in (Waibel and Lee, 1990).

———. "Modularity and scaling in large phonemic time-delay neural networks," *IEEE Transactions on Acoustics, Speech, and Signal Processing*, vol. 37, pp. 1888–1898, Dec. 1989.

WANG, D. L., and J. S. LIM. "The unimportance of phase in speech enhancement," *IEEE Transactions on Acoustics, Speech, and Signal Processing*, vol. 30, pp. 679–681, Aug. 1982.

WATROUS, R., B. LADENDORF, and G. KUHN. "Complete gradient optimization of a recurrent network applied to /b, g, d/ discrimination," *Journal of the Acoustical Society of America*, vol. 37, pp. 1301–1309, Mar. 1990.

WATROUS, R., and L. SHASTRI. "Learning phonetic features using connectionist networks," *Proceedings of the 1st IEEE International Conference on Neural Networks*, San Diego, Calif., vol. IV, pp. 381–388, 1987.

WEINER, N. *Extrapolation, Interpolation, and Smoothing of Stationary Time Series*. Cambridge, Mass.: M.I.T. Press, 1949.

WEINTRAUB, M., H. MURVEIT, M. COHEN et al. "Linguistic constraints in hidden Markov model based speech recognition," *Proceedings of the IEEE International Conference on Acoustics, Speech, and Signal Processing*, Glasgow, Scotland, vol. 2, pp. 699–702, 1989.

WEISS, M. R., and E. ASCHKENASY. "Computerized audio processor," Rome Air Development Center, Final Report RADC-TR-83-109, May 1983.

WEISS, M. R., E. ASCHKENASY, and T. W. PARSONS. "Study and development of the INTEL technique for improving speech intelligibility," Nicolet Scientific Corp., Final Report NSC-FR/4023, Dec. 1974.

WERBOS, P. "Beyond regression: New tools for prediction and analysis in the behavioral sciences." Ph.D. dissertation, Harvard University, 1974.

The Scientific Papers of Sir Charles Wheatstone, London and Westminster Review, vol. 28, 1879. Also *Proceedings of the British Association of Advanced Science Notices*, p. 14, 1835.

WHITE, H. "Learning in artificial neural networks: A statistical perspective," *Neural Computation*, vol. 1, pp. 425–469, 1989.

WIDROW, B., J. R. GROVER, J. M. MCCOOL et al. "Adaptive noise canceling: Principles and applications," *Proceedings of the IEEE*, vol. 63, pp. 1692–1716, Dec. 1975.

WIDROW, B., and M. E. HOFF. "Adaptive switching circuits," *IRE WESCON Convention Record*, pp. 96–104, 1960.

WIDROW, B., J. M. MCCOOL, and M. BALL. "The complex LMS algorithm," *Proceedings of the IEEE*, vol. 63, pp. 719–720, Apr. 1975.

WIDROW, B., J. M. MCCOOL, M. G. LARIMORE et al. "Stationary and nonstationary learning characteristics of the LMS adaptive filter," *Proceedings of the IEEE*, vol. 64, pp. 1151–1162, Aug. 1976.

WIDROW, B., P. MANTEY, L. GRIFFITHS et al. "Adaptive antenna systems," *Proceedings of the IEEE*, vol. 55, pp. 2143–2159, Dec. 1967.

———. "Adaptive filters." In R. Kalman and N. DeClaris, eds., *Aspects of Network and System Theory.* New York: Holt, Rinehart and Winston, pp. 563–587, 1971.

WIDROW, B., and S. D. STEARNS. *Adaptive Signal Processing.* Englewood Cliffs, N.J.: Prentice Hall, 1985.

WILLIAMS, R., and D. ZIPSER. "A learning algorithm for continually running fully recurrent neural networks," ICS Report 8805, University of California at San Diego, 1988.

WILPON, J. G., C.-H. LEE, and L. R. RABINER. "Improvements in connected digit recognition using higher order spectral and energy features," *Proceedings of the IEEE International Conference on Acoustics, Speech, and Signal Processing*, Toronto, Canada, vol. 1, pp. 349–352, 1991.

WILPON, J. G., R. P. MIKKILINENI, D. B. ROE et al. "Speech recognition: From the laboratory to the real world," *AT&T System Technical Journal*, vol. 69, pp. 14–24, Oct. 1990.

WILPON, J. G., L. R. RABINER, and T. B. MARTIN. "An improved word detection algorithm for telephone quality speech incorporating both syntactic and semantic constraints," *AT&T System Technical Journal*, vol. 63, pp. 479–497, 1984.

WINTZ, P. A. "Transform picture coding," *Proceedings of the IEEE*, vol. 60, pp. 880–920, July 1972.

WOLF, J. J., and W. A. WOODS. "The HWIM speech understanding system," *Proceedings of the IEEE International Conference on Acoustics, Speech, and Signal Processing*, Hartford, Conn., vol. 2, pp. 784–787, 1977.

———. "The HWIM speech understanding system." In W. A. Lea, ed., *Trends in Speech Recognition.* Englewood Cliffs, N.J.: Prentice Hall, 1980.

WONG, D. Y., J. D. MARKEL, and A. H. GRAY. "Least squares glottal inverse filtering from the acoustic speech waveform," *IEEE Transactions on Acoustics, Speech, and Signal Processing*, vol. 27, pp. 350–355, Aug. 1979.

WONG, E., and B. HAJEK. *Stochastic Processes in Engineering Systems.* New York: Springer-Verlag, 1984.

WOODS, W. A. "Transition network grammars for natural language analysis," *Communications of the American Association for Computing Machinery*, vol. 13, pp. 591–606, Oct. 1970.

———. "Language processing for speech understanding." Chapter 12 in F. Fallside and W. A. Woods, eds., *Computer Speech Processing*, Englewood Cliffs, N.J.: Prentice Hall, 1983. Reprinted in (Waibel and Lee, 1990).

XUE, Q., Y.-H. HU, and P. MILENKOVIC. "Analysis of the hidden units of the multi-layer perceptron and its application in acoustic-to-articulatory mapping," *Proceedings of the IEEE International Conference on Acoustics, Speech, and Signal Processing*, Albuquerque, N.M., vol. 2, pp. 869–872, 1990.

YONG, M., and A. GERSHO "Vector excitation coding with dynamic bit allocation," *Proceedings of the IEEE GLOBECOM*, pp. 290–294, Dec. 1988.

YOUNGER, D. H. "Recognition and parsing of context-free languages in time n^3," *Information and Control*, vol. 10, pp. 189–208, 1967.

ZELINSKI, R., and P. NOLL. "Adaptive transform coding of speech signals," *IEEE Transactions on Acoustics, Speech, and Signal Processing*, vol. 25, pp. 299–309, Aug. 1977.

ZINSER, R. L., G. MIRCHANDANI, and J. B. EVANS. "Some experimental and theoretical results using a new adaptive filter structure for noise cancellation in the presence of crosstalk," *Proceedings of the IEEE International Conference on Acoustics, Speech, and Signal Processing*, Tampa, Fla., pp. 32.6.1–4, 1985.

ZUE, V. "The use of speech knowledge in automatic speech recognition," *Proceedings of the IEEE*, vol. 73, pp. 1602–1615, Nov. 1985.

ZUE, V., J. GLASS, D. GOODINE et al. "The Voyager speech understanding system: Preliminary development and evaluation," *Proceedings of the IEEE International Conference on Acoustics, Speech, and Signal Processing*, Albuquerque, N.M., vol. 1, pp. 73–76, 1990.

ZUE, V., and M. LAFERRIERE. "Acoustic study of medial /t, d/ in American English," *Journal of the Acoustical Society of America*, vol. 66, pp. 1039–1050, 1979.

ZWICKER, E. "Subdivision of the audible frequency range into critical bands," *Journal of the Acoustical Society of America*, vol. 33, pp. 248, Feb. 1961.

Index